Infectious Diseases of Wild Mammals

THIRD EDITION

Infectious Diseases of Wild Mammals

THIRD EDITION

Edited by Elizabeth S. Williams and Ian K. Barker

Manson Publishing / The Veterinary Press

Elizabeth S. Williams, DVM, PhD, is a professor of veterinary sciences at the University of Wyoming. She has been specializing in the research of infectious diseases of wild animals since 1977. She also serves as a diagnostic pathologist.

Ian K. Barker, DVM, PhD, is a professor of veterinary pathobiology in the Department of Pathology in the Ontario Veterinary College at the University of Guelph. His research focuses on infectious diseases of wild animals, and he is an expert in alimentary tract diseases.

Published by Manson Publishing Ltd, 73 Corringham Road, London NW11 7DL, UK
ISBN 1-84076-005-2

∞ Printed on acid-free paper in the United States of America

First edition, 1970
Second edition, 1981
Third edition, 2001

A CIP catalogue record for this book is available from the British Library.

CONTENTS

PREFACE

Nearly 20 years have passed since the publication of the second edition of *Infectious Diseases of Wild Mammals,* an important resource for, and inspiration to, a generation of wildlife biologists and veterinarians. All those concerned with diseases of wildlife are greatly indebted to the late John Davis, and to Lars Karstad and Dan Trainer, for consolidating knowledge in this field and making it so readily accessible. Their work helped establish the widespread and growing recognition of the influence of disease on populations of wild mammals, consideration of disease in the fields of wildlife management and ecology, and greater appreciation of disease as a component in the relationship among wild mammals, humans, and domestic animals.

To us has fallen the humbling task of regenerating and updating this work, in collaboration with an international field of chapter authors. Our goal has been to summarize knowledge in the field of wildlife diseases relevant to wild mammals, in a form that will be useful to students in wildlife biology or veterinary medicine; wildlife biologists and managers; veterinarians dealing with free-living and captive wildlife; and epidemiologists and public health professionals concerned with wildlife zoonoses.

The format of the third edition follows that of its predecessors, with chapters arranged by taxon of infectious agent. They deal with agents established or suspected as pathogens in wild mammals, and/or transmissible between wild mammals and domestic animals or people. Many are capable of causing devastating diseases, such as tularemia, plague, or rabies, among wild mammals; almost all of these are capable of infecting domestic animals or people, as well. Other zoonotic agents, such as those causing the rodent-borne hemorrhagic fevers and Lyme borreliosis in people, have little or no impact on the wildlife that serve as the reservoir for human infection. Some well-recognized pathogens of domestic animals, such as canine distemper virus and parvoviruses, are emerging in significance or increasingly recognized in free-ranging wildlife. Diseases such as brucellosis, bovine tuberculosis, rinderpest, and foot-and-mouth disease arguably are more strongly associated with domestic animals than with wildlife, but they pose some of the knottiest epidemiologic, management, and ethical problems where wild animals are affected. Some agents, such as the myxoma virus, rabbit calicivirus, and *Salmonella,* have been exploited as biologic agents of vertebrate pest control, with varying degrees of success and safety. Others, such as poxviruses and *Leptospira,* are functional or candidate recombinant vectors for the delivery of antigens to immunize wild mammals. The significance for wild mammals of agents such as *Chlamydia* and its relatives, and the retroviruses and mycoplasmas, seems relatively restricted, taxonomically and/or geographically, while others, such as the lyssaviruses, the poxviruses, *Salmonella,* and *Leptospira,* affect a wide array of wild mammals on most continents.

Knowledge of diseases included in the last edition has greatly expanded, and a number of new agents that fall within the scope of this volume have emerged in the past two decades. Accordingly, we have added major treatments of problems such as rodent-borne hemorrhagic fevers, Lyme borreliosis, transmissible spongiform encephalopathies, and calicivirus infections. Consideration of the implications for wildlife of agents such as the parvoviruses, canine distemper virus, and the rickettsiae has been expanded. And, of necessity, some topics found in the second edition have been truncated, consolidated, or eliminated.

Within the limits of the space allotted, chapter authors provide an entrée to the historical literature on an agent or disease, a summary of current knowledge on the etiology, pathogenesis, immunity, and diagnosis, and discussions of implications of the agent or disease for captive or free-ranging wild mammals, domestic animals, and people. Relevant current information on the biology and epidemiology of pathogens gained by molecular techniques has been incorporated, but the rapid expansion of such knowledge, and the inevitable lag in bringing a work such as this to press, no doubt already will have dated chapters on some very active topics.

Despite the expansion in size of this volume, space limitations forced elimination of most comprehensive reviews, lists of citations, and detailed treatment of clinical signs and comparative diagnosis. Of necessity, many chapters are surveys rather than full reviews of the topic, and readers will need to consult the second edition and other sources cited for fuller historical and current information. The number of illustrations and the number and scope of tables also were limited in deference to textual material. Nonetheless, between these covers is a vast amount of information, covering a wide array of infectious agents affecting virtually all orders of mammals. Reference lists lead to the wider literature, and the index will allow the reader to find and consolidate information affecting particular families or genera of mammals, which may be scattered among several chapters. Walker's *Mammals of the World,* fifth edition, by R.M. Nowak (Johns Hopkins Press, 1991), was used as the authority for mammalian nomenclature, except where another authority is cited by chapter authors.

We gratefully acknowledge the effort that our many authors dedicated to their contributions. Selected on the basis of their expertise and familiarity with the wildlife ramifications of the topic assigned, the authors span the disciplines of wildlife biology, veterinary medicine, epidemiology, and microbiology. Years of experience in the field and in the laboratory, on all continents except Antarctica, have been distilled onto these pages. The editors found it exciting to foster international collaborations on authorship of chapters, often by means of electronic communication, and to participate as they developed and evolved. Like the editors, many authors had to cope with the effects of economic rationalism and downsizing on their workplace. Chapter authorship was an act of dedication, for little reward, often carried out to difficult deadlines under trying circumstances. Sadly, two of our contributors, Drs. Charles Seymour and Werner Heuschele, died during the preparation of their chapters. To the chapter authors goes credit for the quality of the content of this book; the editors assume responsibility for any shortcomings.

We also wish to acknowledge colleagues who read drafts of chapters or who provided technical advice on various topics. Iowa State University Press, in particular Gretchen Van Houten, has been very helpful and extremely accommodating over the course of the preparation of this edition. We greatly appreciate the tolerance and encouragement to carry this project through that were offered by our spouses, Tom Thorne and Susy Carman.

Lastly, we wish to recognize the Wildlife Disease Association and its many dedicated members, who, over the course of nearly 50 years, have shared in the study and dissemination of knowledge of diseases of wildlife, in the interests of animal and human health and welfare, and environmental conservation. The Wildlife Disease Association supported the preparation of this volume and is the beneficiary of the royalties that will accrue from its use.

Infectious Diseases of Wild Mammals

THIRD EDITION

PART I

Viral and Prion Diseases

1 RABIES

CHARLES E. RUPPRECHT, KLAUS STÖHR, AND COURTNEY MEREDITH

Synonyms: al kalabe, beshenstvo, derriengue, hari, hydrophobia (humans), kalevet, kuang cheng, lyssa, mal de calderas, oulo fato, polar madness, rabhas, rage, rabia, rabbia, raiva, thao, tollwut.

INTRODUCTION. Rabies is an acute fatal viral encephalomyelitis. The etiologic agents belong to the genus *Lyssavirus*, in the family *Rhabdovirus*. Described for at least four millennia, it is one of the oldest recognized infectious diseases. A clinical entity indistinguishable from classic rabies has been reported on all inhabited continents. Warm-blooded vertebrates are susceptible to experimental infection, but only mammals are important in the epidemiology of rabies. The etiologic agents are neurotropic viruses that primarily replicate in the central nervous system (CNS) and pass to the salivary glands and saliva. Transmission principally occurs by the bites of infected carnivores and bats. As a major zoonosis, rabies has considerable public health, veterinary, and economic impact, and it should be of concern to all professionals that deal directly with the health of free-ranging or captive mammals. Since the publication of the first edition of this book, significant advances in biotechnology have occurred, particularly in the diagnostic arena with the development of monoclonal antibodies (Mabs) and molecular typing techniques that have shed considerable light on the epidemiology of rabies. In addition, in the vaccine field, there has been substantial application of oral immunization toward wildlife rabies control.

HISTORY AND HOSTS

Africa. Circumstantial evidence indicts Africa as the cradle of lyssavirus evolution, because nowhere else are at least four serotypes or genotypes present (Rupprecht et al. 1991; King et al. 1994). Besides rabies virus, only two European bat lyssaviruses and a newly described Australian virus have become established outside of Africa, perhaps by virtue of bat flight and subsequent zoogeographic spread. Classic serotype/genotype 1 rabies virus, found mainly in dogs *Canis familiaris,* has achieved virtual worldwide distribution, primarily with human assistance (Smith and Seidel 1993), whereas the other viruses by comparison may be viewed somewhat as biologic curiosities.

Arguably, prior to the 20th century, innumerable rabies cycles recognizable in local folklore may have circulated throughout Africa. For most of the continent, however, written records to confirm this assertion hardly exist. For example, by the late 19th and early 20th centuries, numerous apparent human and animal rabies cases derived from wildlife exposure had been observed and sometimes anecdotally reported in South Africa, but actual scientific documentation did not occur until 1928. For much of the rest of Africa, confirmation of the presence of rabies occurred at approximately the same time.

It has been difficult to appraise the primary role of African wildlife realistically in maintaining the disease because of the ubiquitous domestic dog. Many human victims of rabid dogs further obscure the picture of sylvatic rabies. Dog rabies has been known in the Mediterranean littoral region of north Africa since antiquity; in sub-Saharan Africa most, if not all, of the European colonies recorded from their earliest days rabies-like diseases in dogs, other domestic animals, and humans. At this juncture it is impossible to determine whether rabies was imported, always endemic, or if wildlife played any substantial role in its maintenance.

In South Africa, the picture is somewhat clearer. Dr. E. Cluver, an assistant medical officer, published a retrospective clinical account of seven human rabies cases, the earliest of which dated from 1916, in which the biting animal was identified as either a mongoose or a genet (Cluver 1927). Three other cases occurred after dog bites. While investigating these deaths, Cluver found among the indigenous inhabitants a widespread, long-standing belief that the bite of a genet led to a fatal disease some weeks or months later. Today, using Mabs, a distinct rabies variant has been identified that is maintained among members of the subfamily Viverrinae, to which the mongooses, civets, and genets belong. The specific characteristics of this rabies variant suggest that it evolved long ago by adaptation to these particular hosts. The yellow mongoose *Cynictis penicillata* is both common and widely distributed and is probably principally involved in maintaining endemic disease in this particular locale. Although the suggestion of rabies among domestic species existed for centuries (King et al. 1994), it is otherwise difficult to document earlier involvement of African wildlife.

Eurasia. Rabies, at least among dogs, was suggested by the ancient Mesopotamians (ca. 2300 B.C.), Chinese (ca. 782 B.C.), Greeks (ca. 500 B.C.), and Romans alike, but it is uncertain when its existence in wildlife was first appreciated (Wilkinson 1988; Steele and Fernandez 1991; Baer et al. 1996). The disease was prevalent in both domestic animals and wildlife in Western and Central Europe for several centuries throughout the Middle Ages, particularly among foxes and wolves *Canis lupus* (Steck et al. 1968; Wilkinson 1988), although badgers *Meles meles* and bears *Ursus arctos* (ca. 900 A.D.) were also occasionally mentioned (Steele and Fernandez 1991). Many human cases were reported after rabid animal contact, which led to the development of vaccines for postexposure prophylaxis (PEP) in humans at the end of the 19th century (Pasteur et al. 1884; Roux 1887; Calmette 1891). Red fox *Vulpes vulpes* rabies was rampant throughout Europe in the 19th century, largely disappeared for unknown reasons during the first decades of the 20th century, but reemerged after World War II. The disease spread subsequently throughout mainland Europe (Kauker 1966; Wandeler et al. 1974; Bögel et al. 1976; Toma and Andral 1977; Steck and Wandeler 1980; Wachendörfer and Frost 1992; Steele and Fernandez 1991; Blancou et al. 1991). Rabies cases in indigenous terrestrial wildlife have since been reported in all European countries except Sweden, Portugal, Greece, the United Kingdom, Ireland, and the mainland of Norway and Spain; data from Albania are not readily available [World Health Organization (WHO) 1996]. Generally, the rabies situation has improved in Western Europe over the last decade, primarily as a result of large-scale oral vaccination campaigns against fox rabies. With few exceptions, it has been difficult to meet the criteria for rabies elimination (WHO 1990; Office International des Epizooties 1992) despite tremendous progress in rabies control in foxes.

The Americas. Whether rabies existed prior to a European presence is unknown, largely because of the paucity of a written history prior to the late 15th century, but it is likely that the present disease is a mosaic of rabies of the Old and New Worlds. Conceivably, human and livestock deaths described by the original Spanish conquistadores were attributable to indigenous rabid vampire bats *Desmodus rotundus,* but several notable reports throughout the 16th and 17th centuries mention the apparent absence of the disease throughout much of Latin America, at least among dogs (Baer et al. 1996). The apparent significance of insectivorous bats was largely unappreciated until late in the 20th century. In contrast, the ancient Bering land bridge that provided the opportunity for initial human migration may also have enabled the natural introduction of rabies into northern North American latitudes for thousands of years by dispersing Old World carnivore populations (Winkler 1975); maintenance in circumpolar areas cannot be discounted (Crandell 1991). Nevertheless, strong suggestions of the disease did not occur until approximately 200 years after the pivotal European conquest of the Americas; one of the first reports is attributable to a cleric during 1703 in Mexico (Steele and Fernandez 1991), and a prominent Mexican outbreak was recorded in 1709, with the involvement of a wolf with hydrophobia (Baer et al. 1996). In the mid-1700s, reports of rabid foxes and dogs were common throughout the mid-Atlantic British colonies, perhaps complicated by a predilection for red fox hunting and hence widespread translocation, a focus of disease that may have persisted throughout the eastern United States well into the 20th century. One of the first descriptions of the disease in Canada did not occur until 1819, when the Duke of Richmond, then Governor General, was bitten by a pet fox and subsequently died (Steele and Fernandez 1991). By the mid-1800s, skunk rabies was reported from the American prairies and west, spreading northward into Canadian provinces by the next century; it was so prevalent that some proposed the name *Rabies mephitica* to describe skunk rabies (Steele and Fernandez 1991).

Undoubtedly, the unintentional European translocation of infected dogs during typical voyages of 4–6 weeks, which were often shorter than the incubation period of the disease, had a major influence on the global spread of rabies (Smith and Seidel 1993) particularly in the Western Hemisphere.

Another exotic introduction influenced the history and distribution of rabies in the Western Hemisphere. During the mid-1800s, importations began of the small Indian mongoose *Herpestes auropunctatus* for control of rats and snakes on sugarcane plantations in many areas (Everard and Everard 1988, 1992). Consequently, mongoose rabies has been reported on several islands, including Cuba, the Dominican Republic, Grenada, and Puerto Rico. However, the absence of the disease despite the presence of mongoose on other Caribbean islands and Hawaii is somewhat peculiar, a combination perhaps of geographic isolation and historical accident.

Although wildlife rabies may have been present in the far north for centuries (Rosatte 1988; Crandell 1991), documentation of a major outbreak among red and Arctic *Alopex lagopus* foxes did not occur until the mid-20th century, spreading into Alberta, British Columbia, Saskatchewan, Manitoba, Quebec, and Ontario by the 1950s. Remnants from this original Arctic outbreak persist in endemic foci of fox rabies in Alaska, southeastern Canada, and the northeastern United States (Krebs et al. 1997). Rabies virus variants found among gray foxes *Urocyon cinereoargenteus* in Texas and Arizona are distinct (Smith et al. 1995).

Occasional cases of wildlife rabies have been reported from throughout many portions of Latin America since at least the 19th century, but the predominance of dog rabies and its associated human fatalities historically blurred the epidemiologic significance of nondomestic species. In contrast, outbreaks of bovine rabies, and later human cases, of common vampire bat origin were clearly recognized as a fatal ascending paralysis in the late 1800s and early 1900s in Mexico, Brazil, Argentina, Bolivia, Paraguay, Guyana, and Trinidad (Acha and Alba 1988). Today, vampire rabies is widespread from Mexico to Argentina, and it responsible for the deaths of millions of livestock and hundreds of people (Constantine 1988; Schneider et al. 1996). Throughout the 20th century, a number of novel vampire control methods were developed in serious attempts at population reduction and rabies prevention (Lord 1988; Flores-Crespo and Arellano-Sota 1991), but vampire rabies persists.

DISTRIBUTION. With the exception of geographically isolated political units (e.g., the United Kingdom and Japan) that have successfully eliminated rabies through aggressive control measures or those regions that have never reported an indigenous case of the disease (e.g., much of the Caribbean and Pacific Oceania), rabies is widespread on all inhabited continents (Fig. 1.1). At least 48 countries or territories did not report rabies during 1994 (WHO 1996). However, rabies surveillance is passive and biased toward detection in humans and domestic species rather than free-ranging wildlife. It is further complicated by the mobility of infected bats that often escape surveillance efforts.

ETIOLOGY. The viruses that cause rabies belong to the Rhabdoviridae. This family consists of the genera *Vesiculovirus, Ephemerovirus, Lyssavirus,* and many unassigned viruses isolated from a variety of plants, invertebrates, and vertebrates. Viruses in this family are primarily assigned on the basis of their distinctive rod- or bullet-shaped structure, and all are single-stranded, negative-sense, unsegmented RNA viruses.

The genus *Lyssavirus* includes rabies virus and a group of antigenically and genetically related Old World viral species known as serotypes or genotypes with molecular weights of approximately 4.6×10^6 Da (Wunner 1991). Virions are composed of an internal ribonucleoprotein (RNP) core or nucleocapsid, containing the nucleic acid, and an outer lipid envelope covered with transmembrane protein spikes. The viral genome encodes five polypeptides, associated with either the nucleocapsid or the envelope (Tordo and Kouknetzoff 1993). The L (transcriptase), N (nucleoprotein), and the NS (nominal phosphoprotein) proteins comprise, together with the linear viral RNA, the RNP. The μ (matrix) and G (glycoprotein) proteins are associated with the envelope. The N protein comprises the group or genus-specific antigens, whereas the G protein is responsible for the classic serotype (Rupprecht et al. 1991).

Lyssaviruses are adapted to replication in mammalian neural tissue and do not persist in the external environment. Viral susceptibility to chemical and physical degradation partly depends on the source and nature of the infectious material, such as intact brain tissue or a film of saliva, and specific environmental conditions. Although infectious virus may be recovered months later in the frozen carcass of a fox on the Arctic tundra, virus can be inactivated within hours during the summer in the decomposing tissues of a road-killed raccoon *Procyon lotor* in Florida. Being enveloped, the virus is sensitive to many lipid solvents and is rapidly inactivated by exposure to fixatives such as formalin, strong acids and bases, most detergents, and ultraviolet radiation, including direct sunlight. Heat (less than 56° C), drying, and repeated freezing and thawing will usually lead to a rapid inactivation. For example, in rodent carcasses under laboratory conditions, viral infectivity became greatly diminished at 25° C–37° C within 3–4 days, although it was still present as long as 9 days later (Lewis and Thacker 1974).

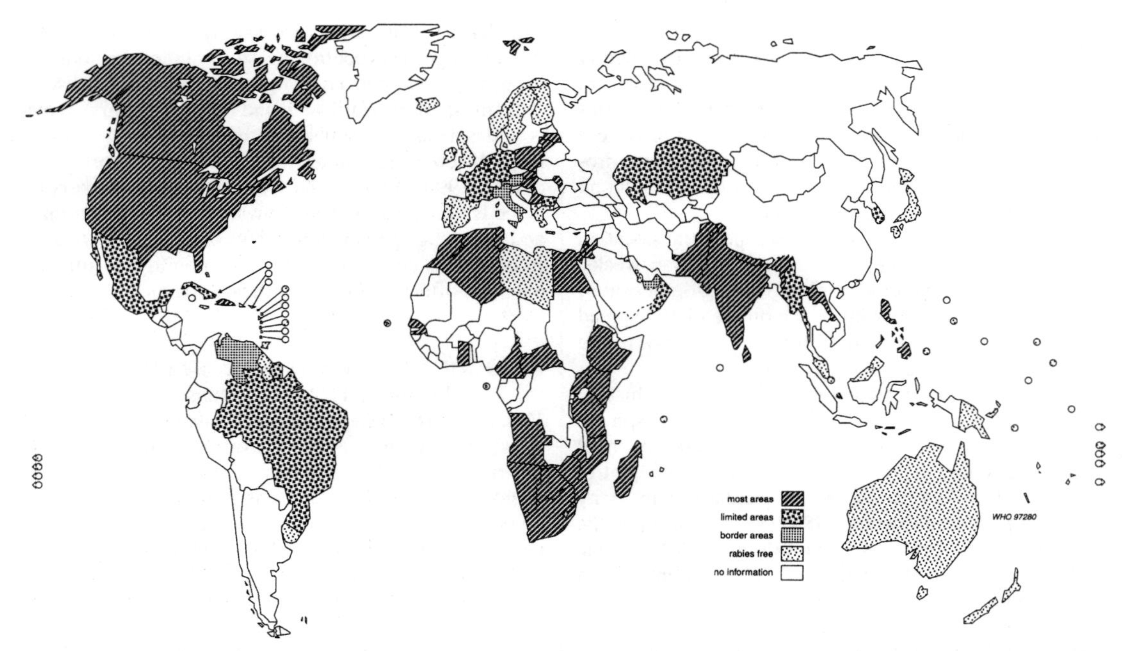

FIG. 1.1—World map depicting geographic distribution of rabies (not to scale). (The designations employed and the presentation of material on this map do not imply the expression of any opinion whatsoever on the part of the World Health Organization concerning the legal status of any country, territory, city, or area of its authorities, or concerning the delimitation of its frontiers or boundaries. Dotted lines represent approximate border lines for which there may not yet be full agreement.)

Previously, rabies and the "rabies-related viruses" were defined on the basis of their morphology, serologic cross-reactivity, and their ability to cause encephalitis when inoculated intracranially into laboratory rodents (Bourhy et al. 1993). Two other rhabdovirus species isolated from insects in Africa—Obodhiang and Kotonkan—that were seemingly aligned with this group on the basis of serology await further genetic characterization. Tentatively, there are at least six other distinct lyssaviruses related to rabies, but they are easily distinguished by their characteristic antigenic and genetic properties. Within each major *Lyssavirus* serotype/genotype are numerous viral variants that exist as *quasi species,* heterogeneous mutant viral populations, subject to Darwinian evolution and punctuated equilibrium (Eigen and Biebricher 1988; Nichol et al. 1993).

***Lyssavirus* Serotype/Genotype 1: Rabies Virus.** Rabies virus is the type species and is the most significant member of the genus as concerns distribution, abundance, and relative veterinary and public health importance. Most of the relevant scientific focus has concentrated on this classic agent. Except where noted otherwise, the principal applied attributes of rabies virus and the rabies-related lyssaviruses are felt to be largely interchangeable in regard to diagnosis, pathobiology, and disease prevention, control, and management.

Besides technical differentiation from other lyssaviruses, isolates of rabies virus have also been informally categorized from the time of Pasteur as either *fixed* laboratory or vaccine strains, adapted by passage in animals or cell culture, or as unadapted street or *wild type* viruses. The use of monoclonal antibodies (Mabs) in the 1970s and the application of molecular techniques in the 1980s, particularly viral nucleic acid detection by reverse transcription and amplification of cDNA by the polymerase chain reaction (RT/PCR) with subsequent analysis of viral nucleotide sequences, have been extremely useful in the identification and differentiation of multiple viral variants originating in major reservoirs throughout the world. Such techniques have been essential in implicating the likely sources of human exposure, when a definitive history of animal bite was unavailable, and in understanding lyssavirus epidemiology and phylogeny. For example, these analyses have provided historical insights into the global dissemination of lyssaviruses and the significance of translocation upon current distribution patterns (Smith and Seidel 1993).

***Lyssavirus* Serotype/Genotype 2: Lagos Bat Virus.** Discovery of the rabies-related lyssaviruses of Africa began in 1956 when Boulger and Porterfield (1958) read reports of rabies virus isolations from bats in the United States. These reports inspired them to investigate the large straw-colored fruit bat *Eidolon helvum* living on Lagos Island, Nigeria. A virus was isolated that the authors named Lagos bat virus and that they concluded, after serologic investigation, was not a rabies virus because it was not neutralized by rabies antiserum. This virus became the prototype of the rabies-related lyssaviruses. The presumed bat host is found on the southern half of the African continent, is remarkably widespread, and may be found anywhere from approximately 16° S to 34° S (Skinner and Smithers 1990).

The second isolation of Lagos bat virus (Sureau et al. 1980) occurred in the Central African Republic from the dwarf epauleted fruit bat *Micropteropus pusillus.* A third bat species was found when an investigation was made into widespread morbidity and mortality among Wahlberg's epauleted fruit bats *Epomophorus wahlbergi* in Natal (Meredith and Standing 1981; Crick et al. 1982).

Eidolon helvum and the epomophorine bats are tree-roosting and gregarious, which may play a role in the epidemiology of Lagos bat virus. Also significant could be the interaction between these species whose breeding grounds overlap, such as during competition for preferred fruits, particularly when in short supply during drought or seasonal variation. Alternatively, several variants may persist in different species.

In the first two recoveries of Lagos bat virus, there was no obvious indication of pathogenicity in the hosts from which they were isolated. In the case of *E. wahlbergi,* however, extensive bat mortality was observed, and the resultant media publicity resulted in recovery of a large number of dead and dying bats from which several viral isolations were made. When inoculated intracerebrally into rodents and vervet monkeys *Chlorocebus pygerythrus,* this virus produced a fatal rabies-like disease with the formation of Negri bodies in the CNS (C. Meredith unpublished). Boulger and Porterfield (1958) were unable to demonstrate Negri bodies in rodents dying of Lagos bat virus infection, but this may have been due to a lack of special stains for the detection of the sometimes rare inclusions.

A further isolation of Lagos bat virus was made from a domestic cat *Felis catus* in Natal (King and Crick 1988). The source of the infection remains unknown. Another cat showing atypical clinical signs, but suspected of being rabid, was found to be infected with Lagos bat virus in Zimbabwe (Foggin 1988). Again the source of the infection could not be determined.

Classification of 115 Ethiopian suspect rabies virus isolates by Mebatsion et al. (1992) revealed that at least one dog had been infected with Lagos bat virus. This is the only record of this viral infection in a canid. No details of this specific clinical case were published.

***Lyssavirus* Serotype/Genotype 3: Mokola Virus.** In 1968, a survey of Nigerian rodents for arboviruses resulted in the isolation of a new virus from viscera of shrews *Crocidura* spp. The virus was later named Mokola from the district of origin and shown to be a rhabdovirus related to rabies virus (Shope et al. 1970; Kemp et al. 1972). This virus became the prototype of *Lyssavirus* serotype/genotype 3. Unlike Lagos bat virus, which has not, as yet, been incriminated in any fatal human case, Mokola virus was reported by Familusi et al. (1972) as the cause of death of a young girl afflicted with a paralytic disease in Nigeria. Later that year, Familusi and Moore (1972) claimed to have recovered Mokola virus from the cerebrospinal fluid (CSF) of a child suffering from a septic meningitis, but the child recovered from the illness without adverse sequelae. The unusual circumstances of this case raised some doubts as to whether Mokola virus was actually the cause of the illness. Mokola virus was recovered from a *Crocidura* shrew in 1974 in Cameroon (le Gonidec et al. 1978) and from the brain of a harsh-furred mouse *Lophuromys sikapusi* caught in the Central African Republic (Saluzzo et al. 1984). Because Mokola virus appears to be widespread among shrews and, possibly, rodents that are common domestic commensals in west and central Africa, such animals may be sources of infection, quite unlike the case for other lyssaviruses.

Foggin (1982, 1983, 1988) described Mokola virus infection in cats and a dog in Zimbabwe. Relatively large amounts of Mokola virus, exceeding 10^3 mouse intracerebral 50% lethal dose, were required to achieve experimental infection of cats, making transmission by bite of a small rodent or shrew unlikely. Possibly, cats may become infected by predation and consumption of infected prey, although Bell and Moore (1971) found cats refractory to this route when using mice dying of serotype/genotype 1 rabies virus.

Because of Foggin's experience in Zimbabwe, all rabies-positive cat isolates in the Republic of South Africa were screened with anti-N Mabs to determine which *Lyssavirus* was involved, and cats from the eastern Cape were shown to be infected with Mokola virus (Meredith et al. 1996). Mebatsion et al. (1992) reported a case of Mokola virus infection in a cat in Ethiopia, a continent away from the South African cases. The role of cats as sentinels for rabies-related lyssaviruses in wildlife is clearly important.

***Lyssavirus* Serotype/Genotype 4: Duvenhage Virus.** The index case occurred in South Africa during 1970, when classic clinical rabies occurred in a man north of Pretoria who had been bitten on the lip by a bat while asleep, possibly Schreibers' long-fingered bat *Miniopterus schreibersi* (Meredith et al. 1971). Brain specimens from the man were negative when originally examined by the fluorescent antibody test (FAT) using a locally prepared conjugate, but Negri bodies were seen in Purkinje cells. Confirmation that this was a rabies-related virus occurred later (Tignor et al. 1977), and the original patient name was selected for viral nomenclature.

Corroboration that the Duvenhage virus originated from an insectivorous bat occurred when it was isolated from the brain of a sick bat. The general description of the bat was suggestive of *Miniopterus* sp. (Schneider et al. 1985).

A third African isolation of Duvenhage virus was made from a common slit-faced bat *Nycteris thebaica* during a survey in Zimbabwe (Foggin 1988). This species is extremely widespread in central, east, and southern Africa. As with Lagos bat virus and Mokola viruses, the epidemiology of Duvenhage virus is not understood.

***Lyssavirus* Serotype/Genotypes 5 and 6: European Bat Virus I and II.** As reviewed by Schneider and Cox (1994), the existence of rabies among European bats was suspected as early as the 1950s. Initial reports suggested an antigenic relationship with Duvenhage virus, but later analyses demonstrated at least two unique *Lyssavirus* variants, termed European bat viruses (EBVs), subtypes I and II. The reservoirs of EBV I and II are believed to be insectivorous bats. Both subtypes are responsible for human fatalities, but no known domestic or wild mammal cases have been confirmed after active disease surveillance. More than 400 total EBV cases have been identified to date, including the first report from Great Britain (Whitby et al. 1996). Considering that many local bat populations are endangered or threatened species and that EBV subtypes appear largely restricted to those portions of Europe where dog and wildlife vaccination is widely practiced, it is hoped that these lyssaviruses will not play a significant role in European public or veterinary health.

***Lyssavirus* Serotype/Genotype 7: Australian Bat Virus.** Although there was no previous evidence of endemic rabies in Australia, a novel virus infection (Fraser et al. 1996) was diagnosed in a black flying fox *Pteropus alecto,* a megachiropteran fruit bat, during the spring of 1996. The bat was submitted to an Australian veterinary laboratory during routine surveillance of bats for another recently discovered virus, the so-called equine morbillivirus (Murray et al. 1995). The virus, hereafter referred to as Australian bat virus (ABV), possessed typical rhabdovirus morphology, was detectable by FAT using standard antirabies diagnostic reagents, and caused a nonsuppurative meningoencephalitis, consistent with lesions induced by other lyssaviruses. The ABV has since been isolated from grey-headed *P. poliocephalus* and little red *P. scapulatus* flying foxes and from a yellow-bellied sheath-tail bat *Saccolaimus flaviventris,* an insectivorous microchiropteran (Tidemann et al. 1997). Infection appears widespread; ABV has been isolated from bats in New South Wales, Queensland, Victoria, and the Northern Territory (Speare et al. 1997). Based on examination of archived fixed tissues, ABV was present in 1995, if not earlier. Antigenically and genetically, ABV isolates are similar to classic rabies viruses, as confirmed by cross-protection studies with commercial human and animal rabies vaccines (Rupprecht et al. 1996a). But they represent a distinct subtype (Gleeson 1997). Provisionally, ABV is considered to represent a new *Lyssavirus:* serotype/genotype 7. The ABV has caused at least one human death, the patient most likely infected while caring for ill bats (Allworth et al. 1996). Previously, no rabies was reported in a west Malaysian survey of 478 bats of 12 different species (Tan et al. 1969) or in a collection of more than 1000 bats in the Philippines (Beran et al. 1972). However, rabies infection was reported during 1978 from a grey-headed flying fox found dead in India (Pal et al. 1980), suggesting that some agents such as ABV may be more prevalent locally among Asian-Pacific bat populations than previously realized. Obviously, these discoveries have raised serious public health concerns (Crerar et al. 1996) on a continent believed free of the disease.

TRANSMISSION AND EPIDEMIOLOGY. A key to understanding the complex epidemiology of wildlife rabies is a basic appreciation of the agent-host-environment triad (Wandeler 1993; Wandeler et al. 1994). Unlike many etiologic agents, a productive infection by a lyssavirus usually kills its host. Thus, propagation to the next susceptible mammal usually only effectively occurs during a relatively short excretion period of the virus during the final stage of the disease, unlike previous beliefs of a long-term clinically normal carrier state. This very small window of infectious opportunity is generally concomitant with illness or in the prior 3–10 days. Virus excretion several weeks before clinical signs is unusual. Many conditions must be met to allow an efficient rabies infectious chain to be maintained without in utero transmission (Steece and Calisher 1989) or other rare transmission events. There are two primary conditions: a pertinent virus-animal interaction (e.g., virus characteristics, species susceptibility, and disease course) and appropriate host natural history (e.g., reproduction, dispersal, social interactions, population demographics, and density). The significance of human influences, predominantly on the latter conditions, should not be overlooked (Rupprecht et al. 1995).

In rabies, the term *reservoir* refers specifically to those species that maintain the disease in nature. Given more than 4000 species in the order Mammalia, all of which are believed susceptible and in theory capable of infecting another mammal, relatively few species qualify as major rabies reservoirs. An oversimplified description of an urban cycle of rabies, maintained by feral domestic dogs, and a sylvatic cycle of rabies, supported by wildlife, fails to communicate the true interactive dynamics of the disease and the frequent overlap of these general categories.

Africa. Probably no country in Africa can validly claim to be truly free of rabies. As in much of the developing world, the story of infectious diseases in Africa, especially regarding zoonoses, is one of inadequate funds, shortage of experienced personnel, lack of equipment, and largely nonexistent modern communications resulting in inadequate surveillance. As a general observation, rabies has spilled over from dogs into wild canids and, as a result, they are now major maintenance hosts in many parts of Africa.

SOUTH AFRICA. Table 1.1 lists the main wildlife species involved in rabies maintenance and transmission in southern Africa. A semi-diagrammatic map (Fig. 1.2), displays the principal rabies vectors. In most African countries, rabies can usually be found wherever there are large human populations with their associated dogs. Prior to the late 1940s, southern Africa was essentially free of dog rabies. Sporadic and isolated cases in dogs, due to wildlife exposure, occurred in South Africa, but the numbers were insignificant (Snyman 1940).

In 1947, dog rabies appeared on the northernmost border of South West Africa (now Namibia) and, by

TABLE 1.1—Wildlife species considered to be likely reservoirs of rabies virus in southern Africa

Family	Common Name	Species
Canidae	Black-backed jackal	*Canis mesomelas*
	Side-striped jackal	*Canis adustus*
	Bat-eared fox	*Otocyon megalotis*
Herpestidae	Yellow mongoose	*Cynictis penicillata*
	Slender mongoose	*Galevella sanguinea*
	Suricate	*Suricata suricatta*
	Water mongoose	*Atilax paludinosus*
Viverridae	Small-spotted genet[a]	*Genetta genetta*
Felidae	African wild cat[a]	*Felis sylvestris*

[a]Rabies in these species is no longer common.

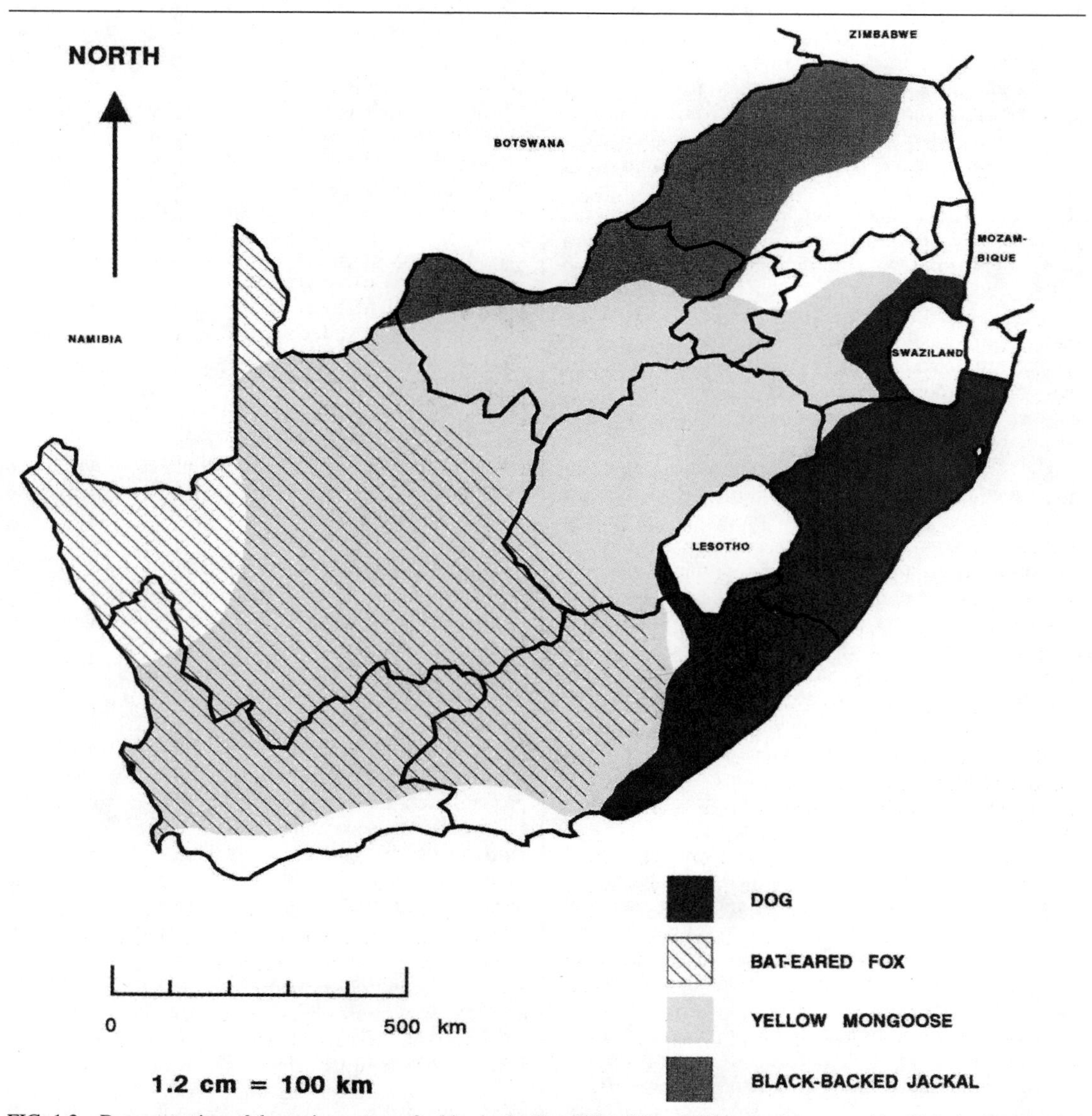

FIG. 1.2—Demonstration of the main vectors of rabies in the Republic of South Africa and the areas in which they predominate (Courtesy of G. Bishop).

1950, it had spread to the northern Transvaal province of South Africa. It reached Natal in 1961. Swanepoel (1994) showed that this epizootic followed the path of greatest human, and therefore domestic dog, population density.

The dog rabies virus variant readily transmits to wild Canidae and Hyaenidae, such as jackals *Canis* spp., bat-eared fox *Otocyon megalotis,* and aardwolf *Proteles cristatus.* Black-backed jackals *Canis mesomelas* and side-striped jackals *Canis adustus,* in Zimbabwe, and *C. mesomelas,*, in northern South Africa, now appear to be major reservoirs of the disease.

A different scenario has emerged in Namibia. As in Zimbabwe and the northern Transvaal, jackal species in the northern woodland savannah soon became major hosts. Southern Namibia is more arid and in this habitat the jackal is largely replaced by the bat-eared fox as a major host.

Sporadic cases of rabies in bat-eared fox in the Cape over the years were shown to be the viverrid variant of rabies virus (Meredith unpublished). This changed in 1970 when cases of what was probably canid strain rabies in bat-eared foxes appeared along the Orange River on the Namibian border. This infiltration continued steadily over the years in the form of a slow-moving epizootic, with substantial bat-eared fox mortality, now confirmed as being due to the canid variant of rabies virus (Thomson and Meredith 1993). Thus, the present situation in the Eastern Cape Province is now complex, with at least two variants of canid virus, that borne by domestic dogs and that derived from the bat-eared fox, superimposed on preexisting mongoose-borne viverrid rabies.

The relative importance of a particular species in any given region appears, in part, proportional to its population density. Thus, the yellow mongoose has a very wide distribution, and rabies occurs wherever it is found. However, it is of greatest importance as a rabies vector in the eastern grassland savannah where it is abundant, and it is progressively less important to the west in the arid, semidesert of the karroo where not only is food less readily available but the sparse, stony, and sandy soil does not favor burrowing. Consequently, colonies are more scattered and isolated.

In South Africa, endemic viverrid rabies is being investigated with anti-N Mabs against field isolates of virus and by nucleotide sequencing of the viral genome. Results to date have shown there to be at least four or five mongoose variants with distribution based, not on species alone, but also on geographic lines (King et al. 1994, Nel et al. 1994, von Teichman et al. 1995). These findings support a theory of genetic drift, that the endemic virus evolved over a long period, isolated in large, stable, nonmigratory mongoose populations where adaptation of both host and virus produced detectable viral variation in definable geographic areas. This situation is somewhat analogous to the phenomenon of compartmentalization seen in fox, skunk *Mephitis* spp., and raccoon *Procyon lotor* rabies in North America (Winkler 1975), but differing from that model in that apparent compartmentalization has also developed and radiated within a single species, based on distribution.

NORTH AFRICA. In Egypt, rabies viruses were reportedly isolated from three gerbils *Gerbillus gerbillus* and a fox *Vulpes* sp. (Botros et al. 1977). In a subsequent study, Botros et al. (1988) used N Mabs in an attempt to classify various rabies isolates from domestic animals and wildlife. Foggin (1988) in Zimbabwe found antibodies to Mokola virus in the sera of 13 of 141 gerbils *Tatera leucogaster* tested. Nowhere else in Africa have gerbils been implicated in serotype/genotype 1 rabies.

In another unusual finding, unreported species of rats were implicated in rabies epidemiology in Algeria (Benelmouffok et al. 1982). Support for this report came from Nigeria (Okolo 1988), where *Rattus* spp. were said to be frequently infected, but these data have not been substantiated.

Rabies virus has been isolated from the Ethiopian wolf or Simien jackal *Canis simensis*, which lives only in the Ethiopian highlands (Sillero-Zubiri et al. 1996). This endangered species consists of a population estimated to be fewer than 1000 and is severely threatened by loss of habitat, interbreeding with domestic dogs, and canine-borne diseases (Ginsberg and MacDonald 1990).

CENTRAL AND WEST AFRICA. A civet *Civettictis civetta* that died of rabies in a Nigerian zoo had been caught 14 days previously in an area reputed to be free of domestic dogs (Enurah et al. 1988). In a large survey, Bula and Mafwala (1988) found rabies in only a few wild species in Zaire: four monkeys, a hyena, and a bat. Apparently these last interesting diagnoses were not investigated further.

EAST AFRICA. The majority of rabies is in domestic dogs, though spillover into black-backed and side-striped jackal populations may have created a wildlife reservoir (Davies 1981). In Kenya, there have been cases in civet and honey badger *Mellivora capensis.* The relatively frequent occurrence of rabies in these two species demonstrates the potential influence of behavior on transmission. For example, the honey badger when molested is quite aggressive and will probably retaliate if attacked by a rabid jackal.

Rabies has also been recorded in spotted hyena *Crocuta crocuta,* bat-eared fox, and white-tailed mongoose *Ichneumia albicauda.* In recent years, African wild dogs *Lycaon pictus* have fallen victim to rabies through contact with domestic canids, with serious implications for survival of this endangered species (Gascoyne et al. 1993a,b). Both Kenyan and Tanzanian populations have been severely diminished by rabies (Kat et al. 1995). Maas (1993) confirmed rabies in bat-eared foxes in Serengeti National Park, Tanzania, in 1986.

WESTERN SOUTH AFRICA. Species involved in rabies are not significantly different than those in South

Africa (Berry 1993). Viverrid rabies, occurring mostly in yellow mongoose, is less common than in more favorable habitats where population densities are much higher. The main rabies reservoir today appears to be the black-backed jackal, as it is in Namibia. Endemic jackal rabies became such a threat to cattle ranching, with losses from a single infected jackal often amounting to ten or more head of cattle, that ranchers were often forced to immunize their livestock against rabies.

Game farming of kudu *Tragelaphus strepsiceros* and, to a lesser extent, eland *Taurotragus oryx* developed in association with cattle ranching. During 1977, high mortality from rabies was noticed in kudu (Barnard and Hassel 1981). Before this outbreak had run its course in 1985, an estimated 50,000 kudu died—approximately 20% of the total population (Hassel 1982; Hubschle 1988). The observation that sick kudu, usually salivating profusely, were the subject of licking and grooming by other animals in the herd led to experiments where it was shown that rabies could be easily transmitted by introducing saliva from infected animals into the buccal cavity (Barnard et al. 1982; Hassel 1982; Hubschle 1988). Control measures were not taken, but this unusual event subsided spontaneously, presumably once the population declined.

Despite the spectacular mammalian biodiversity for which Africa is noted, relatively few studies have implicated many indigenous wildlife species as important in the epidemiology of rabies. Notably, Felidae do not appear important as major reservoirs, although they are obviously competent vectors (Rupprecht and Childs 1996).

Asia and the Middle East. Throughout Asia, the overwhelming reservoir and principal cause of human fatalities is the domestic dog (WHO 1996, 1997). Wildlife, such as mongoose, jackal, fox, raccoon dog *Nyctereutes procynoides,* wolf, and various small mustelids, plays a comparatively much smaller role, at least as currently realized given surveillance limitations (Ahuja et al. 1985; Bahnemann 1985; Koesharyono et al. 1985; Lari 1985; Thongcharoen et al. 1985; Sama et al. 1986; Blancou 1988a). In contrast, with the possible exception of Bahrain and Qatar, wildlife rabies is fairly common in Syria, Jordan, Israel, Saudi Arabia, the United Arab Emirates, Yemen, Oman, Lebanon, Iraq, Iran, and Kuwait (WHO 1996). Rabies was not recorded from Oman and the United Arab Emirates until 1990, when the disease emerged in foxes *V. vulpes arabica* in the northeastern part of Oman and spread subsequently southward toward Salalah and to the north. Data on the number of human, domestic animal, and wildlife rabies cases are difficult to obtain (Fig. 1.3).

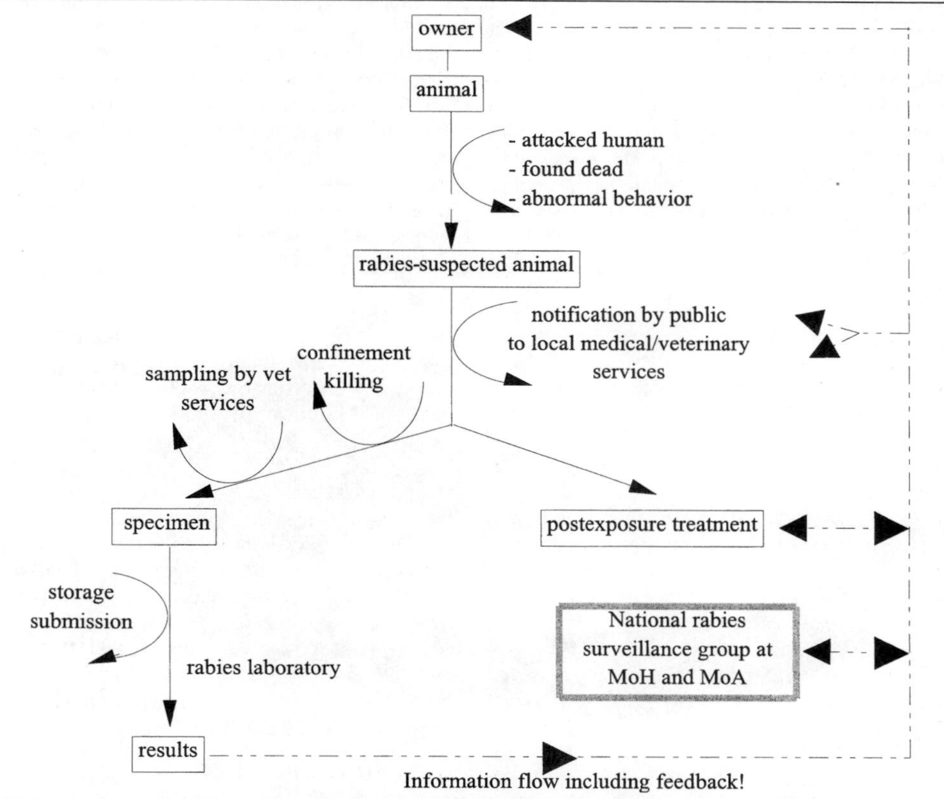

FIG. 1.3—Rabies surveillance and data flow chart. MoH, Ministry of Health; MoA, Ministry of Agriculture.

Wildlife appear to be important rabies sources in most Middle Eastern countries, with the red fox playing a major role. In particular, Israel, Saudi Arabia, Jordan, Oman, Yemen, and the United Arab Emirates reported increasing numbers of rabies cases in red foxes over the last decade. This may be related to improved surveillance, but could also be explained by an increased number and spatial spread of foxes due to habitat alterations and better availability of food and shelter. Most wildlife rabies cases in the Middle East are reported from red fox, wolf *C. lupus arabs*, and white-tailed mongoose. Rabies has also been diagnosed in numerous other wild carnivores, such as the wild cat *Felis silvestris,* fennec fox *Fennecus zerda,* striped hyena *Hyaena hyaena,* and golden jackal *C. aureus* (WHO 1996). Information on bat rabies in the Middle East is largely unavailable.

Fox rabies emerged in the Sultanate of Oman in 1990. Within 1 year, cases in foxes, wolves, mongoose, dogs, and cats were reported from several parts of the country. The disease spread throughout the fox population, which appeared to collapse in 1992, and the few rabies cases reported in foxes after 1992 suggested that the disease may have become endemic (Stöhr 1990, 1995; Novelli and Melankar 1991; Ata et al. 1993). Between 1986 and 1992, most rabies cases diagnosed in Saudi Arabia were from foxes (35%), cats (20%), and dogs (16%). Foxes appear to be a primary reservoir.

A small study of the status and biology of red and sand fox *Vulpes rueppellii* and other small carnivores in desert and semidesert environments was initiated and conducted by the National Wildlife Research Centre, Taif, in a reserve in the western part of the country (Olfermann 1995). The population density per square kilometer of *V. rueppelli* (0.25) and *V. vulpes* (0.07) was smaller than the minimum estimated as necessary to perpetuate rabies in foxes in Europe. However, population turnover was at about the same scale (50%) as found in Europe (66%). Sand foxes had a larger dispersal area and more extensive home range compared to European red fox (Lindsay and MacDonald 1986); consequently, contact may be more frequent between individuals due to overlapping ranges. Knowledge of these and related peculiarities of the fox populations of the Arabian Peninsula may help to explain the persistence of wildlife rabies in the area. Although wolf populations are reported to be high in the northern provinces and a considerable number of baboons *Papio* sp. exist in the eastern part of the country, both appear to be dead-end hosts for rabies.

Europe. As elsewhere, rabies occurs in various domestic and wild mammals in Europe, but there are only two primary reservoirs: terrestrial carnivores and insectivorous bats. Rabies is maintained rather independently in these two compartments, as substantiated by numerous virologic, genetic, and epidemiologic studies and several animal trials. This has been corroborated by the obvious success of oral immunization of foxes in Europe against rabies in areas where bat rabies is prevalent (Baer and Smith 1991; Brass 1994; Schneider and Cox 1994).

Dog rabies was controlled in western and central Europe in the early 1960s, but it still occurs in Turkey. Dogs appear also to be involved in rabies outbreaks in the European part of Russia and Romania, and to a smaller extent in Bulgaria and Lithuania (Koromyslow and Cherkasskiy 1990; Dranseika 1992; WHO 1994a; Cherkasskiy et al. 1995; Toacsen and Moraru 1995; Lalosevic 1996). In contrast, two wild mammalian species serve as major reservoirs for terrestrial rabies in Western Europe: red fox and, to a lesser extent in circumpolar areas, Arctic fox. The red fox is both the most important reservoir and victim of the disease in Europe (Fig. 1.4). Unlike most other wildlife reservoirs, this widespread, opportunistic, highly adaptable canid has

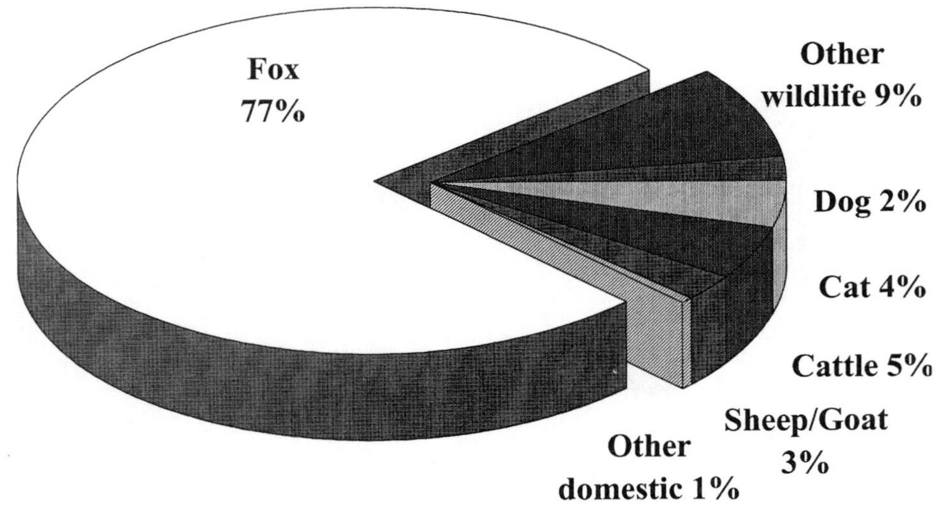

FIG. 1.4—Relative species involvement in rabies in Europe between 1977 and 1996.

been well studied. It possess flexible social structures and population dynamics that promote high conspecific contact and relatively short bursts of intensive, intraspecific transmission, making it an *ideal* rabies host (Winkler 1975; MacDonald 1980; Steck and Wandeler 1980; Anderson et al. 1981; Gessler and Spittler 1982; Blancou et al. 1991; Wandeler 1991; White et al. 1995).

Raccoon dogs are also involved in rabies dynamics, particularly in Eastern Europe and the Baltic, but little is known of the role they play in perpetuation of rabies (Cherkasskiy 1988; Blancou and Wandeler 1989; Westerling 1989; Koromyslow and Cherkasskiy 1990; WHO 1994a; Cherkasskiy et al. 1995). This highly adaptable canid originates from East Asia, but large numbers of the furbearer were released in the western part of the former Soviet Union between 1928 and 1955, and individuals gradually dispersed westward (Lavrov 1971; Nowak 1973). Approximately 100 rabid raccoon dogs were recorded annually over the last decade, representing about 5% of all rabies cases diagnosed in Poland. Similarly, in Estonia, raccoon dogs constituted about 30% of all rabies cases diagnosed in animals between 1991 and 1996. The raccoon dog was the species most frequently involved in a rabies outbreak in Finland during the late 1980s (Reinius 1992). Apparently higher population densities in eastern Russia, Finland, Latvia, Estonia, Lithuania, Ukraine, and Byelorussia, and increasingly more rabies cases diagnosed in raccoon dogs in some Eastern European countries, suggest a separate cycle of raccoon dog rabies, independent of foxes. Only limited data exist on susceptibility of raccoon dogs to the circulating rabies virus(es) in Europe (Blancou 1986, 1988b) and the corresponding susceptibility of foxes.

The first European case of rabies in a bat was reported in 1954, and an additional 14 cases were found in various insectivorous bat species by 1984 (Kappeler 1989; Selimov et al. 1989; Baer and Smith 1991; Brass 1994; Schneider and Cox 1994). Cases unexpectedly increased in 1985, with a peak in 1987, but subsequently declined; only 6–15 cases were reported annually in Europe from 1991 to 1996 (Fig. 1.5). A likely explanation for the sudden increase in bat rabies in Europe in 1985 appeared to be heightened public awareness and attention to bats as potential reservoirs (King 1991; Baer and Smith 1991; Brass 1994; Schneider and Cox 1994). Between 1954 and 1992, more than 85% of the bat rabies cases in Europe were diagnosed in *Eptesicus serotinus.* The reason for the observed higher prevalence of rabies in *E. serotinus* is unknown, but it is speculated to be related to colonial habits and roosting behavior (Brass 1994). The majority of rabid bats have been from Northern Europe, along the coast of the North Sea. Prior to 1996, rabies had not been reported in bats in the United Kingdom despite numerous examinations of bats found dead or ill (King 1991). In May 1996, however, a *Myotis daubentonii,* a species indigenous to the United Kingdom, was found near the south coast of England and was diagnosed as positive for EBV. Although there can be no certainty that the bat had not originated in England, it was suggested that the animal had flown across the English Channel from the mainland (Anonymous 1996a,b).

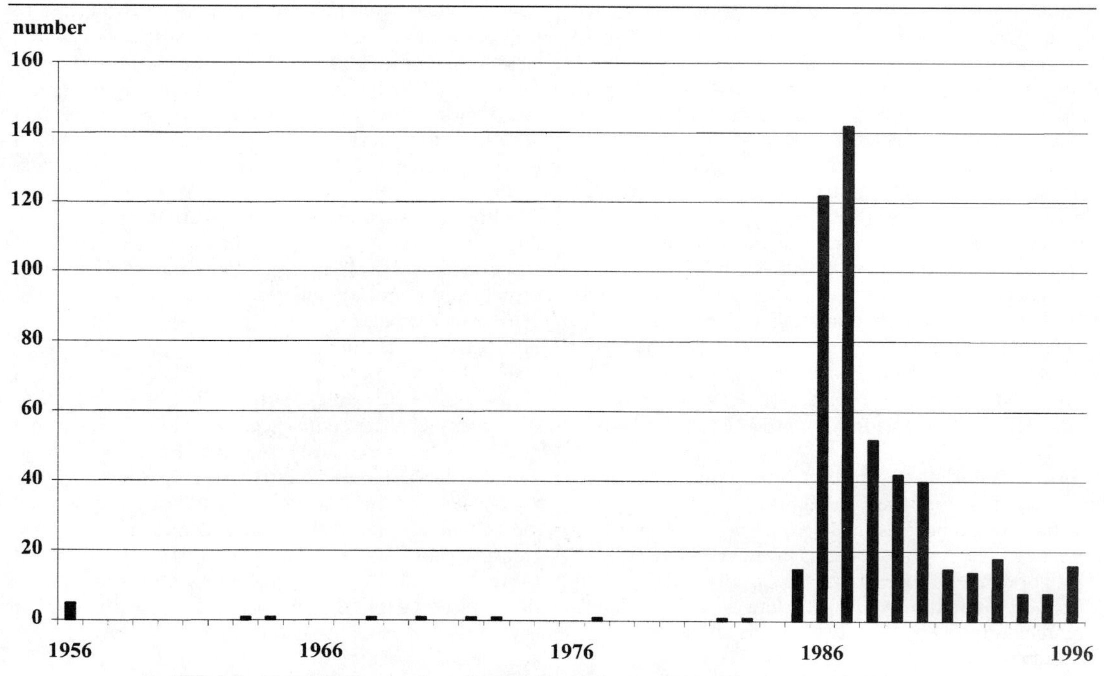

FIG. 1.5—Trends in the number of reported rabies cases in insectivorous bats in Europe.

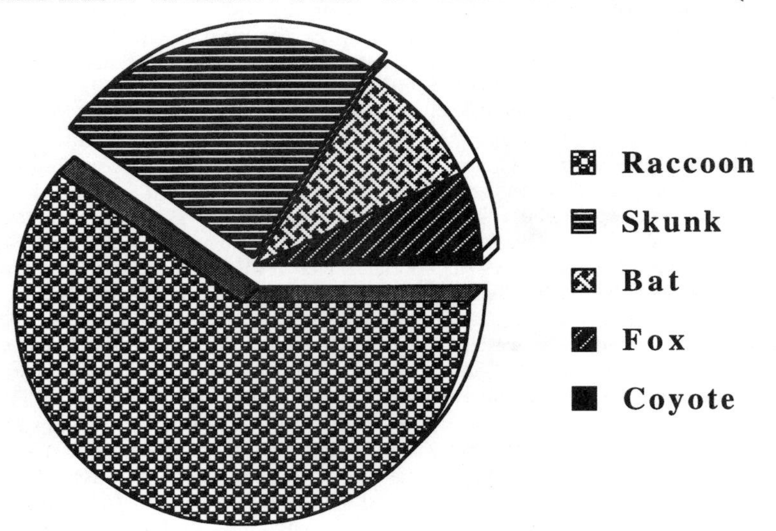

FIG. 1.6—Percent of rabies-positive specimens from various animal species, United States, 1992–96.

The Americas. The epidemiology of rabies in Canada and the United States has changed substantially in the last 100 years (Rosatte 1988; Smith et al. 1992; Rupprecht et al. 1995). Currently, more than 90% of all reported animal rabies cases occur in wildlife (Fig. 1.6), whereas for most of the century the majority of cases were reported in dogs (Krebs et al. 1997). Application of antigenic and molecular techniques has clearly demonstrated that major North American wild reservoirs are raccoons, coyotes *Canis latrans,* skunks, foxes, and bats of several species infected with many different types of viral variants (Smith et al. 1995). Wildlife rabies occurrence today is a multispecies complex, partially related to changes in human demography, animal translocation, environmental alteration, and viral adaptations over time. Epizootics are characterized by viral compartmentalization, usually to a single major host, with fairly discrete geographic boundaries, at least among the Carnivora (Smith 1996). Although many other mammals may be incidentally infected through contact with these reservoir species, such cases are quite sporadic. Once entrenched within a particular niche, viral transmission cycles may occur for decades to centuries (Tinline 1988).

In contrast to most of North America, only individual cases in several species, including fox, skunk, coyote, raccoon, coati *Nasua nasua,* puma *Felis concolor,* lynx *Felis lynx,* and a few others, have been reported throughout Latin America (Acha and Arambulo 1985; Fernandes and Arambulo 1985; Pan American Health Organization 1995), compared to the major reservoir, the domestic dog. More recent antigenic and genetic analyses of Mexican terrestrial rabies isolates demonstrate at least three distinct viral clades suggestive of long-term endemic circulation among local skunk or fox populations (de Mattos et al. 1996a), supporting the contention that a sylvatic rabies cycle may persist beyond dog-to-dog transmission (Loza-Rubio et al. 1996). This may represent a serious threat to the programmatic elimination of dog rabies in the Western Hemisphere (Cifuentes 1988; Pan American Health Organization 1995). For example, Uruguay has not reported a case of terrestrial rabies in some three decades, but the risk of viral acquisition from dogs in surrounding areas or infected bats remains, and wildlife rabies must be considered a possible mechanism for reestablishment of the disease among domestic animals.

Although several species of procyonids range throughout the Americas, endemic rabies has been associated only with raccoons and was initially reported in Florida populations during the late 1940s. This primary outbreak, recognized as a single major viral variant with uncertain origins, gradually spread over the next 30 years throughout the southeastern United States (Rupprecht and Smith 1994). Translocation of infected raccoons from the latter focus to the mid-Atlantic region during the late 1970s (Jenkins and Winkler 1987) led to an intensive raccoon rabies epizootic that continues to the present, stretching east of the Appalachians from New England to Alabama, Georgia, and Florida [Centers for Disease Control and Prevention (CDC) 1997a]. Raccoon rabies is still confined to the United States, and cases typically number between approximately 3000 and 6000 animals per year, usually peaking in the spring (Krebs et al. 1997). A new outbreak reported from eastern Ohio during the spring of 1997 potentially threatens to engulf the entire upper Midwest, unless it can be controlled. No human rabies cases have been documented to date due to the raccoon rabies epizootic despite obvious vector competence (Winkler et al. 1985), but thousands of people receive rabies PEP due to known or presumed exposure to infected animals (Rupprecht et al. 1996b). Undoubtedly, rabies kills substantial numbers of raccoons and other animals, yet its role in population regulation, per se, and its effect on overall survivorship, are poorly understood (Brown et al. 1990).

In the United States and Canada, significant dog rabies control began during the 1940s. A portion of southern Texas bordering Mexico remains one of the few foci of dog rabies. Similar to African reports of outbreaks of jackal rabies from infected domestic dogs, during the late 1980s the number of cases of coyote rabies rose in this southern Texas border zone (Clark et al. 1994), resulting in the majority of coyote rabies reported from the United States between 1991 and 1996 (Krebs et al. 1997). The viral variant in coyotes is indistinguishable from that of infected border dogs (Smith et al. 1995). Previous observations in the United States and Canada during the early part of this century demonstrated that coyotes may support explosive rabies outbreaks (Rosatte 1988). One of the greatest dangers of this particular host is the possibility of continual spillover into the domestic dog population in a region of Texas in which companion animal immunity is less than ideal. In addition, translocation of infected Texas coyotes was thought responsible for at least one case of dog rabies in Alabama during 1993 and seven cases of apparent dog-to-dog rabies transmission in Florida during 1994 (CDC 1995). At least two human fatalities were associated with the outbreak in south Texas during 1991 and 1994, probably due to coyote-dog interactions with subsequent human exposure.

Rabies in skunks has persisted sporadically in the American plains since the last century (Gremillion-Smith and Woolf 1988; Charlton et al. 1991), extending into Canada by the late 1950s. It historically has been associated with human fatalities, perhaps because skunks may be tenacious, aggressive biters when infected. Although initial reports implicated the spotted skunk *Spilogale putorius,* the striped skunk *Mephitis mephitis* has a broader distribution, encompassing most of Canada, the United States, and northern Mexico. Although poorly documented, Greenwood et al. (1997) reported rabies may exert significant effects on skunk population characteristics, such as age structure and productivity. Little is known of the role that hooded *Mephitis macroura* or hognose *Conepatus leuconotus* skunks may play in rabies epidemiology; their northern distribution extends only into the southwestern United States. They may be more important in Mexico, but are infrequently reported rabid.

Between 1961 and 1989, skunks were the most frequently confirmed rabid animal in the United States; during 1996, 23% of the rabid animals reported in the United States were skunks (N = 1656) (Krebs et al. 1997). Annual cases range from 1000 to 4000, with peaks in spring and autumn. At least three major skunk rabies virus variants can be distinguished: north-central plains, lower Midwestern states, and California variants. Reports of rabies in skunks from southeastern Ontario, Quebec, and New England are associated with the rabies virus variant in red foxes; similarly, cases of skunk rabies in the eastern United States appear to be primarily due to spillover infection from raccoons. Skunks experimentally infected with a raccoon variant do not appear to shed virus over prolonged periods and may have decreased opportunity to infect a conspecific, compared to other rabies variants (Charlton et al. 1988).

Compared to the smaller carnivores, reports of rabies in wolves are infrequent, and the epidemiology of rabies in these large canids is poorly understood. Cases occur primarily in circumpolar regions, where human populations are generally sparse. In the United States, most cases have been reported from Alaska, averaging fewer than one case per year from 1980 to date (Krebs et al. 1997). At least in North America, it is unlikely that wolves act as a major rabies reservoir, given their disparate populations. No novel viral variants have been associated with wolves, strongly suggesting that such sporadic cases are spillover infections from rabid Arctic or red fox, as is thought to be the case for other affected taxa within this ecosystem (Taylor et al. 1991). Nevertheless, rabies may seriously alter or limit wolf populations (Weiler et al. 1995; Ballard and Krausman 1997), and rabid wolves are a public health concern. Wolf bites tend to be broad, deep, multiple, and often involve severe wounds to the head. Infected wolves have traditionally been recognized as exceptional in their ability to expose large numbers of people, often associated with significant human mortality, especially in the Old World.

Although other lyssaviruses occur in bats elsewhere, only *Lyssavirus* serotype/genotype 1, rabies virus, has been confirmed in bats from the New World. No doubt related to the prominent notoriety ascribed to vampire bat rabies during the 1930–40s, major interest in insectivorous bat rabies was delayed until the first diagnosis in a bat from Florida during 1953. Consequently, bat rabies was reported throughout the United States and Canada. The history of this early spread was recently reviewed (Brass 1994). Between 600 and 1000 cases of rabid bats are typically diagnosed annually (Krebs et al. 1997). Unlike the disease in carnivores, however, specific geographic boundaries cannot be readily defined for bat rabies. Variants associated with particular bat species often can be found throughout a species migratory range. For example, rabies virus in free-tailed bats *Tadarida brasiliensis* shows minimal variation throughout the southern United States. Most areas of the United States, with the exception of Hawaii (Sasaki et al. 1992), represent a complex mosaic of different bat species affected by distinct rabies variants (Smith et al. 1995). The proportion of rabid bats from those submitted to diagnostic laboratories typically ranges from 5% to 15% (Pybus 1986; Childs et al. 1994; Feller et al.

1997). Additionally, the occurrence of bat rabies appears largely independent of rabies in terrestrial carnivores (Rupprecht et al. 1987; Baer and Smith 1991), although viral spillover to animals besides bats (Daoust et al. 1996) occurs occasionally, such as in cats or gray fox (Smith et al. 1995).

Clearly, the public health significance of insectivorous bats as zoonotic reservoirs is small in comparison to their overall ecological utility, but bats have accounted for an increasing proportion of recent human rabies cases in the United States (CDC 1997c). In Latin America, beyond the well-studied cases in hematophagous bats, rabies has only been infrequently reported from other Chiroptera. However, there is little reason to believe that the tropical viral biodiversity present among Latin American bats will be dissimilar from that noted for North American species, once adequate surveillance has been initiated (Almeida et al. 1994; Diaz et al. 1994; Martorelli et al. 1995, 1996; Uieda et al. 1995; de Mattos et al. 1996b).

CLINICAL SIGNS. There are no definitive or species-specific clinical signs of rabies beyond acute behavioral alterations (Blancou et al. 1991; Charlton et al. 1991; Winkler and Jenkins 1991; Brass 1994); as noted by Sikes (1981) in the previous edition of this book, "the atypical is typical ... ." Severity and variation of signs may be related to the specific site(s) of the primary CNS lesion(s) or to viral strain, dose, and route of infection (Smart and Charlton 1992; Hamir et al. 1996a). At the end of the incubation period, the disease progresses through a short nonspecific prodromal stage to encephalopathy and death within days. The prodromal period usually follows a bite by several weeks; however, periods of less than 10 days to several months are well documented (Charlton 1994). Severe bites to the head and neck, and bites to highly innervated areas, may result in shorter incubation periods. Local cutaneous signs may consist of paresthesia at the site of the exposure, resulting in self-inflicted wounds, probably due to viral excitation of sensory ganglia. Nonspecific clinical signs may include restlessness, inappetance, dysphagia, vomiting, or diarrhea.

An acute neurologic period usually follows the brief prodromal period by 1–2 days. A generalized excitative increase in neurologic activity is observed that is associated with hyperesthesia to auditory, visual, or tactile stimuli and sudden and seemingly unprovoked agitation and extreme aggressive behavior (furious rabies) to animate or inanimate objects. Wildlife may lose their apparent wariness and caution around humans and domestic species, alter their activity cycles, seek solitude, or become more gregarious. Head tilt, head pressing or butting, "stargazing," and altered phonation may be observed. Profound characteristic clinical manifestations in human rabies (Hemachuda 1994), specifically true hydrophobia and aerophobia, which are exaggerated respiratory protective reflexes that result in contractions of the diaphragm and inspiratory accessory muscles triggered by attempts to swallow, the sight or sound of water, or air currents have not been reported in other animals. Cranial nerve manifestations may include facial asymmetry, trismus, choking, a lolling tongue, drooling, drooping of the lower jaw, nictitans prolapse, and anisocoria. There may be additional CNS excitation with hyperactivity, disorientation, confusion, photophobia, pica, incoordination, and convulsive seizures. Autonomic systemic excitation may also result in labile hypertension, hyperventilation, muscle tremors, priapism, altered libido, hypothermia, or hyperthermia.

A paralytic phase (dumb rabies) may follow the agitated or furious phase, or the animal may progress directly to the paralytic phase from the prodromal stage. Increasing prostration, lethargy, paresis, frequent urination or incontinence, tenesmus, constipation, tail flaccidity, or decreased spinal reflexes may be apparent. Ataxia may be observed as knuckling or swaying. Ascending, flaccid, symmetrical or asymmetrical paralysis leads eventually to respiratory and cardiac failure. Hypothalamic and hypophyseal dysfunction may contribute to acute organ wasting, such as cardiomyopathy. Ultimately, rabies most often culminates in coma and generalized multiorgan failure that leads to death 1–10 days (or rarely more) after primary signs or may result in acute death with no premonitory signs.

PATHOGENESIS. Lyssaviruses are the penultimate neurotropic agents (Jackson 1994). For at least two centuries, scholars appreciated that rabies is caused by entry of a novel agent into the body, specifically that saliva contained the infectious agent, and that nerves were the means by which the agent traveled to the CNS (Baer et al. 1996). Nerve destruction by limb amputation, cautery, or surgical sectioning and microtubule-disrupting agents can inhibit axonal transport and prevent disease, indicating that centripetal viral spread from the site of entry to the spinal cord occurs within both motor and sensory axons of peripheral nerves via primary innervation pathways (Baer 1988a).

Lyssavirus replication is believed similar to that of other negative-stranded RNA viruses that have been more intensively studied, but genus-specific peculiarities are likely to be uncovered eventually because of their relatively restricted CNS niche. After exposure, the precise mechanism of entry of virus into the nervous system is controversial. Some data support the assertion that virus binds near the acetylcholine receptor via neuromuscular junctions (Lentz et al. 1983; Spriggs 1985; Baer et al. 1990). These sites are in close proximity to unsheathed nerves at synaptic clefts where virus can gain access to the axoplasm. This hypothesis does not preclude alternative mechanisms, because other protein-based receptors can also bind virus, and acetylcholine receptors are not exclusively found at sites permissive of virus infection (Reagan and Wunner 1985).

Following viral entry, usually through a bite or less frequently via mucous membranes, attachment to cell membranes can occur by conformational changes in the G protein during receptor-mediated fusion (Flamand et al. 1995). After adsorption, particles may penetrate the cytoplasm within an endosome and uncoat to RNP under low-pH conditions (Gaudin et al. 1993). In ribosome-rich locations, such as in the perikaryon or proximal dendrite, the viral RNP may initiate primary transcription. After synthesis, replication of the genomic RNA continues with the production of positive-stranded RNA, acting as a template for progeny negative-stranded RNA. Within the CNS, Lyssavirus replication has been associated almost exclusively with neurons, and there is little evidence to suggest replication in other cells. Viral assembly, and budding of the virion from the infected cell, may occur by the reverse process of initial viral attachment. Notwithstanding that a high proportion of some viral amino acid substitutions may be tolerated while still retaining molecular fidelity, even small alterations at certain antigenic locations may be critical in the determination of function, in part because of molecular influences on secondary protein structure (Coulon et al. 1994). In fact, as little as a single amino acid substitution can have substantial impact on viral virulence (Dietzschold et al. 1983).

The incubation period of rabies, usually 1–3 months following exposure, may vary from as little as several days to several years (Smith et al. 1991), but it is rarely more than 6 months. In nature, this sporadic, unusually slow course of infection in an individual host may assure rabies virus survival in vivo until new susceptible generations gradually reach a density supportive of efficient transmission at the population level. Longer incubation periods may also allow the selection of a specific virus population that can effectively multiply in the salivary glands and be readily excreted in the saliva (Charlton et al. 1984).

The precise location of viral residence during the incubation period is undetermined. Virus may replicate in muscle tissue during early stages of infection, prior to entry into nerves, or it may remain at the site of entry for a prolonged time (Baer 1988a). However, virus replication in muscle cells prior to entry of neurons is not required for development of CNS infection (Shankar et al. 1991). Obviously, these somewhat conflicting observations explain neither the site, the form, nor the mechanism by which virus can exist after exposure during an eclipse phase prior to CNS replication. A number of factors, alone or in combination, may be operative. For example, virus may be sequestered at the original site of entry, undergoing minimal or no replication until a stimulus triggers viral activity in proximity to a peripheral nerve. Virus may persist at the level of the dorsal root ganglia and involve viral RNP, with little replicative activity due to defects in the available biochemical milieu. In contrast, sequestration sites may be nonneural in origin (Ray et al. 1995). Additionally, if overt disease has an immunologic basis, the incubation period may equate to a dynamic stasis perhaps related to delayed immunologic surveillance via antigen-presenting cells. In theory, inocula may also contain defective interfering particles that limit neuronal access by virions until after some variable delay period. Regardless of mechanism, unusually long incubation periods tend to be the exception.

Once CNS neurons become infected, the virus disseminates rapidly along neuronal pathways (Iwasaki 1991; Charlton et al. 1996). Infection spreads to the brainstem, deep cerebellar nuclei, Purkinje cells, and neurons in the cerebral cortex. Involvement of the hippocampus may occur relatively late, primarily infecting pyramidal neurons with little involvement of the dentate gyrus. There also appears to be a predilection for the limbic system, the reticular formation, the pontine tegmentum, and the nuclei of the cranial nerves at the floor of the fourth ventricle. At the height of CNS infection, the virus may traverse back to the peripheral nerves and centrifugally invade highly innervated areas such as the cornea, skin (especially of the head and neck), salivary glands, tonsils, and buccal membranes of the oral cavity (Schneider 1991). There is pronounced secretion of the virus into saliva at a time when agitation and aggressive biting behavior may be present, increasing the odds of viral transmission.

Late in infection, rabies virus may occasionally be found in tissues other than the CNS and salivary glands, such as the adrenal medulla and nasal glands (Balachandran and Charlton 1994). Terminally, practically all innervated organs contain virus (Charlton 1994).

Rabies has one of the highest case-fatality ratios of any infectious disease and is usually considered to be invariably fatal. However, survival from rabies, usually with severe neurologic sequelae, has been documented in a few naturally or experimentally infected cases (Fekadu 1991).

A primary difference in the pathogenesis of rabies between reservoir hosts and dead-end or victim species may not be found. Rather, the explanation for the propensity of particular taxa to fall into these categories may be associated with factors other than the nature of the infection itself and the resultant probability of viral excretion in the individual animal (Wandeler et al. 1994). Some species may be more or less likely to withstand overt aggression by a rabid individual, and small prey species may not easily survive attack by larger predators. Distinct behavioral patterns of small mammal and ungulate species, such as secretive activity or wariness of and ability to escape from rabid predators, may influence the dynamics of rabies. In addition, anatomic differences can affect the ability of an animal to serve as a reservoir, including the effective penetration of virus-laden saliva deep into muscle via the canines and carnassial teeth; the extreme contrast between Carnivora and Edentata is obvious (Leffingwell and Neill 1989).

Rather than simple destruction of cellular elements, select organic dysfunction, owing to pathophysiologic disruption in critical neuronal activities and neurotransmitter imbalance (Tsiang 1993) or programmed cell death as an attempted response to limit infection (Jackson and Rossiter 1997), may ultimately explain the underlying mechanisms of virulence. Also, recognition is needed of the functional interaction between the immune system and the CNS during health and disease. For example, during rabies virus encephalitis, brain receptors for some cytokines may decrease while local concentrations are increasing; moreover, copious amounts of nitric oxide may be produced locally in the CNS in response to viral infection (Koprowski et al. 1993).

PATHOLOGY. No in-depth clinical chemistry investigations have been reported for naturally infected wildlife, and the direct relevance of human rabies data to free-ranging wildlife is questionable. Results of routine clinical laboratory tests are usually nonspecific. Glycosuria may be detected as may prominent lymphocytic pleocytosis of the CSF (Hanlon et al. 1989).

Despite its acute nature, often bizarre clinical manifestations, and virulence, the pathologic findings associated with rabies are usually relatively mild (Rupprecht and Dietzschold 1987). Depending on the disease course, there may be evidence of multiple fresh or healed bite wounds; gross trauma from self-mutilation; foreign bodies in the mouth, esophagus, and stomach; the gastrointestinal track may be devoid of contents; and, in geographic areas where they occur, porcupine quills may be in the muzzle. Results of external examination may be normal, except for an unusual odor (Hubbard 1985), perhaps due to poor hygiene in terminal disease. Typically, few internal gross pathologic changes are visible, other than mild cerebral edema, focal meningeal congestion and thickening, or subtle suggestions of hemorrhagic myelitis.

Nonspecific inflammation of variable severity can be observed in the brain, spinal cord, and ganglia, or it may be lacking (Hamir and Rupprecht 1990a). Severity of infection and, to some extent, the location of pathologic changes may be proportional to the duration of illness. Inflammatory infiltrates, consisting primarily of lymphocytes, macrophages or, rarely, eosinophils may be in the leptomeninges (Hamir and Rupprecht 1990b). Lesions of nonsuppurative encephalomyelitis, including perivascular cuffing, focal to diffuse gliosis, neuronophagia, and rare to modest neuronal necrosis, tend to be sparse in relation to the distribution of viral antigen detected by immunofluorescence (Perl and Good 1991).

Standard light microscopy is focused on identification of Negri bodies which are eosinophilic intracytoplasmic inclusion bodies in infected neurons. Negri bodies are haloed, and may be round, ovoid, ameboid, triangular, or oblong and pink to purple, depending on the stain. For example, Seller's stain produces classic magenta bodies with dark interior basophilic granules arranged in a rosette pattern (Velleca and Forrester 1981). Negri body development is usually directly related to duration of illness. These inclusions are detected in 50%–75% of specimens positive by virus isolation, immunofluorescence, or electron microscopy. Negri bodies are seen most frequently in Purkinje cells in the cerebellum and pyramidal cells in the hippocampus, and less often in medulla oblongata, spinal cord, cerebral cortex, basal ganglia, and peripheral ganglia.

Spongiform lesions of the gray matter, affecting the neuropil and neuronal cell bodies of the thalamus and cerebral cortex, have been documented in experimental and natural rabies cases (Charlton 1984). These should not be confused with lesions of transmissible spongiform encephalopathy (Bundza and Charlton 1988).

In rabies cases, typical rod- or bullet-shaped viral particles are observed most often in CNS and salivary glands (Gosztonyi 1994). Mature virions are approximately 75–80 × 180–200 nm. Negri bodies identified by light microscopy consist of viral nucleocapsids and correspond to matrices seen en masse by ultrastructural microscopy. These matrices correspond to the site of virus replication. Neurons containing viral matrices may have large numbers of viral particles budding from membranes of the rough endoplasmic reticulum and the plasma membrane.

DIAGNOSIS. No clinical signs are pathognomonic for rabies, and postmortem laboratory examination for evidence of lyssaviruses is the only method of definitive diagnosis in animals (Smith 1995). Antemortem methods for detection of virus by skin biopsy, corneal impressions, saliva collection, etc., may be used to confirm a clinical suspicion, but a negative test result does not rule out the possibility of rabies (Blenden et al. 1983).

Given its public health, veterinary, and wildlife management implications, rabies is usually considered a reportable disease, and a rapid, accurate, and inexpensive diagnostic method is highly desirable. Suspect animals should be euthanized in a manner so as not to damage the brain. The head, brain, or carcass of small mammals such as bats should be sent under refrigeration to the diagnostic laboratory. Freezing is not recommended, because it may delay testing until samples thaw. Alternatively, brainstem cores may be sampled in place of the entire organ (Hirose et al. 1991). In field surveillance programs, particularly in tropical regions, samples may be preserved in a 50% saline-glycerine solution.

The FAT for rabies, originally developed by Goldwasser and Kissling (1958) and now refined (Velleca and Forrester 1981; Trimarchi and Debbie 1991), is a rapid, sensitive, and specific diagnostic method used globally. Slide impressions or smears of brainstem, hippocampus, and cerebellum are immersed in fixative (usually cold acetone), before the addition of either fluorescein-labeled monoclonal or polyclonal antirabies virus reagents, and are examined by direct fluorescence microscopy. Typically, reliable results can be obtained in 2–4 hours. Test sensitivity is affected by tissue decomposition and a number of other variables. Use of salivary gland instead of brain to detect viral antigen is not recommended. The FAT on brain can also be used after formalin or other chemical fixation, but it is not

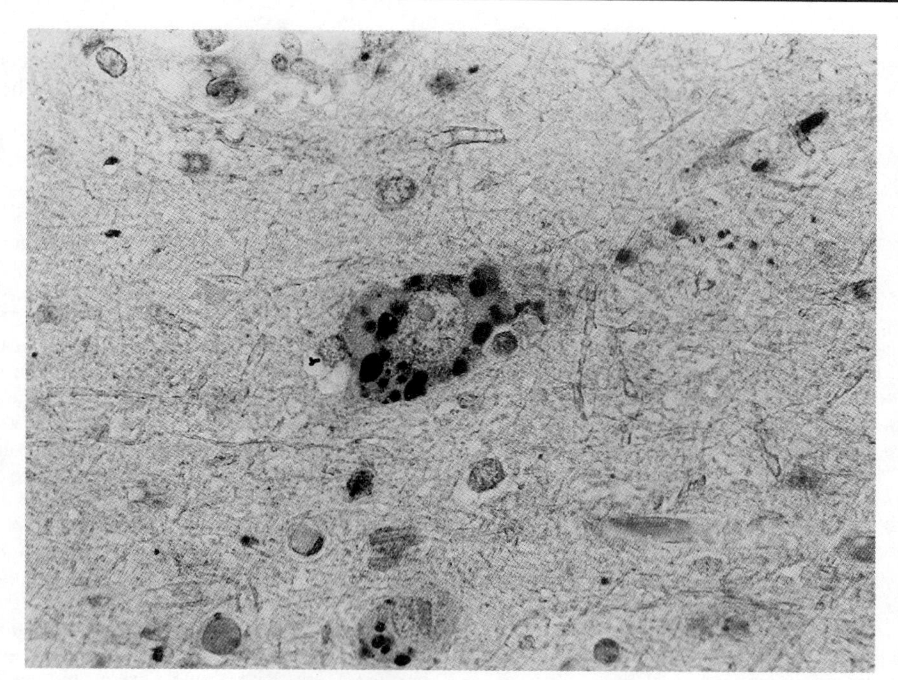

FIG. 1.7—Rabies virus intracytoplasmic inclusions (400×) in a Purkinje cell of an infected striped skunk *Mephitis mephitis,* stained by immunohistochemistry (Courtesy of M. Niezgoda).

routine, and special techniques are necessary for paraffin-embedded sections (Warner et al. 1997). Besides the FAT on formalin-fixed tissue, immunohistochemical methods of diagnosis, employing either a peroxidase or an avidin-biotin complex and Mabs (Fekadu et al. 1988; Hamir et al. 1992, 1996b), can also reliably detect rabies-specific inclusions by light microscopy (Fig. 1.7).

Routine histologic examination of paraffin-embedded brain tissues may demonstrate nonsuppurative encephalitis and Negri bodies. However, microscopic examination for Negri bodies alone may produce false-negative results because not all cases develop inclusion bodies, or false-positive results may be due to nonspecific inclusion bodies (Nietfeld et al. 1989).

Animal inoculation (Koprowski 1996) or cell culture (Rudd and Trimarchi 1987; Crick and King 1988; Sureau et al. 1991) can be used for virus isolation. These techniques are used infrequently.

A variety of serologic procedures are available for detecting antibodies to lyssaviruses (Smith 1991), such as the rapid fluorescent focus inhibition test (Smith et al. 1996) for detection of virus-neutralizing antibodies (VNAs), or enzyme-linked immunosorbent assays (ELISAs) (Barton and Campbell 1988). Demonstration of rabies VNAs in serum only indicates exposure to viral antigen. Antibodies are rarely detected until late in the clinical course of disease. Conversely, induction of VNAs may be evidence of immunity, either from natural exposure or to vaccination. Rabies VNAs are not produced in the CSF from vaccination; their presence in association with suggestive illness supports a diagnosis of rabies. The rare discovery of antibodies in the CSF of a healthy animal is strongly suggestive of prior illness and recovery (Fekadu 1991).

Modern molecular techniques, such as RT/PCR for detection of lyssavirus nucleic acid (Sacramento et al. 1991; Smith et al. 1992), may serve as an adjunct or confirmatory test to routine rabies diagnostics. However, the extreme sensitivity of these techniques greatly increases the probability of false-positive results due to laboratory contamination or technical error. Additionally, because of the diversity of lyssaviruses, false-negative results can occur if initial primer selection is inadequate to compensate for sequence heterogeneity. Moreover, universal primers for all known lyssaviruses have not been identified (Nadin-Davis et al. 1996). Considering these factors, related costs, and the research expertise necessary for proper analysis and interpretation, such molecular techniques are not recommended for routine primary rabies diagnosis at the present time (WHO 1995).

Given the aforementioned diagnostic tools, a case definition of rabies is supported by either virus isolation, detection of viral antigen or nucleic acid, or demonstration of rabies-specific antibodies in the CSF in a clinically suspect animal.

DIFFERENTIAL DIAGNOSES. Rabies should be strongly considered among the differential diagnoses of any suspected mammalian encephalitis, especially in such high-risk taxa as the Carnivora and Chiroptera. Depending on the species in question, rabies may be clinically confused with a number of acquired conditions: viral (e.g., canine distemper, infectious canine hepatitis, pseudorabies, feline infectious peritonitis, feline panleukopenia, malignant catarrhal fever, herpes myelitis, equine encephalitides, or borna disease), bacterial (e.g., botulism, tetanus, or listeriosis), mycotic (e.g., cryptococcosis or blastomycosis), protozoal (e.g., toxoplasmosis or coccidiosis), helminth (e.g., baylisascariasis, parelaphostrongylosis, or verminous encephalitis), or others (e.g., chronic wasting disease and other transmissible spongiform encephalopathies, tick paralysis, intervertebral disc disease, degenerative encephalomyelopathy, or lymphosarcoma). Rabies may be mimicked by a multitude of other diseases whose origins may be traumatic (e.g., gunshot, hit by car, esophageal foreign body, or petrous temporal fracture), toxic (e.g., heavy metals, chlorinated hydrocarbons, or lathyrism), or physiometabolic (e.g., ketosis, polioencephalomalacia, or ischemic encephalopathy).

IMMUNITY. Immunity to rabies involves nonspecific defenses, such as intact, heavily furred, or armored skin that aids in protection against bites, and specific induced responses (Lodmell 1983; Xiang et al. 1995). Viral exposure may or may not lead to a productive viral infection, which may or may not result in detectable immune responses (Niezgoda et al. 1993). The ultimate outcome depends in part on complex interplay of viral and host factors (Nathanson and Gonzalez-Scarano 1991). During the centripetal transport of rabies virus to the brain, especially during long incubation periods, insufficient antigenic mass may be detected by the immune system to induce an appropriate response. In contrast, highly neural invasive strains might be quite immunogenic, but the host response may be slow or inadequate to prevent or clear CNS replication, given an immunologically privileged location of replication. Rabies-specific VNAs may be detected in sera at the onset of illness, but they usually are not detected, if at all, until the terminal stages of disease. These VNAs may interfere with salivary excretion of infectious virus (Charlton et al. 1987). Antibody detection in the CSF of a rabies-suspect animal is considered a reliable indication of present or past CNS infection.

The role of acquired immunity to rabies in naturally exposed wildlife is not known. Red foxes seldom have VNAs, generally less than 4%–7% of surveyed populations (Blancou et al. 1991), which is similar to the seroprevalence of 2%–10% in some insectivorous bats (Trimarchi and Debbie 1977). In epizootic areas, however, presence of VNAs varied from 1% to 3% (Winkler and Jenkins 1991) to more than 20% (Bigler et al. 1983) of raccoons sampled. In Grenada, presence of rabies VNAs ranged between 9% and 55% of surveyed mongoose, and seroprevalence was inversely proportional to the number of rabid mongoose reported. Mongoose with preexisting VNAs responded with higher titer on rabies vaccination (Everard and Everard 1988). The foregoing observations suggest that natural immunity to rabies varies (Rosatte and Gunson 1984; Black and Wiktor 1986; Orr et al. 1988; Follman et al. 1994) from essentially none to a potentially critical mechanism responsible for a robust host-parasite equilibrium (Hill and Beran 1992; Hill et al. 1992, 1993), depending in part on the species and virus variant (Niezgoda et al. 1997).

Induction of protective antiviral immunity is primarily based on response to the G protein. This protein has received considerable study because it forms protrusions that cover the outer surface of the viral envelope and is responsible for reception on the cell membrane and induction of pH-mediated endocytosis (Wunner 1991). It is the only lyssavirus protein known to induce specific VNAs. It also plays a role in eliciting cell-mediated immunity. However, there is not always a clear relationship between level of VNA and resistance to rabies, suggesting that other antigens and immune effector mechanisms are likely involved in protection against lethal virus infection.

Identification of the relative contributions of the complex lyssavirus antigens in humoral and cellular immunity has been slowly delineated and generally based on human immune responses (Celis et al. 1986, 1988; Ertl et al. 1989, 1991; Herzog et al. 1991; Perry and Lodmell 1991). Intact virus and purified viral N, G, NS, and μ proteins can induce high lymphokine secretion. Additionally, T-lymphocyte clones of the helper/inducer class may react not only with classic fixed rabies viruses but also with Duvenhage and Mokola viruses, and many recognize viral RNP determinants. Rabies cytotoxic T-cell responses occur against antigenic determinants of both RNP and G proteins. T-cell clones which react to several lyssaviruses may recognize closely situated epitopes presented in the context of the same major histocompatibility complex molecule.

Certain internal lyssavirus antigens may possess an inherent capacity to enhance immune responsiveness. These have been characterized either as superantigens (Lafon 1994) or as powerful adjuvants (Hooper et al. 1994). In the absence of other antigens, rabies RNP is capable of inducing protection against virus challenge without induced VNAs and can confer immunity against infection by heterologous lyssaviruses (Dietzschold et al. 1987).

Exploiting immunologic advances, modern rabies vaccines maximize potency, purity, safety, and efficacy and minimize cost. Although older modified-live virus vaccines were effective, did not require adjuvant, and induced long-lasting immune responses that involved a full spectrum of immune effectors, they also traditionally carried the risk of vaccine-induced disease especially in the immunocompromised host. Conversely, inactivated animal neural tissue vaccines eliminated the potential for vaccine-associated rabies, but were traditionally hampered by low potency and adverse reactions to myelin basic proteins. Newer vaccines rely on advances in immunobiology to correct these drawbacks. Generation of viral escape mutants under neutralizing Mab pressure in vitro has proven useful in the development of apathogenic rabies virus vaccines that are effective in oral administration (Schumacher et al. 1993). Research during the past several decades has been toward inactivated tissue culture rabies vaccines. In addition, both G and N protein genes have been effectively expressed in a number of prokaryotic and eukaryotic systems for parenteral or oral application (Rupprecht et al. 1986; Taylor and Paoletti 1988; Prehaud et al. 1990; Prevec et al. 1990; Fu et al. 1993; McGarvey et al. 1995; Lodmell et al. 1996; Yarosh et al. 1996). Future expression of viral antigens in transgenic plants (Yusibov et al. 1997) may offer additional opportunities for economic production of vaccines.

CONTROL. In most countries, the rationale for wildlife rabies control stems from public health needs, agricultural concerns, and the often poorly documented perception of direct economic burden caused by rabies. Rabies remains in a population when each individual can infect, on average, one other susceptible member of that population. General approaches to controlling wildlife rabies and its impact are 1) elimination of the reservoir, 2) elimination of rabies in the reservoir, or 3) protection of victim species from infection via a reservoir. These strategies are not mutually exclusive. However, attempting rabies control by true eradication of a particular wildlife host is impractical, expensive, and has ecological consequences, which are usually difficult to predict reliably. In addition, the concept of species eradication is ethically unacceptable in many cultures. The second approach includes wildlife vaccination. The third strategy is used in human and veterinary medicine.

Attempts at controlling wildlife rabies have traditionally centered on population reduction of specific reservoirs below a threshold necessary to maintain the disease. Methods frequently used to attempt this goal in carnivores often involved hunting, including night shooting and trapping and, to a much lesser extent, den gassing with hydrocyanic acid or poisoning with strychnine. In general, hunting, poisoning, or den gassing alone proved inefficient for sustainable population reduction (Toma and Andral 1977; Debbie 1991).

In Europe, fox rabies was projected to be eliminated in an area if 0.3–0.5 foxes/km^2 per year could be removed by constant hunting pressure (Steck and Wandeler 1980). This figure was later confirmed by a deterministic compartmental model (Anderson et al. 1981). However, the usefulness of this figure still remains questionable (Blancou et al. 1991) because hunting intensity is often subject to significant spatial and temporal variability and carrying capacity for the fox may differ significantly in various habitats (Ziemen 1982; Bacon 1985).

In endemic areas, the combined effects of rabies and intensified hunting were usually not efficient enough to reduce the density of foxes below the threshold density. The same applied if single den-gassing campaigns were conducted at the peak of an outbreak (Toma and Andral 1977; Ziemen 1982). Comprehensive den-gassing campaigns were only successful in some areas in combination with intensive hunting or when simultaneous rampant rabies similarly reduced the fox population. In addition, gassing had to be applied on a large enough scale to prevent the rapid repopulation from adjacent areas (Müller 1971, 1992; Wandeler et al. 1974; Westergaard 1982).

Failure to control wildlife rabies can occasionally have direct local agricultural impacts that may even outweigh basic public health considerations (Arellano-Sota 1988; Lord 1992); vampire bat rabies causes substantial livestock mortality in Latin America. The vampire bat is so specialized as to appear vulnerable to technological attack. Factors favoring the feasibility of *Desmodus* elimination include an entirely hematophagous diet and adaptations to domestic animals as preferred prey; a low reproductive rate, favoring population reduction; colonial habits that concentrate individuals; social habits of mutual grooming that favor poisoning by anticoagulants applied to the fur; a high susceptibility to such toxins; apparent minimal ecological sequelae after elimination; and an obvious cost-benefit ratio via institution of control measures (Acha and Alba 1988). However, while it may be possible to suppress local populations significantly in cattle-ranching areas by intense coordinated programs (Lord 1988), true eradication of vampire bats is unlikely because of their great adaptability to a variety of niches, some totally unrelated to livestock. Moreover, even vampire bats possess attributes of direct human medical benefit (Hawkey 1988). Both the pros and cons of a systematic campaign to eliminate vampire bats have been described by Brass (1994).

In contrast, basic logistics of attempts to reduce numbers of small carnivores, such as yellow mongoose in Africa, are substantial, even ignoring ecological consequences of such a project. For many years, animal health departments in South Africa maintained teams that attempted mongoose control on farms where rabies occurred. When Zumpt (1982) showed that repopulation of gassed warrens was initiated within months and was complete within 3 years, the ever-escalating cost of this futile policy was recognized and it was abandoned.

Similarly, jackal-borne rabies control also presents a formidable task. Jackals possess many characteristics that are not seemingly amenable to eradication plans: an omnivorous diet; a high reproductive rate; large home ranges, whereby nonselective poisoning attempts would probably affect nontarget species; solitary habits and adaptability requiring individual targeting; apparent incalculable environmental effects from removal; and enormous projected costs of control campaigns, with cost-benefit ratio that appears quite poor. Thus, a poisoning campaign that was maintained for many years at great expense financially and ecologically in the northern Transvaal bushveld aimed at reducing numbers of *C. mesomelas,* was ultimately stopped when it became clear that control of rabies was not being achieved.

Drawbacks to population reduction alone demanded new solutions, such as the application of oral vaccination of wildlife against rabies (Winkler and Bögel 1992) after its conception at the CDC nearly three decades ago. The idea is based on the induction of herd immunity and is similar to any other vaccination technique: increase resistance in the vaccinee and/or interruption of the infectious chain in a given population (Baer 1988b). However, effectively immunizing free-ranging animals was easier in concept than practice. Wildlife are generally inaccessible by conventional delivery systems; often the knowledge of size, density, spatial distribution, dispersal, and demographics of the target population in the vaccination area is limited; there are uncontrollable environmental impacts on the vaccine and the delivery system; diverse nontarget species, including humans, have potential access to the vaccine; and objective evaluation of efficacy is difficult. Much of the available data on wildlife rabies vaccination were generated through studies of red fox and, more recently, other species.

Parenteral vaccination via manual inoculation or by traps, intranasal vaccination by aerosol trap or den fumigation, and enteric exposure to vaccine in baits were tested on captive and wild foxes. These approaches were not economically or technically viable on a large scale (Wandeler 1991). Oral immunization became possible only with discovery that a modified-live SAD rabies virus strain immunized foxes when given in the mouth (Baer et al. 1971).

The WHO coordinated laboratory research and field application of rabies vaccines for oral wildlife immunization (Wandeler 1991; Winkler 1992). A unique attribute of the evolution of oral vaccination techniques was international cooperation in research and field application (WHO 1989, 1992a; Bögel 1992). In general, vaccines and baits for oral immunization of wildlife against rabies were expected to share a number of characteristics (Wandeler 1988, 1991; Wandeler et al. 1988) and a minimum set of tests were suggested to establish safety and efficacy (WHO 1989, 1992a). Intentional release of modified-live or recombinant viruses is not a trivial undertaking and involves considerable environmental assessment. Such research includes developing models of virus spread and safety in target and nontarget species; study of the potential alteration of host specificity and/or pathogenicity with passage; occurrence of recombination; and possible changes in antigenic properties, due to selective pressure of an immune host population. Data obtained in numerous laboratories and extensive field investigations indicated that recombinant rabies vaccines are safe and efficacious alternatives to modified-live rabies viruses. Studies over the past 14 years have shown that the prototypic vaccinia recombinant vaccine was not pathogenic for more than 50 vertebrate species, including nonhuman primates (Rupprecht et al. 1992a) and immunocompromised animals (Hanlon et al. 1997) and was highly efficient in the field (Aubert 1994; Brochier et al. 1994).

The first field trial with a modified-live rabies vaccine was conducted in Switzerland in 1977. This original experiment showed that the virus would not become established in small mammals in the trial area (Wandeler et al. 1988); free-ranging foxes were orally vaccinated beginning in 1978. The SAD-Bern virus vaccine was placed in sealed plastic blister containers under the skin of chicken heads and was distributed at the entrance of the Rhone Valley in Switzerland, an experiment that stopped the impending rabies outbreak.

The only vaccines used in the field before 1988 were the modified-live SAD-Bern and SAD-B19 strains (Häflinger et al. 1982; Steck et al. 1982; Wandeler et al. 1982; WHO 1980, 1982, 1987, 1988, 1989; Schneider et al. 1983, 1987, 1988, 1989). Five other vaccines have since been applied in the field: a vaccinia-rabies glycoprotein (V-RG) recombinant virus vaccine (Kieny et al. 1984; Wiktor et al. 1984, 1985; Blancou et al. 1986; Rupprecht et al. 1986, 1992b, 1993; WHO 1989, 1991, 1992a; Artois et al. 1990); SAG-1, an attenuated vaccine constructed from a SAD-Bern escape mutant selected in the presence of an anti-G protein Mab (Seif et al. 1985; Prehaud et al. 1988; Tuffereau et al. 1989; Le Blois et al. 1990; Tolson et al. 1990; Benmansour et al. 1991; Guittre et al. 1992; Coulon et al. 1992; Flamand et al. 1993); SAG-2, which is replacing SAG-1 (Schumacher et al. 1993; WHO 1993, 1994b; Lafay et al. 1994); a modified-live vaccine developed in Germany, SADP5/88 (Kintscher et al. 1990; Sinnecker et al. 1990; Stöhr et al. 1990, 1994; WHO 1993, 1994b); and SAD-Vnukovo, another modified-live virus vaccine used extensively in the former USSR (Kovalev et al. 1992; Svrcek et al. 1994, 1996; Süliova et al. 1997; Ondrejka et al. 1997) (Fig. 1.8).

To be effective, vaccine must be brought into contact with oral and/or pharyngeal mucosa in a sufficient number of target animals via a bait. In part, size, texture, shape of bait, and vaccine container, as well as physical and chemical characteristics, determine whether sufficient antigenic mass reaches the immunologically competent regions of the oral cavity. Several systems were developed for vaccine-bait delivery that took into account bait density, distribution method (manual, aerial, or both), sequence and frequency of bait distribution, season, selection of specific baiting areas, strategies for expansion of baiting areas, and the overall duration of vaccination campaigns.

All baits currently used for oral vaccination in Europe are machine manufactured. Today, most European countries give preference to aerial distribution by fixed-wing aircraft or helicopters with manual distribution mainly limited to areas that are difficult to treat by air. In general, fox vaccination campaigns in Europe are conducted at least twice a year in spring and autumn at a bait density of 13–20/km^2. The success of oral immunization campaigns in countries such as Austria, France, Germany, and Switzerland, where most

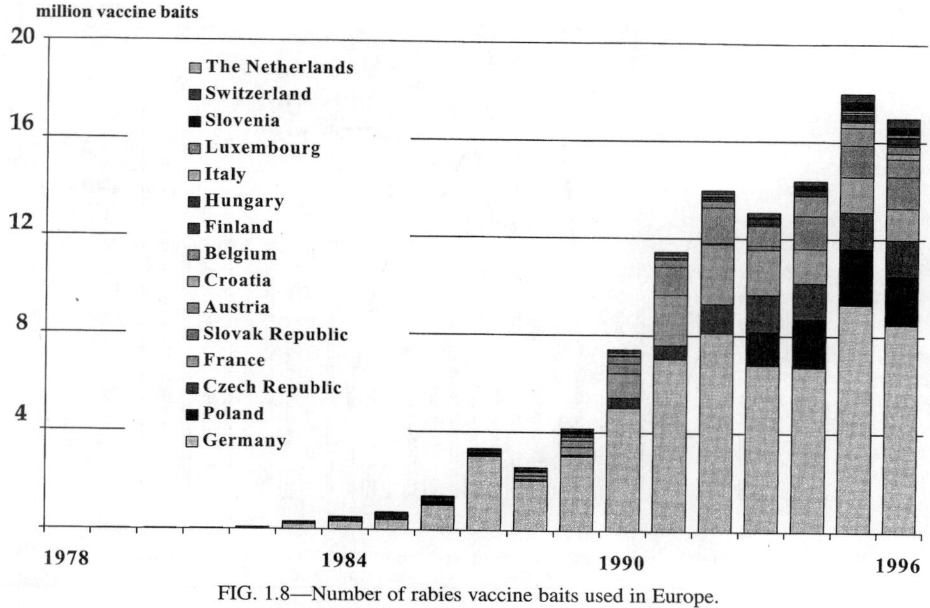

FIG. 1.8—Number of rabies vaccine baits used in Europe.

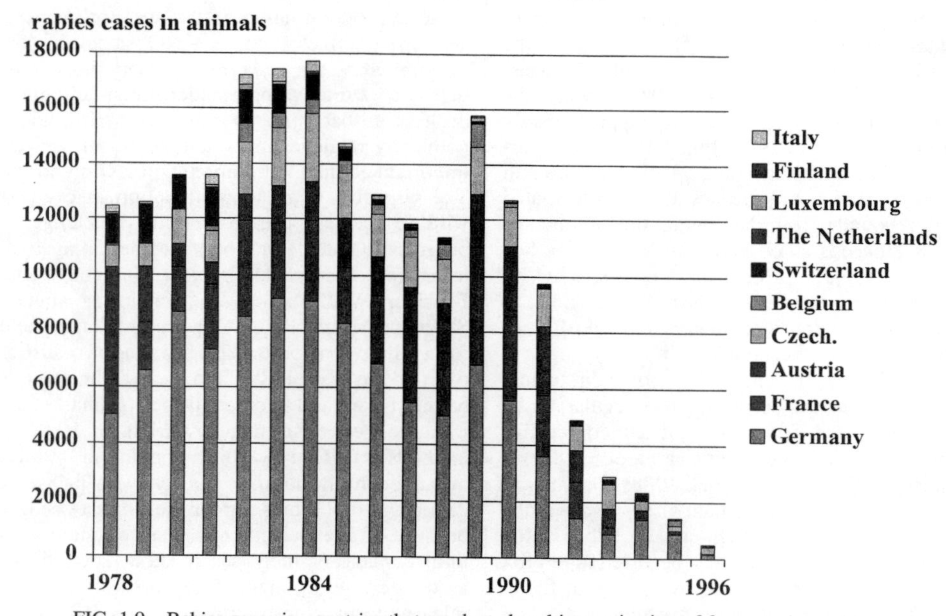

FIG. 1.9—Rabies cases in countries that conducted oral immunization of foxes against rabies.

baits were air dropped, is indirect proof of the efficacy of aerial distribution under the ecological and economic conditions in Europe (Fig. 1.9). Since 1978, more than 110 million baits have been produced in Europe and spread over about 6.1 million km². A turning point toward a goal of eliminating terrestrial rabies from continental Europe occurred in 1989–90 when a number of countries moved toward vaccination over larger connected areas. Almost 17 million baits were distributed in Europe during 1996 (Stöhr unpublished). Costs of vaccine baits have totaled approximately $116 million US since 1978.

Oral immunization greatly reduced the number of rabies cases (Fig. 1.9), but has not yet eliminated the disease. An increase in fox populations following a decrease of rabies was predicted. Based on figures supplied by

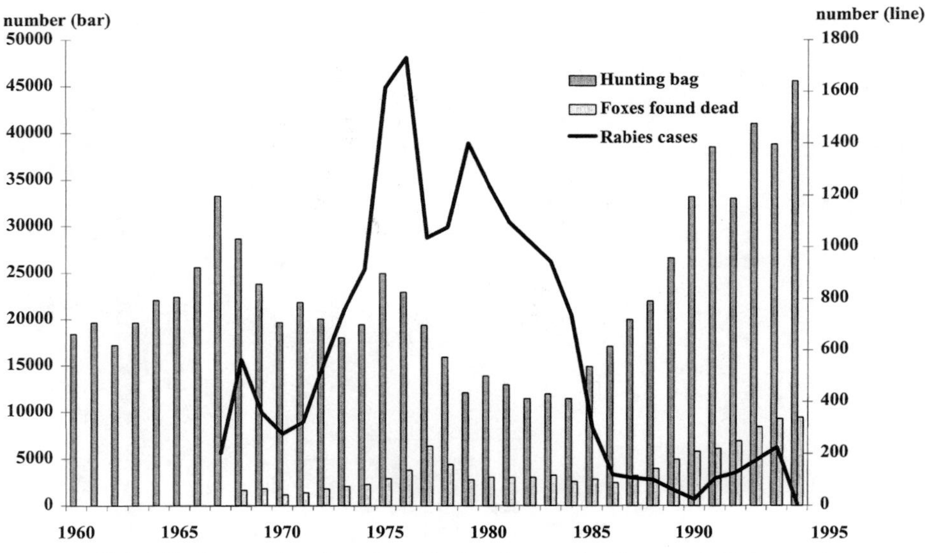

FIG. 1.10—Hunting bag, foxes found dead, and reported rabies cases in Switzerland.

the Federal Directorate of Forestry, Hunting and Wildlife Research Section, Federal Bureau of Environment, Water, and Landscape, Bern, Switzerland, the number of foxes harvested more than tripled between 1981 (N = 12,938) and 1993 (N = 40,993) when the number of rabies cases decreased from approximately 1800 in 1976 to only 6 in 1996 (Fig. 1.10). Thus, coordinated surveillance activities and vaccination are still vital, as rabies incidence decreases and the vaccinated population rebounds. If reduction in the number of rabies cases proceeds as observed during the last few years, large parts of Europe may be expected to become free from rabies in the near future and it is hoped that continental elimination of the disease among foxes will soon become a reality.

North America is the only other continent where wildlife vaccination has occurred on a regular basis, commencing with the first release of a modified-live ERA rabies virus vaccine (Lawson et al. 1987) in Canada during 1985 (Johnston et al. 1988; MacInnes et al. 1988) and the V-RG recombinant virus release in the United States in 1990 (Rupprecht et al. 1992b; Hanlon et al. 1998). Substantial success has occurred in red fox rabies control in southern Ontario by the use of fixed-wing aircraft delivery of baits and parenteral application of inactivated rabies vaccine in urban areas (Rosatte et al. 1993, 1997). By contrast, much slower progress has been observed in the United States (Hanlon and Rupprecht 1998), where the red fox is no longer a major reservoir; field trials have commenced in New Jersey, Massachusetts, New York, Florida, Vermont, and Ohio for raccoon rabies control (Hanlon et al. 1993, 1996; Mitchell and Heilman 1996; Robbins et al. 1996; Roscoe et al. 1996) and in Texas for coyotes and gray foxes (Fearneyhough et al. 1996), but no oral rabies vaccination program has been initiated for skunks, which are largely refractory to the V-RG vaccine.

Eliminating the virus from a reservoir by immunization is an attractive proposition. Such an approach by itself is probably not broadly achievable within economically realistic limits and should be viewed as an important adjunct to traditional rabies control methods. The objective of rabies elimination in even a single wildlife species poses a daunting challenge. Despite progress to date of oral-bait vaccination against rabies in red fox (Blancou et al. 1988; MacInnes et al. 1988; Brochier et al. 1994; Wandeler 1991; Aubert 1994; Stöhr and Meslin 1996), total elimination of the disease from whole continents is probably not a realistic objective (Coyne et al. 1989), and these techniques may not readily be applied to other species (Linhart et al. 1993) in a cost-effective manner (Meltzer 1996). Newer methods of local wildlife population reduction or immunocontraception, in concert with habitat modification and oral rabies immunization, may be important because of unexpected consequences, such as agricultural damage or increases in other zoonotic diseases associated with changes in furbearer populations released from rabies mortality (Fig. 1.10).

PUBLIC HEALTH CONCERNS. Prevention of human rabies is a great public health challenge for most areas of the world. Although many developed countries have successfully controlled domestic animal rabies, the disease remains a major health risk for people in most developing countries (Meslin et al. 1994). There is no effective treatment for clinical rabies, but prompt postexposure prophylaxis (PEP) will almost always prevent disease. With few exceptions, infected animal bites are the ultimate sources of all known human rabies cases. Extremely uncommon nonbite routes of infection include scratches; contamination of mucous membranes; aerosols in laboratory accidents or, rarely, bat cave exposure; corneal transplants; or iatrogenic causes (Fishbein 1991; Hemachuda 1994).

The magnitude of human rabies fatalities worldwide is essentially unknown. Only a fraction of the reported cases of human rabies occur outside the tropic and subtropic regions. Annually, approximately 50,000–100,000 cases of human rabies may occur worldwide, some 5–10 times the number of reported cases (WHO 1996). In the developed world, the economic and emotional impact of evaluating exposure for rabies prophylaxis exacts a much greater toll than the disease itself. Even in the absence of direct mortality, rabies results in numerous expenditures, including expensive human vaccination, a complicated medical infrastructure, and lost wages.

Preexposure rabies prophylaxis is appropriate for high-risk groups, such as wildlife biologists; veterinarians; animal control officers; animal handlers; and rabies research, production, and diagnostic laboratory workers (CDC 1991; WHO 1992b). In addition, any person who is likely to come into contact with potentially infected wildlife, such as spelunkers who explore caves where there are large bat colonies, and persons spending extended periods (longer than 30 days) in direct contact with mammals in Africa, Asia, or Latin America, where medical care may not be readily available, should consider preexposure vaccination.

Modern preexposure vaccination consists of a licensed cell culture rabies vaccine, given intramuscularly or intradermally, on days 0, 7, and 21 or 28. Booster doses are recommended depending on the results of serologic testing (Briggs and Schwenke 1992) and the relative risk of continued exposure. The recommended period between serologic testing is every 2 years for individuals with frequent potential exposure. No routine serologic testing or boostering is needed with an infrequent risk, such as workers in an area of low rabies endemicity. Preexposure vaccination simplifies postexposure management of high-risk groups, without the necessity of costly treatment with expensive rabies immunoglobulin (RIG), and may provide protection against unrecognized exposure.

The issue of rabies PEP is raised when humans are bitten by wild animals. Consultation with local or national public health officials knowledgeable about regional rabies epidemiology is highly encouraged before PEP initiation. Knowledge of the epidemiology of rabies in local reservoirs is important in risk assessment. If exposure is due to an animal species that is essentially rabies free in the wild (e.g., all nonmammals), then the patient should have appropriate local wound care with reassurance that no rabies PEP is required.

Wild mammalian rabies suspects should be euthanized immediately because there is no recommended observation period as with domestic species. Although naturally occurring cases have been described (Dowda and DiSalvo 1984), bites by rodents and lagomorphs almost never require PEP because these taxa are not significant in rabies epidemiology (Winkler 1991; Kulonen and Boldina 1996; Childs et al. 1997). If a wild carnivore or bat that escapes capture and diagnostic evaluation is the cause of human exposure, PEP should be initiated immediately.

When performed promptly, PEP is remarkably successful in preventing human rabies. For example, there have been no documented PEP failures in the United States since new cell culture vaccines and RIGs were licensed in the early 1980s, despite an estimated excess of 40,000 human exposures per year (Rupprecht et al. 1996b). A key feature is local care of bite wounds, which must be washed thoroughly with soap and water to remove any residual saliva that might contain virus and to remove devitalized tissue. Cell culture vaccine, RIG, or other biologicals may be administered using different schedules depending on vaccine status of the person exposed (WHO 1992b).

Without PEP, the chance of developing rabies following the bite of a rabid animal depends in part on the virus dose and the severity and location of the injury. Only a handful of people have survived documented clinical rabies (Alvarez et al. 1994). Each of these patients required prolonged supportive therapy, including mechanical ventilation, and most suffered significant neurologic sequelae. There is no effective antiviral therapy for rabies once the neurologic infection has been established.

Human rabies is quite rare in developed countries that have controlled the disease in dogs. For example, in the United States, human rabies decreased from more than 100 cases in the early 1900s, which were almost all due to rabid dogs, to only 1–2 annually for the last several decades. However, from 1980 to 1996, 17 of the 20 indigenous cases of human rabies resulted from infection with bat rabies virus variants; only one of these cases had a clear history of bite, although many patients had bat contact (CDC 1997c). In 12 of these cases, a silver-haired bat *Lasionycteris noctivagans* or eastern pipistrelle *Pipistrellus subflavus* variant virus was identified. The inability of health professionals to obtain an accurate history on potential human exposures involving bats is problematic. Therefore, PEP may be appropriate when there is reasonable probability that a bat bite occurred, unless the results are negative after diagnostic testing of the bat. Clearly, some bat rabies virus variants may possess peculiar biologic characteristics that support infectious transmission, even under limited bite conditions (Morimoto et al. 1996).

Although not currently recognized as a major zoonotic threat, nonrabies lyssaviruses may pose diagnostic problems and questions about PEP of persons bitten by species not strongly implicated in rabies transmission. Traditional rabies virus vaccines do not provide adequate protection (Dietzschold et al. 1988).

DOMESTIC ANIMAL HEALTH CONCERNS. Rabid wildlife serve as reservoirs of infection for livestock and companion animals. Domestic animals, primarily dogs and domestic cats, forge the link between infected wildlife and humans. Recent advances in vaccinology have resulted in availability of a wide variety of potent, inexpensive, and safe veterinary rabies biologics (Bunn 1991; Chauppis 1995) for at least six domestic species (CDC 1997b). Except in vampire bat rabies areas, it is not common to consider routine livestock vaccination beyond the individual case (Bloch and Diallo 1995).

Minimum prerequisites for domestic animal rabies control include adequate legislation and infrastructure capable of effective administration, public education on the dangers of rabies and the need for control, properly equipped veterinary control authority with full governmental support, sufficient funding for vaccine and vaccination campaigns, and a practical and scientifically sound control program (Held et al. 1991). The minimal ideal vaccine coverage for interruption of disease transmission is 60%–70% of the canine population.

While simple in theory, in practice rabies control programs are not readily achievable in the developing world, with few exceptions (Belotto 1988; Chomel et al. 1988). Many countries that have exchanged a colonial past for democracy find themselves paralyzed when it comes to rabies control, due to financial constraints and the inability to impose often unpopular control measures on their citizens. Opposition stems from a variety of myths, such as the fear that a dog tax will be levied on those attending vaccination centers or the belief that vaccination causes sterility in bitches and loss of stamina in hunting dogs (Meredith unpublished). As a result, vaccination campaigns are often largely ineffective.

MANAGEMENT IMPLICATIONS. Rabies may be the paradigm for many wildlife infectious diseases. Professionals may be faced with a daunting task if they are responsible for management of furbearers and balancing recreational values and public health implications of zoonotic disease. Enzootic rabies may impact endangered species recovery (e.g., Simian jackal, African wild dog, Florida panther, and gray wolf) directly because of overt mortality and impact on social behavior (Gascoyne et al. 1993a,b; MacDonald 1993). If deemed appropriate, commercial inactivated rabies vaccines recommended for domestic species should be used. If individual animal handling is not possible or practical, a commercial oral rabies vaccine and bait currently sanctioned by international authorities may be considered for remote delivery.

Careful consideration should be given to benefits and risks of translocation of nonendangered species from rabies enzootic areas for nuisance animal release, hunting, or trapping (Davidson et al. 1992). Epizootics have been precipitated by inadvertent movement of raccoon, gray fox, and coyote during rabies incubation periods (Krebs et al. 1997).

Many bat populations throughout the world are in decline; the perceived fear of rabies should not be used to support bat population reductions that do not result in significant disease control. Bats may be excluded from human dwellings at appropriate times and seasons, such as after juveniles have left the roost. Creation of artificial bat roosts away from human and domestic animals may decrease the opportunity for bat exposure at human dwellings. When control of hematophagus bats is contemplated, the most species-specific techniques possible should be implemented to reduce ecological impacts (Lord 1988). Clearly, a diverse cadre of wildlife disease professionals have a major responsibility in the management, public education, and applied research related to rabies and its control.

LITERATURE CITED

Acha, P.N., and A.M. Alba. 1988. Economic losses due to *Desmodus rotundus*. In *Natural history of vampire bats*, ed. A.M. Greenhall and U. Schmidt. Boca Raton, FL: CRC, pp. 207–214.

Acha, P.N., and P.V. Arambulo III. 1985. Rabies in the tropics: History and current status. In *Rabies in the tropics*, ed. E. Kuwert, C. Mérieux, H. Koprowski, and K. Bögel. Berlin: Springer-Verlag, pp. 343–359.

Ahuja, S., K.K. Tripathi, S.M. Saha, and S.N. Saxena. 1985. Epidemiology of rabies in India. In *Rabies in the tropics*, ed. E. Kuwert, C. Mérieux, H. Koprowski, and K. Bögel. Berlin: Springer-Verlag, pp. 571–582.

Allworth, A., K. Murray, and J. Morgan. 1996. A human case of encephalitis due to a *Lyssavirus* recently identified in fruit bats. *Communicable Diseases Intelligence* 20:504.

Almeida, M.F., E.A.C. Aauiar, L.F.A. Martorelli, and M.M.S. Silva. 1994. Laboratory diagnosis of rabies in Chiroptera carried out of a metropolitan area of Southeastern Brazil. *Revista de Saude Publica* 28:341–344.

Alvarez, A.L., R. Fajardo, M.E. Lopez, R.R. Pedroza, T. Hemachudha, N. Kamolvarin, C.G. Cortes, and G.M. Baer. 1994. Partial recovery from rabies in a nine-year-old boy. *Pediatric Infectious Disease Journal* 13:1154–1155.

Anderson, R.M., H.C. Jackson, R.M. May, and A.D.M. Smith. 1981. Population dynamics of fox rabies in Europe. *Nature* 289:765–771.

Anonymous. 1996a. Zoonoses—bat rabies case confirmed. *Veterinary Record* 138:630–631.

———. 1996b. Rabies suspected in a bat in New Haven. *Veterinary Record* 138:578.

Arellano-Sota, C. 1988. Biology, ecology, and control of the vampire bat. *Reviews of Infectious Diseases* 10 (Suppl. 4):S615–S619.

Artois, M., K.M. Charlton, N.D. Tolson, G.A. Casey, M.K. Knowles, and J.B. Campbell. 1990. Vaccinia recombinant virus expressing the rabies virus glycoprotein: Safety and efficacy trials in Canadian wildlife. *Canadian Journal of Veterinary Research* 54:504–507.

Ata, F.A., M.H. Tageldin, H.S. al Sumry, and S.I. al-Ismaily. 1993. Rabies in the Sultinate of Oman. *Veterinary Record* 132:68–69.

Aubert, M. 1994. Control of rabies in foxes: What are the appropriate measures? *Veterinary Record* 134:55–59.

Bacon, P.J. 1985. *Population dynamics of rabies in wildlife.* London: Academic, 358 pp.

Baer, G.M. 1988a. Animal models in the pathogenesis and treatment of rabies. *Reviews of Infectious Diseases* 10 (Suppl. 4):S739–S750.

———. 1988b. Oral rabies vaccination: An overview. *Reviews of Infectious Diseases* 10 (Suppl. 4):S644–S648.

Baer, G.M., and J.S. Smith. 1991. Rabies in nonhematophagous bats. In *The natural history of rabies,* ed. G.M. Baer, 2d ed. Boca Raton, FL: CRC, pp. 341–366.

Baer, G.M., M.K. Abelseth, and J.G. Debbie. 1971. Oral vaccination of foxes against rabies. *American Journal of Epidemiology* 93:487–492.

Baer, G.M., J.H. Shaddock, R. Quirion, T.V. Dam, and T.L. Lentz. 1990. Rabies susceptibility and the acetylcholine receptor. *Lancet* 335:664–665.

Baer, G.M., J. Neville, and G.S. Turner. 1996. *Rabbis and rabies: A pictorial history of rabies through the ages.* Cuaulta, Col. Condesa, Mexico, D.F.: de Imprenta Marsella, 134 pp.

Bahnemann, H.G. 1985. Rabies in South East Asia. In *Rabies in the tropics,* ed. E. Kuwert, C. Mérieux, H. Koprowski, and K. Bögel. Berlin: Springer-Verlag, pp. 541–544.

Balachandran, A., and K.M. Charlton. 1994. Experimental rabies infection of non-nervous tissues in skunks (*Mephitis mephitis*) and foxes (*Vulpes vulpes*). *Veterinary Pathology* 31:93–102.

Ballard, W.B., and P.R. Krausman. 1997. Occurrence of rabies in wolves of Alaska. *Journal of Wildlife Diseases* 33:242–245.

Barnard, B.J.H., and R.H. Hassel. 1981. Rabies in kudus (*Tragelaphus strepsiceros*) in South West Africa/Namibia. *Journal of the South African Veterinary Association* 52:309–314.

Barnard, B.J.H., R.H. Hassel, H.J. Geyer, and W.C. de Koker. 1982. Non-bite transmission of rabies in kudu (*Tragelaphus strepsiceros*). *Onderstepoort Journal of Veterinary Research* 49:193–194.

Barton, L.D., and J.B. Campbell. 1988. Measurement of rabies specific antibodies in carnivores by an enzyme-linked immunosorbent assay. *Journal of Wildlife Diseases* 24:246258.

Bell, J.F., and G.J. Moore. 1971. Susceptibility of Carnivora to rabies virus administered orally. *American Journal of Epidemiology* 93:176–182.

Belotto, A.J. 1988. Organization of mass vaccination for dog rabies in Brazil. *Reviews of Infectious Diseases* 10 (Suppl. 4):S693–S696.

Benelmouffok, A., M. Benhassine, and M. Abrous. 1982. La rage en Algerie: Recrudescence et nouvel aspect epizootiologique. *Comparative Immunology of Microbiological and Infectious Diseases* 5:321–326.

Benmansour, A., H. Leblois, P. Coulon, C. Tuffereau, Y. Gaudin, A. Flamand, and F. Lafay. 1991. Antigenicity of rabies virus glycoprotein. *Journal of Virology* 65:4198–4203.

Beran, G.W., A.P. Nocete, O. Elvina, S.B. Gregorio, R.R. Moreno, J.C. Nakao, G.A. Buchett, H.L. Canizares, and F.F. Macasaet. 1972. Epidemiological and control studies on rabies in the Philippines. *Southeast Asian Journal of Tropical Medicine and Public Health* 3:433–445.

Berry, H.H. 1993. Surveillance and control of anthrax and rabies in wild herbivores and carnivores in Namibia. *Revue Scientifique et Technique O.I.E.* 12:137–146.

Bigler, W.J., G.L. Hoff, J.S. Smith, R.G. McLean, H.A. Trevino, and J. Ingwersen. 1983. Persistence of rabies antibody in free-ranging raccoons. *Journal of Infectious Diseases* 148:610.

Black, D., and T.J. Wiktor. 1986. Survey of raccoon hunters for rabies antibody titers: Pilot study. *Journal of the Florida Medical Association* 73:517–520.

Blancou, J. 1986. Epidemiological aspects of virus adaptation to different animal species. In *Viral infections,* ed. W. Marget, W. Land, and E. GablerSandberger. Munich: MMV Medizin Verlag, pp. 134–156.

———. 1988. Ecology and epidemiology of fox rabies. *Reviews of Infectious Diseases* 10 (Suppl. 4):S606–S609.

Blancou, J., and A. Wandeler. 1989. Rabies virus and its vectors in Europe. *Revue Scientifique et Technique O.I.E.* 8:859–861.

Blancou, J., M.P. Kieny, R. Lathe, J.P. Lecocq, P.P. Pastoret, J.P. Soulebot, and P. Desmettre. 1986. Oral vaccination of the fox against rabies using a live recombinant vaccinia virus. *Nature* 322:373–375.

Blancou, J., P.P. Pastoret, B. Brochier, I. Thomas, and K. Bögel. 1988. Vaccinating wild animals against rabies. *Revue Scientifique et Technique O.I.E.* 7:1005–1013.

Blancou, J., M.F.A. Aubert, and M. Artois. 1991. Fox rabies. In *The natural history of rabies,* ed. G.M. Baer, 2d ed. Boca Raton, FL: CRC, pp. 257–290.

Blenden, D.C., J.F. Bell, A.T. Tsao, and J.U. Umoh. 1983. Immunofluorescent examination of the skin of rabies-infected animals as a means of early detection of rabies virus antigen. *Journal of Clinical Microbiology* 18:631–636.

Bloch, N., and I. Diallo. 1995. A probable outbreak of rabies in a group of camels in Niger. *Veterinary Microbiology* 46:281–283.

Bögel, K. 1992. International cooperation in wildlife research and control. In *Wildlife rabies control,* ed. K. Bögel, F.X. Meslin, and M. Kaplan. Kent, UK: Wells Medical, pp. 194–198.

Bögel, K., H. Moegle, F. Knorpp, A.A. Arata, K. Dietz, and P. Diethelm. 1976. Characteristics of the spread of a wildlife rabies epidemic in Europe. *Bulletin of the World Health Organization* 54:433–447.

Botros, B.A.M., R.W. Moch, M. Kerkor, and I. Helmy. 1977. Rabies in the Arab Republic of Egypt: III. Enzootic rabies in wildlife. *Journal of Tropical Medicine and Hygiene* 80:59–62.

Botros, B.A.M., A.W. Salib, P.W. Mellick, J.M. Linn, A.K. Soliman, and R.M. Scott. 1988. Antigenic variation of wild and vaccine rabies strains of Egypt. *Journal of Medical Virology* 24:153–159.

Boulger, L.R., and J.S. Porterfield. 1958. Isolation of a virus from Nigerian fruit bats. *Transactions of the Royal Society of Tropical Medicine and Hygiene* 52:421–424.

Bourhy, H., B. Kissi, and N. Tordo. 1993. Molecular diversity of the *Lyssavirus* genus. *Virology* 194:70–81.

Brass, D. 1994. *Rabies in bats.* Ridgefield, CT: Livia, 335 pp.

Briggs, D.J., and J.R. Schwenke. 1992. Longevity of rabies antibody titre in recipients of human diploid cell rabies vaccine. *Vaccine* 10:125–129.

Brochier, B., D. Boulanger, F. Costy, and P.P. Pastoret. 1994. Towards rabies elimination in Belgium by fox vaccination using a vaccinia-rabies glycoprotein recombinant virus. *Vaccine* 12:1368–1371.

Brown, C.L., C.E. Rupprecht, and W.M. Tzilkowski. 1990. Adult raccoon survival in an enzootic rabies area of Pennsylvania. *Journal of Wildlife Diseases* 26:346–350.

Bula, M., and L. Mafwala. 1988. Le diagnostic de la rage animale a Lumbumbashi, Zaire. *Revue Scientifique et Technique O.I.E.* 7:387–394.

Bundza, A., and K.M. Charlton. 1988. Comparison of spongiform lesions in experimental scrapie and rabies in skunks. *Acta Neuropathologica (Berlin)* 76:275–280.

Bunn, T.O. 1991. Canine and feline vaccines, past and present. In *The natural history of rabies,* ed. G.M. Baer, 2d ed. Boca Raton, FL: CRC, pp. 415–425.

Calmette, A. 1891. Notes sur la rage in Indochine et sur les vaccinations antirabiques practiques a Saigon du 15 avril au ler aout. *Annals of the Institute Pasteur (Paris)* 5:633.

Celis, E., R.W. Miller, T.J. Wiktor, B. Dietzschold, and H. Koprowski. 1986. Isolation and characterization of human T cell lines and clones reactive to rabies virus: Antigen specificity and production of interferon-gamma. *Journal of Immunology* 136:692–697.

Celis, E., D.W. Ou, B. Dietzschold, and H. Koprowski. 1988. Recognition of rabies and rabies-related viruses by T cells derived from human vaccine recipients. *Journal of Virology* 62:3128–3134.

Centers for Disease Control and Prevention (CDC). 1991. Rabies prevention: United States 1991—Recommendations of the Immunization Practices Advisory Committee (ACIP). *Morbidity and Mortality Weekly Report* 40:1–19.

———. 1994. The pathogenesis of rabies and other lyssaviral infections: Recent studies. In *Lyssaviruses,* ed. C.E. Rupprecht, B. Dietzschold, and H. Koprowski. Berlin: Springer-Verlag, pp. 95–120.

———. 1995. Translocation of coyote rabies: Florida 1994. *Morbidity and Mortality Weekly Report* 44:580–587.

———. 1997a. Update: Raccoon rabies epizootic—United States 1996. *Morbidity and Mortality Weekly Report* 45:1117–1120.

———. 1997b. Compendium of animal rabies control 1997: National Association of State Public Health Veterinarians, Inc. *Morbidity and Mortality Weekly Report* 46:1–9.

———. 1997c. Human rabies: Texas and New Jersey 1997. *Morbidity and Mortality Weekly Report* 47:1–5.

Charlton, K.M. 1984. Rabies: Spongiform lesions in the brain. *Acta Neuropathologica (Berlin)* 63:198–202.

Charlton, K.M., G.A. Casey, and W.A. Webster. 1984. Rabies virus in the salivary glands and nasal mucosa of naturally infected skunks. *Canadian Journal of Comparative Medicine* 48:338–339.

Charlton, K.M., G.A. Casey, and J.B. Campbell. 1987. Experimental rabies in skunks: Immune response and salivary gland infection. *Comparative Immunology and Microbiology of Infectious Disease* 10:227–235.

Charlton, K.M., W.A. Webster, G.A. Casey, and C.E. Rupprecht. 1988. Skunk rabies. *Reviews of Infectious Diseases* 10 (Suppl. 4):S626–S628.

Charlton, K.M., W.A. Webster, and G.A. Casey. 1991. Skunk rabies. In *The natural history of rabies,* ed. G.M. Baer, 2d ed. Boca Raton, FL: CRC, pp. 307–324.

Charlton, K.M., G.A. Casey, A.I. Wandeler, and S. Nadin-Davis. 1996. Early events in rabies virus infection of the central nervous system in skunks (*Mephitis mephitis*). *Acta Neuropathologica (Berlin)* 91:89–98.

Chauppis, G.E. 1995. Development of rabies vaccines. In *Rabies in a changing world,* ed. P.H. Beynon and A.T.B. Edney. London: British Small Animal Veterinary Association, pp. 49–59.

Cherkasskiy, B.L. 1988. Roles of the wolf and the raccoon dog in the ecology and epidemiology of rabies in the USSR. *Reviews of Infectious Diseases* 10 (Suppl. 4):S634–S636.

Cherkasskiy, B.L., A.G. Knop, V.A. Vedernikov, V.A. Sedov, A.E. Khairushev, and S.A. Chernichenko. 1995. The epidemiology and epizootiology of rabies on the territory of the former USSR. *Zeitschrift fur Mikrobiology, Epidemiology, und Immunobiology* 1:21–26.

Childs, J.E., C.V. Trimarchi, and J.W. Krebs. 1994. The epidemiology of bat rabies in New York State 1988–92. *Epidemiology and Infection* 113:501–511.

Childs, J.E., L. Colby, J.W. Krebs, T. Strine, M. Feller, D. Noah, C. Drenzek, J.S. Smith, and C.E. Rupprecht. 1997. Surveillance and spatiotemporal associations of rabies in rodents and lagomorphs in the United States 1985–1994. *Journal of Wildlife Diseases* 33:20–27.

Chomel, B., G. Chappuis, F. Bullon, E. Cardenas, T.D. de Beublain, M. Lombard, and E. Giambruno. 1988. Mass vaccination campaign against rabies: Are dogs correctly protected? The Peruvian experience. *Reviews of Infectious Diseases* 10 (Suppl. 4):S697–S702.

Cifuentes, E. 1988. Program for the elimination of urban rabies in Latin America. *Reviews of Infectious Diseases* 10 (Suppl. 4):S689–S692.

Clark, K.A., S.U. Neill, J.S. Smith, P.J. Wilson, V.W. Whadford, and G.W. McKirahan. 1994. Epizootic canine rabies transmitted by coyotes in south Texas. *Journal of the American Veterinary Medical Association* 204:536–540.

Cluver, E. 1927. Rabies in South Africa. *Journal of the Medical Association of South Africa* 1:247–253.

Constantine, D.G. 1988. Transmission of pathogenic microorganisms by vampire bats. In *Natural history of vampire bats,* ed. A.M. Greenhall and U. Schmidt. Boca Raton, FL: CRC, pp. 167–189.

Coulon, P., F. Lafay, H. Leblois, C. Tuffereau, M. Artois, J. Blancou, A. Benmansour, and A. Flamand. 1992. The SAG: A new attenuated oral rabies vaccine. In *Wildlife rabies control,* ed. K. Bögel, F.X. Meslin, and M. Kaplan. Kent, UK: Wells Medical, pp. 105–111.

Coulon, P., F. Lafay, C. Tuffereau, and A. Flamand. 1994. The molecular basis for altered pathogenicity of *Lyssavirus* variants. In *Lyssaviruses,* ed. C.E. Rupprecht, B. Dietzschold, and H. Koprowski. Berlin: Springer-Verlag, pp. 69–84.

Coyne, M.J., G. Smith, and F.E. McAllister. 1989. Mathematic model for the population biology of rabies in raccoons in the mid-Atlantic states. *American Journal of Veterinary Research* 50:2148–2154.

Crandell, R.A. 1991. Arctic fox rabies. In *The natural history of rabies,* ed. G.M. Baer, 2d ed. Boca Raton, FL: CRC, pp. 291–306.

Crerar, S., H. Longbottom, J. Rooney, and P. Thornber. 1996. Human health aspects of a possible *Lyssavirus* in a black flying fox. *Communicable Diseases Intelligence* 20:325.

Crick, J., and A. King. 1988. Culture of rabies virus in vitro. In *Rabies,* ed. J.B. Campbell and K.M. Charlton. Boston: Kluwer Academic, pp. 47–66.

Crick, J., G.H. Tignor, and K. Moreno. 1982. A new isolate of Lagos bat virus from the Republic of South Africa. *Transactions of the Royal Society of Tropical Medicine and Hygiene* 76:211–213.

Daoust, P.-Y., A.I. Wandeler, and G.A. Casey. 1996. Cluster of rabies cases of probable bat origin among red foxes in Prince Edward Island, Canada. *Journal of Wildlife Diseases* 32:403–406.

Davidson, W.R., M.J. Appel, G.L. Doster, O.E. Baker, and J.F. Brown. 1992. Diseases and parasites of red foxes, gray foxes and coyotes from commercial sources selling to fox-chasing enclosures. *Journal of Wildlife Diseases* 28:581–589.

Davies, F.G. 1981. The possible role of wildlife in the natural history of rabies in Kenya. In *Wildlife disease research and economic development,* ed. L. Karstad, B. Nestel, and M. Graham. Ottawa: International Development Research Centre, pp. 28–29.

Debbie, J.G. 1991. Rabies control of terrestrial wildlife by population reduction. In *The natural history of rabies,* ed. G.M. Baer, 2d ed. Boca Raton, FL: CRC, pp. 477–484.

de Mattos, C.C., C.A. de Mattos, A. Aguilar-Setien, E. Loza-Rubio, P.A. Yager, L.A. Orciari, B.I. Osburn, and J.S.

Smith. 1996a. Molecular characterization of rabies viruses obtained in Mexico. In *Proceedings of the 7th annual international meeting: Advances towards rabies control in the Americas,* ed. C.E. Rupprecht. Atlanta: Centers for Disease Control and Prevention, p. 60.

de Mattos, C.C., C.A. de Mattos, J.S. Smith, E.T. Miller, S. Papo, A. Utrera, and B.I. Osburn. 1996b. Genetic characterization of rabies field isolates from Venezuela. *Journal of Clinical Microbiology* 34:1553–1558.

Diaz, A.-M., S. Papo, A. Rodriguez, and J.S. Smith. 1994. Antigenic analysis of rabies-virus isolates from Latin America and the Caribbean. *Journal of Veterinary Medicine Series B* 41:153–160.

Dietzschold, B., W.H. Wunner, T.J. Wiktor, A.D. Lopes, M. Lafon, C.L. Smith, and H. Koprowski. 1983. Characterization of an antigenic determinant of the glycoprotein that correlates with pathogenicity of rabies virus. *Proceedings of the National Academy of Sciences USA* 80:70–74.

Dietzschold, B., H.H. Wang, C.E. Rupprecht, E. Celis, M. Tollis, H. Ertl, E. Heber-Katz, and H. Koprowski. 1987. Induction of protective immunity against rabies by immunization with rabies virus ribonucleoprotein. *Proceedings of the National Academy of Sciences USA* 84:9165–9169.

Dietzschold, B., C.E. Rupprecht, M. Tollis, M. Lafon, J. Mattei, T.J. Wiktor, and H. Koprowski. 1988. Antigenic diversity of the glycoprotein and nucleocapsid proteins of rabies and rabies-related viruses: Implications for epidemiology and control of rabies. *Reviews of Infectious Diseases* 10 (Suppl. 4):S785–S798.

Dowda, H., and A.F. DiSalvo. 1984. Naturally acquired rabies in an eastern chipmunk (*Tamias striatus*). *Journal of Clinical Microbiology* 19:281–282.

Dranseika, A. 1992. Rabies data of Lithuania in 1991. *Rabies Bulletin Europe* 1:15.

Eigen, M., and C.K. Biebricher. 1988. Sequence space and quasispecies distribution. In *RNA genetics.* Vol. 3, *Variability of RNA genomes,* ed. E. Domingo, J.J. Holland, and P. Ahlquist. Boca Raton, FL: CRC, pp. 211–245.

Enurah, L.U., R.A. Ocholi, K.O. Adeniyi, and M.C. Ekwonu. 1988. Rabies in a civet cat (*Civettictis civetta*) in the Jos zoo, Nigeria. *British Veterinary Journal* 144:515–516.

Ertl, H.C., B. Dietzschold, M. Gore, L. Otvos Jr., J.K. Larson, W.H. Wunner, and H. Koprowski. 1989. Induction of rabies virus-specific T-helper cells by synthetic peptides that carry dominant T-helper cell epitopes of the viral ribonucleoprotein. *Journal of Virology* 63:2885–2892.

Ertl, H.C., B. Dietzschold, and L. Otvos, Jr. 1991. T helper cell epitope of rabies virus nucleoprotein defined by tri- and tetrapeptides. *European Journal of Immunology* 21:1–10.

Everard, C.O., and J.D. Everard. 1988. Mongoose rabies. *Reviews of Infectious Diseases* 10 (Suppl. 4):S610–S614.

———. 1992. Mongoose rabies in the Caribbean. *Annals of the New York Academy of Sciences* 653:356–366.

Familusi, J.B., and D.L. Moore. 1972. Isolation of a rabies-related virus from the cerebrospinal fluid of a child with aseptic meningitis. *African Journal of Medical Science* 3:93–96.

Familusi, J.B., B.O. Osunkoya, D.L. Moore, G.E. Kemp, and A. Fabiyi. 1972. A fatal human infection with Mokola virus. *American Journal of Tropical Medicine and Hygiene* 21:959–963.

Fearneyhough, M.G., K.A. Clark, and D.R. Smith. 1996. Oral rabies vaccination program in Texas. In *Proceedings of the 7th annual international meeting: Advances towards rabies control in the Americas,* ed. C.E. Rupprecht. Atlanta: Centers for Disease Control and Prevention, p. 33.

Fekadu, M. 1991. Latency and aborted rabies. In *The natural history of rabies,* ed. G.M. Baer, 2d ed. Boca Raton, FL: CRC, pp. 191–198.

Fekadu, M., P.W. Greer, F.W. Chandler, and D.W. Sanderlin. 1988. Use of the avidin-biotin peroxidase system to detect rabies antigen in formalin-fixed paraffin-embedded tissues. *Journal of Virological Methods* 19:91–96.

Feller, M.J., J.B. Kaneene, and M.G. Stobierski. 1997. Prevalence of rabies in bats in Michigan 1981–1993. *Journal of the American Veterinary Medical Association* 210:195–200.

Fernandes, M.V., and P.V. Arambulo. 1985. Rabies as an international problem. In *World's debt to Pasteur,* ed. H. Koprowski and S.A. Plotkin. New York: Alan R. Liss, pp. 187–218.

Fishbein, D.B. 1991. Rabies in humans. In *The natural history of rabies,* ed. G.M. Baer, 2d ed. Boca Raton, FL: CRC, pp. 519–549.

Flamand, A., P. Coulon, P.F. Lafay, and C. Tuffereau. 1993. Avirulent mutants of rabies virus and their use as live vaccine. *Trends in Microbiology* 1:317–320.

Flamand, A., P. Coulon, Y. Gaudin, F. Lafay, H. Raux, and C. Tuffereau. 1995. Reversible conformational changes of the rabies glycoprotein that mask or expose epitopes involved in virulence. *Journal of Cellular Biochemistry Supplement* 19A:277.

Flores-Crespo, R., and C. Arellano-Sota. 1991. Biology and control of the vampire bat. In *The natural history of rabies,* ed. G.M. Baer, 2d ed. Boca Raton, FL: CRC, pp. 461–476.

Foggin, C.M. 1982. Atypical rabies virus in cats and a dog in Zimbabwe. *Veterinary Record* 110:338.

———. 1983. Mokola virus infection in cats and a dog in Zimbabwe. *Veterinary Record* 113:115.

———. 1988. Rabies and rabies-related viruses in Zimbabwe: Historical, virological and ecological aspects. Ph.D. diss., University of Zimbabwe, Harare, Zimbabwe, 262 pp.

Follmann, E.H., D.G. Ritter, and M. Beller. 1994. Survey of trappers in northern Alaska for rabies antibody. *Epidemiology and Infection* 113:137–141.

Fraser, G.C., P.T. Hooper, R.A. Lunt, A.R. Gould, L.J. Gleeson, A.D. Hyatt, G.M. Russell, and J.A. Kattenbelt. 1996. Encephalitis caused by a *Lyssavirus* in fruit bats in Australia. *Emerging Infectious Diseases* 2:327–331.

Fu, Z.F., C.E. Rupprecht, P. Saikumar, H.S. Niu, I. Babka, W.H. Wunner, H. Koprowski, and B. Dietzschold. 1993. Oral vaccination of raccoons, *Procyon lotor,* with baculovirus-expressed rabies virus glycoprotein. *Vaccine* 11:925–928.

Gascoyne, S.C., M.K. Laurenson, S. Lelo, and M. Borner. 1993a. Rabies in African wild dogs (*Lycaon pictus*) in the Serengeti region, Tanzania. *Journal of Wildlife Diseases* 29:396–402.

Gascoyne, S.C., A.A. King, M.K. Laurenson, M. Borner, B. Schildger, and J. Barrat. 1993b. Aspects of rabies infection and control in the conservation of the African wild dog (*Lycaon pictus*) in the Serengeti Region, Tanzania. *Onderstepoort Journal of Veterinary Research* 60:415–420.

Gaudin, Y., R.W. Ruigrok, M. Knossow, and A. Flamand. 1993. Low-pH conformation changes of rabies virus glycoprotein and their role in membrane fusion. *Journal of Virology* 67:1365–1372.

Gessler, M., and H. Spittler. 1982. Relationship between fox populations and limitation of rabies spread in Rhineland-Palatinate. *Comparative Immunology and Microbiology of Infectious Diseases* 5:293–302.

Ginsberg, J.R., and D.W. MacDonald. 1990. *Foxes, wolves, jackals and dogs: An action plan for the conservation of canids.* Gland, Switzerland: International Union for Conservation of Nature, 116 pp.

Gleeson, L.J. 1997. Australian bat *Lyssavirus:* A newly emerged zoonosis? *Australian Veterinary Journal* 75:188.

Goldwasser, R.A., and R.E. Kissling. 1958. Fluorescent antibody staining of street and fixed rabies virus antigens. *Proceedings of the Society for Experimental Biology and Medicine* 98:219–223.

Gosztonyi, G. 1994. Reproduction of lyssaviruses: Ultrastructural composition of lyssavirus and functional aspects of pathogenesis. In *Lyssaviruses,* ed. C.E. Rupprecht, B. Dietzschold, and H. Koprowski. Berlin: Springer-Verlag, pp. 43–68.

Greenwood, R.J., W.E. Newton, G.L. Pearson, and G.J. Schamber. 1997. Population and movement characteristics of radio-collared striped skunks in North Dakota during an epizootic of rabies. *Journal of Wildlife Diseases* 33:226–241.

Gremillion-Smith, C., and A. Woolf. 1988. Epizootiology of skunk rabies in North America. *Journal of Wildlife Diseases* 24:620–626.

Guittre, C., M. Artois, and A. Flamand. 1992. Use of an avirulent mutant of rabies virus for the oral vaccination of wild animals: Innocuity in several nontarget species. *Annales de Medecine Veterinaire* 136:329–332.

Häflinger, U., P. Bichsel, A. Wandeler, and F. Steck. 1982. Oral Immunization of foxes against rabies: Stabilization and use of bait for the virus. *Zentralblatt Veterinärmedizin Reihe B* 29:604–618.

Hamir, A.N., and C.E. Rupprecht. 1990a. Absence of rabies encephalitis in a raccoon with concurrent rabies and canine distemper infections. *Cornell Veterinarian* 80:197–201.

———. 1990b. Eosinophilic encephalomyelitis in a raccoon experimentally infected with a dog isolate of rabies virus. *Journal of Veterinary Diagnostic Investigation* 2:145–147.

Hamir, A.N., G. Moser, and C.E. Rupprecht. 1992. Morphologic and immunoperoxidase study of neurologic lesions in naturally acquired rabies of raccoons. *Journal of Veterinary Diagnostic Investigation* 4:369–373.

Hamir, A.N., G. Moser, and C.E. Rupprecht. 1996a. Clinicopathologic variation in raccoons infected with different street rabies virus isolates. *Journal of Veterinary Diagnostic Investigation* 8:31–37.

Hamir, A.N., G. Moser, T. Wampler, A. Hattel, B. Dietzschold, and C.E. Rupprecht. 1996b. Use of a single anti-nucleocapsid monoclonal antibody to detect rabies antigen in formalin-fixed, paraffin-embedded tissues. *Veterinary Record* 138:114–115.

Hanlon, C.A., and C.E. Rupprecht. 1998. The reemergence of rabies. In *Emerging infections,* ed. W.M. Scheld, D. Armstrong, and J.M. Hughes. Washington, DC: American Society for Microbiology, pp. 59–80.

Hanlon, C.A., E.L. Ziemer, A.N. Hamir, and C.E. Rupprecht. 1989. Cerebrospinal fluid analysis of rabid and vaccinia-rabies glycoprotein recombinant, orally vaccinated raccoons (*Procyon lotor*). *American Journal of Veterinary Research* 50:364–367.

Hanlon, C.A., J.R. Buchanan, E. Nelson, H.S. Niu, D. Diehl, and C.E. Rupprecht. 1993. A vaccinia-vectored rabies vaccine field trial: Ante- and post-mortem biomarkers. *Revue Scientifique et Technique O.I.E.* 12:99–107.

Hanlon, C.A., A. Willsey, B. Laniewicz, C. Trimarchi, and C.E. Rupprecht. 1996. New York State oral wildlife rabies vaccination: First evaluation for enzootic raccoon rabies control. In *Proceedings of the 7th annual international meeting: Advances towards rabies control in the Americas,* ed. C.E. Rupprecht. Atlanta: Centers for Disease Control and Prevention, p. 26.

Hanlon, C.A., M. Niezgoda, V. Shankar, H.S. Niu, H. Koprowski, and C.E. Rupprecht. 1997. A recombinant vaccinia-rabies virus in the immunocompromised host: Oral innocuity, progressive parenteral infection, and therapeutics. *Vaccine* 15:140–148.

Hanlon, C.A., M. Niezgoda, A.N. Hamir, C. Schumacher, H. Koprowski, and C.E. Rupprecht. 1998. First North American field release of a vaccinia-rabies glycoprotein recombinant virus. *Journal of Wildlife Diseases* 34:228–239.

Hassel, R.H. 1982. Incidence of rabies in kudu in South West Africa/Namibia. *South African Journal of Science* 78:418–421.

Hawkey, C.M. 1988. Salivary antihemostatic factors. In *Natural history of vampire bats,* ed. A.M. Greenhall and U. Schmidt. Boca Raton, FL: CRC, pp. 133–141.

Held, J.R., J.A. Escalante, and W.G. Winkler. 1991. The international management of rabies. In *The natural history of rabies,* ed. G.M. Baer, 2d ed. Boca Raton, FL: CRC, pp. 505–512.

Hemachudha, T. 1994. Human rabies: Clinical aspects, pathogenesis, and potential therapy. In *Lyssaviruses,* ed. C.E. Rupprecht, B. Dietzschold, and H. Koprowski. Berlin: Springer-Verlag, pp. 121–144.

Herzog, M., C. Fritzell, M. Lafage, J.A. Montano Hirose, D. Scott-Algara, and M. Lafon. 1991. T and B cell human responses to European bat *Lyssavirus* after post-exposure rabies vaccination. *Clinical and Experimental Immunology* 85:224–230.

Hill, R.E., Jr., and G.W. Beran. 1992. Experimental inoculation of raccoons, *Procyon lotor,* with rabies virus of skunk origin. *Journal of Wildlife Diseases* 28:51–56.

Hill, R.E., Jr., G.W. Beran, and W.R. Clark. 1992. Demonstration of rabies virus-specific antibody in the sera of free-ranging Iowa raccoons (*Procyon lotor*). *Journal of Wildlife Diseases* 28:377–385.

Hill, R.E., Jr., K.E. Smith, G.W. Beran, and P.D. Beard. 1993. Further studies on the susceptibility of raccoons (*Procyon lotor*) to a rabies virus of skunk origin and comparative susceptibility of striped skunks (*Mephitis mephitis*). *Journal of Wildlife Diseases* 29:475–477.

Hirose, J.A., H. Bourhy, and P. Sureau. 1991. Retro-orbital route for brain specimen collection for rabies diagnosis. *Veterinary Record* 129:291–292.

Hooper, D.C., I. Pierard, A. Modelska, L. Otovos, Z.F. Fu, and H. Koprowski. 1994. Rabies ribonucleocapsid as an oral immunogen and immunological enhancer. *Proceedings of the National Academy of Sciences USA* 91:10,908–10,912.

Hubbard, D.R. 1985. A descriptive epidemiological study of raccoon rabies in a rural environment. *Journal of Wildlife Diseases* 21:105–110.

Hubschle, O.J.B. 1988. Rabies in the kudu antelope (*Tragelaphus strepsiceros*). *Reviews of Infectious Diseases* 10 (Suppl. 4):S629–S633.

Iwasaki, Y. 1991. Spread of virus within the central nervous system. In *The natural history of rabies,* ed. G.M. Baer, 2d ed. Boca Raton, FL: CRC, pp. 121–132.

Jackson, A.C. 1994. Animal models of rabies virus neurovirulence. In *Lyssaviruses,* ed. C.E. Rupprecht, B. Dietzschold, and H. Koprowski. Berlin: Springer-Verlag, pp. 85–93.

Jackson, A.C., and J.P. Rossiter. 1997. Rabies virus infection produces apoptosis. *Neurology* 48:A162.

Jenkins, S.R., and W.G. Winkler. 1987. Descriptive epidemiology from an epizootic of raccoon rabies in the Middle Atlantic States 1982–1983. *American Journal of Epidemiology* 126:429–437.

Johnston, D.H., D.R. Voigt, C.D. MacInnes, P. Bachmann, K.F. Lawson, and C.E. Rupprecht. 1988. An aerial baiting system for the distribution of attenuated or recombinant rabies vaccines for foxes, raccoons, and skunks.

Reviews of Infectious Diseases 10 (Suppl. 4): S660–S664.

Kappeler, A. 1989. Bat rabies surveillance in Europe. *Rabies Bulletin Europe* 13:12–13.

Kat, P.W., K.A. Alexander, J.S. Smith, and L. Munson. 1995. Rabies and African wild dogs in Kenya. *Proceedings of the Royal Society (London) B* 262:229–233.

Kauker, E. 1966. Die Tollwut in Mitteleuropa: Von 1953 bis 1966. *Sitzungsberichte der Heidelberger Akademie der Wissenschafter* 4:123–141.

Kemp, G.E., O.R. Causey, D.L. Moore, A. Odeola, and A. Fabiyi. 1972. Mokola virus: Further studies on IbAn 27377, a new rabies-related etiologic agent of zoonosis in Nigeria. *American Journal of Tropical Medicine and Hygiene* 21:356–359.

Kieny, M.P., R. Lathe, R. Drillien, D. Spehner, S. Skory, D. Schmitt, T.J. Wiktor, H. Koprowski, and J.P. Lecocq. 1984. Expression of rabies virus glycoprotein from a recombinant vaccinia virus. *Nature* 312:163–166.

King, A.A. 1991. Studies of the antigenic relationships of rabies and rabies-related viruses using antinucleoprotein monoclonal antibodies. Ph.D. diss., University of Surrey, Guildford, UK, 143 pp.

King, A.A., and J. Crick. 1988. Rabies-related viruses. In *Rabies,* ed. J.B. Campbell and K.M. Charlton. Boston: Kluwer Academic, pp. 177–199.

King, A.A., C.D. Meredith, and G.E. Thomson. 1994. The biology of southern African *Lyssavirus* variants. In *Lyssaviruses,* ed. C.E. Rupprecht, B. Dietzschold, and H. Koprowski. Berlin: Springer-Verlag, pp. 267–295.

Kintscher, M., J. Bernhard, and I. Lemke. 1990. Investigations of the safety, innocuity and efficacy of the SAD/Potsdam 5/88 strain for rabies vaccination. *Monatshefte für Veterinärmedizin* 45:81–84.

Koesharyono, C., R.J. Theos, and G. Simanjuntak. 1985. The epidemiology of rabies in Indonesia. In *Rabies in the tropics,* ed. E. Kuwert, C. Mérieux, H. Koprowski, and K. Bögel. Berlin: Springer-Verlag, pp. 545–555.

Koprowski, H. 1996. The mouse inoculation test. In *Laboratory techniques in rabies,* ed. F.X. Meslin, M. Kaplan, and H. Koprowski, 4th ed. Geneva: World Health Organization, pp. 80–87.

Koprowski, H., Y.M. Zheng, E. Heber-Katz, L. Rorke, Z.F. Fu, C. Hanlon, and B. Dietzschold. 1993. In vivo expression of inducible nitric oxide synthase in experimentally induced neurologic disease. *Proceedings of the National Academy of Sciences* 90:3024–3027.

Koromyslow, W.A., and B.L. Cherkasskiy. 1990. Rabies in the European part of the Union of Soviet Socialist Republics (SSR) in 1989. *Rabies Bulletin Europe* 3:9–310.

Kovalev, N.A., V.A. Sedov, A.S. Shashenko, D.F. Osidse, and E.V. Ivanovsky. 1992. An attenuated oral vaccine for wild carnivorous animals in the USSR. In *Wildlife rabies control,* ed. K. Bögel, F.X. Meslin, and M. Kaplan. Kent, UK: Wells Medical, pp. 112–114.

Krebs, J.W., J.S. Smith, C.E. Rupprecht, and J.E. Childs. 1997. Rabies surveillance in the United States during 1996. *Journal of the American Veterinary Medical Association* 211:1525–1539.

Kulonen, K., and I. Boldina. 1996. No rabies detected in voles and field mice in a rabies-endemic area. *Journal of Veterinary Medicine Series B* 43:445–447.

Lafay, F., J. Benejean, C. Tuffereau, A. Flamand, and P. Coulon. 1994. Vaccination against rabies: Construction and characterization of SAG2, a double avirulent derivative of SADBern. *Vaccine* 12:317–320.

Lafon, M. 1994. Immunobiology of lyssaviruses: The basis for immunoprotection. In *Lyssaviruses,* ed. C.E. Rupprecht, B. Dietzschold, and H. Koprowski. Berlin: Springer-Verlag, pp. 145–160.

Lalosevic, D., ed. 1996. *Selected papers of rabies prophylaxis and a meeting on rabies on the occasion of the 75th anniversary of Pasteur Institute Novi Sad.* Novi Sad, Yugoslavia: Pasteur Institute Novi Sad, 162 pp.

Lari, F.A. 1985. Epidemiology and prevention of rabies in Pakistan. In *Rabies in the tropics,* ed. E. Kuwert, C. Mérieux, H. Koprowski, and K. Bögel. Berlin: Springer-Verlag, pp. 567–570.

Lavrov, N.P. 1971. The introduction of the raccoon dog (*Nyctereutes procynoides* Gray) in some districts in the USSR. *Trudy Biologii* 29:101–160.

Lawson, K.F., J.G. Black, K.M. Charlton, D.H. Johnston, and A.J. Rhodes. 1987. Safety and immunogenicity of a vaccine bait containing ERA strain of attenuated rabies virus. *Canadian Journal of Veterinary Research* 51:460–464.

Le Blois, H., C. Tuffereau, J. Blancou, M. Artois, A. Aubert, and A. Flamand. 1990. Oral immunization of foxes with avirulent rabies virus mutants. *Veterinary Microbiology* 23:259–266.

Leffingwell, L.M., and S.U. Neill. 1989. Naturally acquired rabies in an armadillo (*Dasypus novemcinctus*) in Texas. *Journal of Clinical Microbiology* 27:174–175.

le Gonidec, G., A. Rickenbach, Y. Robin, and G. Heme. 1978. Isolement d'une souche de virus Mokola au Cameroun. *Annals of Microbiology (Paris)* 129A:245–249.

Lentz, T.L., T.G. Burrage, A.L. Smith, and G.H. Tignor. 1983. The acetylcholine receptor as a cellular receptor for rabies virus. *Yale Journal of Biology and Medicine* 56:315–322.

Lewis, V.J., and W.L. Thacker. 1974. Limitations of deteriorated tissue for rabies diagnosis. *Health Laboratory Science* 11:8–12.

Lindsay, I.M., and D.W. MacDonald. 1986. Behaviour and ecology of the Ruppell's fox, *Vulpes ruppellii,* in Oman. *Mammalia* 50:461–475.

Linhart, S.B., T.E. Creekmore, J.L. Corn, M.D. Whitney, B.D. Snyder, and V.F. Nettles. 1993. Evaluation of baits for oral rabies vaccination of mongooses: Pilot field trials in Antigua West Indies. *Journal of Wildlife Diseases* 29:290–294.

Lodmell, D.L. 1983. Genetic control of resistance to street rabies virus in mice. *Journal of Experimental Medicine* 157:451–460.

Lodmell, D.L., N.B. Ray, and E.C. Ewalt. 1996. DNA immunization against rabies: Strategies for optimization of protective neutralizing antibody responses. In *Proceedings of the 7th annual international meeting: Advances towards rabies control in the Americas,* ed. C.E. Rupprecht. Atlanta: Centers for Disease Control and Prevention, p. 22.

Lord, R.D. 1988. Control of vampire bats. In *Natural history of vampire bats,* ed. A.M. Greenhall and U. Schmidt. Boca Raton, FL: CRC, pp. 215–226.

———. 1992. Seasonal reproduction of vampire bats and its relation to seasonality of bovine rabies. *Journal of Wildlife Diseases* 28:292–294.

Loza-Rubio, E., R. Vargas, E. Hernandez, D. Batalla, and A. Aguilar-Setien. 1996. Investigation of rabies virus strains in Mexico with a panel of monoclonal antibodies used to classify *Lyssavirus. Bulletin of the Pan American Health Organization* 30:31–35.

Maas, B. 1993. Bat-eared fox behavioural ecology and the incidence of rabies in the Serengeti National Park. *Onderstepoort Journal of Veterinary Research* 60:389–393.

MacDonald, D.W. 1980. *Rabies and wildlife: A biologist's perspective.* Oxford: Oxford University Press, 151 pp.

———. 1993. Rabies and wildlife: A conservation problem? *Onderstepoort Journal of Veterinary Research* 60:351–355.

MacInnes, C.D., R.R. Tinline, D.R. Voigt, L.H. Broekhoven, and R.R. Rosatte. 1988. Planning for rabies control in Ontario. *Reviews of Infectious Diseases* 10 (Suppl. 4):S665–S669.

Martorelli, L.F.A., E.A.C. Aguiar, M.F. Almeida, M.M.S. Silva, and E.C.R. Novaes. 1995. Isolation of the rabies virus in a specimen of the insectivorous bat *Myotis nigricans. Revista de Saude Publica* 29:140–141.

Martorelli, L.F.A., E.A.C. Aguiar, M.F. Almeida, M.M.S. Silva, and V.D.F.P. Nunes. 1996. Rabies virus isolation in insectivorous bat *Lasiurus borealis. Revista de Saude Publica* 30:101–102.

McGarvey, P.B., J. Hammond, M.M. Dienelt, D.C. Hooper, Z.F. Fu, B. Dietzschold, H. Koprowski, and F.H. Michaels. 1995. Expression of the rabies virus glycoprotein in transgenic tomatoes. *Bio-Technology (New York)* 13:1484–1487.

Mebatsion, T., J.H. Cox, and J.W. Frost. 1992. Isolation and characterization of 115 street rabies virus isolates from Ethiopia by using monoclonal antibodies: Identification of 2 isolates as Mokola and Lagos bat viruses. *Journal of Infectious Diseases* 166:972–977.

Meltzer, M.I. 1996. Assessing the costs and benefits of an oral vaccine for raccoon rabies: A possible model. *Emerging Infectious Diseases* 2:343–349.

Meredith, C.D., and E. Standing. 1981. Lagos bat virus in South Africa. *Lancet* 1:832–833.

Meredith, C.D., A.P. Rossouw, and H. van Praag Koch. 1971. An unusual case of human rabies thought to be of chiropteran origin. *South African Medical Journal* 45:767–769.

Meredith, C.D., L.H. Nel, and B.F. von Teichman. 1996. Further isolation of Mokola virus in South Africa. *Veterinary Record* 138:119–120.

Meslin, F.-X., D.B. Fishbein, and H.C. Matter 1994. Rationale and prospects for rabies elimination in developing countries. In *Lyssaviruses,* ed. C.E. Rupprecht, B. Dietzschold, and H. Koprowski. Berlin: Springer-Verlag, pp. 1–26.

Mitchell, K.D., and J.P. Heilman. 1996. Raccoon rabies: The urban challenge. In *Proceedings of the 7th annual international meeting: Advances towards rabies control in the Americas,* ed. C.E. Rupprecht. Atlanta: Centers for Disease Control and Prevention, p. 30.

Morimoto, K., M. Patel, S. Corisdeo, D.C. Hooper, Z.F. Fu, C.E. Rupprecht, H. Koprowski, and B. Dietzschold. 1996. Characterization of a unique variant of bat rabies virus responsible for newly emerging human cases in North America. *Proceedings of the National Academy of Sciences USA* 93:5653–5658.

Müller, J. 1971. The effect of fox reduction on the occurrence of rabies: Observations from two outbreaks of rabies in Denmark. *Bulletin Office Internationale des Epizooties* 75:763–776.

———. 1992. Shooting and poisoning of vector species as a means to control wildlife rabies. In *Wildlife rabies control,* ed. K. Bögel, F.X. Meslin, and M. Kaplan. Kent, UK: Wells Medical, pp. 181–182.

Murray, K., R. Rogers, L. Selvey, P. Selleck, A. Hyatt, A. Gould, L. Gleeson, P. Hooper, and H. Westbury. 1995. A novel morbillivirus pneumonia of horses and its transmission to humans. *Emerging Infectious Diseases* 1:31–33.

Nadin-Davis, S.A., W. Huang, and A.I. Wandeler. 1996. The design of strain-specific polymerase chain reactions for discrimination of the raccoon rabies virus strain from indigenous rabies viruses of Ontario. *Journal of Virological Methods* 57:1–14.

Nathanson, N., and F. Gonzalez-Scarano. 1991. Immune response to rabies virus. In *The natural history of rabies,* ed. G.M. Baer, 2d ed. Boca Raton, FL: CRC, pp. 145–161.

Nel, L.H., G.R. Thomson, and B.F. von Teichman. 1994. Molecular epidemiology of rabies virus in South Africa. *Onderstepoort Journal of Veterinary Research* 60:301–306.

Nichol, S.T., J.E. Rowe, and W.M. Fitch. 1993. Punctuated equilibrium and positive Darwinian evolution in vesicular stomatitis virus. *Proceedings of the National Academy of Sciences USA* 90:10,424–10,428.

Nietfeld, J.C., P.M. Rakich, D.E. Tyler, and R.W. Bauer. 1989. Rabies-like inclusions in dogs. *Journal of Veterinary Diagnostic Investigation* 1:333–338.

Niezgoda, M., C.A. Hanlon, and C.E. Rupprecht. 1993. Pathogenesis of experimental rabies virus in raccoons (*Procyon lotor*). In *Proceedings of the 42nd annual Wildlife Disease Association conference.* Guelph, Ontario: Wildlife Disease Association, p. 71.

Niezgoda, M., D.J. Briggs, J. Shaddock, D.W. Dreesen, and C.E. Rupprecht. 1997. Pathogenesis of experimentally induced rabies in domestic ferrets. *American Journal of Veterinary Research* 58:1327–1331.

Novelli, V.M., and P. Malankar. 1991. Epizootic of fox rabies in the Sultanate of Oman. *Transactions of the Royal Society of Tropical Medicine and Hygiene* 85:543.

Nowak, E. 1973. Habitat and distribution of the raccoon dog (*Nyctereutes procynoides* Gray) in Europe. *Beiträge zur Jagd und Wildforschung* 8:351–384.

Office International des Epizooties. 1992. *International animal health code: Mammals, birds and bees.* Paris: Office Internationale des Epizooties, 549 pp.

Okolo, M.I.O. 1988. Studies on dumb rabies in wild rats (*Rattus* species) in eastern Nigeria. *Revue Scientifique et Technique O.I.E.* 7:619–625.

Olfermann, E. 1995. *Small carnivore project at Mahazat as Said (Final Report).* Geneva: World Health Organization, Division of Communicable Diseases, Veterinary Public Health Unit, 15 pp.

Ondrejka, R., A. Durove, S. Svrcek, Z. Benisek, and J. Süliova. 1997. Isolation and identification of *Lyssavirus* strains from an area of Slovakia where oral antirabies vaccine was administered. *Veterinare Medicine Praha* 42:57–60.

Orr, P.H., M.R. Rubin, and F.Y. Aoki. 1988. Naturally acquired serum rabies neutralizing antibody in a Canadian Inuit population. *Arctic Medical Research* 47:699–700.

Pal, S.R., B. Arora, P.N. Chbuttani, S. Broor, S. Choudhury, R. M. Joshi, and S.D. Ray. 1980. Rabies virus infection of a flying fox bat, *Pteropus poliocephalus* in Chandigarh, Northern India. *Tropical and Geographic Medicine* 32:265–267.

Pan American Health Organization. 1995. Vigilancia epidemiologica de la rabia en las Americas 1995. *Boletin de Vigilancia epidemiologica de la Rabia en Las Americas* 27:1–27.

Pasteur, L., C. Chamberland, and E. Roux. 1884. Sur la rage. *Bulletin Academie Medicine* 25:661–664.

Perl, D.P., and P.F. Good. 1991. The pathology of rabies in the central nervous system. In *The natural history of rabies,* ed. G.M. Baer, 2d ed. Boca Raton, FL: CRC, pp. 163–190.

Perry, L.L., and D.L. Lodmell. 1991. Role of CD4+ and CD8+ T cells in murine resistance to street rabies virus. *Journal of Virology* 65:3429–3434.

Prehaud, C., P. Coulon, F. Lafay, C. Thiers, and A. Flamand. 1988. Antigenic site II of the rabies virus glycoprotein: Structure and role in viral virulence. *Journal of Virology* 62:1–7.

Prehaud, C., R.D. Harris, V. Fulop, C.L. Koh, J. Wong, A. Flamand and D.H. Bishop. 1990. Expression, characterization, and purification of a phosphorylated rabies nucleoprotein synthesized in insect cells by baculovirus vectors. *Virology* 178:486–497.

Prevec, L., J.B. Campbell, B.S. Christie, L. Belbeck, and F.L. Graham. 1990. A recombinant human adenovirus vaccine against rabies. *Journal of Infectious Diseases* 161:27–30.

Pybus, M.J. 1986. Rabies in insectivorous bats of western Canada 1979 to 1983. *Journal of Wildlife Diseases* 22:307–313.

Ray, N.B., L.C. Ewalt, and D.L. Lodmell. 1995. Rabies virus replication in primary murine bone marrow macrophages and in human and murine macrophage-like cell lines: Implications for viral persistence. *Journal of Virology* 69:764–772.

Reagan, K.J., and W.H. Wunner. 1985. Rabies virus interaction with various cell lines is independent of the acetylcholine receptor. *Archives of Virology* 84:277–282.

Reinius, S. 1992. Epidemiology of fox/raccoon dog rabies in Finland. In *Wildlife rabies control,* ed. K. Bögel, F.X. Meslin, and M. Kaplan. Kent, UK: Wells Medical, pp. 32–34.

Robbins, A.H., M. Borden, M. Niezgoda, L. Feinstein, B.S. Windmiller, M.W. McGuill, C.E. Rupprecht, and S.L. Rowell. 1996. Oral rabies vaccination of raccoons (*Procyon lotor*) on the Cape Cod isthmus, Massachusetts. In *Proceedings of the 7th annual international meeting: Advances towards rabies control in the Americas,* ed. C.E. Rupprecht. Atlanta: Centers for Disease Control and Prevention, p. 28.

Rosatte, R.C. 1988. Rabies in Canada: History, epidemiology, and control. *Canadian Veterinary Journal* 29:362–365.

Rosatte, R.C., and J.R. Gunson. 1984. Presence of neutralizing antibodies to rabies virus in striped skunks from areas free of skunk rabies in Alberta. *Journal of Wildlife Diseases* 20:171–176.

Rosatte, R.C., C.D. MacInnes, M.J. Power, D.H. Johnston, P. Bachmann, C.P. Nunan, C. Wannop, M. Pedde, and L. Calder. 1993. Tactics for the control of wildlife rabies in Ontario (Canada). *Revue Scientifique et Technique O.I.E.* 12:95–98.

Rosatte, R.C., C.D. MacInnes, R.T. Williams, and O. Williams. 1997. A proactive prevention strategy for raccoon rabies in Ontario, Canada. *Wildlife Society Bulletin* 25:110–116.

Roscoe, D.E., F.E. Sorhage, W.C. Holste, K. Kirk-Pfugh, C. Campbell, and C.E. Rupprecht. 1996. Vaccinia-rabies glycoprotein (V-RG) vaccine field trial, NJ 1992–1995. In *Proceedings of the 7th annual international meeting: Advances towards rabies control in the Americas,* ed. C.E. Rupprecht. Atlanta: Centers for Disease Control and Prevention, p. 29.

Roux, E. 1887. Note sur un moyen de conserver les moelles rabiques ave leur virulence. *Annals of the Institute Pasteur (Paris)* 1:87.

Rudd, R.J., and C.V. Trimarchi. 1987. Comparison of sensitivity of BHK-21 and murine neuroblastoma cells in the isolation of a street strain rabies virus. *Journal of Clinical Microbiology* 25:1456–1458.

Rupprecht, C.E., and J.E. Childs. 1996. Feline rabies. *Feline Practice* 24:15–19.

Rupprecht, C.E., and B. Dietzschold. 1987. Perspectives on rabies virus pathogenesis. *Laboratory Investigation* 57:603–606.

Rupprecht, C.E., and J.S. Smith. 1994. Raccoon rabies: The re-emergence of an epizootic in a densely populated area. *Seminars in Virology* 5:155–164.

Rupprecht, C.E., T.J. Wiktor, D.H. Johnston, A.N. Hamir, B. Dietzschold, W.H. Wunner, L.T. Glickman, and H. Koprowski. 1986. Oral immunization and protection of raccoons (*Procyon lotor*) with a vaccinia-rabies glycoprotein recombinant virus vaccine. *Proceedings of the National Academy of Sciences USA* 83:7947–7950.

Rupprecht, C.E., L.T. Glickman, P.A. Spencer, and T.J. Wiktor. 1987. Epidemiology of rabies virus variants: Differentiation using monoclonal antibodies and discriminant analysis. *American Journal of Epidemiology* 126:298–309.

Rupprecht, C.E., B. Dietzschold, W.H. Wunner, and H. Koprowski. 1991. Antigenic relationships of lyssaviruses. In *The natural history of rabies,* ed. G.M. Baer, 2d ed. Boca Raton, FL: CRC, pp. 69–100.

Rupprecht, C.E., C.A. Hanlon, L.B. Cummins, and H. Koprowski. 1992a. Primate responses to a vaccinia-rabies glycoprotein recombinant virus vaccine. *Vaccine* 10:368–374.

Rupprecht, C.E., C.A. Hanlon, A.N. Hamir, and H. Koprowski. 1992b. Oral wildlife rabies vaccination: Development of a recombinant virus vaccine. *Transactions of the North American Wildlife and Natural Resource Conference* 57:439–452.

Rupprecht, C.E., C.A. Hanlon, M. Niezgoda, J.R. Buchanan, D. Diehl, and H. Koprowski. 1993. Recombinant rabies vaccines: Efficacy assessment in free-ranging animals. *Onderstepoort Journal of Veterinary Research* 60:463–468.

Rupprecht, C.E., J.S. Smith, M. Fekadu, and J.E. Childs. 1995. The ascension of wildlife rabies: A cause for public health concern or intervention? *Emerging Infectious Diseases* 1:107–114.

Rupprecht, C.E., J.S. Smith, P.A. Yager, L.A. Orciari, J.S. Shaddock, D. Sanderlin, M. Niezgoda, S. Whitfield, H. Shoemake, C.K. Warner, L. Gleeson, and K. Murray. 1996a. Preliminary analysis of a new *Lyssavirus* isolated from an Australian fruit bat. In *Proceedings of the 7th annual international meeting: Advances towards rabies control in the Americas,* ed. C.E. Rupprecht. Atlanta: Centers for Disease Control and Prevention, p. 19.

Rupprecht, C.E., J.S. Smith, J. Krebs, M. Niezgoda, and J.E. Childs. 1996b. Current issues in rabies prevention in the United States: Health dilemmas, public coffers, private interests. *Public Health Reports* 111:400–407.

Sacramento, D., H. Bourhy, and N. Tordo. 1991. PCR technique as an alternative method for diagnosis and molecular epidemiology of rabies virus. *Molecular and Cellular Probes* 5:229–240.

Saluzzo, J.-F., P.E. Rollin, C. Dauguet, J.-P. Digoutte, A.-J Georges, and P. Sureau. 1984. Premier isolement du virus Mokola a partir d'un rongeur (*Lophuromys sikapusi*). *Annals of Virology (Paris)* 135E:57–66.

Sama, S.M., S.B. Sharma, and S.N. Saxena. 1986. Surveillance of rabies in animals in North India. *Journal of Communicable Diseases* 18:9–12.

Sasaki, D.M., C.R. Middleton, T.R. Sawa, C.C. Christensen, and G.Y. Kobayashi. 1992. Rabid bat diagnosed in Hawaii. *Hawaii Medical Journal* 51:181–185.

Schneider, L.G. 1991. Spread of virus from the central nervous system. In *The natural history of rabies,* ed. G.M. Baer, 2d ed. Boca Raton, FL: CRC, pp. 133–144.

Schneider, L.G., and J.H. Cox. 1994. Bat lyssaviruses in Europe. In *Lyssaviruses,* ed. C.E. Rupprecht, B. Dietschold, and H. Koprowski. Berlin: SpringerVerlag, pp. 207–218.

Schneider, L.G., G. Wachendörfer, E. Schmittdiel, and J.H. Cox. 1983. A field trial for the oral immunization of foxes against rabies in the Federal Republic of Germany: II. Planning, implementation and evaluation of the field trial. *Tierärztliche Umschau* 38:476–480.

Schneider, L.G., B.J.H. Barnard, and H.P. Schneider. 1985. Application of monoclonal antibodies for epidemiological investigations and oral vaccination studies. 1. African viruses. In *Rabies in the tropics,* ed. E. Kuwert, C. Mérieux, H. Koprowski, and K. Bögel. Berlin: Springer-Verlag, pp. 47–59.

Schneider, L.G., J.H. Cox, W.W. Muller, and K.P. Hohnsbeen. 1987. Field trials for oral immunization of foxes against rabies in Germany: Current situation. *Tierärztliche Umschau* 38:184–198.

———. 1988. Current oral rabies vaccination in Europe: An interim balance. *Revue Infectious Diseases* 10 (Suppl. 4):S654–S659.

Schneider, L.G., U. Wilhelm, and J.H. Cox. 1989. Logistics and execution of the field trial for the oral immunization of foxes against rabies. In *11th International symposium of the World Association of Veterinary Microbiologists, Immunologists and Specialists in Infectious Diseases.* Bologna: Società Editrice Esculapio, pp. 134–139.

Schneider, M.C., C. Santos-Burgoa, J. Aron, B. Munoz, S. Ruiz-Velazco, and W. Uieda. 1996. Potential force of infection of human rabies transmitted by vampire bats in the Amazonian region of Brazil. *American Journal of Tropical Medicine and Hygiene* 55:680–684.

Schumacher, C.L., P. Coulon, F. Lafay, J. Benejean, M.F.A. Aubert, J. Barrat, A. Aubert, and A. Flamand. 1993. SAG-2 oral rabies vaccine. *Onderstepoort Journal of Veterinary Research* 60:459–462.

Seif, I., P. Coulon, P.E. Rollin, and A. Flamand. 1985. Rabies virulence: Effect on pathogenicity and sequence characterization of rabies virus mutations affecting antigenic site III of the glycoprotein. *Journal of Virology* 53:926–934.

Selimov, M.A., A.G. Tatarov, A.D. Botvinkin, E.V. Klueva, L.G. Kulikova, and N.A. Khismatullina. 1989. Rabies-related Yuli virus: Identification with a panel of monoclonal antibodies. *Acta Virologica (Praha)* 33:542–546.

Shankar, V., B. Dietzschold, and H. Koprowski. 1991. Direct entry of rabies virus into the central nervous system without prior local replication. *Journal of Virology* 65:2736–2738.

Shope, R.E., F.A. Murphy, A.K. Harrison, O.R. Causey, G.E. Kemp, D.I.H. Simpson, and D.L. Moore. 1970. Two African viruses serologically and morphologically related to rabies virus. *Journal of Virology* 6:690–692.

Sikes, R.K., Jr. 1981. Rabies. In *Infectious diseases of wild mammals,* ed. J.W. Davis, L.H. Karstad, and D.O. Trainer. Ames: Iowa State University Press, pp. 3–17.

Sillero-Zubiri, C., A.A. King, and D.W. MacDonald. 1996. Rabies and mortality in Ethiopian wolves (*Canis simensis*). *Journal of Wildlife Diseases* 32:80–86.

Sinnecker, H., L. Apitzsch, D. Berndt, C. Schrader, J. Gogolin, and J. Egert. 1990. Development of SAD Potsdam 5/88 rabies live vaccine virus for oral vaccination of foxes, and its characterisation by a mouse model. *Monatshefte für Veterinärmedizin* 45:77–78.

Skinner, J.D., and R.H.N. Smithers. 1990. *The mammals of the southern African subregion.* Pretoria: University of Pretoria, 771 pp.

Smart, N.L., and K.M. Charlton. 1992. The distribution of Challenge virus standard rabies virus versus skunk street rabies virus in the brains of experimentally infected rabid skunks. *Acta Neuropathologica (Berlin)* 84:501–508.

Smith, J.S. 1991. Rabies serology. In *The natural history of rabies,* ed. G.M. Baer, 2d ed. Boca Raton, FL: CRC, pp. 235–252.

———. 1995. Rabies virus. In *Manual of clinical microbiology,* ed. P.R. Murray, E.J. Baron, M.A. Pfaller, F.C. Tenover, and R.H. Yolken, 6th ed. Washington, DC: American Society for Microbiology, pp. 997–1003.

———. 1996. New aspects of rabies with emphasis on epidemiology, diagnosis, and prevention of the disease in the United States. *Clinical Microbiology Reviews* 9:166–176.

Smith, J.S., and H.D. Seidel. 1993. Rabies: A new look at an old disease. *Progress in Medical Virology* 40:82–106.

Smith, J.S., D.B. Fishbein, C.E. Rupprecht, and K. Clark. 1991. Unexplained rabies in three immigrants in the United States: A virologic investigation. *New England Journal of Medicine* 324:205–211.

Smith, J.S., L.A. Orciari, P.A. Yager, H.D. Seidel, and C.K. Warner. 1992. Epidemiologic and historical relationships among 87 rabies virus isolates as determined by limited sequence analysis. *Journal of Infectious Diseases* 166:296–307.

Smith, J.S., L.A. Orciari, and P.A. Yager. 1995. Molecular epidemiology of rabies in the United States. *Seminars in Virology* 6:387–400.

Smith, J.S., P.A. Yager, and G.M. Baer. 1996. A rapid fluorescent focus inhibition test (RFFIT) for determining rabies virus-neutralizing antibody. In *Laboratory techniques in rabies,* ed. F.X. Meslin, M. Kaplan, and H. Koprowski, 4th ed. Geneva: World Health Organization, pp. 181–192.

Snyman, P.S. 1940. The study and control of the vectors of rabies in South Africa. *Onderstepoort Journal of Veterinary Science and Animal Industry* 15:9–140.

Speare, R., L. Skerratt, R. Foster, L. Berger, P. Hooper, R. Lunt, D. Blair, D. Hansman, M. Goulet, and S. Cooper. 1997. Australian bat *Lyssavirus* infection in three fruit bats from north Queensland. *Communicable Diseases Intelligence* 21:117–121.

Spriggs, D.R. 1985. Rabies pathogenesis: Fast times at the neuromuscular junction. *Journal of Infectious Diseases* 152:1362–1363.

Steck, F., and A. Wandeler. 1980. The epidemiology of fox rabies in Europe. *Epidemiologic Reviews* 2:71–96.

Steck, F., P. Addy, E. Schipper, and A. Wandeler. 1968. Der bisherige Verlauf des Tollwutseuchenzuges in der Schweiz. *Schweizer Archiv für Tierheilkunde* 110:597–616.

Steck, F., A. Wandeler, P. Bichsel, S. Capt, and L.G. Schneider. 1982. Oral immunization of foxes against rabies: A field study. *Zentralblatt Veterinärmedizin Reihe B* 29:372–396.

Steece, R.S., and C.H. Calisher. 1989. Evidence for prenatal transfer of rabies virus in the Mexican free-tailed bat (*Tadarida brasiliensis mexicana*). *Journal of Wildlife Diseases* 25:329–334.

Steele, J.H., and P.J. Fernandez. 1991. History of rabies and global aspects. In *The natural history of rabies,* ed. G.M. Baer, 2d ed. Boca Raton, FL: CRC, pp. 1–24.

Stöhr, K. 1990. *Emerging rabies in humans and animals in the Sultanate of Oman: Report of a short-term mission—9–19 December 1990.* Rome: Food and Agriculture Organization, 39 pp.

———. 1995. *Rabies control in Oman: Mission report, 22–25 April 1995.* Geneva: World Health Organization, Division of Communicable Diseases, Veterinary Public Health Unit, 27 pp.

Stöhr, K., and F.X. Meslin. 1996. Progress and setbacks in oral immunization of foxes against rabies in Europe. *Veterinary Record* 139:32–35.

Stöhr, K., E. Karge, H. Gädt, R. Kokles, W. Ehrentraut, W. Wit, and H.G. Fink. 1990. Oral immunization of free-living foxes against rabies: Preparation and implementation of first field trials conducted in East Germany. *Monatshefte für Veterinärmedizin* 45:782–786.

Stöhr, K., P. Stöhr, and T. Müller. 1994. Results and experiences of oral vaccination of foxes against rabies in eastern Germany. *Tierärztliche Umschau* 49:203.

Süliova, J., Z. Benisek, S. Svrcek, A. Durove, and R. Ondrejka. 1997. The effectiveness of inactivated, purified and concentrated experimental rabies vaccine for veterinary use: Immunogenic activity. *Veterinarni Medicina (Praha)* 42:51–56.

Sureau, P., Tignor, G.H., and A.L. Smith. 1980. Antigenic characterization of the Bangui strain (ANCB-672d) of Lagos bat virus. *Annals of Virology* 131:25–32.

Sureau, P., P. Ravisse, and P.E. Rollin. 1991. Rabies diagnosis by animal inoculation, identification of Negri bodies, or ELISA. In *The natural history of rabies,* ed. G.M. Baer, 2d ed. Boca Raton, FL: CRC, pp. 203–217.

Svrcek, S., A. Durove, J. Sokol, J. Süliova, R. Ondrejka, J. Zavadova, Z. Benisek, D. Lani, M. Selimov, J. Vrtiak, D. Feketeova, A. Zurek, B. Lovas, J. Hojsik, O. Borsukova, J. Kuninec, and M. Juska. 1994. Immunoprophylaxis of rabies in wild carnivores. *Slovenska Veterinarna Caseta* 2:50–55.

Svrcek, S., J. Sokol, A. Durove, B. Lovas, J. Zavadova, J. Süliova, R. Ondrejka, Z. Benisek, O. Borsukova, J. Kubinec, and M. Madar. 1996. The current epizootiological situation and immunoprophylaxis of rabies in animals in the Slovak Republic. In *Selected papers on rabies prophylaxis,* ed. D. Lalosevic. Novi Sad, Yugoslavia: Pasteur Institute Novi Sad, pp. 99–107.

Swanepoel, R. 1994. Rabies. In *Infectious diseases of livestock,* ed. J.A.W. Coetzer, G.R. Thomson, and R.C. Tustin. Cape Town, South Africa: Oxford University Press, pp. 493–552.

Tan, D.S.K., A.J. Beck, and M. Omar. 1969. The importance of Malaysian bats in the transmission of viral disease. *Medical Journal of Malaya* 24:32–35.

Taylor, J., and E. Paoletti. 1988. Fowlpox virus as a vector in non-avian species. *Vaccine* 6:466–468.

Taylor, M., B. Elkin, N. Maier, and M. Bradley. 1991. Observation of a polar bear with rabies. *Journal of Wildlife Diseases* 27:337–339.

Thomson, G.R., and C.D. Meredith. 1993. Rabies in bat-eared foxes in South Africa. *Onderstepoort Journal of Veterinary Research* 60:399–403.

Thongcharoen, P., C. Wasi, P. Puthavathana, and L. Chavanich. 1985. Rabies in Thailand. In *Rabies in the tropics,* ed. E. Kuwert, H. Mérieux, H. Koprowski, and K. Bögel. Berlin: Springer-Verlag, pp. 556–565.

Tidemann, C.R, M.J. Vardon, J.E. Nelson, R. Speare, and L.J. Gleeson. 1997. Health and conservation implications of Australian bat *Lyssavirus*. *Australian Zoologist* 30:369–376.

Tignor, G.H., F.A. Murphy, H.F. Clark, R.E. Shope, P. Madore, S.P. Bauer, S.M. Buckley, and C.D. Meredith. 1977. Duvenhage virus: Morphological, biochemical, histopathological and antigenic relationships to the rabies serogroup. *Journal of General Virology* 37:595–611.

Tinline, R.R. 1988. Persistence of rabies in wildlife. In *Rabies,* ed. J.B. Campbell and K.M. Charlton. Boston: Kluwer Academic, pp. 301–322.

Toacsen, E., and S. Moraru. 1995. The evolution of rabies in Romania: The biological characteristics of the strains of the rabies virus isolated in Romania. *Bacteriology, Virusology, Parazitology, Epidemiology* 40:135–140.

Tolson, N.D., K.M. Charlton, R.B. Stewart, G.A. Casey, W.A. Webster, K. MacKenzie, J.B. Campbell, and K.F. Lawson. 1990. Mutants of rabies viruses in skunks: Immune response and pathogenicity. *Canadian Journal of Veterinary Research* 54:178–183.

Toma, B., and L. Andral. 1977. Epidemiology of fox rabies. In *Advances in virus research,* ed. B.A. Lauffer, F.B. Bang, K. Maramorosch, and K.M. Smith. New York: Academic, pp. 1–36.

Tordo, N., and A. Kouknetzoff. 1993. The rabies virus genome: An overview. *Onderstepoort Journal of Veterinary Research* 60:263–269.

Trimarchi, C.V., and J. Debbie. 1977. Naturally occurring rabies virus and neutralizing antibody in two species of insectivorous bats of New York State. *Journal of Wildlife Diseases* 13:366–369.

———. 1991. The fluorescent antibody in rabies. In *The natural history of rabies,* ed. G.M. Baer, 2d ed. Boca Raton, FL: CRC, pp. 219–233.

Tsiang, H. 1993. Pathophysiology of rabies virus infection of the nervous system. *Advances in Virus Research* 42:375–412.

Tuffereau, C., H. Leblois, J. Benejean, P. Coulon, F. Lafay, and A. Flamand. 1989. Arginine or lysine in position 333 of ERA and CVS glycoprotein is necessary for rabies virulence in adult mice. *Virology* 172:206–212.

Uieda, W., N.M.S. Harmani, and M.M.S. Silva. 1995. Rabies in insectivorous bats (Molossidae) of southeastern Brazil. *Revista de Saude Publica* 29:393–397.

Velleca, W.M., and F.T. Forrester. 1981. *Laboratory methods for detecting rabies.* Atlanta: Centers for Disease Control, 159 pp.

von Teichman, B.F., G.R. Thomson, C.D. Meredith, and L.H. Nel. 1995. Molecular epidemiology of rabies virus in South Africa: Evidence for two distinct virus groups. *Journal of General Virology* 76:73–82.

Wachendörfer, G., and J.W. Frost. 1992. Epidemiology of red fox rabies: A review. In *Wildlife rabies control,* ed. K. Bögel, F.X. Meslin, and M. Kaplan. Kent, UK: Wells Medical, pp. 19–31.

Wandeler, A. 1988. Control of wildlife rabies: Europe. In *Rabies,* ed. J.B. Campbell and K.M. Charlton. Boston: Kluwer Academic, pp. 365–380.

———. 1991. Oral immunization of wildlife. In *The natural history of rabies,* ed. G.M. Baer, 2d ed. Boca Raton, FL: CRC, pp. 485–503.

———. 1993. Wildlife rabies in perspective. *Onderstepoort Journal of Veterinary Research* 60:347–350.

Wandeler, A., G. Wachendörfer, U. Förster, H. Krekel, U. Schale, J. Müller, and F. Steck. 1974. Rabies in wild carnivores in central Europe: I. Epidemiological studies. *Zentralblatt Veterinärmedizin Reihe B* 21:735–756.

Wandeler, A., W. Bauder, S. Prochaska, and F. Steck. 1982. Small mammal studies in a SAD baiting area. *Comparative Immunology, Microbiology and Infectious Diseases* 5:173–176.

Wandeler, A., S. Capt, A. Kappeler, and R. Hauser. 1988. Oral immunization of wildlife against rabies: Concept and first field experiments. *Reviews of Infectious Diseases* 10 (Suppl. 4):S649–S653.

Wandeler, A., S.A. Nadin-Davis, R.R. Tinline, and C.E. Rupprecht. 1994. Rabies epidemiology: Some ecological and evolutionary perspectives. In *Lyssaviruses,* ed. C.E. Rupprecht, B. Dietzschold, and H. Koprowski. Berlin: Springer-Verlag, pp. 297–324.

Warner, C.K., S.G. Whitfield, M. Fekadu, and H. Ho. 1997. Procedures for the reproducible detection of rabies virus antigen, mRNA and genome in situ in formalin-fixed tissues. *Journal of Virological Methods* 67:5–12.

Weiler, G.J., G.W. Garner, and D.G. Ritter. 1995. Occurrence of rabies in a wolf population in northeastern Alaska. *Journal of Wildlife Diseases* 31:79–82.

Westergaard, J.M. 1982. Measures applied in Denmark to control the rabies epizootic in 1977–1980. *Comparative Immunology, Microbiology and Infectious Diseases* 5:383–387.

Westerling, B. 1989. A field trial on oral immunization of raccoon dogs and foxes against rabies in Finland 1988–89. *Rabies Bulletin Europe* 2:9–212.

Whitby, J.E., P. Johnstone, G. Parsons, and A.A. King. 1996. Ten-year survey of British bats for the existence of rabies. *Veterinary Record* 139:491–493.

White, P.C.L., S. Harris, and G.C. Smith. 1995. Fox contact behaviour and rabies spread: A model for the estimation of contact probabilities between urban foxes at different population densities and its implications for rabies control in Britain. *Journal of Applied Ecology* 32:693–706.

Wiktor, T.J., R.I. Macfarlan, K.J. Reagan, B. Dietzschold, P.J. Curtis, W.H. Wunner, M.P. Kieny, R. Lathe, J.P. Lecocq, M. Mackett, M. Moss, and H. Koprowski. 1984. Protection from rabies by a vaccinia virus recombinant containing the rabies virus glycoprotein gene. *Proceedings of the National Academy of Sciences* 81:7194–7198.

Wiktor, T.J., R.I. Macfarlan, B. Dietzschold, C. Rupprecht, and W.H. Wunner. 1985. Immunogenic properties of vaccinia recombinant virus expressing the rabies glycoprotein. *Annals of the Institute Pasteur/Virology* 136E:405–411.

Wilkinson, L. 1988. Understanding the nature of rabies: An historical perspective. In *Rabies,* ed. J.B. Campbell and K.M. Charlton. Boston: Kluwer Academic, pp. 1–23.

Winkler, W.G. 1975. Fox rabies. In *The natural history of rabies,* ed. G.M. Baer. New York: Academic, pp. 3–22.

———. 1991. Rodent rabies. In *The natural history of rabies,* ed. G.M. Baer, 2d ed. Boca Raton, FL: CRC, pp. 405–410.

———. 1992. A review of the development of the oral vaccination technique for immunizing wildlife against rabies. In *Wildlife rabies control,* ed. K. Bögel, F.X. Meslin, and M. Kaplan. Kent, UK: Wells Medical, pp. 82–96.

Winkler, W.G., and K. Bögel. 1992. Control of rabies in wildlife. *Scientific American* 266:86–92.

Winkler, W.G., and S.R. Jenkins. 1991. Raccoon rabies. In *The natural history of rabies,* ed. G.M. Baer, 2d ed. Boca Raton, FL: CRC, pp. 325–340.

Winkler, W.G., J.S. Shaddock, and C. Bowman. 1985. Rabies virus in salivary glands of raccoons (*Procyon lotor*). *Journal of Wildlife Diseases* 21:297–298.

World Health Organization (WHO). 1980. *Report of a consultation on rabies prevention and control, Lyon, France, March 1980* (WHO/Rabies/80.188). Geneva: WHO, 16 pp.

———. 1982. *Report on a consultation on oral and enteric mass immunization of wildlife, Geneva, Switzerland, September 1982* (WHO/Rab. Res./82.16). Geneva: WHO, 12 pp.

———. 1987. *Report of workshop on oral immunization of wildlife against rabies in Europe (INTORAL), Tübingen, Germany, October 1986* (WHO/Rab.Res./87.23). Geneva: WHO, 6 pp.

———. 1988. *Report of 2nd international IMVIEssen/WHO symposium on new developments in rabies control, Essen, Germany, July 1988* (WHO/Rab.Res./88.31). Geneva: WHO, 13 pp.

———. 1989. *Report of consultation on requirements and criteria for field trials on oral rabies vaccination of dogs and wild carnivores, Geneva, Switzerland 1–2 March 1989* (WHO/Rab.Res./89.32). Geneva: WHO, 22 pp.

———. 1990. *Report of the seminar on wildlife rabies control, Geneva, Switzerland, July 1990* (WHO/CDS/VPH/90.93). Geneva: WHO, 25 pp.

———. 1991. *Report of the 2nd consultation on oral immunization of dogs against rabies, Geneva, Switzerland, July 1990* (WHO/Rab.Res./91.37). Geneva: WHO, 14 pp.

———. 1992a. *Report of the 3rd consultation on oral immunization of dogs against rabies, Geneva, Switzerland, July 1992* (WHO/Rab.Res./92.38). Geneva: WHO, 19 pp.

———. 1992b. WHO Expert Committee on Rabies, Eighth Report. *WHO Technical Report Series* 824:1–84.

———. 1993. *Report of the 4th consultation on oral Immunization of dogs against rabies, Geneva, Switzerland, June 1993* (WHO/Rab.Res./93.42), Geneva: WHO, 17 pp.

———. 1994a. *Report of a workshop on prevention and control of rabies in Baltic countries, Tallin, Estonia 14–15 February 1994* (WHO/CDC/VPH/94.133). Geneva: WHO, 17 pp.

———. 1994b. *Report of the 5th consultation on oral immunization of dogs against rabies, Geneva, Switzerland, June 1994* (WHO/Rab.Res./94.45). Geneva: WHO, 24 pp.

———. 1995. *WHO workshop on genetic and antigenic molecular epidemiology of lyssaviruses, Niagra Falls, Canada 17 November 1994* (WHO/Rab.Res./94.46). Geneva: WHO, 13 pp.

———. 1996. *World survey of rabies No. 30 for the year 1994* (WHO/EMC/ZOO/96.3). Geneva: WHO, 29 pp.

———. 1997. *Report of the 3rd international symposium on rabies in Asia, Wuhan, China, 11–15 September 1996* (WHO/EMC/ZOO/96.8). Geneva: WHO, 27 pp.

Wunner, W.H. 1991. The chemical composition and molecular structure of rabies viruses. In *The natural history of rabies,* ed. G.M. Baer, 2d ed. Boca Raton, FL: CRC, pp. 31–67.

Xiang, Z.Q., B.B. Knowles, J.W. McCarrick, and H.C.J. Ertl. 1995. Immune effector mechanisms required for protection to rabies virus. *Virology* 214:398–404.

Yarosh, O.K., A.I. Wandeler, F.L. Graham, J.B. Campbell, and L. Prevec. 1996. Human adenovirus type 5 vectors expressing rabies glycoprotein. *Vaccine* 14:1257–1264.

Yusibov, V., A. Modelska, K. Steplewski, M. Agadjanyan, D. Weiner, D.C. Hooper, and H. Koprowski. 1997. Antigens produced in plants by infection with chimeric plant viruses immunize against rabies virus and HIV-1. *Proceedings of the National Academy of Sciences USA* 94:5784–5788.

Ziemen, E. 1982. The effect of rabies on different fox populations in the southwest of the Federal Republic of Germany. *Comparative Immunology, Microbiology and Infectious Diseases* 5:257–289.

Zumpt, I.F. 1982. The yellow mongoose as a rabies vector on the central plateau of South Africa. *South African Journal of Science* 78:417–418.

2 MORBILLIVIRAL DISEASES

RINDERPEST

PAUL ROSSITER

Synonyms: Cattle plague, la peste bovine, steppe murrain.

INTRODUCTION AND HISTORY. Rinderpest has probably had more impact on humans and domestic livestock than any other animal disease. In its worst and most devastating form, *cattle plague,* it can destock whole areas of cattle *Bos taurus,* impoverishing the owners and creating great hardship in rural communities. Indirectly, it has even caused the loss of human life. The Great Ethiopian Famine of 1888–92, in which perhaps a third of the country's human population died, was the result of the loss of most of the region's draft oxen to the pandemic of rinderpest introduced into northeastern Africa in 1887 (Pankhurst 1966). Even in its mildest forms, which are usually seen in long-established endemic foci, the disease causes continual low levels of mortality in young stock, causes poor performance, prohibits stock improvement through introduction of more productive and usually more susceptible breeds, and restricts trade. Perhaps the only positive aspect of rinderpest is that its control was a major stimulus for establishment of veterinary schools in Europe in the 18th century.

Some misinformed wildlife enthusiasts erroneously believe rinderpest is good for Africa and Asia because it continually reduces cattle populations. However, pastoral livestock owners perpetually threatened by rinderpest protect themselves by keeping large herds of cattle in which a few animals will always survive, rather than smaller herds of more productive breeds that would be less destructive to the environment but highly vulnerable to rinderpest.

The rinderpest virus (RPV) can probably infect all cloven-hoofed animals. Rinderpest in wild species has been reviewed by Scott (1981) and Plowright (1968, 1982). The historical and scientific detail in the chapter by Scott (1981) is difficult to improve upon, and this present chapter supplements rather than replaces it. There are comprehensive reviews of all other aspects of rinderpest in domesticated as well as wild animals (Scott 1964; Plowright 1968; Rossiter 1994).

DISTRIBUTION. Rinderpest is present in parts of East Africa, Pakistan and some neighboring areas of southern Asia, parts of the Middle East, and, possibly, in central Asia (Fig. 2.1). Despite occasional setbacks,

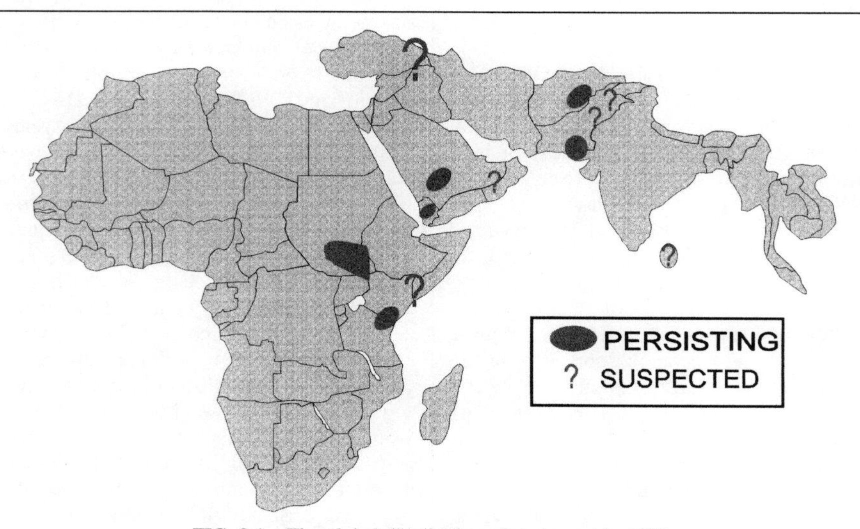

FIG. 2.1—The global distribution of rinderpest in 1997.

the global distribution and annual incidence of rinderpest have decreased steadily during the past 15 years. The Global Rinderpest Eradication Programme (GREP) aims to eradicate rinderpest from the world by the year 2010. As a result, it is increasingly unlikely that rinderpest should ever be responsible for serious mortalities in wildlife populations outside Africa.

HOST RANGE. There is no conclusive new addition to the lists of susceptible species prepared by Scott (1981), though it is possible there are still several species of antelope and other bovids in which rinderpest has not yet been reported, particularly in central and southeast Asia. Both lesser kudu *Tragelaphus imberbis* and greater kudu *Tragelaphus strepsiceros* are naturally affected. Outside Africa and Asia, Cervidae might be expected to be the main group of target species because deer were reported to have been severely affected during the last British epidemic in 1865 (Scott 1981) and white-tailed deer *Odocoileus virginianus* were experimentally infected in North America (Hamdy et al. 1975).

ETIOLOGY. Rinderpest virus is classified in the genus *Morbillivirus,* family *Paramyxoviridae*. The first indication that rinderpest was closely related to other morbilliviruses was shown when dogs living on the Kabete Veterinary Laboratory compound in Kenya that were fed the carcasses of goats *Capra hircus* used for production of caprinized rinderpest vaccine remained unaffected by a severe epidemic of canine distemper (Polding and Simpson 1957). Recent molecular studies determined the entire nucleotide sequence of RPV (Baron and Barrett 1995) and indicate it may be the oldest morbillivirus from which the other members of the genus evolved (Norrby et al. 1985; Chamberlain et al. 1993).

TRANSMISSION AND EPIDEMIOLOGY. Infected animals excrete virus in all secretions and excretions from 24 to 48 hours before the onset of clinical signs either until death or after development of high titers of antibody 7–10 days later (Scott 1964; Plowright 1968). The virus is fragile, highly susceptible to inactivation by physical and chemical means, and usually survives for only a few hours outside the host. Transmission, therefore, requires close contact between infected and susceptible animals. Airborne spread, other than for a few meters inside animal pens, has never been proven to occur outdoors. Recovered animals are solidly immune for the rest of their life and never reexcrete the virus. Indirect transfer of virus by fomites such as vehicles, fodder, and veterinarians is rare and arthropod-borne transmission is unknown.

Consequently, rinderpest is maintained and spread by mixing of infected with susceptible animals. Virtually all new outbreaks can be traced to the introduction of unvaccinated stock, as several recent outbreaks in Africa and Asia have shown (Wamwayi et al. 1992; Rossiter et al. 1998). The last confirmed outbreak of rinderpest in Western Europe was in the Rome Zoo after importation of infected wildlife from Somalia (Cilli and Roetti 1951).

The virus is maintained by continuous cycles of infection in domestic cattle and African buffalo *Syncerus caffer.* Wildlife becomes infected through close contact with infected cattle along the borders of wildlife preserves or at water sources and grazing areas. This is more common in times of drought. Infected wildlife transmit the disease among themselves and can carry it over significant distances to reinfect cattle not epidemiologically linked to the original source of the infection (Carmichael 1938; Plowright 1968; Woodford 1984).

It is widely accepted that wildlife are incapable of persistently maintaining the virus. Evidence for this comes from two sources: empirical observation and serology. The empirical evidence is that rinderpest has been eradicated from areas with significant wildlife populations. For instance, wildlife did not prevent eradication of rinderpest from Europe and southern Africa, or from India and West and Central Africa during the past decade, though the wildlife numbers in these regions are much lower than in East Africa. Although initially high, the wildlife population in southern Africa crashed during the Great Pandemic of 1887–96, probably to levels that could not sustain the virus, and they recovered only after the infection was eradicated from cattle. In East Africa, the debate focused mainly on the Maasai grasslands of southern Tanzania and northern Kenya where high numbers of pastoral cattle range freely among the highest number of species and largest populations of wild ruminants and pigs in the world. Rinderpest was endemic there from at least the 1930s to the early 1960s (Cornell and Reid 1934; Branagan and Hammond 1965). Mortality with clinical disease was noted annually or biannually in both wild animals and domestic stock and acted as a severe constraint on the population size of susceptible species, especially buffalo and wildebeest *Connochaetes taurinus* in which the disease was referred to as "yearling disease" (Talbot and Talbot 1961). When cattle were effectively immunized in the early 1960s (Plowright and Ferris 1962), the virus was eradicated from cattle in the Maasailands. Simultaneously, the disease also disappeared from wildlife even though they were not vaccinated, and their populations began to grow immediately (Sinclair 1973) until finally checked by other limiting factors, such as predation and food supply. Conclusive evidence came when serologic surveys proved that all wildebeest and buffalo born after 1963 were antibody negative (Plowright and McCullough 1967; Taylor and Watson 1967). Thus, the virus was eradicated from both cattle and wildlife in the Maasailands by immunization of cattle alone, and this is excellent evidence that East Africa's wild animals are not a permanent reservoir of RPV.

If the variety and numbers of susceptible species that exist in the Serengeti/Mara ecosystem cannot maintain the infection, then, almost certainly, neither can the increasingly isolated and small wildlife populations elsewhere in Africa. To prove the point, eradication of rinderpest from the Serengeti was repeated again in 1982 when rinderpest recurred in buffalo and other highly susceptible species such as eland *Taurotragus oryx* and giraffe *Giraffa camelopardalis* (Rossiter et al. 1983, 1987; Anderson et al. 1990). Thus, rinderpest has died out twice in Africa's largest wildlife concentrations following eradication of the disease in cattle through vaccination.

Modest prevalences of antibody to RPV in localized wildlife populations in Kenya were reported after eradication of clinical disease from the Serengeti (Rossiter et al. 1983; Wafula et al. 1983; Plowright 1987), but these, together with significant prevalences of antibody in sheep *Ovis aries* and goats (Rossiter et al. 1982a), are now thought to represent unrecognized extensions of infection from nearby domestic cattle that did not cause epidemic outbreaks in the wildlife.

How long wildlife populations can remain infected after an epidemic is uncertain, but evidence from the populations of several hundred thousand wildebeest and up to 60,000 buffalo in the Serengeti/Mara suggests that even in these large populations the longest the infection can probably last before dying out is only 1 or 2 years (Rossiter et al. 1987; Anderson et al. 1990).

The apparent absence of disease in cattle during the recent series of outbreaks of rinderpest in wildlife in Kenya, in the Tsavo National Parks during 1994–95 and Nairobi National Park in late 1996 (Kock et al. 1999), reopened debate on the role of wildlife in the epidemiology of rinderpest. However, detailed investigation of the Nairobi outbreak found evidence of mild rinderpest existing in cattle surrounding the park for 1–2 months before it was confirmed in eland and buffalo, and laboratory-confirmed mild clinical disease was eventually found to be widespread in cattle throughout a large part of southern Kenya between the Tsavo and Nairobi National Parks in late 1996. Both outbreaks affected all ages of buffalo, lesser kudu, and eland (Scott 1964; Plowright 1968), and very high prevalence of clinical disease occurred in infected herds at the peak of infection. A survey of 23 buffalo in Nairobi National Park in 1993 did not detect antibody to RPV (P. Rossiter et al. unpublished), whereas immediately after the 1994–96 outbreaks high titers of antibody occurred in all animals sampled from affected herds. Thus, the outbreaks had all the hallmarks of epidemic and not endemic infection.

Molecular typing (Barrett et al. 1998) indicated that RPV responsible for these two outbreaks in wildlife is unique among the strains circulating in East Africa today and is closely related to the RGK/1 strain isolated from a sick giraffe in eastern Kenya in 1962 (Leiss and Plowright 1964). Currently, the search for the reservoir of this virus is focusing on the cattle of Somali-speaking pastoralists in northeastern Africa. Surveillance is difficult because the disease in zebu cattle is milder than that caused by other circulating strains of the virus (H.W. Wamwayi unpublished). This is not a new phenomenon, because strains isolated from sick eland (Robson et al. 1959) and buffalo (Plowright 1963) in the Maasailands caused mild disease in cattle.

One poorly understood aspect of rinderpest is interspecific transmission. In some wildlife outbreaks, most of the highly susceptible species are involved either simultaneously or sequentially, whereas in other outbreaks, only one or a few species are affected. Clearly, factors influencing both the virulence of the strain and the contact rate between different species are involved. Dobson (1995), using epidemiologic modeling, investigated the latter factor with a view to possibly predicting rinderpest epidemics and their progress in wildlife. The study hinged on observations of the distances regularly measured between different species and the logical assumption that transmission would be most likely to occur between susceptible species that were in close contact with each other. The measurements were made on apparently healthy populations, but during the recent outbreaks in Kenya it was quickly apparent that sick animals had altered behavior, which would facilitate transmission to other species. Sick eland lagged behind or dropped out of their herds, often becoming objects of curiosity to other species. On one occasion, a herd of wildebeest grazed within a few meters of a severely affected eland, and one wildebeest actually walked up to the eland and sniffed at its nose and discharges from less than 30 cm away (P. Rossiter unpublished). Blind kudu left their groups and wandered away from their preferred habitat of confining bush into clearings such as roads and tracks, where they were very accessible to other curious animals. It was frequently reported in the past that sick buffaloes remain around water holes, increasing the chance of transmission to other animals that drank at the same point (Carmichael 1938; Plowright et al. 1964; Branagan 1966). Thus, the altered behavior of the infected host may influence the probability of interspecific transmission. Almost certainly, intrinsic resistance of some species is also involved.

Another aspect of rinderpest in wildlife that needs further clarification is the tendency to regard outbreaks of the disease involving several wild species as one epidemic though this may not be strictly the case. During the recent outbreak in Tsavo National Parks, the distribution and sequence of cases in lesser kudu and buffalo suggested two separate epidemics. The buffalo epidemic was highly visible and spread very rapidly, whereas the lesser kudu epidemic moved much more slowly and at times in different directions than the buffalo epidemic. Similar observations for buffalo and how they spread the disease to warthog *Phacochoerus* spp. and bushbuck *Tragelaphus scriptus* were made by Carmichael (1938).

CLINICAL SIGNS. The clinical signs of classic rinderpest in cattle are well known: depression, fever, ocular and nasal discharges, necrosis of the buccal mucosa, diarrhea, dehydration, and death. Case-mortality rates with virulent strains of virus in highly susceptible cattle populations may exceed 90% (Scott 1964; Plowright 1968).

Many outbreaks in cattle, particularly in endemic areas, however, are less severe either because the animals are resistant to the virus or because the strains of virus have become attenuated. Under these circumstances, clinical signs are less obvious and, with mild strains, one or more of the typical signs may be absent and case-mortality rates less than 10%. In endemic areas where adults are immune, the disease is restricted to the young, usually less than 2 years old, in which the disease frequently goes unnoticed or unreported to veterinary authorities (Scott 1964; Plowright 1968; Rossiter 1994).

Species of wildlife vary in their susceptibility to rinderpest, the disease being most obvious and severe in buffalo, eland, greater and lesser kudu, giraffe, and warthogs. Most reports of rinderpest in wildlife are of disease in African wildlife, but it also causes severe disease and mortality in wild Asian buffalo *Bubalus bubalis,* wild cattle *Bos* spp., wild pigs *Sus scrofa,* deer, gazelle *Gazella* sp., and antelope (Scott 1981), and in Europe was reported to cause high mortality in bison *Bison bonasus* and deer *Cervus* sp. (Scott 1981). Clinical signs seen in susceptible wildlife are essentially the same as those seen in domestic cattle and buffalo, though eye and skin lesions have been reported more frequently, as have behavioral changes such as increased aggression in buffalo (Branagan 1966).

Recent epidemics in wildlife in Kenya have allowed close observation of the disease in wild species. Sick buffalo appeared depressed and had serous discharges from the eyes and nose that often required close observation through binoculars to detect. Corneal opacity and ulceration occurred in a small proportion of animals. In advanced cases, the eyes were so sunken into the orbits that they could hardly be seen, presumably because of severe dehydration caused by diarrhea. The nasal discharges became mucoid to mucopurulent but often were clearly seen only when the animals sneezed. Many animals coughed. Only the most obviously sick animals did not graze or ruminate, but in some infected herds an unusually high proportion of buffalo showed drooling or frothing from the mouth. Younger animals appeared to be more severely affected; most calves under 6 months old disappeared from the herds and those 7–18 months old had, in addition to classic signs, severe skin lesions, such as reddening and rashes in the axilla and perineum, parakeratosis, flaking, and desquamation resembling mange on their flanks and back. Very enlarged parotid lymph nodes were a prominent sign.

Diarrhea, often projectile, was evident in many animals, especially when the epidemics were most severe, but it stopped quickly in animals that recovered. Signs of abortion, including retained placenta in cows without newborn calves, were seen in several affected herds.

FIG. 2.2—A dead young greater kudu with rinderpest on the Rukwa plains of Southern Tanzania in 1941. Note severe ocular discharge. [Photograph believed to have been taken by Thomas and Reid (1944), and kindly provided by Heloise Heyne and Stephan Vogel, Onderstepoort Veterinary Laboratories, Onderstepoort, Republic of South Africa.]

Eland and kudu both showed profuse serous and then mucoid and mucopurulent discharges from their eyes; these discharges stained their faces, making dark streaks (Fig. 2.2) that could be seen from a distance without binoculars. These species also had diarrhea.

Close clinical examination of immobilized animals revealed necrotic epithelial erosions on the lips and gums; on some buccal papillae; on the dorsal, lateral, and ventral aspects of the tongue; on the hard and soft palates; and inside the nares. In eland, oral lesions were usually severe, but lesions in some buffalo were milder. Some buccal lesions in both species resembled ulcers. Young buffalo often had red macular and papular rashes of the skin of the axillae and perineal area, and some buffalo of all ages had small raised nodules, up to 1 cm in diameter, in the skin similar to the lesions of pseudo-lumpy skin disease or even papillomatosis.

The overall impression of the affected buffalo herds in the Tsavo outbreak was of a severe epidemic causing obvious diarrhea and high mortality. In Nairobi, the disease was initially milder with diarrhea less evident; a lower case-mortality rate, especially in adults; and an impression of general unhealthiness in the herd. The animals in convalescent herds were discernibly thin and in poor condition, and often showed raw or healing wounds on their backs and flanks, presumably from being attacked by lions *Panthera leo* (Kock et al. 1999). In fact, increased predation by lions was a feature of the disease in buffalo and perhaps in eland. Convalescence is complicated by the severe immunosuppression caused by the virus, which allows secondary infections (Scott 1964).

PATHOGENESIS. The pathogenesis of rinderpest in wildlife has not been studied but is unlikely to differ significantly from what is known in cattle. An animal is most probably infected via the upper respiratory tract either by direct contact with infected mucus or feces or, more probably, by aerosol. The primary focus of infection is established in draining lymph nodes of the head or throat. The virus infects lymphoblasts and monocytes (Rossiter and Wardley 1985; Rossiter 1994), and spreads via lymph and blood throughout the lymphoid system (Plowright 1964a), giving rise to viremia. Virulent strains then also infect epithelial cells, particularly those of the alimentary tract. In vitro studies showed that RPV grows better in cattle lymphocytes than sheep and goat lymphocytes, whereas the reverse is true for peste des petits ruminants virus (Rossiter and Wardley 1985).

PATHOLOGY. The clinical pathology in wild species has received little study but can be expected to be essentially the same as that seen in domestic species: lymphopenia and hemoconcentration in severe cases with diarrhea.

Gross pathology in wild mammals has received more attention (Wilde 1948; Scott 1981) and is essentially the same as for domestic stock (Maurer et al. 1955) though with notable exceptions for the eyes and skin. The carcass is often emaciated and the external openings soiled with discharges or feces. In cases with severe oral necrosis, the mouth is foul smelling, and the muzzle may be dry and cracked. Mucosal erosions may be seen in the nares and sometimes inside the urogenital openings. Tear staining may be evident on the side of the face. Congestion, hemorrhage, and necrotic erosions can be found to varying degree throughout the alimentary tract though the abomasum and large intestine are often the most severely affected sites. Peyer's patches are often necrotic, frequently congested, and sometimes blackened with hemorrhage.

The longitudinal folds of the large intestine and rectum are often extremely reddened, the so-called tiger stripes, due to congestion and hemorrhage of the submucosal capillaries. Distribution and severity of alimentary tract lesions may vary considerably between animals, and it is important to examine as many carcasses as possible. The carcass and internal lymph nodes are often edematous and congested, sometimes slightly hemorrhagic. In contrast to peste des petits ruminants, the lungs are minimally involved, though congestion and a degree of pneumonia are not uncommon in animals that have been sick for a long time. The gallbladder is frequently full or distended, with congestion and hemorrhages in the mucosa. Necrotic lesions may also be seen in the mucosa of the urogenital tract. In less susceptible wild species, lesions are less severe but consistent with those seen in more highly susceptible species.

Immunosuppression caused by destruction of lymphocytes and monocytes may lead to severe secondary infections and recrudescence of latent infections. As a result, secondary bacterial growth often complicates and exacerbates the lesions. Severe arthritis and synovitis were seen in limb joints of chronically affected lesser kudu (Kock et al. 1999).

In contrast to cattle, a feature of rinderpest in wild species is the high frequency of ocular and dermal lesions. Lesions in giraffe, buffalo, greater and lesser kudu, and eland include corneal ulcers, hypopion, keratitis, uveitis, iridocyclitis, and cataract. Reddening and exudative dermatitis of axilla and inguinal skin and flaky desquamation of skin on the upper surfaces of the body were seen in affected young buffalo. In some buffalo, nodular pocklike lesions were seen throughout the skin, similar perhaps to those reported by Shanthikumar et al. (1985) during the 1980s epidemics in Nigeria.

The histopathology of rinderpest in wildlife was first examined by Thomas and Reid (1944), who investigated rinderpest in buffalo, greater kudu, and eland in the Rukwa area of southern Tanzania during the Second World War. They described lymphoid necrosis and depletion in lymph nodes and spleen, with reticulum cell hyperplasia, giant cell formation, and sinus histiocytosis in some lymph nodes. Epithelial and crypt cell necrosis and giant cells occurred in the alimentary tract. There was edema of the submucosa, and inflammatory cell infiltration, primarily lymphocytic, in the lamina propria of affected gut. Mild periportal subacute hepatitis was recorded in buffalo. In lesser kudu, necrosis occurred in renal tubules, salivary gland, and bile duct epithelia, and in the pancreas. Necrotizing bronchopneumonia was seen in one lesser kudu. Hyperkeratosis and parakeratosis of the eyelid, including the hair follicles, and squamous metaplasia of the sebaceous glands occurred. Lesions in the eyes of both lesser kudu and buffalo were generally more severe than those seen in other organs (N. Kock personal communication; P. Wohlsein personal communication). Immunohistochemistry (Wohlsein et al. 1993) with monoclonal antibodies detected RPV-specific antigen in tissues of cattle with microscopic lesions thereby confirming that these changes were due to the virus rather than secondary infections (P. Wohlsein, personal communication).

DIAGNOSIS. Clinical and gross pathologic signs are sufficient to make a provisional diagnosis in areas where rinderpest is known to occur, but laboratory confirmation is essential. This is carried out either by detection of the virus or by serology. Virus detection can be by isolation of virus in cell culture; through confirmation of specific RNA by polymerase chain reaction (PCR); or by a variety of tests such as agar gel immunodiffusion, immunofluorescence, and immunocapture enzyme-linked immunosorbent assay (ELISA) (Anderson et al. 1997). Immunohistochemistry (Wohlsein et al. 1993) allowed retrospective diagnosis of rinderpest on formalin-fixed tissues collected at the start of the recent Kenyan epidemics. A serologic diagnosis based on antibody detection requires, to be strictly accurate, the demonstration of a rising titer in paired sera. However, collecting paired sera from sick wildlife that may later die and disappear is virtually impossible. In practice, reliable demonstration of high titers of antibody in wild mammals from a population with signs consistent with rinderpest is sufficient evidence for a diagnosis because these animals are not vaccinated (Rossiter et al. 1998).

The reliability of detecting antibodies in recovered animals has allowed serology to become the main tool for studying rinderpest in wild populations. By combining it with careful ageing of sampled animals, it is often possible to determine when rinderpest last infected a wild population. In 1964, Plowright and others (1964) reported that hippopotami *Hippopotamus amphibius* in the Queen Elisabeth National Park in Uganda had been affected up to 30 years before, coinciding very closely with the last reported rinderpest in cattle in that area.

If live animals can be sampled, specimens of whole uncoagulated blood and coagulated blood should be collected together with swabs of ocular and nasal discharges and, if possible, a lymph node biopsy. The swabs should be placed in a minimum amount of phosphate-buffered or normal saline sufficient to prevent them drying out. It is advisable to include an anti-RNAse in the medium if they are to be examined by PCR. The lymph node biopsy sample should be placed in a small amount of phosphate-buffered saline or cell culture medium. These and the uncoagulated blood should then be placed on ice and transported to an appropriate laboratory without delay. Uncoagulated blood must not be frozen, and glycerol, which is a frequent component of virus transport media, must be avoided because it inactivates RPV.

At postmortem examination, specimens of spleen, lymph nodes (peripheral, submaxillary, retropharyngeal, and mesenteric), hemolymph nodes, and sections of alimentary tract that show lesions should be collected aseptically or as cleanly as possible. They should be sent immediately to the laboratory on ice, but if this is not possible they can be kept frozen at –20° C, though there will be some loss of virus titer. Duplicate tissues also should be fixed in 10% buffered formalin for histology and confirmation by immunohistochemistry. The tonsil and eyelid have proved a useful source of antigen in cattle (Brown et al. 1996), including those that have been dead for some time, but eyelids often may be missing from wildlife carcasses that have been scavenged.

Manuals on collection and submission of specimens and on laboratory diagnosis have been prepared by the World Reference Laboratory for Rinderpest, Pirbright Laboratory, United Kingdom, and are available from the Food and Agriculture Organization of the United Nations, Rome.

DIFFERENTIAL DIAGNOSES. Wherever rinderpest is present, any serious outbreak of epidemic lethal disease in wildlife should be treated as rinderpest until proved otherwise. In countries where rinderpest does not occur, it is unlikely that rinderpest will be considered a possibility until the disease is well established.

In Africa, bovine herpesvirus 2 has been shown to cause a rinderpest-like syndrome in buffalo in which it is ubiquitous (Plowright and Jessett 1971; Schieman et al. 1971). Antibodies to bovine virus diarrhea virus and bovine herpesvirus 1 are prevalent in many African (Rweyemamu 1970; Rampton and Jessett 1976; Hedger and Hamblin 1978; Hamblin and Hedger 1979), Eurasian, and American wild species (Nettleton et al. 1988). Bovine virus diarrhea virus was isolated from a giraffe with a rinderpest-like disease (W. Plowright, personal communication). Bovine herpesvirus 1-related viruses were found in wild species, including wildebeest in Africa (Karstad et al. 1974) and cervids throughout the Northern Hemisphere (Nettleton et al. 1988). Peste des petits ruminants virus causes a rinderpest-like disease in wild species (Hamdy and Dardiri 1976; Furley et al. 1987) and should be differentiated from RPV.

Many East African wild bovids carry *Theileria* spp. that under certain circumstances may cause disease. Acute cases in buffalo and eland are rare, though these infections are probably more severe in young animals. Perinatal mortalities from myriad other causes are undoubtedly common in wildlife but are very difficult to study since the carcasses are removed by predators and scavengers before being found by veterinarians or researchers.

Malignant catarrhal fever (MCF) is a recognized rinderpest-like disease and the wildebeest-derived form caused by Alcelaphine herpesvirus 1 (AHV-1) is especially prevalent in East Africa (Plowright et al. 1960). Eland inoculated experimentally with AHV-1 developed typical fatal MCF, whereas, remarkably, buffalo did not (Plowright 1964b). Usually, MCF causes low morbidity in a cattle herd, whereas rinderpest will affect virtually 100% of a fully susceptible herd. Interestingly, a morbillivirus was isolated from a case of MCF in cattle in the United States (Coulter and Storz 1977).

IMMUNITY. When rinderpest was endemic in the Maasailands, the disease in wildlife was nearly always only seen in calves and yearlings, suggesting that these species, like cattle (Brown and Raschid 1965; Plowright 1984), develop lifelong immunity against the disease if they recover. This immunity appears to be absolute and, therefore, outbreaks of rinderpest affecting animals of all ages are considered to be epidemics caused by recently introduced infection. Susceptible animals infected with either attenuated vaccine or virulent virus develop a strong humoral antibody response that usually can be detected by the neutralization test approximately a week later. Neutralizing antibody and probably competitive ELISA antibody (Anderson et al. 1997) remain detectable for life in nearly all animals and correlate strongly with immunity (Plowright and Taylor 1967).

CONTROL AND TREATMENT. In uninfected countries, wildlife are best protected by maintaining the measures that exclude rinderpest from the country. In endemic areas, wildlife are best protected by ensuring that neighboring cattle populations are routinely vaccinated against the disease and by rigidly preventing the incursion of cattle into wildlife sanctuaries. All of the recent epidemics in Kenya were associated with the entry of abnormal numbers of cattle into the National Parks a few weeks before the disease was recognized in wildlife.

Many susceptible wild species have been safely inoculated with the standard live cell culture vaccine (TCRV) (Plowright 1982; Furley et al. 1987; Bengis and Erasmus 1988; Greth et al. 1992; R. Kock personal communication; P. Rossiter and L. Karstad unpublished). No clinical signs were seen, and all animals tested produced neutralizing antibody. Techniques for remote ballistic vaccination could be used in free-ranging wildlife.

Oral immunization is an alternative approach. Deer and gazelle repeatedly given feed that had been sprinkled with freeze-dried TCRV at approximately 50–100 doses per animal appeared to resist infection with peste des petits ruminants virus, though this could not be confirmed by serology (G.R. Scott and S. Hafez personal communication). If trials show this method is successful, it could obviously be very beneficial in immunizing high-risk endangered species in zoological collections and small sanctuaries. However, its potential use for large free-ranging populations requires much more careful evaluation (Rossiter and James 1989).

DOMESTIC ANIMAL HEALTH CONCERNS. In Africa, and to some extent Asia, infected wildlife have long been known to transmit rinderpest across considerable distances. The 1960–62 epidemic of rinderpest in wildlife in Kenya began on the Ethiopian border and spread over 600 km to Garissa with very little recorded involvement of cattle (Stewart 1964, 1968). This ability was reaffirmed recently in Kenya when the disease was confirmed in lesser kudu in the north of Tsavo East National Park in mid-1994 after which infection spread without the aid of domestic cattle to the south of Tsavo West National Park by the end of the year, a distance of 200 km in 6 months. The significance of this particular episode for domestic animals is that the wildlife may have acted as the conduit that transmitted infection from one domestic pastoral ecosystem to another quite unrelated and distant domestic pastoral ecosystem.

This problem will remain in East Africa until rinderpest is eradicated from the continent but is unlikely to be significant elsewhere. If rinderpest were to infect wildlife in Europe or the Americas, appropriate eradication measures to contain the disease, including support by an intensive program of clinical and serologic surveillance in all susceptible species, should be sufficient to eradicate the infection.

MANAGEMENT IMPLICATIONS. All even-toed ungulates from the infected areas of Africa and Asia could, if infected, introduce the disease to new susceptible populations. The most likely manner for this to happen is for animals to be shipped while either incubating the disease or suffering from mild or clinically unrecognizable disease. This can be prevented by adequate quarantine for at least 3 weeks before shipping, with final clinical inspection of all individuals at the point of embarkation. Similar quarantine at the port of disembarkation would seem to be prudent especially if the animals are transported by air. All sick animals without antibodies to the virus should be isolated and released only when they have fully recovered and remain antibody negative. Recovery and development of antibody confirm a diagnosis of rinderpest. At this point most authorities would probably want to destroy the animals, but in fact, since recovered animals never reexcrete virus, the animals should be absolutely virus free after a further quarantine of a month. Most countries stipulate that imported animals must be rinderpest antibody free, but the disease has never been transmitted by fully recovered and immune animals. Therefore, an alternative to risking the importation of antibody-negative but possibly infected animals in the incubation period would be to import quarantined animals that have been certified to have serum antibodies that have been vaccinated 2–4 weeks before. The current development of new vaccines with markers that will enable distinction of vaccine-induced antibodies from those of field strains could make this concept more palatable to traditionally conservative national veterinary authorities.

LITERATURE CITED

Anderson, E.C., M. Jago, T. Mlemgya, C. Timms, A. Payne, and K. Hirji. 1990. A serological survey of rinderpest in wildlife, sheep and goats in Northern Tanzania. *Epidemiology and Infection* 105:203–214.

Anderson, J., T. Barrett, and G.R. Scott. 1997. *Manual on the diagnosis of rinderpest,* 3d ed. Rome: Food and Agriculture Organization.

Baron, M.D., and T. Barrett. 1995. The sequence of the N and L genes of rinderpest virus, and the 5′ and 3′ extra-genic sequences: The completion of the genome sequence of the virus. *Veterinary Microbiology* 44:175–186.

Barrett, T., M.A. Forsyth, K. Inui, H.M. Wamwayi, R. Kock, J. Wambua, J. Mwanzia, and P.B. Rossiter. 1998. Rediscovery of the second African lineage of rinderpest virus: Its epidemiologic significance. *Veterinary Record* 142:669–671.

Bengis, R.G., and J.M. Erasmus. 1988. Wildlife diseases in South Africa: A review. *Revue Scientifique et Technique O.I.E.* 7:807–821.

Branagan, D. 1966. Behaviour of buffalo infected with rinderpest. *Bulletin of Epizootic Diseases of Africa* 14:341–342.

Branagan, D., and J.A. Hammond. 1965. Rinderpest in Tanganyika: A review. *Bulletin of Epizootic Diseases of Africa* 13:225–246.

Brown, R.D., and A. Raschid. 1965. Duration of rinderpest immunity in cattle following vaccination with caprinised rinderpest virus. *Bulletin of Epizootic Diseases of Africa* 13:311–315.

Brown, C.C., L. Ojok, and J.C. Mariner. 1996. Immunohistochemical detection of rinderpest virus: Effects of autolysis and period of fixation. *Research in Veterinary Science* 60:182–184.

Carmichael, J. 1938. Rinderpest in African game. *Journal of Comparative Pathology* 51:264–268.

Chamberlain, R.W., H.W. Wamwayi, E. Hockley, M.S. Shaila, L. Goatley, N.J. Knowles, and T. Barrett. 1993. Evidence for different lineages of rinderpest virus reflecting their geographic isolation. *Journal of General Virology* 74:2775–2780.

Cilli, M.V., and C. Roetti. 1951. l'Epidosio di peste bovina al Giardiono Zoologico di Roma. *Archivio Italiano di Scienze Mediche Tropicali e di Parassitologia* 32:83–94.

Cornell, R.L., and N.R. Reid. 1934. *Rinderpest in wildebeest: Annual report of the Department of Veterinary Science and Animal Husbandry, Tanganyika Territory, 1933.* Dar-es-Salaam: Government Printer.

Coulter, G.R., and J. Storz. 1977. Identification of a cell-associated morbillivirus from cattle affected with malignant catarrhal fever: Antigenic differentiation and cytologic characterization. *American Journal of Veterinary Research* 40:1671–1677.

Dobson, A. 1995. The ecology and epidemiology of rinderpest virus in Serengeti and Ngorongoro conservation area. In *Serengeti II: Dynamics, management, and conservation of an ecosystem,* ed. A.R.E. Sinclair and P. Arcese. Chicago: University of Chicago Press, pp. 485–505.

Furley, C.W., W.P. Taylor, and T.U. Obi. 1987. An outbreak of peste des petits ruminants in a zoological collection. *Veterinary Record* 121:443–447.

Greth, A., D. Calvez, M. Vassart, and P.-C. Lefevre. 1992. Serological survey for bovine bacterial and viral pathogens in captive Arabian oryx (*Oryx leucoryx* Pallas, 1776). *Revue Scientifique et Technique O.I.E.* 11:1163–1168.

Hamdy, F.M., and A.H. Dardiri. 1976. Response of white-tailed deer to infection with peste des petits ruminants virus. *Journal of Wildlife Diseases* 12:191–196.

Hamdy, F.M., A.H. Dardiri, and S.S. Breese Jr. 1975. Experimental infection of white-tailed deer with rinderpest virus. *Journal of Wildlife Diseases* 11:508–515.

Hamblin, C., and R.S. Hedger. 1979. The prevalence of antibodies to bovine diarrhea/mucosal disease virus in African wildlife. *Comparative Immunology and Microbiology of Infectious Diseases* 2:295–303.

Hedger, R.S., and C. Hamblin. 1978. Neutralising antibodies to bovid herpes virus 1 (infectious bovine rhinotracheitis/infectious pustular vulvovaginitis) in African wildlife with special reference to the Cape buffalo (*Syncerus caffer*). *Journal of Comparative Pathology* 88:211–218.

Karstad, L., D.M. Jessett, J.C. Otema, and S. Drevemo. 1974. Vulvovaginitis in wildebeest caused by the virus of infectious bovine rhinotracheitis. *Journal of Wildlife Diseases* 10:392–396.

Kock, R.A., J.M. Wambua, J. Mwanzia, H.W. Wamwayi, E.K. Ndungu, T. Barrett, N.D. Kock, and P. Rossiter. 1999. Rinderpest epidemic in wild ruminants in Kenya, 1993–1997. *Veterinary Record* 145:275–283.

Leiss, B., and W. Plowright. 1964. Studies on the pathogenesis of rinderpest in experimentally infected cattle: I. Correlation of clinical signs, viraemia and virus excretion by various routes. *Journal of Hygiene (London)* 62:81–100.

Maurer, F.D., T.C. Jones, B. Easterday, and D.E. De Tray. 1955. Pathology of rinderpest. *Journal of the American Veterinary Medical Association* 127:512–514.

Nettleton, P.F., E. Thiry, H. Reid, and P.-P. Pastoret. 1988. Herpesvirus infections in Cervidae. *Revue Scientifique et Technique O.I.E.* 7:977–988.

Norrby, E., H. Sheshberadaran, K.C. McCullough, W.C. Carpenter, and C. Örvell. 1985. Is rinderpest the archevirus of the *Morbillivirus* genus? *Intervirology* 23:228–232.

Pankhurst, R. 1966. The Great Ethiopian famine of 1888–1892: A new assessment. *Journal of the History of Medicine and Allied Sciences* 21:95–124.

Plowright, W. 1963. Some properties of strains of rinderpest recently isolated in East Africa. *Research in Veterinary Science* 4:96–108.

———. 1964a. Studies on the pathogenesis of rinderpest in experimental cattle: II. Proliferation of the virus in different tissues following intranasal infection. *Journal of Hygiene (London)* 62:257–281.

———. 1964b. Studies of malignant catarrhal fever in cattle. D.V.Sc. thesis, University of Pretoria.

———. 1968. Rinderpest virus. *Monographs in Virology* 3:25–110.

———. 1982. The effect of rinderpest and rinderpest control on wildlife in Africa. *Symposium of the Zoological Society of London* 50:1–28.

———. 1984. The duration of immunity in cattle following inoculation of rinderpest cell culture vaccine. *Journal of Hygiene (London)* 92:285–296.

———. 1987. Investigations of rinderpest antibody in East African wildlife, 1967–1971. *Revue Scientifique et Technique O.I.E.* 6:497–513.

Plowright, W., and R.D. Ferris. 1962. Studies with rinderpest virus in tissue culture: The use of attenuated virus as a vaccine for cattle. *Research in Veterinary Science* 3:172–182.

Plowright, W., and D.M. Jessett. 1971. Investigations of Allerton-type virus infection in East African game animals and cattle. *Journal of Hygiene (London)* 69:209–222.

Plowright, W., and K.C. McCullough. 1967. Investigations on the incidence of rinderpest virus infection in game animals of N. Tanganyika and S. Kenya 1960/1963. *Journal of Hygiene (London)* 65:343–358.

Plowright, W., and W.P. Taylor. 1967. Long-term studies of the immunity in East African cattle following inoculation with rinderpest culture vaccine. *Research in Veterinary Science* 8:118–128.

Plowright, W., R.D. Ferris, and G.R. Scott. 1960. Blue wildebeest and the aetiological agent of bovine malignant catarrhal fever. *Nature* 188:1167–1169.

Plowright, W., R.M. Laws, and C.S. Rampton. 1964. Serological evidence for the susceptibility of the hippopotamus (*Hippopotamus amphibius* Linnaeus) to natural infection with rinderpest virus. *Journal of Hygiene (London)* 62:329–336.

Polding, J.B., and R.M. Simpson. 1957. A possible immunological relationship between canine distemper and rinderpest. *Veterinary Record* 69:582–584.

Rampton, C.S., and D.M. Jessett. 1976. The prevalence of antibody to infectious rhinotracheitis virus in some game animals in East Africa. *Journal of Wildlife Diseases* 12:2–6.

Robson, J., R.M. Arnold, W. Plowright, and G.R. Scott. 1959. The isolation from an eland of a strain of virus attenuated for cattle. *Bulletin of Epizootic Diseases of Africa* 7:97–102.

Rossiter, P.B. 1994. Rinderpest. In *Infectious diseases of livestock, with special reference to Southern Africa,* ed. J.A.W. Coetzer, G.R. Thomson, and R.C. Tustin. Cape Town: Oxford University Press, pp. 735–757.

Rossiter, P.B., and A.D. James. 1989. An epidemiological model of rinderpest: II. Simulations of the behaviour of the virus in populations. *Tropical Animal Health and Production* 21:69–84.

Rossiter, P.B., and R.C. Wardley. 1985. The differential growth of virulent and avirulent strains of rinderpest

virus in bovine lymphocytes and macrophages. *Journal of General Virology* 66:969–975.
Rossiter, P.B., D.M. Jessett, and W.P. Taylor. 1982a. Neutralising antibodies to rinderpest virus in sheep and goats in western Kenya. *Veterinary Record* 111:504–505.
Rossiter, P.B., L. Karstad, D.M. Jessett, T. Yamamoto, A.H. Dardiri, and E.Z. Mushi. 1982b. Neutralising antibodies to rinderpest virus in wild animal sera collected in Kenya between 1970 and 1981. *Preventive Veterinary Medicine* 1:257–264.
Rossiter, P.B., D.M. Jessett, J.S. Wafula, L. Karstad, S. Chema, W.P. Taylor, L. Rowe, J.C. Nyange, M. Otaru, M. Mumbala, and G.R. Scott. 1983. Re-emergence of rinderpest as a threat in East Africa since 1979. *Veterinary Record* 113:459–461.
Rossiter, P.B., W.P. Taylor, B. Bwangamoi, A.R.H. Ngereza, P.D.S. Moorhouse, J.M. Haresnape, J.S. Wafula, J.F.C. Nyange, and I.D. Gumm. 1987. Continuing presence of rinderpest virus as a threat in East Africa 1983–1985. *Veterinary Record* 120:59–62.
Rossiter, P.B., M. Hussain, R.H. Raja, W. Moghul, Z. Khan, and D.W. Broadbent. 1998. Cattle plague in Shangri-La: Observations on a severe outbreak of rinderpest in northern Pakistan 1994–1995. *Veterinary Record* 143:39–42.
Rweyemamu, M.M. 1970. Probable occurrence of infectious bovine rhinotracheitis virus in Tanzanian wildlife and cattle. *Nature* 225:738–739.
Schieman, B., W. Plowright, and D.M. Jessett. 1971. Allerton-type herpes virus as a cause of lesions of the alimentary tract in a severe disease of Tanzanian buffaloes (*Syncerus caffer*). *Veterinary Record* 89:17–22.
Scott, G.R. 1964. Rinderpest. *Advances in Veterinary Science* 9:113–224.
———. 1981. Rinderpest. In *Infectious diseases of wild mammals,* ed. J.W. Davis, L. Karstad, and D.O. Trainer. Ames: Iowa State University Press, pp. 18–30.
Shanthikumar, S.R., S.A. Malachi, and K.A. Majiyagbe. 1985. Rinderpest outbreak in free-living wildlife in Nigeria. *Veterinary Record* 117:469–470.
Sinclair, A.R.E. 1973. Population increases of buffalo and wildebeest in the Serengeti. *East African Wildlife Journal* 11:93–107.
Stewart, D.R.M. 1964. Rinderpest among wild animals in Kenya, 1960–62. *Bulletin of Epizootic Diseases of Africa* 12:39–42.
———. 1968. Rinderpest among wild animals in Kenya, 1963–66. *Bulletin of Epizootic Diseases of Africa* 16:139–140.
Talbot, L.M., and M.H. Talbot. 1961. Preliminary observations on the population dynamics of the wildebeest in Narok District, Kenya. *East African Agriculture and Forestry Journal* 27:108–116.
Taylor, W.P., and R.M. Watson. 1967. Studies of the epizootiology of rinderpest in blue wildebeest and other game species of northern Tanzania and southern Kenya, 1965–1967. *Journal of Hygiene (London)* 65:537–545.
Thomas, A.D., and N.R. Reid. 1944. Rinderpest in game: A description of an outbreak and an attempt at limiting its spread by means of a bush fence. *Onderstepoort Journal of Veterinary Science and Animal Industry* 20:7–23.
Wafula, J.S., E.Z. Mushi, and L. Karstad. 1983. Antibodies to rinderpest virus in the sera of some wildlife in Kenya. *Bulletin of Animal Health and Production* 30:363–365.
Wamwayi, H.W., D.P. Kariuki, J.S. Wafula, P.B. Rossiter, P.G. Mbuthia, and S.R. Macharia. 1992. Observations on rinderpest in Kenya, 1986–1989. *Revue Scientifique et Technique O.I.E.* 11:769–778.
Wilde, J.K.H. 1948. Rinderpest in some African wild mammals. *Journal of Comparative Pathology* 58:64–72.
Wohlsein, P., G. Trautwein, T.C. Harder, B. Liess, and T. Barrett. 1993. Viral antigen distribution in organs of cattle experimentally infected with rinderpest virus. *Veterinary Pathology* 30:544–554.
Woodford, M.H. 1984. *Rinderpest in wildlife in sub-Saharan Africa: Consultancy report.* Rome: Food and Agriculture Organization, 60 pp.

PESTE DES PETITS RUMINANTS

PAUL ROSSITER

Synonyms: Stomatitis pneumoenteritis complex, kata, pseudorinderpest, goat plague.

INTRODUCTION. Peste des petits ruminants (PPR) was first described in 1942 by Gargadennec and Lalanne, who investigated a syndrome in sheep *Ovis aries* and goats *Capra hircus* in Côte d'Ivoire, West Africa. Because of its clinical and pathologic resemblance to rinderpest, it was called peste des petits ruminants, and it was soon recognized in other French West African countries such as Dahomey (now Benin) and Senegal. Over a decade later, workers in Nigeria began to study a syndrome that occurred primarily in goats and was variously named stomatitis pneumoenteritis complex, pseudorinderpest, and *kata.* Because of its predominance in goats as opposed to sheep in the earlier French descriptions, and because of the crusty labial lesions commonly found in convalescent cases, this syndrome was initially considered to be distinct from PPR. More detailed investigations, however, clearly showed the two diseases were indistinguishable, and it is now universally referred to by its original French name. In those areas where it occurs, it is believed to be the most economically important virus disease of small ruminants.

Although rinderpest virus can infect goats and sheep in Africa, most experimental infections are mild or subclinical. Therefore, it seems probable that some earlier descriptions of severe rinderpest in small ruminants, particularly in Africa, were actually PPR. This was demonstrated by El Hag Ali and Taylor (1984), who used specific antisera and transmission studies to confirm that virus isolates originally believed to be rinderpest virus from goats in Sudan were actually the first isolates of PPR virus in eastern Africa.

To date, PPR has not been reported in free-living wildlife, but undoubtedly it has the potential to cause severe disease in some important wild species. This review focuses on the little data available from experimental infections and outbreaks in zoological collections, supported by standard observations in domestic sheep and goats. Greater detail and extensive references are provided in several reviews (Scott 1981; Taylor 1984; Losos 1986; Rossiter and Taylor 1994).

DISTRIBUTION AND HOST RANGE. In Africa, PPR is widespread throughout West Africa below the Sahara and has been identified in the Sudan and

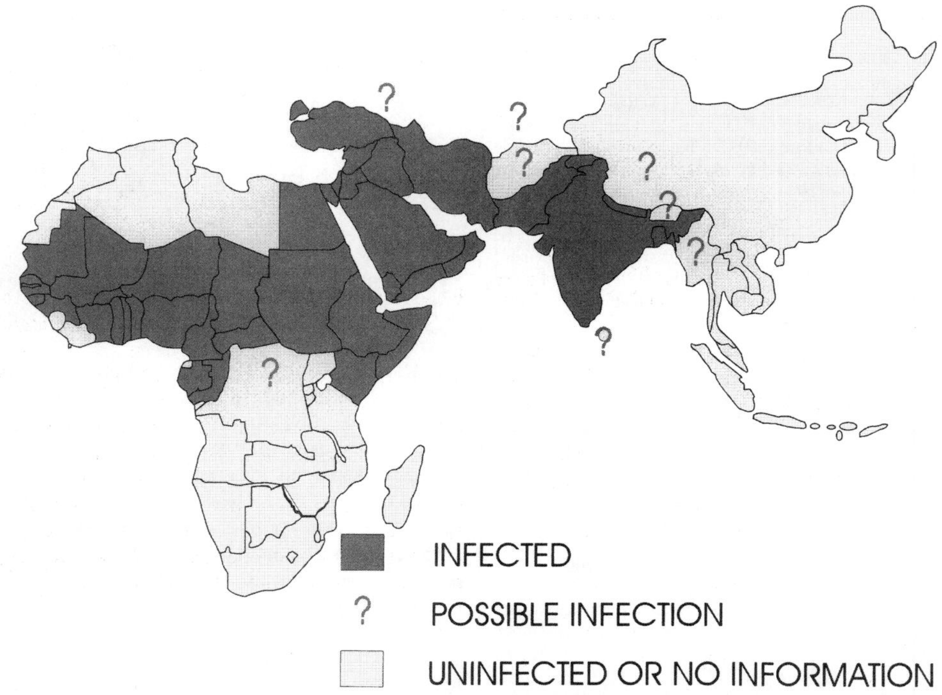

FIG. 2.3—Global distribution of peste des petits ruminants in 1997.

Ethiopia (Fig. 2.3). To date, there is no clinical evidence that PPR exists south of a line from Cameroon to the Sudan. The disease is present in many countries of the Arabian peninsula and Near and Middle East, and it has recently been confirmed in India, Pakistan, Bangladesh, and Iran. There is serologic evidence of its presence in Jordan (Lefevre et al. 1991) and possibly Uganda and Kenya (Wamwayi et al. 1995). The sequence of reports of PPR gives the impression that it is spreading steadily eastward from West Africa. Recent molecular evidence, however, shows that non-African isolates are quite distinct from African isolates, and it may be that PPR was frequently overlooked in areas where rinderpest and contagious caprine pleuropneumonia (CCPP) are common.

In addition to domestic sheep and goats, fatal PPR was confirmed in gemsbok *Oryx gazella,* Dorcas gazelle *Gazella dorcas,* Nubian ibex *Capra ibex nubiana,* and Laristan sheep *Ovis orientalis laristanica* (Furley et al. 1987); fatal and subclinical disease was reported in white-tailed deer *Odocoileus virginianus* (Hamdy and Dardiri 1976); and nilgai *Boselaphus tragocemalus* displayed subclinical infection (Furley et al. 1987). Other wild species are likely susceptible.

ETIOLOGY. Genetic analysis demonstrated that the causative morbillivirus is not a novel variant of rinderpest virus but a distinct member of the genus *Morbillivirus* more distantly related to rinderpest virus in terms of nuclear acid homology than is measles virus of humans (Baron et al. 1994).

TRANSMISSION AND EPIDEMIOLOGY. Clinically affected animals excrete virus in their secretions for approximately 24–48 hours before the onset of clinical signs. Susceptible animals become infected through close contact with infected animals when they ingest or inhale infected material, including aerosol-dispersed virus. Previously uninfected populations are infected through the introduction of live-infected sheep or goats. There is no indirect transmission and no known wildlife reservoir. Recovered animals are immune to reinfection.

In West Africa, clinical disease is perhaps more prevalent in the humid and subhumid regions rather than the drier Sahel. Peak prevalence of PPR occurs following seasonal animal husbandry patterns that allow goats and sheep to be released from their family compounds after the crop harvest (Obi et al. 1983). The epidemiology of PPR in East Africa is less clear. Although the virus has been isolated from sick sheep and goats in Sudan and Ethiopia (Roeder et al. 1994), the pattern of these outbreaks in relation to animal husbandry practices and seasonal marketing is not known. Considering that most of the sheep and goats of Central and East Africa are fully susceptible to PPR, it is a mystery why this infection has not swept through these populations.

Throughout the Middle East, PPR is spread by intensive trading. In Oman, endemic infection is maintained by poorly confined flocks of goats and sheep in both urban and rural situations. The recent descriptions of PPR in south Asia hint that the significant export of small ruminants from this area to the Middle East may also be responsible for some past outbreaks.

In India, sheep and goats frequently show a rinderpest-like syndrome that in the past was usually diagnosed as small-ruminant-adapted rinderpest but has now been shown to be caused more frequently by PPR virus. It is not known whether the disease is a recent introduction, but the fact that molecular genetics clearly distinguishes Asian from African strains suggests this. The epidemiology has not been described, but it is interesting to note that a massive wave of PPR in goats swept through northern India, Bangladesh, and Pakistan during 1993–96 (Amjad et al. 1996; Kulkarni et al. 1996), suggesting that the infection may not have been widespread in these regions before.

The risk to wildlife from PPR is unknown but probably high. In Asia, the virus has reached the major mountain ranges where it must pose a serious threat to the region's endangered wild caprines and ovines. Similarly, gazelle in reserves and elsewhere may be highly susceptible. At present, the Himalayas are a barrier that may prevent the virus affecting some of the endangered bovid species in central Asia. There have been no confirmed reports of PPR in the dwindling wildlife of West Africa. If the infection becomes more prevalent in East African livestock, wild species will inevitably be exposed; probably species most closely related to sheep and goats will be affected.

CLINICAL SIGNS. An incubation period of 3–7 days is followed by 1 or 2 days of fever (40° C–41° C), with increasing dullness and inappetance, after which serous ocular and nasal discharges develop. Within 1–2 days, these become mucopurulent, and the nasal and oral mucosa become red and then covered with small plaques of necrosis. These foci increase in size and coalesce and may involve the whole mouth. The author saw one outbreak in which affected goats had virtually no epithelium visible in the mouth. Dyspnea is common. Diarrhea begins 2–3 days after pyrexia and leads to death 3–7 days later.

Clinical signs in a species may be variable. For instance, after experimental infection of five white-tailed deer, two developed fever, conjunctivitis, mucopurulent nasal discharge, erosive stomatitis, and diarrhea, and they died of the disease. The remaining three had subclinical infection and were immune to subsequent challenge with virulent rinderpest virus (Hamdy and Dardiri 1976).

During an outbreak of PPR in a zoological collection in the United Arab Emirates (Furley et al. 1987), most of the affected animals had a peracute form of the disease and were found dead before signs were seen. In the few animals in which clinical signs were seen, the most common sign was diarrhea, which often contained blood. Clinical illness lasted 13 days in one adult gemsbok and 17 days in a gazelle that reportedly smacked its lips during this time. Outbreaks in zoological collections in Saudi Arabia (Hafez et al. 1987) resulted in severe diarrhea and death in 3 of 31 deer and four of eight gazelle.

PATHOGENESIS. It is assumed that the pathogenesis of PPR is similar to that of rinderpest. Animals are infected by virus crossing the upper respiratory or upper alimentary tract where it infects lymphocytes and monocytes in the draining lymph nodes. Infection then spreads throughout the lymphoid system and is transported via blood to the target epithelium in the alimentary and respiratory tracts. In vitro studies have shown that PPR virus infects more cells and grows to higher titer in lymphoblasts from sheep and goats than in those from cattle, and the reverse is true for rinderpest virus (Rossiter and Wardley 1985).

PATHOLOGY. It seems probable that the lesions in wild species will be the same as those in sheep and goats. Infected goats develop leukopenia that may correlate with the severity of clinical disease. Eosinopenia, monocytosis, and a rise in the packed-cell volume, presumably due to hemoconcentration resulting from diarrhea, have been reported.

The carcass is frequently emaciated and soiled with feces, and the eyelids, nares, and lips encrusted with discharges. The female genital mucosa may be congested and show erosions. Erosions are found throughout the mucosa of the buccal cavity and pharynx and, less frequently, the esophagus. As with rinderpest, erosions in the forestomachs are not always seen but may be found occasionally on the ruminal pillars and the leaves of the omasum. The abomasum and small intestine are usually congested rather than eroded, and hemorrhages have been reported. Peyer's patches may be necrotic, and erosions and congestion can be found on the ileocecal valve. Marked congestion of the apices of the longitudinal folds of the mucosa (zebra striping) may occur in the large intestine and rectum.

The upper respiratory tract mucosa is usually congested, there may be erosions in the nares, and tracheitis has been described. The lungs may be focally or diffusely congested. Changes in the trachea and lungs are more severe if there is concurrent bacterial infection. Lymph nodes may be edematous and friable, but they are rarely enlarged or hemorrhagic. The spleen is often congested and firm.

The reports by Furley et al. (1987) and by Hamdy and Dardiri (1976) provide brief histologic descriptions of erosions that arise from hydropic degeneration and necrosis of epithelial cells in the stratum granulosum but do not penetrate the underlying stratum germinativum.

Multinucleate giant cells, as well as eosinophilic intracytoplasmic and intranuclear inclusions are evident in the epithelium, and inflammatory cells are prominent at the borders of the lesions. The orf-like lesions on the lips contain necrotic debris, fibrin, and mononuclear and polymorphonuclear leukocytes.

Lymphoid tissues, especially germinal centers, are depleted of lymphocytes. Pyknosis and karyorrhexis of lymphocytes are present in the cortex, and there is proliferation of reticuloendothelial cells along the medullary cords and in the sinuses. Necrosis throughout the Peyer's patches has been described, although a mere reduction in the numbers of lymphocytes without necrosis may be all that is evident.

Isolated areas of epithelial necrosis occur in the deep glands of the abomasum, and similar but more widespread changes are seen throughout the intestine together with atrophy of villi and accumulation of debris in the glandular crypts. The lamina propria is infiltrated with lymphocytes, macrophages, and eosinophils.

Areas of necrosis and hyperplasia have been described in the respiratory tract mucosa, and inclusion bodies may be seen in epithelial cells. Interstitial pneumonia, involving infiltration of lymphocytes and neutrophils, and proliferation of pneumocytes are common. A striking feature, not seen in rinderpest, is the presence of giant cells, frequently containing intracytoplasmic and intranuclear viral inclusions, in the alveoli and terminal bronchioles. It seems probable that the virus is responsible for primary pneumonia that then becomes complicated by secondary pathogens. Small foci of hepatocellular necrosis have been described, and there may be evidence of glomerulonephritis.

Ultrastructural studies have shown paramyxovirus-like particles in degenerating cells in the lungs, particularly alveolar macrophages, and in intestinal epithelium. Intranuclear and intracytoplasmic nucleocapsids are observed in intestinal epithelial cells, with apparent extrusion of viral particles from microvilli.

DIAGNOSIS. A presumptive diagnosis of PPR in areas where this disease occurs can be made from the clinical signs and lesions together with epidemiologic evidence such as the introduction of stock from a known infected area. The diagnosis must be confirmed in the laboratory, either by detection of virus-specific antigen or RNA, or by isolation of virus in cell culture. Retrospective confirmation can be made by demonstrating a rise in antibody to PPR virus in paired sera or by serologic evidence in unvaccinated populations.

Specimens should be collected from acutely affected animals with fever and early mucosal lesions. Serum and blood in anticoagulant as well as ocular, nasal, and oral swabs should be collected from live animals. If possible, 20–30 g each of spleen, lymph nodes, lung, and gut mucosa should be collected from freshly dead animals. The specimens should be kept chilled on ice, but not frozen, and sent to the laboratory immediately.

Virus antigen in undiluted swab material or 30% weight/volume tissue suspensions is detected by hyperimmune serum to rinderpest virus in either immunodiffusion or counter-immunoelectrophoresis tests. Hyperimmune serum to PPR virus prepared in goats can also be used. Antigen can also be detected by immunohistochemical staining of tissues (Bundza et al. 1988; Brown et al. 1991).

The virus is isolated in cell culture by inoculating swab material, buffy coat, or 10% tissue suspensions onto young monolayers of primary sheep or goat kidney cells. Vero cells, some lines of primary bovine kidney, and bovine T lymphoblasts are also sensitive (Rossiter et al. 1992). Cytopathic effects that are indistinguishable from rinderpest virus develop after 4 days. The identity of the agent is confirmed as rinderpest virus or PPR virus by serologic tests that use hyperimmune or convalescent sera or monoclonal antibody to either virus. Alternatively, the isolate can be examined with virus-specific monoclonal antibodies or cDNA probes (Diallo et al. 1989; McCullough et al. 1991; Libeau et al. 1994). Polymerase chain reaction (PCR) is now used to distinguish between rinderpest and PPR viruses and for nucleotide sequencing for strain typing (Diallo et al. 1995).

If neither antigen nor infectious virus can be detected in suspect animals, survivors should be bled 2–4 weeks after the first sampling and the paired sera assayed for antibody levels to the virus. A fourfold or greater increase in titer confirms the presumptive diagnosis. Viral antigens may be found in formalin-fixed tissues (Saliki et al. 1994), which are often the only samples collected from autopsies of free-living wildlife.

DIFFERENTIAL DIAGNOSES. Interestingly, PPR usually occurs in countries that are either infected with rinderpest virus or have been in the recent past. In India, one or more strains of rinderpest virus may have become adapted to sheep and goats and, therefore, investigations into PPR-like disease in south and southwest Asia and the Middle East should consider this possibility. In countries that have never reported PPR but do have rinderpest, all outbreaks of rinderpest-like disease in small ruminants should be carefully examined for both PPR and rinderpest viruses. In particular, all morbilliviruses isolated from goats or sheep should be further characterized.

The fact that the respiratory system is frequently involved in PPR increases the range of other conditions that can confuse diagnosis: in particular, CCPP is widespread and prevalent in many parts of sub-Saharan Africa and causes high mortality in goats. In uncomplicated cases of CCPP, there are no mucosal lesions or diarrhea, and sheep are not affected. Concurrent infections with other *Mycoplasma* spp. are common and can obscure the diagnosis of PPR. Adenovirus together with

PPR virus was recovered from the intestines of two goats that died of PPR in Nigeria (Gibbs et al. 1977).

In East Africa, PPR might be confused with Nairobi sheep disease (NSD), but oral lesions are minimal or absent in NSD, its distribution is restricted to areas infested with *Rhipicephalus appendiculatus,* and goats are rarely affected. Foot-and-mouth disease and bluetongue can be ruled out by closer examination of oral lesions, absence of diarrhea, and the possible presence of foot lesions. Nematodiasis is a common cause of diarrhea in sheep and goats, as is bacterial enteritis caused particularly by *Escherichia coli* and *Salmonella* serovars in kids and lambs. In convalescent goats, the labial scabs can be confused with the mouth lesions of orf (contagious ecthyma), but this disease does not have the alimentary and pulmonary tract lesions seen in PPR.

IMMUNITY. Solid immunity develops after infection and recovery with virulent or attenuated strains of PPR and rinderpest viruses. White-tailed deer that survived experimental inoculation with PPR virus all resisted challenge with rinderpest virus (Hamdy and Dardiri 1976). Kids born to dams immunized with rinderpest virus vaccine have protective maternal antibody for up to 3 months (Ata et al. 1989).

CONTROL AND TREATMENT. Affected and in-contact goats and sheep should be isolated from other stock and effective quarantine imposed until at least 1 month after the complete recovery of the last clinical case. Infected premises should be thoroughly cleaned and disinfected with solutions containing lipid solvents and/or agents with high or low pH. Strict control of animal movement is essential but can be difficult to achieve. Consequently, vaccination is very important in the control of PPR.

Since the early observations of cross-immunity between PPR and rinderpest viruses, the use of attenuated rinderpest virus vaccine to protect sheep and goats against PPR (Taylor 1979) is widely practiced (Lefevre and Diallo 1990). This vaccine is safe in pregnant goats and confers clinical immunity against challenge with virulent PPR virus for at least 3 years. Attenuated PPR virus vaccines that protect goats against experimental challenge have been developed (Diallo et al. 1989).

There is no established therapy for affected animals. Hyperimmune serum administered during early clinical disease has been reported to help recovery, but antibiotics have little effect. Fluid replacement should be considered for valuable individuals with diarrhea. Good husbandry and nursing are needed during convalescence.

MANAGEMENT IMPLICATIONS. Wild caprines and ovines from Asia and bovids from Africa could introduce the infection to zoological collections if either in the incubation period or undergoing mild or subclinical infection. Therefore, all new stock should be quarantined for at least 3 weeks before shipping. Conventionally, it is recommended that all animals being moved to a new destination should be seronegative to the virus, but such animals are susceptible, and there is the risk, perhaps small, of them contracting the disease in transit. However, recovered and vaccinated animals have never been shown to reexcrete virus, and it might be more appropriate to vaccinate all stock with attenuated rinderpest vaccine immediately on entry to the quarantine and then 3 weeks later test them to ensure they have high levels of immunity, indicating that they cannot be incubating the infection.

LITERATURE CITED

Amjad, H., Qamar-ul-Islam, M. Forsyth, T. Barrett, and P.B. Rossiter. 1996. Peste des petits ruminants in goats in Pakistan. *Veterinary Record* 139:118–119.

Ata, F.A., H.S. Sumry, G.J. King, S.I. Ismaili, and A.A. Ata. 1989. Duration of maternal immunity to peste des petits ruminants. *Veterinary Record* 124:590–591.

Baron, M.D., L. Goatley, and T. Barrett. 1994. Cloning and sequence analysis of the matrix (M) protein gene of rinderpest virus and evidence for another bovine morbillivirus. *Virology* 200:121–129.

Brown, C.C., J.C. Mariner, and H.J. Olander. 1991. An immunohistochemical study of the pneumonia caused by peste des petits ruminants virus. *Veterinary Pathology* 28:166–170.

Bundza, A., A. Afshar, T.W. Dukes, D.J. Myers, G.C. Dullac, and S.A.W.E. Becker. 1988. Experimental peste des petits ruminants (goat plague) in goats and sheep. *Canadian Journal of Veterinary Research* 52:46–52.

Diallo, A., W.P. Taylor, P.C. Lefevre, and A. Provost. 1989. Attenuation d'une souche de virus de la peste des petits ruminants: Candidat pour un vacci homologue vivant. *Revue des Eleves Medicins Veterinaire des Pays Tropicaux* 42:311–319.

Diallo, A., G. Libeau, E. Couacy-Hymann, and M. Barbron. 1995. Recent developments in the diagnosis of rinderpest and peste des petits ruminants. *Veterinary Microbiology* 44:307–317.

El Hag Ali, B., and W.P. Taylor. 1984. Isolation of peste des petits ruminants virus from the Sudan. *Research in Veterinary Science* 36:1–4.

Furley, C., W.P. Taylor, and T.U. Obi. 1987. An outbreak of peste des petits ruminants in a zoological collection. *Veterinary Record* 121:443–447.

Gargadennec, L., and A. Lalanne. 1942. La peste des petits ruminants. *Bulletin des Services Zootechniques et des Epizooties de l'Afrique Occidentale Française* 5:16–21.

Gibbs, E.P.J., W.P. Taylor, and M.J.P. Lawman. 1997. Isolation of adenoviruses from goats affected with peste des petits ruminants in Nigeria. *Research in Veterinary Science* 23:331–335.

Hafez, S.M., A. Al-Sukayran, D. Dela-Cruz, S.I. Bekairi, and A.I. Radwan. 1987. Serological evidence for the occurrence of peste des petits ruminants among deer and gazelles in Saudi Arabia. In *Proceedings of the first symposium (for) wildlife conservation and development in Saudi Arabia,* ed. A.H. Abu-Zinada, P.D. Goriup, and I.A. Nader. Riyadh: National Commission for Wildlife Conservation and Development; Publication 3.

Hamdy, F.M., and A.H. Dardiri. 1976. Response of white-tailed deer to infection with "peste des petits ruminants" virus. *Journal of Wildlife Diseases* 12:191–196.

Kulkarni, D.D., A.U. Bhikane, M.S. Shaila, P. Varalakshmi, M.P. Apte, and B.W. Narladkar. 1996. Peste des petits

ruminants in goats in India. *Veterinary Record* 138:187–188.
Lefevre, P.C., and A. Diallo. 1990. Peste des petits ruminants. *Revue Scientifique et Technique O.I.E.* 9:951–965.
Lefevre, P.C., A. Diallo, F. Schenkel, S. Hussein, and G. Staak. 1991. Serological evidence of peste des petits ruminants in Jordan. *Veterinary Record* 128:110.
Libeau, G., A. Diallo, F. Colas, and L. Guerre. 1994. Rapid differential diagnosis of rinderpest and peste des petits ruminants using an immunocapture ELISA. *Veterinary Record* 134:300–304.
Losos, G. 1986. Peste des petits ruminants. In *Infectious tropical diseases of domestic animals,* ed. G. Losos. Essex, England: Longman Scientific and Technical.
McCullough, K.C., T.U. Obi, and H. Sheshberadaran. 1991. Identification of epitope(s) on the internal virion proteins of rinderpest virus which are absent from peste des petits ruminants virus. *Veterinary Microbiology* 26:313–321.
Obi, T.U., M.O. Ojo, W.P. Taylor, and L.W. Rowe. 1983. Studies on the epidemiology of peste des petits ruminants in southern Nigeria. *Tropical Veterinarian* 1:209–217.
Roeder, P., G. Abraham, G. Kenfe, and T. Barrett. 1994. Peste des petits ruminants in Ethiopian goats. *Tropical Animal Health and Production* 26:69–73.
Rossiter, P.B., and W.P. Taylor. 1994. Peste des petits ruminants. In *Infectious diseases of livestock, with special reference to southern Africa,* ed. J.A.W. Coetzer, G.R. Thomson, and R.C. Tustin. Cape Town: Oxford University Press, pp. 758–765.
Rossiter, P.B., and R.C. Wardley. 1985. The differential growth of virulent and avirulent strains of rinderpest virus in bovine lymphocytes and macrophages. *Journal of General Virology* 66:969–975.
Rossiter, P.B., K.A.J. Herniman, and H.W. Wamwayi. 1992. Improved isolation of rinderpest virus in transformed bovine T lymphoblast cell lines. *Research in Veterinary Science* 53:11–18.
Saliki, J.T., C.C. Brown, J.A. House, and E.J. Dubovi. 1994. Differential immunohistochemical staining of peste des petits ruminants and rinderpest antigens in formalin-fixed, paraffin-embedded tissues using monoclonal and polyclonal antibodies. *Journal of Veterinary Diagnostic Investigation* 6:96–98.
Scott, G.R. 1981. Rinderpest and peste des petits ruminants. In *Disease monographs.* Vol. 3: *Virus diseases of food animals,* ed. E.P.J. Gibbs. London: Academic, pp. 401–432.
Taylor, W.P. 1979. Protection of goats against peste des petits ruminants with attenuated rinderpest virus. *Research in Veterinary Science* 27:321–324.
———. 1984. The distribution and epidemiology of "peste des petits ruminants." *Preventive Veterinary Medicine* 2:157–166.
Wamwayi, H.M., P.B. Rossiter, D.P. Kariuki, J.S. Wafula, T. Barrett, and J. Anderson. 1995. Peste des petits ruminants antibodies in East Africa. *Veterinary Record* 136:199–200.

CANINE DISTEMPER

ELIZABETH S. WILLIAMS

Synonyms: La maladie de Carré.

INTRODUCTION. Canine distemper (CD), an infectious, contagious viral disease of many species of domestic and wild carnivores, is among the most significant infectious diseases of these species. Canine distemper virus (CDV) is a morbillivirus closely related to measles virus of primates and peste de petits ruminants and rinderpest viruses of ruminants. Because of the importance of CD in domestic dogs *Canis familiaris* and mink *Mustela vison,* pathogenesis, molecular biology, and immunoprophylaxis have been well studied. However, with the exception of a few species, the epidemiology of CD in free-ranging species is poorly understood. Recent emergence of CD in species not previously known to be naturally susceptible, including javelina *Tayassu tajacu,* felids, and marine mammals (see other sections in this chapter), and the significant impact of CD on some endangered species, make CD an infectious disease of major concern for managers of free-ranging and captive carnivores (Montali et al. 1987; Williams and Thorne 1996).

HISTORY. Canine distemper has been known from Europe for at least 200 years. It was reported to be caused by a virus in 1905 (Carré 1905). About that time, CD was recognized in captive nondomestic species; about 50 years later, it was reported in free-ranging wildlife (Helmboldt and Jungherr 1955). The recognized host range of CDV recently expanded with occurrence of epidemics in free-ranging felids and marine mammals.

HOSTS AND DISTRIBUTION. The host range of CDV is wide; species in all families in the order Carnivora (Canidae, Mustelidae, Procyonidae, Hyaenidae, Ursidae, Viverridae, and Felidae) are susceptible to CDV. Budd (1981) provided an extensive list of susceptible species; reports of new hosts continue to appear, and probably all canids, mustelids, and procyonids should be considered susceptible.

Canine distemper occurs worldwide. Coyotes *Canis latrans* and wolves *Canis lupus* are common canid hosts of CDV in North America (Choquette and Kuyt 1974; Stephenson et al. 1982; Guo et al. 1986; Zarnke and Ballard 1987; Gese et al. 1991, 1997; Johnson et al. 1994; Williams et al. 1997, 1998; Cypher et al. 1998). Surveys (Amundson and Yuill 1981) and diagnostic records (Little et al. 1998) demonstrated that red fox *Vulpes vulpes* are susceptible to CD but appear to be more resistant than highly susceptible gray foxes *Urocyon cinereoargenteus* (Hoff et al. 1974; Davidson et al. 1992a,b). Natural CD also occurs in red foxes in Europe (López-Peña et al. 1994; Truyen et al. 1998). A serosurvey of endangered San Joaquin kit foxes *Vulpes macrotis mutica* in California provided evidence of exposure to CDV (McCue and O'Farrell 1988).

Among African canids, most attention has been focused on wild dogs *Lycaon pictus* (Alexander and Appel 1994); black-backed jackals *Canis mesomelas* and side-striped jackals *Canis adustus* are also susceptible (Alexander et al. 1994). Canine distemper is rarely reported in Asian canids though it occurs in free-ranging raccoon dogs *Nyctereutes procyonoides* in Japan (Machida et al. 1993).

Canine distemper has been reported in mustelids in many parts of the world. In North America, black-footed ferrets *Mustela nigripes* are extremely susceptible (Williams et al. 1988); however, species such as striped skunks *Mephitis mephitis* appear to be more resistant (Diters and Nielsen 1978). Reports of CD in European mustelids include epidemics in stone marten *Martes foina,* polecats *Mustela putorius,* European badgers *Meles meles,* and weasels *Mustela* sp. (Kolbl et al. 1990; Alldinger et al. 1993; van Moll et al. 1995).

Canine distemper has been studied in free-ranging raccoons *Procyon lotor* in eastern North America (Hoff et al. 1974; Cranfield et al. 1984; Hable et al. 1992; Roscoe 1993; Schubert et al. 1998). Captive coatis *Nasua nasua* died of CD in Brazil (Cubas 1996). Lesser pandas *Ailurus fulgens* are highly susceptible to CD (Kotani et al. 1989), and there are numerous reports of vaccine-induced disease in this species (Bush et al. 1976; Montali et al. 1987).

Ursids are susceptible to CDV infection, but disease appears to be rare. Serologic evidence of CD was found in captive and free-ranging brown bears *Ursus arctos* in Italy (Marsilio et al. 1997) and in free-ranging black bears *Ursus americanus floridanus* from Florida (Dunbar et al. 1998). There is one report of CD in captive polar bears *Ursus maritimus* and a spectacled bear *Tremarctos ornatus* (von Schöbauer et al. 1984) from Europe. Canine distemper was diagnosed in giant pandas *Ailuropoda melanoleuca* from China, and other pandas with antibodies indicate their susceptibility to infection by CDV (Qui and Mainka 1993; Mainka et al. 1994).

The susceptibility of hyenas and viverrids to CDV has not been well studied, but serologic surveys have shown that at least some of these species are susceptible to CD. Clinical CD was documented in spotted hyenas *Crocuta crocuta* from the Serengeti (Alexander et al. 1995; Haas et al. 1996). Canine distemper was reported in a palm civet *Paguma larvata* from Japan (Machida et al. 1992).

There are few published reports of CD from South America, but it is considered a challenge to management of captive carnivores (Cubas 1996). Approximately 20% of 108 captive maned wolves *Chrysocyon brachyurus* died of CD between 1989 and 1993 in Brazilian zoos (Cubas 1996).

Diagnosis of CD in javelina (Appel et al. 1991) demonstrates the potential of this virus to affect a variety of species. Morbilliviruses suspected of being CDV were isolated from hedgehogs *Erinaceus europaeus* in Europe (Vizoso and Thomas 1981).

ETIOLOGY. Canine distemper virus is in the family *Paramyxoviridae*, subfamily *Paramyxovirinae*, and genus *Morbillivirus* [Pringle 1992; see Lamb and Kolakofsky (1996) for a review of paramyxoviruses]. It is a relatively large (100–700 nm), negative-stranded RNA virus with spherical to filamentous morphology. The lipoprotein envelope is derived from the plasma membrane of the host cell, with membrane (M) proteins on the inside and two surface glycoproteins, hemagglutinin (H) and fusion protein (F), on the outside. The nucleocapsid is helical, has a herringbone pattern, and contains NP, P, and L structural viral proteins.

Canine distemper virus is relatively fragile and quickly inactivated in the environment by ultraviolet light and by heat and drying. Temperatures greater than 50° C rapidly destroy the virus, and it is inactivated in several hours in tissues at 20° C–37° C. It may remain stable for weeks at 4° C and for years when frozen at –65° C (Appel 1987). Common disinfectants readily inactivate CDV (Appel 1987).

There is only one serotype of CDV, but minor antigenic differences are recognized among strains (Blixenkrone-Møller et al. 1992; Alldinger et al. 1993). However, recent sequencing of H and NP genes of CDV isolates from North America, Europe, and Asia show more than one genotype and variation among field and vaccine viruses (Bolt et al. 1997; Haas et al. 1997; Iwatsuki et al. 1997; Yoshida et al. 1998). There is considerable biologic variation among isolates of CDV. For example, some strains of CDV are more neurotropic (Stettler et al. 1997).

TRANSMISSION AND EPIDEMIOLOGY. Transmission of CDV is primarily by aerosol or contact with oral, respiratory, and ocular fluids and exudates containing the virus. Though CDV is shed from skin and in feces and urine, these are probably relatively unimportant in transmission. Transplacental transmission has been documented in dogs (Krakowka et al. 1977); it may occur in wild species but is unlikely to be epidemiologically important. Close association between affected and susceptible animals is necessary due to the relative fragility of CDV in the environment. Viral shedding occurs even if animals are subclinically infected (Appel 1987), and virus may be shed for up to 90 days after infection (Greene and Appel 1998).

Dense populations of susceptible individuals are necessary to sustain epidemics of CD. Epidemiology of CD depends on a variety of factors, such as relative susceptibility of hosts, population density of sympatric susceptible hosts, and intraspecies and interspecies behavior that influence transmission. Susceptibility in some species may be influenced by predisposing factors such as stress and immunosuppression, vaccination, or exposure to a high dose of virus (Greene and Appel 1998).

Observations on the epidemiology of CD in dogs may assist in understanding the disease in free-ranging species. In endemic areas where dog populations are high, clinical disease is mostly seen in pups following loss of maternal antibody at 3–6 months. In isolated populations of dogs, CD is epidemic, and outbreaks may be severe, widespread, and affect all ages (Leighton et al. 1988; Bohm et al. 1989).

The dynamics of CD in eastern North America has been best studied in raccoons (Hoff et al. 1974; Cranfield et al. 1984; Roscoe 1993; Schubert et al. 1998). In New Jersey, epizootics of CD occurred at 4-year intervals (Roscoe 1993), and peaks of prevalence were at the end of the breeding season in March and again in September at the time of kit dispersal. Epizootics of CD were associated with river drainages. Localized raccoon population declines in association with CD have been suspected; however, studies in Ontario, Canada, failed to show clear evidence of population effects (Schubert et al. 1998). Canine distemper was not diagnosed in sympatric red foxes during one epidemic (Cranfield et al. 1984).

On North American short-grass prairies, where densities of carnivores are relatively low, CD occurred as epidemics about every 3–7 years (Williams et al. 1997, 1998). Epidemics occurred during all seasons, and CD was active in study populations for many months. Coyotes were the principle species involved in epidemics, but other susceptible species involved included red fox, swift fox *Vulpes velox,* badgers *Taxidea taxus,* and black-footed ferrets (Williams et al. 1988, 1997, 1998). Behavior probably influences dissemination of the virus. For example, coyotes are relatively gregarious during most of the year, possibly facilitating transmission, in comparison to badgers, which are relatively solitary during most of the year. Canine distemper probably does not have significant effects on coyote populations, though there may be loss of many juveniles during epidemics (Williams et al. 1992).

There has been considerable concern over the potential detrimental effect of CD on populations of African wild dogs (Alexander and Appel 1994). However, study of the epidemiology of CD in Tanzania suggested that African wild dog populations can remain stable and demographically healthy despite 60% seroprevalence to CDV (Creel et al. 1997).

The epidemiology of CD has been studied in stone marten in Europe (Palmer et al. 1983; Steinhagen and Nebel 1985; van Moll et al. 1995). Adults and juvenile marten were affected and prevalence was highest in the summer, coincident with the breeding season. Canine distemper was documented in marten but not in sympatric red foxes, and cases did not increase in dogs in the area (van Moll et al. 1995).

CLINICAL SIGNS. Signs of CD vary depending on species, viral strain, environmental conditions, and host age and immune status. In canids and mink, and probably other species, juveniles appear to be most susceptible (Krakowka and Koestner 1976; Pearson and Gorham 1987). Controlled studies of CD susceptibility have rarely been conducted in nondomestic species, but it is estimated that 25%–75% of susceptible dogs become subclinically infected and clear the virus without developing illness (Greene and Appel 1998). Mortality is variable in adult mink (20%–90%) and is approximately 90% in mink kits; CD is considered 100% fatal in domestic ferrets *Mustela putorius furo* (Budd 1981; Pearson and Gorham 1987). Mortality in experimentally infected raccoons varied from 50% to 100% (Kilham and Herman 1954; Paré 1997). In an experimental study of CD in coyotes, all ten pups died but all seven adults survived (Gier and Ameel 1959); however, antibody status of the adults was not determined prior to infection. Surveys in Wyoming suggested coyote pups may be more susceptible to CD mortality than are yearlings and adults (Williams et al. 1992). Some wild species are so susceptible that few if any individuals exposed to CDV recover; examples are black-footed ferrets (Williams et al. 1988) and gray foxes (Davidson et al. 1992b).

Incubation period ranges from about 1 week to 1 month or more (Appel 1987). Duration of clinical disease is also variable and dependent on many factors but ranges from 1 to 4–6 weeks with resolution and recovery or death. Budd (1981) provides an excellent review of the clinical signs of CD.

Clinical presentation depends somewhat on the species; the classic signs of a canid with CD are depression and mucopurulent oculonasal exudates. Early signs may be subtle or not observed in wild species. They are usually mild serous oculonasal exudates that become mucopurulent. A dry cough may progress to a moist cough. Fever, depression, anorexia, vomiting, and diarrhea frequently occur. Animals that recover from clinical CD or that have a prolonged illness may be in poor body condition.

Central nervous system (CNS) signs may be concurrent or follow systemic disease, usually within 1–5 weeks after recovery but occasionally longer. Neurologic signs depend on the area of brain affected and may include abnormal behavior, convulsions that often take the form of "chewing gum" seizures, cerebellar and vestibular signs, paresis or paralysis, aimless wandering, incoordination, or myoclonus (Greene and Appel 1998). Javelina with CD showed blindness, myoclonus, depression, and circling (Appel et al. 1991).

Canine distemper in procyonids is clinically similar to that described in canids. Raccoons with CNS manifestations are frequently reported as behaving abnormally. Roscoe (1993) reported lethargy as the most commonly observed clinical sign; aggressive behavior was rare. Hyperkeratosis of the foot pads with marked thickening and deep cracks is relatively common in raccoons that have a prolonged clinical course. An early report of CD in raccoons described icterus (Kilham and Herman 1954), but this is uncommon.

In mustelids, the clinical signs first recognized are serous oculonasal exudate, photophobia, and hyperemia and thickening of the eyelids, lips, and anus (Budd 1981; Pearson and Gorham 1987). Secondary bacterial infection of the skin results in pruritus, especially of the face. Other clinical signs include fever, depression, respiratory signs, diarrhea, dehydration, anorexia, behavioral changes, and convulsions.

PATHOGENESIS. Liu and Coffin (1957), Budd (1981), Appel (1987), and Greene and Appel (1998) reviewed pathogenesis of CD infection in mink, ferrets, and dogs. Aerosolized virus enters via epithelium of the upper respiratory tract. Within a few days, CDV multiplies in macrophages and spreads to tonsils, regional lymph nodes, and in low amounts to other lymphoid tissues. By about 1 week, virus may be found in systemic lymphoid tissues, lamina propria of the digestive system, and Kupffer cells in liver. Spread corresponds to development of fever and leukopenia due to viral associated loss of T and B lymphocytes. Damage to the immune system results in immunosuppression (Kaufman et al. 1982).

With viremia and widespread infection of epithelial tissue, virus shedding begins about 1 week after infection (Appel 1987). In wild animals, there probably are no discernible clinical signs at this time. Between 1 and 2 weeks after infection, the host's immune response determines the outcome of infection. If there is a strong antibody response, no clinical illness develops and the virus is cleared. If the animal mounts a weak antibody response, illness may ensue, but by about 3 week after infection, the virus is cleared from most of the body, with the possible exception of lung, skin, and CNS. Virus may be shed by these animals for several months. If there is minimal or no antibody response, severe systemic disease occurs by about 2–3 weeks after infection, with death by 3–4 weeks after infection. If the animal recovers, it may shed virus for 2–3 months. Early development of circulating antibodies is considered crucial to recovery and preventing infection of the CNS.

PATHOLOGY. Changes in hematologic parameters and clinical chemistry in animals with CD are nonspecific. Absolute lymphopenia is common.

Gross external lesions may vary from no obvious lesions to those reflective of clinical signs: oculonasal exudates, hyperkeratosis, diarrhea, and poor body condition. Thymic atrophy is a consistent but nonspecific gross lesion. Interstitial pneumonia to bronchopneumonia, catarrhal to hemorrhagic enteritis, and hyperkeratosis of the nose, lips, eyelids, ears, anus, and foot pads may be present.

Less commonly observed lesions include damage of enamel, dentin, and tooth roots in animals that contract CD prior to eruption of permanent teeth. This lesion is well described in dogs (Dubielzig 1979; Bittegko et al. 1995) and has been suspected in coyotes (E.S. Williams unpublished). Long bone lesions consisting of metaphyseal osteosclerosis occur in dogs (Baumgärtner et al. 1995). Testicular degeneration, orchitis, and epididymitis have been reported in raccoons and dogs (Hamir et al. 1992; Greene and Appel 1998). Degenerative or necrotizing lesions of the retina with scarring and atrophy may occur in dogs (Dungworth 1993) but has not been reported in wild species, though it likely occurs.

Common microscopic lesions are depletion of lymphocytes in paracortical zones and germinal centers of lymph nodes and spleen and thymic atrophy. Interstitial pneumonia and occasionally suppurative bronchopneumonia secondary to bacterial infection are typical. Giant cell pneumonia due to formation of syncytial cells in pulmonary alveoli has been described in raccoons (Karstad and Budd 1964).

Nonsuppurative encephalitis and meningoencephalitis, neuronal necrosis, and focal malacia and demyelination in cerebellar white matter occur (Potgieter and Patton 1984; van Moll et al. 1995). Other typically affected sites are cerebellum and medulla adjacent to the fourth ventricle and optic tracts (Potgieter and Patton 1984; Dungworth 1993). Gliosis is common. Intraocular tissues may be inflamed (Appel 1987; Dungworth 1993).

An important diagnostic feature of CD is the presence of intracytoplasmic and intranuclear eosinophilic inclusion bodies in epithelia, neurons, and astroglia. Inclusion bodies are often in gastric mucosa, enterocytes, pancreatic and biliary duct epithelium, and epithelium of the respiratory and urogenital tract. Transitional epithelium of the urinary bladder is especially useful. Inclusion bodies are 1–5 μm in diameter and may be single or multiple. Inclusion bodies persist 5–6 weeks in lymphoid tissues and urinary tract of dogs.

Secondary bacterial or protozoal infections due to the immunosuppressive effect of CDV may complicate microscopic evaluation of affected animals. Toxoplasmosis is commonly recognized in carnivores (Moller and Nielsen 1964; Reed and Turek 1985; Dubey et al. 1992). Other CD-associated protozoal diseases are sarcocytosis in raccoons (Dubey et al. 1990; Stoffregen and Dubey 1991), encephalitozoonosis in African wild dogs (van Heerden et al. 1989), *Pneumocystis carnii* in dogs (Sukura et al. 1997), and coccidiosis in black-footed ferrets (Williams et al. 1988). Secondary bacterial infections occurred in gray foxes (Black et al. 1996) and Tyzzer's disease was diagnosed in a raccoon (Wojcinski and Barker 1986). Other viral diseases may occur concurrent with CD; parvoviral enteritis has been reported in a coyote (Holzman et al. 1992) and rabies in a raccoon without encephalitis typical of rabies (Hamir and Rupprecht 1990).

DIAGNOSIS. History, clinical observations, and typical gross external lesions, when available, may suggest CD. Abnormal behavior, which might include diurnal activity in nocturnal species, inappropriate interaction with humans, and aggressiveness, suggests CD-associated CNS disease. Some individuals show only mild or atypical signs, and external lesions and additional diagnostic procedures are necessary to confirm CD. Smears of conjunctiva or buffy coats may be examined for inclusion bodies or by fluorescent antibody staining for CDV antigen; however, cytology is not sensitive especially early in CD or in animals with prolonged clinical course. Viral nucleocapsids may be found by negative stain transmission electron microscopy of feces or internal organs such as lung and spleen (Nunamaker and Williams 1986). Typical gross lesions, particularly thymic atrophy, and pneumonia may be suggestive of CD.

Samples to collect at necropsy for diagnostic purposes include fresh chilled and 10% buffered formalin-fixed spleen, lymph nodes, stomach, lung, small intestine, liver, pancreas, urinary bladder, kidney with renal pelvis, and brain. Blood for serology can be collected if it is not extensively hemolyzed. Diagnosis is made by histopathology; demonstration of typical viral inclusion bodies in epithelium, lymphoid tissues, and brain; presence of typical virions in negative-stain electron-microscopic preparations; and detection of viral antigen in tissues by immunofluorescence or immunohistochemistry.

Immunohistochemistry is very useful for detection of CDV antigen in formalin-fixed, paraffin-embedded tissues and is now widely used (Baumgärtner et al. 1989; Palmer et al. 1990; van Moll et al. 1995). Antiserum against CDV or measles virus may be used. Polymerase chain reaction (PCR) and nucleic acid hybridization tests also detect CDV in tissues (Zurbriggen et al. 1993; Haas et al. 1996).

Isolation of CDV is definitive evidence of infection and is useful if epidemiologic study of the virus is desired. It is typically conducted in pulmonary alveolar macrophages or in mitogen-stimulated canine or ferret blood lymphocytes (Appel et al. 1992; J. Cavender personal communication). Vaccine strains of CDV may be isolated on Vero cells (Evans et al. 1991).

Serum neutralization is the standard serologic test for antibodies against CDV (Appel and Robson 1973). Other tests commonly used are enzyme-linked immunosorbent assays (ELISAs), which may enable distinction between immunoglobulin μ (IgM) and IgG. Presence of IgM suggests recent infection or vaccination (Blixenkrone-Møller et al. 1991). Animals that die of CD often do not have antibodies against CDV. In dogs, serum antibody titers tend to vary inversely with the severity of disease (Greene and Appel 1998). Serologic surveys are commonly used to study the presence and epidemiology of CD in wild populations.

DIFFERENTIAL DIAGNOSES. Rabies is the most important differential diagnosis in wild carnivores showing CNS signs and must always be considered because of its public health significance. Canine distemper often is diagnosed in carnivores submitted to diagnostic laboratories for rabies testing (Woolf et al. 1986; K. Mills personal communication). Concurrent rabies and CD occur and must be considered in areas of endemic rabies (Hamir and Rupprecht 1990). Other differential diagnoses that can be distinguished from CD by clinical course and lesions are pseudorabies, caused by Aujeszky's disease herpesvirus (Thawley and Wright 1982); infectious canine hepatitis and fox encephalitis, caused by canine adenovirus 1; parvovirus infections; toxoplasmosis; and intoxications such as lead poisoning.

IMMUNITY, CONTROL, AND TREATMENT. Wild animals that survive CD probably have lifelong immunity to subsequent infection by CDV. Modified-live virus (MLV) vaccines, in species that can be safely vaccinated, induce long-lived immunity.

Immunity to CDV infection involves both humoral and cell-mediated mechanisms. Antibodies clearly play a crucial role in immunity (Winters et al. 1983; Greene and Appel 1998), and animals with serum-neutralizing antibody titers of equal to or more than 1:100 are usually considered protected (Montali et al. 1983). The H and F membrane proteins stimulate protective antibodies against CDV (Norrby et al. 1986). Viral proteins are expressed on the cell surface; thus immune-mediated cytolysis plays a role in clearing the virus from infected animals (Appel et al. 1982).

Many inactivated and MLV CD vaccines have been developed. Canine tissue culture adapted MLV vaccines caused CD in some species (Thomas-Baker 1985; Montali et al. 1987; McInnes et al. 1992). Avian or vero cell-adapted CDV vaccines are safer in sensitive species and dogs than are canid cell culture-adapted vaccines (Appel and Summers 1995). However, some MLV vaccines, even those propagated in avian cell cultures, may induce CD in highly susceptible species, including black-footed ferrets (Carpenter et al. 1976; E.S. Williams unpublished), gray foxes (Halbrooks et al. 1981), fennec foxes *Fennecus zerda* (Montali et al. 1987), lesser pandas (Bush et al. 1976), kinkajous *Potos flavus* (Kazacos et al. 1981), African wild dogs (McCormick 1983; van Heerden et al. 1989; Durchfeld et al. 1990), European mink *Mustela lutreola* (Sutherland-Smith et al. 1997), and long-tailed weasels *Mustela frenata* and ermine *Mustela erminea* (C. Petrini personal communication). Vaccine-induced immunosuppression may have led to death of a lesser panda (Montali et al. 1983).

Relatively few CD vaccines have been tested for safety and efficacy in nondomestic mammals. Some MLV vaccines induced antibodies and were safe in bush dogs, maned wolves, gray foxes (Halbrooks et al. 1981; Montali et al. 1983), badgers (Goodrich et al. 1994), raccoons (Evans 1984; Paré 1997), African wild dogs (Spencer and Burroughs 1992), red wolves *Canis rufus* (Harrenstien et al. 1997), and black-footed ferret × Siberian polecat *Mustela eversmanni* hybrids (List 1994; Williams et al. 1995). However, they induced lymphopenia in badgers and lymphopenia and immunosuppression in hybrid ferrets. African wild dogs, lesser pandas, binturong *Arctictis binturong,* and river otters *Lutra canadensis* failed to develop antibodies following vaccination (van Heerden et al. 1980; Montali et al. 1983, 1987; Hoover et al. 1985).

Thus, use of MLV vaccines in wild species should be approached cautiously and with knowledge of the specific vaccine and sensitivity of the species. Highly susceptible species may be vaccinated with killed CDV vaccines, though protection is not complete (Williams et al. 1995). Vaccination of sick or immunosuppressed individuals of any species should be avoided.

The effect of maternal antibody on MLV vaccination has not been carefully studied in wild mammals; in domestic ferrets, the half-life of maternal antibody was 9.4 days (Appel and Harris 1988) and, in raccoons, it was 10.5 days (Paré 1997). Raccoon kits should be vaccination at 8, 12, and 16 weeks of age (Paré 1997).

Recently developed CD vaccines use purified glycoproteins (Norrby et al. 1986), immune-stimulating complexes (De Vries et al. 1988), and various recombinant poxviruses expressing F and H glycoproteins (Taylor et al. 1992; Pardo et al. 1997; Stephensen et al. 1997). Efficacy of these vaccines for wild species is being tested, and they may offer safe alternatives to MLV vaccines in the future. In preliminary tests, one poxvirus recombinant vaccine was safe in hybrid ferrets and provided protection from challenge with virulent CDV (E.S. Williams and R.J. Montali unpublished). Recombinant vaccines also have potential to be used as oral vaccines in the field.

Treatment of captive wild species with CD is symptomatic, but prognosis is guarded. Antibiotics, fluids, nutrient supplements, and anticonvulsants are used (Greene and Appel 1998). Treatment of free-ranging species with CD is not possible or warranted.

PUBLIC HEALTH CONCERNS. Over the years, there has been concern that CDV could be associated with multiple sclerosis in humans (Rohowsky-Kochan et al. 1995), though this theory remains controversial (Greene and Appel 1998). Paramyxoviruses, especially CDV, have been implicated in Paget's disease of bone in humans, and recently evidence of CDV was found in all of 15 patients tested by in situ-reverse transcriptase PCR (Mee et al. 1998). In any case, domestic dogs would be a more likely source of CDV for humans than would wild carnivores.

DOMESTIC ANIMAL HEALTH CONCERNS. Commercial CDV vaccines provide excellent long-lasting immunity in dogs and furbearers. Because there is such variation in the epidemiology of CD in wild species, the likelihood of them serving as reservoirs of CDV for domestic species probably depends on the location and level of domestic animal vaccination (Appel 1987; Anderson 1995). Transmission of CD between domestic and free-ranging wildlife has not been extensively studied, but the potential exists (van Moll et al. 1995). Arctic foxes *Alopex lagopus* were suspected of spreading CD to sled dogs (Bohm et al. 1989). Pearson and Gorham (1987) suggested wild species were potential sources of CDV for farmed mink. Similarly, free-ranging species such as raccoons could serve as a source of CDV for zoo animals (Cranfield et al. 1984; Paré 1997).

MANAGEMENT IMPLICATIONS. Canine distemper is one of the most important infectious diseases of free-ranging carnivores. It may have significant impact on free-ranging populations of highly susceptible species such as black-footed ferrets (Williams and Thorne 1996) and gray foxes (Davidson et al. 1992b). Canine distemper along with sylvatic plague, caused by *Yersinia pestis,* was responsible for extirpation of black-footed ferrets from the wild (Thorne and Williams 1988). Impacts on populations of other susceptible species are less clear (Creel et al. 1997; Schubert et al. 1998).

Canine distemper is also among the most important infectious diseases of carnivores in captivity and must be considered in design of husbandry protocols, including vaccination, quarantine, and housing arrangements (Williams and Thorne 1996). On more than one occasion, wild carnivores have been translocated or brought into captivity while incubating CD, thus spreading the disease and destroying species-rescue plans, research projects, or disrupting management actions (Thorne and Williams 1988, Williams et al. 1988; Davidson et al. 1992a; D. Hoff and E.S. Williams unpublished).

Because many threatened and endangered carnivores are susceptible to CD, it must be considered in management plans and during restoration efforts. Coyotes are thought to serve as a potential source of CD for endangered San Joaquin kit foxes (Cypher et al. 1998), and there was close homology of CDV isolated from hyenas and sympatric lions *Panthera leo* with CD, suggesting interchange of virus (Haas et al. 1996). Concern has been raised about the potential transmission of CD from domestic dogs in or adjacent to wildlife reserves to native wildlife. This may have occurred in the case of free-ranging African wild dogs (Alexander and Appel 1994) and was though possible in the case of giant pandas in China (Mainka et al. 1994).

LITERATURE CITED

Alexander, K.A., and M.J.G. Appel. 1994. African wild dogs (*Lycaon pictus*) endangered by a canine distemper epizootic among domestic dogs near the Masai Mara National Reserve, Kenya. *Journal of Wildlife Diseases* 30:481–485.

Alexander, K.A., P.W. Kat, R.K. Wayne, and T.K. Fuller. 1994. Serologic survey of selected canine pathogens among free-ranging jackals in Kenya. *Journal of Wildlife Diseases* 30:486–491.

Alexander, K.A., P.W. Kat, L.G. Frank, K.E. Holekamp, L. Smale, C. House, and M.J.G. Appel. 1995. Evidence of canine distemper virus infection among free-ranging spotted hyenas (*Crocuta crocuta*) in the Masai Mara, Kenya. *Journal of Zoo and Wildlife Medicine* 26:201–206.

Alldinger, S., W. Baumgärtner, P. van Moll, and C. Örvell. 1993. In vivo and in vitro expression of canine distemper virus proteins in dogs and nondomestic carnivores. *Archives of Virology* 132:421–428.

Amundson, T.E., and T.M. Yuill. 1981. Prevalence of selected pathogenic microbial agents in the red fox (*Vulpes fulva*) and gray fox (*Urocyon cinereoargenteus*) of southwestern Wisconsin. *Journal of Wildlife Diseases* 17:17–22.

Anderson, E.C. 1995. Morbillivirus infections in wildlife (in relation to their population biology and disease control in domestic animals). *Veterinary Microbiology* 44:319–332.

Appel, M.J.G. 1987. Canine distemper virus. In *Virus infections of carnivores,* ed. M.J.G. Appel. Amsterdam: Elsevier Science, pp. 133–159.

Appel, M.J.G., and W.V. Harris. 1988. Antibody titers in domestic ferret jills and their kits to canine distemper virus vaccine. *Journal of the American Veterinary Medical Association* 193:332–333.

Appel, M.J.G., and D.S. Robson. 1973. A microneutralization test for canine distemper virus. *American Journal of Veterinary Research* 34:1459–1463.

Appel, M.J.G., and B.A. Summers. 1995. Pathogenicity of morbilliviruses for terrestrial carnivores. *Veterinary Microbiology* 44:187–191.

Appel, M.J.G., W.R. Shek, and B.A. Summers. 1982. Lymphocyte-mediated immune cytotoxicity in dogs infected with virulent canine distemper virus. *Infection and Immunity* 37:592–600.

Appel, M.J.G., C. Reggiardo, B.A. Summers, S. Pearce-Kelling, C.J. Maré, T.H. Noon, R.E. Reed, J.N. Shively, and C. Örvell. 1991. Canine distemper virus infection and encephalitis in javelinas (collared peccaries). *Archives of Virology* 119:147–152.

Appel, M.J.G., S. Pearce-Kelling, and B.A. Summers. 1992. Dog lymphocyte cultures facilitate the isolation and growth of virulent canine distemper virus. *Journal of Veterinary Diagnostic Investigation* 4:258–263.

Baumgärtner, W., C. Örvell, and M. Reinacher. 1989. Naturally occurring canine distemper virus encephalitis: Distribution and expression of viral polypeptides in nervous tissue. *Acta Neuropathologica (Berlin)* 78:504–512.

Baumgärtner, W., R.A. Boyce, S. Alldinger, M.K. Axthelm, S.E. Weisbrode, S. Krakowka, and K. Gaedke. 1995. Metaphyseal bone lesions in young dogs with systemic canine distemper infection. *Veterinary Microbiology* 44:201–209.

Bittegeko, B., J. Arnbjerg, R. Nkya, and A. Tevik. 1995. Multiple dental abnormalities following canine distemper infection. *Journal of the American Animal Hospital Association* 31:42–45.

Black, S.S., F.W. Austin, and E. McKinley. 1996. Isolation of *Yersinia pseudotuberculosis* and *Listeria monocytogenes* serotype 4 from a gray fox (*Urocyon cinereoargenteus*) with canine distemper. *Journal of Wildlife Diseases* 32:362–366.

Blixenkrone-Møller, M., I.R. Pedersen, M.J. Appel, and C. Griot. 1991. Detection of IgM antibodies against canine distemper virus in dog and mink sera employing enzyme-linked immunosorbent assay (ELISA). *Journal of Veterinary Diagnostic Investigation* 3:3–9.

Blixenkrone-Møller, M., V. Svansson, M. Appel, J. Krogsrud, P. Have, and C. Örvell. 1992. Antigenic relationships between field isolates of morbilliviruses from different carnivores. *Archives of Virology* 123:279–294.

Bohm, J., M. Blixenkrone-Møller, and E. Lund. 1989. A serious outbreak of canine distemper among sled-dogs in northern Greenland. *Arctic Medical Research* 48:195–203.

Bolt, G., T.D. Jensen, E. Gottschalck, P. Arctander, M.J.G. Appel, R. Buckland, and M. Blixenkrone-Møller. 1997. Genetic diversity of the attachment (H) protein gene of current field isolates of canine distemper virus. *Journal of General Virology* 78:367–372.

Budd, J. 1981. Distemper. In *Infectious diseases of wild mammals,* ed. J.W. Davis, L.H. Karstad, and D.O. Trainer, 2d ed. Ames: Iowa State University Press, pp. 31–44.

Bush, M., R.J. Montali, D. Brownstein, A.E. James, and M.J.G. Appel. 1976. Vaccine-induced canine distemper in a lesser panda. *Journal of the American Veterinary Medical Association* 169:959–960.

Carpenter, J.W., M.J.G. Appel, R.C. Erickson, and M.N. Novilla. 1976. Fatal vaccine-induced canine distemper virus infection in black-footed ferrets. *Journal of the American Veterinary Medical Association* 169:961–964.

Carré, H. 1905. Sur la maladie des jeunes chiens. *Compte Rendue de l'Academie des Sciences [III]* 140:689–690 and 1489–1491.

Choquette, L.P., and E. Kuyt. 1974. Serological indication of canine distemper and of infectious canine hepatitis in wolves (*Canis lupus* L.) in northern Canada. *Journal of Wildlife Diseases* 10:321–324.

Cranfield, M.R., I.K. Barker, K.G. Mehren, and W.A. Rapley. 1984. Canine distemper in wild raccoons (*Procyon lotor*) at the Metropolitan Toronto Zoo. *Canadian Veterinary Journal* 25:63–66.

Creel, S., N.M. Creel, L. Munson, D. Sanderlin, and M.J. Appel. 1997. Serosurvey for selected viral diseases and demography of African wild dogs in Tanzania. *Journal of Wildlife Diseases* 33:823–832.

Cubas, Z.S. 1996. Special challenges of maintaining wild animals in captivity in South America. *Revue Scientific et Technique O.I.E.* 15:267–287.

Cypher, B.L., J.H. Scrivner, K.L. Hammer, and T.P. O'Farrell. 1998. Viral antibodies in coyotes from California. *Journal of Wildlife Diseases* 34:259–264.

Davidson, W.R., M.J. Appel, G.L. Doster, O.E. Baker, and J.F. Brown. 1992a. Diseases and parasites of red foxes, gray foxes, and coyotes from commercial sources selling to fox-chasing enclosures. *Journal of Wildlife Diseases* 28:581–589.

Davidson, W.R., V.F. Nettles, L.E. Hayes, E.W. Howerth, and C.E. Couvillion. 1992b. Diseases diagnosed in gray foxes (*Urocyon cinereoargenteus*) from the southeastern United States. *Journal of Wildlife Diseases* 28:28–33.

De Vries, P., F.G.C. Uytdenhaag, and A.D.M.E. Osterhaus. 1988. Canine distemper virus (CDV) immune stimulating complexes (iscoms), and not measles virus iscoms protect dogs against CDV infection. *Journal of General Virology* 69:536–541.

Diters, R.W., and S.W. Nielsen. 1978. Toxoplasmosis, distemper and herpes virus infection in a skunk (*Mephitis mephitis*). *Journal of Wildlife Diseases* 14:132–136.

Dubey, J.P., A.N. Hamir, C.A. Hanlon, M.J. Topper, and C.E. Rupprecht. 1990. Fatal necrotizing encephalitis in a raccoon associated with a *Sarcocystis*-like protozoa. *Journal of Veterinary Diagnostic Investigation* 2:345–347.

Dubey, J.P., A.N. Hamir, C.A. Hanlon, and C.E. Rupprecht. 1992. Prevalence of *Toxoplasma gondii* in raccoons. *Journal of the American Veterinary Medical Association* 200:534–536.

Dubielzig, R.R. 1979. The effect of canine distemper virus on the ameloblastic layer of the developing tooth. *Veterinary Pathology* 16:268–270.

Dunbar, M.R., M.W. Cunningham, and J.C. Roof. 1998. Seroprevalence of selected disease agents from free-ranging black bears in Florida. *Journal of Wildlife Diseases* 34:612–619.

Dungworth, D.L. 1993. The respiratory system. In *Pathology of domestic animals,* vol. 2, ed. K.V.F. Jubb, P.C. Kennedy, and N. Palmer, 4th ed. New York: Academic, pp. 617–624.

Durchfeld, B., W. Baumgärtner, W. Herbst, and R. Brahm. 1990. Vaccine-associated canine distemper infection in a litter of African hunting dogs (*Lycaon pictus*). *Zentralblatt fur Veterinarmedizin B* 37:203–212.

Evans, R.H. 1984. Studies of a virus in a biological system: Naturally occurring and experimental canine distemper in the raccoon (*Procyon lotor*). M.S. thesis, Southern Illinois University, Carbondale, Illinois, 135 pp.

Evans, M.B., T.O. Bunn, H.T. Hill, and K.B. Platt. 1991. Comparison of in vitro replication and cytopathology caused by strains of canine distemper virus of vaccine

and field origin. *Journal of Veterinary Diagnostic Investigation* 3:127–132.
Gese, E.M., R.D. Schultz, O.J. Rongstad, and D.E. Andersen. 1991. Prevalence of antibodies against canine parvovirus and canine distemper virus in wild coyotes in southeastern Colorado. *Journal of Wildlife Diseases* 27:320–323.
Gese, E.M., R.D. Schultz, M.R. Johnson, E.S. Williams, R.L. Crabtree, and R.L. Ruff. 1997. Serological survey for diseases in free-ranging coyotes (*Canis latrans*) in Yellowstone National Park, Wyoming. *Journal of Wildlife Diseases* 33:47–56.
Gier, H.T., and D.J. Ameel. 1959. Parasites and diseases of Kansas coyotes. *Kansas State University, Agricultural Experiment Station, Technical Bulletin* 91, 34 pp.
Goodrich, J.M., E.S. Williams, and S.W. Buskirk. 1994. Effects of a modified-live virus canine distemper vaccine on captive badgers (*Taxidea taxus*). *Journal of Wildlife Diseases* 30:492–496.
Greene, C.E., and M.J. Appel. 1998. Canine distemper. In *Infectious diseases of the dog and cat,* ed. C.E. Greene, 2d ed. Philadelphia: W.B. Saunders, pp. 9–22.
Guo, W., J.F. Evermann, W.J. Foreyt, F.F. Knowlton, and L.A. Windberg. 1986. Canine distemper virus in coyotes: A serologic survey. *Journal of the American Veterinary Medical Association* 189:1099–1100.
Haas, L., H. Hofer, M. East, P. Wohlsein, B. Liess, and T. Barrett. 1996. Canine distemper virus infection in Serengeti spotted hyaenas. *Veterinary Microbiology* 49:147–152.
Haas, L., W. Martens, I. Greiser-Wilke, L. Mamaev, T. Butina, D. Maack, and T. Barrett. 1997. Analysis of the haemagglutinin gene of current wild-type canine distemper virus isolates from Germany. *Virus Research* 48:165–171.
Hable, C.P., A.N. Hamir, D.E. Snyder, R. Joyner, J. French, V. Nettles, C. Hanlon, and C.E. Rupprecht. 1992. Prerequisites for oral immunization of free-ranging raccoons (*Procyon lotor*) with recombinant rabies virus vaccine: Study site ecology and bait system development. *Journal of Wildlife Diseases* 28:64–79.
Halbrooks, R.D., L.J. Swango, P.R. Schnurrenberger, F.E. Mitchell, and E.P. Hill. 1981. Response of gray foxes to modified live-virus canine distemper vaccines. *Journal of the American Veterinary Medical Association* 179:1170–1174.
Hamir, A.N., and C.E. Rupprecht. 1990. Absence of rabies encephalitis in a raccoon with concurrent rabies and canine distemper infections. *Cornell Veterinarian* 80:197–201.
Hamir, A.N., N. Raju, C. Hable, and C.E. Rupprecht. 1992. Retrospective study of testicular degeneration in raccoons with canine distemper infection. *Journal of Veterinary Diagnostic Investigation* 4:159–163.
Harrenstien, L.A., L. Munson, E.C. Ramsay, C.F. Lucash, S.A. Kania, and L.N. Potgieter. 1997. Antibody responses of red wolves to canine distemper virus and canine parvovirus vaccination. *Journal of Wildlife Diseases* 33:600–605.
Helmbolt, C.F., and E.L. Jungherr. 1955. Distemper complex in wild carnivores simulating rabies. *American Journal of Veterinary Research* 16:463–469.
Hoff, G.L., W.J. Bigler, S.J. Proctor, and L.P. Stallings. 1974. Epizootic of canine distemper virus infection among urban raccoons and gray foxes. *Journal of Wildlife Diseases* 10:423–428.
Holzman, S., M.J. Conroy, and W.R. Davidson. 1992. Diseases, parasites and survival of coyotes in south-central Georgia. *Journal of Wildlife Diseases* 28:572–580.
Hoover, J.P., A.E. Castro, and M.A. Nieves. 1985. Serologic evaluation of vaccinated American river otters. *Journal of the American Veterinary Medical Association* 187:1162–1165.
Iwatsuki, K., N. Miyashita, E. Yoshida, T. Gemma, Y.S. Shin, T. Mori, N. Hirayama, C. Kai, and T. Mikami. 1997. Molecular and phylogenetic analyses of the haemagglutinin (H) proteins of field isolates of canine distemper virus from naturally infected dogs. *Journal of General Virology* 78:373–380.
Johnson, M.R., D.K. Boyd, and D.H. Pletscher. 1994. Serologic investigations of canine parvovirus and canine distemper in relation to wolf (*Canis lupus*) pup mortalities. *Journal of Wildlife Diseases* 30:270–273.
Karstad, L., and J. Budd. 1964. Distemper in raccoons characterized by giant cell pneumonitis. *Canadian Veterinary Journal* 5:326–330.
Kauffman, C.A., A.G. Bergman, and R.P. O'Connor. 1982. Distemper virus infection in ferrets: An animal model of measles induced immunosuppression. *Clinical and Experimental Immunology* 47:617–625.
Kazacos, K.R., H.L. Thacker, H.L. Shivaprasad, and P.P. Burger. 1981. Vaccination-induced distemper in kinkajous. *Journal of the American Veterinary Medical Association* 179:1166–1169.
Kilham, L., and C.M. Herman. 1954. Isolation of an agent causing bilirubinemia and jaundice in raccoons. *Proceedings of the Society for Experimental Biology and Medicine* 85:272–275.
Kolbl, S., H. Schnabel, and M. Mikula. 1990. Distemper as the cause of death in badgers in Austria. *Tierarztliche Praxis* 18:81–84.
Kotani, T., M. Jyo, Y. Odagiri, Y. Sakakibara, and T. Horiuchi. 1989. Canine distemper virus infection in lesser pandas (*Ailurus fulgens*). *Nippon Juigaku Zasshi* 51:1263–1266.
Krakowka, S., and A. Koestner. 1976. Age related susceptibility to canine distemper virus infection in gnotobiotic dogs. *Journal of Infectious Diseases* 134:629–632.
Krakowka, S., E.A. Hoover, A. Koestner, and K. Ketring. 1977. Experimental and natural occurring transplacental transmission of canine distemper virus. *American Journal of Veterinary Research* 38:919–922
Lamb, R.A., and D. Kolakofsky. 1996. *Paramyxoviridae:* The viruses and their replication. In *Fundamental virology,* ed. B.N. Fields, D.M. Knipe, and P.M. Howley. Philadelphia: Lippincott-Raven, pp. 577–604.
Leighton, T., M. Ferguson, A. Gunn, E. Henderson, and G. Stenhouse. 1988. Canine distemper in sled dogs. *Canadian Veterinary Journal* 29:299.
List, K.A. 1994. Investigation of immune function following canine distemper vaccination and challenge in black-footed ferret × Siberian polecat hybrids. M.S. thesis, University of Wyoming, Laramie, Wyoming, 136 pp.
Little, S.E., W.R. Davidson, E.W. Howerth, P.M. Rakich, and V.F. Nettles. 1998. Diseases diagnosed in red foxes from the southeastern United States. *Journal of Wildlife Diseases* 34:620–624.
López-Peña, M., M.I. Quiroga, S. Vázquez, and J.M. Nieto. 1994. Detection of canine distemper viral antigen in foxes (*Vulpes vulpes*) in northwestern Spain. *Journal of Wildlife Diseases* 30:95–98.
Liu, C., and D.L. Coffin. 1957. Studies on canine distemper infection by means of fluorescent-labeled antibody: I. The pathogenesis, pathology, and diagnosis of disease in experimentally infected ferrets. *Virology* 3:115–131.
Machida, N., N. Izumisawa, T. Nakamura, and K. Kiryu. 1992. Canine distemper virus infection in a masked palm civet (*Paguma larvata*). *Journal of Comparative Pathology* 107:439–443.
Machida, N., K. Kiryu, K. Oh-ishi, E. Kanda, N. Izumisawa, and T. Nakamura. 1993. Pathology and epidemiology of canine distemper in raccoon dogs (*Nyctereutes procyonides*). *Journal of Comparative Pathology* 108:383–392.

Mainka, S.A., X. Qui, T. He, and M.J. Appel. 1994. Serologic survey of giant pandas (*Ailuropoda melanoleuca*), and domestic dogs and cats in the Wolong Reserve, China. *Journal of Wildlife Diseases* 30:86–89.

Marsilio, F., P.G. Tiscar, L. Gentile, H.U. Roth, G. Boscagli, M. Tempesta, and A. Gatti. 1997. Serologic survey for selected viral pathogens in brown bears from Italy. *Journal of Wildlife Diseases* 33:304–307.

McCormick, A.E. 1983. Canine distemper in African cape hunting dogs (*Lycaon pictus*): Possibly vaccine induced. *Journal of Zoo Animal Medicine* 14:66–71.

McCue, P.M., and T.P. O'Farrell. 1988. Serological survey for selected diseases of the endangered San Joaquin kit fox (*Vulpes macrotis mutica*). *Journal of Wildlife Diseases* 24:274–281.

McInnes, E.F., R.E. Burroughs, and N.M. Duncan. 1992. Possible vaccine-induced canine distemper in a South American bush dog (*Speothos venaticus*). *Journal of Wildlife Diseases* 28:614–617.

Mee, A.P., J.A. Dixon, J.A. Hoyland, M. Davies, P.L. Selby, and E.B. Mawer. 1998. Detection of canine distemper virus in 100% of Paget's disease samples by in situ-reverse transcriptase-polymerase chain reaction. *Bone* 23:171–175.

Moller, T., and S.W. Nielsen. 1964. Toxoplasmosis in distemper susceptible Carnivora. *Pathologia Veterinaria* 1:189.

Montali, R.J., C.R. Bartz, J.A. Teare, J.T. Allen, M.J.G. Appel, and M. Bush. 1983. Clinical trials with canine distemper vaccines in exotic carnivores. *Journal of the American Veterinary Medical Association* 183: 1163–1167.

Montali, R.J., C.R. Bartz, and M. Bush. 1987. Canine distemper virus. In *Virus infections of carnivores,* ed. M. Appel. Amsterdam: Elsevier Science, pp. 437–443.

Norrby, E., G. Utter, C. Örvell, and M.J.G. Appel. 1986. Protection against canine distemper virus in dogs after immunization with isolated fusion protein. *Journal of Virology* 58:536–541.

Nunamaker, C.E., and E.S. Williams. 1986. TEM as an aid in the diagnosis of canine distemper in wildlife. In *Proceedings of the Electron Microscopy Society of America, Albuquerque, New Mexico,* ed. G.W. Bailey. San Francisco: San Francisco Press, pp. 302–303.

Palmer, D., P. Ossent, A. Valdvogel, and R. Weilenmann. 1983. Distemper encephalitis in stone martens (*Martes foina* Erxleben 1977) in Switzerland. *Schweizer Archiv fur Tierheilkunde* 125:529–536.

Palmer, D., C.R.R. Huxtable, and J.B. Thomas. 1990. Immunohistochemical demonstration of canine distemper virus antigen as an aid to the diagnosis of canine distemper encephalomyelitis. *Research in Veterinary Science* 49:177–181.

Pardo, M.C., J.E. Bauman, and M. Mackowiak. 1997. Protection of dogs against canine distemper by vaccination with a canarypox virus recombinant expressing canine distemper virus fusion and hemagglutinin glycoproteins. *American Journal of Veterinary Research* 58:833–836.

Paré, J.A. 1997. Vaccination of raccoons (*Procyon lotor*) against canine distemper: An experimental study. D.V.Sc. thesis, University of Guelph, Guelph, Ontario, 121 pp.

Pearson, R.C., and J.R. Gorham. 1987. Canine distemper virus. In *Virus infections of carnivores,* ed. M.J.G. Appel. Amsterdam: Elsevier Science, pp. 371–378.

Potgieter, L.N., and C.S. Patton. 1984. Multifocal cerebellar cortical necrosis caused by canine distemper virus infection in a raccoon. *Journal of the American Veterinary Medical Association* 185:1397–1399.

Pringle, C.R. 1992. *Paramyxoviridae.* In *Classification and nomenclature of viruses,* ed. R.I.B. Francki, C.M. Fauguet, D.L. Knudson, and F. Brown. New York: Springer-Verlag, 459 pp.

Qiu, X., and S.A. Mainka. 1993. Review of mortality of the giant panda (*Ailuropoda melanoleuca*). *Journal of Zoo and Wildlife Medicine* 24:425–429.

Reed, W.M., and J.J. Turek. 1985. Concurrent distemper and disseminated toxoplasmosis in a red fox. *Journal of the American Veterinary Medical Association* 187:1264–1265.

Rohowsky-Kochan, C., P.C. Dowling, and S.D. Cook. 1995. Canine distemper virus-specific antibodies in multiple sclerosis. *Neurology* 45:1554–1560.

Roscoe, D.E. 1993. Epizootiology of canine distemper in New Jersey raccoons. *Journal of Wildlife Diseases* 29:390–395.

Schubert, C.A., I.K. Barker, R.C. Rosatte, C.D. MacIness, and T.D. Nudds. 1998. Effect of canine distemper on an urban raccoon population: An experiment. *Ecological Applications* 8:379–387.

Spencer, J., and R. Burroughs. 1992. Antibody response to canine distemper vaccine in African wild dogs. *Journal of Wildlife Diseases* 28:443–444.

Steinhagen, P., and W. Nebel. 1985. Distemper in stone martens (*Martes foina,* Erxl) in Schleswig-Holstein: A contribution to the epidemiology of distemper. *Deutsche Tierarztliche Wochenschrift* 92:178–181.

Stephensen, C.B., J. Welter, S.R. Thaker, J. Taylor, J. Tartaglia, and E. Paoletti. 1997. Canine distemper virus (CDV) infection of ferrets as a model for testing *Morbillivirus* vaccine strategies: NYVAC- and ALVAC-based CDV recombinants protect against symptomatic infection. *Journal of Virology* 71:1506–1513.

Stephenson, R.O., D.G. Ritter, and C.A. Nielsen. 1982. Serologic survey for canine distemper and infectious canine hepatitis in wolves in Alaska. *Journal of Wildlife Diseases* 18:419–424.

Stettler, M., K. Beck, A. Wagner, M. Vandevelde, and A. Zurbriggen. 1997. Determinants of persistence in canine distemper viruses. *Veterinary Microbiology* 57:83–93.

Stoffregen, D.A., and J.P. Dubey. 1991. A *Sarcocystis* sp.-like protozoan and concurrent canine distemper virus infection associated with encephalitis in a raccoon (*Procyon lotor*). *Journal of Wildlife Diseases* 27:688–692.

Sukura, A., J. Laakkonen, and E. Rudback. 1997. Occurrence of *Pneumocystis carinii* in canine distemper. *Acta Veterinaria Scandinavica* 38:201–205.

Sutherland-Smith, M.R., B.A. Rideout, A.B. Mikolon, M.J. Appel, P.J. Morris, A.L. Shima, and D.J. Janssen. 1997. Vaccine-induced canine distemper in European mink, *Mustela lutreola. Journal of Zoo and Wildlife Medicine* 28:312–318.

Taylor, J., R. Weinberg, J. Tartaglia, C. Richardson, G. Alkhatib, D. Briedis, M. Appel, E. Norton, and E. Paoletti. 1992. Nonreplicating viral vectors as potential vaccines: Recombinant canarypox virus expressing measles fusion (F) and hemagglutinin (HA) glycoproteins. *Virology* 187:321–328.

Thawley, D.G., and J.C. Wright. 1982. Pseudorabies virus infection in raccoons: A review. *Journal of Wildlife Diseases* 18:113–116.

Thomas-Baker, B. 1985. Vaccination-induced distemper in maned wolves, vaccination-induced corneal opacity in a maned wolf. In *Proceedings of the American Association of Zoo Veterinarians, Scottsdale, Arizona,* ed. M.S. Silberman and S.D. Silberman. Scottsdale, AZ: American Association of Zoo Veterinarians, p. 53.

Thorne, E.T., and E.S. Williams. 1988. Disease and endangered species: The black-footed ferret as a recent example. *Conservation Biology* 2:66–74.

Truyen, U., T. Muller, R. Heidrich, K. Tackmann, and L.E. Carmichael. 1998. Survey on viral pathogens in wild red foxes (*Vulpes vulpes*) in Germany with emphasis on par-

voviruses and analysis of a DNA sequence from a red fox parvovirus. *Epidemiology and Infection* 121:433–440.
van Heerden, J., W.H. Swart, and D.G. Meltzer. 1980. Serum antibody levels before and after administration of live canine distemper vaccine to the wild dog *Lycaon pictus*. *Journal of South African Veterinary Association* 51:283–284.
van Heerden, J., N. Bainbridge, R.E.J. Burroughs, and N.P.J. Kriek. 1989. Distemper-like disease and encephalitozoonosis in wild dogs (*Lycaon pictus*). *Journal of Wildlife Diseases* 25:70–75.
van Moll, P., A. Alldinger, W. Baumgärtner, and M. Aami. 1995. Distemper in wild carnivores: An epidemiological, histological and immunohistochemical study. *Veterinary Microbiology* 44:193–199.
Vizoso, A.D., and W.E. Thomas. 1981. Paramyxoviruses of the morbilli group in the wild hedgehog *Erinaceus europaeus*. *British Journal of Experimental Pathology* 62:79–86.
von Schönbauer, M., S. Kölbl, and A. Shönbauer-Längle. 1984. Perinatale staupeinfektion bei drei eisbären (*Urusus maritimus*) und bei einem brillenbären (*Tremarctos ornatus*). *Verhanlungsbericht des Internationalen Symposiums über die Erkrankungen der Zootiere* 26:131–136. Abstract.
Williams, E.S., and E.T. Thorne. 1996. Infectious and parasitic diseases of captive carnivores, with special emphasis on the black-footed ferret (*Mustela nigripes*). *Revue Scientifique et Technique O.I.E.* 15:91–114.
Williams, E.S., E.T. Thorne, M.J.G. Appel, and D.W. Belitsky. 1988. Canine distemper in black-footed ferrets (*Mustela nigripes*) from Wyoming. *Journal of Wildlife Diseases* 24:385–398.
Williams, E.S., J. Cavender, C. Lynn, K. Mills, C. Nunamaker, and A. Boerger-Fields. 1992. Survey of coyotes and badgers for diseases in Shirley Basin, Wyoming in 1991. In *Black-footed ferret reintroduction Shirley Basin, Wyoming*, ed. B. Oakleaf, B. Luce, E.T. Thorne, and S. Torbit. Cheyenne: Wyoming Game and Fish Department, pp. 75–106.
Williams, E.S., S.L. Anderson, J. Cavender, C. Lynn, K. List, C. Hearn, and M.J.G. Appel. 1995. Vaccination of black-footed ferrets (*Mustela nigripes*) × Siberian polecat (*M. eversmanni*) hybrids and domestic ferrets (*M. putorius furo*) against canine distemper. *Journal of Wildlife Diseases* 32:417–423.
Williams, E.S., J. Edwards, H. Edwards, V. Welch, S. Dubay, B. Luce, and S. Anderson. 1997. Survey of coyotes for diseases in Shirley Basin, Wyoming in 1996. In *Black-footed ferret reintroduction Shirley Basin, Wyoming*, ed. B. Luce, B. Oakleaf, and E.S. Williams. Cheyenne: Wyoming Game and Fish Department, pp. 34–43
Williams, E.S., J. Edwards, W. Edwards, A. McGuire, S. Dubay, W. Cook, S. Anderson, and P. Jaeger. 1998. Survey of carnivores for diseases in the Conata Basin/Badlands black-footed ferret reintroduction site, 1996–1997. In *Black-footed ferret reintroduction Conata Basin/Badlands, South Dakota*, ed. G. Plumb. Wall, SD: U.S. Forest Service.
Winters, K.A., L.E. Mathes, S. Krakowka, and R.G. Olsen. 1983. Immunoglobulin class response to canine distemper virus in gnotobiotic dogs. *Veterinary Immunology and Immunopathology* 5:209–215.
Wojcinski, Z.W., and I.K. Barker. 1986. Tyzzer's disease as a complication of canine distemper in a raccoon. *Journal of Wildlife Disease* 22:55–59.
Woolf, A., C. Gremillion-Smith, and R.H. Evans. 1986. Evidence of canine distemper virus infection in skunks negative for antibody against rabies virus. *Journal of the American Veterinary Medical Association* 189:1086–1088.
Yoshida, E., K. Iwatsuki, N. Miyashita, T. Gemma, C. Kai, and T. Mikmi. 1998. Molecular analysis of the nucleocapsid protein of recent isolates of canine distemper virus in Japan. *Veterinary Microbiology* 59:237–244.
Zarnke, R.L., and W.B. Ballard. 1987. Serologic survey for selected microbial pathogens of wolves in Alaska, 1975–1982. *Journal of Wildlife Diseases* 23:77–85.
Zurbriggen, A., C. Muller, and M. Vandevelde. 1993. In situ hybridization of virulent canine distemper virus in brain tissue using digoxigenin-labeled probes. *American Journal of Veterinary Research* 54:1457–1461.

FELINE MORBILLIVIRUS INFECTION

LINDA MUNSON

INTRODUCTION. The recent occurrence of several canine distemper (CD) epidemics in wild and captive felids signaled the emergence of a new canine distemper virus (CDV) biotype that is pathogenic in cats (Appel et al. 1994; Roelke-Parker et al. 1996). Before the 1990s, a few individual zoo cats were suspected of having CD (Piat 1950; Blythe et al. 1983; Gould and Fenner 1983), but felids in general were thought to be fairly resistant. Since 1991, however, CDV infections have been reported in five species of free-ranging and captive felids from at least eight discontiguous sites. If this trend continues, CDV may eventually affect all species of felids worldwide.

HISTORY. Canine distemper virus was first suspected to infect felids in 1950 when two young captive lions *Panthera leo* developed typical neurologic signs and lesions (Piat 1950). Three decades later, two tigers *Panthera tigris* and two snow leopards *Panthera uncia* from United States zoos had similar CD-like lesions and signs (Blythe et al. 1983; Gould and Fenner 1983; Fix et al. 1989). Then in 1991 and 1992, three discrete epidemics of CD occurred in captive large felids from United States wildlife parks, including lions, tigers, leopards *Panthera pardus,* and jaguars *Panthera onca* (Appel et al. 1994). During these epizootics, clinical disease was noted in 46 felids and resulted in 21 fatalities. In late 1993 and 1994, a major epidemic of CD occurred among lions and other carnivores in the Serengeti ecosystem in Kenya and Tanzania, causing the death or disappearance of approximately 1000 lions or about one-third of the population (Roelke-Parker et al. 1996). In late 1993, two wild bobcats *Felis rufus* were confirmed to have CDV encephalitis (P.Y. Daoust and S. McBurney personal communication), and CD was reported in a captive lion in Canada (Wood et al. 1995). Since 1994, several other felids in wildlife parks, zoos, and circuses in the United States have been reported with neurologic signs or had lesions compatible with CD (L. Munson unpublished). Retrospective analysis of stored paraffin blocks indicated that CD had caused the death of captive lions and tigers in Europe since at least 1972 (Myers et al. 1997).

DISTRIBUTION AND HOST RANGE. Canine distemper infections have been documented in wild felids in Kenya, Tanzania, and Canada (Roelke-Parker et al. 1996; Kock et al. 1998; P.Y. Daoust and S. McBurney personal communication). In captive felids, CDV disease has been confirmed in the United States and Canada and suspected in Mexico and Europe (Appel et al. 1994; Wood et al. 1995). The global distribution of this virulent felid biotype of CDV currently is unknown, although pathogenic strains of CDV affecting other carnivores occur worldwide.

Susceptibility first appeared to be confined to large felids (Appel et al. 1994). However, recent confirmation of CDV in bobcats (P.Y. Daoust and S. McBurney personal communication) and infection of domestic cats (Harder et al. 1996; L. Munson unpublished) indicates that small felids also are susceptible. The types of felids affected approximate the species composition in zoos.

Most morbilliviral infections have been in captive felids. Captive felids may be more susceptible because of higher animal densities or increased exposure to infected animals, such as raccoons *Procyon lotor,* skunks *Mephitis mephitis,* and domestic dogs *Canis familiaris.* Conversely, infections may simply be detected more often in captive than in wild felids because of closer health monitoring of captive populations. In the Serengeti, lions may have been more susceptible than other felid species because populations were dense at the time of the epidemic and opportunities for infection were greater in prides than in solitary species. During the Serengeti epidemic, CD mortalities also occurred in a leopard, hyenas *Crocuta crocuta,* bat-eared fox *Otocyon megalotis,* and domestic dogs, demonstrating the broad species range of this variant. The virulence of this biotype for striped hyenas *Hyaena hyaena* also was not typical of other CDV strains. In the California epizootics, dogs, fox *Vulpes vulpes,* skunks, and raccoons were affected by the felid virus.

ETIOLOGY. The felid morbillivirus is closely related to CDV strains isolated from canids and other carnivores in the United States and Europe and is currently considered a variant of CDV. Although the H and P genes from Serengeti lion and California leopard viruses had only approximately 91% (H) and 95% (P) nucleotide-sequence homology with other strains of CDV, classifying the felid virus as a unique morbillivirus was not justified (Harder et al. 1996; Roelke-Parker et al. 1996; Carpenter et al. 1998; S. O'Brien unpublished). However, the genetic differences would result in amino acid sequences that were approximately 10% (H) and 15% (P) different than canid isolates (S. O'Brien unpublished). The resulting changes in H gene product (the hemagglutinin envelope protein) may explain the increased pathogenicity of this virus for felids and hyenas, because the H protein mediates host cell attachment and infectivity (Stern et al. 1995).

TRANSMISSION AND EPIDEMIOLOGY. The origin of the CDV that has infected felids was presumed to be other terrestrial carnivores, although the exact mode of transmission to felids is currently unknown. Infectious virus in ocular, oral, or nasal secretions could pass to felids through direct contact or indirectly through aerosol or fomites. Transmission among wild species could occur at congregating sites, such as recently killed prey. Although transplacental transmission of this CDV biotype has been confirmed in hyenas (L. Munson unpublished), transmission by this route in felids has not yet been documented. Significant environmental contamination is unlikely because of the lability of CDV to heat, ultraviolet light, or disinfectants.

During the epidemics, CDV caused high morbidity and mortality in seronegative felids (Appel et al. 1994; Roelke-Parker et al. 1996). Recovered felids developed high serum-neutralizing titers against CDV, which should protect these animals for life (Appel 1987). Nonetheless, if this biotype of CDV persists, CD epidemics will likely recur as populations of seronegative cubs and juveniles emerge. Epidemics should be anticipated where high densities of seronegative felids share habitats with susceptible carnivores.

CLINICAL SIGNS. Affected felids usually had either acute neurologic signs or progressive intestinal and respiratory signs that sometimes progressed to neurologic disease (Appel et al. 1994; Roelke-Parker et al. 1996). The duration of clinical signs ranged from 1 day to several weeks in most felids, although two captive tigers developed progressive neurologic signs over 2 and 14 months, respectively (Blythe et al. 1983; Gould and Fenner 1983).

The neurologic signs were the most conspicuous and included seizures, tremors, disorientation, weakness, ataxia, paraparesis, hyperreflexia, head pressing, and coma. Behavioral changes, such as increased anxiousness and aggression, loss of normal human avoidance, and depression, also occurred. Myoclonus developed in recovered felids but also preceded a rapid decline in neurologic status in another cat (Blythe et al. 1983). Some CDV-infected felids had anorexia, lethargy, dyspnea, mucopurulent ocular and nasal discharge, or diarrhea that was sometimes hemorrhagic. Free-ranging lions also had multiple infected wounds, anemia, lymphadenopathy, emaciation, and rough hair coats (M.E. Roelke-Parker personal communication).

The prognosis for recovery in wild felids is poor. Mortalities occurred from primary CDV infection of the brain or the respiratory or gastrointestinal tracts, or from secondary bacterial or protozoal infections due to CDV-induced immunosuppression. Many CDV-infected free-ranging lions were killed by other predators when moribund or demonstrating unusual behavior (M.E. Roelke-Parker personal communication). The prognosis for recovery is greater in captive felids where medical intervention is feasible.

PATHOGENESIS. The pathogenesis of CD in felids has not been investigated, but is presumed similar to CD in other mammals. In felids, lymphocytes and macrophages were infected and would disseminate virus to other organs. Canine distemper virus was identified by immunohistochemistry in lungs (pneumocytes and rarely bronchial epithelium), brains (neurons and astrocytes), epididymides, and rarely bile ducts (Munson et al. 1995; Roelke-Parker et al. 1996). Neither viral inclusions nor occult antigens were identified in stomachs or urinary bladders, suggesting a more restricted tissue distribution in felids than in canids.

PATHOLOGY. Lymphopenia and a neutrophilic leukocytosis were common hematologic findings. Many CDV-infected felids also had anemia (Roelke-Parker et al. 1996; L. Munson unpublished).

Felids had minimal gross lesions, as is typical in other CDV-infected mammals. Affected free-ranging lions had lymphadenopathy, pulmonary consolidation, ocular and nasal discharge, extensive traumatic wounds, and emaciation (M.E. Roelke-Parker personal communication), whereas most captive felids were in excellent body condition (Appel et al. 1994).

Histologic lesions of CD in felids were subtle in comparison to lesions in CDV-infected canids and did not correlate with the magnitude of clinical signs (Appel et al. 1994; Munson et al. 1995; Roelke-Parker et al. 1996). Felids had only rare inclusion bodies and syncytia, minimal inflammatory reactions, and limited organ involvement. Lymph nodes, spleen, and thymus consistently had either lymphocyte necrosis, depletion, or hyperplasia, and rarely had syncytia and discrete intranuclear or intracytoplasmic inclusion bodies. All felids with neurologic signs had some degree of encephalitis, but lesions were often focal and mild. The hippocampus and parahippocampus were most consistently affected and had neuronal necrosis, microgliosis, astrocytosis, syncytia, and intranuclear or intracytoplasmic inclusions in astrocytes and neurons. Intracytoplasmic inclusions in hippocampal neurons were remarkably similar to the Negri bodies of rabies. Inclusion bodies also were noted in the cerebellum of some cats (Appel et al. 1994). Some felids had nonsuppurative meningoencephalitis with small aggregates of lymphocytes and macrophages around blood vessels and within the neuropil. Some chronic cases had nonsuppurative encephalomyelitis with demyelination. Concurrent toxoplasmosis occurred in three felids (Appel et al. 1994; P.Y. Dauost and S. McBurney personal communication).

Canine distemper virus-infected felids had mild multifocal to diffuse interstitial pneumonia with prominent type-II pneumocyte hyperplasia, alveolar histiocytosis, syncytia and, rarely, intranuclear and intracytoplasmic inclusions. Although bronchial infection with intracytoplasmic inclusions was a prominent feature in captive felids (Appel et al. 1994), the bronchial epithelium was not involved in free-ranging felids (Munson et al. 1995), possibly because they died earlier in the infection. Viral inclusions rarely were identified in the bile duct and epididymal epithelium. Using immunohistochemical methods (Appel et al. 1994; Roelke-Parker et al. 1996), CDV antigens were detected in nuclear and cytoplasmic inclusions and in other sites without lesions or inclusions.

DIAGNOSIS. Canine distemper should be suspected in felids with seizures, myoclonus, or other neurologic signs and/or with respiratory and gastrointestinal signs. In live felids, CDV can be identified by immunofluorescence of conjunctival smears or viral isolation from buffy-coat leukocytes if fresh or fresh-frozen (–70° C) cells are available (Appel et al. 1992). Viral sequences can also be derived from leukocytes or fresh or frozen tissues by reverse transcriptase polymerase chain reaction using oligonucleotide primers based on conserved regions of the CDV phosphoprotein (P) gene (Roelke-Parker et al. 1996).

Many felids with acute infections had no measurable serum-neutralizing (SN) antibodies (Appel et al. 1994). High or rising SN titers occurred after recent infections (Roelke-Parker et al. 1996), but high titers alone did not distinguish recent from previous exposures. Therefore, serology is not the most reliable method of definitive diagnosis during an epidemic.

Definitive diagnosis can be accomplished by histopathology. Identifying the typical lesions of nonsuppurative encephalitis, lymphoid necrosis, or interstitial pneumonia associated with syncytia and viral inclusions provides convincing evidence of CD. Canine distemper virus infection can be confirmed with immunohistochemistry (Myers et al. 1997).

DIFFERENTIAL DIAGNOSES. The neurologic signs and central nervous system lesions in CDV-infected felids were very similar to rabies in distribution and character. In some cases, CDV inclusions in hippocampal neurons were only differentiated from Negri bodies by immunohistochemistry or identification of concurrent intranuclear inclusions. A nonsuppurative encephalitis of unknown cause in zoo and domestic felids (Truyen et al. 1990; Lundgren 1992) also has similar clinical and histologic characteristics as CD, but the lesions from this disease lack immunoreactive CDV antigens or involvement of other organ systems.

CONTROL AND TREATMENT. In zoo felids, CD can be prevented and controlled by preventing contact with indigenous wildlife or domestic pets. On CDV-infected premises, seronegative felids should be isolated from infected animals, and contaminated enclosures and equipment should be cleaned with an appropriate disinfectant. Supportive treatment of CDV-infected felids rarely was beneficial, particularly in neurologic cases. However, recovery has been documented in some zoo felids (Blythe et al. 1983; Appel et al. 1994).

No currently available commercial vaccines are approved for felids, and modified-live virus canine-origin vaccines carry the risk of causing disease (Montali et al. 1991). Therefore, vaccination of felids is not recommended until a safe, efficacious vaccine is available. Felids that recovered from infection developed high CDV-neutralizing antibody titers (Appel et al. 1994; Roelke-Parker et al. 1996) and should be protected for life. In free-ranging lions, the CD epidemic was self-limiting when numbers of susceptible animals diminished.

PUBLIC HEALTH CONCERNS. There are no known cases of transmission of CDV from felids to humans.

DOMESTIC ANIMAL HEALTH CONCERNS. Domestic cats experimentally infected with leopard-origin (Harder et al. 1996) or Serengeti-origin CDV (L. Munson unpublished) developed transient viremia and lymphopenia but not clinically evident neurologic or respiratory disease. Specific pathogen-free domestic cats infected with the Serengeti lion strain also developed anemia (L. Munson unpublished). No spontaneous cases of CD in domestic cats have been reported, but infections could go undetected because clinical signs lack specificity. The Serengeti lion strain of CDV is a definite threat to domestic dogs, as affirmed by mortalities during the Serengeti epidemic (Roelke-Parker et al. 1996).

MANAGEMENT IMPLICATIONS. Managing captive and wild populations to reduce crowding should minimize the risk of a CD epidemic. Any translocation of felids between habitats or movement of other carnivores into felid habitat should include an assessment of the CDV status of indigenous animals and a 30-day quarantine that includes paired SN CDV antibody titers to detect active infection. Keeping the local domestic dog populations vaccinated against CDV and minimizing contact between humans, their domestic pets, and felid populations also would diminish opportunities for epidemics.

LITERATURE CITED

Appel, M.J.G. 1987. Canine distemper virus. In *Virus infections in carnivores,* ed. M.J.G. Appel. Amsterdam: Elsevier Science, pp. 133–160.

Appel, M.J.G., S. PearceKelling, and B.A. Summers. 1992. Dog lymphocyte cultures facilitate the isolation and growth of virulent canine distemper virus. *Journal of Veterinary Diagnostic Investigation* 4:258–263.

Appel, M.J.G., R.A. Yates, G.L. Foley, J.J. Bernstein, S. Santinelli, L.H. Spelman, L.D. Miller, L.H. Arp, M. Anderson, M. Barr, S. Pearce-Kelling, and B.A. Summers. 1994. Canine distemper epizootic in lions, tigers, and leopards in North America. *Journal of Veterinary Diagnostic Investigation* 6:277–288.

Blythe, L.L., J.A. Schmitz, M. Roelke, and S. Skinner. 1983. Chronic encephalomyelitis caused by canine distemper virus in a Bengal tiger. *Journal of the American Veterinary Medical Association* 183:1159–1162.

Carpenter, M.A., M.J. Appel, M.E. Roelke-Parker, L. Munson, H. Hofer, M. East, and S.J. O'Brien. 1998. Genetic characterization of canine distemper virus in Serengeti carnivores. *Veterinary Immunology and Immunopathology* 65:259–266.

Fix, A.S., D.P. Riordan, H.T. Hill, M.A. Gill, and M.B. Evans. 1989. Feline panleukopenia virus and subsequent canine distemper virus infection in two snow leopards (*Panthera uncia*). *Journal of Zoo and Wildlife Medicine* 20:273–281.

Gould, D.H., and W.R. Fenner. 1983. Paramyxovirus-like nucleocapsids associated with encephalitis in a captive Siberian tiger. *Journal of the American Veterinary Medical Association* 183:1319–1322.

Harder, T.C., M. Kenter, H. Vos, K. Siebelink, W. Huisman, G. van Amerongen, C. Örvell, T. Barrett, M.J.G. Appel, and A.D.M.E. Osterhaus. 1996. Canine distemper virus from diseased large felids: Biological properties and phylogenetic relationships. *Journal of General Virology* 77:397–405.

Kock, R., W.S. Chalmers, J. Mwanzia, C. Chillingworth, J. Wambua, P.G. Coleman, and W. Baxendale. 1998. Canine distemper antibodies in lions of the Masai Mara. *Veterinary Record* 142:662–665.

Lundgren, A.L. 1992. Feline nonsuppurative meningoencephalomyelitis: A clinical and pathological study. *Journal of Comparative Pathology* 107:411–425.

Montali, R.J., C.R. Bartz, and M. Bush. 1991. Canine distemper virus. In *Virus infections of carnivores,* ed. M. Appel. Amsterdam: Elsevier Science, pp. 437–443.

Munson, L., M.E. Roelke-Parker, A. Pospischil, G.L.M. Mwamengele, B.A. Summers, R. Kock, and M.J.G. Appel. 1995. The pathology of canine distemper virus (CDV) in East African lions. *Veterinary Pathology* 32:591.

Myers, D.L., A. Zurbriggen, H. Lutz, and A. Pospischil. 1997. Distemper: Not a new disease in lions and tigers. *Clinical Diagnostic Laboratory Immunology* 4:180–184.

Piat, B.L. 1950. Susceptibility of young lions to dog distemper. *Bulletin Service d'Elevage Industrial Animales Afrique Occidental Francais* 3:39–40.

Roelke–Parker, M.E., L. Munson, C. Packer, R. Kock, S. Cleaveland, M. Carpenter, S.J. O'Brien, A. Pospischil, R. HofmannLehmann, H. Lutz, G.L.M. Mwamengele, M.N. Mgasa, G.A. Machange, B.A. Summers, and M.J.G. Appel. 1996. A canine distemper virus epidemic in Serengeti lions (*Panthera leo*). *Nature* 379:441–445.

Stern, L.B., M. Greenberg, J.M. Gershoni, and S. Rozenblatt. 1995. The hemagglutinin envelope protein of canine distemper virus (CDV) confers cell tropism as illustrated by CDV and measles virus complementation analysis. *Journal of Virology* 69:1661–1668.

Truyen, U., N. Stockhofe-Zurwieden, O.R. Kaaden, and J. Pohlenz. 1990. A case report: Encephalitis in lions—Pathological and virological findings. *Deutsche Tierarztliche Wochenschrift* 97:89–91.

Wood, S.L., G.W. Thomson, and D.M. Haines. 1995. Canine distemper virus-like infection in a captive African lioness. *Canadian Veterinary Journal* 36:34–35.

MEASLES

LINDA MUNSON

Synonym: Rubeola.

INTRODUCTION. Measles is a highly contagious systemic viral disease of captive or wild primates in contact with infected human populations [reviewed by Lowenstine (1993)]. Most measles virus (MV) infections have occurred in macaques with high morbidity and low mortality, but infections in colobus monkeys *Colobus guereza* and marmosets *Saguinus* spp. and *Callithrix* spp. were highly lethal (Hime et al. 1975; Albrecht et al. 1980; Lowenstine 1993). All species of primate are presumed susceptible.

ETIOLOGY, TRANSMISSION, AND EPIZOOTIOLOGY. Measles virus is a morbillivirus in the family *Paramyxoviridae*. Humans are the natural host. Measles occurs worldwide but is endemic in regions with no vaccination programs, high human population densities, and inadequate nutrition (Norrby and Oxman 1990). Measles virus is monotypic, but has geographic variation in pathogenicity (Schneider-Schaulies et al. 1995). It is highly contagious and spreads by aerosol from infected humans or nonhuman primates to susceptible primates. A continuous chain of acutely infected animals is necessary for transmission. Only immunologically naive primates become infected, and antibodies acquired from infection afford lifelong protection (McChesney et al. 1989). Immaturity, nutritional deficiencies, stress, crowding, and concurrent infections increase morbidity and mortality in an epidemic (Lowenstine 1993). Measles infections have been reported in the following primates: Gorillas *Gorilla gorilla,* chimpanzees *Pan troglodytes,* orangutan *Pongo pygmaeus,* rhesus *Macaca mulatta* and crab-eating macaques (cynomolgus monkeys) *Macaca fasicularis,* gibbons *Hylobates* spp., baboons *Papio* spp., African green monkeys *Chlorocebus sabaeus,* squirrel monkeys *Saimiri sciureus,* colobus monkeys, leaf monkeys *Presbytis* sp., owl monkeys *Aotus trivirgatus,* hairy saki *Pithecia monachus,* and moustached marmosets *Saguinus mystax* (Hunt et al. 1978; Albrecht et al. 1980; Ott-Joslin 1986; Hastings et al. 1991; Lowenstine 1993). All wild populations that were isolated from humans were seronegative.

CLINICAL SIGNS AND PATHOLOGY. Initial signs are facial erythema and edema, anorexia, fever, and malaise (Lowenstine 1993). In some cases Koplick spots occur on the oral mucosa, and a maculopapular rash appears on the face, chest, and abdomen. Lymphadenopathy, diarrhea, and pneumonia usually follow. Metritis and abortion also can occur. A typical lymphopenia occurs, but otherwise clinical pathologic parameters are unremarkable. Rarely, primates develop encephalitis, which is thought to be caused by a measles-induced autoimmune reaction, although persistent MV infections of the brain have been demonstrated. Measles virus also causes profound, prolonged immunosuppression (McChesney et al. 1989).

Gross lesions include facial edema and erythema, erythematous maculopapular rash, lymphadenopathy, and lobular or patchy consolidation of lungs (Lowenstine 1993). Histopathologic lesions are similar to those of other morbilliviral infections such as canine distemper and include lymphoid necrosis and depletion, ballooning degeneration and necrosis of epithelial cells, syncytia, and intracytoplasmic and/or intranuclear eosinophilic inclusion bodies. Common lesions are giant cell pneumonia with bronchial and bronchiolar necrosis, type-II pneumocyte hyperplasia, and syncytia with inclusion bodies. Warthin-Finkeldey lymphoid or phagocytic giant cells also may occur in lymph nodes, spleen, Peyer's patches, appendix, and tonsils. New World monkeys often have hemorrhagic enteritis with epithelial necrosis and secondary bacterial enteritis. Syncytia also occur in the uterus, urinary bladder, pancreatic ducts, and biliary epithelium (Lowenstine 1993).

In live animals, measles can be confirmed by fluorescent antibody tests of throat swabs or biopsy samples. In tissues, MV antigens can be identified in lesions by immunohistochemical methods. These tests will distinguish MV from other viral infections that cause bronchitis or pneumonia with syncytia or inclusions, such as the paramyxoviruses, respiratory syncytial viruses, herpesviruses, and adenoviruses. Simian immunodeficiency virus can cause pneumonia and enteritis with syncytia, but no inclusions are present (Lowenstine 1993).

CONTROL. Natural infection results in lifelong immunity. Attenuated vaccines produce effective antibody titers in most primates 6 months or older (Ott-Joslin 1986). A 90-day quarantine should be enforced for new imports, and infected animals should be kept in isolation. In the face of an epidemic, human gamma globulin can be used. Zoonotic precautions also should be exercised, especially if contact humans are unvaccinated.

LITERATURE CITED

Albrecht, P., D. Lorenz, M.J. Klutch, J.H. Vickers, and F.A. Ennis. 1980. Fatal measles infection in marmosets pathogenesis and prophylaxis. *Infection and Immunity* 27:969–978.

Hastings, B.E., D. Kenny, L.J. Lowenstine, and J.W. Foster. 1991. Mountain gorillas and measles: Ontogeny of a wildlife vaccination program. In *Proceedings of the American Association of Zoo Veterinarians,* ed. R. Junge, pp. 198–207. Calgary, Alberta: American Association of Zoo Veterinarians.

Hime, J.M., I.F. Keymer, and C.J. Baxter. 1975. Measles in recently imported colobus monkeys (*Colobus guereza*). *Veterinary Record* 97:392.

Hunt, R.D., W.W. Carlton, and N.W. King. 1978. Viral diseases. In *Pathology of laboratory animals,* ed. K. Benirschke, F.M. Garner, and T.C. Jones. New York: SpringerVerlag, pp. 1315–1316.

Lowenstine, L.J. 1993. Measles virus infection, nonhuman primates. In *Nonhuman primates: 1,* ed. T.C. Jones, U. Mohr, and R.D. Hunt. Berlin: SpringerVerlag, pp. 108–118.

McChesney, M.B., R.S. Fujinami, N.W. Lerche, P.A. Marx, and M.B.A. Oldstone. 1989. Virus induced immunosuppression: Infection of peripheral blood mononuclear cells and suppression of immunoglobulin synthesis during natural measles virus infection of rhesus monkeys. *Journal of Infectious Diseases* 159:757–760.

Norrby, E., and M.N. Oxman. 1990. Measles virus. In *Virology,* ed. B.N. Fields and D.M. Knipe, 2d ed. New York: Raven, pp. 1013–1044.

Ott–Joslin, J.E. 1986. Viral diseases in nonhuman primates. In *Zoo and wild animal medicine,* ed. M.E. Fowler, 2d ed. Philadelphia: W.B. Saunders, pp. 674–697.

Schneider–Schaulies, J., L.M. Dunster, S. Schneider–Schaulies, and V. ter Meulen. 1995. Pathogenetic aspects of measles virus infections. *Veterinary Microbiology* 44:113–125.

MORBILLIVIRUS INFECTIONS IN AQUATIC MAMMALS

SEAMUS KENNEDY

Synonym: Distemper.

INTRODUCTION. Morbillivirus infections were unknown in aquatic mammals prior to 1987, but since then there have been five reported epizootics of morbilliviral disease in harbor seals *Phoca vitulina* and gray seals *Halichoerus grypus* in Europe (Kennedy et al. 1988b; Osterhaus and Vedder 1988), Baikal seals *Phoca sibirica* in Siberia (Grachev et al. 1989), striped dolphins *Stenella coeruleoalba* in the Mediterranean Sea (Domingo et al. 1992), and bottlenose dolphins *Tursiops truncatus* along the coast of the United States (Lipscomb et al. 1994a,b; Schulman et al. 1997). Individual cases of distemper and serologic evidence of morbillivirus infection have been reported in these and other species of marine mammal in recent years (Tables 2.1 and 2.2).

HISTORY, DISTRIBUTION, AND HOST RANGE. An epizootic of morbillivirus disease killed approximately 18,000 harbor seals and a few hundred gray seals in northwestern Europe in 1988 (Dietz et al. 1989b; Kennedy 1990). This die-off apparently commenced along the coast of Denmark in April of that year and subsequently spread to the coasts of Sweden, the Netherlands, Norway, Germany, the United Kingdom, and Ireland. It terminated in early 1989, although a small focal outbreak subsequently occurred in northern Norway in late 1989 (Krogsrud et al. 1990). Regional mortality among harbor seals varied from approximately 40% to at least 80%, but the mortality among gray seals was extremely low (Heide-Jorgensen et al. 1992). Discovery of distemper-like inclusions, morbilliviral antigen in tissues (Kennedy et al. 1988b; Bergman et al. 1990), and canine distemper virus (CDV)-neutralizing antibodies in serum of affected seals, followed by the isolation of a morbillivirus (Kennedy et al. 1988b; Osterhaus and Vedder, 1988; Osterhaus et al. 1988) implicated morbillivirus infection as the cause of the seal epizootic. Subsequent studies of this morbillivirus indicated it was a newly recognized virus, which was termed phocine distemper virus (PDV) (Cosby et al. 1988). Distemper has since been diagnosed in a few harbor seals from the eastern coast of the United States, indicating that PDV infection is also present in pinnipeds in this region (Duignan et al. 1993).

TABLE 2.1—Aquatic mammals in which clinical distemper has been reported

Species	Common Name	Location	References
Phoca vitulina	Harbor seal	Northwestern Europe Western Atlantic Canada (captive)	Kennedy 1990 Duignan et al. 1993 Lyons et al. 1993
Phoca sibirica	Baikal seal	Lake Baikal Japan (captive)	Grachev et al. 1989 Nunoya et al. 1990
Phoca groenlandica	Harp seal	Gulf of St. Lawrence	Daoust et al. 1993
Halichoerus grypus	Gray seal	Northwestern Europe	Kennedy 1990; Bergman et al. 1990
Phocoena phocoena	Harbor porpoise	Northwestern Europe	Kennedy et al. 1991, 1992; Visser et al. 1993a
Stenella coeruleoalba	Striped dolphin	Mediterranean Sea	Domingo et al. 1992; Di Guardo et al. 1992; Duignam et al. 1992
Tursiops truncatus	Bottlenose dolphin	Western Atlantic Gulf of Mexico Mediterranean Sea	Lipscomb et al. 1994b; Schulman et al. 1997 Lipscomb et al. 1994a Tsur et al. 1997
Globicephala melas	Pilot whale	Western Atlantic	Duignan et al. 1995b

TABLE 2.2—Serologic evidence of morbillivirus infections in aquatic mammals

Species	Common Name	Location	References
Phoca groenlandica	Harp seal	Barents Sea, West Ice	Markussen and Have 1992; Stuen et al. 1994
		Greenland	Dietz et al. 1989a
		Canada	Henderson et al. 1992
Phoca hispida	Ringed seal	Greenland	Dietz et al. 1989a
		Canada	Henderson et al. 1992
Phoca sibirica	Baikal seal	Lake Baikal	Grachev et al. 1989
Phoca vitulina	Harbor seal	Northwestern Europe	Osterhaus and Vedder 1988; Cornwell et al. 1992; Liess et al. 1989; Carter et al. 1992
		Western Atlantic	Duignan et al. 1995d; Ross et al. 1992
Halichoerus grypus	Gray seal	Northwestern Europe	Carter et al. 1992; Cornwell et al. 1992
		Western Atlantic	Henderson et al. 1992; Ross et al. 1992; Duignan et al. 1995d
Phocoena phocoena	Harbor porpoise	Western Atlantic	Duignan et al. 1995a; Van Bressem et al. 1998a
Hydrurga leptonyx	Leopard seal	Antarctica	Bengston et al. 1991
Lobodon carcinophagus	Crabeater seal	Antarctica	Bengston et al. 1991
Cystophora cristata	Hooded seal	Canada	Ross et al. 1992
		North of Jan Mayen	Stuen et al. 1994
Odobenus rosmarus	Walrus	Northwestern Canada	Duignan et al. 1994
Grampus griseus	Risso's dolphin	Western Atlantic	Duignan et al. 1995a
Delphinus delphis	Common dolphin	Northwestern Europe, Mediterranean Sea	Visser et al. 1993a; Van Bressem et al. 1993, 1998a
		Western Atlantic	Duigan et al. 1995a
Delpinius capensis	Long-beaked dolphin	Southeast Pacific	Van Bressem et al. 1988b
		California	Reidarson et al. 1998
Lagenorhynchus acutus	Atlantic white-sided dolphin	Western Atlantic	Duignan et al. 1995a
Lagenorhynchus albirostris	White-beaked dolphin	Northwestern Europe	Osterhaus et al. 1995; Van Bressem et al.1998a
Lagenodelphis hosei	Fraser's dolphin	Western Atlantic	Duignan et al. 1995a
Stenella coeruleoalba	Striped dolphin	Western Atlantic	Duignan et al. 1995a
		Mediterranean Sea	Visser et al. 1993a
Stenella frontalis	Spotted dolphin	Western Atlantic	Duignan et al. 1995a
Tursiops truncatus	Bottlenose dolphin	Western Atlantic	Geraci 1989
		Southeast Pacific	Van Bressem et al. 1998b
Balaenoptera physalus	Fin whale	Iceland	Blixenkrone-Møller et al. 1996
Balaenoptera acutorostrata	Minke whale	Italy	Di Guardo et al. 1995a
Globicephala macrorhynchus, G. melas	Pilot whales	Western Atlantic	Duignan et al. 1995b; Van Bressem et al. 1998b
Kogia breviceps	Pygmy sperm whale	Western Atlantic	Duignan et al. 1995a
Feresia attenuata	Pygmy killer whale	Western Atlantic	Duignan et al. 1995a
Pseudorca crassidens	False killer whale	Western Atlantic	Duignan et al. 1995a
Trichechus manatus latirostris	Manatee	Florida	Duignan et al. 1995c
Ursus maritimus	Polar bear	Russia and Canada	Follmann et al. 1996

During the 1988 European harbor seal epizootic, it became known that several thousand Baikal seals had died in Lake Baikal from late 1987 to late 1988 (Grachev et al. 1989). Although clinical signs and pathologic alterations in affected animals were similar to those of distemper in European harbor seals, the Baikal seal morbillivirus (BSM) was identified as a strain of CDV (Barrett et al. 1992; Mamaev et al. 1995, 1996). A direct epidemiologic link between the Siberian and European epizootics is therefore unlikely.

Distemper was reported in six harbor porpoises *Phocoena phocoena* found stranded on the coast of Ireland in late 1988 (Kennedy et al. 1988a, 1991). These animals were the first documented examples of morbilliviral infection in a cetacean species. Interspecific transfer of PDV infection was initially suspected because of close geographic and temporal links between these cetaceans and morbillivirus-infected harbor seals. Morbillivirus infection was subsequently found in a few harbor porpoises from the coasts of England, Scotland, and the Netherlands in 1990 (Kennedy et al. 1992; Visser et al. 1993a). Porpoise morbillivirus (PMV) is now known to be a distinct morbillivirus (McCullough et al. 1991; Trudgett et al. 1991; Welsh et al. 1992; Barrett et al. 1993; Visser et al. 1993a).

The third known epizootic of morbillivirus infection in aquatic mammals occurred in striped dolphins in the Mediterranean Sea. It apparently commenced along the coast of Spain in July 1990 (Domingo et al. 1990, 1992; Duignan et al. 1992) and rapidly spread to other regions of the western Mediterranean. It subsided in late 1990 but reemerged in June 1991, spreading eastward to the southern Adriatic and Ionian Seas, Sicilian Channel, southern Tyrrhenian Sea, and coasts of Greece and Turkey before terminating in 1992. Total mortality in the two phases of this epizootic is unknown. The striped dolphin morbillivirus (DMV) has been characterized as another new morbillivirus that is closely related to PMV (Barrett et al. 1993; Visser et al. 1993a; Blixenkrone-Møller et al. 1994; Bolt et al. 1994; Rima et al. 1995).

More than 50% of the in-shore population of Atlantic bottlenose dolphins along the eastern coast of the United States from Virginia to Florida were estimated to have died between June 1987 and May 1988 (*Federal Register* 1993). Brevetoxin produced by the "red tide" marine dinoflagellate, *Ptychodiscus brevis,* was originally implicated as a possible cause of this mortality (Geraci 1989). However, morbillivirus was present in tissues of more than 50% of dolphins examined (Lipscomb et al. 1994b; Schulman et al. 1997). A smaller die-off of this species in the Gulf of Mexico from mid-1993 to mid-1994 was also due to morbillivirus infection (Lipscomb et al. 1994a, 1996). The only other record of morbilliviral disease in bottlenose dolphins is one case from the Mediterranean Sea in 1994 (Tsur et al. 1997). Bottlenose dolphins from the eastern coast of the United States were apparently infected with DMV and PMV, whereas those in the Gulf of Mexico were infected with PMV (Taubenberger et al. 1996).

Lesions of distemper were documented in two of 47 common dolphins *Delphinus delphis ponticus* found stranded on the northern shores of the Black Sea in mid-1994 (Birkun et al. 1999). However, the full extent of mortality due to morbillivirus infection in this incident is unknown. Distemper has also been reported in a long-finned pilot whale *Globicephala melas* calf found stranded on the eastern coast of the United States in late 1989 (Duignan et al. 1995b), in a Pacific white-sided dolphin *Lagenorhynchus obliquidens* found on the coast of Japan in 1998 (Uchida et al. 1999), in a fin whale *Balaenoptera physalus* found on the coast of Belgium in 1997 (Jauniaux et al. 1998), and in a harp seal *Phoca groenlandica* found in the Gulf of St. Lawrence, Canada, in 1991 (Daoust et al. 1993).

More than half the population of approximately 300 Mediterranean monk seals *Monachus monachus* along the coast of Mauritania died in 1997. Although a morbillivirus closely related to DMV was isolated from a few of these animals (Osterhaus et al. 1997; van de Bildt et al. 1999), distemper-like lesions were not apparent in affected seals. A role for morbillivirus infection in this mortality event was not therefore established. A morbillivirus similar to PMV was isolated from a monk seal found on the coast of Greece in 1996 (van de Bildt et al. 1999).

Morbillivirus antigen and nucleic acid were detected in tissues of two hooded seals *Cystophora cristata* from the Mediterranean Sea (Osterhaus et al. 1992), and morbillivirus antigen was found in lung tissue of a white-beaked dolphin *Lagenorhyncus albirostris* from the coast of the Netherlands (Osterhaus et al. 1995). Morbilliviral nucleic acid was identified in tissues of common dolphins that stranded on the coast of southern California from 1995 to 1997 (Reidarson et al. 1998). These animals were seropositive to DMV. Canine distemper virus nucleic acid was detected in brain tissue of a Caspian seal *Phoca caspica* found on the Caspian coast of Azerbaijan in 1997 (Forsyth et al. 1998). However, it is not known whether these viral infections caused disease in these animals.

Serologic evidence of morbillivirus infection occurred in a wide range of pinnipeds and cetaceans from the Northern Hemisphere, southeast Pacific Ocean, and Antarctica; manatees *Trichechus manatus latirostris* from the coast of Florida; and polar bears *Ursus maritimus* from Canada and Russia (Table 2.2).

ETIOLOGY. The morbilliviruses of aquatic mammals are members of the genus *Morbillivirus* in the family *Paramyxoviridae.* They are serologically related, single-stranded RNA viruses that, unlike other paramyxoviruses, lack neuraminidase activity. The nucleocapsids are approximately 16.5–18.5 nm in diameter (Osterhaus et al. 1988; Visser et al. 1990; Van Bressem et al. 1991).

Phocine Distemper Virus. Comparison of the reactivity patterns of PDV and reference strains of CDV, measles virus (MV), and rinderpest virus (RPV) with panels of monoclonal antibodies indicated that PDV has unique epitopes on the nucleoprotein (N), phosphoprotein (P), and hemagglutinin (H) proteins (Cosby et al. 1988; Örvell et al. 1990; Visser et al. 1990; Harder et al. 1991; Örvell and Sheshberadaran 1991; Trudgett et al. 1991; Blixenkrone-Møller et al. 1992a). Sharing of only a few epitopes between the H proteins of CDV and PDV suggests that these viruses evolutionarily diverged in the distant past (Blixenkrone-Møller et al. 1992a). These antigenic differences are reflected in the fact that sera from seals infected with PDV reacted more strongly with RPV, peste des petits ruminants virus (PPRV), and PDV than with CDV (Liess et al. 1989; Bostock et al. 1990; Barrett et al. 1992).

Nucleotide sequence differences between the H, N, and P genes of PDV and the corresponding genes of CDV, MV, and RPV indicate that PDV is more closely related to CDV than to these other morbilliviruses (Curran et al. 1990, 1992b; Blixenkrone-Møller et al. 1992b; Curran and Rima 1992; Rima et al. 1992). However, the sequence differences between PDV and CDV are so great that they should be regarded as separate viral species. Although more highly conserved among morbilliviruses, sequence analysis of the F (fusion) protein of PDV supports this conclusion (Kovamees et al. 1991; Curran et al. 1992a). These data are in agreement with indications from nucleic acid hybridization analysis (Mahy et al. 1988; Bostock et al. 1990; Haas et al. 1991) and studies of the biologic, morphologic, and protein chemical properties of PDV that it represents a newly discovered morbillivirus (Rima et al. 1990; Visser et al. 1990).

Baikal Seal Morbillivirus. Nucleotide-sequence data on the F, H, and P genes of BSM indicate it is more closely related to recent European isolates of CDV than to any other morbillivirus (Visser et al. 1993b; Mamaev et al. 1995, 1996). Phylogenetic analysis of F gene nucleotide sequences reveal a closer evolutionary relationship to CDV than to PDV. This similarity is supported by nucleic acid hybridization data and the demonstration of only minor antigenic differences between BSM and CDV (Visser et al. 1990; Barrett et al. 1992). Collectively, these data indicate that BSM is a strain of CDV.

Cetacean Morbilliviruses. Studies on the cross-reactivities of DMV, PMV, PDV, CDV, and MV with panels of monoclonal antibodies revealed major antigenic differences between DMV and PMV and the other morbilliviruses (McCullough et al. 1991; Trudgett et al. 1991; Osterhaus et al. 1992; Welsh et al. 1992; Visser et al. 1993a). These cetacean viruses are more closely related antigenically to RPV and PPRV of ruminants than to CDV and PDV of carnivores. Furthermore, they differ from each other at only a few epitopes, indicating they are either similar or identical viruses. The findings that titers of neutralizing antibodies to PMV and DMV in sera of several cetacean species are almost identical and higher than those to any other morbillivirus provides further evidence for a close antigenic relationship between these viruses (Van Bressem et al. 1993; Visser et al. 1993a).

Barrett et al. (1993) compared the sequences of the P genes of DMV and PMV with the corresponding sequences of the other known morbilliviruses. The level of nucleotide-sequence homology between the P gene of DMV and PMV is approximately 90% and is similar to that observed between geographically distinct isolates of RPV. The predicted F proteins of DMV and PMV are 94% identical (Bolt et al. 1994). Sequence analysis of the N and H genes indicates that DMV and PMV are almost equidistant from the other morbilliviruses (Blixenkrone-Møller et al. 1994, 1996; Rima et al. 1995).

These genomic and antigenic data indicate that DMV is one of the phylogenetically oldest morbilliviruses known to be in current circulation. Dolphin morbillivirus and PMV may be strains of the same viral species circulating in different cetacean populations (Bolt et al. 1994). From an epidemiologic perspective, these data provide no evidence for linkage between the epizootics in European harbor seals, Baikal seals, and Mediterranean striped dolphins.

TRANSMISSION AND EPIDEMIOLOGY. Studies of the distribution of morbillivirus antigen in tissues of harbor seals (Kennedy et al. 1989; Harder et al. 1992; Munro et al. 1992; Daoust et al. 1993) suggest that viral shedding by the respiratory, urinary, fecal, and ocular routes occurs in pinnipeds. Viral transmission by aerosol and by direct and indirect contact can therefore be postulated. Hauling out and formation of large social groups by pinnipeds almost certainly facilitates lateral spread of infection. Disease due to cetacean morbillivirus infection has not been documented in any pinniped species.

Demonstration of morbillivirus antigen in a wide range of tissues in cetaceans suggests a high potential for viral excretion from all body orifices and skin (Domingo et al. 1992; Kennedy et al. 1991, 1992; Schulman et al. 1997). Detection of virus in the male reproductive tract of harbor porpoises (Kennedy et al. 1992) and a common dolphin (Birkun et al. 1999), and occasionally in the mammary gland of striped dolphins and bottlenose dolphins (Domingo et al. 1992; Schulman et al. 1997), suggests the possibility of venereal and vertical transmission, respectively. It is not known whether transplacental transmission of morbilliviruses occurs in aquatic mammals. Urinary, fecal, ocular, and dermal excretions of virus also seem probable but may be less likely to result in lateral transmission in natural situations because of the dilution and probable destructive effects of seawater on morbilliviruses.

By analogy with distemper in terrestrial animals, transmission by carrier or inapparently infected animals and by contaminated fomites can be postulated for aquatic mammals. The existence of prolonged viremia in morbillivirus-infected pinnipeds (Harder et al. 1992) suggests that the possibility of transmission by blood-sucking arthropods should also be considered.

Epizootics in aquatic mammals probably follow introduction of morbillivirus into immunologically naive populations. Although morbillivirus infections in aquatic mammals date back to at least 1983 (Blixenkrone-Møller et al. 1996), proven morbillivirus epizootics have been recorded only since 1987. Enzootic morbillivirus infection associated with low mortality may be present in some aquatic mammal populations. Alterations in the migratory patterns of these populations may have introduced infection to other unexposed populations, resulting in epizootics.

Harp seals from Greenland have been suggested as vectors responsible for introduction of PDV to harbor seal and gray seal populations in northwestern Europe in 1988 (Dietz et al. 1989a). Differential virus neutralization tests revealed higher titers to DMV and PMV than to other morbilliviruses in sera from Florida manatees (Duignan et al. 1995c) and several odontocete species from the eastern and western Atlantic Ocean and Gulf of Mexico (Duignan et al. 1995b; Blixenkrone-Møller et al. 1996). There may therefore be a link between morbillivirus epizootics in bottlenose dolphins along the coast of the United States (Lipscomb et al. 1994b, 1996) and morbillivirus infections in other cetacean species and manatees in this region. It has been suggested that pilot whales *Globicephala melas* and *G. macrorhynchus* may act as a reservoir of morbillivirus infection for cetacean species in the western Atlantic (Duignan et al. 1995b).

There is strong evidence that the 1987–88 Baikal seal epizootic resulted from CDV transfer from local terrestrial carnivores (Mamaev et al. 1995). This mechanism is also likely to have resulted in CDV infection in Caspian seals (Forsyth et al. 1998). Virus-neutralizing antibodies were present in sera of crabeater seals *Lobodon carcinophagus* from Antarctica in higher titers to CDV than to PDV (Bengston et al. 1991), suggesting that a virus identical or closely related to CDV has also circulated in Antarctic pinnipeds. Sled dogs may have been the source of infection for these seals. In contrast, sera from Atlantic walruses *Odobenus rosmarus* (Duignan et al. 1994) and gray seals and harbor seals from the western Atlantic (Ross et al. 1992; Duignan et al. 1995d) neutralized PDV to higher titers than other morbilliviruses.

It has been suggested that chemical contaminants predisposed pinnipeds and striped dolphins to epizootics of morbillivirus infection. However, there was no evidence of a causal relationship between mortality and tissue concentrations of chemical contaminants in harbor seals and striped dolphins that died in morbillivirus epizootics (Hall et al. 1992; Aguilar and Borrell 1994). Furthermore, dietary supplementation with polychlorinated biphenyls did not increase the severity of disease in harbor seals following experimental challenge with PDV (Harder et al. 1992). It therefore appears that these substances had a minimal, if any, role in these epizootics.

Several epizootics of infectious disease in marine mammals followed periods of increased air temperatures (Lavigne and Schmitz 1990). It is not clear whether this association is coincidental, but climatic change could alter behavior and movement patterns of aquatic mammal populations, thereby influencing the occurrence of morbillivirus epizootics.

CLINICAL SIGNS. Seals naturally infected with morbilliviruses had variable body condition and frequently fever, serous or mucopurulent oculonasal discharge, conjunctivitis, ophthalmitis, keratitis, coughing, dyspnea, diarrhea, lethargy, nervous signs, and abortion (Breuer et al. 1988; Grachev et al. 1989; Kennedy et al. 1989; Bergman et al. 1990; Kennedy 1990; Krogsrud et al. 1990; Nunoya et al. 1990; Have et al. 1991; Heje et al. 1991; Baker 1992; Munro et al. 1992; Duignan et al. 1993). Subcutaneous emphysema of the cervical and thoracic regions resulted in increased buoyancy and consequent inability to submerge (Kennedy et al. 1989; Bergman et al. 1990; Have et al. 1991).

There are few clinical data on cetaceans with distemper. Many affected striped dolphins and bottlenose dolphins were in poor body condition (Geraci 1989; Domingo et al. 1992; Duignan et al. 1992). Loss of body fat stores resulted in poor flotation (Piza 1991). Skin lesions and erosions of the buccal mucosa were also common (Geraci 1989; Domingo et al. 1992). Increased numbers of epizoites were reported on the skin and teeth of morbillivirus-infected striped dolphins, probably reflecting chronic debilitation (Aznar et al. 1994). Piza (1991) reported tachycardia and abnormal respiratory rates in striped dolphins with distemper. Sound emission was weak. Muscle tremors were seen, and some dolphins repeatedly struck their bodies against rocks or breakwaters, possibly as a result of brain damage. Other animals showed no interest in swimming. None of the examined dolphins opened their mouths. Lethargy, incoordination, tachycardia, and tachypnea were reported in common dolphins with distemper (Birkun et al. 1999).

PATHOLOGY. Clinical pathology data on morbillivirus-infected aquatic mammals are scant. Leukopenia and hemoconcentration were reported in striped dolphins and harbor seals with distemper (Piza 1991; Harder et al. 1992). Lymphopenia was described in a common dolphin (Birkun et al. 1999).

Pneumonia was the main gross finding in seals naturally infected with morbilliviruses (Breuer et al. 1988; Kennedy et al. 1989; Bergman et al. 1990; Krogsrud et al. 1990; Nunoya et al. 1990; Have et al. 1991; Heje et al. 1991; Munro et al. 1992; Daoust et al. 1993). Pneumonic lungs frequently failed to collapse and were congested and edematous. Interlobular, subpleural, and mediastinal emphysema were prominent features (Fig. 2.4). Pulmonary abscesses, parasitic granulomas, pleuritis, hydrothorax, and hemothorax were also common (Heje et al. 1991). The airways contained mucopurulent, bloodstained, or frothy exudate. Heavy nematode infections of the lungs were frequently reported (Breuer et al. 1988; Kennedy et al. 1989; Schumacher et al. 1990; Munro et al. 1992). Enlargement of mediastinal, mesenteric, and retroperitoneal lymph nodes was also documented (Schumacher et al. 1990).

Striped dolphins from the 1990–92 Mediterranean epizootic were usually in poor body condition (Domingo et al. 1992). Ulcerative stomatitis was apparent in some animals, but exudative bronchopneumonia was the most prominent necropsy finding. Pneumonic

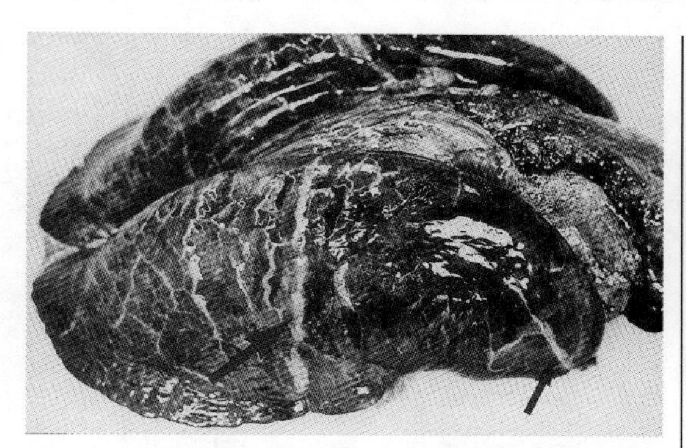

FIG. 2.4—Lungs of a harbor seal with distemper. Note subpleural emphysema (*arrows*).

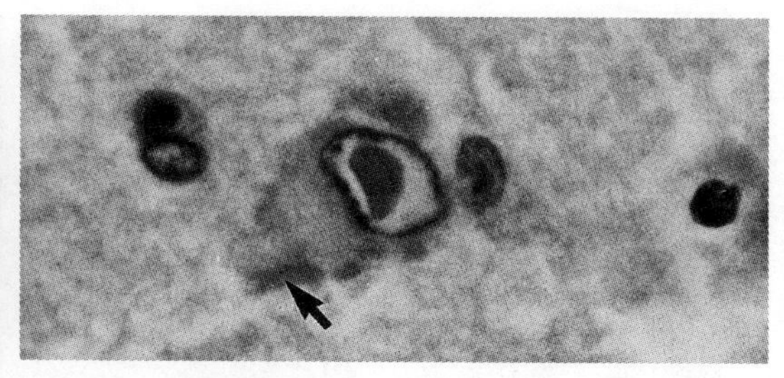

FIG. 2.5—Brain of a harbor seal with distemper. Nuclear and cytoplasmic (*arrow*) inclusions in a neuron. Hematoxylin and eosin.

lungs did not collapse and contained multiple foci of atelectasis and evidence of subpleural emphysema and dilatation of subpleural lymphatics (Di Guardo et al. 1992; Domingo et al. 1992; Duignan et al. 1992). Lung-associated lymph nodes were frequently enlarged and edematous. Variable numbers of nematodes were seen in the airways (Kennedy et al. 1991; Domingo et al. 1992, Duignan et al. 1992). A few striped dolphins had large hemorrhagic necrotic lesions in the cerebral cortex (Domingo et al. 1992).

Pneumonia and ulceration of the skin and buccal mucosa were reported in bottlenose dolphins (Geraci 1989). Many dolphins had evidence of septicemia, including edema of internal organs and accumulation of large quantities of serosanguinous fluid in the pleural and peritoneal cavities. Fibrosis of lung, myocardium, liver, and pancreas was also apparent. In addition, vaginitis and balanoposthitis were documented in morbillivirus-infected harbor porpoises (Kennedy et al. 1991, 1992). Pulmonary emphysema sometimes resulted in formation of large subpleural bullae.

Cardinal histopathologic lesions were present in the respiratory tract, central nervous system (CNS), lymphoid tissues, and epithelia of seals with distemper and were described by many authors (Hofmeister et al. 1988; Kennedy et al. 1989; Bergman et al. 1990; Krogsrud et al. 1990; Nunoya et al. 1990; Schumacher et al. 1990; Heje et al. 1991; Baker 1992; Munro et al. 1992; Daoust et al. 1993; Duignan et al. 1993). Pulmonary changes were characteristic of bronchointerstitial pneumonia. Alveolar and bronchiolar lumina contained serofibrinous exudate, leukocytes, macrophages, and other mononuclear cells. Intra-alveolar hyaline membranes were common and hemorrhage was frequent. Lesions of suppurative bronchopneumonia associated with abscess formation, and bacterial colonization of alveoli and airways, were also present in many animals. Alveolar walls were infiltrated with leukocytes and mononuclear cells and had evidence of fibroplasia in chronic cases. There was proliferation of type-II pneumocytes and formation of intra-alveolar, intrabronchiolar, and intrabronchial syncytia. Acidophilic inclusions were found in cytoplasm and nuclei of nasal, tracheal, and bronchial epithelial cells and in type-II pneumocytes. These structures were single or multiple, approximately 10–20 μm in diameter, and round to ovoid with distinct borders (Nunoya et al. 1990). Pulmonary syncytia and inclusions were frequently obscured by secondary bacterial and parasitic infections.

Lesions of nonsuppurative encephalitis were reported in many harbor seals and in a harp seal. These changes were characterized by necrosis of neurons and glial cells in cerebral cortex, perivascular cuffing, astrocytosis, microglial infiltration, neuronophagia, and focal demyelination. Cerebral cortical necrosis frequently occurred in a laminar pattern. Acidophilic nuclear and cytoplasmic inclusions were frequently seen in astrocytes and neurons (Fig. 2.5).

Marked necrosis and depletion of lymphocytes in lymph nodes, spleen, gut-associated lymphoid tissue, and thymus were commonly found. Cytoplasmic inclusions were occasionally reported in lymph nodes. Damage to lymphoid tissues in morbillivirus-infected animals is likely to have resulted in immunosuppression and consequently increased susceptibility to other infectious agents.

Nuclear and cytoplasmic inclusions were seen in epithelium of the urinary bladder and renal pelvis, biliary and pancreatic ductules, and gastrointestinal tract.

Morbilliviral antigen has been demonstrated by immunoperoxidase staining in sections of trachea, lung (Fig. 2.6), brain, spleen, lymph nodes, thymus, gastric mucosa, epithelium of renal pelvis and bladder, and pancreatic ductules of harbor seals naturally infected with PDV (Kennedy et al. 1989; Munro et al. 1992). Additionally, it has been detected in tonsil and small and large intestines of harbor seals experimentally infected with this virus (Harder et al. 1992; Pohlmeyer et al. 1993).

Microscopic lesions of morbillivirus infection in cetaceans have been reported (Kennedy et al. 1991, 1992; Di Guardo et al. 1992, 1995b; Domingo et al. 1992; Duignan et al. 1992, 1995b; Lipscomb et al. 1994a,b, 1996; Schulman et al. 1997; Jauniaux et al. 1998; Birkun et al. 1999; Uchida et al. 1999). Lung lesions were subacute to chronic bronchointerstitial pneumonia.

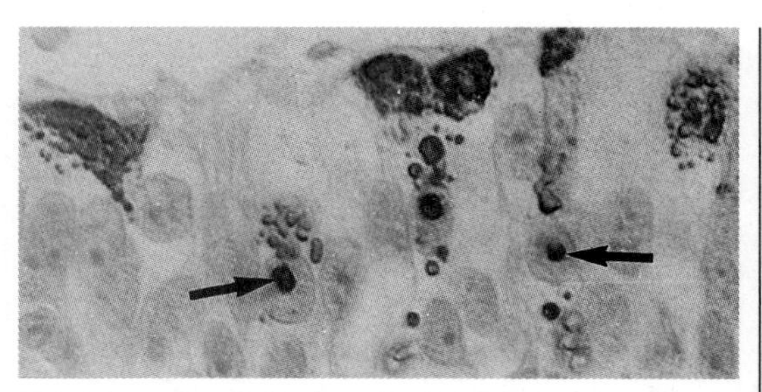

FIG. 2.6—Lung of a harbor seal with distemper. Morbilliviral antigen is apparent in nuclei (*arrows*) and cytoplasm of bronchial epithelial cells. Immunoperoxidase staining using a measles antibody as primary antibody.

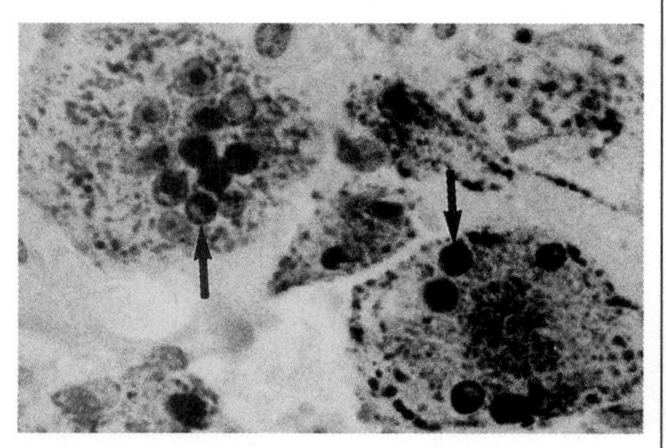

FIG. 2.7—Lung of a harbor porpoise with distemper. Morbilliviral antigen is present in nuclei (*arrows*) and cytoplasm of intra-alveolar syncytia. Immunoperoxidase staining using a monoclonal antibody to phocine distemper virus as primary antibody.

Hyperplasia of type-II pneumocytes and formation of intrabronchial, intrabronchiolar, and intra-alveolar syncytia (Fig. 2.7) were generally more marked than in seals. Nuclear and cytoplasmic inclusions were also more abundant and were commonly seen in bronchial and alveolar epithelium, lung syncytia, and intra-alveolar macrophages. However, syncytia and inclusions were sometimes highly focal or were obscured by severe necrotizing lesions due to opportunistic bacterial and fungal infections. Fibroplasia of alveolar septa was a prominent finding in many striped dolphins, bottlenose dolphins, and harbor porpoises.

The marked neurovirulence of DMV and PMV was reflected in the occurrence of nonsuppurative encephalitis in up to 69% of morbillivirus-infected striped dolphins and in individual harbor porpoises, a common dolphin, and a Pacific striped dolphin. Demyelination, especially of the cerebellum, was reported in approximately 22% of striped dolphins in one study but was not a prominent feature in other striped dolphins or in harbor porpoises. Cytoplasmic and nuclear acidophilic inclusions were common in

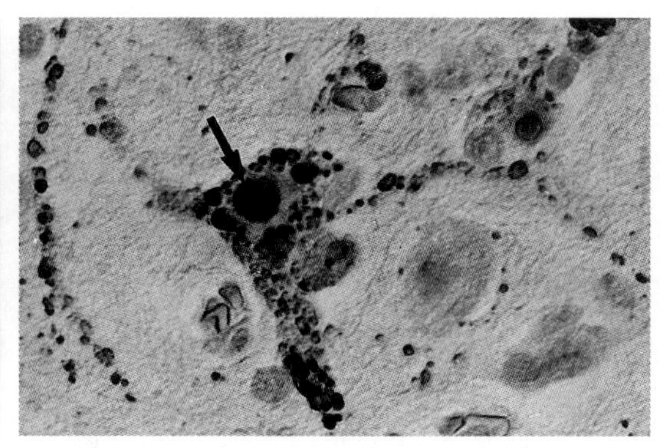

FIG. 2.8—Cerebral cortex of a harbor porpoise with distemper. Morbilliviral antigen is present in nucleus (*arrow*) and in perikaryonal and axonal cytoplasm of neurons. Immunoperoxidase staining using a monoclonal antibody to phocine distemper virus as primary antibody.

degenerate and necrotic neurons, glial cells, and syncytia. Neuronal cytoplasmic inclusions tended to have irregular margins and frequently occupied the entire perikaryon. Abundant morbilliviral antigen was associated with these lesions (Fig. 2.8). Fungal hyphae and *Toxoplasma*-like pseudocysts were seen in necrotic lesions in brains of a few striped dolphins.

Lymphoid cell depletion and necrosis were commonly seen in lymph nodes and spleen. As in measles and rinderpest, but in contrast to PDV infection in seals, syncytia were frequently scattered throughout these tissues. Acidophilic cytoplasmic and nuclear inclusions were found in syncytia, lymphocytes, and reticular cells. Nuclear and cytoplasmic inclusions and/or morbilliviral antigen were detected in a wide variety of epithelia.

Domingo et al. (1995) and Tsur et al. (1997) described nonsuppurative encephalitis associated with the presence of acidophilic nuclear and cytoplasmic inclusions and morbillivirus antigen in brains of a few Mediterranean striped dolphins found stranded in 1994. No lesions attributable to morbillivirus infection nor morbillivirus antigen were found in other tissues of these animals. Restriction of morbillivirus lesions and antigen to the brain may represent a form of chronic infection analogous to subacute sclerosing panencephalitis caused by chronic MV infection in humans. However, further studies are required to determine whether, as in the human disease, this condition is caused by a defective virus.

PATHOGENESIS. Experimental intranasal infection of seronegative harbor seals with a German isolate of PDV resulted in severe clinical disease similar to that reported in natural outbreaks of PDV infection (Harder et al. 1992). Inoculated seals were viremic from days 5–12 after inoculation and occasionally for longer periods. As in natural infections in this species, PDV antigen was detected in a range of tissues, including the respiratory and gastrointestinal tracts, CNS, and lymphoid tissues. Neutralizing antibodies [probably immunoglobulin μ (IgM)] were detected from day 7 after infection, followed by production of IgG from day 11. Unlike dogs *Canis familiaris* recovering from CDV infection, but as in natural outbreaks of distemper in harbor seals (Liess et al. 1989), there was no correlation between PDV-specific antibody titers and recovery from infection. A seronegative gray seal, challenged with PDV, seroconverted to this virus but did not develop signs of disease. These results are in agreement with the finding of low mortality among gray seals during the 1988 European seal morbillivirus epizootic.

Baikal seals experimentally infected with BSM developed mild disease (Titenko et al. 1991). Serum morbillivirus antibodies were detected by enzyme-linked immunosorbent assay (ELISA) 10 days after infection and were maximal at 20 days after infection. Neutralizing serum antibodies were apparent at 16 days after infection and were maximal from 20 days to at least 39 days after inoculation.

DIAGNOSIS. Clinical signs are not sufficiently characteristic for the diagnosis of distemper in aquatic mammals, so laboratory examinations are essential. Examination of smears of conjunctival, nasal, buccal, vaginal, preputial, and urethral mucosae or peripheral leukocytes for the presence of typical inclusions or morbillivirus antigen may be useful. As in dogs infected with CDV, however, the success of this approach probably depends on factors such as the stage of infection and serum antibody titer; negative results should therefore be interpreted with caution.

At necropsy, depletion of lymphocytes in lymphoid tissues, hyperplasia of type-II pneumocytes, and formation of syncytia in lungs, airways, brain, or lymphoid tissues are suggestive of distemper. The presence of acidophilic nuclear and cytoplasmic inclusions in cells of the respiratory, gastrointestinal, and urinary tracts, lymphoid tissues, or CNS can be considered pathognomonic for this disease.

Distemper can be confirmed by immunoperoxidase or immunofluorescence labeling of morbillivirus antigens in paraffin or frozen tissue sections. Antigen-capture ELISA techniques have also been employed to detect morbillivirus antigens in tissues (Visser et al. 1990). Nucleic acid hybridization techniques can be used to detect morbilliviral RNA in tissue extracts (Mahy et al. 1988; Grachev et al. 1989; Haas et al. 1991; Bolt and Blixenkrone-Møller 1994). The sensitivity of these techniques can be increased by reverse transcription of morbilliviral RNA, followed by polymerase chain reaction amplification (RT-PCR) before hybridization. When applied to autolyzed tissues of bottlenose dolphins, RT-PCR was more sensitive than immunoperoxidase staining (Krafft et al. 1995) for detection of morbillivirus infection.

Morbillivirus infection can also be confirmed by virus isolation. Cell cultures used to isolate PDV and BSM from seal tissues include Vero cells and cell lines prepared from seal lung, skin, and kidney (Hofmeister et al. 1988; Kennedy et al. 1988b; Blixenkrone-Møller et al. 1989; Liess et al. 1989; Visser et al. 1990). These viruses have also been isolated by inoculation of tissue extracts into dogs followed by cocultivation of peripheral blood mononuclear cells (PBMCs) from these animals with primary dog lung macrophages (Visser et al. 1990). In vitro isolation of PDV has also been achieved following passage in mink *Mustela vison* (Blixenkrone-Møller et al. 1992a).

Porpoise morbillivirus grew in Vero cells and in bovine fetal lung cells (McCullough et al. 1991; Visser et al. 1993a). When grown in PBMCs of bottlenose dolphin origin, DMV produced syncytia within 7 days (Van Bressem et al. 1993). This virus also was isolated by direct inoculation of tissue homogenates into Vero cells (Van Bressem et al. 1991) and by inoculation into canine kidney epithelial cells followed by cocultivation with Vero cells (Blixenkrone-Møller et al. 1992a).

Virus-neutralizing tests and ELISA techniques have been used to determine morbillivirus antibody titers in sera of aquatic mammals (Visser et al. 1990; Örvell et al. 1990; Blixenkrone-Møller et al. 1992a; Duignan et al. 1995a). Because of the extensive serologic cross-reactions among morbilliviruses, differential virus-neutralizing assays or ELISAs that use specific monoclonal antibodies are required to identify the morbillivirus most likely to have caused the serologic response. Antibody titers indicate exposure to virus but do not imply active infection unless rising titers can be demonstrated.

DIFFERENTIAL DIAGNOSES. The differential diagnosis of distemper in individual aquatic mammals includes diseases that cause respiratory, enteric, and nervous signs; ocular and skin lesions; and abortion. Influenza, salmonellosis, brucellosis, pasteurellosis, and herpesviral, caliciviral, and lungworm infections should be considered. Other possible causes of mass mortality in aquatic mammals include poisoning by algal toxins.

IMMUNITY. The immune response to PDV in harbor seals that died of PDV infection was directed toward the internal (N and P) proteins of this virus (Rima et al. 1990). Antibodies to the external glycoprotein antigens required for immunity to CDV infection in dogs were rarely found in sera of these seals. Unlike dogs infected with CDV, there was no correlation between neutralizing antibody titers and recovery from PDV infection in harbor seals (Liess et al. 1989; Harder et al. 1992). Moderate to high titers of CDV-neutralizing antibodies persisted for at least 6 months in harbor seals recovered from PDV infection and for at least 12 months in gray seal survivors (Cornwell et al. 1992). Unexpectedly low antibody titers in some animals probably resulted from the immunosuppressive effects of morbillivirus infection.

Morbillivirus vaccines have been used experimentally in seals. A candidate CDV immunostimulatory-complex-matrix subunit vaccine and a candidate inactivated whole virus CDV vaccine protected harbor seals against severe clinical disease following challenge with PDV (Visser et al. 1992). However, a few vaccinated animals developed mild respiratory signs indicating that these vaccines permitted PDV infection of the respiratory tract. Vaccination of pregnant gray seals with CDV antigens resulted in clinical protection and passive transfer of antibodies to their pups, but the duration of this immunity is unknown (Carter et al. 1992).

TREATMENT AND CONTROL. There is no specific treatment for distemper in any species. Supportive measures including antibiotic and rehydration therapy may be beneficial. Anthelmintic drug administration for parasitic infections may be indicated. Euthanasia is recommended for animals with severe clinical signs.

Control of distemper in captive aquatic mammals is based on prevention of exposure to morbilliviruses. Previously exposed animals can be identified by serologic testing. All animals to be introduced to a captive group should ideally be of the same serologic status as members of that group and should be quarantined before introduction. Access to terrestrial mammals known to be susceptible to morbillivirus infections should be restricted, and a high standard of hygiene, including disinfection of equipment and instruments, should be practiced, especially following an outbreak of disease. Inactivated CDV vaccines are likely to be useful for control of PDV infection in captive pinnipeds (Visser et al. 1992), but further work is required to determine the range of species that may benefit from vaccination. The use of live morbillivirus vaccines in free-ranging animals is contraindicated.

PUBLIC HEALTH CONCERNS. There are no reports of human infections with aquatic mammal morbilliviruses. However, until the possible but unproven causal relationship between CDV infection and Paget's disease in humans (Gordon et al. 1991) is clarified, suspect cases of morbillivirus infection in aquatic mammals should be handled with caution. Immunosuppression in morbillivirus-infected aquatic mammals may result in an increased susceptibility to infectious diseases, including zoonoses such as salmonellosis, leptospirosis, tuberculosis, brucellosis, vibriosis, influenza, and toxoplasmosis, with consequent health implications for people coming into contact with such animals.

DOMESTIC ANIMAL HEALTH CONCERNS. Transmission of PDV from harbor seals has resulted in outbreaks of distemper in farmed mink (Blixenkrone-Møller et al. 1992a). Specific pathogen-free (SPF) dogs experimentally inoculated with BSM or PDV developed a slight elevation in body temperature, lymphopenia, and mild respiratory signs (Osterhaus et al. 1988, 1989). Challenge of SPF dogs with DMV and PMV resulted in leukopenia, biphasic viremia, and production of neutralizing antibodies but no clinical signs (Visser et al. 1993a). Sheep *Ovis aries,* cattle *Bos taurus,* and goats *Capra hircus* infected with DMV and PMV developed leukopenia, viremia, and neutralizing serum antibodies but no clinical signs except for pyrexia in a goat inoculated with PMV (Visser et al. 1993a).

Natural infections of conventional dogs with aquatic mammal morbilliviruses might result in more severe disease than seen in SPF dogs challenged with tissue culture-adapted virus. Dogs likely to have access to habitats of aquatic mammals should therefore be vaccinated against CDV. Natural infection of ruminants with these aquatic mammal morbilliviruses is improbable, but should it occur, severe disease is unlikely. However, differentiation of antibodies induced by DMV and PMV from those induced by RPV and PPRV could be a concern for regulatory authorities.

MANAGEMENT IMPLICATIONS. There is little that can be done to control morbillivirus epizootics in most species of free-ranging aquatic mammals. The risk of transmission to endangered species may be so great that vaccination (Osterhaus et al. 1992) or translocation to other regions may be considered for those species amenable to such interventions. Rehabilitation or translocation of marine mammals may pose a threat to populations of free-ranging marine mammals. A prior assessment of risks to released or translocated animals and to the aquatic mammal populations, including endangered species, in the area of proposed release is recommended. Development of safe and effective morbillivirus vaccines for use in captive aquatic mammals, especially endangered species, is required.

LITERATURE CITED

Aguilar, A., and A. Borrell. 1994. Abnormally high polychlorinated biphenyl levels in striped dolphins (*Stenella coeruleoalba*) affected by the 1990–92 Mediterranean epizootic. *Science of the Total Environment* 154:237–247.

Aznar, F.J., J.A. Balbuena, and J.A. Raga. 1994. Are epizoites indicators of a western Mediterranean striped dolphin die-off? *Diseases of Aquatic Organisms* 18:159–163.

Baker, J.R. 1992. The pathology of phocine distemper. *Science of the Total Environment* 115:1–7.

Barrett, T., J. Crowther, A.D.M.E. Osterhaus, S.M. Subbarao, J. Groen, I. Haas, I.V. Mamaev, A.M. Titenko, I.K.G. Visser, and C.J. Bostock. 1992. Molecular and serological studies on the recent seal virus epizootics in Europe and Siberia. *Science of the Total Environment* 115:117–132.

Barrett, T., I.K.G. Visser, L. Mamaev, L. Goatley, M.F. van Bressem, and A.D.M.E. Osterhaus. 1993. Dolphin and porpoise morbilliviruses are genetically distinct from phocine distemper virus. *Virology* 193:1010–1012.

Bengston, J.L., P. Boveng, U. Franzen, P. Have, M.-P. Heide-Jorgensen, and T.L. Harkonen. 1991. Antibodies to canine distemper virus in Antarctic seals. *Marine Mammal Science* 7:85–87.

Bergman, A., B. Jarplid, and B.M. Svensson. 1990. Pathological findings indicative of distemper in European seals. *Veterinary Microbiology* 23:331–341.

Birkun, A., Jr., T. Kuiken, S. Krivokhizhin, D.M. Haines, A.D.M.E. Osterhaus, M.W.G. van de Bildt, C.R. Joiris, and U. Siebert. 1999. Epizootic of morbilliviral disease in common dolphins (*Delphinus delphis ponticus*) from the Black Sea. *Veterinary Record* 144:85–92.

Blixenkrone-Møller, M., V. Svansson, P. Have, A. Botner, and J. Nielsen. 1989. Infection studies in mink with seal-derived morbillivirus. *Archives of Virology* 106:165–170.

Blixenkrone-Møller, M., V. Svansson, M. Appel, J. Krogsrud, P. Have, and C. Örvell. 1992a. Antigenic relationships between field isolates of morbilliviruses from different carnivores. *Archives of Virology* 123:279–294.

Blixenkrone-Møller, M., B. Sharma, T.M. Varsanyi, A. Hu, E. Norrby, and J. Kovamees. 1992b. Sequence analysis of the genes encoding the nucleocapsid protein and phosphoprotein (P) of phocid distemper virus, and editing of the P gene transcript. *Journal of General Virology* 73:885–893.

Blixenkrone-Møller, M., G. Bolt, E. Gottschalck, and M. Kenter. 1994. Comparative analysis of the gene encoding the nucleocapsid protein of dolphin morbillivirus reveals its distant evolutionary relationship to measles virus and ruminant morbilliviruses. *Journal of General Virology* 75:2829–2834.

Blixenkrone-Møller, M., G. Bolt, T.D. Jensen, T. Harder, and V. Svansson. 1996. Comparative analysis of the attachment protein gene of dolphin morbillivirus. *Virus Research* 40:47–55.

Bolt, G., and M. Blixenkrone-Møller. 1994. Nucleic acid hybridization analyses confirm the presence of a hitherto unknown morbillivirus in Mediterranean dolphins. *Veterinary Microbiology* 41:363–372.

Bolt, G., M. Blixenkrone-Møller, E. Gottschalck, R.G.A. Wishaupt, M.J. Welsh, J.A.P. Earle, and B.K. Rima. 1994. Nucleotide and deduced amino acid sequences of the matrix (M) and fusion (F) protein genes of cetacean morbilliviruses isolated from a porpoise and a dolphin. *Virus Research* 34:291–304.

Bostock, C.J., T. Barrett, and J.R. Crowther. 1990. Characterization of the European seal morbillivirus. *Veterinary Microbiology* 23:351–360.

Breuer, E.M., R. Ernst, R.J. Hofmeister, J. Hentschke, G. Molle, and H. Ludwig. 1988. First report of canine distemper-like disease and lesions in harbor seals: Pathologic-histologic findings and virus isolation. *Zeitschrift fur Angewandte Zoologie* 75:129–138.

Carter, S.D., D.E. Hughes, V.J. Taylor, and S.C. Bell. 1992. Immune responses in common and gray seals during the seal epizootic. *Science of the Total Environment* 115:83–91.

Cornwell, H.J.C., S.S. Anderson, P.M. Thompson, S.J. Mayer, H.M. Ross, P.P. Pomeroy, and R. Munro. 1992. The serological response of the common seal (*Phoca vitulina*) and the grey seal (*Halichoerus grypus*) to phocine distemper virus as measured by a canine distemper virus neutralisation test. *Science of the Total Environment* 115:99–116.

Cosby, S.L., S. McQuaid, N. Duffy, C. Lyons, B.K. Rima, G.M. Allan, S.J. McCullough, S. Kennedy, J.A. Smyth, F. McNeilly, and C. Craig. 1988. Characterization of a seal morbillivirus. *Nature* 336:115–116.

Curran, M.D., and B.K. Rima. 1992. The genes encoding the phosphoproteins and matrix proteins of phocine distemper virus. *Journal of General Virology* 73:1587–1591.

Curran, M.D., D. O'Loan, S. Kennedy, and B.K. Rima. 1990. Nucleotide sequence analysis of phocine distemper virus reveals its distinctness from canine distemper virus. *Veterinary Record* 127:430–431.

Curran, M.D., Y.J. Lu, and B.K. Rima. 1992a. The fusion protein gene of phocine distemper virus: Nucleotide and deduced amino acid sequences and a comparison of morbillivirus fusion proteins. *Archives of Virology* 126:159–169.

Curran, M.D., D. O'Loan, S. Kennedy, and B.K. Rima. 1992b. Molecular characterization of phocine distemper virus: Gene order and sequence of the gene encoding the attachment (H) protein. *Journal of General Virology* 73:1189–1194.

Daoust, P.Y., D.M. Haines, J. Thorsen, P.J. Duignan, and J.R. Geraci. 1993. Phocine distemper in a harp seal (*Phoca groenlandica*) from the Gulf of St. Lawrence, Canada. *Journal of Wildlife Diseases* 29:114–117.

Department of Commerce, National Oceanic and Atmospheric Administration, National Marine Fisheries Service. 1993. 50 CFR Part 216. Taking and importing marine mammals: Depletion of the coastal-migratory stock of bottlenose dolphins along the Mid-Atlantic Coast. *Federal Register.* 58:17,789–17,791.

Dietz, R., C.T. Hansen, P. Have, and M.-P. Heide-Jorgensen. 1989a. Clue to seal epizootic. *Nature* 338:627.

Dietz, R., M.-P. Heide-Jorgensen, and T. Harkonen. 1989b. Mass deaths of harbor seals (*Phoca vitulina*) in Europe. *Ambio* 18:258–264.

Di Guardo, G., U. Agrimi, D. Amaddeo, M. McAliskey, and S. Kennedy. 1992. Morbillivirus infection in a striped dolphin (*Stenella coeruleoalba*) from the coast of Italy. *Veterinary Record* 130:579–580.

Di Guardo, G., U. Agrimi, L. Morelli, G. Cardeti, G. Terracciano, and S. Kennedy. 1995a. Post-mortem investigations on cetaceans found stranded on the coasts of Italy between 1990 and 1993. *Veterinary Record* 136:439–442.

Di Guardo, G., A. Corradi, U. Agrimi, N. Zizzo, L. Morelli, A. Perillo, C. Kramer, E. Cabassi, and S. Kennedy. 1995b. Neuropathological lesions in cetaceans found stranded from 1991 to 1993 on the coasts of Italy. *European Journal of Veterinary Pathology* 1:47–51.

Domingo, M., L. Ferrer, M. Pumarola, A. Marco, J. Plana, S. Kennedy, M. McAliskey, and B.K. Rima. 1990. Morbillivirus in dolphins. *Nature* 348:21.

Domingo, M., J. Visa, M. Pumarola, A.J. Marco, I. Ferrer, R. Rabanal, and S. Kennedy. 1992. Pathological and immunocytochemical studies of morbillivirus infection in striped dolphins (*Stenella coeruleoalba*). *Veterinary Pathology* 29:1–10.

Domingo, M., M. Vilafranca, J. Visa, N. Prats, A. Trudgett, and I. Visser. 1995. Evidence for chronic morbillivirus infection in the Mediterranean striped dolphin (*Stenella coeruleoalba*). *Veterinary Microbiology* 44:229–239.

Duignan, P.J., J.R. Geraci, J.A. Raga, and N. Calzada. 1992. Pathology of morbillivirus infection in striped dolphins (*Stenella coeruleoalba*) from Valencia and Murcia, Spain. *Canadian Journal of Veterinary Research* 56:242–248.

Duignan, P.J., S. Sadove, J.T. Saliki, and J.R. Geraci. 1993. Phocine distemper in harbor seals (*Phoca vitulina*) from Long Island, New York. *Journal of Wildlife Diseases* 29:465–469.

Duignan, P.J., J.T. Saliki, D.J. St. Aubin, J.A. House, and J.R. Geraci. 1994. Neutralizing antibodies to phocine distemper virus in Atlantic walruses (*Odobenus rosmarus*

rosmarus) from Arctic Canada. *Journal of Wildlife Diseases* 30:90–94.

Duignan, P.J., C. House, J.R. Geraci, N. Duffy, B.K. Rima, M.T. Walsh, G. Early, D.J. St. Aubin, S. Sadove, H. Koopman, and H. Rhinehart. 1995a. Morbillivirus infection in cetaceans of the western Atlantic. *Veterinary Microbiology* 44:241–249.

Duignan, P.J., C. House, J.R. Geraci, G. Early, H.G. Copland, M.T. Walsh, G.D. Bossart, C. Gray, S. Sadove, D.J. St. Aubin, and M. Moore. 1995b. Morbillivirus infection in two species of pilot whales (*Globicephala* sp.) from the western Atlantic. *Marine Mammal Science* 11:150–162.

Duignan, P.J., C. House, M.T. Walsh, T. Campbell, G.D. Bossart, N. Duffy, P.J. Fernandes, B.K. Rima, S. Wright, and J.R. Geraci. 1995c. Morbillivirus infection in manatees. *Marine Mammal Science* 11:441–451.

Duignan, P.J., J.T. Saliki, D.J. St. Aubin, G. Early, S. Sadove, J.A. House, K. Kovacs, and J.R. Geraci. 1995d. Epizootiology of morbillivirus infection in north American harbor seals (*Phoca vitulina*) and gray seals (*Halichoerus grypus*). *Journal of Wildlife Diseases* 31:491–501.

Follmann, E.H., G.W. Garner, J.F. Evermann, and A.J. McKiernan. 1996. Serological evidence of morbillivirus infection in polar bears (*Ursus maritimus*) from Alaska and Russia. *Veterinary Record* 138:615–618.

Forsyth, M.A., S. Kennedy, S. Wilson, T. Eybatov, and T. Barrett. 1998. Canine distemper virus in a Caspian seal. *Veterinary Record* 143:662–664.

Geraci, J.R. 1989. *Clinical investigation of the 1987–88 mass mortality of bottlenose dolphins along the US central and south Atlantic coast: Final report to National Marine Fisheries Service.* US Navy Office of Naval Research, and Marine Mammal Commission. Guelph, Ontario, 63 pp.

Gordon, M.T., D.C. Anderson, and P.T. Sharpe. 1991. Canine distemper virus localized in bone cells of patients with Paget's disease. *Bone* 12:195–201.

Grachev, M.A., V.P. Kumarev, L.V. Mamaev, V.L. Zorin, L.V. Baranova, N.N. Denikina, S.I. Belikov, E.A. Petrov, V.S. Kolesnik, R.S. Kolesnik, V.M. Dorofeev, A.M. Beim, V.N. Kudelin, F.A. Nagieva, and V.N. Sidorov. 1989. Distemper in Baikal seals. *Nature* 338:209–210.

Haas, I., S.M. Subbarao, T. Harder, B. Liess, and T. Barrett. 1991. Detection of phocid distemper virus RNA in seal tissues using slot hybridization and the polymerase chain reaction amplification assay: Genetic evidence that the virus is distinct from canine distemper virus. *Journal of General Virology* 72:825–832.

Hall, A.J., R.J. Law, D.E. Wells, J. Harwood, H.M. Ross, S. Kennedy, C.R. Allchin, C.A. Campbell, and P.P. Pomeroy. 1992. Organochlorine levels in common seals (*Phoca vitulina*) which were victims and survivors of the 1988 phocine distemper epizootic. *Science of the Total Environment* 115:145–162.

Harder, T.C., V. Moennig, I. Greiserwilke, T. Barrett, and B. Liess. 1991. Analysis of antigenic differences between 16 phocine distemper virus isolates and other morbilliviruses. *Archives of Virology* 118:261–268.

Harder, T.C., T. Willhaus, W. Leibold, and B. Liess. 1992. Investigations on course and outcome of phocine distemper virus-infection in harbor seals (*Phoca vitulina*) exposed to polychlorinated biphenyls: Virological and serological investigations. *Journal of Veterinary Medicine Series B* 39:19–31.

Have, P., J. Nielsen, and A. Botner. 1991. The seal death in Danish waters 1988: 2. Virological studies. *Acta Veterinaria Scandinavica* 32:211–219.

Heide-Jorgensen, M.-P., T. Harkonen, R. Dietz, and P.M. Thompson. 1992. Retrospective of the 1988 European seal epizootic. *Diseases of Aquatic Organisms* 13:37–62.

Heje, N.I., P. Henriksen, and B. Aalbaek. 1991. The seal death in Danish waters 1988: 1. Pathological studies. *Acta Veterinaria Scandinavica* 32:205–210.

Henderson, G., A. Trudgett, C. Lyons, and K. Ronald. 1992. Demonstration of antibodies in archival sera from Canadian seals reactive with a European isolate of phocine distemper virus. *Science of the Total Environment* 115:93–98.

Hofmeister, R., E.M. Breuer, R. Ernst, J. Hentschke, G. Molle, and H. Ludwig. 1988. Distemper-like disease in harbor seals: Virus isolation, further pathologic and serologic findings. *Journal of Veterinary Medicine Series B* 35:765–769.

Jauniaux, T., G. Charlier, M. Desmecht, and F. Coignoul. 1998. Lesions of morbillivirus infection in a fin whale (*Balaenoptera physalus*) stranded along the Belgian coast. *Veterinary Record* 143:423–424.

Kennedy, S. 1990. A review of the 1988 European seal morbillivirus epizootic. *Veterinary Record* 127:563–567.

Kennedy, S., J.A. Smyth, P.F. Cush, S.J. McCullough, G.M. Allan, and S. McQuaid. 1988a. Viral distemper now found in porpoises. *Nature* 336:21.

Kennedy, S., J.A. Smyth, S.J. McCullough, G.M. Allan, F. McNeilly, and S. McQuaid. 1988b. Confirmation of cause of recent seal deaths. *Nature* 335:404.

Kennedy, S., J.A. Smyth, P.F. Cush, P. Duignan, M. Platten, S.J. McCullough, and G.M. Allan. 1989. Histopathologic and immunocytochemical studies of distemper in seals. *Veterinary Pathology* 26:97–103.

Kennedy, S., J.A. Smyth, P.F. Cush, M. McAliskey, S.J. McCullough, and B.K. Rima. 1991. Histopathological and immunocytochemical studies of distemper in harbor porpoises. *Veterinary Pathology* 28:1–7.

Kennedy, S., T. Kuiken, H.M. Ross, M. McAliskey, D. Moffett, C.M. McNiven, and M. Carole. 1992. Morbillivirus infection in two common porpoises (*Phocoena phocoena*) from the coasts of England and Scotland. *Veterinary Record* 131:286–290.

Kovamees, J., M. Blixenkrone-Møller, B. Sharma, C. Örvell, and E. Norrby. 1991. The nucleotide sequence and deduced amino acid composition of the hemagglutinin and fusion proteins of the morbillivirus phocid distemper virus. *Journal of General Virology* 72:2959–2966.

Krafft, A., J.H. Lichy, T.P. Lipscomb, B.A. Klaunberg, S. Kennedy, and J.K. Taubenberger. 1995. Postmortem diagnosis of morbillivirus infection in bottle-nosed dolphins (*Tursiops truncatus*) in the Atlantic and Gulf of Mexico epizootics by polymerase chain reaction-based assay. *Journal of Wildlife Diseases* 31:779–785.

Krogsrud, J., O. Evensen, G. Holt, S. Hoie, and N.H. Markussen. 1990. Seal distemper in Norway in 1988 and 1989. *Veterinary Record* 126:460–461.

Lavigne, D.M., and O.J. Schmitz. 1990. Global warming and increasing population densities: A prescription for seal plagues. *Marine Pollution Bulletin* 21:280–284.

Liess, B., H.R. Frey, A. Zaghawa, and M. Stede. 1989. Morbillivirus infection of seals (*Phoca vitulina*) during the 1988 epidemic in the Bay of Heligoland: 1. Mode, frequency and significance of cultural virus isolation and neutralizing antibody detection. *Journal of Veterinary Medicine Series B* 36:601–608.

Lipscomb, T.P., S. Kennedy, D. Moffett, and B.K. Ford. 1994a. Morbilliviral disease in an Atlantic bottle nosed dolphin (*Tursiops truncatus*) from the Gulf of Mexico. *Journal of Wildlife Diseases* 30:572–576.

Lipscomb, T.P., F.Y. Schulman, D. Moffett, and S. Kennedy. 1994b. Morbilliviral disease in Atlantic bottle-nosed dolphins (*Tursiops truncatus*) from the 1987–1988 epizootic. *Journal of Wildlife Diseases* 30:567–571.

Lipscomb, T.P., S. Kennedy, D. Moffett, A. Krafft, B.A. Klaunberg, J.H. Lichy, G.T. Regan, G.A.J. Worthy, and J.K. Taubenberger. 1996. Morbilliviral epizootic in Atlantic bottlenose dolphins (*Tursiops truncatus*) of the Gulf of Mexico. *Journal of Veterinary Diagnostic Investigation* 8:283–290.

Lyons, C., M.J. Welsh, J. Thorsen, K. Ronald, and B.K. Rima. 1993. Canine distemper virus isolated from a captive seal. *Veterinary Record* 132:487–488.

Mahy, B., T. Barrett, S. Evans, E.C. Anderson, and C.J. Bostock. 1988. Characterization of a seal morbillivirus. *Nature* 336:115–116.

Mamaev, L.V., N.N. Denikina, S.I. Belikov, V.E. Volchkov, I.K.G. Visser, M. Fleming, C. Kai, T.C. Harder, B. Liess, A.D.M.E. Osterhaus, and T. Barrett. 1995. Characterisation of morbilliviruses isolated from Lake Baikal seals (*Phoca sibirica*). *Veterinary Microbiology* 44:251–259.

Mamaev, L.V., I.K.G. Visser, S.I. Belikov, N.N. Denikina, T. Harder, L. Goatley, B. Rima, B. Edginton, A.D.M.E. Osterhaus, and T. Barrett. 1996. Canine distemper virus in Lake Baikal seals (*Phoca sibirica*). *Veterinary Record* 138:437–439.

Markussen, N.H., and P. Have. 1992. Phocine distemper virus infection in harp seals (*Phoca groenlandica*). *Marine Mammal Science* 8:19–26.

McCullough, S.J., F. McNeilly, G.M. Allan, S. Kennedy, J.A. Smyth, S.L. Cosby, S. McQuaid, and B.K. Rima. 1991. Isolation and characterization of a porpoise morbillivirus. *Archives of Virology* 118:247–252.

Munro, R., H. Ross, C. Cornwell, and J. Gilmour. 1992. Disease conditions affecting common seals (*Phoca vitulina*) around the Scottish mainland, September–November 1988. *Science of the Total Environment* 115:67–82.

Nunoya, T., M. Tajima, Y. Ishikawa, T. Samejima, H. Ishikawa, and K. Hasegawa. 1990. Occurrence of a canine distemper-like disease in aquarium seals. *Japanese Journal of Veterinary Science* 52:469–477.

Örvell, C., and H. Sheshberadaran. 1991. Phocine distemper virus is phylogenetically related to canine distemper virus. *Veterinary Record* 129:267–269.

Örvell, C., M. Blixenkrone-Møller, V. Svansson, and P. Have. 1990. Immunological relationships between phocid and canine distemper virus studied with monoclonal antibodies. *Journal of General Virology* 71:2085–2092.

Osterhaus, A.D.M.E., and E.J. Vedder. 1988. Identification of virus causing recent seal deaths. *Nature* 335:20.

Osterhaus, A.D.M.E., J. Groen, P. de Vries, F.G.C.M. UytdeHaag, B. Klingeborn, and R. Zarnke. 1988. Canine distemper virus in seals. *Nature* 335:403–404.

Osterhaus, A.D.M.E., J. Groen, F.G.C.M. UytdeHaag, I.K.G. Visser, M.W.G. van de Bildt, A. Bergman, and B. Klingeborn. 1989. Distemper virus in Baikal seals. *Nature* 338:209–210.

Osterhaus, A.D.M.E., I.K.G. Visser, R.L. de Swart, M.F. van Bressem, M.W.G. van de Bildt, C. Örvell, T. Barrett, and J.A. Raga. 1992. Morbillivirus threat to Mediterranean monk seals? *Veterinary Record* 30:141–142.

Osterhaus, A.D.M.E., R.L. de Swart, H.W. Vos, P.S. Ross, M.J.H. Kenter, and T. Barrett. 1995. Morbillivirus infections of aquatic mammals: Newly identified members of the genus. *Veterinary Microbiology* 44:219–227.

Osterhaus, A.D.M.E., J. Groen, H. Niesters, M. van de Bildt, B. Martina, L. Vedder, J. Vos, H. van Egmond, B. Abou Sidi, and M.E.O. Barham. 1997. Morbillivirus in monk seal mass mortality. *Nature* 388:838–839.

Piza, J. 1991. Striped dolphin mortality in the Mediterranean. In *Proceedings of the Mediterranean Striped Dolphin Mortality International Workshop,* ed. X. Pastor and M. Simmonds. Palma de Mallorca, Spain: Greenpeace Mediterranean Sea Project, pp. 93–103.

Pohlmeyer G., J. Pohlenz, and P. Wohlsein. 1993. Intestinal lesions in experimental phocine distemper: Light microscopy, immunohistochemistry and electron microscopy. *Journal of Comparative Pathology* 109:57–69.

Reidarson T.H., J. McBain, C. House, D.P. King, J.L. Stott, A. Krafft, J.K. Taubenberger, J. Heyning, and T.P. Lipscomb. 1998. Morbillivirus infection in stranded common dolphins from the Pacific Ocean. *Journal of Wildlife Diseases* 34:771–776.

Rima B.K., S.L. Cosby, C. Lyons, D. O'Loan, S. Kennedy, S.J. McCullough, J.A. Smyth, and F. McNeilly. 1990. Humoral immune responses in seals infected by phocine distemper virus. *Research in Veterinary Science* 49:114–116.

Rima B.K., M.D. Curran, and S. Kennedy. 1992. Phocine distemper virus, the agent responsible for the 1988 mass mortality of seals. *Science of the Total Environment* 115:45–55.

Rima B.K., R.G.A. Wishaupt, M.J. Welsh, and J.A.P. Earle. 1995. The evolution of morbilliviruses: A comparison of nucleocapsid gene sequences including a porpoise morbillivirus. *Veterinary Microbiology* 44:127–134.

Ross, P.S., I.K.G. Visser, H.W.J. Broeders, M.W.G. van de Bildt, W.D. Bowen, and A.D.M.E. Osterhaus. 1992. Antibodies to phocine distemper virus in Canadian seals. *Veterinary Record* 130:514–516.

Schulman, F.Y., T.P. Lipscomb, D. Moffett, A.E. Krafft, J.H. Lichy, M.M. Tsai, J.K. Taubenberger, and S. Kennedy. 1997. Histologic, immunohistochemical, and polymerase chain reaction studies of bottlenose dolphins from the 1987–1988 United States Atlantic Coast epizootic. *Veterinary Pathology* 34:288–295.

Schumacher, U., H.-P. Horny, G. Heidemann, W. Schultz, and U. Welsch. 1990. Histopathological findings in harbour seals (*Phoca vitulina*) found dead on the German North Sea coast. *Journal of Comparative Pathology* 102:299–309.

Stuen, S., J.M. Arnemo, P. Have, and A.D.M.E. Osterhaus. 1994. Serological investigation of virus infection in harp seals (*Phoca groenlandica*) and hooded seals (*Cystophora cristata*). *Veterinary Record* 134:502–503.

Taubenberger, J.K., M. Tsai, A.E. Krafft, J.H. Lichy, A.H. Reid, F.Y. Schulman, and T.P. Lipscomb. 1996. Two morbilliviruses implicated in bottlenose dolphin epizootics. *Emerging Infectious Diseases* 2:213–216.

Titenko, A.M., T.I. Borisova, A.M. Beim, S.S. Novozhilov, V.N. Kudelin, V.L. Zorin, V.P. Kumarev, and V.S. Kolesnik. 1991. Experimental infection of seal (*Phoca sibirica*) with morbillivirus. *Voprosy Virusologii* 36:511–512.

Trudgett, A., C. Lyons, M.J. Welsh, N. Duffy, S.J. McCullough, and F. McNeilly. 1991. Analysis of a seal and a porpoise morbillivirus using monoclonal antibodies. *Veterinary Record* 128:61.

Tsur, I., O. Goffman, B. Yakobsen, D. Moffett, and S. Kennedy. 1997. Morbillivirus infection in a bottlenose dolphin (*Tursiops truncatus*) from the Mediterranean Sea. *European Journal of Veterinary Pathology* 2:83–85.

Uchida, K., M. Muranaka, Y. Horii, N. Murakami, R. Yamaguchi, and S. Tateyama. 1999. Non-purulent meningoencephalitis of a Pacific striped dolphin (*Lagenorhynchus obliquidens*). *Journal of Veterinary Medical Science* 61:159–162.

Van Bressem, M.F., I.K.G. Visser, M.W.G. van de Bildt, J.S. Teppema, J.A. Raga, and A.D.M.E. Osterhaus. 1991. Morbillivirus infection of Mediterranean striped dolphins (*Stenella coeruleoalba*). *Veterinary Record* 129:471–472.

Van Bressem, M.F., I.K.G. Visser, R.L. de Swart, C. Örvell, L. Stanzani, E. Androukaki, K. Siakavara, and A.D.M.E.

Osterhaus. 1993. Dolphin morbillivirus infection in different parts of the Mediterranean Sea. *Archives of Virology* 129:235–242.

Van Bressem, M.F., P. Jepson, and T. Barrett. 1998a. Further insight on the epidemiology of cetacean morbillivirus in the northeastern Atlantic. *Marine Mammal Science* 14:605–613.

Van Bressem, M.F., K. Van Waerebeek, M. Fleming, and T. Barrett. 1998b. Serological evidence of morbillivirus infection in small cetaceans from the Southeast Pacific. *Veterinary Microbiology* 59:89–98.

van de Bildt, M.W.G., E.J. Vedder, B.E.E. Martina, B.A. Sidi, A.B. Jiddou, M.E.O. Barham, E. Androukaki, A. Komnenou, H.G.M. Neisters, and A.D.M.E. Osterhaus. 1999. Morbilliviruses in Mediterranean monk seals. *Veterinary Microbiology* 69:19–21.

Visser, I.K.G., V.P. Kumarev, C. Örvell, P. de Vries, H.W.J. Broeders, M.W.G. van de Bildt, J. Groen, J.S. Teppema, M.C. Burger, F.G.C.M. UytdeHaag, and A.D.M.E. Osterhaus. 1990. Comparison of two morbilliviruses isolated from seals during outbreaks of distemper in North-West Europe and Siberia. *Archives of Virology* 111:149–164.

Visser, I.K.G., E.J. Vedder, M.W.G. van de Bildt, C. Örvell, T. Barrett, and A.D.M.E Osterhaus. 1992. Canine distemper virus iscoms induce protection in harbor seals (*Phoca vitulina*) against phocid distemper but still allow subsequent infection with phocid distemper virus-1. *Vaccine* 10:435–438.

Visser, I.K.G., M.F. van Bressem, R.L. de Swart, M.W.G. van de Bildt, H.W. Vos, R.W.J. van der Heijden, J.T. Saliki, C. Örvell, P. Kitching, T. Kuiken, T. Barrett, and A.D.M.E. Osterhaus. 1993a. Characterization of morbilliviruses isolated from dolphins and porpoises in Europe. *Journal of General Virology* 74:631–641.

Visser, I.K.G., R.W.J. van der Heijden, M.W.G. van de Bildt, M.J.H. Kenter, C. Örvell, and A.D.M.E. Osterhaus. 1993b. Fusion protein gene nucleotide sequence similarities, shared antigenic sites and phylogenetic analysis suggest that phocid distemper virus type 2 and canine distemper virus belong to the same virus entity. *Journal of General Virology* 74:1989–1994.

Welsh, M.J., C. Lyons, A. Trudgett, B.K. Rima, S.J. McCullough, and C. Örvell. 1992. Characteristics of a cetacean morbillivirus isolated from a porpoise (*Phocoena phocoena*). *Archives of Virology* 125:305–311.

3

BLUETONGUE, EPIZOOTIC HEMORRHAGIC DISEASE, AND OTHER ORBIVIRUS-RELATED DISEASES

ELIZABETH W. HOWERTH, DAVID E. STALLKNECHT, AND PETER D. KIRKLAND

The genus *Orbivirus* in the family *Reoviridae* represents over 120 viral serotypes classified within 14 serogroups (Roy 1996). Most of these double-stranded RNA viruses are vector-borne. Although viruses in several of the orbivirus serogroups have been implicated with human, domestic animal, and wildlife disease, only the bluetongue (BLU) and epizootic hemorrhagic disease (EHD) viruses have been associated with significant disease in wildlife populations.

BLUETONGUE AND EPIZOOTIC HEMORRHAGIC DISEASE VIRAL INFECTIONS

Synonym: Hemorrhagic disease.

Introduction. The diseases caused by both the BLU and EHD viruses are referred to collectively as hemorrhagic disease (HD) (Thomas et al. 1974), for which there are several reviews (Hoff and Hoff 1976; Hoff and Trainer 1978, 1981; Thomas 1981; Gibbs and Greiner 1989). To date, HD has been associated with significant mortality only in North American ungulate populations.

History. The history of HD in North America has been reviewed by Nettles and Stallknecht (1992). Accounts of mortality that are consistent with HD in both white-tailed deer *Odocoileus virginianus* and mule deer *Odocoileus hemionus* in the western United States date back to 1886 and 1901 (Schultz 1979). Unconfirmed HD-like disease and mortality in deer also has been reported in the southeastern (Ruff 1950; Alexander 1954; Trainer 1964; Shope 1967; Nettles and Stallknecht 1992) and the western United States (Shope et al. 1960; Schildman and Hurt 1984) for the period 1908–57. The viral etiology of this disease was confirmed in 1955, when EHD virus serotype 1 was isolated from white-tailed deer in New Jersey (Shope et al. 1960). In 1966, the BLU viruses also were implicated in the etiology of this disease when virus isolations were made from white-tailed deer and desert bighorn sheep *Ovis canadensis* in Texas (Robinson et al. 1967; Stair et al. 1968). Since 1966, HD has been observed annually, especially in white-tailed deer in the southeastern United States (Nettles and Stallknecht 1992). These outbreaks have been associated with both the EHD and BLU viruses.

Distribution. Although the BLU and EHD viruses are widely distributed in temperate and tropical climates worldwide (Gibbs and Greiner 1989), HD in free-ranging wildlife populations has been reported in only the United States and Canada. In the United States, HD has been reported from numerous states generally located in a wide diagonal band extending from the southeastern corner to the northwestern corner of the country, including California (Nettles et al. 1992a,b). In Canada, reports of HD in wild ruminants have been sporadic and have included localized areas in the southern portions of British Columbia, Alberta, and Saskatchewan (Dulac et al. 1992).

Host Range. Hemorrhagic disease primarily affects white-tailed deer and mule deer (Nettles and Stallknecht 1992; Nettles et al. 1992b), but mortality has been reported in pronghorn *Antilocapra americana* during outbreaks in the western United States (Kistner et al. 1975; Thorne et al. 1988). Clinical disease has been reported in desert bighorn sheep with BLU virus, and it was suggested that BLU may have been partially responsible for the disappearance of this species in Texas (Robinson et al. 1967). Other North American ungulates from which the EHD or BLU viruses have been isolated under field, captive, or experimental conditions include bison *Bison bison* (Dulac et al. 1988), elk *Cervus elaphus* (Stott et al. 1982; Jessup 1985; Dulac et al. 1988), bighorn sheep (Robinson et al. 1967; Kistner et al. 1975; Jessup 1985), and mountain goats *Oreamnos americanus* (Dulac et al. 1988). In most of these cases, however, reports of HD-related disease or mortality were not associated with the isolation of these viruses. North American species that have been experimentally infected with EHD viruses but did not develop disease include moose *Alces alces* [Karstad as cited by Hoff and Trainer (1978)], raccoons *Procyon lotor* (Trainer 1964), opossums *Didelphis marsupialis* (Pirtle and Layton 1961), snowshoe hares *Lepus americanus* (Fay et al. 1956), cottontail rabbits *Sylvilagus floridanus* (Fay et al. 1956), red squirrels *Tamiasciurus hudsonicus* (Fay et al. 1956), and snapping turtles *Chelydra serpentina* (Pirtle and Layton 1961).

Clinical disease associated with BLU virus infection has been reported for several ungulate species in zoological collections in North America, including greater kudu *Tragelaphus strepsiceros* and muntjac *Muntiacus reevesi* (Hoff et al. 1973a), and perinatal infections associated with BLU virus serotype 11 have been reported in Grant's gazelle *Gazella granti,* gemsbok *Oryx gazella,* sable antelope *Hippotragus niger,* African buffalo *Syncerus caffer,* ibex *Capra ibex,* hartebeest *Alcelaphus buselaphus,* and addax *Addax nasomaculatus* (Ramsay et al. 1985). One Old World ungulate species, the African buffalo, has been experimentally infected with BLU virus and developed clinical signs (Young 1969). Other species that have been experimentally infected with BLU viruses without disease include blesbok *Damaliscus pygarus* (Neitz 1933) and mountain gazelle *Gazella gazella* (Barzilai and Tadmor 1972).

Although disease has been reported in only one case, antibodies to BLU viruses have been reported from several free-ranging wild ungulate species in Africa, including buffalo, Thomson's gazelle *Gazella thomsonii,* impala *Aepyceros melampus,* hartebeest, eland *Taurotragus oryx,* Grant's gazelle, waterbuck *Kobus ellipsiprymnus,* reedbuck *Redunca fulvorufula,* gemsbok, oribi *Ourebia ourebi,* and wildebeest *Connochaetes taurinus* (Davies and Walker 1974; Simpson 1978, 1979; Hamblin et al. 1990; Alexander et al. 1994). Antibodies to the EHD and BLU viruses have also been reported from deer and antelope exotic to North America, including axis deer *Axis axis,* fallow deer *Dama dama,* sika deer *Cervus nippon,* and blackbuck antelope *Antilope cervicapra* in Texas (Corn et al. 1990). Reported mortality related to naturally acquired BLU virus infection in free-ranging ungulates that are not native to North America is restricted to topi *Damaliscus lunatus* (Wells 1962).

Antibodies to the BLU viruses also have been reported from wild carnivores in Africa, including wild dog *Lycaon pictus,* lion *Panthera leo,* cheetah *Acinonyx jubatus,* spotted hyena *Crocuta crocuta,* and large-spotted genet *Genetta maculata* (Alexander et al. 1994). Since domestic dogs *Canis familiaris* can be infected with the African horse sickness viruses via consumption of contaminated meat (Lubroth 1992), these carnivores may have been infected through predation or scavenging of BLU virus-infected ungulates (Alexander et al. 1994). Bluetongue virus also has been isolated from rodents, the multimammate rat *Mastomys coucha* and the African swamp rat *Otomys irroratus* in South Africa (Du Toit 1955) and white-toothed shrews *Crocidura* spp. in West Africa (Kemp et al. 1974), but the significance of these isolations currently is unknown.

Little is known about wildlife infections with the EHD viruses outside of North America. Old World ungulates that have been experimentally infected with EHD virus without disease include red deer *Cervus elaphus,* roe deer *Capreolus capreolus,* and muntjac *Muntiacus muntjak* (Gibbs and Lawman 1977).

Etiology. Viruses in the BLU and EHD serogroups are 65–80 nm in size, are nonenveloped, and are characterized by a double-capsid structure consisting of an outer capsid or shell and an inner shell, termed the core (Roy 1996). The outer shell is a diffuse layer that obscures the characteristic ring-shaped capsomeres of the core particle (Fig. 3.1) (Monath and Guirakhoo 1996). The genome consists of ten segments of double-stranded RNA, with each segment coding for a major or nonstructural protein. There are seven major proteins (VP1–VP7) and four nonstructural proteins (NS1, NS2, NS3, and NS3A). The outer shell is composed of VP2 and VP5 (GP5), with VP2 representing the major antigenic determinant for serotype-specific neutralization. The inner core is composed of two major proteins (VP3 and VP7) and encloses three minor proteins (VP1, VP4, and VP6) and the double-stranded RNA genome. The major protein on the core's surface is VP7, which represents serogroup-specific antigen. Viral matrices and virus-associated macrotubules formed during replication are composed of NS2 and NS1, respectively (Roy 1996).

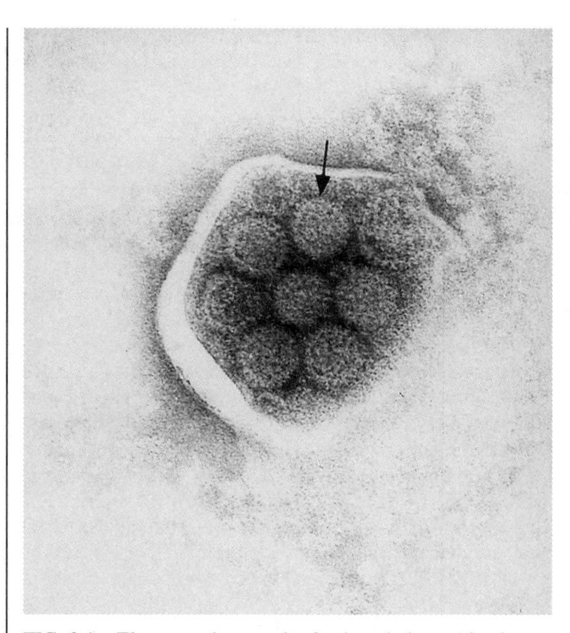

FIG. 3.1—Electron micrograph of epizootic hemorrhagic disease virus. The characteristic ring-shaped capsomeres of the orbiviruses can be seen in some particles (*arrow*). This preparation of virus-infected baby hamster kidney cells was stained with 3% phosphotungstic acid.

There are 24 serotypes (1–24) in the BLU virus serogroup and 10 serotypes (1–9 and Ibaraki) in the EHD virus serogroup worldwide (Monath and Guirakhoo 1996; Roy 1996). However, only BLU virus serotypes 10, 11, 13, and 17, and EHD virus serotypes 1 and 2, have been associated with clinical HD in either experimental or natural infections of wild ungulates (Shope et al. 1960; Thomas et al. 1974; Barber and Jochim 1975; Howerth et al. 1988; Pearson et al. 1992a; Southeastern Cooperative Wildlife Disease Study unpublished).

Transmission and Epidemiology. The BLU and EHD viruses are vectored by various species of *Culicoides* midges (Gibbs and Greiner 1989). Most vector work, to date, has been directed at the BLU viruses, and several reviews are available on this subject (Wirth and Dyce 1985; Gibbs and Greiner 1989; Mellor 1990). Proven *Culicoides* vectors for BLU viruses, worldwide, include *C. imicola* in Africa, *C. variipenis* and *C. insignus* in North America, and *C. fulvus* and *C. actoni* in Australia. Many additional species, however, are regarded as suspect or potential vectors. In North America alone, *C. lahillei, C. stellifer, C. paraenisis, C. obsoletus, C. biguttatus, C. venustus, C. cockerelli, C. freeborni, C. neomontanus, C. hieroglyphicus,* and *C. owhyeensis* have been suggested as possible vectors of the BLU viruses, based on either virus isolations, evidence of virus replication through experimental inoculation, or their observed association with livestock or wildlife in endemic areas.

Much less is known about the vectors of the EHD viruses, but it has been assumed that the list parallels that for the BLU viruses. In North America, EHD virus isolations from insects have been reported only from field-collected *C. variipenis* (Roughton 1975; Jones et al. 1977). The EHD viruses, however, have been shown to replicate in experimentally infected *C. variipennis* (Boorman and Gibbs 1973), *C. venustus* (Jones et al. 1983), and *C. lahillei* (Smith et al. 1996a). Transmission of the EHD viruses to deer has been demonstrated for only *C. variipennis* (Foster et al. 1977; Gibbs and Greiner 1983; Greiner et al. 1985). In field studies, using deer-baited traps, however, *C. variipennis* was not one of the predominant species attacking deer either in endemic areas (Smith et al. 1996b) or during HD epidemics (Smith and Stallknecht 1996), suggesting that other species may be important in EHD virus transmission among white-tailed deer.

Although much remains to be learned regarding the vectors of both the BLU and EHD viruses, the potential for individual species to vector these viruses is extremely difficult to evaluate. Survey techniques can bias field data relating to species abundance, and since animal-based trapping systems are cumbersome and labor intensive, relatively few studies have identified species that commonly feed on wild ungulates in North America (Kramer et al. 1985; Mullen et al. 1985; Gerhardt 1986; Kline and Greiner 1992; Smith and Stallknecht 1996; Smith et al. 1996b). Virus isolation success from field-collected *Culicoides* also may be extremely low, and a very low prevalence of BLU and EHD virus-infected flies has been reported from field-collected *Culicoides* (Greiner et al. 1985, 1989, 1993; Wieser-Schimpf et al. 1993; Mo et al. 1994; Nunamaker et al. 1997). Variation in vector susceptibility also has been reported for both field and experimental studies. Differences in susceptibility to the BLU viruses were reported among species in the *C. variipennis* complex (Tabachnick and Holbrook 1992). In addition, variation in BLU susceptibility also was reported between lab colonies (Jones and Foster 1974; Tabachnick 1991) and field collections (Jones and Foster 1978) of *C. variipennis.*

Hemorrhagic disease occurs in the late summer and early fall (Couvillion et al. 1981), and it has been suggested that this relates to seasonal patterns of vector activity. Variation in both frequency and severity of disease has been reported within the United States (Nettles et al. 1992a). In general, the frequency of HD reports decreases with latitude, whereas the probability that these reports will involve mortality increases. These mortality events, especially those involving large numbers of animals, also are likely to occur sporadically. This pattern includes most of the western United States. In contrast, most reports of the disease in deer in southern parts of the United States do not involve mortality, with numerous animals surviving infection. This is evident by the high antibody prevalence reported from these populations and the detection of HD-related hoof lesions in animals harvested during hunting seasons (Couvillion et al. 1981; Kocan et al. 1982; Stallknecht et al. 1991a). In this "endemic" area, evidence of infection is seen annually or on a 2- to 3-year cycle (Couvillion et al. 1981; Stallknecht et al. 1991a, 1995). In some extreme cases, very high exposure rates with little or no disease are observed. In Texas, antibodies to BLU and EHD viruses have been reported in greater than 80% of surveyed deer (Stallknecht et al. 1996), but HD reports have been infrequent (Nettles et al. 1992a). Differences in antibody prevalence between white-tailed deer populations also have been reported for physiographic regions within the southeastern United States (Stallknecht et al. 1991a) and within individual states (Stallknecht et al. 1995). These differences probably reflect the combined effects of latitude and altitude on climatic variables controlling *Culicoides* populations.

The factors associated with the observed variation in both frequency of reported disease and clinical outcome are not understood but may relate to acquired and innate herd immunity, virulence factors associated with the viruses, and vector species composition and activity patterns. It is interesting that similar unexplained variation in clinical response has been observed in livestock populations in many other areas where these viruses exist (Gibbs 1992).

With regard to acquired herd immunity, it has been reported that previous exposure to a given BLU or EHD serotype will prevent disease associated with homologous virus challenge. This simple relationship, however, is complicated by the diversity of viruses capable of causing HD. It has been shown that multiple serotypes may be involved in HD outbreaks (Thomas et al. 1974) and that these virus serotypes may change in a given area over time (Stott et al. 1981; Stallknecht et al. 1991b, 1995). Differences in the distribution of serotypes over broad geographic areas also have been reported. In the United States, disease associated with the BLU viruses has been reported predominantly from the western states (Johnson 1992). In the southeastern United States, serologic and virus isolation results from white-tailed deer suggest that the EHD viruses, specifically EHD virus serotype 2, predominate (Stallknecht et al. 1995 and unpublished). Potential effects of maternal antibodies also should not be neglected since the 2–3-month persistence of maternal antibodies in white-tailed deer (Hoff et al. 1974) overlaps with the observed seasonality of HD. The potential effects of innate immunity or disease resistance have not been evaluated and currently are supported only by circumstantial evidence.

Specific virulence factors have not been described for the BLU and EHD viruses, and no information suggests that virulence is related to any specific serotype. One concern regarding both virulence and herd immunity relates to genetic recombination. Although this has been reported for the BLU viruses (Roy 1996), the significance of recombination in the epidemiology of HD has not been investigated. Likewise, viral dose, as influenced by variations in vector competence, abundance, and attack rates, also has been suggested as a potential factor affecting clinical outcome (Nettles and Stallknecht 1992), but this has not been further evaluated.

In addition to these many unknowns, there also exists a potential for change. As with many vector-borne diseases, global warming has been suggested as a potential factor affecting the distribution of vectors (Gibbs 1992). Additional problems may be associated with the introduction of exotic BLU and EHD serotypes. Although current information suggests that individual viruses are well adapted to specific vector species and thus to specific geographic areas, this potential problem cannot be dismissed at this time (Gibbs 1992).

Clinical Signs. Clinical signs of BLU and EHD are well described for white-tailed deer and pronghorn and range from sudden death to chronic disease. Infections with these viruses, however, do not always result in disease, and asymptomatic and nonfatal infections may be common. Sick and dead animals often have been found near or in water (Roughton 1975).

Clinical signs in white-tailed deer reported from both natural and experimental BLU and EHD are similar and include hyperemia of the skin and mucous membranes; swelling of the face, conjunctiva, and neck; anorexia; lethargy; weakness and incoordination; excessive salivation, sometimes with blood-tinged saliva; nasal discharge, which can be clear or blood tinged; bloody diarrhea; lameness; hemorrhage at the coronary band; hemorrhages in the skin; hemorrhages on the buccal papillae; hematoma formation at venipuncture sites; conjunctivitis; oral ulcers; necrosis and sloughing of the nasal epithelium; respiratory distress; and recumbency (Pirtle and Layton 1961; Stair et al. 1968; Vosdingh et al. 1968; Thomas and Trainer 1970; Fletch and Karstad 1971; Roughton 1975; Howerth et al. 1988; Fischer et al. 1995). Clinical signs typically progressed from hyperemia to facial and cervical swelling, then lameness followed by hemorrhage, and finally oral ulceration. Icterus has been reported in deer experimentally

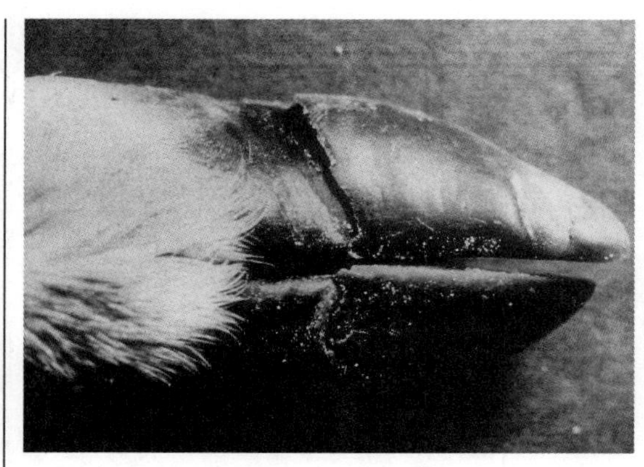

FIG. 3.2—Cracked hooves are a common lesion in animals that have survived hemorrhagic disease.

infected with BLU viruses, most likely due to hemolysis or breakdown of extravasated erythrocytes rather than liver dysfunction (Howerth et al. 1988).

An increased body temperature, generally peaking 5–7 days after inoculation, was reported from experimental studies and paralleled viremia (Fletch and Karstad 1971; Howerth et al. 1988). Subnormal body temperatures also have been observed when death is imminent (Roughton 1975; Quist et al. 1997a). In experimental infections, death occurred 5–15 days after infection (Vosdingh et al. 1968; Fletch and Karstad 1971; Quist et al. 1997a).

Heavy overgrowth of the hooves with an indentation or crack in the wall corresponding to interruption of growth during the period of acute illness has been reported (Stair et al. 1968; Prestwood et al. 1974) (Fig. 3.2). Since the deer hoof grows approximately 0.5 cm per month (Miller et al. 1986), a measurement from the coronary band to this growth interruption can be used to estimate when an animal was initially infected.

As with white-tailed deer, BLU and EHD virus infection in black-tailed deer *Odocoileus hemionus columbianus* and mule deer may be inapparent or result in mortality. In general, however, these species appear to be more resistant to infection with BLU and EHD viruses than are white-tailed deer. Lower mortality in mule deer than white-tailed deer has been reported during EHD outbreaks in areas where these species coexist (Richards 1963; Ditchfield et al. 1964; Hoff et al. 1973b). Black-tailed deer and mule deer did not develop clinical signs when experimentally infected with EHD virus, even though white-tailed deer died when given the same EHD virus inoculum (Pirtle and Layton 1961; Stauber et al. 1977). Similar results were reported for black-tailed deer experimentally infected with BLU virus (Patton et al. 1994). Black-tailed deer naturally infected with BLU and/or EHD viruses were depressed, recumbent, or incoordinated when forced to run (Kistner et al. 1975). Lameness has also been reported in mule deer (Kistner et al. 1975). In addition,

BLU virus has been associated with foot rot in California mule deer (Jessup et al. 1984), and abortion and neonatal deaths have been reported in mule deer infected with BLU virus during the last trimester of pregnancy (Jessup et al. 1984).

Clinical signs in pronghorns infected with the BLU or EHD viruses also have ranged from inapparent to severe disease with death (Jessup et al. 1984; Thorne et al. 1988). Free-ranging pronghorns with BLU were anorexic, incoordinated, reluctant to move when approached, and often recumbent. Some animals without clinical signs died suddenly when they were disturbed (Thorne et al. 1988). Pronghorns experimentally infected with BLU had similar signs, as well as dyspnea by 6 days after inoculation, and some died on postinoculation days 7 and 8 (Hoff and Trainer 1972). Incoordination, convulsions, and "running fits" were described in free-ranging pronghorns with EHD (Richards 1963).

Elk experimentally infected with BLU virus had mild or inapparent clinical signs. Conjunctivitis and diarrhea containing blood and mucus were present by postinoculation days 9 and 10, and the body temperature was transiently elevated between 5 and 9 days after inoculation (Murray and Trainer 1970). A calf from an elk cow experimentally infected with BLU was born weak and latently infected with BLU virus (Stott et al. 1982). Elk experimentally infected with EHD virus did not have clinical signs other than a slight febrile response (Hoff and Trainer 1973).

A desert bighorn sheep naturally with BLU was found weak and staggering, and eventually became recumbent with clear froth in its nostrils (Robinson et al. 1967). Bloody diarrhea was observed in Rocky Mountain bighorn sheep in Colorado and South Dakota with EHD serotype 2 infections (E.S. Williams personal communication). A captive yak *Bos grunniens* with hemorrhagic disease, unconfirmed by virus isolation, was depressed and anorectic, and had bloody feces, hemorrhagic sclera, and sanguineous ocular secretions (Griner and Nelson 1970).

Pathogenesis. Bluetongue viruses and EHD viruses are transmitted to susceptible hosts by *Culicoides* spp. (Monath and Guirakhoo 1996). Studies in cattle *Bos taurus* experimentally infected with BLU virus indicate that initial viral replication probably occurs within macrophages in regional lymph nodes draining the bite site (MacLachlan et al. 1990; Barratt-Boyes and MacLachlan 1994), and that virus then enters the circulation via infected efferent lymph mononuclear cells, causing a low-titer viremia (Barratt-Boyes and MacLachlan 1994). In cattle experimentally infected with BLU virus, the first blood cells from which virus can be isolated are peripheral blood mononuclear cells (PBMCs), and it is most likely these cells that carry virus to secondary sites of replication, principally the spleen (MacLachlan et al. 1990; Barratt-Boyes and MacLachlan 1994). Bluetongue virus released from secondary sites of replication caused a high-titer secondary viremia in experimentally infected cattle which was highly cell associated and apparently disseminated virus to other tissues (Barratt-Boyes and MacLachlan 1994) where it principally infected endothelial cells (MacLachlan et al. 1990). This secondary viremia was nonselectively associated with many cell types in the blood but was primarily associated with red blood cells and platelets (Barratt-Boyes and MacLachlan 1994). Secondary viremia was often prolonged, up to 42 days after inoculation, in cattle experimentally infected with BLU virus, and virus cocirculated with neutralizing antibodies (Barratt-Boyes and MacLachlan 1994). In vitro, BLU virus particles become sequestered in invaginations in cattle red blood cell membranes, and prolonged viremia in cattle has been attributed to this virus-cell association that apparently protects virus from neutralizing antibodies (Brewer and MacLachlan 1992). Protein VP2 has been shown to be solely responsible for the attachment of BLU virus serotype 10 to bovine erythrocytes (Brewer and MacLachlan 1994).

The pathogenesis of these viruses is not as well delineated for wildlife species but is probably similar to that in cattle. In experimental infections of white-tailed deer (Quist et al. 1997a) and elk (Murray and Trainer 1970; Hoff and Trainer 1973) with BLU or EHD viruses and pronghorn (Hoff and Trainer 1972) with BLU virus, viremia generally peaked at day 6 or 7 after infection. Replication of BLU and EHD viruses has been demonstrated in white-tailed deer PBMCs in vitro (Stallknecht et al. 1997) and, as with cattle (Whetter et al. 1989; Barratt-Boyes et al. 1992), replication was primarily associated with monocytes. Virus was isolated from PBMCs from white-tailed deer experimentally infected with EHD virus for a short period during peak viremia (Stallknecht et al. 1997). However, following the detection of precipitating and neutralizing antibodies, virus isolations and the highest viral titers were associated with whole blood and were primarily associated with the erythrocytes (Stallknecht et al. 1997). Periods of viremia for as long as 56 days after inoculation have been reported from white-tailed deer experimentally infected with EHD virus and, in these cases, virus cocirculated with neutralizing antibodies (Quist et al. 1997a). Similar to findings in cattle (Brewer and MacLachlan 1992), both BLU and EHD viruses have been reported to sequester in invaginations in white-tailed deer red blood cell membranes, which may protect virus from neutralizing antibody (Stallknecht et al. 1997). Similar findings were seen in elk experimentally infected with EHD virus where virus was isolated from the white blood cell fraction of blood only during peak viremia, days 4–8 after inoculation but continued to be isolated from the red blood cell fraction up to 30 days after inoculation, despite the development of neutralizing antibodies (Hoff and Trainer 1973).

Both BLU and EHD viruses have been shown to replicate in endothelium in white-tailed deer (Tsai and Karstad 1973; Howerth and Tyler 1988). It has been suggested that the variation in susceptibility to BLU and EHD viruses among the various ruminant species probably reflects the extent of infection and subsequent virus-mediated destruction of the vascular endothelium (MacLachlan et al. 1990). In species that develop severe clinical disease, such as white-tailed deer, viral replication in endothelial cells causes damage that activates the coagulation system, leading to thrombosis (Howerth and Tyler 1988; Howerth et al. 1988). Consumptive coagulation with concurrent fibrinolysis can ensue, culminating in disseminated intravascular coagulation (Howerth et al. 1988). The combination of virus-induced vascular damage and disseminated intravascular coagulation is probably responsible for the hemorrhagic syndrome seen with these viruses (Howerth and Tyler 1988; Howerth et al. 1988). Vascular damage and thrombosis also result in local tissue necrosis, which is typically manifested as muscle necrosis in heart, tongue, abomasum, and skeletal muscle, ulceration of mucosal surfaces, and necrosis in other organs, such as salivary gland and kidney.

Pathology. Leukopenia and lymphopenia, which were often severe, were consistent findings in white-tailed deer during the first week following experimental infection with BLU virus or EHD virus (Howerth et al. 1988; Quist et al. 1997a). Leukocyte counts generally returned to normal by 10 days after inoculation, at which time the lymphocyte counts also rebounded. White-tailed deer experimentally infected with EHD virus that had extremely low lymphocyte counts, generally less than 1000/ml, often had severe clinical signs and died (Quist et al. 1997a). Severely affected white-tailed deer that survived experimental EHD virus infection had decreased hematocrits for as long as 4 weeks after infection (Quist et al. 1997a). Total plasma protein concentration progressively decreased during the first week of experimental BLU virus infection in white-tailed deer (Howerth et al. 1988).

Activation of the coagulation and fibrinolytic systems, which results in disseminated intravascular coagulation, has been demonstrated in both experimental BLU virus and EHD virus infections in white-tailed deer (Debbie et al. 1965; Debbie and Abelseth 1971; Howerth et al. 1988, 1995). Severe prolongation of the activated partial thromboplastin time and a progressive decrease in factor VIII and XII activities developed during the first week of experimental BLU virus and EHD virus infection (Howerth et al. 1988, 1995). One-stage prothrombin time was typically normal or only slightly prolonged, and thrombin time was normal or prolonged during this period (Debbie et al. 1965; Debbie and Abelseth 1971; Howerth et al. 1988, 1995). Factor V activity increased during the first few days of experimental BLU virus infection, probably related to an acute-phase response, but decreased, sometimes below preinfection activity, as the disease progressed probably due to consumptive coagulopathy (Howerth et al. 1988). Fibrinogen concentration was increased, sometimes massively, by day 10 after inoculation and returned to baseline by day 18 following experimental infection with BLU virus and EHD virus (Howerth et al. 1995). Platelet numbers progressively decreased early in experimental BLU virus and EHD virus infection, but deer did not always become thrombocytopenic (Debbie and Abelseth 1971; Howerth et al. 1988, 1995). Prolongation of the Russell viper venom time (Debbie et al. 1965; Debbie and Abelseth 1971) and whole blood clotting time (Debbie et al. 1965) occurred during the 11 days of experimental EHD virus infection. Plasminogen activity decreased and circulating fibrinogen degradation products (FDPs) increased during the first week of BLU and EHD virus infections, indicating fibrinolytic activity (Howerth et al. 1988, 1995). Although abnormalities in coagulation tests and platelet counts returned to baseline by day 18 after inoculation, FDPs were continually detected during recovery, suggesting ongoing lysis of previously formed thrombi (Howerth et al. 1995). Changes in the kallikrein-kinin system, including decreased activity of high molecular weight kininogen and prekallikrein, also occurred in white-tailed deer experimentally infected with BLU or EHD viruses (Howerth et al. 1995).

Hyperbilirubinemia, primarily due to unconjugated bilirubin, was seen in the acute phase of the disease in white-tailed deer experimentally infected with BLU virus and was most likely due to hemolysis or breakdown of extravasated erythrocytes. There were also mild to severe increases in aspartate transaminase and creatine kinase, probably due to skeletal and cardiac muscle necrosis (Howerth et al. 1988).

In elk experimentally infected with BLU virus, leukopenia occurred during peak viremia (Murray and Trainer 1970). Bighorn sheep experimentally infected with BLU virus had decreased platelet counts, but they did not become thrombocytopenic (Robinson et al. 1974). Biphasic leukopenia occurred in mountain gazelle during the first 2 weeks following experimental infection with BLU virus, but clinical disease was not seen (Barzilai and Tadmor 1972).

Gross lesions in natural and experimental BLU and EHD are similar and indistinguishable in white-tailed deer. Lesions are reported to progress with increasing duration of disease from edema, to hemorrhagic diathesis, to ulceration, but deer often have a combination of these lesions. Early in disease, mild to severe pleural, peritoneal, and pericardial effusions commonly are seen in BLU and EHD. Pulmonary edema (Fig. 3.3) and edema in the subcutis, particularly of the head and neck, and along fascial planes (Fay et al. 1956; Shope et al. 1960; Pirtle and Layton 1961; Prestwood et al. 1974; Roughton 1975; Howerth et al. 1988; Fischer et al. 1995) are common. Edema also has been noted in the omentum, mesenteries, joints, and cranium (Roughton 1975).

Petechial, ecchymotic, and suffusive hemorrhages are observed in many tissues as the disease progresses (Fay et al. 1956; Stair et al. 1968; Thomas and Trainer 1970; Fletch and Karstad 1971; Prestwood et al. 1974; Roughton 1975; Howerth et al. 1988). Hemorrhages

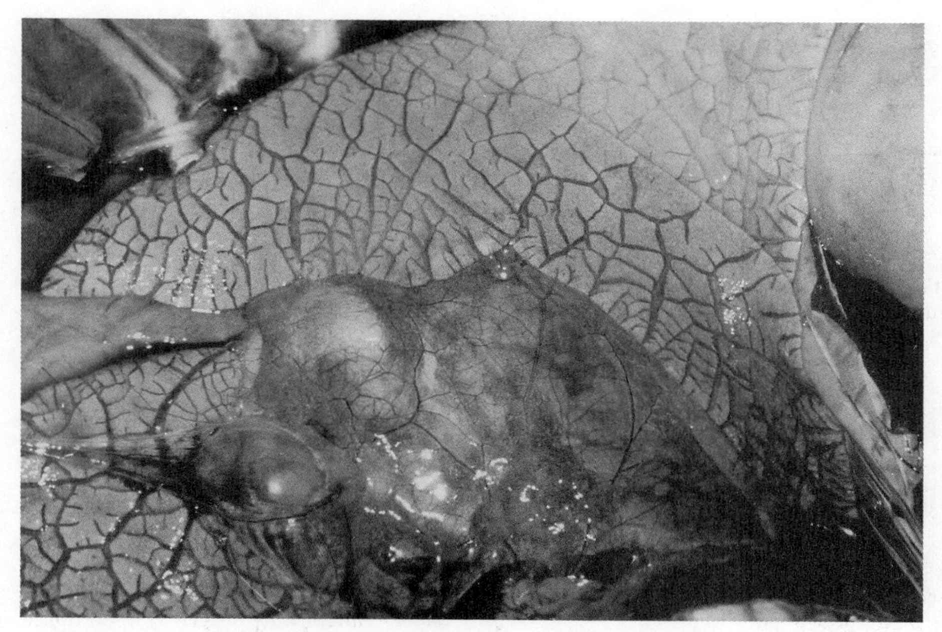

FIG. 3.3—Pulmonary edema is an early finding in white-tailed deer with hemorrhagic disease.

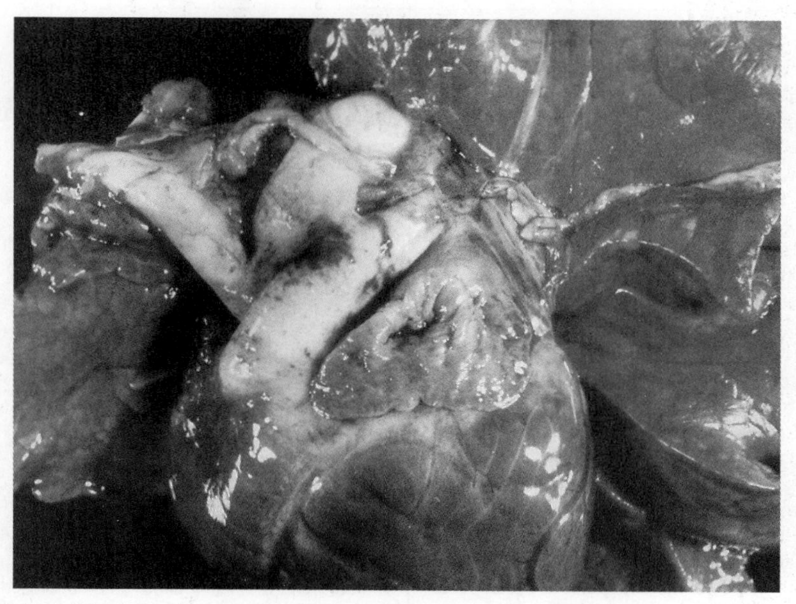

FIG. 3.4—Multiple hemorrhages on the pulmonary artery from a white-tailed deer with hemorrhagic disease. This is a common, if not pathognomonic, lesion.

most frequently involve the gastrointestinal tract (including oral mucosa, abomasum, forestomachs, and intestine) and heart (including epicardium, myocardium, and endocardium). Testicular hemorrhage is common, and hemorrhages have been seen in diaphragm, kidney, urinary bladder, fascia, subcutis, lymph nodes, lung, pleura, peritoneum, thymus, salivary gland, and adrenal. Hemorrhage in the tunica media of the pulmonary artery just distal to the pulmonary valve, which is visible from both the adventitial and intimal surfaces, is a classic, if not pathognomonic, lesion (Fig. 3.4), but hemorrhage over the serosal surface of the pyloric region of the abomasum probably is more commonly seen (Fig. 3.5).

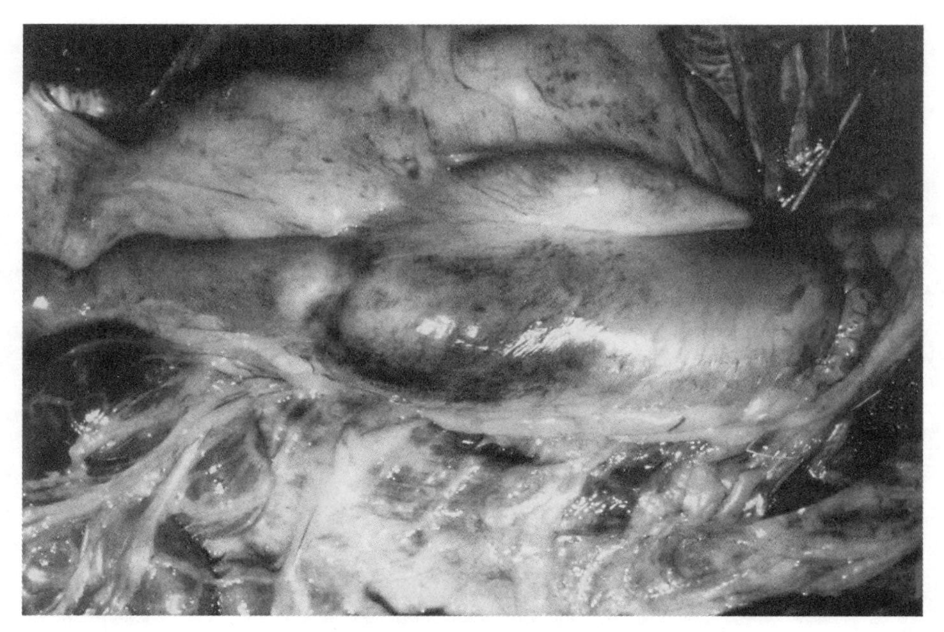

FIG. 3.5—Hemorrhage on the serosa of the pyloric region of the abomasum is common in white-tailed deer with hemorrhagic disease. Note the hemorrhages in the adjacent omentum.

Erosion and ulceration of the buccal papillae, gingiva, palate, tongue, frenulum, forestomachs, and abomasum eventually develops if the animal lives long enough (Stair et al. 1968; Fletch and Karstad 1971; Prestwood et al. 1974; Roughton 1975; Howerth et al. 1988). Buccal papillae, particularly the tips, become hyperemic, hemorrhagic and eventually ulcerated as the disease progresses. These changes are particularly severe over the cheek teeth where friction would be most severe (Fig. 3.6). The tongue is sometimes swollen, but, if the animal lives long enough, necrosis and ulceration may occur. Severe hyperemia of the mucosa of the forestomachs may progress to severe hemorrhage, necrosis, and eventually to ulceration (Fig. 3.7). Rumen hemorrhage involves both the papillae and pillars, and the papillae may mineralize and clump together when hemorrhage and necrosis are extensive (E.W. Howerth unpublished); ruminal contents are sometimes dark brown or red due to blood (Roughton 1975). The abomasal pylorus sometimes has transmural necrosis and mucosal ulceration. Hemorrhagic enteritis and typhlitis with bloody intestinal contents have been observed (Stair et al. 1968; Fischer et al. 1995). Feces range from normal to diarrhetic (Prestwood et al. 1974). The lymph nodes are congested and may ooze serosanguineous fluid (Stair et al. 1968), and splenomegaly may occur (Fischer et al. 1995). Corneal opacities have been described involving one or both eyes (Richards 1963). Foot lesions may range from coronitis and laminitis to complete sloughing of the hoof (Prestwood et al. 1974). Hemorrhage

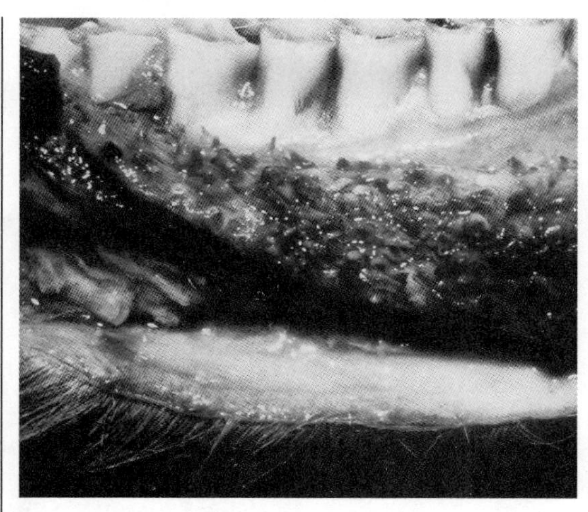

FIG. 3.6—Hemorrhage and necrosis of the buccal papillae is a common finding in white-tailed deer as the disease progresses.

also has been noted at the coronary band (Stair et al. 1968; Fletch and Karstad 1971).

Rumen scarring is not uncommon in animals that survive infection with BLU or EHD viruses (Fay et al. 1956; Prestwood et al. 1974). Cracked or sloughing hooves also may be seen during winter and spring in animals that survive BLU or EHD (Fig. 3.2) (Roughton 1975; Couvillion et al. 1981). Secondary abscessation

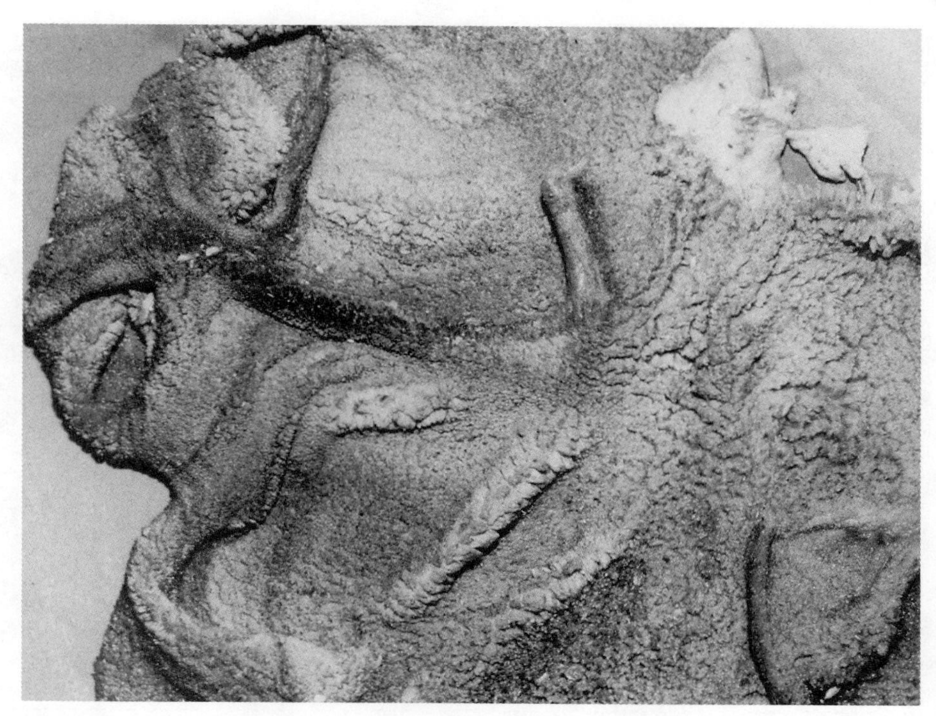

FIG. 3.7—Hemorrhage is common in the forestomachs of white-tailed deer with hemorrhagic disease and may involve both pillars and papillae as seen here.

of the foot sometimes occurs, and animals with foot lesions may be emaciated (Prestwood et al. 1974; Fischer et al. 1995). Deaths also may occur in winter and spring due to complications, such as secondary fungal infection and bacterial abscessation (Roughton 1975).

Black-tailed and mule deer with spontaneous BLU had lymphadenopathy, pleural effusion, cardiac hemorrhage, hemorrhages in the rumen subserosa and papillae, submucosal and subserosal abomasal hemorrhages, and joint effusions (Kistner et al. 1975). Pulmonary edema and areas of oral necrosis were also observed in black-tailed deer naturally infected with BLU virus (Jessup et al. 1984). Mule deer experimentally infected with EHD virus had minimal hemorrhagic lesions and hydropericardium (Pirtle and Layton 1961).

Common findings in pronghorns with naturally occurring BLU were pericardial, subepicardial, and subendocardial hemorrhage, and edema and hemorrhage in the tunica adventitia of the pulmonary artery and dorsal aorta (Thorne et al. 1988). Other common findings included hydrothorax, with fluid that clotted on exposure to air, severe pulmonary edema, edema and/or hemorrhages in the intercartilagenous ligament of the trachea, and subcutaneous tissues in the lower neck, brisket, and hind legs. Hemorrhage occasionally was seen in the lung and subintimally in the pulmonary artery. Hemorrhage was often present in the gastrointestinal tract, particularly on the serosal surface of the rumen and cecum and on the mucosal surface of the abomasum, and was occasionally observed in lymph nodes, urinary bladder, synovial surfaces of joints, conjunctiva, muscles, renal pelvis, thymus, testicle, and under peritoneum and pleura. In some pronghorn, the skeletal muscles were pale and there was splenomegaly. In contrast to white-tailed deer, oral and foot lesions were not observed in pronghorns. Corneal opacity was rarely observed. Pneumonia and hemorrhages were reported in pronghorns experimentally infected with BLU virus (Hoff and Trainer 1972).

Bighorn sheep naturally infected with BLU virus had edema and hemorrhage, particularly intra- and intermuscularly, and blood-tinged pericardial, thoracic, and peritoneal fluid (Kistner et al. 1975). Acute secondary bacterial pneumonia was reported in bighorn sheep from Texas that were naturally infected with BLU virus (Robinson et al. 1967). Naturally infected bighorn sheep with EHD virus had marked congestion and hemorrhage in the forestomachs and small intestines and minimal pulmonary edema (E.S. Williams personal communication).

Captive exotic ruminants, including hartebeest, greater kudu, okapi *Okapia johnstoni*, Reeve's muntjac, and Formosan sika deer, had gross lesions characterized by ecchymotic hemorrhages and transudation during an epidemic of BLU (Hoff et al. 1973a). Captive axis deer, yak, sable antelope, nilgai *Boselaphus tragocamelus,* North

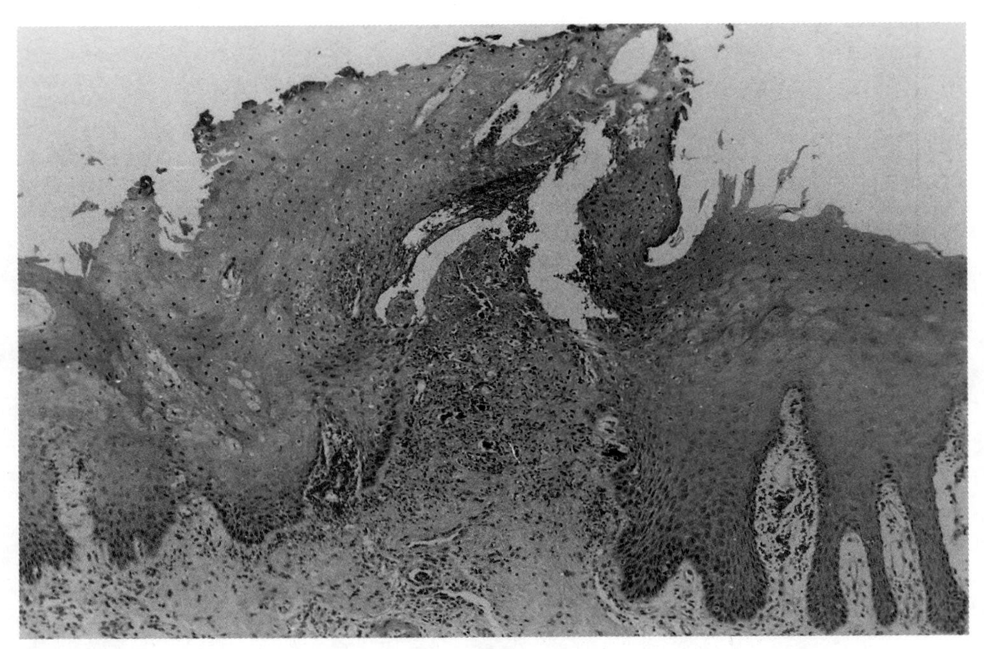

FIG. 3.8—This photomicrograph of tongue from a white-tailed deer has necrosis and inflammation in the lamina propria with necrosis and sloughing of the overlying epithelium, a common lesion. Hematoxylin-eosin stain.

Somali bushbuck *Tragelaphus scriptus,* and European mouflon *Ovis orientalis* also died during an epidemic that was suspected to be hemorrhagic disease but not confirmed by virus isolation (Griner and Nelson 1970). These animals had extensive ecchymosis and petechiation on serosal and mucosal membranes and body muscles, transudation in pericardial, pleural, and peritoneal cavities, swollen pulpy kidneys, swollen livers, and pulmonary congestion and edema.

Congestion, hemorrhage, thrombosis, and necrosis are the primary microscopic findings in white-tailed deer with either BLU or EHD (Shope et al. 1960; Karstad and Trainer 1967; Stair et al. 1968; Fletch and Karstad 1971), but the lesions can be very subtle if animals die early in the course of the disease (E.W. Howerth unpublished). The digestive tract and heart are typically involved, as well as kidneys, major blood vessels, adrenal, thymus, lung, spleen, and lymph nodes. Small vessels underlying the epithelium of the buccal cavity, tongue, and forestomachs initially are lined by swollen endothelium, sometimes with a very mild perivascular infiltration of lymphocytes. Vessels subsequently become dilated, congested, and thrombosed, leading to perivascular hemorrhage and sometimes neutrophilic infiltration and parenchymal necrosis. Changes in the overlying epithelium include spongiosis, leukocyte infiltration, and necrosis, eventually resulting in epithelial sloughing and ulceration (Fig. 3.8). Changes often are striking in the buccal and ruminal papillae where severe hemorrhage and epithelial necrosis may be accompanied by mineralization (E.W. Howerth unpublished). Thrombosis of larger vessels in the muscle of the tongue was associated with hemorrhage and muscle necrosis, and sometimes mineralization (E.W. Howerth unpublished) (Fig. 3.9). Similar thrombosis, hemorrhage, necrosis, and sometimes mineralization can be seen in the myocardium and in the tunica muscularis of the abomasum at the pylorus (E.W. Howerth unpublished). Thrombosis and foci of necrosis can be seen in salivary glands (Stair et al. 1968; Fletch and Karstad 1971). Thrombosis and hemorrhage have been seen in the intestinal wall (Fletch and Karstad 1971), as well as segmental intestinal necrosis due to infarction (Karstad and Trainer 1967). Proximal tubular necrosis and fibrin thrombi in glomeruli have been described in kidney. Thrombosis of capillaries in the tunica media of aorta and pulmonary artery is accompanied by hemorrhage. Congestion, edema, and hemorrhages may be seen in the lung in uncomplicated cases, but these changes can be obscured by inflammation in deer with secondary bacterial pneumonia. There may be thrombosis and hemorrhage in the skin at the coronary band (Fletch and Karstad 1971).

Hemorrhage, lymphoid necrosis with subsequent lymphoid depletion, and erythrophagocytosis and hemosiderosis are seen in spleen and lymph node (Shope et al. 1960; Karstad and Trainer 1967; Stair et al. 1968; E.W. Howerth unpublished). Megakaryocyte hyperplasia with megakaryocyte numbers greater than five times normal has been observed in bone marrow (Howerth et al. 1988).

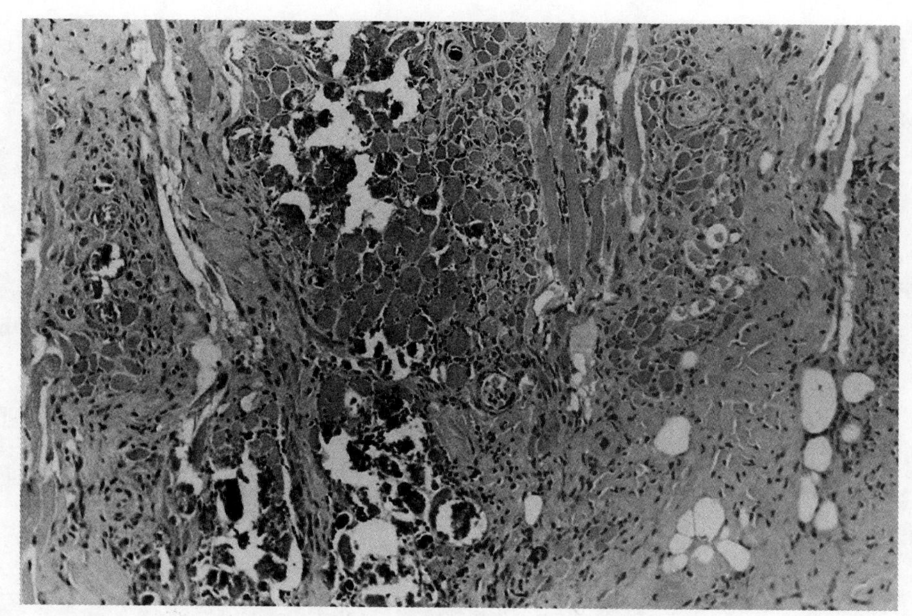

FIG. 3.9—This photomicrograph of tongue from a white-tailed deer has necrosis and mineralization of the musculature, a frequent finding in white-tailed deer with hemorrhagic disease. Hematoxylin-eosin stain.

Additional, less common, lesions that have been described are fibrin deposition in vessels of the central nervous system (Stair et al. 1968), lymphocytic infiltration of interstitium of kidney and adrenal (Stair et al. 1968), centralobular hepatocellular necrosis (Fischer et al. 1995), nonsuppurative meningoencephalitis (E.S. Williams personal communication), interstitial pneumonia, and lymphocytic vasculitis (Fischer et al. 1995).

In pronghorns with BLU, hemorrhages were seen in the heart, abomasum, adrenal cortex, and kidney. Thromboses and fibrinous necrosis of the tunica media of small arteries and arterioles occurred in the digestive tract and perilymphoid connective tissue. There was edema with occasional hemorrhages in the wall of the trachea and bronchi. Eosinophilic exudate filled alveolar spaces, bronchioles, and bronchi; occasionally, fibrin and a few macrophages were present in airspaces. Spleen and lymph nodes had lymphoid depletion. Congestion and focal necrosis with vacuolation of the neuropil and swollen axons were seen in cerebellum, cerebral cortex, ventral brainstem, and thalamus (Thorne et al. 1988).

Bighorn sheep with BLU had hemosiderosis in spleen and lymph node, perivascular hemorrhage in the brainstem, and secondary purulent bacterial pneumonia (Robinson et al. 1967). Hemorrhage in the digestive tract was observed in bighorn sheep with EHD (E.S. Williams personal communication).

Ultrastructural studies have demonstrated that early in infection both BLU and EHD viruses replicate in the cytoplasm of endothelial cells of small vessels (capillaries, arterioles, and venules) of white-tailed deer, with the formation of viral matrices, virus-associated macrotubules, and aggregates of virus particles (Tsai and Karstad 1973; Howerth and Tyler 1988). Subsequently, virus replication was demonstrated in pericytes and vascular smooth muscle cells in BLU-infected deer (Howerth and Tyler 1988). Endothelial cell hypertrophy occurred, as well as endothelial cell degeneration and necrosis, which resulted in denudation of the endothelial lining. Activated platelets and fibrin thrombi were seen in damaged vessels, primarily in areas of endothelial denudation (Fig. 3.10). Vascular damage and thrombosis were associated with vessel rupture, perivascular hemorrhage, and perivascular accumulation of inflammatory cells (Fig. 3.10) (Howerth and Tyler 1988).

Epizootic hemorrhagic disease virus particles were described in platelets of experimentally infected white-tailed deer (Tsai and Karstad 1973) but not in white-tailed deer experimentally infected with BLU virus (Howerth and Tyler 1988). Viral morphogenesis was not demonstrated in bone marrow in white-tailed deer experimentally infected with BLU virus, but megakaryocytes demonstrated massive platelet shedding during the acute phase of the disease and eventually appeared spent, changes attributed to stimulated thrombopoiesis caused by consumption of platelets during thrombosis (Howerth and Tyler 1988). Virus morphogenesis was not demonstrated in epithelium of mucosal surfaces in white-tailed deer experimentally infected with BLU virus, and epithelial changes, characterized by intercellular edema, were considered secondary to vascular changes in the underlying tissues (Howerth and Tyler 1988).

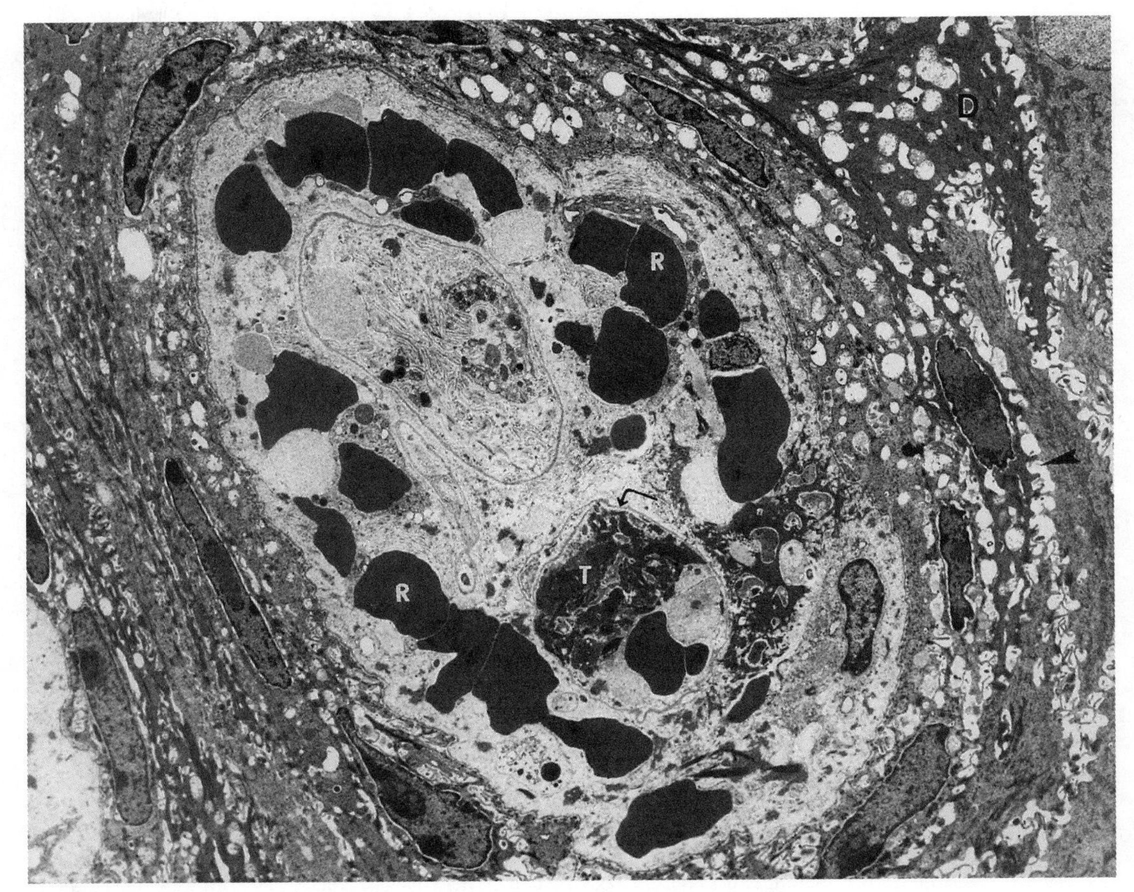

FIG. 3.10—In this electron micrograph of tongue, viral replication has resulted in endothelial necrosis and loss in a capillary (*arrow*). Note the denuded capillary basement membrane and fibrin thrombus (T). Extravasated red blood cells are present in the lamina propria (R). The overlying epithelium has widening of the intercellular spaces (*arrowhead*) and dark degenerate cells (D).

Similar to cattle erythrocytes infected with BLU virus in vitro (Brewer and MacLachlan 1994), virus particles of EHD virus and BLU virus also became sequestered in pits in the white-tailed deer erythrocytic membrane (Stallknecht et al. 1997). Viral replication does not proceed past adherence in bovine erythrocytes (Brewer and MacLachlan 1994) and probably does not in white-tailed deer (Stallknecht et al. 1997). However, this intimate virus-erythrocytic association apparently protects adhered virus particles from neutralizing antibody and is thought to be responsible for the prolonged viremia seen in these viral infections.

Diagnosis. Several serologic tests for antibodies to the BLU and EHD viruses are available, including agar gel immunodiffusion (AGID), serum neutralization (SN) applied to both serum (Pearson et al. 1992b) and blood collected on filter strips (Stallknecht and Davidson 1992), and competitive enzyme-linked immunosorbent assays (cELISAs). All of these techniques can be used effectively in both diagnostic and epidemiologic studies, provided that their advantages and limitations are understood.

The BLU and EHD AGID tests (Pearson and Jochim 1979) have the advantages of simplicity, commercial availability, and application to both the EHD and BLU viruses. These are the only serologic tests for BLU and EHD virus antibodies that can be done with little or no laboratory equipment. The major limitation of these tests involves specificity. Although sensitive, problems with cross-reactions have been reported at the serogroup level (Jochim 1985). Even though seropositive results on either the EHD or BLU AGID give a reliable indication that the animal was exposed to one or more of these viruses, estimates of specific serogroup exposure are inappropriate based on AGID results alone (Stallknecht et al. 1991a).

The SN test is the only serologic test that can provide any indication of previous exposure to a specific BLU or EHD virus serotype. The disadvantages of these tests relate to cross-reactions between serotypes within the serogroup, the need for a laboratory equipped for cell culture-based assays, and the need to run individual tests against all indigenous BLU and EHD virus serotypes. Because of potential cross-reactions at the serotype level, several techniques to aid in the interpretation of SN data are available (Taylor et al. 1985; Stallknecht et al. 1991b, 1996). These techniques have application primarily to population-based studies rather than to diagnostics involving individual animals.

Several ELISAs have been described, especially for BLU viruses. The cELISA, however, is the only ELISA routinely used for serologic testing for antibodies to BLU (Pearson et al. 1992b). Although a cELISA for EHD antibody testing has been developed (Afshar et al. 1997), it is not commercially available at this time. Improved sensitivity has been reported with the BLU cELISA when compared with the AGID (Gustafson et al. 1992), but in most cases these improvements have been negligible. With regard to specificity at the serogroup level, however, these tests represent a major improvement. ELISAs can be adapted to automated systems, results are determined objectively, and the system is very compatible with testing large numbers of samples. cELISAs, however, require more laboratory support than AGID tests and, like AGID results, must be backed by SN tests if information on specific EHD virus or BLU virus serotypes is desired.

It must be emphasized that positive serologic results when used in a clinical setting do not provide confirmatory evidence of infection with the BLU or EHD viruses unless applied to paired sera and the detection of seroconversion. Evidence suggests that these antibodies last for many years in white-tailed deer (Stallknecht et al. 1991a) and elk (E.S. Williams personal communication), and it is apparent from field studies that many healthy animals are seropositive to these viruses.

Virus isolation is the only means by which HD can be confirmed. Both the BLU and EHD viruses can be propagated in embryonated chicken eggs and several cell lines, and these techniques have been described and reviewed by Pearson et al. (1992b). Samples that should be submitted for virus isolation include blood collected in an anticoagulant, such as EDTA or heparin, and spleen. Other tissues that are suitable include lymph node, lung, and bone marrow. All samples should be refrigerated and not frozen. For sample preparation, it is important to lyse red blood cells, since a close association between these viruses and the erythrocyte membrane has been reported for both the BLU (Brewer and MacLachlan 1992) and EHD viruses (Stallknecht et al. 1997). For egg inoculation, an intravenous route is used in 11–13-day-old embryonated chicken eggs. This represents the most sensitive means for isolating BLU viruses. Several tissue culture systems have been used for isolation of EHD and BLU viruses (Wechsler and McHolland 1988), but baby hamster kidney and/or cattle pulmonary artery endothelial (CPAE) cells are most often used. Although CPAE cells are very sensitive for isolation of these viruses, this cell line is persistently infected with bovine viral diarrhea virus, which may influence their use in some laboratories. Viruses can be identified to BLU or EHD serogroup by indirect fluorescent antibody test and to serotype by plaque neutralization (Jochim and Jones 1976) or, in a modification of this technique, in a microtiter-plate format (Quist et al. 1997a).

Several molecular techniques have been applied to BLU and EHD diagnostics, including dot blot (Gonzalez and Knudson 1988), in situ hybridization (Venter et al. 1993; Brown et al. 1996) and polymerase chain reaction (PCR) techniques (Harding et al. 1996; Shad et al. 1997). Of these, reverse transcriptase PCR has shown the most promise, and several techniques have been described. Polymerase chain reaction techniques are available at both the serotype level and the serogroup level. Infections can be confirmed much quicker with PCR than with virus isolation, and the system can be applied to samples that would be unsuitable for virus isolation. Potential problems with these systems include the possibility for contamination, leading to false-positive results, and the need for laboratory equipment support. Another potential problem that has not been evaluated in deer involves the persistence of viral RNA in the host. Positive PCR results could be detected in cattle for up to 160 days and in sheep *Ovis aries* for up to 89 days after infection, long after the virus could be demonstrated by virus isolation, by animal inoculation, or by xenodiagnostic techniques using *Culicoides* (Katz et al. 1994). The biologic significance of this viral RNA persistence is unknown.

Antigen-capture and identification techniques described for these viruses include dot-blot antigen detection, capture ELISA, and immune electron microscopy. These techniques have not been routinely applied to clinical samples but may have application to vector studies (Nunamaker et al. 1997). Bluetongue virus has been identified in tissues of naturally infected white-tailed deer (Stair et al. 1968) and experimentally infected bovine fetuses (Anderson et al. 1989) by using polyclonal antibody and immunofluorescence, and in experimentally infected bovine fetuses (Anderson et al. 1989) and calves (MacLachlan et al. 1990) by using polyclonal or monoclonal antibody and an avidin-biotin complex immunoperoxidase technique.

Differential Diagnoses. In animals that die in the early stages of disease when pulmonary edema, hydrothorax, pericardial effusion, and petechial hemorrhages are the primary lesions, heartwater (cowdriosis) (Dardiri et al. 1987; Okoh et al. 1987), anthrax (Kellogg et al. 1970), sepsis, pasteurellosis (Reed et al. 1976), and clostridial enterotoxemia need to considered in the differential diagnosis. When erosions and ulcers are present, hemorrhagic disease must be differentiated from vesicular stomatitis (Karstad and Hanson 1957), bovine viral diarrhea (Karstad 1981), malignant catarrhal fever (Brown and Bloss 1992), foot-and-mouth disease (McVicar et al. 1974), rinderpest (Hamdy et al. 1975), and peste des petits ruminants

(Hamdy and Dardiri 1976). Severe ruminal ulceration and scarring seen in the more chronic stages could be confused with rumenitis secondary to lactic acidosis. Heavy-metal poisonings cause hemorrhage in the gastrointestinal tract of many species and could be confused with hemorrhagic disease. In mule deer (Woods et al. 1996), adenovirus infection, which is characterized by hemorrhages, pulmonary edema, alimentary tract ulceration, and hemorrhagic enteritis, should be included in the differential diagnosis. Secondary infections, particularly bacterial bronchopneumonia, may make the diagnosis of hemorrhagic disease difficult. Central nervous system signs are prominent in pronghorns, so conditions such as polioencephalomalacia would need to be considered as a differential diagnosis in this species.

Immunity. Both serogroup- and serotype-specific antibodies develop when animals are infected with BLU or EHD viruses. Specificity of serogroup-specific antibody is associated with VP7 (Huismans and Erasmus 1981), although VP3 has been shown to contain group-specific antigenic determinants (Roy 1996). Serotype-specific antibodies are primarily directed to determinants on VP2 (Huismans and Erasmus 1981). Serogroup-specific antibodies may be detected as early as 5–8 days following experimental infection of white-tailed deer with BLU (Thomas and Trainer 1970) or EHD (Quist et al. 1997a) viruses. Serotype-specific neutralizing antibodies can be detected in white-tailed deer as early as 8 days following experimental infection with EHD virus and peak around 2 weeks after inoculation (Quist et al. 1997a). Evidence from serologic surveys also suggests that antibodies to both serogroup- and serotype-specific epitopes are long-lived in white-tailed deer (Stallknecht et al. 1991a). Although viremia rapidly decreases in white-tailed deer following the development of neutralizing antibodies, indicating virus clearance from blood by these antibodies, viremia persists because neutralizing antibody apparently does not affect virus sequestered in pits in red blood cell membranes (Quist et al. 1997a; Stallknecht et al. 1997). Elk experimentally infected with BLU developed neutralizing antibody by week 2 or 3 after exposure, and antibody was still present in sera at 6–7 months after inoculation (Murray and Trainer 1970).

The cell-mediated immune response in wild ruminants to BLU and EHD virus infections is not well defined but has been partially examined in white-tailed deer. In experimental studies, white-tailed deer infected with BLU or EHD viruses had depressed cell-mediated immune responses to nonspecific mitogens during the acute phase of the disease (Quist et al. 1997b). Transient immune deficiency in the early stages of infection may be partially responsible for the secondary bacterial infections often seen with these virus infections. A specific cell-mediated immune response to EHD or BLU viruses, as detected by lymphocyte proliferation assays, was not demonstrated following experimental infection of white-tailed deer with these viruses, but findings did not conclusively rule out a cell-mediated immune response to these viruses (Quist et al. 1997b).

Interferon type I (IFN) is produced early in many viral infections and is thought to be an important nonspecific antiviral response (Johnson et al. 1994), but its role in BLU and EHD virus infections has not been completely elucidated. Interferon levels rose abruptly in white-tailed deer experimentally infected with EHD or BLU viruses, peaked between days 4 and 6 after inoculation, and then fell rapidly so that no IFN was detected after 10 days of infection (Quist et al. 1997a). Peak IFN levels coincided with peak viremia, and both IFN and viremia decreased in concert as precipitating and neutralizing antibodies were first detected in the peripheral blood at 8–10 days after inoculation.

Protective immunity to these viruses is poorly studied in wild ruminants, but studies in sheep indicate that both neutralizing antibody and cell-mediated immunity are important in protection against reinfection (Jeggo et al. 1983, 1986; Stott et al. 1985). There is a good association between virus-neutralizing antibody and protective immunity to homologous virus challenge in sheep (Jeggo et al. 1983). Humoral immunity associated with neutralizing antibody appears to protect only against challenge with the homologous virus serotype, but serial infection with different serotypes may lead to the development of heterotypic neutralizing antibody that may be associated with protection to additional serotypes (Jeggo et al. 1983). Pirtle and Layton (1961) demonstrated that serum from white-tailed deer that survived EHD virus infection protected other deer from a lethal dose of homologous virus; thus, neutralizing antibodies appear to be important in protecting white-tailed deer from reinfection.

Homologous and heterologous protection to BLU viruses has been induced in sheep by "core particles" composed of VP3 and VP7. This protection is probably via cell-mediated immunity because these proteins do not elicit neutralizing antibodies and are known to contain cytotoxic T-cell epitopes (Murray and Eaton 1996).

White-tailed deer surviving either natural or experimental infection with the EHD viruses were protected from disease due to homologous virus challenge (Shope et al. 1960; Quist et al. 1997a). However, whether immunity to one virus will protect deer from infection with a heterologous viruses is less clear. In one study, white-tailed deer experimentally infected with EHD virus serotype 1 were fully susceptible to infection with BLU virus serotype 10 and vice versa (Hoff and Trainer 1974). Quist et al. (1997a) also found that deer initially infected with EHD virus serotype 2 were fully susceptible to BLU virus serotype 10. However, Vosdingh et al. (1968) reported that white-tailed deer that survived BLU virus serotype 10 infection were immune to EHD virus. Pronghorn with natural neutralizing antibody titers to BLU 10 did not succumb to disease when experimentally challenged with a BLU 10 inoculum that caused disease in naive pronghorn, suggesting that previous exposure to the same serotype of BLU virus was protective (Hoff and Trainer 1972).

Control and Treatment. At present, no wildlife management tools or strategies are available to prevent, predict, or minimize potential impacts of this disease.

Public Health Concerns. None of the viruses in the BLU or EHD serogroups have been associated with human disease.

Domestic Animal Concerns. Cattle usually develop subclinical infection when infected with the BLU and EHD viruses, but fever, stomatitis, lameness, and reproductive problems occasionally occur (Hourrigan and Klingsporn 1975; Metcalf et al. 1992). Sheep are very susceptible to infection with BLU viruses, but clinical disease has never been reported in sheep infected with EHD viruses (Luedke et al. 1964; Thompson et al. 1988).

Domestic dogs have been shown to be susceptible to BLU virus infection. Pregnant bitches that were infected with a BLU virus-contaminated vaccine during the last trimester of pregnancy aborted and died due to BLU virus infection (Wilbur et al. 1994). Experimentally, only pregnant bitches developed clinical signs following administration of BLU virus serotype 11 (Brown et al. 1996). Whether there is a potential risk for dogs to acquire BLU infection from eating tissues from infected white-tailed deer or other ruminants is unknown. Dogs were not susceptible to EHD virus when experimentally infected (Shope et al. 1960; E.W. Howerth unpublished).

Management Implications. Although HD is considered the most important viral disease affecting white-tailed deer in the United States (Nettles and Stallknecht 1992), population impacts are poorly documented. There are many reports of high mortality related to HD in deer populations, and it has been suggested that high HD-related mortality combined with the normal hunting kill can result in long-term population declines (Fischer et al. 1995). In many areas, however, repeated HD outbreaks have not represented a limiting factor to deer population growth. The specific risk factors responsible for HD outbreaks in white-tailed deer and other ungulate populations also are poorly defined. Although it is logical to assume that host population density could indirectly affect clinical outcome, there is no evidence that severity of disease is related to population density.

OTHER ORBIVIRUSES OF POSSIBLE WILDLIFE SIGNIFICANCE

Orbivirus Infection in Kangaroos. Epidemics of blindness, most likely due to an orbivirus, occurred in kangaroos in Australia between 1994 and early 1997. However, examination of archival tissues suggested that a similar disease may also have occurred about 20 years previously. Cases of blindness were seen in late summer and autumn of 1994 in western New South Wales and northwestern Victoria, and again in similar areas in the summer and autumn of 1995 with spread into southeastern South Australia. In late 1995 and early 1996, affected kangaroos also were seen in Western Australia. Affected animals have mostly been western gray kangaroos *Macropus fuliginosus,* but eastern gray kangaroos *Macropus giganteus,* red kangaroos *Macropus rufus,* and euros *Macropus robustus* were also affected. Animals of all ages, with the exception of pouch young, appeared to be affected. Estimates of more than 10,000 cases have been made, with 1%–10% of animals in individual populations affected.

Wallal and Warrego viruses of the Wallal and Warrego serogroups, respectively, were isolated from retinal specimens from affected kangaroos from New South Wales, Warrego virus was isolated from a kangaroo from northwestern Victoria, and an unidentified agent has been described in cases from South Australia (Durham et al. 1996). Wallal virus has been detected in retina from affected animals by immunofluorescence using polyclonal antiserum, and nucleic acids from both Wallal and Warrego viruses were also demonstrated in tissue sections by PCR. Subsequently, PCR on material from a case from about 20 years before also tested positive for Wallal virus nucleic acid.

Eye lesions indistinguishable from the lesions seen in field cases have been induced in kangaroos experimentally inoculated with cell culture-derived Wallal virus, and Wallal virus was identified in the retina of these animals by immunofluorescence and PCR. Orbivirus-like particles were observed in degenerating retinal cells by electron microscopy. Although the experimental inoculum was found to contain a low titer of Warrego virus, Wallal virus was considered the most likely cause of the eye lesions.

Cases of blindness were first noted during summer and continued through autumn; cases subsided during winter, suggesting that this disease was caused by a vector-borne agent. The distribution of disease closely followed watercourses where animals had congregated during drought conditions. Both Wallal and Warrego viruses are transmitted by *Culicoides,* and two species, *C. dycei* and *C. austropalpalis,* were shown by PCR to be carrying these viruses during the 1995 outbreak of blindness in kangaroos in New South Wales. Their role as true vectors has not been proven. A limited serologic survey of macropods in New South Wales in 1995 indicated high prevalence of neutralizing antibodies to both Wallal and Warrego viruses.

Behavioral changes were the first indications of disease, with animals being observed in the middle of the day and failing to retreat when approached by humans. Signs of blindness included stumbling, collision with obstacles, exaggerated vertical hopping, and circling. Animals panicked when disturbed, and many kangaroos died from misadventure, starvation, and water deprivation. The only indication of possible abnormalities was a change in the reflectivity of the eyes at night. Multiple white plaques were observed on the surface of the retina during ophthalmologic examination of experimentally induced cases.

Microscopic changes consisted of retinal degeneration, which was so severe in some cases that all retinal tissues were eliminated, and nonsuppurative chorioretinitis, characterized by a severe cellular infiltration and hyperemia. Wallerian degeneration of the optic nerve and tracts occurs, probably secondary to the retinal degeneration. Nonsuppurative optic neuritis and encephalitis also have been observed. Similar lesions in the optic nerves and tracts have been seen in normal kangaroos from populations with a high incidence of disease.

Although both Wallal and Warrego viruses have been known in Australia for approximately 30 years (Doherty et al. 1973), there is only one previous record of sporadic cases of blindness in kangaroos. The critical circumstances that precipitated recent outbreaks of blindness is unknown. However, nucleotide-sequence analysis of recent Wallal virus isolates suggests that there may be genetic diversity from earlier isolates.

African Horse Sickness Viruses. African horse sickness (AHS) is an infectious, noncontagious, insect-borne, orbivirus disease of equids caused by the viruses of the African horse sickness serogroup in which there are ten serotypes (Coetzer and Erasmus 1994a). African horse sickness occurs mainly in southern, eastern, and central Africa, but it has occurred also in North Africa, the Middle East, parts of Asia, Spain, and Portugal (Coetzer and Erasmus 1994a). Horses *Equus caballus* are most susceptible to disease, and typically there is severe morbidity and mortality; mules *Equus caballus* × *Equus asinus* are also susceptible but less so. Donkeys *Equus asinus* are relatively resistant, with most infections being subclinical (Alexander 1948). Burchell's zebra *Equus burchelli* are highly resistant to disease, and only showed a mild fibril response following experimental infection (Erasmus et al. 1978). African horse sickness viruses occasionally cause disease in dogs (Van Rensburg et al. 1981). Although antibodies against AHS viruses have been found in elephants *Loxodonta africana,* spotted hyena, jackal *Canis mesomelas,* and camels *Camelus dromedarius* (Awad et al. 1981; Binepal et al. 1992; Coetzer and Erasmus 1994a), disease has not been reported in these species.

These viruses are all transmitted by *Culicoides* midges (Monath and Guirakhoo 1996). Infection of domestic dogs typically has resulted from the ingestion of infected horse meat (Van Rensburg et al. 1981), and natural insect-borne transmission is probably rare in dogs (McIntosh 1955). The reservoir host of AHS viruses is unknown, but the susceptibility of Burchell's zebras to experimental infection (Erasmus et al. 1978) and the presence of antibodies against AHS viruses in free-living Burchell's zebra (Barnard 1993) suggest that zebras are the reservoir. It was also suspected that the original source of infection for the 1987 outbreak of AHS in Spain was ten zebras imported from Namibia (Lubroth 1988).

There are several reports of human infection with the AHS viruses, but these have been restricted to laboratory workers infected with neurotropic vaccine strains (Van der Meyden et al. 1992). Infection in humans has ranged from subclinical, as documented serologically (Swanepoel et al. 1992), to illnesses characterized by encephalitis and chorioretinitis (Van der Meyden et al. 1992).

Equine Encephalosis Viruses. Equine encephalosis (EE) is an acute insect-borne orbivirus disease caused by viruses of the EE virus serogroup (Coetzer and Erasmus 1994b). Equine encephalosis viruses are endemic in South Africa and Botswana. Although these viruses probably extend beyond southern Africa, no studies have been done to determine their distribution (Coetzer and Erasmus 1994b). Horses of all age groups are susceptible, and these viruses may cause abortions, jaundice, and peracute deaths (Barnard and Paweska 1993; Gerdes and Pieterse 1993). There has been serologic evidence of EE infection in donkeys (Gerdes and Pieterse 1993) and Burchell's zebra (Barnard and Paweska 1993), and EE virus has been isolated from the blood of Burchell's zebra (Gerdes and Pieterse 1993), but disease due to these viruses has never been described. There are at least seven serotypes of EE virus, all of which are transmitted by *Culicoides* midges (Monath and Guirakhoo 1996).

LITERATURE CITED

Afshar, A., J. Anderson, K.H. Nielsen, J.E. Pearson, H.C. Trotter, T.R. Woolhouse, J.A. Thevasagayam, D.E.J. Gall, J. Lapointe, and G.A. Gustafson. 1997. Evaluation of a competitive ELISA for detection of antibodies to epizootic hemorrhagic disease virus of deer. *Journal of Veterinary Diagnostic Investigations* 9:309–311.

Alexander, H.E. 1954. *Deer problems, new to Arkansas—an old story elsewhere.* Little Rock: Arkansas Game and Fish Commission, 23 pp.

Alexander, K.A., N.J. MacLachlan, P.W. Kat, C. House, S.J. O'Brien, N.W. Lerche, M. Sawyer, L.G. Frank, K. Holekamp, L. Smale, J.W. McNutt, M.K. Laurenson, M.G.L. Mills, and B.I. Osburn. 1994. Evidence of natural bluetongue virus infection among African carnivores. *American Journal of Tropical Medicine and Hygiene* 51:568–576.

Alexander, R.A. 1948. The 1944 epizootic of horsesickness in the Middle East. *Onderstepoort Journal of Veterinary Science and Animal Industry* 23:77–92.

Anderson, G.A., D.L. Phillips, A.S. Waldvogel, and B.I. Osburn. 1989. Detection of bluetongue virus in bovine fetuses using the avidin-biotin complex immunoperoxidase method. *Journal of Veterinary Diagnostic Investigations* 1:45–49.

Awad, F.I., M.M. Amin, S.A. Salama, and M.M. Aly. 1981. The incidence of African horse sickness antibodies in animals of various species in Egypt. *Bulletin of Animal Health Production in Africa* 29:285–287.

Barber, T.L., and M.M. Jochim. 1975. Serotyping bluetongue and epizootic hemorrhagic disease virus strains. *Proceedings of the American Association of Veterinary Laboratory Diagnosticians* 18:149–157.

Barnard, B.J.H. 1993. Circulation of African horsesickness virus in zebra (*Equus burchelli*) in the Kruger National Park, South Africa, as measured by the prevalence of type specific antibodies. *Onderstepoort Journal of Veterinary Research* 60:111–117.

Barnard, B.J.H., and J.T. Paweska. 1993. Prevalence of antibodies against some equine viruses in zebra (*Zebra burchelli*) in the Kruger National Park, 1991–1992. *Onderstepoort Journal of Veterinary Research* 60:175–179.

Barratt-Boyes, S.M., and N.J. MacLachlan. 1994. Dynamics of viral spread in bluetongue virus infected calves. *Veterinary Microbiology* 40:361–371.

Barratt-Boyes, S.M., P.V. Rossitto, J.L. Stott, and N.J. MacLachlan. 1992. Flow cytometric analysis of in vitro bluetongue virus infection of bovine blood mononuclear cells. *Journal of General Virology* 73:1953–1960.

Barzilai, E., and A. Tadmor. 1972. Experimental infection of the mountain gazelle (*Gazella gazella*) with bluetongue virus. *Refuah Veterinarith* 29:45–50.

Binepal, V.S., B.N. Wariru, F.G. Davies, R. Soi, and R. Olubayo. 1992. An attempt to define the host range for African horse sickness virus (Orbivirus, Reoviridae) in East Africa, by a serological survey in some Equidae, Camelidae, Loxodontidae and Carnivore. *Veterinary Microbiology* 31:19–23.

Boorman, J., and E.P. Gibbs. 1973. Multiplication of the virus of epizootic haemorrhagic disease of deer in *Culicoides* species (Diptera, Ceratopogonidae). *Archiv für die Gesamte Virusforschung* 41:259–266.

Brewer, A.W., and N.J. MacLachlan. 1992. Ultrastructural characterization of the interaction of bluetongue virus with bovine erythrocytes in vitro. *Veterinary Pathology* 29:356–359.

———. 1994. The pathogenesis of bluetongue virus infection of bovine blood cells in vitro: Ultrastructural characterization. *Archives of Virology* 136:287–298.

Brown, C.C., and L.L. Bloss. 1992. An epizootic of malignant catarrhal fever in a large captive herd of white-tailed deer (*Odocoileus virginianus*). *Journal of Wildlife Diseases* 28:301–305.

Brown, C.C., J.C. Rhyan, M.J. Grubman, and L.A. Wilbur. 1996. Distribution of bluetongue virus in tissues of experimentally infected pregnant dogs as determined by in situ hybridization. *Veterinary Pathology* 33:337–340.

Coetzer, J.A.W., and B.J. Erasmus. 1994a. African horsesickness. In *Infectious diseases of livestock,* ed. J.A.W. Coetzer, G.R. Thomson, and R.C. Tustin. New York: Oxford University Press, pp. 460–475.

———. 1994b. Equine encephalosis. In *Infectious diseases of livestock,* ed. J.A.W. Coetzer, G.R. Thomson, and R.C. Tustin. New York: Oxford University Press, pp. 476–479.

Corn, J.L., D.M. Kavanaugh, D.A. Osborn, S. Denariais, C.M. Milke, and V.F. Nettles. 1990. Survey for diseases and parasites in exotic ruminants in Texas. *Proceedings of the United States Animal Health Association* 94:530–540.

Couvillion, C.E., W.R. Davidson, J.E. Pearson, and G.A. Gustafson. 1981. Hemorrhagic disease among white-tailed deer in the southeast from 1971 through 1980. *Proceedings of the United States Animal Health Association* 85:522–537.

Dardiri, A.H., L.L. Logan, and C.A. Mebus. 1987. Susceptibility of white-tailed deer to experimental heartwater infections. *Journal of Wildlife Diseases* 23:215–219.

Davies, F.G., and A.R. Walker. 1974. The distribution in Kenya of bluetongue virus and antibody, and the *Culicoides* vector. *Journal of Hygiene* 72:265–272.

Debbie, J.G., and M.K. Abelseth. 1971. Pathogenesis of epizootic hemorrhagic disease: I. Blood coagulation during viral infection. *Journal of Infectious Diseases* 124:217–222.

Debbie, J.G., H.C. Rowsell, L.H. Karstad, and J. Ditchfield. 1965. An approach to the pathogenesis of viral hemorrhagic disease of deer. *Transactions of the North American Wildlife and Natural Resources Conference* 30:196–205.

Ditchfield, J., J.G. Debbie, and L.H. Karstad. 1964. The virus of epizootic hemorrhagic disease of deer. *Transactions of the North American Wildlife and Natural Resources Conference* 29:196–201.

Doherty, R.L., J.G. Carley, H.A. Standfast, A.L. Dyce, B.H. Kay, and W.A. Snowdon. 1973. Isolation of arboviruses from mosquitoes, biting midges, sandflies and vertebrates collected in Queensland, 1969 and 1970. *Transactions of the Royal Society of Tropical Medicine and Hygiene* 67:536–543.

Dulac, G.C., C. Dubuc, A. Afshar, D.J. Meyers, A. Bouffard, J. Shapiro, P.T. Shettigara, and D. Ward. 1988. Consecutive outbreaks of epizootic hemorrhagic disease of deer and bluetongue. *Veterinary Record* 122:340.

Dulac, G.C., W.G. Sterritt, C. Dubuc, A. Afshar, D.J. Myers, E.A. Taylor, B.R. Jamieson, and M.W. Martin. 1992. Incursions of orbiviruses in Canada and their serologic monitoring in native animal populations between 1962 and 1991. In *Bluetongue, African horse sickness, and related orbiviruses,* ed. T.E. Walton and B.E. Osburn. Boca Raton, FL: CRC, pp. 120–127.

Durham, P.J.K., J.W. Finnie, D.A. Lawrence, and P. Alexander. 1996. Blindness in South Australian kangaroos. *Australian Veterinary Journal* 73:111–112.

Du Toit, R. 1955. Insect vectors and virus diseases. *Journal of the South African Veterinary Medical Association* 26:263–268.

Erasmus, B.J., E. Young, L.M. Pieterse, and S.T. Boshoff. 1978. The susceptibility of zebra and elephants to African horsesickness. In *Equine infectious disease: Proceedings of the fourth international conference on equine infectious diseases,* ed. J.T. Bryans and H. Gerber. Princeton: Veterinary Publications, pp. 409–413.

Fay, L.D., A.P. Boyce, and W.G. Youatt. 1956. An epizootic in deer in Michigan. *Transactions of the North American wildlife and natural resources conference* 21:173–184.

Fischer, J.R., L.P. Hansen, J.R. Turk, M.A. Miller, W.H. Fales, and H.S. Gosser. 1995. An epizootic of hemorrhagic disease in white-tailed deer (*Odocoileus virginianus*) in Missouri: Necropsy findings and population impact. *Journal of Wildlife Diseases* 31:30–36.

Fletch, A.L., and L.H. Karstad. 1971. Studies on the pathogenesis of experimental epizootic hemorrhagic disease of white-tailed deer. *Canadian Journal of Comparative Medicine* 35:224–229.

Foster, N.M., R.D. Breckon, A.J. Luedke, R.H. Jones, and H.E. Metcalf. 1977. Transmission of two strains of epizootic hemorrhagic disease virus in deer by *Culicoides variipennis. Journal of Wildlife Diseases* 13:9–16.

Gerdes, G.H., and L.M. Pieterse. 1993. The isolation and identification of Potchefstroom virus: A new member of the equine encephalosis group of orbiviruses. *Journal of the South African Veterinary Association* 64:131–132.

Gerhardt, R.R. 1986. *Culicoides* spp. attracted to ruminants in the Great Smoky Mountains National Park, Tennessee. *Journal of Agricultural Entomology* 3:192–197.

Gibbs, E.P.J. 1992. Epidemiology of orbiviruses—bluetongue: Towards 2000 and the search for patterns. In *Bluetongue, African horse sickness, and related orbiviruses,* ed. T.E. Walton and B.I. Osburn. Boca Raton, FL: CRC, pp. 65–75.

Gibbs, E.P.J., and E.C. Greiner. 1983. Bluetongue infections and *Culicoides* species associated with livestock in Florida and the Caribbean region. In *Double-stranded RNA viruses,* ed. R.W. Compans and D.H.L. Bishop. New York: Elsevier Science, pp. 375–382.

———. 1989. Bluetongue and epizootic hemorrhagic disease. In *The arboviruses: Epidemiology and ecology,* vol. 2, ed. T.P. Monath. Boca Raton, FL: CRC, pp. 39–70.

Gibbs, E.P.J., and M.J.P. Lawman. 1977. Infection of British deer and farm animals with epizootic haemorrhagic disease of deer virus. *Journal of Comparative Pathology* 87:335–343.

Gonzalez, H.A., and D.L. Knudson. 1988. Intra- and interserogroup genetic relatedness of orbiviruses: I. Blot hybridization of viruses of Australian serogroups. *Journal of General Virology* 69:125–134.

Greiner, E.C., T.L. Barber, J.E. Pearson, W.L. Kramer, and E.P.J. Gibbs. 1985. Orbiviruses from *Culicoides* in Florida. In *Bluetongue and related orbiviruses,* ed. T.L. Barber, M.M. Jochim, and B.I. Osburn. New York: Alan R. Liss, pp. 195–200.

Greiner, E.C., F.C.M. Alexander, J. Roach, V. Moe, G. Borde, W.P. Taylor, J. Dickinson, and E.P.J. Gibbs. 1989. Bluetongue epidemiology in the Caribbean region: Serological and entomological findings from a pilot sentinel system in Trinidad and Tobago. *Medical and Veterinary Entomology* 3:101–105.

Greiner, E.C., C.L. Mo, E.J. Homan, J. Gonzalez, M. Oviedo, L.H. Thompson, and E.P.J. Gibbs. 1993. Epidemiology of bluetongue in Central America and the Caribbean: Initial entomological findings. *Medical and Veterinary Entomology* 7:309–315.

Griner, L.A., and L.S. Nelson. 1970. Hemorrhagic disease in exotic ruminants in zoo. *Journal of the American Veterinary Medical Association* 157:600–603.

Gustafson, G.A., J.E. Pearson, and K.M. Moser. 1992. A comparison of the bluetongue competitive-ELISA to other serologic tests. In *Bluetongue, African horse sickness, and related orbiviruses,* ed. T.E. Walton and B.I. Osburn. Boca Raton, FL: CRC, pp. 128–139.

Hamblin, C., E.C. Anderson, M. Jago, T. Mlengeya, and K. Hirji. 1990. Antibodies of some pathogenic agents in free-living wild species in Tanzania. *Epidemiology and Infection* 105:585–594.

Hamdy, F.M., and A.H. Dardiri. 1976. Response of white-tailed deer to infection with peste des petits ruminants virus. *Journal of Wildlife Diseases* 12:516–522.

Hamdy, F.M., A.H. Dardiri, D.H. Ferris, and S.S. Breese. 1975. Experimental infection of white-tailed deer with rinderpest virus. *Journal of Wildlife Diseases* 11:508–515.

Harding, M.J., I. Prud'homme, J. Rola, and G.C. Dulac. 1996. Development of PCR-based tests for the identification of North American isolates of epizootic hemorrhagic disease virus. *Canadian Journal of Veterinary Research* 60:59–64.

Hoff, G.L., and D.M. Hoff. 1976. Bluetongue and epizootic hemorrhagic disease: A review of these diseases in nondomestic artiodactyles. *Journal of Zoo Animal Medicine* 7:26–30.

Hoff, G.L., and D.O. Trainer. 1972. Bluetongue virus in pronghorn antelope. *American Journal of Veterinary Research* 33:1013–1016.

———. 1973. Experimental infection in North American elk with epizootic hemorrhagic disease virus. *Journal of Wildlife Diseases* 9:129–132.

———. 1974. Observations on bluetongue and epizootic hemorrhagic disease in white-tailed deer: (1) distribution of virus in the blood (2) cross-challenge. *Journal of Wildlife Diseases* 10:25–31.

———. 1978. Bluetongue and epizootic hemorrhagic disease viruses: Their relationship to wildlife species. *Advances in Veterinary Science and Comparative Medicine* 22:111–132.

———. 1981. Hemorrhagic disease in wild ruminants. In *Infectious diseases of wild mammals,* ed. J.W. Davis, L.H. Karstad, and D.O. Trainer. Ames: Iowa State University Press, pp. 45–53.

Hoff, G.L., L.A. Griner, and D.O. Trainer. 1973a. Bluetongue virus in exotic ruminants. *Journal of the American Veterinary Medical Association* 163:565–567.

Hoff, G.L., S.H. Richards, and D.O. Trainer. 1973b. Epizootic hemorrhagic disease in North Dakota deer. *Journal of Wildlife Management* 37:331–335.

Hoff, G.L., D.O. Trainer, and M.M. Jochim. 1974. Bluetongue virus and white-tailed deer in an enzootic area of Texas. *Journal of Wildlife Diseases* 10:158–163.

Hourrigan, J.L., and A.L. Klingsporn. 1975. Bluetongue: The disease in cattle. *Australian Veterinary Journal* 51:170–175.

Howerth, E.W., and D.E. Tyler. 1988. Experimentally induced bluetongue virus infection in white-tailed deer: Ultrastructural findings. *American Journal of Veterinary Research* 49:1914–1922.

Howerth, E.W., C.E. Greene, and A.K. Prestwood. 1988. Experimentally induced bluetongue virus infection in white-tailed deer: Coagulation, clinical pathologic, and gross pathologic changes. *American Journal of Veterinary Research* 49:1906–1913.

Howerth, E.W., M.E. Dorminy, C.F. Quist, T. Pisell, and D.E. Stallknecht. 1995. Epizootic hemorrhagic disease in white-tailed deer: Changes in the factor XII dependent pathways. In *Proceedings of the 38th annual meeting of the American Association of Veterinary Laboratory Diagnosticians.* Sparks, NV: American Association of Veterinary Laboratory Diagnosticians, p. 38.

Huismans, H., and B.J. Erasmus. 1981. Identification of the serotype-specific and group-specific antigens of bluetongue viruses. *Onderstepoort Journal of Veterinary Research* 48:51–58.

Jeggo, M.H., I.D. Gumm, and W.P. Taylor. 1983. Clinical and serological response of sheep to serial challenge with different bluetongue virus types. *Research in Veterinary Science* 34:205–211.

Jeggo, M.H., R.C. Wardly, J. Brownlie, and A.H. Corteyn. 1986. Serial inoculation of sheep with two bluetongue virus types. *Research in Veterinary Science* 40:386–392.

Jessup, D.A. 1985. Epidemiology of two orbiviruses in California's native wild ruminants: Preliminary report. In *Bluetongue and related orbiviruses,* ed. T.L. Barber, M.M. Jochim, and B.I. Osburn. New York: Alan R. Liss, pp. 53–55.

Jessup, D.A., B.I. Osburn, and W.P. Heuschele. 1984. Bluetongue in California's wild ruminants: Distribution and pathology. *Proceedings of the United States Animal Health Association* 88:616–630.

Jochim, M.M. 1985. An overview of diagnostics for bluetongue. In *Bluetongue and related orbiviruses,* ed. T.L. Barber, M.M. Jochim, and B.I. Osburn. New York: Alan R. Liss, pp. 423–434.

Jochim, M.M., and S.C. Jones. 1976. Plaque neutralization of bluetongue virus and epizootic hemorrhagic disease virus in BHK_{21} cells. *American Journal of Veterinary Research* 37:1345–1347.

Johnson, B.G. 1992. An overview and perspective on orbivirus disease prevalence and occurrence of vectors in North America. In *Bluetongue, African horse sickness, and related orbiviruses,* ed. T.E. Walton and B.I. Osburn. Boca Raton, FL: CRC, pp. 58–64.

Johnson, H.M., F.W. Bazer, B.E. Szente, and M.A. Jarpe. 1994. How interferons fight disease. *Scientific American* 270:68–75.

Jones, R.H., and N.M. Foster. 1974. Oral infection of *Culicoides variipennis* with bluetongue virus: Development of susceptible and resistant lines from a colony population. *Journal of Medical Entomology* 11:316–323.

———. 1978. Heterogeneity of *Culicoides variipennis* field populations to oral infection with bluetongue virus. *American Journal of Tropical Medicine and Hygiene* 27:178–183.

Jones, R.H., R.D. Roughton, N.M. Foster, and B.M. Bando. 1977. *Culicoides,* the vector of epizootic hemorrhagic disease in white-tailed deer in Kentucky in 1971. *Journal of Wildlife Diseases* 13:2–8.

Jones, R.H., E.T. Schmidtmann, and N.M. Foster. 1983. Vector-competence studies for bluetongue and epizootic hemorrhagic disease viruses with *Culicoides venustus. Mosquito News* 43:184–186.

Karstad, L. 1981. Bovine virus diarrhea. In *Infectious disease of wild mammals,* ed. J.W. Davis, L.H. Karstad, and D.O. Trainer. Ames: Iowa State University Press, pp. 209–211.

Karstad, L., and R.P. Hanson. 1957. Vesicular stomatitis in deer. *American Journal of Veterinary Research* 18:162–166.

Karstad, L., and D.O. Trainer. 1967. Histopathology of experimental bluetongue disease of white-tailed deer. *Canadian Veterinary Journal* 8:247–254.

Katz, J., D. Alstad, G. Gustafson, and J. Evermann. 1994. Diagnostic analysis of the prolonged bluetongue virus RNA presence found in the blood of naturally infected cattle and experimentally infected sheep. *Journal of Veterinary Diagnostic Investigations* 6:139–142.

Kellogg, F.E., A.K. Prestwood, and R.E. Noble. 1970. Anthrax epizootic in white-tailed deer. *Journal of Wildlife Diseases* 6:226–228.

Kemp, G.E., O.R. Causey, H.W. Setzer, and D.L. Moore. 1974. Isolation of viruses from wild mammals in West Africa, 1966–1970. *Journal of Wildlife Diseases* 10:279–293.

Kistner, T.P., G.E. Reynolds, L.D. Koller, C.E. Trainer, and D.L. Eastman. 1975. Clinical and serological findings on the distribution of bluetongue and epizootic hemorrhagic disease viruses in Oregon. *American Association of Veterinary Laboratory Diagnosticians* 18:135–148.

Kline, D.L., and E.C. Greiner. 1992. Field observations on the ecology of adult and immature stages of *Culicoides* spp. associated with livestock in Florida. In *Bluetongue, African horse sickness, and related orbiviruses,* ed. T.E. Walton and B.I. Osburn. Boca Raton, FL: CRC, pp. 297–304.

Kocan, A., A.E. Castro, B. Espe, R.T. Doyle, and S.K. Olsen. 1982. Inapparent bluetongue in free-ranging white-tailed deer. *Journal of the American Veterinary Medical Association* 181:1415–1416.

Kramer, W.L., E.C. Greiner, and E.P.J. Gibbs. 1985. A survey of *Culicoides* midges (Diptera: Ceratopogonidae) associated with cattle operations in Florida, USA. *Journal of Medical Entomology* 22:153–162.

Lubroth, J. 1988. African horsesickness and the epizootic in Spain 1987. *Equine Practice* 10:26–33.

Lubroth, J. 1992. The complete epidemiologic cycles of African horse sickness: Our incomplete knowledge. In *Bluetongue, African horse sickness, and related orbiviruses,* ed. T.E. Walton and B.I. Osburn. Boca Raton, FL: CRC, pp. 197–204.

Luedke, A.J., J.G. Bowne, M.M. Jochim, and C. Doyle. 1964. Clinical and pathologic features of bluetongue in sheep. *American Journal of Veterinary Research* 25:963–969.

MacLachlan, N.J., G. Jagels, P.V. Rossitto, P.F. Moore, and H.W. Heidner. 1990. The pathogenesis of experimental bluetongue virus infection of calves. *Veterinary Pathology* 27:223–229.

McIntosh, B.M. 1955. Horsesickness antibodies in the sera of dogs in enzootic areas. *Journal of the South African Veterinary Medical Association* 26:269–272.

McVicar, J.W., P. Sutmoller, D.H. Ferris, and C.H. Campbell. 1974. Foot-and-mouth disease in white-tailed deer: Clinical signs and transmission in the laboratory. *Proceedings of the United States Animal Health Association* 78:169–180.

Mellor, P.S. 1990. The replication of bluetongue virus *Culicoides* vectors. *Current Topics in Microbiology and Immunology* 162:143–161.

Metcalf, H.E., A.J. Luedke, and M.M. Jochim. 1992. Epizootic hemorrhagic disease virus infection in cattle. In *Bluetongue, African horse sickness, and related orbiviruses,* ed. T.E. Walton and B.I. Osburn. Boca Raton, FL: CRC, pp. 222–237.

Miller, K.V., R.L. Marchington, and V.F. Nettles. 1986. The growth rate of hooves of white-tailed deer. *Journal of Wildlife Diseases* 22:129–131.

Mo, C.L., L.H. Thompson, E.J. Homan, M.T. Oviedo, E.G. Greiner, J. Gonzalez, and M.R. Saenz. 1994. Bluetongue virus isolations from vectors and ruminants in Central America and the Caribbean. *American Journal Veterinary Research* 55:211–215.

Monath, T.P., and F. Guirakhoo. 1996. Orbiviruses and coltiviruses. In *Fields virology,* ed. B.N. Fields, D.M. Knipe, and P.M. Howley, 3d ed. Philadelphia: Lippincott-Raven, pp. 1735–1766.

Mullen, G.R., M.E. Hayes, and K.E. Nusbaum. 1985. Potential vectors of bluetongue and epizootic hemorrhagic disease viruses in cattle and white-tailed deer in Alabama. In *Bluetongue and related orbiviruses,* ed. T.L. Barber, M.M. Jochim, and B.I. Osburn. New York: Alan R. Liss, pp. 201–206.

Murray, J.O., and D.O. Trainer. 1970. Bluetongue virus in North American elk. *Journal of Wildlife Diseases* 6:144–148.

Murray, P.K., and B.T. Eaton. 1996. Vaccines for bluetongue. *Australian Veterinary Journal* 73:207–210.

Neitz, W.O. 1933. The blesbuck (*Damaliscus albifrons*) as a carrier of heartwater and blue tongue. *Journal of the South African Veterinary Medical Association* 4:24–26.

Nettles, V.F., and D.E. Stallknecht. 1992. History and progress in the study of hemorrhagic disease of deer. *Transactions of the North American Wildlife and Natural Resources Conference* 57:499–516.

Nettles, V.F., W.R. Davidson, and D.E. Stallknecht. 1992a. Surveillance for hemorrhagic disease in white-tailed deer and other wild ruminants, 1980–1989. *Proceedings of the Annual Conference of the Southeastern Association of Fish and Wildlife Agencies* 46:138–146.

Nettles, V.F., S.A. Hylton, D.E. Stallknecht, and W.R. Davidson. 1992b. Epidemiology of epizootic hemorrhagic disease viruses in wildlife in the USA. In *Bluetongue, African horse sickness, and related orbiviruses,* ed. T.E. Walton and B.I. Osburn. Boca Raton, FL: CRC, pp. 238–248.

Nunamaker, R.A., J.O. Mecham, F.R. Holbrook, and J.A. Lockwood. 1997. Applications of dot-blot, ELISA, and immunoelectron microscopy to field detection of bluetongue virus in *Culicoides variipennis sonorensis:* An ecological perspective. *Journal of Medical Entomology* 34:24–28.

Okoh, A.E.J., I.L. Oyetunde, and J.O. Ibu. 1987. Heartwater infection (cowdriosis) in a sitatunga (*Tragelaphus spekei*) in Nigeria. *Journal of Wildlife Diseases* 23:211–214.

Patton, J.F, T.M. Work, D.A. Jessup, S.K. Hietala, M.N. Oliver, and N.J. MacLachlan. 1994. Serologic detection of bluetongue virus infection of black-tailed deer: Comparison of serum neutralization, agar gel immunodiffusion, and competitive ELISA assays. *Journal of Wildlife Diseases* 30:99–102.

Pearson, J.E., and M.M. Jochim. 1979. Protocol for the immunodiffusion test for bluetongue. *Proceedings of the American Association of Veterinary Laboratory Diagnosticians* 22:463–471.

Pearson, J.E., G.A. Gustafson, A.L. Shafer, and A.D. Alstad. 1992a. Distribution of bluetongue on the United States. In *Bluetongue, African horse sickness, and related orbiviruses,* ed. T.E. Walton and B.I. Osburn. Boca Raton, FL: CRC, pp. 128–139.

———. 1992b. Diagnosis of bluetongue and epizootic hemorrhagic disease. In *Bluetongue, African horse sickness, and related orbiviruses,* ed. T.E. Walton and B.I. Osburn. Boca Raton, FL: CRC, pp. 533–546.

Pirtle, E.C., and J.M. Layton. 1961. Epizootic hemorrhagic disease in white-tailed deer: Characteristics of the South Dakota strain of virus. *American Journal of Veterinary Research* 22:104–108.

Prestwood, A.K., T.P. Kistner, F.E. Kellogg, and F.A. Hayes. 1974. The 1971 outbreak of hemorrhagic disease among white-tailed deer of the southeastern United States. *Journal of Wildlife Diseases* 10:217–224.

Quist, C.F., E.W. Howerth, D.E. Stallknecht, J. Brown, T. Pisell, and V.F. Nettles. 1997a. Host defense responses associated with experimental hemorrhagic disease in white-tailed deer. *Journal of Wildlife Diseases* 33:584–599.

Quist, C.F., E.W. Howerth, D.I. Bounous, and D.E. Stallknecht. 1997b. Cell-mediated immune response and IL-2 production in white-tailed deer infected with hemorrhagic disease viruses. *Veterinary Immunology and Immunopathology* 56:283–297.

Ramsay, E.C., S.J. Rodgers, A.E. Castro, E.L. Stair, and B.M. Baumeister. 1985. Perinatal bluetongue viral infection in exotic ruminants. *Journal of the American Veterinary Medical Association* 187:1249–1251.

Reed, D.E., H. Shave, M.E. Bergeland, and C.E. Gates. 1976. Necropsy and laboratory findings in free-living deer in South Dakota. *Journal of the American Veterinary Medical Association* 169:975–979.

Richards, S.H. 1963. Deer and antelope epizootic in the North Dakota badlands. *Proceedings of the North Dakota Academy of Science* 17:70–71.

Robinson, R.M., T.L. Hailey, C.W. Livingstone, and J.W. Thomas. 1967. Bluetongue in the desert bighorn sheep. *Journal of Wildlife Management* 31:165–168.

Robinson, R.M., T.L. Hailey, R.G. Marburger, and L. Weishuhn. 1974. Vaccination trials in desert bighorn sheep against bluetongue virus. *Journal of Wildlife Diseases* 10:228–231.

Roughton, R.D. 1975. An outbreak of a hemorrhagic disease in white-tailed deer in Kentucky. *Journal of Wildlife Diseases* 11:177–186.

Roy, P. 1996. Orbiviruses and their replication. In *Fields virology,* ed. B.N. Fields, D.M. Knipe, and P.M. Howley, 3d ed. Philadelphia: Lippincott-Raven, pp. 1709–1734.

Ruff, F.J. 1950. What is "blacktongue" among deer. *Wildlife in North Carolina* 14:16–19.

Schildman, G., and J. Hurt. 1984. *Wildlife disease and mortality summary, 1950–1983.* Work Plan S-83, Pittman-Robertson Project W-15-R-40. Lincoln: Nebraska Game and Parks Commission, 8 pp.

Schultz, J.W. 1979. *Floating on the Missouri.* Norman: University of Oklahoma Press, 142 pp.

Shad, G., W.C. Wilson, J.O. Meacham, and J.F. Evermann. 1997. Bluetongue virus detection: A safer reverse-transcriptase polymerase chain reaction for prediction of viremia in sheep. *Journal of Veterinary Diagnostic Investigation* 9:118–124.

Shope, R.E. 1967. The epizootiology of epizootic hemorrhagic disease of deer. *Transactions of the North American Wildlife and Natural Resources Conference* 32:381–386.

Shope, R.E., L.G. MacNamara, and R. Mangold. 1960. A virus-induced epizootic hemorrhagic disease of the Virginia white-tailed deer (*Odocoileus virginianus*). *Journal of Experimental Medicine* 111:155–170.

Simpson, V.R. 1978. Serologic evidence of bluetongue in game animals in Botswana. *Tropical Animal Health Production* 10:55–60.

Simpson, V.R. 1979. Bluetongue antibody in Botswana's domestic and game animals. *Tropical Animal Health Production* 11:43–49.

Smith, K.E., and D.E. Stallknecht. 1996. *Culicoides* (Diptera: Ceratopogonidae) collected during epizootics of hemorrhagic disease among captive white-tailed deer. *Journal of Medical Entomology* 33:507–510.

Smith, K.E., D.E. Stallknecht , and V.F. Nettles. 1996a. Experimental infection of *Culicoides lahillei* (Diptera: Ceratopogonidae) with epizootic hemorrhagic disease virus serotype 2 (Orbivirus: Reoviridae). *Journal of Medical Entomology* 33:117–122.

Smith, K.E., D.E. Stallknecht, C.T. Sewell, E.A. Rollor, G.R. Mullen, and R.R. Anderson. 1996b. Monitoring of *Culicoides* spp. at a site enzootic for hemorrhagic disease in white-tailed deer in Georgia, USA. *Journal of Wildlife Diseases* 32:627–642.

Stair, E.L., R.M. Robinson, and L.P. Jones. 1968. Spontaneous bluetongue in Texas white-tailed deer. *Pathologia Veterinaria* 5:164–173.

Stallknecht, D.E., and W.R. Davidson. 1992. Antibodies to bluetongue and epizootic hemorrhagic disease virus from white-tailed deer blood samples dried on paper strips. *Journal of Wildlife Diseases* 28:306–310.

Stallknecht, D.E., J.L. Blue, E.A. Rollor III, V.F. Nettles, W.R. Davidson, and J. E. Pearson. 1991a. Precipitating antibodies to epizootic hemorrhagic disease and bluetongue viruses in white-tailed deer in the Southeastern United States. *Journal of Wildlife Diseases* 27:238–247.

Stallknecht, D.E., M.L. Kellogg, J.L. Blue, and J.E. Pearson. 1991b. Antibodies to bluetongue and epizootic hemorrhagic disease viruses in a barrier island white-tailed deer population. *Journal of Wildlife Diseases* 27:668–674.

Stallknecht, D.E., V.F. Nettles, E.A. Rollor, and E.W. Howerth. 1995. Epizootic hemorrhagic disease virus and bluetongue virus serotype distribution in white-tailed deer in Georgia. *Journal of Wildlife Diseases* 31:331–338.

Stallknecht, D.E., M.P. Luttrell, K.E. Smith, and V.F. Nettles. 1996. Hemorrhagic disease in white-tailed deer in Texas: A case for enzootic stability. *Journal of Wildlife Diseases* 32:695–700.

Stallknecht, D.E., E.W. Howerth, M.L. Kellogg, C.F. Quist, and T. Pisell. 1997. In vitro replication of epizootic hemorrhagic disease and bluetongue viruses in white-tailed deer peripheral blood mononuclear cells and virus-cell association during in vivo infections. *Journal of Wildlife Diseases* 33:574–583.

Stauber, E.H., R.K. Farrell, and G.R. Spencer. 1977. Nonlethal experimental inoculation of Columbia black-tailed deer (*Odocoileus hemionus columbianus*) with virus of epizootic hemorrhagic deer disease. *American Journal of Veterinary Research* 38:411–412.

Stott, J.L., K.C. Else, B. McGowan, L.K. Wilson, and B.I. Osburn. 1981. Epizootiology of bluetongue virus in the western United States. *Proceedings of the United States Animal Health Association* 85:170–180.

Stott, J.L., L.H. Lauerman, and A.J. Luedke. 1982. Bluetongue virus in pregnant elk and their calves. *American Journal of Veterinary Research* 43:423–428.

Stott, J.L., T.L. Barber, and B.I. Osburn. 1985. Immunologic response of sheep to inactivated and virulent bluetongue virus. *American Journal of Veterinary Research* 46:1043–1049.

Swanepoel, R., B.J. Erasmus, R. Williams, and M.B. Taylor. 1992. Encephalitis and chorioretinitis associated with neurotropic African horsesickness virus infection in laboratory workers: III. Virologic and serologic investigations. *South African Medical Journal* 81:458–461.

Tabachnick, W.J. 1991. Genetic control of oral susceptibility to infection of *Culicoides variipennis* with bluetongue virus. *American Journal of Tropical Medicine and Hygiene* 45:666–671.

Tabachnick, W.J., and F.R. Holbrook. 1992. The *Culicoides variipennis* complex and the distribution of bluetongue viruses in the United States. *Proceedings of the United States Animal Health Association* 95:207–212.

Taylor, W.P., I.D. Gumm, E.P.J. Gibbs, and J. Homan. 1985. The use of serology in bluetongue epidemiology. In *Bluetongue and related orbiviruses,* ed. T.L. Barber, M.M. Jochim, and B.I. Osburn. New York: Alan R. Liss, pp. 461–468.

Thomas, F.C. 1981. Hemorrhagic disease. In *Diseases and parasites of white-tailed deer,* ed. W.R. Davidson, F.A. Hayes, V.F. Nettles, and F.E. Kellogg. Tallahassee, FL: Tall Timbers Research Station, Miscellaneous Publications 7, pp. 87–96.

Thomas, F.C., and D.O. Trainer. 1970. Bluetongue virus in white-tailed deer. *American Journal of Veterinary Research* 31:271–278.

Thomas, F.C., N. Willis, and G. Ruckerbaker. 1974. Identification of the viruses involved in the 1971 outbreak of hemorrhagic disease in southeastern United States white-tailed deer. *Journal of Wildlife Diseases* 10:187–189.

Thompson, L.H., J.O. Mecham, and F.R. Holbrook. 1988. Isolation and characterization of epizootic hemorrhagic disease virus from sheep and cattle in Colorado. *American Journal of Veterinary Research* 49:1050–1052.

Thorne, E.T., E.S. Williams, T.R. Spraker, W. Helms, and T. Segerstrom. 1988. Bluetongue in free-ranging pronghorn antelope (*Antilocapra americana*) in Wyoming: 1976 and 1984. *Journal of Wildlife Diseases* 24:113–119.

Trainer, D.O. 1964. Epizootic hemorrhagic disease of deer. *Journal of Wildlife Management* 28:377–381.

Tsai, K., and L. Karstad. 1973. The pathogenesis of epizootic hemorrhagic disease of deer: An electron microscopic study. *American Journal of Pathology* 70:379–400.

Van der Meyden, C.H., B.J. Erasmus, R. Swanepoel, and O.W. Prozesky. 1992. Encephalitis and chorioretinitis associated with neurotropic African horsesickness virus infection in laboratory workers: I. Clinical and neurological observations. *South African Medical Journal* 81:451–454.

Van Rensburg, I.B.J., J. de Clerk, H.D. Groenewald, and W.S. Botha. 1981. An outbreak of African horse sickness in dogs. *Journal of the South African Veterinary Association* 52:323–325.

Venter, E.H., J.J. van der Lugt, and G.J. Gerdes. 1993. Detection of bluetongue virus RNA in cell cultures and in the central nervous system of experimentally infected mice using in situ hybridization. *Onderstepoort Journal of Veterinary Research* 60:39–45.

Vosdingh, R.A., D.O. Trainer, and B.C. Easterday. 1968. Experimental bluetongue disease in white-tailed deer. *Canadian Journal of Comparative Medicine and Veterinary Science* 32:382–387.

Wechsler, S.J., and L.E. McHolland. 1988. Susceptibilities of 14 cell lines to bluetongue virus infection. *Journal of Clinical Microbiology* 26:2324–2327.

Wells, E.A. 1962. A disease resembling bluetongue occurring in topi (*Damaliscus korrigum ugandae*) in Queen Elizabeth National Park, Uganda. *Veterinary Record* 74:1372–1373.

Whetter, L.E., N.J. MacLachlan, D.H. Gebhard, H.W. Heidner, and P.F. Moore. 1989. Bluetongue virus infection of bovine monocytes. *Journal of General Virology* 70:1663–1676.

Wieser-Schimpf, L., W.C. Wilson, D.D. French, A. Baham, and L.D. Foil. 1993. Bluetongue virus in sheep and cattle and *Culicoides variipennis* and *C. stellifer* (Diptera: Ceratopogonidae) in Louisiana. *Journal of Medical Entomology* 30:719–724.

Wilbur, J.A., J.F. Evermann, R.L. Levings, I.R. Stoll, D.E. Starling, C.A. Spillers, G.A. Gustafson, and A.J. McKeirnan. 1994. Abortion and death in pregnant bitches associated with a canine vaccine contaminated with bluetongue virus. *Journal of the American Veterinary Medical Association* 204:1762–1765.

Wirth, W.W., and A.L. Dyce. 1985. The current taxonomic status of the *Culicoides* vectors of bluetongue viruses. In *Bluetongue and related orbiviruses,* ed. T.L. Barber, M.M. Jochim, and B.I. Osburn. New York: Alan R. Liss, pp. 151–164.

Woods, L.W., P.K. Swift, B.C. Barr, M.C. Horzinek, R.W. Nordhausen, M.H. Stillian, J.F. Patton, M.N. Oliver, K.R. Jones, and N.J. MacLachlan. 1996. Systemic adenovirus infection associated with high mortality in mule deer (*Odocoileus hemionus*) in California. *Veterinary Pathology* 33:125–132.

Young, E. 1969. The significance of infectious diseases in African game populations. *Zoologica Africana* 4:275–281.

4 ARBOVIRUS INFECTIONS

THOMAS M. YUILL AND CHARLES SEYMOUR

INTRODUCTION. Arboviruses (arthropod-borne viruses) replicate in blood-feeding arthropods, which transmit them to vertebrate hosts. The virus again replicates in the infected vertebrate, which develops a viremia and serves as a source of infectious blood meals for biting arthropods, completing the transmission cycle. A variety of blood-feeding arthropods are proven or suspected biologic vectors of these viruses. Viruses, such as poxes, that are transmitted only mechanically to vertebrate hosts by arthropods are excluded from the arboviruses, as are others classified as nonarboviruses by the American Committee on Arthropod-Borne Viruses (Karabatsos 1985).

There is virologic or serologic support for at least 215 arboviruses in eight families [*Togaviridae, Flaviviridae, Bunyaviridae, Rhabdoviridae, Reoviridae, Orthomyxoviridae, Coronaviridae, Herpesviridae,* and several viruses not yet classified to family (Karabatsos 1985)] infecting wild mammals in nature.

The majority of arboviruses infecting wild mammals are not known to cause disease in their hosts, though these hosts may play a role in maintaining the virus in nature. Wild mammals are incidental hosts for other arboviruses, which use avian maintenance hosts, or, in many instances, their status as vertebrate hosts in the epidemiology of these viruses is uncertain. Since few arboviruses, with the exception of bluetongue and epizootic hemorrhagic disease, cause overt disease or significant mortality in wild mammals, little consideration is given to their control or to their implications in wildlife management.

Here we discuss the arboviruses known to cause disease in wild mammals, except bluetongue and epizootic hemorrhagic disease, which are discussed elsewhere. Most viruses discussed here also cause disease in people and domestic animals.

YELLOW FEVER

Synonyms: Yellow jack, fiebre amarilla, fievre jaune, febris flava, black vomit, typhus amaril, typhus icteroide, and *Flavivirus hominis* fever.

History. Yellow fever virus (YFV) has caused devastating epidemics in the tropical Americas at least since the 1600s (Monath 1989, 1991). It was periodically introduced into coastal cities of North America, and even Europe, during the summer. The Yellow Fever Commission was established in the United States in 1900, and subsequently Walter Reed and his colleagues demonstrated that the disease was caused by a virus transmitted between people in urban settings by *Aedes aegypti* mosquitoes. Elucidation of sylvatic (jungle) cycles of zoonotic YFV transmission was begun in the early 1930s by Soper and colleagues.

Although much reduced, YF remains a problem, and the number of cases has increased, particularly in Africa over the past decade, to the greatest number since international reporting began in 1948 (Meegan 1994). Tropical Asia has remained strangely free of the disease, despite the abundance of *Ae. aegypti* and susceptible primates in many areas.

Etiology. YFV is the prototype flavivirus (antigenic group B) of the *Togaviridae.* It is considered a single serotype. There are three genotypes: I (East and Central Africa), IIA (West Africa), and a highly variable IIB (Americas) (Chang et al. 1995).

Transmission and Epidemiology. In tropical America and in Africa, YFV has urban and sylvatic cycles, but the wild mammalian hosts and mosquito vectors differ (Monath 1989, 1991).

In the Americas, in its sylvatic cycle, YFV is transmitted between monkeys in the forest canopy by *Haemagogus* spp. (principally *Hg. janthinomys*), *Sabethes chloropterus,* and *Aedes* (*Ochlerotatus*) *fulvus* mosquitoes. The main monkey hosts for YFV include tamarins (Callitrichidae), howler monkeys *Alouatta* spp., the squirrel monkey *Saimiri sciureus,* the night monkey *Aotus trivirgatus,* spider monkeys *Ateles* spp., and capuchin monkeys *Cebus* spp. (Monath 1989). YF can be fatal in howler monkeys, and finding these monkeys dead in the forest, or a sharp decline in their population, is often the most obvious indication that an epidemic of sylvan YF is occurring.

Local monkey populations are not large enough to sustain virus endemically. Hence, YF in primates occurs as a "wandering epidemic" moving slowly through the population over a large area, returning after affected local populations are replenished with susceptible individuals.

Several wild neotropical mammals could serve as sources of infection for efficient vectors such as *Haemagogus* spp. mosquitoes (Monath 1989). The role of rodents, kinkajous *Potos flavus,* and anteaters *Taman-*

dua tetradactyla and *Cyclopes didactylus* in the maintenance of YFV is unclear, although there is serologic evidence for infection.

In the classic neotropical "woodcutter's disease" cycle, tree cutting brings the mosquitoes to ground level, where mainly male forest workers are bitten. However, *Haemagogus* spp. will come to ground level in small agricultural clearings, infecting women and children (Clarke and Casals 1965). Infected people then transport the virus to populated areas where *Ae. aegypti* are abundant, initiating the urban cycle of transmission.

Transovarial transmission of YFV occurs in both *Ae. aegypti* (Aitkin et al. 1979) and *Haemagogus equinus* (Monath 1991) mosquitoes, which probably accounts for persistence of the virus in endemic areas.

In Africa, *Aedes africanus* transmits YFV to wild primates, especially colobus monkeys *Colobus guereza*, in the forest canopy. In the moist savannah and gallery forest "zones of emergence," *Ae. furcifer-taylori*, *Ae. bromeliae* (*Ae. simpsoni* group), and *Diceromyia* spp. transmit the virus among monkeys and to people (Cordelier 1991; Meegan 1994). Infected monkeys such as grivets *Cercopithecus aethiops* and guenons *Cercopithecus nictitans* come into adjoining agricultural areas, where they are bitten by *A. bromeliae*, an efficient monkey-to-human and person-to-person vector. In drier, brushy areas the YF vector is *Ae. luteocephalus*.

The principal African YFV primate hosts are guenons, grivets, or green monkeys *C. aethiops*, *Cercopithecus ascanius*, and *Cercopithecus mona*; patas or red monkeys *Erythrocebus patas*; baboons *Papio* spp.; colobus monkeys *C. guereza*, *Colobus polykomos*, and *Procolobus badius*; mangabeys *Cercocebus albigena*; bush babies *Galago senegalensis* and *Galago crassicaudatus*; and chimpanzees *Pan* spp. (summarized in Monath et al. 1980, McCrae and Kirya 1982, and Meegan 1994). Unlike neotropical primates, African monkeys infected with YFV seldom die, and epidemics may recur sooner in a given locality than in the Americas.

Though YFV was isolated from an epauleted fruit bat *Epomophorus* sp. in Africa, the role of bats in the epidemiology of YFV is unknown (Andral et al. 1968). Few African wild mammals other than primates develop viremias sufficient to infect vectors (Monath 1989), with the exception of a fat mouse, *Steatomys opimus* (Chippaux-Hyppolite and Chippaux 1969).

Although YFV isolates have been made from larvae and eggs of *Amblyomma variegatum* ticks and transmission accomplished in the laboratory (Germain et al. 1981), the epidemiologic significance of ticks as YFV vectors awaits further study.

Clinical Signs and Pathology. Disease in neotropical monkeys, including howlers, night monkeys, and marmosets, is more severe than in African primates. However, the bush baby may die from experimental YFV (Monath 1989).

Signs of disease in people and monkeys are similar (Bearcroft 1957), although disease may progress more quickly in monkeys (Monath et al. 1981). After a 3- to 6-day incubation period, there is sudden onset of fever and depression, headache, nausea, vomiting, lumbosacral pain, and myalgia in people (Monath 1989). Recovery may begin at this time, or there may be a biphasic course, with a subsequent toxic phase of recurrent fever and bradycardia, conjunctival congestion, gingival hemorrhage, epigastric pain, dehydration, and prostration. In severe cases, there is marked gastrointestinal and uterine hemorrhage, deepening jaundice and, in 20%–50%, renal failure leading to death.

At autopsy, the stomach and intestines may contain blood. The liver is enlarged and the lobular outlines may be obscured. Microscopically, degeneration and necrosis of renal tubular epithelium and midzonal areas of the liver occur. Necrotic liver cells may contain hyaline intracytoplasmic (Councilman) bodies, and hepatocytes commonly undergo fatty change. Lymphocytes are depleted in the spleen and lymph nodes.

Diagnosis and Treatment. Timely diagnosis is essential in control of YF (World Health Organization 1971). Clinical and epidemiologic observations help to establish a tentative diagnosis of YF. Definitive diagnosis is made by isolation of the virus, by detection of circulating antigens, by demonstration of a significant rise in specific YF antibodies, or by observation of characteristic liver lesions in fatal cases.

The virus can be isolated from blood early in the disease or from liver in fatal cases. The virus can be isolated in suckling mice (Karabatsos 1985), in several mammalian and mosquito *Ae. pseudoscutellaris* cell lines, or in intrathoracically inoculated mosquitoes *Toxorhynchites* spp. or *Ae. aegypti* (Monath 1989). Antigen-capture enzyme-linked immunosorbent assay (ELISA) is as sensitive as virus isolation and is faster (Saluzzo et al. 1985).

A variety of serologic tests can be used to detect antibody. However, significant cross-reaction with other flaviviruses may complicate serologic interpretation. An immunoperoxidase monolayer assay is less subject to cross-reaction (Soliman et al. 1997).

Histologic examination of liver may be the only practical diagnostic method for sylvatic YF in remote areas. Viral antigens may be demonstrated in liver sections by immunohistochemistry (Hall et al. 1991).

No specific treatment is available. Fluid and blood replacement is indicated in human patients with significant vomiting or hemorrhage (Clarke and Casals 1965).

Control and Public Health Implications. Human disease can be prevented by the 17D vaccine. Although revaccination is required every 10 years, antibody titers persist at least 30–35 years (Monath 1989). Vaccination programs must have sufficient continuity to maintain a high prevalence of immunity in the population; several YF outbreaks occurred in Africa after mass vaccination programs were discontinued (Monath 1989).

The urban YF cycle in the Americas can be controlled by reduction of *Ae. aegypti* populations. However, *Ae. aegypti* control has stalled in Central and South America. Mosquito control is costly and often competes unsuccessfully with priority social and economic programs in developing countries (Schliessmann and Calheiros 1974).

In the neotropics, although aerial application of ultra-low-volume malathion has been attempted (Monath 1991), elimination of the sylvatic cycle is not feasible. The remoteness from medical attention of people at risk to sylvatic YF and the continued expansion of human populations into forested areas suggest that sylvatic YF will persist (Monath 1991). Control of domestic *Ae. aegypti* vectors through elimination of breeding sites, or by application of adulticides such as malathion, rarely has been sustained over large areas (Monath 1989).

ST. LOUIS ENCEPHALITIS

History. St. Louis encephalitis virus (SLEV), a mosquito-borne flavivirus, causes sporadic but extensive human epidemics in North America. First isolated from a fatal human case in St. Louis, Missouri (Muckenfuss et al. 1933), it has since been found from Canada to Argentina. It also has been isolated from diseased gray foxes *Urocyon cinereoargenteus* in California (Emmons and Lennette 1967).

Transmission and Host Range. *Culex* spp. mosquitoes transmit SLEV in North America: *C. tarsalis* is the vector in the west, *C. pipiens* in the north, C. *pipiens-quinquefasciatus* in the south and central areas, and *C. nigripalpus* in Florida (Luby 1994).

Birds are the chief amplifying hosts throughout North America. However, in the southeastern United States, wild mammals may play an epidemiologic role. In Florida and Mississippi, SLEV and antibody occur in raccoons *Procyon lotor,* Virginia opossums *Didelphis virginiana,* and cotton rats *Sigmodon hispidus* (McLean and Bowen 1980). Antibody has been found in Florida in cotton mice *Peromyscus gossypinus* and armadillos *Dasypus novemcinctus* (Bigler and Hoff 1975; Day et al. 1996), and SLEV antibody increased in prevalence in armadillos following an SLE epidemic in people in Florida, indicating possible involvement in transmission (Day et al. 1995).

In Ohio, antibody was found in big brown bats *Eptesicus fuscus* and little brown bats *Myotis lucifugus.* Experimentally, these bats harbored the virus through hibernation and became viremic on arousal (Herbold et al. 1983). In Ontario, SLEV antibody was detected in coyotes *Canis latrans,* red foxes *Vulpes vulpes,* striped skunks *Mephitis mephitis,* and raccoons (Artsob et al. 1986). Elsewhere in North America, wild mammals are probably insignificant as hosts of SLEV, compared with birds.

On the Pacific coast, other than the aforementioned gray fox isolate, evidence for SLEV infection of wild mammals is only serologic. From Kern County, California, to 52°N in British Columbia, antibodies have been detected in black-tailed jackrabbits *Lepus californicus,* snowshoe hares *Lepus americanus,* Nuttall's cottontails *Sylvilagus nuttallii,* Fresno and Heermann's kangaroo rats *Dipodomys nitratoides* and *Dipodomys heermanni,* southern grasshopper mice *Onychomys torridus,* deer mice *Peromyscus maniculatus,* harvest mice *Reithrodontomys megalotis,* San Joachin antelope ground squirrels *Ammospermophilus nelsoni,* California ground squirrels *Spermophilus beecheyi,* yellow-bellied marmots *Marmota flaviventris,* wood rats *Neotoma fuscipes,* pocket gophers *Thomomys talpoides,* house mice *Mus musculus,* black rats *Rattus rattus,* and Norway rats *Rattus norvegicus* (Hardy et al. 1974; McLean and Bowen 1980).

In Texas, SLEV has been isolated year-round from Mexican free-tailed bats *Tadarida brasiliensis,* and hibernating bats may serve as overwintering hosts (Allen et al. 1970). Antibodies to SLEV also occur in white-tailed deer *Odocoileus virginianus* in Texas (Trainer and Hanson 1969). In north-central North America (Wisconsin, Wyoming, North Dakota, and Alberta), SLEV antibodies have been detected in snowshoe hares, moose *Alces alces,* a pronghorn antelope *Antilocapra americana,* and in both white-tailed and mule deer *Odocoileus hemionus* (Trainer and Hanson 1969; Trainer and Hoff 1971; Hoff et al. 1970, 1973).

In South and Central America, the cycle of SLEV remains unclear. At least seven genera of mosquitoes have yielded SLEV (Karabatsos 1985). As in temperate North America, in the tropics most of the isolates from vertebrates and serologic reactors have been in birds. In Argentina, however, SLEV was isolated from a vesper mouse *Calomys musculinus* and a house mouse (M.S. Sabattini, cited in McLean and Bowen 1980). In Brazil, SLEV was isolated from the grass mouse *Akodon arviculoides,* a rice rat *Oryzomys nigripes, Gigantolaelaps* mites combed from another rice rat *Oryzomys macconnelli* and from a black rat, southern opossums *Didelphis marsupialis,* a howler monkey *Alouatta nigerrina,* a spider monkey *Ateles paniscus,* and a three-toed sloth *Bradypus tridactylus* (Woodall 1967; McLean and Bowen 1980). Three-toed sloths *Bradypus variegatus* and two-toed sloths *Choloepus hoffmanni* in Panama have the highest prevalence of SLEV antibody of all forest birds and mammals tested (Gorgas Memorial Laboratory 1977) and may be significant amplifying hosts (Seymour et al. 1983). Spiny rats *Proechimys semispinosus* in Panama and, in Guatemala, the neotropical fruit bat *Artibeus intermedius,* the Jamaican or Mexican fruit bat *Artibeus jamaicensis,* the great fruit-eating bat *Artibeus lituratus,* the dwarf fruit-eating bat *Artibeus phaeotis,* long-tongued bats *Glossophaga soricina,* and yellow-shouldered bats *Sturnira lilium* also have SLEV antibody (De Rodaniche and Galindo 1961; Ubico and McLean 1995).

Clinical Signs, Pathogenesis, and Pathology. The gray fox from which SLEV was isolated by Emmons and Lennette (1967) was suspected of rabies. A second isolation was made under similar circumstances from another California gray fox, but no histologic studies were made of either animal.

Inoculation of SLEV by peripheral routes has not caused disease in any wild mammals. Of those species known to be naturally infected by SLEV, inoculation has consistently produced viremia and antibody in Virginia opossums, a single gray fox (Kokernot et al. 1969), antelope ground squirrels, Fresno kangaroo rats, both types of sloth (R.G. McLean and Bowen 1980), and big brown and little brown bats (Herbold et al. 1983).

Variable results, in which some, but not all, animals became viremic or developed antibodies, have been reported for Mexican free-tailed bats (Sulkin et al. 1963), cotton rats, and black-tailed jackrabbits (McLean and Bowen 1980). Raccoons did not develop detectable viremia but were not tested for antibody (Kokernot et al. 1969). Black rats and deer mice were refractory to SLEV infection (McLean and Bowen 1980). Of these species, the longest and among the most intense experimental viremias occurred in sloths and free-tailed bats, particularly in hibernating bats.

In people, disease ranges from inapparent infection, through mild febrile illness, to coma and fatal encephalitis. Disease is usually more severe in elderly people than in the young and can leave neurologic sequelae. Histologic changes are limited to the brain and spinal cord and include perivascular cuffing; glial, lymphocytic, and histiocytic nodules; widespread neuronal degeneration; and marked reduction of Purkinje cells (reviewed in Gardner and Reyes 1980).

Diagnosis. Diagnosis depends on virus isolation or on a fourfold or greater increase in antibody titer in paired sera. The virus is most easily isolated by intracranial inoculation of suckling mice. It is identified with specific antisera by using hemagglutination inhibition, virus neutralization, or complement fixation. The same tests are used to detect antibody in serum. Immunoglobulin M (IgM)-capture ELISA is useful for rapid, early diagnosis of SLE in people (Monath et al. 1984). Antigen-capture enzyme immunoassay (Tsai et al. 1987) and reverse transcriptase-polymerase chain reaction (Nawrocki et al. 1996) can be used to detect SLEV antigens or genomic sequences in mosquitoes.

Treatment and Control. There is presently no licensed vaccine, although recombinant vaccinia and baculovirus vaccines (Venugopal et al. 1995) and plasmid-expressed SLEV proteins (Phillpotts et al. 1996) have protected mice. The most effective preventive measures are elimination of mosquito-breeding sites and spraying to control mosquito vector populations (Mitchell et al. 1980).

The appearance of SLEV antibodies in the sera of juvenile birds, such as house sparrows, or of virus in mosquitoes is a signal to begin intensive mosquito control (Lord et al. 1974; Luby 1994), and the public should be warned to wear repellants and attempt to avoid mosquito bites. However, extensive SLEV outbreaks continue to occur in spite of control measures.

OMSK HEMORRHAGIC FEVER

Introduction and Etiology. Omsk hemorrhagic fever virus (OHFV) causes outbreaks of severe hemorrhagic disease in people and muskrats *Ondatra zibethica* in Omsk, Novosibirsk, Kurgan, and Tyumen oblasts in western Siberia (Lvov 1988).

OHFV is a member of the tick-borne encephalitis complex of flaviviruses, with Central European encephalitis, Russian spring-summer encephalitis, louping-ill, Powassan, Negishi, and Kyasanur Forest disease viruses. Clarke (1964) and Kornilova et al. (1970) described two antigenic varieties of OHFV.

Transmission and Host Range. The ticks *Dermacentor reticulatus* and *Ixodes apronophorus* are the principal vectors of OHFV, but others, including *D. marginatus,* also may serve as vectors (Lvov 1988).

In moist areas, OHFV also may be transmitted through rodent urine (Lvov 1988). OHFV was isolated from the urine of wild muskrats and water voles *Arvicola terrestris* and from the urine of experimentally infected muskrats, water voles, root voles *Microtus oeconomus,* and red-cheeked susliks *Spermophilus (Citellus) erythrogenys.* Muskrats and root voles become infected by the oral route. Direct transmission also occurs to hunters and trappers who handle infected muskrats.

Case fatality rates of 80% have been reported in muskrats. However, the true maintenance hosts are probably small rodents and insectivores. Water voles, root voles, and European common shrews *Sorex araneus* have yielded isolates, and sera from 14 species of rodents and shrews have reacted positively against OHF antigen (reviewed in Lvov 1988). Reports of hemagglutination inhibition (HI) antibody in birds are inconclusive, because Eurasian birds are frequently infected by cross-reacting flaviviruses.

Pathogenesis, Pathology, and Clinical Signs. Muskrats, imported into Siberia from North America in 1928, may experience considerable epidemic mortality (Lvov 1988). Inoculation of OHFV into muskrats produces severe hemorrhagic disease, with swelling and necrosis of vascular endothelium and perivascular edema and hemorrhage. Acute lymphadenitis and focal necrosis in the brain are reported. Viremia lasts until death, which may occur up to 23 days after inoculation (Marenko et al. 1974). In another study, 90% of 50 inoculated animals died with clinical signs of weakness, depression, and lethargy. Only seven developed paresis or paralysis, and hemorrhage was reported for a single muskrat (Fedorova 1966).

Infection of red-cheeked susliks produced transient lethargy and an antibody response but no pathologic changes after recovery (Kharitonova 1969). Narrow-skulled voles *Microtus gregalis* developed lethargy, weakness, and spasms, and OHFV was isolated from their brains (Gagarina 1965). Infection with OHFV did not make water voles ill (Dunaev 1976).

Human OHF ranges from a mild febrile illness to fatal hemorrhaging. The case fatality rate is 0.5%–3.0%. Infection is typically marked by leukopenia with a pronounced mononucleosis, low blood pressure, respiratory distress, a biphasic fever, and often hemorrhage. Long-lasting sequelae include limb pain, headache, poor memory and hearing, hair loss, a bleeding tendency, and loss of energy (summarized in Lvov 1998).

Diagnosis. Symptoms and epidemiology may be helpful in differential diagnosis of some human cases. However, definitive diagnosis is based on virus isolation in suckling mice or cell cultures and on identification using specific antisera. Convalescent sera react with OHFV antigen in a variety of serologic tests (Lvov 1988), but results must be interpreted with caution due to possible cross-reactions with related viruses. PCR should be sensitive and specific for detection of OHFV RNA in tissue.

Treatment and Control. Tick avoidance by forest workers, and sanitary precautions by muskrat trappers, may help reduce human infection rates. Although no OHF vaccine is available, tick-borne encephalitis virus vaccines provide some cross-protection and have been used in areas of high OHF risk (Monath and Heinz 1996).

LOUPING-ILL

Introduction and Etiology. Louping-ill virus (LIV), a member of the tick-borne encephalitis complex of flaviviruses, causes an often fatal encephalitis in sheep, and occasional disease in horses, humans, pigs, and cattle (West 1975; Reid 1988).

Originally described in the United Kingdom and Ireland, louping-ill also occurs in Norway (Gao et al. 1993) and perhaps more widely in Europe (Reid 1988).

Transmission and Host Range. LIV is transmitted by the tick *Ixodes ricinus.* The virus has been isolated from the red grouse *Lagopus lagopus,* wood mouse *Apodemus sylvaticus,* European shrew, blue or mountain hare *Lepus timidus,* Old World badger *Meles meles,* red deer *Cervus elaphus,* and roe deer *Capreolus capreolus* (reviewed in Reid 1988). In Scotland, antibody was detected in all of nine hares *Lepus* sp. (Karabatsos 1985) and in 30% of 324 roe deer, red deer, and sika deer *Cervus nippon* (Adam et al. 1977). In England, antibody was found in field voles *Microtus agrestis* and bank voles *Clethrionomys glareolus* (Kaplan et al. 1980). However, only sheep and red grouse appear to be important maintenance hosts in Scotland (Reid 1988), and LIV poses only a minor threat to wild mammals.

Pathology and Pathogenesis. Mild neuronal degeneration accompanied by a minimal nonsuppurative inflammatory response, interpreted as an early response to viral infection, was evident in the brain of a roe deer fawn from which LIV was isolated (Reid et al. 1976). Young red and roe deer experimentally exposed to LIV remained normal clinically, but all developed severe nonsuppurative meningoencephalitis (Reid et al. 1982).

Diagnosis. Definitive diagnosis is made by virus isolation using intracerebral inoculation of suckling mice or cell cultures and subsequent identification with specific antisera (Reid 1988). Viral antigens also can be detected by immunohistochemistry in tissue sections (Krueger and Reid 1994). Louping-ill may be diagnosed presumptively based on a rise in antibody titer in paired sera.

Treatment and Control. Prevention by vaccination may be the only feasible approach in domestic stock (Brotherston and Boyce 1970). Otherwise, control is limited to tick reduction in infested pastures and to prophylactic dipping against ticks. However, there seems to be little justification for attempting control measures in wildlife.

POWASSAN VIRUS

History and Etiology. Powassan virus (POWV) was isolated in 1958 from a fatal human case of encephalitis in Ontario (McLean and Donohue 1959). A member of the tick-borne encephalitis antigenic complex of flaviviruses, it is distributed across northern North America from California and British Columbia to New York, and has been reported from eastern Siberia. A virus similar to POWV, provisionally named deer tick virus, was isolated from *Ixodes scapularis* collected in the northeastern United States (Telford et al. 1997). POWV occasionally causes human encephalitis and was the apparent cause of fatal encephalitis in a wild-caught gray fox (Whitney and Jamnback 1965).

Transmission and Host Range. POWV is transmitted by hard ticks (summarized by Artsob 1988). It has been isolated from *Dermacentor andersoni* in Colorado and in South Dakota. Lvov et al. (1974) reported the isolation of Powassan virus from *Haemaphysalis neumanni, Ixodes persulcatus,* and *Dermacentor silvarum* ticks in far-eastern Russia. Most tick isolates have been from *I. cookei* (parasitic mainly on woodchucks or groundhogs *Marmota monax*) but also from *I. spinipalpus* and *I. marxi* (associated with squirrels) in North America. Powassan virus was isolated from woodchucks and red squirrels *Tamiasciurus hudsonicus* in Ontario (McLean et al. 1967) and from woodchucks and their ticks in New York State (Whitney and Jamnback 1965). *Ixodes scapularis* is an efficient vector in the laboratory, and transovarial transmission of the virus occurs (Costero and Grayson 1996).

Antibody to POWV was found in foxes (not speciated) and raccoons. The virus was isolated from the brains of a fatally ill gray fox in New York and an ill red fox in West Virginia. The virus was isolated from a striped skunk and a long-tailed weasel *Mustela frenata* in Massachusetts, from *I. cookei* collected from them (Main et al. 1979), and from a spotted skunk *Spilogale putorius* in California (Johnson 1987). In South Dakota, Eklund (1963) recovered POWV from deer mice and from *I. spinipalpis* ticks found on them.

Antibody to POWV has been found in a variety of wild mammals across a wide geographic area in North America. In Ontario, antibody against POWV was most prevalent in woodchucks and red squirrels. However, gray squirrels *Sciurus carolinensis,* chipmunks *Tamias striatus,* snowshoe hares, porcupines *Erethizon dorsatum,* coyotes, and striped skunks were also positive. There is serologic evidence for POWV in snowshoe hares in Alberta (Zarnke and Yuill 1981a). Antibody was found in woodland jumping mice *Napaeozapus insignis* in New York. In South Dakota, antibody was detected principally in deer mice but also in meadow voles *Microtus sp.* and chipmunks *Eutamias* sp. In British Columbia, McLean et al. (1970) detected POWV antibody in ground squirrels *Spermophilus columbianus* and *S. lateralis,* chipmunks *Eutamias amoenus,* red squirrels, and yellow-bellied marmots.

Clinical Signs, Pathology, and Pathogenesis. The wild gray fox from which POWV was isolated was twitching involuntarily when captured and died 48 hours later (Whitney and Jamnback 1965). Parenteral inoculation of skunks, red foxes, gray foxes, and woodchucks results in symptomless viremia (Kokernot et al. 1969) whereas, in gray squirrels, it produces viremia in adults and death in neonates (Timoney 1971). Microscopic lesions in POWV-infected wild mammals have not been described. However, the nonsuppurative encephalitis in fatal human cases is indistinguishable from other fatal flavivirus encephalitides (McLean and Donohue 1959). POWV is suspected to cause nonsuppurative encephalitis seen in horses in Ontario, Canada (Little et al. 1985).

Diagnosis. Diagnosis requires virus isolation by intracerebral inoculation of suckling mice or cell cultures, followed by identification using specific antisera. Demonstration of viral RNA by nested reverse transcriptase-PCR provides rapid clinical diagnosis (Meiyu et al. 1997). Demonstration of specific IgM or a fourfold rise in antibody titer in paired sera is also diagnostic of POWV infection.

Treatment, Control, and Public Health Implications. There is no vaccine against POWV for use in people or animals. Control of vector ticks and wild vertebrate host populations is impractical.

KYASANUR FOREST DISEASE

History and Etiology. In 1956, an epidemic of fatal disease occurred in people and monkeys near Kyasanur Forest, Shimoga District, Karanataka (formerly Mysore) State, southern India. In 1957, the etiologic agent was isolated and named Kyasanur Forest disease virus (KFDV) (Work and Trapido 1957). It is a member of the tick-borne encephalitis flavivirus complex.

Though it has been isolated only within a 2500-km^2 area in Karanataka State, KFD has been spreading slowly (Banerjee 1988). The prevalence of disease and location of foci of transmission vary from season to season and year to year. The disease typically appears in early spring and lasts until midsummer (Work et al. 1959); mortality in black-faced or Hanuman langurs *Semnopithecus* (*Presbytis*) *entellus* and bonnet macaques *Macaca radiata* is highest from February to April (Goverdhan et al. 1974).

Transmission. The principal vectors of KFDV are *Haemaphysalis* ticks; many species have yielded KFDV isolates (Boshell et al. 1968b; Banerjee 1988), but the principal vectors to monkeys are *H. spinigera* and *H. turturis* (Trapido et al. 1964; Boshell and Rajagopalan 1968; Boshell et al. 1968b). The virus is transmitted transstadially but not transovarily in these ticks (Banerjee 1988), and it persists in nymphs between outbreaks (Rajagopalan and Anderson 1970).

Ixodes petauristae, as well as four other *Haemaphysalis* spp. found infected in nature can transmit KFDV to vertebrates (Singh et al. 1968; Bhat et al. 1975). *Ornithodoros* sp. ticks collected from the roosts of infected bats during an epidemic in primates and people yielded KFDV (Rajagopalan et al. 1969); the related tick, *O. crossi,* transmitted KFDV experimentally (Bhat and Goverdhan 1973).

Host Range. Hanuman langurs and bonnet monkeys are frequently infected (Goverdhan et al. 1974; Banerjee 1988). Roving bands of monkeys may be important in the expansion of old foci of KFDV transmission and in initiating new foci. However, the natural histories of closely related viruses, combined with the slow rate of turnover of monkey populations, suggest that monkeys are not the primary vertebrate hosts of KFDV.

Small rodents and shrews are the probable maintenance hosts (Banerjee 1988). KFDV has been isolated from the black rat *Rattus rattus,* from the forest rat *Cremnomys (Rattus) blanfordi,* and from house or musk shrews *Suncus murinus.* Antibodies against KFDV have been found in these three species, as well as in striped palm squirrels *Funambulus tristriatus,* large naked-soled gerbils *Tatera indica,* and forest mice *Mus booduga* (Karabatsos 1985). Viremic palm squirrels, shrews, and *C. blanfordi* are infective for ticks (Singh et al. 1968). Shrews and *C. blanfordi* are particularly important maintenance hosts of KFDV, though the other rodents, except black rats, possibly are efficient hosts, as well.

The role of other vertebrates in the epidemiology of KFDV is poorly understood. Although KFDV has been isolated from insectivorous horseshoe bats *Rhinolophus rouxi,* and antibody against KFDV has been detected on 7 of 17 bat species (summarized in Banerjee 1988), their role in KFDV transmission is uncertain (Pavri and Singh 1968; Theiler and Downs 1973). In experimentally infected short-nosed fruit bats *Cynopterus sphinx,* however, KFDV causes viremia and death, and high virus concentrations may be found in salivary glands (Pavri and Singh 1968). The crested porcupine, *Hystrix indica,* harbors immature stages of the tick vectors and develops high-titered viremias, suggesting that it may play a role in virus amplification (Banerjee 1988). Although wild birds are important hosts for vector tick species, KFDV has never been isolated from them (Theiler and Downs 1973). Antibody has been detected rarely in wild birds from endemic areas.

Clinical Signs, Pathology, and Pathogenesis. Epidemic KFD is typically hemorrhagic in monkeys and humans. In people, KFD passes through four week-long clinical stages: a prodrome with fever; hypotension and hepatomegaly; hemorrhage with neurologic signs or bronchopneumonia; and recovery, perhaps with late fever (Adhikari Prabha et al. 1993).

In recently dead Hanuman langurs and bonnet monkeys, there were blood clots at the anus, and swelling and pallor of the adrenal cortex. Microscopically, there was focal hepatocyte necrosis, renal tubular degeneration, and mild nonsuppurative encephalitis. Hemorrhages were present in lungs, kidneys, brain, and adrenals in most animals, and virus was isolated from a variety of tissues (Iyer et al. 1960).

Experimental infections of bonnet monkeys produced diarrhea, bradycardia, hypotension, and death (Webb and Chatterjea 1962) with KFDV-specific gastrointestinal and lymphoid lesions (Kenyon et al. 1992). Marked leukopenia, anemia, and thrombocytopenia may have been mediated immunologically. Other studies (Webb and Burston 1966) showed spinal cord lesions. The paucity of hemorrhages may have been due to the use of a mouse-adapted KFDV.

Inoculation of KFDV into *C. blanfordi, R. rattus, S. murinus,* and *Mus* sp. produced a symptomless viremia; infection of striped palm squirrels caused high viremia and was frequently fatal (Webb 1965; Boshell et al. 1968a). In short-nosed fruit bats, KFDV can cause weakness, paralysis, and death (Pavri and Singh 1968).

Diagnosis. The restricted distribution, seasonality, and hemorrhagic nature of KFD are distinctive. Diagnosis is confirmed by virus isolation in suckling mice or in cell cultures, and identification using specific antisera, or serologically by demonstration of specific antibody in paired sera, is recommended (Banerjee 1988).

Treatment, Control, and Public Health Implications. Treatment in people is supportive; no measures to treat or control the disease are appropriate in wildlife. Repellants such as DEET and pyrethrum have been recommended to reduce the probability of ticks feeding on people (Banerjee 1988). An inactivated viral vaccine reduced the occurrence of KFD in naturally exposed people (Dandawate et al. 1994).

TENSAW VIRUS

History and Etiology. Tensaw virus (TENV) is a bunyavirus belonging to the Bunyamwera antigenic group of the Bunyamwera supergroup. It was first isolated in 1960 from *Anopheles crucians* mosquitoes caught near the Tensaw River, Alabama (Chamberlain et al. 1969), and seems restricted to the southeastern United States. It has been isolated from a sick gray fox in Florida (Bigler and Hoff 1975) and from one sick person (McGowan et al. 1973).

Transmission and Host Range. Although TENV has been isolated many mosquito species, most have been from *An. crucians* (Chamberlain et al. 1969; Wellings et al. 1972). Wild mammals from which TENV has been isolated, other than the gray fox, include marsh rabbits *Sylvilagus palustris* and a sentinel cotton rat, all from Florida (Wellings et al. 1972). Antibody has been found in raccoons in Georgia and Florida, and in marsh rabbits, as well as in people, cattle, and dogs (Chamberlain et al. 1969; Bigler and Hoff 1975).

Clinical Signs, Pathogenesis, and Pathology. The gray fox from which TENV was isolated was a rabies suspect with nervous signs, but no lesions were described (Bigler and Hoff 1975). Inoculation of dogs, New England cottontail rabbits *Sylvilagus transitionalis,* raccoons, and cotton rats produced a symptomless viremia and an antibody response (Sudia et al. 1969). Gray squirrels *Sciurus carolinensis,* chipmunks *Tamias striatus,* and Virginia opossums did not become viremic, and some did not develop antibody. TENV has caused a single reported case of human encephalitis (McGowan et al. 1973).

Diagnosis. Diagnosis depends on virus isolation in suckling mice or cell cultures and identification using group and monotypic reagent antisera, or a fourfold or greater rise in antibody titer in paired sera.

Treatment, Control, and Public Health Implications. Since TENV is not a major human or animal pathogen, there are no vaccines, vector control measures, or specific treatment procedures.

SNOWSHOE HARE VIRUS

Intoduction and Etiology. Snowshoe hare virus (SSHV) is a member of the California encephalitis complex of the Bunyamwera supergroup in the *Bunyaviridae* and is a variant of La Crosse virus. First isolated in Montana in 1959 from a sick snowshoe hare (Burgdorfer et al. 1961), it occurs across northern North America from Alaska, through Canada, to upper New York State (summarized in Artsob 1983, Srihongse et al. 1984, and Calisher et al. 1986b). Occasionally, SSHV causes encephalitis in people in eastern Canada (Artsob 1983) and New York (Srihongse et al. 1984) but is not known to cause disease in domestic animals.

Transmission. SSHV is mosquito-borne. Isolates have come from two species of the genus *Culiseta* and from at least ten different *Aedes* species (Newhouse et al. 1967; Iversen et al. 1969; McLean et al. 1970; McLean 1983; Ritter and Feltz 1974). Of these, *Ae. cinereus* has yielded isolates from the greatest geographic range: Alaska, Yukon, and upper New York State (Whitney et al. 1969; Ritter and Feltz 1974; McLean 1983). In Alaska, Yukon, and Northwest Territories, and in Alberta, SSHV has been isolated from wild adult *Ae. communis* or *communis*-group mosquitoes (Iversen et al. 1969; Hoff et al. 1971; McLean et al. 1973, 1975, 1977c; Ritter and Feltz 1974), and *Ae. communis* is an efficient experimental vector (McLean et al. 1977a).

SSHV probably passes to progeny mosquitoes through the overwintering egg (McLintock et al. 1976; McLean et al. 1977b).

Vertebrate Host Range. Snowshoe hares are the principal vertebrate hosts of SSHV. As well as in Montana, SSHV has been isolated from snowshoe hares in Alberta (Hoff et al. 1971) and in Alaska (Ritter and Feltz 1974), where SSHV also has been isolated from a northern red-backed vole *Clethrionomys rutilus* and from a collared lemming *Dicrostonyx rubicans* (Ritter and Feltz 1974). Antibodies have been detected in snowshoe hares from British Columbia (McLean et al. 1971), Yukon Territory (McLean et al. 1975), Alberta (Zarnke and Yuill 1981b), Northwest Territories (Gaunt et al. 1974), Montana (Newhouse et al. 1967), northern Michigan (Newhouse et al. 1963), and upper New York State (Whitney et al. 1969).

Antibodies to SSHV are less prevalent in other wild mammals and do not occur in birds. Antibody has been found in arctic ground squirrels *Spermophilus parryii* and golden-mantled ground squirrels *Spermophilus lateralis* in the Yukon Territory, British Columbia, and Montana (Newhouse et al. 1967; McLean et al. 1975). In British Columbia, antibodies also have been detected in yellow-bellied marmots and yellow-pine chipmunks *Eutamias amoenus* (McLean et al. 1970). In the Yukon Territory, red squirrels have antibodies (McLean et al. 1975). Zarnke and Yuill (1981b) detected SSHV antibody in Alberta black bears *Ursus americanus,* moose, and bighorn sheep *Ovis canadensis.* In northern New York State, Whitney et al. (1969) reported antibodies in eastern cottontail rabbits *Sylvilagus floridanus,* woodchucks, raccoons, red foxes, a striped skunk, a North American porcupine *Erethizon dorsatum,* a Norway rat, and white-tailed deer.

SSHV was isolated from *Ae. euedes* mosquitoes collected in the upper Ob River, western Siberia, Russia, and prevalence of antibody was high in Old World field mice *Apodemus agrarius,* the northern red-backed vole, the common vole *Microtus arvalis,* and the European common shrew (Mitchell et al. 1993).

Pathology and Pathogenesis. Gross or microscopic abnormalities were not reported in the "emaciated, rather sluggish" Montana hare from which the first isolate originated (Burgdorfer et al. 1961). Inoculation of snowshoe hares and ground squirrels resulted in a symptomless viremia (Newhouse et al. 1967), suggesting that SSHV infection was coincidental in the Montana hare. In Alberta, however, wild hares convalescent from SSHV infection experienced a higher mortality rate than uninfected hares (Yuill et al. 1969).

Treatment, Control, and Public Health Implications. Neither vaccines nor vector control are used, because disease due to SSHV is uncommon in people, unknown in domestic animals, and probably uncommon in wild mammals. Control of the hordes of arctic and boreal *Aedes* mosquito vectors is impractical.

WESTERN EQUINE ENCEPHALITIS

Synonyms: Western encephalitis, western equine encephalomyelitis.

History and Etiology. Western equine encephalitis virus (WEEV) was first isolated from horses in California (Meyer et al. 1931). Although WEEV is distributed across the United States, northward into western Canada and southward into South America, it causes epidemics only in far-western North America, in the prairie states and provinces, and in Argentina and Uruguay (Reisen and Monath 1989).

WEEV is a complex of antigenically related viruses in the *Alphavirus* genus (group A) of the *Togaviridae.* WEEV arose by recombination between an eastern equine encephalitis-like virus and a Sindbis-like virus (Hahn et al. 1988). The WEEV complex is comprised of closely related species from different geographic areas and, in addition to prototype WEEV, include Fort Morgan, Buggy Creek, and Highlands J viruses (Calisher and Karabatsos 1988; Johnson and Peters 1996). Only WEEV and, rarely, Highlands J virus cause disease in people or in animals. This discussion is limited to prototype WEEV.

Transmission and Host Range. The cycles of WEEV vary in western North America and in South America. In the west, *Culex tarsalis* is the principal vector (Hardy et al. 1976). *Aedes melanimon,* a mammal feeder, also is a WEEV vector (Hardy et al. 1974).

Although birds long have been recognized as important vertebrate hosts of WEEV (Holden et al. 1973), wild mammals may play a role in some areas. *Culex tarsalis,* usually a bird feeder, feeds preponderantly on black-tailed jackrabbits in some localities (Nelson et al. 1976), and prevalence of antibody in jackrabbits is high. Jackrabbits infected with some strains of WEEV develop viremias sufficient to infect mosquitoes (Hardy et al. 1974), and transmission involving *Ae. dorsalis* and probably jackrabbits occurred in Utah (Smart et al. 1972).

WEEV has been isolated periodically from western gray squirrels *Sciurus griseus,* California ground squirrels and San Joachin antelope ground squirrels, house mice, a vole *Microtus* sp. (Karabatsos 1985), and a Virginia opossum (Emmons and Lennette 1969). Several of the squirrels were aggressive and had bitten people. The role that these mammals play (if any) in the maintenance of WEEV in the west is not clear, since many of these species suffer fatal infections during the summer transmission period. However, some have been shown to survive normally fatal infection in the winter, with the onset of viremia delayed until the following spring (Hardy et al. 1974).

In the prairies and aspen parklands of western Canada, *C. tarsalis, Ae. melanimon,* and *Culiseta inornata* are WEEV vectors (Burton and McLintock 1970; Hayles et al. 1972). Wild mammals may also be WEEV hosts in the north. Snowshoe hares (Yuill and Hanson 1964; Kiorpes and Yuill 1975) and Richardson's ground squirrels *Spermophilus richardsonii* (Leung et al. 1975, 1976) develop significant viremias. Transmission may occur among Richardson's ground squirrels by cannibalism or via contaminated urine (Leung et al. 1977).

The mechanisms for overwintering of WEEV throughout the temperate zone are unknown. It has been isolated from hibernating snakes and frogs in the west (Gebhardt et al. 1973). Experimentally infected Texas tortoises *Gopherus berlandieri* develop viremias long enough to enable overwintering of virus in warm climates (Bowen 1977). Virus may also overwinter in hibernating infected mosquitoes, by persistent infection of warm-blooded vertebrates, or by reintroduction in migrating birds or bats (Hayes and Wallis 1977).

Sporadic WEE cases have occurred in horses but not in people, in Argentina and Uruguay. In Argentina, WEEV has been isolated from mosquitoes, and antibody prevalence was low in birds but higher (10%) in European brown hares *Lepus europaeus. Aedes albifasciatus* has been implicated as a WEEV vector in northern Argentina (Aviles et al. 1992).

Clinical Signs and Pathology. Richardson's ground squirrels infected with WEEV became weak, depressed, and ataxic as their limbs became paralyzed before they died (Leung et al. 1976, 1977). A few individuals became hyperexcitable. Grossly, vascular congestion and petechial hemorrhages were seen in the cerebral hemispheres. Histologic changes were seen most frequently in the olfactory bulbs, cerebral cortex, and leptomeninges and were marked by severe necrotizing nonsuppurative meningoencephalitis. Terminally, there was marked hemorrhage into neuropil, and edema occurred in the brainstem and cerebrum.

Diagnosis and Treatment. Diagnosis is established by isolation of the virus, by demonstration of viral RNA by reverse transcriptase-PCR (Vodkin et al. 1994), or by development of specific neutralizing antibodies in surviving animals. Isolation can be accomplished in a variety of cell cultures, laboratory mice, or embryonating chicken eggs. Virus isolation from chronically ill animals often is unsuccessful, probably because they have developed neutralizing antibody.

Serologic diagnosis is based on a fourfold or greater rise of antibody titer in paired serum samples. Standard serologic tests are employed (Reisen and Monath 1989). Demonstration of anti-WEEV IgM or its equivalent within a month of illness can also be used to diagnose WEE. However, cross-reaction can occur between WEEV and its relatives within the WEEV antigenic complex, as well as eastern and Venezuelan equine encephalitis viruses. The illness produced by these three viruses in horses is indistinguishable.

Effective vaccines are available for use in horses. Epidemic spread may be prevented or controlled by timely measures to bring vector populations below the transmission threshold and to avoid mosquito bites. House sparrows *Passer domesticus* are preferred hosts of *C. tarsalis* and are effective early indicators of WEEV transmission (Iversen 1994). No specific treatment is available, but symptomatic supportive therapy may be helpful.

EASTERN EQUINE ENCEPHALITIS

Synonyms: Eastern encephalitis, eastern equine encephalomyelitis.

History. Fatal disease compatible with that caused by eastern equine encephalitis virus (EEEV) occurred among horses in the eastern United States in the mid-1800s (Hanson 1973). The virus was first isolated from fatal encephalitis in Maryland horses in 1933 (Giltner and Shahan 1933). EEEV is endemic in eastern North America, and it also has caused epidemics in horses in the Caribbean, and scattered epidemics and cases in Central America and throughout South America (Hanson 1973).

The virus is classified as an *Alphavirus* (group A) of the *Togaviridae.* The North and South American EEEV strains can be differentiated serologically (Casals 1964). EEEV can be divided into three genetic groups by RNA-sequence analysis: a North American-Caribbean clade, an Amazon Basin clade, and a third clade from Panama, Trinidad, Venezuela, Guyana, Ecuador, and Argentina (Weaver et al. 1994).

Transmission and Host Range. In temperate areas, the warm-weather maintenance cycle of EEEV involves wild birds and *Culiseta melanura* mosquitoes (Chamberlain 1968). *Coquillitidea perturbans* spreads the virus from endemic foci to areas where *Ae. vexans* and *Ae. canadensis* can transmit the virus to horses and people (Centers for Disease Control 1990). EEEV was isolated in Florida from the Asian tiger mosquito *Aedes albopictus,* indicating a potential role for this recently introduced exotic mosquito in the epidemiology of EEE (Johnson and Peters 1996).

Antibodies occur in many wild mammals, including Virginia opossums, cotton mice, and cotton rats (Karabatsos 1985; Morris 1988; Day et al. 1996), but only Goldfield et al. (1969) reported frequent EEEV isolations from a mammal, the white-footed mouse *Peromyscus leucopus,* throughout the year. This observation was not confirmed, though, in whitefooted mice from an endemic focus on the eastern shore of Maryland (T.M. Yuill unpublished).

Experimental infection of eight midwestern rodent species produced no mortality following peripheral inoculation, but caused encephalitis and death in gray squirrels and in white-footed mice inoculated intracerebrally. White-footed mice from the endemic focus in Maryland inoculated with high doses of low-passage EEEV from that area developed encephalitis and died, whereas those inoculated with low doses were unaffected (T.M Yuill unpublished). Thus, wild rodents do not appear to play a significant role in EEEV maintenance in the north, and it is unlikely that this virus causes significant mortality in wild mammal populations. The overwintering mechanism for EEEV is unknown.

In neotropical rainforests, EEEV infects both birds and mammals (Theiler and Downs 1973). Virus and antibodies have been found frequently in a wide variety of bird species in both endemic and epidemic situations. However, the differences between North and South American EEEV strains indicate that migrating birds do not mix the virus between the Northern and Southern Hemispheres.

The virus also has been isolated from sentinel mice and hamsters, a sentinel *Cebus* sp. monkey, wild-caught spiny rats *Proechimys* sp. and rice rats *Oryzomys* sp., and a mouse opossum *Marmosa* sp. EEEV has been isolated most frequently from *Culex* (*Melanoconion*) *taeniopus* in the tropics (Morris 1988). In Guatemala, EEEV antibody has been found in big fruit-eating bats *Artibeus intermedius,* the Jamaican fruit bat *Artibeus jamaicensis,* proboscis bats *Rhynchonycteris naso,* and yellow-shouldered bats (Ubico and McLean 1995). However, until the virus-vector-mammal relationships are tested in the laboratory and confirmed in the field, the role of wild mammals in maintenance and amplification of EEEV, or the impact of EEEV on wild mammal populations, will remain unknown.

Diagnosis. A diagnosis of EEEV is established by isolation of the virus by detection of viral RNA, or by serologic methods that demonstrate the development of specific EEEV antibodies in paired serum samples or by IgM capture (Calisher et al. 1986a). Antibodies to EEEV must be differentiated from those to WEE or Venezuelan equine encephalitis viruses, especially in tropical areas where all three may occur. Virus can be isolated in cell cultures, newborn mice, or chicks. A colorimetric dot-blot assay following PCR amplification can be used to detect EEEV RNA in mosquitoes (Armstrong et al. 1995).

Control and Public Health Implications. Inactivated virus vaccines are used in horses. Vector control for EEEV is carried out by local mosquito control districts with increasing reliance on water management and biologic control of larval habitats and less on insecticide application (Gibbs and Tsai 1994).

VENEZUELAN EQUINE ENCEPHALITIS

Synonyms: Peste loca, derrengadera, Venezuelan encephalitis.

History and Etiology. Periodic outbreaks of encephalitis among horses and people have been recognized in northern South America, especially the coastal lowlands of Colombia and Venezuela, since the early 1930s (Groot 1972). The etiologic agent, Venezuelan equine encephalitis virus (VEEV), was isolated from the brain of a sick horse in 1939 (Kubes and Rios 1939). Encephalitis (VEE) appeared in this area again in 1995 after a 22-year absence, causing 75,000 human cases and an estimated 8% mortality rate among the horse population (Rivas et al. 1997).

An epidemic among horses on the high Bogotá altiplano hinted at the ecological versatility of VEEV. Epidemics have occurred as far south as Peru, and an epidemic that began in Guatemala established VEE as one of the major arbovirus problems in the Western Hemisphere. Between 1969 and 1971, it spread over a region 4000 km long, extending northward through Mexico to the lower Rio Grande Valley of Texas, and southward to Costa Rica, causing encephalitis in several thousand people, and killing over 44,000 horses (Groot 1972).

VEE complex viruses are members of the *Alphavirus* genus of the family *Togaviridae* (Johnson and Peters 1996). This complex has six antigenic subtypes: I, VEE virus with variants A/B–F; II, Everglades virus from Florida; III, Mucambo virus with variants A (Mucambo virus from Brazil), B (Tonate virus from French Guiana), and C (71 D-1252 from Peru); IV, Pixuna virus from Brazil; V, Cabassou virus from French Guiana; and VI, AG80-663 virus from Argentina. These subtypes have differing natural histories, virulence, pathology, and public or animal health implications (reviewed in Osorio and Yuill 1994).

Transmission and Host Range. The sporadic appearance of epidemic VEEV subtypes (I-A/B, C and, more recently, E) in horses has not been fully explained. Epidemic VEEV subtype I-C appears to arise from endemic I-D subtype viruses in northern South America (Kinney et al. 1992; Rico-Hesse et al. 1995).

Epidemic VEEV has horses, mules, and donkeys as the main vertebrate hosts and a variety of mosquitoes as vectors. The virus has been isolated from 33 mosquito species in eight genera (summarized in Walton and Grayson 1988). The majority of epidemic VEE virus isolations have come from *Psorophora* (*Ps.*) *confinnis, Ps. discolor, Mansonia* (*M.*) *titillans, M. indubitans, Deinocerites* (*D.*) *pseudes, Aedes thelcter, Ae. sollicitans, Ae. taeniorhynchus, and Ae. scapularis. Aedes aegypti* has been implicated in person-to-person spread, in the absence of infected horses, in Venezuela (Suarez and Bergold 1968).

Other biting arthropods may serve as mechanical or biologic vectors of epidemic VEEV, but their roles have not been assessed adequately. Experimentally, the Asian tiger mosquito, *Ae. albopictus,* recently introduced into the Americas, transmits epidemic variants of VEEV (Beaman and Turell 1991). The black fly *Simulium mexicanum* transmits the virus mechanically, and biologic transmission also may occur (Homan et al. 1985). The tick *Amblyomma cajennense* transmits VEEV with low efficiency (Linthicum et al. 1991).

Although wild mammals may contribute to transmission of epidemic VEEV during an outbreak (Walton and Grayson 1988), they do not sustain virus transmission in the absence of susceptible populations of equids. Intense epidemics of VEEV may cause mortality among some species of wild mammals, especially the younger cohorts.

Several tropical and temperate-zone mammals (including the black-tailed jackrabbit, eastern cottontail rabbit, cotton rat, Virginia opossum, raccoon, striped skunk, big brown bat, little brown bat, Jamaican fruit-eating bat, and great fruit-eating bat) develop a high-titered, sustained viremia with no mortality. Other mammals also have a significant viremia and over half die, including the Mexican ground squirrel *Spermophilus mexicanus,* southern plains wood rat *Neotoma micropus,* northern grasshopper mouse *Onychomys leukogaster,* Mexican pocket mouse *Liomys irroratus,* white-footed mouse, coyote, gray squirrel, and deer mouse (Sudia and Newhouse 1975; Bowen 1976; Seymour et al. 1978a). Direct rodent-to-rodent transmission, without arthropod vectors, has been shown experimentally. Cotton rats infected experimentally transmitted VEEV to in-contact animals (Howard 1974). Lagomorphs and opossums have viremias of lesser magnitude and briefer duration than those of rodents. Although VEEV has been isolated from southern opossums and wooly opossum *Caluromys derbianus* during VEE epidemics, their epidemiologic role is unclear.

Canids develop viremias sufficient to infect mosquitoes, and the virus has been isolated from a naturally infected gray fox. However, populations of foxes and coyotes are relatively low, and they may be more important as disseminators of the virus than as amplifiers.

Bats may play an important part in VEEV epidemiology, as they are susceptible to infection, abundant, and highly mobile (Bowen 1976; Seymour et al. 1978b). Three common species of neotropical bats (the Jamaican fruit-eating bat, the great fruit-eating bat, and the lesser spear-nosed bat *Phyllostomus discolor*) developed high-titered viremias during experimental infection, and VEEV was isolated from a great fruit-eating bat during an outbreak (Sanmartin et al. 1973). Antibodies to VEE subtypes I-A/B or I-C virus have been found in Jamaican fruit-eating bats, short-tailed leaf-nosed bats *Carollia brevicauda,* Hahn's short-tailed bats *C. subrufa,* lesser spear-nosed bats *Phyllostomus discolor,* yellow-shouldered bats, and Anthony's bats *Sturnira ludovici* in Guatemala (Ubico and McLean 1995). VEEV has been isolated from naturally infected common vampire bats *Desmodus rotundus* (Correa-G. et al. 1972), and they can transmit the virus to horses by bite (Seymour and Dickerman 1978).

VEEV has been isolated from commensal rats and house mice during epidemics, but these species seem epidemiologically unimportant. Similarly, the role of wild birds in VEEV epidemiology is unclear (Bowen and McLean 1977), but they appear to be far less important in VEEV epidemiology than in EEEV or WEEV. However, birds, mobile wild mammals, or transported horses and dogs may account for the sudden appearance of VEEV in new locales.

In contrast to epidemic VEEV, the sylvatic subtypes of this virus (variants I-D, usually I-E, and I-F) are maintained in wild mammals, not in horses, and are associated with a more limited group of arthropod vectors. Sylvatic VEEV strains have been associated mainly with species of *Melanoconion* subgenus of *Culex* mosquitoes from northern South America to Mexico and Florida (summarized in Walton and Grayson 1988).

Rodents, including the large-headed rice rat *Oryzomys capito,* spiny rat *Proechimys cayennensis,* and cotton rat, are important hosts of VEEV variant I-D and I-E in the neotropics (Zarate and Scherer 1968; Shope and Woodall 1973). In Guatemala, antibodies have been found in cotton rats, marsh rice rats *Oryzomys palustris,* Salvin's spiny pocket mouse *Liomys salvini,* southern opossum, gray four-eyed opossum *Philander opossum,* and the tropical porcupine *Sphiggerus (Coendou) mexicanus* (Cupp et al. 1986). VEEV variant I-F was isolated from a naturally infected Seba's short-tailed leaf-nosed bat, *Carollia perspicillata,* in Brazil (Calisher et al. 1982). Cotton rats, cotton mice, Virginia opossums, and raccoons are mammalian hosts for VEEV Everglades subtype II in Florida (Lord et al. 1973; Bigler et al. 1974; Day et al. 1996).

Sylvatic VEEV was isolated from southern opossums, gray four-eyed opossums, and South American mouse possum *Marmosa robinsoni* and from dwarf fruit-eating bats, short-tailed leaf-nosed bat *Carollia subrufa,* and Peters' tent-making bat *Uroderma bilobatum* (Scherer et al. 1972; Seymour et al. 1978a). These more mobile mammals may serve to disseminate the sylvatic subtypes and variants to new areas. Birds are not involved in the epidemiology of sylvatic VEEV (Dickerman et al. 1972; Scherer et al. 1976). The impact of sylvatic VEEV on wild mammal populations has not been assessed.

Clinical Signs and Pathology. In equids and in people, VEEV may cause encephalitis or only an undifferentiated fever, or it may be asymptomatic.

The pathologic changes vary with the animal species (Gochenour 1972), but the lesions caused by VEEV in wild mammals have not been studied. In guinea pigs, VEEV results in severe febrile disease with death 2–4 days after inoculation. Encephalitis does not occur; rather, there is massive necrosis of lymphocytes in the spleen and lymph nodes and bone marrow depletion. Mice survive the lymphocyte destruction and bone marrow depletion and develop extensive meningoencephalitis. A few dogs receiving high doses of VEEV died. They had minimal nonsuppurative encephalitis, and the cause of death was not clear.

Disease in horses can vary from mild fever and leukopenia, through severe leukopenia with marked destruction of lymphocytes in lymph nodes and depletion of myeloid elements of the bone marrow, to severe, diffuse necrotizing meningoencephalitis and vasculitis terminating in death. The sylvatic strains usually do not cause disease in horses.

In 88 human cases in Texas, the usual symptoms included headache, myalgia, chills, vomiting, sore throat, and diarrhea. Central nervous system signs were observed in 36% of those under 17 years of age and in 11% of those 17 years of age or older (Bowen et al. 1976). Some Colombian children had neurologic sequelae 1–4 years after VEEV infection (Leon et al. 1975). Transplacental human infection may cause microencephaly and other fetal damage leading to neonatal death (Wenger 1977).

The lesions caused by VEEV in horses are similar to those described for EEEV and WEEV. Case fatality ranges from 38% to 83% (Groot 1972). Sequelae are common in surviving animals, often rendering them unfit for work.

Diagnosis. VEE can be diagnosed by virus isolation, demonstration of the presence of specific viral nucleic acid sequences, development of specific early IgM antibody, or a significant rise in antibody titer to the virus. It may be difficult to isolate VEEV from brain of fatal cases, probably because of the presence of antibody. Virus isolation is more successful from febrile pasturemates. Material possibly containing virus must be handled carefully, since the risk to laboratory personnel of aerosol infection is extremely high.

Viruses can be subtyped with monoclonal antibodies in plaque reduction neutralization (PRN) or ELISAs (Rico-Hesse et al. 1988). For serologic diagnosis, PRN, immunoblots, or the standard hemagglutination-inhibition tests are more specific but less sensitive than the ELISA (Coates et al. 1992). The IgM ELISA can be used for early antibody detection.

Treatment, Control, and Public Health Implications. Epidemic VEEV is prevented by immunization of horses with live attenuated (TC-83) or inactivated vaccines (McKinney 1972). However, maintenance of herd immunity is expensive and logistically difficult in the neotropics. In people, a vaccine with improved safety and potency is under development (Pittman et al. 1996).

Vaccination of horses, combined with vector control, has been effective in controlling epidemics. Human disease is best prevented through control of VEEV in horse populations by vaccination. Avoidance of mosquito bites is recommended during an outbreak or to prevent infection with the sylvatic strains of VEEV. Symptomatic treatment is the only therapeutic approach available.

VESICULAR STOMATITIS

Synonyms: Erosive stomatitis, stomatitis contagiosa of horses, aphthous stomatitis, mal de tierra, and pseudo aftosa.

History and Etiology. A disease resembling vesicular stomatitis (VS) occurred in horses during the U.S. Civil War, in South African horses in the late 1800s, in cattle in the United States in the early 1900s, and in France in horses arriving from the United States in 1915. Two serologically distinct VS viruses were isolated from infected cattle in 1925 and 1926 and were designated VS-Indiana (VSI) and VS-New Jersey (VSNJ) (Hanson 1952). Endemic VSI and VSNJ were recognized in horses, cattle, and swine in South and Central America (Hanson 1975).

Since the recognition of VSI and VSNJ viruses, related viruses of the VS serogroup (*Vesiculovirus* genus of the family *Rhabdoviridae*) have been isolated from domesticated animals, wild mammals, and insects around the world (Yuill 1980; Calisher and Karabatsos 1988; Webb and Holbrook 1989). The vesiculoviruses have been divided into six complexes: VSIV, VSNJV, Piry virus (PIRYV), Chandipura virus (CHPV), Isfahan virus (ISFV), and Bahia Grande (Calisher and Karabatsos 1988). Malpais Spring virus (MSV) has not been assigned to a complex.

Transmission and Epidemiology. Despite decades of study, the epidemiology of VS viruses is incompletely understood. Both VSNJV and VSIV cause periodic epidemics in livestock. Sudden outbreaks, seasonal occurrence, and movements of VSNJ outbreaks in livestock suggest arthropod spread of the virus, and phlebotomine sand flies are increasingly implicated in transmission and maintenance of both VSIV and VSNJV.

Sand flies, principally *Lutzomyia trapidoi,* are naturally infected with VSIV in the American tropics; they transmit the virus transovarially (Tesh et al. 1972) and by bite (Tesh et al. 1971). The black fly *Simulium notatum* also may be a vector (Mead et al. 1997). VSIV has also been isolated from black flies in Colombia, South America (Theiler and Downs 1973), and VSNJV was isolated from *Culicoides* (*Selfia*) sp. in Colorado (Kramer et al. 1990), but it is not known whether these flies are biologic or mechanical vectors.

VSNJV probably overwinters in the sand fly *Lutzomyia shannoni* in an endemic area on Ossabaw Island, Georgia (Comer et al. 1994a). This fly transmits

VSNJV to rodents by bite and transovarially to a small proportion of its progeny (Comer et al. 1990).

How vesiculoviruses persist during interepidemic periods, other than by transovarial transmission in sand-fly populations, is not known. Although VSV may be spread mechanically in dairy cattle and pigs, there is no evidence for direct transmission among domestic stock at pasture (summarized in Webb and Holbrook 1989).

The role of wild vertebrates in the epidemiology of the VS complex viruses is unclear. Many species are infected and develop antibodies, yet very few develop viremias sufficient to infect biting arthropods. For example, on Ossabaw Island, Georgia, sand flies feed preferentially on white-tailed deer and feral swine, and a high proportion of these species develop VSNJV antibodies, though they do not develop viremias and appear not to be amplifying hosts (Fletcher et al. 1991; Stallknecht et al. 1993; Comer et al. 1994b, 1995).

Some suggest that VSIV and VSNJV are plant viruses that become infectious for mammals following passage through arthropods (Hanson 1968; Johnson et al. 1969).

VSNJV is endemic in the southeastern United States, and sporadic epidemics occur in the intermountain west. There is serologic evidence for VSNJV infection of wild mammals in the United States and in Mexico, including neutralizing antibody in white-tailed deer, mule deer, feral swine *Sus scrofa,* pronghorn antelope, bighorn sheep *Ovis canadensis,* elk *Cervus elaphus,* raccoons, bobcats *Felis rufus,* coyotes, gray foxes, black bears *Ursus americanus,* striped skunks, collared peccaries *Tayassu (Dicotyles) tajacu,* Virginia opossums, black-tailed jackrabbits, desert cottontail rabbits *Sylvilagus audubonii,* nutria *Myocastor coypus,* Merriam's kangaroo rats *Dipodomys merriami,* Mexican wood rats *Neotoma mexicanum,* deer mice, white-footed mice, rock mice *Peromyscus difficilis,* northern pygmy mice *Baiomys taylori,* gray squirrels, and a house mouse (Webb et al. 1987; Webb and Holbrook 1989; Aguirre et al. 1992).

Antibody to VSNJV was found in Jamaican fruit bats, dwarf fruit-eating bats, common vampire bats, yellow-shouldered bats, and great stripe-faced bats *Vampyrodes caraccioli* in Guatemala (Ubico and McLean 1995). Prevalence of antibody to VSNJV in the neotropics was highest among several species of rodents, bats, and carnivores (Tesh et al. 1969). It is not known, however, whether these animals contribute to virus maintenance or are dead-end hosts.

Wildlife hosts of VSIV in North America are not well known. In northeastern Mexico, neutralizing antibody occurred in white-tailed deer, bobcats, gray foxes, black bears, striped skunks, collared peccaries, black-tailed jackrabbits, and white-footed mice (Aguirre et al. 1992). In the tropics, antibody prevalence was highest in arboreal and semiarboreal mammals, including two-toed sloths, spider monkeys, and a prehensile-tailed porcupine *Coendou rothschildi* (Srihongse 1969). VSI virus-neutralizing antibodies were found in cotton rats in an endemic area in Costa Rica (Jimenez et al. 1996) and in Jamaican fruit bats and dwarf fruit-eating bats in Guatemala (Ubico and McLean 1995).

The epidemiology of cocal virus (COCV), a member of the VSIV complex, also is unclear. Isolation of this virus from small tropical rodents, their ectoparasitic mites, and *Culex* sp. mosquitoes suggested a rodent-arthropod transmission cycle (Jonkers et al. 1965). However, the viremia in rodents was too low for efficient virus transmission (Jonkers et al. 1964). Although viremias in experimentally infected little brown bats have been sufficient for transmission by *Ae. aegypti* (Donaldson 1970), and COCV antibody has been found in greater spear-nosed bats *Phyllostomus hastatus* in Trinidad, the role of bats in COCV epidemiology is unknown.

PIRYV also has been associated with small mammals. It was isolated from the brown four-eyed opossum *Philander opossum,* and antibodies have been found in southern opossums; the rodents *Proechimys* sp., *Caluromys* sp., and *Oryzomys goeldii;* bats; primates; edentates; pigs; and water buffalo. Experimentally infected marsupials *D. marsupialis* and *Philander* sp. developed viremia and some died (Theiler and Downs 1973).

CHPV has been isolated from *Phlebotomus* sp. sand flies in India (Dhanda et al. 1970) and from African hedgehogs *Atelerix* spp. in Africa (Karabatsos 1985).

ISFV has been isolated from sand flies in Iran, and great gerbils *Rhombomys opinus* and large naked-soled gerbils are infected naturally (Tesh et al. 1977).

Signs. The incubation period of VSI and VSNJ in cattle, swine, horses, and deer varies from 24 hours to several days, terminating with the onset of fever, which may be biphasic. Large vesicles (fluid-filled blisters) may form in the mouth and on the nose, lips, muzzle, coronary band, and teats. During the vesicular stage, animals are often anorexic and depressed and salivate excessively. The vesicle ruptures, leaving a raw erosion, which heals within a few days, barring secondary bacterial or mycotic infection. Animals may become temporarily lame, and teat lesions interfere with milking in cattle (summarized in Reif 1994). Death usually is not associated with VS infection, and its main significance is in its resemblance to foot-and-mouth disease.

Although the signs of VS in domestic animals and experimentally infected deer and pronghorn antelope have been well described, they have not been documented in other wild vertebrates. In white-tailed deer and antelope, the disease generally was milder than in cattle or swine, with the exception of one antelope that died (Jenny et al. 1980; Thorne et al. 1983).

VSIV and VSNJV produce an acute influenza-like disease in people exposed to the virus by infected animals or in the laboratory. Occasionally, vesicles may form in the mouth or on the lips and nose. Infection with PIRYV and CHPV also causes acute, febrile disease in people (summarized in Yuill 1980). Laboratory infections of PIRYV are particularly prevalent. PIRYV, ISFV, and CHPV produce only mild lesions in infected experimental animals (Wilks and House 1985).

Pathogenesis and Pathology. Virus replication begins in the prickle cells of the epithelium. The virus particles mature and bud off the cell surface, accumulating in the intercellular spaces and infecting adjoining cells. The epithelium becomes edematous, and a papule forms. Transudates and polymorphonuclear cells accumulate and form vesicles under the stratum corneum. The overlying epidermis breaks, leaving an erosion, which is quickly repaired (Ribelin 1958; Proctor and Sherman 1975).

Diagnosis and Immunology. VS is the only vesicular disease of horses, but in cattle and swine it cannot be distinguished clinically from foot-and-mouth disease, and in pigs from vesicular exanthema or swine vesicular disease; the diagnosis must be made in the laboratory.

Laboratory diagnosis is based on isolation of the virus, detection and identification of VSV nucleic acid, or demonstration of the specific antigen by complement fixation or the development of specific antibodies. The virus can be isolated in animals, embryonated chicken eggs, or a variety of cell cultures, by using fluid from intact vesicles, the epithelium of ruptured vesicles, or throat swabs (Webb and Holbrook 1989). A reverse transcriptase-PCR and a hemi-nested PCR assay have been developed for detection and identification of VSIV and VSNJV (Rodriguez et al. 1993; Hofner et al. 1994).

Although serologic diagnosis can be made on the basis of significant rise of VSV antibody, greater caution in interpretation must be exercised than with other groups of arboviruses. VSV-neutralizing antibody titers fluctuate widely (Geleta and Holbrook 1961), and an increasing titer could be due to recent infection or to one of the fluctuations that occur many months after original virus exposure. Complement-fixing antibodies fluctuate less (Holbrook 1962), but this test may be problematic with sera from wild mammals. Early, specific antibody may be demonstrated by IgM capture or MAC-ELISA tests, but an ELISA test was not specific enough to differentiate between antibodies to VSIV and VSNJV (Workman et al. 1986).

The population at risk to VSV infection cannot be assessed by presence or absence of antibodies, as it can with other arboviruses. Animals convalescent from VSV infection may become susceptible to infection again with the homologous virus, despite the presence of neutralizing antibody (Casteñeda et al. 1964).

Control and Public Health Implications. Control in livestock or in wildlife is difficult without a clearer understanding of the epidemiology of the disease. The disease must be reported to government veterinary authorities on account of its clinical resemblance to foot-and-mouth disease.

Livestock can be vaccinated with attenuated or inactivated VSNJ or VSI vaccines (Gearhart et al. 1987), and a recombinant vaccinia virus vectored VSIV vaccine expressing the G protein provided cattle with some protection against virus challenge (Mackett et al. 1985).

The aforementioned potential for human disease described must be recognized by those working with VS in the field or laboratory.

ACKNOWLEDGMENT. The assistance of Mrs. Barbara Brown in the preparation of this manuscript is gratefully acknowledged.

LITERATURE CITED

Adam, K.M.G., S.J. Beasley, and D.A. Blewitt. 1977. The occurrence of antibody to *Babesia* and to the virus of louping-ill in deer in Scotland. *Research of Veterinary Science* 23:133–138.

Adhikari Prabha, M.R., M.G. Prabhu, M. Bai, and M.A. Mala. 1993. Clinical study of 100 cases of Kyasanur Forest disease with clinicopathological correlation. *Indian Journal of Medical Science* 47:124–130.

Aguirre, A.A., R.G. McLean, R.S. Cook, and T.J. Quan. 1992. Serologic survey for selected arboviruses and other potential pathogens in wildlife from Mexico. *Journal of Wildlife Diseases* 28:435–442.

Aitkin, T.H.G., R.B. Tesh, B.J. Beaty, and L. Rosen. 1979. Transovarial transmission of yellow fever virus by mosquitoes (*Aedes aegypti*). *American Journal of Tropical Medicine and Hygiene* 28:119–121.

Allen, R., S.K. Taylor, and S.E. Sulkin. 1970. Studies of arthropod-borne virus infections in Chiroptera: VIII. Evidence of natural St. Louis encephalitis virus infection in bats. *American Journal of Tropical Medicine and Hygiene* 19:851–859.

Andral, L., P. Bres, C. Serie, J. Casals, and R. Panthier. 1968. Etudes sur la fièvre jaune en Ethiopie: 3. Etude serologique et virologique de la faune sylvatique. *Bulletin of the World Health Organization* 38:855–861.

Armstrong, P., D. Borovsky, R.E. Shope, C.D. Morris, C.J. Mitchell, N. Karabatsos, N. Komar, and A. Spielman. 1995. Sensitive and specific colorimetric dot assay to detect eastern equine encephalomyelitis viral RNA in mosquitoes (Diptera: Culicidae) after polymerase chain reaction amplification. *Journal of Medical Entomology* 32:42–52.

Artsob, H. 1983. Distribution of California serogroup viruses and virus infection in Canada. In *California serogroup viruses,* ed. C.H. Calisher and W.H. Thompson. New York: Alan R. Liss, pp. 277–292.

———. 1988. Powassan encephalitis. In *The arboviruses: Epidemiology and ecology,* vol. 4, ed. T.P. Monath. Boca Raton, FL: CRC, pp. 29–49.

Artsob, H., L. Spence, C. Th'ng, V. Lampotang, D. Johnson, C. MacInnes, F. Matejka, D. Voigt, and I. Watt. 1986. Arbovirus infections in several Ontario mammals. *Canadian Journal of Veterinary Research* 50:42–46.

Aviles, G., M.S. Sabattini, and C.J. Mitchell. 1992. Transmission of western equine encephalomyelitis virus by Argentine *Aedes albifasciatus* (Diptera: Culicidae). *Journal of Medical Entomology* 29:850–853.

Banerjee, K. 1988. Kyasanur Forest disease. In *The arboviruses: Epidemiology and ecology,* vol. 3, ed. T.P. Monath. Boca Raton, FL: CRC, pp. 93–116.

Beaman, J.R., and M.J. Turell. 1991. Transmission of Venezuelan equine encephalomyelitis virus by strains of *Aedes albopicus* (Diptera: Culicidae) collected in North and South America. *Journal of Medical Entomology* 28:161–173.

Bearcroft, W.G.C. 1957. The histopathology of the liver of yellow fever-infected rhesus monkeys. *Journal of Pathology and Bacteriology* 74:295–303.

Bhat, U.K.M., and M.K. Goverdhan. 1973. Transmission of Kyasanur Forest disease by the soft tick *Ornithodorus crossi*. *Acta Virologica* 17:337–342.

Bhat, H.R., M.A. Sreenivasan, M.K. Goverdhan, and S.V. Naik. 1975. Transmission of Kyasanur Forest disease virus by *Haemaphysalis kyasanurensis* Trapido, Hoogstraal, and Rajagopalan (Acarina: Ixodidae). *Indian Journal of Medical Research* 63:879–887.

Bigler, W.J., and G.L. Hoff. 1975. Arbovirus surveillance in Florida: Wild vertebrate studies, 1965–1974. *Journal of Wildlife Diseases* 11:348–356.

Bigler, W.J., A.K. Ventura, A.L. Lewis, F.M. Wellings, and N.J. Ehrenkranz. 1974. Venezuelan equine encephalomyelitis in Florida: Endemic virus circulation in native rodent populations of Everglades hammocks. *American Journal of Tropical Medicine and Hygiene* 23:513–521.

Boshell, J., and P.K. Rajagopalan. 1968. Observation on the experimental exposure of monkeys, rodents, and shrews to infestations of ticks in forest in Kyasanur Forest disease area. *Indian Journal of Medical Research* 56:573–588.

Boshell, J., M.K. Goverdhan, and P.K. Rajagopalan. 1968a. Preliminary studies on the susceptibility of wild rodents and shrews to KFD virus. *Indian Journal of Medical Research* 56:614–627.

Boshell, J., P.K. Rajagopalan, A.P. Patil, and K.M. Pavri. 1968b. Isolation of Kyasanur Forest disease virus from Ixodid ticks: 1961–1964. *Indian Journal of Medical Research* 56:541–568.

Bowen, G.S. 1976. Experimental infection of North American mammals with epidemic Venezuelan encephalitis virus. *American Journal of Tropical Medicine and Hygiene* 25:891–899.

———. 1977. Prolonged western equine encephalitis viremia in the Texas tortoise *(Gopherus berlandieri)*. *American Journal of Tropical Medicine and Hygiene* 26:171–175.

Bowen, G.S., and R.G. McLean. 1977. Experimental infection of birds with epidemic Venezuelan encephalitis virus. *American Journal of Tropical Medicine and Hygiene* 26:808–814.

Bowen, G.S., T.R. Fashinell, P.B. Dean, and M.B. Gregg. 1976. Clinical aspects of human Venezuelan equine encephalitis in Texas. *Bulletin of the Pan American Health Organization* 10:46–57.

Brotherston, J.G., and J.B. Boyce. 1970. Development of a non-infective protective antigen against louping-ill (Arbovirus group B). *Journal of Comparative Pathology* 80:377–388.

Burgdorfer, W., V.F. Newhouse, and L.A. Thomas. 1961. Isolation of California encephalitis virus from the blood of a snowshoe hare (*Lepus americanus*) in Western Montana. *American Journal of Hygiene* 73:344–349.

Burton, A.N., and J. McLintock. 1970. Further evidence of western encephalitis infection in Saskatchewan mammals and birds and in reindeer in northern Canada. *Canadian Veterinary Journal* 11:232–235.

Calisher, C.H., and N. Karabatsos. 1988. Arbovirus serogroups: Definition and geographic distribution. In *The arboviruses: Epidemiology and ecology,* vol. 1, ed. T.P. Monath. Boca Raton, FL: CRC, pp. 19–59.

Calisher, C.H., R.M. Kinney, O. Lopes de Souza, D.W. Trent, T.P. Monath, and D.B. Francy. 1982. Identification of a new Venezuelan equine encephalitis virus from Brazil. *American Journal of Tropical Medicine and Hygiene* 31:1260–1272.

Calisher, C.H., V.P. Berardi, D.J. Muth, and E.E. Buffy. 1986a. Specificity of immunoglobulin M and G antibody response in humans infected with eastern and western equine encephalitis viruses: Application to rapid serodiagnosis. *Journal of Clinical Microbiology* 23:369–372.

Calisher, C.H., D.B. Francy, G.C. Smith, D.J. Muth, J.S. Lazuick, N. Karabatsos, W.L. Jakob, and R.G. McLean. 1986b. Distribution of Bunyamwera serogroup viruses in North America, 1956–1984. *American Journal of Tropical Medicine and Hygiene* 35:429–443.

Casals, J. 1964. Antigenic variants of eastern equine encephalitis virus. *Journal of Experimental Medicine* 119:547–565.

Casteñeda, G.J., L.H.J. Lauerman, and R.P. Hanson. 1964. Evaluation of virus neutralization tests and association of indices to cattle resistance. *Proceedings of the United States Livestock Sanitary Association* 68:455–468.

Centers for Disease Control. 1990. Arboviral surveillance: United States, 1990. *Morbidity and Mortality Weekly Reports* 39:593–598.

Chamberlain, R.W. 1968. Arboviruses, the arthropod-borne animal viruses. *Current Topics in Microbiology and Immunology* 427:35–58.

Chamberlain, R.W., W.D. Sudia, and P.H. Coleman. 1969. Isolations of an arbovirus of the Bunyamwera group (Tensaw virus) from mosquitoes in the southeastern United States. *American Journal of Tropical Medicine and Hygiene* 18:92–97.

Chang, G.J., B.C. Cropp, R.M. Kinney, D.W. Trent, and D.J. Gubler. 1995. Nucleotide sequence variation of the envelope protein gene identifies two distinct genotypes of yellow fever virus. *Journal of Virology* 69:5573–5780.

Chippaux-Hyppolite, C., and A. Chippaux. 1969. Contribution a l'étude d'un réservoir de virus amaril dans le cycle de certains arbovirus en Centrafrique: I. Etude immunologique chez divers animaux domestiques et sauvages. *Bulletin of Social and Pathological Exotics* 62:1034–1040.

Clarke, D.H. 1964. Further studies on antigenic relationships among the viruses of the group B tick-borne complex. *Bulletin of the World Health Organization* 31:45–56.

Clarke, D.H., and J. Casals. 1965. Arboviruses: Group B. In *Viral and rickettsial infections of man,* ed. F.L. Horsfall and I. Tamm. Philadelphia: Lippincott, pp. 606–658.

Coates, D.M., S.R. Makh, N. Jones, and G. Lloyd. 1992. Assessment of assays for the serodiagnosis of Venezuelan equine encephalitis. *Journal of Infection* 25:279–289.

Comer, J.A., R.B. Tesh, G.B. Modi, J.L. Corn, and V.F. Nettles. 1990. Vesicular stomatitis virus, New Jersey serotype: Replication in and transmission by *Lutzomyia shannoni* (Diptera: Psychodidae). *American Journal of Tropical Medicine and Hygiene* 42:483–490.

Comer, J.A., D.M. Kavanaugh, D.E. Stallknecht, and J.L. Corn. 1994a. Population dynamics of *Lutzomyia shannoni* (Diptera: Psychodidae) in relation to the epizootiology of vesicular stomatitis virus on Ossabaw Island, Georgia. *Journal of Medical Entomology* 31:850–854.

Comer, J.A., W.S. Irby, and D.M. Kavanaugh. 1994b. Hosts of *Lutzomyia shannoni* (Diptera: Psychodidae) in relation to vesicular stomatitis virus on Ossabaw Island, Georgia, U.S.A. *Medical and Veterinary Entomology* 8:325–330.

Comer, J.A., D.E. Stallknecht, and V.F. Nettles. 1995. Incompetence of white-tailed deer as amplifying hosts of vesicular stomatitis virus for *Lutzomyia shannoni* (Diptera: Psychodidae). *Journal of Medical Entomology* 32:738–740.

Cordelier, R. 1991. The epidemiology of yellow fever in Western Africa. *Bulletin of the World Health Organization* 69:73–84.

Correa-G., P., C.H. Calisher, and G.M. Baer. 1972. Epidemic strain of Venezuelan equine encephalomyelitis virus from a vampire bat captured in Oaxaca, Mexico, 1970. *Science* 175:546–547.

Costero A., and M.A. Grayson. 1996. Experimental transmission of Powassan virus (Flaviviridae) by *Ixodes scapu-*

laris ticks (Acari: Ixodidae). *American Journal of Tropical Medicine and Hygiene* 55:536–546.
Cupp, E.W., W.F. Scherer, J.B. Lok, R.J. Brenner, G.M. Dziem, and J.V. Ordonez. 1986. Entomological studies at an enzootic Venezuelan equine encephalitis virus focus in Guatemala, 1977–1980. *American Journal of Tropical Medicine and Hygiene* 35:851–859.
Dandawate, C.N., G.B. Desai, T.R. Achar, and K. Banerjee. 1994. Field evaluation of formalin inactivated Kyasanur Forest disease virus tissue culture vaccine in three districts of Karnataka state. *Indian Journal of Medical Research* 99:152–158.
Day, J.F., E.E. Storrs, L.M. Stark, A.L. Lewis, and S. Williams. 1995. Antibodies to St. Louis encephalitis virus in armadillos from southern Florida. *Journal of Wildlife Diseases* 31:10–14.
Day, J.F., L.M. Stark, J.T. Zhang, A.M. Ramsey, and T.W. Scott. 1996. Antibodies to arthropod-borne encephalitis viruses in small mammals from southern Florida. *Journal of Wildlife Diseases* 32:431–436.
De Rodaniche, E., and P. Galindo. 1961. St. Louis encephalitis in Panama: III. Investigation of local mammals and birds as possible reservoir host. *American Journal of Tropical Medicine and Hygiene* 10:390–392.
Dhanda, V., F.M. Rodrigues, and S.N. Gosh. 1970. Isolation of Chandipura virus from sandflies in Aurangabad. *Indian Journal of Medical Research* 58:179.
Dickerman, R.W., W.F. Scherer, A.S. Moorhouse, E. Toaz, M. E. Essex, and R.E. Steele. 1972. Ecologic studies of Venezuelan encephalitis virus in southeastern Mexico: VI. Infection of wild birds. *American Journal of Tropical Medicine and Hygiene* 21:66–78.
Donaldson, A.I. 1970. Bats as possible maintenance hosts for vesicular stomatitis virus. *American Journal of Epidemiology* 92:132–136.
Dunaev, N.B. 1976. Mixed infection: Tularemia and Omsk hemorrhagic fever in an experiment on *Arvicola terrestris* L. [in Russian]. *Zhurnal po Mikrobiologi, Epidemiologi i Immunobiologi* 8:118–122.
Eklund, C.M. 1963. Role of mammals in maintenance of arboviruses. *Annals of Microbiology* 11:99–105.
Emmons, R.W., and E.H. Lennette. 1967. Isolation of St. Louis encephalitis from a naturally infected gray fox *Urocyon cinereoargenteus*. *Proceedings of the Society of Experimental Biology and Medicine* 125:443–447.
———. 1969. Isolation of western equine encephalomyelitis virus from an opossum. *Science* 163:945–946.
Fedorova, T.N. 1966. Susceptibility and sensibility in muskrats to Omsk hemorrhagic fever virus. In *The muskrat of western Siberia (biocenotic relations, parasite fauna, epizootics and measures of prophylaxis)* [in Russian], ed. A.A. Maksimov and G.I. Netsky. Novosibirsk, Sibirskoe Otdelenie: Biologicheskii Institut, Akademiia Nauk SSSR, pp. 131–135.
Fletcher, W.O., D.E. Stallknecht, M.T. Kearney, and K.A. Eernisse. 1991. Antibodies to vesicular stomatitis New Jersey type virus in white-tailed deer on Ossabaw Island, Georgia, 1985 to 1989. *Journal of Wildlife Diseases* 27:675–680.
Gagarina, A.V. 1965. Transmission of Omsk hemorrhagic fever by ticks. In *Endemic viral infections (hemorrhagic fever with renal syndrome, Crimean hemorrhagic fever, Omsk hemorrhagic fever, and Astrakhan virus) from the* Hyalomma p. plumbeum *tick* [in Russian], vol. 7, ed. M.P. Chumakov. Sbornik Trudy Instituta Poliomielita i Virusnykh Entsefalitov Akademii Meditsinskikh Nauk SSSR, pp. 422–429.
Gao, G.F., W.R. Jiang, M.H. Hussain, K. Venugopal, T.S. Gritsun, H.W. Reid, and E.A. Gould. 1993. Sequencing and antigenic studies of a Norwegian virus isolated from encephalomyelitic sheep confirm the existence of louping ill virus outside Great Britain and Ireland. *Journal of General Virology* 74:109–114.
Gardner, J.J., and M.G. Reyes. 1980. Pathology. In *St. Louis encephalitis,* ed. T.P. Monath. Washington, DC: American Public Health Association, pp. 551–569.
Gaunt, R.A., P.C. Stowe, and C.G. Watson. 1974. Antibody to human pathogens in the wildlife of the Yellowknife, N.W.T., area. *Canadian Journal of Public Health* 65:61.
Gearhart, M.A., P.A. Webb, A.P. Knight, M.D. Salman, J.A. Smith, and G.A. Erickson. 1987. Serum neutralizing antibody titers in dairy cattle administered an inactivated vesicular stomatitis virus vaccine. *Journal of the American Veterinary Medical Association* 191:819–822.
Gebhardt, L.P., S.C. St. Jeor, G.J. Stanton, and D.A. Stringfellow. 1973. Ecology of western encephalitis virus. *Proceedings of the Society of Experimental Biology and Medicine* 142:731–733.
Geleta, J.N., and A.A. Holbrook. 1961. Vesicular stomatitis: Patterns of complement-fixing and serum-neutralizing antibody in serum of convalescent cattle and horses. *American Journal of Veterinary Research* 22:713.
Germain, M., J.F. Saluzzo, J.P. Cornet, R. Cordellier, J.P. Herve, J.P. Digoutte, T.P. Monath, J.J. Salun, V. Deubel, Y. Robin, J. Coz, R. Taufflieb, J.F. Saluzzo, and J.P. Gonzalez. 1981. La fièvre jaune en Afrique: donées récentes et conceptions actuelles. *Médécine Tropicale* 41:31–43.
Gibbs, E.P.J., and T.F. Tasi. 1994. Eastern encephalitis. In *Handbook of zoonoses,* section B: *Viral,* ed. G.W. Beran and J.H. Steele, 2d ed. Boca Raton, FL: CRC, pp. 11–24.
Giltner, L.T., and M.S. Shahan. 1933. The 1933 outbreak of infectious equine encephalomyelitis in the eastern states. *North American Veterinarian* 14:25–27.
Gochenour, W.S., Jr. 1972. The comparative pathology of Venezuelan equine encephalitis virus infection in selected animal hosts. In *Venezuelan encephalitis. Pan American Health Organization Scientific Publication* 243:113–117.
Goldfield, M., M.S. Sussman, R. Altman, and R.P. Kandle. 1969. Eastern equine encephalitis in New Jersey during 1968. *Proceedings of the New Jersey Mosquito Extermination Association* 56:56–63.
Gorgas Memorial Laboratory. 1977. *48th annual report, fiscal year 1976.* 95th Congress, 1st Session, House Document 95-39. Washington, DC: US Government Printing Office.
Goverdhan, M.K., P.K. Rajagopalan, D.P. Narasima Murthy, S. Upadhaya, J. Boshell-M., H. Trapido, and R.T. Ramachandra. 1974. Epizootiology of Kyasanur Forest disease in wild monkeys of Shimoga District, Mysore State (1957–1964). *Indian Journal of Medical Research* 62:497–510.
Groot, H. 1972. The health and economic impact of Venezuelan equine encephalitis (VEE). In *Venezuelan encephalitis. Pan American Health Organization Scientific Publication* 243:7–16.
Hahn, C.S., S. Lustig, E.G. Strauss, and J.H. Strauss. 1988. Western equine encephalitis virus is a recombinant virus. *Proceedings of the National Academy of Sciences USA* 85:5997–6001.
Hall, W.C., T.P. Crowell, D.M. Watts, V.L.R. Barros, H. Kruger, F. Pinheiro, and C.J. Peters. 1991. Demonstration of yellow fever and dengue antigens in formalin-fixed paraffin-embedded human liver by immunohistochemical analysis. *American Journal of Tropical Medicine and Hygiene* 45:408–417.
Hanson, R.P. 1952. The natural history of vesicular stomatitis virus. *Bacteriological Reviews* 16:179–204.
———. 1968. Discussion of the natural history of vesicular stomatitis. *American Journal of Epidemiology* 87:264–266.

———. 1973. Virology and epidemiology of eastern and western arboviral encephalomyelitis of horses. In *II. American arboviral encephalomyelitides of Equidae: Proceedings of the Third International Conference of Equine Infectious Diseases.* Basel: Karger, pp. 200–214.

———. 1975. Vesicular stomatitis. In *Diseases of swine,* ed. H.W. Dunne and A.D. Leman. Ames: Iowa State University Press, pp. 308–324.

Hardy, J.L., W.C. Reeves, R.P. Scrivani, and D.R. Roberts. 1974. Wild mammals as hosts of group A and group B arboviruses in Kern County, California. *American Journal of Tropical Medicine and Hygiene* 23:1165–1177.

Hardy, J.L., W.C. Reeves, and R.D. Sjogren. 1976. Variations in the susceptibility of field and laboratory populations of *Culex tarsalis* to experimental infection with western equine encephalomyelitis virus. *American Journal of Epidemiology* 103:498–505.

Hayes, C.G., and R.C. Wallis. 1977. Ecology of western equine encephalomyelitis in the eastern United States. *Advances in Virus Research* 21:37–83.

Hayles, L.B., J. McLintock, and J.R. Saunders. 1972. Laboratory studies on the transmission of western equine encephalitis virus by Saskatchewan mosquitoes: I. *Culex tarsalis. Canadian Journal of Comparative Medicine* 36:83–88.

Herbold J.R., W.P. Heuschele, and M.A. Parsons. 1983. Reservoir of St. Louis encephalitis virus in Ohio bats. *American Journal of Veterinary Research* 44:1889–1893.

Hoff, G.L., T.M. Yuill, J.O. Iversen, and R.P. Hanson. 1970. Selected microbial agents in snowshoe hares and other vertebrates of Alberta. *Journal of Wildlife Diseases* 6:472–478.

Hoff, G.L., R.O. Anslow, J. Spalatin, and R.P. Hanson. 1971. Isolation of Montana snowshoe hare serotype of California encephalitis virus group from a snowshoe hare and *Aedes* mosquitoes. *Journal of Wildlife Diseases* 7:28–34.

Hoff, G.L., G.J. Issel, D.O. Trainer, and S.H. Richard. 1973. Arbovirus serology in North Dakota mule and white-tailed deer. *Journal of Wildlife Diseases* 9:291–295.

Hofner, M.C., W.C. Carpenter, N.P. Ferris, R.P. Kitching, and F. Ariza Botero. 1994. A hemi-nested PCR assay for the detection and identification of vesicular stomatitis virus nucleic acid. *Journal of Virological Methods* 50:11–20.

Holbrook, A.A. 1962. Duration of immunity and serologic patterns in swine convalescing from vesicular stomatitis. *Journal of the American Veterinary Medicine Association* 141:1463.

Holden, P., R.O. Hayes, C.J. Mitchell, D.B. Francy, J.S. Lazuick, and T.B. Hughes. 1973. House sparrows, *Passer domesticus (L.)* as hosts of arboviruses in Hale County, Texas: I. Field studies, 1965–1969. *American Journal of Tropical Medicine and Hygiene* 22:244–253.

Homan, E.J., F.N. Zuluaga, T.M. Yuill, and H. Lorbacher. 1985. Studies on the transmission of Venezuelan equine encephalitis by Colombian Simuliidae (Diptera). *American Journal of Tropical Medicine and Hygiene* 34:799–804.

Howard, A.T. 1974. Experimental infection and intracage transmission of Venezuelan equine encephalitis virus (subtype IB) among cotton rats, *Sigmodon hispidus* (Say and Ord). *American Journal of Tropical Medicine and Hygiene* 23:1178–1184.

Iversen, J.O. 1994. Western equine encephalitis. In *Handbook of zoonoses,* section B: *Viral,* ed. G.W. Beran and J.H. Steele, 2d ed. Boca Raton, FL: CRC, pp. 25–31.

Iversen, J.O., R.P. Hanson, O. Papadopoulos, C.V. Morris, and G.R. de Foliart. 1969. Isolation of viruses of the California encephalitis virus group from boreal *Aedes* mosquitoes. *American Journal of Tropical Medicine and Hygiene* 18:735–742.

Iyer, C.G.S., T.H. Work, D.P.N. Murthy, H. Trapido, and P. K. Rajagopalan. 1960. Kyasanur Forest disease: VII. Pathological findings in monkeys, *Presbytis entellus* and *Macaca radiata,* found dead in the forest. *Indian Journal of Medical Research* 48:276–286.

Jenny, E.W., G.A. Erickson, W.E. Buisch, W.C. Stewart, and M.A. Mixson. 1980. Surveillance for vesicular stomatitis in the United States 1972 through 1979. *Proceedings, American Association of Veterinary Laboratory Diagnosticians* 23:83–92.

Jimenez, A.E., C. Jimenez, L. Castro, and L. Rodriguez. 1996. Serological survey of small mammals in a vesicular stomatitis virus enzootic area. *Journal of Wildlife Diseases* 32:274–279.

Johnson, H.N. 1987. Isolation of Powassan virus from a spotted skunk in California. *Journal of Wildlife Diseases* 23:152–153.

Johnson, K.M., R.B. Tesh, and P.H. Peralta. 1969. Epidemiology of vesicular stomatitis virus: Some new data and a hypothesis for transmission of the Indiana serotype. *Journal of the American Veterinary Medical Association* 155:2133–2140.

Johnson, R.E., and C.J. Peters. 1996. Alphaviruses. In *Fields virology,* ed. B.N. Fields, D.M. Knipe, and P.M. Howley. Philadelphia: Lippincott-Raven, pp. 843–898.

Jonkers, A.H., L. Spence, C.A. Coakwell, and J.J. Thornton. 1964. Laboratory studies with wild rodents and viruses native to Trinidad. *American Journal of Tropical Medicine and Hygiene* 13:613–619.

Jonkers, A.H., L. Spence, and T.H.G. Aitken. 1965. Cocal virus epizootiology in Bush Bush forest and the Nariva swamp, Trinidad, West Indies: Further studies. *American Journal of Veterinary Research* 26:758–763.

Kaplan, C., T.D. Healing, N. Evans, L. Healing, and A. Prior. 1980. Evidence of infection by viruses in small British field rodents. *Journal of Hygiene* 84:285–294.

Karabatsos, N., ed. 1985. *International catalogue of arboviruses.* San Antonio: American Society of Tropical Medicine and Hygiene, 1147 pp.

Kenyon, R.H., M.K. Rippy, K.T. McKee, P.M. Zack, and C.J. Peters. 1992. Infection of *Macaca radiata* with viruses of the tick-borne encephalitis group. *Microbial Pathogenesis* 13:399–409.

Kharitonova, N.N. 1969. Results of serological investigation of blood from wild and domestic animals in the Omsk hemorrhagic fever focus of northern Kulunda. In *Migratory birds and their role in the disperson of arboviruses* [in Russian], ed. A.I. Cherepanov et al. Novosibirsk, Sibirskoe Otdelenie: Biologicheskii Institut, Akademiia Nauk SSSR, pp. 317–321.

Kinney, R.M., K.R. Tsuchiya, J.M. Sneider, and D.W. Trent. 1992. Genetic evidence that epizootic Venezuelan equine encephalitis may have evolved from enzootic VEE subtype I-D virus. *Virology* 191:569–580.

Kiorpes, A.L., and T.M. Yuill. 1975. Environmental modification of western equine encephalomyelitis infection in the snowshoe hare (*Lepus americanus*). *Infection and Immunity* 11:986–990.

Kokernot, R.H., B. Radivojevic, and R.J. Anderson. 1969. Susceptibility of wild and domesticated animals to four arboviruses. *American Journal of Veterinary Research* 30:2197–2203.

Kornilova, E.A., A.V. Gagarina, and M.P. Chumakov. 1970. Comparative characteristics of Omsk hemorrhagic fever virus isolated from different objects in a natural focus [in Russian]. *Voprosy Virusologii* 2:232–236.

Kramer, W.L., R.H. Jones, F.R. Holbrook, T.E. Walton, and C.H. Calisher. 1990. Isolation of arboviruses from *Culicoides* midges (Diptera: Ceratopogonidae) in Colorado during an epizootic of vesicular stomatitis New Jersey. *Journal of Medical Entomology* 27:487–493.

Krueger, N., and H.W. Reid. 1994. Detection of louping ill virus in formalin-fixed, paraffin wax-embedded tissues of mice, sheep and a pig by the avidin-biotin-complex immunoperoxidase technique. *Veterinary Record* 135:224–225.

Kubes, V., and F.A. Rios. 1939. The causative agent of infectious equine encephalomyelitis in Venezuela. *Science* 90:20–21.

Leon, C.A., R. Jaramillo, S. Martinez, F. Fernandez, H. Tellez, B. Lasso, and R. de Guzman. 1975. Sequelae of Venezuelan equine encephalitis in humans: A four year follow-up. *International Journal of Epidemiology* 4:131–140.

Leung, M.K., A. Burton, J. Iversen, and J. McLintock. 1975. Natural infections of Richardson's ground squirrels with western equine encephalomyelitis virus, Saskatchewan, Canada, 1964–1973. *Canadian Journal of Microbiology* 21:954–958.

Leung, M.K., J. Iversen, J. McLintock, and J.R. Saunders. 1976. Subcutaneous exposure of the Richardson's ground squirrel (*Spermophilus richardsonii* Sabine) to western equine encephalomyelitis virus. *Journal of Wildlife Diseases* 12:237–246.

Leung, M.K., J. McLintock, and J. Iversen. 1977. Intranasal exposure of the Richardson's ground squirrel to western equine encephalomyelitis virus. *Canadian Journal of Comparative Medicine* 42:184–191.

Linthicum, K.J., T.M. Logan, C.L. Bailey, S.W. Gordon, C.J. Peters, T.P. Monath, J.F. Osorio, D.B. Francy, R.G. McLean, and J.W. Leduc. 1991. Venezuelan equine encephalomyelitis virus infection in and transmission by the tick *Amblyomma cajennense* (Arachnide: Ixodidae). *Journal of Medical Entomology* 28:405–412.

Little, P.B., J. Thorsen, W. Moore, and N. Weninger. 1985. Powassan viral encephalitis: A review and experimental studies in the horse and the rabbit. *Veterinary Pathology* 22:500–507.

Lord, R.D., C.H. Calisher, W.D. Sudia, and T.H. Work. 1973. Ecological investigations of vertebrate hosts of Venezuelan equine encephalomyelitis virus in south Florida. *American Journal of Tropical Medicine and Hygiene* 22:116–123.

Lord, R.D., C.H. Calisher, W.A. Chappell, W.R. Metzger, and G.W. Fisher. 1974. Urban St. Louis encephalitis surveillance through wild birds. *American Journal of Epidemiology* 99:360–363.

Luby, J.P. 1994. St. Louis encephalitis. In *Handbook of zoonoses,* section B: *Viral,* ed. G.W. Beran and J.H. Steele, 2d ed. Boca Raton, FL: CRC, Press, pp. 47–58.

Lvov, D.K. 1988. Omsk hemorrhagic fever. In *The arboviruses: Epidemiology and ecology,* vol. 3, ed. T.P. Monath. Boca Raton, FL: CRC, pp. 205–216.

Lvov, D.K., G.N. Leonova, V.L. Gromashevski, N.P. Belikova, L.K. Berezina, A.V. Safronov, O.V. Veselovskaya, Y.P. Gofman, and S.M. Klimenko. 1974. Isolation of Powassan virus from *Haemaphysalis neumanni,* Donitz 1905, ticks in Primorye [in Russian]. *Voprosy Virusologii* 5:538–541.

Mackett, M., T. Yilma, J.K. Rose, and B. Moss. 1985. Vaccinia virus recombinants: Expression of VSV genes and protective immunization of mice and cattle. *Science* 227:433–435.

Main, A.J., A.B. Carey, and W.G. Downs. 1979. Powassan virus in *Ixodes cookei* and Mustelidae in New England. *Journal of Wildlife Diseases* 15:585–591.

Marenko, V.F., N.B. Dunaev, L.S. Egorova, and T.N. Fedorova. 1974. Experimental mixed infection in muskrats (Omsk hemorrhagic fever and tularemia) [in Russian]. *Voprosy Virusologii* 5:545–550.

McCrae, A.W., and B.G. Kirya. 1982. Yellow fever and Zika virus epizootics and enzootics in Uganda. *Transactions of the Royal Society of Tropical Medicine and Hygiene* 76:552–562.

McGowan, J.E., Jr., J.A. Bryan, and M.B. Greggs. 1973. Surveillance of arboviral encephalitis in the United States, 1955–1971. *American Journal of Epidemiology* 97:199–207.

McKinney, R.W. 1972. Inactivated and live VEE vaccines: A review. In *Venezuelan encephalitis. Pan American Health Organization Scientific Publication* 243:369–376.

McLean, D.M. 1983. Yukon isolates of snowshoe hare virus. In *California serogroup viruses,* ed. C.H. Calisher and W.H. Thompson. New York: Alan R. Liss, pp. 247–256.

McLean, D.M., and W.L. Donohue. 1959. Powassan virus: Isolation of virus from a fatal case of encephalitis. *Canadian Medical Association Journal* 80:708–711.

McLean, D.M., C. Cobb, S. Gooderham, C. Smart, A.G. Wilson, and W.E. Wilson. 1967. Powassan virus: Persistence of viral activity during 1966. *Canadian Medical Association Journal* 96:660–664.

McLean, D.M., M.A. Crawford, S.R. Ladyman, R.R. Peers, and K.W. Purvin-Good. 1970. California encephalitis and Powassan virus activity in British Columbia, 1969. *American Journal of Epidemiology* 92:266–272.

McLean, D.M., S.K.A. Bergman, E.J. Goddard, E.A. Graham, and K.W. Purvin-Good. 1971. North-south distribution of arbovirus reservoirs in British Columbia, 1970. *Canadian Journal of Public Health* 62:120–124.

McLean, D.M., A.M. Clarke, E.J. Goddard, E.J. Manes, C.A. Montalbetti, and R.E. Pearson. 1973. California encephalitis virus endemicity in the Yukon Territory, 1972. *Journal of Hygiene* 71:391–402.

McLean, D.M., S.K.A. Bergman, A.P. Gould, P.N. Grass, M.A. Miller, and E.E. Spratt. 1975. California encephalitis virus prevalence throughout the Yukon Territory, 1971–1974. *American Journal of Tropical Medicine and Hygiene* 24:676–684.

McLean, D.M., P.N. Grass, and B.D. Judd. 1977a. California encephalitis virus transmission by arctic and domestic mosquitoes. *Archives of Virology* 55:39–45.

McLean, D.M., P.N. Grass, B.D. Judd, D. Cmiralova, and K. M. Stuart. 1977b. Natural foci of California encephalitis virus activity in the Yukon Territory. *Canadian Journal of Public Health* 68:69–73.

McLean, D.M., P.N. Grass, B.D. Judd, L.V. Ligate, and K.K. Peter. 1977c. Bunyavirus isolations from mosquitoes in the western Canadian arctic. *Journal of Hygiene* 79:61–72.

McLean, R.G., and G.S. Bowen. 1980. Vertebrate hosts. In *St. Louis encephalitis,* ed. T.P. Monath. Washington, DC: American Public Health Association, pp. 381–450.

McLintock, J., P.S. Curry, R.J. Wagner, M.K. Leung, and J.O. Iversen. 1976. Isolation of snowshoe hare virus from *Aedes implicatus* larvae in Saskatchewan. *Mosquito News* 36:233–237.

Mead, D.G., C.J. Maré, and E.W. Cupp. 1997. Vector competence of select black fly species for vesicular stomatitis virus (New Jersey serotype). *American Journal of Tropical Medicine and Hygiene* 57:42–48.

Meegan, J.M. 1994. Yellow Fever. In *Handbook of zoonoses,* section B: *Viral,* ed. G.W. Beran and J.H. Steele, 2d ed. Boca Raton, FL: CRC, pp. 111–124.

Meiyu, F., C. Huosheng, C. Cuihua, T. Xiaodong, J. Lianhua, P. Yifei, C. Weijun, and G. Huiyu. 1997. Detection of flaviviruses by reverse transcriptase-polymerase chain reaction with the universal primer set. *Microbiology and Immunology* 41:209–213.

Meyer, K.F., C.M. Haring, and B. Howitt. 1931. The etiology of epizootic encephalomyelitis in horses in the San Joaquin Valley, 1930. *Science* 74:227–228.

Mitchell, C.J., D.B. Francy, and T.P. Monath. 1980. Arthropod vectors. In *St. Louis encephalitis,* ed. T.P. Monath.

Washington, DC: American Public Health Association, pp. 313–379.

Mitchell, C.J., S.D. Lvov, H.M. Savage, C.H. Calisher, G.C. Smith, D.K. Lvov, and D.J. Gubler. 1993. Vector and host relationships of California serogroup viruses in western Siberia. *American Journal of Tropical Medicine and Hygiene* 49:53–62.

Monath, T.P. 1989. Yellow fever. In *The arboviruses: Epidemiology and ecology,* vol. 5, ed. T.P. Monath. Boca Raton, FL: CRC, pp. 139–231.

Monath, T.P. 1991. Yellow fever: Victor, Victoria? Conqueror, conquest? Epidemics and research of the last forty years and prospects for the future. *American Journal of Tropical Medicine and Hygiene* 45:1–43.

Monath, T.P., and F.X. Heinz. 1996. Flaviviruses. In *Fields virology,* ed. B.N. Fields, D.M. Knipe, and P.M. Howley. Philadelphia: Lippincott-Raven, pp. 961–1034.

Monath, T.P., R.B. Craven, A. Adjukiewicz, M. Germain, D.B. Francy, L. Ferrara, E.M. Samba, H. N'Jie, K. Cham, S. A. Fitzgerald, P.H. Crippen, D.I. Simpson, E.T. Bowen, A. Fabiyi, and J.J. Salaun. 1980. Yellow fever in the Gambia, 1978–1979: Epidemiologic aspects with observations on the occurrence of Orungo virus infections. *American Journal of Tropical Medicine and Hygiene* 29:912–928.

Monath, T.P., K.R. Brinker, F.W. Chandler, G.E. Kemp, and C.B. Cropp. 1981. Pathophysiologic correlations in a rhesus monkey model of yellow fever with special observations on the acute necrosis of B cell areas of lymphatic tissues. *American Journal of Tropical Medicine and Hygiene* 30:431–443.

Monath T.P., R.R. Nystrom, R.E. Bailey, C.H. Calisher, and D.J. Muth. 1984. Immunoglobulin M antibody capture enzyme-linked immunosorbent assay for diagnosis of St. Louis encephalitis. *Journal of Clinical Microbiology* 20:784–790.

Morris, C.D. 1988. Eastern equine encephalitis. In *The arboviruses: Epidemiology and ecology,* vol. 3, ed. T.P. Monath. Boca Raton, FL: CRC, pp. 1–20.

Muckenfuss, R.S., C. Armstrong, and H.A. McCordock. 1933. Encephalitis: Studies on experimental transmission. *Public Health Reports* 48:1341–1343.

Nawrocki, S.J., Y.H. Randle, M.H. Vodkin, J.P. Siegel, and R.J. Novak. 1996. Evaluation of a reverse transcriptase-polymerase chain reaction assay for detecting St. Louis encephalitis virus using field-collected mosquitoes (Diptera: Culicidae). *Journal of Medical Entomology* 33:123–127.

Nelson, R.L., C.H. Tempelis, W.C. Reeves, and M.M. Milby. 1976. Relation of mosquito density to bird: Mammal feeding ratios of *Culex tarsalis* in stable traps. *American Journal of Tropical Medicine and Hygiene* 25:644–654.

Newhouse, V.F., W. Burgdorfer, J.A. McKiel, and J.D. Gregson. 1963. California encephalitis virus: Serologic survey of small wild mammals in northern United States and southern Canada, and isolation of additional strains. *American Journal of Hygiene* 78:123–129.

Newhouse, V.F., W. Burgdorfer, and D. Corwin. 1967. Field and laboratory studies on the hosts and vectors of the snowshoe hare strain of California virus. *Mosquito News* 31:401–408.

Osorio, J.E., and T.M. Yuill. 1994. Venezuelan equine encephalitis. In *Handbook of zoonoses,* section B: *Viral,* ed. G.W. Beran and J.H. Steele, 2d ed. Boca Raton, FL: CRC, pp. 33–46.

Pavri, K.M., and K.R. P. Singh. 1968. Kyasanur Forest disease infection in the frugivorous bats *Cynopterus sphinx. Indian Journal of Medical Research* 56:1202–1204.

Phillpotts, R.J., K. Venugopal, and T. Brooks. 1996. Immunisation with DNA polynucleotides protects mice against lethal challenge with St. Louis encephalitis virus. *Archives of Virology* 141:743–749.

Pittman, P.R., R.S. Makuch, J.A. Mangiafico, T.L. Cannon, P.H. Gibbs, and C.J. Peters. 1996. Long-term duration of detectable neutralizing antibodies after administration of live-attenuated VEE vaccine and following booster vaccination with inactivated VEE vaccine. *Vaccine* 14:337–343.

Proctor, S.J., and K.C. Sherman. 1975. Ultrastructural changes in bovine lingual epithelium infected with vesicular stomatitis virus. *Veterinary Pathology* 12:362–377.

Rajagopalan, P.K., and C.R. Anderson. 1970. Transmonsoonal persistence of Kyasanur Forest disease virus in *Haemaphysalis* nymphs infected in nature. *Indian Journal of Medical Research* 8:1184–1187.

Rajagopalan, P.K., S.D. Paul, and M.A. Sreenivasan. 1969. Involvement of *Rattus blanfordi* (Rodentia: Muridae) in the natural cycle of Kyasanur Forest disease virus. *Indian Journal of Medical Research* 57:999–1002.

Reid, H.W. 1988. Louping-ill. In *The arboviruses: Epidemiology and ecology,* vol. 3, ed. T.P. Monath. Boca Raton, FL: CRC, pp. 117–135.

Reid, H.W., R.M. Barlow, J.B. Boyce, and D.M. Inglis. 1976. Isolation of louping-ill virus from a roe deer *(Capreolus capreolus). Veterinary Record* 98:116.

Reid, H.W., D. Buxton, I. Pow, and J. Finlayson. 1982. Experimental louping-ill virus infection in two species of British deer. *Veterinary Record* 111:61.

Reif, J.S. 1994. Vesicular stomatitis. In *Handbook of zoonoses,* section B: *Viral,* ed. G.W. Beran and J.H. Steele, 2d ed. Boca Raton, FL: CRC, pp. 171–181.

Reisen, W.K., and T.P. Monath. 1989. Western equine encephalitis. In *The arboviruses: Epidemiology and ecology,* vol. 5, ed. T.P. Monath. Boca Raton, FL: CRC, pp. 89–137.

Ribelin, W.E. 1958. The cytopathogenesis of vesicular stomatitis virus infection in cattle. *American Journal of Veterinary Research* 29:66–73.

Rico-Hesse, R., J.T. Roehrig, and R.W. Dickerman. 1988. Monoclonal antibodies define antigenic variation in the ID variety of Venezuelan equine encephalitis virus. *American Journal of Tropical Medicine and Hygiene* 38:187–194.

Rico-Hesse, R., S.C. Weaver, J. de Siger, G. Medina, and R.A. Salas. 1995. Emergence of a new epidemic/epizootic Venezuelan equine encephalitis virus in South America. *Proceedings of the National Academy of Sciences USA* 92:5278–5281.

Ritter, D.G., and E.T. Feltz. 1974. On the natural occurrence of California encephalitis virus and other arboviruses in Alaska. *Canadian Journal of Microbiology* 20:1359–1366.

Rivas, F., L.A. Diaz, V.M. Cardenas, E. Daza, L. Bruzon, A. Alcata, O. de la Hoz, F.M. Caceres, G. Aristizabal, J.W. Martinez, D. Revalo, F. de la Hoz, J. Boshell, T. Camacho, L. Calderon, V.A. Olano, L.I. Villarreal, D. Roselli, G. Alvarez, G. Ludwig, and T. Tsai. 1997. Epidemic Venezuelan equine encephalitis in La Guajira, Colombia, 1995. *Journal of Infectious Diseases* 175:828–832.

Rodriguez, L.L., G.J. Letchworth, C.F. Spiropoulou, and S.T. Nichol. 1993. Rapid detection of vesicular stomatitis virus New Jersey serotype in clinical samples by using polymerase chain reaction. *Journal of Clinical Microbiology* 31:2016–2020.

Saluzzo, J.F., T.P. Monath, M. Cornet, V. Deubel, and J.P. Digoutte. 1985. Comparaison de differentes techniques pour le detection du virus de la fièvre jaune dans les prélèvements humains et les lots de moustiques: Interet d'une méthode rapide de diagnostic par ELISA. *Annals de Institut Pasteur, Virologie* 136E:115–129.

Sanmartin, C., R.B. Mackenzie, H. Trapido, P. Barreto, C. H. Mullenax, E. Gutierrez, and C. Lesmes. 1973. Encefalitis equina venezolana en Colombia, 1967. *Boletin Oficina Sanitaria Panamerica* 74:108–137.

Scherer, W.F, J.V. Ordoñez, P.B. Jahrling, B.A. Pancake, and R.W. Dickerman. 1972. Observations of equines, humans and domestic and wild vertebrates during the 1969 equine epizootic and epidemic of Venezuelan encephalitis in Guatemala. *American Journal of Epidemiology* 95:255–266.

Scherer, W.F., R.W. Dickerman, J.V. Ordoñez, C. Seymour III, L.D. Kramer, P.B. Jahrling, and C.D. Powers. 1976. Ecologic studies of Venezuelan encephalitis virus and isolations of Nepuyo and Patois viruses during 1968–1973 at a marsh habitat near the epicenter of 1969 outbreak in Guatemala. *American Journal of Tropical Medicine and Hygiene* 25:161–162.

Schliessmann, D.J., and L.B. Calheiros. 1974. A review of the status of yellow fever and *Aedes aegypti* eradication programs in the Americas. *Mosquito News* 34:1–9.

Seymour, C., and R.W. Dickerman. 1978. Venezuelan encephalitis virus infection in neotropical bats: III. Experimental studies on virus excretion and nonarthropod transmission. *American Journal of Tropical Medicine and Hygiene* 27:307–312.

Seymour, C., R.W. Dickerman, and M.S. Martin. 1978a. Venezuelan encephalitis virus infection in neotropical bats: I. Natural infection in a Guatemalan enzootic focus. *American Journal of Tropical Medicine and Hygiene* 27:290–296.

———. 1978b. Venezuelan encephalitis virus infection in neotropical bats: II. Experimental infections. *American Journal of Tropical Medicine and Hygiene* 27:297–306.

Seymour, C., P.H. Peralta, and G.G. Montgomery. 1983. Serologic evidence of natural togavirus infections in Panamanian sloths and other vertebrates. *American Journal of Tropical Medicine and Hygiene* 32:854–861.

Shope, R.E., and J.P. Woodall. 1973. Ecological interaction of wildlife, man and a virus of the Venezuelan equine encephalomyelitis complex in a tropical forest. *Journal of Wildlife Diseases* 9:198–203.

Singh, K.R.P., M.K. Goverdhan, and T.R. Rao. 1968. Experimental transmission of Kyasanur Forest disease virus to small mammals by *Ixodes petauristae, I. ceylonensis,* and *Haemaphysalis spinigera. Indian Journal of Medical Research* 56:594–609.

Smart, K.L., R.E. Elbel, R.F.N. Woo, E.R. Kern, G.T. Crane, G.L. Bales, and D.W. Hill. 1972. California and western encephalitis viruses from Bonneville Basin, Utah, in 1965. *Mosquito News* 32:382–390.

Soliman, A.K., D.M. Watts, A.W. Salib, A.E. Shehata, R.R. Arthur, and B.A. Botros. 1997. Application of an immunoperoxidase monolayer assay for the detection of arboviral antibodies. *Journal of Virological Methods* 65:147–151.

Srihongse, S. 1969. Vesicular stomatitis virus infections in Panamanian primates and other vertebrates. *American Journal of Epidemiology* 90:69–82.

Srihongse, S., M.A. Grayson, and R. Diebel. 1984. California serogroup viruses in New York State: The role of subtypes in human infections. *American Journal of Tropical Medicine and Hygiene* 33:1218–1227.

Stallknecht, D.E., D.M. Kavanaugh, J.L. Corn, K.A. Eernisse, J.A. Comer, and V.F. Nettles. 1993. Feral swine as a potential amplifying host for vesicular stomatitis virus New Jersey serotype on Ossabaw Island, Georgia. *Journal of Wildlife Diseases* 29:377–383.

Suarez, O.M., and G.H. Bergold. 1968. Investigations of an outbreak of Venezuelan equine encephalitis in towns of eastern Venezuela. *American Journal of Tropical Medicine and Hygiene* 17:875–880.

Sudia, W.D., and V.F. Newhouse. 1975. Epidemic Venezuelan equine encephalitis in North America: A summary of virus-vector-host relationships. *American Journal of Epidemiology* 101:1–13.

Sudia, W.D., V.F. Newhouse, and W.A. Chappell. 1969. Venezuelan equine encephalitis virus vector studies following a human case in Dade County, Florida, 1968. *Mosquito News* 29:596–600.

Sulkin, S.E., R. Allen, and R. Sims. 1963. Studies of arthropod-borne virus infections in Chiroptera: I. Susceptibility of insectivorous species to experimental infection with Japanese B and St. Louis encephalitis viruses. *American Journal of Tropical Medicine and Hygiene* 12:800–814.

Telford III, S.R., P.M. Armstrong, P. Katavolos, I. Foppa, A.S. Garcia, M.L. Wilson, and A. Spielman. 1997. A new tick-borne encephalitis-like virus infecting New England deer ticks, *Ixodes dammini. Emerging Infectious Diseases* 3:165–170.

Tesh, R.B., P.H. Peralta, and K.M. Johnson. 1969. Ecologic studies of vesicular stomatitis virus: I. Prevalence of infection among animals and humans living in an area of endemic VSV activity. *American Journal of Epidemiology* 90:255–261.

Tesh, R.B., B.N. Chaniotis, and K.M. Johnson. 1971. Vesicular stomatitis virus, Indiana serotype: Multiplication and transmission by experimentally infected phlebotomine sandflies (*Lutzomyia trapidoi*). *American Journal of Epidemiology* 93:491–495.

Tesh, R.B., B.N. Chaniotis, and K.M. Johnson. 1972. Vesicular stomatitis virus (Indiana serotype): Transovarial transmission by phlebotomine sandflies. *Science* 175:1477–1479.

Tesh, R., S. Saidi, E. Javadian, P. Loh, and A. Nadim. 1977. Isfahan virus, a new vesiculovirus. *American Journal of Tropical Medicine and Hygiene* 26:299–306.

Theiler, M., and W.C. Downs. 1973. *The arthropod-borne viruses of vertebrates.* London: Yale University Press.

Thorne, E.T., E.S. Williams, W.J. Adrian, and C.M. Gillin. 1983. Vesicular stomatitis in pronghorn antelope: Serologic survey and artificial infection. *Proceedings of the United States Animal Health Association* 87:638–653.

Timoney, P. 1971. Powassan virus infection in the grey squirrel. *Acta Virologica* 15:429.

Trainer, D.O., and R.P. Hanson 1969. Serologic evidence of arbovirus infections in wild ruminants. *American Journal of Epidemiology* 90:354–358.

Trainer, D.O., and G.L. Hoff. 1971. Serologic evidence of arbovirus activity in a moose population in Alberta. *Journal of Wildlife Diseases* 7:118–119.

Trapido, H., M.K. Goverdhan, P.K. Rajagopalan, and M.J. Rebello. 1964. Ticks ectoparasitic on monkeys in the Kyasanur Forest disease area of Shimoga District, Mysore State, India. *American Journal of Tropical Medicine* 13:763–772.

Tsai, T.F., R.A. Bolin, M. Montoya, R.E. Bailey, D.B. Francy, M. Jozan, and J.T. Roehrig. 1987. Detection of St. Louis encephalitis virus antigen in mosquitoes by capture enzyme immunoassay. *Journal of Clinical Microbiology* 25:370–376.

Ubico, S.R., and R.G. McLean. 1995. Serologic survey of neotropical bats in Guatemala for virus antibodies. *Journal of Wildlife Diseases* 31:1–9.

Venugopal, K., W.R. Jiang, and E.A. Gould. 1995. Immunity to St. Louis encephalitis virus by sequential immunization with recombinant vaccinia and baculovirus derived PrM/E proteins. *Vaccine* 13:1000–1005.

Vodkin, M.H., T. Streit, C.J. Mitchell, G.L. McLaughlin, and R.J. Novak. 1994. PCR-based detection of arboviral

RNA from mosquitoes homogenized in detergent. *Biotechniques* 17:114–116.

Walton, T.E., and M.A. Grayson. 1988. Venezuelan equine encephalitis. In *The arboviruses: Epidemiology and ecology,* vol. 4, ed. T.P. Monath. Boca Raton, FL: CRC, pp 203–231.

Weaver, S.C., A. Hagenbaugh, L.A. Bellew, L. Gousset, V. Mallampalli, J.J. Holland, and T.W. Scott. 1994. Evolution of alphaviruses in the eastern equine encephalomyelitis complex. *Journal of Virology* 68:158–169.

Webb, H.E. 1965. Kyasanur Forest disease virus in three species of rodents. *Transactions of the Royal Society of Tropical Medicine and Hygiene* 59:205–211.

Webb, H.E., and J. Burston. 1966. Clinical and pathological observations with special reference to the nervous system in *Macaca radiata* infected with Kyasanur Forest disease virus. *Transactions of the Royal Society of Tropical Medicine and Hygiene* 60:325–331.

Webb, H.E., and J.B. Chatterjea. 1962. Clinico-pathological observations on monkeys infected with Kyasanur Forest disease virus, with special reference to the hemopoietic system. *British Journal of Haematology* 8:401–413.

Webb, P.A., and F.R. Holbrook. 1989. Vesicular stomatitis. In *The arboviruses: Epidemiology and ecology,* vol. 4, ed. T.P. Monath. Boca Raton, FL: CRC, pp. 1–29.

Webb, P.A., R.G. McLean, G.C. Smith, J.H. Ellenberger, D.B. Francy, T.E. Walton, and T.P. Monath. 1987. Epizootic vesicular stomatitis in Colorado, 1982: Some observations on the possible role of wildlife populations in an enzootic maintenance cycle. *Journal of Wildlife Diseases* 23:192–198.

Wellings, F.M., A.L. Lewis, and L.V. Pierce. 1972. Agents encountered during arboviral ecological studies: Tampa Bay, Florida, 1963 to 1970. *American Journal of Tropical Medicine and Hygiene* 21:201–213.

Wenger, F. 1977. Venezuelan equine encephalitis. *Teratology* 16:359–362.

West, G.P., ed. 1975. *Encyclopedia of animal care,* 11th ed. Baltimore: Williams and Wilkins, pp. 433–434.

Whitney, E., and H. Jamnback. 1965. The first isolation of Powassan virus in New York State. *Proceedings of the Society of Experimental and Biological Medicine* 119:432–435.

Whitney, E., H. Jamnback, R.G. Means, A.P. Roz, and G.A. Rayner. 1969. Isolation and characterization of California encephalitis complex from *Aedes cinereus. American Journal of Tropical Medicine and Hygiene* 18:123–131.

Wilks, C.R., and J.A. House. 1985. Antigenic and pathogenic comparisons of seven vesiculoviruses. In *Veterinary viral diseases: Their significance in Southeast Asia and the Western Pacific,* ed. A.J. Della-Porta. Sydney: Academic, pp. 390–402.

Woodall, J.P. 1967. Virus research in Amazonia. *Atas do Simposio sobre a biota Amazonica (Patalogia)* 6:31–63.

Work, T.H., and H. Trapido. 1957. Summary of preliminary report of investigations of the Virus Research Center on epidemic disease affecting forest villages and wild monkeys of Shimoga District, Mysore. *Indian Journal of Medical Science* 11:340–341.

Work, T.H., F.R. Rodriguez, and P.N. Bhatt. 1959. Virological epidemiology of the 1958 epidemic of Kyasanur Forest disease. *American Journal of Public Health* 49:869–874.

Workman, T., D. Shen, L. Woodard, and T. Yilma. 1986. An enzyme-linked immunosorbent assay for the detection of bovine antibodies to vesicular stomatitis virus. *American Journal of Veterinary Research* 47:1507–1512.

World Health Organization. 1971. Technical guide for a system of yellow fever surveillance. *Weekly Epidemiologic Record* 45:493–500.

Yuill, T.M. 1980. Vesicular stomatitis. In *Handbook of zoonoses,* section B, *Viral zoonoses,* ed. J.H. Steele and G.W. Beran. West Palm Beach, FL: CRC.

Yuill, T.M., and R.P. Hanson. 1964. Serologic evidence of California encephalitis virus and western equine encephalitis virus infection in a population of snowshoe hares. *Zoonoses Research* 3:153–163.

Yuill, T.M., J.O. Iversen, and R.P. Hanson. 1969. Evidence for arbovirus infections in a population of snowshoe hares: A possible mortality factor. *Bulletin of the Wildlife Disease Association* 5:248–253.

Zarate, M.L., and W.F. Scherer. 1968. Contact spread of Venezuelan equine encephalomyelitis virus among cotton rats via urine or feces and the naso- or oropharynx. *American Journal of Tropical Medicine and Hygiene* 17:894–899.

Zarnke, R.L., and T.M. Yuill. 1981a. Powassan virus infection in snowshoe hares (*Lepus americanus*). *Journal of Wildlife Diseases* 17:303–310.

———. 1981b. Serologic survey for selected microbial agents in mammals from Alberta. *Journal of Wildlife Diseases* 17:453–461.

PICORNAVIRUS INFECTIONS

GAVIN R. THOMSON, ROY G. BENGIS, AND CORRIE C. BROWN

The family *Picornaviridae* has five genera, *Aphthovirus, Cardiovirus, Enterovirus, Hepatovirus,* and *Rhinovirus.* Only aphthoviruses, the cause of foot-and-mouth disease in cloven-hoofed animals, and cardioviruses, which cause encephalomyocarditis in many species, are currently recognized as pathogens of wild mammals.

FOOT-AND-MOUTH DISEASE

Synonyms: Aphthous fever, hoof-and-mouth disease.

Introduction and Domestic Animal Health Concerns. Foot-and-mouth disease (FMD) is a highly contagious but usually nonlethal disease of ruminants, Suidae and Camelidae, characterized by vesiculation of the oral mucosa and of the skin of the feet.

In the late 19th century, FMD was widespread throughout the world, but during the early 20th century, the disease was progressively eradicated from areas of intensive livestock husbandry, due to the disease's economic impact. Today, North America and Western Europe are free of FMD, but it persists in the northern half of South America, most African countries, the Middle East, parts of Eastern Europe and most of Asia.

Where FMD was eradicated from livestock, it also generally disappeared from wildlife. In parts of sub-Saharan Africa, however, African buffaloes *Syncerus caffer* are maintenance hosts of FMD virus. Hence, the eradication of FMD from that region is impossible without the destruction of large numbers of African buffaloes, which is untenable.

Countries in which FMD does not occur restrict imports of livestock and their products, and of susceptible wildlife, from countries where the disease is prevalent. Hence, it is a challenge for African nations to maintain their rich wildlife heritage without compromising agricultural development. As well, FMD occasionally can be destructive among wildlife, as occurred in South Africa in the late 19th century and more recently in Israel (Macaulay 1963; Shimshony 1988).

History. In 1545, Fracastorius described FMD in cattle in Italy (Bulloch 1927). The disease was common in France, Germany, and Italy in the 17th and 18th centuries, but it did not spread to Britain until 1839 (Brown 1986). FMD was first reported in the United States in 1870, and a year later it occurred in South America (Brown 1986). The FMD virus was the first filterable agent identified, by Loeffler and Frosch in 1898 (Rueckert 1996).

The early history of FMD in Asia and Africa is not known. However, disease resembling FMD was described from South Africa in 1780, 1850, and 1858, and FMD was officially recorded in 1892 (Thomson 1994).

The last outbreaks of FMD in the United States, Canada, and Mexico occurred in 1929, 1952, and 1953, respectively (Bachrach 1968). FMD eradication was only achieved in 1992 in Western Europe; the disease is still prevalent in the northern half of South America. In Asia, FMD is widespread, with the exception of Japan (eradicated in 1908), Singapore (1935), and Indonesia (1983) (Anonymous 1995). FMD occurs throughout Africa, except in an infection-free zone of South Africa (Anonymous 1994). Australia has been free of FMD since 1872, and it never occurred in New Zealand (Bachrach 1968).

Etiology. FMD viruses are small (approximately 30 nm in diameter), unenveloped, spherical, with icosahedral (T = 1) symmetry. The capsid is comprised of 60 protomeres (poorly defined capsomeres), each consisting of four polypeptides designated VP1–VP4. Three capsid proteins (VP1, VP2, and VP3) are exposed on the surface of the virion, while the fourth (VP4) is associated with the RNA genome. Seven to nine nonstructural virally encoded proteins, involved in virus replication, are produced in infected cells (Porter 1993).

The single plus (messenger sense) strand of RNA, approximately 8400 nucleotides long, is polyadenylated at the 3′ end and contains a 50- to 200-nucleotide poly(C) tract near the 5′ end, within the noncoding region. The genetic organization of picornaviruses is illustrated in Rueckert (1996).

Loops of VP1, each with a tripeptide motif for attaching to cells, protrude from the surface of the virion (Logan et al. 1993). Other attachment sites probably also exist on the virus (Barteling and Vreeswijk 1991). Rueckert (1996) illustrates the structure and mode of replication of picornaviruses.

Seven immunologically distinct types of FMD virus occur: A, O, C, SAT1, SAT2, SAT3, and Asia 1. These form two groups on the basis of RNA hybridization; A, O, C, and Asia 1 in one group, and the three SAT types in the other (Robson et al. 1977). Infection or immunization with one type does not confer protection against another type. The two most prevalent FMD viruses are types O and A, which occur in South America, many parts of central and northern Africa, the Middle East, Eastern Europe, and Asia. Type C, with approximately the same distribution, is less prevalent. The three SAT types are confined to sub-Saharan Africa, and Asia 1 to Asia and the Middle East.

Antigenic variants or subtypes of FMD occur, and many A, O, and C subtypes have specific designations. Variation appears to be greater within the three SAT types, but specific subtypes are not presently recognized (Vosloo et al. 1995).

SAT-type viruses evolve independently by mutation in different regions of southern Africa (Vosloo et al. 1992). The nucleotide heterogeneity of the VP1 genes of such viruses recovered from African buffaloes within Kruger National Park in South Africa may exceed 20% (Vosloo et al. 1995), and the rate at which they mutate in individual persistently infected buffalo has been established (Vosloo et al. 1996). More dramatic mutations may occur by intra- and intertypic recombination between viruses (Krebbs and Marquardt 1992). Deletions or additions within the 5′-untranslated region of the genome have also been demonstrated (Escarmis et al. 1995).

FMD viruses are relatively resistant to environmental conditions but are labile at pH values below 6 and above 9. This is used for cheap and convenient disinfection, for instance by 2% citric acid or 4% washing soda (Na_2CO_3). Washing soda is corrosive but is less liable to neutralization by biologic material such as feces and blood than is citric acid. Other disinfectants, including various acids (usually formulated with detergents), sodium metasilicate, hypochlorites, iodine, and several proprietary commercial disinfectants are effective, whereas phenolic and quaternary ammonium compounds generally are not (Sellers 1968).

Wildlife Host Range. Hedger (1981) tabulated wildlife species that have been recorded with clinical FMD as a result of natural or experimental infection, not always confirmed by virus isolation. These include members of the Bovidae (24 spp.), Cervidae (10 spp.), Suidae (4 spp.), Tayasuidae (1 sp.), and Camelidae (4 spp.) among the Artiodactyla; the Erinaceidae (2 spp.) and Talpidae (1 sp.) among the Insectivora; the Dasypodidae (1 sp.) among the Xenarthra (Edentata); the Leporidae (1 sp.) among the Lagomorpha; the Sciuridae (2 spp.), Bathyergidae (Rhizomyidae) (1 sp.), Muridae (4 spp.), Hystricidae (1 sp.), Hydrochaeridae (1 sp.), Capromyidae (1 sp.), Dasyproctidae (1 sp.), and Chinchillidae (1 sp.) among the Rodentia; Elephantidae (2 spp.) in the Probiscidea; Procaviidae (1 sp.) in the Hyracoidea; Ursidae (1 sp.) among the Carnivora; and various marsupials and monotremes. Since that time, FMD has been reported in the nyala *Tragelaphus angasi* (Bengis 1983) and the mountain gazelle *Gazella gazella* (Shimshony et al. 1986) among the Bovidae, and in the giraffe *Giraffa camelopardalis* (Bengis 1984) in the Giraffidae.

Transmission and Epidemiology. Transmission of FMD usually occurs by close contact between acutely infected and susceptible individuals. High levels of FMD virus occur in respiratory secretions for 1–3 days prior to, and 7–14 days after, the development of lesions (Sellers 1971; Donaldson 1983). Urine and feces contain less virus (Hyslop 1970). FMD occasionally is spread mechanically by contaminated animal products (e.g., milk and meat) or by fomites, vehicles, and people contaminated with the virus. Rarely, FMD has been transmitted over long distances by airborne aerosols (Gloster et al. 1982; Donaldson 1983).

FMD virus excretion by wildlife has been less well studied. Roe deer *Capreolus capreolus,* fallow deer *Dama dama,* sika deer *Cervus nippon,* red deer *Cervus elaphus,* and muntjac *Muntiacus muntjac* excreted FMD virus in approximately the same quantities as sheep and cattle. But deer were considered unlikely to be important in the epidemiology of FMD in the United Kingdom, because they do not contact livestock closely (Gibbs et al. 1975). Routes of virus excretion by African buffalo resembled those of cattle, and it persisted for up to 28 days. Aerosol excretion of low levels of virus was sometimes detected (Gainaru et al. 1986).

There are no reports of the dose of FMD virus required to infect wildlife. In cattle and sheep, the minimal infectious dose by the respiratory route is 25 and 10 cell-culture-infective doses, respectively (Gibson and Donaldson 1986; Donaldson et al. 1987), whereas the dose required for oral infection is about 10,000 times higher (Sellers 1971; Burrows et al. 1981). However, aerosols containing as little as one cell-culture-infective dose established infection in impala *Aepyceros melampus* (R.G. Bengis and G.R. Thomson unpublished). This sensitivity to infection may contribute to the epidemics of FMD in impala described below.

Transmission of FMD by carrier animals has been debated for years. FMD virus persists, often at barely detectable levels (Thomson 1996), in the pharynx of carriers for 4 weeks or longer (Salt 1993), by which time it cannot be detected elsewhere. Carrier status appears to occur only in ruminants (Terpstra 1972). Both cattle and African buffaloes rarely may transmit FMD intermittently for months or years (Thomson 1996), and carrier buffaloes have been proven to transmit SAT viruses to cattle (Dawe et al. 1994; Vosloo et al. 1996).

Other wildlife also may develop a carrier state. Fallow and sika deer regularly developed persistent infection, whereas red deer did so occasionally; roe and muntjac deer, on the other hand, did not (Forman et al. 1974; Gibbs et al. 1975). Persistent infection developed experimentally in kudu *Tragelaphus strepsiceros* (Hedger 1972). Excretion of FMD virus for a little over 4 weeks has also been shown in wildebeest *Connochaetes taurinus* (Anderson et al. 1975) and sable *Hippotragus niger* (Ferris et al. 1989). Among wildlife, however, carrier transmission has been demonstrated only for African buffalo, and it appears to be rare.

In all parts of the world, except sub-Saharan Africa, FMD in free-ranging or captive wildlife appears to have been an extension of the disease in livestock. This has been documented for free-ranging moose *Alces alces* (Magnusson 1939), as well as in fallow (Bartels and Claassen 1936), roe, and red deer in Europe (Cohrs and Weber-Springe 1939). In the former Soviet Union, FMD was described in free-ranging reindeer *Rangifer tarandus* (Kvitkin 1959; Ogryzkov 1963) and saiga *Saiga tatarica* (Khukhorov et al. 1974) whereas, in India, severe clinical signs and mortality were reported in blackbuck *Antilope cervicapra* (Kar et al. 1983). High morbidity and mortality also occurred in free-ranging mountain gazelles in Israel (Shimshony et al. 1986). All these episodes in wildlife occurred during epidemics in cattle. Similarly, outbreaks of FMD in zoological gardens in Paris (Urbain et al. 1938), Zurich (Allenspach 1950), and Buenos Aires (Grosso 1957), coincided with outbreaks in domestic animals. Even in sub-Saharan Africa, where wildlife clearly are involved in the maintenance of FMD, livestock have transmitted the infection to wildlife (Hedger 1976; Thomson et al. 1984; Anderson et al. 1993).

The only locality in which overt FMD has been reported regularly in wildlife over the last 60 years is the Kruger National Park in South Africa, where there have been 31 recorded outbreaks in impala since 1938, and 23 since routine surveillance was introduced in the mid 1960s. Eight (26%) were caused by SAT1, 15 (48%) by SAT2, three (10%) by SAT3, and five (16%) were untyped. Since 1983, however, all nine outbreaks in impala were caused by SAT2.

FMD in impala appears to be density dependent, and because impala depend on water, infection frequently has spread along watercourses. With few exceptions, obvious clinical disease has not occurred in other species in the vicinity of outbreaks in impala (Bengis 1983; Keet et al. 1996). Direct contact between impala inside Kruger National Park and domestic animals outside the park is inhibited by fencing, and livestock are vaccinated. There have been no outbreaks of FMD in livestock in South Africa since 1983.

African buffalo are probably the source of infection for impala, because in interepidemic periods prevalence of antibody to FMD in impala falls (R.G. Bengis and G.R. Thomson unpublished), and persistent infection has not been demonstrated (Anderson et al. 1975). However, FMD epidemics caused by identical viruses have recurred in impala 6–18 months after the original outbreak (Vosloo et al. 1992; Keet et al. 1996). How these viruses survived in these intervening periods remains to be explained. Infection and attack rates have varied, with the latter sometimes much lower than the former, reflecting subclinical infection, as has been seen in impala experimentally (R.G. Bengis and G.R. Thomson unpublished).

Paradoxically, clinical FMD has not been diagnosed in impala elsewhere, although there is serologic evidence of infection in other parts of southern Africa (Anderson et al. 1993).

African buffaloes were recognized as major reservoirs of SAT-type viruses in the 1970s (Hedger et al. 1972). Persistence in individuals for at least 5 years possibly explains why FMD virus can persist for over 20 years in a small isolated group (Condy et al. 1985), since infection probably is not lifelong (Hedger 1976).

How SAT viruses are maintained in buffalo populations in southern Africa is unclear. Infection usually occurs when maternal immunity starts to wane at 2–4 months of age (Condy and Hedger 1978). Calves are not necessarily infected by their dams (Condy and Hedger 1974), and it is presumed that SAT viruses spread mainly during minor epidemics among young animals in breeding herds, with carriers ensuring that the viruses survive interepidemic periods (Thomson et al. 1992). Since most buffalo are born in midsummer, they become susceptible to infection more-or-less synchronously during the dry winter months. Other species likely become exposed while infection is epidemic among buffalo calves, probably around permanent water, where animals congregate.

There is evidence suggesting transmission in both directions between cattle and European hedgehogs *Erinaceus europaeus,* and for latent infection in hibernating hedgehogs (Hulse and Edwards 1937; McLaughlin and Henderson 1947; Macauley 1963). However, these reports should be viewed with caution, because virologic techniques at the time were unreliable, and there is little evidence that hedgehogs have participated in the propagation of FMD in Europe or Africa.

Capybaras *Hydrochaeris hydrochaeris* are susceptible to FMD and speculatively may play a role in the epidemiology of FMD in cattle in South America (Gomes and Rosenberg 1984–85). Although laboratory mice *Mus musculus* and guinea pigs *Cavia porcellus* are highly susceptible to infection with FMD virus, and neonatal mice frequently develop fatal disease (Subak-Sharpe 1961), there is no evidence that mice or other small rodents have been involved in the spread of FMD in the field.

The position of elephants is confusing. Although natural cases of FMD have been reported in captive African elephants *Loxodonta africana* and Asian elephants *Elephas maximus* (Piragino 1970; Pyakural et al. 1976), and African elephants are susceptible to needle inoculation with FMD (Howell et al. 1973), African elephants did not become infected when exposed to artificially infected cohorts or cattle (Howell et al. 1973; Bengis et al. 1984). Furthermore, there was no serologic evidence for infection in elephants culled in Kruger National Park over a 30-year period (Bengis et al. 1984). In southern Africa, African elephants are not considered susceptible to FMD in natural circumstances, and they are not subject to the restrictions on movement imposed on ruminants from FMD-endemic areas.

Although hippopotami *Hippopotamus amphibius* are artiodactyls, which generally are susceptible to FMD, they have not been found with proven FMD, and serology of 877 animals failed to detect evidence of infection (R.G. Bengis unpublished). Rhinoceros *Ceratotherium simum* and *Diceros bicornis* are perissodactyls, which are considered refractory to FMD.

Clinical Signs. FMD in wildlife varies from completely inapparent to acutely lethal. African buffalo mainly develop subclinical infection. Few of over 47,000 buffalo culled in the Kruger National Park, where FMD is endemic, had clinical signs or lesions suggestive of FMD (R.G. Bengis unpublished). However, no FMD virus was recovered from suspect material examined in the laboratory. Experimentally, typical small lesions occurred, particularly on the feet (Anderson et al. 1979; Gainaru et al. 1986). Likewise, impala frequently become infected without showing lesions.

Death due to FMD has been described among mountain gazelles in two Israeli nature reserves (Shimshony et al. 1986; Shimshony 1988), and it also has occurred among impala (Hedger et al. 1972), blackbuck (Kar et al. 1983), saiga (Kindyakov et al. 1972), white-tailed deer (McVicar et al. 1974), and warthogs *Phacochoerus aethiopicus* (R.G. Bengis unpublished). The case fatality rate among mountain gazelles in one outbreak was greater than 50%; at least 1500 animals died. Death was presumed to be due to a combination of heart failure (due to viral myocarditis) in the more acute cases and diabetes mellitus (due to pancreatitis) in cases of longer duration (Shimshony 1988; Perl et al. 1989).

The signs of FMD in wildlife generally are similar to those in domestic animals (Thomson 1994). Some impala may develop severe, although usually nonfatal, FMD, while others remain clinically normal (Thomson et al. 1984; Keet et al. 1996). In the acute stages, animals may develop piloerection, probably due to fever, and locomotor signs relating to foot lesions. These vary from mild "walking on eggs," with arched back and head held low, to severe "carrying leg" lameness. Other signs include licking or shaking of the feet, shifting weight from one leg to the other, holding one hoof off the ground, lagging behind the herd, and lying down with reluctance to rise. Similar signs have been observed in kudu, bushbuck *Tragelaphus scriptus,* nyala, warthogs, and giraffe (R.G. Bengis and D.F. Keet unpublished). In very severe cases, hooves of impala and wild suids may slough (R.G. Bengis unpublished). Secondary bacterial infection of foot lesions is sometimes crippling.

Discontinuity of the skin-hoof junction results in a "break" or fault in the hoof wall as the hoof grows, which is useful for estimating the time since the acute phase of the disease. In impala, it takes 5–6 months for this fault to grow out completely.

Salivation is uncommon in antelope, even in animals with severe mouth lesions.

Unusual signs include progressive emaciation as a result of exocrine and endocrine pancreatic atrophy in mountain gazelles (Perl et al. 1989), loss of horns (Shimshony et al. 1986), erosions at the base of the supernumerary digits in wild suids and kudu (R.G. Bengis unpublished), and lesions on the kneeling pads of warthogs and bushpigs *Potamochoerus porcus* (R.G. Bengis unpublished).

Lesions of the udder or teats have not been documented in wildlife.

Pathogenesis. The pathogenesis of FMD has been studied mainly in cattle and pigs (Burrows et al. 1981; Brown et al. 1995, 1996). The initial site of virus replication is thought to be the respiratory bronchioles of the lung (Brown et al. 1996). The virus then spreads via the bloodstream to Langerhans cells (macrophage-like dendritic cells) in epithelial tissue (Di Girolamo et al. 1995), and all epithelial cells in contact with an infected Langerhans cell become infected (Brown et al. 1995).

The virus is disseminated to many epidermal sites, but lesions develop only in areas subjected to mechanical trauma or physical stress (Gailiunas and Cottral 1966).

A number of mechanisms have been proposed for persistent infection with FMD virus (Salt 1993; Woodbury 1995). The pharynx may be the site of viral persistence, but this has not been confirmed experimentally. Bergmann et al. (1996) found FMD virus-specific genetic sequences in multiple sites, but not in pharyngeal specimens, in cattle up to 2 years after infection. Infection of macrophage-type cells may be central to the pathogenesis of FMD, and virus might persist in these cells in various tissues (Brown et al. 1996).

Pathology. In FMD, vesicles (blisters, or aphthae) develop at multiple sites, generally on the feet and in the mouth (Sutmoller 1992; Barker et al. 1993). Severe lesions occur where there is mechanical stress on infected epithelial surfaces. This varies with the species. Thus, suids, which have a high ratio of body weight to foot size, and which root with the nose, tend to have the most severe lesions on the feet and on the rostrum of the snout. In ruminants, oral lesions can be severe. Foot lesions in cattle usually occur in the interdigital cleft and at the skin-hoof junction of the bulbs of the heel. In impala, as in small domestic ruminants, mouth lesions are usually most severe on the dental pad but may occur elsewhere, especially on the tongue; foot lesions begin as a coronitis, sometimes vesiculating around the entire coronet (Keet et al. 1996). Vesicles at any site rupture early in the course of disease, so that lesions often are eroded by the time that an animal is examined.

In white-tailed deer, FMD was very similar to that seen in cattle, with vesicles on both oral and foot epithelium (McVicar et al. 1974). However, they tended to form preferentially on the bulbs of the heel rather than in the interdigital cleft. White-collared peccaries *Tyassu tajucu* were very susceptible, but the course of the disease was milder and of briefer duration than that in domestic pigs. Vesicles occurred on the snout, tongue, coronary band, and interdigital clefts (Dardiri et al. 1969). Nine-banded armadillos *Dasypus novemcinctus* developed vesicular lesions on foot pads and toes (Wilder et al. 1974).

Young animals of any species may die of myocarditis, which appears grossly as whitened streaklike areas in the myocardium.

Histologically, vesicles begin as clusters of hypereosinophilic degenerating keratinocytes in the stratum spinosum. Intercellular edema fluid accumulates, forming a vesicle that soon ruptures, leaving an eroded surface (Barker et al. 1993). The epithelium in the mouth often regenerates completely within a week, but foot lesions heal more slowly. Myocardial lesions consist of multifocal myocardial degeneration and necrosis with a predominantly lymphocytic cellular response.

Diagnosis. In most countries, suspect cases of FMD must be reported to animal health authorities, to whom falls the responsibility for further investigation and diagnosis. Specialized high-security diagnostic laboratories, of which there are only a few worldwide, are required to meet international standards for safety and competence. Due to the potential for inadvertent spread of the virus, external surfaces of objects to be removed from the site of investigation must be decontaminated. Movement of animals from such sites should be prohibited and that of people controlled until a diagnosis is established.

The diagnosis of FMD in wildlife is more complicated than in domestic stock, because the range and variation in severity of presenting signs is often greater. It is sometimes important to demonstrate infection in wildlife in the absence of clinical signs, either because it tends to be subclinical for the particular species-virus combination or because physical examination of wild animals is difficult to accomplish. Laboratory assistance is indispensable.

FMD should be suspected in animals with compatible orocutaneous or foot lesions or clinical signs. During the acute stage of the infection, tags of vesicular epithelium for virus isolation should be collected into containers containing phosphate-buffered saline (pH 7.4) kept at 4° C. Blood also should be collected into anticoagulant for the detection of viremia. If specimens cannot be delivered to a laboratory within 24 hours, they should be frozen in dry ice or liquid nitrogen, but not in a conventional freezer, since FMD virus does not survive well at –20° C. At necropsy, lymph nodes are useful for virus isolation. Since sudden death, particularly among young animals, may be due to myocarditis, specimens of myocardium should be collected on ice for virus isolation, as well as in 10% formalin for histologic examination in such cases.

Persistent infections in ruminants may be identified by obtaining probang specimens from the pharyngeal region (Kitching and Donaldson 1987). A number of animals need to be sampled because this test is not sensitive enough to detect all persistently infected individuals.

Sera should be collected from as many suspect animals as possible. Positive serologic results may provide conclusive evidence of infection, even though virus may not be recovered. Serology involves either virus neutralization (Esterhuysen et al. 1985) or a variety of enzyme-linked immunosorbent assays (ELISAs), most commonly a blocking ELISA (Hamblin et al. 1986).

Conventionally, FMD viruses in tissue or secretions are detected by isolation in cell cultures (Snowden 1966). Thereafter, virus typing is usually performed by ELISA (Ferris and Dawson 1988; Westbury et al. 1988) or by polymerase chain reaction (PCR) (Locher et al. 1995).

For epidemiologic reasons, it is important to compare the virus involved in an outbreak with other isolates to enable the selection of appropriate vaccine strains, should vaccination be an option as a control strategy. This is accomplished by molecular techniques, while antigenic relationships between the outbreak strain and potential vaccine strains are obtained by cross-neutralization (Pay 1983; Thomson 1994).

Differential Diagnosis. The differential diagnosis of FMD in livestock includes other viral infections that induce vesicular lesions in the mouth and on the feet: vesicular stomatitis, swine vesicular disease, and vesicular exanthema (Thomson 1994); the latter two are diseases of pigs exclusively, and the effect that these infections have on wild ungulates, if any, is unclear. Other infections that, in domestic animals, may produce lesions that may resemble FMD are bovine virus diarrhea, rinderpest, bluetongue, epizootic hemorrhagic disease, malignant catarrhal fever, lumpy skin disease, bovine papular stomatitis, infectious bovine rhinotracheitis, and orf (contagious pustular dermatitis). Fungal infections and photosensitivity rarely may produce skin lesions that could be confused with FMD. To the extent that these or similar conditions may affect wildlife, they must be considered with FMD in a differential diagnosis.

Immunity. The duration of immunity following FMD infection seems to vary among wildlife (Hedger et al. 1972), but there are almost no reliable data. In cattle, antibody concentration reflects immune status (McCullough et al. 1992). Immunity in cattle lasts 1–3, and occasionally up to over 4, years (Bachrach 1968; Brooksby 1982). When immunity is challenged by another viral subtype, the duration of immunity is reduced; the degree of antigenic difference and the duration of immunity are inversely related (Pay 1983).

Although animals that have recovered from infection or been vaccinated rapidly develop virus-neutralizing antibody, protective immunity against FMD virus probably is effected by antibody-dependent phagocytosis by cells of the reticuloendothelial system (McCullough et al. 1992).

Control and Management. Control strategies for FMD in wildlife depend on the locality and the type of livestock husbandry practiced. Because of the potential impact of FMD on the livestock economy, the effect on wildlife often is secondary in the eyes of authorities responsible for animal health. In most countries, FMD is a scheduled or controlled disease and how it is dealt with is stipulated in legislation or animal health regulations.

The first line of defense is to prevent introduction of FMD virus to a susceptible population. In countries where FMD is absent, this is accomplished by prohibition of, or strict controls on, the importation of animals and animal products from FMD-endemic areas; these sanctions extend to wildlife and their products.

In southern Africa, where wildlife are reservoirs of FMD virus, the historic approach has been to separate domestic livestock from wildlife, usually by means of fences (increasingly, double fence lines that preclude direct contact between animals on either side), and to immunize livestock in the vicinity. This generally has been successful in preventing transmission of FMD from wildlife to livestock (Thomson 1995). However, it has been severely criticized by conservationists, because the fences sometimes have blocked migration routes and access of wildlife to water, resulting in ecological disturbance and wildlife mortality (Owen and Owen 1980; Taylor and Martin 1987). Wildlife management may be further complicated by restrictions placed on areas where wildlife may be farmed or ranched; on wildlife translocation; and on the distribution of products derived from wildlife, such as meat, hides, and trophies.

If disease breaks out, there are three basic approaches to the control of FMD in livestock; they sometimes are used in combination. The first is "stamping out," or eradication: infected and in-contact animals are slaughtered quickly, and the carcasses appropriately disposed of, to eliminate the infection completely. This response also may entail attempted elimination of all individuals of susceptible wildlife species within a defined radius of the infected premises to prevent establishment of a wildlife reservoir of infection (Wobeser 1994). Such an approach is usually adopted where the disease is exotic, and where the economic consequences of an outbreak would be severe.

In the second approach, animals in the infected focus are quarantined from those in surrounding areas until the infectious period has passed. This can be done by isolating (and in some cases immunizing) the affected animals, with, ideally, an animal-free zone between the infected and closest susceptible animals.

The third approach involves vaccination of animal populations likely to be exposed to infected individuals. However, immunization does not necessarily protect all animals (Donaldson and Kitching 1989).

For logistical reasons, these measures often are not applicable to FMD outbreaks involving wildlife. Furthermore, with the exception of sub-Saharan Africa, outbreaks of FMD are rare and difficult to anticipate. Therefore, strategies for controlling FMD in wildlife would need to be specifically devised, taking into account the local ecology, epidemiologic features of the outbreak, implications for the livestock population, and logistical and economic realities. Models of the epidemiology of FMD in feral species suggest that, once established, FMD may be difficult or impossible to eradicate in some circumstances (Wobeser 1994).

Public Health Concerns. Although there are reports of the disease in people, it is most unlikely that humans develop FMD. A disease of people named "hand, foot, and mouth disease," caused predominantly by Coxsackie virus A16, but also by other enteroviruses, has clinical features similar to FMD (Melnick 1996) and may account for the reports of FMD in people.

ENCEPHALOMYOCARDITIS

Synonym: Mengovirus infection.

Introduction. Encephalomyocarditis (EMC) virus is thought to produce mainly occult infection in rodents. However, it may cause sporadic outbreaks of myocarditis and sudden death among a wide variety of species, particularly among captive collections of wildlife. It was suspected that EMC was associated with human neurologic disease; however, it now appears that infection of people usually is asymptomatic.

Among domestic animals, EMC virus causes disease most commonly in pigs, in which it was first recognized in Panama (Murnane et al. 1960). In addition to myocarditis and sudden death, it also causes reproductive failure in pigs (Joo 1992). Horses, cattle, dogs, and cats are susceptible to infection but do not develop clinical disease.

Etiology and Distribution. The genus *Cardiovirus* contains two groups: EMC virus and related isolates (Colombia SK, ME, MM, and Mengo viruses), and Theiler's murine encephalomyelitis (TME) virus. Apart from the fact that cardioviruses, like aphthoviruses, have a poly(C) tract in their genome, their structure and replication are typical of the Picornaviridae in general (Rueckert 1996). EMC virus appears to be ubiquitous, and this virus, or antibody to it, has been demonstrated in wild rats and mice worldwide (Grobler et al. 1995).

Transmission, Epidemiology, and Host Range. Experimentally, EMC may be transmitted orally; by intranasal, intratracheal, and aerosol infection of the respiratory tract; and by parenteral injection (Tesh and Wallace 1978; Zimmerman 1994). Transmission by contamination of wounds is efficient in swine (Zimmerman et al. 1993). EMC does not seem very contagious in laboratory rodents and domestic pigs (Littlejohns and Acland 1975).

Rodents are often implicated in outbreaks of EMC. The black rat *Rattus rattus,* the Norway rat *Rattus norvegicus,* and the house mouse *Mus musculus* were readily infected orally but rarely developed clinical disease (Tesh and Wallace 1978). Since these species did not shed virus significantly, they are unlikely reservoirs of infection. Cannibalism of infected animals was required for transmission of EMC among mice (Dick 1953; Vanella et al. 1956) and rats (Kilham et al. 1956a).

However, multimammate rats *Mastomys* sp. were highly susceptible to infection, developed clinical disease, and frequently died (Kilham et al. 1956a); feces contained large amounts of virus. EMC virus also was recovered from the gut and feces of experimentally infected banded mongooses *Mungos mungo,* white-tailed mongooses *Ichneumia albicauda,* and grivet monkeys *Cercopithecus aethiops* (Kilham et al. 1956b). It also has been isolated from feces or intestinal contents of wild squirrels *Sciurus* sp. (Vizoso et al. 1964), cotton rats *Sigmodon hispidus,* and a raccoon *Procyon lotor* (Gainer and Bigler 1967).

EMC virus was recovered from rodents, an opossum *Didelphis virginiana,* and a rabbit *Sylvilagus* sp. during an EMC outbreak (Wells et al. 1989), which appeared to be limited by rodent control. Similarly, the death of an African elephant from EMC at an Australian zoo, during an epidemic in pigs in the region, was associated with a plague of mice (Seaman and Finnie 1987). Hubbard et al. (1992) concluded that rats were the probable source of infection in an EMC epidemic in a colony of baboons *Papio cynocephalus;* rodent control influenced the duration of the outbreak.

Reports of natural outbreaks of EMC among captive wild animals suggest an oral route of infection, by fecal contamination of feed or drinking water, or by consumption of infected carcasses. Lions died after being fed carcasses of herbivores that died of EMC (Simpson et al. 1977; Gaskin et al. 1980). Hay fed to an African elephant that died of EMC had been contaminated by mice (Seaman and Finnie 1987). EMC at the Taronga Zoo, Sydney, Australia, followed a change in feeding practices that made enclosures harder to clean and more attractive to rodents (Reddacliff et al. 1997).

During an EMC epidemic in elephants in Kruger National Park, there was a temporal correlation between a rodent population explosion, a high prevalence of antibody to EMC in multimammate mice *Mastomys natalensis,* and clinical disease in elephants (Grobler et al. 1995). Elephants may have been predisposed to infection by their tendency to uproot and ingest entire tufts of grass, in which rodents sheltered from raptors. EMC does not appear to be transmitted efficiently between elephants. In a recent EMC vaccine trial (Hunter et al. 1998), transmission did not occur between diseased elephants and susceptible contact controls, nor could virus be isolated from the feces of affected animals.

Although rodents are implicated in most outbreaks of EMC, they merely may be a reflection of virus circulating among a wide range of animals in the environment (Zimmerman 1994). The tendency to lump all rodent data together when evaluating surveys, and failure to focus on specific species, may have clouded the issue. It is also unfortunate that in the many surveys other potential wildlife reservoirs of infection have been underrepresented. In Hawaii, neutralizing antibodies to EMC were more prevalent in mongooses, cattle, and pigs than in *Rattus rattus* in the same locality (Tesh and Wallace 1978). Similarly, prevalence of antibody was higher in captive rhesus monkeys than in rats (Dick 1953), and also among horses than in wild rats during an epidemic in Brazil (Causey et al. 1962).

EMC virus also has been isolated from a variety of arthropods, including mosquitoes, ticks, houseflies, and a flea (Dick 1949; Causey et al. 1962). However, experimental attempts to transmit EMC with arthropods have not been successful.

In the final analysis, where and how the virus survives in interepidemic periods is uncertain.

Clinical disease has been documented in a wide variety of captive wildlife. Gaskin et al. (1980) described EMC outbreaks in three zoos in Florida, during which confirmed EMC caused the death of four African elephants, an Asian elephant, a Debrazza monkey *Cercopithecus neglectus,* an African antelope cross, and three chimpanzees *Pan troglodytes.* A further ten African elephants, an Asian elephant, a giraffe, a white rhinoceros *Ceratotherium simum,* a chimpanzee, a Debrazza monkey, and 20 lions also died of suspected EMC. Subsequently, Gaskin et al. (1987) reported mortalities due to EMC in further species: nyala, llama *Lama glama,* two-toed sloth *Choloepus* sp., ringtail lemur *Lemur catta,* ruffed lemur *Varecia variegata,* and an orangutan *Pongo pygmaeus.* Reports from other zoos describe the disease in African elephants (Simpson et al. 1977; Seaman and Finnie 1987), Thomson's gazelles *Gazella thomsoni,* dromedary camels *Camelus dromedarius* (Wells et al. 1989), a Sumatran orangutan (Citino et al. 1988), a squirrel monkey *Saimiri sciureus,* mandrills *Mandrillus sphinx,* a pygmy hippopotamus *Choeropsis liberiensis,* and Goodfellow's tree kangaroos *Dendrolagus goodfellowi* (Reddacliff et al. 1997). An EMC epidemic occurred in a baboon breeding colony (Hubbard et al. 1992). Asymptomatic EMC infection has been reported in a range of rodents, mongooses, marsupials, and birds (Causey et al. 1962; Tesh and Wallace 1978).

Clinical Signs. Most animals affected by EMC are found dead without any prior signs of illness. Rapidly fatal infections have been described in nonhuman primates, porcupines, mongooses, lions, and elephants (Kilham et al. 1956b; Simpson et al. 1977; Gaskin et al. 1980; Hubbard et al. 1992). In less sudden cases, anorexia, listlessness, and moderate to severe dyspnea generally occur. These clinical signs are related to acute congestive heart failure, with pulmonary congestion and edema, hydropericardium, and ascites. Neurologic signs are uncommon, except in smaller primates and some rodents. Abortions and stillbirths, possibly as a result of EMC infection, were reported in baboons (Hubbard et al. 1992).

Many infections in a wide range of species are nonlethal and probably subclinical. For example, antibody to EMC was found in 53% of free-ranging elephants in the Kruger National Park, demonstrating that most survived the infection (Grobler et al. 1995). Of the 64 elephant deaths that could be attributed to EMC, 83% were adult bulls. This gender predilection may be explained by the tendency for testosterone to enhance susceptibility to EMC infection (Friedman et al. 1972).

Three of four unvaccinated juvenile African elephants that were challenged during an EMC vaccine trial developed clinical disease and two died suddenly (Hunter et al. 1998); one developed subclinical disease. Signs included lethargy, depression, anorexia, swinging and flexing movements of the trunk, and sucking the tip of the trunk; a loose stool developed 9–10 days after oral infection or 4–5 days after parenteral infection. Intermittent tremors of the trunk were previously reported by Seaman and Finnie (1987). None of the infected elephants developed fever. Significant electrocardiographic abnormalities occurred during the acute phase of the disease, as has also been reported in a capybara (Wells et al. 1989).

EMC may have contributed to "frothy trunk" syndrome in young free-living elephants in Kruger National Park. This is characterized by acute congestive heart failure with massive pulmonary edema shortly after immobilization. An increase in the incidence of this syndrome from 10% to 25% between 1987 and 1992 was associated with an increase in prevalence of antibody to EMC from 12.5% to 33%. Elephants that survive EMC infection develop moderate to severe myocardial scarring. They may suffer acute congestive heart failure due to the increase in afterload placed on a weakened myocardium that is susceptible to developing ventricular dysrhythmias due to catecholamine release caused by capture drugs (Hattingh et al. 1994).

Pathogenesis. Viremia may occur within 24 hours of infection (Tesh and Wallace 1978). The initial site of replication is unknown. Primary lesions are found in the heart, less frequently in the central nervous system, and rarely in pancreas or skeletal muscle. The virus kills the cells in which it replicates; this, and the associated inflammatory response, are the basis of the tissue damage that occurs.

Pathology. The predominant lesion in EMC is myocardial necrosis (Wells et al. 1989). Consequently, gross lesions may include any of the following sequelae to heart failure: hydrothorax, hydropericardium, pulmonary edema, froth in the trachea and bronchi, and fibrin in the thoracic or abdominal cavities (Hubbard et al. 1992). Grossly, there are often pale streaks within the myocardium, which histologically are areas of myocardial degeneration and necrosis. Inflammatory infiltrates are predominantly lymphocytic, although necrotic myocardial fibers may also contain neutrophils and macrophages. Crystalline arrays of viral particles may be visible in the cytosol of infected cells on electron-microscopic examination but locating such cells is difficult (Hubbard et al. 1992).

Some strains of EMC virus also induce encephalitis, pancreatitis, or myositis (Blanchard et al. 1987; Cronin et al. 1988). In all tissues, there is multifocal necrosis with an associated inflammatory response. There also seems to be some species predilection for the development of lesions in particular organs. For instance, squirrel monkeys developed encephalitis and pancreatitis when inoculated with a strain of EMC virus that caused myocarditis in other species (Blanchard et al. 1987).

Diagnosis. Sudden death or signs of acute heart failure among captive wildlife, particularly primates, should alert clinicians to the possibility of EMC. White necrotic foci or more diffuse pallor of the myocardium, or subepicardial ecchymotic hemorrhages, at necropsy is a further indication, especially if the myocardial necrosis is accompanied by a mononuclear cell infiltration histologically.

Definitive diagnosis requires identification of the virus in the tissues, usually the myocardium, of affected animals. Isolation of EMC virus usually is straightforward in acute cases because it is cytolytic in a variety of cell culture systems. The diagnosis is confirmed by inhibition of infectivity or hemagglutination by antisera specific for EMC virus.

Isolation of EMC virus in more chronic infections usually is not possible, though immunohistochemistry may demonstrate viral antigen in sections of affected tissue, and molecular techniques for detecting the viral genome may be developed. Serology may not be helpful in such cases, because of the prevalence of antibody resulting from previous subclinical infection. Increasing antibody titers in paired sera, or demonstration of an immunoglobulin G response may be helpful, but these are difficult to obtain.

Differential Diagnosis. Conditions which produce cardiac lesions that may be confused with those seen in EMC are FMD in young animals, herpesvirus infection in Asian elephants (Richman et al. 1996), acute capture myopathy involving the myocardium, myocardial necrosis caused by vitamin E/selenium deficiency, infarcts caused by septic emboli; ionophore poisoning, and plant poisoning involving cardiac glycosides. All other causes of acute death among wildlife also should be considered in the differential diagnosis.

Control and Management Implications. Despite the lack of consensus on the role of rodents in the epidemiology of EMC, rodent control was associated with a rapid end to at least two outbreaks (Wells et al. 1989; Hubbard et al. 1992). Hence, it is worth considering in the face of an outbreak. General preventive measures are based on good hygiene, including effective disposal of garbage and rodent proofing of food stores.

Vaccination of highly susceptible primates, elephants, and other species remains a further option, but until recently no safe and effective vaccine was available for use in wild animals. A commercial inactivated vaccine for domestic pigs has been marketed (Oxford Veterinary Laboratories, Worthington, MN, U.S.A.) but its efficacy in wildlife has not been documented.

A beta-propiolactone-inactivated vaccine protected laboratory mice against challenge and produced variable antibody responses in pygmy goats *Capra hircus* and Barbados sheep *Ovis aries,* depending on the adjuvant (Gaskin et al. 1987). A formaldehyde-inactivated vaccine tested during a zoo outbreak was considered safe, but its efficacy was questionable (Wells et al. 1989). An inactivated vaccine derived from a virulent isolate of EMC from a zoo outbreak proved effective in inducing antibody in macaques *Macaca mulatta* (Emerson and Wagner 1996). Osorio et al. (1996) developed a live vaccine attenuated by deletions in the poly(C) tract of the viral genome. Baboons, macaques, and domestic pigs all developed antibody and were protected against challenge with lethal doses of EMC virus.

Following the EMC epidemic in free-ranging elephants in South Africa, an aziradine-inactivated vaccine containing high antigen mass and an oil adjuvant was developed. A single dose elicited high levels of neutralizing antibody in laboratory mice, domestic pigs, and African elephants, and protected mice and elephants against lethal challenge with virulent EMC virus (Hunter et al. 1998). Thus, safe and effective vaccines can be developed, although none aimed specifically at wildlife are as yet commercially available.

Public Health Concerns. Encephalomyocarditis virus has been isolated sporadically from blood, cerebrospinal fluid, and feces of people, particularly children, with aseptic meningitis, encephalomyelitis, and Guillain-Barré syndrome. Rising serum antibody titers also have been reported in some cases. However, Tesh (1978) could find no association between prevalence of antibody to EMC virus in various human populations and the incidence of encephalitis, myocarditis, and diabetes mellitus. The virus does not appear to be highly contagious for people, and surveys among high-risk professions such as veterinarians, zoo workers, farmers, and laboratory personnel have revealed few serologic reactors. Cumulatively, the data indicate that, if infection occurs, a high proportion of EMC infections in people are asymptomatic.

LITERATURE CITED

Allenspach, V. 1950. Foot and mouth disease in the zoological gardens of Zürich [in German]. *Schweizer Archiv für Tierheilkunde* 92:42–47.

Anderson, E.C., J. Anderson, W.J. Doughty, and S. Drevmo. 1975. The pathogenicity of bovine strains of foot-and-mouth disease virus for impala and wildebeest. *Journal of Wildlife Diseases* 11:248–255.

Anderson, E.C., W.J. Doughty, J. Anderson, and R. Paling. 1979. The pathogenesis of foot-and-mouth disease in African buffalo (*Syncerus caffer*) and the role of this species in the epidemiology of the disease in Kenya. *Journal of Comparative Pathology* 89:541–549.

Anderson, E.C., C. Foggin, H. Atkinson, K.J. Sorenson, R.L. Madekurozva, and J. Nqindi. 1993. The role of wild animals other than buffalo in the current epidemiology of foot and mouth disease in Zimbabwe. *Epidemiology and Infection* 111:559–563.

Anonymous. 1994. *International animal health code.* Paris: Office International des Epizooties.

———. 1995. *Animal health year book, 1994.* Rome: Food and Agriculture Organization, Office International des Epizooties, World Health Organization, 250 pp.

Bachrach, H.L. 1968. Foot-and-mouth disease. *Annual Review of Microbiology* 22:201–244.

Barker, I.K., A.A. Van Dreumel, and N. Palmer. 1993. The alimentary system. In *Pathology of domestic animals,* vol. 2, ed. K.V.F. Jubb, P.C. Kennedy, and N. Palmer, 4th ed. San Diego: Academic, pp. 1–318.

Barteling, S.J., and J. Vreeswijk. 1991. Developments in foot-and-mouth disease vaccine. *Vaccine* 9:75–88.

Bartels, and P. Claassen. 1936. Wild game in the etiology of foot and mouth disease [in German]. *Berliner Tierärztliche Wochenschrift* 52:230–233.

Bengis, R.G. 1983. *Annual report 1981–82: State Veterinarian, Skukuza (Kruger National Park).* Pretoria, South Africa: Directorate of Veterinary Services, National Department of Agriculture.

———. 1984. *Annual report 1982–83: State Veterinarian, Skukuza (Kruger National Park).* Pretoria, South Africa: Directorate of Veterinary Services, National Department of Agriculture.

Bengis, R.G., R.S. Hedger, V. de Vos, and L. Hurter. 1984. The role of the African elephant (*Loxodonta africana*) in the epidemiology of foot and mouth disease in the Kruger National Park. In *Proceedings of the 13th World Congress of Diseases in Cattle, World Buiatrics Association, September 17–21.* Durban, South Africa: World Buiatrics Association, pp. 39–44.

Bergmann, I.E., V. Malirat, P.A. de Mello, and I. Gomes. 1996. Detection of foot-and-mouth disease viral sequences in various fluids and tissues during persistence of the virus in cattle. *American Journal of Veterinary Research* 57:134–137.

Blanchard, J.L., A.E. Gutter, K.F. Soike, and G.B. Baskin. 1987. Encephalomyocarditis virus infection in African green and squirrel monkeys: Comparison of pathologic effects. *Laboratory Animal Science* 37:635–639.

Brooksby, J.B. 1982. Portraits of viruses: Foot-and-mouth disease virus. *Intervirology* 18:1–23.

Brown, C.C., H.J. Olander, and R.F. Meyer. 1995. Pathogenesis of foot-and-mouth disease in swine as studied by in-situ hybridization. *Journal of Comparative Pathology* 113:51–58.

Brown, C.C., M.E. Piccone, P.W. Mason, T.St.-C. McKenna, and M.J. Grubman. 1996. Pathogenesis of wild-type and leaderless foot-and-mouth disease virus in cattle. *Journal of Virology* 70:5638–5641.

Brown, F. 1986. Foot-and-mouth disease: One of the remaining great plagues. *Proceedings of the Royal Society of London [B]* 229:215–226.

Bulloch, W. 1927. Foot-and-mouth disease in the 16th century. *Journal of Comparative Pathology and Therapeutics* 40:75–76.

Burrows, R., J.A. Mann, A.J.M. Garland, A. Greig, and D. Goodridge. 1981. The pathogenesis of natural and simulated natural foot-and-mouth disease infection in cattle. *Journal of Comparative Pathology* 91:599–609.

Causey, O.R., R.E. Shope, and H. Laemmert. 1962. Report of an epizootic of encephalomyocarditis virus in Para, Brazil. *Revista Servicio Especial de Saúde Pública* 12:47–50.

Citino, S.B., B.L. Homer, J.M. Gaskin, and D.J. Wickham. 1988. Fatal encephalomyocarditis infection in a Sumatran orangutan (*Pongo pygmaeus abelii*). *Journal of Zoo Animal Medicine* 19:214–218.

Cohrs, P., and W. Weber-Springe. 1939. Maul- und Klauenseuche beim Reh und Hirsch [Foot and mouth disease in roe deer and red deer]. *Deutsche Tierärztliche Wochenschrift* 47:97–103.

Condy, J.B., and R.S. Hedger. 1974. The survival of foot-and-mouth disease virus in African buffalo with non-transference to domestic cattle. *Research in Veterinary Science* 16:182–185.

Condy, J.B., and R.S. Hedger. 1978. Experiences in the establishment of a herd of foot and mouth disease-free African buffalo (*Syncerus caffer*). *South African Journal of Wildlife Research* 8:87–89.

Condy J.B., R.S. Hedger, C. Hamblin, and I.T.R. Barnett. 1985. The duration of the foot-and-mouth disease virus carrier state in African buffalo (i) in the individual animal and (ii) in a free-living herd. *Comparative Immunology, Microbiology and Infectious Diseases* 8:257–265.

Cronin, M.E., L.A. Love, F.W. Miller, P.R. McClintock, and P.H. Plotz. 1988. The natural history of encephalomyocarditis virus-induced myositis and myocarditis in mice: Viral persistence demonstrated by in situ hybridization. *Journal of Experimental Medicine* 168:1639–1648.

Dardiri, A.H., R.J. Yedloutschnig, and W.D. Taylor. 1969. Clinical and serologic response of American white-collared peccaries to African swine fever, foot-and-mouth disease, vesicular stomatitis, vesicular exanthema of swine, hog cholera, and rinderpest viruses. *Proceedings, Annual Meeting of the United States Animal Health Association* 73:437–452.

Dawe, P.S., F.O. Flanagan, R.L. Madekurozwa, K.J. Sorenson, E.C. Anderson, C.M. Foggin, N.P. Ferris, and N.J. Knowles. 1994. Natural transmission of foot-and-mouth disease from African buffalo (*Syncerus caffer*) to cattle in a wildlife area of Zimbabwe. *Veterinary Record* 134:230–232.

Dick, G.W.A. 1949. The relationship of Mengo encephalomyelitis, encephalomyocarditis, Columbia SK and M.M. viruses. *Journal of Immunology* 62:375–386.

———. 1953. Epidemiological notes on some viruses isolated in Uganda. *Transactions of the Royal Society of Tropical Medicine and Hygiene* 47:13–48.

Di Girolamo, W., M. Salas, and R.P. Laguens. 1995. Role of Langerhans cells in the infection of the guinea-pig epidermis with foot-and-mouth disease virus. *Archives of Virology* 83:331–336.

Donaldson, A.I. 1983. Quantitative data on airborne foot-and-mouth disease virus: Its production, carriage and deposition. *Philosophical Transactions of the Royal Society of London [B]* 302:529–534.

Donaldson, A.I., and R.P. Kitching. 1989. Transmission of foot and mouth disease by vaccinated cattle following natural challenge. *Research in Veterinary Science* 46:9–14.

Donaldson, A.I., C.F. Gibson, R. Oliver, C. Hamblin, and R.P. Kitching. 1987. Infection of cattle by airborne foot and mouth disease virus: Minimal doses with O_1 and SAT2 strains. *Research in Veterinary Science* 43:339–346.

Emerson, C.L., and J.L. Wagner. 1996. Antibody responses to two encephalomyocarditis virus vaccines in rhesus macaques (*Macaca mulatta*). *Journal of Medical Primatology* 25:42–45.

Escarmis, C., J. Dopazo, M. Davila, E.L. Palma, and E. Domingo. 1995. Large deletions in the 5′-untranslated region of foot and mouth disease virus of serotype C. *Virus Research* 35:155–167.

Esterhuysen, J.J., G.R. Thomson, J.R.B. Flamand, and R.G. Bengis. 1985. Buffalo in the northern Natal game parks show no serological evidence of infection with foot-and-mouth disease virus. *Onderstepoort Journal of Veterinary Research* 52:63–66.

Ferris, N.P., and M. Dawson. 1988. Routine application of enzyme-linked immunosorbent assay in comparison with complement fixation for the diagnosis of foot-and-mouth and swine vesicular diseases. *Veterinary Microbiology* 16:201–209.

Ferris, N.P., J.B. Condy, I.T.R. Barnett, and R.M. Armstrong. 1989. Experimental infection of eland (*Taurotragus oryx*), sable antelope (*Hippotragus niger*) and buffalo (*Syncerus caffer*) with foot and mouth disease virus. *Journal of Comparative Pathology* 101:307–316.

Forman, A.J., E.P.J. Gibbs, D.J. Baber, K.A.J. Herniman, and I.T. Barnett. 1974. Studies with foot-and-mouth disease in British deer (red, fallow and roe): II. Recovery of virus and serological response. *Journal of Comparative Pathology* 84:221–229.

Friedman, S.B., L.J. Grota, and L.A. Glasgow. 1972. Differential susceptibility of male and female mice to encephalomyocarditis virus: Effects of castration, adrenalectomy and the administration of sex hormones. *Infection and Immunity* 5:637–642.

Gailiunas, P., and G.E. Cottral. 1966. Presence and persistence of foot-and-mouth disease virus in bovine skin. *Journal of Bacteriology* 91:2333–2338.

Gainaru, M.D., G.R. Thomson, R.G. Bengis, J.J. Esterhuysen, W. Bruce, and A. Pini. 1986. Foot-and-mouth disease and the African buffalo (*Syncerus caffer*): II. Virus excretion and transmission during acute infection. *Onderstepoort Journal of Veterinary Research* 53:75–85.

Gainer, J.H., and W.I. Bigler. 1967. Encephalomyocarditis (EMC) virus recovered from two cotton rats and a raccoon. *Bulletin of the Wildlife Disease Association* 3:47–49.

Gaskin, J.M., M.A. Simpson, C.F. Lewis, O.L. Olsen, E.E. Schobert, E.P. Wollenman, C. Marlow, and M.M. Curtis. 1980. The tragedy of encephalomyocarditis infection in zoological parks in Florida. In *Proceedings of the American Association of Zoo Veterinarians.* Washington, DC: American Association of Zoo Veterinarians, pp. 1–7.

Gaskin, J.M., T.L. Andresen, J.H. Olsen, E.E. Schobert, D. Buesse, J.D. Lynch, M. Walsh, S. Citino, and D. Murphy. 1987. Encephalomyocarditis in zoo animals: Recent experiences with the disease and vaccination. In *Proceedings of the First International Conference of Zoological and Avian Medicine, Oahu, Hawaii, September, 1987,* p. 491.

Gibbs, E.P.J., K.A.J. Herniman, and M.J.P. Lawman. 1975. Studies with foot-and-mouth disease in British deer (muntjac and sika). *Journal of Comparative Pathology* 85:361–366.

Gibson, C.F., and A.I. Donaldson. 1986. Exposure of sheep to natural aerosols of foot-and-mouth disease virus. *Research in Veterinary Science* 41:45–49.

Gloster, J., R.F. Sellers, and A.I. Donaldson. 1982. Long distance transport of foot-and-mouth disease virus over the sea. *Veterinary Record* 110:47–52.

Gomes, I., and F.J. Rosenberg. 1984–85. A possible role of capybaras (*Hydrochoaris hydrochoaris hydrochoaris*) in foot-and-mouth disease (FMD) endemicity. *Preventive Veterinary Medicine* 3:197–205.

Grobler, D.J., J.P. Raath, L.E.O. Braack, D.F. Keet, G.H. Gerdes, B.J.H. Barnard, N.P.J. Kriek, J. Jardine, and R. Swanepoel. 1995. An outbreak of encephalomyocarditis virus infection in free ranging African elephants in the Kruger National Park. *Onderstepoort Journal of Veterinary Research* 62:97–108.

Grosso, A.M. 1957. Foot-and-mouth disease in the Buenos Aires Zoo in the past five years [in Spanish]. *Gaceta Veterinaria (Buenos Aires)* 19:54–55.

Hamblin, C., I.T.R. Barnett, and R.S. Hedger. 1986. A new enzyme-linked immunosorbent assay (ELISA) for the detection of antibodies against foot-and-mouth disease virus: I. Development and method of ELISA. *Journal of Immunological Methods* 93:115–121.

Hattingh, J., C.M. Knox, J.P. Raath, and D.F. Keet. 1994. Arterial blood pressure in anaesthetized African elephants. *South African Journal of Wildlife Research* 24:15–17.

Hedger, R.S. 1972. Foot-and-mouth disease and the African buffalo (*Syncerus caffer*). *Journal of Comparative Pathology* 82:19–28.

———. 1976. Foot-and-mouth disease in wildlife with particular reference to the African buffalo (*Syncerus caffer*). In *Wildlife diseases,* ed. L.A. Page. New York: Plenum, pp. 235–244.

———. 1981. Foot-and-mouth disease. In *Infectious diseases of wild mammals,* ed. J.W. Davis, L.H. Karstad, and D.O. Trainer, 2d ed. Ames: Iowa State University Press, pp. 87–96.

Hedger, R.S., J.B. Condy, and S.M. Golding. 1972. Infection of some species of African wildlife with foot-and-mouth disease virus. *Journal of Comparative Pathology* 82:455–461.

Howell, P.G., E. Young, and R.S. Hedger. 1973. Foot and mouth disease in the African elephant (*Loxodonta africana*). *Onderstepoort Journal of Veterinary Research* 40:41–52.

Hubbard, G.B., K.F. Soike, T.M. Butler, K.D. Carey, H. Davis, W.I. Butcher, and C.J. Gauntt. 1992. An encephalomyocarditis epizootic in a baboon colony. *Laboratory Animal Science* 42:233–239.

Hulse, E.C., and J.T. Edwards. 1937. Foot and mouth disease in hibernating hedgehogs. *Journal of Comparative Pathology and Therapeutics* 50:421–430.

Hunter, P., S.P. Swanepoel, J.J. Esterhuysen, J.P. Raath, R.G. Bengis, and J.J. van der Lugt. 1998. The efficacy of an experimental oil-adjuvanted encephalomyocarditis vaccine in elephants, mice and pigs. *Vaccine* 16:55–61.

Hyslop, N.St.G. 1970. The epizootiology and epidemiology of foot and mouth disease. *Advances in Veterinary Science* 14:261–307.

Joo, H.S. 1992. Encephalomyocarditis virus. In *Diseases of swine,* ed. A.D. Leman, B.E. Straw, W.H. Mengling, S. D'Allaire, and J.D. Taylor, 7th ed. Ames: Iowa State University Press, pp. 257–262.

Kar, B.C., N. Hota, and L.N. Acharjyo. 1983. Occurrence of foot-and-mouth disease among some ungulates in captivity. *Indian Veterinary Journal* 60:237–239.

Keet, D.F., P. Hunter, R.G. Bengis, A. Bastos, and G.R. Thomson. 1996. The 1992 foot-and-mouth disease epizootic in the Kruger National Park. *Journal of the South African Veterinary Association* 67:83–87.

Khukhorov, V.M., N.A. Pronina, L.N. Korsun, I.G. Karpenko, and B.A. Kruglikov. 1974. Foot and mouth disease in saiga antelopes [in Russian]. *Veterinariya (Moscow)* 5:60–61.

Kilham, L., P. Mason, and J.N.P. Davies. 1956a. Host-virus relations in encephalomyocarditis (EMC) infection: 1. Infection in wild rats. *American Journal of Tropical Medicine and Hygiene* 5:647–654.

———. 1956b. Host-virus relations in encephalomyocarditis (EMC) infection: 2. Myocarditis in mongooses. *American Journal of Tropical Medicine and Hygiene* 5:655–663.

Kindyakov, V.I., F.M. Nagumanov, and E.S. Tasbulatov. 1972. The epizootiological significance of contact between wild and domestic animals in relation to foot-and-mouth disease [in Russian]. *Voprosy Prirodnoi Ochagovosti Boleznei* 5:63–66.

Kitching, R.P., and A.I. Donaldson. 1987. Collection and transportation of specimens for vesicular virus investigation. *Revue Scientifique et Technique O.I.E.* 6:263–272.

Krebbs, O., and O. Marquardt. 1992. Identification and characterization of foot-and-mouth disease virus O_1, Burwedel/1987 as an intertypic recombinant. *Journal of General Virology* 73:613–619.

Kvitkin, V.P. 1959. Physiopathology of experimental foot-and-mouth disease in reindeer. *Veterinariya (Moscow)* 36:25–28 [Abstract from *Veterinary Bulletin* 30:119, 1960].

Littlejohns, I.R., and H.M. Acland. 1975. Encephalomyocarditis infection of pigs: 2. Experimental disease. *Australian Veterinary Journal* 51:416–422.

Locher, F., V.V.S. Suryanarayana, and J.-D. Tratschin. 1995. Rapid detection and characterization of foot-and-mouth disease virus by restriction enzyme and nucleotide sequence analysis of PCR products. *Journal of Clinical Microbiology* 33:440–444.

Logan, D., R. Abu-Ghazaleh, W. Blakemore, S. Curry, T. Jackson, and A. King. 1993. Structure of a major immunogenic site of foot-and-mouth disease virus. *Nature* 362:566–568.

Macaulay, J.W. 1963. Foot-and-mouth disease in non-domestic animals. *Bulletin of Epizootic Diseases of Africa* 11:143–146.

Magnusson, H. 1939. A case of foot-and-mouth disease in an elk [in German]. *Deutsche Tierärtzliche Wochenschrift* 47:509–511.

McCullough, K.C., F. de Simone, E. Brocchi, L. Capucci, J. R. Crowther, and U. Kihm. 1992. Protective immune response against foot-and-mouth disease. *Journal of Virology* 66:1835–1840.

McLaughlin, J.D., and W.M. Henderson. 1947. The occurrence of foot-and-mouth disease in the hedgehog under natural conditions. *Journal of Hygiene* 45:477–479.

McVicar, J.M., P. Sutmoller, and C.H. Campbell. 1974. Foot-and-mouth disease in white-tailed deer: Clinical signs and transmission in the laboratory. *Proceedings, Annual Meeting of the United States Animal Health Association* 78:169–180.

Melnick, J.L. 1996. Enteroviruses: Polioviruses, coxsackieviruses, echoviruses and newer enteroviruses. In *Fields virology,* vol. 1, ed. B.N. Fields, D.M. Knipe, and P.M. Howley, 3d ed. Philadelphia: Lippincott-Raven, pp. 655–712.

Murnane, T.G., J.E. Craighead, H. Mondragon, and A. Shelokov. 1960. Fatal disease of swine due to encephalomyocarditis virus. *Science* 131:498–499.

Ogryzkov, S.E. 1963. The pathology of foot-and-mouth disease in reindeer [in Russian]. Trudy 2 Vsesojuznoi Konferenzii Patologii i Anatomii Zhivotnych (Izdanie Moskovskoi Veterinarnoi Akademii, Tome 46), pp. 420–425. [English Abstract, *Veterinary Bulletin* 36(2571), 1966.]

Osorio, J.E., G.B. Hubbard, K.F. Soike, M. Girard, S. van der Werf, J.C. Moulin, and A.C. Palmenberg. 1996. Protection of non-murine mammals against encephalomyocarditis virus using a genetically engineered Mengo virus. *Vaccine* 14:155–161.

Owen, M., and D. Owen. 1980. The fences of death. *African Wildlife* 31:25–27.

Pay, T.W.F. 1983. Variation in foot-and-mouth disease: Application to vaccination. *Revue Scientifique et Technique O.I.E.* 2:701–723.

Perl, S., H. Yadin, B. Yakobson, E. Zuckerman, and U. Orgad. 1989. Pathological changes in mountain gazelles challenged with FMD virus, with special reference to pancreatic lesions. *Revue Scientifique et Technique O.I.E.* 8:765–769.

Piragino, S. 1970. An outbreak of foot and mouth disease in circus elephants [in Italian]. *Zooprofilassi* 25:17–22.

Porter, A.G. 1993. Picornavirus nonstructural proteins: Emerging roles in virus replication and inhibition of host cell functions. *Journal of Virology* 67:6917–6921.

Pyakural, S., U. Singh, and N.B. Singh. 1976. An outbreak of foot-and-mouth disease in Indian elephants (*Elaphus maximus*). *Veterinary Record* 99:28–29.

Reddacliff, L.A., P.D. Kirkland, J.W. Hartley, and R. Reece. 1997. Encephalomyocarditis virus infections in an Australian Zoo. *Journal of Zoo and Wildlife Medicine* 28:153–157.

Richman, L.K., R.J. Montali, R.C. Cambre, J. Lehnhardt, M. Kennedy, S. Kania, and L. Potgieter. 1996. Endothelial inclusion body disease: A newly recognized fatal herpes-like infection in Asian elephants. *Proceedings of the Annual Conference of the American Association of Zoo Veterinarians.* Puerto Vallerta: American Association of Zoo Veterinarians, pp. 483–486.

Robson, K.J.H., T.J.R. Harris, and F. Brown. 1977. An assessment by competition hybridization of the sequence homology between the RNSs of the seven serotypes of FMDV. *Journal of General Virology* 37:271–276.

Rueckert, R.R. 1996. Picornaviridae: The viruses and their replication. In *Fields virology,* vol. 1, ed. B.N. Fields, D.M. Knipe, and P.M. Howley, 3d ed. Philadelphia: Lippincott-Raven, pp. 609–654.

Salt, J.S. 1993. The carrier state in foot-and-mouth disease: An immunological review. *British Veterinary Journal* 149:207–223.

Seaman, J.T., and E.P. Finnie. 1987. Acute myocarditis in a captive African elephant (*Loxodonta africana*). *Journal of Wildlife Diseases* 23:170–171.

Sellers, R.F. 1968. The inactivation of foot-and-mouth disease virus by chemicals and inactivants. *Veterinary Record* 83:504–506.

———. 1971. Quantitative aspects of the spread of foot-and-mouth disease. *Veterinary Bulletin* 41:431–439.

Shimshony, A. 1988. Foot and mouth disease in the mountain gazelle in Israel. *Revue Scientifique et Technique O.I.E.* 7:917–923.

Shimshony, A., U. Orgad, D. Baharav, S. Prudovsky, B. Yakobson, B. Bar Moshe, and D. Dagan. 1986. Malignant foot-and-mouth disease in mountain gazelles. *Veterinary Record* 119:175–176.

Simpson, C.F., A.L. Lewis, and J.M. Gaskin. 1977. Encephalomyocarditis virus infection of captive elephants. *Journal of the American Veterinary Medical Association* 171:902–905.

Snowden, W.A. 1966. Growth of foot-and-mouth disease virus in monolayer cultures of calf thyroid cells. *Nature* 210:1079–1080.

Subak-Sharpe, H. 1961. The effect of passage history, route of inoculation, virus strain and host strain on the susceptibility of adult mice to the virus of foot-and-mouth disease. *Archiv für die Gesamte Virusforschung* 11:373–399.

Sutmoller, P. 1992. Vesicular diseases. In *Foreign animal diseases,* ed. W.W. Buisch, J.L. Hyde, and C.A. Mebus. Richmond, VA: US Animal Health Association, pp. 368–382.

Taylor, D.R., and R.B. Martin. 1987. Effects of veterinary fences on wildlife conservation in Zimbabwe. *Environmental Management* 11:327–334.

Terpstra, C. 1972. Pathogenesis of foot-and-mouth disease in experimentally infected pigs. *Bulletin de l'Office International des Epizooties* 77:859–874.

Tesh, R.B. 1978. The prevalence of encephalomyocarditis virus neutralizing antibodies among human populations. *American Journal of Tropical Medicine and Hygiene* 27:144–149.

Tesh, R.B., and G.D. Wallace. 1978. Observations on the natural history of encephalomyocarditis virus. *American Journal of Tropical Medicine and Hygiene* 27:133–143.

Thomson, G.R. 1994. Foot-and-mouth disease. In *Infectious diseases of livestock with special reference to southern Africa,* ed. J.A.W. Coetzer, G.R. Thomson, and R.C. Tustin. Cape Town: Oxford University Press, pp. 825–952.

———. 1995. Overview of foot-and-mouth disease in southern Africa. *Revue Scientifique et Technique O.I.E.* 14:503–520.

———. 1996. The role of carrier animals in the transmission of foot and mouth disease. In *Proceedings of the 64th General Session of the International Committee of the Office International des Epizooties, May 20–24.* Paris: O.I.E., pp. 1–18.

Thomson, G.R., R.G. Bengis, J.J. Esterhuysen, and A. Pini. 1984. Maintenance mechanisms for foot-and-mouth disease in the Kruger National Park and potential avenues for its escape into domestic animal populations. In *Proceedings of the 13th World Congress on Diseases of Cattle, World Buiatrics Association, September 17–21.* Durban, South Africa: World Buiatrics Association, pp. 33–38.

Thomson, G.R., W. Vosloo, J.J. Esterhuysen, and R.G. Bengis. 1992. Maintenance of foot and mouth disease virus in buffalo (*Syncerus caffer* Sparrman, 1979) in Southern Africa. *Revue Scientifique et Technique O.I.E.* 11:1097–1107.

Urbain, A., P. Bullier, and J. Nouvel 1938. A small outbreak of aphthous fever in wild animals in capitivity [in French]. *Bulletin Academie Vétérinaire de France* 11:59–73.

Vanella, J.M., R.E. Kissling, and R.W. Chamberlain. 1956. Transmission studies with encephalomyocarditis virus. *Journal of Infectious Diseases* 98:98–102.

Vizoso, A.D., M.R. Vizoso, and R. Hay. 1964. Isolation of a virus resembling encephalomyocarditis from a red squirrel. *Nature* 201:849–850.

Vosloo, W., N.J. Knowles, and G.R. Thomson. 1992. Genetic relationships between southern African SAT-2 isolates of foot-and-mouth disease virus. *Epidemiology and Infection* 109:547–558.

Vosloo, W., E. Kirkbride, R.G. Bengis, D.F. Keet, and G.R. Thomson. 1995. Genome variation in the SAT types of foot-and-mouth disease viruses prevalent in buffalo (*Syncerus caffer*) in the Kruger National Park and other regions of southern Africa, 1986–1993. *Epidemiology and Infection* 114:203–218.

Vosloo, W., A.D. Bastos, E. Kirkbride, J.J. Esterhuysen, D. Janse van Rensburg, R.G. Bengis, D.F. Keet, and G.R. Thomson. 1996. Persistent infection of African buffalo (*Syncerus caffer*) with SAT-type foot-and-mouth viruses: Rate of fixation of mutations, antigenic change and interspecies transmission. *Journal of General Virology* 77:1457–1467.

Wells, S.K., A.E. Gutter, K.F. Soike, and G.B. Baskin. 1989. Encephalomyocarditis virus: Epizootic in a zoological collection. *Journal of Zoo and Wildlife Medicine* 20:291–296.

Westbury, H.A., W.J. Doughty, A.J. Forman, S. Tangchaitrong, and A. Kongthon. 1988. A comparison of enzyme-linked immunosorbent assay, complement fixation and virus isolation for foot-and-mouth disease diagnosis. *Veterinary Microbiology* 17:21–28.

Wilder, F.W., A.H. Dardiri, and R.J. Yedloutschnig. 1974. Clinical and serologic response of the nine-banded armadillo (*Dasypus novemcinctus*) to viruses of African swine fever, hog cholera, rinderpest, vesicular exanthema of swine, vesicular stomatitis and foot-and-mouth disease. *Proceedings, Annual Meeting of the United States Animal Health Association* 78:188–194.

Wobeser, G.A. 1994. *Investigation and management of disease in wild animals.* New York: Plenum, 250 pp.

Woodbury, E.L. 1995. A review of the possible mechanisms for the persistence of foot-and-mouth disease virus. *Epidemiology and Infection* 114:1–13.

Zimmerman, J.J. 1994. Encephalomyocarditis. In *Handbook of zoonoses,* section B: *Viral,* ed. G.W. Beran and J.H. Steele, 2d ed. Ann Arbor, MI: CRC, pp. 423–436.

Zimmerman, J.J., K. Schwartz, H.T. Hill, M.C. Meetz, R. Simonson, and J.H. Carlson. 1993. Influence of dose and route on transmission of encephalomyocarditis virus in swine. *Journal of Veterinary Diagnostic Investigation* 5:317–321.

6

PARVOVIRUS INFECTIONS

IAN K. BARKER AND COLIN R. PARRISH

Many species of carnivores are susceptible to infection with members of the feline parvovirus subgroup of the family *Parvoviridae* (Siegl et al. 1985). Reviews of these viruses and their associated diseases include those by Appel and Parrish (1987), Pedersen (1987), Pearson and Gorham (1987), Montali et al. (1987), Greene and Scott (1990), Parrish (1990), Pollock and Carmichael (1990a,b), Porter and Larsen (1990), Scott (1990), Kurtzman (1993), and Pollock and Coyne (1993).

Aleutian disease of mink, caused by a distinct parvovirus, which may infect carnivores other than mink, also is considered briefly here.

Other parvoviruses infect primates, including people, and domestic and laboratory animals, causing a wide variety of syndromes (Tijssen 1990); however, there is no information on infections with these or analogous agents in free-ranging wildlife.

FELINE PARVOVIRUS SUBGROUP INFECTIONS

Synonyms: Feline parvovirus (FPV), feline panleukopenia (FP), feline infectious enteritis, feline distemper, spontaneous agranulocytosis, feline ataxia; mink enteritis virus (MEV), mink viral enteritis (MVE), Fort William disease; blue fox parvovirus (BFPV); raccoon parvovirus (RPV); canine parvovirus (CPV), canine parvovirus type 2 (CPV-2); raccoon dog parvovirus (RDPV).

History

FELINE PANLEUKOPENIA. Syndromes now known to be caused by feline parvovirus (FPV) have long been recognized in cats; Verge and Christoforoni (1928) showed the cause to be a filterable virus. Studies during the 1930s and 1940s [see Parrish (1990)] revealed that a number of previously described diseases of domestic cats *Felis catus* (e.g., feline distemper, spontaneous agranulocytosis, feline infectious enteritis, or malignant panleukopenia) were all caused by this agent. During this era, FPV also was implicated in mortality of many species of wild felids in captivity (Torres 1941; Cockburn 1947; Goss 1948; Hyslop 1955). Similar diseases in raccoons *Procyon lotor* and arctic (blue) foxes *Alopex lagopus* were recognized during the 1940s (Waller 1940; Phillips 1943; Goss 1948).

Panleukopenia virus is distributed worldwide (Scott 1990). The virus was isolated in tissue culture first from a domestic cat in the late 1950s (Bolin 1957) and subsequently from a leopard, probably *Panthera pardus* (Johnson 1964), permitting it to be characterized as a parvovirus (Johnson et al. 1974). Panleukopenia was reproduced in cats infected with tissue culture-origin virus (Johnson 1965), and FPV subsequently also was recognized as the cause of feline ataxia in kittens (Johnson et al. 1967). In 1940, formalin-inactivated tissue-based vaccines were introduced commercially for FPV prophylaxis; tissue culture-attenuated live vaccines and more modern inactivated virus vaccines followed (Pollock 1984a).

MINK ENTERITIS VIRUS. In 1947, a new viral gastroenteritis was observed in farmed mink *Mustela vison* near Fort William (now Thunder Bay), Ontario, Canada (Schofield 1949). The history of this disease has been reviewed by Burger and Gorham (1970) and Pearson and Gorham (1987). Initially named Fort William disease and later mink viral enteritis (MVE), within 15 years MVE was observed on mink ranches throughout Canada, the United States, Europe, and Scandinavia (Pearson and Gorham 1987; Parrish 1990), and now it probably occurs wherever mink are farmed (Porter and Larsen 1990).

The relationship between MVE and FP was recognized early (Wills 1952; Wills and Belcher 1956), and subsequent studies revealed that FPV and MEV were closely related biochemically and serologically (Johnson 1967; Johnson et al. 1974; Flagstad 1977). Because MVE was recognized in 1947 as a new syndrome in the well-established Canadian mink industry and then spread rapidly both locally and worldwide, MEV may have adapted to infect mink about that time. It may be an artifact of mink husbandry because it apparently has not been reported from wild mink. Vaccines for MVE were initially comprised of formalin-inactivated tissues from infected mink (Wills and Belcher 1956); later, tissue culture-attenuated live virus or inactivated tissue culture-origin virus were used (Porter and Larsen 1990).

CANINE PARVOVIRUS. During 1978, two new syndromes—myocarditis in pups under about 4 months of age and gastroenteritis in animals of all ages—became epidemic in domestic dogs *Canis familiaris* throughout the world [see Pollock (1984b) and Parrish (1990)]. The enteric disease resembled FP in cats and MVE in mink. A parvovirus was associated with both canine syndromes, and subsequently they were demonstrated to be caused by the same agent (Robinson et al. 1980). It was named canine parvovirus type 2 (CPV-2) to distinguish it from the unrelated minute virus of canines (Pollock and Carmichael 1990a,b). However, use of the term CPV-2 is discouraged, and we refer to this virus as CPV. The close relationship of CPV to FPV was quickly recognized (Appel et al. 1979; Johnson and Spradbrow 1979), and vaccines based on inactivated or attenuated FPV, MEV, and CPV were soon developed (Pollock 1984b).

Canine parvovirus seems to have emerged first in dogs in Europe, and the virus was disseminated rapidly around the world during 1978 (Parrish 1990). The virus also spread quickly among susceptible wild canids. Sera positive for CPV antibodies were first detected in coyotes *Canis latrans* in the United States in 1979, and by 1981 most coyotes examined in Texas, Idaho, Utah (Thomas et. al. 1984), and Ontario (Barker et al. 1983) were seropositive. Positive sera were collected from gray wolves *Canis lupus* in Alaska during 1980 (Zarnke and Ballard 1987). More difficult to rationalize are reports of antibody to a parvovirus in gray wolves in Minnesota as early as 1975–77 (Goyal et al. 1986; Mech et al. 1986), because CPV was unknown in other canids in the United States at that time.

Beginning in 1978, disease attributed to CPV occurred in captive Canidae, including maned wolves *Chrysocyon brachyurus* (Fletcher et al. 1979; Mann et al. 1980), crab-eating foxes *Cerdocyon thous* (Mann et al. 1980), coyotes (Evermann et al. 1980), bush dogs *Speothos venaticus* (Janssen et al. 1982), and dingos *Canis familiaris* (Dietzmann et al. 1987) in zoos and research facilities, and in raccoon dogs *Nyctereutes procyonoides* on fur farms in Finland (Neuvonen et al. 1982).

Etiology. Feline parvovirus subgroup viruses are unenveloped and contain a single-stranded DNA genome about 5100 bases long (Martyn et al. 1990; Parrish 1991). The genome contains promoters for messages for two nonstructural proteins, NS1 and NS2, and for the structural proteins VP1 and VP2. The virus capsid is only about 25 nm in diameter; hence the name, from the Latin *parvus* (small). The capsid is assembled from 60 copies of about 10% VP1 and 90% VP2 molecules. Both FPV and CPV particles are icosahedral with a T = 1 arrangement of capsid proteins. The virus capsids are the primary determinants of host range; differences of only two or three amino acids in three areas on or near the surface determine the ability of CPV and FPV to replicate in dogs or cats or their cultured cells (Tsao et al. 1991; Chang et al. 1992; Truyen and Parrish 1992; Truyen et al. 1994a).

Canine parvovirus probably derived by mutation from FPV or a closely related virus (Truyen et al. 1996). Although CPV and FPV isolates are more than 98% identical in DNA sequence, they can be distinguished readily by antigenic typing with monoclonal antibodies (Parrish and Carmichael 1983), by their characteristic pH and temperature dependence for hemagglutination (Carmichael et al. 1980; Senda et al. 1988), and by host range in cultured cells or animals. The amplification of parvovirus DNA sequences intermediate between FPV and CPV from tissues of red foxes *Vulpes vulpes* from Europe suggests that wildlife may have played a role in interspecies transfer of feline parvovirus subgroup virus to dogs (Truyen et al. 1998).

Phylogenetic analysis, based on gene and VP1 amino acid sequences, defines two major groups based on their relatedness to FPV. As currently understood, the FPV subgroup is rooted in FPV, from which blue fox parvovirus (BFPV) and raccoon parvovirus (RPV) isolates are indistinguishable, and mink enteritis virus (MEV) is a minor variant. Canine parvovirus, although closely related to and probably derived from FPV, is distinguishable on antigenic and genetic grounds and includes raccoon dog parvovirus (RDPV) isolates (Truyen et al. 1995). There is very little antigenic variation among FPV isolates collected decades apart. Although three slightly different antigenic types of MEV coexist, there was no difference in the in vivo cross-protection between them (Parrish et al. 1984).

Since 1978, CPV has undergone sequential antigenic variation. The variant viruses (CPV types 2a and 2b) differ very little from the original CPV, but each replaced the previous virus type in nature within 2 or 3 years (Parrish et al. 1988). Although the antigenic changes were located in each of the two major neutralizing antigenic determinants on the virus (Strassheim et al. 1994), these variations did not influence the efficacy of vaccination (Greenwood et al. 1995).

Host Range. Species from six families (Felidae, Canidae, Procyonidae, Mustelidae, Ursidae, and Viverridae) in the order Carnivora are suspected of being susceptible to parvoviruses of the FP subgroup (Table 6.1). Confirmed clinical disease has not been reported among the Viverridae or Ursidae, and there is no evidence that the Hyaenidae are susceptible. Even within affected families, only certain genera or species have been reported to be susceptible. Syndromes resembling parvovirus infection also have been described in an insectivore (Kranzlin et al. 1993) and a rodent (Frelier et al. 1984).

The natural host ranges of feline subgroup parvoviruses are poorly defined. Many reports are based on descriptions of clinical disease, serologic surveys, or presumptive pathologic diagnoses in natural infections of captive animals, without virus isolation. Well-characterized viruses have been isolated from only a few species, and infection trials to determine host range have involved only some species in the Felidae, Canidae, Mustelidae, and Procyonidae.

By definition, the natural hosts of FPV, MEV, and CPV are cats, mink, and dogs, respectively. Viruses in other species have been named either after a virus from a related host (e.g., CPV from coyotes) or after the host

TABLE 6.1—Relationships and host range of viruses in the feline parvovirus subgroup of the *Parvoviridae,* and the basis for making a host-virus association[a]

	FPV	MEV (= FPV Variant)	CPV	Unknown Parvovirus
Felidae	*A. jubatus*[i,c] *F. bengalensis*[p,c] *F. catus*[e] *F. concolor*[i,p,s] *F. lynx*[c] *F. pardalis*[p,c] *F. rufus*[p,s] *F. serval*[c] *F. silvestris*[i,c] *F. weidii*[p] *F. yagouaroundi*[c] *N. nebulosa*[i] *P. leo*[i,p,s] *P. onca*[p,c] *P. pardus*[i,p,c] *P. tigris*[i,p,c] *P. uncia*[i,p]	*F. catus*[e]	*F. catus*[e] (CPV-2a,b) *F. rufus*[i] (CPV-2b)	
Procyonidae	*P. lotor*[e,i,p] (RPV) *N. nasua*[i]	*P. lotor*[e]		
Mustelidae	*M. vison*[e] *M. putorius*[e] (<3 days)	*M. vison*[e,i]		
Canidae	*A. lagopus*[i,p,s] (BFPV) *C. brachyurus*[i,p] *S. venaticus*[i,p]		*C. familiaris*[e,i,s] *C. latrans*[i,s] *C. lupus*[e,s] *N. procyonoides*[i] (RDPV)	*C. adustus*[s] *C. auratus*[s] *C. mesomelas*[s] *C. thous*[p] *L. pictus*[s] *U. cinereoargenteus*[s] *U. littoralis*[s] *V. macrotis*[s] *V. vulpes*[s]
Ursidae				*U. americanus*[s] *U. arctos*[s] *A. melanoleuca*[s]
Viverridae				"Civet"[c] "Palm civet"[i]

[a] BFPV, blue fox parvovirus; CPV, canine parvovirus; FPV, feline parvovirus; MEV, mink enteritis; RDPV, raccoon dog parvovirus; RPV, raccoon parvovirus. See text for citations.
[e] Experimental infection.
[i] Virus isolation.
[p] Pathologic diagnosis.
[c] Clinical diagnosis.
[s] Serologic reaction.

animal from which they were isolated (e.g., RPV). However, nomenclature of these viruses based on host of origin is potentially misleading and may confuse understanding of their epidemiology. Feline parvovirus seems to infect some species of South American canids, whereas some CPV types may infect cats (see below).

Viral isolates must be characterized biologically, antigenically, and genetically, particularly if new hosts or syndromes are being described. Antibody to a particular parvovirus does not necessarily reflect exposure to that virus, because strong antigenic cross-reactions occur among these agents. This complicates interpretation of serologic surveys of species in which a native parvovirus has not been characterized.

Host ranges of these viruses are complex. Both MEV and FPV infect and replicate in mink and cats, but they seem to be more virulent in the homologous host [FPV in cats and MEV in mink (Burger and Gorham 1970)]. Although MEV caused clinical disease in mink, most infections by feline or raccoon viruses in mink were subclinical (Barker et al. 1983; Parrish et al. 1987). The original 1978 strain of CPV replicated in both canine and feline cells in vitro, and in vivo in dogs but not in cats (Truyen and Parrish 1992). However, CPV types 2a and 2b replicated efficiently in cats, and they have been isolated from cats with natural disease (Truyen et al. 1996). Although FPV replicated in feline cells in vitro and in vivo in cats, it did not replicate in cultured canine cells, though it did replicate in vivo in dogs (Truyen and Parrish 1992; Truyen et al. 1994a).

FELIDAE. All cats are considered susceptible to FPV infection (Fowler 1986), since FP was diagnosed in many species prior to the emergence of CPV. Canine parvovirus types 2a and 2b also infect cats, causing disease, and CPV-2b has been isolated from a captive bobcat *Felis rufus* (C.R. Parrish unpublished). Most reports of clinical disease are in captive animals (Table 6.1), in which FP is considered an important problem (Torres 1941; Cockburn 1947; Goss 1948; Hyslop 1955; Elze et al. 1974; Fowler 1986). Feline panleukopenia virus has been isolated from captive leopards (Johnson 1964; Scott 1990); lions *Panthera leo* (Johnson and Halliwell 1968; Studdert et al. 1973; Scott 1990; Mochizuki et al. 1996); tigers *Panthera tigris* (Johnson and Halliwell 1968; Povey and Davis 1977); snow leopard *Panthera uncia* (Scott 1990); clouded leopard *Neofelis nebulosa* (Cai et al. 1988); cheetah *Acinonyx jubatus* (Valíček et al. 1993); and wild cat *Felis silvestris* (Chappuis and Lernould 1987). Panleukopenia has been diagnosed on the basis of pathologic lesions in a variety of feline species (Table 6.1) (Torres 1941; Vetési and Balsai 1971; Studdert et al. 1973; Woolf and Swart 1974; Singh et al. 1983; Fowler 1986; Wassmer et al. 1988; Fix et al. 1989; Mochizuki et al. 1996).

Among free-ranging populations, serologic surveys indicate that exposure to an FPV-like virus is common in Florida panthers *Felis concolor coryi* in Florida [78% (Roelke et al. 1993)] and mountain lions *F. concolor* in California [93% (Paul-Murphy et al. 1994)], rare to common in bobcats in New York [9%–76%, overall 21% (Fox 1983)], common in lions in some South African national parks [84% (Spencer 1991)] but not others [0% (Spencer and Morkel 1993)], and not detected in Iriomote cats *Felis iriomotensis* in Japan (Mochizuki et al. 1990).

Occasional wild animals may succumb, and FPV has been isolated from a mountain lion from Colorado (C.R. Parrish unpublished) and was implicated in the death of a mountain lion in Wyoming, based on negative-stain electron microscopy and pathology (E.S. Williams unpublished). Panleukopenia was a common cause of sporadic natural mortality in bobcats in California and was associated with a local outbreak in bobcats in Florida (Wassmer et al. 1988).

PROCYONIDAE. The parvovirus of raccoons (Nettles et al. 1980) has been termed RPV, but it is likely FPV; raccoons develop disease after inoculation with MEV or FPV (Barker et al. 1983), and isolates from raccoons do not differ from FPV (Truyen et al. 1995). Raccoons are not susceptible to the 1978 strain of CPV (Appel and Parrish 1982). Along with rabies and canine distemper, parvovirus infection is one of the most important infectious diseases of raccoons. Exposure to parvovirus is moderately common in Ontario [22% (Barker et al. (1983)], though not necessarily throughout its range (Rabinowitz and Potgieter 1984). Although there are few published reports (Waller 1940; Vetési and Balsai 1971; Nettles et al. 1980), parvovirus infection causes occasional local outbreaks of disease or kills individuals, and it may cause high mortality among unvaccinated juveniles in raccoon "orphanages," rehabilitation facilities, and animal shelters (I.K. Barker unpublished), as well as contributing to multiple-etiology gastrointestinal infections (Martin and Zeidner 1992).

Coatis, or coatimundis *Nasua nasua,* are naturally susceptible to FPV, based on mortality and virus isolation (Johnson and Halliwell 1968; Vetési and Balsai 1971). The purported susceptibility of the kinkajou *Poto flavus* (Miller 1961) is unsubstantiated.

MUSTELIDAE. Mink are susceptible to MEV, which causes disease in farmed mink, and to FPV (see above). Canine parvovirus replicated to only very low titers in mink after experimental inoculation (Barker et al. 1983; Parrish et al. 1987).

Domestic ferrets *Mustela putorius furo* become infected by FPV when inoculated in utero or by the intraperitoneal route within 2 days after birth, and they may develop cerebellar hypoplasia (Kilham et al. 1967). However, disease did not occur in ferrets inoculated at 3 days of age or older (Parrish et al. 1987). Adult ferrets developed antibody titers, but not disease, when inoculated with FPV (Johnson et al. 1967). Natural infections of farmed ferrets by related parvoviruses have not been detected serologically (Veijalainen 1986), and parvovirus infection has not been described in nondomestic ferrets (black-footed ferrets *Mustela nigripes,* or polecats *Mustela eversmanni* or *Mustela putorius*).

Wild striped skunks *Mephitis mephitis* had serologic evidence of exposure to parvovirus at very low prevalence (Barker et al. 1983), but virus has not been isolated. Canine parvovirus, FPV, and MEV did not replicate in experimentally exposed skunks, and little or no antibody was elicited (Barker et al. 1983).

Sable *Martes zibellina* resisted challenge with MEV, and vaccination was not considered warranted (Aulova et al. 1989).

CANIDAE. Domestic dogs and all wild members of the genus *Canis* probably are susceptible to CPV (see Table 6.1). Coyotes are exposed commonly to parvovirus, based on serologic surveys in the United States [up to 90% (Thomas et al. 1984), 71% (Gese et al. 1991), 65% (Holzman et al. 1992), 100% (Davidson et al. 1992), 100% (Gese et al. 1997), and 66% (Cypher et al. 1998)] and Canada [85% (Barker et al. 1983)]. Canine parvovirus has been isolated from dead coyotes (Evermann et al. 1980); infection had a marked effect on pup survival in one study (Gese et al. 1997), and there is anecdotal evidence of a population effect (Pence and Windberg 1984).

Gray wolves also have a relatively high prevalence of antibody to parvovirus in Minnesota [more than 50% (Mech et al. 1986; Goyal et al. 1986; Mech and Goyal

1993)], Alaska [31% (Zarnke and Ballard 1987)], Montana [65% (Johnson et al. 1994)], and Wisconsin [53% (Behler-Amass et al. 1995)]. Mortality of wild wolves due to parvovirus has been confirmed, and there may be an effect on recruitment of pups into the population (Mech and Goyal 1993, 1995; Peterson 1995; Mech et al. 1997), perhaps in association with canine distemper (Johnson et al. 1994). A parvovirus was isolated from wolf feces but not further classified (Muneer et al. 1988). However, wolves were susceptible to experimental infection with CPV (Amundson et al. 1986).

Jackals in Kenya also have antibody to an unidentified parvovirus: *Canis aureus,* 56%; *Canis mesomelas,* 19%; *Canis adustus,* 33%; and overall, 34% (Alexander et al. 1994).

The raccoon dog is susceptible to clinical parvovirus gastroenteritis (Neuvonen et al. 1982). Serologic evidence of exposure on fur farms is common [43% (Veijalainen 1986)], and CPV has been isolated and characterized (Veijalainen 1988).

Other Canidae considered susceptible to CPV on the basis of clinical disease and microscopic lesions are the crab-eating fox (Mann et al. 1980), the maned wolf (Fletcher et al. 1979), and the bush dog (Mann et al. 1980). However, CPV infection has not been confirmed from any of these species; assumptions about the susceptibility of bush dogs and maned wolves to CPV seem based on temporal association with the emergence of CPV infections in dogs and are not supported by other studies.

Although Janssen et al. (1982) identified an isolate from bush dogs as CPV, the conditions of hemagglutination (pH 5.8 at 4° C) did not permit discrimination of CPV from FPV. Furthermore, Chappuis and Lernould (1987) confirmed an FPV isolate from bush dogs on the basis of its hemagglutination reaction (strong at pH 6.4 and weak at pH 7.3). Similarly, maned wolves clearly are susceptible to FPV. Prior to the emergence of CPV, there were several reports of mortality due to documented parvovirus infection (Visee et al. 1974; Bieniek et al. 1981), identified as FPV by the latter authors, based on hemagglutination characteristics.

Arctic foxes are infected by a virus that is closely related to MEV and FPV and distinct from CPV (Phillips 1943; Neuvonen et al. 1982; Veijalainen 1988; Mizak 1994; Truyen et al. 1995). Seroprevalence exceeding 20% is reported on Finnish fur farms (Veijalainen 1986).

Red foxes also are susceptible to a similar agent. Parvovirus DNA sequences intermediate between FPV and CPV have been amplified from tissues of red foxes in Europe (Truyen et al. 1998). Antibodies to an undetermined parvovirus were common in wild red foxes in Ontario [79% (Barker et al. 1983)], as they were in Georgia [100% (Davidson et al. 1992)], though the prevalence of antibodies to parvovirus antigen seems lower in red foxes in Germany [13% (Truyen et al. 1998)] and in France [less than 3% (Schwers et. al. 1983)]. Oral inoculation of red foxes with CPV, MEV, or FPV did not elicit clinical disease; although all viruses were shed only in low titer, foxes developed the strongest antibody response to FPV (Barker et al. 1983). In contrast, intravenous inoculation of red foxes with CPV caused leukopenia, virus shedding in feces, and seroconversion but no clinical signs (Buonavoglia et al. 1986).

Exposure to unidentified parvoviruses also seems common, based on serologic surveys, in gray foxes *Urocyon cinereoargenteus* [88% (Davidson et al. 1992)], island foxes *Urocyon littoralis* [up to 59%, but highly variable from island to island (Garcelon et al. 1992)], and kit foxes *Vulpes macrotis* [67%–100% (McCue and O'Farrell 1988)] in the United States.

African hunting dogs *Lycaon pictus* in the Selous Game Reserve in Tanzania had a high prevalence of antibody to an undetermined parvovirus [68% (Creel et al. 1997)], whereas in Kenya prevalence was lower [7% (Alexander et al. 1993)], and in Namibia it was nil in six animals (Laurenson et al. 1997).

VIVERRIDAE. Disease due to parvovirus has not been confirmed in viverrids. No specific histologic lesions were noted, and a virus was not isolated from "civets" with a condition clinically similar to panleukopenia (Nair et al. 1964). Cockburn (1947) listed one "doubtful" case of feline infectious enteritis among civets and genets at the London Zoo, whereas Goss (1948) did not consider viverrids susceptible to FPV. Komolafe (1986) reported a parvovirus in the feces of healthy "palm civets" in a Nigerian zoo.

URSIDAE. Although parvoviruses never have been isolated from bears, and disease is undescribed, 16% of 62 American black bears *Ursus americanus* from Florida (Duncan et al. 1998), 30% of 22 brown bears *Ursus arctos* from Croatia (Madic et al. 1993), and 75% of giant pandas *Ailuropoda melanoleuca* from China (Mainka et al. 1994) had serologic evidence of exposure.

OTHER SPECIES. The report of suspected parvovirus infection in a group of captive North American porcupines *Erethizon dorsatum* (Frelier et al. 1984) is based on a pathologic syndrome (enteritis and myocarditis) similar to that seen in carnivores. However, parvovirus infection was not supported by serologic or virologic evidence.

Parvovirus antigen was demonstrated in tissue of young captive Eurasian hedgehogs *Erinaceus europaeus* with compatible lesions. No virus was isolated, but the authors speculated that FPV may have been contracted by contact with cats (Kranzlin et al. 1993).

Pathogenesis. For citations to the extensive original literature on pathogenesis of FPV, MVE, and CPV, readers are referred to reviews by Pollock (1984a,b), Pollock and Parrish (1985), Appel and Parrish (1987), Pedersen (1987), Pearson and Gorham (1987), Parrish (1990), Pollock and Carmichael (1990a), Porter and Larsen (1990), Scott (1990), Kurtzman (1993), and Barker et al. (1993), and the report by Uttenthal et al. (1990).

Autonomous parvoviruses replicate only in the nucleus of dividing cells, which contributes to differences in the outcome of infections in fetal and neonatal versus older animals. Proliferative lymphoid, hematopoietic, and intestinal epithelial cells are primary targets for virus replication by FPV, MVE, and CPV, and developmentally regulated properties of some other dividing cell populations seem to restrict parvovirus replication.

ANIMALS OLDER THAN FOUR WEEKS. Syndromes seen in wildlife resemble diseases caused by FPV in cats, MEV in mink, and CPV in dogs, which are themselves similar. The virus enters the body and replicates initially in cells of the nasopharynx, the tonsils, or other lymphoid tissues. By 1–3 days after infection, virus is in tonsils, retropharyngeal lymph nodes, thymus, and mesenteric lymph nodes; after 3–4 days, virus also is found in gut-associated lymphoid tissue and Peyer's patches and in intestinal crypt epithelium. Virus probably spreads systemically though a plasma viremia and by traffic of infected lymphoid cells. Infection of dividing cells in lymphoid tissues results in lymphocytolysis, cellular depletion, and, subsequently, tissue regeneration in surviving animals. In FPV and CPV infections, all cell lines in the bone marrow may be severely depleted, although individuals vary. Damage to the myeloid precursor population by FPV probably causes reduced numbers of circulating neutrophils due to reduced recruitment in the face of rapid turnover of mature cells.

All carnivore parvoviruses cause distinctive intestinal lesions, the severity of which strongly influences the clinical outcome. Virus is found in the rapidly dividing epithelial progenitor cells in the crypts of Lieberkühn 4–8 days after infection. Increased epithelial proliferation related to mucosal trauma, age, starvation and refeeding, and concomitant parasitic, bacterial, or viral infections (especially coronavirus in dogs) may increase susceptibility of crypt cells to viral replication and cytolysis, and thereby increase severity of lesions and clinical signs.

Viral killing of proliferative crypt cells, and consequent inability to generate new epithelium, causes villus atrophy as epithelial turnover continues. Overt erosion of the mucosa often occurs. Malabsorption and loss of tissue fluids, plasma protein, and blood into the gut lumen ensue; the resulting diarrhea often contains blood and mucus. Animals may become dehydrated and pyrexic, possibly because of endotoxin uptake from the gut. Severity of clinical disease probably is determined by the degree of damage to the intestine. The size of the virus exposure and the response of individuals to infection vary; infections are often mild or subclinical, because seropositive animals may have no history of disease (Meunier et al. 1980; Parrish et al. 1982). Provided they survive the consequences of intestinal damage, animals recover completely from clinical disease as lymphoid, myeloid, and enteric cell populations regenerate.

FETAL OR NEONATAL INFECTIONS. Enteritis is not observed in very young animals (under about 2–3 weeks of age). Infection of neonates results in infection of the developing cerebellum and other tissues in kittens, or of the heart in CPV-infected puppies. Infection of kittens in utero or shortly after birth can result in viral replication in the external germinal epithelium of the cerebellum, causing cerebellar dysplasia/hypoplasia, which results in ataxia in surviving animals. Infection of neonatal puppies by CPV can cause death from myocarditis, generally between 3 and 8 weeks of age. The age dependence of the syndrome is due to the susceptibility of actively dividing myocardial cells to CPV infection; cell division occurs only until about 15 days of age.

Rarely, generalized prenatal or neonatal infections occur, with virus replication and lesions in many different tissues, perhaps including myocardium. In utero infections of cats by FPV or of arctic foxes by BFPV (Veijalainen and Smeds 1988) may result in fetal death and resorption, abortion, or neonatal death.

Immunity. Antibody in plasma protects against infection and promotes recovery (Meunier et al. 1985a). Antibody absorbed by the offspring from colostrum in the first few days after birth protects against parvovirus infection until it declines naturally to very low concentrations (Parrish et al. 1982; Pollock and Carmichael 1982; Macartney et al. 1988). The amount of colostral antibody passed to the offspring is directly related to the plasma concentration in the dam. Unfortunately, the level of maternal antibody transferred may vary, from dam to dam and within litters, and it cannot be predicted or evaluated readily. In dogs, the half-life of maternal antibody to parvovirus is about 10 days (Parrish et al. 1982; Pollock and Carmichael 1982), and protective levels usually persist until pups are about 8–15 weeks of age (Pollock and Coyne 1993), sometimes longer. The same holds for maternal antibody activity in cats with FPV (Scott et al. 1970) and mink with MVE [see Pearson and Gorham (1987) and Gorham et al. (1988)].

Unfortunately, the concentration of maternal antibody that will interfere with vaccination is less than the amount required to prevent infection with virulent virus (Pollock and Carmichael 1982; Buonavoglia et al. 1992). Hence, there is a period of several days late in the decay of maternal antibody levels when the animal is susceptible to virus in the environment but cannot be vaccinated successfully.

Following natural infection, there is a rapid antibody response, detectable about the time that clinical signs begin, 4–5 days after exposure. As antibody levels rise, extracellular virus is neutralized, so that, by 7–9 days after infection, levels of virus detectable in tissue and feces are markedly reduced (Carman and Povey 1985). The onset and extent of antibody production are important in limiting the magnitude and duration of viremia and, hence, determining the severity of clinical disease (Meunier et al. 1985b). Parvoviruses are potent immunogens, and animals that have recovered from infection typically have complete and persistent (probably lifelong) immunity, which may be boosted by natural reexposure to virus. Whether parvovirus infections cause significant immunocompromise is in debate, but any effect on the immune system is likely to be transient and minor (Pollock and Carmichael 1990a,b).

Transmission and Epidemiology. The infective dose of CPV for a dog may be very low (Pollock and Coyne 1993). Because sick animals can pass billions of infective viruses per gram of feces (Pollock 1982), the potential for contamination of soil, litter, bedding, feeding utensils, clothing, and footwear is great. Most contamination results from virus shed over the period about 4–10 days after infection [see Carman and Povey (1985) and Shen et al. (1986)]. Within about 4 weeks, recovered dogs are not contagious to susceptible contacts (Pollock 1982); some cats can shed virus for several weeks after recovery from FPV (Czisa et al. 1971) and a carrier state persisting for up to a year has been suggested in mink with MVE (Bouillant and Hansen 1965b). Subclinical infections, which occur in some immunized mink and dogs, also may contribute to environmental contamination (Burger and Gorham 1970; Pollock and Coyne 1993).

Parvoviruses are very hardy, surviving for months under cool, moist conditions protected from sunlight, and they are very stable when frozen (Gordon and Angrick 1986). Infectious CPV has persisted in feces held for 6 months at room temperature (Pollock 1982), and MEV may remain viable in the natural environment for 9–12 months (Bouillant and Hanson 1965a; Burger and Gorham 1970). Parvoviruses were detected in 6%–28% of wolf scats collected in Minnesota and the Great Lakes region (Muneer at al. 1988; Behler-Amass et al. 1995). In addition to inanimate fomites, insects such as flies, and birds and rats have been implicated as mechanical carriers of MEV [see Burger and Gorham (1970)], and prepared feed may become contaminated with virus (Henriksen 1987).

Transmission is by the fecal-oral route, probably mainly through ingestion of virus from the environment, rather than by direct contact with infected animals (Reif 1976). Hence, free-ranging wild carnivores, even if solitary, widely dispersed, and at low density, may be exposed at marking sites, latrines, or other areas contaminated by feces deposited by a virus shedder. Introduction of virus into the environment of captive animals on inanimate objects is very difficult to prevent.

Due to the wide host range of some parvoviruses, wildlife may be exposed to virus shed by species with which they share the environment or upon which they may prey, e.g., FPV might be transmitted from domestic cats or raccoons to predatory large felids (Wassmer et al. 1988), or from domestic cats to raccoons in natural habitats or animal shelters.

In naive populations, if exposure is common at sufficient doses over a brief period, an epidemic may occur, with significant mortality in all age classes (Mason et al. 1987), affecting population size (Hesterbeek and Roberts 1995). Such outbreaks have been reported relatively commonly in captive collections of wild carnivores (Studdert et al. 1973; Evermann et al. 1980; Chappuis and Lernould 1987; Mochizuki et al. 1996).

In some populations in which infection is endemic, the high prevalence of antibody (see the section on host range: Felidae and Canidae) suggests that subclinical infection (Bush et al. 1981), or mild disease with recovery, is a common outcome of exposure. A high prevalence of immune animals among the adult population is likely related to the potent immunogenicity of the virus, persistence of antibody, and periodic natural reexposure, boosting antibody titers. However, antibody titers in animals infected in previous years are typically lower than in recently infected animals (Van Rensburg et al. 1987).

In populations with endemic CPV at high prevalence, most new infections will occur among juveniles exposed to CPV following decline of maternal antibody at about 2 months of age (Mason et al. 1987). In such populations, prevalence of clinical cases tends to be cyclic and associated with the relative size of the population of susceptible juveniles (Reif 1976), though not inevitably so (Mason et al. 1987). If breeding is seasonal, the prevalence of clinical parvovirus infections is greatest during the months when juveniles with waning antibody levels (i.e., 2- to 4-month age class) are most common (Reif 1976).

Without knowledge of the prevalence of exposure of susceptible animals, and of case fatality rates, it is difficult, if not impossible, to evaluate the contribution of diseases such as parvovirus infection to mortality at the population level (Dobson and Hudson 1995). Although there is an anecdotal report of epidemic mortality in coyotes soon after the introduction of CPV (Pence and Windberg 1984), when the entire North American population was presumably susceptible, population effects were not obvious as the infection became endemic (Thomas et al. 1984). And African hunting dog populations in which CPV and canine distemper were endemic at moderately high prevalence remained stable and demographically healthy (Creel et al. 1997).

Parvovirus epidemics have been reported only rarely in wild populations [e.g., bobcats (Wassmer et al. 1988) and wolves (Peterson 1995)]. This may reflect a low frequency of introduction of virus into naive populations, though evidence suggests that exposure is common in many populations of susceptible species (see the section on host range). Once the agent becomes endemic in a metapopulation of a susceptible species, naive subpopulations able to support an epidemic are likely to be dispersed and relatively small, and epidemics will have only a transient local impact that is scattered unpredictably in space and time. This, coupled with low intensity of surveillance, may cause sporadic mortalities and small outbreaks to be detected only fortuitously and infrequently.

In addition, the clinical expression of parvovirus infection involves profound depression and lassitude, which are likely to cause an animal to hide rather than become more conspicuous. This may explain failure to find carcasses, despite searches, in a feral cat population in which FPV was assumed on demographic grounds to be causing significant juvenile mortality (Van Rensburg et al. 1987). Notably, fatal FPV infections have been detected in several populations of radio-collared bobcats [see Wassmer et al. (1988)] and in transmitter-implanted coyote pups (Gese et al. 1997), suggesting that long-term active surveillance would enable more accurate and probably higher estimates of parvovirus-associated mortality than have been generated by indirect or passive approaches.

High prevalence of antibody among adult wolves in a population with endemic parvovirus was associated with reduced pup recruitment (Mech and Goyal 1995). No overall decline in wolf numbers was evident, suggesting that pup mortality due to parvovirus at the level occurring in the population studied may not be additive and limiting. However, they suggested the possibility of a limiting effect of CPV infection in that wolf population if it caused high (greater than 50%) pup mortality (Mech and Goyal 1995).

Population decline, associated with poor recruitment, has been observed following the deliberate introduction of FPV into a naive feral cat population that was causing depredations on nesting seabirds on subantarctic Marion Island. Over the 5 years following introduction of FPV, the cat population, which had been expanding at a rate of about 23% per annum, decreased at an average annual rate of 29% (Van Rensburg et al. 1987). Gese et al. (1997) detected a significant impact of parvovirus infection on pup survival in a sample from a Wyoming coyote population in which infection was endemic. Roelke et al. (1993) speculated that endemic FPV may contribute to high mortality in Florida panther cubs exposed as maternal antibody waned, and Fox (1983) suggested that FPV might influence bobcat abundance in New York.

The effect of parvovirus-induced mortality may be more significant if probability of infection is high in populations that are small and/or isolated from sources of immigration. However, immigration from naive populations to endemic areas has the capacity to increase the occurrence of infection (Mech and Goyal 1993, 1995). Johnson et al. (1994) observed pup loss in relatively isolated colonizing wolf packs that had serologic evidence of recent infection by parvovirus and canine distemper.

The insular nature of the Marion Island cat population, in which FPV seemed to have been an effective means of biologic control (Van Rensburg et al. 1987), may have ensured that the effect of presumed FPV-induced juvenile mortality was not compensated for by immigration. Endemic fade-out, or extinction of the virus from the population (Hesterbeek and Roberts 1995), may have occurred in the isolated Isle Royale wolf population when it declined to a very small size in the late 1980s (Peterson 1995).

Clinical signs. Most parvoviruses of this group typically cause gastroenteritis, the syndrome most likely to be encountered in wildlife. The exception to this generalization is BFPV, which is associated mainly with reproductive wastage (Veijalainen and Smeds 1988; Mizak 1994), which would be difficult to detect in a wild population.

PARVOVIRAL GASTROENTERITIS. Clinical presentation varies little with type of virus or host affected (Reynolds 1969; Burger and Gorham 1970; Fletcher et al. 1979; Bieniek et al. 1981; Evermann et al. 1980; Mann et al. 1980; Nettles et al. 1980; Pollock 1984a,b; Mech et al. 1986; Chappuis and Lernould 1987; Pollock and Carmichael 1990a,b; Porter and Larsen 1990; Scott 1990). Depending on the epidemiologic circumstances, the disease may occur as an outbreak or affect only individual animals. Occasional animals may be found moribund or dead without prodromal signs, and animals with prodromal signs may die without having diarrhea. Typically, about day 4 or 5 after exposure, animals develop lethargy, profound depression, and inappetence. Within a day or so, there is an acute onset of fever, vomiting, and diarrhea. The feces may be pasty, porridge-like, or fluid; they are typically foul smelling and may contain mucus or fibrin casts and be blood flecked to overtly hemorrhagic.

Dehydration, acid-base imbalance, hypoproteinemia, and, especially in cats with FPV, leukopenia, occur. Animals that resume eating within 3–4 days of onset of illness are likely to survive, and most animals that are going to die succumb within 4–5 days. Juvenile animals tend to have a higher case fatality rate than do adults (Mason et al. 1987).

PRENATAL AND NEONATAL INFECTIONS. Other than a condition resembling feline ataxia, reported in lion cubs (Leclerc-Cassan 1981); reproductive wastage in blue fox vixens with BFPV due to fetal resorption, abortions, and stillbirths (Veijalainen and Smeds 1988; Mizak 1994); and equivocal reduction in fecundity associated with FPV in bobcats (Wassmer et al. 1988), syndromes due to neonatal or perinatal infection have been reported only in domestic dogs and cats. Consult reviews cited in the Introduction for further information on these syndromes.

Pathology

CLINICAL PATHOLOGY. Panleukopenia is a striking feature of FPV infection in many cats: total white cell counts may fall to 1000–2000/ml or less, and neutrophil counts may decrease to fewer than 200/ml. Lymphocyte numbers decline to a lesser degree, but there is little effect on other leukocytes or red cells unless the animal develops anemia due to blood loss into the gut (Reynolds 1969; Larsen et al. 1976; Hosokawa et al. 1987). Panleukopenia is reported less commonly in CPV infections, although at some point during the clinical course most dogs infected with CPV develop a transient leukopenia with both lymphopenia and neutropenia (Jacobs et al. 1980; Pollock 1982; Macartney et al. 1984; Mason et al. 1987). Similarly, though leukopenia may occur in MVE, it is not a consistent finding (Reynolds 1969; Burger and Gorham 1970).

Anemia and hypoproteinemia, probably due to blood loss into the gut, may occur in animals with parvovirus infections (Jacobs et al. 1980). In addition to changes in the hemogram, electrolyte and blood-gas abnormalities may occur associated with vomiting, diarrhea, and dehydration (Jacobs et al. 1980; Heald et al. 1986).

GROSS AND MICROSCOPIC LESIONS. Lesions in all species are similar. The literature should be consulted for greater detail (Burger and Gorham 1970; Reynolds 1970; Larsen et al. 1976; Fletcher et al. 1979; Mann et al. 1980; Nettles et al. 1980; Barker et al. 1983; Pollock 1984a,b; Pollock and Carmichael 1990a,b; Barker et al. 1993). At necropsy, animals with enteric parvovirus infections are typically dehydrated and, if anemic, pale. In most cases, there are gross lesions in the gastrointestinal tract, but in some, especially in cats and raccoons, they are subtle or not evident, perhaps other than scant fluid feces in the colon. Segmental subserosal to transmural hemorrhage of the small and large intestine, visible from the external aspect, may occur in any species but is most common in dogs with CPV infection and in mink with MVE. Peyer's patches may be prominent.

Gastric contents are usually scant, fluid, and bile stained or bloodstained. Contents of the small intestine are creamy, mucoid, or fluid, perhaps containing fibrin, and in some cases very hemorrhagic. The intestinal and colonic mucosa may appear virtually normal; more commonly, it is segmentally to extensively congested or covered to varying degrees with a fibrinous exudate. Mesenteric lymph nodes often are enlarged, wet, and congested, but they may be reduced in size. The thymus of young animals is consistently atrophic and may be difficult to detect. Bone marrow is pale and gelatinous. The lungs often are congested and edematous.

Microscopically, lesions are found consistently in the small intestine of animals that have died. They reflect the viral insult to the proliferative cells, with ensuing dilation of crypts, atrophy of villi, and erosion and sometimes apparent collapse of the mucosa. Although less consistent and distinctive, involution of lymphoid tissue in the germinal centers of lymph nodes and in the cortex of the thymus is often evident. Marrow typically is hypocellular.

Diagnosis. Parvoviral gastroenteritis in its fulminant form is readily diagnosed clinically. Mild cases of parvovirus enteritis are difficult to differentiate from other causes of transient gastroenteritis, including coronavirus enteritis in dogs and mink, but because recovery is likely, this is of little clinical significance.

Laboratory aids to clinical diagnosis, other than hematology, center on demonstration of virus or the host antibody response. Virus may be detected in feces by an enzyme-linked immunosorbent assay (ELISA) or by hemagglutination, electron microscopy, or virus isolation in tissue culture (Studdert et al. 1983; Pollock and Carmichael 1990a,b). Such tests may yield a high proportion of false-negative results if the samples are taken later than about day 5 or 6 of illness, by which time the antibody response is neutralizing virus (Studdert et al. 1983; Shen et al. 1986). A fourfold rise in antibody titer between the acute and convalescent phases of the disease also supports a diagnosis, but antibody levels may already be high by the time the first sera are collected.

A presumptive diagnosis is readily made post mortem, based on the characteristic suite of gross and microscopic enteric, lymphoid, and myeloid lesions, perhaps with recognition of intranuclear inclusions, or demonstration of viral antigen or nucleic acid by using immune or molecular probes in tissue sections (Jonsson et al. 1988; Waldvogel et al. 1992; Barker et al. 1993; Matsui et al. 1993; Truyen et al. 1994b). Definitive diagnosis commonly rests with isolation and characterization of the type of parvovirus involved or perhaps, in the future, with specific molecular probes.

There are few specific conditions from which parvovirus gastroenteritis must be differentiated in most species; among these are salmonellosis (Eulenberger et al. 1974; Woolf and Swart 1974), which is uncommon in carnivores, and causes of enteric hemorrhage, which in canids include clostridial enteritis and ingestion of anticoagulant rodenticides. In canids, canine coronavirus might be considered (Evermann et al. 1989), as might epizootic catarrhal enteritis in mink, also caused by a coronavirus (Gorham et al. 1990).

Control and Treatment. Control of parvovirus infections is practical only in captive situations, because it rests largely on vaccination to reduce the number of susceptible animals in the population. Environmental contamination also must be minimized, so that during the inevitable window of opportunity for infection of juveniles that occurs as maternal antibody levels wane, the virus challenge is not excessive. Maintaining a naive captive population of susceptible animals requires strict isolation and extreme sanitary measures that are not routinely feasible. If a colostrum-deprived animal, or one that is unvaccinated or of unknown immune status, has been exposed to parvovirus, passive immunization may be considered if the value of the animal or other circumstances warrant (Greene 1990).

VACCINATION. Numerous inactivated-adjuvanted or modified live vaccines (MLVs) are registered for use against parvoviruses in cats, dogs, and mink. These vaccines are recommended in captive wild carnivores that may be susceptible to parvovirus infection (Table 6.1) (Povey and Davis 1977; Fowler 1986). The vaccine selected should be based on similarity of the hosts (e.g., CPV vaccines in coyotes) or the known or probable virus susceptibility of the host to be vaccinated (e.g., FPV vaccine in raccoons).

Use of these biologics in wild carnivores is off-label, and there is no warranty of safety or efficacy. However, there is a long history of use of such vaccines in this context (Gray 1971; Povey and Davis 1977; Fowler 1986) with a low risk of complications (Karesh and Bottomley 1983). The few studies of parvovirus vaccination in wild animals suggest that the response is comparable to that in domestic animals (Bush et al. 1981; Janssen et al. 1982; Green et al. 1984; Spencer and Burroughs 1990, 1991, 1992; Wack et al. 1993), though challenge trials have not been done.

There is debate over vaccine dose in relation to size of species, and over use of inactivated and MLV vaccines. The antigenic efficiency of parvoviruses and use of effective adjuvants mean that it is probably unnecessary to increase the dose, or antigen mass, when using inactivated vaccines, even in large species (Povey and Davis 1977), though this has not been tested. Modified-live virus vaccines, in particular, probably overcome any consideration of size of animal in relation to dose of vaccine (Povey and Davis 1977), but there may be a trade-off between potency of MLV on one hand and safety on the other (Pollock and Coyne 1993). Virus shed in feces after use of MLV may cause clinically insignificant seroconversion in susceptible animals (Pollock and Carmichael 1990b).

Fowler (1986) recommends that MLV vaccines not be used routinely for primary vaccination and never in pregnant females, presumably based on concerns regarding safety. However, because MLV vaccines are more efficacious in the face of waning maternal antibody (Pollock and Coyne 1993), they have been used in zoos as the primary vaccine in offspring from dams that have been well immunized (Montali et al. 1987), and they can be used as a booster vaccination in a previously fully immunized animal (Fowler 1986). Since inactivated vaccines may induce sufficient antibody to interfere with replication of subsequent doses of MLV vaccines, some suggest that MLV vaccines should not be used following inactivated vaccines during the initial immunization regimen.

Interference with primary immunization by residual maternal antibody is the commonest cause of "vaccine failures" in parvovirus immunization of domestic carnivores (Pearson and Gorham 1987; Greene 1990; Pollock and Coyne 1993) and in captive wild carnivores (Janssen et al. 1982; Wack et al. 1993).

Vaccination programs should heed the principles applied to vaccination of domestic carnivores in the face of maternal antibody (Greene 1990; Pollock and Coyne 1993). Current recommendations are the following: minimize potential exposure of offspring under 5 months of age to field virus, begin vaccination at 6–9 weeks of age [or as early as 2 weeks in colostrum-deprived animals (Fowler 1986)], and continue to vaccinate at 2- to 4-week intervals through 20 weeks of age. Annual revaccination of adult animals is recommended but may not be essential.

Alternatively, in heavily contaminated environments, females close to parturition may be hyperimmunized with a killed vaccine, so that high levels of maternal antibody delay the window of susceptibility until offspring are older and better able to withstand the effects of infection. In this circumstance, it will be necessary to delay vaccination until 4–5 months of age or to vaccinate at 2-week intervals until that time.

QUARANTINE. Animals introduced into a facility should be quarantined for 30 days (Montali et al. 1987), or longer if required to complete an immunization series. This will more than encompass the incubation period for parvovirus, should the animal have been exposed prior to arrival. To minimize risk of exposure, species susceptible to these parvoviruses should not be mixed in holdings, and efforts should be made to prevent intrusions by cats, dogs, and raccoons.

SANITATION. Parvoviruses are resistant to many common disinfectants, such as quaternary ammonium compounds (Kennedy et al. 1995) and alcohols, and disinfection requires use of formaldehyde, glutaraldehyde, or chlorine solutions (Scott 1980). Routinely, washable surfaces, equipment, and cage furnishings should be sanitized with effective commercial or generic virucidal agents, among which 0.175% sodium hypochlorite solution is useful if extraneous organic matter does not interfere (Scott 1980; McGavin 1987). Similar disinfectants may be considered for footbaths. Boiling rapidly inactivates parvoviruses, but hot-water washes will likely be ineffective because virus may survive for over 7 hours at 80° C and several days at 56° C (McGavin 1987).

TREATMENT. Animals suspected of parvovirus gastroenteritis should be isolated from other susceptible animals, which should be vaccinated if their immunization status is in question. Therapy is entirely supportive, aimed at mitigating the effects of dehydration and electrolyte imbalance during the phase of intestinal infection [see Greene and Scott (1990), Pollock and Carmichael (1990b), and Pollock and Coyne (1993)]. Prolonged treatment is probably not warranted for animals that have not begun to recover within 5–6 days of onset of signs.

Public Health Concerns. No evidence of human infection with feline subgroup parvoviruses has been found (Lenghaus and Studdert 1980; Binn et al. 1981), and they pose no known human health hazard (McCandlish and Thompson 1990).

Domestic Animal Health Concerns. To the extent that wildlife such as raccoons may pose an incremental risk of exposure of domestic carnivores to parvoviruses, attempts should be made to exclude them or to employ sanitary measures to minimize virus concentrations in the environment.

Management Implications. Mortality due to parvovirus seems rarely to be limiting in free-ranging populations, and it is unlikely to influence management decisions in other than small populations of great value at high risk, when vaccination might be considered (see above). However, FPV has been exploited as an apparently successful agent for biologic control of feral cats in a naive island population (Van Rensburg et al. 1987).

It is impractical to prevent exposure of susceptible carnivores if parvovirus is endemic in other sympatric species. In captive situations, such as zoos, animal shelters, and rehabilitation facilities, good hygiene and a vaccination program should be implemented. Although vaccination of large wild populations is impractical, capture and vaccination of individuals from small, isolated, highly significant, or threatened populations might be considered [e.g., Florida panther (Roelke et al. 1993) or colonizing gray wolf (Mech and Goyal 1995)].

The potential for candidate animals incubating parvovirus infections must be evaluated when translocation programs are contemplated [see Nettles et al. (1980)]. Vaccination and quarantine of translocated animals may mitigate the risk, but fomite-borne introduction of virus into nonendemic localities also must be considered.

ALEUTIAN DISEASE. Aleutian disease (AD) (or infectious plasmacytosis) is caused by persistent infection with a parvovirus distinct from the FP subgroup (Porter and Larsen 1990). Antibody against AD virus has been reported in wild striped skunk, red fox, and raccoon (Ingram and Cho 1974). Marten *Martes americana,* skunk, raccoon dogs, and domestic dogs seroconverted following experimental inoculation, whereas several other species harbored virus for up to 4 weeks (Kenyon et al. 1978; Alexandersen et al. 1985). Raccoons were suspected of playing an epidemiologic role on mink ranches, though lesions were not induced experimentally in raccoons and horizontal transmission seemed unlikely (Oie et al. 1996).

Subclinical infection is common; clinical disease occurs in ranched mink and in domestic ferrets, which have their own strain of AD virus (Porter and Larsen 1990; Welchman et al. 1993). Reproductive inefficiency, weak kits, and wasting associated with renal failure occur in mink (Porter and Larsen 1990), whereas ferrets develop a wasting syndrome or posterior ataxia and paresis (Welchman et al. 1993). The pathogenesis is related to ineffectual overproduction of antibody to AD virus, resulting in hypergammaglobulinemia, associated with plasmacyte infiltrates in a variety of organs. Immune complex-mediated vasculitis and glomerulonephritis are common in mink (Porter and Larsen 1990), although, in ferrets, nonsuppurative meningoencephalomyelitis often occurs (Welchman et al. 1993).

Farmed and feral mink probably are infected with AD virus wherever they occur (Porter and Larsen 1990; Skírnisson et al. 1990), and AD has been reported in farmed and domestic ferrets in North America, the United Kingdom, and New Zealand (Welchman et al. 1993). Therefore, it may be encountered serologically, or perhaps causing lesions and disease, in feral populations of these species, in which its importance is unknown. Aleutian disease has been suspected, but not proven, in a wild European otter *Lutra lutra* in the United Kingdom (Wells et al. 1989). The significance, if any, of AD infection in other species of wild carnivores is unknown.

LITERATURE CITED

Alexander, K.A., P.A. Conrad, I.A. Gardner, C. Parish, M. Appel, M.G. Levy, N. Lerche, and P. Kat. 1993. Serologic survey of selected microbial pathogens in African wild dogs (*Lycaon pictus*) and sympatric domestic dogs (*Canis familiaris*) in Maasai Mara, Kenya. *Journal of Zoo and Wildlife Medicine* 24:140–144.

Alexander, K.A., P.W. Kat, R.K. Wayne, and T.K. Fuller. 1994. Serologic survey of selected canine pathogens among free-ranging jackals in Kenya. *Journal of Wildlife Diseases* 30:486–491.

Alexandersen, S., Å.U. Jensen, M. Hansen, and B. Aasted. 1985. Experimental transmission of Aleutian disease virus (ADV) to different animal species. *Acta Pathologia Microbiologia et Immunologia Scandinavica [B]* 93:195–200.

Amundson, T., J. Zuba, R. Schulz, and N. Thomas. 1986. Experimental infection of gray wolves (*Canis lupus*) with canine parvovirus. *Wildlife Disease Newsletter,* Supplement to *Journal of Wildlife Diseases* 22:4.

Appel, M.J.G., and C.R. Parrish. 1982. Raccoons are not susceptible to canine parvovirus. *Journal of the American Veterinary Medical Association* 181:489.

Appel, M., and C.R. Parrish. 1987. Canine parvovirus type 2. In *Virus infections of carnivores,* ed. M.J. Appel. Amsterdam: Elsevier Science, pp. 69–92.

Appel, M.J.G., F.W. Scott, and L.E. Carmichael. 1979. Isolation and immunisation studies of a canine parvo-like virus from dogs with haemorrhagic enteritis. *Veterinary Record* 105:156–159.

Aulova, S.V., N.S. Bukina, A.K. Kirillov, V.S. Slugin. 1989. Susceptibility of sable (*Martes zibellina*) to mink enteritis parvovirus and to botulism [in Russian, English abstract]. *Veterinariya (Moscow)* 9:28–30.

Barker, I.K., R.C. Povey, and D.R. Voigt. 1983. Response of mink, skunk, red fox and raccoon to inoculation with mink virus enteritis, feline panleukopenia and canine parvovirus and prevalence of antibody to parvovirus in wild carnivores in Ontario. *Canadian Journal of Comparative Medicine* 47:188–197.

Barker, I.K., A.A. van Dreumel, and N. Palmer. 1993. The alimentary system. In *Pathology of domestic animals,* vol. 2, ed. K.V.F. Jubb, P.C. Kennedy, and N. Palmer, 4th ed. San Diego: Academic, pp. 1–318.

Behler-Amass, K., R. Thiel, R. Schultz, A. Wydeven, B. Kohn, S. Schmitt, J. Hamill, and S. Kapil. 1995. Fecal and serologic survey of selected canine pathogens in Great Lakes timber wolves (*Canis lupus lycaon*). In *Proceedings of the Joint Conference of the American Association of Zoo Veterinarians, Wildlife Disease Association and American Association of Wildlife Veterinarians, East Lansing, Michigan,* p. 502.

Bieniek, H.J., W. Encke, R. Gandras, and P. Vogt. 1981. Parvovirus-infecktion beim Mähnenwolf (*Chrysocyon brachyuris,* Illiger 1811). *Kleintierpraxis* 26:291–298.

Binn, L.N., R.H. Marchwicki, E.H. Eckermann, and T.E. Fritz. 1981. Viral antibody studies of laboratory dogs with diarrheal disease. *American Journal of Veterinary Research* 42:1665–1667.

Bolin, V.S. 1957. The cultivation of panleukopenia virus in tissue culture. *Virology* 4:389–390.

Bouillant, A., and R.P. Hanson. 1965a. Epizootiology of mink enteritis: I. Stability of the virus in feces exposed to natural environmental factors. *Canadian Journal of Comparative Medicine and Veterinary Science* 29:125–128.

———. 1965b. Epizootiology of mink enteritis: III. Carrier state in mink. *Canadian Journal of Comparative Medicine and Veterinary Science* 29:183–189.

Buonavoglia, C., P. de Nardo, and A. Fioretti. 1986. Experimental infection of red fox with canine parvovirus. *Journal of Veterinary Medicine B* 33:597–600.

Buonavoglia, C., M. Tollis, D. Buonavoglia, and A. Puccini. 1992. Response of pups with maternal derived antibody to modified-live virus canine parvovirus vaccine. *Comparative Immunology, Microbiology and Infectious Diseases* 15:281–283.

Burger, D., and J.R. Gorham. 1970. Mink virus enteritis. In *Infectious diseases of wild mammals,* ed. J.W. Davis, L.H. Karstad, and D.O. Trainer. Ames: Iowa State University Press, pp. 76–84.

Bush, M., R.C. Povey, and H. Koonse. 1981. Antibody response to an inactivated vaccine for rhinotracheitis, caliciviral disease and panleukopenia in non-domestic felids. *Journal of the American Veterinary Medical Association* 179:1203–1205.

Cai, J.L., Z.X. Zhang, and B.X. Cai. 1988. Studies on viral enteritis in the clouded leopard *Neofelis nebulosa:* Identification of the virus. *Chinese Journal of Veterinary Science and Technology* 4:6–8.

Carman, P.S., and R.C. Povey. 1985. Pathogenesis of canine parvovirus-2 in dogs: Haematology, serology and virus recovery. *Research in Veterinary Science* 38:134–140.

Carmichael, L.E., J.C. Joubert, and R.V.H. Pollock. 1980. Hemagglutination by canine parvovirus: Serologic studies and diagnostic applications. *American Journal of Veterinary Research* 40:784–791.

Chang, S.F., J.Y. Sgro, and C.R. Parrish. 1992. Multiple amino acids in the capsid structure of canine parvovirus coordinately determine the canine host range and specific antigenic and hemagglutination properties. *Journal of Virology* 66:6858–6867.

Chappuis, G., and J.M. Lernould. 1987. Infection à parvovirus félin chez le chien de forêt (*Speothos venaticus*), canide d'Amérique du sud. *Verhandlungsbericht des Internationalen Symposiums über die Erkrankungen der Zootiere* 29:293–297.

Cockburn, A. 1947. Infectious enteritis in the zoological gardens, Regent's Park. *British Veterinary Journal* 103:261–262.

Creel, S., N.M. Creel, L. Munson, D. Sanderlin, and M.J. Appel. 1997. Serosurvey for selected viral diseases and demography of African wild dogs in Tanzania. *Journal of Wildlife Diseases* 33:823–832.

Cypher, B.L., J.H. Scrivener, K.L. Hammer, and T.P. O'Farrell. 1998. Viral antibodies in coyotes from California. *Journal of Wildlife Diseases* 34:259–264.

Czisa, C.K., F.W. Scott, A. de Lahunta, and J.H. Gillespie. 1971. Immune carrier state of feline panleukopenia virus-infected cats. *American Journal of Veterinary Research* 32:419–426.

Davidson, W.R., M.J. Appel, G.L. Doster, O.E. Baker, and J.F. Brown. 1992. Diseases and parasites of red foxes, gray foxes, and coyotes from commercial sources selling to fox-chasing enclosures. *Journal of Wildlife Diseases* 28:581–589.

Dietzmann, U., W. Peter, and B. Beckendorff. 1987. Parvovirusinfektion bei Dingos: Diagnosik, Verlauf und Gedanken zur Bekämpfung. *Verhandlungsbericht des Internationalen Symposiums über die Erkrankungen der Zootiere* 29:299–304.

Dobson, A.P., and P.J. Hudson. 1995. Microparasites: Observed patterns in wild animal populations. In *Ecology of infectious diseases in natural populations,* ed. B.T. Grenfell and A.P. Dobson. Cambridge: Cambridge University Press, pp. 52–89.

Dunbar, M.R., M.W. Cunningham, and J.C. Roof. 1998. Seroprevalence of selected disease agents from free-ranging black bears in Florida. 1998. *Journal of Wildlife Diseases* 34:612–619.

Elze, K., K. Eulenberger, S. Seifert, H. Kronberger, K.F. Schüppel, and U. Schnurrbusch. 1974. Auswertung der Krankengeschichten der Felidenpatienten (1958–1973) des Zoos Leipzig. *Verhandlungsbericht des Internationalen Symposiums über die Erkrankungen der Zootiere* 16:5–18.

Eulenberger, K., K. Elze, S. Seifert, and U. Schnurrbusch. 1974. Zur Differentialdiagnose, Prophylaxe und Therapie der Panleukopenie, Salmonellose und Koliinfection der jungen Grosskatzen (Pantherini). *Verhandlungsbericht des Internationalen Symposiums über die Erkrankungen der Zootiere* 16:55–65.

Evermann, J.F., W. Foreyt, L. Maag-Miller, C.W. Leathers, A.J. McKiernan, and B. LeaMaster. 1980. Acute hemorrhagic enteritis associated with canine coronavirus and parvovirus infections in a captive coyote population. *Journal of the American Veterinary Medical Association* 177:784–786.

Evermann, J.F., A.J. McKiernan, A.K. Eugster, R.F. Solozano, J.K. Collins, J.W. Black, and J.S. Kim. 1989. Update on canine coronavirus infections and interactions with other enteric pathogens of the dog. *Companion Animal Practice* 19:6–12.

Fix, A.S., D.P. Riordan, H.T. Hill, M.A. Gill, and M.B. Evans. 1989. Feline panleukopenia virus and subsequent canine distemper virus infection in two snow leopards (*Panthera uncia*). *Journal of Zoo and Wildlife Medicine* 20:273–281.

Flagstad, A. 1977. Feline panleukopenia virus and mink enteritis virus: A serological study. *Acta Veterinaria Scandinavica* 18:1–9.

Fletcher, K.C., A.K. Eugster, R.E. Schmidt, and G.B. Hubbard. 1979. Parvovirus infection in maned wolves. *Journal of the American Veterinary Medical Association* 175:897–900.

Fowler, M.E. 1986. Carnivora. In *Zoo & Wild Animal Medicine,* ed. M.E. Fowler. Philadelphia: W.B. Saunders, pp. 800–807.

Fox, J.S. 1983. Relationships of diseases and parasites to the distribution and abundance of bobcats in New York. Ph.D. diss., State University of New York, Syracuse, 101 pp.

Frelier, P.F., R.W. Leininger, L.D. Armstrong, P.N. Nation, and R.C. Povey. 1984. Suspected parvovirus infection in porcupines. *Journal of the American Veterinary Medical Association* 185:1291–1294.

Garcelon, D.K., R.K. Wayne, and B.J. Gonzales. 1992. A serologic survey of the island fox (*Urocyon littoralis*) on the Channel Islands, California. *Journal of Wildlife Diseases* 28:223–229.

Gese, E.M., R.D. Schultz, O.J. Rongstad, and D.E. Andersen. 1991. Prevalence of antibodies against canine parvovirus and canine distemper virus in wild coyotes in southeastern Colorado. *Journal of Wildlife Diseases* 27:320–323.

Gese, E.M., R.D. Schultz, M.R. Johnson, E.S. Williams, R.L. Crabtree, and R.L. Ruff. 1997. Serological survey for diseases in fee-ranging coyotes (*Canis latrans*) in Yellowstone National Park, Wyoming. *Journal of Wildlife Diseases* 33:47–56.

Gordon, J.C., and E.J. Angrick. 1986. Canine parvovirus: Environmental effects on infectivity. *American Journal of Veterinary Research* 47:1464–1467.

Gorham, J.R., C. Parrish, D. Shen, and J.A. Baker. 1988. The effect of maternally derived antibody on the immunization of young mink against mink virus enteritis (MVE). *Scientifur* 12:303–304.

Gorham, J.R., J.F. Evermann, A. Ward, R. Pearson, D. Shen, G.R. Hartsough, and C. Leathers. 1990. Detection of coronavirus-like particles from mink with epizootic catarrhal gastroenteritis. *Canadian Journal of Veterinary Research* 54:383–384.

Goss, L.J. 1948. Species susceptibility to the viruses of Carré and feline enteritis. *American Journal of Veterinary Research* 9:65–68.

Goyal, S.M., L.D. Mech, R.A. Rademacher, M.A. Khan, and U.S. Seal. 1986. Antibodies against canine parvovirus in wolves of Minnesota: A serologic study from 1975 through 1985. *Journal of the American Veterinary Medical Association* 189:1092–1094.

Gray, C. 1971. Immunization of the exotic felidae for panleukopenia. *Journal of Zoo Animal Medicine* 3:14–15.

Green, J.F., M.L. Bruss, J.F. Evermann, and P.K. Bergstrom. 1984. Serologic response of captive coyotes (*Canis latrans* Say) to canine parvovirus and accompanying profiles of canine coronavirus titers. *Journal of Wildlife Diseases* 20:6–11.

Greene, C.E. 1990. Immunoprophylaxis and immunotherapy. In *Infectious diseases of the dog and cat,* ed. C.E. Greene. Philadelphia: W.B. Saunders, pp. 21–54.

Greene, C.E., and F.W. Scott. 1990. Feline panleukopenia. In *Infectious diseases of the dog and cat,* ed. C.E. Greene. Philadelphia: W.B. Saunders, pp. 291–299.

Greenwood, N.M., W.S.K. Chalmers, W. Baxendale, and H. Thompson. 1995. Comparison of isolates of canine parvovirus by restriction enzyme analysis, and vaccine efficacy against field strains. *Veterinary Record* 136:63–67.

Heald, R.D., B.D. Jones, and D.A. Schmidt. 1986. Blood gas and electrolyte concentrations in canine parvoviral enteritis. *Journal of the American Animal Hospital Association* 22:745–748.

Henriksen, P. 1987. A case of feed-borne virus transmission in an epidemic outbreak of mink virus enteritis. *Scientifur* 11:146–147.

Hesterbeek, J.A.P., and M.G. Roberts. 1995. Mathematical models for microparasites of wildlife. In *Ecology of infectious diseases in natural populations,* ed. B.T. Grenfell and A.P. Dobson. Cambridge: Cambridge University Press, pp. 90–122.

Holzman, S., M.J. Conroy, and W.R. Davidson. 1992. Diseases, parasites and survival of coyotes in south-central Georgia. *Journal of Wildlife Diseases* 28:572–580.

Hosokawa, S., S. Ichijo, and H. Goto. 1987. Clinical, hematological, and pathological findings in specific pathogen-free cats experimentally infected with feline panleukopenia virus. *Japanese Journal of Veterinary Science* 49:43–50.

Hyslop, N.St.G. 1955. Feline enteritis in the lynx, the cheetah and other wild felidae. *British Veterinary Journal* 111:373–377.

Ingram, D.G., and H.J. Cho. 1974. Aleutian disease in mink: Virology, immunology and pathogenesis. *Journal of Rheumatology* 1:74–92.

Jacobs, R.M., M.G. Weiser, R.L. Hall, and J.J. Kowalski. 1980. Clinicopathologic features of canine parvoviral enteritis. *Journal of the American Animal Hospital Association* 16:809–814.

Janssen, D.L., C.R. Bartz, M. Bush, R.H. Marchwicki, S.J. Grate, and R.J. Montali. 1982. Parvovirus enteritis in vaccinated juvenile bush dogs. *Journal of the American Veterinary Medical Association* 181:1225–1227.

Johnson, M.R., D.K. Boyd, and D.H. Pletscher. 1994. Serologic investigations of canine parvovirus and canine distemper in relation to wolf (*Canis lupus*) pup mortalities. *Journal of Wildlife Diseases* 30:270–273.

Johnson, R.H. 1964. Isolation of a virus from a condition simulating feline panleucopaenia in a leopard. *Veterinary Record* 76:1008–1013.

———. 1965. Feline panleucopaenia: I. Identification of a virus associated with the syndrome. *Research in Veterinary Science* 6:466–471.

———. 1967. Feline panleukopenia virus: In vitro comparison of strains with a mink enteritis virus. *Journal of Small Animal Practice* 8:319–324.

Johnson, R.H., and R.E.W. Halliwell. 1968. Natural susceptibility to feline panleucopaenia of the coati-mundi. *Veterinary Record* 82:582.

Johnson, R.H., and P.B. Spradbrow. 1979. Isolation from dogs with severe enteritis of a parvovirus related to feline panleucopaenia. *Australian Veterinary Journal* 55:151.

Johnson, R.H., G. Margolis, and L. Kilham. 1967. Identity of feline ataxia virus with feline panleucopaenia virus. *Nature* 214:175–177.

Johnson, R.H., G. Siegl, and M. Gautschi. 1974. Characteristics of feline panleucopaenia virus strains enabling definitive classification as parvoviruses. *Archives Gesamte Virusforschung* 46:315–324.

Jonsson, L., C. Magnusson, M. Book, and N. Juntti. 1988. Monoclonal antibodies applied in an immunoperoxidase method for detection of parvovirus in specimens of small intestine from dog and mink. *Acta Veterinaria Scandinavica* 29:263–264.

Karesh, W.B., and G. Bottomley. 1983. Vaccine induced anaphylaxis in a Brazilian jaguar (*Panthera onca plaustrix*). *Journal of Zoo Animal Medicine* 14:133–137.

Kennedy, M.A., V.S. Mellon, G. Caldwell, and L.N.D. Potgieter. 1995. Virucidal efficacy of the newer quaternary ammonium compounds. *Journal of the American Animal Hospital Association* 31:254–258.

Kenyon, A.J., B.J. Kenyon, and E.C. Hahn. 1978. Protides of the Mustelidae: Immunoresponse of mustelids to Aleutian mink disease virus. *American Journal of Veterinary Research* 39:1011–1015.

Kilham, L., G. Margolis, and E.D. Colby. 1967. Congenital infections of cats and ferrets by feline panleukopenia virus manifested by cerebellar hypoplasia. *Laboratory Investigation* 17:465–480.

Komolafe, O.O. 1986. Detection of parvovirus in the faeces of apparently healthy palm civet cats. *Microbios Letters* 31:75–77.

Kranzlin, B., P. Wohlsein, M. Dubberke, and A. Kuczka. 1993. Parvovirusinfektion bei Igeln (*Erinaceus europaeus*). *Kleintierpraxis* 38:675–677.

Kurtzman, G.J. 1993. Feline panleukopenia virus. In *Viruses and bone marrow: Basic research and clinical practice—hematology,* vol. 16, ed. N.S. Young. New York: Marcel Dekker, pp. 119–142.

Larsen, S., A. Flagstad, and B. Aalbak. 1976. Experimental feline panleukopenia in the conventional cat. *Veterinary Pathology* 13:216–240.

Laurenson, K., J. van Heerden, P. Stander, and M.J. van Vuuren. 1997. Seroepidemiological survey of sympatric domestic and wild dogs (*Lycaon pictus*) in Tsumkwe District, north-eastern Namibia. *Onderstepoort Journal of Veterinary Research* 64:313–316.

Leclerc-Cassan, M. 1981. Ataxie cerebelleuse du chaton et maladie des étoiles du lionceu: Note de pathologie comparée. *Receuil Médecine Vétérinaire* 157:741–743.

Lenghaus, C., and M.J. Studdert. 1980. Model for viral myocarditis. *Medical Journal of Australia* 2:42.
Macartney, L., I.A.P. McCandlish, H. Thompson, and H.J.C. Cornwell. 1984. Canine parvovirus enteritis 1: Clinical, haematological and pathological features of experimental infection. *Veterinary Record* 115:201–210.
Macartney, L., H. Thompson, I.A.P. McCandlish, and H.J.C. Cornwell. 1988. Canine parvovirus: Interaction between passive immunity and virulent challenge. *Veterinary Record* 122:573–576.
Madic, J., D. Huber, and B. Lugovic. 1993. Serologic survey for selected viral and rickettsial agents of brown bears (*Ursus arctos*) in Croatia. *Journal of Wildlife Diseases* 29:572–576.
Mainka, S.A., X.M. Qui, T.M. He, and M.J. Appel. 1994. Serologic survey of giant pandas (*Ailuropoda melanoleuca*), and domestic dogs and cats in the Wolong reserve, China. *Journal of Wildlife Diseases* 30:86–89.
Mann, P.C., M. Bush, M.J.G. Appel, B.A. Beehler, and R.J. Montali. 1980. Canine parvovirus infection in South American canids. *Journal of the American Veterinary Medical Association* 177:779–783.
Martin, H.D., and N.S. Zeidner. 1992. Concomitant cryptosporidia, coronavirus and parvovirus infection in a raccoon (*Procyon lotor*). *Journal of Wildlife Diseases* 28:113–115.
Martyn, J.C., B.E. Davidson, and M.J. Studdert. 1990. Nucleotide sequence of feline panleukopenia virus: Comparison with canine parvovirus identifies host-specific differences. *Journal of General Virology* 71:2747–2753.
Mason, M.J., N.A. Gillett, and B.A. Muggenburg. 1987. Clinical, pathological, and epidemiological aspects of canine parvoviral enteritis in an unvaccinated closed beagle colony: 1978–1985. *Journal of the American Animal Hospital Association* 23:183–192.
Matsui, T., J. Matsumoto, T. Kanno, T. Awakura, H. Taniyama, H. Furuoka, and H. Ishikawa. 1993. Intranuclear inclusions in the stratified squamous epithelium of the tongue in dogs and cats with parvovirus infection. *Veterinary Pathology* 30:303–305.
McCandlish, I.A.P., and H. Thompson. 1990. Human parvovirus infections. *Veterinary Record* 127:385.
McCue, P.M., and T.P. O'Farrell. 1988. Serological survey for selected diseases in the endangered San Joaquin kit fox (*Vulpes macrotis mutica*). *Journal of Wildlife Diseases* 24:274–281.
McGavin, D. 1987. Inactivation of canine parvovirus by disinfectants and heat. *Journal of Small Animal Practice* 28:523–535.
Mech, L.D., and S.M. Goyal. 1993. Canine parvovirus effect on wolf population change and pup survival. *Journal of Wildlife Diseases* 29:330–333.
———. 1995. Effects of canine parvovirus on gray wolves in Minnesota. *Journal of Wildlife Management* 59:565–570.
Mech, L.D., S.M. Goyal, C.N. Bota, and U.S. Seal. 1986. Canine parvovirus infection in wolves (*Canis lupus*) from Minnesota. *Journal of Wildlife Diseases* 22:104–106.
Mech, L.D., H.J. Kurtz, and S. Goyal. 1997. Death of a wild wolf from canine parvoviral enteritis. *Journal of Wildlife Diseases* 33:321–322.
Meunier, P.C., L.T. Glickman, M.J.G. Appel, and S.J. Shin. 1980. Canine parvovirus in a commercial kennel: Epidemiologic and pathologic findings. *Cornell Veterinarian* 71:96–110.
Meunier, P.C., B.J. Cooper, M.J.G. Appel, M.E. Lanieu, and D.O. Slauson. 1985a. Pathogenesis of canine parvovirus enteritis: Sequential virus distribution and passive immunization studies. *Veterinary Pathology* 22:617–624.
Meunier, P.C., B.J. Cooper, M.J.G. Appel, and D.O. Slauson. 1985b. Pathogenesis of canine parvovirus enteritis: The importance of viremia. *Veterinary Pathology* 22:60–71.
Miller, R.M. 1961. Distemper in the coati-mundi. *Modern Veterinary Practice* 42:52.
Mizak, B. 1994. Isolation of Polish strains of blue fox parvovirus (BFPV). *Bulletin of the Veterinary Institute in Pulawy* 38:98–104.
Mochizuki, M., M. Akuzawa, and H. Nagatomo. 1990. Serological survey of the Iriomote cat (*Felis iriomotensis*) in Japan. *Journal of Wildlife Diseases* 26:236–245.
Mochizuki, M., H. Hiragi, M. Sueyoshi, Y. Kimoto, S. Takeishi, M. Horiuchi, and R. Yamaguchi. 1996. Antigenic and genomic characteristics of parvovirus isolated from a lion (*Panthera leo*) that died of feline panleukopenia. *Journal of Zoo and Wildlife Medicine* 27:416–420.
Montali, R.J., C.R. Bartz, and M. Bush. 1987. Parvoviruses. In *Virus infections of carnivores,* ed. M.J. Appel. Amsterdam: Elsevier Science, pp. 419–428.
Muneer, M.A., I.O. Farah, K.A. Pomeroy, S.M. Goyal, and L.D. Mech. 1988. Detection of parvoviruses in wolf feces by electron microscopy. *Journal of Wildlife Diseases* 24:170–172.
Nair, K.P.D., R.P. Iyer, and A. Venugopalan. 1964. An outbreak of gastroenteritis in civet cats. *Indian Veterinary Journal* 41:763–765.
Nettles, V.F., J.E. Pearson, G.A. Gustafson, and J.L. Blue. 1980. Parvovirus infection in translocated raccoons. *Journal of the American Veterinary Medical Association* 177:787–789.
Neuvonen, E., P. Veijalainen, and J. Kangas. 1982. Canine parvovirus infection in housed raccoon dogs and foxes in Finland. *Veterinary Record* 110:448–449.
Oie, K.L., G. Durrant, J.B. Wolfinbarger, D. Martin, F. Costello, S. Perryman, D. Hogan, W.J. Hadlow, and M.E. Bloom. 1996. The relationship between capsid protein (VP2) sequence and pathogenicity of Aleutian mink disease parvovirus (ADV): A possible role for raccoons in the transmission of ADV. *Journal of Virology* 70:852–861.
Parrish C.R. 1990. Emergence, natural history, and variation of canine, mink, and feline parvoviruses. *Advances in Virus Research* 38:403–450.
———. 1991. Mapping specific functions in the capsid structure of canine parvovirus and feline panleukopenia virus using infectious plasmid clones. *Virology* 183:195–205.
Parrish, C.R., and L.E. Carmichael. 1983. Antigenic structure and variation of canine parvovirus type-2, feline panleukopenia virus, and mink enteritis virus. *Virology* 129:401–414.
Parrish C.R., R.E. Oliver, and R. McNiven. 1982. Canine parvovirus infection in a colony of dogs. *Veterinary Microbiology* 7:317–324.
Parrish, C.R., J.R. Gorham, T.M. Schwartz, and L.E. Carmichael. 1984. Characterization of antigenic variation among mink enteritis virus isolates. *American Journal of Veterinary Research* 45:2591–2599.
Parrish, C.R., C.W. Leathers, R. Pearson, and J.R. Gorham. 1987. Comparisons of feline panleukopenia virus, canine parvovirus, raccoon parvovirus, and mink enteritis virus and their pathogenicity for mink and ferrets. *American Journal of Veterinary Research* 48:1429–1435.
Parrish, C.R., P. Have, W.J. Foreyt, J.F. Evermann, M. Senda, and L.E. Carmichael. 1988. The global spread and replacement of canine parvovirus strains. *Journal of General Virology* 69:1111–1116.
Paul-Murphy, J., T. Work, D. Hunter, E. McFie, and D. Fjelline. 1994. Serologic survey and serum biochemical reference ranges of the free-ranging mountain lion (*Felis

concolor) in California. *Journal of Wildlife Diseases* 30:205–215.

Pearson, R.C., and J.R. Gorham. 1987. Mink virus enteritis. In *Virus infections of carnivores,* ed. M.J. Appel. Amsterdam: Elsevier Science, pp. 349–360.

Pedersen, N.C. 1987. Feline panleukopenia virus. In *Virus infections of carnivores,* ed. M.J. Appel. Amsterdam: Elsevier Science, pp. 247–254.

Pence, D.P., and L.A. Windberg. 1984. Population dynamics across selected habitat variables of the helminth community in coyotes, *Canis latrans,* from southern Texas. *Journal of Parasitology* 70:735–746.

Peterson, R.O. 1995. *The wolves of Isle Royale: A broken balance.* Minocqua, WI: Willow Creek, 190 pp.

Phillips, C.E. 1943. Haemorrhagic enteritis in the Artic [*sic*] blue fox: Caused by the virus of feline enteritis. *Canadian Journal of Comparative Medicine* 7:33–35.

Pollock, R.V.H. 1982. Experimental canine parvovirus infection in dogs. *Cornell Veterinarian* 72:103–119.

———. 1984a. The parvoviruses: Part I. Feline panleukopenia virus and mink enteritis virus. *Compendium on Continuing Education for the Practicing Veterinarian* 6:227–237.

———. 1984b. The parvoviruses: Part II. Canine parvovirus. *Compendium on Continuing Education for the Practicing Veterinarian* 6:653–664.

Pollock, R.V.H., and L.E. Carmichael. 1982. Maternally derived immunity to canine parvovirus infection: Transfer, decline, and interference with vaccination. *Journal of the American Veterinary Medical Association* 180:37–42.

———. 1990a. The canine parvoviruses. In *CRC handbook of parvoviruses,* vol. 2, ed. P. Tijssen. Boca Raton, FL: CRC, pp. 113–134.

———. 1990b. Canine viral enteritis. In *Infectious diseases of the dog and cat,* ed. C.E. Greene. Philadelphia: W.B. Saunders, pp. 268–287.

Pollock, R.V.H., and M.J. Coyne. 1993. Canine parvovirus. *Veterinary Clinics of North America, Small Animal Practice* 23:555–568.

Pollock, R.V.H., and C.R. Parrish. 1985. Canine parvovirus. In *Comparative pathobiology of viral diseases,* ed. R.G. Olsen, S. Krakowka, and J.R. Blakeslee. Boca Raton, FL: CRC, pp. 145–177.

Porter, D.D., and A.E. Larsen. 1990. Mink parvovirus infections. In *CRC handbook of parvoviruses,* vol. 2, ed. P. Tijssen. Boca Raton, FL: CRC, pp. 87–101.

Povey, R.C., and E.V. Davis. 1977. Panleukopenia and respiratory virus infection in wild felids. In *The world's cats,* vol. 3, no. 3, ed. R.L. Eaton. Seattle: Carnivore Research Institute, Burke Museum, pp. 120–128.

Rabinowitz, A.R., and L.N.D. Potgieter. 1984. Serologic survey for selected viruses in a population of raccoons, *Procyon lotor* (L.), in the Great Smoky Mountains. *Journal of Wildlife Diseases* 20:146–148.

Reif, J.S. 1976. Seasonality, natality and herd immunity in feline panleukopenia. *American Journal of Epidemiology* 103:81–87.

Reynolds, H.A. 1969. Some clinical and hematological features of virus enteritis of mink. *Canadian Journal of Comparative Medicine* 33:155–159.

———. 1970. Pathological changes in virus enteritis of mink. *Canadian Journal of Comparative Medicine* 34:155–163.

Robinson, W.F., G.E. Wilcox, and R.L.P. Flower. 1980. Canine parvoviral disease: Experimental reproduction of the enteric form with a parvovirus isolated from a case of myocarditis. *Veterinary Pathology* 17:589–599.

Roelke, M.E., D.J. Forrester, E.R. Jacobson, G.V. Kollias, F.W. Scott, M.C. Barr, J.F. Evermann, and E.C. Pirtle. 1993. Seroprevalence of infectious disease agents in free-ranging Florida panthers (*Felis concolor coryi*). *Journal of Wildlife Diseases* 29:36–49.

Schofield, F.W. 1949. Virus enteritis in mink. *North American Veterinarian* 30:651–654.

Schwers, A., J. Barrat, J. Blancou, M. Maenhoudt, and P.P. Pastoret. 1983. Récherche d'anticorps envers le parvovirus canin dans des sérums de renards en France. *Annales de Médecine Vétérinaire* 127:544–546.

Scott, F.W. 1980. Virucidal disinfectants and feline viruses. *American Journal of Veterinary Research* 41:410–414.

———. 1990. Feline parvovirus infection. In *CRC handbook of parvoviruses,* vol. 2, ed. P. Tijssen. Boca Raton, FL: CRC, pp. 103–111.

Scott, F.W., C.K. Csiza, and J.H. Gillespie. 1970. Maternally derived immunity to feline panleukopenia. *Journal of the American Veterinary Medical Association* 156:439–453.

Senda, M., N. Hirayama, O. Itoh, and H. Yamamoto. 1988. Canine parvovirus: Strain difference in haemagglutination activity and antigenicity. *Journal of General Virology* 69:349–354

Shen, D.T., A.C.S. Ward, and J.R. Gorham. 1986. Detection of mink enteritis virus in mink feces, using enzyme-linked immunosorbent assay, hemagglutination, and electron microscopy. *American Journal of Veterinary Research* 47:2025–2030.

Siegl, G., R.C. Bates, K.I. Berns, B.J. Carter, D.C. Kelly, E. Kurstak, and P. Tattersall. 1985. Characteristics and taxonomy of *Parvoviridae. Intervirology* 23:61–73.

Singh, B., P. N. Dhingra, N. Singh, and M. Chandra. 1983. Infectious feline enteritis in panther cubs (*Panthera pardus*). *Indian Journal of Animal Sciences* 53:921–924.

Skírnisson, K., E. Gunnarson, and S. Hjartardóttir. 1990. Plasmacytosis-syking í villtum mink á íslandi. *Búvísindi, Icelandic Agricultural Science* 3:113–122.

Spencer, J.A. 1991. Survey of antibodies to feline viruses in free-ranging lions. *South African Journal of Wildlife Research* 21:59–61.

Spencer, J.A., and R.E.J. Burroughs. 1990. Antibody response in wild dogs to canine parvovirus vaccine. *South African Journal of Wildlife Research* 20:14–15.

Spencer, J.A., and R. Burroughs. 1991. Antibody response of captive cheetahs to modified-live feline virus vaccine. *Journal of Wildlife Diseases* 27:578–583.

———. 1992. Decline in maternal immunity and antibody response to vaccine in captive cheetah (*Acinonyx jubatus*) cubs. *Journal of Wildlife Diseases* 28:102–104.

Spencer, J.A., and P. Morkel. 1993. Serological survey of sera from lions in Etosha National Park. *South African Journal of Wildlife Research* 23:60–61.

Strassheim, M.L., A. Gruenberg, P. Veijalainen, J.Y. Sgro, and C.R. Parrish. 1994. Two dominant neutralizing antigenic determinants of canine parvovirus are found on the three-fold spike of the virus capsid. *Virology* 198:175–184.

Studdert, M.J., C.M. Kelly, and K.E. Harrigan. 1973. Isolation of panleucopaenia virus from lions. *Veterinary Record* 93:156–158.

Studdert, M.J., C. Oda, C.A. Riegl, and R.P. Roston. 1983. Aspects of the diagnosis, pathogenesis and epidemiology of canine parvovirus. *Australian Veterinary Journal* 60:197–200.

Thomas, N.J., W.J. Foreyt, J.F. Evermann, L.A. Windberg, and F.F. Knowlton. 1984. Seroprevalence of canine parvovirus in wild coyotes from Texas, Utah, and Idaho (1972 to 1983). *Journal of the American Veterinary Medical Association* 185:1283–1287.

Tijssen, P., ed. 1990. *CRC handbook of parvoviruses,* vol. 2. Boca Raton, FL: CRC, 312 pp.

Torres, S. 1941. Infectious feline gastroenteritis in wild cats. *North American Veterinarian* 22:297–299.

Truyen, U., and C.R. Parrish. 1992. Canine and feline host ranges of canine parvovirus and feline panleukopenia virus: Distinct host cell tropisms of each virus in vitro and in vivo. *Journal of Virology* 66:5399–5408.

Truyen, U., M. Agbandje, and C.R. Parrish. 1994a. Characterization of the feline host range and a specific epitope of feline panleukopenia virus. *Virology* 200:494–503.

Truyen, U., G. Platzer, C.R. Parrish, T. Hanichen, W. Hermanns, and O.R. Kaaden. 1994b. Detection of canine parvovirus DNA in paraffin-embedded tissues by polymerase chain reaction. *Journal of Veterinary Medicine [B]* 41:148–152.

Truyen, U., A. Gruenberg, S.-F. Chang, B. Obermaier, P. Veijalainen, and C.R. Parrish. 1995. Evolution of the feline sub-group parvoviruses and the control of canine host range in vivo. *Journal of Virology* 69:4702–4710.

Truyen, U., J.F. Evermann, E. Vieler, and C.R. Parrish. 1996. Evolution of canine parvovirus involved loss and gain of feline host range. *Virology* 215:186–189.

Truyen, U., T. Muller, R. Heidrich, K. Tackmann, and L.E. Carmichael. 1998. Survey on viral pathogens in wild red foxes (*Vulpes vulpes*) in Germany with emphasis on parvoviruses and analysis of a DNA sequence from a red fox parvovirus. *Epidemiology and Infection* 121:433–440.

Tsao, J., M.S. Chapman, M. Agbandje, W. Keller, K. Smith, H. Wu, M. Luo, T.J. Smith, M.G. Rossmann, R.W. Compans, and C.R. Parrish. 1991. The three-dimensional structure of canine parvovirus and its functional implications. *Science* 251:1456–1464.

Uttenthal, Å., S. Larsen, E. Lund, M.E. Bloom, T. Storgård, and S. Alexandersen. 1990. Analysis of experimental mink enteritis virus infection in mink: In situ hybridization, serology, and histopathology. *Journal of Virology* 64:2768–2779.

Valíček, L., B. Šmid, and J. Váhala. 1993. Demonstration of parvovirus in diarrhoeic African cheetahs (*Acinonyx jubatus jubatus* Schreber, 1775). *Veterinarni Medicina* 38:245–249.

Van Rensburg, P.J.J., J.D. Skinner, and R.J. van Aarde. 1987. Effects of feline panleucopaenia on the population characteristics of feral cats on Marion Island. *Journal of Applied Ecology* 24:63–73.

Veijalainen, P. 1986. A serological survey of enteric parvovirus infections in Finnish fur-bearing animals. *Acta Veterinaria Scandinavica* 27:159–171.

———. 1988. Characterization of biological and antigenic properties of raccoon dog and blue fox parvoviruses: A monoclonal antibody study. *Veterinary Microbiology* 16:219–230.

Veijalainen, P.M.-L., and E. Smeds. 1988. Pathogenesis of blue fox parvovirus on blue fox kits and pregnant vixens. *American Journal of Veterinary Research* 49:1941–1944.

Verge, J., and N. Christoforini. 1928. La gastroenterité infectieuse des chats est elle due à un virus filterable? *Comptes Rendues Séances Societé Biologique (Paris)* 99:312–314.

Vetési, F., and A. Balsai. 1971. Beiträge zur Infektiösen Panleukopenie der Zootiere. *Verhandlungsbericht des Internationalen Symposiums über die Erkrankungen der Zootiere* 13:255–257.

Visee, A.M., P. Zwart, and J. Haagsma. 1974. Zwei Fälle von infektiöser Enteritis bei Mähnenwölfen (*Chrysocyon brachyurus*). *Verhandlungsbericht des Internationalen Symposiums über die Erkrankungen der Zootiere* 16:67–69.

Wack, R.F., L.W. Kramer, W.L. Cupps, S. Clawson, and D.R. Hustead. 1993. The response of cheetahs (*Acinonyx jubatus*) to routine vaccination. *Journal of Zoo and Wildlife Medicine* 24:109–117.

Waldvogel, A.S., S. Hassam, N. Stoerckle, R. Weilenmann, J. D. Tratschin, G. Siegl, and A. Pospischil. 1992. Specific diagnosis of parvovirus enteritis in dogs and cats by in situ hybridization. *Journal of Comparative Pathology* 107:141–146.

Waller, E.F. 1940. Infectious gastroenteritis in raccoons (*Procyon lotor*). *Journal of the American Veterinary Medical Association* 96:266–268.

Wassmer, D. A., D.D. Guenther, and J.N. Layne. 1988. Ecology of the bobcat in south-central Florida. *Bulletin of the Florida State Museum, Biological Sciences* 33:159–228.

Welchman, D. de B., M. Oxenham, and S.H. Done. 1993. Aleutian disease in domestic ferrets: Diagnostic findings and survey results. *Veterinary Record* 132:479–484.

Wells, G.A.H., I. F. Keymer, and K.C. Barnett. 1989. Suspected Aleutian disease in a wild otter (*Lutra lutra*). *Veterinary Record* 125:232–235.

Wills, C.G. 1952. Notes on infectious enteritis of mink and its relationship to feline enteritis. *Canadian Journal of Comparative Medicine* 16:419–420.

Wills, G., and J. Belcher. 1956. The prevention of virus enteritis of mink with commercial feline panleukopenia vaccine. *Journal of the American Veterinary Medical Association* 128:559–560.

Woolf, A., and J. Swart. 1974. An outbreak of feline panleukopenia. *Journal of Zoo Animal Medicine* 5:32–34.

Zarnke, R.L., and W.B. Ballard. 1987. Serologic survey for selected microbial pathogens of wolves in Alaska, 1975–1982. *Journal of Wildlife Diseases* 23:77–85.

7 HERPESVIRUS INFECTIONS

HERPESVIRUSES OF NONHUMAN PRIMATES

NORVAL W. KING

INTRODUCTION. Members of the family *Herpesviridae* have a genome consisting of a linear molecule of double-stranded DNA, enclosed within a 100–110-nm icosahedral nucleocapsid composed of 162 hollow capsomeres, covered by a lipid-containing envelope bearing glycoprotein spikes on its surface. The herpesvirus genome encodes a number of enzymes involved in protein processing, DNA synthesis, and nucleic acid metabolism. Viral DNA synthesis and capsid assembly occur in the nucleus of infected cells, and encapsulated virions acquire their envelope by budding through host cell membranes, usually the inner leaflet of the nuclear membrane. Viral replication causes death of the host cell. Lifelong latent infection and viral persistence occur commonly in the host and may be associated with periodic episodes of reactivation, viral replication, and shedding.

The family consists of three distinct subfamilies: *Alphaherpesvirinae, Betaherpesvirinae,* and *Gammaherpesvirinae* (Table 7.1), which have biologic features that make them distinctive (Roizman 1996).

The alphaherpesviruses have a variable host range; a rapid replicative cycle both in vitro and in vivo; and an affinity for epithelial cells, for which they are highly cytolytic; and they are characterized by lifelong persistence as latent infections, principally in sensory ganglia of infected hosts.

The betaherpesviruses, or cytomegaloviruses, generally have a restricted host range. These agents have a prolonged replicative cycle and are not as rapidly cytolytic as the alphaherpesviruses. In vitro and in vivo, they cause marked enlargement of the nucleus and cytoplasm of infected cells (cytomegaly) and produce large amphoteric intranuclear inclusion bodies and small basophilic cytoplasmic inclusions. The cytomegaloviruses are typically cell associated, with large clusters of enveloped virions accumulating in cytoplasmic vacuoles of infected cells, which accounts for the cytoplasmic inclusions seen. Latent infection with cytomegaloviruses occurs in secretory glands such as the salivary glands, lymphoid organs, kidneys, and other tissues. The subfamily contains only one genus, *Cytomegalovirus,* that affects primate species (Roizman 1996).

TABLE 7.1—Family *Herpesviridae*

Current Scientific Name	Common Name
Alphaherpesvirinae	
Cercopithicine herpesvirus 1	*Herpes simiae* (B virus)
Cercopithicine herpesvirus 6, 7, and 9	Simian varicelloviruses
Human herpesvirus 3	Varicella-zoster
Pongine herpesvirus	
Saimirine herpesvirus 1	*Herpesvirus platyrrhinae* (*H. tamarinus,* herpes T)
Human herpesvirus 1 and 2	*Herpes hominis* (H. simplex)
Cercopithicine herpesvirus 2	*Herpesvirus papionis* (SA-8)
Betaherpesvirinae	
Cytomegaloviruses	
Cercopithicine herpesvirus 3	SA-6
Cercopithicine herpesvirus 4	SA-15
Cercopithicine herpesvirus 5	African green monkey CMV
Cercopithicine herpesvirus 8	Rhesus monkey CMV
Aotine herpesvirus 1, 3, and 4	*Herpes aotus* types 1, 3, and 4
Callitrichine herpesvirus 1 and 2	Marmoset CMV
Cebine herpesvirus 1	Capuchin herpesvirus (AL-5)
Cebine herpesvirus 2	Capuchin herpesvirus (AP-18)
Gammaherpesvirinae	
Rhadinovirus	
Saimiriine herpesvirus 2	*Herpes saimiri*
Ateline herpesvirus 2 and 3	*Herpes ateles*
Lymphocryptovirus	
Epstein-Barr-like viruses	
Pongine herpesvirus 1	Chimpanzee herpes
Pongine herpesvirus papionis	Orangutan herpes
Pongine herpesvirus 3	Gorilla herpes
Cercopithicine herpesvirus 10 and 11	Rhesus leukocyte-associated herpesviruses 1 and 2
Cercopithicine herpesvirus 12	*Herpesvirus papio*
Cercopithicine herpesvirus 14	African green monkey EBV-like virus

CMV, cytomegalovirus; SA, simian agent.

Members of the subfamily *Gammaherpesvirinae* have a restricted host range, generally limited to the family or order to which the natural or reservoir host belongs. In cell culture, all members replicate in lymphoblastoid cells but are specific for cells of either B-cell or T-cell lineage. Some agents in this subfamily are also cytolytic for epithelial or fibroblastic cells. Latent infection with gammaherpesviruses occurs in lymphoid tissues. Two genera comprise this subfamily: *Lymphocryptovirus* and *Rhadinovirus.*

ALPHAHERPESVIRINAE

***Herpesvirus simiae* (B Virus).** B virus (*Herpesvirus simiae,* Cercopithecine herpesvirus 1) occurs commonly as a latent, subclinical infection of Asian macaques. Although not usually associated with disease in the natural host, accidental infection of humans results in a disseminated viral infection characterized by ascending paralysis and a high fatality rate. The increased number of human cases of B-virus infection since 1987 has spurred renewed interest in this zoonotic disease.

HOST RANGE. Asian macaques commonly infected with B virus include the rhesus macaque *Macaca mulatta,* bonnet macaque *Macaca radiata,* Japanese macaque *Macaca fuscata,* stump-tailed macaque *Macaca arctoides,* Formosan rock macaque *Macaca cyclopis,* and cynomolgus monkey *Macaca fascicularis* (Hunt and Blake 1994). Infection of primates other than macaques, including the patas monkey *Erythrocebus patas,* black-and-white colobus monkey *Colobus abyssinicus,* capuchin monkey *Cebus apella,* and common marmoset *Callithrix jacchus,* has reportedly been fatal (Gay and Holden 1933; Loomis et al. 1981; Wilson et al.,1990).

ETIOLOGY. The genome of *H. simiae* is 162 kb long and encodes approximately 23 major proteins. It shares antigenic determinants with the gD and gB glycoproteins of *Herpesvirus hominis* (*simplex*) 1 and 2. Viral replication is rapid, with enveloped nucleocapsids being produced 8–10 hours after infection. In cell culture, multinucleated syncytial cells and Cowdry type-A intranuclear inclusion bodies characterize its cytopathic effect (Whitely 1996).

TRANSMISSION AND EPIDEMIOLOGY. The incidence of infection in immature rhesus macaques is generally low but increases rapidly at or about the time of sexual maturity. In some colonies, 80%–90% of rhesus macaques may be seropositive by 3–4 years of age (Weigler et al. 1993). The proportion of animals with overt oral or labial lesions is considerably lower and, in one large survey, only 2.3% of 14,400 macaques had obvious lesions (Keeble 1960).

Virus transmission occurs venereally, via fomites, or through bite wounds inflicted by infected cagemates. Overcrowding and poor husbandry conditions may result in animals becoming infected at an earlier age and a higher prevalence in the colony. Animals remain infected for life and periodically shed virus in oral and genital secretions. The greatest risk of primary infection occurs during the breeding season in sexually adolescent animals 2–3 years of age (Weigler et al. 1990).

CLINICAL SIGNS. Primary infection of macaques is usually mild and self-limiting. Characteristic transient vesicular lesions, which progress to ulcers, appear on the oral, labial, and genital mucosae; these usually heal within 10–14 days. In those rare cases of disseminated infection, the clinical course can vary from peracute to slowly progressive. In the latter instance, B-virus infection may not be suspected, increasing the risk of human exposure. High morbidity and mortality has occurred in bonnet macaques with a respiratory form of B-virus infection. Animals exhibited signs of coryza, rhinorrhea, cough, and conjunctivitis but no oral lesions (Espana 1973).

PATHOGENESIS. The pathogenesis of B-virus infection in macaques is similar to *H. hominis* (*simplex*) infection in humans. Primary infection leads to an initial round of viral replication within epithelial cells at the site of inoculation, usually the skin or mucous membranes. Subsequently, virions are transported by retrograde axonal flow to the sensory ganglia, where a latent infection is established for the life of the monkey. Centrifugal spread may cause expansion of lesions or, rarely, widespread dissemination of the virus during primary infection.

Factors responsible for reactivation of a latent infection, with recurrence of lesions and shedding of the virus, are poorly understood. Stress, fever, ultraviolet light, tissue or nerve injury, and immunosuppression contribute to reactivation of *H. hominis* (*simplex*) in humans and may also play a role in B-virus infection in macaques. Reactivation of oral lesions in cynomolgus monkeys has been associated with treatment with an immunosuppressive agent (Chellman et al. 1992). However, viral shedding by single-housed seropositive macaques was not enhanced by the stress associated with quarantine, breeding, or parturition (Weir et al. 1993). For reasons that are not understood, reactivation of latent B-virus infection is not common in macaques dying of AIDS following experimental simian immunodeficiency virus (SIV) infection (Lackner 1994).

PATHOLOGY. Histologically, cutaneous and mucosal lesions are characterized by ballooning degeneration of keratinocytes with progression to vesiculation. Multinucleated syncytial cells and eosinophilic to amphophilic intranuclear viral inclusion bodies occur. Inflammatory cells may be present within vesicles, epidermis, and adjacent dermis. Necrosis of endothelial cells within dermal blood vessels with intranuclear viral inclusions may be present in some cases.

Systemic infection of macaques is extremely rare but, when it occurs, it is usually fatal. In these cases, virus becomes widely disseminated throughout major organs, including the lung, liver, spleen, bone marrow, and adrenal cortex (Wilson et al. 1990; Simon et al.

1993). In disseminated disease, there are extensive multifocal areas of necrosis throughout the liver, lung, brain, and lymphoid organs (Espana 1973; Simon et al. 1993).

DIAGNOSIS. B-virus infection should be considered in the diagnosis of any disease of macaques in which there is oral or genital ulceration or multifocal necrotizing hepatitis. Seroconversion follows shortly after primary infection and is associated with resolution of labial or oral ulcers. Antibodies may be detected by enzyme-linked immunosorbent assay (ELISA) and Western blot techniques (Ward and Hilliard 1994). False-negative test results and latently infected, immunologically nonreactive individuals may confound interpretation of test results on individual samples but seem to be relatively uncommon. In addition, because of extensive antigenic cross-reactivity between the two viruses, antibodies to *H. hominis* (*simplex*) can be used immunohistochemically to demonstrate the presence of *H. simiae* antigen in sections of lesions where the latter is suspected, since macaques are not susceptible to *H. hominis*.

CONTROL AND TREATMENT. Actively infected animals shedding virus should not be treated, since this entails considerable risk to the attending personnel. Management measures to reduce risk of exposure are discussed next.

PUBLIC HEALTH CONCERNS. Despite the widespread use of macaques in biomedical research and the high prevalence of B-virus infection, fewer than 40 documented human cases have been described (Holmes et al. 1995). The majority of human infections have resulted from injuries inflicted by macaque monkeys. Rarely, human-to-human contact, respiratory spread, needle-stick injury, laboratory exposure, or unknown source of exposure have been reported (Holmes et al. 1990; Weigler 1992). Although previous infection by *H. hominis* (*simplex*) was thought to afford protection against B-virus infection in people (Hull 1973; Palmer 1987), evidence from recent outbreaks indicates this is not the case (Holmes et al. 1990).

In people, a vesicular dermatitis develops at the site of inoculation as early as 3–5 days or as late as 24 days after exposure. The site of inoculation may be intensely pruritic. Lymphangitis and secondary lymphadenopathy may follow. Neurologic signs develop 3–7 days after the initial cutaneous lesion and consist of an ascending myelitis. Fever, paresthesia, muscle weakness, and conjunctivitis may occur prior to the onset of neurologic signs. In some cases, premonitory signs and a history of exposure to infected macaques have not been documented prior to onset of the ascending myelitis. Early recognition of clinical signs is vital, as treatment with acyclovir or ganciclovir may be beneficial if administered during the initial stages of infection. The case fatality rate in infected people is approximately 70%, death occurring in 10–14 days. Asymptomatic infection and infections characterized by recurrent vesicular rash and respiratory signs have been described in people but these infections are rare (Freifeld et al. 1995).

Recommendations for the prevention and treatment of injuries inflicted by macaque monkeys have been published (Centers for Disease Control and Prevention 1987; Holmes et al. 1995). Proper instruction of animal care and laboratory personnel in the prevention and risks of B-virus infection is imperative (Davenport et al. 1994).

Bite wound kits, along with detailed standard operating procedures, should be available at all institutions where macaques are housed and tissues derived from them are handled. Immediately following exposure, the wound should be washed thoroughly and vigorously with detergent and water for at least 15 minutes. Risk assessment by an infectious disease expert should follow. Early diagnosis of clinical disease and careful clinical follow-up are essential.

MANAGEMENT IMPLICATIONS. Guidelines used to establish B-virus-specific pathogen-free (SPF) colonies have been published (Ward and Hilliard 1994).

Animals are initially screened serologically by titration ELISA and Western blot. Negative animals are kept in single-cage housing or small groups and periodically tested by a modified ELISA for 1 year. Repeated testing is required because animals may be (1) chronically infected and immunologically nonreactive or (2) in the early stages of disease prior to seroconversion. Animals that are repeatedly negative by these criteria can then be moved into larger breeding groups and periodically tested to screen for seroconversion.

Ideally, SPF colonies should be self-sustaining and not require introduction of new animals. Any introductions require appropriate testing and quarantine. Breaks in SPF status may occur from introduction of new animals, contact with fomites contaminated by non-SPF animals, or reactivation of latent infection in seronegative animals.

In smaller facilities, the acquisition of seronegative young animals, subsequently housed individually, reduces the occurrence of primary active infections (Di Giacomo and Shah 1972; Olson et al. 1991) and may be adequate for management of infection in small numbers of animals that are to be kept for short periods.

Simian Varicella Viruses. The simian varicella viruses (SVVs) cause a highly contagious infection of a variety of Old World primates, resulting in high morbidity and mortality. These viruses have been used experimentally in nonhuman primates to investigate aspects of the human varicella-zoster infection.

ETIOLOGY. Closely related simian herpesviruses, including Liverpool vervet virus, patas herpesvirus, Medical Lake macaque virus, and delta herpesvirus cause similar exanthematous disease. Infection of

African green monkeys *Cercopithecus aethiops,* patas monkeys, pigtail macaques *Macaca nemestrina,* Japanese macaques, cynomolgus monkeys, and Formosan rock macaques has been reported.

TRANSMISSION AND EPIDEMIOLOGY. Between 1966 and 1970, explosive outbreaks of SVV occurred in *C. aethiops* at the Liverpool School of Tropical Medicine (Liverpool vervet virus) (Clarkson et al. 1967), in *E. patas* at the Delta Regional Primate Research Center (Delta herpesvirus) (Felsenfeld and Schmidt 1975), and in *M. fascicularis, M. fuscata,* and *M. nemestrina* colonies at the Medical Lake field station of the Washington Regional Primate Research Center (Medical Lake macaque virus) (Blakely et al. 1973; Wenner et al. 1977).

The epidemiology of these outbreaks is poorly understood. On several occasions, they occurred in recently imported animals. Serologic surveys suggested the Medical Lake macaque virus may have originated in Malaysia, and the outbreaks resulted from transmission of the agent from an unidentified reservoir host, possibly the stump-tailed macaque, to more susceptible nonhuman primate hosts.

Reactivation of latent infections in ganglia has been documented in *C. aethiops,* suggesting that, as with other alphaherpesviruses, latently infected animals may periodically shed the virus and infect other animals (Soike et al. 1984; Mahalingam et al. 1992). Once established within a colony, the virus may spread rapidly to susceptible individuals, possibly by the respiratory route.

CLINICAL SIGNS. In natural outbreaks, the clinical course was characterized by the occurrence of a disseminated, vesicular exanthema followed by death within 48 hours. The morbidity and mortality rates were high. The disease in macaques appears to be slightly less severe than in *C. aethiops* and *E. patas.*

PATHOGENESIS. Experimental infection of *C. aethiops* results in a vesicular dermatitis 6–8 days after inoculation (Roberts et al. 1984; Dueland et al. 1992). Viral antigen becomes widely disseminated by 8 days after inoculation and can be demonstrated in the liver, lungs, spleen, adrenal gland, kidney, lymph node, skin, and trigeminal ganglion.

PATHOLOGY. Cutaneous lesions are characterized microscopically by formation of multiple intraepidermal vesicles containing necrotic cell debris, erythrocytes, and rare syncytial cells. Characteristic Cowdry type-A intranuclear inclusion bodies are generally present in keratinocytes adjacent to the vesicle. Within the underlying dermis, there is often a necrotizing vasculitis.

The liver often contains multiple individual to coalescing foci of necrosis, and intranuclear inclusions are commonly present in hepatocytes surrounding areas of necrosis. Similar necrotizing lesions may be present in the lung and gastrointestinal tract. The liver and lung lesions are generally the most severe lesions in SVV infections.

DIAGNOSIS. Clinical laboratory findings in experimentally inoculated animals include a marked neutrophilic leukocytosis, a thrombocytopenia, and elevations in alanine aminotransferase, aspartate aminotransferase, and blood urea nitrogen. Virus isolation is essential to establish a definitive diagnosis.

CONTROL AND TREATMENT. Treatment has not been reported in the naturally occurring outbreaks. Acyclovir and interferon have shown some beneficial effect in experimentally infected animals (Arvin et al. 1983; Soike and Gerone 1995).

PUBLIC HEALTH CONCERNS. There are no reports of the transmission of SVV to human beings.

Chimpanzee Varicella Herpesvirus. A varicella-zoster (chickenpox)-like herpesvirus has been isolated from three juvenile chimpanzees *Pan troglodytes* that developed a mild, self-limiting vesicular dermatitis (McClure and Keeling 1971). This virus was more closely related antigenically to human varicella-zoster than to the SVVs, but whether it represented transmission from an infected human to chimpanzees is not known (Harbour and Caunt 1979). Similar cases have been reported in gorillas *Gorilla gorilla* and orangutans *Pongo pygmaeus* (Heuschele 1960; White et al. 1972).

Herpesvirus platyrrhinae

INTRODUCTION. *Herpesvirus platyrrhinae* (saimirine herpesvirus 1; herpes T; *Herpesvirus tamarinus*) infection has many biologic and clinicopathologic features in common with *H. hominis* (*simplex*) infection of humans. This virus exists commonly as a latent, asymptomatic infection in squirrel monkeys *Saimiri sciureus,* but when transmitted naturally or experimentally to owl monkeys *Aotus* spp. and several species of marmosets *Callithrix* spp. and tamarins *Saguinus* spp., causes an acutely fatal disease (Hunt and Melendez 1966; King et al. 1967).

TRANSMISSION AND EPIDEMIOLOGY. Squirrel monkeys become infected early in life and harbor the virus asymptomatically within sensory ganglia. Periodic reactivation and shedding of infectious virus in oral secretions constitutes the primary source of infection. Serologic surveys demonstrated antibodies to *H. platyrrhinae* in asymptomatic spider monkeys *Ateles* spp., capuchin monkeys *Cebus* spp., and woolly monkeys *Lagothrix* spp., suggesting that these animals may represent additional reservoir hosts in free-ranging populations. Spontaneous infection of laboratory-maintained owl monkeys, marmosets, and tamarins is usually a consequence of accidental exposure to latently infected reservoir hosts. Once introduced into a colony, intraspecies spread may result in an epizootic with high mortality.

CLINICAL SIGNS. In carrier species, infection is generally not associated with clinical signs, and only rarely are oral vesicles and ulcers evident (King et al. 1967). In other susceptible species (owl monkeys, marmosets, and tamarins), inadvertent infection results in an epizootic of high mortality with variable oral, labial, and cutaneous lesions. Clinical signs include pruritus, cutaneous and oral ulceration, anorexia, and depression. Death typically occurs within 24–48 hours.

PATHOGENESIS. Primary infection of squirrel monkeys is characterized by oral and labial vesicles, necrotic plaques, and ulcers that resolve within 10 days. As with *H. hominis* infection in humans, during resolution of the lesions the virus is transported to the trigeminal ganglia, where it persists latently for the life of the animal, with periodic episodes of virus shedding in oral secretions.

In owl monkeys, marmosets, and tamarins, following a 7–10-day incubation period, the virus becomes widely disseminated and causes widespread cytolytic infection of the skin, oral mucosa, and most visceral organs, leading to death.

PATHOLOGY. Cutaneous lesions are characterized by full-thickness necrosis of the epidermis. Adjacent to the necrotic areas, a few viable epithelial cells may remain beneath a coagulum of eosinophilic exudate containing pyknotic debris. Sebaceous glands, hair follicles, and apocrine glands are usually spared. There is mild parakeratosis and intercellular edema within the adjacent epidermis, and scattered multinucleated giant cells containing intranuclear viral inclusions are present adjacent to the necrotic foci. Due to the acutely cytolytic nature of the infection, inflammatory reactions within the dermis may be minimal, consisting only of scattered neutrophils.

Foci of full-thickness necrosis similar to those present in the skin are frequently found in the mucosa of the oral cavity and small and large intestines. In the oral cavity, raised plaques of necrotic epithelium eventually slough, leaving ulcers. Numerous focal areas of necrosis are present in the liver, spleen, lung, kidney, and adrenal gland, and prominent Cowdry type-A intranuclear inclusions are typically present in cells in these organs. When present, encephalitis is usually mild.

DIAGNOSIS. The lesions are essentially identical to those that occur in *H. hominis* (*simplex*) infection in these species; hence, virus isolation and characterization are required to differentiate them.

TREATMENT AND PREVENTION. Contact between susceptible and carrier species should be avoided. A live vaccine is effective in owl monkeys; however, infrequent episodes of vaccine-induced disease have occurred (Daniel et al. 1967). These were characterized by rapidly progressive disseminated infection similar to that seen in the natural disease. In other instances, a slowly progressive ascending myelitis occurred.

Herpesvirus hominis (*H. simplex*)

INTRODUCTION. *Herpesvirus hominis* (*simplex*) infection is a common asymptomatic infection of humans. Inadvertent infection of gibbons *Hylobates lar,* gorillas, tree shrews *Tupaia glis,* and chimpanzees has been described (Smith et al. 1969; Emmons and Lennette 1970; McClure et al. 1972, 1980; Marennikova et al. 1973). In these species, infection usually results in mild self-limiting oral vesicular lesions. Conversely, infection of owl monkeys results in a lethal, disseminated disease similar to that caused by *H. platyrrhinae,* from which it must be differentiated (Hunt and Melendez 1966; Melendez et al. 1969). Although natural infection of callithricids has not been reported (Potkay 1992), intravaginal inoculation of tamarins (*Saguinus oedipus* and *S. fuscicollis*) with *H. simplex* 2 resulted in disseminated disease (Felsburg et al. 1973). Experimental inoculation of capuchin monkeys produced localized disease (Felsburg et al. 1972).

ETIOLOGY. Two distinct subtypes of *H. hominis* (*simplex*) are recognized. *Herpes hominis* (*simplex*) type 1 (HSV-1) is most often responsible for oral lesions and encephalitis in adult people, whereas *H. hominis* (*simplex*) type 2 is responsible for a sexually transmitted disease causing a genital infection in adults and a disseminated infection in infants. Both types are equally pathogenic in experimentally inoculated owl monkeys, and no difference in the clinical disease has been noted (Katzin et al. 1967; Melendez et al. 1969).

An alphaherpesvirus closely related to, but antigenically distinct from HSV-2, has been identified by serology in free-ranging mountain gorillas *G. gorilla beringei* but has not been associated with clinical signs (Eberle 1992).

TRANSMISSION AND EPIDEMIOLOGY. Nonhuman primates are not naturally infected with *H. hominis* (*simplex*) and likely acquire the infection through human contact. Once established in owl monkey colonies, the virus spreads rapidly and results in high morbidity and mortality, though a natural epizootic in a research colony of gibbons was characterized by a more limited spread (Smith et al. 1969; Emmons and Lennette 1970).

PATHOGENESIS. The pathogenesis of *H. hominis* (*simplex*) infection in owl monkeys is essentially identical to that of *H. platyrrhinae* infection in this species, except that encephalitis may be a more frequent sequel.

PATHOLOGY. Multifocal necrotizing and vesicular dermatitis is often most severe on facial skin and accompanied by blepharitis and stomatitis. In gibbons, a multifocal acute meningoencephalitis may be evident in the pons and cerebral cortex. These changes may be accompanied by necrosis, reactive gliosis, and typical Cowdry type-A inclusions.

CLINICAL SIGNS. In gorillas, chimpanzees, and gibbons, infection usually is self-limiting, and clinical signs are restricted to vesiculation and ulceration of mucosal surfaces. During one natural outbreak in gibbons, viral encephalitis developed in a minority of animals well after oral lesions healed (Emmons and Lennette 1970). Generalized disease in susceptible species is identical to that induced by *H. platyrrhinae.*

TREATMENT AND PREVENTION. Protective clothing and face masks should be worn by animal care personnel. A modified-live vaccine proved protective in owl monkeys (Daniel et al. 1978).

Herpesvirus papionis (Simian Agent 8, SA-8)

INTRODUCTION. Herpesvirus papionis (SA-8), originally isolated from African green monkey neural tissue, is antigenically related to *H. simiae* (Malherbe and Harwin 1958). Common in baboons *Papio* spp., it has many features resembling *H. hominis* (*simplex*) 2 infection in humans.

TRANSMISSION AND EPIDEMIOLOGY. *Herpesvirus papionis* is a common asymptomatic infection of baboons (Eichberg et al. 1976). As with other alphaherpesvirus infections, the virus may persist latently within sensory ganglia (Kalter et al. 1978). Recrudescence may occur periodically, often associated with exposure to stress, ultraviolet light, or cold; and the virus is shed in oral and genital secretions. A single outbreak in baboons suggested that the virus was likely transmitted venereally and, as such, may represent a model of genital herpesvirus infection (Levin et al. 1988).

PATHOGENESIS. The pathogenesis of *H. papionis* infection is most similar to *H. hominis* (*simplex*) 2 infection of humans. Skin lesions include hemorrhagic ulcers, vesicles, and pustules.

PATHOLOGY. Following experimental inoculation of infant baboons, a systemic infection occurred with fibrinonecrotic alveolitis and multifocal hepatic necrosis. Intranuclear inclusions identical to those described with *H. simiae* (herpes B) were noted (Eichberg et al. 1976).

CLINICAL SIGNS. Many individuals carry the virus asymptomatically. During primary infection or recrudescence, small vesicles and pustules may be found on the genital and, less frequently, oral mucous membranes (Levin et al. 1988). Genital lesions were occasionally severe, involving the vulvar, penile, or perineal tissues and accompanied by an inguinal lymphadenopathy. Oral lesions were less severe, although a hemorrhagic gingivitis was occasionally noted.

BETAHERPESVIRINAE

Cytomegaloviruses

INTRODUCTION AND HOST RANGE. Cytomegaloviruses (CMVs) cause a common asymptomatic infection of humans and many nonhuman primates. Infection of macaques, capuchin monkeys, woolly monkeys, squirrel monkeys, chimpanzees, baboons, white-lipped tamarins *S. fuscicollis,* and African green monkeys has been demonstrated (Rangan and Chaiban 1980; Davis et al. 1992). Although viruses within this group are generally believed to have a narrow host range, interspecies transmission does occur.

TRANSMISSION AND EPIDEMIOLOGY. Cytomegalovirus infection is common in nonhuman primates and usually is not associated with disease. Transmission is horizontal. The virus is spread in a variety of body secretions, including saliva, blood, urine, milk, and semen and may be shed for extended periods (Asher et al. 1974). Macaques generally become infected within the first year of life (Vogel et al. 1994). Disease occurs in immunocompromised individuals or following intrauterine infection. Rhesus macaques naturally infected with rhesus CMV are susceptible to infection by an antigenically distinct African green monkey CMV and commonly become infected with both strains during captivity (Swack and Hsuing 1982).

PATHOGENESIS. Cytomegalovirus persists as a latent infection and may periodically be shed in bodily secretions. In macaques immunosuppressed by concomitant viral infection (SIV or retrovirus type D) or drug therapy (cyclophosphamide, cortisone, and antithymocyte globulin), reactivation of the virus may be associated with disseminated lesions in the brain, lymph nodes, liver, spleen, kidney, small intestine, nervous system, and arteries.

The pathogenesis of CMV infection in SIV-inoculated macaques shares many similarities with the disease in human patients with AIDS. In SIV-infected macaques, reactivation typically occurs as a terminal opportunistic infection associated with severe CD4+ lymphocyte depletion and is manifest by a necrotizing enteritis, encephalitis, lymphadenitis, and/or pneumonitis (Baskin 1987).

PATHOLOGY. Pulmonary lesions are common and consist of a multifocal to coalescent interstitial pneumonia. Cytomegaly and large, intranuclear Cowdry type-A inclusion bodies may be evident in alveolar septa and septal lining cells. Similar lesions may be found in the liver, spleen, kidney, and testes. Smaller amphophilic, intracytoplasmic inclusions may also be found in some of the cells bearing intranuclear inclusions.

Central nervous system lesions are multifocal, involve primarily the leptomeninges and subjacent neuropil, and are characterized by neutrophilic infiltrates with necrosis and fibrinous exudates. These findings are accompanied by characteristic viral inclusions and often a nonsuppurative, perivascular meningoencephalitis. In the gastrointestinal tract, hemorrhage and, in particular, neutrophilic infiltrates may be prominent. In all locations, these findings may be accompanied by a necrotizing and proliferative vasculitis.

Intrauterine infection of the fetus subsequent to experimental infection of the pregnant female has been reported in rhesus macaques (London et al. 1986) and squirrel monkeys (Ordy et al. 1981).

CLINICAL SIGNS. Infection is usually asymptomatic. Immunosuppressed animals may experience reactivation and dissemination of the virus. In these individuals, clinical signs relate to the anatomic site(s) involved and may include dyspnea, diarrhea, melena, and neurologic signs.

DIAGNOSIS. In situ hybridization often reveals many more cells to be infected than would be anticipated on routine stains and may be useful for diagnosis when only equivocal changes (i.e., mild cytomegaly) are present. It is not routinely available.

GAMMAHERPESVIRINAE

Lymphocryptoviruses. The genus *Lymphocryptovirus* contains a number of species, including Epstein-Barr virus (EBV, human herpesvirus 4), *Herpesvirus pan, H. pongo, H. papio,* and *H. gorilla* (Levy et al. 1971; Falk et al. 1976; Gerber et al. 1977; Rabin et al. 1980). These primate viruses share approximately 35%–45% homology with each other and with EBV (Dillner et al. 1987). Two other herpesviruses, previously referred to as "rhesus leukocyte-associated herpesviruses," currently also are classified as lymphocryptoviruses (Bissell et al. 1973; Frank et al. 1973). These agents occur commonly as latent, asymptomatic infections of rhesus macaque lymphocytes but exhibit no antigenic relationship to human EBV or other known simian herpesviruses.

In humans, EBV infection is usually asymptomatic but may cause infectious mononucleosis. Epstein-Barr virus has also been associated with Burkitt's lymphoma, with nasopharyngeal carcinoma, and with an oral lesion, termed *hairy leukoplakia,* in patients infected with human immunodeficiency virus (Rickinson and Kieff 1996).

HOST RANGE. These viruses apparently lack strict host specificity, since a number of New World primate species are susceptible to infection with human isolates. In one survey, antibodies to EBV were detected in a large number of captive and wild Old World primates (Ishida and Yamamoto 1987). In general, it is accepted that EBV-like viruses are common in Old World primates but not so in New World species and prosimians, though cotton-top tamarins *Saguinus oedipus* are susceptible to experimental infection with human EBV and develop multifocal, large cell lymphoma (Niedobitek et al. 1994).

TRANSMISSION AND EPIDEMIOLOGY. Epidemiology of lymphocryptovirus infections in Old World primates resembles that in people. Infection occurs primarily in the young through contact with infected oral secretions and usually is not associated with clinical signs. In cynomolgus monkeys, infection occurs after the disappearance of maternal immunity; by 1 year of age, virtually all animals harbor the virus (Fujimoto and Honjo 1991). The virus infects B lymphocytes and persists for life, usually in a latent form. Some individuals periodically shed virus and serve as a source of infection for uninfected individuals. A small number of New World species have antibodies to EBV, but the method of transmission and the epidemiology in these species are unknown.

PATHOGENESIS. As members of the Gammaherpesvirinae, lymphocryptoviruses have a particular tropism for lymphocytes. Infection of Old World species is common and usually not associated with disease. Epstein-Barr-like viruses have been associated with malignant lymphoma in baboons (Deinhardt et al. 1978) and macaques (Rangan et al. 1986).

PATHOLOGY. Macaques coinfected with SIV and EBV have been proposed as a model of AIDS-associated Burkitt's lymphoma (Feichtinger et al. 1992). In New World species (*S. oedipus, S. fuscicollis,* and *C. jacchus*), experimental inoculation of some lymphocryptoviruses produces B-cell lymphomas and death, usually 1–2 months after inoculation (Miller et al. 1977).

Proliferative squamous epithelial lesions resembling oral hairy leukoplakia, and containing typical herpesvirus virions, were described in rhesus macaques inoculated with SIV and succumbing to AIDS (Baskin et al. 1995).

Although EBV induction of lymphoma in New World monkeys may closely parallel ateline and saimirine herpes virus infection (see below) (Cameron et al. 1987), its association with lymphoma in Old World species is usually associated with immunodeficient states caused by immunosuppressive drugs or SIV infection, a situation similar to that in human transplant patients and HIV-infected individuals (Feichtinger et al. 1992).

PUBLIC HEALTH CONCERNS. Infection of humans with simian EBV-like agents has not been reported.

Rhadinoviruses (*Herpesvirus saimiri, Herpesvirus ateles*)

INTRODUCTION. While *H. saimiri* and *H. ateles* are common asymptomatic infections of their natural hosts, inoculation into appropriate susceptible hosts results in rapid development of malignant lymphoma or leukemia. Evidence of the ability of these viruses to induce disease in naturally infected hosts is more limited (Hunt et al. 1973).

TRANSMISSION AND EPIDEMIOLOGY. A large proportion of wild squirrel monkeys are infected with *H. saimiri,* and nearly all animals are seropositive by 1.5–2.0 years of age. Similarly, approximately 50% of spider monkeys are seropositive for *H. ateles.* While squirrel and spider monkeys represent the natural reservoir hosts for these two viruses, little is known about the natural history and epidemiologic factors involved in their transmission. Transmission is thought to be horizontal through oral secretions (Falk et al. 1976).

PATHOGENESIS. *Herpesvirus saimiri* and *H. ateles* are both lymphomagenic when inoculated experimentally into appropriate hosts, but *H. saimiri* has been best studied. Recent work focused on genes encoding saimiri transformation-associated protein (STP) and its ability to acutely transform cells in vitro and in vivo (Jung and Desrosiers 1992). Experimental inoculation of *H. saimiri* into owl monkeys *Aotus spp.;* several species of tamarins and marmosets *S. oedipus, S. fuscicollis, S. nigricollis, S. mystax,* and *C. jacchus;* howler monkeys *Alouatta caraya;* and spider monkeys results in lymphoma or lymphocytic leukemia.

Following inoculation, the time to development of lymphoma is variable but may be as soon as 3 weeks. Generally, one of three patterns of disease occurs: (1) survival less than 40 days with development of disseminated lymphoma, often associated with extensive necrosis and replacement of vital organs; (2) survival from 50–150 days with a less aggressive form of lymphoma involving multiple organs and associated with lymphocytic leukemia; or (3) survival over 150 days with localized well-differentiated lymphocytic lymphoma (Hunt et al. 1976).

Limited studies indicate that *H. saimiri* is oncogenic in African green monkeys but not in macaques *M. arctoides* and *M. mulatta* (Melendez et al. 1970).

Experimental inoculation of *H. ateles* into tamarins, marmosets, and owl monkeys induces lymphoma and leukemia. Although the virus has not been studied as extensively, the oncogenic properties of *H. ateles* appear to be similar to *H. saimiri.*

PATHOLOGY. Microscopically, the cells comprising the neoplastic infiltrate vary considerably in their degree of differentiation. In the most aggressive form, the neoplastic cells are large, pleomorphic, and resemble reticulum cells or histiocytes. Many are polygonal to stellate and have a moderate amount of granular cytoplasm. In less aggressive forms, the cells are more differentiated, resembling lymphocytes. Neoplastic cells are polyclonal and are of T-lymphocyte origin.

Ultrastructurally, bizarre convoluted nuclei are seen, and viral particles are not evident. Virus may be recovered in cell culture from tumor explants.

CLINICAL SIGNS. Infection of the natural hosts of *H. saimiri* and *H. ateles* is not associated with clinical signs.

TREATMENT AND PREVENTION. Natural infection of callithricids has been reported infrequently. Exposure to tissue or blood products is probably essential to produce infection in inappropriate hosts.

PUBLIC HEALTH CONCERNS. *Herpesvirus saimiri* induces malignant lymphoma when inoculated into a variety of New World primate species and rabbits (Melendez et al. 1970). The virus also was shown to infect and transform human T lymphocytes in vitro, but there is no evidence that people can be infected or develop tumors as a consequence of exposure to this agent.

Herpesvirus aotus. Several herpesviruses have been isolated from primary owl monkey cell lines. These tissues originated from healthy monkeys, and the animals were asymptomatically and latently infected. *Herpesvirus aotus* type 1 (aotine herpesvirus 1) and *H. aotus* type 3 (ateline herpesvirus 3) belong to the betaherpesvirinae subfamily (Daniel et al. 1973). *Herpesvirus aotus* type 2 (aotine herpesvirus 2) belongs to the gammaherpesvirus subfamily (Barahona et al. 1973). No natural or experimental disease has been associated with them.

Herpesvirus saguinus. A herpesvirus was isolated from marmoset kidney cell cultures but has not been associated with clinical disease (Melendez et al. 1970).

LITERATURE CITED

Arvin, A.M., D.P. Martin, E.A. Gard, and T.C. Merigan. 1983. Interferon prophylaxis against simian varicella in *Erythrocebus patas* monkeys. *Journal of Infectious Diseases* 147:149–154.

Asher, D.M., C.J. Gibbs, D.J. Lang, and D.C. Gajdusek. 1974. Persistent shedding of cytomegalovirus in the urine of healthy rhesus monkeys. *Proceedings of the Society for Experimental Biology* 145:794–801.

Barahona, H.H., L.V. Melendez, N.W. King, M.D. Daniel, C.E.O. Fraser, and A.C. Preville. 1973. *Herpesvirus aotus* type 2: A new viral agent from owl monkeys (*Aotus trivirgatus*). *Journal of Infectious Diseases* 127:171–178.

Baskin, G.B. 1987. Disseminated cytomegalovirus infection in immunodeficient rhesus monkeys. *American Journal of Pathology* 129:345–352.

Baskin, G.B., E.D. Roberts, D. Kuebler, L.N. Martin, B. Blauw, J. Heeney, and C. Zurcher. 1995. Squamous epithelial proliferative lesions associated with rhesus Epstein-Barr virus in simian immunodeficiency virus-infected rhesus macaques. *Journal of Infectious Diseases* 172:535–538.

Bissel, J.A., A.L. Frank, N.R. Dunnick, D.S. Rowe, M.A. Conliffe, P.D. Prakman, and H.M. Meyer Jr. 1973. Rhesus leukocyte-associated herpesvirus: II. Natural and experimental infection. *Journal of Infectious Diseases* 128:630–637.

Blakely, G.A., B. Lourie, and W.G. Morton. 1973. A varicella-like disease in cynomolgus macaque monkeys. *Journal of Infectious Diseases* 127:617–625.

Cameron, K.R., T. Stamminger, M. Craxton, W. Bodemer, R.W. Honess, and B. Fleckenstein. 1987. The 160,000-Mr protein encoded at the right end of the *Herpesvirus saimiri* genome is homologous to the 140,000-Mr membrane antigen encoded at the left end of EBV genome. *Journal of Virology* 61:2063–2070.

Centers for Disease Control and Prevention. 1987. Guidelines for prevention of *Herpesvirus simiae* (B virus) infection in monkey handlers. *Morbidity and Mortality Weekly Report* 36:680–689.

Chellman, G.J., V.S. Lukas, E.M. Eugui, K.P. Altera, S.J. Almquist, and J.K. Hilliard. 1992. Activation of B virus

(*Herpesvirus simiae*) in chronically immunosuppressed cynomolgus monkeys. *Laboratory Animal Science* 42:146–151.

Clarkson, M.J., E. Thorpe, and K. McCarthy. 1967. A virus disease of captive vervet monkeys (*Cercopithecus aethiops*) caused by a new herpesvirus. *Archiv fur die Gesamte Virusforschung* 22:219–234.

Daniel, M.D., A. Karpas, L.V. Melendez, N.W. King, and R.D. Hunt. 1967. Isolation of herpes T virus from spontaneous disease in squirrel monkeys (*Saimiri sciureus*). *Archiv fur die Gesamte Virusforschung* 227:324–331.

Daniel, M.D., L.V. Melendez, N.W. King, H.H. Barahona, C.E.O. Fraser, F.G. Garcia, and D. Silva. 1973. Isolation and characterization of a new virus from owl monkeys: *Herpesvirus aotus* type 3. *American Journal of Physical Anthropology* 38:497–500.

Daniel, M.D., H. Barahona, L.V. Melendez, R.D. Hunt, P.K. Sehgal, B. Marshall, J. Ingalls, and M. Forbes. 1978. Prevention of fatal herpes infections in owl and marmoset monkeys by vaccination. In *Recent advances in primatology: 4. Medicine,* ed. D.J. Chivers and E.H.R. Ford. New York: Academic, pp. 67–69.

Davenport, D.S., D.R. Johnson, G.P. Holmes, D.A. Jewett, S. Ross, and J.K. Hilliard. 1994. Diagnosis and management of human B virus (*Herpes simiae*) infections in Michigan. *Clinical Infectious Disease* 19:33–41.

Davis, K.J., G.B. Hubbard, K.F. Soike, and T.M. Butler. 1992. Fatal necrotizing adenoviral hepatitis in a chimpanzee (*Pan troglodytes*) with disseminated cytomegalovirus infection. *Veterinary Pathology* 29:547–549.

Deinhardt, F., L.G. Falk, A. Wolfe, A. Schudel, and L. Yakovleva. 1978. Susceptibility of marmosets to Epstein-Barr virus-like baboon herpesviruses. *Primate Medicine* 10:163–170.

Di Giacomo, R.F., and K.V. Shah. 1972. Virtual absence of infection with *Herpesvirus simiae* in colony-reared rhesus monkeys (*Macaca mulatta*), with a literature review on antibody prevalence in natural and laboratory rhesus populations. *Laboratory Animal Science* 22:61–67.

Dillner, J., H. Rabin, N. Letvin, G. Henle, and G. Klein. 1987. Nuclear DNA binding proteins determined by the Epstein-Barr virus related simian lymphotropic herpesviruses: *H. gorilla, H. pan, H. pongo,* and *H. papio. Journal of General Virology* 68:1587–1596.

Dueland, A.N., J.R. Martin, M.E. Devlin, M. Wellish, R. Mahalingam, R. Cohrs, K.F. Soike, and D.H. Gilden. 1992. Acute simian varicella infection: Clinical, laboratory, pathologic and virologic features. *Laboratory Investigation* 66:762–773.

Eberle, R. 1992. Evidence for an alphaherpesvirus indigenous to mountain gorillas. *Journal of Medical Primatology* 21:246–251.

Eichberg, J.W., B. McCullough, S.S. Kalter, D.E. Thor, and A.R. Rodriguez. 1976. Clinical, virological and pathological features of herpesvirus SA8 infection in conventional and gnotobiotic infant baboons (*Papio cynocephalus*). *Archives of Virology* 50:255–270.

Emmons, R.W., and E.H. Lennette. 1970. Natural *Herpesvirus hominis* infection of a gibbon (*Hylobates lar*). *Archiv fur die Gesamte Virusforschung* 31:215–218.

Espana, C. 1973. *Herpes simiae* infection in *Macaca radiata. American Journal of Physical Anthropology* 38:447–454.

Falk, L., F. Deinhardt, M. Nonoyama, L.G. Wolfe, C. Berholz, B. Lapin, L. Yakovleva, V. Agrba, G. Henle, and W. Henle. 1976. Properties of a baboon lymphotropic herpes virus related to Epstein-Barr virus. *International Journal of Cancer* 18:798–807.

Feichtinger, H., S. Li, E. Kaaya, P. Putkonen, K. Grunewald, K. Weyrer, D. Bottiger, I. Emberg, A. Linde, G. Biberfeld, and P. Biberfeld. 1992. A monkey model of Epstein-Barr virus associated lymphomagenesis in human acquired immunodeficiency syndrome. *Journal of Experimental Medicine* 176:281–286.

Felsburg, P.J., R.L. Heberling, and S.S. Kalter. 1972. Experimental genital infection of cebus monkeys with oral and genital isolates of *Herpesvirus hominis* types 1 and 2. *Archiv fur die Gesamte Virusforschung* 39:223–227.

Felsburg, P.J., R.L. Heberling, M. Brack, and S.S. Kalter. 1973. Experimental genital infection of the marmoset. *Journal of Medical Primatology* 2:50–60.

Felsenfeld, A.D., and N.J. Schmidt. 1975. Immunological relationship between delta herpesvirus of patas monkeys and varicella-zoster virus of humans. *Infection and Immunity* 12:261–266.

Frank, A.L., J.A. Bissell, D.S. Rowe, N.R. Dunnick, R.E. Mayner, H.E. Hopps, P.D. Parkman, and H.M. Meyer Jr. 1973. Rhesus leukocyte-associated herpesvirus: I. Isolation and characterization of a new herpesvirus recovered from rhesus monkey leukocytes. *Journal of Infectious Diseases* 128:618–629.

Freifeld, A.G., J. Hilliard, J. Southers, M. Murray, B. Savarese, J.M. Schmitt, and S.E. Strauss. 1995. A controlled seroprevalence survey of primate handlers for evidence of asymptomatic herpes B virus infection. *Journal of Infectious Diseases* 171:1031–1034.

Fujimoto, K., and S. Honjo. 1991. Presence of antibody to Cyno-EBV in domestically bred cynomolgus monkeys (*Macaca fascicularis*). *Journal of Medical Primatology* 20:42–45.

Gay, F.P., and M. Holden. 1933. The herpes encephalitis problem. *Journal of Infectious Diseases* 53:287–303.

Gerber, P., S.S. Kalter, G. Schildlovsky, W.D. Peterson, and M.D. Daniel. 1977. Biologic and antigenic characteristics of Epstein-Barr related herpesviruses of chimpanzees and baboons. *International Journal of Cancer* 20:448–459.

Harbour, D.A., and A.E. Caunt. 1979. The serological relationship of varicella-zoster virus to other primate herpesviruses. *Journal of General Virology* 45:469–477.

Heuschele, W.P. 1960. Varicella (chicken pox) in three young anthropoid apes. *Journal of the American Veterinary Medical Association* 136:256–257.

Holmes, G.P., J.K. Hilliard, K.C. Klontz, A.H. Rupert, C.M. Schindler, E. Parrish, G. Griffin, G.S. Ward, N.D. Bernstein, T.W. Bean, M.R. Ball, J.A. Brady, and M.H. Wilder. 1990. B virus (*Herpesvirus simiae*) infection in humans: Epidemiologic investigation of a cluster. *Annals of Internal Medicine* 112:833–839.

Holmes, G.P., L.E. Chapman, J.A. Stewart, J.A. Straus, S.E. Straus, J.K. Hilliard, and D.S. Davenport. 1995. Guidelines for the prevention and treatment of B-virus infections in exposed persons. *Clinical Infectious Disease* 20:421–439.

Hull, R.N. 1973. The simian herpesviruses. In *The herpesviruses,* ed. A.S. Kaplan. New York: Academic, pp. 389–426.

Hunt, R.D., and B. Blake. 1994. Herpesvirus B infection. In *Monographs on the pathology of laboratory animals.* Vol. 1, *Nonhuman primates,* ed. T.C. Jones, U. Mohr, and R.D. Hunt. Berlin: Springer-Verlag, pp. 78–81.

Hunt, R.D., and L.V. Melendez. 1966. Spontaneous herpes-T infection in the owl monkey (*Aotus trivirgatus*). *Veterinary Pathology* 3:1–26.

Hunt, R.D., F.G. Garcia, H.H. Barahona, N.W. King, C.E.O. Fraser, and L.V. Melendez. 1973. Spontaneous *Herpesvirus saimiri* lymphoma in an owl monkey. *Journal of Infectious Diseases* 127:723–725.

Hunt, R.D., B. Blake, and M.D. Daniel. 1976. *Herpesvirus saimiri* lymphoma in owl monkeys (*Aotus trivirgatus*): Susceptibility, latent period, hematologic picture and course. *Theriogenology* 6:139–151.

Ishida, T., and K. Yamamoto. 1987. Survey of nonhuman primates for antibodies reactive with Epstein-Barr virus (EBV) antigens and susceptibility of their lymphocytes for immortalization with EBV. *Journal of Medical Primatology* 16:359–371.

Jung, J.U., and R.C. Desrosiers. 1992. *Herpesvirus saimiri* oncogene STP-488 encodes a phospholipoprotein. *Journal of Virology* 66:1777–1780.

Kalter, S.S., S.A. Weiss, L. Heberling, J.E. Guajardo, and G.C. Smith. 1978. The isolation of herpesvirus from trigeminal ganglia of normal baboons (*Papio cynocephalus*). *Laboratory Animal Science* 28:705–709.

Katzin, D.S., J.D. Connor, L.A. Wilson, and R.S. Sexton. 1967. Experimental *Herpes simplex* infection in the owl monkey. *Proceedings of the Society for Experimental Biology and Medicine* 125:391–398.

Keeble, S.A. 1960. B virus infection in monkeys. *Annals of the New York Academy of Sciences* 85:960–969.

King, N.W., R.D. Hunt, M.D. Daniel, and L.V. Melendez. 1967. Overt herpes-T infection in squirrel monkeys (*Saimiri sciureus*). *Laboratory Animal Care* 17:413–423.

Lackner, A.A. 1994. Pathology of simian immunodeficiency virus induced disease. *Current Topics in Microbiology and Immunology* 188:35–64.

Levin, J.L., J.K. Hilliard, S.L. Lipper, T.M. Butler, and W.J. Goodwin. 1988. A naturally occurring epizootic of simian agent 8 in the baboon. *Laboratory Animal Science* 38:394–397.

Levy, J.A., S.B. Levy, Y. Hirshaut, G. Kafuko, and A. Prince. 1971. Presence of EBV antibodies in sera from wild chimpanzees. *Nature* 233:559–560.

London, W.T., A.J. Martinez, S.A. Houff, W.C. Wallen, B.L. Curfman, R.G. Traub, and J.L. Sever. 1986. Experimental congenital disease with simian cytomegalovirus in rhesus monkeys. *Teratology* 33:323–331.

Loomis, M.R., T. O'Neill, M. Bush, and R.J. Montali. 1981. Fatal herpesvirus infection in patas monkeys and a black and white colobus monkey. *Journal of the American Veterinary Medical Association* 179:1236–1239.

Mahalingam, R., P. Clarke, M. Wellish, A.N. Dueland, K.F. Soike, D.H. Gilden, and R. Cohrs. 1992. Prevalence and distribution of latent simian varicella virus DNA in monkey ganglia. *Virology* 188:193–197.

Malherbe, H., and R. Harwin. 1958. Neurotropic virus in African monkeys. *Lancet* 2:530.

Marennikova, S.S., V.I. Maltseva, E.M. Shelukhina, L.S. Shenkman, and V.I. Korneeva. 1973. A generalized herpetic infection simulating smallpox in a gorilla. *Intervirology* 2:280–286.

McClure, H.M., and M.E. Keeling. 1971. Viral diseases noted at the Yerkes Primate Center colony. *Laboratory Animal Science* 21:1002–1010.

McClure, H.M., M.E. Keeling, B. Olberling, R.D. Hunt, and L.V. Melendez. 1972. Natural *Herpesvirus hominis* infection of tree shrews (*Tupaia glis*). *Laboratory Animal Science* 22:517–521.

McClure, H.M., R.B. Swenson, S.S. Kalter, and T.L. Lester. 1980. Natural genital *Herpesvirus hominis* infection in chimpanzees. *Laboratory Animal Science* 30:895–901.

Melendez, L.V., C. Espana, R.D. Hunt, and M.D. Daniel. 1969. Natural *Herpes simplex* infection in the owl monkey (*Aotus trivirgatus*). *Laboratory Animal Care* 19:38–45.

Melendez, L.V., R.D. Hunt, M.D. Daniel, and B.F. Trum. 1970. New World monkeys, herpes viruses, and cancer. In *Infections and immunosuppression in subhuman primates,* ed. H. Balner and W.I.B. Beveridge. Copenhagen: Munksgaard, pp. 111–117.

Miller, G., T. Shope, D. Coope, L. Waters, J. Pagano, G.W. Bornkamm, and W. Henle. 1977. Lymphoma in cotton-topped marmosets after inoculation with Epstein-Barr virus: Tumor incidence, histologic spectrum, antibody responses, demonstration of viral DNA and characterization of virus. *Journal of Experimental Medicine* 145:948–967.

Niedobitek, G., A. Agathanggelou, S. Finerty, R. Tierney, P. Watkins, E.L. Jones, A. Morgan, L.S. Young, and N. Rooney. 1994. Latent Epstein-Barr virus infection in cottontop tamarins. *American Journal of Pathology* 145:969–978.

Olson, L.C., W.H. Pryor, and J.M. Thomas. 1991. Persistent reduction of B virus (*Herpesvirus simiae*) seropositivity in rhesus macaques acquired for a study of renal allograft tolerance. *Laboratory Animal Science* 41:540–544.

Ordy, J.M., S.R. Rangan, R.H. Wolf, C. Knight, and W.P. Dunlap. 1981. Congenital cytomegalovirus effects on postnatal neurologic development of squirrel monkey (*Saimiri sciureus*) offspring. *Experimental Neurology* 74:728–747.

Palmer, A.E. 1987. B virus, *Herpesvirus simiae:* Historical perspective. *Journal of Medical Primatology* 16:99–130.

Potkay, S. 1992. Diseases of Callitrichidae: A review. *Journal of Medical Primatology* 21:189–236.

Rabin, H., B.C. Strand, R.H. Neubauer, A.M. Brown, R.F. Hopkins, and R.A. Mazur. 1980. Comparisons of nuclear antigens of Epstein-Barr virus (EBV) and EBV-like simian viruses. *Journal of General Virology* 48:265–272.

Rangan, S.R.S., and J. Chaiban. 1980. Isolation and characterization of a cytomegalovirus from the salivary gland of a squirrel monkey (*Saimiri sciureus*). *Laboratory Animal Science* 30:532–540.

Rangan, S.R.S., L.N. Martin, B.E. Bozelka, N. Wang, and B.J. Gormus. 1986. Epstein-Barr virus-related herpesvirus from a rhesus monkey (*Macaca mulatta*) with malignant lymphoma. *International Journal of Cancer* 38:525–432.

Rickinson, A.B., and E. Kieff. 1996. Epstein-Barr virus. In *Fields virology,* vol. 2, ed. B.N. Fields, D.M. Knipe, P.M. Howley, R.M. Chanock, J.L. Melnick, T.P. Monath, B. Roizman, and S.E. Strauss, 3d ed. New York: Raven, pp. 2397–2446.

Roberts, E.D., G.B. Baskin, K. Soike, and S.V. Gibson. 1984. Pathologic changes of experimental simian varicella (Delta herpes) infection in African green monkeys (*Cercopithecus aethiops*). *American Journal of Veterinary Research* 45:523–530.

Roizman, B. 1996. Herpesviridae. In *Fields virology,* vol. 2, ed. B.N. Fields, D.M. Knipe, P.M. Howley, R.M. Chanock, J.L. Melnick, T.P. Monath, B. Roizman, and S.E. Strauss, 3d ed. New York: Raven, pp. 2221–2230.

Simon, M.A., M.D. Daniel, D. Lee-Parritz, N.W. King, and D.J. Ringler. 1993. Disseminated B virus infection in a cynomolgus monkey. *Laboratory Animal Science* 43:545–550.

Smith, P.C., T.M. Yuill, R.D. Buchanan, J.S. Stanton, and V. Chaicumpa. 1969. The gibbon (*Hylobates lar*): A new primate host for *Herpesvirus hominis* I—A natural epizootic in a laboratory colony. *Journal of Infectious Diseases* 120:292–297.

Soike, K.F., and P.J. Gerone. 1995. Acyclovir in the treatment of simian varicella infection of the African green monkey. *American Journal of Medicine* 73:112–117.

Soike, K.F., S.R.S. Rangan, and P.J. Gerone. 1984. Viral disease models of primates. *Advances in Veterinary Science and Comparative Medicine* 28:151–191.

Swack, N.S., and G.D. Hsuing. 1982. Natural and experimental simian cytomegalovirus infections at a primate center. *Journal of Medical Primatology* 11:169–177.

Vogel, P., B.J. Weigler, H. Kerr, A.G. Hendrickx, and P.A. Barry. 1994. Seroepidemiologic studies of CMV infection in a breeding population of rhesus macaques. *Laboratory Animal Science* 44:25–30.

Ward, J.A., and J.K. Hilliard. 1994. B virus-specific pathogen-free (SPF) breeding colonies of macaques: Issues, surveillance, and results in 1992. *Laboratory Animal Science* 44:222–228.

Weigler, B.J. 1992. Biology of B virus in macaques and human hosts: A review. *Clinical Infectious Disease* 14:555–567.

Weigler, B.J., J.A. Robert, D.W. Hird, N.W. Lerche, and J.K. Hilliard. 1990. A cross sectional survey for B virus antibody in a colony of group-housed rhesus macaques. *Laboratory Animal Science* 40:257–261.

Weigler, B.J., D.W. Hird, J.K. Hilliard, N.W. Lerche, J.A. Roberts, and L.M. Scott. 1993. Epidemiology of Cercopithecine herpesvirus 1 (B virus) infection and shedding in a large breeding cohort of rhesus macaques. *Journal of Infectious Diseases* 167:257–263.

Weir, E.C., P.N. Bhatt, R.O. Jacoby, J.K. Hilliard, and S. Morgenstern. 1993. Infrequent shedding and transmission of *Herpesvirus simiae* from seropositive macaques. *Laboratory Animal Science* 43:541–544.

Wenner, H.A., D. Abel, S. Barrick, and P. Seshumurty. 1977. Clinical and pathologenetic studies of Medical Lake virus infections in cynomolgus monkeys (simian varicella). *Journal of Infectious Diseases* 135:611–622.

White, R.J., L. Simmons, and R.B. Wilson. 1972. Chickenpox in young anthropoid apes: Clinical and laboratory findings. *Journal of the American Veterinary Medical Association* 161:690–692.

Whitely, R.J. 1996. Cercopithecine herpesvirus 1 (B virus). In *Fields virology,* vol. 2, ed. B.N. Fields, D.M. Knipe, P.M. Howley, R.M. Chanock, J.L. Melnick, T.P. Monath, B. Roizman, and S.E. Strauss, 3d ed. New York: Raven, pp. 2623–2635.

Wilson, R.B., M.A. Holscher, T. Chang, and J.R. Hodges. 1990. Fatal *Herpesvirus simiae* (B virus) infection in a patas monkey (*Erythrocebus patas*). *Journal of Veterinary Diagnostic Investigation* 2:242–244.

MALIGNANT CATARRHAL FEVER

WERNER P. HEUSCHELE AND HUGH W. REID

Synonyms: MCF, malignant head catarrh, malignant catarrh, wildebeest disease, snotsiekte.

INTRODUCTION. Malignant catarrhal fever (MCF) is a generalized, often fatal lymphoproliferative disease of domestic and captive species of Artiodactyla. It has been described in domestic cattle *Bos taurus,* water buffalo *Bubalus bubalis,* swine *Sus scrofa,* deer, and in many other ungulates belonging to the families Bovidae, Giraffidae, and Cervidae. The disease occurs following infection with either of two herpesviruses: alcelaphine herpesvirus 1 (AHV-1), the natural host of which is wildebeest (*Connochaetes* spp.), or ovine herpesvirus 2 (OHV-2), the natural host of which is domestic sheep *Ovis aries.* These two viruses appear to be the only members of a large group of closely related gammaherpesviruses of the family Bovidae which spread contagiously to species other than the natural host and induce MCF. Neither virus produces detectable clinical disease in its respective natural host. Transmission to other species results in dramatic clinical and pathologic response characterized by high fever, profuse nasal discharge, ophthalmia with corneal opacity, generalized lymphadenopathy, severe inflammation with necrosis and erosions in the mucosae of the alimentary and upper respiratory tracts, and leukopenia. Diarrhea, neurologic signs, skin lesions, and nonsuppurative arthritis may also develop (Heuschele et al. 1984, 1985; Plowright 1986; Heuschele 1988).

HISTORY. The occurrence of MCF in domestic cattle in association with domestic sheep was observed in Europe in the late 1700s. Successful transmission of MCF was first achieved by inoculation of large volumes of blood from bovine cases of sheep-associated MCF into susceptible cattle in 1929 (Götze and Liess 1929). In Africa, MCF was known for centuries to pastoral tribesmen such as the Masai and referred to by them as *wildebeest disease* in recognition of its occurrence when they grazed their cattle in areas where blue wildebeest *Connochaetes taurinus* had been grazing during their calving season (Heuschele 1988).

In the early 19th century, settlers and hunters in southern Africa were familiar with MCF and called it *snotsiekte,* meaning, in Afrikaans, "snotting sickness" (Mettam 1923). Mettam, however, felt that snotsiekte differed from MCF occurring in Europe, because it was more easily transmitted to cattle, had no horn, hoof, or skin involvement, and had consistent marked lymphadenopathy. In the 1930s, following the studies by Götze and Liess (1929), veterinary researchers in East Africa concluded that snotsiekte was the same disease as MCF seen in Europe, based on the similar clinicopathologic manifestations seen in both diseases (Daubney and Hudson 1936). This view was supported by Du Toit and Alexander (1938) and by Wyssman (1938).

The first isolation and identification of AHV-1 that produced MCF when inoculated into cattle were made in 1959 by Plowright from a blue wildebeest (Plowright et al. 1960). The OHV-2, however, has never been isolated, although fragments of DNA identified as specific to the virus have been isolated from lymphoblastoid cells derived from animals with this form of MCF (Baxter et al. 1993).

DISTRIBUTION. Sheep-associated MCF (SA-MCF) occurs worldwide wherever domestic sheep are present. The form derived from wildebeest occurs chiefly in Africa, in their natural habitats (Plowright 1986). However, this form is appearing with increasing frequency in zoos, game parks, and game ranches around the world where wildebeest are kept in captivity in proximity to cattle and other susceptible species (Castro et al. 1982; Ramsay et al. 1982).

HOST RANGE. Evidence that a group of bovid gammaherpesviruses infects a variety of species relies on detection of antibody that cross-reacts in immunologic assays with antigens of AHV-1 or the detection of DNA sequences homologous to those found in the AHV-1 genome. Infection thus seems to be prevalent in African antelope belonging to the subfamilies Bovinae and Hippotraginae, as well species in Caprinae. The AHV-1 infects blue wildebeest and white-tailed wildebeest *Connochaetes gnou* (Plowright et al. 1960; Heuschele et al. 1984). Viruses designated AHV-2 were recovered from hartebeest *Alcelaphus buselaphus* and topi *Damaliscus lunatus,* and hippotragine herpesvirus 1 (HipHV-1) was isolated from roan antelope *Hippotragus equinus.* These three viruses were isolated in tissue culture, at least partially characterized, and shown to produce MCF experimentally. Only AHV-1 has been sequenced, and the entire genomic sequence of the C500 isolate is now available (Ensser et al. 1997). Evidence for related viruses in other species of African antelope relies on detection of serum antibody that cross-reacts with AHV-1 in neutralization and other tests (Plowright 1981, 1986).

Infection of species of sheep and goats *Capra hircus* with related gammaherpesviruses is implied by the consistent detection of antibody to AHV-1 in all species of Caprinae examined. Such antibody may be detected by a variety of serologic tests, but neutralization tests are either negative or detect antibody only at low titer. This implies these viruses are less closely related to AHV-1 than the viruses of African antelope. So far, however, of the subfamily Caprinae, only domestic sheep have been identified as transmitting MCF virus. Further evidence of OHV-2 infection is derived from culture of lymphoblastoid cell lines from cattle and deer with SA-MCF. Such cell lines contain DNA that hybridizes with clones of the unique region of the AHV-1 molecule (Baxter et al. 1993). Thus, from such cell lines, OHV-2 DNA clones have been derived and used to identify a sequence unique to OHV-2. Using polymerase chain reaction (PCR) based on this sequence, it has been possible to identify virus in nasal secretions of lambs, and tissues, including peripheral blood lymphocytes, of ewes and lambs. Thus, although OHV-2 has not been isolated, molecular methodologies are now available for its study and detection (Crawford et al. 1999).

Whereas AHV-1 may spread readily to cattle and other species from periparturient wildebeest herds, transmission of OHV-2 is less predictable. There is a spectrum of susceptibility ranging from the relatively resistant domestic cattle, through the more susceptible deer and water buffalo, to the extremely susceptible banteng *Bos javanicus* and Père David's deer *Elaphurus davidianus.* The reason why disease in domestic pigs *Sus scrofa* is reported only from Norway remains obscure (Loken et al. 1998). The range of host species in which clinical MCF has been diagnosed, serologic evidence of infection has been found, or MCF virus DNA has been demonstrated is summarized in Table 7.2.

TABLE 7.2—Species susceptible to viruses causing malignant catarrhal fever based on virus isolation, polymerase chain reaction, clinical disease, or serologic evidence

Taxon	Common Name
Antilocapridae	
Antilocapra americana	Pronghorn
Bovidae	
Aepyceros melampus	Impala
Antidorcas marsupialis	Springbok
Antilope cervicapra	Blackbuck
Gazella dama ruficollis	Addra gazelle
Gazella dorcas	Dorcas gazelle
Gazella granti	Grant's gazelle
Gazella leptoceros	Slender-horned gazelle
Gazella subgutturosa	Persian gazelle
Neotragus moschatus	Zulu suni
Ourebia ourebi cottoni	Cotton's oribi
Alcelaphus buselaphus	Hartebeest
Connochaetes gnou	Black or white-tailed wildebeest
Connochaetes taurinus	Blue or white-bearded wildebeest, or brindled gnu
Damaliscus lunatus jimela	Topi
Damaliscus pygarus	Blesbok
Addax nasomaculatus	Addax
Hippotragus equinus cottoni	Roan antelope
Hippotragus niger	Sable antelope
Oryx gazella	Gemsbok
Oryx leucoryx	Arabian oryx
Kobus ellipsiprymnus	Waterbuck
Kobus kob thomasi	Uganda kob
Kobus megaceros	Nile lechwe
Boselaphus tragocamelus	Nilgai
Bison bison	American bison, wood bison
Bison bonasus	European bison or wisent
Bos gaurus	Gaur
Bos javanicus	Banteng
Bubalus bubalis	Asian water buffalo
Syncerus caffer	African buffalo
Taurotragus oryx	Eland
Tragelaphus eurycerus	Bongo
Tragelaphus scriptus	Bushbuck
Tragelaphus spekeii	Sitatunga
Tragelaphus strepsiceros	Greater kudu
Tragelaphus angasii	Nyala
Capra aegagrus	Cretan wild goat
Capra falconeri heptneri	Turkomen markhor
Capra ibex	Ibex
Hemitragus hylocrius	Nilgiri tahr
Hemitragus jemlahicus	Himalayan tahr
Ovis aries	Domestic sheep
Ovis vignei	Asiatic urial
Ovis musimon	Mouflon
Ovibos moschatus	Muskox
Rupicapra rupicapra	Chamois
Naemorhedus goral arnouxianus	Goral
Saiga tatarica	Saiga
Cephalophus spp.	Duiker
Camelidae	
Lama glama	Llama

(continued)

TABLE 7.2—*Continued*

Cervidae	
Alces alces	Moose
Capreolus capreolus	Roe deer
Axis axis	Axis deer
Axis porcinus	Indian hog deer
Cervus duvauceli	Barasinga
Cervus elaphus	Red deer, Altai wapiti, elk
Cervus eldii	Eld's deer
Cervus nippon	Sika deer, Formosan sika deer
Cervus timorensis	Sunda sambar
Cervus unicolor malaccensis	Malayan sambar
Dama dama	Fallow deer
Elaphurus davidianus	Père David's deer
Muntiacus reevesi	Reeves' muntjac
Odocoileus hemionus	Mule deer
Odocoileus virginianus	White-tailed deer
Rangifer tarandus	Caribou, reindeer
Tragulidae	
Tragulus javanicus	Mouse deer

Adapted from Lahijani et al. (1994).

Some species appear to be more sensitive to infection and development of clinical disease than others, whereas a few species appear to be refractory to clinical disease, but may become seropositive and have latent infections. Many ruminant species have a relatively high prevalence of infection with MCF viruses without any history of clinical disease. Based on epidemiologic evidence, however, not all species that may be carriers of MCF herpesviruses appear to shed infectious virus into the environment. Many may be *dead-end* hosts the majority of the time, only rarely shedding the virus.

ETIOLOGY

Alcelaphine Herpesviruses. The etiologic agent of wildebeest-origin MCF was first isolated from a blue wildebeest in 1959 (Plowright et al. 1960). This was characterized as a highly cell-associated, lymphotropic herpesvirus, later classified as a member of the family *Herpesviridae*, subfamily *Gammaherpesvirinae* (Plowright et al. 1965). It was at first designated by some as bovine herpesvirus 3. However, when it became evident that wildebeest, not cattle, were the natural hosts for this agent, the name alcelaphine herpesvirus 1 was suggested as more appropriate (Reid and Rowe 1973; Roizman et al. 1981; Roizman 1982). Members of the herpesviridae are characterized by a core containing double-stranded DNA and an icosahedral capsid 100–110 nm in diameter, with 162 hollow capsomers. Symmetrical material called a tegument and an envelope that contains viral glycoprotein spikes on its surface compose the entire virion (Roizman 1996). Due to the lipid envelope that is necessary for infectivity, the herpesviruses are sensitive to lipid solvents and detergents. Virions of AHV-1 are formed in the nucleus of the cell and bud through the nuclear membrane into vesicles on the plasma membrane (Castro et al. 1982). The wildebeest virus AHV-1 can readily be isolated and propagated in primary or secondary and higher-passaged cell cultures of a number of bovine cell types, especially thyroid (BTh) (Plowright et al. 1960), kidney (BK and BEK) (Plowright 1986), calf testis (BT) (Plowright 1981, 1986), and in nondomestic ovine kidney cells, such as from fetal aoudad *Ammotragus lervia* (FAK) and fetal mouflon sheep *Ovis musimon* (FMSK) (Heuschele and Fletcher 1984; Seal et al. 1988, 1989a,b). Isolation of virus from either MCF-affected animals or latently infected animals can be achieved only through cocultivation of cells, because the virus is strictly cell associated. Initially, cytopathic effects (CPEs) are characterized by the formation of microsyncytia, and infectivity is largely cell associated. Following serial passage, infectivity becomes increasingly cell free and is associated with a change in CPEs from macrosyncytia to refractile cells, rounding, and loss of the ability of certain isolates upon serial passage to induce disease following experimental inoculation.

The AHV-2 viruses isolated from topi and hartebeest induce MCF following experimental inoculation of cattle or rabbits *Oryctolagus cuniculus* but normally do not appear to spread and cause MCF. In 1984, a herpesvirus resembling AHV-1 was isolated from a hartebeest that had died of clinical MCF at the San Diego Wild Animal Park. However, it was later found that two alcelaphine herpesviruses had, in fact, been originally isolated. When these were studied by restriction endonuclease (RE) analysis and comparative cell culture growth, one was found to grow both in bovine embryonic lung cells (BELs) and FAK cells producing CPEs and cell-free virus, whereas the other isolate produced CPEs and cell-free virus in only the FAK ovine kidney cells. Restriction endonuclease analysis and DNA hybridization of these isolates indicated that one isolate was AHV-1 and the other was homologous with AHV-2 (Seal et al. 1989a,b).

A related virus that was capable of producing MCF on inoculation into rabbits was isolated from lymph node cell cultures of a normal roan antelope, though the lesions were distinct from those produced by AHV-1 or OHV-2. Viral DNA of HipHV-1 was distinct from AHV-1 DNA: though it hybridized strongly with several clones of AHV-1, it hybridized only weakly with two clones of OHV-2, indicating a closer relationship with AHV-1 than with OHV-2. This molecular evidence is suggestive of possible infections occurring in susceptible hosts with two related strains of AHV-1.

Ovine Herpesvirus 2. The etiologic agent of SA-MCF remains elusive. Despite numerous isolation attempts and reports, no virus that can be regarded as the causative agent has been isolated either from cases of this form of MCF or from sheep. Evidence of a virus antigenically related to AHV-1 being involved in SA-MCF is based on detection of antibody that reacted with AHV-1 in an indirect immunofluorescence antibody (IIFA) test in virtually every sheep tested (Rossiter 1981a) and detection of antibody in sheep serum that reacted with the major structural proteins of AHV-1 in immunoblots (Herring et al. 1989). Finally, hamsters experimentally infected with SA-MCF virus from cattle and deer developed antibody to AHV-1 (Reid et al. 1989).

Despite consistent failure to isolate an etiologic virus from cases of SA-MCF, it has been possible to generate lymphoblastoid cell lines from cattle and deer with this form of the disease, some of which transmitted MCF to rabbits following experimental inoculation. The DNA from such cell lines hybridized with a number of DNA clones of the unique region of the AHV-1 molecule, and subsequent screening of a genomic library from these cells identified a number of viral clones that have been established to be OHV-2. Sequence data from one of these clones form the basis of a PCR enabling detection of the virus in sheep and MCF-affected cattle, deer, water buffalo, moose *Alces alces,* and bison *Bison bison* (Baxter et al. 1993; Hua et al. 1999).

TRANSMISSION AND EPIDEMIOLOGY. Transmission of AHV-1 in wildebeest occurs both vertically and horizontally (Plowright 1965a,b, 1984; Rweyemamu et al. 1974; Mushi et al. 1980a,b). Vertical transmission of AHV-1 by transplacental infection has been proved by the recovery of virus from fetuses or the blood of wildebeest calves estimated to be 1 week of age or less (Plowright 1965a). Transmission of AHV-1 from wildebeest occurs primarily from newborn calves to cohorts, cattle, or other susceptible species as a result of inhalation of aerosol droplets or ingestion of food or forage contaminated with AHV-1 in nasal secretions, tears, or feces (Rweyemamu et al. 1974; Mushi et al. 1980a,b; Heuschele et al. 1984). Presumably, this shedding ceases at 3–4 months of age after the calf has developed an adequate immune response to the virus (Mushi et al. 1980a,b,). In adult wildebeest, AHV-1 is highly cell associated and not readily transmissible. Shedding of cell-free virus in nasal secretions has, however, been demonstrated in adult wildebeest following stress and corticosteroid administration (Rweyemamu et al. 1974; Heuschele et al. 1985).

As with wildebeest, most free-living adult hartebeest and topi examined have neutralizing antibody to AHV-1; the only seronegative animals detected are calves that either failed to ingest colostral antibody or in which maternal antibody had decayed. Therefore, AHV-2 appears to be transmitted efficiently through contact or milk, ensuring that all calves become infected during the first few months of life. Despite close proximity of cattle and parturient herds of topi and hartebeest, there is no evidence that AHV-2 can spread to cattle. Similarly, of the gammaherpesviruses that infect the Hippotraginae, there would appear to be highly efficient intraspecies transmission but no spread to other species.

Transmission of OHV-2 in sheep probably follows a similar pattern to that of AHV-1 in wildebeest. Baxter and others (1997) found that all of a group of nine lambs excreted OHV-2 DNA in nasal secretion within 2 months of birth and that virus was distributed in a variety of lymphoid and other tissues. On the other hand, after using a combination of PCR and competitive inhibition enzyme-linked immunosorbent assay (CI-ELISA), Li et al. (1998) suggested that lambs did not normally become infected until they were more than 2.5 months of age and did not begin to seroconvert until more than 6 months of age. This apparent discrepancy is probably due to differences in tissues examined in the two studies and the high specificity of the CI-ELISA, which did not detect early antibody responses.

Factors governing transmission to other species are obscure. It is presumed that, following infection, sheep experience periodic reactivation episodes. The normal sporadic occurrence of MCF in individual animals among groups of apparently equally susceptible cohorts is not readily understood. However, when extremely susceptible species such as Père David's deer and banteng are exposed to sheep, all will develop MCF. Occasionally, MCF may occur in larger numbers of individuals not normally considered highly susceptible. When a particular sheep flock is implicated in such cases, it tends to remain highly contagious for some years. One proposed explanation of this phenomenon is that OHV-2 may mutate to become more infectious to species other than sheep. Infection of other species in the subfamily Caprinae is presumed to be similar, but their role as possible sources of MCF has not been established.

The evidence suggests that animals developing MCF due to AHV-1 or OHV-2 are not contagious and may be dead-end hosts. This is assumed to be due to very limited viral replication in these animals, such that no cell-free virus is produced on epithelial surfaces. Virus infectivity can, however, be recovered from AHV-1-affected animals by employing techniques that retain cell viability. Transmission of the OHV-2 form of MCF can also be achieved using large volumes of blood or cell suspensions from affected tissue. This has, however, been inconsistent, though it would appear to be more readily achieved from highly susceptible species such as deer and banteng.

Latent or asymptomatic persistence of OHV-2 in cattle has been shown to occur with greater frequency than previously believed (O'Toole et al. 1995, 1997). This has been based on demonstration of serum antibodies to MCF viruses by CI-ELISA (Li et al. 1996a,b) and by PCR demonstration of OHV-2 DNA amplicons from peripheral lymphocyte DNA of domestic and wild cattle.

CLINICAL SIGNS. Clinical MCF in cattle has been arbitrarily divided into four forms (Plowright 1981, 1986; Heuschele et al. 1985). There is considerable variation and overlap among these artificial categories, and their usage may be confusing. Incubation periods in natural cases are not known, but epidemiologic evidence indicates it may be as long as 200 days. Experimentally, the incubation period has varied from 9–77 days (Heuschele and Seal 1992).

Peracute form: In this form, severe inflammation of the oral and nasal mucosa and hemorrhagic gastroenteritis occur with a course of 1–3 days.

Intestinal form: Pyrexia, diarrhea, hyperemia of oral and nasal mucosa with accompanying discharges, and lymphadenopathy with a clinical course of 4–9 days is descriptive of this form.

Head-and-eye form: This is the typical syndrome of MCF with pyrexia, nasal, and ocular discharges progressing from serous to mucopurulent. Encrustation of the muzzle and nares occurs in later stages, causing obstruction of the nostrils and dyspnea, open-mouthed breathing, and drooling. There is intense hyperemia and multifocal or diffuse necrosis of the oral mucosa, usually on the lips, gums, hard and soft palate, and buccal mucosa. Sloughing of the tips of buccal papillae, leaving them reddened and blunted, is a common finding.

Ocular signs include lacrimation progressing to purulent exudation, photophobia, hyperemia, edema of the palpebral conjunctiva, and injection of scleral vessels. Corneal opacity, starting peripherally and progressing centripetally, results in partial or complete blindness. Corneal opacity is usually bilateral but occasionally is unilateral. Hypopyon may also occur.

Pyrexia is common and usually high (40° C–42° C) until the animal becomes moribund, at which time the body temperature drops. Increased thirst accompanies the pyrexia, and anorexia is seen in the late clinical stages. Constipation is common in this form of MCF, but terminal diarrhea is sometimes observed. Though not frequently seen, nervous signs may include trembling or shivering, incoordination, and terminal nystagmus.

Necrotic skin lesions occasionally are seen, and horn as well as hoof wall may be loosened or sloughed. The course of the head-and-eye form, which is invariably fatal, is usually 7–18 days.

Mild forms: These are syndromes caused by experimental infection of cattle with attenuated viruses and are usually not fatal.

While the manifestations of the head-and-eye form of MCF are considered the typical syndrome of MCF in cattle, clinical signs in exotic ruminants are less dramatic and not specifically diagnostic. Malignant catarrhal fever is usually manifested by conjunctivitis, photophobia, moderate corneal clouding that may be unilateral, fever, depression, variable lymphadenopathy, occasionally diarrhea, and usually mild serous nasal discharge. Death may be sudden, following a brief course of hemorrhagic diarrhea. Inflammation of the oral and nasal cavity is usually less severe than in cattle and only occasionally progresses to mucosal erosions.

PATHOLOGY. Gross lesions vary considerably, depending on the course of disease. Animals that die peracutely may have few lesions other than hemorrhagic enterocolitis. In the more protracted intestinal and head-and-eye forms, the carcass may be normal, dehydrated, or emaciated. The muzzle is often encrusted and raw. Cutaneous lesions sometimes occur as generalized exanthema with exudation of lymph, crusting, and matting of the hair. Hyperemia is apparent in unpigmented skin. These lesions are frequently seen in the ventral thorax and abdomen, inguinal region, perineum and loins, and sometimes on the head. Generalized lymphadenopathy is common.

Lesions in the respiratory system range from mild to severe. When the clinical course is short, there is slight serous nasal discharge and hyperemia of the nasal mucosa. Later, the discharge becomes more copious and mucopurulent accompanied by intense nasal mucosal hyperemia, edema, and small erosions. Occasionally, croupous pseudomembrane formation is seen. Lesions in the nasal passages may extend to the frontal sinuses. Pharyngeal and laryngeal mucosae are hyperemic and edematous and later develop multiple erosions that are often covered with gray-yellow pseudomembranes. Inflammation and sometimes petechiation and ulceration are seen in the tracheobronchial mucosa. Lungs are often edematous and sometimes emphysematous but may appear normal. Bronchopneumonia may complicate chronic cases.

The alimentary tract may appear normal in peracute cases. When the course is longer, alimentary lesions are commensurately more severe and include mild to severe mucosal hyperemia and edema, and erosions and ulcerations especially on the dental pad and gingival surfaces, palate, tongue, and buccal papillae. Mucosal inflammation, hemorrhage, and erosions may also be found in the rest of the digestive tract, including the esophagus, abomasum, small intestine, colon, and rectum. Feces are usually scant, dry, pasty, or bloodstained.

Urinary tract lesions include hyperemia and sometimes marked distention and prominence of bladder mucosal vessels and mucosal edema. Occasionally, petechial to severe hemorrhage and mucosal erosion and ulceration may occur. Kidneys may be normal or mottled with patches of beige raised areas. Petechia or ecchymoses may occur in the renal pelvis and uterus. The liver is usually slightly enlarged and has a prominent reticular pattern. Hemorrhages and erosions may occur in the mucosa of the gallbladder.

Angiitis typified by microscopic lesions of extensive vasculitis, perivasculitis, and lymphoreticular proliferation in lymphoid organs with mononuclear infiltrations in kidney, liver, adrenal glands, and central nervous system are strongly suggestive of MCF. Typical herpesvirus inclusion bodies are not present.

DIAGNOSIS. A history indicating contact with parturient sheep or wildebeest and typical clinical features provide a basis for tentative diagnosis of MCF. Gross lesions of inflammation and erosions in nasal passages, alimentary tract, and urinary bladder; enlarged lymph nodes; corneal opacity; and prominent tortuous small arteries in the subcutis, thorax, and abdomen provide evidence for a presumptive diagnosis of MCF. Typical microscopic lesions are strongly suggestive, and some consider pathognomonic, for MCF. Tissues for histopathology should be fixed in 10% neutral buffered formalin and include lung, kidney, liver, adrenal gland, lymph nodes, eyes, oral epithelium, esophagus, Peyer's patches in ileum, urinary bladder, brain, carotid rete, thyroid, heart, and skin lesions.

Virus isolation is conducted on fetal bovine or ovine thyroid, kidney, lung, or spleen cells, or by intraperitoneal or intravenous inoculation of domestic rabbits or cattle. Buffy-coat leukocytes from unclotted blood provide the best source for virus isolation. Fusion of leukocytes with suspended tissue culture cells by using polyethylene glycol (PEG; mol. wt. 1000–4000) enhances the sensitivity of virus isolation. Refrigerated, but not frozen, tissues for virus isolation should include spleen, lung, lymph nodes, adrenal gland, and thyroid gland. These tissues should be collected as soon after death as possible because the virus rapidly loses infectivity in an animal dead more than 1 hour. Development of characteristic herpesvirus CPEs in inoculated cell cultures may require several subculture passages of intact cells with fresh susceptible cells. Development of cell-free virus is enhanced by incubation of cultures at 33° C–34° C. However, an inhibitory factor in tissues may prevent successful primary isolation of AHV-1 (Plowright 1986). Cowdry type-A intranuclear inclusion bodies are formed in infected cell cultures. Virus isolates are identified by typical herpesvirus CPEs, by immunofluorescence or immunoperoxidase reaction, by virus neutralization, and/or by electron microscopy.

Antibodies against MCF viruses may be detected by IIFA test, complement fixation (CF), CI-ELISA, and virus neutralization (Rossiter and Jessett 1980; Rossiter 1981a,b; Wan et al. 1988b; Seal et al. 1989a; Li et al. 1998). The standardized virus neutralization test is at present the most reliable means of detecting potential MCF virus carriers, because it is highly specific for MCF and does not cross-react with other herpesviruses as do the CI-ELISA, IIFA, and CF test. The CI-ELISA shows promise and may become the serologic test of choice. It must be noted that many animals dying acutely or peracutely of MCF fail to develop detectable MCF antibodies. Samples for serology should be paired when possible and taken first during the acute phase of disease and then during convalescence 3–4 weeks later or at death.

With the advent of PCR and related molecular diagnostic procedures in recent years, specific assays for MCF have been developed (Hsu et al. 1990; Seal et al. 1990; Bridgen and Reid 1991; Katz et al. 1991; Reid and Bridgen 1991; Murphy et al. 1994; Lahijani et al. 1994; Crawford et al. 1999; Hua et al. 1999). These diagnostic methods are relatively rapid (24–48 hours) and can have high sensitivity and specificity. The assay for OHV-2 is an amplification in a nested PCR with primers derived from a genome fragment that is homologous to AHV-1 but contains OHV-2-specific sequence. For routine PCR assays, unfixed tissues or peripheral blood leukocytes are preferred, but formalin- or ethanol-fixed tissue held less than 3–4 weeks can be used (Crawford et al. 1999). A competitive PCR has been shown to detect genomic OHV-2 in moose, deer, bison, cattle, and sheep (Hua et al. 1999). If non-SA-MCF is suspected, primers for AHV-1 should also be used. These PCR procedures could prove very useful in epidemiologic studies.

DIFFERENTIAL DIAGNOSES. Clinical MCF must be distinguished from other diseases that produce inflammation, erosions, and ulcerations of the nasal and alimentary tracts. Some of these include bovine viral diarrhea-mucosal disease, bluetongue, epizootic hemorrhagic disease, rinderpest, vesicular diseases such as foot-and-mouth disease and vesicular stomatitis, ingested caustics, and some poisonous plants and mycotoxins.

IMMUNITY. Transplacental infection of the fetus occurs in wildebeest; thus, the annual occurrence of MCF in cattle in the Masai Reserve in Kenya and Tanzania during wildebeest calving (Plowright 1990). Cattle and experimentally infected rabbits that recover from MCF have solid immunity against all strains of MCF virus. An effective vaccine is not available for MCF and, although some viral strains have undergone attenuation in cell cultures, no efficacious vaccines have yet been developed. This may be due to the highly cell-associated nature of the herpesvirus and the potential for latency, which may not be conducive to attack by humoral antibodies. Experimentally, inactivated vaccines have been inconsistent in inducing protection against virulent virus challenge (Heuschele et al. 1985). At present, vaccination does not appear to be a viable means of MCF prevention.

CONTROL AND IMPLICATIONS. Cattle exposed to MCF do not routinely respond to treatments, although supportive therapy has been used (Plowright 1981). The recent availability of derivatives of acyclovir compounds that inhibit replication of herpesviruses shows promise in potential treatment regimens. A report of inhibition of the replication of the alcelaphine herpesvirus by using recombinant interferons (Wan et al. 1998a) could be considered in development of a treatment regimen for valuable hoofstock.

Since a vaccine does not currently exist for MCF, the most effective control is by prevention of contact of susceptible species with reservoir hosts (i.e., wildebeest, other alcelaphine species, sheep, goats, and other seropositive species). Since transmission may occur on common pastures or when susceptible animals are in close proximity to carrier hosts, separation of these species is imperative. Carrier species should never be in contact with susceptible animals during calving or when carriers are nursing their young. Additionally, the introduction of carrier species into game farms or zoological parks should require a quarantine period wherein extensive molecular and serologic testing can be conducted. Care should be taken that water, food, and bedding materials cannot serve to transmit infection among housed species. There is no evidence of contact transmission between susceptible hosts, nor is there evidence that isolates of MCF are infectious to humans (Heuschele and Seal 1992).

LITERATURE CITED

Baxter S.I., I. Pow, A. Bridgen, and H.W. Reid. 1993. PCR detection of the sheep–associated agent of malignant catarrhal fever. *Archives of Virology* 132:145–159.

Baxter S.I., A. Wiyono, I. Pow, and H.W. Reid. 1997. Identification of ovine herpesvirus–2 infection in sheep. *Archives of Virology* 142:823–831.

Bridgen A., and H.W. Reid. 1991. Derivation of a DNA clone corresponding to the viral agent of sheep–associated malignant catarrhal fever. *Research in Veterinary Science* 50:38–44.

Castro, A.E., G.G. Daley, M.A. Zimmer, D.L. Whitenack, and J. Jensen. 1982. Malignant catarrhal fever in an Indian gaur and greater kudu: Experimental transmission, isolation, and identification of a herpesvirus. *American Journal of Veterinary Research* 43:5–11.

Crawford T.B., H. Li, and D. O'Toole. 1999. Diagnosis of malignant catarrhal fever by PCR using formalin–fixed, paraffin–embedded tissues. *Journal of Veterinary Diagnostic Investigation* 11:111–116.

Daubney, R., and J.R. Hudson. 1936. Transmission experiments with bovine malignant catarrh. *Journal of Comparative Pathology* 49:63–89.

Du Toit, P.J., and R.A. Alexander. 1938. Malignant catarrhal fever and similar diseases. In *Proceedings of the 13th International Veterinary Congress, Zurich, Switzerland,* vol. 1, pp. 553–559.

Ensser, A., R. Pflanz, and B. Fleckenstein. 1997. Primary structure of the alcelaphine herpesvirus 1 genome. *Journal of Virology* 71:6517–6525.

Götze, R., and J. Liess. 1929. Erfolgreiche Ubertragungsversuche des bösartigen Katarrhalfiebers von Rind zu Rind: Indentität mit der Südafrikanischen Snotsiekte. *Deutsche Tierarztliche Wochenschrift* 37:433–437.

Herring A., H. Reid, N. Inglis, and I. Pow. 1989. Immunoblotting analysis of the reaction of wildebeest, sheep and cattle sera with the structural antigens of alcelaphine herpesvirus–1 (malignant catarrhal fever virus). *Veterinary Microbiology* 19:205–215.

Heuschele, W.P. 1988. Malignant catarrhal fever: A review of a serious hazard for exotic and domestic ruminants. *Zoological Garten* 58:123.

Heuschele, W.P., and H.R. Fletcher. 1984. Improved methods for the diagnosis of malignant catarrhal fever. *Proceedings of the American Association of Veterinary Laboratory Diagnosticians* 27:137–150.

Heuschele, W.P., and B.S. Seal. 1992. Malignant catarrhal fever. In *Veterinary diagnostic virology,* ed. A.E. Castro and W.P. Heuschele. Philadelphia: Mosby Year Book, pp. 108–112.

Heuschele, W.P., B.S. Seal, J. Oosterhuis, D. Janssen, and P.T. Robinson. 1984. Epidemiological aspects of malignant catarrhal fever in the USA. *Proceedings of the United States Animal Health Association* 88:640–651.

Heuschele, W.P., N.O. Nielsen, J.E. Oosterhuis, and A.E. Castro. 1985. Dexamethasone–induced recrudescence of malignant catarrhal fever and associated lymphosarcoma and granulomatous disease in a Formosan sika deer (*Cervus nippon taiouanus*). *American Journal of Veterinary Research* 46:1578–1583.

Hsu, D., L.M. Shih, A.E. Castro, and Y.C. Zee. 1990. A diagnostic method to detect alcelaphine herpesvirus-1 of malignant catarrhal fever using polymerase chain reaction. *Archives of Virology* 114:259–263.

Hua Y., H. Li, and T.B. Crawford. 1999. Quantitation of sheep–associated malignant catarrhal fever viral DNA by competitive polymerase chain reaction. *Journal of Veterinary Diagnostic Investigation* 11:117–121.

Katz, J., B. Seal, and J. Ridpath. 1991. Molecular diagnosis of alcelaphine herpesvirus (malignant catarrhal fever) infections by nested amplification of viral DNA in bovine blood buffy coat specimens. *Journal of Veterinary Diagnostic Investigation* 3:193–198.

Lahijani R.S., S.M. Sutton, R.B. Klieforth, M.F. Murphy, and W.P. Heuschele. 1994. Application of polymerase chain reaction to detect animals latently infected with agents of malignant catarrhal fever. *Journal of Veterinary Diagnostic Investigation* 6:403–409.

Li, H., D.T. Shen, D.A. Jessup, D.P. Knowles, J.R. Gorham, T. Thorne, D. O'Toole, and T.B. Crawford. 1996a. Prevalence of antibodies to malignant catarrhal fever virus in wild and domestic ruminants by competitive-inhibition ELISA. *Journal of Wildlife Diseases* 32:437–443.

Li, H., D.T. Shen, D. O'Toole, W.C. Davis, D.P. Knowles, J.R. Gorham, and T.B. Crawford. 1996b. Malignant catarrhal fever virus: Characterization of a United States isolate and development of diagnostic assays. *Annals of the New York Academy of Sciences* 791:198–210.

Li, H., G. Snowder, D. O'Toole, and T.B. Crawford. 1998. Transmission of ovine herpesvirus 2 in lambs. *Journal of Clinical Microbiology* 36:223–226.

Loken, T., M. Aleksandersen, H. Reid, and I. Pow. 1998. Malignant catarrhal fever caused by ovine herpesvirus-2 in pigs in Norway. *Veterinary Record* 143:464–467.

Mettam, R.W.M. 1923. Snotsiekte in cattle. *Director Veterinary Education and Research Union South Africa, Report* 9 and 10:395–432.

Murphy M.F., R.B. Klieforth, R.S. Lahijani, and W.P. Heuschele. 1994. Diagnosis of malignant catarrhal fever by polymerase chain reaction amplification of alcelaphine herpesvirus 1 sequence. *Journal of Wildlife Diseases* 30:377–382.

Mushi E.Z., L. Karstad, and D.M. Jessett. 1980a. Isolation of bovine malignant catarrhal fever virus from ocular and nasal secretions of wildebeest calves. *Research in Veterinary Science* 29:168–171.

Mushi E.Z., P.B. Rossiter, L. Karstad, and D.M. Jessett. 1980b. The demonstration of cell–free malignant catarrhal fever herpesvirus in wildebeest nasal secretions. *Journal of Hygiene* 85:175–179.

O'Toole, D., H. Li, S. Roberts, J. Rovnak, J. De Martini, J. Cavender, B. Williams, and T. Crawford. 1995. Chronic generalized obliterative arteriopathy in cattle: A sequel to sheep-associated malignant catarrhal fever. *Journal of Veterinary Diagnostic Investigation* 7:108–121.

O'Toole, D., H. Li, W.R. Miller, and T.B. Crawford. 1997. Chronic and recovered cases of sheep-associated malignant catarrhal fever in cattle. *Veterinary Record* 140:519–524.

Plowright, W. 1965a. Malignant catarrhal fever in East Africa: II. Observations on wildebeest calves at the laboratory and contact transmission of the infection to cattle. *Research in Veterinary Science* 6:69–83.

———. 1965b. Malignant catarrhal fever in East Africa: III. Neutralizing antibody in freeliving wildebeest. *Research in Veterinary Science* 8:129–136.

———. 1981. Herpesviruses of wild ungulates, including malignant catarrhal fever virus. In *Infectious diseases of wild mammals,* ed. J.W. Davis, L.H. Karstad, and D.O. Trainer, 2d ed. Ames: Iowa State University Press, pp. 126–146.

———. 1984. Malignant catarrhal fever virus: A lymphotropic herpesvirus of ruminants. In *Latent herpesviruses in veterinary medicine,* ed. G. Wittmann, R.M. Gaskell, and H.J. Rziha. The Hague: Martinus Nijhoff, pp. 279–305.

———. 1986. Malignant catarrhal fever. *Revue Scientifique et Technique O.I.E.* 5:897–918.

———. 1990. Malignant catarrhal fever virus. In *Virus infections of ruminants,* vol. 3, ed. Z. Dinter and B. Morein. Amsterdam: Elsevier Science, pp. 123–150.

Plowright, W., R.D. Ferris, and G.R. Scott. 1960. Blue wildebeest and the aetiological agent of malignant catarrhal fever. *Nature* 188:1167–1169.

Plowright, W., R.F. Macadam, and J.A. Armstrong. 1965. Growth and characterisation of the virus of malignant catarrhal fever in East Africa. *Journal of General Microbiology* 39:253–266.

Ramsey, E.R., A.E. Castro, and B.M. Banmeister. 1982. Investigations of malignant catarrhal fever in ruminants at the Oklahoma City Zoo. *Proceedings of the United States Animal Health Association* 86:571–582.

Reid, H.W., and A. Bridgen. 1991. Recovery of a herpesvirus from an roan antelope (*Hippotragus equinus*). *Veterinary Microbiology* 28:269–278.

Reid, H.W., and L. Rowe. 1973. The attenuation of a herpesvirus (malignant catarrhal fever virus) isolated from hartebeest (*Alcelaphus buselaphus cokei,* Gunther). *Research in Veterinary Science* 15:144–146.

Reid, H.W., I. Pow, and D. Buxton. 1989. Antibody to alcelaphine herpesvirus-1 (AHV-1) in hamsters experimentally infected with AHV-1 and the "sheep-associated" agent of malignant catarrhal fever. *Research in Veterinary Science* 47:383–386.

Roizman, B. 1982. The Herpesviridae: General description, taxonomy, and classification. In *The herpesviruses,* vol. 1, ed. B. Roizman. New York: Plenum, p. 1.

———. 1996. Herpesviridae. In *Fields virology,* vol. 2, ed. B.N. Fields, D.M. Knipe, P.M. Howley, R.M. Chanock, J.L. Melnick, T.P. Monath, B. Roizman, and S.E. Straus, 3d ed. Philadelphia: Lippincott-Raven, pp. 2221–2230.

Roizman, B., L.E. Carmichael, F. Deinhardt F, G. de–The, A.J. Nahmias, W. Plowright, F. Rapp, P. Sheldrick, M. Takahashi, K. Wolf. 1981. Herpesviridae: Definition, provisional nomenclature, and taxonomy. The Herpesvirus Study Group, the International Committee on Taxonomy of Viruses. *Intervirology* 16:201–217.

Rossiter, P.B. 1981a. Antibodies to malignant catarrhal fever virus in sheep sera. *Journal of Comparative Pathology* 91:303–311.

———. 1981b. Immunofluorescence and immunoperoxidase techniques for detecting antibodies to malignant catarrhal fever in infected cattle. *Tropical Animal Health and Production* 13:189–192.

Rossiter, P.B., and D.M. Jessett. 1980. A complement fixation test for antigens of and antibodies to malignant catarrhal fever virus. *Research in Veterinary Sciences* 28:228–233.

Rweyemanu, M.M., L. Karstad, E.Z. Mushi, J. Otema, D.M. Jessett, L. Rowe, S. Drevemo, and J.G. Grootenhuis. 1974. Malignant catarrhal fever virus in nasal secretions of wildebeest: A probable mechanism for virus transmission. *Journal of Wildlife Diseases* 10:478–487.

Seal, B.S., R.B. Kleiforth, A.E. Castro, and W.P. Heuschele. 1988. Replication of alcelaphine herpesviruses in various cell culture systems and subsequent purification of virus. *Journal of Tissue Culture Methods* 11:49–56.

Seal, B.S., W.P. Heuschele, and R.B. Klieforth. 1989a. Prevalence of antibodies to alcelaphine herpesvirus–1 and nucleic acid hybridization analysis of viruses isolated from captive exotic ruminants. *American Journal of Veterinary Research* 9:1447–1453.

Seal, B.S., W.P. Heuschele, W.H. Welch, and W.P. Heuschele. 1989b. Alcelaphine herpesviruses 1 and 2 SDS–PAGE analysis of virion polypeptides, restriction endonuclease analysis of genomic DNA and virus replication restriction in different cell types. *Archives of Virology* 106:301–320.

Seal, B.S., R.B. Kleiforth, and W.P. Heuschele. 1990. Restriction endonuclease analysis of alcelaphine herpesvirus 1 DNA and molecular cloning of virus genomic DNA for potential diagnostic use. *Journal of Veterinary Diagnostic Investigation* 2:92–102.

Wan S.K., A.E. Castro, and R.W. Fulton. 1988a. Effect of interferons on the replication of alcelaphine herpesvirus–1 of malignant catarrhal fever. *Journal of Wildlife Diseases* 24:484–490.

Wan S.K., A.E. Castro, W.P. Heuschele, and E.C. Ramsay. 1988b. Enzyme linked immunosorbent assay for detecting antibodies to alcelaphine herpesvirus of malignant catarrhal fever in exotic ruminants. *American Journal of Veterinary Research* 49:164–168.

Wyssman, E. 1938. Bosartiges Katarrhalfieber und ahnliche Krankheiten. In *Proceedings of the 13th International Veterinary Congress, Zurich, Switzerland* 1:560–569.

PSEUDORABIES (AUJESZKY'S DISEASE)

DAVID E. STALLKNECHT AND ELIZABETH W. HOWERTH

Synonyms: Mad itch, infectious bulbar paralysis, suid herpesvirus 1, *Herpes suis.*

INTRODUCTION. Pseudorabies (Aujeszky's disease), caused by suid herpesvirus 1, which is also called pseudorabies virus (PRV), is an important disease of domestic swine *Sus scrofa.* Although the virus is present in wild boar *Sus scrofa* and feral swine populations in both Europe and the United States, disease associated with these naturally occurring infections has not been reported. Pseudorabies, however, has been documented in many other domestic and wildlife species where it has been associated with fatal pruritus.

HISTORY. The history of pseudorabies has been reviewed (Galloway 1938; Hanson 1954; Whittmann and Rziha 1989). The disease was first described by Aujeszky (1902) in cattle *Bos taurus,* dogs *Canis familiaris,* and cats *Felis catus,* and accounts of a pseudorabies-like illness were reported from the United States as early as 1813 (Hanson 1954). In 1910, a viral etiology was demonstrated, and the relationship between mad itch and pseudorabies in swine was established in 1931 (Shope 1931). The virus was cultivated initially in rabbit tissue (Traub 1953), followed by embryonated chicken eggs (Moril and Graham 1941) and tissue culture (Scherer 1953). Since the 1960s, pseudorabies has emerged as a significant domestic swine disease (Gustafson 1986).

DISTRIBUTION AND HOST RANGE. With some exceptions (Canada, Norway, Finland, England, Switzerland, and Australia), it is believed that domestic swine are infected with suid herpesvirus 1 worldwide (Van Oirschot 1992). Several countries, however, including the United States, are in the process of eliminating this virus from their domestic swine populations (Kluge et al. 1992). Feral and wild swine represent significant viral reservoirs, and serologic results suggest that most of the feral swine populations in the United States are infected (Clarke et al. 1983; Nettles and Erickson 1984; Corn et al. 1986; Nettles 1989; Van der Leek et al. 1993). The virus also has been documented by isolation and serology in wild swine populations in Europe (Ordas et al. 1983; Dahle et al. 1993; Capua et al. 1997a,b). The distribution of the disease in nonporcine domestic and wildlife species depends on direct or indirect contact with infected domestic, feral, or wild swine.

Domestic and wild swine are the reservoir hosts for suid herpesvirus 1. Disease associated with swine infections, however, has not been reported in these populations, suggesting that an efficient host-parasite relationship exists under natural conditions. Pseudorabies could not be experimentally produced in European wild boar (Tozzini et al. 1982). In contrast, disease was produced in feral swine that were experimentally infected with a domestic swine isolate but not with a virus originally isolated from feral swine (Hahn et al. 1997).

Other than swine, suid herpesvirus 1 can infect and cause disease in a diversity of wild vertebrate hosts, but reports of natural infections are not common. Pseudorabies has been reported in a Florida panther *Felis concolor coryi* (Glass et al. 1994), rats *Rattus* spp. (Von Becker and Herrmann 1963), raccoons *Procyon lotor* (Kirkpatrick et al. 1980), and a roe deer *Capreolus capreolus* (Nikolitsch 1954). In most of these cases, contact with swine was either known or suspected. Wild rodents have been suggested as a potential reservoir of suid herpesvirus 1 (Shope 1935), but, while susceptible, a carrier status has not been demonstrated (Maes et al. 1979). Raccoons also have been suggested as a potential source of infection for domestic swine, but their potential involvement in the maintenance or transmission cycle is short term and at best limited (Kirkpatrick et al. 1980). Pseudorabies was reported in captive brown bears *Ursus arctos* (Xenon et al. 1997) and a black bear *Ursus americanus* (Schultze et al. 1986). Antibodies to suid herpesvirus 1 also were reported in a free-living black bear from Florida that perhaps contacted infected feral swine (Pirtle et al. 1986). Three captive coyotes *Canis latrans* developed pseudorabies after they were fed offal (Raymond et al. 1997). Pseudorabies has been associated with several outbreaks on mink *Mustela vison* and fox *Vulpes vulpes* farms in Europe, and these outbreaks have been attributed to the feeding of swine offal (Konrad and Blazek 1958; Ljubashenko et al. 1958; Ugorski 1957; Vanek et al. 1962; Lapcevic 1964).

Species that are susceptible under experimental conditions include fox, skunk *Mephitis mephitis,* muskrat *Ondatra zibethica,* cottontail *Sylvilagus floridanus,* raccoon, badger *Taxidea taxus,* woodchuck *Marmota monax,* opossum *Didelphis marsupialis,* and white-tailed deer *Odocoileus virginianus* (Trainer and Karstad 1963). Galloway (1938) and Shahan et al. (1947) also reported that polecats *Mustela* sp., hedgehogs, porcupine, jackal *Canis* sp., ferret *Mustela putorius furo,* sable *Martes zibellina,* marmoset, rhesus monkey *Macaca mulatta,* sparrow hawk *Falco* sp., pigeon *Columbia livia,* goose, duck, and buzzard also were susceptible to experimental infections. Naturally acquired infections in birds have not been reported, but dead birds have been observed in association with suid herpesvirus 1-infected domestic swine herds (Gustafson 1986). Poikilothermic animals are reported to be refractory to experimental infection (Beran 1992).

ETIOLOGY. Suid herpesvirus 1 is included in the subfamily *Alphaherpesvirus* of *Herpesviridae.* The virus has a lytic replication cycle with cytopathic effects in cell culture similar to other mammalian herpesviruses. The virion is enveloped, contains linear double-stranded DNA (approximately 150 kb), and ranges from 150 to 180 nm in size. The genome encodes at least 50 proteins. The envelope consists of at least nine proteins, eight of which are glycoproteins. The nucleocapsid is 105–110 nm in size and consists of eight proteins. One serotype is known. Strain differences, however, can be detected by using panels of monoclonal antibodies or restriction endonuclease analyses (Kluge et al. 1992). The virus is sensitive to lipid solvents and detergents (Trainer 1981).

TRANSMISSION AND EPIDEMIOLOGY. Suid herpesvirus 1 is maintained in swine populations through the establishment of latent infection primarily in sensory nervous tissues (Gustafson 1986). During initial infection or reactivation of latent infections, virus can be isolated from nasal and tonsillar swabs and swabs of the genital tract. In domestic swine, virus may be excreted from the nose and mouth for up to 17 days (McFerran and Dow 1964), and aerosol and contact transmission involving naso-oral fluids is considered most important. In contrast, transmission in feral swine populations is believed to be associated primarily with reproduction and viral shedding via the genital tract (Romero et al. 1997).

Transmission also can occur by the oral route through ingestion of infected animal carcasses. Oral transmission has been demonstrated in feral pigs (Hahn et al. 1997), rats (Maes et al. 1979), raccoons (Kirkpatrick et al. 1980) and fox (Trainer and Karstad 1963). Viral persistence in the environment is pH and temperature dependent (Davies and Beran 1981), and virus has been reported to persist in the environment for up to 18 days in soils, 2–7 days in nonchlorinated water, and 2 days in swine sewage lagoon effluent (Beran 1992).

Suid herpesvirus 1 is maintained in swine populations, with feral and wild swine populations representing the only recognized wildlife reservoirs. All other wildlife species can be regarded as a dead-end hosts in the maintenance cycle. In feral swine populations, infection generally is associated with the adult age classes, and transmission occurs annually (Pirtle et al. 1989). Transmission is associated with food-dependent peaks in breeding activity (D.E. Stallknecht unpublished data). Under natural conditions of viral shedding (D.E. Stallknecht unpublished data) and steroid-induced viral shedding (Romero et al. 1997) in feral swine, suid herpesvirus 1 can be routinely recovered from swabs of the genital tract but not from nasal swabs. The probability of virus isolation is greater in swabs collected from boars. It is not understood whether this variation between domestic and feral swine represents tropism variation associated with wild or domestic swine viruses or reflects differences in the primary site of infection. However, when feral swine were experimentally infected with both domestic and feral swine isolates via the nasal route, most isolations were made from nasal swabs (Hahn et al. 1997).

CLINICAL SIGNS. Raccoons naturally infected with pseudorabies had pruritus and signs of central nervous system (CNS) dysfunction (Goyal et al. 1986). Raccoons died following experimental exposure to PRV, with depression and pruritus being the most common clinical signs (Trainer and Karstad 1963; Wright and Thawley 1980). Respiratory signs, salivation, grinding of teeth, disorientation, incoordination, paralysis, and convulsions also were observed in experimental infections (Trainer and Karstad 1963; Wright and Thawley 1980; Platt et al. 1983). A black bear with pseudorabies had clinical signs characterized by lethargy, depression, anorexia, staggering, moaning, and shaking, but the signs may have been partially attributable to concurrent intestinal volvulus (Schultze et al. 1986). Four captive brown bears that developed pseudorabies after the consumption of raw pork had depression, anorexia, and hypersalivation, followed by severe pruritus with self-mutilation. Respiratory distress and paralysis eventually led to death within 24 hours of the onset of clinical signs (Xenon et al. 1997). Anorexia, severe depression, hypersalivation, nervous excitability, dyspnea, and coma were seen in captive mink with pseudorabies (Kimman and van Oirschot 1986). Similar signs, and sometimes pruritus, were seen in mink experimentally infected with PRV via oral and subcutaneous inoculation (Goto et al. 1968; Quiroga et al. 1997).

Although PRV infections are well documented in feral swine, no clinical signs have been observed. In domestic swine, clinical signs depend on the strain of virus, the infectious dose, and age of the infected swine. Adult swine may have respiratory signs, and pregnant females may abort, but mortality is low. Neonatal pigs typically have neurologic signs and usually die. Slightly older pigs may have respiratory signs in addition to neurologic signs, but the mortality is lower than among neonatal pigs. Respiratory signs with high morbidity and low mortality are seen in grower/finishing swine (Kluge et al. 1992).

Anorexia and pruritus at the site of inoculation occurred in a white-tailed deer experimentally infected with PRV (Trainer and Karstad 1963). The deer died within 48 hours of inoculation, following a period of recumbency, excessive salivation, grinding of teeth, and clonic spasms. Red fox, skunk, badgers, opossums, cottontails, muskrats, woodchucks (Trainer and Karstad 1963), arctic fox *Alopex lagopus* (Quiroga et al. 1995), and brown bats *Eptesicus fuscus* (Reagan et al. 1953) died following experimental inoculation with PRV. Clinical signs in these animals included anorexia, depression, excessive salivation, paralysis, clonic spasms, convulsions, and coma. A captive coyote with pseudorabies was anorexic and had abnormal vocalization (Raymond et al. 1997).

PATHOGENESIS. The pathogenesis of pseudorabies in species other than swine is obscure (Kluge et al. 1992). The pathogenesis in swine varies depending on viral strain, age of the pig, size of the inoculum, and route of infection (Kluge et al. 1992). In domestic swine the oral-nasal route of infection is most common, whereas in feral swine venereal transmission appears to be the principal route of infection (Romero et al. 1997). Following oral-nasal infection, primary replication occurs in the epithelium of the nasopharynx and tonsil and spreads via lymphatics to regional lymph nodes or via nerves to the CNS. Widespread virus distribution in the body may occur with highly virulent strains. As with other herpesviruses, latent infections develop with virus persisting in ganglia, such as the trigeminal ganglia.

Pseudorabies virus infection is typically fatal in species other than swine. Carnivores apparently become infected after either direct or indirect contact with infected swine by eating virus-containing tissues (Trainer and Karstad 1963; Glass et al. 1994; Xenon et al. 1997). It has been suggested that infections may also occur via the aerogenous route in animals living in close contact with infected swine (Kimman and van Oirschot 1986). Unlike other species, mink develop predominantly vascular lesions after infection with PRV, and viremia and endotheliotropism appear to be important factors in the pathogenesis of the disease in this species (Kimman and van Oirschot 1986).

PATHOLOGY. Gross lesions in clinically affected animals are often minimal or nonexistent. Raccoons, red fox, skunk, badger, opossum, cottontail, muskrat, and woodchuck experimentally infected with PRV had minimal lesions that were confined to inflamed turbinates, generalized congestion, petechiation of the heart, congestion of the meningeal vessels, pulmonary edema, and cutaneous self-trauma secondary to pruritus (Trainer and Karstad 1963). Severe self-mutilation was the primary finding in naturally infected brown bears (Xenon et al. 1997). Affected mink had widespread petechia and ecchymoses (Kimman and van Oirschot 1986; Quiroga et al. 1997), black gastric and intestinal contents (Quiroga et al. 1997), acute hemorrhagic enteritis with fibrin (Kimman and van Oirschot 1986), and pulmonary congestion and edema (Kimman and van Oirschot 1986). Captive coyotes had alopecia and hyperkeratosis of the submandibular skin, with underlying edema and hemorrhage, and intestinal hemorrhage (Raymond et al. 1997). In an experimentally infected white-tailed deer, evidence of pruritus was the only gross finding (Trainer and Karstad 1963). Experimentally infected arctic fox had petechia, ecchymoses, scratches in the head region, and reactive lymphoid tissue (Quiroga et al. 1995).

Focal nonsuppurative encephalitis with perivascular cuffing and glial nodules in the brainstem was seen in a naturally infected raccoon (Goyal et al. 1986). Raccoons experimentally infected with PRV had congestion and hemorrhages in a variety of organs, patchy myocardial necrosis, necrosis of lymphoid tissues, degeneration in acinar and islet cells of the pancreas, with large intranuclear inclusion bodies in these cells as well as pancreatic ganglion cells, and glial nodules in the medulla oblongata (Trainer and Karstad 1963). Coyotes had nonsuppurative meningoencephalitis, with intranuclear inclusion bodies in a few neurons (Raymond et al. 1997). A black bear had splenic lymphoid depletion and multifocal hepatic and adrenal necrosis, with intranuclear inclusion bodies (Schultze et al. 1986). Hyalin and fibrinoid degeneration and necrosis of blood vessels in heart, brain, gastrointestinal tract, and occasionally elsewhere in the body with associated hemorrhages were seen in naturally infected mink (Kimman and van Oirschot 1986). Focal malacia was associated with angiopathy in the brain. Additional lesions in mink were pulmonary hemorrhage, lymphoid necrosis, and hemoglobinuric nephrosis. Nonsuppurative encephalitis confined to the brainstem and fibrinoid degeneration of vessel walls in pharynx, larynx, and myocardium were observed in mink experimentally infected with PRV (Quiroga et al. 1997). Other than cutaneous lesions attributable to pruritus, an experimentally infected white-tailed deer had no significant lesions (Trainer and Karstad 1963). Microscopic lesions in experimentally infected red fox, badger, cottontail, muskrat, and woodchuck were limited and nonspecific, primarily consisting of degeneration of kidney and liver, hemorrhages in multiple organs, congestion, and lymphoid necrosis or depletion (Trainer and Karstad 1963). Experimentally infected opossums had lesions in turbinates, myocardium, kidneys, spleen, and adrenals, which consisted of necrosis and/or neutrophilic infiltration with intranuclear inclusion bodies in cells in spleen, adrenal, and para-adrenal ganglia (Trainer and Karstad 1963). Experimentally infected skunks had necrosis and inflammation in the adrenal and liver, with intranuclear inclusion bodies in these organs (Trainer and Karstad 1963). Nonsuppurative meningoencephalitis with neuronal necrosis, gliosis, neuronophagia, and perivascular cuffing was seen in experimentally infected arctic fox (Quiroga et al. 1995).

Although disease due to PRV infection has not been reported in feral swine, if it occurred, microscopic lesions similar to those in domestic pigs probably would be present. These lesions are well documented (Kluge et al. 1992) and include nonsuppurative meningoencephalitis and ganglioneuritis; necrosis in the tonsil and intestine, with intranuclear inclusions in crypt epithelium; necrotic bronchitis, bronchiolitis, and alveolitis, with associated intranuclear inclusion bodies; focal necrosis in spleen, liver, lymph node, and adrenal, with intranuclear inclusion bodies in cells surrounding the necrotic areas; and periorchitis, endometritis, vaginitis, and necrotic placentitis.

DIAGNOSIS. Tissues for virus isolation (tonsil, brainstem, cerebrum, spleen, or lung) and swab samples (nasal or genital) should be refrigerated or frozen on dry ice (Gustafson 1986; Kluge et al. 1992). Tissue culture medium or a balanced salt solution can be used as transport medium but should not contain heparin, agar, or other sulfated polyanions, because these block viral adsorption to cells. Although PRV replicates in most cell lines (e.g., Vero cells), most laboratories use swine kidney cells (Romero et al. 1997). In cell culture, cytopathic effect is usually evident in 2–5 days. The presence of suid herpesvirus 1 can be confirmed by indirect flourescent antibody (IFA) technique.

Virus can be isolated from latently infected swine by reactivation of infection through steroid treatment (Gutekunst et al. 1980; Thawley et al. 1984) and by direct isolation using explant cultures of tonsil, cervical lymph nodes, nasal mucosa, and trigeminal ganglion (Sabo and Rajcani 1976; Beran et al. 1980; Brockmeier et al. 1993). In domestic swine, however, the trigeminal ganglion is believed to be the major site of viral latency and, therefore, is recommended for isolation attempts (Wheeler and Osorio 1991). Suid herpesvirus 1 also has been isolated from genital swabs of dexamethasone-treated feral swine (Romero et al. 1997), but isolation from explant cultures from latently infected feral or wild swine has not been reported.

Virus can be detected directly in tissues by fluorescent antibody procedures, but, although rapid, they may not be as sensitive as virus isolation (Kluge et al. 1992). Viral antigens in tissues can be detected also by using immunoperoxidase methods and enzyme immune assays (Osorio 1992). Several in situ hybridization and polymerase chain reaction (PCR) techniques for detection of viral DNA have been described (McFarlane and Thawley 1985; Belak et al. 1989; Brown et al. 1990; Jestin et al. 1990; Maes et al. 1990; Wheeler and Osorio 1991; Brockmeier et al. 1993). A PCR technique to detect RNA associated with latency transcripts also is available (Cheung 1995). These PCR techniques are ideally suited for confirmation of latent infections.

A wide variety of serologic tests are available (Gustafson 1986; Kluge et al. 1992; Osorio 1992), but virus neutralization, enzyme-linked immunosorbent assay (ELISA), and latex agglutination (LA) tests are the most used. Virus neutralization is regarded as the benchmark serologic test, although titers are often low. The LA test is commercially available and can be done in the field. Several ELISAs also are commercially available and are ideal for testing large numbers of samples. Differential ELISAs to distinguish between antibodies to wild-type viruses and gene-deletion vaccine viruses are also in current use.

DIFFERENTIAL DIAGNOSES. The major differential diagnoses for pseudorabies are rabies and canine distemper in species susceptible to canine distemper virus. In feral swine, a variety of other viral diseases, including Teschen and Talfan diseases, hog cholera, African swine fever, vomiting and wasting disease, swine vesicular disease, and encephalomyocarditis virus, may cause clinical signs and microscopic lesions that could be confused with pseudorabies. In other species, the presence of pruritus should help differentiate pseudorabies from other diseases. Other viral infections, in particular infection with other herpesviruses, could be confused with pseudorabies. For example, there is evidence that raccoons are infected with another herpesvirus that causes lesions that have to be differentiated from pseudorabies (Hamir et al. 1995). In canids, canine herpesvirus and canine adenovirus infections also could cause an encephalitis that could be confused with pseudorabies, while feline immunodeficiency virus and perhaps feline infectious peritonitis virus infections would have to be ruled out in cases of viral encephalitis in felids.

IMMUNITY. Information on the immune response to infection with suid herpesvirus 1 is limited to domestic swine. In domestic pigs, an immunoglobulin μ (IgM) antibody response is present within 6–8 days after infection, followed by an IgA and IgG response 8–10 days after infection (Kimman et al. 1992). After primary infection, pigs are immune to reinfection for at least 3 months. Maternal-derived immunity also is incomplete and persists for up to 4 months. It is likely that the immune responses seen in feral and wild swine are similar. Antibody prevalence in feral swine populations may be high, resulting in a high prevalence of maternally derived antibodies in pigs less than 2 months of age. Because clinical pseudorabies in domestic herds is often in young pigs, this high prevalence of maternal antibodies in infected feral swine populations may explain the lack of reported disease in wild populations. Information on antibodies in other wildlife species is sparse due to the fatal nature of this disease in nonporcine hosts.

CONTROL AND TREATMENT. Contact between domestic swine and feral swine or wild boar should be prevented. Control of pseudorabies in feral swine and wild boar populations currently is not feasible either through vaccination or population manipulations. Although vaccines are available, vaccination does not prevent infection, and in a single field trial was not effective in reducing antibody prevalence in a feral swine population (D.E. Stallknecht unpublished data). It is unknown whether maintenance of the virus in these populations is population dependent. Movement and establishment of new feral swine populations should be discouraged. Pseudorabies in other wildlife species is not treatable but can be prevented if direct contact with swine or their tissues can be eliminated.

PUBLIC HEALTH CONCERNS. There have been several illnesses in humans suspected to be pseudorabies virus infections, but none have been confirmed by virus isolation.

DOMESTIC ANIMAL HEALTH CONCERNS. Concerns will increase about PRV in feral swine as this virus is eradicated from domestic swine populations. Free-ranging feral swine and wild boar populations will continue to represent a reservoir of PRV and a potential source of infection to domestic animals.

MANAGEMENT IMPLICATIONS. There is no evidence that suid herpesvirus 1 infections adversely affect feral or wild swine populations. However, if attempts to eradicate this virus from domestic swine herds are successful, more attention will be devoted to the control of this virus in wild populations. Populations of feral and wild swine harboring PRV may impact endangered carnivores that prey on them (Capua et al. 1997b). Infections in black bears and panthers in Florida have been attributed to contact with infected feral swine.

LITERATURE CITED

Aujeszky, A. 1902. Ueber eine neue Infektionskrankheit bei Hautieren. *Zentralblatt Bakteriologie (Orig.)* 32:353–357.

Belak, S., A. Ballagipordany, J. Flensburg, and A. Virtanen. 1989. Detection of pseudorabies virus DNA sequences by the polymerase chain reaction. *Archives of Virology* 108:279–286.

Beran, G.W. 1992. Transmission of Aujeszky's disease virus. In *First international symposium on eradication of pseudorabies (Aujeszky's) virus,* ed. R.B. Morrison. St. Paul: University of Minnesota, College of Veterinary Medicine, pp. 93–112.

Beran, G.W., E.B. Davies, P.V. Arambulo, L.A. Will, H.T. Hill, and D.L. Rock. 1980. Persistence of pseudorabies virus in infected swine. *Journal of the American Veterinary Medical Association* 176:998–1000.

Brockmeier, S.L., K.M. Lager, and W.L. Mengeling. 1993. Comparison of in vivo reactivation, in vitro replication, and polymerase chain reaction for detection of latent pseudorabies virus infection in swine. *Journal of Veterinary Diagnostic Investigation* 5:505–509.

Brown, T.M., F.A. Osorio, and D.L. Rock. 1990. Diagnosis of latent pseudorabies virus infection using in situ hybridization. *Veterinary Microbiology* 24:273–280.

Capua, I., C. Casaccia, G. Calzetta, and V. Caporale. 1997a. Characterization of Aujeszky's disease viruses isolated form domestic animals and from wild boar (*Sus scrofa*) in Italy between 1972 and 1995. *Veterinary Microbiology* 57:143–149.

Capua, I., R. Fico, M. Banks, M. Tamba, and G. Calzetta. 1997b. Isolation and characterization of an Aujeszky's disease virus naturally infecting a wild boar (*Sus scrofa*). *Veterinary Microbiology* 55:141–146.

Cheung, A.K. 1995. Investigation of pseudorabies virus DNA and RNA in trigeminal ganglia and tonsil tissues of latently infected swine. *American Journal of Veterinary Research* 56:45–50.

Clarke, R.K., D.A. Jessup, D.W. Hurd, R. Ruppanner, and M.E. Meyer. 1983. Serologic survey of California wild hogs for antibodies against zoonotic disease agents. *Journal of the American Veterinary Medical Association* 183:1248–1251.

Corn, J.L., P.K. Swiderek, B.O. Blackburn, G.A. Erickson, A.B. Thiermann, and V.F. Nettles. 1986. Survey of selected diseases in wild swine in Texas. *Journal of the American Veterinary Medical Association* 189:1029–1032.

Dahle, J., T. Patzelt, G. Schagemann, and B. Liess. 1993. Antibody prevalence of hog cholera, bovine viral diarrhoea and Aujeszky's disease virus in wild boars in northern Germany. *Deutsche Tierarztliche Wochenschrift* 100:330–333.

Davies E.B., and G.W. Beran. 1981. Influence of environmental factors upon the survival of Aujeszky's disease virus. *Research in Veterinary Science* 31:32–36.

Galloway, I.A. 1938. Aujeszky's disease. *Veterinary Record* 50:745–763.

Glass, C.M., R.G. McLean, J.B. Katz, D.S. Maehr, C.B. Cropp, L.J. Kirk, A.J. McKeirnan, and J.F. Evermann. 1994. Isolation of pseudorabies (Aujeszky's disease) virus from a Florida panther. *Journal of Wildlife Diseases* 30:180–184.

Goto H., J.R. Gorham, and K.W. Hagen. 1968. Clinical observation of experimental pseudorabies in mink and ferrets. *Japanese Journal of Veterinary Science* 30:257–263.

Goyal S.M, R. Drolet, and P. King. 1986. Pseudorabies in free-ranging raccoons. *Journal of the American Veterinary Medical Association* 189:1163–1164.

Gustafson, D.P. 1986. Pseudorabies. In *Diseases of swine,* ed. A.D. Leman, B.E. Straw, R.D. Glock, W.L. Mengeling, R.H.C. Penny, and E. Scholl, 6th ed. Ames: Iowa State University Press, pp. 274–289.

Gutekunst, D.E., E.C. Pirtle, L.D. Miller, and W.C. Stewart. 1980. Isolation of pseudorabies virus from trigeminal ganglia of a latently infected sow. *American Journal of Veterinary Research* 41:1315–1316.

Hahn E.C., G.R. Page, P.S. Hahn, K.D. Gillis, C. Romero, J.A. Annelli, and E.P.J. Gibbs. 1997. Mechanisms of transmission of Aujeszky's disease virus originating from feral swine in the USA. *Veterinary Microbiology* 55:123–130.

Hamir A.N., G. Moser, M. Kao, N. Raju, and C.E. Rupprecht. 1995. Herpesvirus-like infection in a raccoon (*Procyon lotor*). *Journal of Wildlife Diseases* 31:420–423.

Hanson, R.P. 1954. The history of pseudorabies in the United States. *Journal of the American Veterinary Medical Association.* 124: 259–261.

Jestin, A., T. Foulon, B. Pertuiset, P. Blanchard, and M. Labourdet. 1990. Rapid detection of pseudorabies virus genomic sequences in biological samples from infected pigs using polymerase chain reaction DNA amplification. *Veterinary Microbiology* 23:317–318.

Kimman T.G., and J.T. van Oirschot. 1986. Pathology of Aujeszky's disease in mink. *Veterinary Pathology* 23:303–309.

Kimman T.G., T.G., A.T.J. Bianchi, and D. van Zaane. 1992. The immune system and the response to pseudorabies virus infection. In *First international symposium on eradication of pseudorabies (Aujeszky's) virus,* ed. R.B. Morrison. St. Paul: University of Minnesota, College of Veterinary Medicine, pp. 49–62.

Kirkpatrick, C.M., C.L. Kanitz, and S.M. McCrocklin. 1980. Possible role of wild mammals in the transmission of pseudorabies to swine. *Journal of Wildlife Diseases* 16:601–614.

Kluge, J.P., G.W. Beran, H.T. Hill, and K.B. Platt. 1992. Pseudorabies (Aujeszky's disease). In *Diseases of swine,* ed. A.D. Leman, B.E. Straw, W.L. Mengeling, S. D'Allaire, and D.J. Taylor, 7th ed. Ames: Iowa State University Press, pp. 312–323.

Konrad, J., and K. Blazek. 1958. Aujeszky's disease in mink. *Veterinary Bulletin* 29:307. Abstract 1755.

Lapcevic, E. 1964. Aujeszky's disease in mink. *Veterinary Bulletin* 34:587–588. Abstract 3673.

Ljubashenko, S.Y., A.F. Tyulpanova, and V.M. Grishin. 1958. Aujeszky's disease in mink, arctic fox, and silver fox. *Veterinary Bulletin* 34:244. Abstract 1386.

Maes, R.K., C.L. Kanitz, and D.P. Gustafson. 1979. Pseudorabies virus infections in wild and laboratory rats. *American Journal of Veterinary Research* 40:393–396.

Maes, R.K., C.E. Beisel, S.J. Spatz, and B.J. Thacker. 1990. Polymerase chain reaction amplification of pseudorabies virus DNA from acutely and latently infected cells. *Veterinary Microbiology* 24:281–295.

McFarlane, R.G., and D.G. Thawley. 1985. DNA hybridization procedure to detect latent pseudorabies virus DNA in swine tissue. *American Journal of Veterinary Research* 46:1133–1136.

McFerran, J.B., and C. Dow. 1964. The excretion of Aujeszky's disease virus by experimentally infected pigs. *Research in Veterinary Science* 5:405–410.

Moril, C.C., and R. Graham. 1941. An outbreak of bovine pseudorabies, or "mad itch." *American Journal of Veterinary Research* 2:35–40.

Nettles, V.F. 1989. Diseases of wild swine. In *Proceedings of the feral pig symposium, Orlando, Florida.* Madison, WI: Livestock Conservation Institute, pp. 16–18.

Nettles, V.F., and G.A. Erickson. 1984. Pseudorabies in wild swine. *Proceedings of the United States Animal Health Association* 88:505–506.

Nikolitsch, M. 1954. Die Aujeszkysch Krankheit beim Reh. *Wiener Tierarztliche Monatsschrift* 41:603–605.

Ordas, A., C. Sanchez-Botija, and S. Diaz. 1983. *Epidemiological Studies on African Swine Fever in Spain: Report EUR 8466.* EN Luxembourg and Commission of the European Communities, pp. 67–73.

Osorio, F.A. 1992. Diagnosis of pseudorabies (Aujeszky's disease) virus infections. In *First international symposium on eradication of pseudorabies (Aujeszky's) virus,* ed.

R.B. Morrison. St. Paul: University of Minnesota, College of Veterinary Medicine, pp. 17–32.
Pirtle, E.C., M.E. Roelke, and J. Brady. 1986. Antibodies against pseudorabies virus in serum of a Florida black bear cub. *Journal of the American Veterinary Medical Association* 189:1164.
Pirtle, E.C., J.M. Sacks, V.F. Nettles, and E.A. Rollor III. 1989. Prevalence and transmission of pseudorabies virus in an isolated population of feral swine. *Journal of Wildlife Diseases* 25:605–607.
Platt K.B., D.L. Graham, and R.A. Faaborg. 1983. Pseudorabies: Experimental studies in raccoons with different virus strains. *Journal of Wildlife Diseases* 19:297–301.
Quiroga M.I., S. Vazquez, M. Lopez-Pena, F. Guerroro, and J.M. Niets. 1995. Experimental Aujeszky's disease in blue foxes (*Alopex lagopus*). *Zentralblatt für Veterinarmedizin [A]* 42:649–657.
Quiroga M.I., M. Lopez-Pena, S. Vazquez, and J.M. Nieto. 1997. Distribution of Aujeszky's disease virus in experimentally infected mink (*Mustela vison*). *Deutsche Tierarztliche Wochenschrift* 104:147–150.
Raymond, J.T., R.G. Gillespie, M. Woodruff, and E.B. Janovitz. 1997. Pseudorabies in captive coyotes. *Journal of Wildlife Diseases* 34:916–918.
Reagan R.L., W.C. Day, R.T. Marley, and A.L. Brueckner. 1953. Effect of pseudorabies virus (Aujeszky stain) in the large brown bat (*Eptesicus fuscus*). *American Journal of Veterinary Research* 51:331–332.
Romero, C.H., P. Meade, J. Santagata, K. Gillis, G. Lollis, E.C. Hahn, and E.P.J. Gibbs. 1997. Genital infection and transmission of pseudorabies virus in feral swine in Florida, USA. *Veterinary Microbiology* 55:131–139.
Sabo, A., and J. Rajcani. 1976. Latent pseudorabies virus infection in pigs. *Acta Virologica (Praha)* 20:208–214.
Scherer, W.F. 1953. The utilization of a pure strain of mammalian cells (Earle) for the cultivation of viruses in vitro: I. Multiplication of pseudorabies and herpes simplex viruses. *American Journal of Pathology* 29:113–137.
Schultze, A.E., R.K. Maes, and D.C. Taylor. 1986. Pseudorabies and volvulus in a black bear. *Journal of the American Veterinary Medical Association* 189:1165–1186.
Shahan, M.S., R.L. Knudson, H.R. Seibold, and C.N. Dale. 1947. Aujeszky's disease (pseudorabies): A review, with notes on two strains of the virus. *North American Veterinarian* 28:440–449.
Shope, R.E. 1931. An experimental study of "mad itch" with especial reference to its relationship to pseudorabies. *Journal of Experimental Medicine* 54:233–248.
———. 1935. Experiments on the epidemiology of pseudorabies: II. Prevalence of the disease among middle western swine and the possible role of rats in herd-to-herd infections. *Journal of Experimental Medicine* 62:101–107.
Thawley, D.G., R.F. Solorzano, and M.E. Johnson. 1984. Confirmation of pseudorabies virus infection using virus recrudescence by dexamethasone treatment and in vitro lymphocyte stimulation. *American Journal of Veterinary Research* 45:981–983.
Tozzini, F., A. Poli, and G.D. Croce. 1982. Experimental infection of European wild swine (*Sus scrofa*) with pseudorabies virus. *Journal of Wildlife Diseases* 18:425–428.
Trainer, D.O. 1981. Pseudorabies. In *Infectious diseases of wild mammals,* ed. J.W. Davis, L.H. Karstad, and D.O. Trainer, 2d ed. Ames: Iowa State University Press, pp. 102–107.
Trainer, D., and L. Karstad. 1963. Experimental pseudorabies in some wild North American mammals. *Zoonoses Research* 2:135–151.
Traub, E. 1953. Cultivation of pseudorabies virus. *Journal of Experimental Medicine* 58:663–681.
Ugorski, L. 1957. Aujeszky's disease in silver foxes. *Veterinary Bulletin* 29:307. Abstract 1756.
Van der Leek, M.L., H.N. Becker, E.C. Pirtle, P. Humphrey, C.L. Adams, B.P. All, G.A. Erickson, R.C. Belden, W.B. Frankenberger, and E.P. Gibbs. 1993. Prevalence of pseudorabies (Aujeszky's disease) virus antibodies in feral swine in Florida. *Journal of Wildlife Diseases* 29:403–409.
Vanek, J., Groch, L., and A. Sanda. 1962. Aujeszky's disease in mink. *Veterinary Bulletin* 32:364. Abstract 1853.
Van Oirschot, J.T. 1992. Why pseudorabies is a candidate for eradication. In *First international symposium on eradication of pseudorabies (Aujeszky's) virus,* ed. R.B. Morrison. St. Paul: University of Minnesota, College of Veterinary Medicine, pp. 85–92.
Von Becker, C.H., and H.J. Herrmann. 1963. Zur ubertragbarkeit des Aujeszky-virus durch die ratte. *Montatshefte für Veterinaermedizin* 18:181–184.
Wheeler, J.G., and F.A. Osorio. 1991. Investigation of sites of pseudorabies virus latency, using polymerase chain reaction. *American Journal of Veterinary Research* 52:1799–1803.
Whittmann, G., and J.J. Rziha. 1989. Herpesvirus diseases of cattle, horses, and pigs. In *Developments in veterinary virology.* Boston: Kluwer Academic, pp. 230–325.
Wright J.C., and D.G. Thawley. 1980. Role of the raccoon in the transmission of pseudorabies: A field and laboratory investigation. *American Journal of Veterinary Research* 41:581–583
Xenon, E., I. Capua, C. Cassaccia, A. Zuin, and A. Moresco. 1997. Isolation and characterization of Aujeszky's disease virus in captive brown bears from Italy. *Journal of Wildlife Diseases* 33:632–634.

ELEPHANT HERPESVIRUS INFECTIONS

LAURA K. RICHMAN AND RICHARD J. MONTALI

Synonyms: Elephant endotheliotropic inclusion body disease, loxodontal herpesvirus-induced pulmonary nodules, loxodontal elephantid herpesvirus, herpes-like virus-induced cutaneous papillomatosis of African elephants, African elephant herpesvirus, Asian elephant herpesvirus.

INTRODUCTION. Several herpesvirus diseases of elephants have been described in the literature. Since the 1981 review of elephant herpesvirus (Plowright 1981), new herpesviruses that cause a highly fatal systemic disease in African *Loxodonta africana* and Asian *Elaphus maximus* elephants have been identified. The original elephant loxodontal herpesvirus was described as an incidental finding in pulmonary nodules from asymptomatic elephants culled in South Africa (Basson et al. 1971; McCully et al. 1971). The virus was never characterized beyond its ultrastructural features; a claim of virus isolation (cited by Plowright 1981) could not be confirmed. The relationship of the loxodontal herpesvirus to the newly described elephant herpesviruses is unknown. The most recently characterized elephant herpesvirus disease manifests as an acute, generalized hemorrhagic diathesis primarily involving heart, liver, intestine, and tongue.

HISTORY. Ten elephants in eight North American zoos had a highly fatal disease with common clinical and postmortem findings attributed to endotheliotropic herpesviruses (Richman et al. 1999). Eight were Asian elephants, one of which survived after a course of the antiherpesvirus drug, famciclovir (Schmitt and Hardy 1998). The lesions in the African elephants were nearly identical to those in the Asian elephants; the herpesvirus appears to be a distinct virus but related to the virus that caused disseminated disease in the Asian species.

DISTRIBUTION. As of July 1998, there were ten confirmed cases of disseminated herpesvirus disease in elephants in North America. The cases have been sporadic and scattered; a distinct epidemiologic pattern has not yet been elucidated. Cases in other countries include a young Asian circus elephant in Switzerland (Ossent et al. 1990), three cases in Asian elephants elsewhere in Europe, and a calf from a zoo in Israel that died with lesions attributed to this disease. The status of the disease in free-ranging elephants in Africa and Asia is undetermined.

HOST RANGE. To date, all investigations strongly suggest the disease is limited to elephants.

ETIOLOGY. Polymerase chain reaction (PCR) on DNA extracted from tissues with lesions from seven Asian elephants and from blood of the elephant that survived yielded products with identical protein sequences in the terminase gene region of the herpesvirus. In contrast, nucleotide sequences in this region from the African elephant virus were 20% different. Sequence comparison for the herpesvirus DNA polymerase region showed 76% protein identity and 35% nucleotide difference between the viruses detected in the Asian and African elephants (Richman et al. 1999). These viruses are unlike any known herpesvirus in the two gene regions so far studied and may represent a new subfamily of herpesviruses. To date, virus isolation has been unsuccessful.

TRANSMISSION AND EPIZOOTIOLOGY. The index case for acute, disseminated disease occurred in 1995: a 16-month-old Asian elephant born at the National Zoological Park in Washington, D.C. Recent cases have included an 11-month-old African elephant and a 17-month-old Asian elephant. In retrospective studies, six additional Asian elephant cases and an African elephant case occurring between 1983 and 1993 were identified in North America. Eight of ten elephants with confirmed disease ranged in age from 18 months to 7 years. There was no sex or seasonal predilection, and affected elephants were widely distributed across the continent. Further losses from the herpesvirus disease did not occur in elephant calves raised at three of the facilities.

Between 1983 and 1996, a total of 34 Asian elephants were born in North America, eight of which died with lesions attributed to the endotheliotropic herpesvirus disease. Only seven African elephants were born in North America during the same period, and two died of this disease.

CLINICAL SIGNS. Onset of disease is sudden, with a course of 1–7 days. The first signs typically include lethargy, anorexia, mild colic, and edematous swelling of the head, neck, proboscis, and thoracic limbs. Cyanosis occurs at the tip of the tongue and spreads caudally at any stage of the disease. Disease is usually fatal.

PATHOGENESIS. One of the hallmarks of the endotheliotropic herpesviruses is hemorrhagic diathesis, particularly in heart, liver, and tongue. Once the elephant becomes viremic, productive infection of capillary endothelial cells by the herpesvirus is cytocidal, leading to capillary leakage and widespread hemorrhage. Severe diffuse myocardial hemorrhage is likely associated with circulatory shock.

PATHOLOGY. Pertinent laboratory findings are derived from three elephants, which variably showed leukopenia, thrombocytopenia, and a low erythrocyte count. Anemia persisted for 1 week in the one Asian elephant survivor.

Gross lesions typically include pericardial effusion, with extensive petechial to ecchymotic hemorrhages involving the entire myocardium. Petechia may be found in the visceral and parietal peritoneal surfaces and, variably, there is cyanosis of the tongue. In addition, hepatomegaly may be present. Oral, laryngeal, and intestinal ulcers often occur.

Plowright (1981) described a high prevalence of lymphoid nodules (3–30-mm diameter) in lungs of African elephants with elephant herpesvirus infection. Some nodules were encapsulated and contained cavities, but these were not thought to cause clinical disease.

Salient microscopic findings consist of extensive microhemorrhages throughout the heart and tongue associated with edema and mild infiltrates of lymphocytes, monocytes, and a few neutrophils. The liver shows interstitial edema with similar inflammatory cell types and minimal hepatocellular degenerative changes. Ulcers in the larynx and digestive tract are acute, with necrotic surface epithelial cells still intact in some areas. Capillary endothelial cells of the myocardium, tongue, and sinusoids of the liver contain amphophilic to basophilic viral inclusion bodies in the nuclei (Fig. 7.1). The endothelial cells containing inclusion bodies are often closely associated with microhemorrhages in heart and tongue. Inclusion bodies are less frequent in the capillary endothelial cells of the inner submucosa and smooth muscle layers of the intestinal tract and have not been seen in ulcers or in blood vessels larger than capillaries.

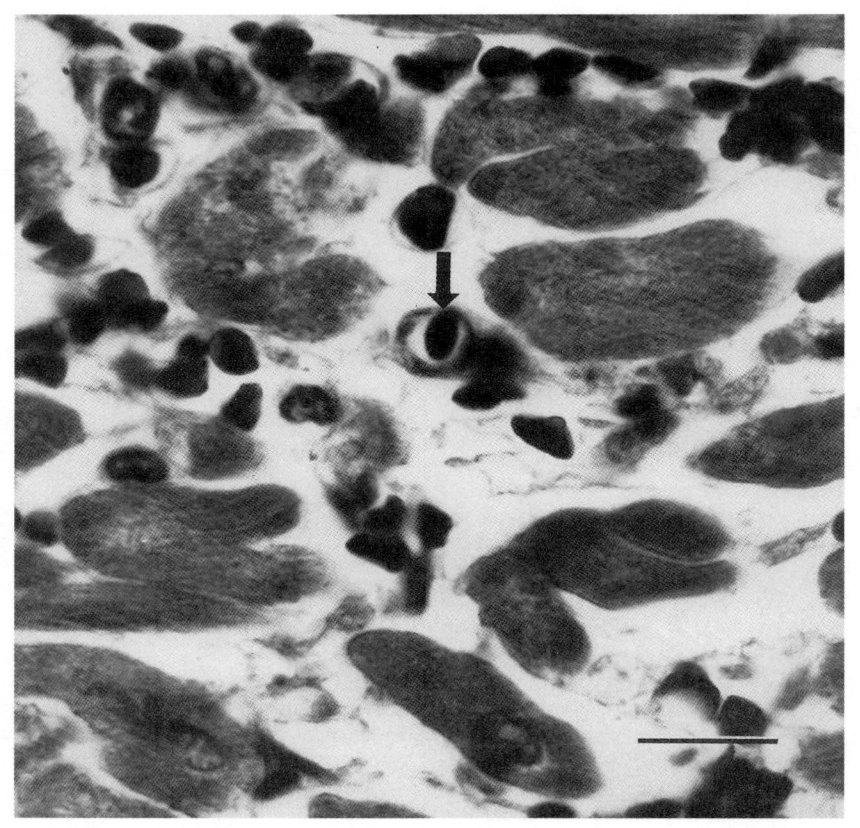

FIG. 7.1—Photomicrograph of the heart from a young Asian elephant that died from endotheliotropic herpesvirus infection. The intranuclear inclusion body is in a capillary endothelial cell (*arrow*). Similar inclusion bodies may be found in hepatic sinusoidal lining cells, as well as capillary endothelial cells within the tongue, intestinal tract, spleen, and less frequently, in other organs. Bar = 25 μm.

Lymphoid nodules in lungs of African elephants commonly contained germinal centers and metaplastic or hyperplastic epithelium with intranuclear inclusion bodies. Syncytia were common, and eosinophilic bodies were in the cytoplasm (Plowright 1981).

By electron microscopy, the inclusion bodies within the capillary or sinusoidal endothelial cells contained 80–92-nm-diameter viral nucleocapsids morphologically consistent with herpesvirions. Viral capsids are limited to the nucleus; no viruses were seen within the cytoplasm or intercellular regions.

DIAGNOSIS. During the early, viremic stage of the disseminated disease, the virus can be detected in peripheral blood by PCR (Richman et al. 1999). The herpesvirus was detectable in diminishing amounts in peripheral blood of the surviving Asian elephant for 7 weeks. A presumptive postmortem diagnosis can be made by light and electron microscopy, and heart, liver, tongue, and intestine can be tested by PCR for genomic sequences of elephant herpesvirus. Herpesvirus has not yet been isolated in tissue culture from any of the cases. Serologic evidence of herpesvirus infection of Asian elephants has been reported by virus neutralization and immunoblotting (Metzler et al. 1990).

DIFFERENTIAL DIAGNOSES. Where appropriate, encephalomyocarditis virus infection should be ruled out. Other diseases of elephants that may have an acute onset and a rapid death include clostridial enterotoxemia, leptospirosis, salmonellosis, other bacterial septicemias, toxicoses, and micronutrient deficiencies such as vitamin-E deficiency.

IMMUNITY. The fact that a majority of elephants affected by the disseminated form of the disease were relatively young suggests that adults may develop immunity. However, the immune responses of elephants to infection with these viruses have not yet been studied.

CONTROL AND TREATMENT. The one Asian elephant survivor was diagnosed early in the course of the disease by PCR on peripheral blood. A 3-week course of treatment with the human antiviral drug, famciclovir (Famvir) (Schmitt and Hardy 1998), appeared to decrease viral load in serial peripheral blood samples.

PUBLIC HEALTH AND DOMESTIC ANIMAL CONCERNS. There are no currently identified public health concerns.

MANAGEMENT IMPLICATIONS. Specific management practices as they apply to endotheliotropic herpesvirus disease in elephants have not yet been developed and await more epidemiologic information about this disease.

LITERATURE CITED

Basson, P.A., R.M. McCully, V. De Vos, E. Young, and S.P. Kruger. 1971. Some parasitic and other natural diseases of the African elephant in the Kruger National Park. *Onderstepoort Journal of Veterinary Research* 38:239–254.

McCully, R.M., P.A. Basson, J.G. Pienaar, B.J.E. Erasmus, and E. Young. 1971. Herpes nodules in the lung of the African elephant in the Kruger National Park. *Onderstepoort Journal of Veterinary Research* 38:239–254.

Metzler A.E., P. Ossent, F. Guscetti, A. Rubel, and E.M. Lang. 1990. Serological evidence of herpesvirus infection in captive Asian elephants (*Elephas maximus*). *Journal of Wildlife Diseases* 26:41–49.

Ossent P., F. Guscetti, A.E. Metzler, E.M. Lang, A. Rubel, and B. Hauser. 1990. Acute and fatal herpesvirus infection in a young Asian elephant (*Elephas maximus*). *Veterinary Pathology* 27:131–133.

Plowright, W. 1981. Herpesvirus of wild ungulates, including malignant catarrhal fever virus. In *Infectious diseases of wild mammals,* ed. J.W. Davis, L.H. Karstad, and D.O. Trainer, 2d ed. Ames: Iowa State University Press, pp. 126–146.

Richman, L.K., R.J. Montali, R.L. Garber, M.A. Kennedy, J. Lehnhardt, T. Hidebrandt, D. Schmitt, D. Hardy, D.J. Alcendor, and G.S. Hayward. 1999. Novel endotheliotropic herpesvirus fatal for Asian and African elephants. *Science* 283:1171–1176.

Schmitt, D.L., and D.A. Hardy. 1998. Use of famciclouir for the treatment of herpes virus in an Asian elephant. *Journal of Elephant Managers Association* 9:103–104.

CAPRINE HERPESVIRUS

MONIKA ENGELS

Synonyms: Caprine herpesvirus 1 (formerly caprine herpesvirus 2, bovid herpesvirus 6).

INTRODUCTION. Caprine herpesvirus 1 (CapHV-1) infections in domestic animals have been reviewed (Koptopoulos 1992). They cause severe generalized disease in newborn goat kids (Saito et al. 1974; Mettler et al. 1979). In adult animals, subclinical infections are most common, but respiratory and genital tract infections have been observed (Horner et al. 1982; Buddle et al. 1990); abortions may occur (Berrios et al. 1975; Waldvogel et al. 1981; Williams et al. 1997).

HISTORY AND DISTRIBUTION. Caprine herpesvirus 1 was isolated first in California in 1974 from newborn goat kids suffering from enteritis (Saito et al. 1974). Five years later, CapHV-1 was isolated in Switzerland from young kids with a similar disease (Mettler et al. 1979). The virus has been isolated from adult goats in Europe, North America, New Zealand, and Australia (Horner et al. 1982; Grewal and Wells 1986; Williams et al. 1997). The distribution of CapHV-1 is probably worldwide (Plebani et al. 1983; Kao et al. 1985; Koptopoulos et al. 1988).

HOST RANGE. The natural host range of CapHV-1 may not be strictly restricted to goats. A few serologic surveys of wild ruminants (Plebani et al. 1983; Nixon et al. 1988) have detected antibodies against CapHV-1 only in ibex *Capra ibex* from France (Thiry et al. 1988). Nevertheless, since domestic calves and lambs are susceptible to experimental infection (Berrios et al. 1975; Engels et al. 1992; Papanastasopoulou et al. 1991), other ruminants might also be susceptible. In addition, CapHV-1 shares genetic and antigenic relationship with bovine herpesvirus 1 (BHV-1) and other ruminant alphaherpesviruses (Brake and Studdert 1985; Ackermann et al. 1986; Engels et al. 1987; Nixon et al. 1988). Antibodies against BHV-1 have been found in a number of wild ungulates in the families Bovidae, Cervidae, Giraffidae, Hippopotamidae, and Suidae [reviewed in Wyler et al. (1989)]. In many cases, though, the sera were not tested for antibodies against CapHV-1. Thus, further seroepidemiologic studies using both viruses are necessary to determine the host range of CapHV-1 among wild ungulates.

ETIOLOGY. Based on its biologic and genetic properties, CapHV-1 is a typical alphaherpesvirus. Although cells of ruminant origin are preferable, CapHV-1 replicates in a broad spectrum of cells. It induces a cytopathic effect characterized by cell rounding and swelling, followed by detachment in clumps (Engels et al. 1983). Caprine herpesvirus 1 has a group-D genome that is similar to the BHV-1 genome (Engels et al. 1983, 1987). Minor differences in the restriction patterns of different CapHV-1 strains have been observed (Engels et al. 1987; Rimstad et al. 1992).

Caprine herpesvirus 1 is antigenically related to BHV-1 and other ruminant alphaherpesviruses (Engels et al. 1983; Kao et al. 1985; Nixon et al. 1988; Martin et al. 1990). Typically there is serologic one-way cross-reactivity, e.g., antibodies against BHV-1 neutralize

CapHV-1 to a much higher extent than CapHV-1 antibodies neutralize BHV-1, allowing for possible differentiation based on serologic responses (Hasler and Engels 1986).

TRANSMISSION AND EPIZOOTIOLOGY. Acutely and latently infected animals are the main source of infection. Acutely infected animals excrete virus with ocular, nasal, and genital secretions (Mettler et al. 1979; Tarigan et al. 1987). Thus, direct contact with virus by ingestion or inhalation is considered the main route of transmission.

Based on serologic findings in goat herds, virus apparently was mainly spread during the breeding season. Venereal transmission may be responsible for introduction into and maintenance of the infection within herds (Tarigan et al. 1990).

CLINICAL SIGNS. Primary signs in young goat kids are progressive weakness and abdominal pain. Conjunctivitis, serous to purulent nasal discharge, erosions in the oral cavity, and petechia in the skin are also observed. Most kids die within 1–4 days, but some may survive.

In animals with genital lesions, clinical signs resemble those of BHV-1-induced vulvovaginitis and balanoposthitis in cattle. Hyperemia and edema occur in the genital mucous membranes, with multiple papules, vesicles, and ulcers. The lesions disappear within 4–6 days (Horner et al. 1982; Grewal and Wells 1986).

Acute pneumonia associated with CapHV-1 and *Mannheimia (Pasteurella) haemolytica* was described in goats (Buddle et al. 1990). Recent studies have shown that abortions occur in association with CapHV-1 infections (Williams et al. 1997).

PATHOGENESIS. The pathogenesis of CapHV-1 infections is similar to that of other alphaherpesvirus infections (Engels and Ackermann 1996). Virus spread within the host may be initiated by leukocyte-associated viremia, leading to generalized disease or abortion (Smith 1997). Similar to other herpesviruses, CapHV-1 can establish latent infections (Koptopoulos et al. 1988); natural reactivation may occur at times of stress or treatment with high levels of dexamethasone (Buonavoglia et al. 1996).

PATHOLOGY. Lesions are found throughout the digestive tract. The most extensive lesions are in cecum and colon, where there is extensive thickening of the mucosa. In the genital tract, there may be petechial hemorrhages and numerous small erosions and ulcerations on the mucosa (Tarigan et al. 1990). The anterior lobes of the lungs may be consolidated and, occasionally, there is fibrinous pleuritis (Buddle et al. 1990). No gross lesions occur in fetuses aborted due to CapHV-1 infection (Williams et al. 1997).

Microscopic lesions are focal or extensive epithelial necrosis and ulceration in the digestive tract, urinary bladder, genital mucosa, and skin. Ballooning degeneration of epithelial cells is observed in skin vesicles (Waldvogel et al. 1981). Intranuclear, acidophilic inclusion bodies are in the epithelium in areas of ulceration and necrosis (Saito et al. 1974; Mettler et al. 1979; Tarigan et al. 1990). Furthermore, edema of myocardium, gliosis, and meningitis (Papanastasopoulou et al. 1991), as well as nonsuppurative interstitial pneumonia, occur (Waldvogel et al. 1981). Focal necrosis occurs in liver, lungs, and spleen of fetuses aborted due to CapHV-1 infection; hepatocytes adjacent to foci of necrosis contain eosinophilic intranuclear inclusion bodies (Williams et al. 1997).

DIAGNOSIS. Diagnosis is based on isolation of virus in primary bovine, lamb, or goat cell cultures. Viruses can be identified by immune electron microscopy, immunofluorescence, or serum neutralization. Viral antigen can be demonstrated by immunohistochemistry in infected tissues. Serum neutralization tests may be useful for diagnostics and seroepidemiology (Kao et al. 1985; Hasler and Engels 1986).

DIFFERENTIAL DIAGNOSIS. Chlamydial and mycoplasmal infections must be considered as differential diagnoses of generalized CapHV-1 disease (Saito et al. 1974). In cases of abortion, chlamydiosis, leptospirosis, toxoplasmosis, and brucellosis, as well as bovine virus diarrhea virus, bluetongue virus, and BHV-1 infections, must be considered (Koptopoulos et al. 1988; Williams et al. 1997).

IMMUNITY. Humoral antibodies can detected 1–2 weeks after primary infection (Berrios et al. 1975; Ackermann et al. 1986; Papanastasopoulou et al. 1991). Antibodies have lifelong persistence due to booster reactions following reactivation. The role of cell-mediated immunity in CapHV-1 infection is unknown.

CONTROL AND TREATMENT. Due to direct-contact transmission, seropositive captive animals can be culled or separated from seronegative members of the herd. No vaccines are available.

PUBLIC HEALTH CONCERNS. There are no known human health concerns associated with CapHV-1.

DOMESTIC ANIMAL HEALTH CONCERNS. This is a disease of domestic goats, and a virus reservoir in wildlife is not known to exist.

LITERATURE CITED

Ackermann, M., A.E. Metzler, H. McDonagh, L. Bruckner, H.K. Müller, and U. Kihm. 1986. Stellen nichtbovine Paarhufer ein IBR-Virus-Reservoir dar? I. BHV-1 und CapHV-1-Infektions- und Reaktivierungsversuche an Ziegen, Virustyp-Spezifität der humoralen Antikörper und Charakterisierung der viralen Antigene. *Schweizer Archiv für Tierheilkunde* 128:557–573.

Berrios, P.E., D.G. McKercher, and H.D. Knight. 1975. Pathogenicity of a caprine herpesvirus. *American Journal of Veterinary Research* 36:1763–1769.

Brake, F., and M.J. Studdert. 1985. Molecular epidemiology and pathogenesis of ruminant herpesviruses including bovine, buffalo and caprine herpesvirus 1 and bovine encephalitis herpesvirus. *Australian Veterinary Journal* 62:331–334.

Buddle, B.M., A. Pfeffer, D.J.W. Cole, H.D. Pulford, and M.J. Ralston. 1990. A caprine pneumonia outbreak associated with caprine herpesvirus and *Pasteurella haemolytica* respiratory infections. *New Zealand Veterinary Journal* 38:28–31.

Buonavoglia, C., M. Tempesta, A. Cavalli, V. Voigt, D. Buonavoglia, A. Conserva, and M. Corrente. 1996. Reactivation of caprine herpesvirus 1 in latently infected goats. *Comparative Immunology, Microbiology and Infectious Diseases* 19:275–281.

Engels, M., and M. Ackermann. 1996. Pathogenesis of ruminant herpesvirus infections. *Veterinary Microbiology* 53:3–15.

Engels, M., H. Gelderblom, G. Darai, and H. Ludwig. 1983. Goat herpesviruses: Biological and physicochemical properties. *Journal of General Virology* 64:2237–2247.

Engels, M., E. Loepfe, P. Wild, E. Schraner, and R. Wyler. 1987. The genome of caprine herpesvirus 1: Genome structure and relatedness to bovine herpesvirus 1. *Journal of General Virology* 68:2019–2023.

Engels, M., M. Palatini, A.E. Metzler, U. Probst, U. Kihm, and M. Ackermann. 1992. Interactions of bovine and caprine herpesviruses with the natural and the foreign host. *Veterinary Microbiology* 33:69–78.

Grewal, A.S., and R. Wells. 1986. Vulvovaginitis of goats due to a herpesvirus. *Australian Veterinary Journal* 63:79–82.

Hasler, J., and M. Engels. 1986. Stellen nichtbovine Paarhufer ein IBR-Virus-Reservoir dar? II. Seroepidemiologische Untersuchungen an Ziegen, Schafen, Schweinen und Wildpaarhufern in der Schweiz. *Schweizer Archiv für Tierheilkunde* 128:575–585.

Horner, G.W., R. Hunter, and A.M. Day. 1982. An outbreak of vulvovaginitis in goats caused by a caprine herpesvirus. *New Zealand Veterinary Journal* 30:150–152.

Kao, M., T. Leiskau, G. Koptopoulos, O. Papadopoulos, G.W. Horner, B. Hyllseth, M. Fadel, A.H. Gedi, O.C. Straub, and H. Ludwig. 1985. Goat herpesvirus infection: A survey of specific antibodies in different countries. In *Immunity to herpesvirus infections of domestic animals: Report EUR 9737 EN,* ed. P.-P. Pastoret, E. Thiry, and J. Saliki. C.E.C. Agriculture, pp. 93–97.

Koptopoulos, G. 1992. Goat herpesvirus 1 infection: A review. *Veterinary Bulletin* 62:77–84.

Koptopoulos, G., M. Papanastasopoulou, O. Papadopoulos, and H. Ludwig. 1988. The epizootiology of caprine herpesvirus (BHV-6) infections in goat populations in Greece. *Comparative Immunology, Microbiology and Infectious Diseases* 11:199–205.

Martin, W.B., G. Castrucci, F. Frigeri, and M.A. Ferrari. 1990. Serological comparison of some animal herpesviruses. *Comparative Immunology, Microbiology and Infectious Diseases* 13:75–84.

Mettler, F., M. Engels, P. Wild, and A. Bivetti. 1979. Herpesvirus-Infektion bei Zicklein in der Schweiz. *Schweizer Archiv für Tierheilkunde* 121:655–662.

Nixon, P., S. Edwards, and H. White. 1988. Serological comparisons of antigenically related herpesviruses in cattle, red deer and goats. *Veterinary Research Communications* 12:355–362.

Papanastasopoulou, M., G. Koptopoulos, S. Lekkas, O. Papadopoulos, and H. Ludwig. 1991. An experimental study on the pathogenicity of the caprine herpesvirus type 1 (CHV-1). *Comparative Immunology, Microbiology and Infectious Diseases* 14:47–53.

Plebani, G.F., M. Engels, A.E. Metzler, and R. Wyler. 1983. Caprines Herpesvirus in der Schweiz: Verbreitung, Häufigkeit und Latenz der Infektion. *Schweizer Archiv für Tierheilkunde* 125:395–411.

Rimstad, E., R. Krona, and B. Hyllseth. 1992. Comparison of herpesviruses isolated from reindeer, goats, and cattle by restriction endonuclease analysis. *Archives of Virology* 123:389–997.

Saito, J.K., D.H. Gribble, P.E. Berrios, H.D. Knight, and D.G. McKercher. 1974. A new herpesvirus isolate from goats: Preliminary report. *American Journal of Veterinary Research* 35:847–848.

Smith, K.C. 1997. Herpesviral abortion in domestic animals. *Veterinary Journal* 153:253–268.

Tarigan, S., R.F. Webb, and D. Kirkland. 1987. Caprine herpesvirus from balanoposthitis. *Australian Veterinary Journal* 64:321.

Tarigan, S., P.W. Ladds, and R.A. Foster. 1990. Genital pathology of feral male goats. *Australian Veterinary Journal* 67:286–290.

Thiry, E., M. Vercouter, J. Dubuisson, J. Barrat, C. Sepulchre, C. Gerardy, C. Meersschaert, B. Collin, J. Blancou, and P.-P. Pastoret. 1988. Serological survey of herpesvirus infections in wild ruminants of France and Belgium. *Journal of Wildlife Diseases* 24:268–273.

Waldvogel, A., M. Engels, P. Wild, H. Stünzi, and R. Wyler. 1981. Caprine herpesvirus infection in Switzerland: Some aspects of its pathogenicity. *Zentralblatt für Veterinarmedizin [B]* 28:612–623.

Williams, N.M., M.L. Vickers, R.R. Tramontin, M.B. Petrites-Murphy, and G.P. Allen. 1997. Multiple abortions associated with caprine herpesvirus infection in a goat herd. *Journal of the American Veterinary Medical Association* 211:89–91.

Wyler, R., M. Engels, and M. Schwyzer. 1989. Infectious bovine rhinotracheitis/vulvo-vaginitis. In *Herpesvirus diseases of cattle, horses and pigs,* ed. G. Wittmann. Boston: Kluwer Academic, pp. 1–72.

OTHER HERPESVIRUSES

ANTHONY E. CASTRO

BOVINE HERPESVIRUS 1

Synonyms: Infectious bovine rhinotracheitis, IBR, infectious pustular vulvovaginitis, IPV, balanoposthitis, red nose.

Introduction. Bovine herpesvirus 1 (BHV-1), an alphaherpesvirus, is a pathogen primarily of cattle *Bos taurus,* producing an upper respiratory infection, viral pneumonia, vaginitis, and balanoposthitis. It has been associated with keratoconjunctivitis, encephalitis, and abortions of cattle. The herpesvirus has been isolated from goats *Capra hircus* and swine *Sus scrofa* with similar disease syndromes as cattle. The virus is primarily found in feedlot and confined animals, affecting primarily those over 6 months of age (Plowright 1981; Radostits et al. 1994). Infectious bovine rhinotracheitis (IBR) virus contains a core of double-stranded DNA, is 100–110 nm in diameter with 162 capsomeres, and has an envelope that has virus glycoprotein spikes on its surface. The presence of a lipid envelope is necessary for its infectivity in the field; thus, the herpesvirus is sensitive to lipid solvents and certain detergents (Roizman 1996).

Distribution and Host Range. Plowright (1981) reviewed IBR in wild species. Although primarily a virus of domestic cattle (Ludwig 1983), the herpesvirus has been isolated or antibodies detected in a large number of wild species. From North America and Europe, these include white-tailed deer *Odocoileus virginianus,* caribou *Rangifer tarandus,* red deer *Cervus elaphus,* mule deer *Odocoileus hemionus,* pronghorn *Antilocapra americana,* American bison *Bison bison,* and Peninsula bighorn sheep *Ovis canadensis cremnobates* (Chow and Davis 1964; Elazhary et al. 1981; Plowright 1981; Nixon et al. 1988; Lamontague et al. 1989; Sadi et al. 1991; Clark et al. 1993; Taylor et al. 1997). The virus is widespread in African wildlife; BHV-1 or antibodies have been found in African buffalo *Syncerus caffer,* eland *Taurotragus oryx,* wildebeest *Connochaetes taurinus,* topi *Damaliscus lunatus,* hartebeest *Alcelaphus buselaphus,* sable antelope *Hippotragus niger,* roan antelope *Hippotragus equinus,* impala *Aepyceros melampus,* kob *Kobus kob,* lechwe *Kobus leche,* waterbuck *Kobus ellipsiprymnus,* reedbuck *Redunca redunca,* Thomson's gazelle *Gazella thomsonii,* and hippopotamus *Hippopotamus amphibius* (Hedger and Hamblin 1978; Plowright 1981; Anderson and Rowe 1998). In Africa, wildlife may be latent carriers for this virus; recrudescence of latent virus in wildebeest followed administration of corticosteroids (Plowright 1981).

Clinical Signs and Transmission. Wild animals seldom display clinical signs associated with BHV-1 infection. Pustular vulvovaginitis occurred in wildebeest treated with corticosteroids (Plowright 1981). The virus is transmitted via semen, excretions and secretions, and fetal fluids. Several genotypes of the virus exist, which may explain its varied clinical expression in different individuals and species. Evidence suggests that herpesvirus may become latent and reside in the trigeminal ganglia (Radostits et al. 1994). Incubation periods vary from 3–7 days to as long as 20 days. The virus primarily replicates in the epithelial cells of nasal mucosal membranes and genital tract (Ludwig 1983). Replication is accompanied by a fever that may be as high as 42.2° C. Secondary bacterial pneumonia is common.

Diagnosis, Treatment, and Control. Serology and virus isolation are used for diagnosis of BHV-1 infection. By restriction endonuclease (RE) analysis, field isolates can be differentiated from attenuated vaccine strains (Whetstone et al. 1986). Additionally, isolates of herpesviruses from cases of IPV and IBR demonstrated differences by RE analysis (Engels et al. 1981).

Treatment of BHV-1 infection usually includes supportive treatment and antibiotics. Vaccination with attenuated or inactivated vaccines may be warranted in confined and valuable susceptible ruminants. However, pregnant hosts should not be given attenuated virus vaccines.

BOVINE HERPESVIRUS 2. Bovine herpesvirus 2 (BHV-2 or bovine herpes mammillitis virus, Allerton virus, pseudo-lumpy skin disease virus) is classified in the family *Alphaherpesvirinae* with morphology identical to other herpesviruses. All the strains are serologically similar, but BHV-2 shares glycoprotein antigens involved with host immunity that are common with herpes hominis (simplex) HSV-1 and 2 (Plowright 1981; Osorio 1992).

The virus primarily affects cattle. It was first isolated in Africa and has been reported in African buffalo (Schiemann et al. 1971). It has been isolated from cattle in Africa, North America, Europe, and Australia. Virus-neutralizing antibodies have been found in wildebeest, oryx *Oryx gazella,* bushbuck *Tragelaphus scriptus,* waterbuck, impala, giraffe *Giraffa camelopardalis,* eland, and hippopotamus (Plowright 1981). Wild species may be subclinical reservoirs (Osorio 1992).

Infection results in skin lesions on the udder and teats and may affect the perineum or vulvovaginal mucosa (Plowright 1981). Evidence suggests that biting insects and mechanical trauma to the teats may be the mode of transmission. Incubation varies by the route of inoculation, but lesions typically appear in 1–3 days, with generalized skin lesions seen within 4–8 days (Kahrs 1981). Latency is a characteristic of BHV-2; thus, stress or treatment with corticosteroids may cause reactivation. Lower body temperatures in the target tissues appear to play a role in the productive infection. The virus produces vesicular lesions on the udder, which should be examined for virus to rule out rinderpest and foot-and-mouth disease. Infected animals are usually diagnosed by serology.

Natural infections usually confer solid immunity in the host. Control of insect vectors is also a management method for decreasing exposure of susceptible animals (Kahrs 1981). New antiviral agents used for local and systemic treatment of herpesviruses [vidarabine, a DNA base analogue, and phosphonoacetic acid, an inhibitor of viral DNA polymerase (Osorio 1992)] appear to be effective in controlling the disease. Due to the self-limiting nature of the infection, only supportive therapy may be warranted in affected animals.

BOVINE HERPESVIRUS 4. Bovine herpesvirus 4 (BHV-4), or bovine cytomegalovirus, belongs to the

Gammaherpesvirinae and was formerly referred to as bovine herpesvirus 3 (BHV-3). There appear to be some serologic cross-reactions of these cytomegaloviruses with alcelaphine herpesvirus 1 (Rossiter et al. 1988) and other bovine herpesviruses. The virus produces a variety of disease syndromes (Metzler 1992). Some isolates cause respiratory disease (Potgieter and Aldridge 1977); metritis and vaginitis have been reported in affected cattle. A cytomegalovirus isolate was obtained from an African buffalo (Rossiter et al. 1988).

This herpesvirus has been detected in buffalo and infection rates vary from 2% to 70% as determined by serology. It consistently establishes persistent infections but rarely produces clinical disease (Metzler 1992). By restriction endonuclease analysis, BHV-4 isolates are different from other bovine herpesviruses. The level of virus-neutralizing antibodies in wild ungulates is unknown, but BHV-4 infection appears to be self-limiting; thus, vaccinations are not warranted (Metzler 1992).

CERVID HERPESVIRUSES. Many serologic surveys of free-ranging cervids have show evidence of exposure to BHV-1 or related viruses. It has been recognized that herpesviruses occur naturally in cervid populations. These are alphaherpesviruses, which are closely related to BHV-1 (Ek-Kommonen et al. 1982, 1986; Lyaku et al. 1992a,b; Vanderplasschen et al. 1993) and designated cervid herpesvirus 1 (CerHV-1), first isolated from red deer (Inglis et al. 1983), and reindeer herpesvirus (CerHV-2), first isolated from reindeer (*Rangifer tarandus*) (Ek-Kommonen et al. 1986). Serologic cross-reactions occur between these viruses, but deer sera reacted most strongly with homologous cervid virus (Nixon et al. 1988; Lyaku et al. 1992a). Thus, unless both bovine and cervid viruses are used in neutralization tests, there could be confusion as to which virus is circulating in a given population.

Cervid herpesviruses have only been reported from Europe (Inglis et al. 1983; Ek-Kummonen et al. 1986; Thiry et al. 1988); however, they may be more widespread than currently recognized and should be considered when antibodies against alphaherpesviruses are detected in cervids. They do not appear to be significant pathogens except in unusual circumstances when they may cause keratoconjunctivitis (Inglis et al. 1983; Reid et al. 1986).

MACROPODID HERPESVIRUSES. Two alphaherpesviruses (Roizman 1996) occur in Australian macropods and have been associated with morbidity and mortality of captive wallabies and kangaroos. Macropodid herpesvirus 1 was first isolated from tissues of parma wallabies *Macropus parma* that died of herpesvirus infection (Finnie et al. 1976). Another herpesvirus (Macropodid herpesvirus 2) was isolated from tissues of dorcopsis wallaby *Dorcopsis muelleri* (Wilks et al. 1981). Both viruses have been characterized by restriction mapping (Johnson and Whalley 1987, 1990).

Several epizootics in captive macropods resulted in mortality in parma wallaby, quokka *Setonix brachyurus,* dorcopsis wallaby, rat kangaroos *Bettongia penicillata* and *Aepyprymnus refescens,* western gray kangaroo *Macropus fuliginosus* (Finnie et al. 1976; Callinan and Kefford 1981; Wilks et al. 1981), and red kangaroo *Macropus rufus* (Britt et al. 1994). Serologic surveys for antibodies against macropodid herpesviruses demonstrated high seroprevalence in free-ranging and captive macropods, including red kangaroo, eastern gray kangaroo *Macropus giganteus,* western gray kangaroo, tammar wallaby *Macropus eugenii,* Tasmanian red-necked wallaby *Macropus rufogriseus,* Goodfellows tree kangaroo *Dendrolagus matschiei,* and parma wallaby (Webber and Whalley 1978; Kerr et al. 1981; Wilks et al. 1981).

Stress appeared to be a factor in several outbreaks of herpesviral disease in captive macropods (Callinan and Kefford 1981). Clinical signs of herpesviral infection included fever, vesicles and ulcers on mucous membranes, conjunctivitis, keratitis, rhinitis, and incoordination (Callinan and Kefford 1981). Lesions occurred in the epithelium of the digestive tract, including the glandular stomach, mucous membranes, skin, respiratory tract, liver, and adrenal cortex. Microscopic lesions were characterized by vesiculation and ulceration of mucous membranes, hepatocellular necrosis, and bronchiolar epithelial hyperplasia; intranuclear inclusion bodies were common (Acland 1981; Callinan and Kefford 1981).

These diseases may be important in management of captive macropods throughout the world; recrudescence of latent virus may follow stress and crowding. Several antiherpesvirus compounds have shown activity against these viruses in vitro (Smith 1996; Smith and Whalley 1998).

LITERATURE CITED

Acland, H.M. 1981. Parma wallaby herpesvirus infection. *Journal of Wildlife Diseases* 17:471–477.

Anderson, E.C., and L.W. Rowe. 1998. The prevalence of antibody to the viruses of bovine virus diarrhea, bovine herpes virus 1, Rift Valley fever, ephemeral fever and bluetongue and to *Leptospira* sp. in free-ranging wildlife in Zimbabwe. *Epidemiology and Infection* 121:441–449.

Britt, J.O., Jr., D.F. Frost, and J.M. Cockrill. 1994. Fatal herpesviral hepatitis in red kangaroo (*Macropus rufus*). *Journal of Zoo and Wildlife Medicine* 25:580–584.

Callinan, R.B., and B. Kefford. 1981. Mortalities associated with herpesvirus infection in captive macropods. *Journal of Wildlife Diseases* 17:311–317.

Chow, T.L., and R.W. Davis. 1964. The susceptibility of mule deer to infectious bovine rhinotracheitis. *American Journal of Veterinary Research* 25:518–519.

Clark, R.K., C.A. Whetstone, A.E. Castro, M.H. Jorgensen, J.F. Jensen, and D.A. Jessup. 1993. Restriction endonuclease analysis of herpesviruses isolated from two Peninsular bighorn sheep (*Ovis canadensis cremnobates*). *Journal of Wildlife Diseases* 29:50–56.

Ek-Kommonen, C., P. Veijalainen, M. Rantala, and E. Neuvonen. 1982. Neutralizing antibodies to bovine herpesvirus 1 in reindeer. *Acta Veterinaria Scandinavica* 23:565–569.

Ek-Kommonen, C., C., S. Pelkonen, and P.F. Nettleton. 1986. Isolation of a herpesvirus serologically related to bovine herpesvirus 1 from a reindeer (*Rangifer tarandus*). *Acta Veterinaria Scandinavica* 27:299–301.

Elazhary, M.A.S.Y., J.L. Frechette, A. Silim, and R.S. Roy. 1981. Serological evidence of some bovine viruses in the caribou (*Rangifer tarandus caribou*) in Quebec. *Journal of Wildlife Diseases* 17:609–612.

Engels, M., F. Stech, and R. Wyler. 1981. Comparison of the genomes of infectious bovine rhinotracheitis and infectious pustular vulvovaginitis virus strains by restriction endonuclease analysis. *Archives of Virology* 67:169–174.

Finnie, E.P., I.R. Littlejohns, and H.M. Acland. 1976. Mortalities in parma wallabies (*Macropus parma*) associated with probable herpesvirus. *Australian Veterinary Journal* 52:294.

Hedger, R.S., and C. Hamblin. 1978. Neutralizing antibodies to bovid herpes virus 1 (infectious bovine rhinotracheitis/infectious pustular vulvo-vaginitis) in African wildlife with special reference to the cape buffalo (*Syncerus caffer*). *Journal of Comparative Pathology* 88:211–218.

Inglis, D.M, J.M. Bowie, M.J. Allan, and P.F. Nettleton. 1983. Ocular disease in red deer calves associated with a herpesvirus infection. *Veterinary Record* 113:182–183.

Johnson, M.A., and J.M. Whalley. 1987. Restriction enzyme maps of the macropodid herpesvirus 2 genome. *Archives of Virology* 96:153–168.

———. 1990. Structure and physical map of the genome of parma wallaby herpesvirus. *Virus Research* 18:41–48.

Kahrs, R.F. 1981. Herpes mammillitis. In *Viral diseases of cattle.* Ames: Iowa State University Press, pp. 127–133.

Kerr, A., J.M. Whalley, and W.E. Poole. 1981. Herpesvirus neutralising antibody in grey kangaroos. *Australian Veterinary Journal* 57:347–348.

Lamontague, L., L. Sadi, and R. Joyal. 1989. Serological evidence of bovine herpes 1-related virus infection in the white-tailed deer population on Anacosti Island, Quebec. *Journal of Wildlife Diseases* 25:202–205.

Ludwig, H. 1983. Bovine herpesviruses. In *The herpesviruses,* vol. 2, ed. B. Roizman. New York: Plenum, pp. 135–214.

Lyaku, J.R., P.F. Nettleton, and H. Marsden. 1992a. A comparison of serological relationships among five ruminant alphaherpesviruses by ELISA. *Archives of Virology* 124:333–341.

Lyaku, J.R., J.A. Sinclair, P.F. Nettleton, and H.S. Marsden. 1992b. Production and characterization of monoclonal antibodies to cervine herpesvirus-1. *Veterinary Microbiology* 32:229–239.

Metzler, A.E. 1992. Cytomegeloviruses. In *Veterinary diagnostic virology: A practitioner's guide,* ed. A.E. Castro and W.P. Heuschele. St. Louis: Mosby Year Book, pp. 99–100.

Nixon, P., S. Edwards, and H. White. 1988. Serological comparisons of antigenically related herpesvirus in cattle, red deer, and goats. *Veterinary Research Communications* 12:355–362.

Osorio, F.A. 1992. Bovine mammillitis. In *Veterinary diagnostic virology: A practitioner's guide,* ed. A.E. Castro and W.P. Heuschele. St. Louis: Mosby Yearbook, pp. 83–85.

Plowright, W. 1981. Herpesviruses of wild ungulates, including malignant catarrhal fever virus. In *Infectious diseases of wild mammals,* ed. J.W. Davis, L.H. Karstad, and D.O. Trainer, 2d ed. Ames: Iowa State University Press, pp. 138–142.

Potgieter, L.N.D., and P.L. Aldridge. 1977. Frequency of occurrence of viruses associated with respiratory tract disease of cattle in Oklahoma: Serologic survey for bovine herpesvirus DN599. *American Journal of Veterinary Research* 38:1243–1245.

Radostits, O.M., D.C. Blood, and C.C. Gay. 1994. Infectious bovine rhinotracheitis In *Veterinary medicine: A textbook of the diseases of cattle, sheep, pig, goats and horses,* ed. O.M. Radostits, D.C. Blood, and C.C. Gay, 8th ed. Philadelphia: Baillière Tindall. pp. 1061–1070.

Reid, H.W., P.F. Nettleton, I. Pow, and J.A. Sinclair. 1986. Experimental infection of red deer (*Cervus elaphus*) and cattle with a herpesvirus isolated from red deer. *Veterinary Record* 118:156–158.

Roizman, B. 1996. Herpesviridae. In *Fields virology,* ed. B.N. Fields, D.M. Knipe, P.M. Hawley, et al., 3d ed. Philadelphia: Lippincott-Raven, pp. 2221–2230.

Rossiter, P.B., I.D. Grimm, and P.K. Mirangi. 1988. Immunological relationships between malignant catarrhal fever virus (alcelaphine herpesvirus 1) and bovine cytomegalovirus (bovine herpesvirus 3). *Veterinary Microbiology* 16:211–218.

Sadi, L., R. Joyal, M. St-Georges, and L. Lamontagne. 1991. Serologic survey of white-tailed deer on Anacosti Island, Quebec for bovine herpesvirus 1, bovine viral diarrhea, and parainfluenza 3. *Journal of Wildlife Diseases* 27:569–577.

Schiemann, B., W. Plowright, and D.M. Jessett. 1971. Allerton-type herpesvirus as a cause of lesions of the alimentary tract in a severe disease of Tanzanian buffaloes (*Syncerus caffer*). *Veterinary Record* 89:17–22.

Smith, G. 1996. In vitro sensitivity of macropodid herpesvirus 2 to selected anti-herpetic compounds. *Journal of Wildlife Diseases* 32:117–120.

Smith, G., and J.M. Whalley. 1998. (E)-5-(2'-bromovinyl)-2'-deoxyuidine inhibition of macropodid herpesvirus 1 in vitro. *Journal of Zoo and Wildlife Medicine* 29:157–159.

Taylor, S.K., V.M. Lane, D.L. Hunter, K.G. Eyre, S. Kaufman, S. Frye, and M.R. Johnson. 1997. Serologic survey for infectious pathogens in free-ranging American bison. *Journal of Wildlife Diseases* 33:308–311.

Thiry, E., M. Vercouter, J. Dubuisson, J. Barrat, C. Sepulchre, C. Gerardy, C. Meersschaert, B. Collin, J. Blancou, and P.-P. Pastoret. 1988. Serological survey of herpesvirus infections in wild ruminants of France and Belgium. *Journal of Wildlife Diseases* 24:268–273.

Vanderplasschen, A., M. Bublot, P.-P. Pastoret, and E. Thiry. 1993. Restriction maps of the DNA of cervid herpesvirus 1 and cervid herpesvirus 2, two viruses related to bovine herpesvirus 1. *Archives of Virology* 128:379–388.

Webber, C.E., and J.M. Whalley. 1978. Widespread occurrence in Australian marsupials of neutralizing antibodies to a herpesvirus from a parma wallaby. *Australian Journal of Experimental Biology and Medical Science* 56:351–357.

Whetstone, C.A., J.G. Wheeler, and D.E. Reed. 1986. Investigation of possible vaccine-induced epizootics of infectious bovine rhinotracheitis, using restriction endonuclease analysis of viral DNA. *American Journal of Veterinary Research* 47:1789–1795.

Wilks, C.R., B. Kefford, and R.B. Callinan. 1981. Herpesvirus as a cause of fatal disease in Australian wallabies. *Journal of Comparative Pathology* 91:461–465.

8

POXVIRUS INFECTIONS

ANTHONY J. ROBINSON AND PETER J. KERR

CLASSIFICATION. Poxviruses (the Poxviridae) comprise two subfamilies, Entomopoxvirinae and Chordopoxvirinae, viruses of insects and vertebrates, respectively.

The Chordopoxvirinae comprise eight genera, including the avipoxviruses, which infect birds, plus some unclassified viruses of reptiles and other species. The seven genera of mammalian poxviruses are defined by cross-protection in animals, neutralization assays, and cross-hybridization of genomic DNA. Within a genus, the viruses are named for the species infected or the disease caused (see Table 8.1) (Murphy et al. 1995).

STRUCTURE. Most poxviruses resemble orthopoxviruses; they are brick shaped, 200–400 nm long, and covered with irregular tubular elements. However, parapoxvirus particles are oval rather than brick shaped, and the tubular elements are arranged in a more regular spiral fashion, so the virion resembles a roll of yarn. Two forms of particles occur in negatively stained preparations. The mulberry or "M" form has the appearance just described, whereas the capsular or "C" form, on which surface tubules are not apparent, appears empty.

Intracellular mature virions (IMVs), found in the cytoplasm, predominate in negatively stained material. Extracellular enveloped virions (EEVs) are surrounded by an envelope containing virally encoded proteins. EEVs, which are actively released from infected cells, appear to play a role in spread of virus within the host. Both IMVs and EEVs are infectious.

A dumbbell-shaped core is evident within virions in sections under the electron microscope. The core contains the linear double-stranded DNA genome, structural proteins, and virally encoded enzymes that synthesize viral mRNA during the early stages of infection.

Immature forms also may be seen ultrastructurally, within an amorphous cytoplasmic matrix, often termed a "virus factory." By light microscopy, these may appear as irregular amorphous B-type inclusion bodies, which stain basophilic or amphophilic with hematoxylin and eosin. Some, but not all, poxviruses produce A-type inclusion bodies, which consist of a more discrete acidophilic matrix containing mature virions only.

TABLE 8.1—Poxviruses of mammals

Genus	Species
Orthopoxvirus	Buffalopox virus[a], camelpox virus, cowpox virus, ectromelia virus (mousepox), monkeypox virus, rabbit pox virus[a], raccoonpox virus, skunkpox virus[b], taterapox virus, Uasin Gishu virus[b], horsepox virus, vaccinia virus[c], variola virus (smallpox virus), volepox virus
Parapoxvirus	Auzdyk virus[b], bovine papular stomatitis virus, chamois contagious ecthyma virus[b], orf virus[c], pseudocowpox virus, red deer parapoxvirus, red squirrelpox virus[b], sealpox virus[b]
Capripoxvirus	Goatpox virus, lumpy skin disease virus, sheeppox virus[c]
Leporipoxvirus	Hare fibroma virus, myxoma virus[c], rabbit fibroma (Shope fibroma) virus, squirrel fibroma virus
Suipoxvirus	Swinepox virus[c]
Molluscipoxvirus	Molluscum contagiosum virus
Yatapoxvirus	Tanapox virus, yaba monkey tumor virus[c]
Unassigned	Cetacean pox virus, cotia virus, gray kangaroo pox virus, red kangaroo pox virus, quokka pox virus, molluscum-like viruses of horse, donkey, and chimpanzee, mule deer pox virus, marmoset pox virus

[a] Rabbitpox virus and buffalopox virus are regarded as variants of vaccinia virus.
[b] Tentative members of the genus.
[c] Prototypical virus.

REPLICATION. Poxviruses replicate within the cytoplasm of infected cells, passing through stages of cell entry, uncoating, early gene expression, DNA replication, late gene expression, virion assembly, and release from the cell. Cellular DNA, RNA, and protein synthesis are inhibited to varying degrees, depending on the virus (Moss 1996).

PATHOGENESIS. Poxviruses use many routes of infection, e.g., respiratory (variola); percutaneous inoculation by insect vectors (myxoma virus); and penetration of skin or oral mucosa through microabrasions (parapoxviruses). Poxviruses can cause localized infections, usually of the epidermis and dermis, e.g., parapoxviruses in most animals, and fibroma virus in cottontail rabbits *Sylvilagus floridanus.* Such infections typically resolve fully with minimal detrimental effect on the host.

They may also cause generalized disease. The epithelium of the respiratory tract, skin, or oral mucosa is initially infected, and there the virus replicates, forming a primary lesion that may not be obvious. Infection of the draining lymphoid tissue occurs, and then the virus spreads systemically via the bloodstream (viremia) to set up secondary infections at skin or mucosal surfaces remote from the primary site. In ectromelia and some other poxviruses, replication also occurs in the liver and spleen (Fenner et al. 1989). Systemic poxvirus infections are characterized by a generalized skin rash and high mortality rate, e.g., capripox of sheep, and ectromelia of mice.

The same virus can cause localized infection in one host and generalized disease in another. Myxoma virus causes a benign fibroma in the jungle rabbit *Sylvilagus brasiliensis,* whereas in the European rabbit *Oryctolagus cuniculus* it causes the lethal disease myxomatosis. Similarly, cowpox virus, probably a virus of rodents, causes systemic disease in large felids but only a localized skin lesion in cattle.

The basis for this variation in pathogenicity in different hosts is poorly understood. However, two basic mechanisms are apparent. Firstly, poxviruses encode proteins that suppress or subvert the immune response. These include interferon-binding and tumor necrosis factor-binding proteins, inhibitors of interleukin 1 and complement, inhibitors of apoptosis, and inflammatory inhibitors (Moss 1996). Secondly, the ability to replicate in lymphocytes and spread systemically may be critical (Strayer 1992). Poxviruses also contain genes for homologues of epidermal or endothelial growth factors (Moss 1996).

Recovery from poxvirus infection and subsequent immunity are probably largely cell mediated, although antibody can play a role (Fenner 1996).

DIAGNOSIS. Many poxviruses and their wildlife hosts seem extremely well adapted to each other. Poxviruses often are detected only when they spill over from their natural hosts into susceptible indicator hosts, such as domestic animals or people. Classic examples include myxoma virus and monkeypox virus. Disease may become apparent only when animals are intensively managed, e.g., parapoxvirus infection in red deer *Cervus elaphus.*

Diagnosis of poxvirus infection often is based on clinical signs and gross lesions alone. Most present with single or multiple raised papular, pustular, or scabby, sometimes ulcerated, lesions on the skin, or as punctate ulcers on mucous membranes. There are exceptions, such as the fibromatous or myxomatous swellings caused by the leporipoxviruses, or the "tattoo" lesions seen in dolphins. Depending on the host species and the virus, there may be signs of systemic dissemination, such as fever, loss of appetite, and multiple lesions over much of the surface of the animal and in deeper tissues, such as lung or liver.

Microscopic lesions of cutaneous and mucosal surfaces usually involve initial epithelial cell swelling and hyperplasia, with ensuing necrosis of infected epithelium, associated with dermal and intraepithelial mixed inflammatory cell infiltrates. If epithelial necrosis is severe, ulceration, with healing by granulation, may ensue. There also may be hypertrophy of connective tissue cells in the dermis or submucosa, and Feulgen-positive cytoplasmic inclusions may be evident in affected epithelial or stromal cells. Some poxviruses induce proliferative skin lesions that adopt a papillomatous (wartlike), or characteristic "molluscum contagiosum," pattern (Yager and Scott 1993), whereas others induce formation of nodular dermal "fibromas," or edematous myxoma-like dermal thickening.

To confirm a diagnosis of poxvirus infection, it is usually sufficient to demonstrate poxvirus particles in negatively stained preparations of lesion material (a rapid technique) or in thin sections under the electron microscope. Virions usually are numerous in lesion material. Identification of the virus requires DNA-DNA hybridization, polymerase chain reaction (PCR), restriction endonuclease analysis, and serology, often available only in specialized laboratories. Animal inoculation to determine host range is seldom used.

A detailed and still relevant review of the morphology, biology, clinicopathologic syndromes and diagnosis of animal poxviruses is available (Tripathy et al. 1981). For more recent molecular diagnostic techniques, see Esposito and Knight (1985) and Ropp et al. (1995).

POXVIRUS INFECTIONS OF WILD MAMMALS. Very little is known about many poxviruses of wild mammals under natural conditions. Much has been inferred from laboratory studies or from field surveys for lesions or antibodies. Often, observations of poxvirus infection are incidental, opportunistic, or accidental. Many poxviruses of wild mammals have not been isolated and characterized; those that have been characterized are summarized in Table 8.1.

This discussion is arranged by taxonomy of affected hosts. If a particular genus or species of mammal is affected by more than one type of virus, the information is arranged by genus of poxvirus.

MARSUPIALIA. Poxvirus lesions have been described in four macropodid marsupials: quokka *Setonix brachyurus* (Papadimitriou and Ashman 1972), red kangaroo *Macropus rufus* (Bagnall and Wilson 1974), eastern gray kangaroo *Macropus giganteus* (McKenzie et al. 1979), and western gray kangaroo *Macropus fuliginosus* (Presidente 1978; Rothwell et al. 1984).

The lesions on the tail of the quokka grossly and microscopically resembled papillomas, ranging from a few millimeters to 4–5 cm in diameter. Eosinophilic cytoplasmic inclusions containing poxvirus particles were obvious in epithelium.

The proliferative cutaneous lesions on the young red kangaroos, one western gray kangaroo, and an eastern gray kangaroo were 1–2 cm in diameter and have been considered to resemble molluscum contagiosum of humans, though Rothwell et al. (1984) considered the lesion in another western gray kangaroo to be papillomatous. Typical poxvirus particles were present in the large eosinophilic inclusion bodies.

PRIMATES

Monkeypox Virus

HISTORY. Monkeypox is caused by an orthopoxvirus with a wide host range. First observed in cynomolgus monkeys *Macaca fascicularis* (von Magnus et al. 1959), it also causes a disease resembling smallpox in people (Fenner et al. 1989). It seems to occur naturally only in the tropical rainforests of Africa (Arita et al. 1972).

HOST RANGE. Outbreaks of monkeypox in captive animals have been recorded in giant anteaters *Myrmecophaga tridactyla,* cynomolgus monkeys, rhesus monkeys *Macaca mulatta,* pig-tailed macaques *Macaca nemestrina,* orangutans *Pongo pygmaeus,* gorillas *Gorilla gorilla,* chimpanzees *Pan troglodytes,* gibbons *Hylobates lar,* squirrel monkeys *Saimiri sciureus,* owl-faced monkeys *Cercopithecus hamlyni,* African green (grivet) monkeys *Cercopithecus aethiops* (subclinical), marmosets *Callithrix jacchus,* langurs *Semnopithecus (Presbytus) entellus* and *Trachypithecus* (*Presbytus*) *cristatus,* and baboons *Papio cynocephalus* (experimental).

Antibodies have been demonstrated in the following primates in the wild: *C. aethiops, Cercopithecus petaurista, Cercopithecus nictitans, Cercopithecus ascanius, Cercopithecus pogonias, Cercopithecus mona, Cercocebus galeritus, Piliocolobus badius, Allenopithecus nigroviridis,* and *P. troglodytes.* In addition, antibodies have been found in four species of African squirrel: *Heliosciurus rufobrachium, Heliosciurus gambianus, Funisciurus anerythrus,* and *Funisciurus lemniscatus.* Experimentally, the common African rodent *Mastomys natalensis* is susceptible to infection, as are laboratory rabbits and mice (von Magnus et al. 1959; Peters 1966; Espana 1971; Fenner et al. 1989; Fenner 1994b).

TRANSMISSION AND EPIDEMIOLOGY. Humans appear to become infected by direct contact with infected animals. The virus may circulate among arboreal mammals in tropical forests in equatorial Africa (Fenner et al. 1989). The main hosts of monkeypox virus appear to be squirrels, particularly *H. rufobrachium* and *F. anerythrus* (Fenner 1994b), and monkeypox virus has been isolated from a squirrel with generalized skin lesions (Khodakevich et al. 1986).

CLINICAL SIGNS AND PATHOLOGY. Monkeypox is a systemic disease characterized by fever and rash. Morbidity and mortality in primate colonies may be relatively low (Espana 1971). Clinical signs and outcome vary with the species infected. In cynomolgus monkeys, the disease typically presents as a generalized petechial rash that develops into a maculopapular to pustular eruption over the entire body. Lesions gradually scab over and resolve, leaving distinct scars. The general health of the animals is relatively unaffected (von Magnus et al. 1959).

In contrast, a high mortality rate occurred among orangutans at the Rotterdam Zoo. Animals had difficulty in eating, associated with yellow-white plaques on the oral mucosa, followed by inapparent thickening of the facial skin. Some died at this stage with severe dyspnea. Multiple pox eruptions occurred over the body of surviving animals (Peters 1966).

DIAGNOSIS. Diagnosis is based on conventional criteria described in the introductory section. Monkeypox virus is cultured on chorioallantoic membranes of embryonated hen eggs or on a variety of cell lines. Serologic diagnosis is complicated by the strong cross-reactivity with other orthopoxviruses (Gispen et al. 1976; Marennikova et al. 1981).

CONTROL AND TREATMENT. Monkeypox can be prevented in primate colonies by vaccination with vaccinia virus or by quarantine (Espana 1971). There is no specific treatment. There have been no outbreaks in primate colonies since 1968 (Fenner et al. 1989).

PUBLIC HEALTH IMPLICATIONS. Monkeypox virus is a zoonosis, clinically virtually indistinguishable from smallpox. Human monkeypox has been restricted to Liberia, Sierra Leone, Zaire, Nigeria, Cameroon, and the Central African Republic, where cases still occur. Most appear to be due to contact with infected animals; however, person-to-person transmission occurs (Fenner et al. 1989; Aplogan et al. 1997).

MANAGEMENT IMPLICATIONS. Monkeypox virus was introduced into primate colonies and zoos with wild-caught monkeys, especially *C. aethiops* (Fenner et al. 1989). The wide host range, including people, and the significant mortality rate dictate that any species of possible carriers from the endemic region must be adequately quarantined.

Yatapoxviruses. The genus *Yatapoxvirus* contains two species, Tanapox virus and Yaba monkey tumor virus, which show some serologic and genetic relatedness; a third virus, Yaba-like disease virus, is now regarded as a strain of Tanapox virus (Knight et al. 1989).

TANAPOX VIRUS

HISTORY. Tanapox virus originally was isolated from outbreaks of skin lesions in people in the Tana valley of Kenya and subsequently from outbreaks in primate colonies in the United States. Disease has occurred in

Kenya and Zaire (Downie et al. 1971; Jezek et al. 1985).

TRANSMISSION, DISTRIBUTION, AND HOST RANGE. Tanapox virus appears to be endemic in equatorial Africa. Although disease due to Tanapox virus has not been detected in wild primates, many African and Asian primates, including *M. mulatta, Macaca speciosa* (probably *M. arctoides*), *M. nemestrina, Macaca radiata, M. fascicularis, Macaca nigra,* and *Semnopithecus entellus,* have been infected naturally or experimentally in captivity (Hall and McNulty 1967; McNulty et al. 1968; Crandell et al. 1969; Downie et al. 1971; Espana 1971). Natural reservoirs of the virus have not been identified, but antibodies have been found in vervet monkeys *Cercopithecus pygerythrus* from Ethiopia and Kenya (Downie and Espana 1973).

Nonprimate species seem refractory to infection (Downie et al. 1971). Animal handlers can be infected through scratches, but in Africa insect vectors are probably important. There is no evidence for direct-contact transmission (Jezek et al. 1985).

CLINICAL SIGNS, PATHOLOGY, AND DIAGNOSIS. A brief fever is followed by the development of firm maculopapular nodules, usually on the face or limbs. Papules developed 3–5 days after infection in rhesus monkeys or human volunteers. Histologically, the epidermis is hyperplastic, with some cells containing inclusions. There is little inflammatory infiltrate in the underlying dermis; however, ulceration and secondary infection may occur.

Diagnosis is based on histology, serology, and virus isolation in CV-1 cells (Knight et al. 1989).

PUBLIC HEALTH IMPLICATIONS. Humans are susceptible to Tanapox virus in natural outbreaks of the disease and by contact with infected captive animals.

Yaba Monkey Tumor Virus

TRANSMISSION AND HOST RANGE. Yaba monkey tumor virus originally was isolated from an outbreak of cutaneous histiocytomas in rhesus monkeys kept outdoors at Yaba, Nigeria (Bearcroft and Jamieson 1958). Yaba monkey tumor virus disease has not been observed in free-ranging animals. Based on serology, however, the virus occurs in wild African and Southeast Asian primates (Downie 1974). A range of African and Asian primates are susceptible: *M. mulata, M. nemestrina, M. fascicularis, M. speciosa (?arctoides), Papio hamadryas,* and *Papio papio,* but *Cercopithecus tantalus, Cercocebus torquatus, Erythrocebus patas,* and *Cebus apella* were not susceptible (Espana 1971).

CLINICAL SIGNS. Tumors occur on the face, palms, and interdigital areas and on the mucosal surfaces of the nose, sinuses, lips, and palate. Affected animals show no other signs. The tumors may be dramatically protuberant and extensive in rhesus monkeys (Bearcroft and Jamieson 1958).

PATHOGENESIS AND PATHOLOGY. Tumors develop within 5 days and, in most species, progress to 2.5–4.5 cm in diameter over about 6 weeks. Although the draining lymph node is reactive and contains cells similar to those of the tumor, spread of the virus to internal organs is rare. However, multiple tumors may arise along the lymphatic vessels draining the skin around a tumor (Bearcroft and Jamieson 1958). Tumors regress spontaneously over 2–3 months. Animals are immune to superinfection while tumors are present but, following regression, immunity is lost, and reinfection can occur.

The tumors consist of masses of pleomorphic mononuclear cells with abundant cytoplasm, derived from histiocytes. Many contain large eosinophilic intracytoplasmic inclusions. There may also be reactive fibrosis with some inflammation (Bearcroft and Jamieson 1958). In comparison, Tanapox virus lesions are largely confined to the epidermis.

DIAGNOSIS. Diagnosis of *Yatapoxvirus* infections is based on histology and serology and on isolation of virus in a restricted range of cell types (BSC-1 and JINET) (Rouhandeh 1988).

PUBLIC HEALTH IMPLICATIONS. Humans are susceptible to accidental and experimental infection (Grace and Mirand 1963).

XENARTHA. Pox infection has been described in captive anteaters. See the sections on primates (monkeypox virus) and Rodentia (cowpox virus).

LAGOMORPHA

Rabbitpox Virus. Rabbitpox virus, which is derived from vaccinia virus, is an acute infection of laboratory rabbits *Oryctolagus cuniculus,* with no known implications for wild mammals (Fenner 1994d).

Leporipoxviruses. Members of the genus *Leporipoxvirus* infect rabbits, hares, and squirrels (Table 8.1). The type species is myxoma virus. The genus was defined on the basis of antigenic cross-reactivity and cross-neutralization tests (Fenner 1965; Woodroofe and Fenner 1965). The virions resemble orthopoxvirus particles.

Myxomatosis and Myxoma Virus

NATURAL HISTORY AND HOST RANGE. Myxomatosis, the disease caused in European rabbits by myxoma virus, was first observed in laboratory rabbits in Uruguay in 1896. Myxomatosis in the Americas has subsequently been described in *O. cuniculus* in Brazil, Argentina, Colombia, Panama, California, Oregon, and Baja California (Fenner and Ross 1994).

The natural hosts of myxoma virus are the jungle rabbit or tapeti *Sylvilagus brasiliensis* in South and Central America, and the brush rabbit *Sylvilagus bachmani* in California. In these rabbits, myxoma virus causes a cutaneous fibroma, not a systemic disease. Virus isolates from California and South America are closely related, but the two virus types can be distinguished antigenically and by DNA restriction endonuclease profiles (Fenner and Ratcliffe 1965; Woodroofe and Fenner 1965; P.J. Kerr unpublished).

Although both myxoma virus strains cause lethal disease in *O. cuniculus, S. bachmani* is resistant to Brazilian virus, and *S. brasiliensis* to Californian (Regnery and Marshall 1971).

CLINICAL SIGNS IN *ORYCTOLAGUS*. Clinical signs vary with the strain of myxoma virus and the nature of the host, i.e., virulent or attenuated, South American or Californian virus, in domestic laboratory rabbit and susceptible or resistant wild rabbit (Fenner and Ratcliffe 1965).

In wild *Oryctolagus* infected with Brazilian strains, the rabbit is bedraggled, the eyes are partially or fully closed due to swelling of the eyelids, and there is mucopurulent discharge from the conjunctiva and nose. The ears are edematous, and the head is swollen. Nodules 0.1–1.0 cm in diameter may be found on the eyelids, face, and ears and, frequently, over the remainder of the body. The anogenital region is often severely swollen. Scrotal edema is common, and the testes may be enlarged or shrunken. Severely affected animals appear blind, and they often exhibit respiratory distress. In chronic cases, rabbits may be severely emaciated and have large secondary myxomatous swellings over the legs and body.

Californian strains of myxoma virus are rapidly lethal in domestic rabbits but may not cause classic myxomatosis. Often, infected animals are simply found dead with conjunctival erythema, slightly swollen eyelids, and a bloody discharge from the mouth or anus. Nervous signs may be observed prior to death (Fenner and Ratcliffe 1965). Domestic rabbits with more classic myxomatosis have been described from California, Oregon, and Mexico.

CLINICAL SIGNS IN *SYLVILAGUS* AND *LEPUS*. Jungle rabbits develop a localized fibroma, up to several centimeters in diameter, at the site of inoculation with Brazilian strains of myxoma virus. This may persist for 10–40 days. There are no signs of systemic disease in adult rabbits; however, there may be the potential for systemic disease in animals with immature immune systems (Fenner and Ratcliffe 1965).

A localized fibroma, usually at the base of the ears, on the lips, or on the feet, also occurs in *S. bachmani* infected with Californian strains of myxoma virus. This fibroma eventually scabs and regresses over some weeks or months (Marshall and Regnery 1960; Regnery and Miller 1972). Systemic disease or prolonged persistence of tumors is not seen, even in young rabbits (Grodhaus et al. 1963).

Infection of other species of *Sylvilagus* and *Brachylagus* rabbits with myxoma virus has not been seen in the wild. Experimentally, *Brachylagus (Sylvilagus) idahoensis, Sylvilagus audubonii, Sylvilagus nuttallii,* and *Sylvilagus floridanus* can be infected with Californian myxoma virus, but the levels of virus in the subsequent lesions are too low for mosquito transmission (Regnery and Marshall 1971). *Sylvilagus audubonii* developed fibromas and *S. nuttallii* developed fatal myxomatosis when inoculated with Brazilian myxoma virus (Regnery 1971), but *Sylvilagus transitionalis* was resistant to such infection (Hyde and Gardner 1939).

Naturally infected European hares *Lepus europaeus* and mountain hares *Lepus timidus* can develop clinical myxomatosis, but few animals show disease experimentally (Fenner and Ratcliffe 1965). Black-tailed jack rabbits *Lepus californicus* were resistant to infection with Californian and South American viruses (Regnery and Marshall 1971).

TRANSMISSION. Myxoma virus transmission involving all susceptible species is similar (Fenner and Ratcliffe 1965). The virus is passively transmitted by blood-feeding arthropods, such as mosquitoes or fleas. Virus particles adhere to the mouthparts of the vector as it probes through the epidermis overlying a lesion and are then inoculated into a new host when the vector feeds. The virus does not replicate within the vector, so there is no virus-vector specificity as occurs in arboviruses; the importance of a particular vector depends on its prevalence and behavior (Myers et al. 1994; Rogers et al. 1994).

Virus also is shed in discharges (Mykytowycz 1958) and can be transmitted by close contact and indirectly via thorns and thistles. Some strains undergo respiratory transmission (Fenner and Ross 1994).

PATHOGENESIS. In the natural *Sylvilagus* host, the virus replicates at the site of inoculation within the epidermis and dermis, and transmission occurs from this site. There does not appear to be a systemic phase of infection.

In contrast, in *Oryctolagus,* the virus replicates at the inoculation site and spreads to the draining lymph node, where it replicates to high titer. Subsequently, it is found in spleen, other lymph nodes, skin away from the inoculation site, mucosal surfaces such as conjunctiva, and in the testes, lungs, and liver. Virus is present in leukocytes, rather than plasma, and presumably it is in these cells that the virus is transmitted around the body (Fenner and Woodroofe 1953).

The cause of death in acutely affected rabbits is unclear, although secondary bacterial infections probably cause death in chronic cases.

PATHOLOGY IN *ORYCTOLAGUS*. Animals have loss of body fat. The spleen may be markedly swollen, and lymph nodes are enlarged, edematous, and often hemorrhagic. Hemorrhages have been described in the body cavity, organs, and skin, especially in European rabbits infected with Californian strains of myxoma virus (Fenner and Woodroofe 1953). The conjunctiva and nasal mucosa are swollen, with extensive mucopurulent discharge and crusting. Patchy pulmonary consolidation or hemorrhage may be evident.

The histopathology has been described in detail elsewhere (Hurst 1937; Yuill 1981). Lesions are most obvious in skin and mucocutaneous surfaces, such as the nasal mucosa and conjunctiva, as well as in lymph nodes draining affected skin, and in spleen.

The epidermis overlying cutaneous tumors is proliferative and thickened, and cytoplasmic eosinophilic inclusions may be present. Necrosis of epithelium is evident, and intraepithelial pustules occur. The dermis is edematous, and breakdown of fibrillar elements is evident. Large elongate or stellate basophilic stromal cells, termed myxoma cells, within which cytoplasmic inclusions may be seen, are particularly associated with the walls of small vessels. Local hemorrhage and acute inflammation are common in the dermis and superficial muscle, but lesions do not spread through the underlying fascial planes.

In the lymph nodes, acute inflammation and lymphocyte destruction occur together with proliferation of reticulum cells, but foci of normal lymphoid tissue may persist. The node may be depleted of lymphocytes. Vascular changes similar to those in the dermis occur, and hemorrhage is common. Microscopic changes in other lymphoid tissues, such as spleen, peribronchiolar lymphoid tissue, thymus, and cecal tonsil, resemble those in the lymph nodes but are generally less severe (Hurst 1937). There may be acute orchitis, with myxoma cells in the interstitium, and myxoma cells may be present in other organs, including bone marrow.

DIAGNOSIS IN *ORYCTOLAGUS*. Diagnosis is based on the characteristic clinical signs and pathologic findings, together with a history of myxomatosis in the area.

Electron microscopy of lesion material may reveal typical poxvirus particles, while fluorescent antibody tests on frozen tissue sections can provide rapid confirmation of the diagnosis. Polymerase chain reaction is rapid and reasonably sensitive and can be used on extracts of infected tissues (Fountain et al. 1997). Virus can be isolated easily from infected tissues or conjunctival swabs by culture in a variety of cell lines, or by inoculation of rabbits or the chorioallantoic membrane of fertile hen eggs. Antibodies can be measured using enzyme-linked immunosorbent assay (ELISA), virus neutralization assays, or complement fixation assays (Fenner et al. 1953; Kerr 1997).

DIFFERENTIAL DIAGNOSIS. Pasteurellosis and keratoconjunctivitis are conditions to be differentiated from myxomatosis. Rabbit hemorrhagic disease should be considered where high mortality rates are occurring in rabbit populations, but it is easily distinguished from myxomatosis.

IMMUNITY. European rabbits recovered from myxomatosis are considered immune for life. However, there are reports of reinfection (Fenner and Ross 1994) or of possible recrudescent infections (Williams et al. 1973).

Antibodies can be detected from 8 to 10 days after infection; these persist for at least 2 years and probably for the life of the rabbit (Kerr 1997). Maternal antibodies cross the placenta and persist in the young for 6–8 weeks; they may provide limited protection from mosquito-transmitted myxomatosis (Fenner and Marshall 1954; Kerr 1997).

Immunity has not been investigated extensively in *S. brasiliensis*. In *S. bachmani*, complement-fixing and neutralizing antibodies have been detected, but both are relatively short-lived (Marshall et al. 1963).

CONTROL AND TREATMENT. Control of myxomatosis in wild European rabbits has different implications in different countries. In Australia, where myxomatosis is still regarded as an important adjunct to wild rabbit control, no control measures are undertaken, and vaccination of domestic rabbits against myxomatosis is illegal. In France, where the rabbit is regarded as an important game animal, large-scale vaccination has been used in both domestic and wild rabbits (Fenner and Ross 1994). Vaccination can be by either live attenuated virus vaccines or by inoculation with rabbit fibroma virus, which is innocuous in European rabbits, though the immunity obtained may be rather short-lived.

The only other control measures for domestic and laboratory rabbits are quarantine, and exclusion of vectors by insect-proof screening and insecticide treatment.

There is no specific treatment for myxomatosis.

PUBLIC HEALTH IMPLICATIONS. Myxomatosis has no known public health implications. People in close contact with rabbits during an epidemic of myxomatosis did not develop neutralizing antibodies to the virus (Jackson et al. 1966), and ill effects have not been reported following human inoculation with myxoma virus (Burnet 1968).

MANAGEMENT IMPLICATIONS. The potential for biologic control of wild *O. cuniculus* by using myxoma virus was recognized very early. In 1950, it was successfully released into Australia (Fenner and Ratcliffe 1965), where introduced European rabbits had spread across the continent in plague proportions, causing serious economic and environmental damage. Although myxomatosis was initially highly effective at reducing the rabbit population, the subsequent emergence of attenuated strains of virus and genetically resistant rabbits reduced its impact (Fenner 1983). Nevertheless, it remains a major factor in controlling wild rabbit populations.

In 1952, myxoma virus was introduced into France to control rabbits and subsequently spread to continental Europe and Britain. A similar, although slower, reduction in impact occurred (Fenner and Ross 1994; Kerr and Best 1998). Myxoma virus has also been released on subantarctic islands, in Chile and Argentina and, unsuccessfully, in New Zealand.

Estimates of the impact of myxomatosis on infected rabbits in endemic areas range from 30%–40% mortality in Australia to 47%–69% in Britain (Ross et al. 1989). When the effect of myxomatosis has been reduced, either by vaccination or by removal of vectors, the rabbit population has increased significantly.

Epidemic myxomatosis may interfere with hunting of *O. cuniculus,* due to the collapse of the rabbit population (Fenner and Ratcliffe 1965; Rogers et al. 1994). In addition, there may be effects on predators dependent on rabbits, and on vegetation grazed by rabbits. In North America, *S. audubonii* and *S. nuttallii* have the potential to support Brazilian strains of myxoma virus, and this could adversely affect populations of *S. nuttallii* (Regnery 1971).

Rabbit Fibroma Virus

NATURAL HISTORY. Rabbit fibroma virus or Shope's fibroma virus (Shope 1932) occurs in eastern cottontail rabbits *S. floridanus* in eastern and midwestern North America. In these animals, it causes a benign cutaneous "fibroma" at the site of infection. A similar lesion occurs in experimentally infected European rabbits.

The natural disease is relatively common, a pattern of low-grade endemicity being punctuated by sporadic minor epidemics (Herman et al. 1956). Rabbits with fibromas were most common in the summer and autumn but also were found in winter.

Rabbit fibroma virus can be considered an eastern American variant of myxoma virus (Fenner 1965) with a similar mode of transmission. Though it resembles myxoma virus genetically, there are differences in specific genes. Amino acid and nucleotide-sequence homology are high; however, the restriction profiles of rabbit fibroma virus and myxoma virus DNA differ (Delange et al. 1984; Russell and Robbins 1989).

CLINICAL SIGNS. In naturally infected cottontails, infection is seen as a cutaneous "tumor" several centimeters in diameter, most commonly on the legs, especially the dorsal surface of the hind feet, but also on the face and elsewhere. The surface of the tumor usually is smooth, although there may be some scabbing associated with trauma (Kilham and Fisher 1954).

TRANSMISSION. Transmission in the field is presumed to be by biting arthropods, since fleas and mosquitoes transmit the virus experimentally (Kilham and Woke 1953; Kilham and Dalmat 1955). The virus does not replicate in the vector (Day et al. 1956).

Persistence and infectivity of the fibromas are probably critical for overwintering of the virus in northern America, if mosquitoes are the main vector. In cottontails, approximately 1 month old, fibromas experimentally induced by mosquito feeding remain infectious for up to 10 months (Kilham and Dalmat 1955). In older cottontails, fibromas regress after 2–5 months (Kilham and Fisher 1954; Yuill and Hanson 1964).

HOST RANGE. Eastern cottontails are the natural hosts. Natural infections of wild European rabbits have not been reported, but domestic European rabbits can be naturally and experimentally infected. Jackrabbits, presumably *L. californicus,* and snowshoe hares *Lepus americanus,* have been experimentally infected (Hyde 1936; Yuill 1981). No lesions were produced in *S. audubonii, S. nuttallii, S. bachmani, L. europaeus,* or a range of wild and laboratory animals (Herman et al. 1956; Fenner and Ratcliffe 1965).

PATHOGENESIS. In adult cottontails, fibroma virus is confined to the dermis and epidermis surrounding the inoculation site. Skin thickenings in cottontails were first seen 7–8 days after inoculation, and progressed to large smooth surfaced fibromas within several weeks. High titers of virus persisted in the fibroma up to 142 days after inoculation. Tumors persisted for up to 150 days and then disappeared over a few weeks. Viremia was not detected at any stage (Kilham and Fisher 1954).

The pathogenesis is different in very young cottontails, which develop viremia and systemic sites of virus proliferation, dying about 4 weeks after inoculation. In slightly older cottontails, the virus is not lethal, but the tumors become very large (Yuill and Hanson 1964).

PATHOLOGY. The tumors on eastern cottontails are about 1.5–2.0 cm in diameter and firm, white, and moist on cut surface. Microscopically, the overlying epithelium is thickened, with pegs of epithelial cells projecting into the mass of the tumor. Many of the epidermal cells are enlarged and may contain an eosinophilic inclusion. In the underlying dermal mass, most of the stromal cells are elongate with scant cytoplasm and thin spindle-shaped nuclei. However, eosinophilic inclusions may be found in stromal cells with large round or oval nuclei and abundant cytoplasm, especially in earlier lesions. Cell arrangement is compact and irregular with abundant collagen fibers. Mitotic figures are rare. Some blood vessels are cuffed with lymphocytes and plasma cells, and densely packed lymphocytes are present at the base of the tumor mass (Shope 1932).

In adult European rabbits, rabbit fibroma virus causes only a transient fibroma, which is rapidly cleared. The pathology and pathogenesis of rabbit fibroma virus in this species are described in Hurst (1938), Fenner and Ratcliffe (1965), and Strayer and Sell (1983).

DIAGNOSIS. In cottontails in eastern North America, diagnosis is by clinical signs and characteristic microscopic appearance. It can be confirmed by the demonstration of poxviruses by electron microscopy, by transmission to susceptible domestic or cottontail rabbits, by isolation on RK13 cell monolayers, or by molecular techniques.

Differential diagnosis in cottontails could include nonviral fibromas or viral papillomas, or granulomas (all very rare).

IMMUNITY. Cottontails become resistant to reinoculation between 4 and 6 days following primary inoculation. Serum-neutralizing antibodies appear weeks later. These antibodies persist for the life of the fibroma and then wane fairly rapidly (Kilham and Fisher 1954).

CONTROL AND TREATMENT. Control or treatment is unlikely to be necessary or justified.

PUBLIC HEALTH IMPLICATIONS. There is no evidence of transfer of the infection to people through either handling or consumption of rabbits with fibromas (Herman et al. 1956).

DOMESTIC ANIMAL IMPLICATIONS. Since it cross-protects, rabbit fibroma virus has been used in domestic European rabbits as a vaccine against myxomatosis. Farmed European rabbits in Texas and Ohio have become naturally infected with rabbit fibroma virus, causing mortality in suckling kittens (Joiner et al. 1971; Raflo et al. 1973).

MANAGEMENT IMPLICATIONS. There is no evidence that rabbit fibroma virus has had any effect on the population dynamics of cottontails (Yuill and Hanson 1964).

HARE FIBROMA VIRUS. A *Leporipoxvirus* of European hares *L. europaeus* causes solitary or multiple skin tumors 1–3 cm in diameter, mostly on the head, ears, or legs. Epidemics occurred in late summer and autumn in Germany, Italy, and France; however, the disease apparently has not been reported since 1964 (Fenner 1994a). Microscopically, the lesions resembled those of rabbit fibroma virus in European rabbits. Black-tailed jackrabbits proved susceptible to experimental infection (Regnery and Marshall 1971).

AFRICAN HARE FIBROMA VIRUS. An unclassified poxvirus was observed in lesions from hares *Lepus capensis* in Kenya in 1976 (Karstad et al. 1977). Grossly, lesions were small sessile tumors 0.5–1.0 cm in diameter; microscopically, the lesions resembled rabbit fibroma virus in cottontails. Typical poxvirus particles were observed by electron microscopy, but no virus was isolated. It is unclear whether this virus should be differentiated from hare fibroma virus.

RODENTIA

Mousepox Virus

Synonym: Infectious ectromelia virus.

NATURAL HISTORY AND HOST RANGE. Mouse pox was first described in laboratory mice *Mus musculus* in 1930 (Marchal 1930). Infections often begin on the feet and can result in swelling, gangrene, and the loss of the lower limb; hence, the term ectromelia. Fatal systemic spread is common, resulting in lesions on any skin surface (Buller and Wallace 1988; Fenner 1994c). The origin of the virus is unknown. A wildlife rodent reservoir is suspected but not proven (Groppel 1962).

There is serologic evidence for orthopoxvirus infection in foxes, cats, and wild boars in Germany (Mayr et al. 1995; Muller et al. 1996). Although this could be due to exposure to any orthopoxvirus, the most likely candidates are ectromelia-like or cowpox-like viruses (see below). An orthopoxvirus closely related, if not identical, to ectromelia virus caused congenital ectromelia in farmed silver foxes *Vulpes vulpes* in the Czech republic and Russia (Czerny and Mahnel 1990; Mahnel et al. 1993).

The host range of the ectromelia virus of laboratory mice is restricted to mice of the subgenus *Mus* in the genus *Mus,* and to *Mus minutoides* in the subgenus *Nannomys.* Rabbits and guinea pigs are only mildly susceptible, and humans are not susceptible (Fenner et al. 1989).

Cowpox Virus

Synonyms: Carnivorepox, catpox, elephantpox, ratpox, and vaccinia viruses.

HISTORY. Originally, the terms smallpox vaccine, cowpox virus, and vaccinia virus were used interchangeably (Baxby et al. 1982; Fenner et al. 1989). In 1939, an isolate of cowpox virus was shown to be distinct from vaccinia virus (Downie 1939a,b) and is now known to be distinct from all other orthopoxviruses.

Since 1960, viruses closely related to but distinct from the U.K. strain of cowpox virus have been isolated from serious outbreaks of disease in many species of captive wild animals in Europe and the United Kingdom (see below). Subsequently, cowpox virus has been reported in domestic cats (Thomsett et al. 1978; Gaskell et al. 1987).

It is now apparent that the reservoir or maintenance hosts for these viruses are wild rodents found in the United Kingdom, continental Europe, and western and central Asia (Baxby 1977; Marennikova et al. 1978; Pilaski and RosenWolff 1988; Fenner et al. 1989). It is likely that there are a number of distinct cowpox-like viruses maintained in different hosts in different regions. The diseases are usually described in terms of the animal affected, together with the suffix "pox," e.g., catpox and elephantpox.

HOST RANGE. The host range of these viruses is broad and includes Asian elephants *Elephas maximus,* African elephants *Loxodonta africana,* and rhinoceroses *Ceratotherium simum* and *Diceros bicornis* (Pilaski and RosenWolff 1988); okapis *Okapia johnstoni* (Zwart et al. 1971; Pilaski and RosenWolff 1988); a range of Felidae, including lion *Panthera leo,* cheetah *Acinonyx jubatus,* black panther/jaguar *Panthera onca,* puma *Felis concolor,* ocelot *Felis pardalis,* far-eastern cat *Felis bengalensis,* and European wild cat *Felis silvestris* (Marennikova et al. 1977); the genet *Genetta* sp.; and the giant anteater *Myrmecophaga tridactyla* (Marennikova et al. 1976, 1977).

Vaccinia virus and cowpox virus will infect most mammals, including humans, and it is likely that this is true for the cowpox virus variants isolated from zoo mammals. Cowpox is a relatively severe but uncommon infection in people (Baxby et al. 1994), but human cases have been associated with cats in the United Kingdom (Bennett et al. 1990).

ETIOLOGY. Cowpox and cowpox-like viruses are members of the genus *Orthopoxvirus* (Murphy et al. 1995), distinguished from related viruses by biologic properties, antigenicity, and molecular characteristics (Mackett and Archard 1979; Esposito and Knight 1985; Fenner et al. 1989; Czerny and Mahnel 1990; Ropp et al. 1995).

TRANSMISSION AND EPIDEMIOLOGY. Poxviruses generally are resistant and would survive for many weeks in a laboratory or zoo environment.

The origin and hence the mechanism of transmission are not known for most outbreaks, but the infection presumably is contracted from a wild rodent reservoir host. Only in an outbreak of carnivorepox in the Moscow Zoo was the source traced, to white rats used as food for the zoo animals (Marennikova and Shelukhina 1976; Marennikova et al. 1978). Antibody to orthopoxvirus was detected in great gerbils *Rhombomys opimus* and large-toothed yellow susliks *Spermophilus (Citellus) fulvus* (Marennikova et al. 1978). The viruses isolated from a rat in the white rat colony, a puma from the zoo outbreak, and the great gerbil were indistinguishable, but clearly different from a U.K. strain of cowpox and from ectromelia. Gerbils and susliks probably were the wildlife reservoir for the virus that caused infection in the white rat colony and subsequently the outbreak in felids at the Moscow Zoo.

The reservoir for the U.K. strain of cowpox has not been determined unequivocally, but orthopoxvirus antibody has been detected in wood mice *Apodemus sylvaticus,* field voles *Microtus agrestis,* and bank voles *Clethrionomys glareolus* in England and Wales (Kaplan et al. 1980; Crouch et al. 1995), and these animals are susceptible to infection (Bennett et al. 1997). In Scandinavia, bank voles and Norway lemmings *Lemmus lemmus* may be reservoirs (Tryland et al. 1998) and, in Germany, orthopoxvirus antibody has been found in voles *Microtus arvalis,* bank voles, yellow-necked mice *Apodemus flavicollis,* and the brown rat *Rattus norvegicus* (Pilaski and Jacoby 1993).

CLINICAL SIGNS. The diseases caused by the cowpox-like viruses range from mild skin lesions to severe, fatal, systemic disease. In the outbreak in cats in the Moscow Zoo (Marennikova et al. 1977), both dermal and pulmonary forms of the disease occurred. The dermal form ranged from crusty rashes to confluent pocks, with the death of an ocelot. In the pulmonary form of the disease, clinical signs were anorexia, lethargy, fever, increased respiratory rate, coughing, and cyanosis. No skin lesions developed, and all of the affected animals died. A similar picture was seen in cheetahs in zoos in the United Kingdom (Baxby et al. 1979, 1982).

Severe dermal lesions, but with respiratory involvement, were also seen in two giant anteaters (Marennikova et al. 1976). At the Royal Rotterdam Zoo, a cowpox-like virus affected all five okapis (Zwart et al. 1971). A 2-month-old calf ultimately developed severe skin and oral lesions, and died. Two adults showed similar clinical signs but recovered, while another two developed a few pox lesions and recovered uneventfully.

Most of the 67 cases of cowpox-like disease in Asian elephants in zoos and circuses in continental Europe between 1960 and 1986 were severe, and seven deaths occurred (Pilaski and RosenWolff 1988). Two of 13 affected African elephants also died. Lesions appeared on the skin of the feet, legs, perineum, head, and trunk. Lesions in the mouth and tongue led to difficulty chewing and swallowing. Conjunctivitis and swelling of the head also were seen (Gehring et al. 1972).

PATHOGENESIS. The pathogenesis of cowpox-like disease has not been studied in detail but can be reasonably expected to follow that of other systemic orthopoxvirus infections (Fenner et al. 1989). Experimental inoculation of yellow susliks and great gerbils with the white rat virus caused 100% mortality in the susliks. Some great gerbils survived intranasal and cutaneous inoculation, but all succumbed to oral inoculation. The white rat virus also infects Norway rats *Rattus norvegicus* (Maiboroda 1982). In all animals, systemic disease occurred, the course of which mimicked the disease in the zoo cats.

PATHOLOGY. Skin lesions, except for those in domestic cats, appear as typical pox lesions, evolving through papular, vesicular, pustular, and scabbing phases. In domestic cats, the lesions tend to evolve from glistening red hairless areas to ulcers and scabs, possibly because cats lick the lesions. In the pulmonary form of the disease in lions, cheetahs, and black panthers, changes typical of viral pneumonia were seen, with consolidated lung lobes, fibrin in the bronchi, and fluid in the pleural cavity (Marennikova et al. 1977). Oral and esophageal erosions were seen in some of the cats and in the anteaters.

In the zoo cats, typical type-A and type-B cytoplasmic inclusions were found in epithelial cells in skin, bronchi, alveoli, tongue, and pharynx (Marennikova et al. 1977), and similar inclusions were present in skin sections from a cheetah from the outbreak in the Whipsnade Zoo (Baxby et al. 1982) and in the okapis (Zwart et al. 1971).

DIAGNOSIS. A tentative diagnosis of cowpox-like virus infection can be based on the species affected, clinical signs, and gross and microscopic lesions. Orthopoxvirus particles can be demonstrated readily in negatively stained unfixed lesion material, and PCR can be attempted on the same material. Cultivation of the virus in cell cultures, and further biologic and molecular tests, can be used to determine the type of orthopoxvirus involved. Demonstration of rising titers to orthopoxvirus antigens could be considered if appropriately paired sera were available.

DIFFERENTIAL DIAGNOSIS. Other poxvirus infections theoretically could produce signs similar to cowpox-like viruses in zoo animals. The clinical and pathologic picture in elephants and zoo cats is distinctive, but the disease in anteaters should be differentiated from that caused by monkeypox (Gispen et al. 1967).

IMMUNITY. Recovered animals should be immune to reinfection with any orthopoxvirus.

CONTROL AND TREATMENT. Monitoring of the food supply to eliminate possible animal sources of infection is appropriate in potentially endemic areas. However, the source of infection in most endemic areas has not been identified, which makes recommendations for control more difficult. Rodent control can be recommended but is difficult to carry out effectively.

Outbreaks in zoo elephants have been controlled by isolation of sick animals and vaccination of contacts with the MVA strain of vaccinia virus (Pilaski and RosenWolff 1988; Fenner et al. 1989). Rhinoceroses have also been vaccinated with the MVA strain without ill effect. Since the report by Pilaski and Rosen-Wolff (1988), vaccination of 75 Asian and 39 African elephants held in zoos and circuses in Germany with the Lister (Elstree) strain of vaccinia virus has been carried out without adverse consequences, and it is now the recommended vaccine strain (Pilaski and Zhou 1991). In cheetahs, vaccination with the Lister strain seemed to induce neither a lesion nor an immune response (Baxby et al. 1982). It may be difficult to protect zoo cats unless a suitable strain of vaccinia virus is found.

There is no treatment for pox infection other than supportive therapy. Antivaccinia gammaglobulin did not alter the course of disease in cheetahs (Baxby et al. 1982) but does not seem to have been tried in other species.

PUBLIC HEALTH CONCERNS. Cowpox and cowpox-like viruses can infect people. Gloves and eye protection should be worn by staff handling infected animals. Smallpox vaccine protects against these viruses, and consideration should be given to vaccinating those at risk during an outbreak. However, smallpox vaccine is no longer readily available.

DOMESTIC ANIMAL HEALTH CONCERNS. Domestic animals in endemic areas are probably continually exposed to the unknown wildlife reservoir of the virus, so control measures are difficult to recommend. Cowpox will continue to occur sporadically in cattle and cats in the United Kingdom until the source of infection is identified and rational means of control devised. Vaccination of domestic animals is not warranted.

Volepox Virus. Volepox virus was isolated from a skin lesion on a vole *Microtus californicus* trapped in San Mateo County, California (Regnery 1987). It has also been isolated from a scab on a piñon mouse *Peromyscus truei* trapped in the same area (Knight et al. 1992). Inoculation of voles caused a localized lesion and seroconversion. Based on serology, volepox virus is endemic in voles from the San Francisco Bay area.

It is an orthopoxvirus more closely related genetically to the other North American orthopoxviruses, raccoonpox virus and skunkpox virus, than to orthopoxviruses such as vaccinia and cowpox viruses. It forms irregular plaques on Vero cell monolayers, with large syncytia on the periphery of the plaques. A-type inclusion bodies occur (Knight et al. 1992).

Taterapox Virus (Gerbilpox Virus). Taterapox virus was isolated from a naked sole gerbil *Tatera valida* in Dahomey, West Africa, in 1968 (Lourie et al. 1975). Taterapox virus has a restriction endonuclease map that is distinct from both ectromelia and variola viruses (Esposito and Knight 1985).

The virus has been termed "gerbilpox virus" but needs to be differentiated from the cowpox-like virus isolated from great gerbils in Turkmenia (Marennikova et al. 1978; Fenner 1994e). Common gerbils *Meriones unguiculatus* were not susceptible to infection (Lourie et al. 1975).

Parapoxvirus of Red and Gray Squirrels. A morphologically typical parapoxvirus caused eyelid lesions in a red squirrel *Sciurus vulgaris* in the United Kingdom (Scott et al. 1981), and a similar virus was isolated from a lip lesion in a red squirrel (Sands et al. 1984). It failed to cause lesions in colostrum-deprived lambs and was not antigenically related to contagious ecthyma virus. The infection usually is fatal in red squirrels (Sainsbury et al. 1997). A parapoxvirus also has been reported in the United Kingdom in a gray squirrel *Sciurus carolinensis* (Duff et al. 1996). There is speculation that the gray squirrel is the reservoir host for the virus and that the virus is contributing to the decline in the red squirrel population. Generally, the prevalence of antibody in the red squirrel population is low, whereas in the British gray squirrel population it is high (Sainsbury et al. 1997). However, parapoxvirus infection is unknown in gray squirrels in North America, where the species is native.

Squirrel Fibroma Virus

HISTORY AND HOST RANGE. Skin tumors on gray squirrels *S. carolinensis* caused by a leporipoxvirus related to rabbit fibroma virus (Fenner 1965; Woodroofe and Fenner 1965) were first described in 1953 (Kilham et al. 1953), and there have been sporadic reports subsequently (Novilla et al. 1981). Similar tumors in squirrels, diagnosed as fibromas, were recorded from 1936 to 1950, and there is a record of a fibroma in a porcupine *Erethizon dorsatum* in 1938 and a fox squirrel *Sciurus niger* in 1940 (Herman and

Reilly 1955). However, there are no virologic data to confirm whether these species are susceptible to poxviral fibromas. A poxvirus tentatively classified as a *Leporipoxvirus* also has been reported in western gray squirrels *Sciurus griseus* from California (Regnery 1975).

While natural disease associated with the leporipoxvirus of gray squirrels has been confirmed only in squirrels, the virus also grew well in experimentally infected woodchucks *Marmota monax.* Based on the high titers of virus found in lesions in young woodchucks, and the severe lesions and systemic dissemination in young squirrels, Kilham (1955) suggested that the woodchuck, rather than the squirrel, may be the natural host. Domestic rabbits can be infected, especially by inoculation with woodchuck-passaged virus. Attempts to infect guinea pigs, mice, chinchillas, and golden hamsters were unsuccessful (Kirschstein et al. 1958; Hirth et al. 1969).

DISTRIBUTION. Fibromas have been reported on gray squirrels from Maryland, North Carolina, New York State, Connecticut, and Virginia in the eastern United States, and Ontario in Canada.

CLINICAL SIGNS. Firm elevated cutaneous nodules from several millimeters up to several centimeters in diameter vary in form from flat plaques up to a centimeter or so thick, to nearly pedunculated masses. The overlying skin may be thick and corrugated, or smooth and hairless or partially haired; lesions may ulcerate severely. Tumors can involve most parts of the body, including mucocutaneous junctions such as lips, genital openings, and eyelids, where the most obvious sign may be a prominent thickening of the epidermis rather than a nodule. Tumors may be solitary but can be extremely numerous. In some cases, squirrels may be in poor body condition or have concurrent infections (Novilla et al. 1981).

TRANSMISSION. Transmission is presumed to be mechanical, by biting arthropods, such as mosquitoes or the squirrel flea *Orchopeus howardi.* Experimentally, squirrel fibroma virus has been transmitted by mosquitoes *Aedes aegypti* and *Anopheles quadrimaculata.* The virus was most readily transmitted to suckling squirrels, which developed large single or multiple fibromas that persisted for a month or more and were infectious by mosquito transfer. Adult squirrels were more difficult to infect, did not develop large tumors, and did not transmit by mosquito (Kilham 1955).

PATHOGENESIS. Virus replicates in dermis or epidermis. In a 4-week-old squirrel, fibromas appeared in 12–14 days; by 4 weeks, they were 2 cm in diameter, at which stage central necrosis appeared. By 32 days, more than 60 small tumors were distributed over the body. Multiple tumors are probably due to a viremia, and this is supported by the development of lesions in internal organs such as lungs. There was extreme variation in outcome among littermates, with some animals developing multiple tumors and significant disease while others had only small primary lesions (Kilham 1955).

PATHOLOGY. The gross lesions vary with the stage of disease, severity, and involvement of internal organs. Skin lesions range from large tumors to marked thickening of the epidermis around eyes and ears. Varying degrees of hyperkeratosis of the overlying skin have been described (Kilham et al. 1953; Hirth et al. 1969; King et al. 1972). Severe involvement of internal organs has been described, the lungs containing white/gray focal firm masses, and multiple pale areas present in the liver. The spleen may be enlarged with prominent germinal centers (Kilham 1955; King et al. 1972).

Microscopically, the tumors resemble those due to rabbit fibroma virus in cottontails. Hyperplastic, hyperkeratotic epithelium, in which cells are swollen and vacuolate, covers a mass of plump spindle-shaped fibroblastic cells (Hirth et al. 1969). Perivascular infiltration of lymphocytes, plasma cells, and histiocytes involves blood vessels deep in the tumor (Kilham et al. 1953). Eosinophilic intracytoplasmic inclusions are prominent in the enlarged epithelial cells and are often seen in the underlying connective tissue cells.

Multiple discrete areas of adenomatoid proliferation of alveolar lining cells are evident in squirrels with lung lesions, and there is proliferation of loose interstitial connective tissue. Intracytoplasmic inclusions are evident in affected pulmonary epithelium and stroma. Similar stromal lesions may occur in lymph nodes, kidneys, and liver. Electron microscopy of skin and lung lesions revealed poxvirus particles in almost every cell (King et al. 1972).

DIAGNOSIS. Diagnosis is based on the clinical appearance and histology. The virus forms small clear plaques on rabbit embryo fibroblasts, and foci in rabbit kidney cells, but no lesions in chick embryo fibroblasts (Woodroofe and Fenner 1965).

Parapoxvirus infection of red squirrels produces scabbing lesions, rather than nodules, and the viruses differ in morphology in negatively stained lesion material.

IMMUNITY. There is very little information on the immunity following infection with squirrel fibroma virus. Recovered squirrels were resistant to challenge and had circulating neutralizing antibodies, but nothing is known about persistence of antibody. In the only serologic survey published, one of 120 gray squirrels had neutralizing antibodies; it was the only one to have lesions (Kilham 1955).

DOMESTIC ANIMAL IMPLICATIONS. Squirrel fibroma virus causes a small transient fibroma at the site of inoculation in domestic European rabbits (Fenner 1965). There are no implications for other domestic animals.

MANAGEMENT IMPLICATIONS. Though the disease is relatively uncommonly described in the literature, it is commonly encountered in many areas and can occur locally in epidemic proportions. However, it probably has little long-term impact on gray squirrel populations. Control measures are not available and likely not warranted. Treatment of affected squirrels that come into care is supportive, but severely affected animals should be euthanatized.

Squirrelpox Virus. See the section on primates (monkeypox virus).

Cotia Virus. Cotia virus is an unclassified poxvirus isolated from arbovirus sentinel laboratory mice in South America. A natural host has not been identified, but the virus is presumably insect transmitted, possibly from a wildlife host. It seems more closely related to the leporipoxviruses and swinepox virus than to other poxviruses (Ueda et al. 1995).

CETACEA

Poxviruses of Dolphins and Porpoises. Poxvirus infections in cetaceans manifest themselves as skin lesions variously called "targets," "watered-silk," "ring," "pinhole," "circle," or "tattoo" lesions (Simpson and Gardner 1972; Greenwood et al. 1974; Ridgeway 1975; Geraci et al. 1979). They appear as slightly raised irregular areas that are gray, black, or yellowish, with sharply demarcated edges. On close examination, the lesions are composed of numerous small pigmented dots.

Lesions were commonly found on dusky dolphins *Lagenorhynchus obscurus,* long-beaked common dolphins *Delphinus delphis,* Burmeister's porpoises *Phocoena spinipinnis,* and bottlenose dolphins *Tursiops truncatus,* among 339 small cetaceans captured off the coast of Peru (Van Bressem et al. 1993; Van Bressem and van Waerebeek 1996).

The presence of orthopoxvirus-like particles in these lesions has been confirmed (Flom and Houk 1979; Geraci et al. 1979; Smith et al. 1983; Van Bressem et al. 1993). Poxvirus also has been confirmed in "tattoo" lesions in the Atlantic white-sided dolphin *Lagenorhynchus acutus* (Geraci et al. 1979), and "tattoo" lesions have been recorded in the white-beaked dolphin *Lagenorhynchus albirostris,* the striped dolphin *Stenella coeruleoalba,* the long-finned pilot whale *Globicephala melaena,* and the harbor porpoise *Phocoena phocoena* (Baker 1992; Baker and Martin 1992).

Histologically, there is a sharp transition between normal and involved skin (Geraci et al. 1979). The stratum intermedium within the central involved zone is pale, with marked cytoplasmic vacuolation and a prominent reticular pattern of keratinaceous fibers. In the adjacent peripheral transitional zone, cells of the stratum intermedium contain small pale eosinophilic intracytoplasmic inclusions, which contain poxvirus particles. The uninfected epidermis around the lesion is compressed.

Lesions persist for many months without resolution, but biopsy may precipitate an accelerated resolution of the remaining lesion. Natural resolution may be heralded by a change in the appearance of the lesion, from flat to raised and edematous (Smith et al. 1983). Antibody directed toward the virus has been detected by immune electron microscopy (Smith et al. 1983).

CARNIVORA

Catpox Virus and Carnivorepox Virus. See the section on Rodentia (cowpox virus).

Foxpox Virus. See the section on Rodentia (mousepox virus).

Raccoonpox Virus. Raccoonpox virus was isolated from the upper respiratory tracts of two asymptomatic raccoons *Procyon lotor* from Maryland in 1962 (Alexander et al. 1972). It appeared to be widespread, as 22 of the 92 raccoons were seropositive. Inoculation of a single raccoon did not produce a lesion, but the animal did seroconvert (Thomas et al. 1974).

Raccoonpox virus is an orthopoxvirus more closely related to volepox virus than to vaccinia or cowpox viruses. Type-A inclusions are present in infected cells, and it forms syncytia in cell culture. It has been developed as a vaccine vector for rabies and panleukopenia (Hu et al. 1997).

Skunk Pox Virus. A poxvirus closely related to raccoonpox virus and volepox virus was isolated from the lungs of a striped skunk *Mephitis mephitis* shot in Colfax, Washington (Cavallaro and Esposito 1992).

PINNIPEDIA

Sealpox Viruses

Host Range and Distribution. Poxvirus infections have been described in pinnipeds from North America, South America, the United Kingdom, and the Netherlands. Poxvirus particles have been described in nodular skin lesions in six species of pinnipeds, both free-living and captive: California sea lions *Zalophus californianus* (Wilson et al. 1969, 1972a; Hastings et al. 1989); South American sea lions *Otaria flavescens* (Wilson and PoglayenNeuwall 1971; Okada and Fujimoto 1984); harbor seals *Phoca vitulina* (Wilson et al. 1972b; Hastings et al. 1989); northern fur seals *Callorhinus ursinus* (Hadlow et al. 1980); gray seals *Halichoerus grypus* (Hicks and Worthy 1987; Osterhaus et al. 1990, 1994; Simpson et al. 1994; Nettleton et al. 1995); and northern elephant seals *Mirounga angustirostris* (Hastings et al. 1989).

They appear to be parapoxviruses, except for a report on South American sea lions (Wilson and PoglayenNeuwall 1971) and one on gray seals (Osterhaus et al. 1990), in which the virions were orthopoxvirus-like. The gray seal isolate reacted with ectromelia antiserum; both parapoxvirus-like and orthopoxvirus-like particles were seen in this case (Osterhaus et al. 1990).

Nodular skin disease, possibly viral, also has been reported in captive Steller sea lions *Eumetopias jubatus* and elephant seals *Mirounga* sp. (Wilson et al. 1972). Antibodies to a gray seal parapoxvirus have been found in a ringed seal *Phoca hispida* and Baikal seals *Phoca sibirica* (Osterhaus et al. 1994).

The species specificity of these viruses among pinnipeds is unknown, and analogous poxviruses may occur in other species of pinnipeds. Seal parapoxvirus can infect people. Persons handling infected gray seals developed typical parapoxvirus lesions on the fingers (Hicks and Worthy 1987). In both cases, the scabs detached easily after 35–36 days, but in one, lesions recurred at the initial site over the next 6 months. Healing occurred with little scar formation. Natural disease has not been reported in other species, and attempts to infect specific pathogen-free lambs with a gray seal isolate were unsuccessful (Nettleton et al. 1995).

TRANSMISSION AND EPIDEMIOLOGY. The epidemiology of seal pox in the wild is undescribed, but transmission of the parapoxvirus has been reported in captive harbor seals (Wilson et al. 1972c; Hastings et al. 1989) and gray seals (Hicks and Worthy 1987). Gray seals held in dry enclosures remained asymptomatic, while most held in cement saltwater tanks developed lesions. The rough tanks and wet environment appeared to promote virus transmission, and circumstantial evidence suggests that the virus remained viable for over 10 months.

Transmission also seems to have occurred between harbor seals and elephant seals housed separately, perhaps via waste material or arthropods (Hastings et al. 1989).

CLINICAL SIGNS. The disease manifests as obvious nodular, sometimes suppurative, lesions anywhere on the skin, but commonly on the face, neck, and flippers. There are no obvious systemic signs. Deaths of seals with the disease generally have been attributed to other causes.

PATHOGENESIS AND PATHOLOGY. The incubation period of natural cases in harbor seals was 3–5 weeks. Within 1 week, lesions increased in thickness and reached a diameter of 2–3 cm, ultimately taking at least 15 weeks to resolve (Wilson et al. 1972c).

A mild to moderate transient neutrophilia has been described in affected harbor seals, associated with an elevated total protein (Hastings et al. 1989).

In all species, small lesions (0.5 cm) were discrete oval or round, smooth, firm nodules. Larger lesions (1.5–3.0 cm) appeared gray-white and granular or fissured. Hair was lost or reduced over lesions, and they were suppurative. In the South American sea lion and northern fur seal, larger nodules were sometimes umbilicate. Alopecia or shortened hair was seen over healed lesions.

Histologically, there is epithelial proliferation and acanthosis, with intracytoplasmic eosinophilic inclusions in superficial or follicular epithelium or, in northern fur seals, in nodules of epithelial cells seemingly isolated in the dermis (Hadlow et al. 1980). In gray seals (Hicks and Worthy 1987) and harbor seals (Hastings et al. 1989), a significant dermal mixed inflammatory response was seen.

DIAGNOSIS. The gross and microscopic appearance is typical, and the etiology can be confirmed by demonstrating poxvirus particles in negatively stained unfixed lesion material or by attempts to isolate virus.

Since an orthopoxvirus has been found in cases of seal pox (Osterhaus et al. 1990), the identity of the virus involved should be confirmed by negative-staining electron microscopy and virus isolation, if possible. *Dermatophilus congolensis* and other causes of skin nodules, such as neoplasms, parasites, or trauma, would need to be considered.

IMMUNITY. Seroconversion to a gray seal isolate was demonstrated during an outbreak in that species (Osterhaus et al. 1994). Recurrent infection was suspected in a harbor seal, which suggests that immunity may not be long-lived (Hastings et al. 1989).

CONTROL AND TREATMENT. In a captive seal colony, the affected animal should be isolated in a dry area, and animals at risk also might be segregated in a relatively dry environment. If feasible, contaminated tanks could be drained and treated with suitable virucidal compounds. Control in wild populations is neither indicated nor feasible.

PUBLIC HEALTH CONCERNS. Parapoxvirus infection of seals is zoonotic. Staff working with seals should be informed about the disease and the fact that the virus is transmissible to humans. Protective gloves might be worn when handling infected animals.

MANAGEMENT IMPLICATIONS. Recently captured animals should be quarantined in a relatively dry environment and examined for lesions over a 6-week period. Ideally, animals should be examined prior to shipment and affected animals refused.

PROBOSCIDEA

Elephantpox Virus. See the section on Rodentia (cowpox virus).

PERISSODACTYLA

Horsepox Virus. Horsepox, or "grease," may have been caused by cowpox virus (Fenner et al. 1989) and, hence, have been transmitted from the same unidentified wild rodent reservoir. However, horsepox has not been seen recently in Europe, whereas cowpox-like viruses have been reported relatively commonly. Horsepox has not been recognized in wild or feral horses.

Uasin Gishu Disease Virus. This skin disease of horses was recognized on the Uasin Gishu Plateau in Kenya in 1934, and a similar disease occurred in Zaire (Bugyaki 1959), Rwanda, and Burundi (Dekeyser et al. 1960). First considered to be fungal, it is now known to be caused by a poxvirus (Kaminjolo et al. 1974a,b) thought to have a wildlife reservoir host.

ARTIODACTYLA

Cowpox and Vaccinia Viruses. See the section on Rodentia.

Buffalopox Virus. Buffalopox virus is derived from vaccinia virus, presumably originally spread from humans immunized against smallpox. It occurs in buffaloes *Bubalis bubalis* in India and in Indonesia, Italy, Pakistan, Russia, and Egypt (Yager and Scott 1993). At least in India, buffalopox virus must have established an independent transmission cycle, possibly involving a wild rodent reservoir, analogous to cowpox virus (Fenner et al. 1989). The virus produces cutaneous lesions and keratoconjunctivitis in milking buffaloes (Mallick et al. 1990). The broad host range of vaccinia virus suggests that buffalopox virus could spread into wildlife. It also causes pocks, lymphadenitis, and fever in people (Ramanan 1996).

Camelpox Virus. Camelpox, caused by an orthopoxvirus, is a serious disease of dromedary camels *Camelus dromedarius* throughout their African and Asian range (Davies et al. 1975; Kriz 1982; Fenner et al. 1989). It is one of two poxvirus diseases of camels, the other being caused by a parapoxvirus (see below).

In Somalia, the disease, characterized by fever and a typical pox rash, varied from mild to severe and fatal and occurred mainly in young camels (Kriz 1982; Jezek et al. 1983). Following experimental inoculation, the disease followed the typical course of generalized poxvirus infection (Baxby et al. 1975).

Camelpox virus is distinguishable from other orthopoxviruses on biologic and molecular grounds (Marennikova et al. 1974; Esposito and Knight 1985; Ropp et al. 1995). In nature, it probably only affects dromedary camels, and humans generally appear to be refractory (Fenner et al. 1989). Vaccination is not normally practiced.

Contagious Ecthyma Virus. Contagious ecthyma is a common disease of farmed sheep and goats that is caused by the parapoxvirus orf virus (Robinson and Balassu 1981; Robinson and Lyttle 1992). In wild artiodactyls, the disease usually is also called contagious ecthyma, modified by the common name of the species involved.

DISTRIBUTION AND HOST RANGE. Contagious ecthyma occurs where sheep and goats are raised, and it also infects a broad range of wild artiodactyls. It is widespread in wild bighorn sheep *Ovis canadensis* in the Rocky Mountains from Alaska to California (Connell 1954; Blood 1971; Samuel et al. 1975; Lance 1979; Lance et al. 1981; Turner and Payson 1982; Elliott et al. 1994; l'Heureux et al. 1996). It has been found in thinhorn or Dall's sheep *Ovis dalli* in Alaska (Dieterich et al. 1981; Smith et al. 1982; Zarnke et al. 1983) and also appears to be endemic in herds of Rocky Mountain goat *Oreamnos americanus* in Canada (Samuel et al. 1975; Hebert et al. 1977).

The disease occurs in the managed herds of musk oxen *Ovibos moschatus* and caribou (reindeer) *Rangifer tarandus* in Alaska (Bell 1931; Dieterich et al. 1981; Zarnke et al. 1983) and in musk oxen and reindeer in Norway (Kummeneje 1979), but a serologic survey of free-ranging musk oxen in northeastern Greenland failed to reveal antibody (Clausen and Hjort 1986).

The disease is prevalent in Japanese serow *Capricornis crispus* (Suzuki et al. 1986, 1993). Contagious ecthyma has been described also in chamois *Rupicapra rupicapra* in Europe (Carrara 1959), whereas, in New Zealand, it is endemic in the introduced chamois and Himalayan tahr *Hemitragus jemlahicus* in the Southern Alps of the South Island (Kater and Hansen 1962; Daniel and Christie 1963). It has also been reported in steenbok *Raphicerus campestris* (Guarda 1959), in an alpaca *Lama pacos* in the Hanover Zoo (Hartung 1980), and in west Caucasian tur *Capra caucasica* at the Toronto Zoo (Smith et al. 1984). A case of contagious ecthyma occurred in a gazelle *Gazella gazella* introduced into a herd of sheep and goats in Israel (Yeruham et al. 1964).

In the case of Rocky Mountain bighorns, Dall sheep, Rocky Mountain goats, Himalayan tahr, and musk oxen, the virus has been designated as orf virus by successful transmission to domestic sheep or goats. In the other species, the virus has not been identified beyond its description as a parapoxvirus, but it is assumed to be orf virus.

Other species of artiodactyls are susceptible to orf virus experimentally. A moose calf *Alces alces* and a caribou fawn were susceptible to virus derived from a Dall sheep, but only mild lesions were seen (Zarnke et al. 1983). Young white-tailed deer *Odocoileus virginianus,* mule deer *Odocoileus hemionus,* wapiti *Cervus elaphus,* and pronghorns *Antilocapra americana* were susceptible to virus derived from a Rocky Mountain bighorn sheep (Lance et al. 1983), though lesions were produced in only one of three sites of inoculation (oral mucosa) in the white-tailed deer and wapiti. Of these four species, contagious ecthyma has been found only in caribou in the wild.

There is little information on the prevalence or distribution of contagious ecthyma in other wild ungulates, but it, or closely related viruses, probably will be detected in additional species as they are more closely observed or managed.

Human infections also are reported regularly (Robinson and Lyttle 1992). There are two reports of suspected contagious ecthyma in dogs (Wilkinson et al. 1970; Hartung 1980). Experimentally, contagious ecthyma has induced lesions in calves, a foal, and monkeys, but other species tested have appeared refractory to infection (Robinson and Balassu 1981).

TRANSMISSION AND EPIDEMIOLOGY. Transmission is by contact with affected animals or with contaminated objects or surfaces (fomites). Infection occurs mainly in the skin, through disruption of the outer keratinized layer. Lesions on the lips and nares in sheep and goats often follow exposure to thistles or rough feed, and in sheep, teat and udder lesions occur following suckling by clinically affected lambs. Transmission probably follows a similar pattern in wild ungulates.

The virus can survive for years in scabs stored at 4° C in the dark and for many months at room temperature. In nature, scab material lying in sheltered locations where animals habitually camp, or at watering points or salt licks, likely acts as the main source of infection for successive crops of young. In captive situations, substrate in the pen, feeders, and instruments such as ear punches may be contaminated. There is no evidence that orf virus causes latent infections, but lesions may persist for many months on the heads of rams (McKeever 1984), and successive udder infections can occur in ewes (Schmidt 1967). Persistent and recurrent infections probably also play a part in the epidemiology of the disease in wild ungulates.

The proportion of seropositive animals in herds of bighorn sheep (Turner and Payson 1982) and serow (Suzuki et al. 1986, 1993) fluctuates markedly from year to year. The prevalence of disease in any particular year likely depends on the number of immunes in the population and on herd density.

CLINICAL SIGNS. The scabby lesions of contagious ecthyma are most commonly found on the lips and on the skin of the face and/or the coronet, as well as the udder, vulva, pizzle, and oral mucosa, but they can occur elsewhere on the skin or mucous membrane. Lesions can range from a few difficult-to-detect crusts to thick, hard, coalescing scabs that cover the entire face or lower limb.

If mouth lesions are severe enough, animals will not feed, and lameness may occur due to foot lesions. In both cases, animals will lose condition and may even die of starvation. Bighorn lambs with clinical signs of contagious ecthyma gained less weight than those without signs and were lighter as yearlings (l'Heureux et al. 1996).

Extensive proliferative lesions on the gums, oral mucocutaneous junctions, and sometimes skin have been reported in lambs, musk oxen, reindeer, and Japanese serow. In musk oxen, lesions have been described as multiple, multilobulated papillomas, often large, on lips, muzzle, and nostrils and to a lesser extent on the eyelids, neck, chest, and perianal region. Papillomatous, rather than overtly scabby, lesions have also been described in reindeer and Japanese serow; and in young lambs and kids, lesions on the gingiva are often multilobulated "cauliflower-like" papillomas.

Generally, affected animals are afebrile and recover uneventfully, but severe outbreaks, sometimes with deaths, mainly among young animals, have been reported in farmed sheep, Rocky Mountain goats (Samuel et al. 1975), and Dall sheep (Dieterich et al. 1981) in the wild and in managed musk oxen (Dieterich et al. 1981). In farmed sheep, generalized disease has been reported rarely, with lesions found throughout the stratified epithelium of the alimentary tract (Robinson and Balassu 1981). This appears to be due to multiple sites of infection rather than systemic spread of the virus, and it might occur in wild artiodactyls.

PATHOGENESIS. In sheep, small pustules develop by 5 days after infection and, by day 7, scabs are forming. More pustules and scab formation are seen over the next 10 days, and the enlarging lesion is encircled by reddened skin. By 3 weeks, the lesions are resolving and, by 4 weeks, scabs are detaching and easily removed. Typically, lesions heal without scarring (Robinson and Balassu 1981; Robinson and Lyttle 1992).

PATHOLOGY. The microscopic lesions of contagious ecthyma in the skin or upper alimentary tract of *Ovis* and *Capra* species and their close allies are typical of poxvirus infection (Yager and Scott 1993). Cytoplasmic eosinophilic inclusion bodies are usually evident in keratinocytes.

In musk oxen (Dieterich et al. 1981) and Japanese serow (Okada et al. 1984), the proliferative lesions were papillomatous with extensive development of rete pegs and very little swelling and necrosis of epidermal cells. Whether this is due to a different host response to infection or a function of virus strain is not clear.

DIAGNOSIS. Contagious ecthyma usually is diagnosed on the basis of the species involved and the appearance of the lesions in the absence of signs of systemic disease, perhaps in association with demonstration of parapox particles in negatively stained preparations, and histopathology. Animal transmission experiments, or the application of molecular techniques that can differentiate related parapoxviruses (Gassmann et al. 1985; Robinson and Mercer 1995), may be warranted to reveal hitherto undescribed poxviruses of this genus from wild animals.

Past infection and prevalence of exposure in a population can be detected by a range of serologic techniques, including serum neutralization, complement fixation, immunodiffusion, or ELISA.

DIFFERENTIAL DIAGNOSIS. Contagious ecthyma can be confused with dermatitis caused by *Dermatophilus congolensis,* and the two diseases can occur concurrently (Cooper et al. 1970). Sheeppox and goatpox viruses, and other poorly characterized orthopoxviruses, might cause scabby lesions in wild artiodactyles, and these lesions could be confused with contagious ecthyma. The demonstration of parapoxvirus particles in lesion material, and the different clinicopathologic syndrome, should distinguish contagious ecthyma from sheep or goat pox.

IMMUNITY. The duration of immunity in wild animals is probably similar to that seen in sheep. Immunity to reinfection on the mouth or feet persists for up to 5 months following recovery from natural disease (Schmidt 1967). Subsequently, short-lived lesions of little consequence may occur on the mouth or feet. Lesions can occur on the udder in animals immune to infection on the mouth (Schmidt 1967), and this also may occur in wild animals. Maternal antibody, passed to neonates via the colostrum, is not protective (Buddle and Pulford 1984). Protective immunity is probably entirely cell mediated (Robinson and Lyttle 1992).

CONTROL AND TREATMENT. Control of the disease in free-ranging wild animals is problematic, and in most situations is not warranted. Although the disease might become extinct in small isolated flocks or herds, the chances of it being reintroduced with immigrants is high, and the consequences will depend on the level of herd immunity. If a flock or herd has been maintained free of contagious ecthyma, the possibility of introducing the disease through introductions of, or contact with, infected wildlife or domestic sheep and goats must be considered. This should also be borne in mind when introducing artiodactyls into zoos, since populations of susceptible artiodactyls bred in isolation may be naive to the infection.

Vaccination in managed herds could be contemplated, but current vaccines consist of fully virulent virus and, unless the disease is already endemic, vaccination is contraindicated. There are no effective antiviral treatments, but antibiotics could be administered where secondary infection is suspected. Large scabs may be removed if they interfere with feeding. Oral and/or parenteral fluids are indicated if animals are not eating or drinking.

PUBLIC HEALTH CONCERNS. Orf virus infection is an occupational hazard for those who handle sheep, goats, or their allies (Robinson and Balassu 1981; Robinson and Petersen 1983; Robinson and Lyttle 1992). Human disease has been acquired from Rocky Mountain goats (Carr 1968) and from muskox and reindeer (Falk 1978).

In people, the disease is seldom serious. Draining lymph nodes become swollen and painful, and mild pyrexia may occur, but the cutaneous lesions, which resemble those in animals, usually resolve within 6 weeks without extensive scarring. Occasionally, particularly in immunocompromised individuals, the lesion can undergo massive proliferation, to form a so-called "giant orf" that persists for many months. Other serious but rare complications have been described (Robinson and Lyttle 1992).

DOMESTIC ANIMAL HEALTH CONCERNS. Transmission of the virus from wild animals to domestic animals usually is not of concern, unless the sheep or goat flock were one of the minority in which contagious ecthyma was not endemic.

Contagious Ecthyma Virus of Camels. A parapoxvirus causes contagious ecthyma in dromedary camels *C. dromedarius*. The local name of the disease in Khazakstan has been variously transliterated as "auzdik" (Buchnev et al. 1969), "auzduk" (Roslyakov 1972), and "auzdyk" (Munz et al. 1986), and in Mongolia the disease is known as "amru" (Dashtseren et al. 1984). It also has been reported in Kenya (Munz et al. 1986) and Somalia (Moallin and Zessin 1988), and is likely distributed over the range of the dromedary.

Morbidity can approach 100%, but, although disease might be severe, the mortality rate was less than 1%. In many cases, the picture resembled that of contagious ecthyma in sheep, with lesions on the lips and occasionally nostrils. However, unlike the disease in sheep, generalized (not necessarily systemic) disease, with lesions all over the body, was seen in Mongolia, and swelling of the entire head occurred in some animals. The disease must be distinguished from camelpox and papilloma virus of camels.

Mild lesions were induced experimentally in lambs, and mild lesions were also induced in camels with contagious ecthyma virus of sheep. Previous infection with contagious ecthyma virus of sheep afforded some protection to challenge with the camel parapoxvirus, but vaccinia virus did not. There are no reports of infection in humans.

The virus has been cultivated on hen eggs, and an egg-adapted camel virus vaccine has been produced (Dashtseren et al. 1984).

Parapoxvirus of Red Deer. A pustular dermatitis occurred in farmed red deer *Cervus elaphus* in New Zealand (Horner et al. 1987). Lesions were on the muzzle, face, ears and, in some cases, on the legs and neck. In stags, lesions were commonly seen on the velvet antler. The mortality rate was high on one farm, but whether this was due to the parapoxvirus infection was unclear. Histologically, the lesions were characteristic of a cutaneous poxvirus infection. Typical parapoxvirus particles were seen, and the virus was grown in sheep testis cells. The deer virus is a previously undescribed parapoxvirus, based on comparison with orf virus and the viruses of pseudocowpox and bovine papular stomatitis (Robinson and Mercer 1995).

Only mild lesions were induced in sheep. Similarly, orf virus produced only mild lesions in red deer. The red deer virus has not been implicated in human infections, though this might reflect a lack of reporting, if lesions were assumed to be orf. Though red deer were introduced into New Zealand from Europe, this disease has not been reported in red deer in Eurasia, suggesting that red deer may not be the natural host of this virus.

Capripoxviruses. Poxviruses in the genus *Capripoxvirus* fall into three groups: sheeppox virus, goatpox virus, and lumpy skin disease virus (Murphy et al. 1995). Although the three viruses have been given separate names, there is cross-infection among sheep, goats, and cattle. Hence, the three diseases are often referred to simply as capripox (Carn 1993).

The capripoxviruses cause serious systemic infections in their hosts. Sheeppox and goatpox are currently prevalent throughout the Near and Middle East, India, Bangladesh, and North and Central Africa. Lumpy skin disease has been reported in all sub-Saharan African countries, in Egypt, and in Israel.

Sheeppox and goatpox have not been described in wild or captive sheeplike or goatlike animals, although there seems to be no reason why they could not occasionally be affected.

Attempts have been made to find a wildlife reservoir for the lumpy skin disease virus in Africa. A young giraffe *Giraffa camelopardalis* and a young impala *Aepyceros melampus* were experimentally infected and died with typical lumpy skin disease. However, buffalo *Syncerus caffer* calves and two adult black wildebeest *Connochaetes gnou* were unaffected and did not seroconvert (Young et al. 1970). Antibodies to lumpy skin disease virus have been found in 6 of 44 wildlife species in Africa: African buffalo, greater kudu *Tragelaphus strepsiceros,* waterbuck *Kobus ellipsiprymnus,* reedbuck *Redunca arundinum,* impala, springbok *Antidorcas marsupialis,* and giraffe (Hedger and Hamblin 1983). However, the prevalence of reactors was low, and the authors concluded that African wildlife are unlikely to be an important reservoir of lumpy skin disease virus.

Clinical lumpy skin disease occurred in a single captive-bred Arabian oryx (*Oryx leucoryx*) in a closed herd in Saudi Arabia (Greth et al. 1992). Lumps were present on the skin over the whole animal, but it eventually recovered. The virus was not isolated, but typical poxvirus particles were seen, and a rising antibody titer to sheeppox virus was recorded. The virus was thought to be transmitted either by direct contact with sheep at the perimeter fence or insect-borne.

Swinepox Virus. Swinepox virus infection causes sporadic disease in domestic pigs *Sus scrofa* and is likely to spread to wild or feral swine where these are in close contact with domestic pigs. The virus is transmitted by the pig louse, *Haematopinus suis,* and presumably by other arthropod vectors. The disease is typical of a low-grade poxvirus infection with papules on abdomen and legs that become pustular and then scab and crust. Histologically, lesions are confined to the epidermis and dermis, with some involvement of the draining lymph nodes (Yager and Scott 1993).

Poxvirus Infections of Cervids. Poxviruses with orthopoxvirus morphology have been found in mule deer *Odocoileus hemionus* (Williams et al. 1985) and black-tailed deer *O. hemionus* (Patton et al. 1996) in the United States, and in a herd of reindeer in the Toronto Zoo (Barker et al. 1980).

Lesions in mule deer occurred a year apart in two fawns in Wyoming. Both were weak, with keratoconjunctivitis and crusting lesions on the face and nose. A poxvirus was isolated in cell cultures from both cases. Since the mule deer poxvirus reacted with a probe of capripoxvirus DNA (J. Esposito personal communication), a member of the capripoxvirus genus distinct from the African/Asian viruses may be present in North America.

The poxvirus from *Odocoileus hemionus* was serendipitously isolated in testicular cell culture from an animal that had died during an epidemic of adenovirus infection in California.

The disease in reindeer at the Toronto Zoo first appeared as a keratoconjunctivitis and later involved other areas of the body. Raised crusting lesions were found on the haired areas of the nose, legs, and vulva, and an ulcer was found on the hard palate. Poxvirus particles resembling orthopox were seen, but virus isolation in eggs and tissue culture was unsuccessful. Disease was not seen in white-tailed deer, wapiti *Cervus elaphus,* or moose held in separate enclosures nearby.

LITERATURE CITED

Alexander, A.D., V. Flyger, Y.F. Herman, S.J. McConnell, N. Rothstein, and R.H. Yager. 1972. Survey of wild mammals in a Chesapeake Bay area for selected zoonoses. *Journal of Wildlife Diseases* 8:119–126.

Aplogan, A., V. Mangindula, P.T. Muamba, G.N. Mwema, L. Okito, R.G. Pebody, C.E. Roth, L.S. Shongo, M. Szczeniowski, and K.F. Tschioko. 1997. Human Monkeypox–Kasai Oriental, Democratic Republic of Congo, February 1996–October 1997. *Morbidity and Mortality Weekly Report* 46:1168–1171.

Arita, I., R. Gispen, S.S. Kalter, L.T. Wah, S.S. Marennikova, R. Netter, and I. Tagaya. 1972. Outbreaks of monkeypox and serological surveys in nonhuman primates. *Bulletin of the World Health Organization* 46:625–631.

Bagnall, B.G., and G.R. Wilson. 1974. Molluscum contagiosum in a red kangaroo. *Australian Journal of Dermatology* 15:115–120.

Baker, J.R. 1992. Causes of mortality and parasites and incidental lesions in dolphins and whales from British waters. *Veterinary Record* 130:569–572.

Baker, J.R., and A.R. Martin. 1992. Causes of mortality and parasites and incidental lesions in harbour porpoises (*Phocoena phocoena*) from British waters. *Veterinary Record* 130:554–448.

Barker, I.K., K.G. Mehren, W.A. Rapley, and A.N. Gagnon. 1980. Kerato-conjunctivitis and oral/cutaneous lesions associated with poxvirus infection in reindeer (*Rangifer tarandus tarandus*). In *Proceedings of the Symposium on the Comparative Pathology of Zoo Animals, Washington, October 1978,* ed. R.J. Montali and G. Migaki. Washington, DC: Smithsonian Institution, pp. 171–177.

Baxby, D. 1977. Is cowpox misnamed? A review of 10 human cases. *British Medical Journal* 1:1379–1381.

Baxby, D., H. Ramyar, M. Hessami, and B. Ghaboosi. 1975. Response of camels to intradermal inoculation with camelpox and smallpox viruses. *Infection and Immunity* 11:617–621.

Baxby, D., D.G. Ashton, D. Jones, L.R. Thomsett, and E.M. Denham. 1979. Cowpox virus infection in unusual hosts. *Veterinary Record* 109:175.

Baxby, D., D.G. Ashton, D.M. Jones, and L.R. Thomsett. 1982. An outbreak of cowpox in captive cheetahs: Virological and epidemiological studies. *Journal of Hygiene* 89:365–372.

Baxby, D., M. Bennett, and B. Getty. 1994. Human cowpox 1969–93: A review based on 54 cases. *British Journal of Dermatology* 131:598–607.

Bearcroft, W.G.C., and M.F. Jamieson. 1958. An outbreak of subcutaneous tumours in rhesus monkeys. *Nature* 182:195–196.

Bell, W. D. 1931. Experiments in establishing musk oxen in Alaska. *Journal of Mammology* 12:292–297.

Bennett, M., C.J. Gaskell, D. Baxby, R.M. Gaskell, D.F. Kelly, and J. Naido. 1990. Feline cowpox infection. *Journal of Small Animal Practice* 31:167–173.

Bennett, M., A.J. Crouch, M. Begon, B. Duffy, S. Feore, R. M. Gaskell, D.F. Kelly, C.M. McCracken, L. Vicary, and D. Baxby. 1997. Cowpox in British voles and mice. *Journal of Comparative Pathology* 116:35–44.

Blood, D.A. 1971. Contagious ecthyma in Rocky Mountain big horn sheep. *Journal of Wildlife Management* 35:270–275.

Buchnev, K.N., R.G. Sadykov, S.Z. Tulepbayev, and A.A. Roslyakov. 1969. Smallpoxlike disease of camels "Auzdik" [in Russian]. *Trudy Almaatinskogo Zootekhnicheskogo Instituta* 16:36–47.

Buddle, B.M., and H.D. Pulford. 1984. Effect of passively acquired antibodies and vaccination on the immune response to contagious ecthyma virus. *Veterinary Microbiology* 9:515–522.

Bugyaki, L. 1959. Dermatose contagieuse des ruminants et du cheval. *Bulletin de l'Office International des Epizooties* 51:237–249.

Buller, R.M.L., and G.D. Wallace. 1988. Ectromelia (mousepox) virus. In *Virus diseases in laboratory and captive animals,* ed. G. Darai. Boston: Martinus Nijhoff, pp. 63–82.

Burnet, M. 1968. *Changing patterns,* 1st ed. Melbourne: Heinemann, 282 pp.

Carn, V.M. 1993. Control of capripoxvirus infections. *Vaccine* 11:1275–1279.

Carr, R.W. 1968. A case of orf (ecthyma contagiosum; contagious pustular dermatitis) contracted by a human from a wild Alaskan mountain goat. *Alaska Medicine* June: 75–77.

Carrara, O. 1959. Reperti di patologia spontanea nel camoscio (*Rupricapra rupicapra* L.). *Atti della Societa Italiana delle Scienze Veterinarie* 13:460–464.

Cavallaro, K.F., and J.J. Esposito. 1992. Sequences of the raccoon poxvirus hemagglutinin protein. *Virology* 190:434–439.

Clausen, B., and P. Hjort. 1986. Survey for antibodies against various infectious disease agents in muskoxen (*Ovibos moschatus*) from Jamesonland, northeast Greenland. *Journal of Wildlife Diseases* 22:264–266.

Connell, R. 1954. Contagious ecthyma in Rocky Mountain bighorn sheep. *Canadian Journal of Comparative Medicine* 18:59–60.

Cooper, B.S., R.E. Lynch, and P.M. Marshall. 1970. An outbreak of contagious pustular dermatitis associated with *Dermatophilus congolensis* infection. *New Zealand Veterinary Journal* 18:199–201.

Crandell, R.A., H.W. Casey, and W.B. Brumlow. 1969. Studies of a newly recognized poxvirus of monkeys. *Journal of Infectious Diseases* 119:80–88.

Crouch, A.C., D. Baxby, C.M. McCracken, R.M. Gaskell, and M. Bennett. 1995. Serological evidence for the reservoir hosts of cowpox virus in British wildlife. *Epidemiology and Infection* 115:185–191.

Czerny, C.P., and H. Mahnel. 1990. Structural and functional analysis of orthopoxvirus epitopes with neutralizing monoclonal antibodies. *Journal of General Virology* 71:2341–2352.

Daniel, M.J., and A.H.C. Christie. 1963. Untersuchungen über Krankheiten der Gemse (*Rupricapra rupicapra* L.) und des Thars (*Hemitragus jemlaicus* Smith) in den Sudalpen von Neuseeland. *Schweizer Archiv für Tierheilkunde* 105:399–411.

Dashtseren, T., B.V. Solovyev, F. Varejka, and A. Khokhoo. 1984. Camel contagious ecthyma (pustular dermatitis). *Acta Virologica (Praha)* 28:122–127.

Davies, F.G., J.N. Mungai, and T. Shaw. 1975. Characteristics of a Kenyan camelpox virus. *Journal of Hygiene* 75:381–385.

Day, M.F., F. Fenner, G.M. Woodroofe, and G.A. McIntyre. 1956. Further studies on the mechanism of mosquito transmission of myxomatosis in the European rabbit. *Journal of Hygiene* 54:258–283.

Dekeyser, J., L. Delcambe, and D. Thienpont. 1960. Activité thérapeutique de l'iturine et du chinisol sur la teigne du cheval à *Microsporum equinum* Bodin. *Epizootic Diseases of Africa* 8:279–288.

Delange, A.M., C. Macaulay, W. Block, T. Mueller, and G. McFadden. 1984. Tumorigenic poxviruses: Construction of the physical map of the Shope fibroma virus genome. *Journal of Virology* 50:408–416.

Dieterich, R.A., G.R. Spencer, D. Burger, A.M. Gallina, and J. VanderSchalie. 1981. Contagious ecthyma in Alaskan muskoxen and Dall sheep. *Journal of the American Veterinary Medical Association* 179:1140–1143.

Downie, A.W. 1939a. The immunological relationship of the virus of spontaneous cowpox to vaccinia virus. *British Journal of Experimental Pathology* 20:158–176.

———. 1939b. A study of the lesions produced experimentally by cowpox virus. *Journal of Pathology and Bacteriology* 48:361–379.

———. 1974. Serological evidence of infection with Tana and Yaba pox viruses among several species of monkey. *Journal of Hygiene* 72:245–250.

Downie, A.W., and C. Espana. 1973. A comparative study of Tanapox and Yaba viruses. *Journal of General Virology* 19:37–49.

Downie, A.W., C.H. Taylor-Robinson, A.E. Caunt, G.S. Nelson, P.E.C. Manson-Bahr, and T.C.H. Matthews. 1971. Tanapox: A new disease caused by a poxvirus. *British Medical Journal* 1:363–368.

Duff, J.P., A. Scott, and I.F. Keymer. 1996. Parapoxvirus infection of the grey squirrel. *Veterinary Record* 138:527.

Elliott, L.F., W.M. Boyce, R.K. Clark, and D.A. Jessup. 1994. Geographic analysis of pathogen exposure in bighorn sheep (*Ovis canadensis*). *Journal of Wildlife Diseases* 30:315–318.

Espana, C. 1971. Review of some outbreaks of viral disease in captive nonhuman primates. *Laboratory Animal Science* 21:1023–1031.

Esposito, J.J., and J.C. Knight. 1985. Orthopoxvirus DNA: A comparison of restriction profiles and maps. *Virology* 143:230–251.

Falk, E.S. 1978. Parapoxvirus infections of reindeer and muskox associated with unusual human infections. *British Journal of Dermatology* 99:647–654.

Fenner, F. 1965. Viruses of the myxomafibroma subgroup of the poxviruses: II. Comparison of soluble antigens by gel diffusion tests, and a general discussion of the subgroup. *Australian Journal of Experimental Biology and Medical Science* 43:143–156.

———. 1983. Biological control as exemplified by smallpox eradication and myxomatosis. *Proceedings of the Royal Society of London [B]* 218:259–285.

———. 1994a. Hare fibroma virus. In *Virus infections of rodents and lagomorphs,* ed. A.D.M.E. Osterhaus. Amsterdam: Elsevier, pp. 77–79.

———. 1994b. Monkeypox virus. In *Virus infections of rodents and lagomorphs,* ed. A.D.M.E. Osterhaus. Amsterdam: Elsevier, pp. 33–35.

———. 1994c. Mousepox (ectromelia). In *Virus infections of rodents and lagomorphs,* ed. A.D.M.E. Osterhaus. Amsterdam: Elsevier, pp. 5–25.

———. 1994d. Rabbitpox virus. In *Virus infections of rodents and lagomorphs,* ed. A.D.M.E. Osterhaus. Amsterdam: Elsevier, pp. 51–57.

———. 1994e. Tatera poxvirus. In *Virus infections of rodents and lagomorphs,* ed. A.D.M.E. Osterhaus. Amsterdam: Elsevier, pp. 37–38.

———. 1996. Poxviruses. In *Fields virology,* ed. F.N. Fields, D.M. Knipe, P.M. Howley, R.M. Chanock, J.L. Melnick, T.P. Monath, B. Roizman, and S.E. Straus. Philadelphia: Lippincott-Raven, pp. 2673–2702.

Fenner, F., and I.D. Marshall. 1954. Passive immunity in myxomatosis of the European rabbit (*Oryctolagus cuniculus*): The protection conferred on kittens by immune does. *Journal of Hygiene* 52:321–336.

Fenner, F., and F.N. Ratcliffe. 1965. *Myxomatosis.* Cambridge: Cambridge University Press, 379 pp.

Fenner, F., and J. Ross. 1994. Myxomatosis. In *The European rabbit: The history and biology of a successful colonizer,* ed. H.V. Thompson and C.M. King. Oxford: Oxford University Press, pp. 205–240.

Fenner, F., and G.M. Woodroofe. 1953. The pathogenesis of infectious myxomatosis: The mechanism of infection and the immunological response in the European rabbit (*Oryctolagus cuniculus*). *British Journal of Experimental Pathology* 34:400–410.

Fenner, F., I.D. Marshall, and G.M. Woodroofe. 1953. Studies in the epidemiology of infectious myxomatosis of rabbits: I. Recovery of Australian wild rabbits (*Oryctolagus cuniculus*) from myxomatosis under field conditions. *Journal of Hygiene* 51:225–244.

Fenner, F., R. Wittek, and K.R. Dumbell. 1989. *The orthopoxviruses.* San Diego: Academic, 432 pp.

Flom, J.O., and E.J. Houk. 1979. Morphologic evidence of poxvirus in "tattoo" lesions from captive bottlenosed dolphins. *Journal of Wildlife Diseases* 15:593–596.

Fountain, S., M.K. Holland, L.A. Hinds, P.A. Janssens, and P.J. Kerr. 1997. Interstitial orchitis with impaired steroidogenesis and spermatogenesis in the testes of rabbits infected with an attenuated strain of myxoma virus. *Journal of Reproduction and Fertility* 110:161–169.

Gaskell, R.M., D. Baxby, and M. Bennett. 1987. Poxviruses. In *Virus infections of carnivores,* ed. M. J. Appel. Amsterdam: Elsevier, pp. 217–226.

Gassmann, U., R. Wyler, and R. Wittek. 1985. Analysis of parapoxvirus genomes. *Archives of Virology* 83:17–31.

Gehring, H., H. Mahnel, and H. Mayer. 1972. Elephantenpocken. *Zentralblatt für Veterinärmedizin [B]* 19:258–261.

Geraci, J.R., B.D. Hicks, and D.J. St. Aubin. 1979. Dolphinpox: A skin disease of cetaceans. *Canadian Journal of Comparative Medicine* 43:399–404.

Gispen, R., J. Verlinde, and P. Zwart. 1967. Histopathological and virological studies on monkeypox. *Archiv Gesamte für Virusforschung* 21:205–216.

Gispen, R., B. BrandSaathof, and A.C. Hekker. 1976. Monkeypox-specific antibodies in human and simian sera from the Ivory Coast and Nigeria. *Bulletin of the World Health Organization* 53:355–360.

Grace, J.T., and E.A. Mirand. 1963. Human susceptibility to a simian tumor virus. *Annals of the New York Academy of Sciences* 108:1123–1128.

Greenwood, A.G., R.J. Harrison, and H.W. Whitting. 1974. Functional and pathological aspects of the skin of marine mammals. In *Functional anatomy of marine mammals,* ed. R.J. Harrison. London: Academic, pp. 73–111.

Greth, A., J.M. Gourreau, M. Vassart, Nguyen-Ba-Vy, M. Wyers, and P.C. Lefevre. 1992. Capripoxvirus disease in an Arabian oryx (*Oryx leucoryx*) from Saudi Arabia. *Journal of Wildlife Diseases* 28:295–300.

Grodhaus, G., D.C. Regnery, and I.D. Marshall. 1963. Studies in the epidemiology of myxomatosis in California: II. The experimental transmission of myxomatosis between brush rabbits (*Sylvilagus bachmani*) by several species of mosquitoes. *American Journal of Hygiene* 77:205–212.

Groppel, K.H. 1962. Über das Vorkommen von Ektromelie (Mausepocken) unter Wildmausen. *Archiv für Experimentelle Veterinärmedizin* 16:243–278.

Guarda, F. 1959. Contributo allo studio anatomoisto-patologico dell'ectima contagioso dei camosci e stambecchi. *Annali della Facolta di Medicina Veterniaria dell'Universita de Torino* 9:37–52.

Hadlow, W.J., N.F. Cheville, and W.L. Jellison. 1980. Occurrence of pox in a northern fur seal on the Pribilof Islands in 1951. *Journal of Wildlife Diseases* 16:305–312.

Hall, A.J., and W.P. McNulty. 1967. A contagious pox disease in monkeys. *Journal of the American Veterinary Medical Association* 151:833–838.

Hartung, J. 1980. Lippengrind des Schafes. *Tierarztliche Praxis* 8:435–438.

Hastings, B.E., L.J. Lowenstine, L.J. Gage, and R.J. Munn. 1989. An epizootic of seal pox in pinnipeds at a rehabilitation center. *Journal of Zoo and Wildlife Medicine* 20:282–290.

Hebert, D.M., W.M. Samuel, and G.W. Smith. 1977. Contagious ecthyma in mountain goat of coastal British Columbia. *Journal of Wildlife Diseases* 13:135–136.

Hedger, R.S., and C. Hamblin. 1983. Neutralising antibodies to lumpy skin disease virus in African wildlife. *Comparative Immunology, Microbiology and Infectious Diseases* 6:209–213.

Herman, C.M., and J.R. Reilly. 1955. Skin tumors on squirrels. *Journal of Wildlife Management* 19:402–403.

Herman, C.M., L. Kilham, and O. Warbach. 1956. Incidence of Shope's rabbit fibroma at the Patuxent research refuge. *Journal of Wildlife Management* 20:85–90.

Hicks, B.D., and G.A.J. Worthy. 1987. Sealpox in captive grey seals (*Halichoerus grypus*). *Journal of Wildlife Diseases* 23:1–6.

Hirth, R.S., D.S. Wyand, A.D. Osborne, and C.N. Burke. 1969. Epidermal changes caused by squirrel poxvirus. *Journal of the American Veterinary Medical Association* 155:1120–1125.

Horner, G.W., A.J. Robinson, R. Hunter, B.T. Cox, and R. Smith. 1987. Parapoxvirus infections in New Zealand farmed red deer. *New Zealand Veterinary Journal* 35:41–45.

Hu, L., C. Ngichbe, C.V. Trimarchi, J.J. Esposito, and F.W. Scott. 1997. Raccoon poxvirus live recombinant feline panleukopaenia virus VP2 and rabies virus glycoprotein bivalent vaccine. *Vaccine* 15:1466–1472.

Hurst, E.W. 1937. Myxoma and the Shope fibroma: I. The histology of myxoma. *British Journal of Experimental Pathology* 18:1–15.

———. 1938. Myxoma and the Shope fibroma: IV. The histology of Shope fibroma. *Australian Journal of Experimental Biology and Medical Science* 16:53–64.

Hyde, R.R. 1936. The relationship between the viruses of infectious myxomatosis and the Shope fibroma of rabbits. *American Journal of Hygiene* 23:278–297.
Hyde, R.R., and R.E. Gardner. 1939. Transmission experiments with the fibroma (Shope) and myxoma (Sanarelli). *American Journal of Hygiene* 30:57–63.
Jackson, E.W., C.R. Dorn, J.K. Saito, and D.G. McKercher. 1966. Absence of serological evidence of myxoma virus infection in humans exposed during an outbreak of myxomatosis. *Nature* 211:313–314.
Jezek, Z., B. Kriz, and V. Rothbauer. 1983. Camelpox and its risk to the human population. *Journal of Hygiene, Epidemiology, Microbiology and Immunology* 27:29–42.
Jezek, Z., I. Arita, M. Szczeniowski, K.M. Paluka, K. Ruti, and J. Nakano. 1985. Human Tanapox in Zaire: Clinical and epidemiological observations on cases confirmed by laboratory studies. *Bulletin of the World Health Organization* 63:1027–1035.
Joiner, G.N., J.H. Jardine, and C.A. Gleiser. 1971. An epizootic of Shope fibromatosis in a commercial rabbitry. *Journal of the American Veterinary Medical Association* 159:1583–1587.
Kaminjolo, J.S., L.W. Johnson, H. Frank, and J.N. Gicho. 1974a. Vaccinia-like poxvirus identified in a horse with a skin disease. *Zentralblatt für Veterinärmedizin [B]* 21:202–206.
Kaminjolo, J.S., P.N. Nyaga, and J.N. Gicho. 1974b. Isolation, cultivation and characterisation of a poxvirus from some horses in Kenya. *Zentralblatt für Veterinärmedizin [B]* 21:592–601.
Kaplan, C., T.D. Healing, N. Evans, L. Healing, and A. Prior. 1980. Evidence for infection by viruses in small British field rodents. *Journal of Hygiene* 84:285–294.
Karstad, L., J. Thorsen, G. Davies, and J.S. Kaminjolo. 1977. Poxvirus fibromas on African hares. *Journal of Wildlife Diseases* 13:245–247.
Kater, J.C., and N.F. Hansen. 1962. Contagious ecthyma in wild thar in the South Island. *New Zealand Veterinary Journal* 10:116–117.
Kerr, P.J. 1997. An ELISA for epidemiological studies of myxomatosis: Persistence of antibodies to myxoma virus in European rabbits (*Oryctolagus cuniculus*). *Wildlife Research* 24:53–65.
Kerr, P.J., and S.M. Best. 1998. Myxoma virus in rabbits. *Revue Scientifique et Technique Office International des Epizooties* 17:256–268.
Khodakevich, L., Z. Jezek, and K. Kinzanzka. 1986. Isolation of monkeypox virus from wild squirrel infected in nature. *Lancet* 1:98–99.
Kilham, L. 1955. Metastasizing viral fibromas of gray squirrels: Pathogenesis and mosquito transmission. *American Journal of Hygiene* 61:55–63.
Kilham, L., and H.T. Dalmat. 1955. Host-virus-mosquito relations of Shope fibromas in cottontail rabbits. *American Journal of Hygiene* 61:45–54.
Kilham, L., and E.R. Fisher. 1954. Pathogenesis of fibromas in cottontail rabbits. *American Journal of Hygiene* 61:45–54.
Kilham, L., and P.A. Woke. 1953. Laboratory transmission of fibromas (Shope) in cottontail rabbits by means of fleas and mosquitoes. *Proceedings of the Society for Experimental Biology and Medicine* 83:296–301.
Kilham, L., C.M. Herman, and E.R. Fisher. 1953. Naturally occurring fibromas of grey squirrels related to Shope's rabbit fibroma. *Proceedings of the Society for Experimental Biology and Medicine* 82:298–301.
King, J.M., A. Woolf, and J.N. Shively. 1972. Naturally occurring squirrel fibroma virus with involvement of internal organs. *Journal of Wildlife Diseases* 8:321–324.
Kirschstein, R.L., A.S. Rabson, and L. Kilham. 1958. Pulmonary lesions produced by fibroma viruses in squirrels and rabbits. *Cancer Research* 18:1340–1344.
Knight, J.C., F.J. Novembre, D.R. Brown, C.S. Goldsmith, and J.J. Esposito. 1989. Studies on Tanapox virus. *Virology* 172:116–124.
Knight, J.C., C.S. Goldsmith, A. Tamin, R.L. Regnery, D.C. Regnery, and J.J. Esposito. 1992. Further studies on the orthopoxviruses volepox virus and raccoon poxvirus. *Virology* 190:423–433.
Kriz, B. 1982. A study of camelpox in Somalia. *Journal of Comparative Pathology* 92:1–8.
Kummeneje, K. 1979. Contagious ecthyma (orf) in reindeer (*Rangifer t. tarandus*). *Veterinary Record* 105:60–61.
Lance, W. 1979. A review of contagious ecthyma in wild ruminants. In *Proceedings of the Annual Conference of the American Association of Zoo Veterinarians,* pp. 130–131.
Lance, W., W. Adrian, and B. Widhalm. 1981. An epizootic of contagious ecthyma in rocky mountain bighorn sheep in Colorado. *Journal of Wildlife Diseases* 17:601–603.
Lance, W.R., C.P. Hibler, and J. de Martini. 1983. Experimental contagious ecthyma in mule deer, whitetailed deer, pronghorn and wapiti. *Journal of Wildlife Diseases* 19:165–169.
l'Heureux, N., M. Festa-Bianchet, and J.T. Jorgenson. 1996. Effects of visible signs of contagious ecthyma on mass and survival of bighorn lambs. *Journal of Wildlife Diseases* 32:286–292.
Lourie, B., J.H. Nakano, G.E. Kemp, and H.W. Setzer. 1975. Isolation of a poxvirus from an African rodent. *Journal of Infectious Diseases* 132:677–681.
Mackett, M., and L.C. Archard. 1979. Conservation and variation in orthopoxvirus genome structure. *Journal of General Virology* 45:683–701.
Mahnel, H., J. Holejsovsky, P. Bartak, and C.P. Czerny. 1993. Kongenitale »Ektromelie« bei Pelztieren durch *Orthopoxvirus muris. Tierarztliche Praxis* 21:469–472.
Maiboroda, A.D. 1982. Experimental infection of Norvegian [*sic*] rats (*Rattus norvegicus*) with ratpox virus. *Acta Virologica (Praha)* 26:288–291.
Mallick, K.P., V.S. Rawany, and C.S. Celly. 1990. A report on buffalo pox outbreak in Pathalgaon block of district Raigarh (Madya Pradesh). *Indian Veterinary Journal* 67:1173–1174.
Marchal, J. 1930. Infectious ectromelia: A hitherto undescribed virus disease of mice. *Journal of Pathology* 33:713–728.
Marennikova, S.S., and E.M. Shelukhina. 1976. White rats as source of pox infection in Carnivora of the family Felidae. *Acta Virologica (Praha)* 20:442.
Marennikova, S.S., L.S. Shenkman, E.M. Shelukhina, and N.N. Maltseva. 1974. Isolation of a camelpox virus and investigation of its properties. *Acta Virologica (Praha)* 18:423–428.
Marennikova, S.S., N.N. Maltseva, and V.I. Korneeva. 1976. Pox in giant anteaters due to agent similar to cowpox virus. *British Veterinary Journal* 132:182–186.
Marennikova, S.S., N.N. Maltseva, V.I. Korneeva, and N.M. Garanina. 1977. Outbreak of pox disease among Carnivora (Felidae) and Edentata. *Journal of Infectious Diseases* 135:358–366.
Marennikova, S.S., I.D. Ladnyj, Z.I. Ogorodnikova, E.M. Shelukhina, and N.N. Maltseva. 1978. Identification and study of a poxvirus isolated from wild rodents in Turkmenia. *Archives of Virology* 56:7–14.
Marennikova, S.S., N.N. Malceva, and N.A. Habahpaseva. 1981. ELISA: A simple test for detecting and differentiating antibodies to closely related orthopoxviruses. *Bulletin of the World Health Organization* 59:365–369.

Marshall, I.D., and D.C. Regnery. 1960. Myxomatosis in a Californian brush rabbit. *Nature* 188:73–74.
Marshall, I.D., D.C. Regnery, and G. Grodhaus. 1963. Studies in the epidemiology of myxomatosis in California: I. Observations on two outbreaks of myxomatosis in coastal California and the recovery of myxoma virus from a brush rabbit (*Sylvilagus bachmani*). *American Journal of Hygiene* 77:195–204.
Mayr, A., J. Lauer, and C.P. Czerny. 1995. Neue Fakten über die Verbreitung von Orthopokenvirusinfektion bei Katzen, Füchsen und Wildschweinen sowie über die Entwicklung einer Schutzimpfung gegen die Katzenpocken. *Praktische Tierärzt* 76:961–967.
McKeever, D. 1984. Persistent orf. *Veterinary Record* 115:334–335.
McKenzie, R.A., F.R. Fay, and C. Prior. 1979. Poxvirus infection in the skin of an eastern grey kangaroo. *Australian Veterinary Journal* 55:31–43.
McNulty, W.P., W.C. Lobitz, F. Hu, C.A. Maruffo, and A.S. Hall. 1968. A pox disease in monkeys transmitted to man. *Archives of Dermatology* 97:286–293.
Moallin, A.S., and K.H. Zessin. 1988. Outbreak of camel contagious ecthyma in central Somalia. *Tropical Animal Health and Production* 20:185–186.
Moss, B. 1996. Poxviridae: The viruses and their replication. In *Fields virology,* ed. F.N. Fields, D.M. Knipe, P.M. Howley, R.M. Chanock, J.L. Melnick, T.P. Monath, B. Roizman, and S.E. Straus. Philadelphia: Lippincott-Raven, pp. 2637–2671.
Muller, T., K. Henning, M. Kramer, C.P. Czerny, H. Meyer, and K. Ziedler. 1996. Seroprevalence of orthopox virus specific antibodies in red foxes (*Vulpes vulpes*) in the Federal State Brandenburg, Germany. *Journal of Wildlife Diseases* 32:348–353.
Munz, E., D. Schillinger, M. Reimann, and H. Mahnel. 1986. Electron microscopical diagnosis of ecthyma contagiosum in camels (*Camelus dromedarius*): First report of the disease in Kenya. *Zentralblatt für Veterinärmedizin [B]* 33:73–77.
Murphy, F.A., C.M. Fauquet, D.H.L. Bishop, S.A. Ghabrial, A.W. Jarvis, G.P. Martelli, M.A. Mayo, and M.D. Summers. 1995. *Virus taxonomy: Classification and nomenclature of viruses—Sixth report of the International Committee on Taxonomy of Viruses.* Vienna: Springer-Verlag, 586 pp.
Myers, K., I. Parer, D. Wood, and BD. Cooke. 1994. The rabbit in Australia. In *The European rabbit: The history and biology of a successful colonizer,* ed. H.V. Thompson and C.M. King. Oxford: Oxford University Press, pp. 108–157.
Mykytowycz, R. 1958. Contact transmission of infectious myxomatosis of the rabbit *Oryctolagus cuniculus* (L.). *C.S.I.R.O. Wildlife Research* 3:1–6.
Nettleton, P.F., R. Munro, I. Pow, J. Gilray, E.W. Gray, and H.W. Reid. 1995. Isolation of a parapoxvirus from a grey seal (*Halichoerus grypus*). *Veterinary Record* 137:562–564.
Novilla, M.N., V. Flyger, E.R. Jacobson, S.K. Dutta, and E.M. Sacchi. 1981. Systemic phycomycosis and multiple fibromas in a gray squirrel (*Sciurus carolinensis*). *Journal of Wildlife Diseases* 17:89–95.
Okada, K., and Y. Fujimoto. 1984. The fine structure of cytoplasmic and intranuclear inclusions of seal pox. *Japanese Journal of Veterinary Science* 46:401–404.
Okada, H.M., K. Okada, S. Numakunai, and K. Ohshima. 1984. Histopathologic studies on mucosal and cutaneous lesions in contagious papular dermatitis of Japanese serow (*Capricornis crispus*). *Japanese Journal of Veterinary Science* 46:257–264.
Osterhaus, A.D., H.W. Broeders, I.K. Visser, J.S. Teppema, and E.J. Vedder. 1990. Isolation of an orthopoxvirus from pox-like lesions of a grey seal (*Halichoerus grypus*). *Veterinary Record* 127:91–92.
Osterhaus, A.D.M. E., H.W.J. Broeders, I.K.G. Visser, J. S.B. Teppema, and T. Kuiken. 1994. Isolation of a parapoxvirus from pox-like lesions in grey seals. *Veterinary Record* 135:601–602.
Papadimitriou, J.M., and R.B. Ashman. 1972. A poxvirus in a marsupial papilloma. *Journal of General Virology* 16:87–89.
Patton, J.F., R.W. Nordhausen, L.W. Woods, and N.J. MacLachlan. 1996. Isolation of a poxvirus from a black-tailed deer (*Odocoileus hemionus columbianus*). *Journal of Wildlife Diseases* 32:531–533.
Peters, J.C. 1966. An epizootic of monkey pox at Rotterdam zoo. *International Zoo Yearbook* 6:274–275.
Pilaski, J., and F. Jacoby. 1993. Die Kuhpocken-erkrankungen der Zootiere. *Erkrankungen der Zootiere* 35:39–50.
Pilaski, J., and A. RosenWolff. 1988. Poxvirus infection in zoo-kept mammals. In *Virus diseases in laboratory and captive animals,* ed. G. Darai. Boston: Martinus Nijhoff, pp. 83–100.
Pilaski, J., and X. Zhou. 1991. Die Pockenimpfung der Elefanten. *Erkrankungen der Zootiere* 33:203–211.
Presidente, P.J.A. 1978. Diseases seen in free-ranging marsupials and those held in captivity. In *The University of Sydney Postgraduate Committee in veterinary science course for veterinarians: Fauna, part B.* Sydney: University of Sydney, pp. 457–471.
Raflo, C.P., R.G. Olsen, S.P. Pakes, and W.S. Webster. 1973. Characterization of a fibroma virus isolated from naturally occurring skin tumors in domestic rabbits. *Laboratory Animal Science* 23:525–532.
Ramanan, C. 1996. Buffalopox. *International Journal of Dermatology* 35:128–130.
Regnery, D.C. 1971. The epidemic potential of Brazilian myxoma virus (Lausanne strain) for three species of North American cottontails. *American Journal of Epidemiology* 94:514–519.
———. 1987. Isolation and partial characterization of an orthopoxvirus from a California vole (*Microtus californicus*). *Archives of Virology* 94:159–162.
Regnery, D.C., and I.D. Marshall. 1971. Studies in the epidemiology of myxomatosis in California: V. The susceptibility of six leporid species to Californian myxoma virus and the relative infectivity of their tumours for mosquitoes. *American Journal of Epidemiology* 94:508–513.
Regnery, D.C., and J.H. Miller. 1972. A myxoma virus epizootic in a brush rabbit population. *Journal of Wildlife Diseases* 8:327–331.
Regnery, R.L. 1975. Preliminary studies on an unusual poxvirus of the western grey squirrel (*Sciurus griseus griseus*) of North America. *Intervirology* 5:364–366.
Ridgeway, S.H. 1975. Common diseases of small cetaceans. *Journal of the American Veterinary Medical Association* 167:533–539.
Robinson, A.J., and T.C. Balassu. 1981. Contagious pustular dermatitis (orf). *Veterinary Bulletin* 51:771–781.
Robinson, A.J., and D.J. Lyttle. 1992. Parapoxviruses: Their biology and potential as recombinant vaccines. In *Recombinant poxviruses,* ed. M. Binns and G. Smith. Boca Raton, FL: CRC, pp. 285–327.
Robinson, A.J., and A.A. Mercer. 1995. Parapoxvirus of red deer: Evidence for its inclusion as a new member of the genus *Parapoxvirus. Virology* 208:812–815.
Robinson, A.J., and G.V. Petersen. 1983. Orf virus infection of workers in the meat industry. *New Zealand Medical Journal* 96:81–85.

Rogers, P.M., C.P. Arthur, and R.C. Soriguer. 1994. The rabbit in continental Europe. In *The European rabbit: The history and biology of a successful colonizer,* ed. H.V. Thompson and C.M. King. Oxford: Oxford University Press, pp. 22–63.

Ropp, S.L., Q. Jin, J.C. Knight, R.F. Massung, and J.J. Esposito. 1995. PCR strategy for identification and differentiation of smallpox and other orthopoxviruses. *Journal of Clinical Microbiology* 33:2069–2076.

Roslyakov, A.A. 1972. Comparative ultrastructure of viruses of camelpox, poxlike disease of camels (Auzduk) and contagious ecthyma of sheep [in Russian]. *Voprosy Virusologii* 17:26–30.

Ross, J., A.M. Tittensor, A.P. Fox, and M.F. Sanders. 1989. Myxomatosis in farmland populations in England and Wales. *Epidemiology and Infection* 103:333–357.

Rothwell. T.L.W., J.M. Keep, F.-N. Xu, and D.J. Middleton. 1984. Poxvirus in marsupial skin lesions. *Australian Veterinary Journal* 61:409–410.

Rouhandeh, H. 1988. Yaba virus. In *Virus diseases in laboratory and captive animals,* ed. G. Darai. Boston: Martinus Nijhoff, pp. 1–15.

Russell, R.J., and S.J. Robbins. 1989. Cloning and molecular characterization of the myxoma virus genome. *Virology* 170:147–159.

Sainsbury, A.W., P. Nettleton, and J. Gurnell. 1997. Recent developments in the study of parapoxvirus in red and grey squirrels. In *The conservation of red squirrels,* Sciurus vulgaris, ed. L.J. Gurnell and P. Lurz. London: People's Trust for Endangered Species, pp. 105–108.

Samuel, W.M., G.A. Chalmers, J.G. Stelfox, A. Loewen, and J.J. Thomsen. 1975. Contagious ecthyma in bighorn sheep and mountain goat in western Canada. *Journal of Wildlife Diseases* 11:26–31.

Sands, J.J., A.C. Scott, and J.W. Harkness. 1984. Isolation in cell culture of a poxvirus from the red squirrel (*Sciurus vulgaris*). *Veterinary Record* 114:117–118.

Schmidt, D. 1967. Experimentelle Beitrage zur Kenntnis der Dermatitis pustulosa des Schafes: V. Untersuchungen über die Ausbildung der Immunitat gegen das Virus der Dermatitis pustulosa an verschiedenen Stellen der Korperoberflache. *Archiv für Experimentelle Veterinärmedizin* 21:937–945.

Scott, A.C., I.F. Keymer, and J. Labram. 1981. Parapoxvirus infection of the red squirrel. *Veterinary Record* 109:202.

Shope, R.E. 1932. A filtrable virus causing a tumour-like condition in rabbits and its relationship to virus myxomatosum. *Journal of Experimental Medicine* 56:803–822.

Simpson, J.G., and M.B. Gardner. 1972. Comparative microscopic anatomy of selected marine mammals. In *Mammals of the sea,* ed. S.H. Ridgeway. Springfield, IL: Charles C. Thomas, pp. 363–377.

Simpson, V.R., N.C. Stuart, M.J. Stack, H.A. Ross, and J.C.H. Head. 1994. Parapox infection in grey seals (*Halichoerus grypus*) in Cornwall. *Veterinary Record* 134:292–296.

Smith, A.W., D.E. Skilling, S.H. Ridgway, and C.A. Fenner. 1983. Regression of cetacean tattoo lesions concurrent with conversion of precipitin antibody against a poxvirus. *Journal of the American Veterinary Medical Association* 183:1219–1222.

Smith, D.A., I.K. Barker, K.G. Mehren, and G.J. Crawshaw. 1984. Poxvirus infections in ruminants at the Metropolitan Toronto Zoo. In *Proceedings of the Annual Conference of the American Association of Zoo Veterinarians,* pp. 97–98.

Smith, T.C., W.E. Heimer, and W.J. Foreyt. 1982. Contagious ecthyma in an adult Dall sheep (*Ovis dalli dalli*) in Alaska. *Journal of Wildlife Diseases* 18:111–112.

Strayer, D.S. 1992. Determinants of virus-related suppression of immune responses as observed during infection with an oncogenic poxvirus. In *Progress in medical virology,* ed. J.L. Melnick. Basel: Karger, pp. 228–255.

Strayer, D.S., and S. Sell. 1983. Immunohistology of malignant rabbit fibroma virus: A comparative study with rabbit myxoma virus. *Journal of the National Cancer Institute* 71:106–116.

Suzuki, T., N. Minamoto, M. Sugiyama, T. Kinjo, Y. Suzuki, M. Sugimura, and Y. Atoji. 1993. Isolation and antibody prevalence of a parapoxvirus in wild Japanese serows (*Capricornis crispus*). *Journal of Wildlife Diseases* 29:384–389.

Suzuki, Y., M. Sugimura, Y. Atoji, N. Minamoto, and T. Kinjo. 1986. Widespread parapox infection in wild Japanese serows, *Capricornis crispus. Japanese Journal of Veterinary Science* 48:1279–1282.

Thomas, E.K., E.L. Palmer, J.F. Obijeski, and J.H. Nakano. 1974. Further characterization of raccoonpox virus. *Archives of Virology* 49:217–227.

Thomsett, L.R., D. Baxby, and E.M.H. Denham. 1978. Cowpox in the domestic cat. *Veterinary Record* 108:567.

Tripathy, D.K., L.E. Hanson, and R.A. Crandall. 1981. Poxviruses of veterinary importance: Diagnosis of infections. In *Comparative diagnosis of viral diseases,* ed. E. Kurstak and C. Kurstak. New York: Academic, pp. 267–346.

Tryland, M., T. Sandvik, R. Mehl, M. Bennett, T. Traavik, and O. Olsvik. 1998. Serosurvey for orthopoxviruses in rodents and shrews from Norway. *Journal of Wildlife Diseases* 34:240–250.

Turner, J.C., and J.B. Payson. 1982. Prevalence of antibodies of selected infectious disease agents in the peninsular desert bighorn sheep (*Ovis canadensis cremnobates*) of the Santa Rosa Mountains, California. *Journal of Wildlife Diseases* 18:243–245.

Ueda, Y., S. Morikawa, and T. Watanabe. 1995. Unclassified poxvirus: Characterization and physical mapping of Cotia virus DNA and location of a sequence capable of encoding a thymidine kinase. *Virology* 210:67–72.

Van Bressem, M.F., and K. Van Waerebeek. 1996. Epidemiology of poxvirus in small cetaceans from the eastern South Pacific. *Marine Mammal Science* 12:371–382.

Van Bressem, M.F., K. Van Waerebeek, J.C. Reyes, D. Dekegel, and P.P. Pastoret. 1993. Evidence of poxvirus in dusky dolphin (*Lagenorhynchus obscurus*) and Burmeister's porpoise (*Phocoena spinipinnis*) from coastal Peru. *Journal of Wildlife Diseases* 29:109–113.

Von Magnus, D., D.K. Anderson, K.B. Peterson, and A. Birch-Anderson. 1959. A poxlike disease in cynomolgus monkeys. *Acta Pathologia Scandinavica* 46:156–176.

Wilkinson, G.T., J. Prydie, and J. Scarnell. 1970. Possible "orf" (contagious pustular dermatitis, contagious ecthyma of sheep) infection in the dog. *Veterinary Record* 87:766–767.

Williams, E.S., V.M. Becerra, E.T. Thorne, T.J. Graham, M.J. Owens, and C.E. Nunamaker. 1985. Spontaneous poxviral dermatitis and keratoconjunctivitis in free-ranging mule deer (*Odocoileus hemionus*) in Wyoming. *Journal of Wildlife Diseases* 21:430–433.

Williams, R.T., P.J. Fullagar, C. Kogon, and C. Davey. 1973. Observations on a naturally occurring winter epizootic of myxomatosis at Canberra, Australia, in the presence of rabbit fleas (*Spilopsyllus cuniculi* Dale) and virulent myxoma virus. *Journal of Applied Ecology* 10:417–427.

Wilson, T.M., and I. Poglayen-Neuwall. 1971. Pox in South American sea lions (*Otaria byronia*). *Canadian Journal of Comparative Medicine* 35:174–177.

Wilson, T.M., N.F. Cheville, and L. Karstad. 1969. Seal pox. *Bulletin of the Wildlife Diseases Association* 5:412–418.

Wilson, T.M., A.D. Boothe, and N.F. Cheville. 1972a. Sealpox field survey. *Journal of Wildlife Diseases* 8:158–160.

Wilson, T.M., N.F. Cheville, and A.D. Boothe. 1972b. Sealpox questionnaire survey. *Journal of Wildlife Diseases* 8:155–157.

Wilson, T.M., R.W. Dykes, and K.S. Tsai. 1972c. Pox in young, captive harbor seals. *Journal of the American Veterinary Medical Association* 161:611–617.

Woodroofe, G.W., and F. Fenner. 1965. Viruses of the myxoma-fibroma subgroup of the poxviruses: I. Plaque production in cultured cells, plaque-reduction tests and cross-protection tests in rabbits. *Australian Journal of Experimental Biology and Medical Science* 43:123–142.

Yager, J.A., and D.W. Scott. 1993. Viral diseases of skin. In *Pathology of domestic animals,* ed. K.V.F. Jubb, P.C. Kennedy, and N. Palmer. San Diego: Academic, pp. 628–644.

Yeruham, I., A. Nyska, and A. Abraham. 1964. Parapox infection in a gazelle kid (*Gazella gazella*). *Journal of Wildlife Diseases* 30:260–262.

Young, E., P.A. Basson, and K.E. Weiss. 1970. Experimental infection of game animals with lumpy skin disease virus (prototype strain Neethling). *Onderstepoort Journal of Veterinary Research* 37:79–88.

Yuill, T.M. 1981. Myxomatosis and fibromatosis. In *Infectious diseases of wild mammals,* ed. J.W. Davis, L.H. Karstad, and D.O. Trainer, 2d ed. Ames: Iowa State University Press, pp. 154–177.

Yuill, T.M., and R.P. Hanson. 1964. Infection of suckling cottontail rabbits with Shope's fibroma virus. *Proceedings of the Society of Experimental Biology and Medicine* 117:376–380.

Zarnke, R.L., R.A. Dieterich, K.A. Neiland, and G. Ranglack. 1983. Serologic and experimental investigations of contagious ecthyma in Alaska. *Journal of Wildlife Diseases* 19:170–174.

Zwart, P., R. Gispen, and J.C. Peters. 1971. Cowpox in okapis (*Okapia johnstoni*) at Rotterdam zoo. *British Veterinary Journal* 127:20–24.

9 ADENOVIRAL DISEASES

LESLIE W. WOODS

Adenoviruses cause disease in humans and in a wide range of animal species, both domestic and nondomestic. Some adenoviruses are capable of causing epidemics resulting in high mortality. However, clinical disease associated with adenoviral infection is usually sporadic and limited to neonates or immunologically compromised individuals (Fenner et al. 1993). Diseases in domestic mammals caused by adenoviruses include respiratory or enteric disease in cattle *Bos taurus* caused by bovine adenoviruses types 1–10 (Horner et al. 1989; Bürki 1990; Smyth et al. 1996) and in sheep *Ovis aries* caused by ovine adenoviruses 1–6 (Belak 1990), and hepatitis in dogs *Canis familiaris* caused by canine adenovirus type 1 (CAV-1) (Rubarth 1947). Some adenoviruses that commonly infect domestic animals may also infect nondomestic species of the same families. For example, infectious canine hepatitis (ICH) virus of dogs also affects wild members of the Canidae, including coyotes *Canis latrans* (Marler et al. 1976) and wolves *Canis lupus* (Choquette and Kuyt 1974). Important nondomestic mammalian adenoviral diseases that can manifest as epidemics include fox encephalitis caused by CAV-1, originally described in silver fox *Vulpes vulpes* (Green 1925), and the newly described adenovirus hemorrhagic disease of deer *Odocoileus* spp. (Woods et al. 1996).

Adenoviruses were first classified as distinct viruses in 1953 when a transmissible agent was identified in tonsil and adenoid cultures from tissues surgically removed from children (Horne 1973; Horwitz 1990). Adenoviruses are nonenveloped, icosahedral particles, 60–90 nm in diameter, which are composed of a capsid surrounding a core of DNA and arginine-rich proteins (Palmer and Martin 1988; Fenner et al. 1993). The capsid is made up of 252 capsomeres, 240 hexons, and 12 pentons. The virions have 20 triangular surfaces and 12 vertices from which variable-length fibers project, depending on the serotype of the virus. The adenovirus genome is a single linear molecule of double-stranded DNA with a molecular weight between 20 and 25 kDa.

The virion uncoats after infecting the cell. Nonviral polypeptides are synthesized first from viral DNA, followed by DNA replication and synthesis of viral DNA and structural polypeptides. Virions are assembled in the nucleus, shutting off host cell DNA, RNA, and protein synthesis, and are released when the damaged cell ruptures. Adenoviruses are highly host specific and are typically tropic for the epithelium or endothelium in the respiratory or alimentary tracts.

INFECTIOUS CANINE HEPATITIS

Synonyms: Rubarth's disease, fox encephalitis, enzootic fox encephalitis, epizootic fox encephalitis, fox distemper, hepatitis contagiosa canis, encephalitis infectiosa vulpis, canine endotheliitis.

History. Infectious canine hepatitis, caused by CAV-1, was first described in silver fox (Green 1925). Initial work by Green determined the disease was caused by a filterable agent that also caused hepatitis in experimentally infected dogs (Green et al. 1930; Green and Shillinger 1934). Natural disease in dogs was first described in the 1930s and 1940s (Cowdry and Scott 1930; Rubarth 1947). Infectious canine hepatitis virus was shown to be antigenically related to fox encephalitis virus in 1949 (Siedentopf and Carlson 1949), and it was finally classified as an adenovirus in 1962 (Rowe and Hartley 1962).

In 1950, Chaddock and Carlson reported a probable, but unconfirmed, case of ICH in a captive polar bear *Ursus maritimus* that recovered after administration of antiserum to ICH virus (Chaddock and Carlson 1950). The caretaker for the bear had cared for a gray fox *Urocyon cinereoargenteus* diagnosed 3 weeks previously with ICH. Confirmation of the susceptibility of bears occurred in 1979 when CAV-1 was isolated from tissues of two 3-month-old unvaccinated American black bear *Ursus americanus* cubs that died with neurologic signs (Pursell et al. 1983). Two unvaccinated wolf pups handled by the same caretaker developed ataxia and convulsions, and had neutralizing antibody to both CAV-1 and to the adenovirus isolated from the bear cubs. In 1983, an epizootic of ICH in American black bears resulted in 24 deaths of 148 bears in a wildlife park in South Dakota (Collins et al. 1984).

Distribution. Serologic evidence of infection with CAV-1 or reports of disease associated with this virus have been documented throughout the world, including the United States, Sweden, Norway, Poland, and Canada (Green 1925; Cowdry and Scott 1930; Rubarth 1947; Kummeneje 1971; Choquette and Kuyt 1974; Cabasso 1981).

Host Range. Members of the Canidae, Mustelidae, and Ursidae are susceptible to CAV-1 infection. Reports of clinical disease in wildlife associated with natural infection are limited and most involved captive animals. Natural and experimental infections with variable degrees of susceptibility have been reported in foxes, wolves, coyotes, striped skunks *Mephitis mephitis,* raccoons *Procyon lotor,* mink *Mustela vison,* ferrets *Mustela putorius furo,* and bears (Green 1925; Green et al. 1934, 1943; Pursell et al. 1983). Canine adenovirus 1 was isolated from young striped skunks with hepatitis (Karstad et al. 1975). Disease associated with unconfirmed infection with canine adenovirus was reported in a captive polar bear (Chaddock and Carlson 1950). Canine adenovirus 1 was isolated from captive and free-living black bears that died of encephalitis and hepatitis (Pursell et al. 1983; Collins et al. 1984).

Serologic surveys have provided additional information on the prevalence of infection in various wildlife populations. Seroprevalence was low in free-ranging black bears (1 of 33) and a grizzly bear *Ursus arctos horribilis* (0 of 1) tested in northeastern Washington (Foreyt et al. 1986) and 12% (90 of 725) in grizzly bears from Alaska (Zarnke and Evans 1989). Greater than 50% of coyotes from west Texas were positive for exposure to CAV-1 (Trainer and Knowlton 1968), and 62% of skunks from the eastern United States were seropositive (Alexander et al. 1972). Seroprevalence was 12% in free-living wolves in a small population sampled in Canada (Choquette and Kuyt 1974) and 81% (72 of 87) in Alaska (Zarnke and Ballard 1987).

Etiology. Infectious canine hepatitis is caused by CAV-1. The virus is made up of double-stranded DNA surrounded by a capsid with 252 capsomeres. Negatively stained preparations are icosahedral and are approximately 82 nm in diameter. The virus is related to human adenovirus type 5 and 7 (Darbyshire and Pereira 1964) and can be propagated in dog kidney, testicular, lung, liver or spleen tissues, pig kidney, ferret kidney, and raccoon kidney cultures (Cabasso 1981). The adenovirus isolated from fatally infected bears was antigenically related to CAV-1, and restriction endonuclease digestion patterns were also similar (Collins et al. 1984).

Transmission. Infectious canine hepatitis is highly contagious. Viral shedding occurs during disease in urine, nasal, and conjunctival secretions and in feces. The virus persists in the kidneys and may be shed in the urine for several months after recovery (Carmichael 1977; Bürki 1990). The virus can be transmitted through direct contact with infected animals or contaminated fomites (Cabasso 1981). Adenoviruses are typically stable in the environment (Cabasso 1981; Bürki 1990).

Clinical Signs. In foxes, anorexia and rhinitis may develop after a 2- to 6-day incubation period, followed by mucoid or bloody diarrhea, hyperexcitability, seizures, paralysis, coma, and death (Cabasso 1981). Death may occur after a short clinical course or may occur suddenly without prior clinical signs. Keratitis occasionally develops in nonfatal cases. In skunks, a 2-day course of ocular discharge, diarrhea, inappetence, and depression, resulting in death, developed 6 days following exposure (Karstad et al. 1975). Both adult and young American black bears were reported to be affected. Most died within 12 hours of first showing clinical signs (Pursell et al. 1983; Collins et al. 1984). Anorexia, lethargy, ataxia, abdominal pain, excessive salivation, vomiting, paralysis, periodic nystagmus, and convulsions were reported prior to death in captive and free-living black bears infected with CAV-1. Two of four surviving bears developed unilateral transient corneal opacity.

Pathogenesis. Infectivity studies in dogs indicate initial infection can occur through the oropharynx, resulting in viremia with distribution to the viscera and central nervous system. Virus has been isolated from the blood, spleen, spinal cord, and brain of foxes infected with CAV-1 (Cabasso 1981).

Pathology. Gross lesions in dogs are typically distinct and include ascites, and serosal petechial and ecchymotic hemorrhages; edema and hemorrhage of the superficial lymph nodes; gallbladder edema; and fine, uniform, yellowish mottling in the liver, which is turgid and friable (Kelly 1993). Gross lesions described in foxes are less distinct and are typically nonspecific; generalized congestion is the only apparent change noted (Green et al. 1934; Cabasso 1981). The liver, spleen, and adrenal glands may be mildly enlarged. Enlarged, pale mesenteric lymph nodes, congested spleens, ascites, diffuse hyperemia of the gastric mucosa, and disseminated petechial hemorrhages in thymus, lymph nodes, urinary bladder, skeletal muscle, and epicardium were described in fatally infected bears (Pursell et al. 1983; Collins et al. 1984).

Periacinar necrosis with intranuclear inclusion bodies in hepatocytes and Kupffer cells is typically described in fatally infected dogs. Microscopic lesions in other organs are due to vascular injury (Kelly 1993). Primary vascular damage is the basis for alterations in many tissues of foxes that die of CAV-1 infection. Vasculitis and disseminated hemorrhages are seen in the brain (Summers et al. 1995). In the liver and kidneys, however, necrosis results from direct viral infection of parenchymal epithelium. Acidophilic or basophilic intranuclear inclusion bodies are found in hepatocytes and biliary and adrenal epithelium, as well as endothelial cells in various tissues. Mononuclear inflammatory cell infiltrates may be present in the periportal regions of the liver, interstitium of the lungs, and within meninges. In the reports of black bear with fatal CAV-1 infection, significant findings consisted of mild neutrophilic meningoencephalitis with minute hemorrhages, gliosis, vascular endothelial hypertrophy, and intranuclear inclusion bodies in the endothelium of the brain (Pursell et al. 1983; Collins et al. 1984). Multifocal, necrotizing, lymphocytic hepatitis with intranuclear inclusion bodies in hepatocytes and intranuclear inclusion bodies in endothelial cells in renal glomeruli and urinary bladder were extraneural lesions described in bears dying of CAV-1 infection.

Diagnosis. A presumptive diagnosis of adenoviral infection is based on the presence of characteristic intranuclear inclusion bodies in affected epithelial or endothelial cells. The diagnosis of infection with CAV-1 is confirmed by immunohistochemistry using fluorescein- or immunoperoxidase-labeled CAV-1-specific antibody or virus isolation. Canine adenovirus 1-neutralizing antibody titers determined by microtiter serum neutralization test are helpful to determine exposure or serologic prevalence of CAV-1 infection in wildlife populations.

Differential Diagnoses. Rabies and canine distemper should be considered in the differential diagnosis of ICH in susceptible species exhibiting neurologic signs. Absence of Negri bodies and/or rabies antigen by fluorescent antibody test aids in ruling out rabies. All of the families susceptible to canine adenovirus are also susceptible to canine distemper virus (Habermann et al. 1958; Fowler 1986). Microscopically, canine distemper virus typically elicits neuronal necrosis, gliosis, and associated lymphoplasmacytic inflammation in the gray matter of the central nervous system, with demyelination in the white matter, in contrast to vasculitis with little associated gliosis in animals infected with CAV-1 (Summers et al. 1995). Canine distemper virus intranuclear and intracytoplasmic inclusions are present in the neurons and astroglia, whereas intranuclear inclusions are seen in vascular endothelial cells with adenovirus infection. Diagnostic methods distinguishing the viruses, such as transmission electron microscopy, immunohistochemistry, and virus isolation, may be used.

Immunity. Adenoviruses are typically potent antigens, and antibodies produced against homologous strains of adenovirus provide protection (Bürki 1990). Green reported a high degree of immunity in foxes that were inoculated a second time after surviving experimental infection (Green et al. 1930).

Control and Treatment. Successful results were reported with administration of hyperimmune serum and vaccination programs initiated during outbreaks (Cabasso 1981). Supportive care for captive wildlife, including administration of broad-spectrum systemic antibiotics to combat secondary bacterial infections, intravenous fluid support, and whole blood transfusions, may aid recovery in less severely affected animals (Ettinger 1983).

Public Health Concerns. Canine adenovirus does not cause disease in humans, but antibodies have been detected by virus neutralization tests (Horwitz 1990).

Domestic Animal Concerns and Wildlife Management Implications. Dogs experimentally infected with CAV-1 isolated from foxes dying of fox encephalitis developed focal hepatic necrosis (Green et al. 1934). Transmission among species is well documented (Cabasso 1981; Pursell et al. 1983), and it is possible that unvaccinated domestic dogs could contract infection from wild species dying of adenovirus infection. Infection of susceptible wildlife by infected domestic animals, however, has more far-reaching implications (Foreyt et al. 1986). Therefore, it is generally not recommended to mix species or allow contact between wild and domestic species. Wildlife reintroduced into the wild following rehabilitation could infect resident wild populations. Therefore, vaccination of susceptible captive carnivores with killed CAV-1 vaccine should be considered in rehabilitation programs. Canine adenovirus 1 causes little impact on populations of free-ranging species.

ADENOVIRUS HEMORRHAGIC DISEASE OF DEER

History. During the latter half of 1993, an epizootic of hemorrhagic disease caused by a previously unrecognized adenovirus was believed to be responsible for the deaths of thousands of mule deer *Odocoileus hemionus* throughout California (Woods et al. 1996). Most affected animals were black-tailed deer *Odocoileus hemionus columbianus;* Rocky Mountain mule deer *Odocoileus hemionus hemionus* were less frequently affected. During the disease investigation, adenovirus was demonstrated in archived tissues of two deer that died during a similar epizootic that occurred in northern California in 1987. Although neither bluetongue (BT) virus nor epizootic hemorrhagic disease (EHD) virus was isolated in either case, BT or EHD virus was the presumed etiology in both cases. It now seems likely that some previous unconfirmed epizootics of presumptive BT or EHD in California deer may have been caused by this previously unrecognized adenovirus.

Distribution. California is the only state in which adenovirus hemorrhagic disease (AHD) has been reported. The 1993 epizootic in California affected herds in over 17 counties, from Yosemite National Park to the Oregon border. Outbreaks in rehabilitation centers involving small numbers of deer were diagnosed infrequently during the subsequent 3 years in California. Seroprevalence of adenovirus infection in mule deer herds throughout California and the western states has not yet been determined. Hemorrhagic disease due to BT or EHD viruses has been reported throughout the country in mule deer (Jessup et al. 1984) and white-tailed deer *Odocoileus virginianus* (Hoff and Trainer 1981). The striking similarities seen on necropsy in animals that die of AHD and those infected with BT or EHD viruses make the likelihood of a misdiagnosis great if based solely on gross pathology. Therefore, it is likely that additional cases of AHD will soon be recognized in other regions of the United States where mule deer species are present.

Host Range. Subspecies of mule deer, particularly black-tailed deer, are most susceptible to clinical disease due to AHD virus infection. Fawns appear to be more susceptible to fatal infection than adults. White-tailed deer may also be susceptible to infection and clinical disease. Additionally, similar focal lesions have been observed in captive pronghorn *Antilocapra americana* in which adenovirus infection has been suspected but not proven (L.W. Woods unpublished).

Etiology. The AHD virus of deer is characteristic of the Adenoviridae (Horne 1973). Negatively stained virions are typically 77.6–79.6 nm in diameter, with triangular facets (Fig. 9.1, inset) and four capsomeric structures between vertices (Woods et al. 1996). The AHD virus has been shown to be intimately associated with lesions responsible for morbidity and mortality both in naturally and experimentally infected deer (Woods et al. 1996, 1997, 1999). Coinfection with a helper virus, as is sometimes required with adenovirus-associated disease in other species, does not appear to be necessary to reproduce the disease, and none has been found in naturally infected deer.

In the laboratory, the virus can be propagated in black-tailed deer pulmonary artery endothelial cells (Woods et al. 1996). The virus does not produce obvious cytopathic effects in cultured endothelial cells. Virus isolation attempts in Vero cells, baby hamster kidney, bovine turbinate, rabbit kidney, white-tailed deer carotid artery, and black-tailed deer testicular cells have been unsuccessful. Antigenic similarities between the deer adenovirus and bovine adenovirus type 5 have been demonstrated. Cross-reactions with porcine, equine, canine, and bovine adenovirus type 3 does not occur.

Clinical Signs. Two forms of adenoviral disease are recognized: a systemic form and a localized form. Clinical signs in deer with the systemic form may include ptyalism, diarrhea, regurgitation of rumen contents, seizures, and recumbency prior to death (Woods et al. 1996). With the localized form of disease, deer may develop swollen muzzles and oral abscesses, with subsequent anorexia and ptyalism followed by emaciation and death.

Pathogenesis. In the systemic form, endothelial cells in most organ systems may be infected, but vasculitis occurs mostly in lungs and alimentary tract, resulting in pulmonary edema and hemorrhagic enteropathy (Woods et al. 1996, 1997). Endothelial cell necrosis may trigger disseminated intravascular coagulopathy

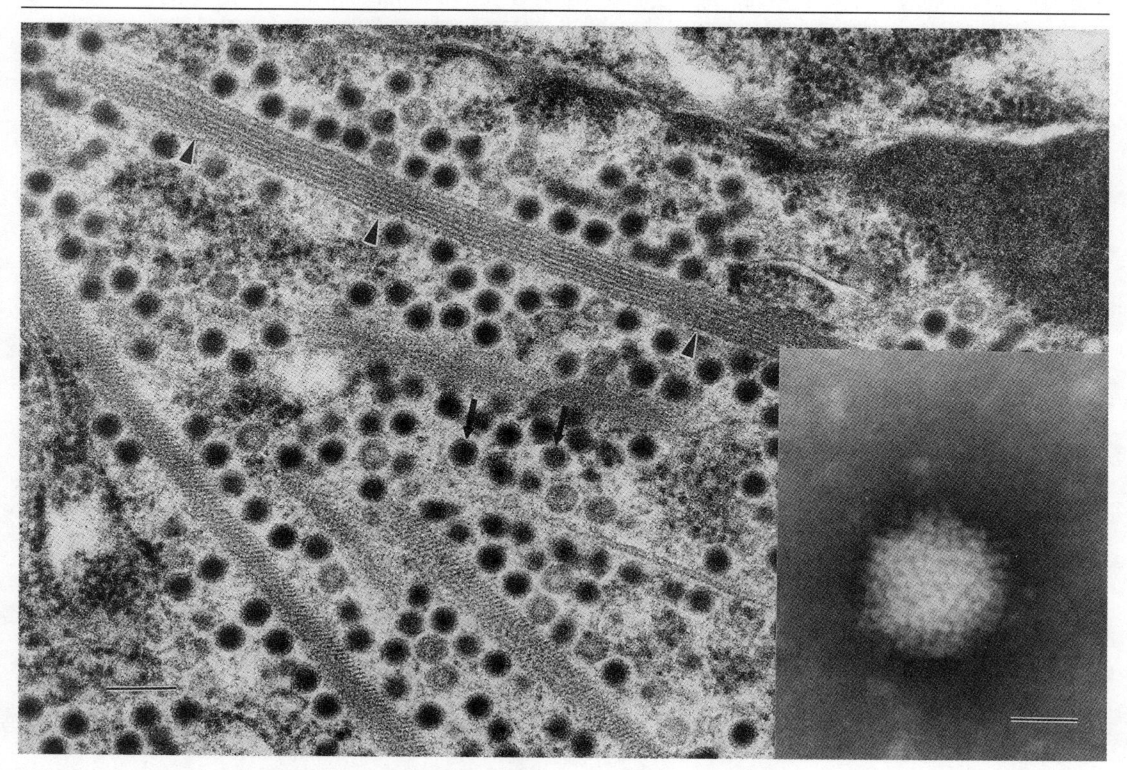

FIG. 9.1—Transmission electron photomicrograph of a pulmonary vascular endothelial cell: black-tailed deer with systemic adenovirus hemorrhagic disease. The nucleus contains numerous adenovirus particles (*arrows*) associated with long granular protein paracrystals (*arrowheads*). Lead citrate and uranyl acetate. Bar = 150 nm. *Inset:* Negatively stained black-tailed deer pulmonary artery endothelial cells infected with adenovirus demonstrate a virion, 78 nm in diameter, with triangular facets. Phosphotungstic acid. Bar = 40 nm.

as seen in the orbivirus hemorrhagic diseases (Fletch and Karstad 1971; Howerth et al. 1988). In the localized form of the disease, vascular damage is seen predominantly in the upper alimentary tract. Endothelial cell necrosis with subsequent thrombosis of vessels in the mouth results in ischemic necrosis and secondary aerobic and anaerobic bacterial infection. Chronic wasting due to inability to eat precedes death. The role that humoral or cell-mediated immunity plays in the development of systemic verses localized infection has not been determined.

Pathology. Pulmonary edema is a consistent finding (Fig. 9.2), and hemorrhagic enteropathy (Fig. 9.3) is less frequently present in animals that die of natural disease (Woods et al. 1996). Some deer with systemic disease also have erosions and ulcers of the upper alimentary tract. Animals with localized disease have large focal or solitary ulcers with deep necrosis of the underlying tissue in the oral cavity (Fig. 9.4), particularly in the lip, pharynx, tongue and, less frequently, the forestomachs (Fig. 9.5). Osteomyelitis is occasionally present. Perirenal and epicardial fat may be absent or minimal and gelatinous.

Acute to subacute panvasculitis is seen primarily in the lungs and alimentary tract, and less frequently in the brain, pulmonary artery, spleen, kidney, uterus, urinary bladder, and lymph nodes in animals that die of natural and experimental infection (Woods et al. 1996, 1997, 1999). Vessels exhibit a range of0dhanges from endothelial hypertrophy to endothelial cell necrosis, fibrinoid necrosis of the tunica intima and media, and intramural inflammatory cell infiltrates. Indistinct eosinophilic or amphophilic intranuclear inclusion bodies may be present in endothelial cells in arteries, veins, arterioles, venules and sometimes capillaries in the lungs and alimentary tract (Fig. 9.6), and less frequently brain, spleen, urinary bladder, kidney, and uterus. Variable degrees of lymphocytic necrosis and depletion are found in retropharyngeal, mandibular, parotid, and cranial cervical lymph nodes. Additional significant histologic changes are consistent with primary vascular damage, including pulmonary edema, ulceration in the upper alimentary tract, and diffuse mucosal hemorrhage in the small and large intestines.

Microscopic changes in deer with localized natural and experimental infection consist of extensive focal necrosis in the upper alimentary tract subtended by subacute to chronic inflammation and vasculitis. Endothelial intranuclear inclusion bodies are rarely present. As such, the lesions seen in the localized form appear often as rather nonspecific focal necroulcerative lesions or "abscesses" in the mouth, particularly in the pharynx. Vasculitis characterized by fibrinoid necrosis of the tunica media, luminal fibrinocellular thrombi, or inflammatory cell infiltrates in the tunica media or intima is seen in vessels in the subjacent viable tissue. Endothelial intranuclear inclusions, present in vessels in the underlying viable tissue, are transient. Therefore, although inclusions have been demonstrated in experimental infection (Woods et al. 1997), they can rarely be found in animals that die of the localized natural infection due to the chronicity of the lesions (Woods et al. 1996). Systemic changes are typically absent in these animals.

Transmission electron microscopy demonstrates variable degrees of endothelial cell degeneration ranging from dilatation of the endoplasmic reticulum and mitochondria to severe nuclear and cytoplasmic fragmentation (Woods et al. 1996, 1997). Adenovirus particles are loosely dispersed in nuclei often associated with long needle-like protein paracrystals (Fig. 9.1). Icosahedral viral nucleocapsids are 68–72 nm in diameter with electron-dense cores.

Diagnosis and Differential Diagnoses. Characteristic gross changes in deer lead to a presumptive diagnosis of AHD, EHD, or BT. Identification of endothelial intranuclear inclusions in association with vascular lesions confirms the diagnosis of AHD. Immunohistochemistry using fluorescein- or immunoperoxidase-labeled antibody directed against bovine adenovirus type 5 or the deer adenovirus can provide a definitive diagnosis. Transmission electron-microscopic examination of affected tissue or virus isolation may also be used. Identical gross lesions between BT, EHD, and AHD make confirmation of a diagnosis by histologic examination and antigen detection crucial. Endothelial intranuclear inclusion bodies may sometimes be indistinct or absent in some animals dying of acute disease and are most times absent in affected tissues of animals with localized infection. Adenovirus antigen detection in animals with chronic localized infection is often unrewarding, and distinction between the various causes of hemorrhagic disease then becomes difficult due to the microscopic similarities seen with these diseases (Karstad and Trainer 1967; Hoff and Trainer 1978, 1981). Although serologic tests are not currently available on a diagnostic basis for this adenovirus, negative results of BT virus- and EHD virus-specific polymerase chain reaction (PCR) or serologic tests for BT virus and EHD virus may be helpful to rule out the orbivirus hemorrhagic diseases in a herd outbreak (Akita et al. 1993; Aradaib et al. 1994; Patton et al. 1994). Positive serology or PCR results for BT or EHD viruses must be evaluated with caution, since these may represent transient, asymptomatic infections with these orbiviruses (Stauber et al. 1977; Work et al. 1992; L.W. Woods unpublished).

Necrotic stomatitis in cervids, caused by *Fusobacterium necrophorum,* may present with similar gross lesions as AHD.

Public Health Concerns. Risk of AHD virus infection to humans has not been determined. However, since adenoviruses are typically host specific, human infection is unlikely.

Management Implications. Susceptibility of domestic animals or farmed cervids to AHD has not yet been

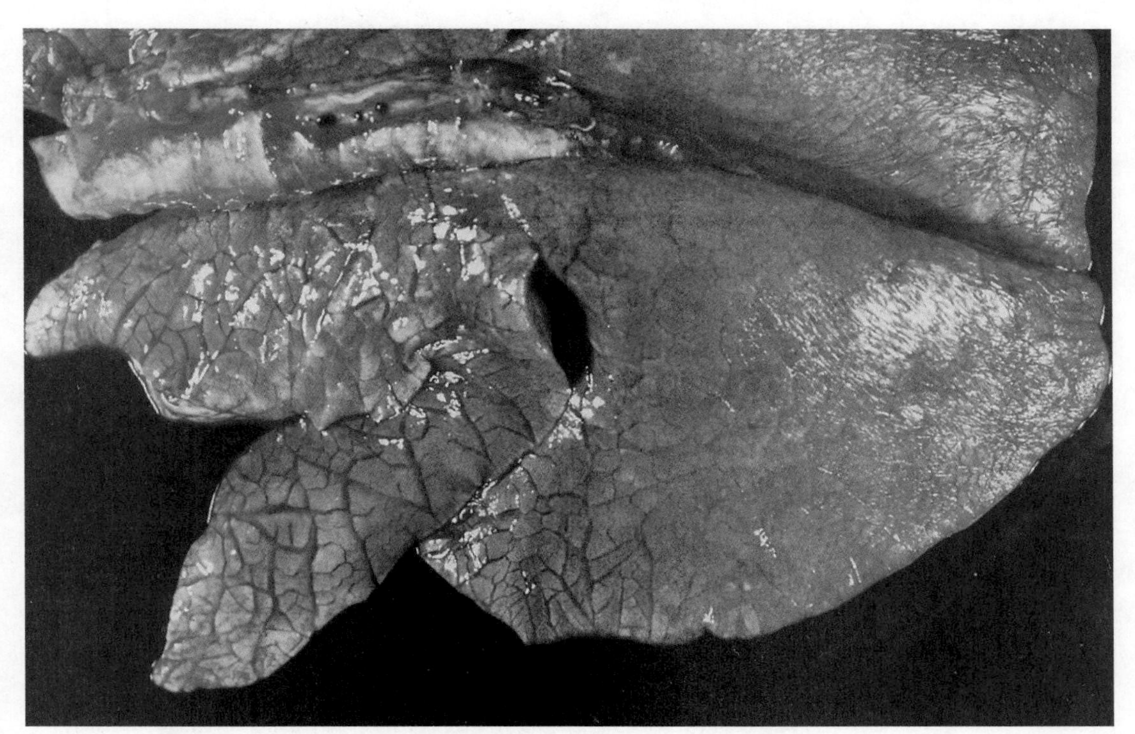

FIG. 9.2—Lungs: black-tailed deer with systemic adenovirus hemorrhagic disease. Pulmonary edema is most prominent in the cranial, middle, and anterior portions of the caudal lung lobes. Note the wide separation of lobules by interlobular septal edema.

FIG. 9.3—Colon: black-tailed deer with systemic adenovirus hemorrhagic disease. The lumen of the spiral colon is exposed, and hemorrhagic contents are demonstrated by *arrows.*

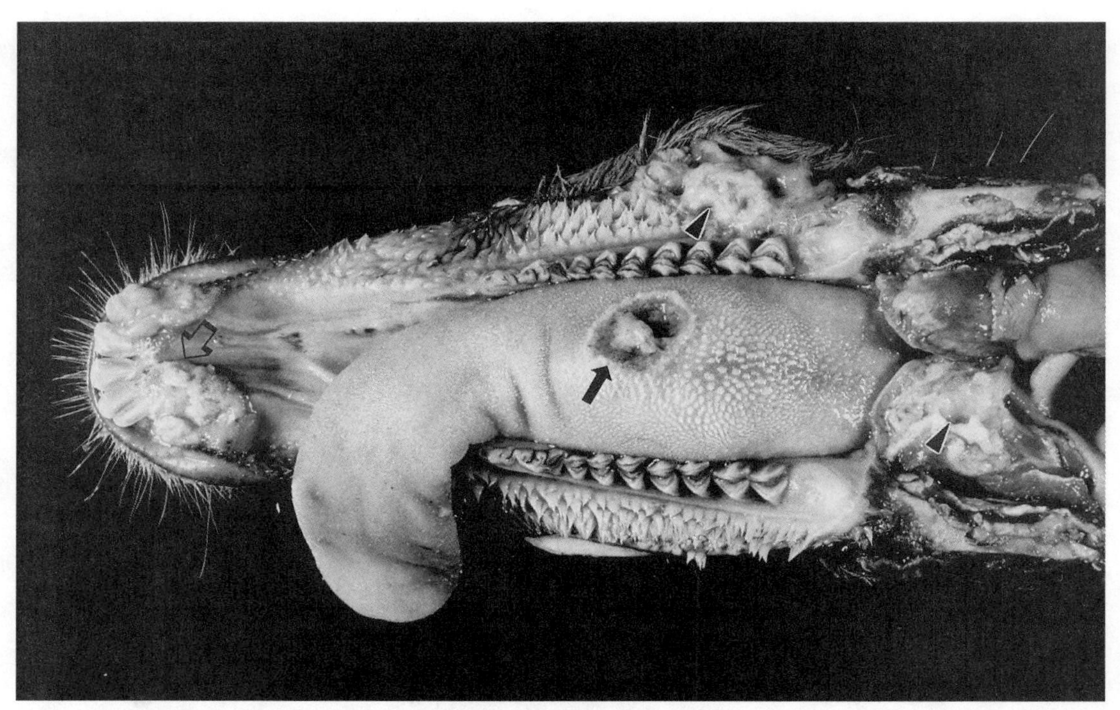

FIG. 9.4—Mandible: black-tailed deer with localized adenovirus infection. The dorsal surface of the tongue has a large, deep ulcer (*black arrow*). There is necrosis at the tip of the mandible (*open arrow*) and necrosis of the buccal mucosa and pharynx (*arrowheads*).

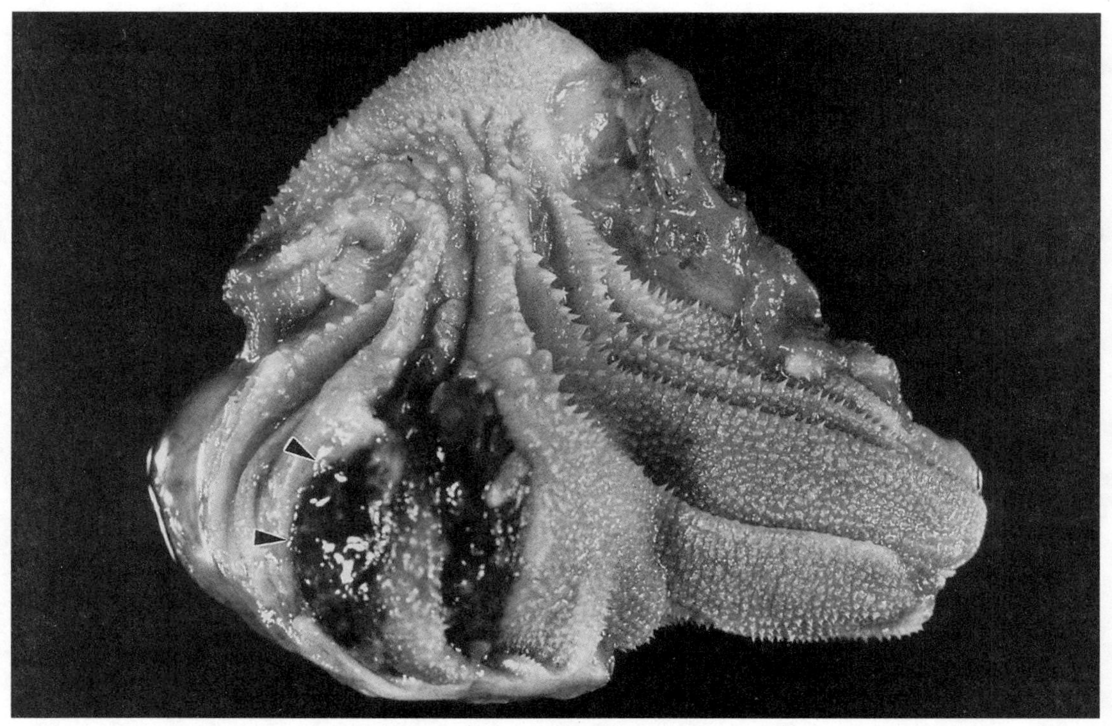

FIG. 9.5—Omasum: black-tailed deer with localized adenovirus infection. *Arrowheads* demonstrate the local area of necrosis and hemorrhage in the omasal mucosa.

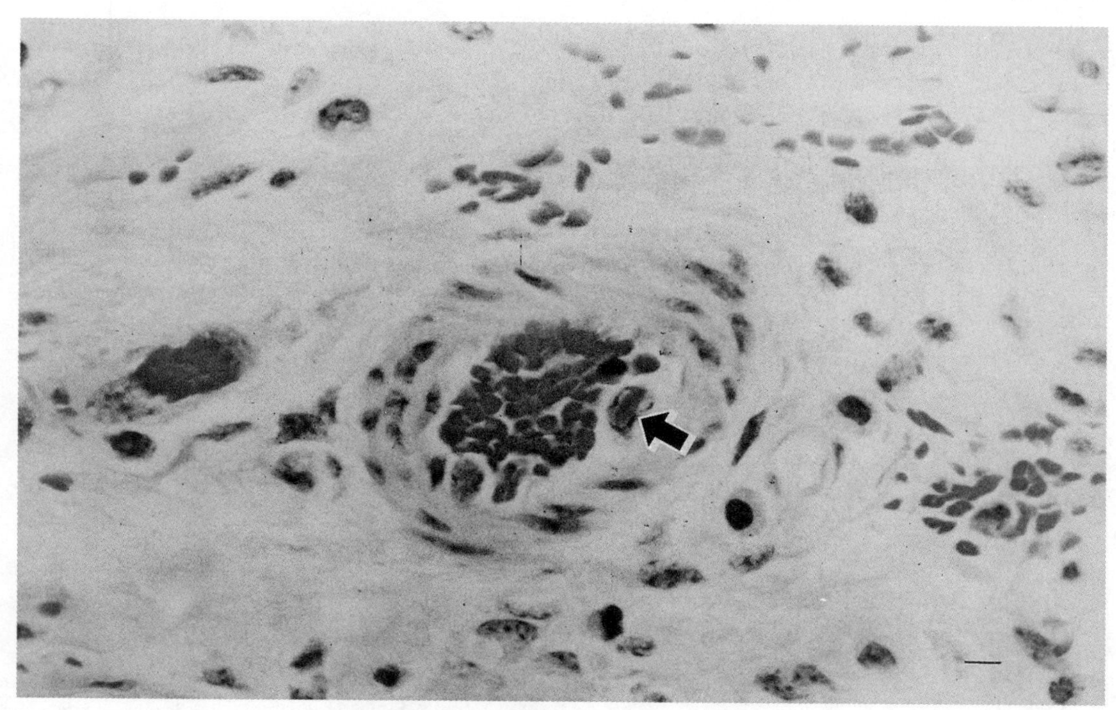

FIG. 9.6—Submucosal vessel in the small intestines: black-tailed deer with systemic adenovirus hemorrhagic disease. The hypertrophic endothelial cell has a intranuclear inclusion body (*arrow*). Hematoxylin-eosin. Bar = 10 μm.

determined. Close monitoring for AHD, particularly in mule deer, should be maintained and stringent policies established for movement of animals from affected locations and release of animals from rehabilitation centers. Diagnosticians should be aware of the similarity of AHD and the orbiviral hemorrhagic diseases, and appropriate tests should conducted to distinguish these diseases.

MISCELLANEOUS ADENOVIRAL DISEASES. A sporadic form of adenovirus infection has previously been reported in deer. Adenovirus-like particles were identified in bronchiolar epithelium associated with a bronchiolitis in a red deer *Cervus elaphus* in New Zealand (Horner and Read 1982). Bovine adenovirus type 6 was isolated from the lung of a fallow deer *Dama dama* with bronchiolitis from the Budapest Zoological and Botanic Gardens (Boros et al. 1985). Adenoviruses apparently unrelated to AHD virus have been observed in renal epithelium from captive mule deer and Rocky Mountain elk *Cervus elaphus nelsoni* in Colorado and Wyoming and from a case of necrotizing hepatitis in an elk (E.S. Williams personal communication).

In a serologic survey for selected viral infections of Rocky Mountain bighorn sheep *Ovis canadensis,* 62% had group specific adenoviral antibodies. It was suggested in this report that adenovirus may be involved in the pneumonia complex of bighorn sheep (Parks and England 1974). Adenovirus has also been identified in gazelles *Gazella gazella* [D.E. Skilling unpublished data cited by Smith et al. (1987)].

There are several reports of adenovirus infection in marine mammals. Acute hepatic necrosis was described in California sea lions *Zalophus californianus* in two separate reports (Britt et al. 1979; Dierauf et al. 1981). Clinical signs included diarrhea, anorexia, abdominal pain, posterior paralysis, polydipsia, and photophobia. In one report, hepatic necrosis was distributed primarily in the central and periportal regions of the lobules. Intranuclear inclusion bodies were seen in Kupffer cells, and vasculitis was present in the hepatic arteries and central veins. In another report, intranuclear inclusion bodies were seen in hepatocytes associated with random necrosis. Additional changes noted were suppurative pneumonia and lymphoid depletion in lymph nodes.

Adenoviral infections were reported in sei *Balaenoptera borealis* and bowhead *Balaena mysticetus* whales. The adenovirus isolated in 1977 from the sei whale was propagated in Vero cells and was reportedly the second virus isolated from whales (Smith and Skilling 1979). Two untyped adenoviruses were isolated from colon samples of two different bowhead whales. However, neutralizing antibody to the adenoviruses could not be demonstrated (Smith et al. 1987).

Adenovirus was isolated in an autoculture of renal medullary cells isolated from a Franklin's ground squirrel *Spermophilus franklini* with karyomegaly and intranuclear inclusions in sloughed renal medullary epithelial cells (Durham et al. 1988). An outbreak of adenovirus infection in tree shrew *Tupaia glis* infants resulted in the death of 12 (80%) of 15 animals. Multifocal necrosis with intranuclear inclusion bodies was present in the stomach and in small and large intestines. One infant had necrotizing cholangiohepatitis (Schoeb and DaRif 1984). In a serologic survey of 231 Australian brush-tailed possums *Trichosurus vulpecula* in New Zealand, 3.5% had precipitating antibodies to the group-specific antigen of mammalian adenoviruses (Rice et al. 1991).

Adenovirus was isolated from the blood of a white-footed deer mouse *Peromyscus maniculatus* in Kern County, California (Reeves et al. 1967). No complement-fixing antibody was demonstrated in this mouse. The virus was easily propagated in hamster cells, suggesting the possible ease with which the virus may cross species lines in nature. None of 90 sera of *P. maniculatus* captured from the same area of Kern county had complement-fixing antibody to the adenovirus isolate (E-20308) or adenovirus type 2. Virus was not neutralized by rabbit antisera to human adenoviruses 1–30 or mouse adenovirus.

Simian adenovirus caused necrotizing alveolitis and bronchiolitis in young laboratory-reared simian primates (Boyce et al. 1978). Most fatal infections occurred in primates under 5 weeks of age. A juvenile rhesus monkey *Macaca mulatta* died of necrotizing pancreatitis associated with adenoviral inclusions in necrotic pancreatic acinar cells (Chandler et al. 1974). Adenovirus was isolated from 36 monkeys in an extensive study that involved virus isolation attempts from approximately 450 langur *Presbytis entellus* and bonnet *Macaca radiata* monkeys in India (Bhatt et al. 1966).

LITERATURE CITED

Akita, G.Y., J. Glenn, A.E. Castro, and B.I. Osburn. 1993. Detection of bluetongue virus in clinical samples by polymerase chain reaction. *Journal of Veterinary Diagnostic Investigation* 5:154–158.

Alexander, A.D., V. Flyger, Y.F. Herman, S.J. McConnell, N. Rothstein, and R.H. Yager. 1972. Survey of wild mammals in a Chesapeake Bay area for selected zoonoses. *Journal of Wildlife Diseases* 8:119–126.

Aradaib, I.E., G.Y. Akita, and B.I. Osburn. 1994. Detection of epizootic hemorrhagic disease virus serotypes 1 and 2 in cell culture and clinical samples using polymerase chain reaction. *Journal of Veterinary Diagnostic Investigation* 6:143–147.

Belak, S. 1990. Ovine adenoviruses. In *Virus infections of ruminants,* vol. 3, ed. Z. Dinter and B. Morein. Amsterdam: Elsevier Science, pp. 171–185.

Bhatt, P.N., M.K. Goverdhan, M.F. Shaffer, M.F., C.D. Brandt, and J.P. Fox. 1966. Viral infections of monkeys in their natural habitat in southern India: I. Some properties of cytopathic agents isolated from bonnet and langur monkeys. *American Journal of Tropical Medicine and Hygiene* 15:551–560.

Boros, G., Z. Graf, M. Benkö, and A. Bartha. 1985. Isolation of bovine adenovirus from fallow deer (*Dama dama*). *Acta Veterinaria Hungarica* 33:119–123.

Boyce, J.T., W.E. Giddens, and M. Valerio. 1978. Simian adenoviral pneumonia. *American Journal of Pathology* 91:259–270.

Britt, J.O., A.Z. Nagy, and E.B. Howard. 1979. Acute viral hepatitis in California sea lions. *Journal of the American Veterinary Medical Association* 175:921–923.

Bürki, F. 1990. Bovine adenoviruses. In *Virus infections of ruminants,* vol. 3, ed. Z. Dinter and B. Morein. Amsterdam: Elsevier Science, pp. 161–169.

Cabasso, V.J. 1981. Infectious canine hepatitis. In *Infectious diseases of wild mammals,* ed. J.W. Davis, L.H. Karstad, and D.O. Trainer, 2d ed. Ames: Iowa State University Press, pp. 191–195.

Carmichael, L.E. 1977. Canine adenovirus infection. In *Current veterinary therapy,* ed. R.W. Kirk, 6th ed. Philadelphia: W.B. Saunders, pp. 1303–1305.

Chaddock, T.T., and W.E. Carlson. 1950. Fox encephalitis (infectious canine hepatitis) in the dog. In *The North American veterinarian,* vol. 31, ed. J.V. Lacroix, H.P. Hoskins, and W.W. Armistead. Evanston, IL: American Veterinary, pp. 35–41.

Chandler, F.W., C.S. Callaway, and S.R. Adams. 1974. Pancreatitis associated with an adenovirus in a rhesus monkey. *Veterinary Pathology* 11:165–171.

Choquette, L.P.E., and E. Kuyt. 1974. Serological indication of canine distemper and of infectious canine hepatitis in wolves (*Canis lupus* L.) in Northern Canada. *Journal of Wildlife Diseases* 10:321–324.

Collins, J.E., P. Leslie, D. Johnson, D. Nelson, W. Peden, R. Boswell, and H. Draayer. 1984. Epizootic of adenovirus infection in American black bears. *Journal of the American Veterinary Medical Association* 185:1430–1432.

Cowdry, E.V., and G.H. Scott. 1930. A comparison of certain intranuclear inclusions found in the livers of dogs without history of infection with intranuclear inclusions characteristic of the action of filtrable viruses. *Archives of Pathology* 9:1184–1196.

Darbyshire, J.H., and H.G. Pereira. 1964. An adenovirus precipitating antibody present in some sera of different animal species and its association with bovine respiratory disease. *Nature* 201:895–897.

Dierauf, L.A., L.J. Lowenstine, and C. Jerome. 1981. Viral hepatitis (adenovirus) in a California sea lion. *Journal of the American Veterinary Medical Association* 179:1194–1197.

Durham, P.J., F.A. Leighton, and G.A. Wobeser. 1988. An adenovirus infection of the kidney of Franklin's ground squirrels (*Spermophilus franklini*) in Saskatchewan, Canada. *Journal of Wildlife Diseases* 24:636–641.

Ettinger, S.J. 1983. Bacterial, viral and other infectious problems. In *Textbook of veterinary internal medicine: Diseases of the cat and dog,* ed. S.J. Ettinger, 2d ed. Philadelphia: W.B. Saunders, pp. 273–276.

Fenner, F.J., E.P.J. Gibbs, F.A. Murphy, R. Rott, M.J. Studdert, and D.O. White. 1993. Adenoviridae. In *Veterinary virology,* 2d ed. New York: Academic, pp. 329–336.

Fletch, A.L., and L.H. Karstad. 1971. Studies on the pathogenesis of experimental epizootic hemorrhagic disease of white-tailed deer. *Canadian Journal of Comparative Medicine* 35:224–229.

Foreyt, W.J., J.F. Evermann, and J. Hickman. 1986. Serologic survey for adenovirus infection in wild bears in Washington. *Journal of Wildlife Management* 50:273–274.

Fowler, M.E. 1986. Carnivora. In *Zoo and wild animal medicine,* ed. M.E. Fowler, 2d ed. Philadelphia: W.B. Saunders, pp. 800–807.

Green R.G. 1925. Distemper in the silver fox (*Vulpes vulpes*). *Proceedings of the Society of Experimental Biology and Medicine* 22:546–548.

Green R.G., and J.E. Shillinger. 1934. Epizootic fox encephalitis: VI. A description of the experimental infection in dogs. *American Journal of Hygiene* 19:362–391.

Green R.G., N.R. Ziegler, B.B. Green, and E.T. Dewey. 1930. Epizootic fox encephalitis: I. General description. *American Journal of Hygiene* 12:109–129.

Green R.G., N.R. Ziegler, W.E. Carlson, J.E. Shillinger, S.H. Tyler, and E.T. Dewey. 1934. Epizootic fox encephalitis: V. General and pathogenic properties of the virus. *American Journal of Hygiene* 18:343–361.

Green R.G., C.A. Evans, and H.Y. Yanamura. 1943. Susceptibility of the raccoon to fox encephalitis. *Proceedings of the Society for Experimental Biology and Medicine* 53:186–187.

Habermann, R.T., C.M. Herman, and F.P. Williams. 1958. Distemper in raccoons and foxes suspected of having rabies. *Journal of the American Veterinary Medical Association* 132:31–35.

Hoff, G.L., and D.O. Trainer. 1978. Bluetongue and epizootic hemorrhagic disease viruses: Their relationship to wildlife species. *Advances in Veterinary Science and Comparative Medicine* 22:111–130.

———. 1981. Hemorrhagic diseases of wild ruminants. In *Infectious diseases of wild mammals,* ed. J.W. Davis, L.H. Karstad, and D.O. Trainer, 2d ed. Ames: Iowa State University Press, pp. 45–53.

Horne, R.W. 1973. The structure of adenovirus particles and their components. In *Ultrastructure of animal viruses and bacteriophages: An atlas,* vol. 5, ed. A.J. Dalton and F. Haguenau. New York: Academic, pp. 67–81.

Horner, G.W., and D.H. Read. 1982. Presumptive adenoviral bronchiolitis in a red deer (*Cervus elaphus*). *Journal of Comparative Pathology* 92:631–634.

Horner, G.W., R. Hunter, A. Bartha, and M. Benkö. 1989. A new subgroup 2 bovine adenovirus proposed as the prototype strain 10. *Archives in Virology* 109:121–124.

Horwitz, M.S. 1990. Adenoviridae and their replication. In *Virology,* vol. 2, ed. B.N. Fields and D.M. Knipe. New York: Raven, pp. 1679–1682.

Howerth, E.W., C.E. Greene, and A.K. Prestwood. 1988. Experimentally induced bluetongue virus infection in white-tailed deer: Coagulation, clinical pathologic, and gross pathologic changes. *American Journal of Veterinary Research* 49:1906–1913.

Jessup, D.A., B.I. Osburn, and W.P. Heuschele. 1984. Bluetongue in California's wild ruminants: Distribution and pathology. *Proceedings of the United States Animal Health Association* 88:616–630.

Karstad, L., and D.O. Trainer. 1967. Histopathology of experimental bluetongue disease of white-tailed deer. *Canadian Veterinary Journal* 8:247–254.

Karstad, L., R. Ramsden, T.J. Berry, and L.N. Binn. 1975. Hepatitis in skunks caused by the virus of infectious canine hepatitis. *Journal of Wildlife Diseases* 11:494–496.

Kelly, W.R. 1993. The liver and biliary system. In *Pathology of domestic animals,* vol. 2, ed. K.V.F. Jubb, P.C. Kennedy, and N. Palmer, 4th ed. San Diego: Academic, pp. 364–366.

Kummeneje, K. 1971. Encephalitis infectiosa vulpis (*Hepatitis contagiosa canis*) in a blue fox farm. *Nordisk Veterinaermedicin* 23:352–360.

Marler, R.J., S.M. Kruckenberg, J.E. Cook, and C.M. O'Keefe. 1976. Encephalitis in a confined coyote. *Journal of the American Veterinary Medical Association* 169:964–965.

Palmer, E.L., and M.L. Martin. 1988. Adenoviridae. In *Electron microscopy in viral diagnostics.* Boca Raton, FL: CRC, pp. 159–164.

Parks, J.B., and J.J. England. 1974. A serological survey for selected viral infections of Rocky Mountain bighorn sheep. *Journal of Wildlife Diseases* 10:107–110.

Patton, J.F., T.M. Work, D.A. Jessup, S.K. Hietala, M.N. Oliver, and N.J. MacLachlan. 1994. Serologic detection of bluetongue virus infection of black-tailed deer: Comparison of serum neutralization, agar gel immunodiffusion, and competitive ELISA assays. *Journal of Wildlife Diseases* 30:99–102.

Pursell, A.R., B.P. Stuart, and E. Styer. 1983. Isolation of an adenovirus from black bear cubs. *Journal of Wildlife Diseases* 19:269–271.

Reeves, W.C., R.P. Scrivani, W.E. Pugh, and W.P. Rowe. 1967. Recovery of an adenovirus from a feral rodent *Peromyscus maniculatus. Proceedings of the Society of Experimental Biology and Medicine* 124:1173–1175.

Rice, M., C.R. Wilks, D. Pfeiffer, and R. Jackson. 1991. Adenovirus precipitating antibodies in the sera of brush-tailed possums in New Zealand. *New Zealand Veterinary* Journal 39:58–60.

Rowe, W.P., and J.W. Hartley. 1962. A general review of the adenoviruses. *Annals of the New York Academy of Sciences* 101:466–474.

Rubarth, S. 1947. An acute virus disease with liver lesions in dogs (hepatitis contagiosa canis): A pathologico-anatomical and etiological investigation. *Acta Pathologica Microbiologica Scandinavica, Supplement* 69:9–207.

Schoeb, T.R., and C.A. DaRif. 1984. Adenoviral infection in infant tree shrews. *Journal of the American Veterinary Medical Association* 185:1363–1366.

Siedentopf, H.A., and W.E. Carlson. 1949. A comparative study of the fox encephalitis virus and the virus of infectious canine hepatitis. *Journal of the American Veterinary Medical Association* 115:109–111.

Smith, A.W., and D.E. Skilling. 1979. Viruses and virus diseases of marine mammals. *Journal of the American Veterinary Medical Association* 175:918–920.

Smith, A.W., D.E. Skilling, K. Benirschke, T.F. Albert, and J.E. Barlough. 1987. Serology and virology of the bowhead whale (*Balaena mysticetus* L.). *Journal of Wildlife Diseases* 23:92–98.

Smyth, J.A., M. Benkö, D.A. Moffett, and B. Harrach. 1996. Bovine adenovirus type 10 identified in fatal cases of adenovirus-associated enteric disease in cattle by in situ hybridization. *Journal of Clinical Microbiology* 34:1270–1274.

Stauber, E.H., R.K. Farrell, and G.R. Spencer. 1977. Nonlethal experimental inoculation of Columbia black-tailed deer (*Odocoileus hemionus columbianus*) with virus of epizootic hemorrhagic deer disease. *American Journal of Veterinary Research* 38:411–412.

Summers, B.A., J.F. Cummings, and A. de Lahunta. 1995. Inflammatory diseases of the central nervous system. In *Veterinary neuropathology,* ed. B.A. Summers, J.F. Cummings, and A. de Lahunta. St. Louis: Mosby-Year Book, pp. 102–117.

Trainer, D.O., and F.F. Knowlton. 1968. Serologic evidence of diseases in Texas coyotes. *Journal of Wildlife Management* 32:981–983.

Woods, L.W., P.K. Swift, B.C. Barr, M.C. Horzinek, R.W. Nordhausen, M.H. Stillian, J.F. Patton, M.N. Oliver, K.R. Jones, and N.J. MacLachlan. 1996. Systemic adenovirus infection associated with high mortality in mule deer (*Odocoileus hemionus*) in California. *Veterinary Pathology* 33:125–132.

Woods, L.W., R.S. Hanley, P.H.W. Chiu, M. Burd, R.W. Nordhausen, M.H. Stillian, and P.K. Swift. 1997. Experimental adenovirus hemorrhagic disease in yearling black-tailed deer. *Journal of Wildlife Diseases* 33:801–811.

Woods, L.W., R.S. Hanley, P.H.W. Chiu, H. Lehmkuhl, R. Nordhausen, and M. Stillian. 1999. Lesions and transmission of experimental adenovirus hemorrhagic disease in black-tailed deer fawns. *Veterinary Pathology* 36:100–110.

Work, T.M., D.A. Jessup, and M.M. Sawyer. 1992. Experimental bluetongue and epizootic hemorrhagic disease virus infection in California black-tailed deer. *Journal of Wildlife Diseases* 28:623–628.

Zarnke, R.L., and W.B. Ballard. 1987. Serologic survey for selected microbial pathogens of wolves in Alaska, 1975–1982. *Journal of Wildlife Diseases* 23:77–85.

Zarnke, R.L., and M.B. Evans. 1989. Serologic survey for infectious canine hepatitis virus in grizzly bears *(Ursus arctos)* from Alaska, 1973 to 1987. *Journal of Wildlife Diseases* 25:568–573.

10

RETROVIRUS INFECTIONS

MICHAEL WORLEY

INTRODUCTION. The family *Retroviridae* is a large group of RNA viruses characterized by a linear, single-stranded nonsegmented RNA genome, and a replicative strategy that includes reverse transcription of the viral RNA into linear double-stranded DNA and the subsequent integration of this DNA into the genome of the cell. A recent classification of the *Retroviridae* and an extensive discussion of their molecular organization and replication are found in Coffin (1996).

Retrovirus-like DNA sequences with which no infectious virus has been associated are found in the genomes of most eukaryotes, causing no known problems. In a broad range of species, however, exogenous retroviruses are associated with a variety of disease syndromes, including rapid and long-latency malignant tumors, wasting diseases, neurologic disorders, and immunodeficiencies. Retrovirus infection can also lead to lifelong viremia without any obvious ill effects.

Information on retroviral infections of nondomestic cats, nonhuman primates, and several other wildlife species is summarized in this chapter.

FELINE LEUKEMIA VIRUS INFECTION

Etiology. Feline leukemia virus (FeLV) belongs to the mammalian type-C oncovirus serogroup of the mammalian type-C retroviruses (unclassified genus).

Host Range and Distribution. Although FeLV is a major pathogen in domestic cats throughout the world, infection and resulting disease are rare in nondomestic felids. However, FeLV antigen has been detected in both a free-ranging and a captive-raised cougar (puma) *Felis concolor* (Meric 1984; Jessup et al. 1993), a captive-raised clouded leopard *Neofelis nebulosa* (Citino 1986), a captive-raised cheetah *Acinonyx jubatus* (Briggs and Ott 1986), and a free-ranging European wildcat *Felis silvestris* (Boid et al. 1991). Virus was isolated from only the free-ranging cougar and the European wildcat. In addition, FeLV has been isolated from a cell line derived from a leopard cat *Felis bengalensis* (Rasheed and Gardner 1981).

Transmission and Epidemiology. There are two basic routes of infection: horizontal transmission via excretions and secretions from infected to susceptible cats, and vertical (in utero) transmission from infected mothers to their fetuses (Pedersen 1988; Hoover and Mullins 1991). Although prolonged intimate contact between cats usually is required for horizontal spread of FeLV, transmission through bite wounds is more efficient because a larger amount of virus can be injected directly into the body.

Clinical Signs. Infection in nondomestic cats has been associated with lethargy, anorexia, emaciation, and dehydration (Meric 1984; Briggs and Ott 1986; Jessup et al. 1993). Additional signs have included pale mucous membranes, with petechiae and ecchymoses (Meric 1984), and enlarged peripheral lymph nodes (Citino 1986).

Pathology. Of the nondomestic cats described with FeLV infection, only the two cougars died or were euthanized, and a complete necropsy was performed only on the free-ranging animal (Jessup et al. 1993). That animal had generalized benign lymphoproliferative disease and evidence of *Escherichia coli* septicemia. Virus was isolated from peripheral blood mononuclear cells. Lymphoid hyperplasia and anemia were believed to be direct effects of the FeLV infection.

Diagnosis. Termination of viremia has been associated with the appearance of virus-specific antibodies. Hence, FeLV traditionally has been diagnosed using the immunofluorescent antibody (IFA) test for cell-associated viral antigens in blood smears or by enzyme-linked immunosorbent assay (ELISA), which detects soluble group-specific viral antigens in the plasma or serum of infected cats (Hardy and Zuckerman 1991). Viral components in peripheral blood and paraffin-embedded tissues have been detected by polymerase chain reaction (PCR) (Jackson et al. 1996).

Control. A number of commercial FeLV vaccines have been marketed, including combination products. The relative safety and efficacy of each product are discussed in Loar (1993). Although rarely found in wild felids, the potential effects of FeLV established in captive populations of endangered felids could be devastating. A surveillance program, consisting of periodic testing for FeLV antigen by IFA or ELISA, should be part of all management plans for captive groups of endangered cats.

Human Health Considerations. Feline leukemia virus infections appear to be limited to felids. There is no scientific support for any role of FeLV in human disease.

FELINE IMMUNODEFICIENCY VIRUS INFECTION

Etiology. The feline immunodeficiency virus (FIV) is a member of the genus *Lentivirus* of the family *Retroviridae.* Viruses known in nondomestic felids have been termed FIV because of their antigenic and molecular relatedness to domestic cat isolates.

History. Pedersen et al. (1987) first isolated FIV from domestic cats with an immunodeficiency syndrome, and infection of nondomestic felids subsequently was reported by Barr et al. (1989).

Host Range. Members of at least 18 of the 37 species in the family Felidae possess serum antibodies that react with FIV antigens (Carpenter and O'Brien 1995). Partial or complete nucleotide sequences have been obtained from FIV isolated from pumas (Langley et al. 1994), lions *Panthera leo* (Brown et al. 1994), and Pallas' cat *Felis manul* (Barr et al. 1995).

Distribution. Exposure to FIV has been documented in wild pumas in Florida (24% antibody positive), the western, southwestern, and intermountain United States (32% antibody positive), and British Columbia and Alaska (40% antibody positive), and in captive animals throughout the Americas (14% antibody positive) (Carpenter et al. 1996). Virus exposure has been detected in free-ranging lions in Kenya and Tanzania (81% antibody positive), Botswana (25% antibody positive), and South Africa (83%–91% antibody positive) but not in Namibia or India; in cheetahs in Kenya-Tanzania (22% antibody positive) and Botswana (25% antibody positive); and in leopards *Panthera pardus* in Botswana (16% antibody positive) and South Africa (71% antibody positive) (Olmsted et al. 1992; Spencer et al. 1992; Brown et al. 1993; Osofsky et al. 1994). Virus-exposed captive-raised nondomestic felids have been identified in zoos throughout the United States and Europe (Lutz et al. 1992; Brown et al. 1993).

Transmission. In domestic cats, transmission of FIV appears to occur via the saliva during biting (Yamamoto et al. 1988). Infection also can be transmitted experimentally by artificial insemination with fresh semen from chronically infected asymptomatic males (Jordan et al. 1996) and to kittens via milk from queens experimentally infected immediately postpartum (Sellon et al. 1994). In nondomestic species, the routes of transmission have not been established.

Clinical Signs. Domestic cats naturally infected with FIV develop an immune deficiency manifested as an array of disease processes caused by secondary or opportunistic pathogens, such as generalized cowpox virus infection, feline calicivirus, generalized demodectic and notoedric mange mite infestations, toxoplasmosis, candidiasis, cryptococcosis, and atypical mycobacteriosis (Sparger 1993).

In nondomestic felids, however, there is no clear correlation between virus infection and disease. Although inverted CD4:CD8 (helper-inducer:cytotoxic-suppressor) T-cell ratios were reported in 6 of 11 FIV-seropositive captive lions (Kennedy-Stoskopf et al. 1994), there was no relationship between seropositivity, inverted T-cell ratios, and clinical disease.

Pathogenesis. The pathogenesis of FIV infection has been studied in experimentally infected domestic cats (Sparger 1993). Some stages of infection have been delineated (Yamamoto et al. 1988), and the progressive immune dysfunction has been investigated (Ackley et al. 1990; Torten et al. 1991).

Pathology. Because of the lack of an association between FIV infection and clinical disease in nondomestic felids, there are no specific laboratory findings or lesions that correlate with infection. Approximately half of the domestic cats with naturally occurring infection showed multiple hematologic abnormalities (Sparkes et al. 1993), and cytopenia was more common in cats with advanced clinical signs of disease. Pathologic changes involving both naturally and experimentally infected domestic cats have been described by Sparger (1993). The most common intestinal tract changes included small intestinal villus blunting, loss of villi, and crypt dilatation; ulceration and necrotizing pyogranulomatous inflammation of the large intestinal tract; and ulcerative stomatitis. Lymph node changes included follicular hyperplasia, follicular involution, and a mixture of hyperplasia and involution in the same lymph node. Central nervous system lesions have included choroid plexus fibrosis, white-matter vacuolation, and perivascular cuffing by macrophages and lymphocytes.

Diagnosis. General principles of retrovirus immunodetection tests have been reviewed by Hardy (1991). Routine methods for diagnosis of FIV in all species of cats involve detection of serum antibody. Specimens that are repeatedly reactive by a commercial ELISA should be confirmed with a Western blot, which detects the presence of antibody to specific viral proteins. In the future, PCR to detect FIV infection in animals that are antibody negative may be possible, using primers derived from the nucleotide sequence of a homologous region of the reverse transcriptase domain of the polymerase gene from puma, African lion, and Pallas' cat strains of FIV.

Immunity. Sera of nondomestic cats screened for FIV antibody by Western blot react most strongly with the major capsid and core proteins (Olmsted et al. 1992;

Brown et al. 1993). The type and activity of antibodies, as well as the cellular immune responses produced after natural infection in these species, have not been studied.

Control. Spread of FIV among zoos may be difficult to control using test and isolation methods, since current tests are unable to detect the various viral strains that may be present. In addition, lack of a clear association between virus exposure/ infection and clinical disease may result in limited enthusiasm among some institutions for a surveillance program.

Although not commercially available, several types of experimental vaccines have been tested in domestic cats. Immunization with whole inactivated virus vaccines appeared to provide the best protection against challenge with either homologous or heterologous FIV strains (Yamamoto et al. 1993; Hosie and Flynn 1996).

Public Health Concerns. There is no evidence linking FIV infection to any human disease. Studies have failed to identify FIV antibodies in people that have been bitten by or otherwise had contact with FIV-infected cats (Sparger 1993).

Management Implications. The control of FIV transmission within captive groups of endangered cats is a high priority for the cheetah and lion Species Survival Plans and the Felid Taxon Advisory Group of the American Zoo and Aquarium Association. Because insufficient data are available to eliminate the possibility of an association with clinical disease, it is prudent to test all captive cats for FIV-specific antibodies, and seropositive animals should be housed separately from uninfected animals. Serologic testing should also be considered when planning movement of captives and translocation of wild cats.

SIMIAN RETROVIRUS INFECTION

Etiology. Simian retrovirus (SRV), also known as simian type-D virus, is a member of the type-D retroviruses, an unclassified genus of the family *Retroviridae.* Five different serotypes have been described, all associated with immunodeficiency disease (Lowenstine 1993a).

History. Type-D viruses, serologically related to a previously described prototype Mason-Pfizer monkey virus, were detected between 1983 and 1985 at the New England, California, Oregon, and Washington Primate Centers in several species of macaques *Macaca* spp. with a fatal immunosuppressive disease (Gardner 1996).

Host Range. Evidence of infection with exogenous type-D retroviruses has been found almost exclusively in members of the genus *Macaca* and is widespread in macaque populations both in captivity and the wild. Serum antibodies reactive with type-D retrovirus proteins have been detected in captive talapoins *Miopithecus talapoin* and in African green monkeys *Cercopithecus aethiops,* although virus has not been isolated (Lowenstine 1993b).

Distribution. Serologic evidence of infection has been found in bonnet monkeys *Macaca radiata* in southern India, rhesus macaques *Macaca mulatta* in southwestern China, and pig-tailed macaques *Macaca nemestrina* and cynomolgus macaques *Macaca fascicularis* in Indonesia (Lerche 1993).

Transmission and Epidemiology. Natural transmission requires close physical contact. The virus is shed in saliva, and social interactions such as grooming, licking, and biting probably facilitate its spread (Lowenstine 1993a). Vertical transmission (mother to infant), both transplacentally and postnatally, has been documented (Lerche 1993). An inapparent carrier state can occur. These animals are healthy, antibody negative, and may shed virus for years before eventually becoming clinically ill (Lerche et al. 1986).

Disease may be sporadic, endemic, or epidemic (Lerche 1993). Sporadic disease occurs in individually housed chronic carriers. Epidemics can occur in groups of naive animals into which virus-positive animals have been introduced, or when naive animals of the same or different species are introduced into a group in which the disease is endemic. In an outdoor-housed population of rhesus macaques with endemic type-D virus infection, the mortality rate among introduced juveniles was 85% in the first 9 months (Lerche et al. 1987). Endemic disease is seen in large breeding groups, often following an epidemic. There is high infant and juvenile mortality; about 45% of each birth cohort can die before reaching sexual maturity (Lerche et al. 1987).

Clinical Signs. Disease in infected rhesus macaques may present as three syndromes. Severe acquired immunodeficiency syndrome (SAIDS) has either an acute course of 1–5 months or a more protracted course of 7 or more months. Persistent lymphadenopathy (referred to as SAIDS-related complex) usually progresses over a period of 2 or more years to overt lymphadenopathy. Fibroproliferative disorders often develop coincident with SAIDS or its related complex (Lowenstine 1993a).

Animals that are fatally infected develop a spectrum of immunologic and hematologic abnormalities that may have a disease course ranging from 1 month to several years. A case definition has been developed for simian AIDS caused by type-D retrovirus in rhesus macaques. Generalized lymphadenopathy and/or splenomegaly is combined with four or more of the following: weight loss (more than 10% of body weight); anemia (packed cell volume of less than 30%) or neutropenia (less than 1.7×10^9/L) in the presence of bone marrow hyperplasia; persistent lymphopenia (less than 1.6×10^9/L); persistent or intermittent diarrhea poorly responsive to therapy; bacterial infections of the skin, oral cavity, or other sites unresponsive to therapy; or opportunistic infection, such as cytomegalovirus infection, cryptosporidiosis, or oral and esophageal candidiasis (Henrickson et al. 1984).

Pathogenesis. Type-D retroviruses have a broad tropism in vitro, including B cells, CD4 and CD8 T cells, macrophages, and epithelial cells from certain organs (Lowenstine 1993a). Neutropenia often occurs during acute infection and predisposes animals to infection with pyogenic bacteria. Macrophages and other antigen-presenting cells are infected, but macrophage function is not diminished. Lymphopenia occurs in many infected animals, especially in the terminal stages of disease, and is due to reduction in the number of CD4 and CD8 T cells and the number of B cells. Profound lymphoid depletion, suggesting lymphoid necrosis, is often seen in the lymph nodes and spleens of these terminal animals (Lowenstine 1993a).

Pathology. No lesions are pathognomonic for SRV infection, except perhaps the fibroproliferative disorders of retroperitoneal and subcutaneous fibromatosis. Infection with SRV must be suspected in susceptible species dying of compatible clinicopathologic syndromes, often involving opportunistic or recrudescent latent infectious agents. The gross and microscopic lesions in various phases of SAIDS have been reviewed by Lowenstine (1993a).

Diagnosis. Demonstration of SRV-specific antibodies by Western blots, or isolation of the virus, confirms SRV infection in animals with compatible syndromes. Detection of inapparent carriers requires virus isolation or the detection of proviral DNA using PCR (Lowenstine and Lerche 1988). These procedures are performed only in specialized reference laboratories or in the laboratories of investigators conducting research on this virus.

Differential Diagnosis. Infection with the simian immunodeficiency virus (SIV), a lentivirus, should be included in the differential diagnosis for macaques with opportunistic infections consistent with immune deficiency. Type-D retrovirus and SIV infections in rhesus macaques can be distinguished on the basis of clinical signs and lesions, in the absence of virus isolation or serology (Lowenstine 1993a).

Immunity. Nearly all macaques infected with type-D virus develop antibodies to the viral proteins. Many animals develop a strong neutralizing antibody response, eliminate the virus, and recover from the infection. However, in vitro blastogenesis responses of macaque peripheral blood mononuclear cells to a number of mitogens are diminished as early as 4 weeks after infection (Lowenstine 1993a).

Control. Type-D virus infections can be eliminated from a group of animals by removing individuals found to be antibody or virus positive. Because new infections may become apparent as late as 6 months after initial testing, the test and removal program must be performed at monthly intervals for perhaps as long as a year after the introduction of new animals (Lowenstine 1993b). Experimental vaccination with either formalin-killed whole SRV or recombinant vaccinia virus expressing SRV envelope glycoproteins confers protection against challenge infection (Marx et al. 1986; Brody et al. 1992).

Public Health Concerns. Simian type-D viruses do not appear to have any zoonotic potential. Serologic testing of individuals exposed to infected nonhuman primates or with known accidental parenteral exposure has not provided any evidence that these viruses are communicable to humans (Lowenstine and Lerche 1993).

Management Implications. Infection with type-D retroviruses represents a significant ongoing problem for animal management, particularly in primate centers. Since inapparent healthy carriers occur, it is imperative that virus isolation or PCR testing and serologic screening of both resident and newly acquired animals be undertaken prior to formation of any new groups.

SIMIAN IMMUNODEFICIENCY VIRUS INFECTION

Etiology. The eight different SIV isolate types are all classified in the genus *Lentivirus* of the family *Retroviridae.* As a result of their genetic relationship to the human immunodeficiency viruses (HIVs), these viruses of nonhuman primates have been designated simian immunodeficiency viruses, even though only four have been associated with immunodeficiency disorders (King 1993). The different SIV isolate types include SIVMAC from rhesus macaques, SIVSMM from sooty mangabey monkeys *Cercocebus atys,* SIVMNE from pig-tailed macaques *Macaca nemestrina,* SIVSTM from stump-tailed macaques *Macaca arctoides,* SIVAGM from African green monkeys, SIVMND from mandrills *Mandrillus sphinx,* SIVCMZ from chimpanzees *Pan troglodytes,* and SIVSYK from Sykes' monkeys *Cercopithecus mitis* (King 1993).

History. The prototype simian AIDS lentivirus, SIVMAC, was first identified in 1985 at the New England Regional Primate Research Center (NERPRC) (Gardner 1996). Retrospective evidence indicates that SIVMAC first isolated at the NERPRC almost certainly originated in 1969 in rhesus macaques at the California Regional Primate Research Center that were inadvertently infected with SIV, either directly (e.g., by biting) or indirectly (e.g., by needle) from asymptomatic sooty mangabeys (Mansfield et al. 1995). In a review of the history of simian AIDS, the possible origins of SIVMNE and SIVSTM also were discussed (Gardner 1996).

Host Range. Infection with the SIVs appears to be endemic, lifelong, and clinically inapparent in most species of African primates. Species in which SIV infection or exposure have been detected were summarized in Lowenstine (1993b). Antibody- or virus-positive captive, but not free-ranging, macaques have been identified.

Distribution. African green monkeys comprise a phenotypically diverse group of primates inhabiting virtually all of sub-Saharan Africa. Considering the diversity of other African primate species in which SIV infection has been detected, it is reasonable to assume that the distribution of the virus correlates with the distribution of primates on the continent.

Transmission and Epidemiology. The principal mode of transmission of SIV among free-ranging populations and within captive colonies has not been established. Since SIV is readily transmissible experimentally by infected blood products, contamination of bite wounds with blood from infected animals is the most likely possibility. Transplacental transmission has been documented in a rhesus monkey with naturally acquired infection (Daniel et al. 1988). The virus also has been experimentally transmitted via the genital mucosa (Miller et al. 1989).

Clinical Signs. The clinical signs and symptoms of SIVMAC infection in macaques vary and depend on the nature and organ specificity of the opportunistic infections. Early in the course of infection, lymphadenopathy of the axillary and inguinal lymph nodes is a frequent finding, as is a maculopapular rash that is most obvious on the less-haired portions of the body (King 1993). Diarrhea leading to dehydration and weight loss is also a frequent finding.

Pathogenesis. Like HIV, the SIVs have a marked tropism for cells that express the CD4 molecule on their surface. These cells include helper T cells and antigen-presenting cells of monocyte-macrophage origin. In species of monkeys in which the virus causes fatal disease, there is a profound depletion of CD4 lymphocytes, leading to severe immune dysfunction and death from opportunistic infections or lymphoma (King 1993). Macrophages may be responsible for dissemination of virus to nonlymphoid tissue, such as the brain, because of their mobility.

Pathology. Hematologic alterations in virus-infected monkeys are variable. The most consistent finding is a reduction in the number of circulating CD4 helper/inducer lymphocytes in the peripheral blood. Measurement of absolute numbers of circulating CD4 lymphocytes in SIVMAC-infected rhesus monkeys is an extremely useful prognostic test (King 1993).

Lymphoid organs may be hypertrophied, normal in size, or atrophic, depending on the duration of the disease. The findings in other visceral organs reflect the nature and extent of the various opportunistic infections to which infected animals usually succumb. Primary lesions in SIV-infected animals—including six distinct microscopic patterns of tissue reaction in lymph nodes and spleen, meningoencephalomyelitis, glomerulonephritis, severe blunting and atrophy of small intestinal villi, and a recurrent maculopapular rash with perivascular infiltrates of lymphocytes in superficial dermal capillaries—have been described in detail (King 1993).

Diagnosis. For screening large colonies, an ELISA using HIV-2 as the antigen is useful. This assay may not detect animals infected with more distantly related strains, such as SIVAGM, SIVMND, SIVSYK, and SIVCMZ. In such cases, immunoblotting or immune precipitation with electrophoresis is more accurate than ELISA (Lowenstine and Lerche 1993). As with SRV, these procedures are performed only in reference laboratories.

Infection may precede seroconversion by several weeks, and sera should be tested at acquisition and again in 3 months. However, this schedule may not be sufficient to detect all infected animals, and additional specialized techniques such as PCR may be needed to certify that an animal is free of SIV infection (Lowenstine and Lerche 1993).

Differential Diagnosis. Other conditions that cause immune dysfunction leading to death from opportunistic infections and/or lymphoma must be considered. Principal among these is infection with type-D retroviruses, which occurs with greater frequency than SIV in many primate colonies. These two retrovirus infections can be differentiated readily by serologic and virus isolation procedures and, in the absence of these, on the basis of clinicopathologic syndrome (Lowenstine 1993a).

Immunity. In naturally infected African species, SIVs appear to cause persistent viremia in the presence of an antibody response, but no overt disease. In Asian macaques, however, infection leads to viremia and an AIDS-like disease with a variable antibody response on the part of the host. The humoral immune response to SIVMAC, since it reflects CD4 lymphocyte function, can be used to predict the clinical course of an infected animal's disease. Those animals that survive the longest develop high-titer antibody responses to viral envelope and capsid/core proteins, whereas those animals that develop low-titer antibody responses to these antigens usually have a short clinical course (King 1993).

Control. Several experimental approaches to vaccination have been attempted in macaques. Animals immunized with recombinant vaccinia virus expressing the major envelope glycoprotein, gp160, of SIVMNE and boosted with gp160 produced in baculovirus-infected cells were protected against an intravenous challenge of the homologous virus (Hu et al. 1992). Rhesus macaques inoculated with an attenuated macrophage-tropic recombinant of a primary SIV isolate produced cross-reactive neutralizing antibodies that were associated with protective responses to the homologous, as well as heterologous, primary isolate (Clements et al. 1995). Rhesus monkeys vaccinated with a live attenuated deletion mutant had protection that increased with the time since vaccination (Wyand et al. 1996).

Public Health Concerns. Although the zoonotic potential of SIV is low, two laboratory workers have developed antibodies to SIV following a needle stick or exposure to infected blood. Neither showed any evidence of clinical disease (Centers for Disease Control 1992).

Management Implications. Transmission of SIV between African and Asian primates dictates that these species should not be housed in direct contact. Since infected African primates may live their entire lives without adverse effects, and occasional chronic carrier macaques exist who remain healthy for six or more years, a persistent surveillance program is required, consisting of serologic screening and virus isolation.

SIMIAN T-CELL LYMPHOTROPIC VIRUS TYPE 1

Etiology. Simian T-cell lymphotropic virus (STLV) is a member of the "BLV-HTLV retroviruses," an unclassified group that also includes bovine leukemia virus and human T-lymphotropic viruses 1 and 2 (HTLV-1 and HTLV-2). These are type-C retroviruses distinct from the "mammalian type-C retroviruses." Simian T-cell lymphotropic virus-1 shares 90%–95% nucleotide homology with HTLV-1.

History. Evidence that nonhuman primates are infected with HTLV-1-like viruses was first obtained from serologic surveys of captive and free-living monkeys (Lowenstine 1993b). The first isolation of STLV-1 was from Japanese macaques *Macaca fuscata* from an area endemic for HTLV-1 (Miyoshi et al. 1983b).

Host Range. Studies have demonstrated that antibodies to STLV-1 are widespread in both wild and captive populations of Asian and African primates, including the African apes and cercopithecine monkeys, but not colobines or prosimians (Lowenstine and Lerche 1993). Infection appears to be rare in gibbons *Hylobates* spp. and orangutans *Pongo pygmaeus.* Infection rates vary among populations within the same species. Although surveys of New World primates have not been exhaustive, there is no evidence of infection in these species (Lowenstine and Lerche 1993).

Distribution. Virus infection seems to be most prevalent in sub-Saharan Africa, Japan, Southeast Asia, and Indonesia.

Transmission and Epidemiology. The exact mode of transmission of STLV-1 is unknown, but the sexual route and maternal transmission to offspring in milk have been postulated (Miyoshi et al. 1983a). In primate centers, inoculation of individuals with blood and other biologic material from infected animals may establish infection (Voevodin et al. 1996).

Prevalence of antibodies was as high as 60% in wild olive baboons *Papio anubis,* 50% in Japanese macaques, and 45% in hamadryas baboons *Papio hamadryas* (Hubbard et al. 1993). The two large outbreaks of malignant lymphoma known in captive baboons may have been the result of interspecies transmission of macaque STLV-1, one outbreak possibly resulting from the importation of Vietnamese rhesus macaques in the early 1960s (Voevodin et al. 1996). Therefore, it is possible STLV-1 can cross genus barriers and exhibit a higher degree of oncogenicity in the new host.

Clinical Signs. Although infection with STLV-1 is widespread, other than the outbreaks of lymphoma in captive baboons, clinical illness in nonhuman primates infected with STLV-1 seems limited.

Persistent lymphocytosis and abnormal circulating lymphocytes have been reported in antibody-positive African green monkeys, and T-cell lymphomas and leukemias associated with STLV-1 infection have occurred in several species (Lowenstine and Lerche 1993). A chronic wasting syndrome was described in three captive lowland gorillas seropositive for STLV-1. Although lymphoid depletion was noted in two of the three animals, STLV-1 was not proven as the cause of this syndrome (Blakeslee et al. 1987).

In one outbreak of STLV-1-associated non-Hodgkin's lymphoma (NHL) in baboons, radiography revealed multifocal, variably sized opacities consistent with either neoplastic or fungal disease in the lungs of 17 animals, and nodular skin lesions were found in two individuals (Hubbard et al. 1993).

Pathology. In an outbreak of NHL in baboons, necropsy usually revealed generalized lymphadenopathy but, in some cases, there was only regional node enlargement and, in others, none (Hubbard et al. 1993). Histologic findings in lymph node biopsy specimens were diagnostically and morphologically consistent with NHL. There was increased lung weight, patchy pulmonary consolidation, and pleural thickening with fibrin tags. Microscopic lung lesions varied from neoplastic lymphocytic infiltrates in perivascular locations and around airways, to diffuse infiltration of lung parenchyma by neoplastic and inflammatory cells, edema, and necrosis.

Diagnosis. Screening for antibodies to STLV-1 can be done by commercial ELISA using HTLV-1 as the antigen. Positive samples should be confirmed by Western blots. A definitive diagnosis depends on virus isolation in cultured T cells.

Immunity. Infected animals develop antibodies to the structural proteins of the virus. This can be detected by Western blotting or radioimmunoprecipitation analysis. Virus continues to be expressed in the presence of antibody (Lowenstine 1993b).

Control. There are no vaccines against STLV-1 infection. This may be due to the low prevalence of disease in HTLV-1-seropositive humans and, therefore, the lack of urgency in developing animal models for vaccine production.

Public Health Concerns. Although it is assumed that there is no direct transmission of STLV-1 to humans, the transmission of rhesus macaque STLV-1 to baboons, with a higher degree of oncogenicity in the new host, suggests that a potential biohazard for humans cannot be discounted (Voevodin et al. 1996).

Management Implications. As with SIV, the risk of cross-species transmission of STLV-1 dictates that African and Asian primates should not be housed together. Because some cases of T-cell lymphomas and leukemias have been reported in African primates, colonies of animals should be monitored for evidence of antibody.

GIBBON APE LEUKEMIA VIRUS. Like FeLV, gibbon ape leukemia virus (GaLV) belongs to the mammalian type-C oncovirus serogroup of the mammalian type-C retroviruses. Both white-handed gibbons *Hylobates lar* and white-cheeked gibbons *Hylobates concolor* have been identified as hosts (Lowenstine and Lerche 1988). Virus has been isolated from white-handed gibbons that had hemolymphatic malignancies including lymphosarcoma, lymphoblastic leukemia, and myelogenous leukemia, as well as from apparently healthy individuals. A serologic survey of captive gibbons determined that seropositivity was highest in colonies in which gibbons with leukemia or lymphoma were present (Kawakami et al. 1973).

Antibody-positive gibbons are generally virus culture negative (Kawakami et al. 1972), but it has not been determined by newer techniques such as PCR whether all virus culture-negative individuals are truly virus negative. Virus-positive, antibody-negative animals can act as carriers and, although GaLV is shed in urine and feces, the exact mode of natural transmission remains uncertain (Kawakami et al. 1977).

Although GaLV can be grown in human lymphocyte cell lines, there is no evidence that the virus is transmissible to humans. Transmissibility to other species of nonhuman primates has not been studied. Testing for GaLV either by serologic methods, virus isolation, or PCR is not readily available.

KOALA RETROVIRUS. Lymphoid neoplasia is the most common form of tumor in free-ranging and captive koalas *Phascolarctos cinereus* (Spencer and Canfield 1996). Virus particles morphologically similar to that of oncoviruses (mammalian type-C viruses) have been detected in the bone marrow of a free-ranging koala with leukemia (Canfield et al. 1988). Retrovirus infection has been detected in a captive colony of koalas in which lymphoma/leukemia, osteochondroma, nonregenerative anemia, and opportunistic infections have been documented (Worley et al. 1993).

Virus was detected in short-term lymphocyte cultures from animals with disease, as well as from clinically normal animals. Further work is necessary to confirm the role of retrovirus infection in these syndromes in koalas.

RETROVIRUS IN SWEDISH MOOSE. Since 1985, more than 1000 wild Swedish moose *Alces alces* have developed a syndrome characterized by severe diarrhea, dehydration, emaciation, excessive salivation, licking, and teeth grinding. In addition, animals showed neurologic disturbances, with lowering of the head, drooping ears, hind-limb ataxia, impaired vision, and lack of fear of humans (Rehbinder et al. 1991). A striking feature is the abnormally low lymphocyte counts ($0.5–1.7 \times 10^9$/L) observed in approximately 44% of the diseased moose (Merza et al. 1994).

Cocultivation of a moose fetal kidney cell line with moose leukocytes resulted in cytopathic effect and the expression of retrovirus particles. They resemble mammalian type-C retroviruses morphologically, and polyclonal antibodies to several retroviruses reacted with purified moose retrovirus in an ELISA (Merza et al. 1994).

LITERATURE CITED

Ackley, C.D., J.K. Yamamoto, N. Levy, N.C. Pedersen, and M.D. Cooper. 1990. Immunologic abnormalities in pathogen-free cats experimentally infected with feline immunodeficiency virus. *Journal of Virology* 64:5652–5655.

Barr, M.C., P.P. Calle, M.E. Roelke, and F.W. Scott. 1989. Feline immunodeficiency virus infection in nondomestic felids. *Journal of Zoo and Wildlife Medicine* 20:265–272.

Barr, M.C., L. Zou, D.L. Holzschu, L. Phillips, F.W. Scott, J.W. Casey and R.J. Avery. 1995. Isolation of a highly cytopathic lentivirus from a nondomestic cat. *Journal of Virology* 69:7371–7374.

Blakeslee, J.R., H.M. McClure, D.C. Anderson, R.M. Bauer, L.Y. Huff, and R.G. Olsen. 1987. Chronic fatal disease in gorillas seropositive for simian T-lymphotropic virus I antibodies. *Cancer Letters* 37:1–6.

Boid, R., S. McOrist, T.W. Jones, N. Easterbee, A.L. Hubbard, and O. Jarrett. 1991. Isolation of FeLV from a wild felid (*Felis silvestris*). *Veterinary Record* 128:256.

Briggs, M.B., and R.L. Ott. 1986. Feline leukemia virus infection in a captive cheetah and the clinical and antibody response of six captive cheetahs to vaccination with a subunit feline leukemia virus vaccine. *Journal of the American Veterinary Medical Association* 189:1197–199.

Brody, B.A., E. Hunter, J.D. Kluge, R. Lasarow, M. Gardner, and P.A. Marx. 1992. Protection of macaques against infection with simian type-D retrovirus (SRV-1) by immunization with recombinant vaccinia virus expressing the envelope glycoproteins of either SRV-1 or Mason-Pfizer monkey virus (SRV-3). *Journal of Virology* 66:3950–3954.

Brown, E.W., S. Miththapala, and S.J. O'Brien. 1993. Prevalence of exposure to feline immunodeficiency virus in exotic felid species. *Journal of Zoo and Wildlife Medicine* 24:357–364.

Brown, E.W., N. Yuhki, C. Packer, and S.J. O'Brien. 1994. A lion lentivirus related to feline immunodeficiency virus: Epidemiologic and phylogenetic aspects. *Journal of Virology* 68:5953–5968.

Canfield, P.J., J.M. Sabine, and D.N. Love. 1988. Virus particles associated with leukemia in a koala. *Australian Veterinary Journal* 65:327–328.

Carpenter, M.A., and S.J. O'Brien. 1995. Coadaptation and immunodeficiency virus: Lessons from the Felidae. *Current Opinion in Genetics and Development* 5:739–745.

Carpenter, M.A., E.W. Brown, M. Culver, W.E. Johnson, J. Pecon-Slattery, D. Brousset, and S.J. O'Brien. 1996. Genetic and phylogenetic divergence of feline immunodeficiency virus in the puma (*Puma concolor*). *Journal of Virology* 70:6682–6693.

Centers for Disease Control. 1992. Seroconversion to simian immunodeficiency virus in two laboratory workers. *Morbidity Mortality Weekly Report* 41:678–681.

Citino, S.B. 1986. Transient FeLV viremia in a clouded leopard. *Journal of Zoo Animal Medicine* 17:5–7.

Clements, J.E., R.C. Montelaro, M.C. Zink, A.M. Amedee, S. Miller, A.M. Trichel, B. Jagerski, D. Hauer, L.N. Martin, R.P. Bohm, and M. Murphey-Corb. 1995. Cross-protective immune responses induced in rhesus macaques by immunization with attenuated macrophage-tropic simian immunodeficiency virus. *Journal of Virology* 69:2737–2744.

Coffin, J.M. 1996. Retroviruses: The viruses and their replication. In *Fields virology,* ed. B.N. Fields, D.M. Knipe, P.M. Howley, R.M. Chanock, J.L. Melnick, T.P. Monath, B. Roizman, and S.E. Straus. Philadelphia: Lippincott-Raven, pp. 1767–1848.

Daniel, M.D., N.L. Letvin, P.K. Sehgal, D.K. Schmidt, D.P. Silva, K.R. Solomon, F.S. Hodi, D.J. Ringler, R. Hunt, N.W. King, and R.C. Desrosiers. 1988. Prevalence of antibodies to three retroviruses in a captive colony of macaque monkeys. *International Journal of Cancer* 41:601–608.

Gardner, M.B. 1996. The history of simian AIDS. *Journal of Medical Primatology* 25:148–157.

Hardy, W.D. 1991. General principles of retrovirus immunodetection tests. *Journal of the American Veterinary Medical Association* 199:1282–1287.

Hardy, W.D., and E.E. Zuckerman. 1991. Ten-year study comparing enzyme-linked immunosorbent assay with the immunofluorescent antibody test for detection of feline leukemia virus infection in cats. *Journal of the American Veterinary Medical Association* 199:1365–1373.

Henrickson, R.V., D.H. Maul, N.W. Lerche, K.G. Osborn, L.J. Lowenstine, S. Prahalada, J.L. Sever, D.L. Madden, and M.B. Gardner. 1984. Clinical features of simian acquired immunodeficiency syndrome (SAIDS) in rhesus monkeys. *Laboratory Animal Science* 34:140–145.

Hoover, E.A., and J.I. Mullins. 1991. Feline leukemia virus infection and diseases. *Journal of the American Veterinary Medical Association* 199:1287–1297.

Hosie, M.J., and J.N. Flynn. 1996. Feline immunodeficiency virus vaccination: Characterization of the immune correlates of protection. *Journal of Virology* 70:7561–7568.

Hu, S.-L., K. Abrams, G.N. Barber, P. Moran, J.M. Zarling, A.J. Langlois, L. Kuller, W.R. Morton, and R.E. Benveniste. 1992. Protection of macaques against SIV infection by subunit vaccines of SIV envelope glycoprotein gp160. *Science* 255:456–459.

Hubbard, G.B., J.P. Moné, J.S. Allan, K.J. Davis III, M.M. Leland, P.M. Banks, and B. Smir. 1993. Spontaneously generated non-Hodgkin's lymphoma in twenty-seven simian T-cell leukemia virus type 1 antibody-positive baboons (*Papio* species). *Laboratory Animal Science* 43:301–309.

Jackson, M.L., D.M. Haines, S.M. Taylor, and V. Misra. 1996. Feline leukemia virus detection by ELISA and PCR in peripheral blood from 68 cats with high, moderate, or low suspicion of having FeLV-related disease. *Journal of Veterinary Diagnostic Investigation* 8:25–30.

Jessup, D.A., C. Pettan, L.J. Lowenstine, and N.C. Pedersen. 1993. Feline leukemia virus infection and renal spirochetosis in a free-ranging cougar (*Felis concolor*). *Journal of Zoo and Wildlife Medicine* 24:73–79.

Jordan, H.L., J. Howard, R.K. Sellon, D.E. Wildt, W.A. Tompkins, and S. Kennedy-Stoskopf. 1996. Transmission of feline immunodeficiency virus in domestic cats via artificial insemination. *Journal of Virology* 70:8224–8228.

Kawakami, T.G., S.D. Huff, P.M. Buckley, D.L. Dungworth, and S.P. Snyder. 1972. C-type virus associated with gibbon lymphosarcoma. *Nature New Biology* 235:170–171.

Kawakami, T.G., P.M. Buckley, and T.S. McDowell. 1973. Antibodies to simian type-C virus antigen in sera of gibbons (*Hylobates lar*). *Nature New Biology* 246:105–107.

Kawakami, T.G., L. Sun, and T.S. McDowell. 1977. Infectious C-type virus shed by healthy gibbons. *Nature* 268:448–449.

Kennedy-Stoskopf, S., D.H. Gebhard, R.V. English, L.H. Spelman, and M. Briggs. 1994. Clinical implications of feline immunodeficiency virus infection in African lions (*Panthera leo*): Preliminary findings. In *Proceedings of the American Association of Zoo Veterinarians.* Pittsburgh: American Association of Zoo Veterinarians, pp. 345–346.

King, N.W. 1993. Simian immunodeficiency virus infections. In *Nonhuman primates I,* ed. T.C. Jones, U. Mohr, and R.D. Hunt. New York: Springer-Verlag, pp. 5–20.

Langley, R.J., V.M. Hirsch, S.J. O'Brien, D. Adger-Johnson, R.M. Goeken, and R.A. Olmsted. 1994. Nucleotide sequence analysis of puma lentivirus (PLV-14): Genomic organization and relationship to other lentiviruses. *Virology* 202:853–864.

Lerche, N.W. 1993. Emerging viral diseases of nonhuman primates in the wild. In *Zoo and wild animal medicine: Current Therapy 3,* ed. M.E. Fowler. Philadelphia: W.B. Saunders, pp. 340–344.

Lerche, N.W., K.G. Osborn, P.A. Marx, S. Prahalada, D.H. Maul, L.J. Lowenstine, R.J. Munn, M.L. Bryant, R.V. Henrickson, L.O. Arthur, R.V. Gilden, C.S. Barker, E. Hunter, and M.B. Gardner. 1986. Inapparent carriers of simian AIDS type D retrovirus and disease transmission with saliva. *Journal of the National Cancer Institute* 77:489–496.

Lerche, N.W., P.A. Marx, K.G. Osborn, D.H. Maul, L.J. Lowenstine, M.L. Bleviss, P. Moody, R.V. Henrickson, and M.B. Gardner. 1987. Natural history of endemic type D retrovirus infection and acquired immune deficiency syndrome in group-housed rhesus monkeys. *Journal of the National Cancer Institute* 79:847–854.

Loar, A.S. 1993. Feline leukemia virus: Immunization and prevention. *Veterinary Clinics of North America: Small Animal Practice* 23:193–212.

Lowenstine, L.J. 1993a. Type D retrovirus infection, macaques. In *Nonhuman primates I,* ed. T.C. Jones, U. Mohr, and R.D. Hunt. New York: Springer-Verlag, pp. 20–32.

———. 1993b. Lymphotropic and immunosuppressive retroviruses of nonhuman primates: A review and update. In *Proceedings of the annual meeting of the American Association of Zoo Veterinarians.* St. Louis: American Association of Zoo Veterinarians, pp. 51–60.

Lowenstine, L.J. and N.W. Lerche. 1988. Retrovirus infections of nonhuman primates: A review. *Journal of Zoo Animal Medicine* 19:168–187.

———. 1993. Nonhuman primate retroviruses and simian acquired immunodeficiency syndrome. In *Zoo and wild animal medicine: Current Therapy 3,* ed. M.E. Fowler. Philadelphia: W.B. Saunders, pp. 373–378.

Lutz, H., E. Isenbügel, R. Lehmann, R. H. Sabapara, and C. Wolfensberger. 1992. Retrovirus infections in nondomestic felids: Serological studies and attempts to isolate a lentivirus. *Veterinary Immunology and Immunopathology* 35:215–224.

Mansfield, K.G., N.W. Lerche, M.B. Gardner, and A.A. Lackner. 1995. Origins of simian immunodeficiency virus infection in macaques at the New England Regional Primate Research Center. *Journal of Medical Primatology* 24:116–122.

Marx, P.A., N.C. Pedersen, N.W. Lerche, K.G. Osborn, L.J. Lowenstine, A.A. Lackner, D.H. Maul, H.-S. Kwang, J.D. Kluge, C.P. Zaiss, V. Sharpe, A.P. Spinner, A.C. Allison, and M.B. Gardner. 1986. Prevention of simian acquired immune deficiency syndrome with a formalin-inactivated type D retrovirus vaccine. *Journal of Virology* 60:431–435.

Meric, S.M. 1984. Suspected feline leukemia virus infection and pancytopenia in a western cougar. *Journal of the American Veterinary Medical Association* 185:1390–1391.

Merza, M.E. Larsson, M. Stéen, and B. Morein. 1994. Association of a retrovirus with a wasting condition in the Swedish moose. *Virology* 202:956–961.

Miller, C.J., N.J. Alexander, S. Sutjipto, A.A. Lackner, A. Getti, A.A. Hendrickx, L.J. Lowenstine, M. Jennings, and P.A. Marx. 1989. Genital mucosal transmission of simian immunodeficiency virus: Animal model for heterosexual transmission of human immunodeficiency virus. *Journal of Virology* 63:4277–4284.

Miyoshi, L., M. Fujishita, and H. Taguchi. 1983a. Horizontal transmission of adult T-cell leukemia virus from male to female Japanese monkey. *Lancet* 2:241–242.

Miyoshi, I., S. Yoshimoto, M. Fujishita, Y. Ohtsuki, H. Taguchi, Y. Shiraishi, T. Agaki, and M. Minezawa. 1983b. Isolation in culture of a type C virus from a Japanese monkey seropositive to adult T-cell leukemia-associated antigens. *Gann* 74:323–326.

Olmsted, R.A., R. Langley, M.E. Roelke, R.M. Goeken, D. Adger-Johnson, J.P. Goff, J.P. Albert, C. Packer, M.K. Laurenson, T.M. Caro, L. Scheepers, D.E. Wildt, M. Bush, J.S. Martenson, and S.J. O'Brien. 1992. Worldwide prevalence of lentivirus infection in wild feline species: Epidemiologic and phylogenetic aspects. *Journal of Virology* 66:6008–6018.

Osofsky, S.A., W.D. Hardy, and K.J. Hirsch. 1994. Serologic evaluation of free-ranging lions (*Panthera leo*), leopards (*Panthera pardus*) and cheetahs (*Acinonyx jubatus*) for feline lentivirus and feline leukemia virus in Botswana. In *Proceedings of the American Association of Zoo Veterinarians.* Pittsburgh: American Association of Zoo Veterinarians, pp. 398–402.

Pedersen, N.C. 1988. Feline leukemia virus infection. In *Feline infectious diseases,* ed. N.C. Pedersen. Goleta, CA: American Veterinary, pp. 83–106.

Pedersen, N.C., E.W. Ho, M.L. Brown, and J.K. Yamamoto. 1987. Isolation of a T-lymphotropic virus from domestic cats with an immunodeficiency-like syndrome. *Science* 235:790–793.

Rasheed, S., and M.B. Gardner. 1981. Isolation of feline leukemia virus from a leopard cat cell line and search for retrovirus in wild felidae. *Journal of the National Cancer Institute* 67:929–933.

Rehbinder, C., E. Gimeno, K. Belák, S. Belák, M. Stéen, E. Rivera, and T. Nikkilä. 1991. A bovine viral diarrhea/mucosal disease-like syndrome in moose (*Alces alces*): Investigations on the central nervous system. *Veterinary Record* 129:552–554.

Sellon, R.K., H.L. Jordan, S. Kennedy-Stoskopf, M.B. Tompkins, and W.A.F. Tompkins. 1994. Feline immunodeficiency virus can be experimentally transmitted via milk during acute maternal infection. *Journal of Virology* 68:3380–3385.

Sparger, E.E. 1993. Current thoughts on feline immunodeficiency virus infection. In *Feline infectious diseases. Veterinary Clinics of North America: Small Animal Practice,* ed. J.D. Hoskins and A.S. Loar. Philadelphia: W.B. Saunders, pp. 173–191.

Sparkes, A.H., C.D. Hopper, W.G. Millard, T.J. Gruffydd-Jones, and D.A. Harbour. 1993. Feline immunodeficiency virus infection: Clinicopathologic findings in 90 naturally occurring cases. *Journal of Veterinary Internal Medicine* 7:85–90.

Spencer, A.J., and P.J. Canfield. 1996. Lymphoid neoplasia in the koala (*Phascolarctos cinereus*): A review and classification of 31 cases. *Journal of Zoo and Wildlife Medicine* 27:303–314.

Spencer, J.A., A.A. van Dijk, M.C. Horzinek, H.F. Egberink, R.G. Bengis, D.F. Keet, S. Morikawa, and D.H.L. Bishop. 1992. Incidence of feline immunodeficiency virus reactive antibodies in free-ranging lions of the Kruger National Park and the Etosha National Park in southern Africa detected by recombinant FIV p24 antigen. *Onderstepoort Journal of Veterinary Research* 59:315–322.

Torten, M., M. Franchini, J.E. Barlough, J.W. George, E. Mozes, H. Lutz, and N.C. Pedersen. 1991. Progressive immune dysfunction in cats experimentally infected with feline immunodeficiency virus. *Journal of Virology* 65:2225–2230.

Voevodin, A., E. Samilchuk, H. Schatzl, E. Boeri, and G. Franchini. 1996. Interspecies transmission of macaque simian T-cell leukemia/lymphoma virus type 1 in baboons resulted in an outbreak of malignant lymphoma. *Journal of Virology* 70:1633–1639.

Worley, M., B. Rideout, A. Shima, and D. Janssen. 1993. Opportunistic infections, cancer and hematologic disorders associated with retrovirus infection in the koala. In *Proceedings of the American Association of Zoo Veterinarians.* St. Louis: American Association of Zoo Veterinarians, pp. 181–182.

Wyand, M.S., K.H. Manson, M. Garcia-Moll, D. Montefiori, and R.C. Desrosiers. 1996. Vaccine protection by a triple deletion mutant of simian immunodeficiency virus. *Journal of Virology* 70:3724–3733.

Yamamoto, J.K., E. Sparger, E.W. Ho, P.R. Andersen, T.P. O'Connor, C.P. Mandell, L. Lowenstine, R. Munn, and N.C. Pedersen. 1988. Pathogenesis of experimentally induced feline immunodeficiency virus infection in cats. *American Journal of Veterinary Research* 49:1246–1258.

Yamamoto, J.K., T. Hohdatsu, R.A. Olmsted, R. Pu, H. Louie, H.A. Zochlinski, V. Azevedo, H.M. Johnson, G.A. Soulds, and M.B. Gardner. 1993. Experimental vaccine protection against homologous and heterologous strains of feline immunodeficiency virus. *Journal of Virology* 67:601–605.

11

PAPILLOMAVIRUS INFECTIONS

JOHN P. SUNDBERG, MARC VAN RANST, AND A. BENNETT JENSON

Synonyms: Warts, papillomas, fibromas, fibropapillomas, squamous cell carcinomas, adenosquamous carcinomas, and sebaceous carcinomas.

INTRODUCTION. Papillomas or warts are easily recognized skin or mucous membrane tumors. They are common in wild species and are particularly well known in deer (Cosgrove and Fay 1981; Sundberg and Nielson 1981). Most of these tumors are caused by papillomaviruses (PVs). Tumors can become quite large and numerous and persist for prolonged periods, but they usually regress, providing the individual with long-term immunity to future infections. Occasionally, the tumors will progress to squamous cell carcinoma, adenosquamous carcinoma, or sebaceous carcinoma that will eventually kill the host if not treated. Specific mucosotropic PV types and those infecting immunodeficient individuals are prone to induce malignancies. Virtually all mammalian species are hosts for one or more PV. This chapter provides an overview of PV infections in mammals.

HISTORY. Papillomas of the skin in animals are easily recognized and therefore have been described for centuries. The stablemaster for the Caliph of Baghdad described warts in the horse *Equus caballus* in the 9th century (Erk 1976). McFadyean and Hobday (1898) successfully transmitted canine oral papillomatosis in 1898, and Cadeac (1901) transmitted warts from horse to horse in 1901. James Watson, codiscoverer of the structure of DNA, was the first to study the DNA of the cottontail rabbit PV (CRPV), determining that this virus contained double-stranded DNA (Watson and Littlefield 1960). Crawford and Crawford (1963) reported the first characterization of a human PV (HPV) in its supercoiled, circular, and linear forms. Through the 1960s and early 1970s, it was still assumed that all wartlike lesions in humans and animals were caused by a single, essentially species-specific PV in which the clinical features were determined by the anatomic location of the lesion (Rowson and Mahy 1967). Through advances in molecular techniques, including DNA cloning methods with restriction endonucleases, ligases, and plasmid vectors, the secrets of these viruses were slowly revealed. The HPV diversity has changed from five types when Law et al. (1979) talked about the "remarkable plurality of the HPVs" to the over 80 now known for this host species alone (Van Ranst et al. 1996). The same observations have been described for domestic and wild animals within the last decade (Olson 1987; Sundberg 1987; Sundberg et al. 1996b, 1997a).

DISTRIBUTION. Papillomaviruses have a worldwide distribution. Specific viral types and subtypes in human populations have been traced and used as a method to determine population migration patterns (Ho et al. 1993; Bernard et al. 1994; Van Ranst et al. 1995).

HOST RANGE. At least 50 mammalian species have been confirmed to be infected by species-specific PVs (Table 11.1). Papillomavirus divergence appears to

Table 11.1—Phylogenetic relationship of mammalian hosts infected with papillomaviruses

Order/Family	Number of Papillomaviruses	Types Cloned and Characterized
Marsupialia		
Didelphidae	1	0
Edentata		
Dasypodidae	1	0
Primates		
Cercopithecidae	3	3
Pongidae	2	2
Hominidae	80+	80+
Carnivora		
Canidae	2+	2
Ursidae	1	0
Procyonidae	1	0
Felidae	8	3
Cetacea		
Physeteridae	2	0
Proboscidea		
Elephantidae	1	0
Perissodactyla		
Equidae	2	1
Artiodacyla		
Cervidae	6	5
Giraffidae	1	0
Antilocapridae	1	0
Bovidae	9	6
Rodentia		
Castroidae	1	0
Muridae	2	2
Lagomorpha		
Leporidae	2	2
Total	125+	106+

Reprinted and expanded from Sundberg et al. 1997.

have followed phylogenetic evolution of the host species. For example, although some families, notably Felidae, have few reports of PV infections, within the last few years over eight new feline PVs have been identified (Carney et al. 1990; Sundberg et al. 1996a,b; J.P. Sundberg unpublished). Studies are in progress to compare divergence of the genome of these viruses relative to the well-characterized evolution of cats (O'Brien et al. 1985, 1997).

ETIOLOGY. Papillomaviruses are double-stranded DNA viruses that consist of a nonenveloped icosahedral capsid structure that is 50–55 nm in diameter (Pfister 1987) (Fig. 11.1). Their genome is 7.4–8.6 kb in size, depending on the species. The genome consists of the long control region, formerly called the noncoding region and later the upstream regulatory region. Early genes are responsible for viral DNA replication, transcription control, and cellular transformation (Banks and Matlashewski 1996; Crook and Vousden 1996; Doorbar 1996; Thierry 1996). The late genes code for the major and minor capsid proteins, including the various antigenic epitopes.

Nomenclature for nonhuman PVs has been used loosely. Several PVs have been studied for over 50 years, and their designations are well established in the literature. These include PVs that infect cattle *Bos taurus* (bovine PVs, BPVs) and cottontail rabbits *Sylvilagus floridanus* (CRPV). Since Richard Shope published the first report of CRPV (Shope and Hurst 1933), CRPV is often referred to as Shope PV. Another case that falls into this group is the canine oral PV, designated COPV. In addition to having been in common use for many years, COPV has been found to infect two closely related canine species: domestic dogs *Canis familiaris* and coyotes *Canis latrans* (Samuel et al. 1978; Sundberg et al. 1991).

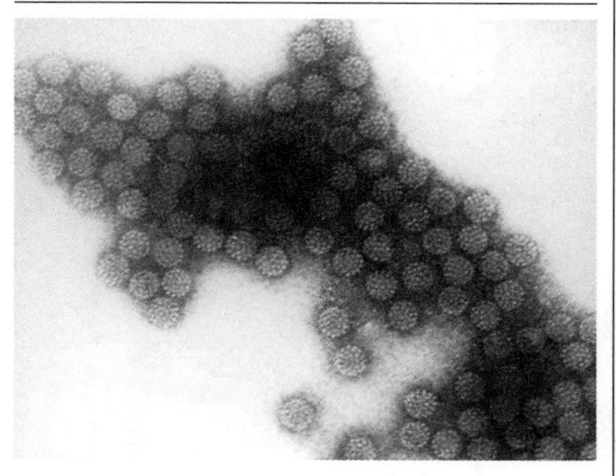

FIG 11.1—Purified white-tailed deer papillomaviruses in a negative-stained preparation are round, nonenveloped, and 55 nm in diameter.

With their work on the multimammate rat *Mastomys natalensis* PV, Muller and Gissmann (1978) set a precedent for naming PVs infecting uncommon hosts. They named the virus after the scientific name of the host species, thus MnPV. A year later, Coggin and zur Hausen (1979) proposed use of the first two letters of the common name of the host species in a similar manner. Based on these papers, the nonhuman PV nomenclature can be unified by using the scientific or common name of the host for the viral designation (Sundberg et al. 1996b). For example, the European harvest mouse *Micromys minutus* PV (Sundberg et al. 1987, 1988; O'Banion et al. 1988) is designated MmPV because it was reported first, and the rhesus macaque *Macaca mulatta* PV (Kloster et al. 1988) is designated RhPV based on its common name. Once molecular criteria (less than 50% homology) have been met that differentiate two distinct PVs that infect a single host species (Coggin and zur Hausen 1979), the viruses are numbered sequentially in the chronological order in which they were identified.

TRANSMISSION AND EPIDEMIOLOGY. Direct contact with lesions or cornified cells embedded on commonly used scratching posts is considered to be the common mode of transmission. Friend (1967) speculated that sparring among white-tailed deer *Odocoileus virginianus* bucks during the rut served to spread PV among deer. Infected cornified cells containing millions of virions become embedded in small skin wounds. Enzymes in the skin break down the cells, releasing the virions that attach to epithelial cells, probably basal cells, via surface receptors thought to be for normal cytokines.

Transmission usually is among individuals of the same species or phylogenetically closely related species. The best example of cross-species transmission is that of CRPV, which causes mild, productive, benign papillomas in cottontail rabbits, the natural host, but when injected into domestic rabbits *Oryctolagus cuniculus* this virus causes malignant, nonproductive, squamous cell carcinomas (Sundberg and O'Banion 1989). The few known exceptions to intraspecific transmission are listed in Table 11.2. The PVs isolated from nonhuman primates, although different, are remarkably similar to HPVs at the molecular level (Reszka et al. 1991; Sundberg et al. 1992; Van Ranst et al. 1992). As more nonhuman primate PVs are identified and studied (Sundberg and Reichmann 1993), particularly from those hosts phylogenetically closely related to humans, such as pygmy chimpanzees *Pan paniscus* (Sundberg et al. 1992; Van Ranst et al. 1991, 1992), they might prove to be infectious for humans.

CLINICAL SIGNS. Lesions caused by PVs are evident on the surface of the skin and mucous membranes. Animals may become weak and debilitated when the tumor burden becomes large or when the tumors interfere with vision or glutition based on anatomic location.

TABLE 11.2—Papillomaviruses that infect more than one species

Natural Host	Virus	Experimentally Infected Host	References
Domestic cattle *Bos taurus*	BPV-1, BPV-2	Horse *Equus caballus*	Segre et al. 1955
		Hamster *Mesocrietus auratus*	Freidman et al. 1963; Koller and Olson 1972
		Pica *Onchotono rufescense*	Puget et al. 1975
		Mice *Mus musculus*	Boiron et al. 1964; Freidman et al. 1965
Cottontail rabbit *Sylvilagus floridans*	CRPV	Domestic rabbit *Oryctolagus cuniculus*	Rous and Beard 1935
Domestic rabbit	OcPV	Cottontail rabbit	Parsons and Kidd 1936
		Hamster	Sundberg et al. 1985c
White-tailed deer *Odocoileus virginianus*	OvPV	Hamster	Koller and Olson 1972
Moose *Alces alces*	AaPV	Hamster	Stenlund et al. 1983
Domestic sheep *Ovis aries*	OaPV	Hamster	Gibbs et al. 1975
Coyote *Canis latrans*	COPV	Dog *Canis familiaris*	Sundberg et al. 1991
Human	HPV-?	Baboon *Papio papio*	Atanasiu 1948

From Sundberg et al. 1997.

PATHOLOGY. Tumors caused by PVs can be exophytic (above the level of the skin or mucous membrane) or endophytic (invaginate below the surface); smooth surfaced or verrucated (convoluted surface similar to that of a cauliflower head); unpigmented or of various colors ranging to black; and firm in consistency (Figs. 11.2–11.5). Size ranges vary dramatically from very small and difficult to identify on the tongues of rabbits or penises of sperm whales *Physeter catodon* (Lambertsen et al. 1987) to large, numerous, and sometimes confluent as seen on ruminants with immunodeficiencies (Sundberg and Nielsen 1980, 1981).

The PVs induce both lytic and proliferative changes in cells, most notably in keratinocytes. Tumors commonly arise on the haired and glabrous skin and mucous membranes, all of which are covered by stratified squamous epithelium. Benign tumors are classified as papillomas (Figs. 11.6 and 11.7), fibropapillomas (Fig. 11.8), and fibromas (Fig. 11.9), depending on the ratio of fibrous connective tissue to hyperplastic epithelium. It is notable that ruminants are the primary group that develop fibropapillomas and fibromas. Malignant tumors are primarily squamous cell carcinomas. Sebaceous carcinomas are associated with infection in several species, primarily of cattle and wild mice. The morphologic features of these tumors vary in different host species (Sundberg 1987).

Productive infections, in which virions are assembled within infected cells, are characterized by lytic changes. The subtle cytopathology of PV infection varies with each specific virus type, but the general features of lytic changes include (1) cytoplasmic swelling, (2) loss of cytoplasmic staining, (3) cytoplasmic keratohyalin-like granules that are often large and hyperchromatic in the stratum granulosum, and (4) solitary or multiple, intranuclear, amphophilic inclusions. These cells may be solitary or in small clusters and are limited to the upper stratum spinosum and throughout the stratum granulosum. Cells exhibiting this type of cytopathology have been called koilocytes (hollow

FIG 11.2—Black, exophytic fibromas on a white-tailed deer [Reprinted with permission from Sundberg (1987)].

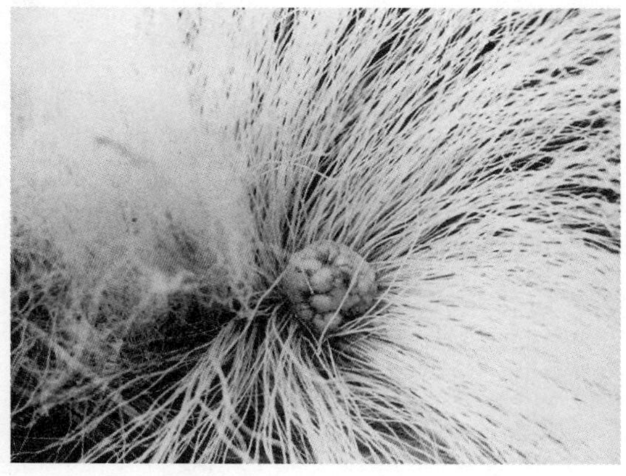

FIG 11.3—White fibroma in the unpigmented tail hair of a white-tailed deer [Reprinted with permission from Sundberg (1987)].

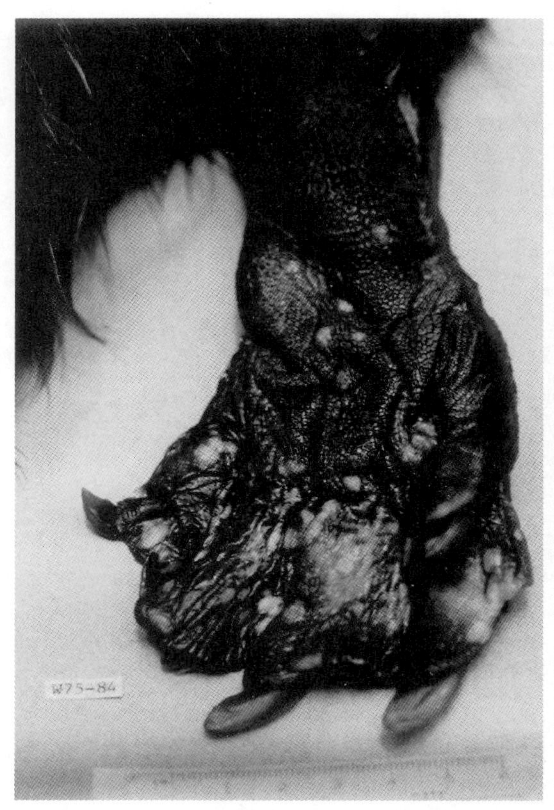

FIG 11.4—Sessile papillomas on the foot of a beaver [Reprinted by permission from Sundberg (1987)].

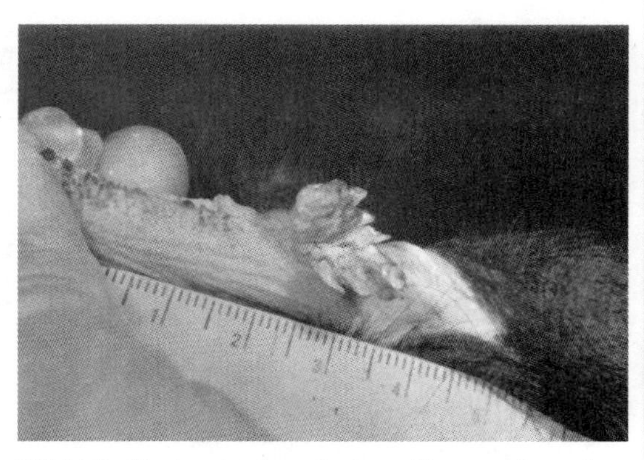

FIG 11.5—Unpigmented exophytic papilloma on the penis of a colobus monkey *Colobus guereza* [Reprinted with permission from Sundberg (1987), courtesy of Dr. Shima].

cells), pale cells, or clear cells (Fig. 11.7) (Koss 1987; Sundberg 1987). Immunohistochemistry is useful for identification of PV antigen in hyperplastic and neoplastic lesions (Sundberg et al. 1984; Lim et al. 1990).

Lesion regression is associated with infiltration of lymphocytes that begins around blood vessels, leading

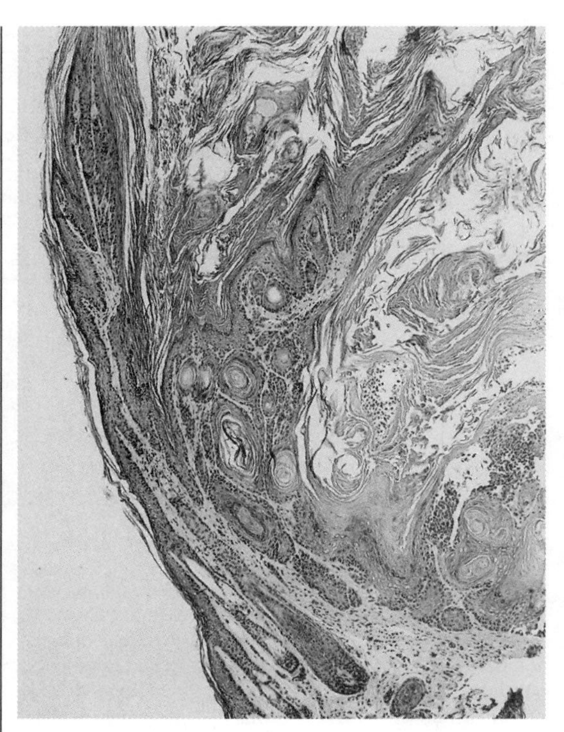

FIG 11.6—Cutaneous papilloma with marked cornification forming a horn in a European harvest mouse [Reprinted with permission from Sundberg et al. (1987)].

to hyalinization of the connective tissue and sloughing of tumors (Sundberg et al. 1985a).

Ultrastructural features vary with the host and viral type involved. However, the common feature to all productive infections of diagnostic importance is finding clusters of uniformly sized virus-like particles in and around the nucleolus in early infections that progress to formation of large crystalline arrays of virions in late infections (Fig. 11.10). These findings are limited to the upper layers of the stratum spinosum, stratum granulosum, and stratum corneum. Virions are never found in the cytoplasm of normal keratinocytes. Infected cells in the stratum granulosum, the koilocytes, undergo degenerative changes and develop abnormal keratohyalin-like granules, the specific changes of which vary with virus type (Sundberg et al. 1985b).

DIAGNOSIS. The gross and histologic diagnosis is usually relatively straightforward. However, determination of whether the lesion is associated with PV infection, of whether the infection is productive, of whether virus or viral DNA are present, and of which type is present requires specialized tests not generally available to diagnosticians.

A large number of monoclonal and polyclonal antibodies have been generated to study the PVs (Lim et al. 1990; Jenson et al. 1997). Polyclonal antibodies are available commercially (DAKO Corporation, Carpin-

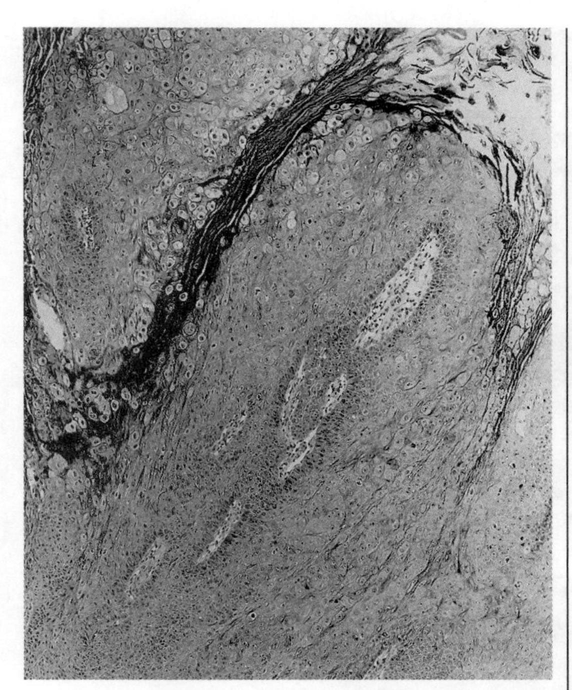

FIG 11.7—Oral papilloma in a dog exhibiting koilocytotic atypia beneath the stratum corneum [Reprinted with permission from Sundberg (1987)].

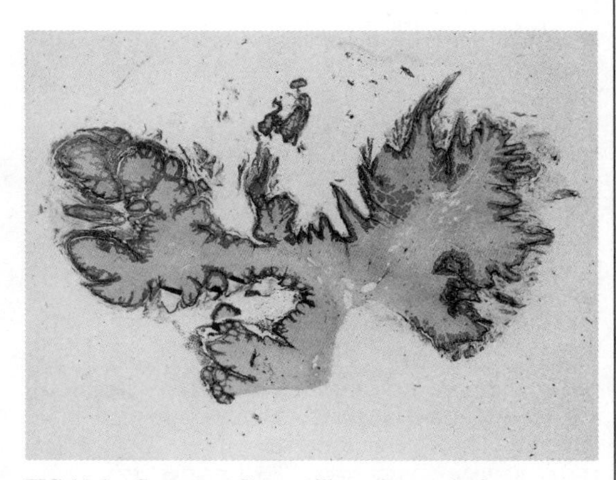

FIG 11.8—Cutaneous fibropapilloma in a mule deer *Odocoileus heminus* [Reprinted with permission from Sundberg and Lancaster (1988)].

teria, California) and are routinely used to screen tissue sections for evidence of productive infections. Monoclonal antibodies can be used in a panel to evaluate and partially type papillomas by immunohistochemistry (Fig. 11.11) (Lim et al. 1990; Sundberg et al. 1996b). This can be useful for specimens for which no frozen tissues are available for molecular studies.

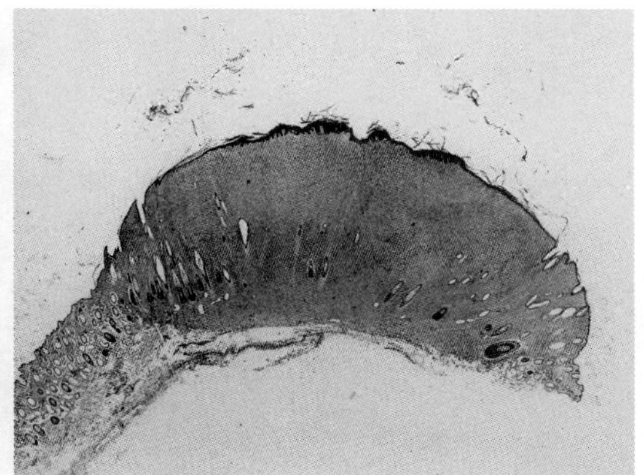

FIG 11.9—Cutaneous fibroma in a white-tailed deer [Reprinted with permission from Sundberg and Lancaster 1988)].

Tumors containing epithelium are the best sources of viral DNA for molecular typing and cloning of the viral genomes of novel PVs. Frozen tumors provide the best starting materials. Papillomaviruses are very stable in the environment; therefore, –20° C household freezers are adequate to preserve the viruses if –80° C freezers are unavailable. For nonhuman PVs, viral typing is limited primarily to research laboratories

DIFFERENTIAL DIAGNOSES. The tumor type is a morphologic, not an etiologic, diagnosis. Histopathology is required to arrive at this diagnosis. Based on cytologic features supported by immunohistochemistry and molecular studies, the etiology can be determined. The lesions discussed here may be caused by a number of other agents, including poxviruses, herpesviruses, adenoviruses, rhabdoviruses, retroviruses, anaerobic bacteria, chemical carcinogens, and genetic factors (Sundberg et al. 1996b, 1997b).

IMMUNITY. Once an animal is infected with a PV, the lesions persist for weeks to months but eventually 75%–80% regress, leaving the host immune to future infections. Regression of lesions is associated with development of virus-neutralizing antibodies that can prevent infection (Ghim et al. 2000). Antibodies also can be detected by enzyme-linked immunosorbent assay or hemagglutination inhibition in convalescent animals. Unfortunately, lesions linger in some individuals, and some carcinogenic PVs may persist in a latent state.

CONTROL AND TREATMENT. Most cases of PV infections are self-limiting and will eventually disappear. Thus, control and treatment are not needed in free-ranging populations. However, the disease and PVs tend to persist for weeks to months or longer in affected animals. Presence of viruses with known oncogenic potential in a particular species may justify

FIG 11.10—Crystalline array of virus-like particles in a mule deer fibroma [Reprinted with permission from Sundberg et al. (1985b)].

FIG 11.11—Papillomavirus capsid proteins can be identified by immunohistochemistry within the nucleus of koilocytes in the stratum granulosum of a dog oral papilloma [Reprinted with permission from Sundberg (1987)].

more aggressive intervention when early lesions advance to premalignant states.

Surgery, either in a classic sense by removal as a biopsy, laser ablation, or debulking tumors mechanically (removing warts, use of radon seeds, liquid nitrogen, or irritant chemicals) has been used with various degrees of success in domestic animals for decades. Scarring and depigmentation can be untoward side effects.

Prevention of PV-induced tumors by vaccination has been attempted with various degrees of success in domestic animals (Olson et al. 1968; Barthold et al. 1976; Bell et al. 1994; Ghim et al. 1995). However, vaccination during the early stages of infection may actually have prolonged fibropapillomatosis in cattle (Olson and Skidmore 1959).

PUBLIC HEALTH CONCERNS. Papillomas and fibromas on nonprimate wild species may be unsightly but are of little concern to humans. Because these tumors are on the skin, they are removed when a harvested animal is skinned and do not affect the quality of meat. Although PVs are thought to be species specific and therefore not a concern to clinicians or animal handlers, some of these viruses are transmissible between closely related species (Table 11.2). This raises a concern for people handling nonhuman primates in particular (Sundberg et al. 1992; Van Ranst et al. 1992).

DOMESTIC ANIMAL HEALTH CONCERNS. Papillomavirus infections may be transmitted between closely related species. Of concern is that the virus might be more virulent in the aberrant host than in the natural host. The COPV infects both domestic and wild canids, and some cases in dogs progress to squamous cell carcinoma (Sundberg et al. 1991; Sundberg and O'Banion 1989). The same situation occurs with CRPV transmitted between two species of rabbits (Sundberg and O'Banion 1989). Wild felids brought into captivity with PV infections pose the potential problem of infecting resident or feral cats (Sundberg et al. 1996a).

MANAGEMENT IMPLICATIONS. Because PV-induced warts and fibromas are usually self-limiting and will regress, they seldom cause significant problems to individual animals. If the location of the lesion is in a place where essential functions of the animal are compromised (unusually large tumors, tumors on the eylids that interfere with sight, or tumors on the mouth that interfere with feeding), the tumor may cause significant illness or death. As with many infectious diseases, separation of susceptible animals from infectious animals or contaminated premises may be effective in preventing disease, although the premises may remain contaminated for many years due to the environmental resistance of these viruses. Development of vaccines in the future could be effective in controlling disease in captive wild animals should a PV problem be recognized.

LITERATURE CITED

Atanasiu, P. 1948. Transmission de la verrue commune au singe cynocephale (*Papio papio*). *Annales de l'Institut Pasteur* 74:246–248.

Banks, L., and G. Matlashewski. 1996. Biochemical and biological activities of the HPV E5 proteins. In *Papillomavirus reviews: Current research on papillomaviruses,* ed. C. Lacey. Leeds, UK: Leeds University Press, pp. 39–45.

Barthold, S.W., C. Olson, and L.L. Larson. 1976. Precipitin response of cattle to commercial wart vaccine. *American Journal of Veterinary Research* 37:449–451.

Bell, J., J.P. Sundberg, S.-J. Ghim, J. Newsome, A.B. Jenson, and R. Schelgel. 1994. A formalin-inactivated vaccine protects against mucosal papillomavirus infection: A canine model. *Pathobiology* 62:194–198.

Bernard, H.-U., S.-Y. Chan, and H. Delius. 1994. Evolution of papillomaviruses. *Current Topics in Microbiology and Immunology* 186:33–54.

Boiron, M., J.P. Levy, M. Thomas, J.C. Friedman, and J.C. Bernard. 1964. Some properties of bovine papilloma virus. *Nature* 201:423–424.

Cadeac, M. 1901. Sur la transmission experimentale des papillomes des diverses especes. *Bulletin Societe des Sciences Veterinaires et de Medecine Comparee de Lyon* 4:280–286.

Carney, H.C., J.J. England, E.C. Hodgin, H.E. Whiteley, D.L. Adkison, and J.P. Sundberg. 1990. Papillomavirus infection of aged Persian cats. *Journal of Veterinary Diagnostic Investigation* 2:294–299.

Coggin, J.R., Jr., and H. zur Hausen. 1979. Workshop on papillomaviruses and cancer. *Cancer Research* 39:545–546.

Cosgrove, G.E., and L.D. Fay. 1981. Viral tumors. In *Infectious diseases of wild mammals,* ed. J.W. Davis, L.H. Karstad, and D.O. Trainer, 2d ed. Ames: Iowa State University Press, pp. 424–426.

Crawford, L.V., and E.M. Crawford. 1963. A comparative study of polyoma and papilloma viruses. *Virology* 21:258–263.

Crook, T., and K.H. Vousden. 1996. HPV oncoprotein function. In *Papillomavirus reviews: Current research on papillomaviruses,* ed. C. Lacey. Leeds, UK: Leeds University Press, pp. 55–60.

Doorbar, J. 1996. The E4 proteins and their role in the viral life cycle. In *Papillomavirus reviews: Current research on papillomaviruses,* ed. C. Lacey. Leeds, UK: Leeds University Press, pp. 31–38.

Erk, N. 1976. A study of Kitab al-Hail wal-Baitara by Muhammed Ibu ahi Hazam. *Historia Medicinae Veterinariae* 1:101–104.

Freidman, J.C., J.P. Levy, J. Lasneret, M. Thomas. M.M. Boiron, and J. Bernard. 1963. Induction de fibromes sous-cutanes chez le hamster dore par inoculation d'extraits acellulaires de papillomes bovins. *Comptes Rendus des Seances de l'Academie des Sciences [III]* 257:2328–2331.

Freidman, J.C., J. Lasneret, L. Gibeaux, and M. Boiron. 1965. Developement de fibromes proliferatifs chez la souris a l'aide d'extraits acellulares de papillomes bovins et leur transformation maligne greffes isologues. *Revue Medecine Veterinaire* 141:115–122.

Friend, M. 1967. Skin tumors in New York deer. *Bulletin of the Wildlife Disease Association* 3:102–104.

Ghim, S.-J., J. Suzich, J. Tamura, J.A. Bell, W. White, J. Newsome, F. Hill, P. Warrener, J. Sundberg, A.B. Jenson, and R. Schelgel. 1995. Formalin-inactivated oral papilloma extracts and recombinant L1 vaccines protect completely against mucosal papillomavirus infection: A canine model. In *Vaccines95: Molecular approaches to control of infectious disease,* ed. S.-J. Ghim, J. Newsom, A.B. Jenson, and R. Schlegel. Cold Spring Harbor, NY: Cold Spring Harbor, pp. 373–380.

Ghim, S.-J., J. Newsome, J. Bell, J.P. Sundberg, R. Schlegel, and A.B. Jenson. 2000. Spontaneously regressing oral papillomas induce systemic antibodies that neutralize canine oral papillomavirus. *Experimental and Molecular Pathology* 68:147–151.

Gibbs, E.P.J., C.J. Smale, and M.J.P. Lawman. 1975. Warts in sheep. *Journal of Comparative Pathology* 85:327–334.

Ho, L., S.Y. Chan, R.D. Burk, B.C. Das, K. Fujinaga, J.P. Icenogle, T. Kahn, N. Kiviat, W. Lancaster, P.P. Mavromara-Nazos, and M.V. Bernard. 1993. The genetic drift of human papillomavirus type 16 is a means of reconstructing viral spread and the movement of ancient human populations. *Journal of Virology* 67:6413–6423.

Jenson, A.B., M.C. Jenson, L. Cowsert, S.-J. Ghim, and J.P. Sundberg. 1997. Multiplicity of uses of monoclonal antibodies that define papillomavirus linear immunodominant epitopes. *Immunologic Research* 16:115–119.

Kloster, B.E., D.A. Manias, R.S. Ostrow, M.K. Shaver, S.W. McPherson, S.R.S. Rangen, and A.J. Faras. 1988. Molecular cloning and characterization of the DNA of two papillomaviruses from monkeys. *Virology* 166:30–40.

Koller, L.D., and C. Olson. 1972. Attempted transmission of warts from man, cattle, and horses, and of deer fibroma to selected hosts. *Journal of Investigative Dermatology* 58:366–368.

Koss, L.G. 1987. Cytologic and histologic manifestations of human papillomavirus infection of the female genital tract and their clinical significance. *Cancer* 60:1942–1950.

Lambertsen, R.H., B.A. Kohn, J.P. Sundberg, and C.D. Buergelt. 1987. Genital papillomatosis in sperm whale bulls. *Journal of Wildlife Diseases* 23:361–367.

Law, M.F., W.D. Lancaster, and P.M. Howley. 1979. Conserved polynucleotide sequences among the genomes of papillomaviruses. *Journal of Virology* 32:199–207.

Lim, P.S., A.B. Jenson, L. Cowsert, Y. Nakai, L.Y. Lim, X.W. Jin, and J.P. Sundberg 1990. Distribution and specific identification of papillomavirus major capsid protein epitopes by immunocytochemistry and epitope scanning of synthetic peptides. *Journal of Infectious Diseases* 162:1263–1269.

McFadyean, J., and F. Hobday. 1898. Note on the experimental transmission of warts in the dog. *Journal of Comparative Pathology and Therapeutics* 11:341–344.

Muller, H., and L. Gissmann 1978. *Mastomys natalensis* papillomavirus (MnPV), the causative agent of epithelial proliferations: Characterization of the virus particle. *Journal of General Virology* 41:315–323.

O'Banion, M.K., M.E. Reichmann, and J.P. Sundberg. 1988. Cloning and characterization of a papillomavirus associated with papillomas and carcinomas in the European harvest mouse (*Micromys minutus*). *Journal of Virology* 62:226–233.

O'Brien, S.J., M.E. Roelke, L. Marker, A. Newman, C.A. Winkler, D. Meltzer, L. Colly, J.P. Evermann, M. Bush, and D.E. Wildt. 1985. Genetic basis for species vulnerability in the cheetah. *Science* 227:1428–1434.

O'Brien, S.J., J. Wienbert, and L.A. Lyons. 1997. Comparative genomics: Lessons from cats. *Trends in Genetics* 13:393–399.

Olson, C. 1987. Animal papillomas: Historical perspectives. In *The papovaviridae.* Vol. 2, *The papillomaviruses,* ed. N.P. Salzman and P.M. Howley. New York: Plenum, pp. 39–66.

Olson, C., C., and L.V. Skidmore. 1959. Therapy of experimentally induced bovine cutaneous papillomatosis with vaccines and excision. *Journal of the American Veterinary Medical Association* 135:339–343.

Olson, C., M.G. Robl, and L.L. Larson 1968. Cutaneous and penile bovine fibropapillomatosis and its control. *Journal of the American Veterinary Medical Association* 153:1189–1194.

Parsons, R.J., and J.G. Kidd. 1936. A virus causing oral papillomatosis in rabbits. *Proceedings of the Society for Experimental Biology and Medicine* 35:441–443.

Pfister, H. 1987. Papillomaviruses: General description, taxonomy, and classification. In *The papovaviridae.* Vol. 2, *The papillomaviruses,* ed. N.P. Salzman and P.M. Howley. New York: Plenum, pp. 1–38.

Puget, A., M. Favre, and G. Orth. 1975. Induction de tumeurs fibroblastiques cutanees ou sous-cutanees chez l'Ochotone afghan (*Ochotona rufescens rufescens*) par inoculation du virus du papillome bovin. *Comptes Rendus Hebdomadaires des Sceanes de l'Academie des Sciences. D: Sciences Naturelles* 280:2813–2816.

Reszka, A.A., J.P. Sundberg, and M.E. Reichmann. 1991. In vitro transformation and molecular characterization of colobus monkey venereal papillomavirus DNA. *Virology* 181:787–792.

Rous, P., and J.W. Beard. 1935. The progression to carcinoma of virus-induced rabbit papillomas (Shope). *Journal of Experimental Medicine* 62:523–548.

Rowson, K.E.K., and B.W.J. Mahy. 1967. Human papova (wart) virus. *Bacteriological Reviews* 31:110–131.

Samuel, W.M., G.A. Chalmers, and J.R. Gunson. 1978. Oral papillomatosis in coyotes (*Canis latrans*) and wolves (*Canis lupus*) of Alberta. *Journal of Wildlife Diseases* 14:165–169.

Segre, D., C. Olson, and A.B. Hoerlein. 1955. Neutralization of bovine papillomavirus with serums from cattle and horses with experimental papillomas. *American Journal of Veterinary Research* 16:517–520.

Shope, R.E., and E.W. Hurst. 1933. Infectious papillomatosis of rabbits. *Journal of Experimental Medicine* 58:607–624.

Stenlund, A., J. Moreno-Lopez, H. Ahola, and U. Pettersson. 1983. European elk papillomavirus: Characterization of the genome, induction of tumors in animals, and transformation in vitro. *Journal of General Virology* 48:370–376.

Sundberg, J.P. 1987. Papillomavirus infections in animals. In *Papillomaviruses and human disease,* ed. K. Syrjanen, L. Gissmann, and L.G. Koss. Heidelberg: Springer-Verlag, pp. 40–103.

Sundberg, J.P., and W.D. Lancaster. 1988. Deer papillomaviruses. In *Virus diseases of laboratory and captive animals,* ed. G. Darai. Boston: Martinus Nijhoff, pp. 279–291.

Sundberg, J.P., and S.W. Nielsen. 1980. Neoplastic diseases. In *Bovine medicine and surgery,* ed. H.E. Amstutz. Santa Barbara, CA: American Veterinary, pp. 615–647.

———. 1981. Deer fibroma: A review. *Canadian Veterinary Journal* 22:385–388.

Sundberg, J.P., and M.K. O'Banion. 1989. Animal papillomaviruses associated with malignant tumors. *Advances in Viral Oncology* 8:55–71.

Sundberg, J.P., and M.E. Reichmann. 1993. Papillomavirus infections. In *Non-human primates: II. Monographs on pathology of laboratory animals,* ed. T.C. Jones, U. Mohr, and R.D. Hunt. Heidelberg: Springer-Verlag, pp. 1–8.

Sundberg, J.P., R.E. Junge, and W.D. Lancaster. 1984. Immunoperoxidase localization of papillomaviruses in hyperplastic and neoplastic epithelial lesions of animals. *American Journal of Veterinary Research* 45:1441–1446.

Sundberg, J.P., R.J. Chiodini, and S.W. Nielsen. 1985a. Transmission of the white-tailed deer cutaneous fibroma. *American Journal of Veterinary Research* 46:1150–1154.

Sundberg, J.P., D.L. Hill, E.S. Williams, and S.W. Nielsen. 1985b. Light and electron microscopic comparisons of cutaneous fibromas in white-tailed and mule deer. *American Journal of Veterinary Research* 46:2200–2206.

Sundberg, J.P., R.E. Junge, and M.O. El Shazly. 1985c. Oral papillomatosis in New Zealand white rabbits. *American Journal of Veterinary Research* 46:664–668.

Sundberg, J.P., M.K. O'Banion, and M.E. Reichmann. 1987. Mouse papillomavirus: Pathology and characterization of the virus. *Cancer Cells* 5:373–379.

Sundberg, J.P., M.K. O'Banion, A. Shima, and M.E. Reichmann. 1988. Papillomas and carcinomas associated with a papillomavirus in European harvest mice (*Micromys minutus*). *Veterinary Pathology* 25:356–361.

Sundberg, J.P., A. Reszka, E. Williams, and M.E. Reichmann. 1991. An oral papillomavirus that infected one coyote and three dogs. *Veterinary Pathology* 28:87–88.

Sundberg, J.P., A.L. Shima, and D.L. Adkison. 1992. Oral papillomavirus infection in a pygmy chimpanzee (*Pan paniscus*). *Journal of Veterinary Laboratory Investigation* 4:70–74.

Sundberg, J.P., R.J. Montali, M. Bush, L.G. Phillips, S.J. O'Brien, A.B. Jenson, R.D. Burk, and M. van Ranst. 1996a. Papillomavirus-associated focal oral hyperplasia in wild and captive Asian lions (*Panthera leo persica*). *Journal of Zoo and Wildlife Medicine* 27:61–70.

Sundberg, J.P., M. Van Ranst, R.D. Burk, and A.B. Jenson. 1996b. The nonhuman (animal) papillomaviruses: Host range, epitope conservation, and molecular diversity. In *Human papillomavirus infections in dermatology and venereology,* ed. G. Gross and G. von Krogh. Boca Raton, FL: CRC, pp. 47–68.

Sundberg, J.P., S.-J. Ghim, M. Van Ranst, and A.B. Jenson. 1997a. Nonhuman papillomaviruses: Host range, pathology, epitope conservation, and new vaccine approaches. In *Spontaneous animal tumors: A survey,* ed. L. Rossi, R. Richardson, and J. Harshbarger. Milan: Press Point di Abbiategrasso, pp. 33–40.

Sundberg, J.P., B.A. Sundberg, and W.G. Beamer. 1997b. Comparison of chemical carcinogen skin tumor induction efficiency in inbred, mutant, and hybrid strains of mice: Morphologic variations of induced tumors and absence of a papillomavirus co-carcinogen. *Molecular Carcinogenesis* 20:19–32.

Thierry, F. 1996. HPV proteins in the control of HPV transcription. In *Papillomavirus reviews: Current research on papillomaviruses,* ed. C. Lacey. Leeds, UK: Leeds University Press, pp. 21–29.

Van Ranst, M., A. Fuse, H. Sobis, W. de Meurichy, S.M. Syrjanen, A. Billiau, and G. Opdenakker. 1991. A papillomavirus related to HPV type 13 in oral focal epithelial hyperplasia in the pygmy chimpanzee. *Journal of Oral Pathology and Medicine* 20:325–331.

Van Ranst, M., A. Fuse, P. Fiten, E. Beuken, H. Pfister, R.D. Burk, and G. Opdenakker. 1992. Human papillomavirus type 13 and pygmy chimpanzee papillomavirus type 1: Comparison of the genome organizations. *Virology* 190:587–596.

Van Ranst, M., J.B. Kaplan, J.P. Sundberg, and R.D. Burk. 1995. Molecular evolution of papillomaviruses. In *Molecular basis of virus evolution,* ed. A.J. Gibbs, C.H. Calisher, and F. García-Arenal. Cambridge: Cambridge University Press, pp. 455–476.

Van Ranst, M., R. Tachezy, and R.D. Burk. 1996. Human papillomaviruses: A never ending story? In *Papillomavirus reviews: Current research on papillomaviruses,* ed. C. Lacey. Leeds, UK: Leeds University Press, pp. 1–19.

Watson, J.D., and J.W. Littlefield. 1960. Some properties of DNA from Shope papillomavirus. *Journal of Molecular Biology* 2:161–165.

12

PESTIVIRUS INFECTIONS

HANA VAN CAMPEN, KAI FRÖLICH, AND MARTIN HOFMANN

Pestiviruses primarily infect ungulates belonging to the order Artiodactyla, within which there are 11 species of pigs and 173 species of ruminants (Nettleton 1990). Bovine viral diarrhea virus (BVDV), border disease virus (BDV), and classic swine fever virus (CSFV) are classified in the genus *Pestivirus* within the family *Flaviviridae,* based on the organization and expression of their genomes (Horzinek 1991; Rice 1996; Thiel et al. 1996). Bovine viral diarrhea virus and BDV, and/or closely related viruses, infect a wide variety of wild ungulates (Nettleton 1990; Loken 1995a). The impact of BVDV- or BDV-associated diseases on the health of free-ranging wild ruminant populations is currently unknown (Nettleton 1990; Depner et al. 1991); there are few confirmed cases of pestivirus-caused disease in these species and no evidence that these viruses have significant population impacts. However, these viruses are major pathogens of cattle *Bos taurus* and sheep *Ovis aries* worldwide, with significant economic impact on livestock production (Duffell and Harkness 1985; Brownlie et al. 1987; Thiel et al. 1996). Classic swine fever virus, also known as hog cholera virus, is a highly contagious pestivirus affecting Suidae, i.e., domestic pigs and wild boars *Sus scrofa.* Swine are the only species that develop clinical disease after CSFV infection. The virus can also experimentally be transmitted to domestic and wild ruminants, albeit without causing clinical signs.

In addition to various ruminant and pig species, only camels *Camelus dromedarius* and guanaco *Lama guanicoe* (Doyle and Heuschele 1983), rabbits *Oryctolagus cuniculus* (Baker et al. 1954; Fernelius et al. 1969; Frölich and Streich 1998), and wallaby *Macropus rufogriseus* (Munday 1966) appear to be susceptible to pestivirus infections (Table 12.1).

BOVINE VIRAL DIARRHEA VIRUS AND BORDER DISEASE VIRUS INFECTIONS

Synonyms: Bovine viral diarrhea (BVD), mucosal disease (MD); border disease (BD), hairy shaker, fuzzy lamb.

Distribution and Host Range. Evidence for BVDV infections of free-ranging populations include serologic surveys and virus isolations (Table 12.1). Serologic surveys in free-ranging and captive populations demonstrated prior infection with BVDV or related pestiviruses in more than 40 species (Nettleton 1990) in North America (Riemann et al. 1979; Doyle and Heuschele 1983), Africa (Hamblin and Hedger 1979), Australia (McKenzie et al. 1985), and Europe (Frölich 1993; Frölich and Flach 1998).

Reports of disease in wild species lacking viral isolation and characterization must be viewed skeptically because similar clinical signs are described for other infectious diseases and, therefore, cannot be attributed with certainty to pestivirus infection. For example, early reports of disease and pathologic findings in white-tailed deer *Odocoileus virginianus* (Fay and Boyce 1955; Shope et al. 1955; Richards et al. 1956) are more compatible with epizootic hemorrhagic disease of deer than with BVDV infections.

Isolations of BVDV from wild ruminants are few (Table 12.1). Some viral isolates can be ascribed to infection with BVDV of domestic livestock origin by history (Romvary 1965). Analysis of other viruses by monoclonal antibody panels or nucleic acid sequence data confirmed that these viruses are BVDV (Edwards et al. 1988; Cay et al. 1989; J.F. Ridpath personal communication). Antigenic and sequence differences from BVDV have been described for a pestivirus isolated from red deer *Cervus elaphus* by Baradel et al. (1988). Cytopathogenic pestiviruses were isolated from two seronegative roe deer *Capreolus capreolus* from northern Germany (Frölich and Hofmann 1995). Sequence analyses showed that these isolates (SH9/11) represent a new strain within the BVDV group (Fischer et al. 1998).

In the Netherlands, Peters (1966) described clinical signs suggestive of BVD in captive Père David's deer *Elaphurus davidianus,* banteng *Bos javanicus,* and yak *Bos grunniens.* In captive ungulates in Germany, lesions typical of BVD were found in banteng, axis deer *Axis axis,* and fallow deer *Dama dama* (Mehring 1965) and in alpaca *Lama pacos* (Kast and Kraus 1968), reindeer *Rangifer tarandus* (Steger 1973), and fallow deer (Weber et al. 1982). Brass et al. (1966) observed a BVD-like disease in 11 ruminants in the Hannover Zoo, including seven gazelles *Gazella dorcas, Gazella rufifrons, Gazella granti,* one muntjac *Muntiacus* sp., one gaur *Bos gaurus,* and two banteng. In Poland, Sosnowski (1977) reported BVD in captive bison *Bison bonasus.*

TABLE 12.1—Free-living or captive wild mammals with evidence of pestivirus infection

Species	Continent; Captive (C) or Free-Living (FL)	Evidence for Infection: Antibody (AB), Virus Isolation (VI), or Lesions (L)	References
Cervidae			
Muntjac *Muntiacus reevesi*	North America (C)	VI	Doyle and Heuschele 1983
Chinese water deer *Hydropotes inermis*	Europe (FL)	AB	Lawman et al. 1978; Frölich and Flach 1998
Axis deer *Axis axis*	North America (C) Europe (C)	AB/VI/L	Mehring 1965; Riemann et al. 1979; Doyle and Heuschele 1983
Barasingha *Cervus duvaucelii*	North America (C)	VI	Doyle and Heuschele 1983
Red deer *Cervus elaphus*	Europe (C,FL)	AB/VI	McMartin et al. 1977; Lawman et al. 1978; Weber et al. 1978; Nettleton et al. 1980; McKenzie et al. 1985; Baradel et al. 1988; Dedek et al. 1988; Frölich 1993
Wapiti/elk *Cervus elaphus nelsoni*	North America (FL)	AB	Kingscote et al. 1987; Aguirre et al. 1995
Sika deer *Cervus nippon*	Europe (FL)	AB	Lawman et al. 1978
Fallow deer *Dama dama*	Europe (C,FL)	AB/VI/L	Mehring 1965; McDiarmid 1975; Lawman et al. 1978; Neumann 1980; Weber et al. 1982; Edwards et al. 1988; Giovanni et al. 1988; Diaz et al. 1988; Frölich 1993
	Australia (FL)	AB	Munday 1972; English 1982
Père David's deer *Elaphurus davidianus*	Europe (C)	AB/L	Peters 1966; Frölich and Flach 1998
Mule deer *Odocoileus hemionus*	North America (FL)	AB	Richards et al. 1956; Couvillion et al. 1977; Stauber et al. 1977; Aguirre et al. 1995
White-tailed deer *Odocoileus virginianus*	North America (FL)	AB/L	Shope et al. 1955a; Fay and Boyce 1955a; Richards et al. 1956a; Kahrs et al. 1964; Friend and Halterman 1967
Roe deer *Capreolus capreolus*	Europe (FL)	AB/VI/L	Romvary 1965; Schellner 1977; Weber et al. 1978; Feinstein et al. 1987; Baradel et al. 1988; Dedek et al. 1988; Frölich and Hofmann 1995; Frölich 1995; Fischer et al. 1998
Moose *Alces alces*	North America (FL) Europe (FL)	AB/L	Kocan et al. 1986; Feinstein et al. 1987; Rehbinder et al. 1991; Merza et al. 1994; Cedersmyg et al. unpublished
Caribou/reindeer *Rangifer tarandus*	Europe (C,FL)	AB/L	Steger 1973; Rehbinder et al. 1991; Stuen et al. 1993
	North America (FL)	AB	Elazhary et al. 1981; Zarnke 1983
Giraffidae			
Giraffe *Giraffa camelopardalis*	Africa (FL)	AB/VI	Plowright 1969; Hamblin and Hedger 1979; Depner et al. 1991
Antilocapridae			
Pronghorn antelope *Antilocapra americana*	North America (C, FL)	AB	Barrett and Chalmers 1975; Stauber et al. 1980; Doyle and Heuschele 1983
Bovidae			
Duiker *Sylvicapra grimmia*	Africa (FL)	AB	Hamblin and Hedger 1979
Dik-dik *Madoqua kirkii*	North America (C)	AB	Doyle and Heuschele 1983
Eland *Taurotragus* spp.	Africa (FL)	AB	Hamblin and Hedger 1979; Depner et al. 1991
	North America (C)	AB	Doyle and Heuschele 1983

(continued)

TABLE 12.1—*Continued*

Species	Continent; Captive (C) or Free-Living (FL)	Evidence for Infection: Antibody (AB), Virus Isolation (VI), or Lesions (L)	References
Greater kudu *Trageaphus strepsieeros*	Africa (FL)	AB	Hamlin and Hedger 1979; Depner et al. 1991
	North America (C)	AB	Doyle and Heuschele 1983
Nyala *Tragelaphus angasii*	Africa (FL)	AB	Hamblin and Hedger 1979
Nilgai *Boselaphus tragocamelus*	North America (C)	VI	Doyle and Heuschele 1983
Oryx/gemsbok *Oryx gazella*	Africa (FL)	AB	Hamblin and Hedger 1979; Depner et al. 1991
	North America (C)	AB	Doyle and Heuschele 1983
Scimitar-horned oryx *Oryx dammah*	Europe (C)	AB	Frölich and Flach 1998
Sable antelope *Hippotragus niger*	Africa (FL)	AB	Hamblin and Hedger 1979; Depner et al. 1991
Roan antelope *Hippotragus equinus*	Africa (FL)	AB	Depner et al. 1991
Waterbuck *Kobus ellipsiprymnus*	Africa (FL)	AB	Hamblin and Hedger 1979;
	North America (C)	VI/AB	Doyle and Heuschele 1983
Lechwe *Kobus leche*	Africa (FL)	AB	Hamblin and Hedger 1979
Reedbuck *Redunca redunca*	Africa (FL)	AB	Provost et al. 1967; Hamblin and Hedger 1979
Hartebeest *Alcelaphus buselaphus*	Africa (FL)	AB	Provost et al. 1967; Hamblin and Hedger 1979
Tsessebe/topi *Damaliscus lunatus*	Africa (FL)	AB	Hamblin and Hedger 1979; Hyera et al. 1992
Blesbok/bontebok *Damaliscus pygargus*	North America (C)	AB	Doyle and Heuschele 1983
Impala *Aepyceros melampus*	Africa (FL)	AB	Hamblin and Hedger 1979
Gazelle *Gazella rufifrons*	Africa (FL)	AB	Provost et al. 1967
	Europe (C)	L	Brass et al. 1966
Dorcas gazelle *Gazella dorcas*	Europe (C)	L	Brass et al. 1966
Grant's gazelle *Gazella granti*	Europe (C)	L	Brass et al. 1966
Springbuck *Antidorcas marsupialis*	Africa (FL)	AB	Hamblin and Hedger 1979
	North America (C)	AB	Doyle and Heuschele 1983
Serow *Capricornis sumatraensis*	North America (C)	VI	Doyle and Heuschele 1983
Chamois *Rupicapra rupicapra*	Europe (FL)	AB	Baradel et al. 1988
Mountain goat *Oreamnos americanus*	North America (C)	AB	Doyle and Heuschele 1983
Barbary sheep *Ammotragus lervia*	North America (C)	AB	Doyle and Heuschele 1983
Ibex *Capra ibex*	Europe (FL)	AB	Baradel et al. 1988
Bezoar *Capra aegagrus aegagrus*	North America (C)	AB	Doyle and Heuschele 1983
Pygmy goat *Capra hircus*	North America (C)	VI	Doyle and Heuschele 1983
Urial *Ovis vignei*	North America (C)	AB	Doyle and Heuschele 1983
Rocky Mountain bighorn sheep *Ovis canadensis*	North America (FL)	AB	Parks and England 1974; Clark et al. 1985
African buffalo *Syncerus caffer*	Africa (FL)	AB/VI	Provost et al. 1967; Hamblin and Hedger 1979; Hyera 1989
Bison *Bison bison*	North America (FL)	AB	Williams et al. 1993; Taylor et al. 1997
	Europe (C)	L	Sosnowski 1977
European bison *Bison bonasus*	Europe (C)	AB	Frölich and Flach 1998
Yak *Bos grunniens*	Europe (C)	L	Peters 1966
Banteng *Bos javanicus*	Europe (C)	L	Mehring 1965; Brass et al. 1966; Peters 1966

(continued)

TABLE 12.1—*Continued*

Species	Continent; Captive (C) or Free-Living (FL)	Evidence for Infection: Antibody (AB), Virus Isolation (VI), or Lesions (L)	References
Gaur *Bos gaurus*	North America (C)	AB	Doyle and Heuschele 1983
	Europe (C)	L	Brass et al. 1966
Camelidae			
Dromedary *Camelus dromedarius*	North America (C)	AB	Doyle and Heuschele 1983
Llama *Lama glama*	North America (C)	AB	Doyle and Heuschele 1983
Alpaca *Lama pacos*	Europe (C)	L	Kast and Kraus 1968
Suidae			
Wart hog *Phacochoerus aethiopicus*	Africa (FL)	AB	Hamblin and Hedger 1979
Wild boar *Sus scrofa*	Europe (FL)	AB	Dahle et al. 1993; Oslage 1993
Leporidae			
Rabbit *Oryctolagus cuniculus*	Europe (FL)	AB/VI	Baker et al. 1954; Fernelius et al. 1969; Frölich and Streich 1998
Macropodidae			
Wallaby *Macropus rufogriseus*	Australia (FL)	AB	Munday 1966

[a]Diagnosis is questionable.
Modified from Nettleton (1990).

Etiology. Bovine viral diarrhea virus is a positive-sense, single-stranded RNA virus with a single open reading frame encoding four structural and six non-structural proteins (Donis 1995). The virions are pleomorphic, spherical structures 50–60 nm in diameter, with a bilaminar envelope of cellular origin surrounding a semidense core of 20- to 25-mm diameter (Bielefeldt-Ohmann 1990). Virions mature within intracytoplasmic membranes, and virus is liberated by exocytosis of virus-containing membrane vesicles (Thiel et al. 1996). Infectivity of pestiviruses is lost at elevated temperatures and by treatment with detergents and lipid solvents. The viruses are able to withstand a relatively broad pH range (Depner et al. 1992; Thiel et al. 1996). Bovine viral diarrhea virus and BDV exist in two biotypes: noncytopathic (ncp) and cytopathic (cp) for cultured cells. Most isolates are ncp, but cp viruses are recovered from animals with a specific BVDV-related affliction called mucosal disease.

Transmission and Epidemiology. The principal reservoirs of BVDV and BDV are persistently infected (PI) cattle and sheep by virtue of the high titer of virus shed in their secretions (Brownlie 1990; Traven et al. 1991; Houe 1995). Virus is also present in large amounts in aborted fetuses, fetal membranes, and uterocervical fluids (McGowan et al. 1993). Inquisitive behavior can lead to infection from these sources. In comparison, transmission of pestiviruses by acutely infected animals is inefficient (McGowan et al. 1993; H. Van Campen et al. unpublished). Transmission by mechanical and insect vectors (*Stomoxys calcitrans* or *Haematopota phivialis*) has been reported (Meyling et al. 1990; Tarry et al. 1991).

The role of pestiviruses in wild ruminant populations and the interactions between wild ungulates and domestic livestock are not well understood (Nettleton 1990; Aguirre et al. 1995; Frölich 1995). Suspected sources of the virus for wild animals include direct contact with infected livestock, shared feed and watering areas, or the presence of pestivirus-infected individuals within wild populations (Van Campen and Williams 1996). Romvary (1965) diagnosed BVD in roe deer living adjacent to a cattle farm where BVD had previously caused severe losses. High seroprevalences in free-ranging populations have been explained by proximity to livestock (Thorsen and Henderson 1971; Barrett and Chalmers 1975; Stauber et al. 1977; Aguirre et al. 1995).

Several investigators have speculated that an independent cycle occurs among wild ruminants with BVDV infection (Weber et al. 1982; Liebermann et al. 1989). Frölich (1995) found no significant difference in antibody prevalence among deer in habitats with high, intermediate, or low density of cattle. It is likely that the BVD isolates from roe deer (Frölich and Hofmann 1995) represent a new strain within the BVDV group (Fischer et al. 1998), indicating that unique pestiviruses circulate in wild ruminants. If this is the case, then contact with cattle is not essential for the appearance of BVDV-like infections in free-ranging wildlife. Factors necessary for the maintenance of pestivirus infections include a population size sufficient to provide a continuous source of susceptible animals, herd behavior allowing interaction between infected and susceptible animals, and timing of reproduction and survivorship of the young. The high seroprevalence in free-ranging caribou *Rangifer tarandus* in Canada [60%–70% (Elazhary et al. 1981)] with no direct domestic ruminant contact for 25 years supports this contention. The large size of caribou herds may be a significant requirement for the maintenance of pestivirus infections.

Neumann et al. (1980) suggested a causal relationship between the spread of BVDV in cattle and its occurrence in deer. Wild ruminants have been speculated to serve as a reservoir of BVDV for cattle (Meyling et al. 1990); however, evidence for this proposal is lacking. Pastoret et al. (1988) suggested that wild species do not play a determinant role in transmitting infection to domestic cattle.

Clinical Signs, Pathogenesis, and Pathology. Clinical signs of pestivirus infections in wild ruminants like those in cattle depend on the virulence of the isolate, the immune status of the animal, and the route of transmission. Relatively little is known about disease caused by pestiviruses in wild ruminants; thus, some discussion of these diseases in domestic livestock is warranted.

Acute infections in cattle are usually subclinical (Wilhelmsen et al. 1990); however, some North American type-2 BVDVs cause peracute death and hemorrhagic disease (Carman et al. 1994). In cattle and sheep, fetal pestivirus infections may lead to abortion, fetal malformations, stillbirths, and weakened neonates (Done et al. 1980; Loken 1995b; Moennig and Liess 1995). Early fetal losses are manifested as infertility and prolonged breeding and birthing intervals. Abortions can occur at any stage of gestation, depending on when the dam is infected. Transplacental infections have not been reported in wild animals; however, mummified fetuses and stillborn and normal healthy fawns were observed in white-tailed deer experimentally inoculated with a cp BVDV originally isolated from a white-tailed deer (A.W. McClurkin personal communication). Bovine fetuses infected with ncp BVDV during the first trimester of gestation may become PI with the virus (Brownlie et al. 1984; McClurkin et al. 1984). These animals are often unthrifty but may be of normal size and development. It is not known whether persistent infections occur in wild ruminants.

Persistently infected cattle eventually develop MD with severe diarrhea, dehydration, and fever, which has low morbidity within a group but causes high mortality. Pairs of cp and ncp viruses are isolated from MD cases; the cp virus is thought to arise by spontaneous mutation from the endogenous ncp virus or be introduced by vaccination (Bolin et al. 1985; Brownlie 1990; Moenning et al. 1990).

Doyle and Heuschele (1983) described fever, cloudy corneas, and depression, signs compatible with malignant catarrhal fever, in nilgai *Boselaphus tragocamelus,* axis deer, and barasingha *Cervus duvaucelii* from which BVDV was isolated in addition to a herpesvirus. In Sweden, lesions characteristic of BVD were observed in captive fallow deer (Diaz et al. 1988), as well as free-ranging moose *Alces alces* and roe deer (Feinstein et al. 1987). In addition to copper deficiency and a retrovirus (*Alces* leukotrophic oncovirus), a BVDV-like pestivirus is suspected in the wasting syndrome (Älvsborg disease) of free-ranging moose. The pathologic changes in moose resemble those of BVD (Rehbinder et al. 1991; Merza et al. 1994) and are supported by serologic findings (Cedersmyg et al. unpublished). Älvsborg disease, associated with death of more than 1000 moose, is currently regarded as a multifactorial disease that includes BVDV as one of at least three important components.

Experimental inoculation of red deer, white-tailed deer, and mule deer *Odocoileus hemionus* with ncp BVDV resulted in subclinical infections (McMartin et al. 1977; Van Campen et al. 1997). Infection with cp BVDV (Singer) resulted in transient mild diarrhea, coronitis, and laminitis in reindeer (Morton et al. 1990). Severe disease has been reported in free-ranging roe deer (Romvary 1965). Clinical signs included weakness, lack of fear, impaired hearing and vision, dehydration, and emaciation. Lesions included erosion and ulceration of the oral mucosa, hemorrhagic enteritis, and general physical impairment. Pyrexia, anorexia, salivation, and nasal discharge have been described; some animals had skin lesions and may have been lame due to interdigital ulceration and inflammation of the coronary bands (Romvary 1965; Wiesner 1987; Neumann et al. 1980; Morton et al. 1990; Nettleton 1994).

Diagnosis. The clinical and pathologic diagnosis is confirmed by virus isolation, demonstration of viral antigen in tissues by immunofluorescent antibody staining or immunohistochemistry, or detection of viral RNA. Virus should be isolated from buffy-coat cells, plasma, serum, or nasal secretions collected early in infection (days 2–15) from acutely infected or suspected PI animals. Samples of thymus, spleen, lung, liver, mesenteric lymph nodes, tonsils, intestines, and kidney should be collected at necropsy for culture (Potgeiter 1992; Brock 1995; Thiel et al. 1996). Primary bovine embryonic testicle or kidney cells, or bovine turbinate cells, are preferred for virus isolation. Most isolates are ncp; therefore, inoculated cells should be examined for viral antigen by direct (FA) or indirect (IFA) fluorescent antibody tests using specific monoclonal or polyclonal antibodies. Cell cultures and fetal bovine sera used in culture media should be free of contaminating BVDV (Hassan and Scott 1986). Prior to the mid-1980s, culture media potentially contaminated with BVDV caused considerable difficulty in confirming virus isolations.

Viral antigen can be detected in acetone-fixed frozen tissue sections or impression smears of respiratory epithelium with immunofluorescent antibody reagents (Miller and Wilkie 1979) or in formalin-fixed paraffin-embedded tissues by immunohistochemical techniques (Edwards et al. 1988; Haines et al. 1992).

Several enzyme-linked immunosorbent assay (ELISA) protocols have been developed for demonstration of pestiviral antigen in serum or plasma (Meyling 1983; Sandvik and Krogsrud 1995) and leukocytes (Frey et al. 1991; Mignon et al. 1991; Bitsch and Ronsholt 1995; Brock 1995) and are particularly useful in screening large numbers of animals. Laboratory diagnosis may be made by detection of the highly conserved 5′-untranslated region of the viral genome in tissues or serum by using reverse transcriptase-polymerase chain reaction (RT-PCR) (Hooft van Iddekinge et al. 1992; Alansari et al. 1993; Frölich and Hofmann 1995; Thiel et al. 1996). Subsequent nucleotide sequencing or a second round of amplification of the respective region enables discrimination among the different pestivirus groups (De Moerlooze et al. 1993; Ridpath et al. 1994).

Evidence of recent infection can also be provided by finding a greater than fourfold increase in virus-neutralizing (VN) titers between acute and convalescent serum samples following suspected postnatal infection (Nettleton 1994). Antibody titers persist for years, probably for life, in recovered cattle (Kahrs et al. 1966). Little is known about the kinetics of antibody titers in wild animals. Serology is of limited use in the diagnosis of abortions, because infection and seroconversion may occur weeks to months prior to abortion. Presuckle serum samples of newborns with antibody titers indicate exposure of the fetus to virus in the last trimester of pregnancy. Colostral antibodies will obscure serologic diagnosis of infection in neonates and mask the detection of viremia in PIs. Standard VN assays are usually performed using a reference BVDV strain (type 1) (Carbrey et al. 1971). Due to antigenic variation among virus isolates, however, sera from wild species should be examined in VN assays using type-1 and type-2 BVDV and BDV. Although some cross-reactivity is found among antibodies to these viruses, individual animals may develop VN titers to the infecting pestivirus with little or no cross-reactivity to others; therefore, serologic surveys against one reference virus may not detect antibody titers in animals infected with antigenically dissimilar pestiviruses (Van Campen and Williams 1996). Enzyme-linked immunosorbent assays to detect anti-BVDV antibodies have been developed for cattle (Straver et al. 1987; Durham and Hassard 1990) but require anti-species conjugated antibody reagents that may be difficult to acquire for some wild species.

Differential Diagnoses. Similarities in clinical signs and lesions require differentiation of BVDV infection from malignant catarrhal fever, rinderpest, foot-and-mouth disease, parapoxvirus disease (bovine papular stomatitis), herpesvirus infections (cervid herpesvirus 1, bovine herpesvirus 1, caprine herpesvirus 1, and rangifer herpesvirus 1), salmonellosis, enzootic keratoconjunctivitis, bluetongue, epizootic hemorrhagic disease, and enteritis caused by gastrointestinal parasites.

Immunity. Development of antibodies to BVDV similar to that described for cattle have been reported by Morton et al. (1990) in reindeer, Van Campen et al. (1997) in mule deer and white-tailed deer, and Hyera et al. (1993) in wildebeest *Connochaetes* sp. In naive animals, antibodies are detected 8–15 days after infection. More rapid increases may occur in animals with preexisting antibody titers to BVDV (H. Van Campen et al. unpublished data). Preexisting antibodies probably protect against development of severe disease caused by acute infection; however, they are unlikely to protect against fetal infection.

Control, Management Implications, and Domestic Animal Concerns. Control of BVD and BD in captivity requires separation of wild animals from livestock and avoiding contact with contaminated biologicals. New arrivals should be quarantined. Where quarantine and separation from livestock are not feasible, and there is concern about BVDV infection, animals may be vaccinated and boosted with inactivated BVDV vaccines to prevent severe disease due to acute infection. Managers should be cognizant that vaccination has not been shown to protect against fetal infections if dams are exposed to antigenically diverse pestiviruses, and thus, there is a potential for production of PIs.

Control of virus in free-ranging populations is more problematic where wild species share range with cattle or sheep. Pestiviruses have not been successfully eliminated from livestock, and domestic animals may serve as a nidus of infection for other ruminants. In small populations or species that form groups for only limited periods, the chance for viral transmission to susceptible animals is low, and the virus infection may be self-limiting. For these species, physical separation from livestock may result in elimination of pestivirus infections and a seronegative population (Sadi et al. 1991). The ability of pestiviruses to survive in wild populations may also be hampered by low recruitment of young, including any PI females, into the reproducing population. Thus, the potential to create the crucial reservoir of these viruses, the PI animal, is low. In species that form large herds (e.g., caribou), the virus may continue to circulate regardless of contact with domestic livestock.

CLASSIC SWINE FEVER VIRUS (HOG CHOLERA VIRUS) INFECTIONS

Synonyms: Classic swine fever (CSF), hog cholera

Distribution and Host Range. Classic swine fever in domestic pigs was once endemic worldwide. The virus has been eradicated in North America, Australia, and most European countries but is still present in Germany, Italy, Austria, as well as other European countries and many areas in Asia and South America (Fig. 12.1). In the recent past, outbreaks of CSF in wild boars were reported from France, Germany, Italy, and Austria [Office International des Epizooties (OIE) 1994, 1995].

Transmission and Epidemiology. Transmission of CSFV occurs mainly by direct contact with infected

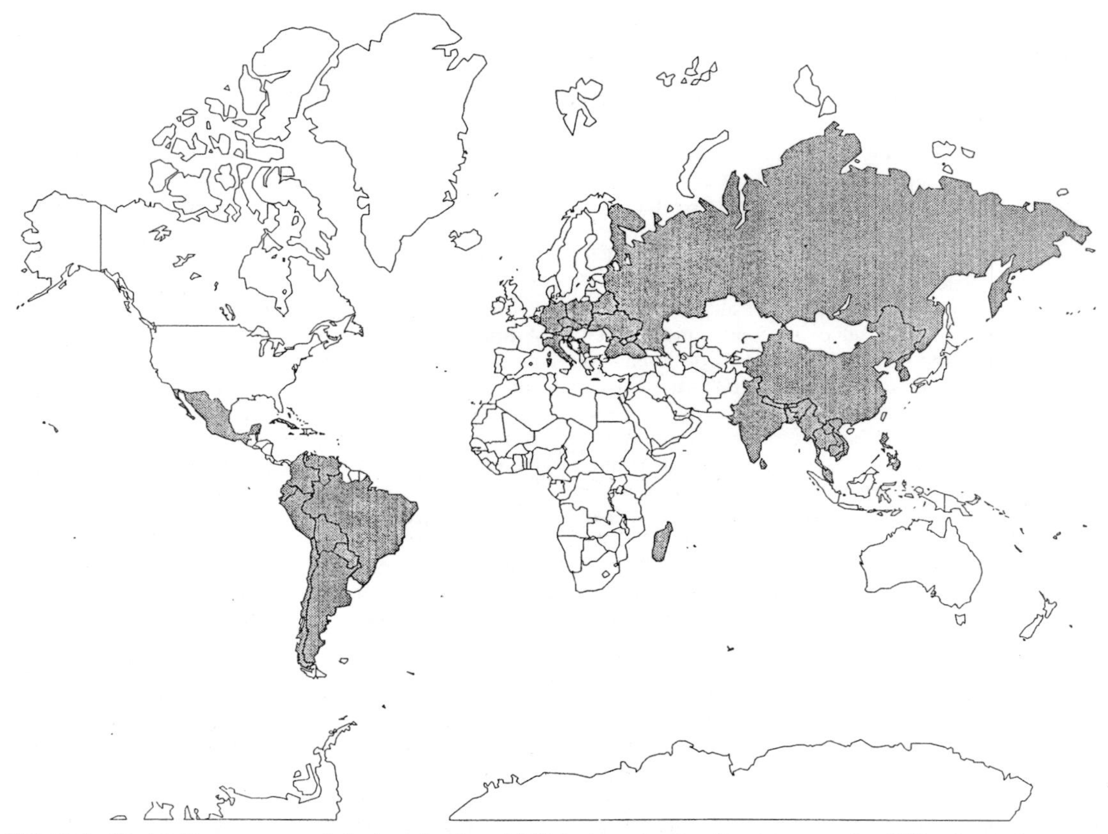

FIG. 12.1—Worldwide occurrence of classic swine fever (CSF) in domestic pigs. Countries reporting CSF outbreaks between 1994 and 1995 are *shaded.* Data were taken from the Office International des Epizooties Bulletins. Since data on the incidence of CSF in wild boars are scarce, it must be assumed that the disease can affect them wherever CSF in domestic pigs occurs. Wild boars exist in Europe, Asia, and North and South America, as well as in Australia.

pigs, contaminated feces, or feedstuffs (Fig. 12.2). The feeding of swill containing meat from CSFV-infected animals or feeding virus-containing offal of wild boar carcasses is of particular importance. Disposal of dead pigs by leaving or burying them in the woods resulted in CSF outbreaks in native wild boar populations (Dahle and Liess 1992; Aubert et al. 1994).

Another epidemiologic factor is infection of pregnant sows resulting in birth of PI piglets (Depner et al. 1995). The disease is more likely to spread to wild boars in areas where domestic pigs are kept on free range and they share the same habitat (Laddomada et al. 1994). Recent data suggest that the appearance of seropositive wild boars might be related to vaccination of domestic pigs with an avirulent CSFV vaccine strain. The vaccine, which is no longer in use, was thought to have spread to wild boars through consumption of offal from vaccinated pigs (Dahle et al. 1993).

At present, the epidemiologic role of wild boars as a reservoir for CSF remains highly questionable. Classic swine fever virus is unlikely to persist within the wild boar population without being repeatedly reintroduced from domestic pigs (Nettles et al. 1989; Picard et al. 1993; Ahl 1994; Aubert et al. 1994). This hypothesis is supported by the absence of CSF in wild boar populations in countries where CSF has been eradicated from domestic pigs (Aubert et al. 1994). Wild boar populations in Europe are rapidly expanding, however, and the higher animal density has increased the likelihood that the disease could persist in wild populations (Oslage et al. 1994).

Clinical Signs, Pathogenesis, and Pathology. Classic swine fever has been studied extensively in experimental infections of domestic pigs, as well as under field conditions [see Liess (1987), Van Oirschot (1988), and Dahle and Liess (1992)], whereas systematic studies in wild boars have only recently been conducted (Depner et al. 1995). These investigations have confirmed that the course of disease in wild boars is similar to the one observed in domestic pigs (Loepelmann and Dedek 1987). Manifestations of disease in pigs are influenced

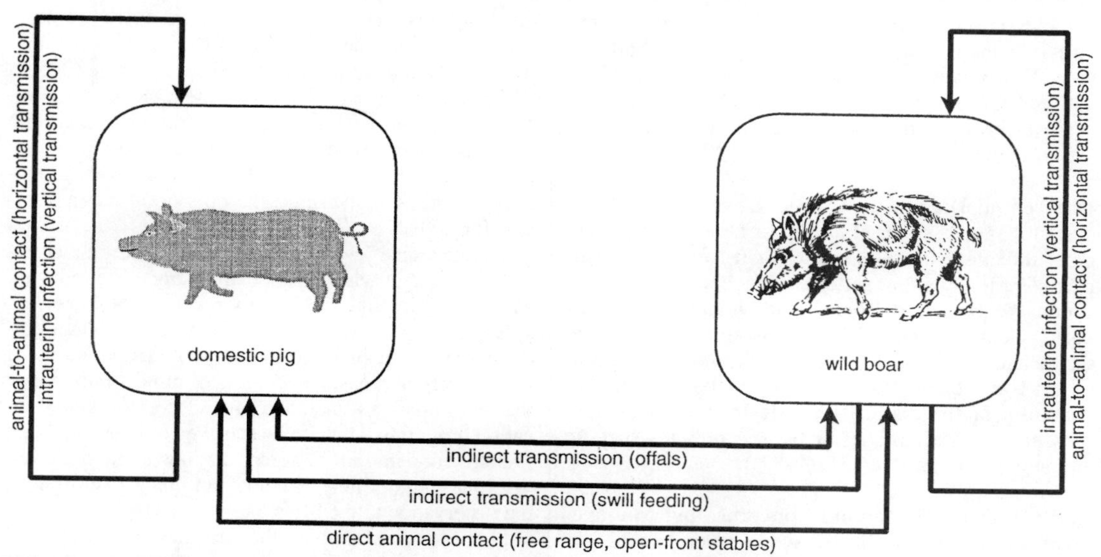

FIG. 12.2—Transmission of classic swine fever virus. The virus can be transmitted within and between wild boars and domestic pigs directly via animal-to-animal contact and by indirect contacts.

by several factors, such as the virulence and dose of the virus, and the genetic background of the host, as well as environmental factors. The peracute form of CSF causes high mortality in which pigs die without typical signs 2–5 days after exposure to the virus. The so-called "typical acute form" is characterized by febrile disease 6–8 days after infection, with leukopenia, obstipation followed by diarrhea, petechial hemorrhages, cyanosis of the skin, and neurologic signs, such as staggering and posterior paresis. Subclinical and chronic infections occur, the latter resulting in signs similar to but milder than acute CSF. Although the animals initially seem to recover, they may eventually die. Wild boars infected with CSFV exhibit altered behaviors, such as loss of natural shyness, approaching or even entering farm buildings, or leaving the forests during daylight (Loepelmann and Dedek 1987). In addition to postnatal infections, pregnant sows can become infected by a virus strain of low virulence resulting in the birth of mummified or stillborn fetuses, as well as healthy piglets. Piglets born alive, however, are PI with CSFV; their growth is retarded, and they invariably die by several months of age. Chronic and persistently infected pigs shed CSFV and are, therefore, important in the epidemiology of CSF.

Under natural conditions, CSFV invades the host via the oronasal route. The epithelial cells of the tonsillar crypts are the initial target tissue for virus replication. Classic swine fever virus infects primarily monocytes, macrophages, and endothelial cells, and it was therefore proposed that CSF is a disease of the immune system (Trautwein 1988).

Classic swine fever is a typical hemorrhagic disease. A consistent finding in acute CSF is severe thrombocytopenia occurring 2–3 days after infection and preceding other clinical signs by several days. Erythema, cyanosis, and petechial hemorrhages occur in the skin. Damaged endothelia and generalized thrombosis lead to petechial hemorrhages and hemorrhagic diatheses in most organs, most prominently in kidneys, urinary bladder, larynx, and lymph nodes. Infarcts along the spleen margin are considered highly suggestive of CSF but are not present in all cases. Highly virulent CSFV isolates can infect endothelium of the central nervous system, resulting in severe neurologic damage (Kamolsiriprichaiporn et al. 1992). In chronic forms of the disease, necrosis of the mucosa in the large intestine ("button ulcers") can often be found. Histopathologically, nonsuppurative encephalitis with severe vasculitis is found in fatal CSF cases.

Diagnosis. Conventional methods such as virus isolation on cell culture, antigen detection by immunofluorescence or immunohistochemistry on tissue sections (Carbrey 1988), and molecular biologic techniques, including amplification of viral RNA by PCR (Hofmann et al. 1994), can be used to diagnose CSFV infections. Classic swine fever virus can be isolated from blood or organ suspensions of acutely diseased animals. Tonsils, kidney, spleen, and lymph nodes are the most suitable organs to collect for virus isolation. It is important to differentiate between CSFV and ruminant pestiviruses that are capable of infecting wild boars (Dahle et al. 1993). Classic swine fever is a notifiable disease controlled by eradication in many countries (see below), whereas no regulatory actions are taken in the case of ruminant pestivirus infections in pigs. Accurate typing of the isolated CSFV strain by molecular methods is required for epidemiologic tracing of the virus. Antibodies can be detected by ELISA and VN tests. Newly developed tests enable accurate discrimination between antibodies against CSFV and those against other pestiviruses (Wensvoort et al. 1988; Moser et al. 1996).

Differential Diagnoses. Classic swine fever is difficult to diagnose either clinically or at necropsy. Many other infectious diseases, such as African swine fever, pseudorabies, or salmonellosis, as well as some intoxications (e.g., salt poisoning and vitamin-K antagonists), induce similar clinical signs and lesions. Since the course of CSF in wild boars is far more difficult to observe and clinical signs are often less prominent, diagnosis of the disease is made by virus isolation, presence of viral antigen in tissues, or virus-specific antibody detection [see OIE (1992)].

Control, Domestic Animal Concerns, and Management Implications. Classic swine fever is a highly contagious animal disease and therefore is included on "List A" of the Office International des Epizooties. In many countries, the disease is notifiable, and infected pig herds must be depopulated to eradicate the disease. Classic swine fever virus-infected wild boars must be destroyed, as well. Because wild boars pose a certain threat for reintroducing the disease into the domestic pig population, control strategies include reduction of wild boar density by increased hunting activity. This strategy has often yielded unsatisfactory results because of the high reproduction rate of wild boars and the increased risk of spreading CSF by chasing wild boars out of their native habitats (Aubert et al. 1994). Vaccination of wild boars with an attenuated CSFV strain has been considered in France, Italy, and Germany. Vaccination is presently being carried out in an experimental setting in two CSF-endemic wild boar populations in the eastern part of Germany. There are two major drawbacks to this approach: first, although initial results were promising, it was not possible to vaccinate an adequate percentage of the wild boar population to prevent spread of the wild-type virus. Secondly, without marker vaccines, a serologic diagnosis is impossible, since infected and vaccinated animals can no longer be differentiated. Control measures should focus on strict separation of wild boars from domestic pigs. This can be achieved by confining wild boars to certain areas bordered either by natural obstacles (lakes, mountains, etc.) or man-made obstacles (highways, etc.). Concurrently, domestic pigs must be prevented from coming in direct contact with wild boars by confining pigs in closed stables and not allowing free-ranging swine management.

A very important aspect of the control of CSF in wild boars is collaboration between farmers and hunters. It is crucial that hunters do not leave offal of wild boars in the forest if CSF is present in an area nor should they bring any wild boar products onto pig farms. Vice versa, farmers must properly dispose of dead pigs. Hunters and farmers need to be properly instructed about the danger involved in swill feeding, particularly since infectious CSFV can survive for several months in frozen wild boar meat (Krassnig et al. 1995). In countries where CSF is present in domestic pigs, mandatory virologic and/or serologic examination of all hunter-killed wild boars would enable regulatory officials to determine their importance as a reservoir for CSF.

Classic swine fever is considered the most economically important viral disease in many pig-producing countries. Although wild boars may appear to be less susceptible to the disease, they play an important epidemiologic role in maintaining an endemic status of CSF in certain areas. Hence, it is of utmost importance that wild boars are included in CSF control and eradication programs. This can be achieved only by close and mutual cooperation between diagnostic laboratories, veterinary authorities, farmers, and hunters. In light of the observation that CSF is likely to disappear from the wild boar population once the virus has been eradicated in domestic pigs, it is conceivable that if a worldwide eradication of CSF can be achieved, this devastating disease can also be eliminated from wild boars.

LITERATURE CITED

Aguirre, A.A., D.E. Hansen, E.E. Starkey, and R.G. McLean. 1995. Serologic survey of wild cervids for potential disease agents in selected national parks in the United States. *Preventive Veterinary Medicine* 21:313–322.

Ahl, R. 1994. Zur Schweinepestsituation in Deutschland in den Jahren 1992 bis 1993. *Deutsche Tierärzteblatt* 4:314–316.

Alansari, H., K.V. Brock, and L.N.D. Potgieter. 1993. Single and double polymerase chain reaction for detection of bovine viral diarrhea virus in tissue culture and sera. *Journal of Veterinary Diagnostic Investigation* 5:148–153.

Aubert, M., M. Picard, E. Fouquet, J. Conde, C. Cruciere, R. Ferry, E. Albina, J. Barrat, and F. Vedeau. 1994. Classical swine fever in European wild boar. *Annales de Medecine Veterinaire* 138:239–247.

Baker, J.A., C.J. York, J.H. Gillespie, and G.B. Mitchell. 1954. Viral diarrhea in cattle. *American Journal of Veterinary Research* 15:525–531.

Baradel, J.M., J. Barrat, J. Blancou, J.M. Boutin, C. Chastel, G. Dannacher, D. Delorme, Y.L. Gerard, J.M. Gourreau, U. Kihm, B. Larenaudie, C. le Goff, P.-P. Pastoret, P. Perreau, A. Schwers, E. Thiry, D. Trap, G. Eilenberg, and P. Vannier. 1988. Results of a serological survey of wild mammals in France. *Revue Scientifique et Technique O.I.E.* 7:873–883.

Barrett, M., and G. Chalmers. 1975. A serologic survey of pronghorns in Alberta and Saskatchewan. *Journal of Wildlife Diseases* 11:157–163.

Bielefeldt-Ohmann, H. 1990. Electron microscopy of bovine virus diarrhoea virus. *Revue Scientifique et Technique O.I.E.* 9:61–73.

Bitsch, V., and L. Ronsholt. 1995. Control of bovine viral diarrhea virus infection without application of vaccines. In *The veterinary clinics of North America,* ed. J.C. Baker and H. Houe. Philadelphia: W.B. Saunders, pp. 627–640.

Bolin, S.R., A.W. McClurkin, R.C. Cutlip, and M.F. Coria. 1985. Response of cattle persistently infected with noncytopathic bovine viral diarrhea virus to vaccination for bovine viral diarrhea and to subsequent challenge exposure with cytopathic bovine viral diarrhea virus. *American Journal of Veterinary Research* 46:2467–2470.

Brass, W., L.C. Schulze, and S. Ueberschär. 1966. Über das Auftreten von mucosal Disease ähnlichen Erscheinungen bei Zoowiederkäuern. *Deutsche Tierärztliche Wochenschrift* 73:155–158.

Brock, K.V. 1995. Diagnosis of bovine viral diarrhea virus infections. In *The veterinary clinics of North America,* ed. J.C. Baker and H. Houe. Philadelphia: W.B. Saunders, pp. 549–562.

Brownlie, J. 1990. Pathogenesis of mucosal disease and molecular aspects of bovine viral diarrhea virus. *Veterinary Microbiology* 23:371–382.

Brownlie, J., M.C. Clarke, and C.J. Howard. 1984. Experimental production of fatal mucosal disease in cattle. *Veterinary Record* 114:535–536.

Brownlie, J., M.C. Clarke, C.J. Howard, and D.H. Popock. 1987. Pathogenesis and epidemiology of bovine virus diarrhoea virus infection of cattle. *Annales de Recherches Veterinaires* 18:157–166.

Carbrey, E.A. 1988. Diagnostic procedures. In *Classical swine fever and related infections,* ed. B. Liess. Boston: Martinus Nijhoff, pp. 99–114.

Carbrey, E.A., L.N. Brown, T.L. Chow, R.F. Kahrs, D.G. McKercher, L.K. Smithies, and T.W. Tamoglia. 1971. Recommended standard laboratory techniques for diagnosing infectious bovine rhinotracheitis, bovine viral diarrhea and shipping fever (parainfluenza-3). *Proceedings of the United States Animal Health Association* 75:629–648.

Carman, S., T. van Dreumel, and R. Trembley. 1994. Severe acute bovine viral diarrhea (BVD) in Ontario in 1993. In *Proceedings of the 37th annual meeting of the American Association of Veterinary Laboratory Diagnosticians, Grand Rapids, Michigan,* p. 19.

Cay, B., G. Chappuis, C. Coulibaly, Z. Dinter, S. Edwards, I. Greiser-Wilke, M. Gunn, P. Have, G. Hess, N. Juntti, B. Liess, A. Mateo, P. McHugh, V. Moennig, P. Nettleton, and G. Wesvoort. 1989. Comparative analysis of monoclonal antibodies against pestiviruses: Report of an international workshop. *Veterinary Microbiology* 20:123–129.

Clark, R.K., D.A. Jessup, M.D. Kock, and R.A. Weaver. 1985. Survey of desert bighorn sheep in California for exposure to selected infectious disease. *Journal of the American Veterinary Medical Association* 11:1175–1179.

Couvillion, C.E., E.W. Jenney, J.E. Pearson, and M.E. Coker. 1977. Survey for antibodies to viruses of bovine viral diarrhea, bluetongue and epizootic hemorrhagic disease in hunter-killed mule deer in New Mexico. *Journal of the American Veterinary Medical Association* 177:790–791.

Dahle, J., and B. Liess. 1992. A review on classical swine fever infections in pigs: Epizootiology, clinical disease and pathology. *Comparative Immunology, Microbiology and Infectious Diseases* 15:203–211.

Dahle, J., T. Patzelt, G. Schageman, and B. Liess. 1993. Antibody prevalence of hog cholera, bovine viral diarrhoea and Aujeszky's disease virus in wild boars in Northern Germany. *Deutsche Tierärztliche Wochenschrift* 100:330–333.

Dedek, J., H. Loepelmann, R. Kokles, C. Kretschmar, M. Müller, and H. Bergmann. 1988. Ergebnisse serologischer Untersuchungen auf Antikörper gegen das Virus der BVD/MD beim Rot-, Reh-, Dam- und Muffelwild. *Monatshefte für Veterinärmedizin* 43:63–65.

De Moerlooze, L., C. Lecomte, S. Brown-Shimmer, D. Schmetz, C. Guiot, D. Vandenbergh, D. Allaer, M. Rossius, G. Chappuis, D. Dina, A. Renard, and J.A. Martial. 1993. Nucleotide sequence of the bovine viral diarrhoea virus Osloss strain: Comparison with related viruses and identification of specific DNA probes in the 5′ untranslated region. *Journal of General Virology* 74:1433–1438.

Depner, K., O.J.B. Hübschle, and B. Liess. 1991. Prevalence of ruminant pestivirus infections in Namibia. *Onderstepoort Journal of Veterinary Research* 58:107–109.

Depner, K., T. Bauer, and B. Liess. 1992. Thermal and pH stability of pestiviruses. *Revue Scientifique et Technique O.I.E.* 11:885–893.

Depner, K., A. Müller, A. Gruber, A. Rodrigues, K. Bickhardt, and B. Liess. 1995. Classical swine fever in wild boar (*Sus scrofa*): Experimental infections and viral persistence. *Deutsche Tierärztliche Wochenschrift* 102:381–384.

Diaz, R., M. Steen, C. Rehbinder, and S. Alenius. 1988. An outbreak of a disease in farmed fallow deer (*Dama dama*) resembling bovine viral diarrhea/mucosal disease. *Acta Veterinaria Scandinavia* 29:369–376.

Done, J.T., S. Terlecki, C. Richardson, J.W. Harkness, J.J. Sands, D.S.P. Patterson, D. Sweasey, I.G. Shaw, C.E. Winkler, and S.J. Duffell. 1980. Bovine virus diarrhoea-mucosal disease virus: Pathogenicity for the fetal calf following maternal infection. *Veterinary Record* 106:473–479.

Donis, R.O. 1995. Molecular biology of bovine viral diarrhea virus and its interactions with the host. In *The veterinary clinics of North America,* ed. J.C. Baker and H. Houe. Philadelphia: W.B. Saunders, pp. 393–423.

Doyle, L.G., and W.P. Heuschele. 1983. Bovine viral diarrhea infection in captive exotic ruminants. *Journal of the American Veterinary Medical Association* 183:1257–1259.

Duffell, S.J., and J.W. Harkness. 1985. Bovine virus diarrhoea-mucosal disease infection in cattle. *Veterinary Record* 117:240–245.

Durham, P.J.K., and L.E. Hassard. 1990. An enzyme-linked immunosorbent assay (ELISA) for antibodies to bovine viral diarrhea virus. *Veterinary Microbiology* 22:1–10.

Edwards, S., J.J. Sands, and J.W. Harkness. 1988. The application of monoclonal antibody panels to characterise pestivirus isolates from ruminants in Great Britain. *Archives of Virology* 102:197–206.

Elazhary, M.A.S.Y., J.L. Frechette, A. Silim, and R.S. Roy. 1981. Serological evidence of some bovine viruses in caribou in Quebec. *Journal of Wildlife Diseases* 17:609–613.

English, A.W. 1982. Serological survey of wild fallow deer in New South Wales. *Veterinary Record* 110:153–154.

Fay, L.D., and A.P. Boyce. 1955. A new disease of deer? In *17th midwest wildlife conference* [cited by Richards et al. (1956)]

Feinstein, R., C. Rehbinder, E. Rivera, T. Nikkila, and M. Steen. 1987. Intracytoplasmic inclusion bodies associated with vesicular, ulcerative and necrotizing lesions of the digestive mucosa of a roe deer (*Capreolus capreolus*) and a moose (*Alces alces*). *Acta Veterinaria Scandinavia* 28:197–200.

Fernelius, A.L., G. Lambert, and R.A. Packer. 1969. Bovine viral diarrhea virus-host cell interactions: Adaptation, propagation, modification and detection of virus in rabbits. *American Journal of Veterinary Research* 30:1541–1550.

Fischer, S., E. Weiland, and K. Frölich. 1998. Characterization of a bovine viral diarrhea virus isolated from roe deer in Germany. *Journal of Wildlife Diseases* 31:47–55.

Frey, H.R., K. Depner, C.C. Gelfert, and B. Liess. 1991. BVD virus isolation techniques for use in cattle herds with or without previous BVD history. *Archives of Virology* (Supplement) 3:257–260.

Friend, M., and L.G. Halterman. 1967. Serologic survey of two deer herds in New York State. *Bulletin of the Wildlife Disease Association* 3:32–34.

Frölich, K. 1993. Bovine Virus diarrhoe/mucosal Disease (BVD/MD) bei Cerviden in unterschiedlichen Freiland- und Gehegepopulationen: Seroepizootiologie und Virusisolierung. Ph.D. diss., Berlin Freie Universität, Berlin, 102 pp.

———. 1995. Bovine viral diarrhea and mucosal disease in free-ranging and captive deer (Cervidae) in Germany. *Journal of Wildlife Diseases* 31:247–250.

Frölich, K., and E.J. Flach. 1998. Long-term viral serology of semi-free-living and captive ungulates. *Journal of Zoo and Wildlife Medicine* 29:165–170.

Frölich, K., and M. Hofmann. 1995. Isolation of bovine viral diarrhea virus-like pestivirus from roe deer (*Capreolus capreolus*). *Journal of Wildlife Diseases* 31:243–246.

Frölich, K., and W.J. Streich. 1998. Serologic evidence of bovine viral diarrhea virus in free-ranging rabbits from Germany. *Journal of Wildlife Diseases* 34:173–178.

Giovanni, A., F.M. Cancellotti, C. Turilli, and E. Randi. 1988. Serological investigations of some bacterial and viral pathogens in fallow deer (*Cervus dama*) and wild boar (*Sus scrofa*) of the San Rossore Preserve, Tuscany, Italy. *Journal of Wildlife Diseases* 24:127–132.

Haines, D.M., E.G. Clark, and E.J. Dubovi. 1992. Monoclonal antibody-based immunohistochemical detection of bovine viral diarrhea virus in formalin fixed paraffin embedded tissues. *Veterinary Pathology* 29:27–32.

Hamblin, C., and R.S. Hedger. 1979. The prevalence of antibodies to bovine viral diarrhea/mucosal disease virus in African wildlife. *Comparative Immunology, Microbiology and Infectious Diseases* 2:295–303.

Hassan, A.K.M., and G.R. Scott. 1986. A technique to obviate the risk of inadvertent infection of cell cultures with bovine viral diarrhoea virus. *Journal of Comparative Pathology* 96:241–246.

Hofmann, M.A., K. Brechtbühl, and N. Stäuber. 1994. Rapid characterization of new pestivirus strains by direct sequencing of PCR-amplified cDNA from the 5′ noncoding region. *Archives of Virology* 139:217–229.

Hooft van Iddekinge, B.J., J.L. van Wamel, H.G. van Gennip, and R.J. Moormann. 1992. Application of the polymerase chain reaction to the detection of bovine viral diarrhea virus infections in cattle. *Veterinary Microbiology* 30:21–34.

Horzinek, M.C. 1991. Pestiviruses—taxonomic perspectives. *Archives of Virology* (*Supplement*) 3:1–5.

Houe, H. 1995. Epidemiology of bovine viral diarrhea virus. In *The veterinary clinics of North America,* ed. J.C. Baker and H. Houe. Philadelphia: W.B. Saunders, pp. 521–548.

Hyera, J.M.K. 1989. Bovine viral diarrhoea (BVD) in domestic and wild ruminants in northern Tanzania. Ph.D. diss., Hannover Veterinary School, Hannover, Germany, 218 pp.

Hyera, J.M.K., B. Liess, E. Anderson, and K.N. Hirji. 1992. Prevalence of antibodies to bovine viral diarrhoea virus in some wild ruminants in northern Tanzania. *Bulletin of Animal Health and Production in Africa* 40:143–151.

Hyera, J.M.K., T. Mlengeya, H.R. Frey, W. Heuschele, and B. Liess. 1993. Virological and serological observations in wildebeests inoculated intranasally with a non-cytopathogenic bovine viral diarrhoea (BVD) virus isolate from east Africa. *Bulletin of Animal Health and Production in Africa* 41:329–330.

Kahrs, R., G. Atkinson, J.A. Baker, L. Carmichael, L. Coggins, J. Gillespie, P. Langer, V. Marshall, D. Robso, and B. Sheffy. 1964. Serological studies on the incidence of bovine viral diarrhea, infectious bovine rhinotracheitis, bovine myxovirus parainfluenza-3 and *Leptospira pomona* in New York State. *Cornell Veterinarian* 54:360–369.

Kahrs, R., D.S. Robson, and J.A. Baker. 1966. Epidemiological considerations for the control of bovine virus diarrhea. *Proceedings of the United States Livestock Sanitary Association* 70:145–153.

Kamolsiriprichaiporn, S., P.T. Hooper, C.J. Morrissy, and H.A. Westbury. 1992. A comparison of the pathogenicity of 2 strains of hog cholera virus: 1. Clinical and pathological studies. *Australian Veterinary Journal* 69:240–244.

Kast, A., and M. Kraus. 1968. MKS-Impfung als auslösender Faktor einer latenten MD im Zoo. *Verhandlungsbericht Erkrankung der Zootiere* 10:175–179.

Kingscote, B.F., W.D.G. Yates, and G.B. Tiffin. 1987. Diseases of wapiti utilizing cattle range in southwestern Alberta. *Journal of Wildlife Diseases* 23:86–91.

Kocan, A., A.W. Franzmann, K.A. Waldrup, and G.J. Kubat. 1986. Serological studies of selected infectious diseases of moose (*Alces alces*) from Alaska. *Journal of Wildlife Diseases* 22:418–511.

Krassnig, R., W. Schuller, J. Heinirich, F. Werfring, P. Kalaus, and M. Fruhwirth. 1995. Isolierung des Erregers der Europäischen Schweinepest (ESP) aus importiertem gefrorenem Wildschweinfleisch. *Deutsche Tierärztliche Wochenschrift* 102:56.

Laddomada, A., C. Patta, A. Oggiano, A. Caccia, A. Ruiu, P. Cossu, and A. Firinu. 1994. Epidemiology of classical swine fever in Sardinia: A serological survey of wild boar and comparison with African swine fever. *Veterinary Record* 134:183–187.

Lawman, M.J.P., D. Evens, E.P.J. Gibbs, A. McDiarmid, and L. Rowe. 1978. A preliminary survey of British deer for antibody to some virus diseases of farm animals. *British Veterinary Journal* 134:85–91.

Liebermann, H.D., J. Tabbaa, H. Dedek, H. Loepelmann, J. Stubbe, and H.J. Selbitz. 1989. Serologische Untersuchungen auf ausgewählte Virusinfektionen bei Wildwiederkäuern in der DDR. *Monatshefte für Veterinärmedizin* 44:380–382.

Liess, B. 1987. Pathogenesis and epidemiology of hog cholera. *Annales des Recherches Veterinaires* 18:139–145.

Loepelmann, H., and J. Dedek. 1987. Erfahrungen bei der Bekkämpfung der Schweinepest beim Schwarzwild in einem Beobachtungsgebiet der DDR. *Monatshefte für Veterinärmedizin* 42:313–316.

Loken, T. 1995a. Ruminant pestivirus infections in animals other than cattle and sheep. In *The veterinary clinics of North America,* ed. J.C. Baker and H. Houe. Philadelphia: W.B. Saunders, pp. 597–614.

———. 1995b. Border disease in sheep. In *The veterinary clinics of North America,* ed. J.C. Baker and H. Houe. Philadelphia: W.B. Saunders, pp. 579–595.

McClurkin, A.W., E.T. Littledike, and R.C. Cutlip. 1984. Production of cattle immunotolerant to bovine viral diarrhea virus (BVDV). *Canadian Journal of Comparative Medicine* 48:156–161.

McDiarmid, A. 1975. Some diseases of wild deer in the United Kingdom. *Veterinary Record* 97:6–9.

McGowan, M.R., P.D. Kirkland, S.G. Richards, and I.R. Littlejohns. 1993. Increased reproductive losses in cattle infected with bovine pestivirus around the time of insemination. *Veterinary Record* 133:39–43.

McKenzie, R.A., P.E. Green, A.M. Thornton, Y.S. Chung, and A.R. McKenzie. 1985. Diseases of deer in south eastern Queensland. *Australian Veterinary Journal* 62:424.

McMartin, D.A., D.R. Snodgrass, and W. Gonjall. 1977. Bovine virus diarrhoea antibody in a Scottish red deer. *Veterinary Record* 100:85–91.

Mehring, M. 1965. Beitrag zur Pathologie der Säugetiere zoologischer Gärten. Ph.D. diss., Hannover Veterinary School, Hannover, Germany, 51 pp.

Merza, M., E. Larsson, M. Steen, and B. Morein. 1994. Association of a retrovirus with a wasting condition in the Swedish moose. *Virology* 202:956–961.

Meyling, A. 1983. An immunoperoxidase (PO) technique for detection of BVD virus in serum of clinically and subclinically infected cattle. *Proceedings of the International Symposium for Veterinary Laboratory Diagnostics* 3:179–184.

Meyling, A., H. Houe, and A.M. Jensen. 1990. Epidemiology of bovine virus diarrhoea virus. *Revue Scientifique et Technique O.I.E.* 9:75–93.

Mignon, B., J. Dubuisson, E. Baranowski, I. Koromyslov, E. Ernst, D. Boulanger, S. Waxweiler, and P.-P. Pastoret. 1991. A monoclonal ELISA for bovine viral diarrhoea pestivirus antigen detection in persistently infected cattle. *Journal of Virological Methods* 35:177–188.

Miller, R.B., and B.N. Wilkie. 1979. Indirect fluorescent antibody test as a method for detecting antibodies in aborted fetuses (cattle). *Canadian Journal of Comparative Medicine* 43:255–261.

Moennig, V., and B. Liess. 1995. Pathogenesis of intrauterine infections with BVD virus. In *The veterinary clinics of North America,* ed. J.C. Baker and H. Houe. Philadelphia: W.B. Saunders, pp. 477–485.

Moennig, V., H.R. Frey, E. Liebler, J. Pohlenz, and B. Liess. 1990. Reproduction of mucosal disease with cytopathogenic bovine viral diarrhoea virus selected in vitro. *Veterinary Record* 127:200–203.

Morton, J.K., J.F. Evermann, and R.A. Dietrich. 1990. Experimental infection of reindeer with bovine viral diarrhea virus. *Rangifer* 10:75–77.

Moser, C., N. Ruggli, J.D. Tratschin, and M.A. Hofmann. 1996. Detection of antibodies against classical swine fever virus in swine sera by indirect ELISA using recombinant envelope glycoprotein E2. *Veterinary Microbiology* 51:41–53.

Munday, B.L. 1966. *Diseases of Tasmania's free-living animals.* Research Bulletin 5. Tasmania Department of Agriculture.

———. 1972. A serological study of some infectious diseases of Tasmanian wildlife. *Journal of Wildlife Diseases* 8:169–175.

Nettles, V.F., J.L. Corn, G.A., Erickson, and D.A. Jessup. 1989. A survey of wild swine in the United States for evidence of hog cholera. *Journal of Wildlife Diseases* 25:61–65.

Nettleton, P.F. 1990. Pestivirus infections in ruminants other than cattle. *Revue Scientifique et Technique O.I.E.* 9:131–150.

———. 1994. Mucosal disease (bovine virus diarrhoea). In *Management and diseases of deer,* ed. T.L. Alexander and D. Buxton, 2d ed. Long, UK: Veterinary Deer Society, pp. 124–125.

Nettleton, P.F., J.A. Herring, and W. Corrigal. 1980. Isolation of bovine virus diarrhea virus from a Scottish red deer. *Veterinary Record* 107:425–426.

Neumann, W., J. Buitkamp, G. Bechmann, and W. Plöger. 1980. BVD/MD Infektion bei einem Damhirsch. *Deutsche Tierärztliche Wochenschrift* 87:94.

Office International des Epizooties (OIE). 1992. *Manual of standards for diagnostic tests and vaccines.* Paris: OIE, pp. 109–116.

———. 1994. *Bulletin 106.* Paris: OIE.

———. 1995. *Bulletin 107.* Paris: OIE.

Oslage, U. 1993. Erhebung zur Prävalenz von Antikörpern gegen das Virus der Europäischen Schweinepest (ESP) in den Wildschweinpopulationen der Bundesländer Sachsen-Anhalt und Brandenburg. Ph.D. diss., Hannover Tierärztliche Hochschule, Hannover, Germany, 147 pp.

Oslage, U., J. Dahle, T. Müller, M. Kramer, D. Beier, and B. Liess. 1994. Prävalenz von Antikörpern gegen die Viren der Europäischen Schweinepest, der Aujeszky'schen Krankheit und des "Porcinen reproduktive and respiratory syndrome" (PRRS) bei Wildschweinen in den Bundesländern Sachsen-Anhalt und Brandenburg. *Deutsche Tierärztliche Wochenschrift* 101:33–38.

Parks, J.B., and J.J. England. 1974. A serological survey for selected viral infections of Rocky Mountain bighorn sheep. *Journal of Wildlife Diseases* 10:107–110.

Pastoret, P.-P., E. Thiry, B. Brochier, A. Schwers, I. Thomas, and J. Dubuisson. 1988. Diseases of wild animals transmissible to domestic animals. *Revue Scientifique et Technique O.I.E.* 7:705–736.

Peters, J.C. 1966. Mucosal Disease im Rotterdamer Zoologischen Garten. *Verhandlungsbericht Erkrankung der Zootiere* 7:329–333.

Picard, M., C. Burger, E. Plateau, and C. Cruciere. 1993. La peste porcine classique chez les sangliers: Un visage epidemiologique nouveau de la maladie. *Bulletin de la Societe des Veterinaires Pratiques de France* 77:81–92.

Plowright, W. 1969. Other diseases in relation to J.P. 15 Programme. In *1st annual meeting, Joint Campaign against Rinderpest (J.P.15), Mogadishu, Somalia.*

Potgeiter, L.N.D. 1992. Bovine viral diarrhea. In *Veterinary diagnostic virology,* ed. A.E. Castro and W.P. Heuschele. St. Louis: Mosby-Year Book, pp. 88–92.

Provost, A, K. Bögel, C. Borredon, and Y. Maurice. 1967. La maladie des muqueuses en Afrique centrale observations cliniques et epizootiologiques. *Revuė d'Elevage et de Medecine Veterinaire des pays Tropicaux* 20:27–49.

Rehbinder, C., E. Gimeno, K. Belak, M. Steen, E. Rivera, and T. Nikkilä. 1991. A bovine viral diarrhoea/mucosal disease-like syndrome in moose (*Alces alces*): Investigations on the central nervous system. *Veterinary Record* 129:552–554.

Rice, C.M. 1996. Flaviviridae: The viruses and their replication. In *Fields virology,* vol. 1, ed. B.N. Fields, D.M. Knipe, and P.M. Hawley. Philadelphia: Lippincott-Raven, pp. 931–959.

Richards, S.H., I.A. Schippers, D.F. Eveleth, and R.F. Shumard. 1956. Mucosal disease of deer. *Veterinary Medicine* 51:358–362.

Ridpath, J.F., S.R. Bolin, and E.J. Dubovi. 1994. Segregation of bovine viral diarrhea virus into genotypes. *Virology* 205:66–74.

Riemann, H.P., R. Ruppauer, P. Willeberg, C.E. Franti, W.H. Elliot, R.A. Fisher, O.A. Brunetti, J.A. Aho, J.A. Howarth, and D.E. Behymer. 1979. Serological profile of exotic deer at Point Reyes National Seashore. *Journal of American Veterinary Medical Association* 175:911–913.

Romvary, J. 1965. Incidence of virus diarrhea among roes. *Acta Veterinaria Hungaria* 15:451–455.

Sadi, L., R. Loyel, M. St. George, and L. Lamontagu. 1991. Serologic survey of white-tailed deer on Anticosti Island, Quebec for bovine herpesvirus, bovine viral diarrhea, and parainfluenza 3. *Journal of Wildlife Diseases* 27:569–577.

Sandvik, T., and J. Krogsrud. 1995. Evaluation of an antigen-capture ELISA for detection of bovine viral diarrhea virus in cattle blood samples. *Journal of Veterinary Diagnostic Investigation* 7:65–71.

Schellner, H.P. 1977. Untersuchungsergebnisse von Fallwild und anderen ausgewählten Tierarten von 1973–1976 in Bayern. *Tierärztliche Umschau* 32:225–229.

Shope, R.E., L.G. MacNamara, and R. Mangold. 1955. Deer mortality/epizootic hemorrhagic disease of deer. *New Jersey Outdoors* 6:16 [cited by Richards et al. (1956)].

Sosnowski, A. 1977. Infektionskrankheiten im Zoologischen Garten Lodz in den Jahren 1961 bis 1975. *Verhandlungsbericht Erkrankung der Zootiere* 19:147–151.

Stauber, E., C.H. Nellis, R.A. Magonigle, and H.W. Vaughn. 1977. Prevalence of selected livestock pathogens in Idaho mule deer. *Journal Wildlife Management* 41:515–519.

Stauber, E., R. Authenrieth, D.O. Markham, and V. Withebeck. 1980. A seroepidemiologic survey of three pronghorn populations in southeastern Idaho 1975–1977. *Journal of Wildlife Diseases* 16:109–115.

Steger, G. 1973. Untersuchungsergebnisse an 1500 Objekten aus der Gruppe wildlebender Wiederkäuer. *Verhandlungsbericht Erkrankung der Zootiere* 15:25–34.

Straver, P.J., W.G.J. Middel, F. Westenbrink, and P.W. de Leeuw. 1987. An ELISA for BVD virus serology. In *Pestivirus infections of ruminants,* ed. J.W. Harkness. Luxembourg: Office for Official Publications of the European Communities, pp. 81–85.

Stuen, S., J. Krogsrud, B. Hyllseth, and N.J.C. Tyler. 1993. Serosurvey of three virus infections in reindeer in northern Norway and Svalbard. *Rangifer* 13:215–219.

Tarry, D.W., L. Bernal, and S. Edwards. 1991. Transmission of bovine virus diarrhea virus by blood feeding flies. *Veterinary Record* 128:82–84.

Taylor, S.K., V.M. Lane, D.L. Hunter, K.G. Eyre, S. Kaufman, S. Frey, and M.R. Johnson. 1997. Serologic survey for infectious pathogens in free-ranging American bison. *Journal of Wildlife Diseases* 33:308–311.

Thiel, H.J., P.G.W. Plagemann, and V. Moennig. 1996. Pestiviruses. In *Fields virology,* ed. B.N. Fields, D.M. Knipe, P.M. Howley, R.M. Chanock, J.L. Melnick, T.P. Monath, B. Roizman, and S.E. Straus, 3d ed. Philadelphia: Lippincott-Raven, pp. 1059–1073.

Thorsen, J., and J.P. Henderson. 1971. Survey for antibody to infectious bovine rhinotracheitis (IBR), bovine virus diarrhea (BVD) and parainfluenza-3 (PI3) in moose sera. *Journal of Wildlife Diseases* 7:93–95.

Trautwein, G. 1988. Pathology and pathogenesis of the disease. In *Classical swine fever and related infections,* ed. B. Liess. Boston: Martinus Nijhoff, pp. 27–54.

Traven, M., S. Alenius, C. Fossum, and B. Larsson. 1991. Primary bovine viral diarrhoea virus infection in calves following direct contact with a persistently viraemic calf. *Journal of Veterinary Medicine [B].* 38:453–462.

Van Campen, H., and E.S. Williams. 1996. Wildlife and bovine viral diarrhea virus. In *Proceedings of the international symposium, Bovine viral diarrhea virus, a 50 year review.* Ithaca: Cornell University, pp. 167–175.

Van Campen, H., E.S. Williams, J. Edwards, W. Cook, and G. Stout. 1997. Experimental infection of deer (*Odocoileus* spp.) with bovine viral diarrhea virus. *Journal of Wildlife Diseases* 33:567–573.

Van Oirschot, J.T. 1988. Description of the virus infection. In *Classical swine fever and related infections,* ed. B. Liess. Boston: Martinus Nijhoff, pp. 1–25.

Weber, A., J. Paulsen, and H. Krauss. 1978. Seroepidemiologische Untersuchungen zum Vorkommen von Infektionskrankheiten bei einheimischem Schalenwild. *Praktische Tierarzt* 59:353–358.

Weber, A., K.P. Hürter, and C. Commicau. 1982. Über das Vorkommen des Virus diarrhoe/mucosal Disease-Virus bei Cerviden in Rheinland-Pfalz. *Deutsche Tierärztliche Wochenschrift* 89:1–3.

Wensvoort, G., M. Bloemraad, and C. Terpstra. 1988. An enzyme immunoassay employing monoclonal antibodies and detecting specifically antibodies to classical swine fever virus. *Veterinary Microbiology* 17:129–140.

Wiesner, H. 1987. Das Reh. In *Krankheiten der Wildtiere,* ed. K. Gabrisch and P. Zwart. Hannover: Schlütersche Verlagsanstalt, pp. 467–493.

Wilhelmsen, C.L., S.R. Bolin, J.F. Ridpath, N.F. Cheville, and J.P. Kluge. 1990. Experimental primary postnatal bovine viral diarrhea viral infections in six-month-old calves. *Veterinary Pathology* 27:235–243.

Williams, E.S., E.T., Thorne, S. Anderson, and J.D. Herriges Jr. 1993. Brucellosis in free-ranging bison (*Bison bison*) from Teton County, Wyoming. *Journal of Wildlife Diseases* 29:118–122.

Zarnke, R.L. 1983. Serologic survey for selected microbial pathogens in Alaskan wildlife. *Journal of Wildlife Diseases* 19:324–329.

13

CORONAVIRAL INFECTIONS

JAMES F. EVERMANN AND DAVID A. BENFIELD

Synonyms: Coronavirus enteritis, coronavirus pneumonia, coronavirus encephalitis, feline infectious peritonitis, canine coronavirus, bovine coronavirus, porcine coronavirus, elk coronavirus

INTRODUCTION. The coronaviruses have emerged over the past 30 years to become one of the most widely studied virus groups affecting animals (Saif and Heckert 1990; Holmes and Lai 1996). The viruses were initially named based on the disease syndrome they were isolated from, and subsequently shown to cause following experimental inoculation (Pensaert et al. 1970; Stair et al. 1972; Mebus et al. 1975). Coronaviruses are currently placed in the family *Coronaviridae,* and together with the family *Arteriviridae,* comprise a new order: *Nidovirales* (Cavanagh 1997).

Virtually every animal species that has been studied has been shown to be infected by a coronavirus. These viruses are well documented to cause diarrhea and respiratory disease in domestic ungulates, including sheep *Ovis aries,* cattle *Bos taurus,* swine *Sus scrofa,* and horses *Equus caballus.* The coronaviruses infect predominantly neonatal animals, but older cattle and swine are also infected (Collins et al. 1987; Saif and Heckert 1990). The role of coronavirus in causing morbidity and mortality in wild animals is just beginning to be recognized. Notable among the coronavirus infections of wild mammals are feline coronavirus/feline infectious peritonitis (FIP) of large felids, primarily cheetahs *Acinonyx jubatus,* and enteric coronaviral infections of captive wild ruminants and swine (Chasey et al. 1984; Evermann et al. 1988; Heeney et al. 1990; Tsunemitsu et al. 1995; Majhdi et al. 1997). Due to the sparsity of information on coronaviral infections of wildlife, a comparative approach has been taken in this chapter. Important coronaviral infections of ungulates and carnivores are discussed and, whenever applicable, correlated with wildlife species.

HISTORY. The actual prevalence of coronavirus in wild populations has been underestimated due to studies that rely on serologic results to assess prevalence of infection (Gardner et al. 1996). Coronaviral infections are generally limited to the mucosal surfaces of either the respiratory tract or the gastrointestinal tract and therefore may not stimulate high levels of circulating antibodies (Saif and Heckert 1990; Holmes and Lai 1996). A 1978 serologic survey of caribou *Rangifer tarandus* from the George River area in Northern Quebec, Canada, indicated that 13% had antibodies that neutralized bovine coronaviruses (Tsunemitsu et al. 1995). Lower seroprevalence in wild mammals contrasts with domestic herds, where prevalence of adult animals with antibodies to coronavirus is usually greater than 80% (Rodak et al. 1982). True assessment of coronavirus prevalence would require more frequent testing and/or more sensitive assays, such as polymerase chain reaction (PCR), that detect low levels of virus in feces or respiratory secretions (Shockley et al. 1987; Benfield et al. 1991; Herrewegh et al. 1995). Alternatively, coronavirus infections of wild mammals may be rare, or the animals are inherently resistant to infection and sufficient virus is not shed to infect large numbers of animals in the population at risk.

Coronavirus infections of domestic animals were initially recognized in 1951 and were primarily associated with enteric disease (Barker et al. 1993). The viruses were difficult to culture and, as a result, were primarily detected by histopathology and fluorescent antibody staining of gut sections (Mebus et al. 1975; Langpap et al. 1979). In 1968, the first reports of coronavirus associated with calf scours were noted. The enteric coronaviruses were frequently associated with concurrent infection with rotavirus and enterotoxigenic strains of *Escherichia coli.* It was during this time that electron microscopy (EM) was beginning to be used in the routine diagnosis of viral scours in domestic animals (Pass et al. 1982). Coronaviral infection was regarded as one of the primary causes of calf scours. In 1971, coronaviral infections were noted in domestic canids and felids. Enteric disease was the primary clinical and pathologic form of the canine disease. However, the disease that attracted primary attention in cats was an immune-mediated disorder: FIP. Based on EM of fixed lesions, this disease was considered to be caused by a coronavirus (Barker 1993). In subsequent years, FIP virus was cultured, and Koch's postulates were confirmed. Later, another coronavirus of cats was detected and referred to as feline enteric coronavirus. The first report of a bovine-like coronavirus in wild captive mammals was in sitatunga *Tragelaphus spekei* in 1984 (Chasey et al. 1984).

A parallel group of coronaviruses, murine hepatitis virus (MHV), was being studied in mice in the 1960s.

TABLE 13.1—Coronavirus infections of mammals

	Enteric	Respiratory	Neurologic	Other Manifestations
Bovidae	Bovine coronavirus	Bovine respiratory coronavirus	NR[a]	NR
	Ovine coronavirus	NR	NR	NR
Felidae	Feline enteric coronavirus	Feline infectious peritonitis	Feline infectious peritonitis	Feline infectious peritonitis (ocular)
Canidae	Canine coronavirus	NR	Canine infectious peritonitis[b]	NR
Muridae	Diarrhea virus of infant mice	NR	Murine hepatitis virus	Murine hepatitis virus (ocular)
	Sialodacryoadenitis virus	Sialodacryoadenitis virus	NR	NR
Leporidae	Rabbit enteric coronavirus	NR	NR	Rabbit pleuritis virus
Suidae	Transmissible gastroenteritis virus	Porcine respiratory coronavirus	Hemagglutinating encephalomyelitis virus	NR
Equidae	Equine coronavirus	NR	NR	NR
Primate	Simian coronavirus	NR	NR	NR
Mustelidae	Mink enteric coronavirus	NR	NR	NR

[a]NR, not reported.
[b]Following vaccination.

These viruses were recognized initially in mice held in captivity that had other viral infections, such as murine leukemia virus, and the studies were later expanded to include wild meadow voles *Microtus pennsylvanicus* (Descoteaux and Mihok 1986). The MHV group constitutes a wide range of pathogens from avirulent to virulent (Compton et al. 1993). The virulence of the strains is host genotype specific, suggesting that the host range of coronavirus-induced disease is related to the host genetics (Barthold et al. 1993).

The coronaviruses occupy a wide ecological niche in nature. Table 13.1 list the common coronaviral infections of animals. Although the diseases caused by coronaviruses were initially described in domesticated mammals, it became apparent that wildlife were susceptible to infection and, in some cases, diseases associated with coronavirus (Evermann et al. 1980; Foreyt and Evermann 1985; Roelke et al. 1993; Tsunemitsu et al. 1995).

DISTRIBUTION AND HOSTS. Occurrence of coronaviruses in mammals is widespread. The viruses are enveloped and, as such, are highly labile outside the host (Tennant et al. 1994). Coronaviruses primarily persist in hosts as subclinical infections of the mucosal surfaces of adult animals (Collins et al. 1987). The viruses are intermittently shed in body secretions (saliva, aerosol, etc.) and excretions (feces) throughout life. Viral transmission is unusually high during periods of pregnancy and from young animals that acquire the infection and progress onto disease. It is for this reason that coronaviral diseases are noted in areas of high animal density and during times of parturition, when neonatal animals are at risk.

The host range of the coronaviruses is generally restricted to single or closely related animal species. Interspecies transmission has been reported for canine coronavirus between dogs *Canis familiaris* and cats *Felis catus,* bovine coronavirus between cattle and elk *Cervus elaphus,* porcine coronavirus from pigs to dogs and foxes, etc. (McArdle et al. 1992). The host range of the coronaviruses is primarily restricted due to receptors on the surface of host mammalian cells (Holmes and Lai 1996).

ETIOLOGY. The coronaviruses constitute a genus within the family *Coronaviridae* (Cavanagh 1997). Coronaviruses are large, enveloped, positive-sense RNA viruses. The coronaviruses also have the largest genome (27–32 kb) of RNA viruses. The presence of the lipid envelope imparts pleomorphism to size and shape of the virion (Fig. 13.1). The virions mature by budding into intracellular membranes such as the rough endoplasmic reticulum and Golgi apparatus to acquire the lipid envelope containing inserted viral proteins and glycoproteins (Compton et al. 1993). Most coronavirions contain one row of club-shaped peplomers or surface (S) projections approximately 12–15 nm in length, and others contain a second row of short spikes that compose the hemagglutinin-esterase (HE) glycoprotein on the envelope. The coronaviruses have a unique method of replication producing six to seven subgenomic messenger RNAs (mRNAs) with common 3′ ends and a 5′ leader. The genomic RNA, like the mRNAs, contains a 5′ cap and 3′ polyadenylated tail. The viral genome encodes for 3–4 structural proteins and several nonstructural proteins. Most of the genome (20 kb) consists of two overlapping open reading frames ORF1a and ORF1b that encode the viral RNA-dependent-RNA polymerase, proteases, and other unrecognized proteins. The remaining 7–12 kb encode for the structural proteins. The coding sequence of the structural proteins is highly conserved in most coronaviruses with the 5′-pol-S-M-N-3′. The unique replication method of a coronaviruses imparts a high rate of mutation due to recombination (Lai 1996). Although a

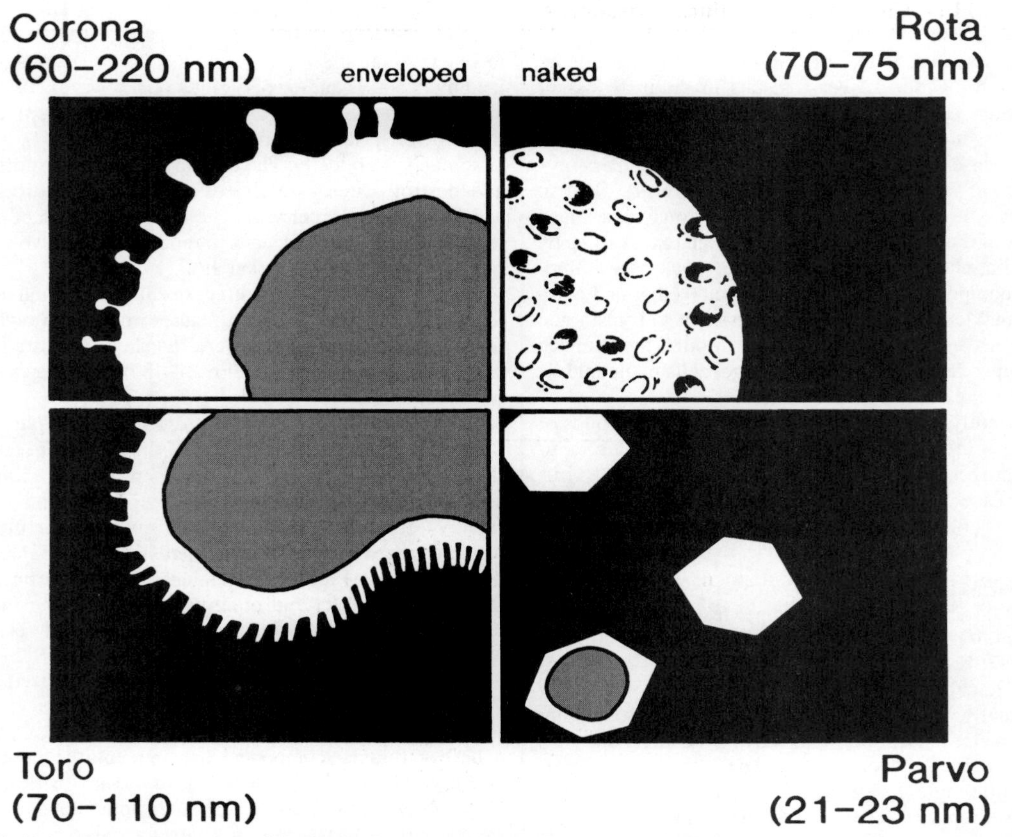

FIG. 13.1— Schematic diagram of the four major enteric viruses of mammals: coronavirus, rotavirus, torovirus, and parvovirus.

majority of the recombinations or mutations are silent, a rare escape mutant may occur and result in altered virulence. It is likely that feline enteric coronavirus mutates to the lethal FIP virus (Evermann et al. 1995; Poland et al. 1996). Once mutation occurs, the susceptibility of the host to disease is regulated by the host genotype and subsequent immune response (Foley and Pedersen 1996).

The mature coronavirus virion includes the nucleocapsid or N protein of 50–60 kDa. The N protein is a phosphoprotein that interacts with viral RNA to form an icosahedral ribonucleoprotein complex, and may also elicit cell-mediated immunity. The glycoprotein M (20–35 kDa) is a membrane-spanning glycoprotein that penetrates the lipid bilayer of the virion envelope three times. The M glycoprotein has a single accessible glycosylation site that is either N- or O-glycosylated, depending on the coronavirus. Antibody to the external domain of the M glycoprotein neutralizes virus in the presence of complement. The M protein may also function to bind the nucleocapsid to the viral envelope during virus budding. The M protein of some coronaviruses can also induce interferon-alpha. The S glycoprotein (90–180 kDa) is the structural protein of the peplomers on the surface of the virion. Functions attributed to the S glycoprotein include cell attachment, membrane fusion to mediate entry of the nucleocapsid, and induction of complement-independent neutralizing antibodies. The HE glycoprotein is primarily restricted to some group-II coronaviruses. The protein is a 130–140-kDa disulfide-linked dimmer of a 65–70-kDa protein that forms short spikes. Coronaviruses that express the HE bind to 9-O-acetylated neuraminic acid reside on glycoproteins or glycolipids and cause hemagglutination and hemadsorption. The HE also contains acetylesterase activity that cleaves acetyl groups from the substrate, potentially eluting adsorbed virions and destroying the HE-binding activity of the glycans on the cell membrane. The HE glycoproteins permit initial adsorption of the virus to cell membranes, but subsequent interaction of the S glycoprotein with its glycoprotein receptor may be required for fusion of the viral envelope with cell membranes. The HE is not required for infectivity in vitro.

Coronaviruses are not exceptionally stable in the environment (Holmes and Lai 1996). These viruses are thermolabile and highly photosensitive. Storage at refrigerator or room temperature will result in loss of infectivity over days or months while storage at –20°C–80° C for 12 years results in minimal loss of virus titer. The more common occurrence of coronavirus infections in the winter months may relate to the fact that these viruses are best preserved by lower temperatures and lower ultraviolet light levels that are prevalent in winter. The lipid envelope of the coronaviruses also makes virions susceptible to chemical inactivation by formalin, phenol, beta-propiolactone, quaternary ammonium compounds, and the lipid solvents ether and chloroform. Most coronaviruses are resistant to trypsin and low pH, which allows for passage through the stomach and upper small intestine to the target cells in the middle to lower small intestine and colon.

Currently, there are three distinct antigenic groups of coronaviruses. Most of the related viruses share common antigenic epitopes on the nucleocapsid of the virus and nucleocapsid gene sequences. There is also cross-reactivity observed for the S and M structural proteins.

TRANSMISSION. Coronaviruses are shed in mucosal secretions from the upper respiratory tract and in excretions from the gastrointestinal tract (Collins et al. 1987; Kapil and Goyal 1995). Transmission is generally regarded as horizontal from parent to offspring postnatally. It may also occur from one adult to another adult in close proximity. This may be the likely scenario with cattle, elk, deer, and muskox *Ovibos moschatus* that commingle. Evidence for vertical transmission has not been reported for the coronavirus family.

EPIDEMIOLOGY. There have been limited prevalence studies for coronaviral infections of wild mammals. Coronavirus infections of domestic cattle, pigs, dogs, and cats are regarded as endemic, with greater than 80% of the populations seropositive by 1 year of age (Barker et al. 1993). In wild populations, several factors might limit coronavirus infection. These include low animal density, limited interspecies transmission, no insect vectors, high lability of the virus outside the host, and restricted host range due to specific viral receptors. This is generally reflected in seroprevalence studies of wild animals, such as canids to canine coronavirus (1.7%), felids to felid coronavirus (2%), and various bovids to bovine coronavirus (range, 6.6%–13.3%) (Evermann et al. 1980, 1988; Foreyt and Evermann 1985; Tsunemitsu et al. 1995). There are exceptions, and one may argue that when wild mammals are managed to any extent, such as on common winter feeding grounds, the risk of infection increases accordingly. The cheetah's exposure to feline coronavirus varies according to the habitat and may reflect incursion by domestic cats or dietary exposure to cross-reacting coronaviruses of feral swine (Evermann et al. 1988; Heeney et al. 1990).

CLINICAL SIGNS. Coronavirus infections of mammals result in at least three major disease manifestations. The first, and most common, is enteritis, followed by respiratory dysfunction ranging from rhinitis to pneumonia, and then systematic disease characterized by hepatitis and/or peritonitis (Barker et al. 1993).

The hallmarks of enteric coronavirus infections are tropism for gastrointestinal epithelial cells and failure to spread systemically. The enteric coronaviruses infect and destroy enterocytes, resulting in villous atrophy and fusion of adjacent villi. The loss of function of the mature absorptive cells leads to reduced absorptive surfaces in the intestine (Barker et al. 1993).

The clinical signs are a direct result of intestinal cell damage and manifested as a malabsorptive, maldigestive diarrhea. In case of severe diarrhea, dehydration occurs and death ensues within 24–48 hours after onset of clinical signs.

The respiratory coronaviruses are unique in that the viruses may have adapted to entry via the upper respiratory mucosa to ensure persistence in the host (Kapil and Goyal 1995). Porcine respiratory coronavirus (PRCV), for example, is a deletion mutant of the more virulent enteric coronavirus, transmissible gastroenteritis virus. Thus, PRCV isolates have lost their tropism for the enteric tract and preferentially replicate in the respiratory tract (Rasschaert et al. 1990; Wesley et al. 1990; Sanchez et al. 1992).

The systemic coronaviruses are best characterized by virulent strains of MHV and FIP virus (Evermann et al. 1988; Barthold et al. 1993). Both of these diseases appear to have an immune component that augments the disease. These viruses have a propensity to infect and persist in macrophages. In the case of FIP and captive felids such as cheetahs, the disease is characterized by a fatal immune-mediated vasculitis. Other large felids, such as the lion *Panthera leo,* do not appear susceptible to disease, although evidence of infection has been reported, based on serologic studies (Heeney et al. 1990).

PATHOGENESIS AND PATHOLOGY. The enteric coronaviruses infect enterocytes throughout the length of the villi and the length of the small intestine (Saif and Heckert 1990; Holmes and Lai 1996). The lesions are a direct result of the cytolytic nature of the virus (Barker et al. 1993). Absorptive epithelial cells, which line the small intestinal villi, are destroyed by the coronavirus and exfoliate. Epithelial cells on villi are constantly being replaced by cells that originate in the crypts and migrate up the sides of the villi. The turnover rate of these cells is slower in immature animals, leading to less rapid repair of villous atrophy. Loss of virus-infected cells results in marked shortening of villi, reduced absorptive capacity of the small intestine, and malabsorptive diarrhea. Lesions and consequences are most severe in young animals. Bovine enteric coronaviruses produce a persistent infection of villous enterocytes throughout the distal portion of the small intestine and colon.

Gross lesions include milk- or bile-stained fluid in the stomach. The small intestine is usually thin walled, flaccid, and contains yellow fluid with flecks of mucus. There is an absence of fat absorption in the mesenteric lymphatics. The colon and cecum are often filled with watery fluid. Microscopically, the principal lesion is marked shortening or atrophy of the villi due to the exfoliation of the absorptive epithelial cells. Villi appear stumpy and club shaped, and fusion between villi is common. The virus does not replicate in crypt cells, which provide the replacement cells for the villi. Crypt epithelium is usually hyperplastic, indicating increased mitotic activity. In bovids, the colon may contain exfoliated, flattened, squamous epithelium and mild inflammation in colonic glands (Barker et al. 1993).

Lesions of FIP are markedly different than those described for enteric coronavirus infections. At necropsy, these cats are in poor to emaciated body condition and have abdominal distension due to fluid accumulation. Peritonitis occurs in most but not all animals with FIP. Serosal surfaces are often covered with fibrin, giving them a granular appearance, and granulomas in liver, spleen, kidney, and small intestine are common. Abdominal and thoracic lymph nodes may be enlarged. In some cases, lesions are restricted to inflammation in the eyes and nervous system. The characteristic microscopic lesion is generalized vasculitis and perivasculitis especially of venules. Neutrophils, lymphocytes, plasma cells, and macrophages accumulate in and around affected vessels. Lesions in the various organs result primarily from vascular damage (Barker 1993).

DIAGNOSIS AND DIFFERENTIAL DIAGNOSES. Methods for diagnosis of coronavirus infections in wild mammals are similar to those used to detect viral infections in domestic animals (Benfield and Saif 1990; Crouch et al. 1984; Gorham et al. 1990; Deeb et al. 1993). Diagnosis is based on clinical signs; detection of virus, viral antigen, or viral nucleic acid; serology; and microscopic lesions. Clinical signs are of little diagnostic value, because coronaviral infections cause signs that mimic other enteric infections. Virus isolation is often unsuccessful, because coronaviruses are difficult to adapt to cell culture and are present in excretions and secretions that contain bacteria and other compounds cytotoxic to cell cultures (Benfield and Saif 1990). Detection of coronavirus particles by EM and immunoelectron microscopy (IEM) of fecal material continues to be the "gold standard" for diagnosis of enteric coronavirus infection in domestic mammals (Stair et al. 1972; Langpap et al. 1979; Heckert et al. 1989) and wild mammals (Chasey et al. 1984; Tsunemitsu et al. 1995) (Fig. 13.2). Detection of viral antigens in the cytoplasm of infected cells in frozen intestinal or fixed sections by immunofluorescence (IF) or immunohistochemistry (IHC) is also economical and reliable for diagnosis of coronavirus infections (Pensaert et al. 1970; Mebus et al. 1975; Shoup et al. 1996). Other techniques such as enzyme-linked immunosorbent assays (ELISA) (Crouch et al. 1984; Reynolds et al. 1984; Smith et al. 1996) and cDNA probes (Shockley

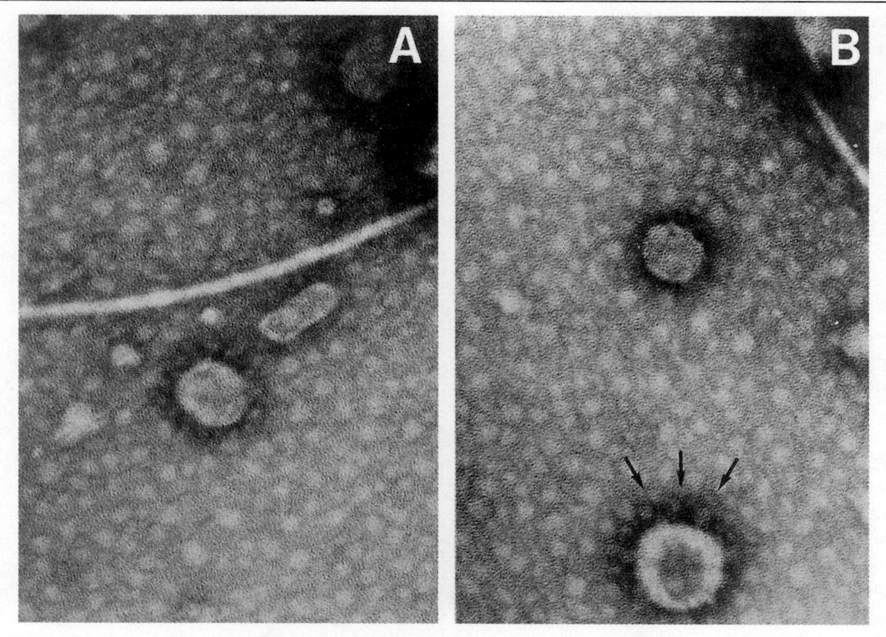

FIG. 13.2—Cheetah coronavirus particles. Size ranges from 120 nm (**A**) to 150 nm (**B**). Note peplomers extending from the intact virion (*arrows*). The virus particles are in a fecal sample obtained from clinically normal cheetah. ×210,000.

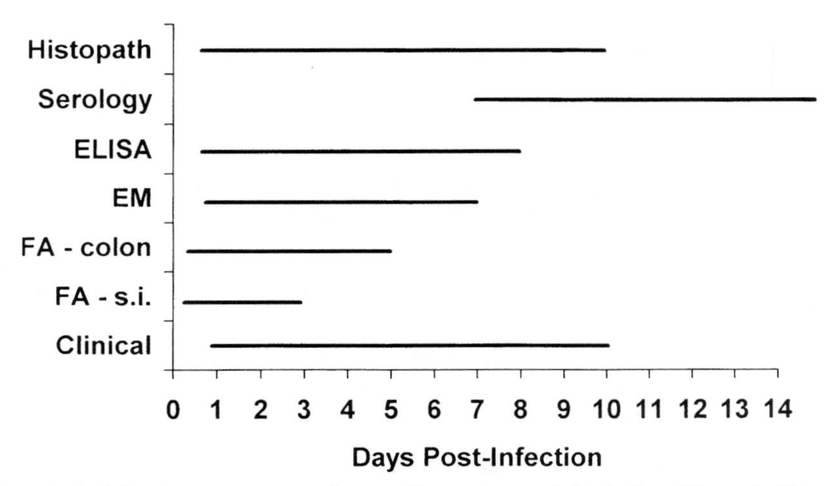

FIG. 13.3—Schematic depicting the most opportune times to diagnose coronaviral infections: Histopath, histopathology; ELISA, enzyme-linked immunosorbent assay; EM, direct or immunoelectron microscopy; FA, fluorescent antibody; s.i., small intestine; and *clinical* means signs of diarrhea, vomiting, respiratory dysfunction. Days after infection indicates the days following exposure.

et al. 1987; Benfield et al. 1991), have not been as reliable as EM and IEM for detection of coronavirus particles in fecal material. Microscopic lesions of villous atrophy are not specific for coronavirus infections and need to be confirmed by additional tests such as IF or IHC to detect the presence of coronaviral antigens in the remaining enterocytes.

Each of the various diagnostic assays mentioned have specific windows of sensitivity for detection of virus and viral antigens (Fig. 13.3). Coronaviruses are cytolytic, and the exfoliation of infected cells into the intestinal lumen limits the usefulness of techniques such as IF and IHC too early in the infection, whereas EM and ELISA can detect virions or viral antigens in fecal material for longer periods. Loss of epithelial cells especially narrows the window of opportunity for detection of coronaviruses by techniques such as IF and IHC that require the structural integrity of the complete cell for identification of viral antigen.

There is a need to develop more sensitive and reliable assays for detection of coronaviruses in the excretions and secretions in which the virus is shed in nature. Recent use of PCR technology, such as that used to detect feline coronavirus in body fluids of cats, may offer a possible method of viral detection (Herrewegh et al. 1995). Serology (neutralization and hemagglutination-inhibition assays) are useful only for retrospective diagnosis and epidemiologic surveys. Serologic surveys have been most commonly used to detect the presence of coronaviruses in wild mammals, such as caribou (Elazhary et al. 1981).

Coronaviruses produce few clinical signs or lesions that are specific to these viruses only (Barker et al. 1993). Differential diagnosis includes other enteropathogenic viruses (rotavirus, adenovirus, torovirus, parvovirus, and bovine viral diarrheal virus), bacteria (*Campylobacter, Clostridium,* enterotoxigenic and enterohemorrhagic *Escherichia coli, Salmonella* spp., and *Serpulina* spp.), parasites (various nematodes and trematodes), and protozoa (coccidia and *Cryptosporidium*) that induce diarrhea (Evermann et al. 1980; Saif and Heckert 1990; Martin and Zeidner 1992; Koopmans and Horzinek 1995).

IMMUNITY. Localized immunity is critical to minimizing the impact of coronaviral infections at the respiratory and gastrointestinal mucosal surfaces (Gustafsson et al. 1996). During the first few weeks of life, neonatal mammals depend on colostral immunoglobulin G (IgG) for passive immunity (Kapil et al. 1994). This form of protection has been referred to as *lactogenic immunity* and persists for several weeks after colostral immunoglobulins have waned. The predominant immunoglobulin in milk is IgA in species with simple stomachs and IgG1 in ruminants (Lamm et al. 1996). Eventually, secretory IgA is generated by the host in the form of active immunity. This form of immunity is antigen dependent and is constantly in stages of reinfection, restimulation, and localized protection (El-Kanawati et al. 1996; Lamm et al. 1996).

TREATMENT AND CONTROL. The control of coronaviruses depends heavily on adequate intake of colostral antibody and maintaining the neonate on the dam for sustained periods. Neonates born to first-lactation animals are more prone to coronavirus-induced diarrhea due to lack of protective antibody.

Treatment of coronaviral diarrhea is usually symptomatic, with fluid rehydration, electrolyte therapy, and provision of a warm, dry environment (Barker et al. 1991; Saif and Heckert 1990). Modified-live vaccines have been used in commercial bovine and porcine herds with limited success.

Biosecurity is the main defense against coronavirus infections in domestic herds. The missing link in the epidemiology of coronaviral infection is where the virus "overwinters" during warmer months of the year, when the prevalence of infection is lower (Gulland 1996). Subclinically infected adult animals are suspected carriers (Collins et al. 1987; Tennant et al. 1994; Storz et al. 1996). High animal density and commingling with domestic species should be avoided with captive wild mammals.

PUBLIC AND DOMESTIC ANIMAL HEALTH CONCERNS. Although there are human strains of coronavirus, these are regarded as host specific. There is no recognized zoonotic potential of the animal coronaviruses in humans.

The potential for interspecies transmission of coronaviruses among domestic and wild animals is possible (Evermann et al. 1980; Ballou 1993; Cunningham 1996). The coronaviruses of animals are usually very species specific due to receptor specificity, resulting in cross-infection between closely related species such as wild felids and domestic cats, wild canids and domestic dogs, and less closely related species such as cattle and elk. The potential for interspecies transmission is minimal unless common range or habitat is utilized, since the coronaviruses are extremely labile outside the host animal (Tennant et al. 1994).

MANAGEMENT IMPLICATIONS. It is important to recognize the host range of the respective coronaviruses in order to take appropriate management steps when wild mammals are winter fed, captured for translocation, held captive for research purposes, maintained in zoologic collections, or farmed (Spalding and Forrester 1993). Coronaviruses are not known to be significant pathogens in free-ranging wildlife populations.

LITERATURE CITED

Ballou, J.D. 1993. Assessing the risks of infectious disease in captive breeding and reintroduction programs. *Journal of Zoo and Wildlife Medicine* 24:327–335.

Barker, I.K. 1993. The peritoneum and retroperitoneum. In *Pathology of domestic animals,* ed. K.V.F. Jubb, P.C. Kennedy, and N. Palmer, 4th ed. San Diego: Academic, pp. 438–441.

Barker, I.K., A.A. van Dreumel, and N. Palmer. 1993. The alimentary system. In *Pathology of domestic animals,* ed. K.V.F. Jubb, P.C. Kennedy, and N. Palmer, 4th ed. San Diego: Academic, pp. 184–190.

Barthold, S.W., D.S. Beck, and A.L. Smith. 1993. Enterotropic coronavirus (mouse hepatitis virus) in mice: Influence of host age and strain an infection and disease. *Laboratory Animal Science* 43:276–284.

Benfield, D.A., and L.J. Saif. 1990. Cell culture propagation of a coronavirus isolated from cows with winter dysentery. *Journal of Clinical Microbiology* 28:1454–1457.

Benfield, D.A., D.J. Jackwood, I. Bae, L.J. Saif, and R.D. Wesley. 1991. Detection of transmissible gastroenteritis virus using cDNA probes. *Archives of Virology* 116:91–106.

Cavanagh, D. 1997. Nidovirales: A new order comprising *Coronaviridae* and *Arteriviridae. Archives of Virology* 142:629–633.

Chasey, D., D.J. Reynolds, J.C. Bridger, T.G. Debney, and A.C. Scott. 1984. Identification of coronaviruses in exotic species of Bovidae. *Veterinary Record* 115:602–603.

Collins, J.K., C.A. Ringel, J.D. Olson, and A. Fountain. 1987. Shedding of enteric coronavirus in adult cattle. *American Journal of Veterinary Research* 48:361–365.

Compton, S.R., S.W. Barthold, and A.L. Smith. 1993. The cellular and molecular pathogenesis of coronaviruses. *Laboratory Animal Science* 43:15–28.

Crouch, C.F., T.J.G. Raybould, and S.D. Acres. 1984. Monoclonal antibody capture enzyme-linked immunosorbent assay for detection of bovine enteric coronavirus. *Journal of Clinical Microbiology* 19:388–393.

Cunningham, A.A. 1996. Disease risks of wildlife translocation. *Conservation Biology* 10:349–353.

Deeb, B.J., R.F. Di Giacomo, J.F. Evermann, and M.E. Thouless. 1993. Prevalence of coronavirus antibodies in rabbits. *Laboratory Animal Science* 43:431–433.

Descoteaux, J.P., and S. Mihok. 1986. Serologic study on prevalence of murine viruses in a population of wild meadow voles (*Microtus pennsylvanicus*). *Journal of Wildlife Diseases* 22:314–319.

Elazhary, M.A.S.Y., J.L. Frechette, A. Silim, and R.S. Roy. 1981. Serological evidence of some bovine viruses in the caribou (*Rangifer tarandus caribou*) in Quebec. *Journal of Wildlife Diseases* 17:609–612.

El-Kanawati, Z.R., H. Tsunemitsu, D.R. Smith, and L.J. Saif. 1996. Infection and cross-protection studies of winter dysentery and calf diarrhea bovine coronavirus strains in colostrum-deprived and gnotobiotic calves. *American Journal of Veterinary Research* 57:48–53.

Evermann, J.F., W. Foreyt, L. Maag-Miller, C.W. Leathers, A.J. McKeirnan, and B. Leamaster. 1980. Acute hemorrhagic enteritis associated with canine coronavirus and parvovirus infections in a captive coyote population. *Journal of American Medical Association* 177:784–786.

Evermann, J.F., J.L. Heeney, M.E. Roelke, A.J. McKeirnan, and S.J. O'Brien. 1988. Biological and pathological consequences of feline infectious peritonitis virus infection in the cheetah. *Archives of Virology* 102:155–171.

Evermann, J.F., C.J. Henry, and S.L. Marks. 1995. Feline infectious peritonitis. *Journal of the American Veterinary Medical Association* 206:1130–1134.

Foley, J.E., and N.C. Pedersen. 1996. The inheritance of susceptibility to feline infectious peritonitis in purebred catteries. *Feline Practice* 24:14–22.

Foreyt, W.J., and J.F. Evermann. 1985. Serologic survey of canine coronavirus in wild coyotes in the western United States. *Journal of Wildlife Diseases* 21:428–430.

Gardner, I.A., S. Hietala, and W.M. Boyce. 1996. Validity of using serological tests for diagnosis of diseases in wild animals. *Revue Scientifique et Technique O.I.E.* 15:323–335.

Gorham, J.R., J.F. Evermann, A. Ward, R. Pearson, D. Shen, G.R. Harsough, and C.W. Leathers. 1990. Detection of coronavirus-like particles from mink with epizootic catarrhal gastroenteritis. *Canadian Journal of Veterinary Research* 54:383–384.

Gulland, F.M.D. 1996. Impact of infectious diseases in wild animal populations: A review. In *Ecology of infectious diseases in natural populations,* ed. B.T. Grenfell and A.P. Dobson. Cambridge: Cambridge University Press, pp. 20–51.

Gustafsson, E., G. Blomquist, A. Bellman, R. Holmdahl, A. Mattsson, and R. Mattsson. 1996. Maternal antibodies protect immunoglobulin deficient neonatal mice from mouse hepatitis virus (MHV) B associated wasting syndrome. *American Journal of Reproductive Immunology* 36:33–39.

Heckert, R.A., L.J. Saif, and G.N. Myers. 1989. Development of protein A-gold immunoelectron microscopy for detection of bovine coronavirus in calves: Comparison with ELISA and direct immunofluorescence of nasal epithelial cells. *Veterinary Microbiology* 19:217–231.

Heeney, J.L., J.F. Evermann, A.J. McKeirnan, L. Marker-Kraus, M.E. Roelke, M. Bush, D.E. Wild, G. Meltzer, L. Colly, J. Lukas, V.J. Manton, T. Caro, and S.J. O'Brien. 1990. Prevalence and implications of feline coronavirus infections of captive and free-ranging cheetahs (*Acinonyx jubatus*). *Journal of Virology* 64:1964–1972.

Herrewegh, A.A.P.M., R.J. de Groot, A. Cepica, H.F. Egberink, M.C. Horzinek, and P.J. Rottier. 1995. Detection of feline coronavirus RNA in feces, tissues, and body fluids of naturally infected cats by reverse transcriptase PCR. *Journal of Clinical Microbiology* 33:684–689.

Holmes, K.V., and M.M.C. Lai. 1996. Coronaviridae: The viruses and their replication. In *Virology,* ed. B.N. Fields, D.M. Knipe, and P.M. Howley, 3d ed. Philadelphia: Lippincott-Raven, pp. 1075–1093.

Kapil, S., and S.M. Goyal. 1995. Bovine coronavirus-associated respiratory disease. *Compendium of Continuing Education for Practicing Veterinarians* 17:1179–1181.

Kapil, S., A.M. Trent, and S.M. Goyal. 1994. Antibody responses in spiral colon, ileum, and jejunum of bovine coronavirus-infected neonatal calves. *Comparative Immunology and Microbiology and Infectious Diseases* 17:139–149.

Koopmans, M., and M.C. Horzinek. 1995. The pathogenesis of torovirus infections in animals and humans. In *The Coronaviridae,* ed. S.G. Siddell. New York: Plenum, pp. 403–413.

Lai, M.M.C. 1996. Recombination in large RNA viruses: Coronaviruses. *Seminars in Virology* 7:381–388.

Lamm, M.E., J.G. Nedrud, C.S. Kaetzel, and M.B. Mazanec. 1996. New insights into epithelial cell function in mucosal immunity: Neutralization of intracellular pathogens and excretion of antigens by IgA. In *Essentials of mucosal immunity,* ed. M.F. Kagnoff and H. Kiyono. New York: Academic, pp. 141–150.

Langpap, T.J., M.E. Bergeland, and D.E. Reed. 1979. Coronaviral enteritis of young calves: Virologic and pathologic findings in naturally occurring infections. *American Journal of Veterinary Research* 40:1476–1478.

Majhdi, F., H.C. Minocha, and S. Kapil. 1997. Isolation and characterization of a coronavirus from elk calves with diarrhea. *Journal of Clinical Microbiology* 35:2937–2942.

Martin, H.D., and N.S. Zeidner. 1992. Concomitant cryptosporidia, coronavirus, and parvovirus infection in a raccoon (*Procyon lotor*). *Journal of Wildlife Diseases* 28:113–115.

McArdle, F., M. Bennett, R.M. Gaskell, B. Tennant, D.F. Kelly, and C.J. Gaskell. 1992. Induction and enhancement of feline infectious peritonitis by canine coronavirus. *American Journal of Veterinary Research* 53:1500–1506.

Mebus, C.A., L.E. Newman, and E.L. Stair. 1975. Scanning electron, light, and immunofluorescent microscopy of the intestine of a gnotobiotic calf infected with calf diarrheal coronavirus. *American Journal of Veterinary Research* 36:1719–1725.

Pass, J.A., W.J. Penhale, G.E. Wilcox, and R.G. Batey. 1982. Intestinal coronavirus-like particles in sheep with diarrhea. *Veterinary Record* 111:106–107.

Pensaert, M.B., E.O. Haelterman, and T. Burnstein. 1970. Transmissible gastroenteritis of swine: Virus intestinal cell interactions: I. Immunofluorescence, histopathology, and virus production in the small intestine through the course of infection. *Archiv für die gesamte Virusforschung* [*Archives of Virology*] 31:321–334.

Poland, A.M., H. Vennema, J.E. Foley, and N.C. Pedersen. 1996. Two related strains of feline infectious peritonitis virus isolated from immunocompromised cats infected with a feline enteric coronavirus. *Journal of Clinical Microbiology* 34:3180–3184.

Rasschaert, D., M. Duarte, and H. Laude. 1990. Porcine respiratory coronavirus differs from transmissible gastroenteritis virus by a few genomic deletions. *Journal of General Virology* 71:2599–2607.

Reynolds, D.J., D. Chasey, A.C. Scott, and J.C. Bridger. 1984. Evaluation of ELISA and electron microscopy for the detection of bovine coronavirus and rotavirus in bovine feces. *Veterinary Record* 114:397–401.

Rodak, L., L.A. Babiuk, and S.D. Acres. 1982. Detection by radioimmunoassay and ELISA of coronavirus antibodies in bovine serum and lacteal secretions. *Journal of Clinical Microbiology* 16:34–40.

Roelke, M.E., D.J. Forrester, E.R. Jacobson, G.V. Kolias, F.W. Scott, M.C. Barr, J.F. Evermann, and E.C. Pirtle. 1993. Seroprevalence of infectious disease agents in free-ranging Florida panthers (*Felis concolor coryi*). *Journal of Wildlife Diseases* 29:36–49.

Saif, L.J., and R.A. Heckert. 1990. Enteropathogenic coronaviruses. In *Viral diarrheas of man and animals,* ed. L.J. Saif and K.W. Theil. Boca Raton, FL: CRC, pp. 187–252.

Sanchez, C.M., F. Gebauer, C. Sune, A. Mendez, J. Dopazo, and L. Enjuanes. 1992. Genetic evolution and tropism of transmissible gastroenteritis coronaviruses. *Virology* 190:92–105.

Shockley, L.J., P.A. Kapke, W. Lapps, D.A. Brian, L.N.D. Potgeiter, and R. Woods. 1987. Diagnosis of porcine and bovine enteric coronavirus infections using cDNA probes. *Journal of Clinical Microbiology* 25:1591–1596.

Shoup, D.I., D.E. Swayne, D.J. Jackwood, and L.J. Saif. 1996. Immunohistochemistry of transmissible gastroenteritis virus antigens in fixed paraffin-embedded tissues. *Journal of Veterinary Diagnostic Investigation* 8:161–167.

Smith, D.R., H. Tsunemitsu, R.A. Heckert, and L.J. Saif. 1996. Evaluation of two antigen-capture ELISAs using polyclonal or monoclonal antibodies for the detection of bovine coronavirus. *Journal of Veterinary Diagnostic Investigation* 8:99–105.

Spalding, M.G., and D.J. Forrester. 1993. Disease monitoring of free-ranging and released wildlife. *Journal of Zoo and Wildlife Medicine* 24:271–280.

Stair, E.L., M.B. Rhodes, R.G. White, and C.A. Mebus. 1972. Neonatal calf diarrhea: Purification and electron microscopy of a coronavirus-like agent. *American Journal of Veterinary Research* 33:1147–1152.

Storz, J., L. Stine, A. Liem, and G.A. Anderson. 1996. Coronavirus isolation from nasal swab samples in cattle with signs of respiratory tract disease after shipping. *Journal of American Veterinary Medical Association* 208:1452–1455.

Tennant, B.J., R.M. Gaskell, and C.J. Gaskell. 1994. Studies on the survival of canine coronavirus under different environmental conditions. *Veterinary Microbiology* 42:255–259.

Tsunemitsu, H., Z.A. el-Kanawati, D.R. Smith, H.H. Reed, and L.J. Saif. 1995. Isolation of coronaviruses antigenically indistinguishable from bovine coronavirus from wild ruminants with diarrhea. *Journal of Clinical Microbiology* 33:3264–3269.

Wesley, R.D., R.D. Woods, and A.K. Cheung. 1990. Genetic basis for the pathogenesis of transmissible gastroenteritis virus. *Journal of Virology* 64:4761–4766.

14 RODENT-BORNE HEMORRHAGIC FEVER VIRUSES

JAMES N. MILLS AND JAMES E. CHILDS

INTRODUCTION. The rodent-borne hemorrhagic fevers are caused by two groups of RNA viruses: the family *Arenaviridae,* and the genus *Hantavirus,* of the family *Bunyaviridae.* Although the two groups are phylogenetically distinct, they share many characteristics. Typically, each unique virus type is maintained in nature by a single rodent host species in which it causes a chronic infection that may be completely asymptomatic.

This chronic infection and long-term shedding of virus into the environment are key to the maintenance of the virus in reservoir populations and to transmission to humans. Human infection most likely results from the inhalation of infectious virus in aerosols of rodent secreta or excreta, and epidemics of hantavirus- and arenavirus-related disease among humans have been associated with high densities of reservoir populations.

The name *hemorrhagic fever* derives from the tendency of these viruses to cause microvascular damage and changes in vascular permeability. This effect may be manifest to varying degrees, ranging from conjunctival injection and flushing to generalized mucous membrane hemorrhage and shock. Due to the seriousness of the diseases and the potential for aerosol spread, research with most of these agents in their natural hosts or other experimental animals can be conducted only in maximum containment laboratories.

HANTAVIRUSES

Synonyms: Hemorrhagic fever with renal syndrome: Korean hemorrhagic fever, nephropathia epidemica, Songo fever, hemorrhagic nephrosonephritis, epidemic hemorrhagic fever.
Hantavirus pulmonary syndrome: None in common usage.

History. Hemorrhagic fever with renal syndrome (HFRS) came to the attention of western science during the Korean conflict, when over 3000 cases of Korean hemorrhagic fever were diagnosed among United Nations troops. The etiologic agent and prototype *Hantavirus* (Hantaan virus, HTNV) was isolated in 1976.

Old World hantaviruses cause both severe and mild forms of HFRS. Hosts for these viruses are found among two groups of rodents of the family Muridae: the Old World rats and mice [subfamily Murinae (Table 14.1A)] and the voles [subfamily Arvicolinae (Table 14.1B)].

In 1993, an outbreak of severe pulmonary disease among people in the southwestern United States was shown to be caused by a hantavirus. Knowledge of the characteristics of Old World hantaviruses enabled rapid serologic and genetic analyses that linked the virus, subsequently named Sin Nombre virus (SNV), to the deer mouse, *Peromyscus maniculatus* (Nichol et al. 1993). The occurrence of human cases outside the geographic range of *P. maniculatus* led to the identification of hantaviruses associated with other species of sigmodontine rodents (Table 14.1C).

At least 13 distinct hantaviruses have been identified through viral isolation, and at least 13 others have been characterized (Table 14.1) on the basis of genetic analyses of RNA sequences amplified by reverse transcriptase-polymerase chain reaction (RT-PCR). This chapter focuses on those viruses that are demonstrated agents of human disease.

Distribution and Host Range. The distribution of HFRS and hantavirus pulmonary syndrome (HPS) generally coincides with the distribution of the host species (Table 14.1 and Figs. 14.1 and 14.2). The primary host of Puumala virus (PUUV), *Clethrionomys glareolus,* is common and widespread in various forest habitats in Western Europe. Although human disease associated with Seoul virus (SEOV) occurs most commonly in urban areas of China and Korea (Lee et al. 1982a; Chen and Qiu 1993), the virus was recently associated with disease in the United States (Glass et al. 1994) and Brazil. Norway rats are distributed throughout the world (Fig. 14.1), and antibody reactive with SEOV has been detected in rats from every continent except Antarctica (LeDuc et al. 1986).

Although most HPS has occurred in the Western United States, cases have been confirmed from New York, Pennsylvania, Virginia, West Virginia, North Carolina, Florida, Louisiana, Iowa, Wisconsin, Illinois, and Indiana. As of December 1997, there were 200 confirmed cases of HPS in 29 U.S. states plus the three western provinces of Canada. Whereas the primary reservoir of SNV, the deer mouse, is one of the most common and widespread rodents in North America, the ranges of other host species associated with HPS in North America are more restricted (Fig. 14.2).

Hantaviruses associated with HPS have been identified from human tissue in Paraguay (Williams et al. 1997), Uruguay, Brazil, Chile, and Argentina (Lopez et

TABLE 14.1—Hantaviruses associated with hosts of the rodent family Muridae and with the insectivore, *Suncus murinus:* their known distribution and disease associations

Virus	Host	Distribution of Virus	Disease[c]	Reference
A. Order Rodentia, family Muridae, subfamily Murinae				
Hantaan[a]	*Apodemus agrarius*	Far East, Russia, Northern Asia, Balkans	Severe HFRS	Lee et al. 1978
Seoul[a]	*Rattus norvegicus*	Worldwide	Mild/Moderate HFRS	Lee et al. 1982a
Dobrava[a]	*Apodemus flavicollis*	Balkans	Severe HFRS	Avsic-Zupanc et al. 1992
Thai[a]	*Bandicota indica*	India	None recognized	Elwell et al. 1985
B. Order Rodentia, family Muridae, subfamily Arvicolinae				
Puumala[a]	*Clethrionomys glareolus*	Europe, Scandinavia, Russia, Balkans	Mild HFRS (NE)[b]	Brummer Korvenkontio et al. 1980
Prospect Hill[a]	*Microtus pennsylvanicus*	North America	None recognized	Lee et al. 1982b
Bloodland Lake	*Microtus ochrogaster*	North America	None recognized	Hjelle et al. 1995b
Isla Vista	*Microtus californicus*	Western U.S.A., Mexico	None recognized	Song et al. 1995
Tula	*Microtus arvalis/ M. rossiaemeridionalis*	Russia, Slovakia	None recognized	Plyusnin et al. 1994
Khabarovsk	*Microtus fortis*	Far Eastern Russia	None recognized	Horling et al. 1996
Topografov	*Lemmus sibericus*	Siberia	None recognized	Plyusnin et al. 1996
C. Subfamily Sigmodontinae				
Sin Nombre[a]	*Peromyscus maniculatus*	U.S.A. and Canada	HPS	Childs et al. 1994
New York[a]	*Peromyscus leucopus*	East and Central U.S.A.	HPS	Hjelle et al. 1995c
Black Creek Canal[a]	*Sigmodon hispidus*	Southeastern U.S.A.	HPS	Rollin et al. 1995
Bayou[a]	*Oryzomys palustris*	Southeastern U.S.A.	HPS	Morzunov et al. 1995
El Moro Canyon[a]	*Reithrodontomys megalotis*	Western U.S.A., Mexico	None recognized	Hjelle et al. 1994
Rio Segundo	*Reithrodontomys mexicanus*	Costa Rica	None recognized	Hjelle et al. 1995a
Cano Delgadito[a]	*Sigmodon alstoni*	Venezuela	None recognized	C.F. Fulhorst et al., personal communication
Juquitiba	Unknown	Brazil	HPS	S.T. Nichol et al., personal commmunication
Rio Mamore	*Oligoryzomys microtis*	Bolivia	HPS	Hjelle et al. 1996
Laguna Negra[a]	*Calomys laucha*	Western Paraguay	HPS	Williams et al. 1997
Andes	*Oligoryzomys longicaudatus*	Southwestern Argentina/Chile	HPS	Levis et al. 1998
Lechiguanas	*Oligoryzomys flavescens*	Central Argentina	HPS	Levis et al. 1998
Bermejo	*Oligoryzomys chacoensis*	Northwestern Argentina	None recognized	Levis et al. 1998
Oran	*Oligoryzomys longicaudatus?*	Northwestern Argentina	HPS	Levis et al. 1998
Maciel	*Bolomys obscurus*	Central Argentina	None recognized	Levis et al. 1998
Pergamino	*Akodon azarae*	Central Argentina	None recognized	Levis et al. 1998
D. Order Insectivora, Family Soricidae				
Thottapalayam[a]	*Suncus murinus*	India	None recognized	Carey et al. 1971

[a]Virus isolated in cell culture; others identified from genetic sequences.
[b]NE, nephropathia epidemica.
[c]HFRS, hemorrhagic fever with renal syndrome; HPS, hantavirus pulmonary syndrome.

al. 1997). Hantaviruses are associated with *Akodon azarae, Bolomys obscurus, Oligoryzomys flavescens, O. longicaudatus,* and *O. chacoensis* from Argentina (Levis et al. 1998); *Calomys laucha* and *Akodon toba* from Paraguay (Williams et al. 1997); and *O. longicaudatus,* and *Akodon olivaceus* from Chile (J.N. Mills et al. unpublished).

Thottapalayam virus was isolated from an insectivore (Table 14.1). Reports of hantavirus isolates from bats (Kim et al. 1994) and birds (Slonova et al. 1992)

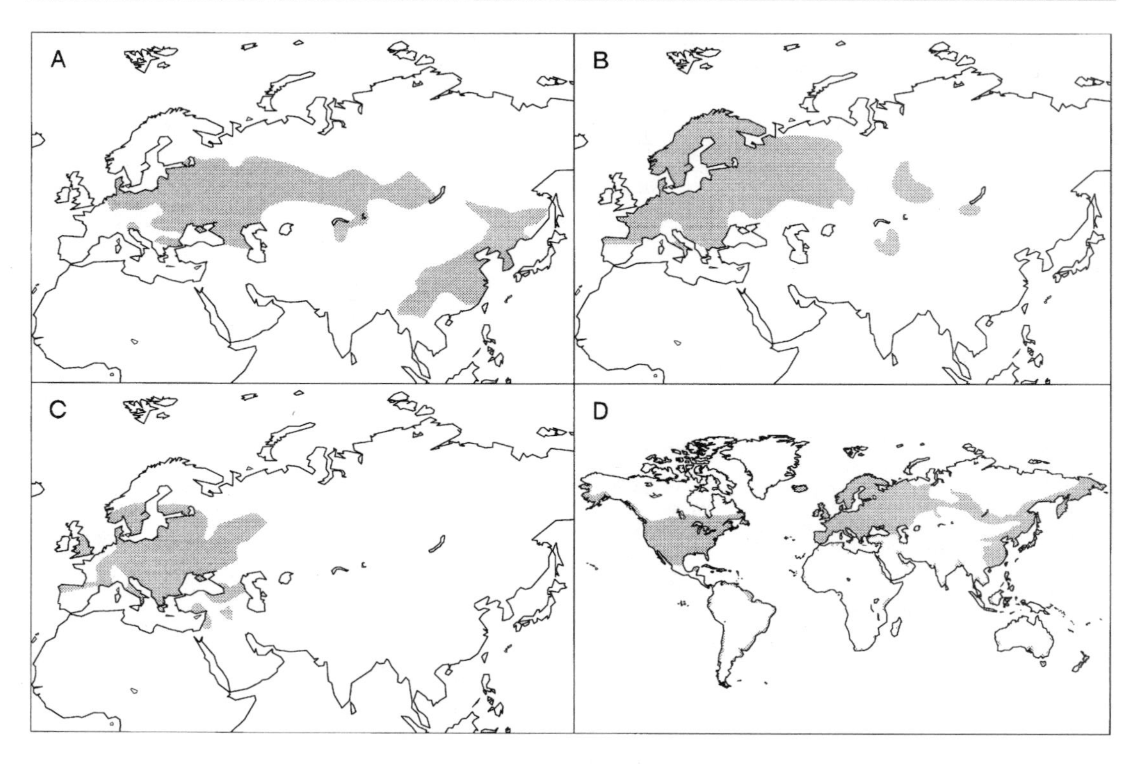

FIG. 14.1—Geographic range of the rodent hosts for Old World hantaviruses: (A) *Apodemus agrarius,* (B) *Clethrionomys glareolus,* (C) *Apodemus flavicollis,* and (D) *Rattus norvegicus.* [A, B, and C adapted from Corbet (1978); D adapted from Brooks and Rowe (1987)].

exist. Carnivores that feed on infected rodents may become infected, as evidenced by antibody (Nowotny et al. 1994) and at least one viral isolate (Zhao-zhaung et al. 1985) from domestic cats. It is not unreasonable to expect that any mammal that shares habitat with a hantavirus reservoir species might be occasionally infected and therefore show antibody. Only the specific host, however, would be likely to develop persistent infection and shed large quantities of infectious virus into the environment.

Etiology. The hantaviruses constitute a genus of antigenically and genetically related viruses in the family Bunyaviridae. Virions are lipid enveloped, spherical to oval, average 90–100 nm in diameter, and contain a trisegmented RNA genome of negative sense that codes for a nucleocapsid and a nonstructural protein (S segment), two envelope glycoproteins, and a nonstructural protein (M), and a replicase/transcriptase (L) (Schmaljohn et al. 1985).

Transmission. Hantavirus infection in the natural host results in chronic, asymptomatic infection, with dissemination of virus in multiple organs. Infectious doses of HTNV in *Apodemus agrarius* (Lee et al. 1981) and PUUV in *C. glareolus* (Yanagihara et al. 1985) are shed for extended, although variable, periods in urine, feces, and saliva. Transmission from rodents to humans is believed to be primarily via fine-particle aerosols of virus-contaminated rodent excreta (Tsai 1987). Laboratory rats are susceptible to infection by HTNV, SEOV, and PUUV by either aerosol or intramuscular (IM) routes (Nuzum et al. 1988), although IM infectious doses are approximately 100-fold lower than those required to infect rats by aerosol. Vertical transmission by infected female rodents has not been convincingly demonstrated. Rat pups born to SEOV-infected females typically possess maternal antibody that is protective (Dohmae et al. 1993). Field studies indicate that transmission of hantaviruses within reservoir populations occurs primarily horizontally (vide infra). Person-to-person transmission of hantavirus is extremely rare but has been documented for Andes virus in southern Argentina (Wells et al. 1997).

Epidemiology. Although prevalence data are limited, many Old World rodents and insectivores may carry hantavirus or hantavirus antigen (Table 14.2). At least 25 New World rodents and a lagomorph have antibody reactive to New World hantaviruses (Table 14.3). All species that have both a high prevalence and a large number of samples tested are arvicolines [hosts of Prospect Hill virus (PHV) and PHV-like viruses] or sigmodontines (hosts of SNV and SNV-like viruses).

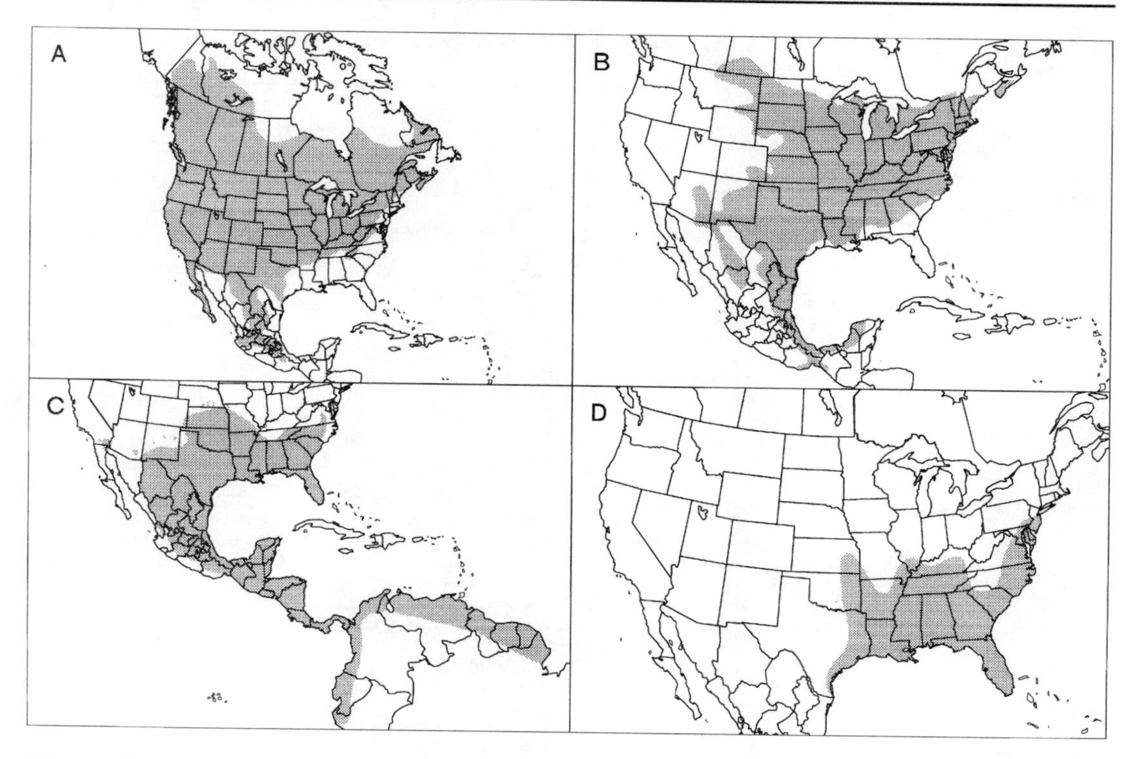

FIG. 14.2—Geographic range of the rodent hosts for New World hantaviruses: (A) *Peromyscus maniculatus* [after Carleton (1989)], (B) *Peromyscus leucopus* [after Carleton (1989)], (C) *Sigmodon hispidus* [after Hall and Kelson (1959) and Hershkovitz (1955)], and (D) *Oryzomys palustris* [Adapted from Wolfe (1982)].

Systematic studies of the dynamics of hantavirus infection in Old World hosts are rare. No longitudinal data exist for HTNV and *A. agrarius,* but prevalence of infection can be greater than 13% (N = 817) in disease-endemic areas of Korea (Lee et al. 1981). The prevalence of antibody to PUUV among Swedish populations of *C. glareolus* peaked in the spring and correlated with vole density the previous fall and spring. Antibody prevalence (overall 19% in 2544 *C. glareolus*) also appeared to be positively correlated with the cyclic, multiannual variation in vole density. Voles showed an age-specific pattern in antibody prevalence indicative of horizontal transmission (Niklasson et al. 1995). The year-to-year incidence of human disease has been correlated with the density of voles in Scandinavia and in Russia (Niklasson et al. 1995).

Norway rats infected with SEOV in Baltimore showed age-related acquisition of antibody, with over 70% of the largest animals seropositive, suggesting horizontal transmission (Childs et al. 1985). The incidence of seroconversion correlated with the acquisition of scars or wounds, suggesting virus transmission through biting (Glass et al. 1988).

SNV infection in reservoir populations is more common in older/larger rodents, and males are typically more frequently infected than females. Infected *Peromyscus* spp. were found in all habitats from desert to alpine tundra in the southwestern United States, although prevalence was greatest (16%–20%) at middle-altitude habitats, including pinyon-juniper, where the majority of human HPS cases occurred (Mills et al. 1997). Antibody prevalence was 30% among deer mice during an investigation of the initial HPS outbreak (Childs et al. 1994), but recent studies in the same geographic area found a prevalence of 11% (Mills et al. 1997; T.G. Ksiazek et al. personal communication).

In New York State, 22% of 405 white-footed mice from Long Island had antibody reactive with SNV, whereas only 2% of 404 from the rest of the state had antibody (White et al. 1996); 18% of 90 *Sigmodon hispidus* captured during an investigation of the HPS case in Dade County, Florida, had antibody reactive with Black Creek Canal virus [BCCV (Rollin et al. 1995)]; 5.5% of 165 *Oryzomys palustris* tested from seven states in the southeastern United States had antibody reactive with SNV and Bayou virus antigen (Ksiazek et al. 1997).

Clinical Disease. Although antigen is readily demonstrable in many organs, hantavirus infection in the rodent host appears asymptomatic. SEOV-infected rats are as large and fecund, and survive as long as uninfected animals (Childs et al. 1985).

In humans, severe HFRS can progress through five stages after a 2- to 3-week incubation period: a febrile

TABLE 14.2—Old World mammal species and locations where hantavirus or hantavirus antigen-positive animals have been reported

Order Family/Subfamily	Species	Countries
Rodentia		
Muridae/Murinae	*Apodemus agrarius*	Korea, China, Russia, Czechoslovakia, former Yugoslavia
	Apodemus flavicollis	European Russia, Czechoslovakia, former Yugoslavia, Greece
	Apodemus peninsulae	Far Eastern Russia, China
	Apodemus speciosus	China, Japan
	Apodemus sylvaticus	European Russia, Belgium, France, Norway, former Yugoslavia
	Bandicota bengalensis	Thailand, Burma
	Bandicota indica	Thailand
	Micromys minutus	European Russia
	Mus musculus	Russia, China
	Rattus confucianus	China
	Rattus losea	China
	Rattus nitidus	China
	Rattus norvegicus	Korea, Japan, China, Russia, Egypt
	Rattus rattus	Korea
	Rattus tanezumi	China
Muridae/Arvicolinae	*Clethrionomys glareolus*	Finland, Sweden, Norway, France, Belgium, Czechoslovakia, former Yugoslavia, European Russia
	Clethrionomys rufocanus	Sweden, Norway, Far Eastern Russia, China, Japan
	Clethrionomys rutilus	Norway, Russia
	Microtus agrestis	European Russia
	Microtus arvalis	European Russia, China
	Microtus fortis	Far Eastern Russia, China
	Microtus montebellii	Japan
	Microtus oeconomus	European Russia
	Microtus rossiaemeridionalis	Russia
Muridae/Cricetinae	*Cricetulus barabensis*	Far Eastern Russia, China
Muridae/Sciuridae	*Tamias sibiricus*	Far Eastern Russia
Insectivora		
Soricidae	*Neomys fodiens*	European Russia
	Sorex araneus	European Russia
	Sorex minutus	European Russia
	Suncus murinus	China
Talpidae	*Talpa europaea*	European Russia
Carnivora		
Felidae	*Felis catus*	China, Austria, England?

Modified from Yanagihara and Gajdusek (1987).

phase of 2–7 days, a hypotensive phase of 2 hours to 3 days, an oliguric phase of 3–7 days, a diuretic phase of several days to weeks, and a convalescent stage that may last weeks to months. The first three phases may be associated with hemorrhagic complications and disseminated intravascular coagulation (McKee et al. 1991). The mortality rate from infection with HTNV may be 10%–15% (McKee et al. 1991) and that from Dobrava virus (DOBV) may be 15%–30% (LeDuc et al. 1992).

The milder forms of HFRS caused by PUUV or SEOV have a shorter clinical course, and the stages are not clearly differentiated. Mortality rate is less than 1% (Park 1986). Antibodies to SEOV have been found in patients with hypertensive renal disease in the United States (Glass et al. 1993). Inapparent infection is common; the ratio of clinical cases to PUUV infections was estimated at 1:10 (Niklasson and LeDuc 1987).

HPS is characterized by a nonspecific prodrome of fever and myalgia, often accompanied by headache, nausea and vomiting, cough, and dyspnea. Within hours or days, a dramatic respiratory distress syndrome can develop (Duchin et al. 1994). Consistent hematologic features include thrombocytopenia, hemoconcentration, left-shifted neutrophilic leukocytosis, and reactive lymphocytes (Zaki et al. 1995). Radiography shows early, prominent interstitial edema and nonperipheral distribution of initial airspace. The overall HPS mortality rate in the United States is 45%.

Pathogenesis and Pathology. Laboratory studies conducted with the natural hosts of HTNV (Lee et al. 1981), PUUV (Yanagihara et al. 1985), and BCCV (Hutchinson et al. 1998) have shown a common pattern. Following an initial, brief viremia, antigen is found in lung, spleen, kidney, and other organs for extended periods, in spite of the presence of high-titered antibody in serum. Endothelial cells are primary targets, though SEOV has been isolated from macrophages and granulocytes (Nagai et al. 1985).

White-footed mice (*Peromyscus leucopus*) naturally infected with the NY-1 strain of SNV had virus particles in pulmonary endothelial cells, and alveolar septa

TABLE 14.3—Small mammals with antibody reactive to Sin Nombre, Prospect Hill, or Black Creek Canal viruses in North America[a]. Taxonomy follows Wilson and Reeder (1993); only species found antibody-positive are listed[b]

Order Family/Subfamily	Species	Number Tested	Percent Positive
Rodentia			
Sciuridae	*Ammospermophilus leucurus*	48	2.1
	Spermophilus variegatus	131	0.8
	Tamias dorsalis	223	2.7
	T. minimus	185	1.1
	T. quadrivittatus	65	6.1
Heteromyidae	*Perognathus fasciatus*	2	50.0
	P. parvus	151	0.7
Muridae/Arvicolinae	*Clethrionomys gapperi*	124	2.4
	Microtus mexicanus	25	12.0
	M. montanus	133	6.0
	M. ochrogaster	5	20.0
Muridae/Murinae	*Mus musculus*	273	0.7
	Rattus norvegicus	102	1.0
	Rattus rattus	303	1.6
Muridae/Sigmodontinae	*Neotoma albigula*	283	0.7
	Onychomys leucogaster	31	3.2
	Oryzomys palustris	205	4.9
	Peromyscus boylii	420	17.9
	P. gossypinus	140	0.7
	P. leucopus	1421	7.5
	P. maniculatus	3527	14.9
	P. nasutus	14	7.1
	P. truei	467	13.0
	Reithrodontomys megalotis	97	19.6
	Sigmodon hispidus	1176	11.4
Lagomorpha			
Leporidae	*Sylvilagus audubonii*	14	7.1

[a]As reported by Childs et al. (1994), G.E. Glass et al. (personal communication), Kaufman et al. (1994), Ksiazek et al. (1997), Mills et al. (1997, 1998), Otteson et al. (1996), Rollin et al. (1995), and White et al. (1996).

[b]Species tested and found to be negative include (number tested in parentheses): **Insectivora**/Soricidae: *Blarina brevicauda* (5) *Sorex monticolus* (1), *S. nanus* (1), *S. vagrans* (1); **Carnivora**/Mustelidae: *Mephitis mephitis* (1), *Mustela frenata* (2); **Rodentia**/Sciuridae: *Ammospermophilus harrisii* (30), A. *interpres* (1), *Cynomys gunnisoni* (1), *Glaucomys volans* (1), *Marmota flaviventris* (32), *Sciurus carolinensis* (1), *Spermophilus armatus* (2), *S. beecheyi* (16), *S. lateralis* (47), *S. spilosoma* (12), *S. tridecemlineatus* (10), *Tamias canipes* (5), *T. cinereicollis* (9), *T. palmeri* (6), *T. quadrimaculatus* (7), *T. rufus* (3), *T. senex* (38), *T. striatus* (32), *T. umbrinus* (68), *Tamiasciurus hudsonicus* (15); Geomyidae: *Thomomys bottae* (4), *Thomomys talpoides* (2); Heteromyidae: *Chaetodipus baileyi* (14), *C. formosus* (21), *C. hispidus* (12), *C. intermedius* (50), *C. penicillatus* (151), *Dipodomys deserti* (14), *D. merriami* (167), *D. microps* (14), *D. ordii* (234), *D. panamintinus* (9), *D. spectabilis* (4), *Microdipodops pallidus* (6), *Perognathus amplus* (27), *P. flavescens* (3), *P. flavus* (22), *P. longimembris* (1); Dipodidae: *Napaeozapus insignis* (3), *Zapus hudsonicus* (20), *Z. princeps* (53); Muridae/Arvicolinae: *Lemmiscus curtatus* (4), *Microtus longicaudus* (45), *M. pennsylvanicus* (44), *Synaptomys cooperi* (1); Muridae/Sigmodontinae: *Neotoma albigula* (1), *N. cinerea* (16), *N. floridana* (13), *N. fuscipes* (3), *N. lepida* (67), *N. mexicana* (61), *N. micropus* (8), *N. stephensi* (14), *Ochrotomys nuttalli* (7), *Onychomys arenicola* (2), *Onychomys torridus* (11), *Peromyscus attwateri* (46), *P. crinitus* (3), *P. difficilis* (13), *P. eremicus* (109), *Reithrodontomys montanus* (3), *Sigmodon arizonae* (22), *S. hispidus* (14); **Lagomorpha**/Leporidae: *Sylvilagus floridanus* (2)

were edematous with hyperplasia of type-1 pneumocytes. Lymphohistiocytic infiltrates were observed in hepatic portal zones, and slightly increased numbers of immunoblasts occurred in the spleen (Llyubsky et al. 1996). Pulmonary histopathology in white-footed mice was similar to that seen in humans but much less severe (Llyubsky et al. 1996), and deer mice infected with SNV show no diminution of respiratory function (O'Connor and Hayes 1997).

In human HFRS, the kidney is primarily affected. Histopathology includes retroperitoneal edema and hemorrhagic necrosis of the renal medulla, the anterior pituitary, and the cardiac right atrium (Hullinghorst and Steer 1953). Pulmonary edema and pleural effusions are seen in some patients with HFRS caused by HTNV (Kim 1965) and by PUUV (Linderholm et al. 1992).

HPS in humans causes interstitial pneumonitis with a variable mononuclear cell infiltrate, edema, and focal hyaline membrane formation. There is marked viral tropism for capillary endothelial cells, especially in the lungs. Antigen is present in follicular dendritic cells in spleen and lymph nodes; macrophages in lung, spleen, and bone marrow; and peripheral blood mononuclear cells. Morbidity appears to be caused by endothelial damage leading to hemoconcentration, pulmonary edema, and shock (Zaki et al. 1995).

Laboratory Diagnosis. Hantaviral infection of animals or humans is usually diagnosed by serology (Lee et al. 1989). Other methods are now being used or assessed, including RT-PCR and immunohistochemistry (Ksiazek et al. 1995). The indirect fluorescent antibody (IFA) test and enzyme immunoassay (EIA) are the most common serologic assays. Past or current infection in rodents may be indicated by the presence of immunoglobulin G (IgG). IgM or significantly increased IgG titers between acute- and convalescent-phase samples indicate recent infection in humans. Cross-reactivity in the immune response to hantavirus infection makes the use of several antigens helpful in diagnosis, even when the agent may be a novel virus (Ksiazek et al. 1995). Western blotting using recombinant proteins has enabled epitope mapping of SNV (Jenison et al. 1994). Antigen detection has not been useful for diagnosis of HTNV infection (LeDuc et al. 1990).

The plaque reduction neutralization test (PRNT) has been used to differentiate between immune responses to antigenically similar agents and to type hantaviruses (Chu et al. 1994). This technique requires containment facilities to grow and handle virus.

Although hantaviruses are difficult to isolate (LeDuc and Lee 1989), the combination of RT-PCR and gene expression provides homologous antigens without isolating virus (Feldmann et al. 1993). Molecular genetic analyses were used in the detection, identification, and subsequent elucidation of the epidemiology of SNV infection and HPS in North America (Nichol et al. 1993). The same methods demonstrate genetic variability in hantaviruses that cause mild HFRS in Europe (Pilaski et al. 1994).

Treatment. Care of patients with hantavirus disease is supportive (Lee and van der Groen 1989; Levy and Simpson 1994). Attention to fluid and electrolyte balance, hemodynamics, and renal or pulmonary function is critical. Intravenous ribavirin has been effective for severe HFRS (Huggins et al. 1991). Although ribavirin inhibits SNV in vitro, its efficacy in treating patients has not been established.

ARENAVIRUSES

Synonyms: Lymphocytic choriomeningitis: Callitrichid hepatitis (in zoo monkeys).
Lassa fever: None.
Argentine hemorrhagic fever: Fiebre hemorragica argentina, mal de los rastrojos.
Bolivian hemorrhagic fever: Fiebre hemorragica boliviana.
Venezuelan hemorrhagic fever: Fiebre hemorragica venezolana.
Sabia virus infection: Brazilian hemorrhagic fever.

History. The Old World arenaviruses are associated with murid rodents (Table 14.4A). The prototype arenavirus—lymphocytic choriomeningitis virus (LCMV)—was isolated in 1933 (Armstrong and Lillie 1934) and subsequently was shown to be a cause of aseptic meningitis in humans, as well as frequent asymptomatic infections in laboratory mice (*Mus musculus*).

Lassa virus (LASV) was isolated in 1969 from patients with severe viral hemorrhagic fever in the town of Lassa, Nigeria (Buckley and Casals 1970). Lassa fever (LF) is common in West Africa, and case-fatality ratios are about 15%–16%. Three "Lassa-like" viruses have been isolated from murid rodents in other parts of sub-Saharan Africa (Table 14.4A).

The New World arenaviruses are associated with sigmodontine rodents (Table 14.4B). The first South American hemorrhagic fever viruses isolated were Junin virus (JUNV) and Machupo virus (MACV), etiologic agents of Argentine hemorrhagic fever (AHF) and Bolivian hemorrhagic fever (BHF) (Table 14.4B). The association of LCMV, JUNV, LASV, and MACV with chronic infection in specific rodent hosts, and the morphologic and antigenic relationships among these viruses, led to the recognition of the family Arenaviridae (Rowe et al. 1970). In 1989, an outbreak of hemorrhagic fever in Venezuela led to the isolation of Guanarito virus (GUAV), and a single fatal case of hemorrhagic fever in Sao Paulo, Brazil, resulted in the isolation of Sabia virus (SABV). Several arenaviruses not associated with human disease also have been identified from the Americas (Table 14.4B). Only the important human pathogens are discussed in this chapter.

Distribution and Host Range. In the Old World, LCMV is widely distributed in association with its cosmopolitan host, the house mouse (*Mus musculus*). LASV has been isolated from humans or rodents in Nigeria, Sierra Leone, Guinea, and Liberia, but serologic surveys show that LASV or Lassa-like viruses are present in at least ten other African countries (Peters 1991). LASV is carried by mice of the *Mastomys natalensis* species complex. At least eight species occur in Africa south of the Sahara, and their systematics and distribution are complex and poorly understood (Robbins and Van der Straeten 1989).

In the New World, the AHF-endemic area, which encompassed an area of approximately 10,000 km^2 in the 1950s, has been progressively expanding and now encompasses more than 150,000 km^2 in four provinces (Maiztegui et al. 1986). The AHF-endemic area includes only a small part of the geographic range of the primary host species, *Calomys musculinus,* which is common and widely distributed in central and northwestern Argentina (Redford and Eisenberg 1992). Infection has not been found in *C. musculinus* outside of the disease-endemic area.

The BHF-endemic area is restricted to the department of Beni, Bolivia [Pan American Health Organization (PAHO) 1982]. The range of the host, *Calomys callosus,* includes northern Argentina and Paraguay, through eastern Bolivia to east-central Brazil (Musser and Carleton 1993).

Human cases of VHF are restricted to a 9000-km^2 area in southwestern Portuguesa State and adjoining areas of Barinas State on the Venezuelan llanos (De Manzione et al. 1998). Rodents (principally *Zygodontomys brevicauda* and *Sigmodon alstoni*) with antibody reactive to GUAV are captured throughout much of

TABLE 14.4—Arenaviruses associated with rodent hosts of the family Muridae

Virus	Host	Distribution of Virus	Disease	Reference
A. Subfamily Murinae				
Lymphocytic choriomeningitis	*Mus musculus*	Europe, Americas, perhaps worldwide	Lymphocytic choriomeningitis	Armstrong and Lillie 1934
Lassa	*Mastomys* spp.	West Africa	Lassa fever	Frame et al. 1970
Ippy	*Arvicanthus* spp.?	Central African Republic	None recognized	Swanepoel et al. 1985
Mopeia	*Mastomys natalensis*	Mozambique, Zimbabwe	None recognized	Wulff et al. 1977
Mobala	*Praomys jacksoni*	Central African Republic	None recognized	Gonzalez et al. 1983
B. Subfamily Sigmodontinae				
Junin	*Calomys musculinus*	Central Argentina	Argentine hemorrhagic fever (AHF)	Parodi et al. 1958
Machupo	*Calomys callosus*	Beni Department, Bolivia	Bolivian hemorrhagic fever (BHF)	Johnson et al. 1965
Guanarito	*Zygodontomys brevicauda*	Central llanos, Venezuela	Venezuelan hemorrhagic fever (VHF)	Salas et al. 1991
Sabia	Unknown	Near Sao Paulo, Brazil	Not named	Coimbra et al. 1994
Amapari	*Neacomys guianae*	Ampa Territory, Brazil	None recognized	Pinheiro et al. 1966
Flexal	*Oryzomys* spp.?	Para State, Brazil	None recognized[a]	Pinheiro et al. 1977
Latino	*Calomys callosus*	Beni Department, Bolivia	None recognized	Webb et al. 1973
Oliveros	*Bolomys obscurus*	Central Argentina	None recognized	Mills et al. 1996
Parana	*Oryzomys buccinatus*?	Missiones Province, Paraguay	None recognized	Webb et al. 1970
Pichinde	*Oryzomys albigularis*	Columbia	None recognized	Trapido and Sanmartin 1971
Pirital	*Sigmodon alstoni*	Central llanos, Venezuela	None recognized	Fulhorst et al. 1997
Tacaribe	Unknown[b]	Trinidad	None recognized	Downs et al. 1963
Tamiami	*Sigmodon hispidus*	South Florida	None recognized	Calisher et al. 1970
Whitewater Arroyo	*Neotoma albigula*	Southwestern U.S.A.	None recognized	Fulhorst et al. 1996; Kosoy et al. 1996

[a]One documented laboratory infection.
[b]Original report lists bats of the genus *Artibeus* as the reservoir; subsequent attempts to isolate virus from *Artibeus* have been unsuccessful.

Portuguesa State and adjacent areas. *Zygodontomys brevicauda,* the probable reservoir of GUAV, is a savanna species found from southeastern Costa Rica through Brazil north of the Amazon (Musser and Carleton 1993). *Sigmodon alstoni* occurs in savanna habitats in Colombia, Venezuela, Guyana, Surinam, and northern Brazil (Musser and Carleton 1993).

The disease-endemic area and animal reservoir of SABV are unknown.

Etiology. Arenaviruses are lipid-enveloped, single-stranded, bisegmented RNA viruses containing two genome segments of ambisense polarity (Southern 1996). The roughly spherical virions vary from 60 to more than 300 nm in diameter and have a "sandy" (Latin, *arena* = sand) appearance due to the incorporation of host cell ribosomes. The S RNA segment encodes the nucleocapsid protein and the envelope glycoproteins; the L segment encodes the polymerase and the zinc-binding proteins.

Transmission. Human infection is thought to occur principally via inhalation of aerosolized virus shed in excretions or secretions of persistently infected rodents. Contaminated food or fomites may also be an important source of infection in those cases in which the host is commensal or peridomestic (e.g., *C. callosus* and MACV or *Mastomys* species and LASV). Risk for AHF is high among farmworkers, especially tractor drivers, and epidemics peak during the fall.

Transmission of LCMV and LASV within reservoir populations is thought to be primarily vertical. All pups born to chronically infected females are themselves chronically infected [perhaps transovarially (Mims 1975)]. In contrast, inoculation of adult hosts with LCMV or LASV results in an immunizing infection without significant shedding of infectious virus. The offspring of these animals are also protected by maternal antibody.

Both MACV and JUNV can have deleterious effects when vertically transmitted among rodents, and horizontal transmission is likely important in viral maintenance. Infection with JUNV among wild *C. musculinus* was more frequent among males than females and was positively correlated with age and the presence of wounds or scars (Mills et al. 1994). These data suggest that aggressive encounters among males may be an important mechanism of transmission.

Epidemiology. In Germany, populations of house mice had a low overall prevalence of LCMV antibody (3.6%), and most infected animals came from the north (Ackermann 1973). In the United States, infection within *M. musculus* populations is highly focal; antibody prevalences ranged from 0 (of 753) to 27% (214 of 594) on four farms in California (Gardner et al. 1973). In inner-city Baltimore infections in mice were clustered within residential blocks (Childs et al. 1991).

Human infection with LCMV reflected local infection in house mouse populations in Europe (Blumenthal et al. 1968), England (Smithard and Macrae 1951), and the United States (Armstrong and Sweet 1939). Incidence of human disease peaks in fall-winter (Adair et al. 1953), presumably with the seasonal invasion of houses by mice. Outbreaks of human LCM also have been traced to pet Syrian hamsters (*Mesocricetus auratus*) in Germany (Hitchcock 1994) and the United States (Gregg 1975).

Two species within the *M. natalensis* species complex are frequently associated with LASV infection. The 32-chromosome species, *M. huberti,* tends to be found in houses, whereas the 38-chromosome species, *M. erythroleucus,* is found in the surrounding bush areas (Robbins et al. 1983; Happold 1987). Both species are frequently infected [ca. 30% prevalence (McCormick et al. 1987b)]. Risk for LF is associated with large populations of household rodents, poor sanitation, and catching, preparing, and consuming rodents (McCormick and Fisher-Hock 1992). Infection in *Mastomys* populations is focal, and antibody prevalence among villages in Sierra Leone varied from less than 1% to more than 80% (McCormick et al. 1987b).

In the AHF-endemic area, *C. musculinus* occur in crop fields and adjacent weedy borders. JUNV infection in *C. musculinus* populations varied from 0 (of 195) to 8.4% (20 of 237) among sites within 30 km of each other. At a given site, month-to-month prevalences varied from 0 to greater than 50% (Mills et al. 1994). Populations of *C. musculinus* peak in the fall when humans enter the fields for the harvest. When favorable climatic conditions result in increased rodent populations, epidemics of AHF can occur (Mills et al. 1992).

The reservoir of MACV, *C. callosus,* is primarily a species of grassland/forest ecotones but also lives in villages on high ground in seasonally inundated areas of the Beni (Kuns 1965). Although sporadic rural cases of BHF primarily affect men, large outbreaks in villages affect both genders and all age groups. Cases may be clustered and associated with houses that were infested with rodents (Johnson 1989). MACV was isolated from 13 of 17 rodents trapped in or near homes during a BHF outbreak (Johnson 1965). Infection in *C. callosus* varied from 35% in sites where human disease was present to 5% in localities where no disease had occurred for at least 2 years (Johnson 1985).

In VHF-endemic areas, 47% of 40 *S. alstoni* yielded viral isolates compared with 11% of *Z. brevicauda* (Tesh et al. 1993). Genetic studies have shown that viral isolates from *Z. brevicauda* were identical to GUAV isolated from human patients, whereas isolates from *S. alstoni* were a distinct arenavirus now called Pirital virus (Fulhorst et al. 1997).

Clinical Disease. The outcome of arenaviral infection in the rodent host is variable. Laboratory experiments indicate this is influenced by five factors: age and immune status of the host, genetic susceptibility of the host, route of exposure, dose of virus, and strain of virus (Childs and Peters 1993). A vast literature is available concerning laboratory investigations involving most of these variables in the case of LCMV and *M. musculus* [see reviews in Childs and Peters (1993) and in Peters et al. (1996)]. Studies indicate that natural infection of LCMV in *M. musculus* is chronic and asymptomatic.

Nonhuman primates may be "naturally" infected with LCMV. The virus caused severe disease among caged colonies of rhesus macaques (Peters 1994), and outbreaks of fatal hepatitis (callitrichid hepatitis) among zoo populations of marmosets likely resulted from eating naturally infected wild mice or mice supplied by keepers as a protein supplement (Stephensen et al. 1991).

Laboratory studies with JUNV have shown that *C. musculinus* inoculated intranasally at birth developed chronic infection, grew more slowly, and had higher mortality and lower reproductive success than uninfected mice (Vitullo et al. 1987). Although 50% of adult *C. musculinus* inoculated in a similar experiment showed viral persistence, infection did not affect reproductive success (Vitullo and Merani 1990). Laboratory experiments with MACV (Johnson 1985) have shown that females infected at birth are sterile. Field studies showed that JUNV infection had no affect on growth rates, body mass, movement patterns, or longevity of subadult and adult *C. musculinus* (Mills et al. 1992).

LCMV infection in humans, which is probably underdiagnosed (Jahrling and Peters 1992), usually produces a syndrome of fever, myalgia, and leukopenia, which is rarely serious. However, LCMV infection during pregnancy has been associated with hydrocephalus, chorioretinitis, psychomotor retardation, and neonatal death (Larsen et al. 1993; Peters et al. 1996). Other syndromes caused by LCMV infection may include parotitis, pancreatitis, and orchitis.

The arenaviral hemorrhagic fevers have an insidious prodrome of fever, malaise, muscle aches, and retroorbital headache. This may be followed by hypotension, conjunctival injection, bleeding from the gums, and petechiae on the palate, throat, chest, and axillae. Neurologic signs such as tremors, dizziness, and coordination problems are common with the South American hemorrhagic fevers but less frequent with LF. Severe disease can include extensive petechial hemorrhages, bleeding from mucous membranes, hypotension and shock, coma, and convulsions (Peters et al. 1996). Convalescence is frequently protracted. Unilateral or bilateral deafness is a common sequela (29% of prospectively followed cohort) with LF (Cummins et al. 1990). The mortality rate is 10%–33% among patients with arenaviral hemorrhagic fevers, although treatment with immune plasma has significantly decreased mortality from Argentine hemorrhagic fever (vide infra).

Pathogenesis and Pathology. Few data are available concerning the pathogenesis of JUNV, MACV, LASV, or GUAV in their natural hosts. T-cell-mediated immunopathology plays a central role in the pathogenesis of fatal LCMV infection involving the choroid plexus of mice (Cole and Nathanson 1974). The basis for chronicity of arenavirus infections in reservoir species has been attributed to the minimal direct cytopathic effects of viral replication and the potential for immunosuppressive or immunopathologic effects of infection in lymphoreticular tissue (Murphy et al. 1976; Peters et al. 1996). LCMV infection of the anterior pituitary of mice results in a decrease in growth hormone production and stunting (Oldstone et al. 1985). Stunting is also seen in laboratory infections of *C. musculinus* with JUNV (Vitullo et al. 1987).

The LCMV-mouse model of immunopathogenesis does not mirror events in arenaviral hemorrhagic fevers. A fatal infection model using Pichinde virus (which is not pathogenic for humans) in strain-13 guinea pigs indicates an initial viral tropism for macrophages and epithelial cells. Minor histopathologic damage includes focal necrosis of the liver, adrenal glands, and spleen, interstitial pneumonitis, and an intestinal lesion associated with infection of mucosal macrophages. Death of the animal results from a dramatic decrease in cardiac output brought about by the secretion of soluble mediators that decrease heart function, including leukotrienes, platelet-activating factor, and endorphins (Peters et al. 1996).

In human arenaviral hemorrhagic fevers, leukopenia and thrombocytopenia are common, although in LF leukocyte counts can be normal or even elevated (Fisher-Hoch et al. 1988). Liver enzymes, aspartate aminotransferase (AST), and alanine aminotransferase (ALT) can be elevated and indicate a poor prognosis. The pathophysiology of hemorrhage is poorly understood. Studies in primate models suggest platelet function may be hampered by decreased production, increased consumption, and impaired platelet activity. Diminished clotting-factor activity likely results from depressed factor synthesis due to hepatocellular dysfunction. These coagulation abnormalities combined with viral-induced damage or dysfunction of the vascular endothelium result in bleeding or hemorrhagic diathesis (Peters et al. 1987).

Diagnosis. Laboratory diagnosis of arenaviral infections in rodents, humans, or other animals can be made by viral isolation, EIA, IFA, PRNT, or RT-PCR (Peters et al. 1996). The most useful test for diagnosis of recent arenavirus infection in humans is the IgM-capture EIA. Virus in acutely or chronically infected rodent hosts may be isolated from blood, oral swabs, or homogenates of most organs on monolayers of Vero or Vero E-6 cells. The arenaviruses are broadly cross-reactive by IFA and EIA but may be distinguished by PRNT of virus in Vero cells (this requires containment facilities). Immunofluorescent antibody is detectable 1–3 weeks after infection of adult animals; this period may be delayed in neonates or chronically infected animals (Peters 1994). Antigen-capture EIA was used to determine infection status of rodents during a long-term study in the AHF-endemic area and was highly correlated with the simultaneous presence of immunofluorescent antibody (Mills et al. 1994). Sequencing of viral RNA amplified by RT-PCR was useful for distinguishing among related arenaviruses (Bowen et al. 1996).

Treatment. Aggressive supportive treatment is important in the management of patients with arenaviral hemorrhagic fevers (Peters 1991). Immune plasma is effective in the treatment of patients with AHF (Maiztegui et al. 1979). This treatment has been used for LF and BHF, but its efficacy is unproven (Peters et al. 1987). Intravenous ribavirin was effective in the treatment of LF (McCormick et al. 1986), and results for AHF were equivocal but encouraging (Enria and Maiztegui 1994). Recent experience with ribavirin treatment of a laboratory worker exposed to SABV also was encouraging (Barry et al. 1995).

PREVENTION AND CONTROL OF RODENT-BORNE HEMORRHAGIC FEVER VIRUSES.

Although prevention of infection by hantaviruses and arenaviruses in natural populations of mice is impractical, prevention of such infections in commercial and laboratory rodent colonies is possible. Introduction of LCMV into mouse and hamster colonies has been documented (Skinner and Knight 1971). Hantavirus infections in laboratory rodent colonies have been the source of extensive outbreaks of human disease (Kulagin et al. 1962). Laboratory colonies should be screened periodically by serologic testing, because infection may be completely asymptomatic. In addition, abnormalities in colony production (e.g., decreased fertility, runting) or unusual experimental results, especially in immunologic studies, should be cause for suspicion and investigation of possible arenaviral infection (Peters 1994).

Precautions when initiating laboratory colonies from wild rodents caught in areas endemic for hantaviruses or arenaviruses should include quarantine and serologic screening upon capture and again after 30 days (Mills et al. 1995b). Captive primates should not be fed baby mice from unscreened sources, and rodent control around their enclosures is important.

Risk factors for human hantaviral disease include occupations such as farming, forestry, and activities that place humans in contact with rodents or their infectious excretions or secretions (Ellis et al. 1995).

Recommendations for preventing hantaviral disease in the United States through reducing rodent contact have been published [Centers for Disease Control and Prevention (CDC) 1993]. Elimination of rodents from human habitations and "rodent proofing" of rural homes are achievable (Glass et al. 1997).

Briefly, recommendations for risk reduction include eliminating rodents from the home environment through using spring-loaded snap-traps; preventing entry of rodents by sealing holes, cracks, and other openings with screen, caulking, or steel wool; reducing the availability of food and nesting sites; and modifying the habitat within 30 m of the home to eliminate rodent harborage.

Rodent contamination of in buildings should be eliminated as follows. Open doors and windows and ventilate the area for at least 30 min. Soak dead rodents, droppings, nesting materials, and other contaminated objects with general household disinfectant or hypochlorite solution (5% household bleach in water). Wearing rubber gloves, place these materials in double plastic bags and dispose of by burning, burying, or through the regular trash removal system. Using a solution of water, detergent, and disinfectant, mop floors and disinfect countertops, drawers, and other surfaces (carpets and upholstery may be steam cleaned or shampooed). Disinfect and then remove the rubber gloves and wash hands with soap and water. Cleanup of homes of persons with hantavirus infection or buildings with heavy rodent infestations require special precautions, and responsible public health officials should be contacted (CDC 1993).

Rodent reduction was successful in the control of an epidemic of BHF in San Joaquin, Bolivia (Kuns 1965). An ongoing program of rodent trapping in villages in the BHF-endemic area may be, at least partly, responsible for the scarcity of cases since 1974 (PAHO 1982). These methods would not be practical against reservoir populations associated with open habitat or agricultural areas, such as *C. musculinus* or *Z. brevicauda,* but, because of the peridomestic affinities of some *Mastomys* species, they could be a useful prophylaxis for LF. Education programs aimed at improving sanitation and finding alternatives to human consumption of *Mastomys* might also reduce the incidence of LF.

Although recommendations for reducing the incidence of AHF by habitat modification or crop replacement have been published (Kravetz 1977), the habitat and competitive interactions among species are complex, and such modifications should be carefully evaluated before being instituted.

Personnel who handle wild or laboratory rodents or rodent tissues or cell lines may be at risk. Investigators conducting field studies involving potential reservoir species in hantavirus or arenavirus disease-endemic areas should use standard precautions (Mills et al. 1995a,b). Traps containing captured animals should be handled wearing rubber gloves and in a manner so as to prevent contact with infectious rodent excretions and secretions or the creation and inhalation of aerosols of these materials. Rodents in traps, or contaminated traps, should be transported in double plastic bags to contain potentially infectious aerosols. These bags should be opened and rodents handled only in an isolated outdoor area or containment laboratory by personnel wearing protective equipment (latex gloves, gowns or coveralls, respirators with high-efficiency-particulate air filters, and goggles). Sampling should be conducted on anesthetized animals to prevent bites, and use of sharp instruments such as needles and scalpels should be avoided or minimized.

Instruments, working surfaces, and traps should be decontaminated using a disinfectant (such as 5% hospital-strength Lysol), and contaminated gloves, gowns, and wastes should be double bagged and autoclaved or incinerated. Rodent carcasses can be rendered noninfectious by fixing in 10% formalin for 48 hours (Mills et al. 1995a). Frozen tissues and blood should be considered infectious. Guidelines describing safe laboratory practices for working with these agents have been published [CDC and National Institutes of Health (NIH) 1993; CDC 1994].

Vaccines for HTNV and SEOV have been developed and used in China and in North and South Korea. More than 1,200,000 people have been vaccinated with an inactivated HFRS vaccine in North Korea, and a decreased incidence of HFRS has been reported (Schmaljohn 1994). Vaccines for HTNV or SEOV would probably not protect against SNV or related viruses, and it is uncertain what populations in North America would be targeted if a vaccine were available.

A live-attenuated vaccine was highly effective in preventing AHF (World Health Organization 1993). This vaccine may be cross-protective against MACV (P.B. Jahrling personal communication). Recombinant vaccinia virus vaccines against LF are being studied in animal models (Auperin 1993).

LASV has been associated with extensive outbreaks in hospital settings where virus is transmitted during surgical procedures or through the use of nonsterile instruments and needles. These outbreaks can be controlled by implementation of appropriate barrier nursing precautions, including the use of gowns, gloves, masks, and sterile instruments.

PUBLIC HEALTH CONCERNS FOR RODENT-BORNE HEMORRHAGIC FEVER VIRUSES.

The rodent-borne hemorrhagic fevers have serious epidemic potential. An estimated 100,000–200,000 cases of HFRS occur annually in Europe and Asia (Lee 1989). LF causes 100,000–300,000 cases and 5000 deaths annually (McCormick et al. 1987a). Although the number of cases of the HPS and the South American hemorrhagic fevers is small by comparison, their severity causes concern among public health officials as well as the general public in disease-endemic areas.

Imported LF has been reported in Europe (Emond 1986; CDC 1988), the United States (Holmes et al.

1990), Israel (Schlaeffer et al. 1988), and Japan (Hirabayashi et al. 1988), where the concern is for possible secondary transmission before the nature of the disease is recognized.

Professionals such as wildlife biologists, pest control workers, and laboratorians or animal caretakers working with wild rodents or rodent-derived tissues may be at increased risk of infection and should take special precautions.

LITERATURE CITED

Ackermann, R. 1973. Epidemiological aspects of lymphocytic choriomeningitis in man. In *Lymphocytic choriomeningitis virus and other arenaviruses,* ed. F. Lehmann-Grube. New York: Elsevier, pp. 234–237.

Adair, C.V., R.L. Gauld, and J.E. Smadel. 1953. Aseptic meningitis, a disease of diverse etiology: Clinical and etiologic studies on 854 cases. *Annals of Internal Medicine* 39:675–704.

Armstrong, C., and R. Lillie. 1934. Experimental lymphocytic choriomeningitis of monkeys and mice produced by a virus encountered in studies of the 1933 St. Louis encephalitis epidemic. *Public Health Reports* 49:1019–1027.

Armstrong, C., and L.K. Sweet. 1939. Lymphocytic choriomeningitis. *Public Health Reports* 54:673–684.

Auperin, D.D. 1993. Construction and evaluation of recombinant virus vaccines for Lassa fever. In *The Arenaviridae,* ed. M.S. Salvato. New York: Plenum, pp. 259–280.

Avsic-Zupanc, T., S.Y. Xiao, R. Stojanovic, A. Gligic, G. Vandergroen, and J.W. LeDuc. 1992. Characterization of Dobrava virus: A hantavirus from Slovenia, Yugoslavia. *Journal of Medical Virology* 38:132–137.

Barry, M.M. Russi, L. Armstrong, D. Geller, R. Tesh. L. Dembry, J.P. Gonzalez, A.S. Khan, and C.J. Peters. 1995. Treatment of a laboratory-acquired Sabia virus infection [Brief report]. *New England Journal of Medicine* 333:294–296.

Blumenthal, W., R. Ackermann, and W. Scheid. 1968. Distribution of the lymphocytic choriomeningitis virus in an endemic area [in German]. *Deutsche Medizinische Wochenschrift* 93:944–948.

Bowen, M.D., C.J. Peters, J.N. Mills, and S.T. Nichol. 1996. Oliveros virus: A novel arenavirus from Argentina. *Virology* 217:362–366.

Brooks, J.E., and F.P. Rowe. 1987. *Commensal rodent control.* Geneva: World Health Organization, Vector Biology and Control Division; WHO/VBC/87.949.

Brummer-Korvenkontio, M., A. Vaheri, T. Hovi, C.-H. von Bonsdorff, J. Vuorimies, T. Manni, K. Penttinen, N. Oker-Blom, and J. Lähdevirta. 1980. Nephropathia epidemica: Detection of antigen in bank voles and serologic diagnosis of human infection. *Journal of Infectious Diseases* 141:131–134.

Buckley, S.M., and J. Casals. 1970. Lassa fever, a new virus disease of man from West Africa: III. Isolation and characterization of the virus. *American Journal of Tropical Medicine and Hygiene* 19:680–691.

Calisher, C.H., T. Tzianabos, R.D. Lord, and P.H. Coleman. 1970. Tamiami virus: A new member of the Tacaribe group. *American Journal of Tropical Medicine and Hygiene* 19:520–526.

Carey, D.E., R. Reuben, K.N. Panicker, R.E. Shope, and R.M. Myers. 1971. Thottapalayam virus: A presumptive arbovirus isolated from a shrew in India. *Indian Journal of Medical Research* 59:1758–1760.

Carleton, M.D. 1989. Systematics and evolution. In *Advances in the study of* Peromyscus *(Rodentia),* ed. G.L. Kirkland and J.N. Layne. Lubbock: Texas Tech University Press, pp. 7–141.

Centers for Disease Control and Prevention (CDC). 1988. Management of patients with suspected viral hemorrhagic fever. *Morbidity and Mortality Weekly Report* 37:1–16.

———. 1993. Hantavirus infection—Southwestern United States: Interim recommendations for risk reduction. *Morbidity and Mortality Weekly Report* 42:1–13.

———. 1994. Laboratory management of agents associated with hantavirus pulmonary syndrome: Interim biosafety guidelines. *Morbidity and Mortality Weekly Report* 43:1–7.

Centers for Disease Control and Prevention (CDC) and National Institutes of Health (NIH). 1993. Biosafety in microbiological and biomedical laboratories, 3d ed. Washington, DC: US Government Printing Office, 177 pp.

Chen, H.-X., and F.-X. Qiu. 1993. Epidemiologic surveillance on the hemorrhagic fever with renal syndrome in China. *Chinese Journal of Medicine* 106:857–863.

Childs, J.E., and C.J. Peters 1993. Ecology and epidemiology of arenaviruses and their hosts. In *The Arenaviridae,* ed. M.S. Salvato. New York: Plenum, pp. 331–384.

Childs, J.E., G.W. Korch, G.A. Smith, A.D. Terry, and J.W. LeDuc. 1985. Geographical distribution and age related prevalence of antibody to Hantaan-like virus in rat populations of Baltimore, Maryland, USA. *American Journal of Tropical Medicine and Hygiene* 34:385–387.

Childs, J.E., G.E. Glass, T.G. Ksiazek, C.A. Rossi, J.G. Oro, and J.W. LeDuc. 1991. Human-rodent contact and infection with lymphocytic choriomeningitis and Seoul viruses in an inner-city population. *American Journal of Tropical Medicine and Hygiene* 44:117–121.

Childs, J.E., T.G. Ksiazek, C.F. Spiropoulou, J.W. Krebs, S. Morzunov, G.O. Maupin, P.E. Rollin, J. Sarisky, and R.E. Enscore. 1994. Serologic and genetic identification of *Peromyscus maniculatus* as the primary rodent reservoir for a new hantavirus in the southwestern United States. *Journal of Infectious Diseases* 169:1271–1280.

Chu, Y.K., C. Rossi, J.W. LeDuc, H.W. Lee, C.S. Schmaljohn, and J.M. Dalrymple. 1994. Serological relationships among viruses in the *Hantavirus* genus, family Bunyaviridae. *Virology* 198:196–204.

Coimbra, T.L., E.S. Nassar, M.N. Burattini, L.T. de Souza, I.B. Ferreira, I.M. Rocco, A.P. Travassos da Rosa, P.F.C. Vasconcelos, F.P. Pinheiro, J.W. LeDuc, R. Rico-Hesse, J. Gonzalez, P.B. Jahrling, and R.B. Tesh. 1994. New arenavirus isolated in Brazil. *Lancet* 343:391–392.

Cole, G.A., and N. Nathanson. 1974. Lymphocytic choriomeningitis. *Progress in Medical Virology* 18:94–110.

Corbet, G.B. 1978. *The mammals of the palaearctic region: A taxonomic review.* Ithaca: Cornell University Press, 314 pp.

Cummins, D., J.B. McCormick, D. Bennett, J.A. Samba, B. Farrar, S.J. Machin, and S.P. Fisher Hoch. 1990. Acute sensorineural deafness in Lassa fever. *Journal of the American Medical Association* 264:2093–2096.

De Manzione, N., R.A. Salas, H. Paredes, O. Godoy, L. Rojas, F. Araoz, C.F. Fulhorst, T.G. Ksiazek, J.N. Mills, B.A. Ellis, C.J. Peters, and R.B. Tesh. 1998. Venezuelan hemorrhagic fever: Clinical and epidemiological studies of 165 cases. *Clinical Infectious Diseases* 26:308–313.

Dohmae, K., U. Koshimizu, and Y. Nishimune. 1993. In utero and mammary transfer of hantavirus antibody from dams to infant rats. *Laboratory Animal Science* 43:557–561.

Downs, W.G., C.R. Anderson, L. Spence, T.H.G. Aitken, and A.H. Greenhall. 1963. Tacaribe virus: A new agent isolated from *Artibeus* bats and mosquitoes in Trinidad, West Indies. *American Journal of Tropical Medicine and Hygiene* 12:640–646.

Duchin, J.S., F.T. Koster, C.J. Peters, G.L. Simpson, B. Tempest, S.R. Zaki, P.E. Rollin, S. Nichol, and E.T. Umland.

1994. Hantavirus pulmonary syndrome: A clinical description of 17 patients with a newly recognized disease. The Hantavirus Study Group. *New England Journal of Medicine* 330:949–955.

Ellis, B.A., J.N. Mills, and J.E. Childs. 1995. Rodent-borne hemorrhagic fever viruses of importance to agricultural workers. *Journal of Agromedicine* 2:7–44.

Elwell, M.R., G.S. Ward, M. Tingpalapong, and J.W. LeDuc. 1985. Serologic evidence of Hantaan-like virus in rodents and man in Thailand. *Southeast Asian Journal of Tropical Medicine and Public Health* 16:349–354.

Emond, R.T. 1986. Viral haemorrhagic fevers. *Journal of Infection* 13:103–106.

Enria, D.A., and J.I. Maiztegui. 1994. Antiviral treatment of Argentine hemorrhagic fever. *Antiviral Research* 23:23–31.

Feldmann, H., A. Sanchez, S. Morzunov, C.F. Spiropoulou, P.E. Rollin, T.G. Ksiazek, C.J. Peters, and S.T. Nichol. 1993. Utilization of autopsy RNA for the synthesis of the nucleocapsid antigen of a newly recognized virus associated with hantavirus pulmonary syndrome. *Virus Research* 30:351–367.

Fisher-Hoch, S., J.B. McCormick, D. Sasso, and R.B. Craven. 1988. Hematologic dysfunction in Lassa fever. *Journal of Medical Virology* 26:127–135.

Frame, J.D., J.M. Baldwin Jr., D.J. Gocke, and J. Troup. 1970. Lassa fever: A new virus disease of man from west Africa: I. Clinical description and pathological findings. *American Journal of Tropical Medicine and Hygiene* 19:670–676.

Fulhorst, C.F., M.D. Bowen, T.G. Ksiazek, P.E. Rollin, S.T. Nichol, M.Y. Kosoy, and C.J. Peters. 1996. Isolation and characterization of Whitewater Arroyo virus, a novel North American arenavirus. *Virology* 224:114–120.

Fulhorst, C.F., M.D. Bowen, R.A. Salas, N. de Manzione, G. Duno, A. Utrera, T.G. Ksiazek, C.J. Peters, S.T. Nichol, and R.B. Tesh. 1997. Isolation and characterization of Pirital virus, a newly discovered South American arenavirus. *American Journal of Tropical Medicine and Hygiene* 56:548–553.

Gardner, M.B., B.E. Henderson, J.D. Estes, H. Menck, J.C. Parker, and R.J. Huebner. 1973. Unusually high incidence of spontaneous lymphomas in wild house mice. *Journal of the National Cancer Institute* 50:1571–1579.

Glass, G.E., J.E. Childs, G.W. Korch, and J.W. LeDuc. 1988. Association of intraspecific wounding with hantaviral infection in wild rats (*Rattus norvegicus*). *Epidemiology and Infection* 101:459–472.

Glass, G.E., A.J. Watson, J.W. LeDuc, G.D. Kelen, T.C. Quinn, and J.E. Childs. 1993. Infection with a ratborne hantavirus in US residents is consistently associated with hypertensive renal disease. *Journal of Infectious Diseases* 167:614–620.

Glass, G.E., A.J. Watson, J.W. LeDuc, and J.E. Childs. 1994. Domestic cases of hemorrhagic fever with renal syndrome in the United States. *Nephron* 68:48–51.

Glass, G.E., J.S. Johnson, G.A. Hodenbach, C.L.J. DiSalvo, C.J. Peters, J.E. Childs, and J.N. Mills. 1997. Experimental evaluation of rodent exclusion methods to reduce hantavirus transmission to humans in rural housing. *American Journal of Tropical Medicine and Hygiene* 56:359–364.

Gonzalez, J.P., J.B. McCormick, J.F. Saluzzo, J.P. Herve, A.J. Georges, and K.M. Johnson. 1983. An arenavirus isolated from wild-caught rodents (*Praomys* species) in the Central African Republic. *Intervirology* 19:105–112.

Gregg, M.B. 1975. Recent outbreaks of lymphocytic choriomeningitis in the United States of America. *Bulletin of the World Health Organization* 52:549–554.

Hall, E.R., and K.R. Kelson. 1959. *The mammals of North America.* New York: Ronald, 1083 pp.

Happold, D.C.D. 1987. *The mammals of Nigeria.* New York: Clarendon, 402 pp.

Hershkovitz, P. 1955. South American marsh rats, genus *Holochilus,* with a summary of sigmodont rodents. *Fieldiana Zoology* 37:639–673.

Hirabayashi, Y., S. Oka, H. Goto, K. Shimada, T. Kurata, S.P. Fisher Hoch, and J.B. McCormick. 1988. An imported case of Lassa fever with late appearance of polyserositis. *Journal of Infectious Diseases* 158:872–875.

Hitchcock, J.C. 1994. The European ferret, *Mustela putorius* (family Mustelidae), its public health, wildlife and agricultural significance. In *Proceedings 16th Vertebrate Pest Conference,* ed. W.S. Halverson and A.C. Crabb. Davis: University of California, pp. 207–212.

Hjelle, B., F. Chavez-Giles, N. Torrez-Martinez, T. Yates, J. Sarisky, J. Webb, and M. Ascher. 1994. Genetic identification of a novel hantavirus of the harvest mouse *Reithrodontomys megalotis. Journal of Virology* 68:6751–6754.

Hjelle, B., B. Anderson, N. Torrez-Martinez, W. Song, W. Gannon, and T. Yates. 1995a. Prevalence and geographic genetic variation of hantaviruses of New World harvest mice (*Reithrodontomys*): Identification of a divergent genotype from a Costa Rican *Reithrodontomys mexicanus. Virology* 207:452–459.

Hjelle, B., S.A. Jenison, D.E. Goade, W.B. Green, R.M. Feddersen, and A. Scott. 1995b. Hantaviruses: Clinical, microbiologic, and epidemiologic aspects. *Critical Reviews in Clinical Laboratory Sciences* 32:469–508.

Hjelle, B., J. Krolikowski, N. Torrez-Martinez, F. Chavez-Giles, C. Vanner, and E. Laposata. 1995c. Phylogenetically distinct hantavirus implicated in a case of hantavirus pulmonary syndrome in the northeastern United States. *Journal of Medical Virology* 46:21–27.

Hjelle, B., N. Torrez-Martinez, and F.T. Koster. 1996. Hantavirus pulmonary syndrome-related virus from Bolivia [Letter]. *Lancet* 347:57.

Holmes, G.P., J.B. McCormick, S.C. Trock, R.A. Chase, S.M. Lewis, C.A. Mason, P.A. Hall, et al. 1990. Lassa fever in the United States: Investigation of a case and new guidelines for management. *New England Journal of Medicine* 323:1120–1123.

Horling, J., V. Chizhikov, A. Lundkvist, M. Jonsson, L. Ivanov, A. Dekonenko, B. Niklasson, T. Dzagurova, C.J. Peters, E. Tkachenko, and S. Nichol. 1996. Khabarovsk virus: A phylogenetically and serologically distinct *Hantavirus* isolated from *Microtus fortis* trapped in far-east Russia. *Journal of General Virology* 77:687–694.

Huggins, J.W., C.M. Hsiang, T.M. Cosgriff, M.Y. Guang, J.I. Smith, Z.O. Wu, J.W. LeDuc, et al. 1991. Prospective, double-blind, concurrent, placebo-controlled clinical trial of intravenous ribavirin therapy of hemorrhagic fever with renal syndrome. *Journal of Infectious Diseases* 164:1119–1127.

Hullinghorst, R.L., and A. Steer. 1953. Pathology of epidemic hemorrhagic fever. *Annals of Internal Medicine* 38:77–101.

Hutchinson, K.L., P.E. Rollin, and C.J. Peters. 1998. Pathogenesis of a North American hantavirus, Black Creek Canal virus, in experimentally infected *Sigmodon hispidus. American Journal of Tropical Medicine and Hygiene* 59:58–65.

Jahrling, P.B., and C.J. Peters. 1992. Lymphocytic choriomeningitis virus: A neglected pathogen of man. *Archives of Pathology and Laboratory Medicine* 116:486–488.

Jenison, S., T. Yamada, C. Morris, B. Anderson, N. Torrez-Martinez, and N. Keller. 1994. Characterization of human antibody responses to Four Corners hantavirus infections among patients with hantavirus pulmonary syndrome. *Journal of Virology* 68:3000–3006.

Johnson, K.M. 1965. Epidemiology of Machupo virus infection: III. Significance of virological observations in man and animals. *American Journal of Tropical Medicine and Hygiene* 14:816–818.

———. 1985. Arenaviruses. In *Virology,* ed. B.N. Fields and K.N. Knipe. New York: Raven, pp. 1033–1053.

———. 1989. Arenaviruses. In *Viral infections of humans,* ed. A.S. Evans, 3d ed. New York: Plenum, pp. 133–152.

Johnson, K.M., N.H. Wiebenga, R.B. Mackenzie, M.L. Kuns, N.M. Tauraso, A. Shelokov, P.A. Webb, G. Justines, and H.K. Beye. 1965. Virus isolations from human cases of hemorrhagic fever in Bolivia. *Proceedings of the Society of Experimental Biology and Medicine* 118:113–118.

Kaufman, G.A., D.W. Kaufman, B.R. McMillan, and D.E. Brillhart. 1994. Prevalence of hantavirus antibodies in natural populations of deer mice in north central Kansas. *Prairie Naturalist* 26:209–216.

Kim, D. 1965. Clinical analysis of 111 fatal cases of epidemic hemorrhagic fever. *American Journal of Medicine* 39:218–220.

Kim, G.R., Y.T. Lee, and C.H. Park. 1994. A new natural reservoir of hantavirus: Isolation of hantaviruses from lung tissues of bats. *Archives of Virology* 134:85–95.

Kosoy, M.Y., L.H. Elliott, T.G. Ksiazek, C.F. Fulhorst, P.E. Rollin, J.E. Childs, J.N. Mills, G.O. Maupin, and C.J. Peters. 1996. Prevalence of antibodies to arenaviruses in rodents from the southern and western United States: Evidence for an arenavirus associated with the genus *Neotoma. American Journal of Tropical Medicine and Hygiene* 54:570–576.

Kravetz, F.O. 1977. Ecologia y control de reservorios. *Ciencia e Investigacion* 33:235–242.

Ksiazek, T.G., C.J. Peters, P.E. Rollin, S. Zaki, S. Nichol, C. Spiropoulou, S. Morzunov, H. Feldmann, A. Sanchez, A.S. Khan, B.W.J. Mahy, K. Wachsmuth, and J.C. Butler. 1995. Identification of a new North American hantavirus that causes acute pulmonary insufficiency. *American Journal of Tropical Medicine and Hygiene* 52:117–123.

Ksiazek, T.G., S.T. Nichol, J.N. Mills, M.G. Groves, A. Wozniak, S. McAdams, M.C. Monroe, A.M. Johnson, M.L. Martin, C.J. Peters, and P.E. Rollin. 1997. Isolation, genetic diversity, and geographic distribution of Bayou virus (Bunyaviridae: *Hantavirus*). *American Journal of Tropical Medicine and Hygiene* 57:445–448.

Kulagin, S.M., N.I. Fedorova, and E.S. Ketiladze. 1962. Laboratory outbreak of hemorrhagic fever with renal syndrome [in Russian]. *Journal of Microbiology, Epidemiology and Immunology* 33:121–126.

Kuns, M.L. 1965. Epidemiology of Machupo virus infection: II. Ecological and control studies of hemorrhagic fever. *American Journal of Tropical Medicine and Hygiene* 14:813–816.

Larsen, P.D., S.A. Chartrand, K.M. Tomashek, L.G. Hauser, and T. G. Ksiazek. 1993. Hydrocephalus complicating lymphocytic choriomeningitis virus infection. *Pediatric Infectious Disease Journal* 12:528–531.

LeDuc, J.W., and H.W. Lee 1989. Virus isolation and identification. In *Manual of hemorrhagic fever with renal syndrome,* ed. H.W. Lee and J.M. Dalrymple. Seoul: Korea University, pp. 61–66.

LeDuc, J.W., G.A. Smith, J.E. Childs, F.P. Pinheiro, J.I. Maiztegui, B. Niklasson, A. Antoniades, et al. 1986. Global survey of antibody to Hantaan-related viruses among peridomestic rodents. *Bulletin of the World Health Organization* 64:139–144.

LeDuc, J W., T.G. Ksiazek, C.A. Rossi, and J.M. Dalrymple. 1990. A retrospective analysis of sera collected by the Hemorrhagic Fever Commission during the Korean Conflict. *Journal of Infectious Diseases* 162:1182–1184.

LeDuc, J.W., J.E. Childs, and G.E. Glass. 1992. The hantaviruses, etiologic agents of hemorrhagic fever with renal syndrome: A possible cause of hypertension and chronic renal disease in the United States. *Annual Review of Public Health* 13:79–98.

Lee, H.W. 1989. WHO collaborating center for virus reference and research. In *Manual of hemorrhagic fever with renal syndrome,* ed. H.W. Lee and J.M. Dalrymple. Seoul: Korea University, pp. 36–38.

Lee, H.W., and G. van der Groen. 1989. Hemorrhagic fever with renal syndrome. *Progress in Medical Virology* 36:62–102.

Lee, H.W., P.W. Lee, and K.M. Johnson. 1978. Isolation of the etiologic agent of Korean hemorrhagic fever. *Journal of Infectious Diseases* 137:298–308.

Lee, H.W., G.R. French, P.W. Lee, L.J. Baek, K. Tsuchiya, and R.S. Foulke. 1981. Observations on natural and laboratory infection of rodents with the etiologic agent of Korean hemorrhagic fever. *American Journal of Tropical Medicine and Hygiene* 30:477–482.

Lee, H.W., L.J. Baek, and K.M. Johnson. 1982a. Isolation of Hantaan virus, the etiologic agent of Korean hemorrhagic fever, from wild urban rats. *Journal of Infectious Diseases* 146:638–644.

Lee, P.-W., H.L. Amyx, D.C. Gajdusek, R.T. Yanagihara, D. Goldgaber, and C.J. Gibbs. 1982b. New haemorrhagic fever with renal syndrome-related virus in indigenous wild rodents in United States. *Lancet* 2:1405–1406.

Lee, P.W., J.M. Meegan, J.W. LeDuc, E.A. Tkachenko, A.P. Tvanov, G.V. Rezapkin, S.G. Drozdov, T. Kitamura, T. F. Tsai, and J.M. Dalrymple 1989. Serologic techniques for detection of Hantaan virus infection, related antigens and antibodies. In *Manual of hemorrhagic fever with renal syndrome,* ed. H.W. Lee and J.M. Dalrymple. Seoul: Korea University, pp. 75–106.

Levis, S., S.P. Morzunov, J.E. Rowe, D. Enria, N. Pini, G. Calderon, M. Sabattini, and S.C. St. Jeor. 1998. Genetic diversity and epidemiology of hantaviruses in Argentina. *Journal of Infectious Diseases* 177:529–538.

Levy, H., and S.Q. Simpson. 1994. Hantavirus pulmonary syndrome. *American Journal of Respiratory Critical Care Medicine* 149:1710–1713.

Linderholm, M., A. Billstrom, B. Settergren, and A. Tarnvik. 1992. Pulmonary involvement in nephropathia epidemica as demonstrated by computed tomography. *Infection* 20:263–266.

Llyubsky, S., I. Gavrilovskaya, B. Luft, and E. Mackow. 1996. Histopathology of *Peromyscus leucopus* naturally infected with pathogenic NY-1 hantaviruses: Pathologic markers of HPS infection in mice. *Laboratory Investigation* 74:627–633.

Lopez, N., P. Padula, C. Rossi, S. Miguel, A. Edelstein, E. Ramirez, and M. T. Franze-Fernandez. 1997. Genetic characterization and phylogeny of Andes virus and variants from Argentina and Chile. *Virus Research* 50:77–84.

Maiztegui, J.I., N.J. Fernandez, and A.J. de Damilano. 1979. Efficacy of immune plasma in treatment of Argentine haemorrhagic fever and association between treatment and a late neurological syndrome. *Lancet* 2:1216–1217.

Maiztegui, J.I., A. Briggiler, D. Enria, and M.R. Feuillade. 1986. Progressive extension of the endemic area and changing incidence of Argentine hemorrhagic fever. *Medical Microbiology and Immunology* 175:149–152.

McCormick, J.B., and S.P. Fisher-Hock 1992. Arenaviruses and other hemorrhagic fever viruses. In *Infectious diseases,* ed. S.L. Gorbach, J.G. Bartlett, and N.R. Blacklow. Philadelphia: W.B. Saunders, pp. 1842–1853.

McCormick, J.B., I.J. King, P.A. Webb, C.L. Scribner, R.B. Craven, K.M. Johnson, L.H. Elliott, and R. Belmont Williams. 1986. Lassa fever: Effective therapy with ribavirin. *New England Journal of Medicine* 314:20–26.

McCormick, J.B., I.J. King, P.A. Webb, E.S. Smith, S. Trippel, and T.C. Tong. 1987a. A case-control study of the clinical diagnosis and course of Lassa fever. *Journal of Infectious Diseases* 155:445–455.

McCormick, J.B., P.A. Webb, J.W. Krebs, K.M. Johnson, and E.S. Smith. 1987b. A prospective study of the epidemiology and ecology of Lassa fever. *Journal of Infectious Diseases* 155:437–444.

McKee, K.T., Jr., J.W. LeDuc, and C.J. Peters 1991. Hantaviruses. In *Textbook of human virology,* ed. R.B. Belshe. St. Louis: Mosby Year Book, pp. 615–632.

Mills, J.N., B.A. Ellis, K.T. McKee, G.E. Calderon, J.I. Maiztegui, G.O. Nelson, T.G. Ksiazek, C.J. Peters, and J.E. Childs. 1992. A longitudinal study of Junin virus activity in the rodent reservoir of Argentine hemorrhagic fever. *American Journal of Tropical Medicine and Hygiene* 47:749–763.

Mills, J.N., B.A. Ellis, J.E. Childs, K.T. McKee Jr., J.I. Maiztegui, C.J. Peters, T.G. Ksiazek, and P.B. Jahrling. 1994. Prevalence of infection with Junin virus in rodent populations in the epidemic area of Argentine hemorrhagic fever. *American Journal of Tropical Medicine and Hygiene* 51:554–562.

Mills, J.N., J.E. Childs, T.G. Ksiazek, C.J. Peters, and W.M. Velleca. 1995a. *Methods for trapping and sampling small mammals for virologic testing.* Atlanta: US Department of Health and Human Services, 61 pp.

Mills, J.N., T.L. Yates, J.E. Childs, R.R. Parmenter, T.G. Ksiazek, P.E. Rollin, and C.J. Peters. 1995b. Guidelines for working with rodents potentially infected with hantavirus. *Journal of Mammalogy* 76:716–722.

Mills, J.N., J.G. Barrera Oro, D.S. Bressler, J.E. Childs, R.B. Tesh, J.F. Smith, D.A. Enria, T.W. Geisbert, K.T. McKee Jr., M.D. Bowen, C.J. Peters, and P.B. Jahrling. 1996. Characterization of Oliveros virus, a new member of the Tacaribe complex (Arenaviridae: *Arenavirus*). *American Journal of Tropical Medicine and Hygiene* 54:399–404.

Mills, J.N., T.G. Ksiazek, B.A. Ellis, P.E. Rollin, S.T. Nichol, T.L. Yates, W.L. Gannon, C.E. Levy, D.M. Engelthaler, T. Davis, D.T. Tanda, W. Frampton, C.R. Nichols, C.J. Peters, and J.E. Childs. 1997. Patterns of association with host and habitat: Antibody reactive with Sin Nombre virus in small mammals in the major biotic communities of the Southwestern United States. *American Journal of Tropical Medicine and Hygiene* 56:273–284.

Mills, J.N., J.M. Johnson, T.G. Ksiazek, B.A. Ellis, P.E. Rollin, T.L. Yates, M.O. Mann, M.R. Johnson, M.L. Campbell, J. Miyashiro, M. Patrick, M. Zyzak, D. Lavender, M.G. Novak, K. Schmidt, C.J. Peters, and J.E. Childs. 1998. A survey of hantavirus antibody in small-mammal populations in selected U.S. national parks. *American Journal of Tropical Medicine and Hygiene* 58:525–532.

Mims, C.A. 1975. The meaning of persistent infections in nature. *Bulletin of the World Health Organization* 52:747–751.

Morzunov, S.P., H. Feldmann, C.F. Spiropoulou, V.A. Semenova, P.E. Rollin, T.G. Ksiazek, C.J. Peters, and S.T. Nichol. 1995. A newly recognized virus associated with a fatal case of hantavirus pulmonary syndrome in Louisiana. *Journal of Virology* 69:1980–1983.

Murphy, F.A., W.C. Winn, D.H. Walker, M.R. Flemister, and S.G. Whitfield. 1976. Early lymphoreticular viral tropism and antigen persistence: Tamiami virus infection in the cotton rat. *Laboratory Investigation* 34:125–140.

Musser, G.G., and M.D. Carleton. 1993. Family Muridae. In *Mammal species of the world: A taxonomic and geographic reference,* ed. D.E. Wilson and D.M. Reeder, 2d ed. Washington, DC: Smithsonian Institution, pp. 501–755.

Nagai, T., O. Tanishita, Y. Takahashi, T. Yamanouchi, K. Domae, K. Kondo, J.R.J. Dantas, M. Takahashi, and K. Yamanishi. 1985. Isolation of haemorrhagic fever with renal syndrome virus from leukocytes of rats and virus replication in cultures of rat and human macrophages. *Journal of General Virology* 66:1271–1278.

Nichol, S.T., C.F. Spiropoulou, S. Morzunov, P.E. Rollin, T.G. Ksiazek, H. Feldmann, A. Sanchez, J.E. Childs, S. Zaki, and C.J. Peters. 1993. Genetic identification of a hantavirus associated with an outbreak of acute respiratory illness. *Science* 262:914–917.

Niklasson, B., and J.W. LeDuc. 1987. Epidemiology of nephropathia epidemica in Sweden. *Journal of Infectious Diseases* 155:269–276.

Niklasson, B., B. Hornfeldt, A. Lundkvist, S. Bjorsten, and J. LeDuc. 1995. Temporal dynamics of Puumala virus antibody prevalence in voles and of nephropathia epidemica incidence in humans. *American Journal of Tropical Medicine and Hygiene* 53:134–140.

Nowotny, N., H. Weissenboeck, S. Aberle, and F. Hinterdorfer. 1994. Hantavirus infection in the domestic cat. *Journal of the American Medical Association* 272:1100–1101.

Nuzum, E.O., C.A. Rossi, E.H. Stephenson, and J.W. LeDuc. 1988. Aerosol transmission of Hantaan and related viruses to laboratory rats. *American Journal of Tropical Medicine and Hygiene* 38:636–640.

O'Connor, C.S., and J.P. Hayes. 1997. Sin Nombre virus does not impair respiratory function of wild deer mice. *Journal of Mammalogy* 78:661–668.

Oldstone, M.B.A., R. Ahmed, J. Byrne, M.J. Buchmeier, Y. Riviere, and P. Southern. 1985. Virus and immune responses: Lymphocytic choriomeningitis virus as a prototype model of viral pathogenesis. *British Medical Bulletin* 44:70–74.

Otteson, E.W., J. Riolo, J.E. Rowe, S.T. Nichol, T.G. Ksiazek, P.E. Rollin, and S.C. St. Jeor. 1996. Occurrence of hantavirus within the rodent population of northeastern California and Nevada. *American Journal of Tropical Medicine and Hygiene* 54:127–133.

Pan American Health Organization (PAHO). 1982. Bolivian hemorrhagic fever. *Epidemiological Bulletin Pan American Health Organization* 3:15–16.

Park, J.G. 1986. Prevalence of antibodies to Hantaan virus among house rats in the harbor area of Manila and normal adults in the Philippines. *Korean University Medical Journal* 23:55–56. Abstract.

Parodi, A.S., D.J. Greenway, H.R. Ruggiero, E. Rivero, M.J. Frigerio, N. Mettler, F. Garzon, M. Boxaca, L.B. de Guerrero, and R. Nota. 1958. Sobre la etiologia del brote epidemica de Junin. *Dia Medico* 30:2300–2302.

Peters, C.J. 1991. Arenaviruses. In *Textbook of human virology,* ed. R.B. Belshe, 2d ed. St. Louis: Mosby Year Book, pp. 541–570.

———. 1994. Arenaviridae. In *Viral infections of rodents and lagomorphs,* ed. A.D.M.E. Osterhaus. New York: Elsevier, pp. 317–341.

Peters, C.J., P.B. Jahrling, C.T. Liu, R.H. Kenyon, K.T.J. McKee, and J.G. Barrera Oro. 1987. Experimental studies of arenaviral hemorrhagic fevers. *Current Topics in Microbiology and Immunology* 134:5–68.

Peters, C.J., M. Buchmeier, P.E. Rollin, and T.G. Ksiazek 1996. Arenaviruses. In *Virology,* ed. B.N. Fields, D.M. Knipe, and P.M. Howley. Philadelphia: Lippincott-Raven, pp. 1521–1551.

Pilaski, J., H. Feldmann, S. Morzunov, P.E. Rollin, S.L. Ruo, B. Lauer, C.J. Peters, and S.T. Nichol. 1994. Genetic identification of a new Puumala virus strain causing severe hemorrhagic fever with renal syndrome in Germany. *Journal of Infectious Diseases* 170:1456–1462.

Pinheiro, F.P., R.E. Shope, A.H. Paes de Andrade, G. Bensabath, G.V. Cacios, and J. Casals. 1966. Amapari, a new virus of the Tacaribe group from rodents and mites of Amapa territory, Brazil. *Proceedings of the Society for Experimental Biology and Medicine* 122:531–535.

Pinheiro, F.P., J.P. Woodall, A.P.A. Travassos da Rosa, and J.F. Travassos de Rosa. 1977. Studies on arenaviruses in Brazil. *Medicina (Buenos Aires)* 37:175–181.

Plyusnin, A., O. Vapalahti, H. Lankinen, H. Lehvaslaiho, N. Apekina, Y. Myasnikov, H. Kallio-Kokko, H. Henttonen, A. Lundkvist, M. Brummer-Korvenkontio, I. Gavrilovskaya, and A. Vaheri. 1994. Tula virus: A newly detected hantavirus carried by European common voles. *Journal of Virology* 68:7833–7839.

Plyusnin, A., O. Vapalahti, A. Lundkvist, H. Henttonen, and A. Vaheri. 1996. Newly recognized hantavirus in Siberian lemmings. *Lancet* 347:1835–1836.

Redford, K.H., and J.F. Eisenberg. 1992. *Mammals of the neotropics: The southern cone.* Chicago: University of Chicago Press, 430 pp.

Robbins, C.B., and E. van der Straeten. 1989. Comments on the systematics of *Mastomys* Thomas 1915 with the description of a new west African species. *Senckenbergiana Biologica* 69:1–14.

Robbins, C.B., J.W. Krebs, and K.M. Johnson. 1983. *Mastomys* (Rodentia: Muridae) species distinguished by hemoglobin differences. *American Journal of Tropical Medicine and Hygiene* 32:624–630.

Rollin, P.E., T.G. Ksiazek, L.H. Elliott, E.V. Ravkov, M.L. Martin, S. Morzunov, W. Livingstone, M. Monroe, G. Glass, S. Ruo, A.S. Khan, J.E. Childs, S. Nichol, and C.J. Peters. 1995. Isolation of Black Creek Canal virus, a new hantavirus from *Sigmodon hispidus* in Florida. *Journal of Medical Virology* 46:35–39.

Rowe, W.P., F.A. Murphy, G.H. Bergold, J. Casals, J. Hotchin, K.M. Johnson, F. Lehmann-Grube, C.A. Mims, E. Trub, and P.A. Webb. 1970. Arenoviruses: Proposed name for a newly defined virus group. *Journal of Virology* 5:651–652.

Salas, R., N. Manzione, R.B. Tesh, R. Rico-Hesse, R.E. Shope, A. Betancourt, O. Godoy, R. Bruzual, M.E. Pacheco, B. Ramos, M.E. Taibo, J.G. Tamayo, E. Jaimes, C. Vasquez, F. Araoz, and J. Querales. 1991. Venezuelan haemorrhagic fever. *Lancet* 338:1033–1036.

Schlaeffer, F., Y. Bar Lavie, E. Sikuler, M. Alkan, and A. Keynan. 1988. Evidence against high contagiousness of Lassa fever. *Transactions of the Royal Society of Tropical Medicine and Hygiene* 82:311.

Schmaljohn, C. 1994. Prospects for vaccines to control viruses in the family Bunyaviridae. *Reviews in Medical Virology* 4:185–196.

Schmaljohn, C.S., S.E. Hasty, J.M. Dalrymple, J.W. LeDuc, H.W. Lee, C.H. von Bonsdorff, M. Brummer Korvenkontio, A. Vaheri, T.F. Tsai, H.L. Regnery, D. Goldgaber, and P.W. Lee. 1985. Antigenic and genetic properties of viruses linked to hemorrhagic fever with renal syndrome. *Science* 227:1041–1044.

Skinner, H.H., and E.H. Knight. 1971. Monitoring mouse stocks for lymphocytic choriomeningitis virus: A human pathogen. *Laboratory Animals* 5:73–87.

Slonova, R.A., E.A. Tkachenko, E.L. Kushnarev, T.K. Dzagurova, and T.I. Astakova. 1992. Hantavirus isolation from birds. *Acta Virologica (Praha)* 36:493.

Smithard, E.H.R., and A.D. Macrae. 1951. Lymphocytic choriomeningitis associated human and mouse infection. *British Medical Journal* 51:1299–1300.

Song, W., N. Torrez-Martinez, W. Irwin, F.J. Harrison, R. Davis, M. Ascher, M. Jay, and B. Hjelle. 1995. Isla Vista virus: A genetically novel hantavirus of the California vole *Microtus californicus. Journal of General Virology* 76:3195–3199.

Southern, P.J. 1996. Arenaviridae: The viruses and their replication. In *Fields virology,* ed. B.N. Fields, D.M. Knipe, and P.M. Howley, 3d ed. Philadelphia: Lippincott-Raven, pp. 1505–1520.

Stephensen, C.B., J.R. Jacob, R.J. Montali, K.V. Holmes, E. Muchmore, R.W. Compans, E.D. Arms, M.J. Buchmeier, and R.E. Lanford. 1991. Isolation of an arenavirus from a marmoset with callitrichid hepatitis and its serologic association with disease. *Journal of Virology* 65:3995–4000.

Swanepoel, R., P.A. Leman, A.J. Shepherd, S.P. Shepherd, M.P. Kiley, and J.B. McCormick. 1985. Identification of Ippy virus as a Lassa-fever-related virus. *Lancet* 1:639.

Tesh, R.B., M.L. Wilson, R. Salas, N.M. de Manzione, D. Tovar, T.G. Ksiazek, and C.J. Peters. 1993. Field studies on the epidemiology of Venezuelan hemorrhagic fever: Implication of the cotton rat *Sigmodon alstoni* as the probable rodent reservoir. *American Journal of Tropical Medicine and Hygiene* 49:227–235.

Trapido, H., and C. Sanmartin. 1971. Pichinde virus: A new virus of the Tacaribe group from Columbia. *American Journal of Tropical Medicine and Hygiene* 20:631–641.

Tsai, T.F. 1987. Hemorrhagic fever with renal syndrome: Mode of transmission to humans. *Laboratory Animal Science* 37:428–430.

Vitullo, A.D., and M.S. Merani. 1990. Vertical transmission of Junin virus in experimentally infected adult *Calomys musculinus. Intervirology* 31:339–344.

Vitullo, A.D., V.L. Hodara, and M.S. Merani. 1987. Effect of persistent infection with Junin virus on growth and reproduction of its natural reservoir, *Calomys musculinus. American Journal of Tropical Medicine and Hygiene* 37:663–669.

Webb, P.A., K.M. Johnson, J.B. Hibbs, and M.L. Kuns. 1970. Parana, a new Tacaribe complex virus from Paraguay. *Archiv für die gesamte Virusforschung* 32:379–388.

Webb, P.A., K.M. Johnson, C.J. Peters, and G. Justines. 1973. Behavior of Machupo and Latino viruses in *Calomys callosus* from two geographic areas of Bolivia. In *Lymphocytic choriomeningitis virus and other arenaviruses,* ed. F. Lehmann-Grube. Berlin: Springer-Verlag, pp. 313–322.

Wells, R.M., S.S. Estani, Z.E. Yadon, D. Enria, P. Padula, J.N. Mills, C.J. Peters, and E.L. Segura. 1997. An unusual hantavirus outbreak in southern Argentina: Person-to-person transmission? *Emerging Infectious Diseases* 3:1–4.

White, D.J., R.G. Means, G.S. Birkhead, E.M. Bosler, L.J. Grady, N. Chatterjee, J. Woodall, B. Hjelle, P.E. Rollin, T.G. Ksiazek, and D.L. Morse. 1996. Human and rodent hantavirus infection in New York State. *Archives of Internal Medicine* 156:722–726.

Williams, R.J., R.T. Bryan, J.N. Mills, E. Palma, I. Bera, F. de Velasquez, E. Baez, W.E. Schmidt, R.E. Figueroa, C.J. Peters, S.R. Zaki, A.S. Khan, and T.G. Ksiazek. 1997. An outbreak of hantavirus pulmonary syndrome in western Paraguay. *American Journal of Tropical Medicine and Hygiene* 57:274–282.

Wilson, D.E., and D.M. Reeder. 1993. *Mammal species of the world,* 2d ed. Washington, DC: Smithsonian Institution, 1207 pp.

Wolfe, J.L. 1982. *Oryzomys palustris. Mammalian Species* 176:1–5.

World Health Organization (WHO). 1993. Vaccination against Argentine haemorrhagic fever. *Weekly Epidemiological Record* 68:233–234.

Wulff, H., B.M. McIntosh, D.B Hamner, and K.M. Johnson. 1977. Isolation of an arenavirus closely related to Lassa virus from *Mastomys natalensis* in south-east Africa. *Bulletin of the World Health Organization* 55:441–444.

Yanagihara, R., and D.C. Gajdusek. 1987. Hemorrhagic fever with renal syndrome: Global epidemiology and ecology of hantavirus infections. In *Medical virology VI,* ed. L.M. de la Maza and E.M. Peterson. New York: Elsevier, pp. 171–214.

Yanagihara, R., H.L. Amyx, and D.C. Gajdusek. 1985. Experimental infection with Puumala virus, the etiologic agent of nephropathia epidemica, in bank voles (*Clethrionomys glareolus*). *Journal of Virology* 55:34–38.

Zaki, S.R., P.W. Greer, L.M. Coffield, C.S. Goldsmith, K.B. Nolte, K. Foucar, R.M. Feddersen, R.E. Zumwalt, G.L. Miller, A.S. Khan, P.E. Rollin, T.G. Ksiazek, S.T. Nichol, B.W.J. Mahy, and C.J. Peters. 1995. Hantavirus pulmonary syndrome: Pathogenesis of an emerging infectious disease. *American Journal of Pathology* 146:552–579.

Zhao-zhaung, L., H. Hao, and L. Yue-xian. 1985. Characteristics of EHF virus isolated from cats in endemic area of Anhui province. *Chinese Journal of Microbiology and Immunology* 5:81–83.

15 ORTHOMYXOVIRUS AND PARAMYXOVIRUS INFECTIONS

HANA VAN CAMPEN AND GREG EARLY

INFLUENZA A VIRUS INFECTIONS

Introduction. Influenza A viruses infect a wide variety of species, including birds, humans, pigs *Sus scrofa,* horses *Equus caballus* (Easterday 1975), and harbor seals *Phoca vitulina* (Geraci et al. 1982). Although influenza A viruses have been periodically isolated from dogs *Canis familiaris,* domestic cats *Felis catus,* monkeys, and cattle *Bos taurus* (Easterday 1975), the natural reservoir for the virus is hypothesized to be aquatic birds (Hinshaw et al. 1978; Murphy and Webster 1996), in which the virus may be maintained from year to year through direct contact with virus-laden aquatic habitats (Ito et al. 1995). Direct transmission among avian species is common (Easterday 1975), but, despite antigenic and genetic relatedness of avian and mammalian influenza viruses, transmission is not as well documented between species other than birds. Interspecies transmission has been strongly suggested between pigs and humans (Wentworth et al. 1994), birds and pigs (Bean et al. 1992), birds and marine mammals (Hinshaw et al. 1984), and seals and humans (Webster et al. 1992). A regular association of avian viruses with influenza outbreaks in seals suggests that transmission of these viruses to seals occurs frequently and may be an important link in the evolution of new mammalian strains. Because transmission of an influenza virus from seals to humans has been reported (Webster et al. 1981a,b), this link's role in the appearance of new pandemic strains in humans should be carefully considered.

Etiology. Influenza A viruses are members of the *Orthomyxoviridae.* The virions are pleomorphic and covered with a lipid envelope derived from the plasma membrane of the host cell. The genome consists of eight negative-sense, single-stranded RNA segments encoding three polymerases (PA, PB1, and PB2), the hemagglutinin (HA) and neuraminidase (NA) glycoproteins, nucleoprotein (NP), matrix proteins (M_1 and M_2), and nonstructural proteins (NS_1 and NS_2). The virus replicates by clipping the m^5GpppXm caps from the 5' end of cellular mRNAs. The caps serve as primers for the transcription of mRNA by the viral polymerase complex from the genomic segment. The resulting complementary RNA also serves as the template for the production of progeny negative-sense RNA segments. Transcription occurs in the nucleus of infected cells, and transcripts are transported to the cytoplasm, where translation of viral proteins occurs (Lamb and Krug 1996).

Influenza A viruses are defined as subtypes by the serologic identification of the HA and NA glycoproteins. The greatest variety of influenza subtypes have been isolated from avian species (Hinshaw et al. 1978). Fourteen serologically distinct HA and nine NA subtypes have been identified to date. The segmented nature of the genome is important in the natural history of the virus. Coinfection of a host cell with two different strains of influenza viruses can result in progeny viruses possessing gene segments from both parental viruses. The result of the reassortment of viral segments is the emergence of new subtypes in the progeny viruses that is called antigenic shift. Low fidelity of the viral polymerases gives rise to considerable genetic variability within a single gene segment. Point mutations in the HA and NA genes accumulate over time in response to selection by populational immunity, and this is termed antigenic drift (Murphy and Webster 1996).

Host Range. Prior to their association with clinical disease in free-ranging marine mammals, influenza viruses were incidentally identified in Pacific minke whales *Balaenoptera acutorostrata* (Lvov et al. 1978). Low levels of antibodies to a human serotype of influenza virus were also reported in the Pacific fur seal *Callorhinus ursinus* (Sazanov et al. 1976).

The first association between influenza viruses and disease in free-ranging marine mammal populations was made during a short but devastating epidemic of pneumonia in harbor seals *Phoca vitulina* along the New England coast between December 1979 and October 1980. The outbreak was characterized by sudden and dramatic increase in mortality among a wintering aggregation of seals on Billingsgate Shoal, Cape Cod, Massachusetts. Within the first 6 months of the outbreak, over 400 dead seals were recovered along the New England coast, more than ten times the number of dead animals that had been recovered historically from the same geographic area. Among the seals tightly clustered together on Billingsgate Shoal, mortality may have been as high as 25% within the first 2 months of the outbreak (Geraci et al. 1982). Mortality slowed as aggregates dispersed and seals began seasonal movements north during the spring. By October 1980, the disease appeared to have run its course. Conservative estimates, based on dead animals recovered, indicated that at least 3%–5% of the 10,000–14,000 harbor seals in New England at the time had died (Geraci et al. 1982).

An influenza A virus A/Seal/Mass/1/80 (H7N7) was isolated from the lungs and brains of dead seals (Lang et al. 1981; Webster et al. 1981b; Geraci et al. 1982). An assortment of aerobic and anaerobic bacteria and an unidentified mycoplasma were isolated from lungs of many affected seals; however, none were considered to be the primary pathogen (Geraci et al. 1982). The influenza virus was genetically similar to fowl plague virus Dutch/27 (H7N7) but did not replicate in avian hosts (Webster et al. 1981b). A/Seal/Mass/1/80 replicated in pigs, cats, and domestic ferrets *Mustela putorius furo* with shedding 5–8 days after inoculation and in guinea pigs *Cavia porcellus* and suckling mice *Mus musculus* for 1–3 days after intracerebral inoculation without causing clinical disease (Webster et al. 1981b). The virus replicated in experimentally infected harbor seals (Webster et al. 1981b) and harp seals *Phoca groenlandica* (Geraci et al. 1984) but did not replicate in gray seals *Halichoerus grypus* (Geraci et al. 1984). Harbor seals inoculated by intratracheal and intranasal routes developed pneumonitis, conjunctivitis, and mucopurulent nasal discharge (Webster et al. 1981b). In another study, moderate pneumonia occurred in harbor seals, mild pneumonia was described in harp seals, and gray seals did not develop clinical signs of infection (Geraci et al. 1984). Failure of the virus to induce severe disease experimentally implicates other factors, such as crowding (Geraci et al. 1984), shifts in host range (Truyen et al. 1995), or cumulative effects of infection with a variety of organisms as contributors to acute mortality. Isolation of virus from brains of seals during the epizootic indicated that the virus can spread systemically under some conditions, which is similar to the nervous form of fowl plague (Webster et al. 1981b).

A second virus, A/Seal/MA/133/82 (H4N5), isolated from a harbor seal stranded in the summer of 1982 (Hinshaw et al. 1984), was genetically distinct from the previously isolated seal virus. The A/Seal/MA/133/82 was genetically and antigenically similar to avian viruses; unlike A/Seal/Mass/1/80, however, it replicated well in experimentally infected ducks, harbor seals, harp seals, and ringed seals *Phoca hispida*. The virus did not cause severe disease experimentally (Hinshaw et al. 1984). From January to March 1983, though, an increase in seal mortality of roughly 3–4 times expected was reported along the New England coast, and similar viruses were isolated from lungs, lymph nodes, and brains of dead seals. Because it is unlikely that the virus was introduced twice into the same population, the time lag between the first isolation of the virus and the observed mortality event indicated that the virus could both be maintained by and produce severe disease in free-ranging seal populations (Hinshaw et al. 1984).

Five additional influenza viruses have been isolated from harbor seals since 1989, although there has been no pattern of acute mortality reminiscent of earlier epizootics. In January 1991, two H4 influenza viruses, A/Seal/MA/3807/91 and A/Seal/MA/3810/91, were isolated. The N6 of influenza virus A/Seal/MA/ 3807/91 and A/Seal/MA/3810/91 (H4N6) isolates differs from the NA of previous seal isolates and is likely of avian origin (Callan et al. 1995). Between January and February 1992, three H3N3 viruses (A/Seal/ MA3911/92, A/Seal/MA/3984/92, and A/Seal/MA/ 4007/92) were isolated from harbor seal carcasses recovered along the coast of Cape Cod, Massachusetts. These viruses were the first H3 subtypes isolated from seals, although viruses of this subtype are known to infect other mammalian hosts, including humans, pigs, and horses. Analysis of the H3 gene sequences of seal isolates showed them to be most closely related to H3 viruses isolated from North American birds, and they are the first isolates recovered from a mammalian host associated with this phylogenetic branch (Callan et al. 1995). Infected seals had lesions consistent with those in earlier epidemics, but infection of seals with these viruses did not cause severe mortality (Callan et al. 1995).

Influenza viruses of the subtypes H13N9 and H13N2 were isolated from a moribund pilot whale *Globicephala melaena* (Hinshaw et al. 1986). The H13N9 virus was closely related to avian viruses and replicated in experimentally infected birds without inducing disease (Hinshaw et al. 1986). Isolates of an H1N3 influenza A virus were made from minke whales (Lvov et al. 1978).

Transmission and Epidemiology. Shorebirds are suspected as the source of influenza A viruses for marine mammals, based on close association of these species in nature. However, transfer of virus between marine species has not been excluded. Most influenza infections in seal populations do not result in significant mortality, and survivors appear to be repeatedly infected. Transmission of avian viruses to whales is suggested by the nature of the isolates made from a whale (Hinshaw et al. 1986); however, the behavior of the viruses within cetaceans is unknown. Susceptibility of other marine mammals to influenza A viral infections is also unknown. The paucity of reports regarding other mammals may be a result of low mortality associated with most influenza virus infections and, thus, lack of recognition of these infections. Episodes of high mortality in seals may be due to the introduction of particularly pathogenic strains and cumulative action of several pathogens and stresses, including adverse environmental conditions. Seals may be important links in the emergence of new virus strains, with the possibility of genetic reassortment of viruses and the potential for interspecies transmission.

Seals may be infected by direct aerosol transmission either via respiratory or ocular infection. Virus apparently spreads quickly geographically and across species. Shortly after the initial influenza epidemic in seals, antibody titers to A/Seal/Mass/1/80 were found in gray seals more than 800 km removed and in a geographically isolated location (Sable Island, Nova Scotia) from the original outbreak (Geraci et al. 1982).

Clinical Signs and Pathology. Severely affected seals exhibit signs of respiratory distress and weak, uncoordinated movements. Seals often have a frothy white to bloody discharge from the nose and mouth, and occasionally shake their heads to clear the discharge. Characteristically, severely affected seals have swollen necks and subcutaneous emphysema resulting from the leakage of air from damaged airways through the thoracic inlet and into subcutaneous tissues of the neck. These seals are unable to dive and, in some cases, seals are so buoyant that they drift with tide and wind. During the 1979–80 epidemic, disease developed in approximately 3 days, with some seals dying acutely within hours of feeding.

Influenza A virus causes acute interstitial pneumonia characterized by necrotizing bronchitis and bronchiolitis, and hemorrhagic alveolitis (Geraci et al. 1982). Hilar lymph nodes were grossly enlarged and hemorrhagic; microscopically large numbers of macrophages and erythrocytes were in medullary sinuses. Lungs were generally uniformly deep red with bloody or froth-filled airways. Often lungs had pockets of emphysema and prominent separation of the interlobular septa by gas and fluid. Trapped air, leaking from damaged alveoli, formed characteristic emphysema in tissues surrounding the neck, thoracic inlet, pericardium, and, in extreme cases, along the dorsal musculature and kidneys. Small perivascular hemorrhages and mild degenerative changes such as spongiosis and swelling of axons were documented in the brains of some animals.

Diagnosis and Differential Diagnoses. Influenza A virus infection can be diagnosed by inoculation of embryonated chicken eggs with fluids from nasal swabs or tracheal washings of live animals, and lung and tracheal homogenates collected at necropsy. Swabs should be placed in transport media in leak-proof vials, and all specimens should be shipped on ice to the laboratory for virus isolation. Influenza A viruses are typed by hemagglutination inhibition (HI) and neuraminidase inhibition (NI) using reference antisera (Kendal and Pereira 1982). Another viral etiology of pneumonia in seals is phocine distemper virus. Mycoplasmas and other bacterial species should be considered and appropriate cultures prepared. Serologic testing may also be useful in making a diagnosis and in surveillance (De Boer et al. 1990).

Treatment and Control. There is no specific treatment for influenza viral infection or pneumonia. Affected seals captured during epidemics were given supportive symptomatic treatment. Experimentally infected harp seals appeared to develop protective immunity (Geraci et al. 1984), suggesting that vaccination, although logistically difficult or impossible, would be technically feasible.

Public Health Significance. A seal experimentally inoculated with A/Seal/Mass/1/80 (H7N7) and known to be shedding virus sneezed in the face of an investigator. The investigator developed purulent conjunctivitis with intense periorbital swelling and pain in the right eye within 40 hours of the episode. Samples from the eye taken 2 days after exposure contained high titers of influenza A virus identical to the seal virus (Webster et al. 1981a) and were negative for mycoplasma. During the 1979–80 epidemic, four workers involved in seal autopsies developed a similar conjunctivitis within 2 days of known exposure to their eyes (Webster et al. 1981a). In all cases, conjunctivitis lasted 5–7 days, and recovery was uneventful. Antibodies to the seal virus were not detected in lacrimal secretions or serum of the individuals. Indications are that infection might occur in the eye without general systemic response (Webster et al. 1981b). Although respiratory disease associated with influenza A viral infection did not occur in any of the affected individuals, the frequent recovery of virus from the brains of affected seals was of concern, since it indicated systemic infection (Webster et al. 1981b). The ability of the virus to infect humans underscores the potential role of seals in the transfer of influenza viruses from avian reservoirs to mammalian species.

RESPIRATORY SYNCYTIAL VIRUS AND PARAINFLUENZA VIRUS

Introduction. *Paramyxoviridae* includes viruses that cause respiratory disease in a wide range of mammals. This family is divided into the genera *Paramyxovirus, Morbillivirus, Rubulavirus,* and *Pneumovirus,* of which parainfluenza viruses and respiratory syncytial viruses are discussed in this chapter. Parainfluenza viruses (PIVs) infect subhuman primates, wild ruminants, dogs, domestic cattle, swine, sheep *Ovis aries,* goats *Capra hircus,* rabbits *Oryctolagus cuniculus,* and rodents. Respiratory syncytial viruses (RSVs) infect primates, wild ruminants, and domestic sheep and cattle. Uncomplicated PIV infections are usually subclinical but can cause interstitial pneumonia. In general, RSV infections cause more severe disease in the form of interstitial pneumonia than do PIVs. Defense mechanisms of the respiratory tract may be compromised by PIV or RSV infection predisposing to secondary bacterial infections. These viruses are characteristically highly infectious and transmissible by aerosols. Seroprevalence to PIV is nearly ubiquitous in wild ruminants and in some primate populations.

Etiology. The *Paramyxoviridae* are pleomorphic (150–300 nm), single-stranded RNA viruses with a long helical nucleocapsid surrounded by a lipid envelope [reviewed by Lamb and Kolakofsky (1996)]. The 15- to 16-kb negative-sense RNA genome encodes six or more proteins and glycoproteins. The genomic RNA is bound to nucleoproteins to form a helical nucleocapsid from which mRNAs and full-length, positive-sense RNA replicative templates are transcribed. In general, the genes are ordered sequentially starting with that encoding the nucleoprotein (N/NP) at the 3' end, followed by the genes for phosphoprotein (P), matrix (M), viral attachment (G) (RSV), fusion (F_o), and hemagglutinin-neuraminidase (HN) (PIV), and ending with the gene for the RNA-dependent RNA polymerase (L). The lipid envelope is derived from the plasma membrane of the infected cell and contains the HN/G glycoprotein and the F_0 protein. The HN glycoprotein is the viral attachment protein that binds to receptors on cell surfaces. The F glycoprotein is involved in fusion and release of the nucleocapsid into the cytoplasm and is responsible for the fusion of infected cells to others to form multinucleated syncytia, a hallmark of these virus infections.

Distribution and Host Range. Serologic and virologic evidence of PIV and RSV infections has been found in a variety of free-ranging ruminant species over all continents (Table 15.1). In addition, some serologic evidence for PIV infections has been reported in free-living subhuman primates in Asia and captive primates originating in Africa.

Transmission and Epidemiology. Despite their labile nature, PIV and RSV are highly infectious and efficiently transmitted by aerosol and contact with nasal secretions [reviewed by Black (1991)]. As such, PIVs infect a large proportion of a population, as judged by seroprevalences of 82%–100% for pronghorn *Antilocapra americana* (Barrett and Chalmers 1975; Stauber et al. 1980), 82%–84% of white-tailed deer *Odocoileus virginianus* (Sadi et al. 1991), and 95% in mule deer *Odocoileus hemionus* (Stauber et al. 1977). Prevalence of antibodies to RSV is also reported to be high for mule deer (89%) and elk *Cervus elaphus nelsoni* (84%) (Aguirre et al. 1995).

The preponderance of initial infections occurs in the young (Barrett and Chalmers 1975; Stauber et al. 1977, 1980). Crowding, mixing groups of animals, and low air exchanges in captivity increase the probability of exposure to an infectious viral dose. In humans and domestic species, RSV outbreaks occur in the fall and winter. Stressors such as handling, shipment, environmental conditions, and secondary bacterial infections are associated with severe respiratory disease in both RSV and PIV infections.

Livestock have been suspected of serving as a source of PIV infection of wild ruminant populations where

TABLE 15.1—Serologic and virologic evidence for parainfluenza virus 3 (PIV-3), simian virus 5 (SV5), and respiratory syncytial virus (RSV) infection in wild mammals

Species	Continent; Captive or Free-Living[a]	Evidence for Infection Antibody Virus Isolation Lesions (L)	References
ARTIODACTYLA			
Cervidae			
Axis deer *Axis axis*	North America (FL)	PIV-3(Ab)	Riemann et al. 1979
Elk *Cervus elaphus nelsoni*	North America (FL)	PIV-3(Ab) RSV(Ab)	Kingscote et al. 1987; Aguirre et al. 1995
Fallow deer *Dama dama*	North America (FL)	PIV-3(Ab)	Riemann et al. 1979
	North America (C)	PIV-3(VI)	Aguirre et al. 1995
Mule deer *Odocoileus hemionus*	North America (FL)	PIV-3(Ab)	Stauber et al. 1977; Aguirre et al. 1995
	North America (C)	PIV-3(VI)	Thorsen et al. 1977
	North America (FL)	RSV(Ab)	Johnson et al.1986b; Aguirre et al. 1995
White-tailed deer *Odocoileus virginianus*	North America (FL)	PIV-3(Ab) RSV(Ab)	Shah et al. 1965; Ingebrigtsen et al. 1986; Sadi et al. 1991 Johnson et al. 1986b
Roe deer *Capreolus capreolus*	Europe (FL)	PIV-3(Ab)	Baradel et al. 1988
Moose *Alces alces*	North America (FL)	PIV-3(Ab)	Thorsen and Henderson 1971; Johnson et al. 1973; Dieterich 1981
Caribou/reindeer *Rangifer tarandus*	Europe (FL) North America (FL, C)	PIV-3(Ab)	Rehbinder et al. 1991 Dieterich 1981
Antilocapridae			
Pronghorn antelope *Antilocapra americana*	North America (FL)	PIV-3(Ab)	Barrett and Chalmers 1975; Stauber et al. 1980; Johnson et al. 1986a; Kingscote and Bohac 1986
		PIV-3(VI) RSV (Ab)	Thorsen et al. 1977

(continued)

TABLE 15.1—*Continued*

Species	Continent; Captive or Free-Living[a]	Evidence for Infection Antibody Virus Isolation Lesions (L)	References
Bovidae			
Eland *Taurotragus* spp.	Africa (FL)	PIV-3 (Ab)	Hamblin and Hedger 1978
Greater kudu *Tragelaphus strepsiceros*	Africa (FL)	PIV-3 (Ab)	Hamblin and Hedger 1978
Oryx/gemsbok *Oryx gazella*	Africa (FL)	PI3-V (Ab)	Hamblin and Hedger 1978
Sable antelope *Hippotragus niger*	Africa (FL)	PI3-V (Ab)	Hamblin and Hedger 1978
Roan antelope *Hippotragus equinus*	Africa (FL)	PIV-3 (Ab)	Hamblin and Hedger 1978
Waterbuck *Kobus ellipsiprymnus*	Africa (FL)	PIV-3 (Ab)	Hamblin and Hedger 1978
Lechwe *Kobus leche*	Africa (FL)	PIV-3 (Ab)	Hamblin and Hedger 1978
Reedbuck *Redunca redunca*	Africa (FL)	PIV-3 (Ab)	Hamblin and Hedger 1978
Hartebeest *Alcelaphus buselaphus*	Africa (FL)	PIV-3 (Ab)	Hamblin and Hedger 1978
Topi/tsessebe *Damaliscus* lunatus	Africa (FL)	PIV-3 (Ab)	Hamblin and Hedger 1978
Blue wildebeest *Connochaetes taurinus*	Africa (FL)	PIV-3 (Ab)	Hamblin and Hedger 1978
Impala *Aepyceros melampus*	Africa (FL)	PIV-3 (Ab)	Hamblin and Hedger 1978
Springbuck *Antidorcas marsupialis*	Africa (FL)	PIV-3 (Ab)	Hamblin and Hedger 1978
Cape buffalo *Syncerus caffer*	Africa (FL)	PIV-3 (Ab)	Hamblin et al 1980.
Bison *Bison bison*	North America (FL)	PIV-3(Ab)	Zarnke 1983; Zarnke and Erickson 1990; Williams et al. 1993; Taylor et al. 1997
Dall sheep Ovis dalli	North America (FL)	PIV-3 (Ab)	Dieterich 1981
	North America (FL)	RSV (Ab)	Baradel et al. 1988
Bighorn sheep *Ovis canadensis*	North America (C)	PIV-3(Ab)	Howe et al. 1966; Parks and England 1974; Spraker et al. 1984; Clark et al. 1985; Spraker and Collins 1986
	North America (FL)	PIV-3(VI)	Parks et al. 1972; Spraker 1979; Clark et al. 1985
		RSV(Ab)	Dunbar et al. 1985; Clark et al. 1985; Spraker and Collins 1986
		RSV(VI)	Spraker and Collins 1986
Mountain goat Rocky *Oreamnos americanus*	North America (FL)	RSV (Ab)	Dunbar and Foreyt 1986
Ibex *Capra ibex*	Europe (FL)	PIV-3(Ab)	Baradel et al. 1988
Chamois *Rupricapra rupricapra*	Europe (FL)	PIV-3 (Ab)	Baradel et al. 1988
Suidae			
Wart hog *Phacochoerus aethiopicus*	Africa (FL)	PIV-3 (Ab)	Hamblin and Hedger 1978
Bush pig *Potamochoerus porcus*	Africa (FL)	PIV-3 (Ab)	Hamblin and Hedger 1978
PRIMATES			
Cercopithecidae			
Patas monkey *Erythrocebus patas*	Europe (C)	PIV (Ab, VI, L)	Churchill 1963
Rhesus macaque *Macaca mulatta*	Asia (FL)	SV5 (Ab)	Bhatt et al. 1966
Bonnet monkey *Macaca radiata*	Asia (FL)	SV5 (Ab)	Bhatt et al. 1966
Hunuman langur *Semnopithecus entellus*	Asia (FL)	SV5 (Ab)	Bhatt et al. 1966

[a]C, captive; FL, free-living.
[b]Ab, antibody; VI, virus isolation; L, lesions.

they share range (Thorsen and Henderson 1971; Stauber et al. 1977; Clark et al. 1985; Kingscote et al. 1987). However, studies of white-tailed deer geographically isolated from livestock indicate that large wild ruminant populations can maintain PIV (Sadi et al. 1991).

Clinical Signs, Pathogenesis, and Pathology. Most PIV infections of wild ruminant species are subclinical (Thorsen et al. 1977; Hamblin and Hedger 1978; Clark et al. 1985; Bryson et al. 1988). However, PIV type 3 (PIV-3) infections and fatal pneumonia have been reported in captive Rocky Mountain bighorn sheep *Ovis canadensis,* with 100% mortality within the group (Parks et al. 1972). At autopsy, the diaphragmatic, cardiac, and apical lung lobes were reported to be consolidated, with gray nodules and pleural adhesions.

Parainfluenza virus-3 viral pneumonia in captive patas monkeys *Erythrocebus patas* resulted in losses of up to 75%. Lesions of bronchopneumonia with fibrinous pleuritis, pericarditis, and peritonitis were described, and PIV-3 was isolated from clinically normal animals, as well as lung samples from fatal cases. Pneumococci were also isolated from lung samples, underscoring the importance of secondary infections in the severity of disease. Experimental inoculation resulted in fever (39° C–41° C) 5–6 days after infection, with slight nasal discharge. Viral shedding was detected in throat swabs for 7–11 days after infection, and seroconversion occurred 14 days after inoculation. Stress due to shipment of the animals was implicated in the occurrence of fatal disease (Churchill 1963).

Respiratory syncytial virus infections of wild ruminants may be subclinical or result in pneumonia (Spraker and Collins 1986; Foreyt and Evermann 1988). Moderate suppurative rhinitis, tracheitis, and subacute suppurative bronchopneumonia were observed in a bighorn sheep from which RSV was isolated (Spraker and Collins 1986). Respiratory syncytial viral infections in ruminants may be biphasic with fever and depression 2–4 days after infection, followed by a short period of recovery and then by more severe respiratory signs. Experimental inoculation of white-tailed deer fawns with an ovine isolate of RSV resulted in mild rhinitis, consolidation of approximately 10% of the lung mass, intracytoplasmic viral inclusions in the epithelial cells of alveolar walls, and perivascular and peribronchiolar lymphocytic accumulations (Bryson et al. 1988). Reinfections with either PIV or RSV are common and tend to be subclinical.

Diagnosis and Differential Diagnoses. The characteristic histopathologic findings of syncytia formed from the respiratory epithelium is helpful in diagnosis of PIV and RSV infection in cases of interstitial pneumonia. Eosinophilic cytoplasmic inclusions in respiratory epithelium may also be observed. Diagnosis can be strengthened by demonstrating viral antigen by using immunofluorescent antibody (FA) staining on frozen tissue sections or by immunohistochemical techniques. Respiratory syncytial viral antigen can be detected in nasal swabs and tracheal washes by enzyme-linked immunosorbent assay.

Both viruses are difficult to isolate from field cases. Lung samples from animals in the acute stages of infection should be immediately processed after collection, and virus isolation attempted using primary bovine embryonic kidney or lung cells. Following the development of cytopathic changes, including syncytia formation, the viruses are identified by FA.

Serum neutralizing antibody (SN) titers to bovine reference viruses (Carbrey et al. 1971) indicate past infection with PIV or RSV, and a fourfold or greater rise in SN titer is indicative of recent infection in living animals. Paired sera, of course, are seldom available from free-ranging animals.

Differential diagnoses for viral respiratory infections of wild ruminants should include infections by adenoviruses, bluetongue and epizootic hemorrhagic disease viruses, bovine herpesviruses, respiratory coronaviruses, enteroviruses, pestiviruses, and rhinoviruses. For primates, the differential includes adenoviruses, coronaviruses, influenza A viruses, morbillivirus, and rhinoviruses.

Control, Management Implications, and Domestic Animal Health Concerns. Few studies have been made on the efficacy of vaccination in wild mammals. Bighorn sheep vaccinated with a modified live bovine RSV vaccine developed neutralizing antibody titers (Foreyt and Evermann 1988); however, as is so often the case in experimental infections, clinical signs were not observed either in vaccinated or in unvaccinated bighorn following challenge. Immunity due to natural infection is short-lived, and reinfections are common. Maternally derived antibody does not prevent infection but reduces severity and prevalence of disease in domestic calves (Kimman et al. 1988). Vaccination of captive ruminants with modified-live BRSV/PIV-3 vaccines may be used to reduce severity of disease if these viruses pose a problem in a particular situation.

Domestic ruminants are a suspected source of PIV and RSV for wild ruminants because these viruses commonly circulate in domestic animal populations. Proximity to livestock is associated with declining or static bighorn sheep populations (Clark et al. 1985). Conversely, the risk of viral transmission from wild ruminants to domestic species is unknown. Separation of wild ruminants from contact with domestic species is prudent where possible in controlling PIV-3 and RSV infections in wild species.

Similarly, primates should be considered to be susceptible to human strains of PIV-3 and RSV. Therefore, contact with humans acutely ill with upper respiratory signs should be prevented. Subhuman primates have served as models of PIV and RSV infections in humans. Immunization with formalized antigens should be avoided because these vaccines have been implicated in an increase in the severity of disease upon natural infection. Passive immunity, on the other hand, is helpful, and RSV infections have been successfully treated with human immunoglobulin preparations.

PARAMYXOVIRUSES OF FLYING FOXES. In 1994 and 1995, a previously unrecognized viral disease caused the death of domestic horses and humans in Australia (Murray et al. 1995; O'Sullivan et al. 1997). The virus was initially thought to be a morbillivirus and designated "equine morbillivirus." However, subsequent study indicated the virus, now called Hendra virus, could not be classified in the genus *Morbillivirus* and should be a new genus in the subfamily *Paramyxovirinae* (Yu et al. 1998a,b). Extensive serologic surveys of 46 wild species in areas where these outbreaks occurred demonstrated virus-neutralizing antibodies only in flying foxes, bats of the suborder Megachiroptera (Young et al. 1996). Overall seroprevalence was approximately 9%, and spectacled fruit bats *Pteropus conspicillatus,* black fruit bats *Pteropus alecto,* little red fruit bats *Pteropus scapulatus,* and gray-headed fruit bats *Pteropus poliocephalus* had antibodies against Hendra virus, suggesting they may be natural host of this virus.

Evidence that flying foxes are hosts for another recently identified paramyxovirus, designated Menangle virus, is also based on serology. In these studies, antibodies were found in gray-headed fruit bats, black fruit bats, and spectacled fruit bats (Philbey et al. 1998). This virus was implicated in illness in pigs and humans.

LITERATURE CITED

Aguirre, A.A., D.E. Hansen, E.E. Starkey, and R.G. McLean. 1995. Serologic survey of wild cervids for potential disease agents in selected national parks in the United States. *Preventive Veterinary Medicine* 21:313–322.

Baradel, J.M., J. Barrat, J. Blancou, J.M. Boutin, C. Chastel, G. Dannacher, D. Delmore, Y. Gerard, J.M. Gourreau, V. Kihm, B. Larenaudie, C. Legoff, P.P. Pastoret, P. Perreau, A. Schwers, E. Thiry, D. Trap, G. Uilenberg, and P. Vannier. 1988. Results of a serological survey of wild mammals in France. *Revue Scientifique et Technique O.I.E.* 7:873–883.

Barrett, M., and G. Chalmers. 1975. A serologic survey of pronghorns in Alberta and Saskatchewan. *Journal of Wildlife Diseases* 11:157–163.

Bean, W.J., M. Schell, J. Katz, Y. Kawaoka, C. Naeve, O. Gorman, and R.G. Webster. 1992. Evolution of the H3 influenza virus hemagglutinin from human and nonhuman hosts. *Journal of Virology* 66:1129–1138.

Bhatt, P.N., C.D. Brandt, R.A. Weiss, J.P. Fox, and M.F. Shaffer. 1966. Viral infections of monkeys in their natural habitat in southern India. *American Journal of Tropical Medicine and Hygiene* 15:561–566.

Black, F.L. 1991. Epidemiology of Paramyxoviridae. In *The paramyxoviruses,* ed. D.W. Kingsbury. New York: Plenum, pp. 509–536.

Bryson, D.G., J.F. Evermann, H.D. Liggitt, W.J. Foreyt, and R.G. Breeze. 1988. Studies on the pathogenesis and interspecies transmission of respiratory syncytial virus isolated from sheep. *American Journal of Veterinary Research* 49:1424–1430.

Callan, R.J., G. Early, H. Kida, and V.S. Hinshaw. 1995. The appearance of H3 influenza viruses in seals. *Journal of General Virology* 76:199–203.

Carbrey, E.A., L.N. Brown, T.L. Chow, R.F. Kahrs, D.G. McKercher, L.K. Smithies, and T.W. Tamoglia. 1971. Recommended standard laboratory techniques for diagnosing infectious bovine rhinotracheitis, bovine virus diarrhea and shipping fever (parainfluenza-3). *Proceedings of the United States Animal Health Association* 75:629–648.

Churchill, A.E. 1963. The isolation of parainfluenza-3 virus from fatal cases of pneumonia in *Erythrocebus patas* monkeys. *British Journal of Experimental Pathology* 45:529–537.

Clark, R.K., D.A. Jessup, M.D. Kock, and R.A. Weaver. 1985. Survey of desert bighorn sheep in California for exposure to selected infectious disease. *Journal of the American Veterinary Medical Association* 11:1175–1179.

De Boer, G.F., W. Buck, and A.D. Osterhaus. 1990. An ELISA for detecting antibodies against influenza A nucleoprotein in humans and various animal species. *Archives of Virology* 115:47–61.

Dieterich, R.A. 1981. Respiratory viruses. In *Alaska wildlife diseases,* ed. R.A. Dieterich. Fairbanks: University of Alaska Press, pp. 28–29.

Dunbar, M.R., and W.J. Foreyt. 1986. Serologic evidence of respiratory syncytial virus infection in free-ranging mountain goats (*Oreamnos americanus*). *Journal of Wildlife Diseases* 22:415–416.

Dunbar, M.R., D.A. Jessup, J.F. Evermann, and W.J. Foreyt. 1985. Seroprevalence of respiratory syncytial virus in free-ranging bighorn sheep. *Journal of the American Veterinary Association* 187:1173–1174.

Easterday, B.C. 1975. Animal influenza. In *The influenza viruses and influenza,* ed. E.D. Kilbourne. New York: Academic, pp. 449–481.

Foreyt, W.J., and J.F. Evermann. 1988. Response of vaccinated and unvaccinated bighorn sheep (*Ovis canadensis canadensis*) to experimental respiratory syncytial virus challenge. *Journal of Wildlife Diseases* 24:356–359.

Geraci, J.R., D.J. St. Aubin, I.K. Barker, R.G. Webster, V.S. Hinshaw, W.J. Bean, H.L. Ruhnke, J.H. Prescott, G. Early, A.S. Baker, S. Madoff, and R.T. Schooley. 1982. Mass mortality of harbor seals: Pneumonia associated with influenza A virus. *Science* 215:1129–1131.

Geraci, J.R., D.J. St. Aubin, I.K. Barker, V.S. Hinshaw, R.G. Webster, and H.L. Ruhnke. 1984. Susceptibility of grey (*Halichoerus grypus*) and harp seals (*Phoca groenlandica*) to the influenza virus and mycoplasma of epizootic pneumonia of harbor seals (*Phoca vitulina*). *Canadian Journal of Fish and Aquatic Science* 41:151–156.

Hamblin, C., and R.S. Hedger. 1978. Neutralising antibodies to parainfluenza 3 virus in African wildlife, with special reference to the Cape buffalo (*Syncerus caffer*). *Journal of Wildlife Diseases* 14:378–388.

Hamblin, C., Hedger, R.S., and J.B. Condy. 1980. The isolation of parainfluenza-3 virus from free-living African buffalo (*Syncerus caffer*). *Veterinary Record* 107:18.

Hinshaw, V.S., R.G. Webster, and B. Turner. 1978. Novel influenza A viruses isolated from Canadian feral ducks: Including strains antigenically related to swine influenza (HSW1N1) viruses. *Journal of General Virology* 41:115–127.

Hinshaw, V.S., W.J. Bean, R.G. Webster, J.E. Rehg, P. Fiorelli, G. Early, J.R. Geraci, and D.J. St. Aubin. 1984. Are seals frequently infected with avian influenza viruses? *Journal of Virology* 51:863–865.

Hinshaw, V.S., W.J. Bean, J.R. Geraci, P. Fiorelli, G. Early, and R.G. Webster. 1986. Characterization of two

influenza A viruses from a pilot whale. *Journal of Virology* 58:655–656.
Howe, D.L., G.T. Woods, and G. Marquis. 1966. Infection of bighorn sheep (*Ovis canadensis*) with *Myxovirus parainfluenza*-3 and other respiratory viruses: Results of serologic tests and culture of nasal swabs and lung tissue. *Bulletin of the Wildlife Disease Association* 2:34–37.
Ingebrigtsen, D.K., J.R. Ludwig, and A.W. McClurkin. 1986. Occurrence of antibodies to the etiologic agents of infectious bovine rhinotracheitis, parainfluenza 3, leptospirosis, and brucellosis in white-tailed deer in Minnesota. *Journal of Wildlife Diseases* 22:83–86.
Ito, T., K. Okazaki, Y. Kawaoka, A. Takada, R.G. Webster, and H. Kida. 1995. Perpetuation of influenza A viruses in Alaskan waterfowl reservoirs. *Archives of Virology* 140:1163–1172.
Johnson, D.W., M. Wheelock, B. Kolstad, and P.D. Karns. 1973. Serologic survey of Minnesota moose for infectious bovine rhinotracheitis (IBR), bovine virus diarrhea (BVD), parainfluenza 3 (PI3), and bluetongue (BT) antibodies. *Proceedings of the United States Animal Health Association* 76:702–710.
Johnson, J.L., T.L. Barber, M.L. Frey, and G. Nason. 1986a. Serosurvey for selected pathogens in hunter-killed pronghorns in western Nebraska. *Journal of Wildlife Diseases* 22:87–90.
———. 1986b. Serologic survey of selected pathogens in white-tailed and mule deer in western Nebraska. *Journal of Wildlife Diseases* 22:515–519.
Kendal, A.P., and M.S. Pereira. 1982. *Concepts and procedures for laboratory-based influenza surveillance.* Washington, DC: U.S. Department of Health and Human Services.
Kimman, T.G., G.M. Zimmer, F. Westenbrink, J. Mars, and E. van Leeuwen. 1988. Epidemiological study of bovine respiratory syncytial virus infections in calves: Influence of maternal antibodies on the outcome of disease. *Veterinary Record* 123:104–109.
Kingscote, B.F., and J.G. Bohac. 1986. Antibodies to bovine bacterial and viral pathogens in pronghorns in Alberta, 1983. *Journal of Wildlife Diseases* 22:511–514.
Kingscote, B.F., W.D.G. Yates, and G.B. Tiffin. 1987. Diseases of wapiti utilizing cattle range in southwestern Alberta. *Journal of Wildlife Diseases* 23:86–91.
Lamb, R.A., and D. Kolakofsky. 1996. Paramyxoviridae: The viruses and their replication. In *Fundamental virology,* ed. B.N. Fields, D.M. Knipe, and P.M. Howley, 3d ed. Philadelphia: Lippincott-Raven, pp. 1177–1204.
Lamb, R.A., and R.M. Krug. 1996. Orthomyxoviridae: The viruses and their replication. In *Fundamental virology,* ed. B.N. Fields, D.M. Knipe, and P.M. Howley, 3d ed. Philadelphia: Lippincott-Raven, pp. 1353–1395.
Lang, G., A. Gagnon, and J.R. Geraci. 1981. Isolation of influenza A virus from seals. *Archives of Virology* 68:189–195.
Lvov, D.K., V.M. Zhdanov, A.A. Sazonov, N.A. Braude, E.A. Vladimirtceva, L.V. Agafonva, E.I. Skljanskara, N.V. Kaverin, V.I. Reznik, T.V. Podcernjaeva, S.M. Klimenko, V.P. Andrejev, and M.A. Yakhno. 1978. Comparison of influenza viruses isolated from man and from whales. *Bulletin of the World Health Organization* 56:923–930.
Murphy, B.R., and R.G. Webster. 1996. Orthomyxoviruses. In *Fundamental virology,* ed. B.N. Fields, D.M. Knipe, and P.M. Howley, 3d ed. Philadelphia: Lippincott-Raven, pp. 1397–1445.
Murray, K., P. Selleck, P. Hooper, A. Hyatt, A. Gould, L. Gleeson, H. Westbury, L. Hiley, L. Selvey, B. Rodwell, and P. Ketterer. 1995. A morbillivirus that caused fatal disease in horses and humans. *Science* 268:94–97.
O'Sullivan, J.D., A.M. Allworth, D.L. Paterson, T.M. Snow, R. Boots, L.J. Gleeson, A.R. Gould, A.D. Hyatt, and J. Bradfield. 1997. Fatal encephalitis due to novel paramyxovirus transmitted from horses. *Lancet* 349:93–95.
Parks, J.B., and J.J. England. 1974. A serological survey for selected viral infections of Rocky Mountain bighorn sheep. *Journal of Wildlife Diseases* 10:107–110.
Parks, J.B., G. Post, T. Thorne, and P. Nash. 1972. Parainfluenza 3 virus infection in Rocky Mountain bighorn sheep. *Journal of the American Veterinary Medical Association* 161:669–672.
Philbey, A.W., P.D. Kirkland, A.D. Ross, R.J. Davis, A.B. Gleeson, R.J. Love, P.W. Daniels, A.R. Gould, and A.D. Hyatt. 1998. An apparently new virus (family Paramyxoviridae) infectious for pigs, humans, and fruit bats. *Emerging Infectious Diseases* 4:269–271 (www.cdc.gov/ncidod/eid/vol4no2/philbey.htm).
Rehbinder, C., S. Belak, and M. Nordkvist. 1991. A serological, retrospective study in reindeer on five different viruses. *Rangifer* 12:191–195.
Riemann, H.P., R. Ruppauer, P. Willeberg, C.E. Franti, W.H. Elliot, R.A. Fisher, O.A. Brunetti, J.A. Aho, J.A. Howarth, and D.E. Behymer. 1979. Serological profile of exotic deer at Point Reyes National Seashore. *Journal of the American Veterinary Medical Association* 175:911–913.
Sadi, L., R. Loyel, M. St. George, and L. Lamontagu. 1991. Serologic survey of white-tailed deer on Anticosti Island, Quebec for bovine herpesviral, bovine viral diarrhea, and parainfluenza 3. *Journal of Wildlife Diseases* 27:569–577.
Sazanov, A.A., D.K. Lvov, N.A. Broude, N.V. Portyanko, S.S. Yamicova, A.A. Timofeeva, T.V. Sokolava, and A.G. Pogrebenko. 1976. Data on virological and serological examination of seabirds and fur seals in Tyuleryi Island of Sakhalinsk region. In *Ecology of viruses,* ed. D.D. Lvov. *D.I. Ivanonovsky Institute of Virology* 4:157–160.
Shah, D.V., G.B. Shaller, V. Flyger, and C.M. Herman. 1965. Antibodies to *Myxovirus parainfluenza* 3 in sera of wild deer. *Bulletin of the Wildlife Disease Association* 2:31–32.
Spraker, T.R., and J.K. Collins. 1986. Isolation and serologic evidence of a respiratory syncytial virus in bighorn sheep from Colorado. *Journal of Wildlife Diseases* 22:416–418.
Spraker, T.R., C.P. Hibler, G.G. Schoonveld, and W.S. Adney. 1984. Pathologic changes and microorganisms found in bighorn sheep during a stress-related die-off. *Journal of Wildlife Diseases* 20:319–327.
Stauber, E.H., C.H. Nellis, R.A. Magonigle, and H.W. Vaughn. 1977. Prevalence of reactors to selected livestock pathogens in Idaho mule deer. *Journal of Wildlife Management* 41:515–519.
Stauber, E.H., R. Authenrieth, D.O. Markham, and V. Withebeck. 1980. A seroepidemiologic of three pronghorn (*Antilocapra americana*) populations in southeastern Idaho 1975–1977. *Journal of Wildlife Diseases* 16:109–115.
Taylor, S.K., V.M. Lane, D.L. Hunter, K.G. Eyre, S. Kaufman, S. Frye, and M.R. Johnson. 1997. Serologic survey for infectious pathogens in free-ranging American bison. *Journal of Wildlife Diseases* 33:308–311.
Thorsen, J., and J.P. Henderson. 1971. Survey for antibody to infectious IBR, BVD and parainfluenza-3 in moose sera. *Journal of Wildlife Diseases* 7:93–95.
Thorsen, J., L. Karstad, M.W. Barrett, and G.A. Chalmers. 1977. Viruses isolated from captive and free-ranging wild ruminants in Alberta. *Journal of Wildlife Diseases* 13:74–79.

Truyen, U., C.R. Parrish, T.C. Harder, and O.R. Kaaden. 1995. There is nothing permanent except change: The emergence of new virus diseases. *Veterinary Microbiology* 43:103–122.

Webster, R.G., J.R. Geraci, and G. Petersson. 1981a. Conjunctivitis in human beings caused by influenza virus of seals. *New England Journal of Medicine* 304:911.

Webster, R.G., V.S. Hinshaw, W.J. Bean, K.L. van Wyke, J.R. Geraci, D.J. St. Aubin, and G. Petersson. 1981b. Characterization of an influenza A virus from seals. *Virology* 113:712–724.

Webster, R.G., W.J. Bean, O.T. Gorman, T.M. Chambers, and Y. Kawaoka. 1992. Evolution and ecology of influenza A viruses. *Microbiology Reviews* 56:152–179.

Wentworth, D.E., B.L. Thompson, X.Y. Xu, H.L. Regnery, A.J. Cooley, M.W. Macgregor, N.J. Cox, and V.S. Hinshaw. 1994. An influenza A (H1N1) virus closely related to swine influenza virus responsible for a fatal case of human influenza. *Journal of Virology* 68:2051–2058.

Williams, E.S., E.T. Thorne, S. Anderson, and J.D. Herriges Jr. 1993. Brucellosis in free-ranging bison *(Bison bison)* from Teton County, Wyoming. *Journal of Wildlife Diseases* 29:118–122.

Young, P.L., K. Halpin, P.W. Selleck, H. Field, J.L. Gravel, M.A. Kelly, and J.S. MacKenzie. 1996. Serologic evidence for the presence in *Pteropus* bats of a paramyxovirus related to equine morbillivirus. *Emerging Infectious Diseases* 2:239–240. (www.cdc.gov/ncidod/eid/vol2no3/young.htm)

Yu, M., E. Hansson, B. Shiell, W. Michalski, B.T. Eaton, and L.F. Wang. 1998a. Sequence analysis of the Hendra virus nucleoprotein gene: Comparison with other members of the subfamily *Paramyxovirinae. Journal of General Virology* 79:1775–1780.

Yu, M., E. Hansson, J.P. Langedijk, E.T. Eaton, and L.F. Wang. 1998b. The attachment protein of Hendra virus has high structural similarity but limited primary sequence homology compared with viruses in the genus *Paramyxovirus. Virology* 251:227–233.

Zarnke, R.L. 1983. Serologic survey for selected microbial pathogens in Alaskan wildlife. *Journal of Wildlife Diseases* 19:324–329.

Zarnke, R.L., and G.A. Erickson. 1990. Serum antibody prevalence of parainfluenza 3 virus in a free-ranging bison (*Bison bison*) herd from Alaska. *Journal of Wildlife Diseases* 26:416–419.

16

CALICIVIRUS INFECTIONS

COR LENGHAUS, MICHAEL J. STUDDERT, AND DOLORES GAVIER-WIDÉN

INTRODUCTION. The family *Caliciviridae* contains four genera: *Vesivirus,* including vesicular exanthema of swine virus (VESV), San Miguel sea lion virus (SMSV), and feline calicivirus (FCV); *Lagovirus,* which includes rabbit hemorrhagic disease virus (RHDV) and the related European brown hare syndrome virus (EBHSV); and two unnamed genera that contain Norwalk-like and Sapporo-like viruses, members of which cause human gastroenteritis (Green et al. 2000). Other viruses with typical calicivirus morphology have been identified, but usually not cultivated, often from enteric infections, in cattle, swine, dogs, rabbits, chickens, reptiles, amphibians, and insects.

The extent of infection and disease in wildlife caused by caliciviruses is poorly defined. SMSV, which infects a wide range of marine species, almost certainly emerged from the sea in the form of VESV (Smith et al. 1998c). FCV causes disease in cheetahs, lions, and tigers, and probably infects all species of cats. RHD and EBHS have emerged recently as major pandemic diseases of wild rabbits and hares, respectively (Morisse 1991).

PROPERTIES OF CALICIVIRUSES. Caliciviruses are so called because of the cup-shaped surface depressions (*calix* is Latin for cup) that give the virion its unique appearance, though some caliciviruses lack the characteristic morphology. Virions are typically about 40 nm in diameter. The 32 cup-shaped surface structures comprising 90 archlike capsomers are arranged in T = 3 icosahedral symmetry. Individual capsomers are dimers of a large 56- to 76-kDa capsid protein. A minor polypeptide (15 kDa) comprises less than 2% of total capsid protein, and a third protein (10–15 kDa) is covalently linked to the RNA of the virion (Clarke and Lambden 1997). Caliciviruses are relatively resistant to heat and intermediate in pH stability (greater than 99% inactivated at pH 3).

Calicivirus genomes are approximately 7.5-kb, positive-sense, polyadenylated, ss-RNA. The sequences of the genomes of FCV, RHDV, and EBHSV have been determined (Clarke and Lambden 1997). The isolated virion RNA is infectious (Ohlinger et al. 1993).

Caliciviruses replicate in the cytoplasm, where they are found as scattered single particles, as characteristic linear arrays along microfibrils or within membranous structures, or as paracrystalline arrays (Studdert 1978). VESV, SMSV, and FCV grow readily, with obvious cytopathic effect, in cell cultures derived from pig and feline tissues, respectively. VESV and SMSV also grow in Vero cells. RHDV and EBHSV so far have proved noncultivable in cell culture, although Konig et al. (1998) detected RHDV-specific polypeptides in isolated rabbit hepatocyte cultures infected with purified RHDV.

The host range of FCV and human caliciviruses appears to be restricted to a primary host and perhaps some related species. FCV has shown no evidence of host range expansion, despite its high prevalence, worldwide distribution, and the daily exposure of many nonfeline species, including humans, to the virus. On the other hand, the emergence of VESV in 1932, almost certainly from SMSV in pinnipeds (Smith and Boyt 1990), and the recent emergence of RHDV, possibly by transfer from a related lagomorph (Lenghaus et al. 1994) or by evolution from an apathogenic calicivirus strain in rabbits (Capucci et al. 1996), support the contention that caliciviruses may have a greater capacity to host switch than do many other viruses.

The potential host range of caliciviruses became topical when RHDV was being considered as a biological control agent for wild rabbits in Australia and New Zealand [see Capucci and Lavazza (1998), Mead (1998), and Smith et al. 1998b,c)].

Single-stranded RNA viruses, such as caliciviruses, have high mutation rates, since replication relies on a virion RNA-dependent RNA polymerase, for which there is no proofreading mechanism. Such high rates of mutation suggest high genetic instability, stimulating speculation about the capacity of caliciviruses to expand their host range.

Both VESV and FCV have a high degree of antigenic heterogeneity (Smith and Boyt 1990; Kreutz et al. 1998). The mutability of regions of the genome coding for epitopes involved in serum neutralization has enhanced the view that caliciviruses are genetically highly unstable and therefore also likely to undergo shifts in host range. Whether the regions of the calicivirus genome that are involved in antigenicity are involved in host specificity is not known. However, the apparent "plasticity" of VESV host range and the apparent "rigidity" of the FCV host range suggest the linking of antigenic heterogeneity with diversity of host ranges as too simplistic. The rate of change of RHDV in this new virus-host relationship is relatively small (Nowotny et al. 1997).

This debate currently cannot reach definitive conclusions, since there is no precise understanding, in molecular terms, of the basis for calicivirus host range. The variability of viruses as they coevolve with their hosts should not obscure the reality of the overall stability of virus-host relationships. There currently is no hard evidence proving either of the opposing allegations: high risk of host switching by caliciviruses generally, or no significant risk of that event.

RABBIT HEMORRHAGIC DISEASE

Synonyms: Rabbit calicivirus disease, hemorrhagic pneumonia of rabbits, hemorrhagic tracheitis of rabbits, infectious hepatitis of rabbits, necrotic hepatitis of rabbits, rabbit plague (*tuen* in Chinese), rabbit viral sudden death, rabbit X disease.

Introduction and History. Rabbit hemorrhagic disease virus was first recognized in the European rabbit *Oryctolagus cuniculus* in the mid-1980s. Infection by RHDV in susceptible adult rabbits causes an acute, fulminating, and often fatal disease (Chasey 1997). The graphic synonyms used to describe RHD reflect its potential virulence.

Many of the synonyms for RHD were applied when the disease first emerged in farmed domestic rabbits in China in 1984, before the etiology was identified (Liu et al. 1984; Chen 1991). It is difficult to trace the early spread of RHD from China to Europe. In 1986, it suddenly emerged in Italy, where millions of farmed rabbits died in a few months. It impacted in France, Hungary, the Czech and Slovak Republics, and the former U.S.S.R. in 1987; in Spain, Portugal, Switzerland, and Egypt in 1988; and subsequently in progressively more areas of the Old World. It was first recognized in the United Kingdom in 1992, after rabbit owners returned from a rabbit show in continental Europe (Xu and Chen 1989; Chasey 1994).

In its first incursion into the Americas, RHD occurred in domestic rabbits in Mexico in 1988 after the importation of contaminated frozen rabbit carcasses. It was subsequently eradicated by test and slaughter (Heneidi Zeckua et al. 1997).

Inevitably, RHD spread into wild rabbits in Europe, with an equally dramatic impact (Marchandeau et al. 1998). Organized groups of recreational shooters were able to exert pressure to have research on RHD funded further, to preserve wild rabbit populations for the use of hunters.

Recognition that European brown hare syndrome, of which hunters had been aware since the 1970s, was also caused by a calicivirus led to speculation that the two diseases might have an identical cause. There is 53%–71% homology between the genomes of EBHSV and RHDV isolates, with greater than 90% homology within each group (Le Gall et al. 1996; Nowotny et al. 1997). These viruses probably share a common ancestor rather than one being a recent derivative of the other.

The origins of these viruses are difficult to trace. They may have arisen by mutation from an avirulent calicivirus of Eurasian lagomorphs (Rodák et al. 1990; Capucci et al. 1996), or by introduction of a new virus, avirulent in its native host, possibly South American rabbits *Sylvilagus* sp. or hares *Lepus* sp., imported into Europe in large numbers for recreational hunting in the 1970s and 1980s (Cancellotti and Renzi 1991).

Authorities in Australia and New Zealand became interested in RHDV as a potential tool for the biological control of wild European rabbits, introduced into both countries during the mid-1800s, and an intractable pest and threat to the environment since.

Work began in 1991 to determine host range by inoculating RHDV strain CAPM V-351 from the Czech Republic into laboratory and wild rabbits and a range of nontarget mammals and birds; to determine susceptibility of cell culture systems away from the rabbit host; to develop diagnostic tests for assay of the virus and its antibody; to assess whether RHD-like viruses already existed by serologic surveys of wild lagomorphs in Australia and New Zealand; and to investigate effects on rabbits of coinfection with RHDV and myxomatosis virus, already in use as a biological control agent (see Chapter 8).

Following review of available information (Munro and Williams 1994), field investigation of RHD under contained conditions was begun on Wardang Island, Spencer Gulf, South Australia. Single-warren colonies of wild rabbits isolated behind rabbit-proof fences were infected and observed. There was little spread of virus from directly inoculated rabbits during initial studies in early 1995. Lack of intimate contact between rabbits and the low level of activity of flying insects may have been important in the limited spread of disease at this time (C. Lenghaus unpublished).

During subsequent experiments, however, it became increasingly difficult to contain RHDV to specific sites. By September 1995, it had escaped from the study compound and, in early October 1995, RHD was diagnosed on the adjacent mainland. Early spread of the virus was roughly northward, suggesting insect spread via the prevailing wind. During 1996, the virus progressively appeared in disparate regions of Australia, even against prevailing winds (Kovaliski 1998). People, particularly pastoralists and farmers, actively disseminated the virus at this time, probably by moving rabbits recently dead of RHD.

Impact on rabbit numbers was immediate and dramatic, with declines in spotlight rabbit counts of 65%–95% commonly reported, particularly in drier areas (Mutze et al. 1998). Impact in temperate climate areas was generally less dramatic, due in part to the higher prevalence of young nonsusceptible rabbits.

In 1996, dissemination of RHD was legalized in Australia, and efforts began to "fill in" areas not yet affected, by deliberately releasing rabbits infected with RHDV. RHD reached Tasmania in January 1997, before release was sanctioned. In New Zealand, surreptitious introduction of RHD into the South Island was detected in August 1997, and release of RHDV subsequently was legalized.

Etiology. RHD is caused by a calicivirus, based on the morphology and physicochemical properties of the virus, its molecular genetics, and the infectivity of purified viral RNA after intrahepatic inoculation of susceptible rabbits. Preliminary reports that it was a parvovirus were not substantiated (Xu et al. 1988; Gregg and House 1989). Early studies on RHDV and the related European brown hare syndrome virus are summarized in Morisse (1991) and in Ohlinger et al. (1993).

RHDV has a major structural protein of 60 kDa and an RNA genome of 7437 nucleotides, within which there are two open reading frames coding for polypeptides of 2344 and 117 amino acids in length (Meyers et al. 1991; Gould et al. 1997). It differs from a recently described, related but nonpathogenic, calicivirus of European rabbits, which has been termed rabbit calicivirus (RCV) by Capucci et al. (1996).

Transmission, Epidemiology, and Host Range. In the Australian context, involving potential dissemination of the virus for biological control of wild rabbits, safety as well as efficacy was important. Species specificity with respect to domestic animals and wildlife was of prime concern, due to the risk that virus could infect and cause disease in nontarget animals, or be attenuated in the course of such infections, rendering it less virulent for rabbits.

Prior to 1993, there was no evidence that RHDV produced virus-productive infections in any species other than rabbits; laboratory rodents and domestic poultry had been deliberately inoculated (see Chen 1991), and there had also been ample opportunity for natural contact exposure of humans, birds, and grazing animals by infected rabbits in Eurasia. Lavazza et al. (1996) confirmed earlier observations that inoculation of European hares and rabbits with doses of RHDV and EBHSV that are fatal in the homologous host produces no disease and only a low antibody response in the heterologous host.

Using protocols to evaluate host specificity that involved intramuscular inoculation of young adult animals with 1000 rabbit LD_{50} of RHDV, followed by systematic serologic, virologic, and pathologic observations (Lenghaus et al. 1994; Collins et al. 1995; Gould et al. 1997), no evidence of a productive viral infection was detected in any of 31 domestic, feral, or Australian native animal and bird species (Table 16.1). Low-level antibody production in white mice was considered a response to antigenic load. In New Zealand, kiwis *Apteryx australis* seroconverted following RHDV inoculation but without productive viral infection, and short-tailed bats *Mystacina tuberculata* did not develop an antibody response or produce virus (Buddle et al. 1997). In Europe, red foxes *Vulpes vulpes* produced specific antibody when fed livers of rabbits infected with RHDV (Leighton et al. 1995), and free-ranging foxes may develop antibodies to RHDV (Frölich et al. 1998). This probably reflects a response to ingested antigen that has reached the peripheral circulation, rather than productive viral infection.

TABLE 16.1—Nontarget species tested and found refractory to infection with rabbit hemorrhagic disease at the Australian Animal Health Laboratory

Domestic animals	Horse, cow, sheep, fallow deer *Dama dama,* goat, pig, dog, cat, and chicken.
Feral animals	Red fox, European brown hare, ferret *Mustela putorius,* rat *Rattus norvegicus,* and house mouse *Mus musculus.*
Australian native mammals	Bush rat *Rattus fuscipes,* spinifex hopping mouse *Notomys alexis,* plains rat *Pseudomys australis,* fat-tailed dunnart *Sminthopsis crassicaudata,* northern brown bandicoot *Isoodon macrourus,* brush-tailed bettong *Bettongia penicillata,* tammar wallaby *Macropus eugenii,* common brushtail possum *Trichosurus vulpecula,* common wombat *Vombatus ursinus,* short-beaked echidna *Tachyglossus aculeatus,* and koala *Phascolarctos cinereus.*
Birds	Long-billed corella *Cacatua tenuirostris,* feral pigeon *Columba livia,* silver gull *Larus novaehollandiae,* brown falcon *Falco berigora,* and emu *Dromaius novaehollandiae.*
Reptiles	Common blue-tongued lizard *Tiliqua scincoides.*

Inability to culture RHD in vitro has made transmission and epidemiologic studies difficult. Direct rabbit-to-rabbit transmission of RHD occurs, without the need for arthropod vectors, although fleas, flies, and mosquitoes may act as mechanical carriers of RHDV. However, a fully susceptible rabbit could remain in immediate contact for up to 6 hours with one that had recently died of RHD, and not become infected (C. Lenghaus unpublished).

Rabbits over about 8 weeks of age were as susceptible as adults, but only about 50% of rabbits 5–6 weeks old were. Neonatal rabbits were resistant to disease but did support some viral replication (Lenghaus et al. 1994). Such age-related resistance is compatible with the epidemiology of the natural disease.

Infectious virus did not persist for more than a month in the environment of the experimental facility, or in rabbit carcasses, but virus suspensions retained infectivity for at least 9 months at 4° C (C. Lenghaus and B.J. Collins unpublished). In contrast to reports that suggested that the oral-fecal route was important in transmission (see Morisse 1991), Collins et al. (1996) could not demonstrate RHDV in feces of infected rabbits and found that rabbit urine caused a rapid degradation of virus.

Spread of virus on fomites, by people contaminated from infected rabbits, or in rabbit products and tissues, is important in the dissemination of RHD. Fleas and mosquitoes transmit RHDV mechanically (Lenghaus et al. 1994), and the oral and anal excretions of flies ("fly-spots") also are infectious (Asgari et al. 1998). Scavenger birds feeding on carcasses of infected dead rabbits also may have spread the disease over long distances, including over expanses of water, such as to the United Kingdom or Ireland (Graham et al. 1996).

Clinical Signs. The disease often is peracute, with adult rabbits found dead or dying usually within a day of developing signs of inappetence, depression, and fever. Rabbits become comatose a few hours before death and usually die quietly. Subacute clinical infections also occur, with survival for periods of a week or more and, in some cases, recovery. Many rabbits younger than about 6 weeks of age will survive infection without apparent disease, although a mortality rate approaching 100% may occur in rabbits older than 8 weeks.

Pathogenesis. RHDV has a predilection for hepatocytes and for splenic histiocytes, in which it induces apoptotic cell death (Alonso et al. 1998). Procoagulant substances released by damaged cells presumably initiate the disseminated intravascular coagulation (DIC) that, with fulminant liver failure, causes rapid death in the rabbits that succumb. The reduced virulence of the virus for rabbits progressively younger than 8 weeks may reflect reduced capacity of younger rabbits to support productive viral infection.

Pathology. Dead rabbits usually are in good body condition, although often moderately dehydrated. Those that have been dead for some time may have blood-stained froth at the nares. The mucosa of the turbinates and trachea usually is deeply congested. The lungs are wet, congested, and may contain hemorrhagic foci. The liver is pale and swollen, with an accentuated lobular pattern. The spleen is markedly swollen and engorged with blood, and the kidneys also may be dark and swollen. Fat depots may be slightly icteric.

Microscopically, fibrin thrombi are common in the lungs, spleen, liver, kidneys, heart, and occasionally other organs of freshly dead rabbits. There is often severe pulmonary congestion and edema. There is characteristically coagulation necrosis of periportal hepatocytes, which may extend to midzonal areas or involve most of the classic liver lobule. The sinusoidal architecture of the spleen usually is obliterated by a coagulum of fibrin, cell debris, and erythrocytes. Splenic lymphoid follicles often are spared, although they can be severely depleted of lymphocytes. Occasionally, sequelae of ischemia may occur, such as small malacic foci in the brain or acute myocardial necrosis. There are no residual lesions in rabbits that recover from RHD. Chronic respiratory, splenic, or hepatic disease are not recognized.

Diagnosis. Sudden death in adult European rabbits should arouse suspicion of RHD. Specific immunohistochemical stains demonstrate RHDV antigen in hepatocytes and splenic histiocytes. Typical calicivirus particles can be demonstrated by electron microscopy in the supernatant of liver from rabbits that have died of RHD, and virions aggregate in homologous antiserum. The diagnosis can be confirmed by rapid antigen-detection enzyme-linked immunosorbent assays (ELISAs) using liver from disease suspects or by polymerase chain reaction (Ros Bascuñana et al. 1997). The virus agglutinates human erythrocytes (groups O and B at least), and this has been the basis of virologic and serologic assays in the past (Chasey 1997). Antibody-detection ELISAs are available for serologic investigation.

Differential diagnosis includes acute septicemia, or rapidly acting poisons used for rabbit population control.

Immunity. Active immunity in recovered rabbits is apparently lifelong. Rabbit kittens acquire passive immunity from their dams, and it persists for up to 12 weeks after birth (C. Lenghaus unpublished). However, the insusceptibility of young rabbits to RHDV is not antibody dependent.

Commercial inactivated virus vaccines, based on liver homogenates from infected rabbits, can provide protection for up to 12 months after an initial dose. Subunit vaccines, based on viral capsid protein VP 60, are protective experimentally (Nagesha et al. 1995).

Control and Treatment. Control of the disease in wild rabbits has not been seriously addressed.

Infection does not spread rapidly through a rabbitry if rabbits are physically separated by at least half a meter, although contaminated insects will disseminate virus in these circumstances. To provide short-term protection, hyperimmune serum might be given to valuable rabbits in the face of an outbreak.

Prevention of RHD in domestic rabbits should be based on a rational vaccination program. Otherwise, in countries where the disease is endemic, there is a danger of introducing the infection onto clean premises, either through people who have been in contact with infected rabbits or on fomites. Commercial or research colonies and, desirably, pet rabbits, should be maintained on a sterilized, pelleted ration. Flying insects should be excluded by screening doors and windows.

Treatment of clinically affected rabbits is likely to be ineffective.

Public Health and Domestic Animal Concerns. Discussion in Australia after the escape of RHD from Wardang Island in 1995 centered largely on the possibility that the virus might cross species barriers. The broad range of nontarget animals that were demonstrably refractory to infection with RHDV (Table 16.1) assuaged, but did not eliminate, such concerns, though there is no convincing evidence of host switching by RHDV in the 14 years since RHD was first described. The amount of deliberate nontarget species testing done with RHDV exceeds that for most other animal viruses in general circulation. Considerable scientific credibility was lost with some sections of the community, in the short term at least, because of the escape of RHDV. This may in future inhibit application of other potentially useful biological control agents.

To date, there is no firm indication that RHDV infects people (Capucci and Lavazza 1998; Mead 1998), although there is debate about the interpretation of some data (Smith et al. 1998b). Carman et al. (1998) found no clinical or serologic evidence of RHDV infection in 259 people occupationally exposed to the virus.

In endemic areas, there is a need to protect pet rabbits, laboratory rabbits, and commercial rabbitries from the disease.

Management Implications. RHD has proven moderately effective in rabbit control in Australia, at least in the short term, though not quite so spectacularly as myxomatosis was, because of its inability to eliminate young rabbits. This is a serious drawback—rabbits are sexually mature in 3–4 months and, under ideal conditions, rabbit does can have a litter of five or more young at monthly intervals. To "breed like rabbits" is a truism. The impact of RHD has been greatest in arid regions and less in areas with higher rainfall, reflecting the capacity of young survivors, in the more benign environment, to breed replacements rapidly for numbers lost. The unexpected release and often uncontrolled dissemination of RHD also meant that integrated rabbit control programs were not in place, and the full potential of this agent to assist in reducing rabbit populations probably has not been realized yet.

Local rabbit eradication has traditionally involved use of poisoned baits, fumigation of warrens, and destruction of rabbit harbor (warrens, rock piles, etc.). Rabbits also have been excluded from specific areas by the construction of expensive, rabbit-proof, mesh fences.

Infectious agents of biological control may act on a broader geographic and time scale. Myxomatosis depends on temperate conditions to support the arthropods that spread it, and now has a variable impact because of mutual selection pressure that has yielded attenuated virus and resistant rabbits (Williams et al. 1995). The same selection pressures operate on RHD. A shifting equilibrium will evolve between virulence of RHDV and host resistance in regional areas; RHD alone will not eradicate rabbits in Australasia (Barlow and Kean 1998).

The impact of RHD would be maximized if the disease occurred when young rabbits were not present, as at the end of a prolonged dry spell that inhibited rabbit breeding. Concomitant use of myxomatosis and RHD may be contraindicated, since some strains of myxoma virus inhibit the full expression of RHD (C. Lenghaus unpublished). However, this needs to be weighed against the relative susceptibility of young rabbits to myxomatosis (Fenner and Ratcliffe 1965). The situation is multifactorial, involving the natural resistance of young rabbits to RHD, the persistence of protective maternal antibodies, and the timing of rabbit breeding affecting the emergence of susceptible rabbits in sufficient numbers, as well as the seasonality of RHD and myxomatosis.

Once rabbit numbers are reduced by disease, it is imperative to destroy as much rabbit habitat as possible, by ripping warrens, etc., so that rabbits surviving the epidemic are further exposed to the debilitating effects of the elements and of predators. However, this is not feasible in vast tracts of the more remote areas of Australia because of cost and lack of personnel.

By contrast, in Europe, wild rabbit conservation is of prime concern as part of overall game management. Hunters should be encouraged to limit their activities in known infected areas until after an epidemic has passed, to try and minimize spread to other areas. Hunters should be encouraged to skin and eviscerate their catch on site, rather than en route and remote from the area where the rabbit was caught. Translocation of rabbits from known infected areas should be prevented, if possible. Rabbit numbers had not recovered to pre-RHD levels in Spain after 6 years (Villafuerte et al. 1995).

EUROPEAN BROWN HARE SYNDROME

Synonyms: Leporid caliciviral hepatitis, viral hepatitis of hares, fältharesjuka, acute hepatosis.

Introduction and History. European brown hare syndrome (EBHS) is a caliciviral infection of free-living and farmed hares characterized by acute necrotic hepatitis.

Epidemic disease with acute liver necrosis was observed in wild European brown hares *Lepus europaeus* in Sweden in the early 1980s. The disease was called *fältharesjuka,* later translated to EBHS. A similar syndrome had been observed in Denmark and later was reported from other European countries. The cause was unknown. Toxins, selenium deficiency, and oilseed rape were thought possibly to be involved, until RHD appeared in Europe and its similarity to EBHS was recognized. Lavazza and Vecchi (1989) first observed viral particles resembling RHDV in the livers of hares that had succumbed of EBHS.

EBHS occurred in Europe many years before the appearance of RHD. Outbreaks with postmortem findings consistent with EBHS have been described from England since 1976 (Duff et al. 1994) and were known by hunters in Scandinavia in the early 1970s. Antibodies to EBHS were found in the sera of 45 of 133 hares from England archived since 1962, but it is not known whether they had the disease (Duff et al. 1997). The earliest confirmed case was from 1980 in Sweden (Gavier-Widén and Mörner 1991).

Distribution and Host Range. By 1999, EBHS had been reported from most Western and Eastern European countries, but it was not known outside Europe. The distribution of the disease appears to correspond with the distribution of the European brown hare; it is not present where only other species of hares occur, such as in Norway.

Wild and farmed European brown hare and varying, mountain, or northern hare *Lepus timidus* are susceptible. No other species are known to be affected, either naturally or experimentally. There is no evidence of natural infection of European rabbits with EBHSV, even in highly endemic areas where hares and rabbits coexist. As noted in the discussion of RHD, experimental inoculation of rabbits and hares with the reciprocal virus results in no virus replication and stimulates only a slight, and nonprotective, antibody response (Lavazza et al. 1996).

Etiology. The EBHSV is a calicivirus related to RHDV. In its morphology, hemagglutinating properties, and inability to grow in cell culture, it resembles RHDV, and the viruses are antigenically related. They are also closely related genetically but distant from other caliciviruses (Wirblich et al. 1994; Le Gall et al. 1996; Nowotny et al. 1997).

Transmission and Epidemiology. EBHSV is highly contagious. The virus is transmitted directly or indirectly, mainly by orofecal and respiratory routes. Vectors or reservoir hosts have not been identified. The persistence of the virus in the environment has not been specifically studied but presumably is similar to RHDV.

Only adult hares appear to succumb, and EBHS has not been observed in hares younger than approximately 50 days. Hares of 2–3 months of age may contract infection but usually do not develop clinical disease.

Even though a decline in hare numbers can be attributed to outbreaks following introduction of EBHS, areas where EBHS is endemic appear to maintain a stable hare population in which most hares are immune and mortality rates are low. In highly endemic areas in Italy, 95% of the free-ranging hares have antibodies to EBHSV (Scicluna et al. 1994).

Mortality is highest in the fall, when the population is most dense and the young of the year become susceptible. At that time, hunters commonly report finding dead hares in groups of approximately 10–15 animals, and such mortality is observed in the same area in consecutive years.

In captivity, mortality rates are variable and range from 10% to 100% (Henriksen et al. 1989; Zanni et al. 1993). In successive years, up to 10% of the hares on known infected farms may die of EBHS. In Denmark, new hares that were introduced to repopulate farms 3–4 weeks after the beginning of the mortality all died of EBHS 1–2 weeks later (Henriksen et al. 1989).

Clinical Signs. In the peracute form, hares die without showing signs. In acute and subacute forms, affected hares are depressed, anorexic and, in some cases, develop nervous signs and abnormal behavior, probably due to hepatic encephalopathy (Henriksen et al. 1989; Zanni et al. 1993). Affected free-ranging hares are typically described as "crazy hares" and lose their fear of people and dogs, run in circles, and adopt abnormal postures. Death occurs after 2–7 days.

Pathogenesis. The pathogenesis has not been studied fully experimentally in hares, but it is probably similar to RHD. The liver is the target organ: the virus specifically infects and causes necrosis of hepatocytes. Liver insufficiency due to widespread hepatocellular necrosis, in association with hemorrhage and hepatic encephalopathy, leads to death. Terminal DIC is observed less frequently in hares with EBHS than in rabbits with RHD. Hares that follow a longer disease course often develop jaundice and chronic hepatitis. Hares that recover may have liver fibrosis and remain in poor body condition.

Pathology. In experimentally infected hares, leukopenia, sometimes severe, occurred 24 hours after infection. This was associated with transient lymphopenia. Aspartate aminotransferase, alanine aminotransferase, and bilirubin increased abruptly at day 4 after inoculation, shortly before death. Hematologic and biochemical abnormalities were less marked in hares that survived and were nearly normal by 10 days after infection (Gavier-Widén 1993).

Hares that die acutely may have a pale, fatty, yellow liver, with accentuated lobular pattern and hemorrhages, but gross lesions in the liver can be difficult to identify. Signs of circulatory collapse often are observed. These include an enlarged, congested spleen, dark red tracheal and nasal mucosa, pulmonary edema and hemorrhage, petechial serosal hemorrhages, and bloodstained intestinal contents. These findings are common but not specific to EBHS; similar signs are observed in terminal stages of other diseases, such as acute infections. A full stomach, mucoid colitis, and jaundice sometimes are observed (Marcato et al. 1989; Poli et al. 1991; Gavier-Widén 1992).

Histologically, the characteristic liver lesion is single-cell necrosis of hepatocytes in periportal areas. Necrosis varies in extent and also may affect midzonal areas or the whole lobule. Inflammatory cell infiltrates may be absent or scant in acute cases, but there may be portal infiltration by macrophages, lymphocytes, and a few heterophils, and granulocytes may infiltrate the parenchyma. Fatty change of hepatocytes is common (Gavier-Widén 1992). Mitochondrial mineralization is frequently observed in hepatocytes undergoing necrosis (Gavier-Widén 1994).

Livers of hares with a clinical course longer than approximately 3–4 days have less extensive necrosis, more inflammatory changes, more pronounced fatty change, and bile duct proliferation. At this stage, the virus is mainly located in macrophages (Gavier-Widén 1994). EBHSV antigen may be found in livers of hares with chronic hepatitis, indicating that chronic disease with persistence of virus also occurs (Gavier-Widén 1992). In the kidneys, tubular cell necrosis with mineralization is common. Changes in other organs are related to terminal circulatory failure and accompanying hemorrhage.

Diagnosis. A tentative clinical diagnosis of EBHS can be made in hares with abnormal behavior. Gross changes specific to EBHS may be difficult to identify or may be absent at autopsy. The diagnosis therefore is based on the demonstration of the virus and recognition of the typical microscopic liver lesions. The antigen can be demonstrated in formalin-fixed liver by immunohistochemistry. Virus may be identified in liver homogenates or other tissues by hemagglutination, several types of ELISA, negative-staining electron microscopy, immune electron microscopy, and PCR. Serologic tests include hemagglutination inhibition and ELISA (Gavier-Widén and Mörner 1991; Capucci et al. 1991; Ros Bascuñana et al. 1997).

Differential Diagnosis. Toxoplasmosis often is confused with EBHS clinically and at postmortem examination. Acute bacterial infections such as pseudotuberculosis, tularemia, pasteurellosis, and listeriosis may clinically resemble EBHS. Chronic forms of EBHS are observed incidentally in hares that die of other causes, such as trauma.

Immunity. Newborn hares may have antibodies against EBHSV, indicating that there is colostral or transplacental passive transfer of immunity (Gavier-Widén and Mörner 1991). Hares that survive infection develop long-lasting, protective immunity.

Treatment and Control. There is no specific treatment for EBHS. Antiserum given during the incubation period may prevent death.

There are no commercial vaccines for hares, but when EBHS mortality begins in a farm, autogenous killed virus vaccines can be produced from livers of the hares that die first, to immunize the remaining apparently unaffected hares (D. Gavier-Widén unpublished).

In the wild, following introduction of EBHSV, spread of the disease to the naive population in contiguous areas is inevitable. Eradication of EBHS in a wild population of hares is not feasible. Translocation of hares from infected to free areas should be avoided.

On hare farms, prevention is based on quarantine, serologic testing of new hares to be introduced, and elimination of contact with wild hares by fencing. Opportunity for indirect infection through grass or hay contaminated with excreta from wild hares, or through contaminated clothes or shoes, should be eliminated.

Human and Domestic Animal Health Concerns. The narrow host range of EBHSV mitigates against threats to the health of people and domestic animals other than farmed hares.

Management Implications. The large outbreaks of mortality and associated "disappearance" of local populations of wild hares in Europe during the 1980s was of concern to hunters. As the disease became endemic and populations recovered, mortality was sporadic and impact on populations low in most of Europe. In Sweden, by 1993, there was no association between occurrence of EBHS and annual hunter harvest of hares (Gavier-Widén and Mörner 1993).

Translocation of hares from Europe to areas free of EBHS should be avoided, since infection is prevalent and often unrecognized. If translocation is imperative, seronegative imports should be quarantined with in-contact susceptible sentinel hares, to detect incubating or carrier animals.

SAN MIGUEL SEA LION VIRUS AND VESICULAR EXANTHEMA OF SWINE VIRUS

Synonyms: Pinniped caliciviruses, marine caliciviruses.

History. In 1932, a vesicular disease, diagnosed as foot-and-mouth disease, occurred in pigs in southern California; it was eradicated by quarantine and slaughter. When it recurred in 1933 and 1934, it was realized that the disease was new, and it was called vesicular exanthema of swine (VES). Between 1932 and 1951, VES occurred intermittently in southern California, affecting millions of swine. There was a clear association with the feeding of garbage, but outbreaks continued, despite laws requiring that all garbage fed to pigs be cooked.

In 1952, VES was diagnosed outside California for the first time and, by September 1953, it had occurred in over 40 states. Federal enforcement of garbage cooking and a slaughter program resulted in a rapid decline in the occurrence of disease, although small isolated outbreaks of VES associated with garbage feeding occurred in Hawaii and Iceland. The last outbreak of VES occurred in 1956, and the disease was declared eradicated from the United States in 1959. It has not recurred (Smith and Boyt 1990).

In 1972, a calicivirus designated San Miguel sea lion virus (SMSV) was isolated during an outbreak of abortion among California sea lions *Zalophus californianus* on San Miguel Island, off the southern Californian coast. It closely resembled VESV and was transmissible to pigs. Subsequently, a similar virus, which was virulent in swine, was isolated from opal eye perch *Girella nigricans* from the west coast of the United States (Smith and Boyt 1990).

Etiology. SMSV and VESV have numerous antigenic types. More than one type may be isolated during a single outbreak of VES and even from a single pig. Although named as separate viruses for regulatory reasons (VES has been declared officially eradicated in the United States), SMSV and VESV are best considered part of a single species or genotype of calicivirus with many serotypes (at least 14 VESV and 17 SMSV or similar) (Neill et al. 1995; Smith et al. 1998c).

Distribution and Host Range. Caliciviruses in the SMSV group are endemic in pinnipeds, and apparently in some cetacean species, along the western seaboard of North America. They have been isolated from California sea lion (California); northern fur seal *Callorhinus ursinus* (California and Pribiloff Islands, Alaska); northern elephant seal *Mirounga angustirostris* (California); walrus *Odobenus rosmarus* (Chukchi Sea and Bering Sea); Atlantic bottlenosed dolphin *Tursiops truncatus* (captive animals, San Diego, California); and Stellar sea lion *Eumetopias jubatus* (Oregon). Viruses in this group also have been isolated from opal eye perch and sea lion liver fluke (*Zalophatrema* sp.).

Antibody to viruses in this group has been demonstrated in many marine mammals: California sea lions, northern fur seals, Stellar sea lions, walrus, northern Elephant seals, Hawaiian monk seals *Monachus schauinslandi,* Pacific bottlenosed dolphins *Tursiops gillii,* bowhead whales *Balaena mysticetus,* gray whales *Eschrictius robustus,* fin whales *Balaenoptera physalus,* sei whales *Balaenoptera borealis,* and sperm whales *Physeter catodon.* Among free-ranging terrestrial mammals, antibody has been detected in feral swine *Sus scrofa,* feral donkeys *Equus asinus,* and gray foxes *Urocyon cinereoargenteus,* as well as several species of captive and domestic ruminants, and mink *Mustela vison* (Smith and Boyt 1990).

Transmission and Epidemiology. The crowding of pinnipeds on rookeries or breeding sites provides ample opportunity for the virus to spread by direct contact, or through contamination of water. SMSV may persist in an infected individual, and vertical transmission also has been suggested. The extent to which pinnipeds and other marine mammals share their caliciviruses has not been fully defined, nor has the role of fish and invertebrates such as the sea lion liver fluke and the fur seal lungworm *Parafilaroides* sp. in the marine cycle of infection (Smith and Boyt 1990; Smith et al. 1998c).

VESV initially became established in pig herds by transfer from the marine environment, probably by feeding uncooked garbage containing fish scraps or tissues of stranded sea lions. Within a pig herd, pig-to-pig transmission of VESV occurred when vesicles ruptured, shedding large quantities of virus into the environment, promoting infection by contact or via fomites. Significant outbreaks of VES in wild pig populations have not been recognized, although there is serologic evidence for infection of wild pigs along the California coast (Smith and Latham 1978).

Clinical Signs. In naturally infected pinnipeds, signs associated with caliciviruses include abortion and vesicular lesions (fluid-filled blisters) up to 3 cm in diameter on the unhaired part of the dorsal and ventral aspects of the flippers. Vesicles usually rupture, leaving an erosion, but they apparently also may regress without rupturing (Smith and Boyt 1990).

The extent to which caliciviruses may contribute to the die-offs that occur among pinnipeds on the California coast has not been well defined, though leptospirosis, which causes abortion, too, clearly also is involved (Gulland et al. 1996). SMSV has been associated with endemic disease in California sea lions, but epidemic disease may occur in northern fur seals on the Pribiloff Islands, where up to about 2% of fur seals may have skin lesions (Smith and Boyt 1990).

VES is an acute, febrile contagious disease characterized by the formation of vesicles on the snout, within the oral cavity, and on the feet. It is clinically indistinguishable from the other three vesicular diseases that affect pigs: foot-and-mouth disease, vesicular stomatitis, and swine vesicular disease. Vesicles easily rupture, leaving raw, bleeding, exceedingly painful ulcers that subsequently become covered with a fibrinous pseudomembrane. There is severe four-footed lameness. Morbidity may be high, but mortality is low and, in uncomplicated cases, recovery occurs after 1–2 weeks. Pregnant sows may abort, and lactating sows may cease milk production (Smith and Boyt 1990).

Pathogenesis. The pathogenesis of skin lesions due to calicivirus infection in pinnipeds is identical to VESV lesions in swine (Smith and Boyt 1990). In pigs, VESV gains entry via abrasions, usually around the snout and mouth or on the feet. There is local viral replication and viremia. Secondary vesicles, mainly on the feet, may occur as a result of direct local spread or following viremia. The vesicular lesions follow replication of the virus in the epidermal cells, leading to cell lysis, inflammatory cell infiltration, and a separation of the epidermal layers by fluid transudate.

Diagnosis. In pinnipeds, diagnosis of SMSV infection depends on isolation and identification of the agent from vesicular fluid, tissue, or feces. The virus produces cytopathology in a wide range of cultured cells; most commonly Vero cells are used.

Clinically, VES cannot be distinguished from the other three known vesicular diseases, and where any of these diseases is suspected in pigs, animal health authorities must be informed.

Immunity. Little information is available on immunity to caliciviruses in pinnipeds. Recovered pigs are immune to the particular antigenic type of VESV with which they were infected but not to other types. Since eradication was used to control the disease, long-term immunity and vaccine development were not at issue.

Control and Treatment. VES is now an extinct disease, although the virus is still present in marine mammals and could well reappear were conditions for transmission to swine favorable. Control would be based on early diagnosis and eradication by test and slaughter. Premises would be destocked and thoroughly cleaned and disinfected, traditionally with 2% sodium hydroxide.

Calicivirus infections in marine mammals are not amenable to control, and there seems to be no indication for such a policy.

Public Health Concerns. Viruses in this group are not known to be highly infectious for humans. There is a single report of human infection with SMSV 5 associated with skin lesions (Smith et al. 1998a).

Domestic Animal Health Concerns. Suspicion of a vesicular disease in an ungulate in most countries requires mandatory notification of appropriate veterinary authorities. Affected properties or regions would be subject to quarantine regulations until the disease was declared eradicated. Cooking or total prohibition of garbage fed to swine is an essential element in control of a number of infectious diseases in addition to VES.

Management Implications. There are no obvious management implications related to SMSV in marine mammals.

FELINE CALICIVIRUS

Introduction. The first feline calicivirus was isolated in 1957. Subsequently, FCV was recognized as one of the two major causes of viral upper respiratory tract disease in cats, the other being feline herpesvirus 1 (feline rhinotracheitis virus).

FCV causes a common disease of domestic cats throughout the world. Economic losses caused by the cost of treatment, and by death of valuable cats, are substantial. FCV also causes clinical disease in captive collections of wild cats and circulates in some populations of free-ranging wild Felidae.

Etiology. FCV is a member of the genus *Vesivirus* and has properties typical of caliciviruses.

Host Range and Distribution. FCV occurs worldwide in domestic cats, and all members of the Felidae are probably susceptible. Among wild felids, FCV has been isolated from captive cheetah *Acinonyx jubatus* (Love 1975) and lions *Panthera leo* (Kadoi et al. 1997); clinical disease occurred in both these species and in captive Siberian tigers *Panthera tigris* (Kadoi et al. 1997).

Although antibody to FCV was not detected in free-ranging lions in two national parks in South Africa (Spencer 1991; Spencer and Morkel 1993), it was highly prevalent (70% of 311) in lions from national parks in Tanzania, though regional prevalence varied markedly (Hofmann-Lehmann et al. 1996). It also was present in wild panthers *Felis concolor* from Florida [56% of 38 (Roelke et al. 1993)] and from California [17% of 58 (Paul-Murphy et al. 1994)]. Free-ranging Iriomote cats *Felis iriomotensis* in Japan also have antibody to FCV [58% of 19 (Mochizuki et al. 1990)]. The extent of exposure to FCV in other free-ranging populations of wild cats has not been defined, nor has disease been described in free-ranging wild felids.

Transmission and Epidemiology. Acute upper respiratory tract disease is common in domestic cats, and FCV can be recovered from about 50% of such cases. It is transmitted by contact, and particularly by sneezing, when cats are closely confined. By 1 year of age, virtually all cats have antibodies to the virus, and clinical disease is rare in cats over this age. Recovered cats remain persistently infected for many months or years and shed virus from the oropharynx, presumably as a consequence of low-grade tonsillar infection.

Infection in wild felids probably is epidemiologically similar.

Pathogenesis and Clinical Signs. FCV produces vesicular lesions on the muzzle, within the oral cavity, and in the respiratory tract. These tend to rupture quickly.

FCV infection in domestic cats may produce subclinical, acute, or subacute disease, usually characterized by conjunctivitis, rhinitis, and tracheitis, pneumonia (mainly in young kittens), and by vesiculation/erosion/ulceration of the epithelium of the oral cavity and muzzle. Lesions also occur occasionally on the inner surface of the forelimb. There is fever, anorexia, lethargy, stiff gait, and usually a profuse ocular and nasal discharge. Morbidity is high; mortality may reach 30% in very young kittens, and recovery is followed by a prolonged carrier state (Gaskell and Dawson 1994).

Signs described in captive cheetahs, lions, and Siberian tigers resemble those in domestic cats (Love 1975; Kadoi et al. 1997).

Diagnosis. Clinically, the disease caused by FCV cannot be differentiated from feline rhinotracheitis. Diagnosis depends on laboratory tests. Both viruses can be isolated readily in cultures of feline cells and may be differentiated rapidly by electron microscopy or by lipid-solvent sensitivity. PCR and a full range of serologic tests are available for the diagnosis of FCV infections.

Immunity. Cats recovered from FCV infection or immunized with FCV vaccine appear to remain relatively free of disease when further exposed, despite the considerable antigenic heterogeneity among FCV isolates. Annual revaccination is recommended.

Although apparently antigenically heterologous in some serologic systems, FCV strains probably are related as a single antigenic type with minor conformational variations in the two neutralizing epitopes, which correspond with hypervariable regions in the genome of the capsid protein (Geissler et al. 1997; Tohya et al. 1997). Such observations paved the way for the development of monotypic vaccines, although there is some suggestion that heterogeneity of the genome in these regions may be associated with vaccine failure (Radford et al. 1997). As well, in persistently infected cats, under the selection pressure of neutralizing antibody, there may be evolution of this component of the FCV genome, in order to effect antigenic change and immune evasion (Radford et al. 1998).

Cell-mediated immune responses, including the generation of cytotoxic T lymphocytes able to lyse autologous cells, have been described for FCV (Tham and Studdert 1987).

Control and Treatment. FCV infection is most difficult to control in large, open cat populations and, even if desired, control would not be feasible among free-ranging wild felids.

Attenuated live virus and inactivated FCV vaccines are widely used in domestic cats, usually in combination with feline herpesvirus 1 vaccine. Similar vaccines are recommended in captive wild felids (Bittle 1993). Cheetahs vaccinated with commercial modified live virus vaccines against FCV responded better than did those vaccinated with inactivated vaccines (Spencer and Burroughs 1991, 1992). Modified live vaccine strains of FCV are not innocuous, however, and if taken orally, for instance by licking a vaccination site, they can cause acute and persistent disease (Pedersen and Hawkins 1995), suggesting that only inactivated vaccines should be used in captive felids.

Similar principles apply to control of the disease by management in populations of domestic and captive wild felids. Clinically ill cats should be isolated, and incoming cats of uncertain status should be quarantined for at least a week before being introduced into the general colony. Although recovered cats remain persistently infected, the amount of virus shed usually is not large, so they pose less of a threat to in-contact cats than cats with obvious clinical disease.

Public Health Concerns. There appear to be no major human public health implications, since FCV is not known to infect humans.

ACKNOWLEDGMENTS. Cor Lenghaus acknowledges the role of Harvey Westbury in initiating the RHD Project at AAHL and records his appreciation to Belinda Collins and Nagaratnam Ratnamohan, who were coworkers with him on the RHD Project.

LITERATURE CITED

Alonso, C., J.M. Oviedo, J.M. Martín-Alonso, E. Díaz, J.A. Boga, and F. Parra. 1998. Programmed cell death in the pathogenesis of rabbit hemorrhagic disease. *Archives of Virology* 143:321–332.

Asgari, S., J.R.E. Hardy, R.G. Sinclair, and B.D. Cooke. 1998. Field evidence for mechanical transmission of rabbit haemorrhagic disease virus (RHDV) by flies (Diptera, Calliphoridae) among wild rabbits in Australia. *Virus Research* 54:123–132.

Barlow, N.D., and J.M. Kean. 1998. Simple model for the impact of rabbit calicivirus disease on Australasian rabbits. *Ecological Modeling* 109:225–241.

Bittle, J.L. 1993. Use of vaccines in exotic animals. *Journal of Zoo and Wildlife Medicine* 24:352–356.

Buddle, B.M., G.W. Delisle, K. McColl, B.J. Collins, C. Morrissy, and H.A. Westbury. 1997. Response of the northern brown kiwi, *Apteryx australis* Mantelli and the lesser short-tailed bat, *Mystacina tuberculata* to a measured dose of rabbit haemorrhagic disease virus. *New Zealand Veterinary Journal* 45:109–113.

Cancellotti, F.M., and M. Renzi. 1991. Epidemiology and current situation of viral haemorrhagic disease of rabbits and the European brown hare syndrome in Italy. *Scientific and Technical Review of the Office International des Épizooties* 10:409–420.

Capucci, L., and A. Lavazza. 1998. A brief update on rabbit hemorrhagic disease virus. *Emerging Infectious Diseases* 4:343–344.

Capucci, L., M.T. Scicluna, and A. Lavazza. 1991. Diagnosis of viral haemorrhagic disease of rabbits and the European brown hare syndrome. *Scientific and Technical Review of the Office International des Épizooties* 10:347–370.

Capucci, L., P. Fusi, A. Lavazza, M.L. Pacciarini, and C. Rossi. 1996. Detection and preliminary characterization of a new rabbit calicivirus related to rabbit hemorrhagic disease virus but nonpathogenic. *Journal of Virology* 70:8614–8623.

Carman, J.A., M.G. Garner, M.G. Catton, S. Thomas, H.A. Westbury, R.M. Cannon, B.J. Collins, and I.G. Tribe. 1998. Viral haemorrhagic disease of rabbits and human health. *Epidemiology and Infection* 121:409–418.

Chasey, D. 1994. Possible origin of rabbit haemorrhagic disease in the United Kingdom. *Veterinary Record* 135:496–499.

———. 1997. Rabbit hemorrhagic disease: The new scourge of *Oryctolagus cuniculus*. *Laboratory Animals* 31:33–44.

Chen, L., ed. 1991. *International symposium on rabbit haemorrhagic disease.* Beijing: Chinese Association of Animal and Veterinary Sciences, 427 pp.

Clarke, I.N., and P.R. Lambden. 1997. The molecular biology of caliciviruses. *Journal of General Virology* 78:291–301.

Collins, B.J., J.R. White, C. Lenghaus, V. Boyd, and H.A. Westbury. 1995. A competition ELISA for the detection of antibodies to rabbit haemorrhagic disease virus. *Veterinary Microbiology* 43:85–96.

Collins, B.J., J.R. White, C. Lenghaus, C.J. Morrissy, and H.A. Westbury. 1996. Presence of rabbit haemorrhagic disease virus antigen in rabbit tissues as revealed by a monoclonal antibody dependent capture ELISA. *Journal of Virological Methods* 58:145–154.

Duff, J.P., D. Chasey, R. Munro, and M. Wooldridge. 1994. European brown hare syndrome in England. *Veterinary Record* 134:669–673.

Duff, J.P., K. Whitwell, and D. Chasey. 1997. The emergence and epidemiology of European brown hare syndrome in the UK. In *Proceedings first international symposium on caliciviruses,* ed. D. Chasey, R.M. Gaskell, and I.N. Clarke. Addlestone, UK: European Society for Veterinary Virology, pp. 176–181.

Fenner, F., and F.N. Ratcliffe. 1965. *Myxomatosis.* Cambridge: Cambridge University Press, 379 pp.

Frölich, K., F. Klima, and J. Dedek. 1998. Antibodies against rabbit hemorrhagic disease virus in free-ranging red foxes from Germany. *Journal of Wildlife Diseases* 34:436–442.

Gaskell, R.M., and S. Dawson. 1994. Viral-induced upper respiratory tract disease. In *Feline medicine and therapeutics,* ed. E.A Chandler, C.J. Gaskell, and R.M. Gaskell, 2d ed. Oxford: Blackwell, pp. 453–472.

Gavier-Widén, D. 1992. Epidemiology, pathology and pathogenesis of two related viral hepatites of leporids. Ph.D. diss., University of California, Davis, 135 pp.

———. 1993. Viral hepatitis of rabbits and hares in Scandinavia. In *Zoo and wild animal medicine: Current Therapy 3,* ed. M.E. Fowler. Philadelphia: W.B. Saunders, pp. 322–325.

———. 1994. Morphologic and immunohistochemical characterization of the hepatic lesions associated with European brown hare syndrome. *Veterinary Pathology* 31:327–334.

Gavier-Widén, D., and Mörner, T. 1991. Epidemiology and diagnosis of European brown hare syndrome in Scandinavian countries: A review. *Scientific and Technical Review of the Office International des Épizooties* 10:453–458.

———. 1993. Descriptive epizootiological study of European brown hare syndrome in Sweden. *Journal of Wildlife Diseases* 29:15–20.

Geissler, K., K. Schneider, G. Platzer, B. Truyen, O.R. Kaaden, and U. Truyen. 1997. Genetic and antigenic heterogeneity among feline calicivirus isolates from distinct disease manifestations. *Virus Research* 48:193–206.

Gould, A.R., J.A. Kattenbelt, C. Lenghaus, C.J. Morrissy, T. Chamberlain, B.J. Collins, and H.A. Westbury. 1997. The complete nucleotide sequence of rabbit haemorrhagic disease virus (Czech strain V351): Use of the polymerase chain reaction to detect replication in Australian vertebrates and analysis of viral population sequence variation. *Virus Research* 47:7–17.

Graham, D.A., J. Cassidy, N. Beggs, W.L. Curran, I.E. McLaren, T.J. Connor, and S. Kennedy. 1996. Rabbit viral haemorrhagic disease in Northern Ireland. *Veterinary Record* 138:47.

Gregg, D.A., and C. House. 1989. Necrotic hepatitis of rabbits in Mexico: A parvovirus. *Veterinary Record* 126:603–604.

Green, K.Y., T. Ando, M.S. Balayan, T. Berke, I.N. Clarke, M.K. Estes, D.O. Matson, S. Nakata, J.D. Neill, M.J. Studdert, and H.-J. Thiel. 2000. Taxonomy of the caliciviruses. *Journal of Infectious Diseases* 181(Suppl. 2): S322–S330.

Gulland, F.M. D., M. Koski, L.J. Lowenstine, A. Colagross, L. Morgan, and T. Spraker. 1996. Leptospirosis in California sea lions (*Zalophus californianus*) stranded along the central California coast, 1981–1994. *Journal Of Wildlife Diseases* 32:572–580.

Heneidi Zeckua, A., C. Zepeda Sein, A. Mateos Poumian, and G. Velázquez. 1997. Model for evaluating the risk of introducing rabbit viral hemorrhagic disease based on experience in Mexico [in Spanish]. *Scientific and Technical Review of the Office International des Épizooties* 16:91–103.

Henriksen, P., D. Gavier, and F. Elling. 1989. Acute necrotising hepatitis in Danish farmed hares. *Veterinary Record* 125:486–487.

Hoffmann-Lehmann, R., D. Fehr, M. Grob, M. Eglizoli, C. Packer, J.S. Martinson, S.J. O'Brien, and H. Lutz. 1996. Prevalence of antibodies to feline parvovirus, calicivirus, herpesvirus, coronavirus, and immunodeficiency virus and of feline leukemia virus antigen and the interrelationship of these viral infections in free-ranging lions in east Africa. *Clinical and Diagnostic Laboratory Immunology* 3:554–562.

Kadoi, K., M. Kiryu, M. Iwabuchi, H. Kamata, M. Yukawa, and Y. Inaba. 1997. A strain of calicivirus isolated from lions with vesicular lesions on tongue and snout. *New Microbiologica* 20:141–148.

Konig, M., H.-J. Thiel, and G. Meyers. 1998. Detection of viral proteins after infection of cultured hepatocytes with rabbit hemorrhagic disease virus. *Journal of Virology* 72:4492–4497.

Kovaliski, J. 1998. Monitoring the spread of rabbit hemorrhagic disease virus as a new biological agent for control of wild European rabbits in Australia. *Journal of Wildlife Diseases* 34:421–428.

Kreutz, L.C., R.P. Johnson, and B.S. Seal. 1998. Phenotypic and genotypic variation of feline calicivirus during persistent infection in cats. *Veterinary Microbiology* 59:229–236.

Lavazza, A., and G. Vecchi. 1989. Osservazioni zu alcuni episodi di mortalità nella lepre: Evidenziazione al microscopio elettronico di una particella virale—Nota preliminare. *Selezione Veterinaria* 30:461–468.

Lavazza, A., M.T. Scicluna, and L. Capucci. 1996. Susceptibility of hares and rabbits to the European brown hare syndrome virus (EBHSV) and rabbit hemorrhagic disease virus (RHDV) under experimental conditions. *Zentralblatt für Veterinärmedizin [B]* 43:401–410.

Le Gall, G., S. Huguet, P. Vende, J.F. Vautherot, and D. Rasschaert. 1996. European brown hare syndrome virus B molecular cloning and the sequencing of the genome. *Journal of General Virology* 77:1693–1697.

Leighton, F.A., M. Artois, L. Capucci, D. Gavier-Widén, and J.-P. Morisse. 1995. Antibody response to rabbit viral hemorrhagic disease virus in red foxes (*Vulpes vulpes*) consuming livers of infected rabbits (*Oryctolagus cuniculus*). *Journal of Wildlife Diseases* 31:541–544.

Lenghaus, C., H. Westbury, B. Collins, N. Ratnamohan, and C. Morrissy. 1994. Overview of the RHD project in Australia. In *Rabbit haemorrhagic disease: Issues in assessment for biological control,* ed. R.K. Munro and R.T. Williams. Canberra: Bureau of Resource Sciences, pp. 104–129.

Liu, S.J., H.P. Xue, B.Q. Pu, and N.H. Qian. 1984. A new viral disease of rabbits [in Chinese]. *Animal Husbandry and Veterinary Medicine* 16:253–255.

Love, D.N. 1975. Pathogenicity of a strain of feline calicivirus for domestic kittens. *Australian Veterinary Journal* 51:541–546.

Marcato, P.S., C. Benazzi, M. Galeotti, and L. della Salda. 1989. l'Epatite necrotica infettiva dei leporidi. *Rivista di Conigliocoltura* 8:41–50.

Marchandeau, S., J. Chantal, Y. Portejoie, S. Barraud, and Y. Chaval. 1998. Impact of viral hemorrhagic disease on a wild population of European rabbits in France. *Journal of Wildlife Diseases* 34:429–435.

Mead, C. 1998. Rabbit hemorrhagic disease. *Emerging Infectious Diseases* 4:344–345.

Meyers, G., C. Wirblich, and H.-J. Thiel. 1991. Rabbit hemorrhagic disease virus: Molecular cloning and nucleotide sequencing of a calicivirus genome. *Virology* 184:664–676.

Mochizuki, M., M. Akuzawa, and H. Nagatomo. 1990. Serological survey of the Iriomote cat (*Felis iriomotensis*) in Japan. *Journal of Wildlife Diseases* 26:236–245.

Morisse, J.-P., ed. 1991. Viral haemorrhagic disease of rabbits and the European brown hare syndrome. *Scientific and Technical Review of the Office International des Épizooties* 10:263–524.

Munro, R.K., and R.T. Williams, eds. 1994. *Rabbit haemorrhagic disease: Issues in assessment for biological control.* Canberra: Bureau of Resource Sciences, 168 pp.

Mutze, G., B. Cooke, and P. Alexander. 1998. The initial impact of rabbit hemorrhagic disease on European rabbit populations in South Australia. *Journal of Wildlife Diseases* 34:221–227.

Nagesha, H.S., L.F. Wang, A.D. Hyatt, C.J. Morrissy, C. Lenghaus, and H.A. Westbury. 1995. Self-assembly, antigenicity, and immunogenicity of the rabbit haemorrhagic disease virus (Czechoslovakian strain V-351) capsid protein expressed in baculovirus. *Archives of Virology* 140:1095–1108.

Neill, J.D., R.F. Meyer, and B.S. Seal. 1995. Genetic relatedness of the caliciviruses: San Miguel sea lion and vesicular exanthema of swine viruses constitute a single genotype within the Caliciviridae. *Journal of Virology* 69:4484–4488.

Nowotny, N., C. Ros Bascuñana, A. Ballagi-Pordány, D. Gavier-Widén, M. Uhlen, and S.I. Belák. 1997. Phylogenetic analysis of rabbit haemorrhagic disease and European brown hare syndrome viruses by comparison of sequences from the capsid protein gene. *Archives of Virology* 142:657–673.

Ohlinger, V.F., B. Haas, and H.-J. Thiel. 1993. Rabbit hemorrhagic disease (RHD): Characterization of the causative calicivirus. *Veterinary Research* 24:103–116.

Paul-Murphy, J., T. Work, D. Hunter, E. McFie, and D. Fjelline. 1994. Serological survey and serum biochemical reference ranges of the free-ranging mountain lion (*Felis concolor*) in California. *Journal of Wildlife Diseases* 30:205–215.

Pedersen, N.C., and K.F. Hawkins. 1995. Mechanisms of persistence of acute and chronic feline calicivirus infection in the face of vaccination. *Veterinary Microbiology* 47:141–156.

Poli, A., M. Nigro, D. Gallazi, G. Sironi, A. Lavazza, and D. Gelmetti. 1991. Acute hepatosis in the European brown hare (*Lepus europaeus*) in Italy. *Journal of Wildlife Diseases* 27:621–629.

Radford, A.D., M. Bennett, F. McCardle, S. Dawson, P.C. Turner, M.A. Glenn, and R.M. Gaskell. 1997. The use of sequence analysis of a feline calicivirus (FCV) hypervariable region in the epidemiological investigation of FCV related disease and vaccine failures. *Vaccine* 15:1451–1458.

Radford, A.D., P.C. Turner, M. Bennett, F. McCardle, S. Dawson, M.A. Glenn, R.A. Williams, and R.M. Gaskell. 1998. Quasispecies evolution of a hypervariable region of the feline calicivirus capsid gene in cell culture and in persistently infected cats. *Journal of General Virology* 79(Pt. 1):1–10.

Rodák, L., B. Smíd, B. Valícek, T. Veselý, J. Stepánek, J. Hampl, and E. Jurák. 1990. Enzyme-linked immunosorbent assay of antibodies to rabbit haemorrhagic disease virus and determination of its major structural proteins. *Journal of General Virology* 71:1075–1080.

Roelke, M.E., D.J. Forrester, E.R. Jacobson, G.V. Kollias, F.W. Scott, M.C. Barr, J.F. Evermann, and E.C. Pirtle. 1993. Seroprevalence of infectious disease agents in free-ranging Florida panthers (*Felis concolor coryi*). *Journal of Wildlife Diseases* 29:36–49.

Ros Bascuñana, C., N. Nowotny, and S. Belák. 1997. Detection and differentiation of rabbit hemorrhagic disease and European brown hare syndrome viruses by amplification of VP60 genomic sequences from fresh and fixed tissue specimens. *Journal of Clinical Microbiology* 35:2492–2495.

Scicluna, M.T., A. Lavazza, and L. Capucci. 1994. European brown hare syndrome in northern Italy: Results of a virological and serological survey. *Scientific and Technical Review of the Office International des Épizooties* 13:893–904.

Smith, A.W., and P.M. Boyt. 1990. Caliciviruses of ocean origin: A review. *Journal of Zoo and Wildlife Medicine* 21:3–23.

Smith, A.W., and A.B. Latham. 1978. Prevalence of vesicular exanthema of swine antibodies in feral animals associated with the southern California coastal zones. *American Journal of Veterinary Research* 39:291–296.

Smith, A.W., E.S. Berry, D.E. Skilling, J.E. Barlough, S.E. Poet, T. Berke, J. Mead, and D.O. Matson. 1998a. In vitro isolation and characterization of a calicivirus causing a vesicular disease of the hands and feet. *Clinical Infectious Diseases* 26:434–439.

Smith, A.W., N.J. Cherry, and D.O. Matson. 1998b. Reply to Drs. Capucci, Lavazza and Mead. *Emerging Infectious Diseases* 4:345–346.

Smith, A.W., D.E. Skilling, N. Cherry, J.H. Mead, and D.O. Matson. 1998c. Calicivirus emergence from ocean reservoirs: Zoonotic and interspecies movements. *Emerging Infectious Diseases* 4:13–20.

Spencer, J.A. 1991. Survey of antibodies to feline viruses in free-ranging lions. *South African Journal of Wildlife Research* 21:59–61.

Spencer, J.A., and R. Burroughs. 1991. Antibody response of captive cheetahs to modified-live feline virus vaccine. *Journal of Wildlife Diseases* 27:578–583.

———. 1992. Decline in maternal immunity and antibody response to vaccine in captive cheetah (*Acinonyx jubatus*) cubs. *Journal of Wildlife Diseases* 28:102–104.

Spencer, J.A., and P. Morkel. 1993. Serological survey of sera from lions in Etosha National Park. *South African Journal of Wildlife Research* 23:60–61.

Studdert, M.J. 1978. Caliciviruses [Brief review]. *Archives of Virology* 58:157–191.

Tham, K.M., and M.J. Studdert. 1987. Antibody and cell-mediated immune responses to feline calicivirus following inactivated vaccine and challenge. *Journal of Veterinary Medicine [B]* 34:640–654.

Tohya, Y., N. Yokoyama, K. Maeda, Y. Kawaguchi, and T. Mikami. 1997. Mapping of antigenic sites involved in neutralization on the capsid protein of caliciviruses. *Journal of General Virology* 78(Pt. 2):303–305.

Villafuerte, R., C. Calvete, J.C. Blanco, and J. Lucientes. 1995. Incidence of viral hemorrhagic disease in wild rabbit populations in Spain. *Mammalia* 59:651–659.

Williams, C.K., I. Parer, B.J. Coman, J. Burley, and M.L. Braysher. 1995. *Managing vertebrate pests: Rabbits.* Canberra: Australian Government Publishing Service, 284 pp.

Wirblich, C., G. Meyers, V.F. Ohlinger, L. Capucci, U. Eskens, B. Haas, and H.-J. Thiel. 1994. European brown hare syndrome virus: Relationship to rabbit hemorrhagic disease virus and other caliciviruses. *Journal of Virology* 68:5164–5173.

Xu, Z.J., and W.X. Chen. 1989. Viral haemorrhagic disease in rabbits: A review. *Veterinary Research Communications* 13:205–212.

Xu, W.Y., N.X. Du, and S.J. Liu. 1988. A new virus isolated from haemorrhagic disease in rabbits. In *Proceedings of the 4th World Rabbit Science Association Congress, Budapest, Hungary,* pp. 456–461.

Zanni, M.L., M.C. Benassi, A. Lavazza, and L. Capucci. 1993. Clinical evolution and diagnosis of an outbreak of European brown hare syndrome in hares reared in captivity. *Scientific and Technical Review of the Office Internationale des Épizooties* 12:931–940.

17

TRANSMISSIBLE SPONGIFORM ENCEPHALOPATHIES

ELIZABETH S. WILLIAMS, JAMES K. KIRKWOOD,
AND MICHAEL W. MILLER

INTRODUCTION. The transmissible spongiform encephalopathies (TSEs) comprise an unusual group of neurologic diseases of humans and animals. They are apparently caused by proteinaceous agents called *prions* that are devoid of nucleic acids (Prusiner 1982). Although debate continues (Chesebro 1998; Farquhar et al. 1998), there is now a great deal of evidence supporting the hypothesis that TSEs are caused by abnormal, protease-resistant forms (PrP^{res}) of cellular proteins (PrP^{c}) coded for and normally synthesized in central nervous system (CNS) and lymphoid tissues (Prusiner 1991). It is thought that these abnormal proteins arise through posttranslational modifications in tertiary structure of PrP^{c}, resulting in decreased α-helical content and increased amounts of β-sheet (Prusiner 1997). In humans, PrP^{res} may arise sporadically through somatic mutations or spontaneous conversion of PrP^{c} to $PrP^{res;}$ as a result of germline mutations in the PrP gene resulting in familial disease; or they may be acquired by infection (Prusiner 1997). In animals, TSEs are infectious; spontaneous and familial forms have not been identified, though they may occur.

On entering a susceptible host by some natural or experimental process, PrP^{res} promotes production of species-specific PrP^{res} from PrP^{c} in lymphoid and CNS tissues. The finding that PrP^{res} catalyzes production of PrP^{res} from PrP^{c} in vitro added weight to the hypothesis that this is its mode of action in vivo (Kocisko et al. 1994; Raymond et al. 1997).

Thus, although the TSEs behave like infectious diseases, the agents appear to have no inherent genetic identity, and if this is so, the disease is more correctly perceived and classified as a special type of toxicity. Prions have remarkable resistance to environmental conditions and a range of treatments that typically kill or inactivate conventional infectious agents (Millison et al. 1976; Taylor et al. 1995).

Prior to 1980, naturally occurring TSEs had been reported in four species: scrapie in domestic sheep *Ovis aries* and goats *Capra hircus* (Dickinson 1976); transmissible mink encephalopathy (TME) in mink *Mustela vison* (Hartsough and Burger 1965); and kuru, Creutzfeldt-Jakob disease (CJD), and Gerstmann-Sträussler-Scheinker syndrome of humans (Prusiner and Hadlow 1979; Collinge and Palmer 1997). More recently, chronic wasting disease (CWD) was reported in deer *Odocoileus* spp. and Rocky Mountain elk *Cervus elaphus nelsoni* in the United States (Williams and Young 1980, 1982). Bovine spongiform encephalopathy (BSE) was diagnosed in cattle *Bos taurus* (Wells et al. 1987), in domestic cats *Felis catus* (Pearson et al. 1992), and in wild mammals in or from Great Britain (Jeffrey and Wells 1988; Kirkwood and Cunningham 1994a) or in France (Bons et al. 1996, 1999). Bovine spongiform encephalopathy was associated with a variant of CJD (vCJD) in a few humans beginning in 1996 (Will et al. 1996).

A disease indistinguishable from scrapie occurred in mouflon *Ovis musimon* in the United Kingdom (Wood et al. 1992). In addition, several suspect cases of TSE were reported in albino tigers *Panthera tigris* (Kelly et al. 1980) and ostriches *Struthio camelus* (Schoon et al. 1991), but these were not confirmed and probably were not prion diseases.

Recent studies draw somewhat conflicting conclusions about the pathogenesis of the TSEs. In part, this may be due to variation in the different agents, different doses and routes of exposure, and different animal models used to study these diseases; natural hosts are seldom employed in these studies.

Substantial evidence exists for genetic variation in susceptibility to some prion diseases among and within species. For example, there are differences in susceptibility to scrapie among breeds of sheep (Hunter et al. 1992; O'Rourke et al. 1997b) and differences in incubation period associated with genotype in mice (Bruce et al. 1994). Genetic variation in susceptibility to sporadic and iatrogenic prion disease in humans is recognized (Collinge and Palmer 1994). In contrast, there is no evidence for variation in susceptibility to BSE among cattle (Wilesmith 1994).

Studies of the pathogenesis of scrapie after intragastric inoculation of mice suggested neural spread of the agent from the gastrointestinal tract to thoracic spinal cord via the sympathetic nervous system (Kimberlin and Walker 1989). In hamsters orally infected with scrapie, the route to the CNS was hypothesized to be the vagus nerve to the parasympathetic vagal nucleus (dorsal motor nucleus of the vagus) in the medulla oblongata, the initial site of detection of PrP^{res} in the CNS (Beekes et al. 1998). Evidence of infectivity in cattle orally infected with large doses of BSE agent was found first in the CNS in thoracic and lumbar spinal cord (Wells et al. 1998). Neuroinvasion in scrapie-infected mice was linked to B lymphocytes (Klein et al. 1997). There is no known immune response to TSE agents in affected hosts; however, the lymphoreticular system plays a role in pathogenesis of disease in rodent models.

Histopathologic changes in animals and humans with TSEs are qualitatively similar and confined to the CNS. Lesions include vacuolation of neuronal perikarya and neurites, neuronal degeneration and loss, gliosis (mainly astrocytic), and accumulation of PrP^{res} (Wells and McGill 1992; McGill and Wells 1993). The pathogenetic mechanisms of neurodegeneration are not understood but are under study (Sakaguchi et al. 1996; Tobler et al. 1996; Jeffrey et al. 1997; Williams et al. 1997; Hegde et al. 1998). Scrapie-associated fibrils (SAFs), which are fibrillar aggregates of PrP^{res}, may be revealed by electron-microscopic examination of detergent extracts of brain from affected animals (Merz et al. 1984; Hope et al. 1988; Wells and McGill 1992).

With the importance of BSE and scrapie in domestic livestock, and the heightened concern about the relationship of these diseases and human health, more attention will certainly be focused on the TSEs in the future.

CHRONIC WASTING DISEASE

History and Distribution. Chronic wasting disease (CWD) was first recognized in 1967 as a clinical syndrome of unknown etiology among captive mule deer *Odocoileus hemionus* at wildlife research facilities in Colorado (Williams and Young 1992). The disease was diagnosed in 1978 as a spongiform encephalopathy by histopathologic examination of CNS from affected animals. Shortly afterward CWD was recognized among captive deer in Wyoming (Williams and Young 1980). Diagnosis of CWD in Rocky Mountain elk from these same facilities quickly followed (Williams and Young 1982). Deer and elk in a few zoological gardens in the United States and Canada were identified with CWD in subsequent years (Williams and Young 1992). Apparently it did not persist in these locations. Chronic wasting disease has recently become a concern to the game farm industry following its diagnosis in elk in Saskatchewan, Canada, and in South Dakota, Nebraska, Montana, Colorado, and Oklahoma.

In 1981, CWD was recognized in a free-ranging elk in Colorado (Spraker et al. 1997). Subsequently, it was found in free-ranging elk in Wyoming, and in free-ranging mule deer [1985 (M.W. Miller unpublished)] and white-tailed deer *Odocoileus virginianus* [1990 (E.S. Williams unpublished)] in both states. The known distribution of CWD currently includes captive and free-ranging cervids in southeast Wyoming and north-central and northeast Colorado (Miller et al. 2000) and several game farms in the United States and Canada.

Host Range. Only three species of Cervidae are known to be naturally susceptible to CWD: mule deer, white-tailed deer, and Rocky Mountain elk. Subspecies of these cervids probably are also naturally susceptible.

Pronghorn *Antilocapra americanus,* Rocky Mountain bighorn sheep *Ovis canadensis,* mouflon, mountain goats *Oreamnos americana,* moose *Alces alces,* and a blackbuck *Antilope cervicapra* have been in contact with CWD-affected deer and elk or resided in premises where CWD had occurred but have not developed the disease. Domestic livestock are not known to be naturally susceptible to CWD, and a few cattle, sheep, and goats have resided in research facilities with CWD for prolonged periods without developing the disease.

Many species are experimentally susceptible to CWD by intracerebral inoculation, an unnatural but commonly used route for the study of prion disease. Mink, domestic ferret *Mustela putorius furo,* squirrel monkey *Saimiri sciureus,* mule deer, domestic goat (Williams and Young 1992), and laboratory mice (Bruce et al. 1997) are susceptible to CWD by this route on primary passage.

Etiology. The origin of CWD is not known. Spontaneous development of PrP^{res} might have occurred in deer, with subsequent transmission to other deer and elk. An alternate explanation is that CWD is actually scrapie occurring in cervids. Chronic wasting disease could also have originated by infection with an as-yet-unrecognized prion.

Characteristics of the agent causing CWD are poorly understood, but the agent is presumed to be a prion. Based on mouse strain typing, it appears to differ from the BSE agent (Bruce et al. 1997), many strains of scrapie, and the TME agent (M.E. Bruce personal communication). The marked similarity of CNS lesions and epidemiology strongly suggests CWD agent is the same in captive and free-ranging deer and elk.

Transmission and Epidemiology. The mode of transmission of CWD is unknown. There is no evidence that CWD is a food-borne disease associated with rendered ruminant meat and bonemeal as was the case in BSE (Wilesmith et al. 1988). Occurrence of the disease among captive deer and elk, many of which were acquired as neonates, fawns, or adults, provides strong evidence of lateral transmission (Williams and Young 1992; Miller et al. 1998; Miller et al. 2000). Maternal transmission may also occur; however, this has not been definitively determined. It is likely transmission occurred from mule deer to elk.

The scrapie agent is found in many lymphoid tissues, including those of the digestive tract (Hadlow et al. 1980, 1982), suggesting the agent may be shed through the alimentary tract. Lymphoid tissues of affected deer and elk contain PrP^{res}; thus, alimentary tract shedding may also occur in CWD. The TSE agents are extremely resistant in the environment (Brown and Gajdusek 1991); pasture contamination has been suspected of being the source of scrapie agent in some outbreaks of sheep scrapie (Greig 1940; Pálsson 1979). Concentration of deer and elk in captivity or by artificial feeding may increase the likelihood of transmission between individuals.

The youngest animal diagnosed with natural CWD was 17 months of age, suggesting this as an approximate minimum incubation period; however, without knowledge of when the animal was infected, it is impossible to accurately determine the incubation period. Maximum incubation periods are not known. Most cases of CWD among deer and elk residing in facilities with a long history of CWD are in 3–7-year-old animals. The age of onset of clinical signs is variable in animals brought into facilities as adults or among animals in herds newly recognized to have CWD. For example, one elk in a presumed newly infected herd was more than 15 years old. It is not known when during the course of infection an animal may be infectious.

In one study, more than 90% of mule deer residing on a premises for more than 2 years died or were euthanized due to CWD (Williams and Young 1980). Chronic wasting disease was the primary cause of adult mortality [5 (71%) of 7 and 4 (23%) of 23] in two captive elk herds (Miller et al. 1998).

Relatively little is known about the epidemiology of CWD in free-ranging cervids. In addition to necropsy and examination of brains from animals showing clinical signs suggestive of CWD to determine its distribution (targeted surveillance), brains from deer and elk harvested by hunters in the CWD-endemic area have been used to estimate prevalence. Within endemic areas, prevalence of preclinical CWD, based on histopathology and/or immunohistochemistry for PrP^{res}, is estimated at less than 1%–8% (Miller et al. 2000). Chronic wasting disease has not been found in cervids outside the endemic areas.

Preliminary modeling suggested lateral transmission is necessary to maintain CWD at the prevalence observed in surveillance programs. Maternal transmission may occur, but in the model this route of transmission alone was not adequate to maintain the disease at observed levels (Miller et al. 2000).

Clinical Signs. The most striking clinical features of CWD in deer and elk are loss of body condition and changes in behavior. Clinical signs of CWD may be more subtle and prolonged in elk than in mule deer. Affected animals may increase or decrease their interaction with handlers or other members of the herd. They may show repetitive behaviors, such as walking set patterns in their pens or pastures, show periods of somnolence or depression from which they are easily roused, and may carry their head and ears lowered. Affected animals continue to eat, but they consume reduced amounts of feed, leading to gradual loss of body condition. As the disease progresses, many affected animals display polydipsia and polyuria; increased salivation with resultant slobbering or drooling; and incoordination, particularly posterior ataxia, fine head tremors, and wide-based stance. Esophageal dilatation, hyperexcitability, and syncope are rarely seen. Death is inevitable.

In captive herds newly experiencing CWD, sporadic cases of prime-aged animals losing condition, being unresponsive to symptomatic treatment, and death from pneumonia are common. Aspiration pneumonia, presumably from difficulty swallowing and hypersalivation, may lead to misdiagnosis of the condition if the brain is not examined.

The clinical course of CWD varies from a few days to a year, with most animals surviving a few weeks to 3–4 months. Although a protracted clinical disease is typical, occasionally acute death may occur in white-tailed deer (M.W. Miller unpublished). Caretakers familiar with individual animals often recognize subtle changes in behavior well before those not familiar with the particular animal detect abnormalities or serious weight loss occurs. Also, those who have seen clinically affected animals are more astute at detecting early behavioral changes than naive observers.

The clinical course of CWD in free-ranging deer and elk is probably shorter than in captivity. Wild cervids must forage, find water, and are susceptible to predation, all factors affecting longevity of sick animals in the wild.

Pathogenesis. The pathogenesis of CWD is not specifically known, though considerable research is currently under way to better understand the dynamics of the disease in deer and elk. Based on similarities in clinical course, neuropathology, and distribution of PrP^{res}, pathogenesis of CWD is likely similar to scrapie (Hadlow et al. 1980, 1982) The CWD agent probably enters the animal by ingestion, perhaps from environmental contamination or direct interaction with animals shedding the agent. In mule deer fawns experimentally infected with CWD, PrP^{res} was detected in retropharyngeal and ileocecal lymph nodes, tonsil, and Peyer's patches by 42 days after inoculation (Sigurdson et al. 1999).

The parasympathetic vagal nucleus in the medulla oblongata is the site of the most severe and consistent lesions in deer (Williams and Young 1993) and is the site of PrP^{res} accumulation, prior to development of spongiform changes (E.S. Williams unpublished; T.R. Spraker personal communication). Distribution of lesions in the brain (Williams and Young 1993) may explain clinical signs. Emaciation may be associated with hypothalamic damage, and polydipsia may reflect damage to the paraventricular and supraoptic nuclei and subsequent diabetes insipidus (Williams and Young 1992).

Pathology. Alterations in clinical chemistry and hematology may occur in CWD-affected animals, but the alterations are not diagnostic. In captive deer, low urine specific gravity (1.002–1.010) reflects polydipsia and possibly inability to concentrate urine (Williams and Young 1980). In free-ranging animals, urine specific gravity may not be as low because they may not have ready access to water and may be dehydrated at death. Other nonspecific changes in clinical pathology reflect emaciation or intercurrent diseases.

The gross lesions of CWD are nonspecific. Carcasses may be in poor nutritional state or emaciated, but may be in fair condition if the animal died of aspiration pneumonia or after only a short clinical course. Aspiration pneumonia with or without fibrinous pleuritis may be present in some animals. Rumen contents contain excessive water in those animals displaying polydipsia; sometimes the contents appear frothy. Sand and gravel are often abundant in the forestomachs.

Microscopic lesions of CWD have been described in mule deer and elk (Williams and Young 1993; Hadlow 1996). The lesions are qualitatively typical of TSEs. Distribution of lesions is similar in deer and elk, with some minor differences in degree. In all cases of clinical CWD, lesions are in the parasympathetic vagal nucleus in the dorsal portion of the medulla oblongata at the obex, in hypothalamus and thalamus, and in olfactory tracts and cortex. Other regions of the brain, in particular, thalamus and cerebellum, show typical spongiform changes with varying degrees of severity. Lesions are usually mild in the cerebral cortex, hippocampus, and basal ganglia.

Plaques composed of PrP^{res} can be appreciated on routine hematoxylin-eosin staining in most clinically affected white-tailed deer and in a few mule deer but are not obvious in elk (Bahmanyar et al. 1985; Williams and Young 1992). In white-tailed deer, plaques are often surrounded by vacuoles in the neuropil, which allows them to be easily visualized. The plaques stain strongly on immunohistochemistry for PrP^{res} by using a variety of polyclonal and monoclonal antibodies (Guiroy et al. 1991a,b; Williams and Young 1992; Liberski et al. 1993; O'Rourke et al. 1998b). Patterns of immunostaining in CWD include granularity and amorphous clumps on neuronal membranes, perivascular aggregates, and large, apparently extracellular accumulations of PrP^{res}.

Scrapie-associated fibrils are found in brains and spleen of deer and elk with CWD (Williams and Young 1992; Spraker et al. 1997). The ultrastructural lesions of CWD are typical of lesions found in the other TSEs (Guiroy et al. 1993, 1994; Liberski et al. 1993).

Diagnosis. Clinical signs of CWD are not specific, and currently diagnosis is based on examination of the brain for spongiform lesions and/or accumulation of PrP^{res}. The parasympathetic vagal nucleus in the dorsal portion of the medulla oblongata at the obex is the most important site to be examined for diagnosis of CWD (Williams and Young 1993) and should be submitted for histopathologic examination on every animal suspected of having CWD. The whole head or whole brain can be submitted to the diagnostic laboratory to ensure that the correct portion of the brain is examined. Supplemental tests include negative-stain electron microscopy for SAF or Western blotting for detection of PrP^{res} in brain (Williams and Young 1992; Spraker et al. 1997).

Demonstration of PrP^{res} in lymph nodes, tonsil, and conjunctival lymphoid tissues is useful in antemortem diagnosis of sheep scrapie (Ikegami et al. 1991; Schreuder et al. 1996, 1998; O'Rourke et al. 1998a). These techniques are currently being tested in deer and elk to determine their sensitivity and specificity.

Differential Diagnoses. Differential diagnoses of CWD in deer and elk include a wide variety of diseases that cause CNS disease and emaciation. Animals with brain abscesses, traumatic injuries, encephalitis, meningitis, peritonitis, pneumonia, arthritis, starvation, nutritional deficiencies, and dental attrition have been submitted to laboratories as CWD suspects. Aspiration pneumonia is often seen as a terminal event in deer and elk with CWD and, when it is recognized in a prime-aged cervid, CWD should be considered.

Immunity. There is no known immune response to the CWD agent. In sheep and mice, PrP genotype plays a major role in development of scrapie. There is marked homology between mule deer, white-tailed deer, and elk PrP gene sequences (Cérvenakova et al. 1997; K. O'Rourke personal communication). Polymorphism was detected in mule deer (codon 138, serine or asparagine) (Cérvenakova et al. 1997; O'Rourke et al. 1997a), white-tailed deer (K. O'Rourke personal communication), and elk [codon 132 (129), methionine or leucine] (Cérvenakova et al. 1997; Schatzl et al. 1997; O'Rourke et al. 1998b). It is not yet known whether particular PrP genotypes confer resistance or increase susceptibility to CWD; however, codon 132 methionine homozygotes were overrepresented in free-ranging and captive CWD-affected elk when compared to unaffected elk (O'Rourke et al. 1999).

Control and Treatment. There is no known treatment for animals affected with CWD, and it is considered 100% fatal once clinical signs develop. If an affected animal develops pneumonia, treatment with antibiotics might prolong the course of illness but will not alter the fatal outcome.

Control of CWD is problematic. In the face of long incubation periods, subtle early clinical signs, absence of reliable antemortem diagnostic tests, extremely resistant infectious agent, possible environmental contamination, and lack of understanding of transmission, designing methods for control or eradication of CWD is extremely difficult. Management currently involves quarantine or depopulation of CWD-affected herds. Two attempts to eradicate CWD from captive cervid facilities failed, though the cause of the failure was not determined; residual environmental contamination following facility cleanup was possible (Williams and Young 1992).

Management of CWD in free-ranging animals is even more problematic. Long-term active surveillance to determine distribution and prevalence of CWD has been instituted to assist in evaluating changes over time and effect of management intervention. Translocation and artificially feeding cervids in the endemic areas has been banned in an attempt to limit range expansion and to decrease transmission of CWD. Localized population reduction in areas of high CWD prevalence is being considered.

Public Health Concerns. No cases of human disease have been associated with CWD. There is a long history of human exposure to scrapie through handling and eating sheep tissues, including brain, yet there is no evidence that this presents a risk to human health. However, in the absence of complete information and in consideration of the similarities of animal and human TSEs, hunters harvesting deer and elk in the endemic areas or meat processors and taxidermists handling cervid carcasses should take some commonsense measures to avoid exposure to the agent and to other zoonotic pathogens. Sick animals should not be harvested for consumption; hunters, game-meat processors, and taxidermists should wear latex or rubber gloves when dressing a deer or elk from these areas; and the brain, spinal cord, lymph nodes, spleen, tonsils, and eyes should be discarded and not consumed, because these organs probably contain the greatest amount of CWD agent. Since TSE agents have never been demonstrated in skeletal muscle, boning game meat is an effective way to reduce the potential for exposure.

Management Implications. The presence of CWD in captive and free-ranging cervids is a serious management problem. Captive populations are quarantined, which limits usefulness and value of the animals for research or commerce. Indemnity for depopulated cervids currently is not available. Guidelines for management of captive herds with CWD are being developed by federal, state, and provincial animal health officials.

Implications for free-ranging populations of deer and elk are significant. Deer and elk are not translocated from CWD-endemic areas, surveillance programs are expensive for wildlife management agencies, and the impact of the disease on the population dynamics of deer and elk is not currently known. Preliminary modeling suggests that CWD could detrimentally affect populations in endemic areas (M.W. Miller unpublished). Public and agency concerns and perceptions about human health risks associated with all the TSEs may ultimately influence management of herds of free-ranging cervids in the endemic areas.

BOVINE SPONGIFORM ENCEPHALOPATHY IN NONDOMESTIC SPECIES

Distribution and Host Range. Cases of TSE, now recognized as caused by the BSE agent, were diagnosed in ten species of Bovidae and Felidae (Table 17.1) at or from zoological collections in the British Isles (Kirkwood and Cunningham 1994a). Cases occurred in cheetah *Acinonyx jubatus* exported to Australia (Peet and Curran 1992) and France (Baron et al. 1997). A possible case of TSE associated with BSE agent was reported in a rhesus macaque *Macaca mulatta* (Bons et al. 1996); however, this diagnosis has been questioned (Baker et al. 1996). Recently, spongiform encephalopathy associated with oral exposure to the BSE agent was confirmed in captive brown lemurs *Eulemur fluvus* and

TABLE 17.1—Wild mammals reported with naturally occurring transmissible spongiform encephalopathies

Scientific name	Common Name	Disease[a]	References
Bovidae			
Taurotragus oryx	Eland[b]	BSE	Fleetwood and Furley 1990; Kirkwood and Cunningham 1994a
Tragelaphus strepsiceros	Greater kudu[b]	BSE	Kirkwood et al. 1990; Kirkwood and and Cunningham 1994a
Tragelaphus angasii	Nyala[b]	BSE	Jeffrey and Wells 1988
Oryx dammah	Scimitar-horned oryx[b]	BSE	Kirkwood and Cunningham 1994a
Oryx gazella	Gemsbok[b]	BSE	Jeffrey and Wells 1988
Oryx leucoryx	Arabian oryx[b]	BSE	Kirkwood et al. 1990
Bison bison	Bison	BSE	R. Bradley, personal communication
Ovis musimon	Mouflon[b]	Scrapie	Wood et al. 1992
Cervidae			
Odocoileus hemionus	Mule deer[b,c]	CWD	Williams and Young 1980
Odocoileus virginianus	White-tailed deer[b,c]	CWD	Spraker et al. 1997
Cervus elaphus nelsoni	Rocky Mountain elk[b,c]	CWD	Williams and Young 1982
Felidae			
Felis concolor	Cougar[b]	BSE	Willoughby et al. 1992
Felis pardalis	Ocelot[b]	BSE	Kirkwood and Cunningham 1994b
Panthera tigris	Tiger[b]	BSE	Kirkwood and Cunningham 1999
Acinonyx jubatus	Cheetah[b]	BSE	Peet and Curran 1992; Kirkwood and Cunningham 1994b; Kirkwood et al. 1995; Baron et al. 1997
Mustelidae			
Mustela vison	Mink[b]	TME	Hartsough and Burger 1965; Hadlow and Karstad 1968; Hartung et al. 1970

[a]BSE, bovine spongiform encephalopathy; CWD, chronic wasting disease; TME, transmissible mink encepalopathy.
[b]Captive animals.
[c]Free-ranging animals.

a mongoose lemur *Eulemur mongoz* in France (Bons et al. 1999).

Transmission and Epidemiology. The epidemiology of BSE in zoo animals in Great Britain is similar to that of BSE in cattle. The epidemic in cattle arose through the practice of including ruminant-derived protein in cattle feeds (Wilesmith et al. 1988, 1991). It was thought that changes in rendering procedures used in preparing this material resulted in failure to inactivate the agent, which was hypothesized to be a strain of scrapie from sheep. The first clinical cases were diagnosed in cattle in 1986. Subsequent analysis of the epidemic in cattle revealed there must have been widespread exposure to the agent via proprietary feeds starting during the winter of 1980–81 (Wilesmith et al. 1988, 1991).

The cases in zoo animals are thought to have been caused by the BSE agent for three reasons: their temporal and geographic coincidence with the BSE epidemic in cattle; affected zoo animals were either known, or suspected, to have been exposed to contaminated feeds; and the pathology and incubation period of the disease in various strains of mice inoculated with brain homogenates from an affected nyala *Tragelaphus angasii* and a greater kudu *Tragelaphus strepsiceros* were nearly identical to those occurring when mice were inoculated with BSE from cattle (Jeffrey et al. 1992; Bruce et al. 1994). The ungulates were exposed to feeds containing ruminant-derived protein, and the carnivores were exposed to tissues, probably including CNS, from cattle incubating BSE that were considered unfit for human consumption (Kirkwood and Cunningham 1994a,b).

The question of whether BSE is laterally or maternally transmissible in cattle has received vigorous investigation. At present, there is some indication that it is transmissible vertically or by other routes at a low rate (Donnelly et al. 1997a; Wilesmith et al. 1997). The occurrence of cases in greater kudu that were born after the July 1988 ban on inclusion of ruminant-derived protein in ruminant feeds [Her Majesty's Stationery Office (HMSO) 1988] and that were not thought to have been exposed via the diet raised the possibility that lateral transmission might have occurred in this species (Kirkwood et al. 1992, 1994; Cunningham et al. 1993; Kirkwood and Cunningham 1994a). However, the pattern of the epidemic in cattle has since revealed that some degree of feed contamination was present for a considerable period after the July 1988 ban, and the possibility that the kudu were exposed to these feeds cannot be excluded. Because of this, and the fact that no further cases have occurred in this species since 1992, which exceeds the apparent average incubation period of 31 months, it is possible that all the kudu cases could, as in cattle, have been due to ingestion of contaminated feeds.

It seems reasonable that the cases in eland *Taurotragus oryx* (Fleetwood and Furley 1990) and scimitar-horned oryx *Oryx dammah,* born, like some of the kudu, quite long after the July 1988 feed ban, were due to exposure to contaminated feeds. Measures to ensure the exclusion of ruminant-derived protein from feeds were subsequently tightened in the United Kingdom, and there has been a marked decline in the number of cases among cattle, indicating the efficacy of these measures (Donnelly et al. 1997b). Decline in the number of new cases in zoo ungulates during recent years supports this and, although no firm conclusions can be drawn at this stage, provides no evidence for natural transmission between antelope.

Because dates of infection of affected zoo animals were not known, incubation periods of the disease could not be determined precisely. However, data on age at death suggest that incubation periods vary between species, and they are clearly longer in Felidae (62–84 months) than in Bovidae (28–48 months).

Clinical Signs. Clinical signs in zoo animals have been reviewed by Kirkwood and Cunningham (1994a), and a detailed description of clinical signs in one greater kudu has been published (Kirkwood et al. 1994). These include various signs of CNS dysfunction, including ataxia, abnormal head and ear posture, fine muscle tremors, myoclonus, dullness, behavioral changes, excessive lip and tongue movements, and weight loss. In most cases, the disease progressed over weeks, and there was gradual progression of severity but, in some cases, the disease appeared to have a rapid onset and a course of only a few days.

Pathogenesis. The specific pathogenesis of BSE in zoo animals has not been studied. The route of spread of the agent after oral exposure to central nervous and other tissues remains unclear. Although infectivity has been detected in several tissues other than CNS in sheep with scrapie and cattle with BSE, lesions have been observed only in the CNS.

Pathology. The comparative pathology of BSE and the recent cases of spongiform encephalopathy in greater kudu and domestic cats have been reviewed (Wells et al. 1993). In cheetah, spongiform changes involved the entire brain axis, and vacuolation of the neuropil was the most prominent feature (Kirkwood et al. 1995). All the zoo animals that were examined for SAF were positive (Kirkwood and Cunningham 1994a).

Diagnosis. Clinical signs are not specific, but the disease may be strongly suspected in animals that show progressive behavioral changes or ataxia, postural abnormalities, and abnormal muscle movements; the suspect animals reside in or were imported from the United Kingdom or other European countries with endemic BSE; and where there is potential exposure to BSE-contaminated feeds. The disease cannot be confirmed during life, and diagnosis depends on detection of characteristic histopathologic changes and other analyses of CNS material collected at postmortem examination. In addition to detection of SAF, these analyses include immunostaining and immunoblotting techniques for PrP^{res} (Scott et al. 1990). Further confirmation and, possibly, some information about strain type can be obtained by inoculating brain homogenates into panels of various genotypes of mice and studying the incubation periods and lesion profiles (Jeffrey et al. 1992; Bruce et al. 1994, 1997).

Immunity. There is no known immune response to TSE agents. Patchiness of the distribution of cases among taxa in zoo animals suggested variation in susceptibility to the BSE agent among species (Kirkwood and Cunningham 1994a; Kirkwood et al. 1995). However, this remains to be confirmed.

Treatment and Control. No treatment is available to halt, reverse, or delay the development of these diseases. Control of BSE in zoo animals has been discussed (Cunningham 1991; Kirkwood and Cunningham 1992, 1994a). Measures to prevent inclusion of rendered products in feeds (HMSO 1988) for zoo ruminants should effectively control the disease unless vertical or horizontal transmission occurs.

Since September 1990, there has been a statutory ban in the United Kingdom on feeding specified offal (brain, spinal cord, spleen, thymus, tonsils, and intestines) from cattle older than 6 months to any animals (HMSO 1990). This legislation did not preclude feeding tissues from zoo ungulates, but Kirkwood and Cunningham (1994a) considered it advisable not to feed to animals the offals of any species that could have been exposed to BSE. Furthermore, in the absence of information about tissue distribution of the agent in zoo animals, they considered that it would be prudent to avoid using any tissues of zoo animals in BSE-endemic countries as food for others.

Public Health Concerns. There is no known human health risk from zoo ungulates or felids with BSE, because these animals are not part of the human food chain. Contact with clinically affected animals is not considered a health risk, but appropriate protective measures should be taken during postmortem examinations.

Management Concerns. Management implications depend on whether BSE is naturally transmissible among zoo animals. If it is, then introduction of an incubating animal into a population of captive or free-living wild animals is likely to have serious consequences (Cunningham 1991; Kirkwood and Cunningham 1994a). The disease may therefore severely compromise movements between zoological collections for breeding management or for reintroduction to the wild. Animals that could have been exposed to the BSE agent, their offspring, or contacts should not be moved into populations that have not been exposed, unless the damage that this would cause to a conservation breeding program outweighs the risk of introduction of a TSE (Kirkwood and Cunningham 1994a). However, even if there is no risk of spread to conspecifics during life, tissues from affected or carrier animals could present a risk if eaten by other animals. For this reason, it has been recommended that no animal that could have been exposed to the BSE or other TSE agent should be used in reintroduction programs (Cunningham 1991; Kirkwood and Cunningham 1994a).

TRANSMISSIBLE MINK ENCEPHALOPATHY.

Transmissible mink encephalopathy is a rare TSE of ranched mink (Marsh and Hanson 1979); it has never been diagnosed in free-ranging mink. Only a few outbreaks have occurred in North America and Europe (Marsh 1976). The disease is thought to be associated with inadvertently incorporating sheep with scrapie into mink feed (Marsh and Hanson 1979); however, several TME outbreaks were associated with feeding cattle and not sheep (Marsh et al. 1991). This has led to the hypothesis that an unidentified spongiform encephalopathy may be circulating in cattle in the United States (Marsh et al. 1991; Robinson 1996). Neither BSE nor any other bovine TSE has been identified in the United States.

Transmissible mink encephalopathy causes 60%–100% morbidity within a population and 100% mortality of affected mink during outbreaks (Robinson 1996). Animals show behavioral changes and become aggressive, ataxic, and carry their tail over their backs, until they become somnolent, moribund, and die. The disease is not transmissible among affected animals except occasionally by bite wounds or cannibalism (Marsh and Hanson 1979). The microscopic lesions are qualitatively typical of the TSEs, but the lesions tend to be more severe in the rostral portions of the brain in comparison to the distribution of lesions in ruminants (Eckroade et al. 1979). Transmissible mink encephalopathy has been experimentally transmitted by intracerebral inoculation to cattle (Robinson et al. 1995) and to sheep, goats, and a variety of laboratory species, including primates (Marsh 1976; Hadlow et al. 1987). Striped skunks *Mephitis mephitis* and raccoons *Procyon lotor* were also experimentally susceptible to TME (Eckroade et al. 1973). Transmissible mink encephalopathy may be considered of greatest importance as a model of the TSEs, primarily through the carefully crafted studies of Marsh, Hadlow, and colleagues, rather than as a significant disease of domestic animals or humans. It is of potential management concern to those raising mink but is of no known concern to free-ranging species.

LITERATURE CITED

Bahmanyar, S., E.S. Williams, F.B. Johnson, S. Young, and D.C. Gajdusek. 1985. Amyloid plaques in spongiform encephalopathy of mule deer. *Journal of Comparative Pathology* 95:1–5.

Baker, H.F., R.M. Ridley, G.A. Wells, and J.W. Ironside. 1996. Spontaneous spongiform encephalopathy in a monkey. *Lancet* 348:955–956.

Baron, T., P. Belli, J.Y. Madec, F. Moutou, C. Vitaud, and M. Savey. 1997. Spongiform encephalopathy in an imported cheetah in France. *Veterinary Record* 141:270–271.

Beekes, M., P.A. McBride, and E. Baldauf. 1998. Cerebral targeting indicates vagal spread of infection in hamsters fed with scrapie. *Journal of General Virology* 79:601–607.

Bons, N., N. Mestre-Frances, Y. Charnay, and F. Tagliavini. 1996. Spontaneous spongiform encephalopathy in a young adult rhesus monkey. *Lancet* 348:55.

Bons, N., N. Mestre-Frances, P. Belli, F. Cathala, D.C. Gajdusek, and P. Brown. 1999. Natural and experimental oral infection of nonhuman primates by bovine spongiform encephalopathy agents. *Proceedings of the National Academy of Sciences USA* 96:4046–4051.

Brown, P., and D.C. Gajdusek. 1991. Survival of scrapie virus after 3 years' interment. *Lancet* 337:269–70.

Bruce, M.E., A. Chree, I. McConnell, J. Foster, G. Pearson, and H. Fraser. 1994. Transmission of bovine spongiform encephalopathy and scrapie to mice: Strain variation and species barrier. *Philosophical Transactions of the Royal Society of London [B]* 343:405–411.

Bruce, M.E., R.G. Will, J.W. Ironside, I. McConnell, D. Drummond, A. Suttie, L. McCardle, A. Chree, J. Hope, C. Birkett, S. Cousens, H. Fraser, and C.J. Bostock. 1997. Transmission to mice indicate that 'new variant' CJD is caused by the BSE agent. *Nature* 389:498–501.

Cérvenakova, L., R. Rohwer, E.S. Williams, P. Brown, and D.C. Gajdusek. 1997. High sequence homology of the PrP gene in mule deer and Rocky Mountain elk. *Lancet* 350:219–220.

Chesebro, B. 1998. BSE and prions: Uncertainties about the agent. *Science* 279:42–43.

Collinge, J., and M.S. Palmer. 1994. Molecular genetics of human prion diseases. *Philosophical Transactions of the Royal Society of London [B]* 343:371–378.

———. 1997. Human prion diseases. In *Prion diseases,* ed. J. Collinge and M.S. Palmer. New York: Oxford University Press, pp. 18–56.

Cunningham, A.A. 1991. Bovine spongiform encephalopathy and British zoos. *Journal of Zoo and Wildlife Medicine* 11:605–634.

Cunningham, A.A., G.A.H. Wells, A.C. Scott, J.K. Kirkwood, and J.E.F. Barnett. 1993. Transmissible spongiform encephalopathy in greater kudu (*Tragelaphus strepsiceros*). *Veterinary Record* 132:68.

Dickinson, A.G. 1976. Scrapie in sheep and goats. In *Slow virus diseases of animals and man,* ed. R.H. Kimberlin. Amsterdam: North Holland, pp. 209–241.

Donnelly, C.A., N.M. Ferguson, A.C. Ghani, J.W. Wilesmith, and R.M. Anderson. 1997a. Analysis of dam-calf pairs of BSE cases: Confirmation of a maternal risk enhancement. *Proceedings of the Royal Society of London [B]* 264:1647–1656.

Donnelly, C.A., A.C. Ghani, N.M. Ferguson, and R.M. Anderson. 1997b. Recent trends in the BSE epidemic. *Nature* 389:903.

Eckroade, R.J., G.M. Zu Rhein, and R.P. Hanson. 1973. Transmissible mink encephalopathy in carnivores: Clinical, light and electron microscopic studies in raccoons, skunks and ferrets. *Journal of Wildlife Diseases* 9:229–240.

———. 1979. Experimental transmissible mink encephalopathy: Brain lesions and their sequential development. In *Slow transmissible diseases of the nervous system,* vol. 1, ed. S.B. Prusiner and W.J. Hadlow. New York: Academic, pp. 409–449.

Farquhar, C.F., R.A. Somerville, and M.E. Bruce. 1998. Straining the prion hypothesis. *Nature* 391:345–346.

Fleetwood, A.J., and C.W. Furley. 1990. Spongiform encephalopathy in an eland. *Veterinary Record* 126:408–409.

Greig, J.R. 1940. Scrapie: Observations on the transmission of the disease by mediate contact. *Veterinary Journal* 96:203–206.

Guiroy, D.C., E.S. Williams, R. Yanagihara, and D.C. Gajdusek. 1991a. Immunolocalization of scrapie amyloid (PrP27–30) in chronic wasting disease of Rocky Mountain elk and hybrids of captive mule deer and white-tailed deer. *Neuroscience Letters* 126:195–198.

———. 1991b. Topographic distribution of scrapie amyloid-immunoreactive plaques in chronic wasting disease in captive mule deer (*Odocoileus hemionus hemionus*). *Acta Neuropathologica (Berlin)* 81:475–478.

Guiroy, D.C., E.S. Williams, P.P. Liberski, I. Wakayama, and D.C. Gajdusek. 1993. Ultrastructural neuropathology of chronic wasting disease in captive mule deer. *Acta Neuropathologica (Berlin)* 85:437–444.

Guiroy, D.C., P.P. Liberski, E.S. Williams, and D.C. Gajdusek. 1994. Electron microscopic findings in brain of Rocky Mountain elk with chronic wasting disease. *Folia Neuropathology* 32:171–173.

Hadlow, W.J. 1996. Differing neurohistologic images of scrapie, transmissible spongiform encephalopathy, and chronic wasting disease of mule deer and elk. In *Bovine spongiform encephalopathy: The BSE dilemma,* ed. C.J. Gibbs Jr. New York: Springer-Verlag, pp. 122–137.

Hadlow, W.J., and L. Karstad. 1968. Transmissible encephalopathy of mink in Ontario. *Canadian Veterinary Journal* 9:193–196.

Hadlow, W.J., R.C. Kennedy, R.E. Race, and C.M. Eklund. 1980. Virologic and neurohistologic findings in dairy goats affected with natural scrapie. *Veterinary Pathology* 17:187–199.

Hadlow, W.J., R.C. Kennedy, and R.E. Race. 1982. Natural infection of Suffolk sheep with scrapie virus. *Journal of Infectious Disease* 146:657–664.

Hadlow, W.J., R.E. Race, and R.C. Kennedy. 1987. Experimental infection of sheep and goats with transmissible mink encephalopathy virus. *Canadian Journal of Veterinary Research* 51:135–144.

Hartsough, G.R., and D. Burger. 1965. Encephalopathy of mink: I. Epizootiologic and clinical observations. *Journal of Infectious Diseases* 115:387–392.

Hartung, J., H. Zimmermann, and U. Johannsen. 1970. Infectious encephalopathy in the mink: I. Clinical, epizootiological and experimental studies. *Monatshefte für Veterinärmedizin* 15:385–388.

Hegde, R.S., J.A. Mastrianni, M.R. Scott, K.A. de Fea, P. Tremblay, M. Torchia, S.J. de Armond, S.B. Prusiner, and V.R. Lingappa. 1998. A transmembrane form of the prion protein in neurodegenerative disease. *Science* 279:827–834.

Her Majesty's Stationery Office (HMSO). 1988. *The bovine spongiform encephalopathy order 1988.* Statutory Instrument Number 1039. London: HMSO.

———. 1990. *Bovine spongiform encephalopathy (No. 2).* Amendment Order 1990. London: HMSO.

Hope, J., L.J. Reekie, N. Hunter, G. Multhaup, K. Beyreuther, H. White, A.C. Scott, M.J. Stack, M. Dawson, and G.A. Wells. 1988. Fibrils from brains of cows with new cattle disease contain scrapie-associated protein. *Nature* 336:390–392.

Hunter, N., J.D. Foster, and J. Hope. 1992. Natural scrapie in British sheep: Breeds, ages and PrP gene polymorphisms. *Veterinary Record* 130:389–392.

Ikegami, Y., M. Ito, H. Isomura, E. Momotani, K. Sasaki, Y. Muramatsu, N. Ishiguro, and M. Shinagawa. 1991. Preclinical and clinical diagnosis of scrapie by detection of PrP protein in tissues of sheep. *Veterinary Record* 128:271–275.

Jeffrey, M., and G.A.H. Wells. 1988. Spongiform encephalopathy in a nyala (*Tragelaphus angasi*). *Veterinary Pathology* 25:398–399.

Jeffrey, M., J.R. Scott, A. Williams, and H. Fraser. 1992. Ultrastructural features of spongiform encephalopathy transmitted to mice from three species of Bovidae. *Acta Neuropathologica (Berlin)* 84:559–569.

Jeffrey, M., C.M. Goodsir, M.E. Bruce, P.A. McBride, and J.R. Fraser. 1997. In vivo toxicity of prion protein in murine scrapie: Ultrastructural and immunogold studies. *Neuropathology and Applied Neurobiology* 23:93–101.

Kelly, D.F., H. Pearson, A.I. Wright, and L.W. Greenham. 1980. Morbidity in captive white tigers. In *The comparative pathology of zoo animals,* ed. R.J. Montali and G. Migaki. Front Royal, VA: Smithsonian Institution, pp. 183–188.

Kimberlin, R.H., and C.A. Walker. 1989. Pathogenesis of scrapie in mice after intragastric infection. *Virus Research* 12:213–220.

Kirkwood, J.K., and A.A. Cunningham. 1992. Spongiform encephalopathy in zoo ungulates: Implications for translocation and reintroduction. In *Proceedings of the conference of the American Association of Zoo Veterinarians and the American Association of Wildlife Veterinarians, Oakland, California,* ed. R.E. Junge. Oakland, CA: American Association of Zoo Veterinarians, pp. 26–27.

———. 1994a. Epidemiological observations on spongiform encephalopathies in captive wild animals in the British Isles. *Veterinary Record* 135:296–303.

———. 1994b. Patterns of incidence of spongiform encephalopathy in captive wild animals in the British Isles. In *Wildlife Disease Association, European Division, Symposium, Paris,* pp. 28.

———. 1999. Scrapie-like spongiform encephalopathies (prion diseases) in nondomesticated species. In *Zoo and wild animal medicine,* ed. M.E. Fowler and R.E. Miller. Philadelphia: W.B. Saunders, pp. 662–668.

Kirkwood, J.K., G.A.H. Wells, J.W. Wilesmith, A.A. Cunningham, and S.I. Jackson. 1990. Spongiform encephalopathy in an Arabian oryx (*Oryx leucoryx*) and a greater kudu (*Tragelaphus strepsiceros*). *Veterinary Record* 127:418–420.

Kirkwood, J.K., G.A.H. Wells, A. A. Cunningham, S.I. Jackson, A.C. Scott, M. Dawson, and J.W. Wilesmith. 1992. Scrapie-like encephalopathy in a greater kudu (*Tragelaphus strepsiceros*) which had not been fed ruminant-derived protein. *Veterinary Record* 130:365–367.

Kirkwood, J.K., A.A. Cunningham, A.R. Austin, G.A.H. Wells, and A.W. Sainsbury. 1994. Spongiform encephalopathy in a greater kudu (*Tragelaphus strepsiceros*) introduced into an affected group. *Veterinary Record* 134:167–168.

Kirkwood, J.K., A.A. Cunningham, E.J. Flach, S.M. Thornton, and G.A.H. Wells. 1995. Spongiform encephalopathy in another captive cheetah (*Acinonyx jubatus*): Evidence for variation in susceptibility or incubation periods between species? *Journal of Zoo and Wildlife Medicine* 26:577–582.

Klein, M.A., R. Frigg, E. Flechsig, A.J. Raeber, U. Kalinke, H. Bluethmann, F. Bootz, M. Suter, R.M. Zinkernagel, and A. Aguzzi. 1997. A crucial role for B cells in neuroinvasive scrapie. *Nature* 390:687–690.

Kocisko, D.A., J.H. Come, S.A. Priola, B. Chesebro, G.J. Raymond, P.T. Lansbury, and B. Caughley. 1994. Cell-free formation of protease-resistant prion protein. *Nature* 370:471–474.

Liberski, P.P., D.C. Guiroy, E.S. Williams, R. Yangihara, P. Brown, and D.C. Gajdusek. 1993. The amyloid plaque. In *Light and microscopic neuropathology of slow virus disorders,* ed. P.P. Liberski. Boca Raton, FL: CRC, pp. 295–347.

Marsh, R.F. 1976. The subacute spongiform encephalopathies. *Frontiers of Biology* 44:359–380.

Marsh, R.F., and R.P. Hanson. 1979. On the origin of transmissible mink encephalopathy. In *Slow transmissible diseases of the nervous system,* vol. 1, ed. S.B. Prusiner and W.J. Hadlow. New York: Academic, pp. 451–460.

Marsh, R.F., R.A. Bessen, S. Lehmann, and G.R. Hartsough. 1991. Epidemiological and experimental studies on a new incident of transmissible mink encephalopathy. *Journal of General Virology* 72:589–594.

McGill, I.S., and G.A. Wells. 1993. Neuropathological findings in cattle with clinically suspect but histologically unconfirmed bovine spongiform encephalopathy (BSE). *Journal of Comparative Pathology* 108:241–260.

Merz, P.A., R.G. Rohwer, K. Kascsak, H.M. Wisniewiski, R.A. Somerville, C.J. Gibbs Jr., and D.C. Gajdusek. 1984. Infection-specific particle from the unconventional slow virus diseases. *Science* 225:437–440.

Miller, M.W., M.A. Wild, and E.S. Williams. 1998. Epidemiology of chronic wasting disease in Rocky Mountain elk. *Journal of Wildlife Diseases* 34:532–538.

Miller, M.W., E.S. Williams, C.W. McCarty, T.R. Spraker, T.J. Kreeger, C.T. Larsen, and E.T. Thorne. 2000. Epidemiology of chronic wasting disease in free-ranging cervids. *Journal of Wildlife Diseases.* 36:676–690.

Millison, G.C., G.D. Hunter, and R.H. Kimberlin. 1976. The physico-chemical nature of the scrapie agent. In *Slow virus diseases of animals and man,* ed. R.H. Kimberlin. Amsterdam: North Holland, pp. 243–266.

O'Rourke, K., T.R. Spraker, M.W. Miller, and E.S. Williams. 1997a. *Three alleles of the prion protein gene in mule deer (*Odocoileus hemionus hemionus*) with chronic wasting disease.* GenBank, Accession nos. 2213811, 2213938, 2213936, http://www.ncbi.nlm.gov/Entrez/protein.html.

O'Rourke, K., G.R. Holyoak, W.W. Clark, J.R. Mickelson, S. Wang, R.P. Melco, T.E. Besser, and W.C. Foote. 1997b. PrP genotypes and experimental scrapie in orally inoculated Suffolk sheep in the United States. *Journal of General Virology* 78:975–978.

O'Rourke, K., T.V. Baszler, S.M. Parish, and D.P. Knowles. 1998a. Preclinical detection of PrPsc in nictitating membrane lymphoid tissue of sheep. *Veterinary Record* 142:489–491.

O'Rourke, K., T.V. Baszler, J.M. Miller, T.R. Spraker, I. Sadler-Riggleman, and D.P. Knowles. 1998b. Monoclonal antibody F89/160.1.5 defines a conserved epitope on the ruminant prion protein. *Journal of Clinical Microbiology* 36:1750–1755.

O'Rourke, K.I., T.E. Besser, M.W. Miller, T.F. Cline, T.R. Spraker, A.L. Jenny, M.A. Wild, G.L. Zebarth, and E.S. Williams. 1999. PrP genotypes of captive and free-ranging Rocky Mountain elk (*Cervus elaphus nelsoni*) with chronic wasting disease. *Journal of General Virology* 80:2765–2769.

Pálsson, P.A. 1979. Rida (scrapie) in Iceland and its epidemiology. In *Slow transmissible diseases of the nervous system,* vol. 1, ed. S.B. Prusiner and W.J. Hadlow. New York: Academic, pp. 357–366.

Pearson, G.R., J.M. Wyatt, T.J. Gruffydd-Jones, J. Hope, A. Chong, R.J. Higgins, A.C. Scott, and G.A. Wells. 1992. Feline spongiform encephalopathy: Fibril and PrP studies. *Veterinary Record* 131:307–310.

Peet, R.L., and J.M. Curran. 1992. Spongiform encephalopathy in an imported cheetah (*Acinonyx jubatus*). *Australian Veterinary Journal* 69:117.

Prusiner, S.B. 1982. Novel proteinaceous infectious particles cause scrapie. *Science* 216:136–144.

———. 1991. Molecular biology of prion diseases. *Science* 252:1515–1522.

———. 1997. Prion diseases and the BSE crisis. *Science* 278:245–251.

Prusiner, S.B., and W.J. Hadlow, eds. 1979. *Slow transmissible diseases of the nervous system,* vol. 1. New York: Academic, 472 pp.

Raymond, G.J., J. Hope, D.A. Kocisko, S.A. Priola, L.D. Raymond, A. Bossers, J. Ironside, R.G. Will, S.G. Chen, R.B. Petersen, P. Gambetti, R. Rubenstein, M.A. Smits, P.T. Lansbury Jr., and B. Caughey. 1997. Molecular assessment of the potential transmissibilities of BSE and scrapie to humans. *Nature* 388:285–288.

Robinson, M.M. 1996. An assessment of transmissible mink encephalopathy as an indicator of bovine scrapie in U.S. cattle. In *Bovine spongiform encephalopathy: The BSE dilemma,* ed. C.J. Gibbs Jr. New York: Springer-Verlag, pp. 97–107.

Robinson, M.M., W.J. Hadlow, D.P. Knowles, T.P. Huff, P.A. Lacy, R.F. Marsh, and J.R. Gorham. 1995. Infection of cattle with the agents of TME and scrapie. *Journal of Comparative Pathology* 113:241–251.

Sakaguchi, S., S. Katamine, N. Nishida, R. Moriuchi, K. Shigematsu, T. Sugimoto, A. Nakatani, Y. Kataoka, T. Houtani, S. Shirabe, H. Okada, S. Hasegawa, T. Miyamoto, and T. Noda. 1996. Loss of cerebellar Purkinje cells in aged mice homozygous for a disrupted PrP gene. *Nature* 380:528–531.

Schatzl, H.M., F. Wopfner, S. Gilch, A. von Brunn, and G. Jager. 1997. Is codon 129 of prion protein polymorphic in human beings but not in animals? *Lancet* 349:219–220.

Schoon, H.-A., D. Brunckhorst, and J. Pohlenz. 1991. Spongiform encephalopathy in a red-necked ostrich [in German]. *Tierärztliche Praxis* 19:263–265. Abstract.

Schreuder, B.E.C., L.J.M. van Keulen, M.E.W. Vromans, J.P.M. Langeveld, and M.A. Smits. 1996. Preclinical test for prion diseases. *Nature* 381:563.

———. 1998. Tonsillar biopsy and PrPsc detection in the preclinical diagnosis of scrapie. *Veterinary Record* 142:564–568.

Scott, A.C., G.A.H. Wells, M.J. Stack, H. White, and M. Dawson. 1990. Bovine spongiform encephalopathy: Detection and quantitation of fibrils, fibril protein (PrP) and vacuolation in brain. *Veterinary Microbiology* 23:295–305.

Sigurdson, C.J., E.S. Williams, M.W. Miller, T.R. Spraker, K.I. O'Rourke, and E.A. Hoover. 1999. Oral transmission and early lymphoid tropism of chronic wasting disease PrP^{res} in mule deer fawns (*Odocoileus hemionus*). *Journal of General Virology* 80:2757–2764.

Spraker, T.R., M.W. Miller, E.S. Williams, D.M. Getzy, W.J. Adrian, G.G. Schoonveld, R.A. Spowart, K.I. O'Rourke, J.M. Miller, and P.A. Merz. 1997. Spongiform encephalopathy in free-ranging mule deer (*Odocoileus hemionus*), white-tailed deer (*Odocoileus virginianus*), and Rocky Mountain elk (*Cervus elaphus nelsoni*) in northcentral Colorado. *Journal of Wildlife Diseases* 33:1–6.

Taylor, D.M., S.L. Woodgate, and M.J. Atkinson. 1995. Inactivation of the bovine spongiform encephalopathy agent by rendering procedures. *Veterinary Record* 137:605–610.

Tobler, I., S.E. Gaus, T. Deboer, P. Achermann, M. Fischer, T. Rulicke, M. Moser, B. Oesch, P.A. McBride, and J.C. Manson. 1996. Altered circadian activity rhythms and sleep in mice devoid of prion protein. *Nature* 380:639–642.

Wells, G.A., and I.S. McGill. 1992. Recently described scrapie-like encephalopathies of animals: Case definitions. *Research in Veterinary Science* 53:1–10.

Wells, G.A., A.C. Scott, C.T. Johnson, R.F. Gunning, R.D. Hancock, M. Jeffrey, M. Dawson, and R. Bradley. 1987. A novel progressive spongiform encephalopathy in cattle. *Veterinary Record* 121:419–420.

Wells, G.A., S.A.C. Hawkins, A.A. Cunningham, W.H. Blamire, J.W. Wilesmith, A.R. Sayers, and P. Harris. 1993. Comparative pathology of the new transmissible spongiform encephalopathies. In *Transmissible spongiform encephalopathies,* ed. R. Bradley and B. Marchant. Brussels: European Commission, pp. 327–345.

Wells, G.A., S.A.C. Hawkins, R.B. Green, A.R. Austin, I. Dexter, Y.I. Spencer, M.J. Chaplin, M.J. Stack, and M. Dawson. 1998. Preliminary observations on the pathogenesis of experimental bovine spongiform encephalopathy (BSE): An update. *Veterinary Record* 142:103–106.

Wilesmith, J.W. 1994. An epidemiologist's view of bovine spongiform encephalopathy. *Philosophical Transactions of the Royal Society of London [B]* 343:357–361.

Wilesmith, J.W., G.A.H. Wells, M.P. Cranwell, and J.B.M. Ryan. 1988. Bovine spongiform encephalopathy: Epidemiological studies. *Veterinary Record* 123:638–644.

Wilesmith, J.W., J.B.M. Ryan, and M.J. Atkinson. 1991. Bovine spongiform encephalopathy: Epidemiological studies on the origin. *Veterinary Record* 128:199–203.

Wilesmith, J.W., G.A. Wells, J.B. Ryan, D. Gavier-Widen, and M.M. Simmons. 1997. A cohort study to examine maternally-associated risk factors for bovine spongiform encephalopathy. *Veterinary Record* 141:239–43.

Will, R.G., J.W. Ironside, M. Zeidler, S.N. Cousens, K. Estibeiro, A. Alperovitch, S. Poser, M. Pocchiari, A. Hofman, and P.G. Smith. 1996. A new variant of Creutzfeldt-Jakob disease in the UK. *Lancet* 347:921–925.

Williams, A., P.J. Lucassen, D. Ritchie, and M. Bruce. 1997. PrP deposition, microglial activation, and neuronal apoptosis in murine scrapie. *Experimental Neurology* 144:433–438.

Williams, E.S., and S. Young. 1980. Chronic wasting disease of captive mule deer: A spongiform encephalopathy. *Journal of Wildlife Diseases* 16:89–98.

———. 1982. Spongiform encephalopathy of Rocky Mountain elk. *Journal of Wildlife Diseases* 18:465–471.

———. 1992. Spongiform encephalopathies of Cervidae. *Scientific and Technical Review Office of International Epizootics* 11:551–567.

———. 1993. Neuropathology of chronic wasting disease of mule deer (*Odocoileus hemionus*) and elk (*Cervus elaphus nelsoni*). *Veterinary Pathology* 30:36–45.

Willoughby, K., D.F. Kelly, D.G. Lyon, and G.A.H. Wells. 1992. Spongiform encephalopathy in a captive puma (*Felis concolor*). *Veterinary Record* 131:431–434.

Wood, J.L.N., L.J. Lund, and S.H. Done. 1992. The natural occurrence of scrapie in moufflon. *Veterinary Record* 130:25–27.

PART II Bacterial and Mycotic Diseases

18 TULAREMIA

TORSTEN MÖRNER AND EDWARD ADDISON

Synonyms: Tularemia, rabbit fever, hare plague, lemming fever, deerfly fever, Ohara's disease, yato-byo.

HISTORY. While seeking foci of plague (*Yersinia pestis*) in ground squirrels *Spermophilus beecheyi* in Tulare County, California, McCoy (1911) described "A plague-like disease of rodents." The causative organism was isolated in 1912 (McCoy and Chapin 1912) and named *Bacterium tularense* (now *Francisella tularensis*) for the county in which the disease was recognized. Francis (1921) proposed the name tularemia for the disease.

Although tularemia was first reported in the United States in 1911, it probably was observed elsewhere prior to that time. Pollitzer (1963) cites authors who, in the late 19th century, reported outbreaks in Western Siberia of what was believed to be anthrax but most probably was tularemia, whereas in Europe, Horne (1911) reported a disease in lemmings in Norway that almost certainly was tularemia. In 1820 in Japan, a disease was described in people who had eaten rabbit meat. It was later identified as tularemia and named Ohara's disease (Ohara 1954).

See Hopla (1974), Jellison (1974), and Bell (1980) for more detailed historical accounts of tularemia, and Hopla and Hopla (1994) for further references.

DISTRIBUTION AND HOST RANGE. The adaptation of *F. tularensis* to different species of lagomorphs and rodents, and the distribution of the disease, indicate that it may have coevolved with lagomorphs and rodents of the Northern Hemisphere (Olsuf'ev 1959).

The disease is known throughout the continental United States, as well as in Alaska, Canada, and Mexico (Bell and Riley 1981). Tularemia has been reported in almost all parts of the former USSR (Pfahler-Jung 1989). In Europe, tularemia is reported to have spread westward from the former USSR in three waves—the first up to 1928, the second between 1929 and 1938, and the third after 1940 (Jusatz 1952)—although this is difficult to reconcile with the view that tularemia has been established in most of the Northern Hemisphere for a very long time (Hopla 1974). Tularemia has been reported in most countries in Central Europe but apparently does not occur naturally in Spain, Portugal, Ireland, and Great Britain (Pfahler-Jung 1989). The disease has been diagnosed only in people in the Netherlands and Denmark.

In Asia, tularemia is known in Turkey, Burma, China and Japan, Iran, Afghanistan, and Lebanon (Bell and Riley 1981; Pfahler-Jung 1989). In Africa, tularemia has been reported from Tunisia and Senegal (Jusatz 1961).

Francisella tularensis has one of the broadest host ranges of all bacteria, having been reported from 190 species of mammals, 23 species of birds, 3 species of amphibians, and 88 species of invertebrates up to 1989 (Hopla 1974; Pfahler-Jung 1989). Most species of mammals known to be infected up to 1980 are tabulated by Bell and Riley (1981).

Despite the broad host range, tularemia is primarily a disease of lagomorphs and rodents (Hopla and Hopla 1994). These species possess a combination of qualities that predispose them to the disease, principally high susceptibility and moderate to high sensitivity to infection with *F. tularensis* (see the section "Pathogenesis"), often combined with an adapted tick fauna that vectors transmission of the agent (Hopla and Hopla 1994).

Among lagomorphs, hares *Lepus* spp. and New World rabbits *Sylvilagus* spp. are the most important hosts for *F. tularensis,* whereas the Old World rabbit *Oryctolagus cuniculus* is relatively resistant to infection. Important hosts among rodents are water voles *Arvicola* spp., muskrat *Ondatra zibithecus,* lemmings *Lemmus* spp., voles *Microtus* spp., hamsters *Cricetus cricetus,* and red-backed voles *Clethrionomys* spp. (Hopla and Hopla 1994).

In North America, tularemia in lagomorphs most commonly involves cottontail rabbits *Sylvilagus* sp., black-tailed jackrabbits *Lepus californicus,* and snowshoe hares *Lepus americanus* (Jellison 1974; Hopla and Hopla 1994), whereas among rodents, the disease is most frequently found in beaver *Castor canadensis* and muskrat (Jellison 1974).

In Central Europe, the disease is most common in European brown hares *Lepus europaeus* (Kemenes 1976; Sterba and Krul 1985) or in small rodents (Jusatz 1961; Borcic et al. 1978). In France, tularemia has been reported as the cause of death in 3% of the European brown hares (Lamarque et al. 1996). Tularemia in Northern Europe and Fennoscandia occurs most frequently among varying hares *Lepus timidus* (Mörner et al. 1988a), although it is seen occasionally in many other species, such as European brown hares (Mörner 1994), squirrels *Sciurus vulgaris,* lemmings *Lemmus lemmus* (Olin 1942), and other small rodents (Myrbäck et al. 1968; Borg et al. 1969; Omland et al. 1977).

In the former USSR, foci of tularemia are maintained by a few commonly infected species, namely water voles, common voles *Microtus arvalis,* hares *L. timidus, L. europaeus, L. tolai,* house mice *Mus musculus,* muskrats, hamsters, and common shrews *Sorex araneus* (Olsuf'ev 1974).

Although wild mammals other than rodents and lagomorphs are relatively insignificant in the epidemiology of tularemia, an outbreak has been reported in wild gray foxes *Urocyon cinereoargenteus* (Schlotthauer et al. 1935), and people have acquired infection from wild ursids (Chase et al. 1980), canids (Anonymous 1980), and cervids (Emmons et al. 1976). Birds are relatively resistant to infection with *F. tularensis,* and their epidemiologic role generally is minor. However, they may act as important hosts for some of the major tick vectors, and particularly gallinaceous species such as grouse may be a source of human infection (Hopla and Hopla 1994).

ETIOLOGY. *Francisella tularensis* is a Gram-negative, pleomorphic rod, 0.2 × 0.2–0.7 μm in size (Eigelsbach and McGann 1984). At various times, it has been referred to the genera *Bacterium, Pasteurella, Brucella,* and others (Jellison 1974).

At least two subspecies are recognized: *Francisella tularensis* biovar *tularensis* (syn. *nearctica;* also known as type A) and *Francisella tularensis* biovar *palaearctica* (syn. *holarctica;* type B), as proposed by Olsuf'ev et al. (1959). *Francisella t. tularensis,* which ferments glycerol and is citrulline ureidase positive, is highly virulent for humans and domestic rabbits, and, until recently, was considered to be present only in North America. *Francisella t. palaearctica,* which does not ferment glycerol and is citrulline ureidase negative, is less virulent for humans and domestic rabbits and is found in both Eurasia and North America (Jellison 1974).

Two additional subdivisions of *F. t. palaearctica* have been proposed (Olsuf'ev 1970): *F. t. palaearctica japonica* for the organism from Japan, and *F. t. palaearctica mediaasiatica* for an organism from Central Asia. More recently, these have been demonstrated to share significant genetic homology with *F. t. tularensis,* rather than with *F. t. palaearctica* (Forsman et al. 1990), implying that the type-A organism is not limited to North America (Hopla and Hopla 1994).

Francisella tularensis is an obligate intracellular parasite (Hahn and Kaufmann 1981) and seems not to produce potent exotoxins (Skrodzki 1968). The presence of endotoxin is uncertain but has been reported (Finegold et al. 1969).

TRANSMISSION AND EPIDEMIOLOGY. *Francisella tularensis* is a highly infectious agent that can enter the body in several ways: inoculated by blood-feeding arthropod vectors directly into the blood or tissues of the host; by contact with blood or tissues of infected animals, through intact or lacerated skin, or through ocular mucous membranes; by inhalation of infected aerosols or particles; or by ingestion of contaminated water or meat (Jellison 1974).

Blood-feeding arthropods that may act as vectors include mosquitoes, fleas, tabanid flies (mainly deerflies *Chrysops* spp.), and ticks (Hopla 1974; Jellison 1974; Hopla and Hopla 1994).

The species of tick vectors vary geographically. In North America, these are principally *Dermacentor vari-*

abilis, D. andersoni, D. parumpertis and *D. occidentalis; Haemaphysalis leporis-palustris; Amblyomma americanum;* and *Ixodes dentatus;* though others may be implicated (Hopla 1974; Jellison 1974). *Haemaphysalis leporis-palustris* and *I. dentatus* are highly associated with lagomorphs and birds, and probably play a major role in transmission among wildlife (Bell 1980). However, they are not anthropophilic, so they are rarely transmit infection to people. In Eurasia, *D. pictus, D. marginatus, I. ricinus,* and *Rhipicephalus rossica* are important, while in Japan, *H. flavis, I. japonensis,* and *I. nipponensis* are vectors of tularemia (Hopla 1974).

Transstadial transmission of *F. tularensis* is required for vector competence, and nymphs, infected while feeding as larvae, are significant in the vector role. Transovarial transmission may occur in some species of ticks, resulting in infected unfed larvae and enhancing vector capacity, but it seems to be exceptional (Hopla and Hopla 1994).

The bacteria can be transmitted among voles *Microtus subarvalis* via consumption of infected corpses (Olsuf'ev and Shlygina 1979; Shlygina and Olsuf'ev 1982), with potential for some voles to survive and be carriers (Olsuf'ev and Shlygina 1979). Large numbers of bacteria can be excreted in the urine of infected voles (Bell and Stewart 1975, 1983; Mörner 1994), and voles may be reservoirs for maintenance of tularemia between epidemics (Shlygina et al. 1987). People have acquired tularemia from contaminated water (Jellison et al. 1950), and *F. tularensis* has been isolated from water and mud in endemic areas, even when there is no evidence of disease in animals (Parker et. al. 1951; Ul'yanova et al. 1982).

The extent to which rodents, mud, or water are sources for initiation of epidemics, and media for the survival of bacteria in the wild between epidemics, remains unclear. Previous studies have demonstrated experimentally and, to a limited extent, in the wild, that some of these scenarios are possible. Survival of some highly sensitive voles *Microtus rossiaemeridionalis* following experimental infection, with bacteriuria in some surviving voles (Olsuf'ev et al. 1984), and evidence of prior exposure and survival of wild cottontail rabbits (Lepitzki et al. 1990; Woolf et al. 1993; Shoemaker et al. 1997) have prompted broader perspectives on the fate of exposed susceptible wildlife.

Some species, such as voles, which are susceptible to infection but which do not succumb quickly to disease, may be important reservoirs of infection in nature. The presence of *F. tularensis* in the ecosystem becomes evident following transmission to more sensitive species, probably mainly by contamination of water. Increased prevalence of *F. tularensis* infection in voles, largely unnoticed, may underlie the outbreaks of disease in species such as muskrats and beaver (Hopla and Hopla 1994).

Airborne infection may occur as the result of aerosolization of infected droplets or from inhalation of dust contaminated with rodent tissues, urine, or feces (see Bell 1980 and Syrjälä et al. 1985).

The epidemiology of tularemia varies worldwide, depending on which *Francisella* subspecies and host animal species are involved and the mode of transmission within the local ecosystem.

In North America, two main systems of tularemia are recognized. The first is caused by *F. t. tularensis* (type A) and is closely associated with the cottontail rabbit, ticks, and biting flies (Jellison 1974). Tularemia may act as a regulatory mechanism in rabbit populations (Woolf et al. 1993), though infection need not be fatal (Lepitzki et al. 1990; Shoemaker et al. 1997). This system of tularemia is responsible for about 70% of human cases in the United States but does not seem to occur in Canada (Jellison 1974). *Francisella t. tularensis* is highly virulent, with a mortality rate of about 5%–7% in untreated people. Human cases usually occur during summer from tick bites (Taylor et al. 1991) or from handling rabbits during the fall and winter rabbit-hunting season (Yeatter and Thompson 1952).

The second system is caused by *F. t. palaearctica* (type B), which occurs mainly in the northern United States and in Canada (Jellison 1974). The species most affected are muskrat and beaver. Transmission is presumably waterborne, and the disease may be seen as either widespread epidemics or localized focal outbreaks. Human cases occur during winter trapping seasons, usually in association with handling carcasses of infected animals. Mortality among untreated people is rare (Jellison 1974).

Despite these general differences in geographic and host distributions, and in transmission, *F. tularensis* types A and B can occur simultaneously in the same ecosystem (Markowitz et al. 1985), and both can be tick transmitted (Ditchfield et al. 1960; Markowitz et al. 1985). In Canada, human tularemia associated with handling "rabbits" is probably mainly type B, circulating in a system involving *H. leporis-palustris* and snowshoe hares (Jellison 1974).

The incidence of tularemia in North America varies both spatially and temporally at large scales. This is reflected in reports of a disease in beaver and muskrat that, in the case of the former species at least, was probably tularemia. Jellison (1974) referred to "disastrous epidemics in the reservoir animals." The naturalist Ernest Thomson Seton, probably referring to tularemia in beaver, wrote "Some kind of distemper was prevailing among these animals, which destroyed them in vast numbers. . . Since that year the Beaver have never been so plentiful in the country of the Red River and Hudson Bay as they used formerly to be." From 1940 to 1959, tularemia occurred much more commonly in humans in the southeastern third of the United States than further west and north (Bell 1965), whereas there was an increased incidence in the west and northwest during the 1960s (Jellison 1974). Temporally, there were many more human cases of tularemia during the 1930s and early 1940s than previously and more recently (Bell 1965; Jellison 1974).

In Ontario, tularemia epidemics in beaver seem to have occurred twice during the past 50 years, in the late

1940s–early 1950s (Labzoffsky and Sprent 1952) and the early 1980s (unpublished). Similarly, the last extensive epidemics of tularemia in muskrats in Ontario were in 1955 (Fyvie et al. 1959) and 1960 (Ditchfield et al. 1960). Otherwise, most cases of tularemia among people and wildlife have been sporadic, with little indication that the disease had spread widely among furbearers (Walker and Moore 1971; Artsob et al. 1984; Addison unpublished).

The epidemiology of tularemia in the former USSR varies geographically. In many systems, the principal reservoir host of the organism is the water vole or water rat *Arvicola terrestris,* but many other rodents and hares can be involved. Transmission among wildlife is probably mediated mainly by ticks, by waterborne infection, and by ingestion of fodder contaminated by rodent excreta (Bell 1980).

Three different forms of human tularemia occur in the former USSR (Pollitzer 1963). The most common form occurs in the spring and is related to massive outbreaks of tularemia among water rats. People then contract tularemia either by direct contact with infected animals or from infected water. The second form of human tularemia occurs in the autumn and is vectorborne, normally transmitted by mosquitoes. The third form occurs during winter and results from contact with infected small rodents. Tularemia seems to occur much more frequently in the former USSR than in North America. In some areas, the disease is sufficiently prevalent that there is extensive preexposure vaccination of people (Olsuf'ev 1959; Dobrokhotov et al. 1987).

In Northern Europe and Scandinavia, tularemia is most frequently observed in late summer and early autumn as large epidemics among varying hares (Mörner 1986) and also occasionally as epidemics among people (Christenson 1984). Olin (1942) showed that human cases at this time of the year were mosquito-transmitted. Similarly, tularemia among hares in Scandinavia to a large extent follows the same seasonal pattern and is believed to be transmitted by mosquitoes (Mörner 1994). Epidemics in farmers have also occurred following inhalation of dust from hay contaminated with rodents or their excreta (Syrjälä et al. 1985).

In Central Europe, tularemia is normally spread to people by contact during skinning of infected animals or by the processing of their meat (Jusatz 1961). Consequently, human tularemia is most often found in winter during the hunting season for hares (Girard 1950).

In Japan, tularemia involves hares, in which it occurs mainly during winter and late spring; people usually acquire the infection from handling dead hares or from eating infected meat (Ohara et al. 1991).

CLINICAL SIGNS. Clinical signs in wild animals are poorly documented, mainly because of the acute character of tularemia in most species. In more sensitive animals, clinical signs of brief, severe apathy are followed by fatal septicemia. The course of the disease lasts approximately 2–10 days in sensitive species, and animals are usually presented for diagnosis dead. Experimentally infected hares demonstrated only minor clinical signs before becoming moribund (Borg et al. 1969; Mörner 1994).

In less sensitive animals, clinical signs associated with the route of infection might be observed, with local inflammation or ulceration at a portal of entry and enlargement of lymph nodes draining the affected area. However, nonspecific signs—depression and elevated body temperature—are always present, and these may be the only clinical signs evident. Most domestic mammals usually do not manifest any signs of tularemia, even though they do develop specific antibodies to the organism following infection.

In people, the more common ulceroglandular and glandular syndromes, and the less common oculoglandular form, are typified by fever and swollen lymph nodes, usually those draining the site where infection occurred—the skin, in which ulceration at a primary site of infection may be evident, or the eye, in which conjunctivitis occurs. Affected lymph nodes may become abscessed. Such syndromes typically follow an arthropod bite or other cutaneous exposure, or contamination of the conjunctiva. The oropharyngeal form, which follows consumption of contaminated water or meat, is typified by pharyngitis and tonsillitis and swelling of lymph nodes in the neck. Typhoidal tularemia presents as an influenza-like illness, with persistent fever but no localizing signs. It is often accompanied by radiographically evident bronchopneumonia and by hilar lymph node enlargement. Pneumonia is a sequel to inhalation exposure but also may occur in a minority of ulceroglandular cases.

PATHOGENESIS. Tularemia is in many species an acute infectious disease, although the outcome is affected by factors such as innate host susceptibility or acquired immunity, and dose and virulence of the infecting bacteria (Moe et al. 1975). Olsuf'ev (1959) described three distinct groups of mammals based on their tularemia susceptibility (to infection) and sensitivity (severity of clinical response): group I, highly susceptible and highly sensitive; group II, susceptible but with low sensitivity; and group III, low susceptibility and practically insensitive. Most rodents and some insectivores were in group I, some rodents and insectivores were in group II, and all carnivores tested were in group III.

After the organism enters the body or bloodstream, it multiplies locally, where it may cause necrosis and ulceration. It then invades small vessels, spreads along the superficial and deep lymphatics, and causes inflammation and necrosis in lymph nodes and scattered foci of coagulation necrosis in liver, spleen, bone marrow, and lungs (Lillie and Francis 1937; Meyer 1965).

PATHOLOGY. Infections with *F. tularensis* elicit gross and microscopic lesions similar to those in other acute infections with intracellular Gram-negative bacteria (Kitamura et al. 1956; Bell and Riley 1981; Syrjälä 1985).

At necropsy, animals dying from acute tularemia are usually in good body condition. There are signs of septicemia characterized by pale foci of necrosis randomly distributed in the liver, in the bone marrow, and in the spleen, which is usually enlarged. Necrotic foci vary in size, but, in some cases, may be barely visible to the naked eye. The lungs are usually congested and edematous, and there may be areas of consolidation and fibrinous pneumonia or pleuritis. Strands of fibrin may be present in the abdominal cavity. Foci of caseous necrosis often are present in one or more lymph nodes, especially in the abdominal cavity. Blood vessels are congested and thrombotic.

In less sensitive species, the gross findings at necropsy are consistent with a more chronic course of disease, with loss of body condition, and with lesions somewhat reminiscent of tuberculosis: granulomas in lymph nodes, liver, spleen, lungs, and kidneys (Bell and Riley 1981; Mörner 1994).

Microscopically, the most common finding in sensitive species is characteristic multifocal coagulation necrosis in the spleen, liver, lymph nodes, bone marrow, and lungs. These lesions are initially characterized by karyolysis, cytolysis, and absence of obvious neutrophils or macrophages. In less acute cases, the lesions are granulomatous, with central necrosis surrounded by vacuolated macrophages, epithelioid cells, and giant cells. The gross and microscopic picture may vary from species to species, but, as a general rule, the microscopic lesions are more or less constant within a species, highly susceptible species having acute lesions, and less susceptible species more chronic lesions (Bell and Riley 1981).

DIAGNOSIS. The postmortem diagnosis of tularemia is based on demonstrating the causative agent, even in cases where the microscopic picture is typical of tularemia [Office International des Epizooties (OIE) 1996].

The organism is normally demonstrated in impression smears of liver, spleen, bone marrow, kidney, or lung, or in tissue sections. Gram staining of smears reveals Gram-negative bacteria, which normally are numerous but which may be overlooked because of their very small size and resemblance to stain precipitate. They can be demonstrated more specifically by direct or indirect fluorescent antibody staining, a safe and rapid diagnostic tool (Karlsson et al. 1970; Mörner and Mattsson 1988). Bacteria also can be demonstrated in tissue sections by using immunofluorescent or immunohistochemical methods (Mörner 1981; Mörner et al. 1988a). Large numbers of bacteria usually are seen in necrotic foci and in blood.

Francisella tularensis can be identified by culture. Due to the highly infectious nature of the organism, culture should be attempted only in laboratories with a high level of containment (level 2 or 3). Isolation can be a problem since *F. tularensis* grows poorly on most ordinary bacterial culture media, although it occasionally may be isolated on blood agar. However, it grows on glucose cystine blood agar or other media containing sufficient cystine or cysteine, such as Francis medium, McCoy and Chapin medium, modified Thayer-Martin agar, glucose cysteine agar with thiamine (Eigelsbach and McGann 1984; OIE 1996), or chocolate agar (Reary and Klotz 1988). Heart blood, liver, spleen, and bone marrow should be cultured. *Francisella tularensis* is slow growing and also often is outnumbered by *Escherichia coli* and other bacteria in carcasses, making it more difficult to demonstrate in culture (Mörner et al. 1988a). The bacteria are nonmotile, non-spore-forming, bipolar-staining uniform coccobacilli in young cultures but pleomorphic in older cultures.

The two types of *F. tularensis* can be distinguished on the basis of cultural characteristics and virulence. *Francisella t. tularensis* (type A) ferments glycerol and is highly virulent for humans and domestic rabbits, whereas *F. t. palaearctica* (type B) does not ferment glycerol and is less virulent for humans and domestic rabbits (Eigelsbach and McGann 1984).

If *F. tularensis* is difficult to isolate on primary culture, it may be isolated (under appropriate animal utilization protocols and conditions of high biosafety) following inoculation of tissue suspensions from suspect cases into laboratory animals, such as mice or guinea pigs, which are highly sensitive to infection (Weaver and Hollis 1980; OIE 1996). The organism then can be cultured from tissues such as spleen and liver, in which it will be numerous.

Criteria for the identification of *F. tularensis* include growth on special media, distinctive cellular morphology, and specific fluorescent antibody (OIE 1996) and slide agglutination reactions. Type A can be distinguished from type B by fermentation of glycerol.

The bacteria also can be identified by hybridization with probes specific to 16S rRNA of the two *F. tularensis* types (Forsman et al. 1990), by polymerase chain reaction (PCR) (Long et al. 1993; Fulop et al. 1996), or by immunoelectron microscopy (Geisbert et al. 1993).

Serologic tests currently are used for diagnosis in people but have limited use in sensitive species of wildlife, which usually die before specific antibodies develop. However, sera or saline extracts of lung tissue (Carlsson et al. 1979; Mörner and Mattsson 1988; Mörner et al. 1988b) can be employed for seroepidemiologic surveys of species that are more resistant to infection, in order to detect tularemia activity in a locality (Gese et al. 1997; Shoemaker et al. 1997).

The standard serologic test is the tube-agglutination test, using as antigen an inactivated suspension of *F. tularensis* (OIE 1996). Possible cross-reactions with *Brucella abortus* and *B. melitensis* have to be considered. The enzyme-linked immunosorbent assay enables early serodiagnosis of tularemia (Carlsson et al. 1979), including differentiation of immunoglobulin isotypes (Shoemaker et al. 1997).

DIFFERENTIAL DIAGNOSIS. The differential diagnosis of tularemia will vary with the species and geographic locality but should include any disease producing necrotizing, suppurative, or granulomatous lymphadenitis, multifocal hepatitis and splenitis, and/or autopsy findings compatible with Gram-negative bacterial septicemia. These include plague due to *Yersinia pestis;* pseudotuberculosis due to *Y. pseudotuberculosis;* mycobacterial infections; staphylococcal infections; salmonellosis; Tyzzer's disease; systemic herpesvirus infections; and parasites, such as *Capillaria hepatica,* ascarid nematodes, or larval cestodes, which may encyst in the liver.

IMMUNITY. Following human infection with *F. tularensis,* an antibody response is measurable by the second week (Carlsson et al. 1979; Ohara et al. 1984). Immunoglobulin A (IgA) and IgM serum antibodies are most prominent at 14 days after infection; serum IgG increases a week later (Sandström et al. 1995) to produce higher IgG-IgM ratios (Carlsson et al. 1979). Serum antibody is highest during the second month, after which it declines gradually, though many reports document titers for 10 years or more after infection (Tärnvik 1989; Ericsson et al. 1994). A strong antibody response is indicative of prior infection, but absence of specific antibody does not indicate that there has been no exposure to *F. tularensis.*

Antibody is ineffective by itself in protecting against challenge by more virulent strains of *F. tularensis* (Tärnvik 1989). Thus, questions remain as to the protective significance of the antibody-mediated arm of the immune response.

In people, cell-mediated immunity stimulated by polypeptides in the bacterial cell wall of *F. tularensis* (Sandström et al. 1987) is measurable during the first week after exposure and strong during the second week (Syrjälä et al. 1984), and remains identifiable for at least 9–25 years (Koskela and Herva 1980; Tärnvik et al. 1985; Ericsson et al. 1994).

Neutrophils are the predominant leukocyte at points of entry of *F. tularensis* within the first 24 hours after infection, and in mice neutrophils are required for protection against primary infection (Sjöstedt et al. 1994). However, the mechanism of action of the neutrophil remains unclear. Neutrophils might produce cytokines that attract macrophages and T cells to the point of entry of the infection (Sjöstedt et al. 1994). Reaction of antigen with serum antibodies also may promote production of cytokines and subsequent activation of T cells (Rhinehart-Jones et al. 1994).

Tärnvik (1989)recommended that future studies on the mechanism of protective immunity to *F. tularensis* should focus on the antigenic membrane polypeptides.

CONTROL. In most ecosystems, tularemia cannot be controlled. In general, protective measures must be undertaken by individual outdoor recreationists or workers, or rural residents, who are most at risk of exposure. An effective public health education system must be in place to raise public awareness of that risk.

However, steps have been taken to ameliorate the impact of tularemia on people and livestock, most commonly in the former USSR, where prevalence of infection in vectors and reservoir hosts has been monitored in foci of tularemia transmission. Preexposure vaccination of people at high risk, and reduction in the principal reservoir hosts and vectors have been the common preventive practices employed in the former USSR.

In the past, reservoir hosts were controlled with poisoning (Olsuf'ev 1959). Hunting also was used to reduce numbers of water rats, other game rodents, and hares. Among vectors, greatest emphasis has been placed on control of ticks by treating vegetation and cattle *Bos taurus* with acaricides (Olsuf'ev 1959). Agricultural practices have been adapted, by timely and efficient harvesting to reduce lagomorph and rodent food and by spring and autumn plowing to reduce habitat. Some of the practices employed in the former USSR might not be considered environmentally acceptable in other jurisdictions.

To prevent contact transmission, rubber gloves should be worn by trappers or hunters when skinning muskrat or beaver, or while dressing lagomorphs, in areas where these species are possibly infected with *F. tularensis.* Meat from potentially infected animals should be well cooked to destroy bacteria, and water from sources that may be contaminated by *F. tularensis* may be boiled, filtered, or chlorinated by outdoorspersons or residents, depending on the volume and feasibility.

In areas where tularemia is predominantly tick-transmitted, avoiding activities that result in encounters with tick-infested habitat or potentially infected mammals will reduce risk of disease. Other preventive measures include use of tick repellants such as DEET or permethrin and vigilance at removing ticks from clothes and the person.

A phenol-killed vaccine was administered prophylactically to people in the United States from 1932 to about 1960 (Jellison 1974). An attenuated live vaccine was developed and put into widespread use among people at high risk of infection in the former USSR (Olsuf'ev 1959), as well as in some laboratory workers at high risk in other parts of the world. Although it has been much more efficient than the killed vaccine (Burke 1977) and is the best vaccine available, it does not always provide complete protection (Tärnvik 1989).

TREATMENT. Streptomycin remains the drug of choice in treatment of tularemia (Jellison 1974; Enderlin et al. 1994). It is both bacteriostatic and bactericidal and has a very high rate of cure, with no relapses if used appropriately (Enderlin et al. 1994). Gentamicin is considered a reasonable alternative if streptomycin is not available (Enderlin et al. 1994). Relapses occur most frequently when people are treated with bacteriostatic drugs (Miller et al. 1989).

DOMESTIC ANIMAL CONCERNS. Although tularemia is mainly a disease of wild lagomorphs and rodents, it occurs sporadically among domestic animals and in zoological gardens. Domestic animals normally contract the disease from infected food (e.g., dead rabbits), or from infected vectors such as ticks, mosquitoes, or deerflies (Jellison 1974). Cats *Felis domesticus* and sheep *Ovis aries* seem to be more frequently infected than other domestic animals in North America [see Jellison (1974), Capellan and Fong (1993), and Woods et al. (1998)], although tularemia is also known in dogs *Canis familiaris* and horses *Equus caballus* (Jellison 1974; Gustafson and DeBowes 1996).

Tularemia has been transmitted from free-living wildlife to primates in zoos on several occasions, probably by fleas or by ingestion of infected wildlife (Nayar et al. 1979; Calle et al. 1993). Tularemia also has occurred in ranch beaver (Bell et al. 1962) and in ranch mink *Mustela vison* fed meat from infected animals (Henson et al. 1978) in North America.

MANAGEMENT IMPLICATIONS. Tularemia should be recognized as a natural component of many ecosystems. For example, in a northern boreal ecosystem, Hörnfeldt (1978) identified relationships among the populations of voles *Clethrionomys* spp., mountain hare, tetraonid birds, red fox *Vulpes vulpes,* owls, and the prevalence of tularemia. Tularemia is a natural factor influencing lagomorph and rodent populations that themselves can impact on habitat (Jellison 1974).

The threat of tularemia increased with increased densities of lagomorphs (*S. floridanus, S. aquaticus,* and *L. townsendii*) in a lagomorph-tick system in the central United States, and Yeatter and Thompson (1952) recommended delaying the opening of the rabbit-hunting season until colder late autumn-early winter weather reduced tick densities and, hence, the likelihood of tick-transmitted tularemia in people (Yeatter and Thompson 1952).

Within a waterborne tularemia system, trappers need to be aware of their heightened risk of exposure, be aware of means to reduce that risk by simple hygiene, have knowledge of the symptoms of tularemia, and realize the need to identify possible exposure to tularemia when seeking medical attention. In some jurisdictions, trappers are required to harvest a quota of beaver or muskrat. Wildlife managers must be flexible and adjust the harvest quotas if there has been significant loss of furbearers due to tularemia.

In a 7-year period during the 1920s, over 270,000 lagomorphs were translocated into Pennsylvania from areas where tularemia was known to have occurred (Jellison 1974). Hundreds of thousands of cottontails were moved from U.S. midwestern states where tularemia was endemic to other eastern and northeastern states. Translocation of potential reservoirs of tularemia from endemic ecosystems should be avoided or done only after thorough consideration of the implications of translocation of tularemia and other diseases.

LITERATURE CITED

Anonymous. 1980. Tularemia from a fox bite: Colorado. *Veterinary Public Health Notes* August 1980:6–7.

Artsob, H., L. Spence, G. Surgeoner, J. McCreadie, J. Thorsen, C. Th'ng, and V. Lampotang. 1984. Isolation of *Francisella tularensis* and Powassan virus from ticks (Acari: Ixodidae) in Ontario, Canada. *Journal of Medical Entomology* 21:165–168.

Bell, J.F. 1965. Ecology of tularemia in North America. *Journal of Jinsen Medicine* 11:33–44.

———. 1980. Tularemia. In *CRC handbook series in zoonoses,* section A: *Bacterial, rickettsial and mycotic diseases,* vol. 2, ed. J.H. Steele. Boca Raton, FL: CRC, pp. 161–193.

Bell, J.F., and J.R. Reilly. 1981. Tularemia. In *Infectious diseases of wild mammals,* ed. J.W. Davis, L.H. Karstad, and D.O. Trainer, 2d ed. Ames: Iowa State University Press, pp. 213–231.

Bell, J.F., and S.J. Stewart. 1975. Chronic shedding tularemia nephritis in rodents: Possible relation to occurrence of *Francisella tularensis* in lotic waters. *Journal of Wildlife Diseases* 11:421–430.

———. 1983. Quantum differences in oral susceptibility of voles, *Microtus pennsylvanicus,* to virulent *Francisella tularensis* type B, in drinking water: Implications to epidemiology. *Ecology of Disease* 2:151–156.

Bell, J.F., C.R. Owen, W.L. Jellison, G.J. Moore, and E.O. Buker. 1962. Epizootic tularemia in pen-raised beavers and field trials of vaccines. *American Journal of Veterinary Research* 23:884–888.

Borcic, B., B. Aleraj, M. Zutic, and D. Mikacic. 1978. The role of ticks (Ixodidae) in the maintenance of tularemia natural focus in central Posavina [in Croatian]. *Veterinarski Archiv* 48:277–283.

Borg, K., E. Hanko, T. Krunajevic, N.-G. Nilsson, and P.O. Nilsson. 1969. On tularemia in the varying hare (*Lepus timidus* L.). *Nordisk Veterinärmedicin* 21:95–104.

Burke, D.S. 1977. Immunization against tularemia: Analysis of the effectiveness of live *Francisella tularensis* vaccine in prevention of laboratory-acquired tularemia. *Journal of Infectious Diseases* 135:55–60.

Calle, P.P., D.L. Bowerman, and W.J. Pape. 1993. Nonhuman primate tularemia (*Francisella tularensis*) epizootic in a zoological park. *Journal of Zoo and Wildlife Medicine* 24:459–468.

Capellan, J., and I.W. Fong. 1993. Tularemia from a cat bite: Case report and review of feline-associated tularemia. *Clinical Infectious Diseases* 16:472–475.

Carlsson, H.E., A.A. Lindberg, G. Lindberg, B. Hederstedt, K.-A. Karlsson, and B.O. Agell. 1979. Enzyme-linked immunosorbent assay for immunological diagnosis of human tularemia. *Journal of Clinical Microbiology* 10:615–621.

Chase, D., H.H. Handsfield, J. Allard, and J. Taylor. 1980. Tularemia acquired from a bear: Washington. *Morbidity and Mortality Weekly Report* 29:57.

Christenson, B. 1984. An outbreak of tularemia in the northern part of central Sweden. *Scandinavian Journal of Infectious Diseases* 16:285–290.

Ditchfield, J., E.B. Meads, and R.J. Julian. 1960. Tularemia of muskrats in eastern Ontario. *Canadian Journal of Public Health* 51:474–478.

Dobrokhotov, B.P., P.M. Baranovskii, T.N. Demidova, and I.S. Meshcheryakova. 1987. Changes in biocenoses of meadow-field tularemia natural foci and their stability conditioned by anthropogenic factors in the European part of the USSR [in Russian]. *Zoologicheskii Zhurnal* 66:1430–1434.

Eigelsbach, H.T., and V.G. McGann. 1984. Gram-negative aerobic cocci, genus *Francisella* Dorofe'ev 1947. In *Bergey's manual of systematic bacteriology,* vol. 1. ed. N.R. Krieg and J.G. Holts. Baltimore: Williams and Wilkins, pp. 394–399.

Emmons, R.W., J. Ruskin, M.L. Bissett, D.A. Uyeda, R.M. Wood, and C.L. Lear. 1976. Tularemia in a mule deer. *Journal of Wildlife Diseases* 12:459–463.

Enderlin, G., L. Morales, R.F. Jacobs, and J.T. Cross. 1994. Streptomycin and alternative agents for the treatment of tularemia: Review of the literature. *Clinical Infectious Diseases* 19:42–47.

Ericsson, M., G. Sandström, A. Sjöstedt, and A. Tärnvik. 1994. Persistence of cell-mediated immunity and decline of humoral immunity to the intracellular bacterium *Francisella tularensis* 25 years after natural infection. *Journal of Infectious Diseases* 170:110–114.

Finegold, M.J., J.D. Pulliam, M.E. Landay, and G.G. Wright. 1969. Pathological changes in rabbits injected with *Pasteurella tularensis* killed by ionizing radiation. *Journal of Infectious Diseases* 119:635–640.

Forsman, M., G. Sandström, and B. Jaurin. 1990. Identification of *Francisella* species and discrimination of type A and type B strains of *F. tularensis* by 16S rRNA analysis. *Applied and Environmental Microbiology* 56:949–955.

Francis, E. 1921. The occurrence of tularemia in nature as a disease of man. *Public Health Reports* 36:1731–1738.

Fulop, M., D. Leslie, and R. Titball. 1996. A rapid, highly sensitive method for the detection of *Francisella tularensis* in clinical samples using the polymerase chain reaction. *American Journal of Tropical Medical Hygiene* 54:364–366.

Fyvie, A., W.G. Ross, and N.A. Labzoffsky. 1959. Tularemia among muskrats on Walpole Island, Lake St. Clair, Ontario. *Canadian Journal of Comparative Medicine* 23:153–156.

Geisbert, T.W., P.B. Jahrling, and J.W. Ezzell Jr. 1993. Use of immunoelectron microscopy to demonstrate *Francisella tularensis. Journal of Clinical Microbiology* 31:1936–1939.

Gese, E.M., R.D. Schultz, M.B. Johnson, E.S. Williams, R.L. Crabtree, and R.L. Ruff. 1997. Serological survey for diseases in free-ranging coyotes (*Canis latrans*) in Yellowstone National Park, Wyoming. *Journal of Wildlife Diseases* 33:47–56.

Girard, G. 1950. Aspects épidemiologiques de la tularémie en France. *Presse Médical* 66:568–570.

Gustafson, B.W., and L.J. DeBowes. 1996. Tularemia in a dog. *Journal of the American Animal Hospital Association* 32:339–341.

Hahn, H., and S.H.E. Kaufmann. 1981. The role of cell mediated immunity in bacterial infections. *Review of Infectious Diseases* 3:1221–1250.

Henson, J.B., J.R. Gorham, and D.T. Shen. 1978. An outbreak of tularemia in mink. *Cornell Veterinarian* 68:78–83.

Hopla, C.E. 1974. The ecology of tularemia. *Advances in Veterinary Science and Comparative Medicine* 18:25–53.

Hopla, C.E., and A.K. Hopla. 1994. Tularemia. In *Handbook of zoonoses,* ed. G. W. Beran, 2d ed. Boca Raton, FL: CRC, pp. 113–125.

Horne, H. 1911. En lemaen- og marsvinpest. *Nordisk Veterinär Tidskrift* 23:16–33.

Hörnfeldt, B. 1978. Synchronous population fluctuations in voles, small game, owls, and tularemia in northern Sweden. *Oecologia (Berlin)* 32:141–152.

Jellison, W.L. 1974. *Tularemia in North America 1930–1974.* Missoula: University of Montana, 276 pp.

Jellison, W.L., D.C. Epler, E. Kuhns, and G.M. Kohls. 1950. Tularemia in man from a domestic rural water supply. *Public Health Reports* 65:1219–1226.

Jusatz, H.J. 1952. Tularämie in Europa 1926–1951. In *Welt-Seuchen Atlas I–III,* ed. E. Rodenwaldt. Hamburg: Falk Verlag, pp. 34–38.

———. 1961. The geographical distribution of tularemia throughout the world, 1911–1959. In *World atlas of epidemic diseases,* ed. E. Rodenwaldt. Hamburg: Falk Verlag, pp. 23–28.

Karlsson, K.A., S. Dahlstrand, E. Hanko, and O. Soderlind. 1970. Demonstration of *Francisella tularensis* in sylvan animals with the aid of fluorescent antibodies. *Acta Pathologica et Microbiologica Scandinavia [B]* 78:647–651.

Kemenes F. 1976. Die Epizootologie der Hasentularämie in Ungarn. In *Proceedings 2 Internationales Arbeitskolloqium über "Naturherde von Infektionskrankheiten in Zentraleuropa."* Graz: University Institute of Hygiene, pp. 285–289.

Kitamura, S., M. Fukada, H. Takeda, S. Ouchi, S. Nakano, and T. Unagami. 1956. Pathology of tularemia. *Acta Pathologica Japonica* 6:719–764.

Koskela, P., and E. Herva. 1980. Cell-mediated immunity against *Francisella tularensis* after natural infection. *Scandinavian Journal of Infectious Diseases* 12:281–287.

Labzoffsky, N.A., and J.F.A. Sprent. 1952. Tularemia among beaver and muskrat in Ontario. *Canadian Journal of Medical Sciences* 30:250–255.

Lamarque, F., J. Barrat, and F. Moutou. 1996. Principal diagnoses for determining causes of mortality in the European brown hare (*Lepus europaeus*) found dead in France between 1986 and 1994. *Gibier Faune Sauvage* 13:53–72.

Lepitzki, D.A., A. Woolf, and M. Cooper. 1990. Serological prevalence of tularemia in cottontail rabbits in southern Illinois. *Journal of Wildlife Diseases* 26:279–282.

Lillie, R.D., and E. Francis. 1937. The pathology of tularaemia in the Belgian hare (*Oryctolagus cuniculus*); the pathology of tularaemia in the black-tailed rabbit (*Lepus* sp.); the pathology of tularaemia in the cottontail rabbit (*Sylvilagus floridanus*). In *The pathology of tularaemia.* Washington, DC: National Institute of Health; U.S. Treasury Department, Bulletin 167:83–126.

Long, G.W., J.J. Oprandy, R.B. Narayanan, A.H. Fortier, K.R. Porter, and C.A. Nacy. 1993. Detection of *Francisella tularensis* in blood by polymerase chain reaction. *Journal of Clinical Microbiology* 31:152–154.

Markowitz, L.E., N.A. Hynes, P. de la Cruz, E. Campos, J.M. Barbaree, B.D. Plikaytis, D. Mosier, and A.F. Kaufmann. 1985. Tick-borne tularemia: An outbreak of lymphadenopathy in children. *Journal of the American Medical Association* 254:2922–2925.

McCoy, G.W. 1911. A plague-like disease of rodents. *Public Health Bulletin* 43:53–71 and plates.

McCoy, G.W., and C.W. Chapin. 1912. *Bacterium tularense* the cause of a plague-like disease of rodents. *U.S. Public Health and Marine Hospital Bulletin* 53:17–23.

Meyer, K.F. 1965. *Pasteurella* and *Francisella.* In *Bacteria and mycotic infections of man,* ed. R.J. Dubos and J.G. Hersch, 4th ed. Philadelphia: Lippincott, pp. 681–697.

Miller, S.D., M.B. Snyder, M. Kleerekoper, and C.H. Grossman. 1989. Ulceroglandular tularemia: A typical case of relapse. *Henry Ford Hospital Medical Journal* 37:73–75.

Moe, J.B., P.G. Canonico, J.L. Stookey, M.C. Powanda, and G.L. Cockerell. 1975. Pathogenesis of tularemia in immune and nonimmune rats. *American Journal of Veterinary Research* 36:1505–1510.

Mörner, T. 1981. The use of FA technique for detecting *Francisella tularensis* in formalin fixed material: A method useful in routine post mortem work. *Acta Veterinaria Scandinavia* 22:296–306.

———. 1986. The occurrence of tularemia in Sweden. In *Verhandlungsbericht des Internationalen Symposiums über die Erkrankungen der Zootiere* 28:327–331.

———. 1994. Tularemia in hares in Sweden with special reference to identification of *Francisella tularensis.* Ph.D. diss., University of Agricultural Sciences, Uppsala, Sweden, 154 pp.

Mörner, T., and R. Mattsson. 1988. Experimental infection of five species of raptors and of hooded crows with *Francisella tularensis* biovar *palaearctica. Journal of Wildlife Diseases* 24:15–21.

Mörner, T., G. Sandström, R. Mattsson, and P.-O. Nilsson. 1988a. Infections with *Francisella tularensis* biovar *palaearctica* in hares (*Lepus timidus, Lepus europaeus*) from Sweden. *Journal of Wildlife Diseases* 24:422–433.

Mörner, T., G. Sandström, and R. Mattsson. 1988b. Comparison of serum and lung extracts for surveys of wild animals for antibodies to *Francisella tularensis* biovar *palaearctica. Journal of Wildlife Diseases* 24:10–14.

Myrbäck, K.E., O. Ringertz, and S. Dahlstrand. 1968. An epidemic of tularemia in Sweden during the summer of 1967. *Acta Pathologica et Microbiologica Scandinavia* 72:463–464.

Nayar, G.P., G.J. Crawshaw, and J.L. Neufeld. 1979. Tularemia in a group of nonhuman primates. *Journal of the American Veterinary Medical Association* 175:962–963.

Office International des Epizooties (OIE). 1996. Tularemia. In *Manual of standards for diagnostic tests and vaccines.* Paris: Office International des Epizooties, pp. 584–588.

Ohara, S. 1954. Studies on Yato-byo (Ohara's disease, tularemia in Japan): Report I. *Japanese Journal of Experimental Medicine* 24:69–79.

Ohara, S., T. Sato, and M. Homma. 1984. Time course of the production of immunoglobulins in patients with tularemia. *Japanese Journal of Bacteriology* 39:103–106.

Ohara, Y., T. Sato, H. Fujita, T. Ueno, and M. Homma. 1991. Clinical manifestations of tularemia in Japan: Analysis of 1,355 cases observed between 1924 and 1987. *Infections* 19:14–17.

Olin, G. 1942. The occurrence and mode of transmission of tularemia in Sweden. *Acta Pathologica et Microbiologica Scandinavica* 19:220–247.

Olsuf'ev, N. 1959. Tularemia. In *Human diseases with natural foci,* ed. Y.N. Pavlovsky. Moscow: Foreign Language Publishing House, pp. 219–281.

———. 1970. Taxonomy and characteristics of the genus *Francisella* Dorofe'ev, 1947 [in Russian]. *Journal of Hygiene, Epidemiology, Microbiology and Immunology* 14:67–74.

———. 1974. *Tularemia: WHO inter-regional travelling seminar on natural foci of zoonoses* [in Russian]. Moscow: World Health Organization, 28 pp.

Olsuf'ev, N.G., and K.N. Shlygina. 1979. Role of cannibalism in the development of tularemia with nonlethal outcome in common voles highly sensitive to tularemia [in Russian]. *Zoologicheskii Zhurnal* 58:933–936.

Olsuf'ev, N.G., O.S. Emelyanova, and T.N. Dunayeva. 1959. Comparative study of strains of *B. tularense:* II. Evaluation of criteria of virulence of *Bacterium tularense* in the old and the new world and their taxonomy [in Russian]. *Journal of Hygiene, Epidemiology, Microbiology and Immunology* 3:138–149.

Olsuf'ev, N.G., K.N. Shlygina, and E.V. Ananova. 1984. Persistence of *Francisella tularensis* tularemia agent in the organism of highly sensitive rodents after oral infection. *Journal of Hygiene, Epidemiology, Microbiology and Immunology* 28:441–454.

Omland, T., E. Christiansen, B. Jonsson, G. Kapperud, and R. Wiger. 1977. A survey of tularemia in wild mammals from Fennoscandia. *Journal of Wildlife Diseases* 13:393–399.

Parker, R.R., E.A. Steinhaus, G.M. Kohls, and W.L. Jellison. 1951. *Contamination of natural waters and mud with* Pasteurella tularensis *and tularemia in beavers and muskrats in the northwestern United States.* Washington, DC: U.S. National Institutes of Health, 61 pp; Bulletin 193.

Pfahler-Jung, K. 1989. Die globale Verbreitungen der Tularemie. Ph.D. diss., Justus-Liebig University, Giessen, Germany, 244 pp.

Pollitzer, R. 1963. *History and incidence of tularemia in the Soviet Union: A review.* New York: Institute of Contemporary Russian Studies, Fordham University, 366 pp.

Reary, B.W., and S.A. Klotz. 1988. Enhancing recovery of *Francisella tularensis* from blood. *Diagnostic Microbiology and Infectious Disease* 11:117–119.

Rhinehart-Jones, T.R., A.H. Fortier, and K.L. Elkins. 1994. Transfer of immunity against lethal murine *Francisella* infection by specific antibody depends on host gamma interferon and T cells. *Infection and Immunity* 62:3129–3137.

Sandström, G., A. Tärnvik, and H. Wolf-Watz. 1987. Immunospecific T-lymphocyte stimulation by membrane proteins from *Francisella tularensis. Journal of Clinical Microbiology* 25:641–644.

Sandström, G., D.M. Waag, A. Sjöstedt, B. Segerstedt, and J.C. Williams. 1995. Immunization against tularemia: Analysis of the immunospecific response in man after vaccination with a new lot of the live vaccine strain of *Francisella tularensis. Immunology and Infectious Diseases (Oxford)* 5:108–114.

Schlotthauer, C.F., L. Thompson, and C. Olson. 1935. Tularemia in wild grey foxes: Report of an epizootic. *Journal of Infectious Diseases* 56:28–30.

Shlygina, K.N., and N.G. Olsuf'ev. 1982. Latent tularemia in common voles (*Microtis subarvalis*) in an experiment [in Russian]. *Zhurnal Microbiologii, Epidemiologii i Immunologii* no. 4:101–104.

Shlygina, K.N., P.M. Baranovskii, E.V. Ananova, and N.G. Olsuf'ev. 1987. On the possibility of the atypical course of tularemia (persistence) in common voles, *Microtis arvalis* Pall [in Russian]. *Zhurnal Microbiologii, Epidemiologii i Immunologii* no. 3:26–29.

Shoemaker, D., A. Woolf, M. Cooper, and R. Kirkpatrick. 1997. The humoral response of cottontail rabbits (*Sylvilagus floridanus*) naturally infected with *Francisella tularensis* in southern Illinois. *Journal of Wildlife Diseases* 33:733–737.

Sjöstedt, A., J.W. Conlan, and R.J. North. 1994. Neutrophils are critical for host defense against primary infection with the facultative intracellular bacterium *Francisella tularensis* in mice and participate in defense against reinfection. *Infection and Immunity* 62:2779–2783.

Skrodzki, E. 1968. Investigations of the pathogenesis of tularemia: VII. Attempts to discover *F. tularensis* toxins. *Biuletyn Instytutu Medycyny Morskiej w Gdansku* 19:69–76.

Sterba, F., and J. Krul. 1985. Pathologish-morphologische Untersuchungen zur Differentialdiagnose von Bruzellose, Tularämie und Pseudotuberkulose beim Hasen (*Lepus europaeus* L.). In *Proceedings of the 17th Congress of the International Union of Game Biologists,* ed. S.A. de Crombrugghe. Brussels: International Union of Game Biologists, pp. 763–771.

Syrjälä, H. 1985. Laboratory diagnosis of different clinical forms and severities of human tularemia caused by *Francisella tularensis* biovar *palaearctica.* Ph.D. diss., University of Oulu, Oulu, Finland, 55 pp.

Syrjälä, H., E. Herva, J. Ilonen, K. Saukkonen, and A. Salminen. 1984. A whole-blood lymphocyte stimulation test for the diagnosis of human tularemia. *Journal of Infectious Diseases* 150:912–915.

Syrjälä, H., P. Kujala, V. Myllyla, and A. Salminen. 1985. Airborne transmission of tularemia in farmers. *Scandinavian Journal of Infectious Diseases* 17:371–375.

Tärnvik, A. 1989. Nature of protective immunity to *Francisella tularensis. Review of Infectious Diseases* 11:440–451.

Tärnvik, A., M.-L. Lofgren, S. Lofgren, G. Sandström, and H. Wolf-Watz. 1985. Long-lasting cell-mediated immunity induced by a live *Francisella tularensis* vaccine. *Journal of Clinical Microbiology* 22:527–530.

Taylor, J.P., G.R. Istre, T.C. McChesney, F.T. Satalowich, R.L. Parker, and L.M. McFarland. 1991. Epidemiologic characteristics of human tularemia in the southwest-central United States, 1981–1987. *American Journal of Epidemiology* 133:1032–1038.

Ul'yanova, N.I., M.A. Bessonova, L.N. Panasik, V.N. Svimonishvili, and L.S. Grishina. 1982. Results of the prolonged study of the backwater foci of tularemia and its prevention (based on the materials obtained in the Leningrad oblast, Russian SFSR, USSR) [in Russian]. *Zhurnal Microbiologii, Epidemiologii i Immunologii* no. 2:104–107.

Walker, W.J., and C.A. Moore. 1971. Tularemia: Experience in the Hamilton area. *Canadian Medical Association Journal* 105:390–393 and 396.

Weaver, R.E., and D. Hollis. 1980. Gram-negative fermentative bacteria and *Francisella tularensis.* In *Manual of clinical microbiology,* ed. E. Lenette, A. Balows, W. Hausler Jr., and J. P. Traunt. Washington, DC: American Society for Microbiology, pp. 242–262.

Woods, J.P., M.A. Crystal, R.J. Morton, and R.J. Panciera. 1998. Tularemia in two cats. *Journal of the American Veterinary Medical Association* 212:81–83.

Woolf, A., D.R. Shoemaker, and M. Cooper. 1993. Evidence of tularemia regulating a semi-isolated cottontail rabbit population. *Journal of Wildlife Management* 57:144–157.

Yeatter, R.E., and D.H. Thompson. 1952. Tularemia, weather, and rabbit populations. *Bulletin of the Illinois Natural History Survey* 25:351–382.

19

PLAGUE AND YERSINIOSIS

PETER W. GASPER AND ROWENA P. WATSON

Emphasis in this chapter is given to sylvatic plague caused by *Yersinia pestis,* because this infamous and reportable zoonotic disease is the most life-threatening and important pathogenic yersiniae of wild mammals. Of the 11 species of *Yersinia,* family Enterobacteriaceae, only four are considered to be primary pathogens: *Y. pestis, Y. enterocolitica, Y. pseudotuberculosis,* and *Y. ruckeri.* The diseases they cause are plague in mammals, yersiniosis due to either *Y. pseudotuberculosis* or *Y. enterocolitica* in mammals and birds, and red mouth in fish, respectively.

Yersinia are Gram-negative, non-spore-forming, facultatively anaerobic, small (0.5–0.8 × 1–3 μm) bacteria exhibiting pleomorphism varying from coccobacilli to rods and sometimes forming small chains. Like all Enterobacteriaceae, they ferment glucose, do not liquefy alginate, are usually nitrate reductase positive (Brubaker 1991) and oxidase negative. They express the enterobacterial common antigen on the outer cell surface. The type species for this genus is *Y. pestis* (Bercovier and Mollaret 1984).

The primary definition of *plague* is an acute infectious disease with a high fatality rate in many mammal species caused by *Y. pestis.* The secondary definition is "any epidemic or pestilence." It is in this sense that names such as cattle plague and fowl plague are used but describe diseases not caused by *Y. pestis.*

Although *Y. pestis* is renowned for causing three human pandemics and is estimated to have killed approximately 200 million people during the course of recorded history (McEvedy 1988), its life cycle and permanent reservoirs occur in wild mammals. There are enzootic sylvatic foci of this bacterium throughout the world where the organism is carried and maintained in a complex cycle within indigenous wild rodent and flea species. Epizootic plague represents transmission from the usual rodent hosts into other species, most commonly other rodents but occasionally other mammals, including humans, that are highly susceptible to *Y. pestis* infection. Plague epidemics in people are associated with commensal or urban rats and their fleas or other humans.

Yersiniosis is the preferred term for infections caused by *Y. enterocolitica* and *Y. pseudotuberculosis.* In contrast to the acute and often fatal outcome of *Y. pestis* infections, *Y. enterocolitica* and *Y. pseudotuberculosis* generally cause prolonged infections. Both of these bacteria may cause gastroenteritis, which may appear as acute, subacute, latent, or chronic disease. Formerly, these infections were called *pseudotuberculosis* because of the tubercle-like abscesses found in some affected animals; this name is still sometimes used.

Investigations of the pathogenic, structural, and metabolic properties of *Y. pestis, Y. enterocolitica,* and *Y. pseudotuberculosis* have attempted to determine why the diseases they induce are so dissimilar and why there is such a dramatic difference in their virulence. The emerging answer is that these bacteria have multiple virulence factors, only some of which are shared. Two general factors make these *Yersinia* spp. successful parasites: invasion and evasion. These bacteria show a predilection for lymphoid tissue and an ability to avoid annihilation by the host immune system.

In addition to the bacterial chromosome, these three yersiniae have plasmids containing important genetic information. All have the low-calcium-response (Lcr) virulence plasmid. This is an approximately 70-kb plasmid that encodes several established virulence factors, among them the V antigen and a set of *Yersinia* outer-membrane proteins known as *Yops.* These antigens facilitate various antihost activities such as preventing phagocytosis, causing lysis of phagocytic cells, and moderating the immune response (Cornelius 1994; Perry and Fetherston 1997). The DNA sequences of the Lcr plasmids are largely homologous in all three species (Portnoy et al. 1984). All pathogenic yersiniae have a shared chromosomal virulence determinant, the *inv* gene. The *inv* gene encodes an outer-membrane host cell invasin protein that allows penetration of nonprofessional phagocytes (Brubaker 1991). The chromosome also contains genes for fimbriae on the outer surface of the three species, thought to enhance adhesion to mucous membrane surfaces (Carniel 1995). *Yersinia pestis* has lost many integral metabolic pathways present in *Y. enterocolitica* and *Y. pseudotuberculosis.* In so doing, it has acquired specialized structures and pathways. It is the only member of the genus that requires an animal host in its natural environment, and it expresses different virulence factors depending on whether its host is mammalian or arthropod.

SYLVATIC PLAGUE

Synonyms: Plague, bubonic plague, yersiniosis, pseudotuberculosis.

Introduction. Plague is a flea-transmitted disease affecting and perpetuated by rodents. Persistence of *Y. pestis* is the result of an elaborate interaction of at least four elements: (1) the agent, (2) fleas, (3) vertebrate hosts, and (4) the environment. Because *Y. pestis* is endemic in certain rodent populations on five continents, plague is and will remain a threat to human health worldwide (Butler 1989; Dennis 1994; Gage and Dennis 1995; Dennis and Hughes 1997; Galimand et al. 1997). Plague, human cholera, and yellow fever are the only international quarantinable infectious diseases of people.

History. "For centuries, plague repeatedly had a major impact on human history by altering and upsetting the political structures and population dynamics of communities and countries swept by enigmatic epidemics with high rates of morbidity and mortality. Because the mode of transmission was neither known nor suspected, epidemic plague caused mass hysteria, fear, and panic. Explanations for the death and destruction were attributed to a variety of factors, including cosmological alignment, divine punishment, or poisoning by ethnic groups" (Poland et al. 1994).

Of all the infectious diseases of wild mammals, plague best illustrates how a disease of wildlife can affect the course of human history. Although not all historical plagues have been documented to be due to *Y. pestis,* there is good evidence that *Y. pestis* has been killing humans since the 12th century B.C., when a plague epidemic befell the Philistines. Since then, there have been three documented pandemics; from 541 to 750 A.D., from the 8th century to 14th century, and from 1855 until today. The discovery of the etiologic agent, its vector, and the epidemiology of plague has been elucidated only in the last 100 years of this 3000-year history.

The first pandemic, termed the *Justinian plague,* affected the Byzantine Empire. It began in Pelusium Egypt and continued through the Middle East and Mediterranean Basin and into Mediterranean Europe. As in each pandemic, epidemics occurred in 8- to 12-year cycles and eventually affected North Africa, Europe, Central and Southern Asia, and Arabia. Records of morbidity and mortality are poor. However, it is estimated that from 541 to 700 A.D. there was a human population loss of 50%–60%. The second pandemic, called the *Black Death,* occurred in the 14th century (Perry and Fetherston 1997). This outbreak of plague is said to have killed one of every four Europeans. With subsequent smaller epidemics raising the toll to one in three, an estimated 75 million people, most from the poorer classes, died. This disease had lasting impact on rural society and English law by decreasing the number of peasants relative to the number of land owners, which in turn increased the value and bargaining power of the surviving serfs. Also, the preeminence of the Church was shaken when clergy were unable to stop the carnage (Poos 1991; Palmer 1993).

The third pandemic is thought to have started in 1855 in the Chinese province of Yüannan, where it quickly spread to China's southern coast to Guangzhou and Hong Kong. Plague was disseminated from here via merchant steamships to the Philippines, Japan, Australia, Hawaii, North America, South America, Africa, Europe, the Middle East, and India. India was particularly devastated, losing a million people per year by 1903, with an estimated total of 12.5 million deaths between 1898 and 1918. It was in Hong Kong in 1894 that a young Swiss physician, Dr. Alexandre E.J. Yersin, and a Japanese microbiologist, Dr. Shibasaburo Kitasato, independently discovered the etiologic agent of plague. Kitasato called the bacteria he found the "bacillus of bubonic plague." Yersin described the same plague organism, adding that it was Gram negative. Although the organisms they discovered were identical, Kitasato maintained that his and Yersin's bacteria were different. Yersin later moved to Vietnam and established a branch of the Pasteur Institute in Indochina. Kitasato went on to a distinguished and productive scientific career. Unfortunately, Kitasato's contention that the agent was Gram positive, likely due to laboratory contamination, clouded the issue of who discovered the agent of bubonic plague. Since 1970, the official name of the plague organism (previously known as *Pasteurella pestis*) has been *Yersinia pestis* (Gregg 1978; Perry and Fetherston 1997).

Folklore suggested the connection between wild mammals and plague epidemics, but the scientific basis was not known until the beginning of this century. The Bible mentions the involvement of rats with plague, and in the Hindu 12th-century work, the Bhagavata Purana, people were advised to leave their homes when rats began dying and falling from the ceilings: hence the term *rat-fall.* The fairy tale of the Pied Piper is thought to recall the tragedy of the death of a village's children after they contracted *Y. pestis* from fleas on dead rats that they were instructed to throw into the river. The collective wisdom of the centuries and the observed correlation between a rat epizootic and the incidence of plague among people was misinterpreted by the British in Hong Kong and India in the last years of the 19th century. They concluded rats caught the disease from humans, rather than the other way around. In Bombay, in 1897, fire engines were used to pump phenol solution onto the walls and floors of infected houses, and millions of gallons were pumped into the sewers each day. The *Y. pestis*-infected rats fled. The result was a more rapid and extensive spread of the infection. In 1908, the work of the French scientist Dr. Paul Louis Simmond documented that fleas transmitted *Y. pestis* from rats to people (Gregg 1978; Perry and Fetherston 1997).

The third pandemic has extended into this century. Between 1965 and 1970, there were over 25,000 human plague cases reported in South Vietnam, which represented the largest plague epidemic in the second half of the 20th century; the actual number of cases was estimated to be between 100,000 and 250,000 (Gregg 1978).

The history of plague among humans is impressive and has been given a focus here to illustrate the ramifications an infectious disease of wild mammals can have on human populations and vice versa. Although little or no documentation is available for how these human pandemics have affected wild mammal populations, the natural history of certain mammals was probably altered concomitantly and irrevocably because of *Y. pestis.*

Distribution. *Yersinia pestis* is thought to have originated in Central Asia or Africa and to have been disseminated to each continent by infected rats on ships. Plague is maintained in wild rodent populations in semiarid areas on every continent except Australia and Antarctica [World Health Organization (WHO) 1980; Gage and Dennis 1995]. The climates of particular ecological niches are thought to be an important reason for the centuries-long survival of *Y. pestis.* Other, more humid environments where plague is found are likely foci with an unnatural set of conditions that support the maintenance of temporary rodent host-flea populations. Plague may circulate for periods varying from a single season to several decades in humid environments; however, such foci will likely cease to persist in the long term.

Plague in North America illustrates how *Y. pestis* persists under various conditions. From 1900 to 1925, there were five large human epidemics of plague in North America at Pacific and Gulf Coast ports. The major plague outbreaks in the eastern United States (New Orleans and New York City) certainly infected local urban rodent populations but failed to become enzootic. The 1925 Los Angeles epidemic was the last urban rat-borne, pneumonic plague outbreak in the United States. Since 1944, there have been almost 400 documented cases of human plague in the United States (Craven et al. 1993), all traced to endemic foci in the semiarid western states. Whether these plague foci originated from steamship transport or arrived via suitable vertebrate hosts crossing the Bering Strait thousands of years ago is unclear. It is clear that *Y. pestis* is slowly spreading eastward in North America; surveillance for plague in rodent and rodent-consuming carnivore populations during the 1990s indicates that *Y. pestis* has spread to counties in eastern Montana, western Nebraska, western North Dakota, and eastern Texas that were believed to be free of this disease since widespread animal surveillance began in the 1930s [Craven et al. 1993; Centers for Disease Control and Prevention (CDC) 1994b; Gage and Dennis 1995].

Host Range. Survival of *Y. pestis* in semiarid areas is accomplished through a complex series of transmission cycles involving susceptible wild rodents and their fleas. Plague has been found in association with about a third of rodent families or subfamilies and has been reported to naturally infect the majority of rodent genera, including more than 200 species of rodents throughout the world. However, only 30–40 species serve as permanent reservoirs of plague, with the remainder being temporary or epizootic hosts. Birds, lagomorphs, carnivores, and primates are not involved as principals in enzootic cycles, although they may play an occasional role in these cycles and in disseminating infection by transporting infective fleas or prey.

In reconciling the virulence of plague in some mammals with its persistence, it is helpful to divide hosts into different epidemiologically important categories. Poland and colleagues (1994) divided *Y. pestis* hosts as follows: (1) enzootic or maintenance rodent hosts, (2) epizootic or amplification rodent hosts, (3) resistant nonrodent hosts, and (4) susceptible nonrodent hosts.

ENZOOTIC, MAINTENANCE, OR RESERVOIR RODENT HOSTS. Certain features are characteristic of enzootic or maintenance hosts found within permanent plague foci: "1. Maintenance of *Y. pestis* by only a few genera of relatively resistant rodent hosts. 2. A broad heterogenic response of reservoir hosts to *Y. pestis* challenge; mortality is low and rarely becomes apparent except by careful studies. 3. A long multiestrus breeding season with successive multiple litters and high reproductive potential. 4. A short natural life expectancy and a high replacement rate of individuals. 5. Vector flea activity throughout the year. 6. Human infections rarely resulting from exposure in permanent foci. 7. Antibody detection in host populations, varying from a small proportion of trapped rodents to essentially 100% of captured animals" (Poland et al. 1994).

Enzootic, reservoir hosts do not suffer 100% mortality from plague. Resistant species, or resistant individuals, are potentially infective to fleas because these species or individuals typically experience a transient, nonfatal, bacteremia. These animals perpetuate the enzootic foci of *Y. pestis.*

Experimental studies were conducted on California voles *Microtus californicus* from two populations. The voles from a population without plague were susceptible to plague, whereas the other voles from a population with enzootic plague were highly resistant (Quan and Kartman 1962). Further investigation of genetically inherited resistance was done using these voles. The F_1 generation showed resistance to plague similar to the resistant parent and was not influenced by gender (Hubbert and Goldenberg 1970).

Heritable resistance to plague was also examined in the grasshopper mouse *Onychomys leucogaster.* First-generation offspring from populations in Colorado, where an epizootic of plague had recently occurred, were 12,500 times more resistant to plague than first-generation mice from a nonenzootic area of Oklahoma (Thomas et al. 1988; Gage and Montenieri 1994). Grasshopper mice are also of interest because they are omnivorous and may receive relatively high exposures to *Y. pestis* via consumption of diseased prey. In addition, grasshopper mice are host for an unusually large number of flea species, many of which are known plague vectors and/or have other rodent hosts involved in the plague cycle (Thomas 1988; Thomas et al. 1988; Gage and Montenieri 1994).

Quan et al. (1985) studied plague in rock squirrels *Spermophilus variegatus,* which are known to have some genetically selected resistance in different populations (Marchette et al. 1962; Quan et al. 1985). Some squirrels withstood experimental exposure to large numbers of bacteria, whereas others died after minimal exposure. In most locations, however, ground squirrels are considered epizootic hosts.

EPIZOOTIC OR AMPLIFICATION RODENT HOSTS. Poland et al. (1994) identified certain features that are characteristic of epizootic or amplification plague hosts found within permanent plague foci: "1. Rodents with low to moderate resistance to plague morbidity and mortality. 2. Relatively little heterogeneity among highly susceptible rodents in response to challenge with *Y. pestis.* 3. Since death rate is great, the available host population density must be high for the outbreak to be sustained. 4. High mortality favors spread of plague, amplifies the intensity of the epizootic and increases the risk of human infection."

Prairie dogs *Cynomys* spp. in the western United States are exemplary epizootic hosts. Plague was first observed in Gunnison's prairie dogs *Cynomys gunnisoni* in 1932 in Arizona, and investigations of numerous epizootics since have documented that prairie dogs are exquisitely sensitive to naturally occurring plague. Mortality may be 100% in affected black-tailed *Cynomys ludovicianus* and Gunnison's prairie dog colonies (Rayor 1985). Mortality reported for white-tailed prairie dogs *Cynomys leucurus* is less, possibly as a consequence of lower density of prairie dogs per hectare (20–30/ha for black-tailed and Gunnison's versus 3–4/ha for white-tailed) (Barnes 1993; Cully 1993; Fitzgerald 1993) and possible differences in flea species and environmental conditions. The 50% lethal dose (LD_{50}) is approximately 50 *Y. pestis* in white-tailed prairie dogs (E.S. Williams personal communication).

Two recent publications have confirmed the complexity of interrelations among species of fleas, small mammals, *Y. pestis,* and susceptible Gunnison's and white-tailed prairie dogs in New Mexico and Wyoming (Anderson and Williams 1997; Cully et al. 1997). However, the precipitating factors necessary for epizootics and the role of fleas and the particular cohabitating mammals in perpetuating the survival of *Y. pestis* remain unknown (Heller 1991; Anderson and Williams 1997; Cully et al. 1997).

High mortality (or a "rat fall") of ground squirrels or prairie dogs often occurs prior to *Y. pestis* infections in susceptible nonrodent hosts. Risk for plague in these hosts is greatest when epizootics in susceptible rodents force infected fleas to seek alternative hosts. It is possible that these episodes of high mortality among epizootic rodent plague hosts do not contribute to the maintenance of *Y. pestis* in nature, but alternatively, these apparent breaches of enzootic cycles may play a role by contributing to the dissemination of *Y. pestis* to new niches by incidental hosts such as carnivores or lagomorphs.

RESISTANT NONRODENT PLAGUE HOSTS. Most ungulates (Gordon et al. 1979; Christie et al. 1980; Christie 1987; Thorne et al. 1987; Jessup et al. 1989; Gage and Montenieri 1994) and rodent-consuming carnivores (McCoy 1911; Rust et al. 1971a,b; Kilonzo 1980; Barnes 1982; Zielinski 1984; Tabor and Thomas 1986; Williams et al. 1991; Orloski and Eidson 1995) studied for exposure and susceptibility to plague mount a robust immune response to *Y. pestis* infection that is usually sufficient to prevent development of signs. The list of animals examined that have tested positive for anti-F_1 *Y. pestis* antibodies is long. Some of these species become ill but rarely die of plague.

Infrequently, clinical plague is seen in domestic dogs *Canis familiaris* (Orloski and Eidson 1995) and coyotes *Canis latrans* (Von Reyn et al. 1976a). Although deeper are not usually considered susceptible to *Y. pestis* illness, Jessup et al. (1989) reported severe ocular infection resulting in the death of a black-tailed deer *Odocoileus hemionus columbianus.* Similar ocular plague has been diagnosed in mule deer *Odocoileus hemionus* in Colorado (M.W. Miller unpublished). Thorne et al. (1987) reported a single case of fatal systemic plague infection in a young mule deer.

SUSCEPTIBLE NONRODENT PLAGUE HOSTS. Certain nonrodent hosts, most notably primates including humans, domestic cats *Felis catus* (Eidson et al. 1991; Gasper et al. 1993), wild cats such as bobcats *Felis rufus* and mountain lions *Felis concolor* (Marchette et al. 1962; Tabor and Thomas 1986; Paul et al. 1994), and black-footed ferrets *Mustela nigripes* (Williams et al. 1994), exhibit high morbidity and mortality when infected with *Y. pestis.* It is in these hosts that the classifications of bubonic, pneumonic, and septicemic plague are used. The unifying feature in these hosts is their permissiveness to the rapid proliferation of *Y. pestis.* Cats, in particular, are an important link in transmission of plague from rodent hosts to humans.

Etiology. Three biovars of *Y. pestis* are biochemically distinguishable based on their ability to ferment glycerol and reduce nitrates to nitrites but have basically identical virulence. Biovar *orientalis* originated in China, and is probably responsible for the sylvatic plague existing in the United States today. Biovar *antiqua* is purported to be the oldest biotype that came out of Central Asia and is now isolated in Africa. *Mediaevalis* is believed to have slowly arisen from the *antiqua* biotype and was the cause of the famous European pandemics (Brubaker 1991; Perry and Fetherston 1997).

Yersinia pestis requires an animal host for its long-term survival, although it may be maintained in culture under proper laboratory conditions. *Yersinia pestis* is killed by desiccation and temperatures above 40° C. Routine disinfectants are effective in killing *Y. pestis*.

Temperature and ionic calcium concentration are two major regulators of virulence factors in *Yersinia* spp. Some are expressed in vivo following a rise in temperature from 26° C to 37° C. Other environmental cues, such as nucleotide and Mg^{2+} concentrations and pH also induce expression of antigens, depending on the microenvironment (cell type) and macroenvironment (host type).

Yersinia pestis growing at low temperature inside a flea makes little or none of the F_1 capsular antigen or the LCR antigens; thus, upon entering a mammalian host from a flea, the bacilli are easily phagocytosed by the polymorphonuclear cells and monocytes (Perry and Fetherston 1997). A certain amount of time is needed for multiplication at the warmer temperatures within a mammal for the bacteria to become endowed with the glycoprotein protective coat and to start producing a variety of virulence factors (Price et al. 1991).

Molecular Biology. The genome of *Y. pestis* is approximately 4380 kb (Lucier and Brubaker 1992). Plasmid and chromosomal encoded virulence factors for *Y. pestis* have been the subject of much research. No specific factor alone is associated with conferring ultimate virulence.

Plasmids are genetic material in bacteria that are separate from chromosomal genes (Brubaker 1991). In pathogenic yersiniae, distinct plasmids encode for virulence factors and are found consistently in wild-type isolates. All three pathogenic yersiniae have the Lcr plasmid. The low-calcium-response stimulon (LCRS) codes for many virulence factors, including V antigen, Yops, Lcr proteins, specific Yop chaperones (Syc), and Yop secretion (Ysc), that control the expression and secretion of the Yops and LcrV (Perry and Fetherston 1997). The V antigen, encoded by the LcrV, is a secreted protein. It is a regulator of the LCRS operon as well as serving in a protective manner to prevent phagocytosis (Price et al. 1991). Recent studies link the LcrV with immunosuppression, possibly inhibiting the production of cytokines and thereby dampening the inflammatory response (Nakajima et al. 1995; Perry and Fetherston 1997). The W antigen is antiphagocytic (Lecker et al. 1989; Brubaker 1991).

Most of the 11 Yop proteins expressed by *Y. pestis* are considered to be virulence factors and function in multiple antiphagocytic and anti-inflammatory capacities (Perry and Fetherston 1997). YopM functions in an immunologically similar manner to a human platelet surface protein and thus may work to inhibit the aggregation of platelets that occurs during the inflammatory response in a mammalian host (Leung and Straley 1989; Reisner and Straley 1992). Studies have shown that Yop E is required for virulence and growth in the liver and spleen. It also has cytotoxic activity (Rosqvist et al. 1994). A more detailed review of *Yersinia* outer-membrane protein molecules can be found in Perry and Fetherston (1997).

Yersinia pestis has two additional virulence-encoding plasmids not found in the other yersiniae. These are the 9.5-kb pesticin (Pst) and the 110-kb exotoxin (Tox) (Protsenko et al. 1983). The Pst plasmid encodes the bacteriocin pesticin, an outer-membrane peptide named plasminogen activator (Pla protease), and a presumed pesticin immunity protein. The Pla protease enables fibrinolytic activity to occur that may facilitate in vivo dissemination of the bacteria (Brubaker 1991). The Pla protease also hydrolyzes many Yop proteins following their synthesis and secretion. It has been postulated that secretion and breakdown of *Y. pestis* Yops facilitates evasion of the host immune system. Loss of the Pst plasmid has been shown to prevent dissemination of the bacteria from peripheral sites (Brubaker et al. 1965).

The Tox plasmid is the largest of the *Y. pestis* plasmids (Perry and Fetherston 1997). Products of the Tox plasmid—murine toxin and the F_1 capsular antigen—enhance the acute nature of plague infection (Brubaker 1991). Friedlander et al. (1995) and others determined that the F_1 capsular protein aids the bacteria in avoidance of phagocytosis. The F_1 proteins are temperature regulated, expressed at 37° C but not at 26° C, and polymerize into a gel-like envelope or capsule around the bacteria (Brubaker 1991). F_1 is the classic antigen used in tests to detect antibodies to *Y. pestis* and was considered at one time to be the definitive virulence factor. However, a recent study established that certain F_1-negative strains were fully virulent (Davis et al. 1996). The murine exotoxin consists of two forms of a protein toxic in mice and rats but is not as toxic in most other species. It is associated with the cell envelope or membrane until cell death (Perry and Fetherston 1997).

The chromosomally encoded fimbriae common to the three pathogenic yersiniae is called pH6 antigen in *Y. pestis* (Lindler et al. 1990; Lindler and Tall 1993; Price et al. 1995). The ability to express fimbriae is considered important because a mutation in this gene results in a 200-fold increase in the LD_{50} of the mutant strain (Carniel 1995).

The chromosomal pigmentation (pgm) locus contains the locus endowing hemin storage ability, termed hms. This is considered an indirect virulence factor because it does not affect the mammalian host but rather influences the ability of the bacteria to sufficiently block the flea (Hinnebusch et al. 1996).

Transmission and Epidemiology. The principal mode of transmission of *Y. pestis* in mammals is via flea bite. Over 1500 flea species have been described as potential vectors, although only a fraction of these are considered epidemiologically important. Other arthropods may carry *Y. pestis* (Poland and Barnes 1979), but no other arthropod has proven to be a natural vector (Pollitzer and Meyer 1961). Federov (1960) experimentally infected camels *Camelus* sp. by using ticks as vectors, but ticks are not considered significant natural vectors of plague.

A single flea bite can transmit up to 15,000 bacteria, which is more than sufficient to effect transmission (Burroughs 1947). In 1914, the role of fleas in plague transmission was elucidated with the observation of *blocking* (Bacot and Martin 1914). Essentially, the flea takes its blood meal from a *Y. pestis*-infected host, introducing the bacteria into its stomach. Here the bacilli respond to changes in temperature and chemical composition. After approximately 2 days in fleas that have not cleared the infection, brown-speckled clusters of bacilli combined with fibrinoid material, and probably hemin, accumulate in the stomach (Bacot and Martin 1914; Cavanaugh 1971; Perry and Fetherston 1997). The bacterial masses may extend into the proventriculus and esophagus. The bacteria infect only the alimentary canal and have not been found in other organs or tissues of the flea (Perry and Fetherston 1997). By 3–9 days following ingestion or sometimes as long as 25 days, depending on a variety of factors, the digestive entrance of the flea is sufficiently blocked. When the starving flea tries to feed, incoming blood causes the esophagus to expand due to the blocked proventriculus, the blood mixes with the bacilli from the plug, and the infected blood is regurgitated back into the host, although the flea remains blocked. This process continues until the flea clears the infection or dies of dehydration and starvation.

Not all blocked fleas will transmit bacteria (Perry and Fetherston 1997). The ability of the bacilli to block a flea is temperature dependent and does not occur above 28° C. Furthermore, blocked fleas will clear an infection if kept at ambient temperatures above the blocking minimum (Cavanaugh 1971). It has been theorized that a flea could also transmit plague by fecal contamination of a bite, because *Y. pestis* has been isolated from flea feces.

Ranking flea species for their vector potential is a complicated evaluation involving host specificity and range, geographic distribution, seasonal prevalence, and life cycle, such as nesting versus body fleas. Some fleas will do what is termed *straggling,* meaning if their primary host is not available they will use other mammalian substitutes. This could contribute to amplification of infection during plague epizootics.

Additionally, the degree of bacteremia in the host and whether or not antibodies are present at the time the blood meal is taken will affect the efficacy of the flea as a vector. Host animal behavior, social structure (large or small groups, solitary), and feeding habits are additional components that affect transmission of plague by fleas. In some species, exposure to plague will be determined by burrowing habits, because infected fleas may remain in burrows for prolonged periods. Lagomorphs and insectivores are known to share habitats with sylvatic rodents, which may enhance exposure to their plague-transmitting fleas (Clover et al. 1989).

Foci of enzootic plague are maintained by certain rodent and flea associations (Table 19.1) (Poland and Barnes 1979). Anderson and Williams (1997) reported that flea species changed on white-tailed prairie dogs from spring to late summer.

A less common mode of transmission than flea bite is ingestion of, or exposure to, another mammal infected with *Y. pestis.* Transmission of the infection in this manner differs significantly from flea-borne transmission for two reasons. First, the numbers of bacteria are much greater in the carcass of an infected mammal compared with a flea bite. Secondly, plague bacilli ingested in prey or inhaled from mammalian tissues or fluids are coming from a mammalian host environment and are producing virulence factors to assist in evading the immune response. Thus, the virulence of the organism is greater from these sources than from fleas. During ingestion, the greatest hazard occurs as the bones or other sharp objects in the prey's bacteria-laden carcass traverse the predator or scavenger's mouth and oropharynx. Any new or preexisting puncture wounds or abrasions can provide the window needed for entrance into the host.

Smith (1994) noted a correlation between the feeding habits of carnivores and omnivores and their exposure to plague. The number of prey species consumed as a percentage of the diet, and whether these were reservoir or susceptible plague hosts, seemed to have an impact on the occurrence of plague antibodies in carnivores. In California, bobcats and foxes *Vulpes* sp. eat cottontail rabbits *Sylvilagus* spp., wood rats *Neotoma* spp. and, less frequently, ground squirrels *Spermophilus* spp. Thus, they are exposed to enzootic hosts (wood rats) and epizootic hosts (ground squirrels), and the likelihood of infection near enzootic foci is high. Mountain lions usually eat deer but will eat wild rodents. Smith (1994) noted that mountain lions are most often exposed to plague when deer populations are low. In regions where plague exists, omnivores tend to have lower seroprevalence than carnivores that principally eat *Y. pestis*-vulnerable species.

In long-term studies in an enzootic area of plague in prairie dogs in Wyoming, coyotes and badgers *Taxidea taxus* had consistently high antibody titers and seroprevalence (greater than 90%) to *Y. pestis.* Prairie dogs presumably made up a large percentage of their diet. In fact, a prairie dog that had died of plague was found in the stomach of a coyote from this area (Williams et al. 1992).

Inhalation of aerosolized bacteria by a mammal in close proximity to an animal with pneumonic plague is a rare but effective mode of transmission. Transmission from domestic cats to humans via scratches, bites, and contact with draining abscesses has been reported (California Department of Health Services 1980; Weniger et al. 1984; Werner et al. 1984; Doll et al. 1994). Close contact with recently dead animals, such as skinning an animal that died of plague, has been shown to cause infection in humans (Von Reyn et al. 1976a,b; Christie et al. 1980; Poland et al. 1973, 1994).

Viable *Y. pestis* can be isolated from soft tissues of carcasses for approximately 1 week and from more protected tissues such as bone marrow over longer

TABLE 19.1—Host-flea complexes involved in *Yersinia pestis* epizootic amplification in western North America, by geographic region

States and Regions	Rodent Species (Hosts)	Flea Species (Vectors)
Arizona, New Mexico, southern Colorado, southern Utah	Rock squirrels Gunnison's prairie dogs	*Oropsylla montana, Hoplopsyllus anomalus* *Oropsylla hirsuta, Oropsylla tuberculata, Oropsylla labis*
Colorado		
Eastern high plains	Black-tailed prairie dogs	*Oropsylla hirsuta, O. tuberculata*
Montane areas	Wyoming ground squirrels	*O. labis, Oropsylla idahoensis*
Denver, Fort Collins, etc.	Fox squirrels	*Orchopeas howardi*
California		
Sierra Nevada	Chipmunks	*Eumolplanus eumolpi,Monopsyllus ciliatus, O. idahoensis, O. montana*
Valleys, foothills, montane	Rock squirrels	*O. montana, H. anomalus*
Mountain meadows	Ground squirrels	*O. idahoensis, Oropsylla rockwoodi, O. montana*
Idaho		
Southern Great Basin	Ground squirrels	Unknown
Nevada, Arizona, New Mexico	Antelope ground squirrels	*Oropsylla bacchi*
Wyoming		
Cheyenne	Fox squirrels	*Orchopeas howardi, O. labis*
High plains	Wyoming ground squirrels	*O. bacchi, O. tuberculata*
High plains	White-tailed prairie dogs	*O. tuberculata, O. labis, O. idahoensis*
Western United States	Wood rats	*Orchopeas sexdentatus, Orchopeas neotomae*

Adapted from Poland et al. (1994) and Anderson and Williams (1997).

intervals. In frozen animal tissues, the bacteria may survive for years (Poland et al. 1994). Transmission of infection from contaminated inanimate matter or long-dead organic materials seems to be negligible.

Clinical Signs, Pathogenesis, and Pathology. The lesions seen in *Y. pestis*-infected mammals vary greatly according to the mode of transmission of the organism and susceptibility of the host. The virulence of *Y. pestis* is derived from its ability to proliferate rapidly and massively in susceptible hosts. Highly susceptible rodents, like prairie dogs and mice, often die so quickly after infection that lesions may be subtle. Murine toxin produced by *Y. pestis* is believed to be a beta-adrenergic antagonist that blocks the receptor for this molecule, thereby leading to circulatory collapse (Montie 1981). This could account for the rapid death in these species.

The three classic clinical manifestations of *Y. pestis* infection include bubonic, pneumonic, and septicemic plague and are seen primarily in susceptible nonrodent species. These forms have been best documented and described in people. An individual may have any combination of them or only one.

Bubonic plague gets its name from the Latin root *bubo,* meaning "swelling," and is descriptive of the lymphadenomegaly that occurs. Acute local inflammation is triggered in the area where the bacteria are introduced, usually by flea-bite inoculation. The bacterium's first destination in the new host will be within the polymorphonuclear cells and the mononuclear cells of the skin and subcutaneous tissues. As they replicate in this intracellular environment, the bacteria express virulence factors that render them less susceptible to phagocytosis when released. If the bacteria can be stopped and contained at the level of the lymph nodes that drain the inoculation site, bubonic plague results. Inguinal and axillary lymph nodes are most frequently involved. Infections following ingestion will frequently produce lymphadenopathy of the submandibular, anterior cervical, medial retropharyngeal, and mesenteric lymph nodes. Hemorrhage and necrosis may occur diffusely in the lymph nodes within a period of several days. In some animals, lymph nodes rupture through the integument and drain spontaneously, as has been seen in cats (Macy and Gasper 1990; Gasper et al. 1993; Gasper 1997) and black-footed ferret × Siberian polecat *Mustela eversmanni* hybrids (E.S. Williams personal communication). Typically highly susceptible wild rodents such as tree and ground squirrels seldom develop buboes, though they may occasionally be present in prairie dogs.

Primary septicemic plague is defined as bacteremia without the presence of palpable buboes. This can occur by direct inoculation of *Y. pestis* into the bloodstream following ingestion of an infected meal by a carnivore or intravascular injection via a bite of an infected animal. Secondary septicemic plague occurs when bacteria in the lymphatics break through affected buboes into the bloodstream.

The bacteria infect many organs via the blood, although liver and spleen are frequently the first to develop lesions. If the organism succeeds in evading the animal's defense mechanisms, fulminating septicemic plague will, in some species, induce disseminated intravascular coagulation and endotoxic shock with widespread hemorrhages and thrombi resulting in extravasation of blood causing a dark reddish-black discoloration of affected tissues, especially the subcutis. The descriptive term *black death* originates from this pathologic lesion (Lichtenberg 1989; Gasper et al. 1993; Gasper 1997). Humans, cats, black-footed ferrets, and prairie dogs may exhibit "black death" lesions.

Primary pneumonic plague may result from aspiration of aerosolized droplets containing *Y. pestis* from an animal with pneumonic plague directly into the airways. Fulminating bacteremic and pneumonic forms of plague are much more severe manifestations of plague than is the bubonic form and have a poor prognosis. Hematogenous spread of the bacteria into the lungs in septicemic plague results in secondary pneumonic plague. The extensive surface area of the pulmonary capillary system may successfully filter and confine *Y. pestis* in the lungs.

The length of time from inoculation to death is variable. White laboratory mice die within 1–3 days with subtle lesions of splenomegaly, hemorrhagic lymphadenitis, and pulmonary edema. Gross lesions of plague in wild rodents, in particular prairie dogs and ground squirrels, vary depending on the duration of clinical disease. Usually death is peracute or acute; gross lesions are subtle and similar to those already mentioned. In addition, marked congestion of subcutaneous vessels is frequently observed. If the duration of clinical disease is longer, pulmonary edema and hemorrhage, serous to fibrinous pleural effusions, splenomegaly, and multifocal splenic and hepatic necrosis may be present. Hemorrhagic lymphadenitis (bubo) is found in sites draining flea bites and may include internal inguinal nodes in addition to axillary and external inguinal lymph nodes. If prairie dogs contract the infection via cannibalism, hemorrhagic lymphadenitis may occur in mesenteric nodes and nodes of the head.

Studies of plague in domestic cats (Eidson et al. 1991; Gasper et al. 1993) and hybrid ferrets (E.S. Williams personal communication) allowed for close observation of clinical signs and pathology in susceptible, nonrodent hosts. Of the symptomatic cats, 54% were bubonic, 15% were pneumonic, and 92% had *Y. pestis* cultured from at least one blood culture. This transient bacteremia was not classified as septicemic plague. Cats exhibited malaise, myalgia, anorexia, and swollen lymph nodes 2–3 days after eating a plague-infected mouse. *Yersinia pestis* induced a rapid febrile response and neutrophilia. These peaked in 3–6 days. Clinical biochemical changes were consistent with a glucocorticoid stress response, dehydration, and blood loss. Consistent findings included increased levels of blood glucose, blood urea nitrogen, potassium, and chloride. Albumin levels fell to below normal levels. Spontaneous draining of buboes, abatement of fever, and neutrophilia correlated with recovery (Eidson et al. 1991; Gasper et al. 1993). In a prospective study of 16 cats fed a plague-infected mouse to mimic the natural route of infection, 38% died within 4–9 days (mean, 5.7 days), 44% developed transient illness and recovered in 14 days, and 19% showed no signs of illness (Gasper et al. 1993).

Lymph node enlargement and abscessation are seen in nodes draining the area of oral inoculation or flea bite. Sublingual abscesses and buboes in submandibular and cervical lymph nodes were seen in the majority of cats with naturally occurring plague, and logistic regression analysis/case-control studies of feline cases in New Mexico indicated that ingestion is a more important route of transmission in cats than flea bites (Eidson et al. 1991). This is probably also true of wild carnivores. Lymph nodes may be enlarged ten times normal size or more. They are soft, pulpy, and may be plum-colored due to hemorrhage. Complete infarction or rupture through the skin may occur. Occasionally, cats and other mammals with plague have abscesses of the face, gluteal muscles, and hind limbs or forelimbs (Eidson et al. 1991). These may be cases of flea-bite-contracted plague. Alternatively, *Y. pestis* could have been transferred from the oropharynx during self-grooming or injected via bite wounds from a plague-infected rodent or cat.

Histologically, lesions of plague teem with *Y. pestis*. Marked tissue swelling early in infection is due to protein and polysaccharide-rich effusion, with few inflammatory cells. This is followed by necrosis of tissues and blood vessels, with hemorrhage and thrombosis. Finally, neutrophilic infiltrates progressively accumulate in zones around the necrotic areas. Normal nodal architecture is totally replaced by hemorrhage, necrosis, and large clusters of bacteria surrounded by a mantle of neutrophils and early fibrosis.

Pneumonic lesions in cats are multifocal areas of bronchopneumonia, with microscopic semblance to buboes: a central area of bacteria, hemorrhage, and necrotic debris with dense neutrophilic infiltrations in surrounding alveoli and small bronchioles (Eidson et al. 1991; Gasper et al. 1993).

Death within 3–7 days was typical in hybrid and black-footed ferrets exposed to *Y. pestis* either orally or via subcutaneous injection. The animals rapidly became febrile, anorexic, and depressed. Lesions were characterized by hemorrhagic and necrotizing regional lymphadenopathy. Severe pulmonary edema and occasional perivascular hemorrhage occurred in most affected animals. Hemorrhage was present also in the mucosa of the distal colon in many affected ferrets (Williams et al. 1994 and personal communication). Gross and microscopic lesions are similar in white-tailed prairie dogs, except that subcutaneous vascular congestion and hemorrhage are relatively common (E.S. Williams personal communication).

One mule deer with systemic plague had bronchopneumonia and lymphadenitis (Thorne et al. 1987).

Ocular plague has been documented in mule deer in California and Colorado (Jessup et al. 1989; M.W. Miller unpublished). Necrosis and suppurative inflammation occurred in the eyes, and numerous *Y. pestis* were isolated from the vitreous humor, retina, and conjunctiva.

Diagnosis. A history of rapid population decline of colonial rodents such as prairie dogs and ground squirrels is suggestive plague. Confirmation requires collection of dead rodents, using appropriate personal protection, and submission to a diagnostic laboratory for evaluation. Individual wild rodents or other susceptible mammals may be found dead; plague should be considered a differential diagnosis in areas of enzootic plague. Observation of clinically affected wild mammals is unlikely.

A suspect case of plague can usually be confirmed quickly because of the large numbers of organisms present in infected tissues. Provisional evidence is presence of bipolar-staining, safety-pin-like, Gram-negative coccobacilli in lymph node aspirates with Wayson or Giemsa stain. Lymph node or abscess aspirates or tracheal washes can be tested with fluorescent antibody (FA) staining, the most sensitive and specific rapid diagnostic test.

Gross and microscopic lesions may be suggestive of plague, but they are not definitive. Immunohistochemistry may be of assistance in making a diagnosis on sections of fixed tissues (Williams et al. 1994). Culture of *Y. pestis* is necessary for definitive diagnosis. Affected lymph nodes, blood, spleen, liver, and lung are preferred tissues to be submitted for culture. The organism is grown on blood agar and colonies identified by morphology, Gram stain, biochemical reactions, and phage lysis (Quan et al. 1979). In rare cases, such as heavy contamination, inoculation of suspect material into laboratory mice, with subsequent cultural identification, is used for diagnosis.

Freshly dead animals are always the best for diagnostic purposes; however, even extremely decomposed carcasses and washes of the marrow cavity of dried bones may be successfully tested by the FA test, though the sensitivity is probably low.

Animals surviving infection with *Y. pestis* develop serum antibodies that can be used for diagnosis of exposure. Historically, passive hemagglutination (PHA) and the hemagglutination inhibition test were used for serologic diagnosis of exposure to *Y. pestis* (Hudson and Kartman 1967). A PHA antibody titer 1:32 is usually considered evidence of exposure to *Y. pestis*. An enzyme linked immunosorbent assay is currently being validated by the U.S. Centers for Disease Control (M. Chu personal communication).

Serologic surveillance of carnivores to monitor plague activity has been used in and near enzootic foci. The presence of antibodies to *Y. pestis* indicates exposure and existence of plague within the animal's home range. Detectable antibodies may remain for long periods, as evidenced by a survey done in badgers where PHA antibodies were detected 300 days after seroconversion (WHO 1970). Traditional sentinel species in surveys in North America have been domestic dogs, coyotes, and badgers (Messick et al. 1983; Gage and Montenieri 1994). In California, feral pigs *Sus scrofa* were used in addition to wild carnivores (Smith 1994). Other wild carnivores used in surveillance include black bears *Ursus americanus,* mountain lions, and foxes. Blood can be obtained from these species when handled for management purposes, during predator reduction programs, or from animals that have been harvested by hunters (Clark et al. 1983; Clover et al. 1989; Gage and Montenieri 1994; Smith 1994). Serologic evidence of exposure in rodent-eating carnivores gives an indication of plague activity that would have required hundreds of rodent samplings to make evident (Gage and Montenieri 1994).

Surveillance may also be accomplished by examination of fleas collected in areas of suspected plague activity. Fleas are collected on flannel squares, identified, pooled, titurated, and injected into mice to detect the presence of *Y. pestis*. Inoculated mice are observed for at least 21 days for evidence of disease. Tissues from mice that die are cultured for *Y. pestis*. Direct culture of fleas is not very sensitive, and mouse passage greatly improves detection of the organism. Newer techniques, such as polymerase chain reaction and in situ hybridization, have been developed (Trebesius et al. 1998) and may replace mouse inoculation as these techniques are refined.

Differential Diagnoses. Acute bacterial infections may mimic sylvatic plague. Tularemia, caused by *Francisella tularensis,* shares several key features with plague, and the geographic distributions overlap (Kaufman 1990). Both infections cause an acute, severe febrile disease in some species, and pneumonia or septicemia can ensue. Although colonial rodents are susceptible to tularemia and it may be found during surveillance for plague (E.S. Williams personal communication), it does not appear to induce the high mortality typical of plague. Septicemic or pneumonic pasteurellosis caused by *Pasteurella* spp. may cause individual mortality in wild rodents and occasional localized epizootics. In addition to causing death of rodents, *Pasteurella* spp. may cause lymphadenitis in carnivores that mimics plague buboes. Culture will differentiate these conditions.

Gross lesions of plague, such as foci of necrosis in liver and spleen, may be caused by a variety of other septicemic bacterial infections. Yersiniosis due to *Y. pseudotuberculosis* or *Y. enterocolitica* may induce gross and microscopic lesions similar to plague that must be differentiated by culture. Some parasites or parasite migration tracts may cause pale foci in the liver, but these are usually easily distinguished from lesions of plague.

In addition to plague, acute population declines of colonial rodents may be caused by poisoning (zinc phosphide, strychnine, etc.) as part of pest control programs, shooting for recreational or pest control purposes, and localized environmental catastrophes, such as thunderstorms and flooding.

Immunity. It remains to be elucidated why some rodents and carnivores are resistant and some are susceptible and whether and how resistant individuals arise. Antibodies of maternal origin were detected in experimental studies in the Norway rat *Rattus norvegicus* (Williams et al. 1974, 1977), but antibodies are not always sufficient to protect against fatal *Y. pestis* infection. The difference between a resistant individual and a susceptible individual is, therefore, a manifestation of its innate immune response, its current immune status, and the dose and route of *Y. pestis* transmission.

Control and Treatment. Treatment of wild species for plague is seldom possible or appropriate. On the rare occasion when the decision is made to treat a wild animal, antibiotic therapy and supportive treatment should be instituted. All live plague suspects should be handled with gloves while wearing eye protection, gowns, and high-density surgical masks or respirators. Plague suspects should be dusted with carbamates (Orsted et al. 1998) or pyrethrins (Beard et al. 1992) to kill fleas.

Although *Y. pestis* is susceptible to a wide variety of antimicrobial agents in vitro, it is resistant to drugs like penicillin and ampicillin in vivo. Tetracyclines and trimethoprim-sulfamethoxazole appear to be the most effective antibiotics in uncomplicated cases (Eidson et al. 1991; Gasper 1997).

Vaccination of free-ranging wild mammals is not possible at this time. The possibility that *Y. pseudotuberculosis* infection may provide immunity to *Y. pestis* infection is an interesting tenet worthy of study in the context of management and immunoprotection of wild mammals (Mollaret 1995).

For decades, scientists from many disciplines have been working to develop a vaccine to protect humans from plague. In some circumstances, there may be a need for plague vaccines for animals either because they are important in exposing humans to plague or because they are a susceptible endangered species. The currently licensed vaccine for animals in the United Kingdom and United States is a human vaccine [plague vaccine, USP (CDC 1996)] that is a formalin-killed whole cell preparation. It relies on the generation of an antibody response against the F_1 capsular antigen. There has been indirect evidence that this vaccine provides significant protection against bubonic plague although not against pneumonic plague. It does not protect domestic cats against challenge with *Y. pestis* (Gasper 1997). An attenuated vaccine provides protection for humans, but severe side effects also occur and thus it is not widely used (Anderson et al. 1996). Bioterrorism has reemerged as a focus in human warfare, and the need for a plague vaccine has become even more pressing.

In some circumstances, either for protection of human health or related to management of endangered species, treatment of the environment to kill fleas has been conducted. Historically, this has been accomplished with the carbamate insecticide carbaryl (Barnes et al. 1972). White-tailed prairie dog towns in Wyoming were dusted with carbaryl in an attempt to break an epizootic of sylvatic plague and protect the prey base of endangered black-footed ferrets (Thorne and Williams 1988). Similarly, black-tailed prairie dog colonies in Montana were treated with permethrins to kill fleas in an attempt to protect reintroduction sites for black-footed ferrets. A significant problem with widespread use of these products is killing nontarget insect species, with associated ramifications to the ecosystem.

Attempts to eliminate sylvatic plague by eradicating rodent hosts have been made in the past but are no longer considered appropriate in the wild. However, control of wild rodents within or adjacent to human habitation is appropriate and is performed for sanitary reasons and because of a variety of rodent-borne pathogens, not only plague. It is important to recognize that killing rodent hosts during an epizootic, without also applying flea controls, could lead to hungry fleas switching from normal rodent hosts to domestic pets and humans. Food sources and rodent habitats, such as rock and brush piles, should be kept away from houses. It is also important to treat pet dogs and cats frequently for fleas if they are allowed to roam outside.

Public Health Concerns. Persons working with wildlife and in veterinary practices in plague-endemic areas should be aware of the signs and symptoms of plague and the risks associated with handling plague-infected animals; appropriate personal protections should be taken to prevent exposure and infection (CDC 1994b). Such protection may include wearing gloves, eye protection, and appropriate respiratory precautions when handling animals suspected of having plague (CDC 1994b). Some high-risk individuals may wish to consider vaccination or prophylactic antibiotic treatment (Gage et al. 1995). Depending on the situation, those at risk should consult with knowledgeable physicians for medical advice.

In collaboration with local public health authorities, wildlife managers may provide public education about plague in endemic areas, may restrict activities such as camping in areas with active plague, or may issue warnings to hunters, trappers, and outdoor recreationists about ways to reduce the risk of exposure to plague.

Acute onset of fever, chills, swollen and painful lymph nodes (buboes), and malaise characterizes bubonic plague in humans. Septicemic and pneumonic plague in humans are extremely serious, life-threatening conditions and clinically are characterized by fever, chills, prostration, coughing, respiratory difficulties. Shock, hemorrhage, and death may follow. Those working with wildlife should seek medical advice if illness follows possible plague exposure.

Before the advent of antibiotics, mortality from bubonic plague in humans varied between 50% and

90%, and pneumonic plague and septicemic plague were invariably fatal. Today, with appropriate antibiotic therapy, mortality among people is about 5%–20%. In human cases, chloramphenicol and aminoglycoside antibiotics are believed to be the most effective in severe infections. The efficacy of antibiotics depends greatly on timing. When given after 24 hours of pneumonic or septicemic plague, little benefit may be noted.

Antimicrobial drug resistance to many bacteria, including *Y. pestis,* is an increasingly important public health concern (Ostroff 1995; Dennis and Hughes 1997). Of particular concern is the recent report of an antibiotic-resistant strain of *Y. pestis* in a human with plague (Galimand et al. 1997).

Worldwide there are 1000–2000 human cases each year (Gage and Dennis 1995). Twenty-four countries reported human plague cases between 1980 and 1994. In 1994, a short but significant outbreak of pneumonic or bubonic plague occurred in India (CDC 1994a; Dennis 1994; Campbell and Hughes 1995; Jayaraman 1995). The outbreak followed a typical pattern, with death of rats just prior to the human epidemic. An earthquake in Latur in October 1993 is theorized to have caused a disruption of the burrows of a variety of wild rodents with endemic *Y. pestis.* These rodents commingled with the more susceptible domestic rats. Dead rats were documented in August 1994; shortly thereafter, approximately 10% of the people in the village contracted bubonic plague. One person traveled from there to Surat; subsequently, Surat announced a number of cases of pneumonic plague, and 600,000 residents fled in a panic that fortunately did not result in a more extensive epidemic. This episode highlights how human plague is usually preceded by an environmental change that disrupts wild rodents with endemic *Y. pestis,* leading to infection of urban hosts. In addition, it is theorized that stopping nationwide malaria insect control programs enabled reemergence of the flea population (CDC 1994a; Dennis 1994; Campbell and Hughes 1995; Gage and Dennis 1995; Jayaraman 1995).

As humans continue to encroach upon traditional animal habitats, plague may result. Further investigations and more detailed knowledge of the complex biology and ecology of *Y. pestis* can hopefully prevent reenactment of the devastating effect that plague has had on large numbers of humans and domestic and wild mammals.

Management Implications. A plague epizootic can decimate localized populations of colonial rodents and may be important if these species are considered threatened or endangered. Endangered prairie dog species, such as the Utah prairie dog *Cynomys parvidens,* are at risk due to sylvatic plague.

In addition to canine distemper, plague epidemics have been devastating to the North American black-footed ferret, which is an endangered species. Previous studies had suggested that these animals probably were resistant because taxonomically close species, domestic ferrets *Mustela putorius furo* and Siberian polecats, were not clinically affected following experimental infection (Williams et al. 1991). However, subsequently it was found that black-footed ferrets are highly susceptible to plague mortality. One report of plague in a captive black-footed ferret described acute mortality within days of exposure (Williams et al. 1994), and more than 20 captive black-footed ferrets died of plague following exposure to prairie dogs with plague (D.E. Biggins and D. Garell unpublished). Black-footed ferrets are susceptible to effects of plague directly by fatal infection and indirectly because plague kills prairie dogs, which are the prey upon which they depend. Sylvatic plague is considered the most significant impediment to recovery of free-ranging black-footed ferret populations.

In addition, prairie dogs are considered a keystone species because their extensive burrows create an ecosystem that supports at least 170 wild species, some of which are being considered for listing as threatened or endangered. Loss of prairie dogs due to plague may significantly impact these systems.

Domestic Animal Concerns. The susceptibility of domestic cats to plague is the most significant domestic animal concern. The disease in cats has been well documented by Eidson et al. (1991) and Gasper et al. (1993). Domestic dogs rarely contract plague (Rust 1971a,b); however, they are susceptible to infection. Domestic pets that spend time outside in plague areas should be treated for fleas because of the possibility they could transport these vectors into homes. Plague in other domestic species is not an important consideration.

YERSINIOSIS. Yersiniosis is a zoonotic disease caused by *Y. pseudotuberculosis* or *Y. enterocolitica* in a wide variety of animals around the world. Conceptually, yersiniosis is similar to the disease seen with pathogenic strains of *Escherichia coli. Yersinia pseudotuberculosis* and *Y. enterocolitica* are a heterogeneous group of bacterial strains that survive in both animal and environmental reservoirs. Distinct strains or serogroups are considered pathogenic, whereas others are avirulent. Though primarily a gastrointestinal disease, extraintestinal yersiniosis may be observed.

Both *Y. pseudotuberculosis* and *Y. enterocolitica* are endowed with a large repertoire of chromosomal and plasmid-encoded determinants that enable their survival under varied circumstances. They may survive and even replicate in soil and aquatic environments outside of a host for months to years because of their minimal nutritional requirements and tolerance of temperature extremes from 5° C–42° C (Brubaker 1991; Bottone 1997). The pathogenic variants of *Y. pseudotuberculosis* and *Y. enterocolitica* have a *Y. pestis*-like predilection for lymphoid tissue, and they survive in intracellular and extracellular locations within a host.

Yersinia pseudotuberculosis is more closely related to *Y. pestis* than it is to *Y. enterocolitica* (Bercovier et al. 1980) and is more often associated with yersiniosis in wildlife. It has been postulated that *Y. pseudotuberculosis* is a mutant *Y. pestis* that arose in Europe during the second plague pandemic, which provided rodents with sufficient immunity to *Y. pestis* so that Europe was protected from the third plague pandemic (Mollaret 1995). Yersiniosis in humans invariably refers to infections with *Y. enterocolitica* and is considered an "emerging infection" (Institute of Medicine 1992).

History and Distribution. *Yersinia pseudotuberculosis* was originally described as *Bacillus pseudotuberculosis* in 1883 when it was isolated from a guinea pig *Cavia porcellus* inoculated with samples from affected human tissue. The distinguishing feature of this bacterium at the time was "its attitude to infect animals slowly until an intercurrent factor lets the disease occur" (Mollaret 1995). In 1929, it was named *Pasteurella pseudotuberculosis* (Kapperud 1994). It was also called *Shigella pseudotuberculosis* for a time prior to 1935 (Kapperud 1994). In an excellent review, Mollaret (1995) states "Europe was undoubtedly the birthplace of *Y. pseudotuberculosis.* Although, after the first and particularly after the second world wars, *Y. pseudotuberculosis* was progressively isolated outside Europe, this initial and exclusive localization in Europe was largely proved by the countless bacteriological controls carried out from the beginning of the century by the harbor laboratories for plague and rat control. Rats captured and examined postmortem from boats as well as on land enabled the isolation of *Y. pseudotuberculosis* outside of Europe." *Yersinia pseudotuberculosis* received its current name in 1965.

The first reference to *Y. enterocolitica* described a Gram-negative coccobacillus isolated from facial abscesses of a farmer with cervical lymphadenitis (McIver and Pike 1934). It was initially called *Bacterium enterocoliticum* and then renamed *Pasteurella* X and finally *Yersinia enterocolitica* in the 1960s (Butler 1983). It was isolated from humans with gastroenteritis in Western Europe, North and South America, Africa, Asia, and the Middle East (Mollaret 1995). Worldwide surveillance data suggest a marked increase in human yersiniosis in the last two decades (Ostroff 1995). Only after being recognized as a human pathogen did these bacteria begin to be studied as agents of disease in veterinary medicine.

Yersinia pseudotuberculosis and *Y. enterocolitica* have now been found on every continent except Antarctica. They are widely distributed in humans, animals, soil, water, and food (milk and meat products; vegetables) in numerous countries throughout the world (Ostroff 1995; Bottone 1997). Yersiniosis has been noted in wildlife worldwide (Welsh et al. 1992).

Host Range. *Yersinia pseudotuberculosis* and *Y. enterocolitica* have been isolated from artiodactyls, carnivores, rodents, marsupials, swine, and primates (Mair 1973; Mingrone and Fantasia 1988). The presence of these bacteria in both cold-blooded vertebrates and invertebrates rarely has been reported (Mair 1973; Hurvell 1981; Shayegani et al. 1986; Cover and Aber 1989; Kwaga and Iversen 1993). Species vary in their susceptibility to yersiniosis. In contrast to *Y. pestis,* there do not appear to be particular enzootic and epizootic hosts.

Yersiniosis is considered to be one of the most significant infectious diseases of farmed cervids in New Zealand, the United States, and Australia (Henderson 1983; Haigh and Hudson 1993). Frequently, animals younger than 1 year of age are affected (Sanford 1995). *Yersinia pseudotuberculosis* type III was the cause of yersiniosis outbreaks in Ontario, which included red deer *Cervus elaphus elaphus,* elk *Cervus elaphus,* and their hybrids, and fallow deer *Dama dama.* In addition to gastrointestinal disease, abortions and mastitis occur in captive hoofed stock due to either pathogenic species (Welsh et al. 1992; Welsh and Stair 1993; Sanford 1995; Hum et al. 1997).

Outbreaks in captive ungulates seldom result in high mortality. An epizootic of *Y. pseudotuberculosis* type III killed blackbuck antelope *Antilope cervicapra* and addax *Addax nasomaculatus* in a wildlife park (Welsh et al. 1992). A winter outbreak at the National Zoo in Washington, DC, resulted in acute death of one dik-dik *Madoqua kirkii,* three blesbok *Damaliscus pygargus,* and one giant anteater *Myrmecophaga tridactyla* due to *Y. pseudotuberculosis* type IA (Baskin et al. 1977).

New and Old World monkeys have succumbed to yersiniosis due to both enteric pathogens, although more often from *Y. pseudotuberculosis* (McClure et al. 1971; Chang et al. 1980; Rosenberg et al. 1980). These outbreaks occurred in animals that were raised in outdoor colonies, as well as those reared in cages.

Disease caused by *Y. pseudotuberculosis* affected over 100 free-ranging muskox *Ovibos moschatus* (Blake et al. 1991). Sporadic cases have been documented in foxes (Mair 1973; Balck et al. 1996). European hares *Lepus europaeus* are highly susceptible to yersiniosis from both enteric species (Mair 1973; Wuthe et al. 1995). Sporadic cases have been reported in beavers *Castor canadensis* and raccoons *Procyon lotor* (Hacking and Sileo 1974).

Etiology. Wild-type *Y. pseudotuberculosis* and *Y. pestis* strains are closely related (Bercovier et al. 1980; Brubaker 1991). Plasmid contents represent the primary structural differences between *Y. pestis* and these species. Changes in temperature alter bacterial gene expression, which arms these microbes with special features for their survival under varying conditions. Some of these virulence factors (Miller et al. 1988) are expressed at 37° C but not at 25° C. One virulence factor, YadA, may be associated with promoting chronic rather than acute disease, thereby facilitating fecal dissemination (Brubaker 1991).

Serotyping and biotyping of these organisms are based on somatic antigens and biochemical reactions, respectively. Somatic and flagellar antigens are diverse and numerous. The somatic O antigen, or cell wall antigen, is a polymer of repeating oligosaccharide units of three or four monosaccharides. This is the most important antigen for serotyping. Further antigenic typing of some O antigen groups may be done using the flagellar H antigens, which differ due to amino acid sequence (Aleksic et al. 1986). Identifying *Y. pseudotuberculosis* and *Y. enterocolitica* at the level of specific biotypes and serotypes is a critical component of proper diagnosis (Quan 1979). Common pathogenic strains in mammals in North America are IA, IB, and III (Mair 1973; Baskin et al. 1977).

Transmission and Epidemiology. The two routes of infection in animals are through fecally contaminated water and food sources and ingestion of infected prey by carnivores (Mair et al. 1967). Birds and rats are reported to be reservoirs of infection for *Y. pseudotuberculosis* (Obwolo 1976). Food contaminated by rodent or bird feces has been suspected as the original source of the bacteria in some epidemics. Pigeons *Columba livia* and rats trapped in areas experiencing an outbreak of yersiniosis were culture positive for *Y. pseudotuberculosis* (Baskin et al. 1977). Some birds, primarily raptors and migratory species, likely contribute to transmission.

Outbreaks of yersiniosis may be precipitated by stressors such as cold and wet weather, decreases or changes in food availability, overcrowding, or capture. There is a seasonal occurrence, with increased incidence associated with the colder temperatures of late autumn, winter, and early spring (Mair 1973; Zwart 1993). Affected animals may be carriers of the bacteria that break with disease due to the physiologic effects of stress. Yersiniosis may also occur as a consequence of exposure to the bacteria concomitant with stressors or simply as a sporadic infection.

Clinical Signs. The signs of *Y. pseudotuberculosis* and *Y. enterocolitica* infection are very similar and usually are a mild to severe gastroenteritis, possibly accompanied by mesenteric lymphadenitis. Septicemia follows in the most severe cases. Clinical signs include lethargy, anorexia, diarrhea, respiratory distress, incoordination, and emaciation. Acute, subacute, and chronic infections may occur.

Pathology and Pathogenesis. The bacteria gain entry to the host by ingestion and invade the intestinal epithelium usually in the jejunum or ileum. They attach and penetrate the mucus layer overlying mucosal epithelial cells and ultimately adhere to and colonize intestinal cell brush border membranes. From this site, the bacteria are transported into the lamina propria by dome (M cell) epithelium overlying Peyer's patches (Hanski et al. 1989). In elegant experimental studies, colonies of *Y. enterocolitica* could be observed beneath, but not within, the intact epithelium of Peyer's patches (Hanski et al. 1989). Following an intracellular stage in which different virulence factors are expressed, pathogenic strains may be disseminated throughout the host. They have a propensity to infiltrate the liver and spleen; lungs, kidneys, and bone marrow may also be involved, although less commonly (Zwart 1993).

Gross lesions include 1–3-cm-diameter, gray to yellow, granulomatous nodules throughout liver, spleen, and, sometimes, the lungs. Mesenteric lymph node infection and necrotic gastrointestinal lesions are frequent. Occasionally, the spleen is enlarged, and serofibrinous pneumonia and peritonitis are seen.

Microscopically, the nodules consist of areas of central caseation to liquefaction, usually with bacterial colonies in the necrotic debris, and a surrounding zone of lymphocytes and macrophages. There is usually little encapsulation and no giant cells. Histologic lesions may mimic plague and tularemia and must be differentiated by FA and/or culture.

Diagnosis and Differential Diagnoses. Because of the shared clinical and pathologic features of infection due to *Y. enterocolitica* or *Y. pseudotuberculosis,* diagnosis depends on culture. Feces and tissues are used for cultures. Excellent reviews are available that discuss techniques for culture and identification (Quan 1979; Bottone 1997).

Many conditions resemble yersiniosis, including clostridial enterotoxemias (Welsh et al. 1992) and salmonellosis. In addition, other types of bacterial septicemias may resemble yersiniosis, including pasteurellosis, plague, salmonellosis, and tularemia.

Serologic cross-reactions have been documented between *Salmonella* groups B and D, some strains of *E. coli, Enterobacter cloacae,* and *Y. pseudotuberculosis* serogroups II, IV, IVA, and VI (Bercovier and Mollaret 1984). Similarly, antibodies induced by *Yersinia enterocolitica* serogroup O:9 cross-react with some *Brucella* spp., and serogroup O:12 cross-reacts with *Salmonella* factor O:47 (Bercovier and Mollaret 1984). Antigenic similarity exists between *Y. enterocolitica* and *Vibrio cholerae,* as well.

Domestic Animal Concerns. Yersiniosis is not considered a serious disease of domestic species, though morbidity and mortality may occur.

Public Health Concerns. Yersiniae are ubiquitous in the environment. Appropriate hygienic measures should be employed to prevent human exposure during handling of an animal suspected of having yersiniosis. Many cases of yersiniosis in humans may not be true zoonoses, because strains of the bacteria historically associated with human disease are not those commonly isolated from nonhuman animal sources (Kapperud 1994).

Yersinia enterocolitica is the most frequently isolated yersiniae from people and is considered by some to be an emerging infectious disease (Bissett et al. 1990; Ostroff 1995; Bottone 1997). It causes abdominal symptoms in humans similar to those of appendicitis. The severity of illness in humans ranges from self-limited enterocolitis to fatal systemic infection (Cover and Aber 1989; Bottone 1997). Extraintestinal symptoms include bacteremia, arthritis, pharyngitis, pyomyositis, and erythema nodosum. Human infection with *Y. pseudotuberculosis* is rare.

Sources of human infection include ingestion of contaminated food and water, contact with infected animals, and blood transfusions (Hacking and Sileo 1974; Cover and Aber 1989; Bissett et al. 1990; Bottone 1997). Disease following consumption of water fecally contaminated by wild animals has been reported (Bissett et al. 1990). Consumption of wild hares in the winter months has been associated with human yersiniosis in Belgium (Hacking and Sileo 1974).

Management Implications. Yersiniosis usually affects only individual or a few animals and is of little concern in populations of free-ranging species. It is of more concern in animals that are concentrated or stressed and may be an important disease of farmed cervids.

LITERATURE CITED

Aleksic, S., J. Bockemuehl, and F. Lange. 1986. Studies on the serology of flagellar antigens of *Yersinia enterocolitica* and related *Yersinia* species. *Zentrablatt fur Bakteriologie, Mikrobiologie, und Hygiene [A]* 261:299–310.

Anderson, G.W., Jr., S.E. Leary, E.D. Williamson, R.W. Titball, S.L. Welkos, P.L. Worsham, and A.M. Friedlander. 1996. Recombinant V antigen protects mice against pneumonic and bubonic plague caused by F_1-capsule-positive and -negative strains of *Yersinia pestis. Infection and Immunity* 64:4580–4585.

Anderson, S.H., and E.S. Williams. 1997. Plague in a complex of white-tailed prairie dogs and associated small mammals in Wyoming. *Journal of Wildlife Diseases* 33:720–732.

Bacot, A.W., and C.J. Martin. 1914. Observations on the mechanism of the transmission of plague by fleas. *Journal of Hygiene* 13:423–439.

Balck, S.S., F.W. Austin, and E. McKinley. 1996. *Yersinia pseudotuberculosis* and *Listeria monocytogenes* serotype 4 from a gray fox (*Urocyon cinereoargenteus*) with canine distemper. *Journal of Wildlife Diseases* 32:362–366.

Barnes, A.M. 1982. Surveillance and control of bubonic plague in the United States. *Symposium of the Zoological Society of London* 50:237–270.

———. 1993. A review of plague and its relevance to prairie dog populations and the black-footed ferret. In *Proceedings of the symposium on the management of prairie dog complexes for the reintroduction of the black-footed ferret,* ed. J.L. Oldemeyer, D.E. Biggins, B.J. Miller, and R. Crete. Washington, DC: U.S. Fish and Wildlife Service, pp. 28–37.

Barnes, A.M., L.J. Ogden, and E.G. Campos. 1972. Control of the plague vector *Opisocrostis hirsutus,* by treatment of prairie dog (*Cynomys ludovicianus*) burrows with 2 percent carbaryl dust. *Journal of Medical Entomology* 9:330–333.

Baskin, G.B., R.J. Montali, and M. Bush. 1977. Yersiniosis in captive exotic mammals. *Journal of the American Veterinary Medical Association* 171:908–912.

Beard, M.L., S.T. Rose, A.M. Barnes, and J.A. Montenieri. 1992. Control of *Oropsylla hirsuta,* a plague vector, by treatment of prairie dog burrows with 0.5% permethrin dust. *Journal of Medical Entomology* 29:25–29.

Bercovier, H., and H.H. Mollaret. 1984. Genus XIV: *Yersinia.* In *Bergey's manual of systematic bacteriology,* ed. N.R. Krieg. Baltimore: Williams and Wilkins, pp. 498–506.

Bercovier, H., H.H. Mollaret, J.M. Alonso, J. Brault, G.R. Fanning, A.G. Steigerwalt, and D.J. Brenner. 1980. Intra- and interspecies relatedness of *Yersinia pestis* by DNA hybridization and its relationship to *Yersinia pseudotuberculosis. Current Microbiology* 4:225–229.

Bissett, M.A., C. Powers, S.L. Abbott, and J.M. Janda. 1990. Epidemiologic investigations of *Yersinia enterocolitica* and related species: Sources, frequency, and serogroup distribution. *Journal of Clinical Microbiology* 28:910–912.

Blake, J.E., B.D. McLean, and A. Gunn. 1991. Yersiniosis in free-ranging muskoxen on Banks Island, Northwest Territories, Canada. *Journal of Wildlife Diseases* 27:527–533.

Bottone, E.J. 1997. *Yersinia enterocolitica:* The charisma continues. *Clinical Microbiology Reviews* 10:257–276.

Brubaker, R.R. 1991. Factors promoting acute and chronic diseases caused by yersiniae. *Clinical Microbiology Reviews* 4:309–324.

Brubaker, R.R., E.D. Beeskey, and M.J. Surgalla. 1965. *Pasteurella pestis:* Role of pesticin I and iron in experimental plague. *Science* 149:422–424.

Burroughs, A.L. 1947. Sylvatic plague studies: The vector efficiency of nine species of fleas compared with *Xenopsylla cheopis. Journal of Hygiene* 45:371–396.

Butler, T. 1983. *Plague and other* Yersinia *infections.* New York: Plenum.

———. 1989. The black death past and present: 1. Plague in the 1980s. *Transactions of the Royal Society of Tropical Medicine and Hygiene* 83:458–460.

California Department of Health Services. 1980. Death from primary plague pneumonia at South Lake Tahoe. *California Morbidity Weekly Report* 48:1.

Campbell, G.L., and J.M. Hughes. 1995. Plague in India: A new warning from an old nemesis. *Annals of Internal Medicine* 112:151–153.

Carniel, E. 1995. Chromosomal virulence factors of *Yersinia. Contributions to Microbiology and Immunology* 13:218–224.

Cavanaugh, D.C. 1971. The specific effect of temperature upon the transmission of the plague bacillus by the oriental rat flea (*Xenopsylla cheopis*). *American Journal of Tropical Medical and Hygiene* 20:264–272.

Centers for Disease Control and Prevention (CDC). 1994a. Update: Human plague—India, 1994. *Morbidity and Mortality Weekly Report* 43:722–723.

———. 1994b. Human plague: United States, 1993–1994. *Morbidity and Mortality Weekly Report* 43:242–246.

———. 1996. Prevention of plague: Recommendations of the Advisory Committee on Immunization Practices (ACIP). *Morbidity and Mortality Weekly Report* 45:1–15.

Chang, J., J.L. Wagner, and R.W. Kornegay. 1980. Fatal *Yersinia pseudotuberculosis* infection in captive bushbabies. *Journal of the American Veterinary Medical Association* 177:820–821.

Christie, A.B. 1987. Plague: Tularemia. In *Infectious diseases: Epidemiology and clinical practice.* Edinburgh: Churchill Livingstone, pp. 1036–1069.

Christie, A.B., T.H. Chen, and S.S. Elberg. 1980. Plague in camels and goats: Their role in human epidemics. *Journal of Infectious Diseases* 141:724–726.

Clark, R.K., D.A. Jessup, D.W. Hird, R. Rupanner, and M.E. Meyer. 1983. Serologic survey of California wild hogs for antibodies against selected zoonotic disease agents. *Journal of the American Veterinary Medical Association* 183:1248–1251.

Clover, J.R., T.D. Hofstra, B.G. Kuluris, M.T. Schroeder, B.C. Nelson, A.M. Barnes, and R.G. Boltzer. 1989. Serologic evidence of *Yersinia pestis* infection in small mammals and bears from a temperate rainforest in North Coastal California. *Journal of Wildlife Diseases* 25:52–60.

Cornelius, G.R. 1994. *Yersinia* pathogenicity factors. *Current Topics in Microbiology and Immunology* 192:245–263.

Cover, T.L., and R.C. Aber. 1989. *Yersinia enterocolitica. New England Journal of Medicine* 321:16–24.

Craven, R.B., G.O. Maupin, M.L. Beard, T.J. Quan, and A.M. Barnes. 1993. Reported cases of human plague infections in the United States, 1970–1991. *Journal of Medical Entomology* 30:758–761.

Cully, J.F., Jr. 1993. Plague, prairie dogs and black-footed ferrets. In *Proceedings of the symposium on the management of prairie dog complexes for the reintroduction of the black-footed ferret,* ed. J.L. Oldemeyer, D.E. Biggins, B.J. Miller, and R. Crete. Washington, DC: U.S. Fish and Wildlife Service, pp. 38–49.

Cully, J.F., Jr., A.M. Barnes, T.J. Quan, and G. Maupin. 1997. Dynamics of plague in a Gunnison's prairie dog colony complex from New Mexico. *Journal of Wildlife Diseases* 33:706–719.

Davis, K.J., D.L. Fritz, M.L. Pitt, S.L. Welkos, P.L. Worsham, and A.M. Friedlander. 1996. Pathology of experimental pneumonic plague produced by fraction 1-positive and fraction 1-negative *Yersinia pestis* in African green monkeys (*Cercopithecus aethiops*). *Archives of Pathology and Laboratory Medicine* 120:156–163.

Dennis, D.T. 1994. Plague in India: Lessons for public health everywhere. *British Medical Journal* 309:893–894.

Dennis, D.T., and J.M. Hughes. 1997. Multidrug resistance in plague. *New England Journal of Medicine* 337:702–704.

Doll, J.M., P.S. Zeitz, P. Ettestad, A.L. Bucholtz, T. Davis, and K. Gage. 1994. Cat-transmitted fatal pneumonic plague in a person who traveled from Colorado to Arizona. *American Journal of Tropical Medicine and Hygiene* 51:109–114.

Eidson, M., J.P. Thilsted, and O.J. Rollag. 1991. Clinical, clinicopathologic, and pathologic features of plague in cats: 119 cases (1977–1988). *Journal of the American Veterinary Medical Association* 199:1191–1197.

Federov, V.N. 1960. Plague in camels and its prevention in the USSR. *Bulletin of the World Health Organization* 34:911–918.

Fitzgerald, J.P. 1993. The ecology of plague in Gunnison's prairie dogs and suggestions for the recovery of black-footed ferrets. In *Proceedings of the symposium on the management of prairie dog complexes for the reintroduction of the black-footed ferret,* ed. J.L. Oldemeyer, D.E. Biggins, B.J. Miller, and R. Crete. Washington, DC: U.S. Fish and Wildlife Service, pp. 50–59.

Friedlander, A.M., S.L. Welkos, P.L. Worsham, G.P. Andrews, D.G. Heath, G.W. Anderson Jr., M.L. Pitt, J. Estep, and K. Davis. 1995. Relationship between virulence and immunity as revealed in recent studies of the F_1 capsule of *Yersinia pestis. Clinical Infectious Diseases* 21(Suppl. 2):S178–S181.

Gage, K.L., and D.T. Dennis. 1995. *Centers for Disease Control and Prevention, Division of Vector-Borne Infectious Diseases Plague Surveillance Summary* 5:1–21.

Gage, K.L., and J.A. Montenieri. 1994. The role of predators in the ecology, epidemiology, and surveillance of plague in the United States. In *16th vertebrate pest conference,* ed. W.S. Halverson and A.C.H. Crabb. Davis: University of California, pp. 200–206.

Gage, K.L., R.S. Ostfeld, and J.G. Olson. 1995. Nonviral vector-borne zoonoses associated with mammals in the United States. *Journal of Mammalogy* 76:695–715.

Galimand, M., Guiyoule, A., Gerbaud, G., Rosoamanana, B., Chanteau, S., Carniel, E., and Courvalin, P. 1997. Multidrug resistance in *Yersinia pestis* mediated by a transferable plasmid. *New England Journal of Medicine* 337:677–680.

Gasper, P.W. 1997. Plague. In *Consultations in feline internal medicine 3,* ed. J.R. August. Philadelphia: W.B. Saunders, pp. 12–22.

Gasper, P.W., A.M. Barnes, T.J. Quan, J.P. Benziger, L.G. Carter, M.L. Beard, and G.O. Maupin. 1993. Plague (*Yersinia pestis*) in cats: Description of experimentally induced disease. *Journal of Medical Entomology* 30:20–26.

Gordon, D.H., M. Isaacson, and P. Taylor. 1979. Plague antibody in large African animals. *Infection and Immunity* 26:767–769.

Gregg, C.T. 1978. *Plague! The shocking story of a dread disease in America today.* New York: Charles Scribner's Sons.

Hacking, M.A., and L. Sileo. 1974. *Yersinia enterocolitica* and *Yersinia pseudotuberculosis* from wildlife in Ontario. *Journal of Wildlife Diseases* 10:452–457.

Haigh, J.C., and R.J. Hudson. 1993. *Farming wapiti and red deer.* St. Louis: C.V. Mosby, 369 pp.

Hanski, C., U. Kutschka, H.P. Schmoranzer, M. Naumann, A. Stallmach, H. Hahn, H. Menge, and E.O. Riecken. 1989. Immunohistochemical and electron microscopic study of interaction of *Yersinia enterocolitica* serotype O8 with intestinal mucosa during experimental enteritis. *Infection and Immunity* 58:673–678.

Heller, G. 1991. *The dynamics of plague in a white-tailed prairie dog complex in Wyoming.* Master's thesis. Laramie: University of Wyoming, 153 pp.

Henderson, T.G. 1983. Yersiniosis in deer from the Otago-Southland region of New Zealand. *New Zealand Veterinary Journal* 31:221.

Hinnebusch, B.J., R.D. Perry, and T.G. Schwan. 1996. Role of the *Yersinia pestis* hemin storage (hms) locus in the transmission of plague by fleas. *Science* 273:367–370.

Hubbert, W.T., and M.I. Goldenberg. 1970. Natural resistance to plague: Genetic basis in the vole (*Microtus californicus*). *American Journal of Tropical Medicine and Hygiene* 19:1015–1019.

Hudson, B.W., and L. Kartman. 1967. The use of the passive hemagglutination test in epidemiologic investigations of sylvatic plague in the United States. *Bulletin of the Wildlife Disease Association* 3:50–59.

Hum, S., S. Slattery, and S.C.J. Love. 1997. Enteritis associated with *Yersinia pseudotuberculosis* infection in a buffalo. *Australian Veterinary Journal* 75:95–97.

Hurvell, B. 1981. Zoonotic *Yersinia enterocolitica* infection: Host range, clinical manifestations, and transmission between animals and man. In *Yersinia enterocolitica,* ed. E.J. Bottone. Boca Raton, FL: CRC, pp. 152–159.

Institute of Medicine. 1992. *Emerging infections: Microbial threats to health in the United States.* Washington, DC: National Academy Press.

Jayaraman, K.S. 1995. India confirms identity of plague. *Nature* 373:650.

Jessup, D.A., C.J. Murphy, N. Kock, S. Jang, and L. Hoefler. 1989. Ocular lesions of plague (*Yersinia pestis*) in a black-tailed deer (*Odocoileus hemionus columbianus*). *Journal of Zoo and Wildlife Medicine* 20:360–363.

Kapperud, G. 1994. *Yersinia enterocolitica* infection. In *CRC Handbook series in zoonoses,* section A: *Bacterial, rickettsial, and mycotic diseases,* ed. J.H. Steele. Boca Raton, FL: CRC, pp. 343–353.

Kaufman, A.F. 1990. Tularemia. In *Clinical microbiology: Infectious diseases of the dog and cat,* ed. C.E. Greene. Philadelphia: W.B. Saunders, pp. 628–631.

Kilonzo, B.S. 1980. Studies determining the involvement of domestic animals in plague epidemiology in Tanzania. *Tanzanian Veterinary Bulletin* 2:37–44.

Kwaga, J., and J.O. Iversen. 1993. Isolation of *Yersinia enterocolitica* (O:5,27 biotype 2) from a common garter snake. *Journal of Wildlife Diseases* 29:127–129.

Lecker, S., R. Lill, T. Ziegelhoffer, C. Georgopoulos, P.J. Bassford Jr., C.A. Kumamoto, and W. Wickner. 1989. Three pure chaperone proteins of *Escherichia coli*-Sec B, trigger factor, and GroEL-form soluble complexes with precursor proteins in vitro. *EMBO Journal* 8:2703–2709.

Leung, K.Y., and S.C. Straley. 1989. The yopM gene of *Yersinia pestis* encodes a released protein having homology with the human platelet surface protein GPIb alpha. *Journal of Bacteriology* 171:4623–4632.

Lichtenberg, F. 1989. Infectious disease. In *Robbins pathologic basis of disease,* ed. R.S. Cotran, V. Kumar, and S.L. Robbins. Philadelphia: W.B. Saunders, pp. 362–364.

Lindler, L.E., and B.D. Tall. 1993. *Yersinia pestis* pH 6 antigen forms fimbriae and is induced by intracellular association with macrophages. *Molecular Microbiology* 8:311–324.

Lindler, L.E., M.S. Klempner, and S.C. Straley. 1990. *Yersinia pestis* pH 6 antigen: Genetic, biochemical, and virulence characterization of a protein involved in the pathogenesis of bubonic plague. *Infection and Immunity* 58:2563–2577.

Lucier, T.S., and R.R. Brubaker. 1992. Determination of genome size, macrorestriction pattern polymorphism, nonpigmentation-specific deletion in *Yersinia pestis* by pulsed-field gel electrophoresis. *Journal of Bacteriology* 174:2078–2086.

Macy, D.W., and P.W. Gasper. 1990. Plague. In *Infectious diseases of the dog and cat,* ed. C.E. Greene. Philadelphia: W.B. Saunders, pp. 621–627.

Mair, N.S. 1973. Yersiniosis in wildlife and its public health implications. *Journal of Wildlife Diseases* 9:64–71.

Mair, N.S., J.F. Harboune, M.T. Greenwood, and G. White. 1967. *Pasteurella pseudotuberculosis* infection in the cat: Two cases. *Veterinary Record* 81:461–462.

Marchette, N.J., D.L. Lundgren, P.S. Nicholes, J.B. Bushman, and D. Vest. 1962. Studies on infectious diseases in wild animals in Utah: II. Susceptibility of wild mammals to experimental plague. *Zoonoses Research* 1:225–250.

McClure, H.M., R.E. Weaver, and A.F. Kaufmann. 1971. Pseudotuberculosis in nonhuman primates: Infection with organisms of the *Yersinia enterocolitica* group. *Laboratory Animal Science* 21:376–382.

McCoy, G.W. 1911. The susceptibility to plague of the weasel, the chipmunk, and the pocket gopher. *Journal of Infectious Diseases* 8:42–46.

McEvedy, C. 1988. The bubonic plague. *Scientific American* 258:118–123.

McIver, M.A., and R.M. Pike. 1934. Chronic glanders-like infection of face caused by an organism resembling *Flavobacterium pseudomallei.* In *Clinical miscellany,* ed. Whitmore. Cooperstown, NY: Charles C. Thomas.

Messick, J.P., G.W. Smith, and A.M. Barnes. 1983. Serologic testing of badgers to monitor plague in southwestern Idaho. *Journal of Wildlife Diseases* 19:1–6.

Miller, V.L., B.B. Finlay, and S. Falkow. 1988. Factors essential for the penetration of mammalian cells by *Yersinia. Current Topics in Microbiology and Immunology* 138:15–39.

Mingrone, M.G., and M. Fantasia. 1988. Characteristics of *Yersinia* spp. isolated from wild and zoo animals. *Journal of Wildlife Diseases* 24:25–29.

Mollaret, H.H. 1995. Fifteen centuries of yersiniosis. *Contributions to Microbiology and Immunology* 13:1–4.

Montie, T.C. 1981. Properties and pharmacological action of plague murine toxin. *Pharmacology and Therapeutics* 12:491–499.

Nakajima, R., V.L. Motin, and R.R. Brubaker. 1995. Suppression of cytokines in mice by protein A-V antigen fusion peptide and restoration of synthesis by active immunization. *Infection and Immunity* 6:36021–3029.

Obwolo, M.J. 1976. A review of yersiniosis (*Yersinia pseudotuberculosis* infection). *Veterinary Bulletin* 46:167–171.

Orloski, K.A., and M. Eidson. 1995. *Yersinia pestis* infection in three dogs. *Journal of the American Veterinary Medical Association* 207:316–318.

Orsted, K.M., S.A. Dubay, M.F. Raisbeck, R.S. Siemion, D.A. Sanchez, and E.S. Williams. 1998. Lack of relay toxicity in ferret hybrids fed carbaryl-treated prairie dogs. *Journal of Wildlife Diseases* 34:362–364.

Ostroff, S. 1995. *Yersinia* as an emerging infection: Epidemiologic aspects of yersiniosis. *Contributions to Microbiology and Immunology* 13:5–10.

Palmer, R.C. 1993. *English law in the age of the black death, 1348–1381.* Chapel Hill: University of North Carolina Press.

Paul, M.J., T. Work, D. Hunter, E. McFie, and D. Fjelline. 1994. Serologic survey and serum biochemical reference ranges of the free-ranging mountain lion (*Felis concolor*) in California. *Journal of Wildlife Diseases* 30:205–215.

Perry, R.D., and J.D. Fetherston. 1997. *Yersinia pestis:* Etiologic agent of plague. *Clinical Microbiology Reviews* 10:35–66.

Poland, J.D., and A.M. Barnes. 1979. Plague. In *CRC Handbook series in zoonoses,* section A: *Bacterial, rickettsial, and mycotic diseases,* ed. J.H. Steele. Boca Raton, FL: CRC, pp. 515–597.

Poland, J.D., A.M. Barnes, and J.J. Herman. 1973. Human bubonic plague from exposure to a naturally infected wild carnivore. *American Journal of Epidemiology* 97:332–337.

Poland, J.D., T.J. Quan, and A.M. Barnes. 1994. Plague. In *CRC Handbook series in zoonoses,* section A: *Bacterial, rickettsial, and mycotic diseases,* ed. J.H. Steele. Boca Raton, FL: CRC, pp. 93–112.

Pollitzer, R., and K.F. Meyer. 1961. The ecology of plague. In *Studies in disease ecology,* ed. J.M. May. New York: Hafner, pp. 433–490.

Poos, L.R. 1991. A rural society after the black death: Essex 1350–1525. Cambridge: Cambridge University Press.

Portnoy, D.A., H. Wolf-Watz, I. Bolin, A.B. Beeder, and S. Falkow. 1984. Characterization of common virulence plasmids in *Yersinia* species and their role in the expression of outer membrane proteins. *Infection and Immunity* 43:108–114.

Price, S.B., C. Cowan, R.D. Perry, and S.C. Straley. 1991. The *Yersinia pestis* V antigen is a regulatory protein necessary for Ca^{2+}-dependent growth and maximal expression of low-Ca^{2+} response virulence genes. *Journal of Bacteriology* 173:2649–2657.

Price, S.B., and M.D. Freeman, and K. Yeh. 1995. Transcriptional analysis of the *Yersinia pestis* pH 6 antigen gene. *Journal of Bacteriology* 177:5997–6000.

Protsenko, O.A., P.I. Anisimov, O.T. Masarov, N.P. Donnov, Y.A. Popov, and A.M. Kokushkin. 1983. Detection and characterization of *Yersinia pestis* plasmids determining pesticin I, fraction I antigen and mouse toxin synthesis. *Genetika* 19:1081–1090.

Quan, T.J. 1979. Biotypic and serotypic profiles of 367 *Yersinia enterocolitica* cultures of human and environmental origin in the United States. *Contributions to Microbiology and Immunology* 9:41–47.

Quan, S.F., and L. Kartman. 1962. Ecologic studies of wild rodent plague in the San Francisco Bay area of California: VIII. *Zoonoses Research* 1:121–144.

Quan, T.J., K.R. Tsuchiya, and L.G. Carter. 1979. Isolation of pathogens other than *Yersinia pestis* during plague investigations. *Journal of Wildlife Diseases* 15:505–510.

Quan, T.J., A.M. Barnes, L.G. Carter, and K.R. Tsuchiya. 1985. Experimental plague in rock squirrels, *Spermophilus variegatus* (Erxleben). *Journal of Wildlife Diseases* 21:205–210.

Rayor, L.S. 1985. Dynamics of a plague outbreak in Gunnison's prairie dog. *Journal of Mammalogy* 66:194–196.

Reisner, B.S., and S.C. Straley. 1992. *Yersinia pestis* YopM: Thrombin binding and overexpression. *Infection and Immunity* 60:5242–5252.

Rosenberg, D.P., N.W. Lerche, and R.V. Henrickson. 1980. *Yersinia pseudotuberculosis* infection in a group of *Macaca fascicularis. Journal of the American Veterinary Medical Association* 177:818–819.

Rosqvist, R., K.E. Magnusson, and H. Wolf-Watz. 1994. Target cell contact triggers expression and polarized transfer of *Yersinia* YopE cytotoxin into mammalian cells. *EMBO Journal* 13:964–972.

Rust, J.H., D.C. Cavanaugh, R. O'Shita, and J.D. Marshall Jr. 1971a. The role of domestic animals in the epidemiology of plague: I. Experimental infection in dogs and cats. *Journal of Infectious Diseases* 124:522–526.

Rust, J.H., B.E. Miller, M. Bahmanyar, J.D. Marshall Jr., S. Purnaveja, D.C. Cavanaugh, and U.S. Hla. 1971b. The role of domestic animals in the epidemiology of plague: II. Antibody of *Yersinia pestis* in sera of dogs and cats. *Journal of Infectious Diseases* 124:527–531.

Sanford, S.E. 1995. Outbreaks of yersiniosis caused by *Yersinia pseudotuberculosis* in farmed cervids. *Journal of Veterinary Diagnostic Investigation* 7:78–81.

Shayegani, M., W.B. Stone, I. DeForge, T. Root, L.M. Parsons, and P. Maupin. 1986. *Yersinia enterocolitica* and related species isolated from wildlife in New York State. *Applied and Environmental Microbiology* 52:420–424.

Smith, C.R. 1994. Wild carnivores as plague indicators in California: A cooperative interagency disease surveillance program. In *16th vertebrate pest conference,* ed. W.S. Halverson and A.C. Crabb. Davis: University of California, pp. 192–199.

Tabor, S.P., and R.E. Thomas. 1986. The occurrence of plague (*Yersinia pestis*) in a bobcat from the Trans-Pecos area of Texas. *Southwest Naturalist* 31:135—136.

Thomas, R.E. 1988. A review of flea collection records from *Onychomys leucogaster* with observations on the role of grasshopper mice in the epizootiology of wild rodent plague. *Great Basin Naturalist* 48:83–95.

Thomas, R.E., A.M. Barnes, T.J. Quan, M.L. Beard, L.G. Carter, and C.E. Hopla. 1988. Susceptibility to *Yersinia pestis* in the northern grasshopper mouse (*Onychomys leucogaster*). *Journal of Wildlife Diseases* 24:327–333.

Thorne, E.T., and E.S. Williams. 1988. Disease and endangered species: The black-footed ferret as a recent example. *Conservation Biology* 2:66–73.

Thorne, E.T., T.J. Quan, E.S. Williams, T.J. Walthall, and D. Daniels. 1987. Plague in a free-ranging mule deer from Wyoming. *Journal of Wildlife Diseases* 23:155–159.

Trebesius, K., D. Harmsen, A. Rakin, J. Schmelz, and J. Heesemann. 1998. Development of rRNA-targeted PCR and in situ hybridization with fluorescently labelled oligonucleotides for detection of *Yersinia* species. *Journal of Clinical Microbiology* 36:2557–2564.

Von Reyn, C.F., A.M. Barnes, S.N. Weber, T.J. Quan, and W.J. Dean. 1976a. Bubonic plague from direct exposure to a naturally infected wild coyote. *American Journal of Tropical Medicine and Hygiene* 25:626–629.

Von Reyn, C.F., A.M. Barnes, S.N. Weber, and U.G. Hodgin. 1976b. Bubonic plague from exposure to a rabbit: A documented case in the United States. *American Journal of Epidemiology* 104:84–87.

Welsh, R.D., and E.L. Stair. 1993. *Yersinia pseudotuberculosis* bovine abortion. *Journal of Veterinary Diagnostic Investigation* 5:109–111.

Welsh, R.D., R.W. Ely, and R.J. Holland. 1992. Epizootic of *Yersinia pseudotuberculosis* in a wildlife park. *Journal of the American Veterinary Medical Association* 201:142–144.

Weniger, B.G., A.J. Warren, V. Forseth, G.W. Shipps, T. Creelman, J. Gorton, and A.M. Barnes. 1984. Human bubonic plague transmitted by a domestic cat scratch. *Journal of the American Veterinary Medical Association* 251:927–928.

Werner, S.B., C.E. Weidmer, B.C. Nelson, G.S. Nygaard, R.M. Goethals, and J.D. Poland. 1984. Primary plague pneumonia contracted from a domestic cat in south Lake Tahoe, California. *Journal of the American Veterinary Medical Association* 251:929–931.

Williams, E.S., E.T. Thorne, T.J. Quan, and S.L. Anderson. 1991. Experimental infection of domestic ferrets (*Mustela putorius furo*) and Siberian polecats (*Mustela eversmanni*) with *Yersinia pestis. Journal of Wildlife Diseases* 27:441–445.

Williams, E.S., J. Cavender, C. Lynn, K. Mills, C. Nunamaker, and A. Boerger-Fields. 1992. Survey of coyotes and badgers for diseases in Shirley Basin, Wyoming in 1991. In *Black-footed ferret reintroduction, Shirley Basin, Wyoming,* ed. B. Oakleaf, B. Luce, E.T. Thorne, and S. Torbit. Cheyenne: Wyoming Game and Fish Department, pp. 75–106.

Williams, E.S., K. Mills, D.R. Kwiatkowski, E.T. Thorne, and F.A. Boerger-Fields. 1994. Plague in a black-footed ferret (*Mustela nigripes*). *Journal of Wildlife Diseases* 30:581–585.

Williams, J.E., J.D. Marshall Jr., D.M. Schaberg, R.F. Huntley, D.N. Harrison, and D.C. Cavanaugh. 1974. Antibody and resistance to infection with *Yersinia pestis* in the progeny of immunized rats. *Journal of Infectious Diseases* 129(Suppl.):572–577.

Williams, J.E., G.H. Eisenberg Jr., and D. C. Cavanaugh. 1977. Decline of maternal antibodies to plague in Norway rats. *Journal of Hygiene* 78:27–31.

World Health Organization (WHO). 1970. Passive hemagglutination test. *World Health Organization, Technical Report Series* 447:23–25.

———. 1980. Human plague in 1979. *World Health Organization Weekly Epidemiology Record* 32:241–244.

Wuthe, H.-H., S. Aleksic, and S. Kwapil. 1995. *Yersinia* in the European brown hare of northern Germany. In *Yersiniosis: Present and future,* ed. G. Ravagnan and C. Chiesa. Basel: Karger, pp. 51–54.

Zielinski, W.J. 1984. Plague in pine martens and the fleas associated with its occurrence. *Great Basin Naturalist* 44:170–175.

Zwart, P. 1993. Yersiniosis in nondomestic birds and mammals. In *Zoo and wild animal medicine: Current therapy 3,* ed. M.E. Fowler. Philadelphia: W.B. Saunders, pp. 52–53.

20

PASTEURELLOSIS

MICHAEL W. MILLER

INTRODUCTION. *Pasteurellosis* refers to a variety of localized and systemic infections caused by bacteria in the genera *Pasteurella* and *Mannheimia.* These bacteria are widely distributed among wild and domestic mammals. Although generally regarded as opportunistic pathogens, some species or strains of *Pasteurella* and *Mannheimia* can serve as primary pathogens in wild mammals. Most cases of pasteurellosis have little or no appreciable impact on wildlife population dynamics, but epidemics of pneumonic or septicemic pasteurellosis can cause significant mortality in susceptible species.

HISTORY. The evolutionary history of *Pasteurella* spp. infections in wild mammals remains unknown. Documented cases and epidemics in free-ranging populations likely represent only a small fraction of those that have occurred. The earliest report of pasteurellosis in wild mammals was apparently made in Europe in 1878 by Bollinger, who described lesions later associated with septicemic pasteurellosis in deer, wild boar *Sus scrofa,* and cattle *Bos taurus* [see references in Rosen (1981b) and Carter and de Alwis (1989)]. Subsequent reports of pasteurellosis in domestic animals, as well as a variety of free-ranging and captive wild mammals, followed over the next 120 years as syndromes like *hemorrhagic septicemia* were recognized and bacteriologic techniques were developed and standardized. Occurrence of epidemic pasteurellosis in native North American wild ruminants generally followed settlement and establishment of domestic livestock grazing throughout the continent, and may reflect historical introduction of novel *Pasteurella* or *Mannheimia* strains into naive wildlife populations in the late 1800s (Grinnell 1928; Skinner 1928).

DISTRIBUTION. *Pasteurella* and *Mannheimia* spp. infect domestic, feral, and free-ranging mammals worldwide. Captive mammals are probably exposed to and infected with these bacteria wherever they are held. Infected wild mammals have been reported from Europe, Africa, Asia, North America, and Australia. Pasteurellosis epidemics in free-ranging populations have been described in Europe and North America.

HOST RANGE. *Pasteurella* and *Mannheimia* spp. infect most mammalian families, as well as a wide range of nonmammalian species (Mutters et al. 1989). Many wild mammals carry one or more strains as commensal flora (Brogden and Rhoades 1983; Quan et al. 1986). Previous summaries of host range (Biberstein 1981; Rosen 1981b) acknowledge that most, if not all, vertebrate species are probably susceptible to infection with one of the many species and strains of these bacteria.

Cases of localized or systemic disease associated with *Pasteurella* and *Mannheimia* spp. infections have been reported for a remarkable variety of domesticated, captive, or free-ranging terrestrial, aquatic, and marine mammal species (Rosen 1981b; Wallach and Boever 1983; Fowler 1986). Such cases seem to occur sporadically and appear to be of little or no consequence to host populations.

Epidemic pasteurellosis has been reported from a relatively small subset of susceptible wild mammals but may have more significant implications for affected populations and species. Hemorrhagic septicemia, a distinct disease entity, causes periodic widespread mortality among domesticated water buffalo *Bubalus bubalis* and cattle in southeast Asia (Carter and de Alwis 1989); other sympatric mammalian species occasionally succumb to hemorrhagic septicemia during epidemics (De Alwis 1992, 1995) but are not central to its epidemiology. Three epidemics of septicemic pasteurellosis occurred in bison *Bison bison* during 1911–1965 (Mohler and Eichhorn 1912–13; Gochenour 1924; Heddleston et al. 1967). Although market hunting and wholesale slaughter were largely responsible for precipitous declines of North America's bison numbers in the late 1800s, the possible role of pasteurellosis (and other infectious diseases) in that decline probably should not be discounted entirely. Periodic food-borne pasteurellosis epidemics affecting silver fox *Vulpes vulpes,* mink *Mustela vison,* and coypu *Myocastor coypus* farmed in the former Soviet Union have caused considerable economic losses (Lyubashenko and Dukur 1983). Pasteurellosis has concurrently affected wild and farmed European brown hares *Lepus europaeus* (Devriese et al. 1991) in Belgium, and pasteurellosis is a common cause of mortality in hares elsewhere in Europe. Epidemics occasionally affect captive pinnipeds and cetaceans (Sweeney 1986; Dunn 1990). Pneumonic or septicemic pasteurellosis epidemics have also affected free-ranging mule deer *Odocoileus hemionus* (Quortrup 1942), white-tailed deer *Odocoileus virginianus* (Rosen 1981a), wapiti *Cervus elaphus nelsoni* (Franson and Smith 1988), captive fallow deer *Dama dama* (Jones and Hussaina 1982; Carrigan et al. 1991), and reindeer *Rangifer tarandus* (Nordkvist and Karlsson 1962).

Pasteurellosis has played a significant role in declines of bighorn sheep *Ovis canadensis* populations throughout western North America (Post 1962; Hobbs and Miller 1992). Epidemic pasteurellosis was first diagnosed in free-ranging bighorns in 1935 (Potts 1937), but earlier unconfirmed epidemics (Rush 1927; Grinnell 1928; Spencer 1943; Buechner 1960) seem likely (Post 1962). The most extensive documented bighorn pasteurellosis epidemic occurred in 1981–84 and involved free-ranging populations in British Columbia and Alberta, Canada, and in Montana (Onderka and Wishart 1984).

ETIOLOGY. Pasteurellosis is caused by infection with bacteria of the genera *Pasteurella* and *Mannheimia.* Phenotypically, *Pasteurella* spp. are small, pleomorphic, Gram-negative rods or coccobacilli. These bacteria are nonmotile, facultatively anaerobic or micro-aerophilic, produce oxidase and/or alkaline phosphatase, ferment glucose, and reduce nitrate (Kilian and Frederiksen 1981; Mutters et al. 1989). Fresh isolates often display bipolar staining, particularly with Giemsa or Wright's stain.

There is tremendous heterogeneity both among and within species in the genus *Pasteurella.* This heterogeneity has fostered a rather dynamic and somewhat confusing taxonomic classification scheme. At least 17 species of *Pasteurella* are currently distinguished on the basis of phenotypic traits (Bisgaard 1995); six additional species have been proposed since 1989 (Bisgaard 1995) and others will likely be identified. Although DNA hybridization studies excluded 11 new or existing species from *Pasteurella sensu stricto* (Mutters et al. 1989; Bisgaard 1995), these presently remain classified in the genus *Pasteurella. Pasteurella haemolytica* was recently reclassified as *Mannheimia haemolytica* (Angen et al. 1999). Adding to this complexity, strains within a species may be further distinguished by serotyping, biotyping, and biogrouping (Kilian and Frederiksen 1981; Bisgaard and Mutters 1986; Rimler and Rhoades 1989; Adlam 1989). Of the many *Pasteurella* spp. identified, few have been associated with disease in wild mammals. Only three of these—*P. multocida, P. haemolytica* (now *Mannheimia haemolytica*), and *P. trehalosi*—appear potentially important in free-ranging population dynamics (Table 20.1).

Pasteurella multocida is the most ubiquitous of these organisms and has been associated with epidemic disease in both domestic and wild mammals (Table 20.1). Pathogenic strains are often encapsulated and appear as smooth, iridescent colonies on culture. Five capsular serogroups and 16 somatic serotypes are commonly used to characterize and differentiate *P. multocida* isolates (Carter 1955; Heddleston et al. 1972; Rimler and Rhoades 1987). These serologic classifications are conventionally reported by capsular group letter and somatic type number (e.g., B:2). Wild mammals commonly yield untypeable or cross-reacting isolates (e.g., A:3,4). *Pasteurella multocida* serotypes B:2 and E:2 (= Namioka 6:B and 6:E) have been consistently associated with hemorrhagic septicemia in water buffalo and cattle (Carter and de Alwis 1989; De Alwis 1995); *P. multocida* serotype B:2 was isolated from an epidemic in bison in 1922 (Rhoades and Rimler 1992). Serotypes A:2, A:3,4, B:1, and B:3,4 also have caused epidemics of septicemic pasteurellosis in wild ruminants (Rimler et al. 1987; Carrigan et al. 1991; Rhoades and Rimler 1992; Rimler and Wilson 1994) and may be endemic in some free-ranging populations (Wilson et al. 1995). Epidemics in rabbits *Oryctolagus cuniculus* are commonly caused by serotype A or D (Deeb 1997). Untyped strains of *P. multocida* have also been linked to epidemics of septicemia or hemorrhagic enteritis in fox, mink, coypu, sea lions, and dolphins (Lyubashenko and Dukur 1983; Sweeney 1986; Dunn 1990).

Mannheimia haemolytica and *P. trehalosi* primarily affect wild and domestic ruminants. These two species are typically distinguished from other *Pasteurella* spp. by formation of a narrow zone of β-hemolysis on 7% ovine or bovine blood agar (Adlam 1989); however, *M. haemolytica* and *P. trehalosi* isolated from bighorn

TABLE 20.1—Of the 23 or more recognized species of *Pasteurella,* only three have been associated with epidemic disease in wild mammals

Species	Serotypes	Diseases	Affected Host Families[a]
Pasteurella multocida	Capsular: A, B, D, E, F Somatic: 1-16	Pneumonia, enteritis, and/or septicemia	Bovidae, Cervidae, Leporidae, Capromyidae, Mustelidae, Canidae, Delphinidae, Otariidae
	Capsular: B, E Somatic: 2	Hemorrhagic septicemia	Bovidae
Mannheimia (Pasteurella) haemolytica	1, 2, 5-9, 11-14, 16	Pneumonia, septicemia	Bovidae, Leporidae, Delphinidae
Pasteurella trehalosi[b]	3, 4, 10, 15	Pneumonia, septicemia	Bovidae

[a] See text for species and references.
[b] Formerly *Pasteurella haemolytica,* biotype T.

sheep and other wild mammals often show little or no hemolysis (Onderka et al. 1988; Wild and Miller 1991, 1994). The recent proposal to differentiate *P. trehalosi* from *P. haemolytica* (*M. haemolytica*) (Sneath and Stevens 1990) stemmed from early recognition of two biotypes (A and T) of *P. haemolytica* that differed in their abilities to ferment arabinose and trehalose (Smith 1961). Under the new speciation scheme, arabinose-fermenting strains (formerly biotype A) are classified as *M. haemolytica,* whereas those fermenting trehalose (formerly biotype T) are now classified as *P. trehalosi.* Although the new classification has some biologic significance with respect to pasteurellosis of domestic sheep *Ovis aries* (Gilmour and Gilmour 1989), it seems less meaningful in the epidemiology of pasteurellosis in wild mammals. Sixteen serotypes of *M. haemolytica* and *P. trehalosi* (Table 20.1) have been recognized based on presence of soluble surface antigens detected by passive hemagglutination or rapid plate agglutination tests (Biberstein et al. 1960; Frank and Wessman 1978). As with *P. multocida,* wild mammals sometimes yield untypeable or cross-reacting isolates. Among wild ruminants, *M. haemolytica* and *P. trehalosi* have been most intensively studied in bighorn sheep. Both species can cause pneumonia or septicemia in bighorns, and at least three serotypes—*M. haemolytica* serotypes 1 and 2 and *P. trehalosi* serotype 10 (formerly *P. haemolytica* T10)—can be primary pathogens in bighorns (Onderka et al. 1988; Foreyt et al. 1994; Kraabel et al. 1998). *Mannheimia haemolytica* or *M. haemolytica*-like bacteria also have caused epidemics in European hares (Devriese et al. 1991) and dolphins (Sweeney 1986).

TRANSMISSION AND EPIDEMIOLOGY. The commensal nature, ubiquitous distribution, and heterogeneity of *Pasteurella* and *Mannheimia* spp. among mammalian hosts cloud several features of their transmission and epidemiology. It is likely that the young of many host species are colonized with commensal strains essentially at birth. Transmission undoubtedly occurs during intense postpartum interactions between dam and offspring. Increased abundance of *M. haemolytica* in oropharyngeal swabs of postpartum bighorn ewes (M.W. Miller unpublished) suggests parturition itself may facilitate such transmission. Both intraspecific and interspecific transmissions also occur during other interactions, via either direct contact or aerosolization. Although these bacteria are relatively vulnerable in the environment, contaminated excreta, exudates, water, feed, and other fomites can be sources of infection for hours to weeks. Transmission, especially from carnivores and rodents, also occurs via bites and scratches. Food-borne transmission has been implicated in pasteurellosis epidemics of farmed canids and mustelids (Lyubashenko and Dukur 1983). Fleas and ticks have been mentioned as potential vectors (Shewen and Conlon 1993).

Pasteurella spp. most commonly function as endemic, opportunistic pathogens. Consequently, predisposing factors like trauma, stress, or intercurrent disease often play integral roles in both isolated and epidemic pasteurellosis. Bite wounds or other trauma often precede abscesses caused by *P. multocida.* Severe weather events, crowding, or other environmental stressors have been linked to pneumonic and septicemic pasteurellosis in both captive and free-ranging wild mammals (Quortrup 1942; Spraker et al. 1984; Franson and Smith 1988; Carter and De Alwis 1989). Stress-induced suppression of host immunity is believed to be the underlying mechanism triggering these cases. Seasonal weather patterns appear to precipitate or perpetuate some epidemics (Nordkvist and Karlsson 1962; Carter and De Alwis 1989). Septicemic pasteurellosis in captive pinnipeds also showed a seasonal trend, although wild waterfowl or some other host species were speculated as a source of infection (Dunn 1990). Prior or concurrent infections with parainfluenza-3 virus, respiratory syncytial virus, Sendai virus, or other viruses may predispose some host species to pasteurellosis, as may infections with mycoplasma or chlamydia. Heavy parasitism also has been regarded as a predisposing factor for pneumonic pasteurellosis, particularly among wild ruminants. Lungworms *Protostrongylus* spp. or *Müellerius* spp. appear to be integral components of some respiratory disease epidemics in bighorn sheep (Spraker and Hibler 1982). Population-level immunity to specific *Pasteurella* spp. strains also appears to influence the frequency and severity of epidemics (Miller et al. 1991; De Alwis 1992; Hobbs and Miller 1992).

In some cases, *P. multocida, M. haemolytica,* and *P. trehalosi* may cause epidemic disease in free-ranging mammals in the absence of significant predisposing factors. In these situations, individuals carrying a pathogenic and perhaps novel bacterial strain are probably responsible for its introduction or reintroduction into a susceptible host population (Miller et al. 1991). A similar process has been suggested for some hemorrhagic septicemia epidemics in water buffalo and cattle (Carter and De Alwis 1989). Whether highly pathogenic strains are "normal" flora seems questionable in some wildlife species. Stressors could still play a role in precipitating shedding or clinical disease in carriers but are not necessary to sustain such epidemics. Some *M. haemolytica* and *P. trehalosi* strains carried as normal commensal flora by healthy domestic sheep are highly pathogenic in bighorn sheep and Dall sheep *Ovis dalli dalli* (Foreyt and Jessup 1982; Onderka and Wishart 1988; Onderka et al. 1988; Foreyt 1989; Foreyt et al. 1996). Once introduced, however, pathogenic strains may become endemic and continue cycling in affected populations for decades (Miller et al. 1991; Hobbs and Miller 1992; M.W. Miller unpublished). Whether similar processes apply to a *M. haemolytica*-like strain that appears to be a primary pathogen of European hares (Devriese et al. 1991) remains undetermined.

PATHOGENESIS

Agent Factors. Pathogenic mechanisms vary among the species and strains of *Pasteurella* (Shewen and Conlon 1993) and also may be influenced by transmission routes and host factors. Because these mechanisms have been studied mainly in domestic species, some unique features of pathogenic interactions between *Pasteurella* spp. and wild mammal hosts likely remain undescribed. A lipopolysaccharide (endotoxin) equivalent in activity to that of other Gram-negative bacteria is common to all *Pasteurella* spp. Both conserved and unique antigenic epitopes occur, and their expression may be influenced by growth phase, microenvironmental conditions, and host factors. Lipopolysaccharide diversity may contribute to observed differences in host specificity and pathogenicity among species and strains (Fenwick 1995). Some species and strains also produce polysaccharide capsules and/or cell-associated outer-membrane proteins that help resist phagocytosis.

Fimbriae, enzymes (hyaluronidase and neuraminidase), outer-membrane proteins (transferrin binding proteins), and/or heat-labile exotoxins may be produced to aid in colonization, self-defense, and/or nutrient release or acquisition from host tissues. Fimbriae that facilitate adhesion of virulent *P. multocida* type-A strains to mucosal epithelium appear to be important in the pathogenesis of pasteurellosis in rabbits. Although endotoxin and capsules have central roles in diseases caused by *P. multocida,* the ability to produce hyaluronidase may contribute to pathogenicity of the highly virulent type-B strains that cause hemorrhagic septicemia (De Alwis 1995). Some *P. multocida* and *M. haemolytica* strains possess host-specific transferrin-binding proteins that acquire iron needed for colonization and replication (Kirby et al. 1995). Pathogenic *M. haemolytica* and *P. trehalosi* strains produce a soluble, heat-labile leukotoxin that impairs phagocytosis and lymphocyte proliferation at low concentrations and lyses leukocytes at higher concentrations. This leukotoxin shows high specificity for ruminant leukocytes (Shewen and Wilke 1983; Confer et al. 1990; Silflow and Foreyt 1994) and is closely related to the alpha-hemolysin of *Escherichia coli* (Strathdee and Lo 1987).

Contributing Host and Environmental Factors. Diseases of wild mammalian hosts caused by *Pasteurella* spp. generally fall into one of three categories: localized abscesses, respiratory disease, or septicemia. Abscesses caused by *Pasteurella* spp. most commonly arise from bite or scratch wounds but also can arise as a sequela of respiratory or systemic pasteurellosis.

Pasteurella multocida, M. haemolytica, and *P. trehalosi* also cause both pneumonic and septicemic pasteurellosis in wild mammals. After exposure via inhalation or ingestion, initial colonization usually occurs in the upper respiratory tract, particularly the tonsillar region. Exposure dose, strain virulence, proliferation rate, and host immune responses probably affect the ultimate outcome of infection. Environmental stressors or intercurrent infections are widely regarded as prerequisite to development of disease in domestic animals (Shewen and Conlon 1993) and may play some role in many wild mammals, as well. Stressful conditions may foster bacterial replication (Grey and Thomson 1971) and suppress immune responses (Kelley 1988); simulated stress increased leukotoxin-mediated phagocyte death rates in bighorn sheep (Kraabel and Miller 1997). Intercurrent infections may compromise primary host defenses, thereby fostering deeper colonization (Jakab 1981).

In pneumonic pasteurellosis, bacteria inhaled deep into dependent portions of the lung continue colonization and propagation. Capsules and fimbriae probably aid some strains in resisting phagocytosis. Pulmonary infections with *M. haemolytica* may be facilitated by leukotoxin-mediated lysis of alveolar macrophages, neutrophils, and platelets. Chemotactic effects of both leukotoxin and released cellular enzymes seem to augment local tissue damage by helping recruit additional phagocytes into the area (Slocombe et al. 1985). In this regard, lungworm larvae or other foreign material [e.g., dust (Spraker et al. 1984)] trapped in dependent parts of the lungs might exacerbate pneumonic pasteurellosis in bighorn sheep by increasing the numbers of alveolar macrophages, neutrophils, and/or eosinophils exposed to leukotoxin. Bighorn and Dall sheep phagocytes are highly susceptible to damage induced by *M. haemolytica* leukotoxin (Silflow and Foreyt 1994; Foreyt et al. 1996); this and other differences in immune responses (Silflow et al. 1989, 1991) may contribute to these species' vulnerability to pasteurellosis. However, prior recovery from clinical or subclinical disease may provide bighorns with immunity to subsequent challenges (Miller et al. 1991).

Development of septicemic pasteurellosis seems dependent on penetration of mucosal surfaces by virulent organisms (De Alwis 1995). Encapsulation probably aids in resisting phagocytosis once tissue entry is gained. Bacteria multiply locally and then enter the bloodstream as emboli that lodge in capillary beds. Damage is primarily attributable to endotoxin-mediated effects (Collins 1977), which may be augmented by leukotoxin in septicemias caused by *M. haemolytica.*

CLINICAL SIGNS. Clinical signs of pasteurellosis in wild mammals vary with site, form, and duration of disease. Painful localized swelling, sometimes accompanied by fever or depression, may result from infections of bite or scratch wounds. Recurrent purulent nasal discharge ("snuffles") is the principal sign of chronic enzootic pasteurellosis in rabbits, although depression, exertional dyspnea, ocular discharge, head tilt, and subcutaneous or genital swelling also may occur (DiGiacomo et al. 1991; Deeb 1997). With peracute or acute septicemic or pneumonic pasteurellosis, death is often the first or only sign reported.

Depression, anorexia, mucopurulent or blood-tinged nasal discharge, drooped ears, distinctive breath odor (like a *Pasteurella* spp. culture), coughing, and/or respiratory distress have been observed in wild ruminants suffering from pneumonic pasteurellosis. Various combinations of depression, anorexia, ataxia, vomiting and diarrhea, respiratory distress, cyanosis, and subcutaneous edema have been observed during epidemics in farmed furbearers (Lyubashenko and Dukur 1983). Emaciation, serous to purulent ocular and nasal discharge, and joint swelling may also occur in individuals suffering from chronic pasteurellosis. Only subtle behavioral changes, anorexia, and/or abdominal distress have been observed preceding acute deaths in affected pinnipeds and cetaceans (Dunn 1990). Hemorrhagic septicemia may produce a short and somewhat unique clinical course in water buffalo that includes fever and anorexia, followed by respiratory distress with profuse salivation and nasal discharge, and then recumbency and death via terminal septicemia (De Alwis 1995).

PATHOLOGY. Clinical pathologic changes associated with pasteurellosis in wild mammals are rarely measured or reported, particularly in free-ranging cases. Complete blood count responses to infection are nonspecific and likely indistinguishable from typical host species' responses to other bacterial infections (Benjamin 1978; Fowler 1986); in wild ruminants, elevated serum fibrinogen concentrations are likely. Distribution and duration of infection may influence these responses. Changes in clinical chemistry parameters reflect organ-specific damage (Spraker et al. 1984) and should be minimal in localized infections. Bipolar-staining pleomorphic coccobacilli, free or phagocytosed, may be observed in Wright's or Giemsa-stained purulent exudate from abscesses caused by *P. multocida.* Occasionally, bacteria can be demonstrated in blood smears or organ impression smears from animals with systemic infections. These are not diagnostic.

Gross lesions caused by pasteurellosis vary with site, form, and duration of disease, and are not diagnostic. Abscesses caused by *Pasteurella* spp. infections are typically suppurative. They often contain substantial amounts of purulent, foul-smelling fluid and necrotic tissue debris. In rabbits, abscesses usually are well encapsulated and contain thick white exudate (Deeb 1997). Hemorrhage or necrosis may be seen in surrounding tissues.

Lesions associated with pneumonic pasteurellosis occur primarily in the thoracic cavity (Spraker and Hibler 1982; Deeb 1997). In peracute cases, lungs may be heavy and appear congested and edematous with petechial or ecchymotic hemorrhages visible beneath the pleura; airways may contain mucoid or fibrinous exudate. Acute lesions are more typical of bronchopneumonia. In fresh carcasses, the odor typical of cultured *Pasteurella* spp. may be detectable. Affected portions of lung are firm and dark red-gray. In most cases, an obvious line demarcates affected from normal tissue. Cranioventral aspects of the lungs are most commonly affected, particularly in ruminants. Nasal passages, trachea, and bronchi may be reddened. Airways often contain frothy exudate or serosanguinous fluid. Serosanguinous or serogelatinous fluid is also often present in the thoracic cavity. Fibrin is typically present on the pleural surface, and adhesions may be present. Tonsils and cervical and thoracic lymph nodes appear enlarged and may be edematous or hemorrhagic on cut surface. Otitis media is sometimes encountered in rabbits and in bighorn lambs. Emaciation, pulmonary abscesses and fibrous adhesions, sinusitis, arthritis, subcutaneous abscesses, pyometra, and orchitis in various combinations can be encountered in subacute and chronic cases.

Lesions of septicemic pasteurellosis occur throughout affected carcasses (Jones and Hussaini 1982; Lyubashenko and Dukur 1983; Carter and De Alwis 1989; Deeb 1997). Those most typically encountered include subcutaneous hemorrhage and edema, petechial and ecchymotic subserosal hemorrhages, and serofibrinous pericarditis, pleuropneumonia, and peritonitis with or without adhesions. Pulmonary and/or gastrointestinal congestion or hemorrhage also occur. Catarrhal or hemorrhagic enteritis has been observed in affected furbearers and marine mammals. Generalized lymphadenopathy usually occurs, cut surfaces are typically edematous and reddened or darkened with multiple hemorrhages, and splenic enlargement may occur in some species.

Histopathologic lesions of pasteurellosis are generally characterized by necrosis, hemorrhage, and fibrinopurulent changes; they are not diagnostic. Lesions of fibrinopurulent bronchopneumonia (Spraker and Hibler 1982; Onderka et al. 1988; Foreyt et al. 1994) often include proteinaceous bronchiolar and alveolar hemorrhage and/or edema, accompanied by combinations of neutrophils, macrophages, cell debris, fibrin, bacterial colonies, and coagulative necrosis that vary with duration of disease and bacterial strain. *Oat cells* have been observed in some bighorn lesions (Onderka et al. 1988). Fibrin also is usually evident on pleural surfaces and in interlobular septae, sometimes admixed with inflammatory cells and debris. Rhinitis, tracheitis, and bronchitis may be observed.

Microscopic lesions of peracute-acute septicemic pasteurellosis (Franson and Smith 1988; Carter and De Alwis 1989; Carrigan et al. 1991) include hyperemia and hemorrhage in numerous tissues, fibrinohemorrhagic interstitial pneumonia, and widespread bacterial embolism. Hemorrhagic gastroenteritis, splenitis, hepatitis, nephritis, and myocarditis may be seen; cloudy swelling in kidney and liver is sometimes reported.

DIAGNOSIS. Diagnosing pasteurellosis requires isolation and identification of *Pasteurella* or *Mannheimia* spp. from representative lesions or carcasses in sufficient quantities to assure involvement in the disease process in question. Modified Cary and Blair media or other special transport media may aid in recovering isolates from field samples (Wild and Miller 1991, 1994; Foreyt and Lagerquist 1994). Selective media also may be used to enhance recovery of *Pasteurella* spp. (Ward et al. 1986). Because *Pasteurella* spp. are common commensals, isolating organisms in the absence of disease may be of limited significance. Moreover, because many *Pasteurella* spp. are opportunistic pathogens, the possible involvement of other agents should not be discounted. Serotyping, biotyping, biogrouping, genomic fingerprinting, and polymerase chain reaction-based assays can be used to further characterize isolates in epidemiologic investigations (Jaworski et al. 1993, 1997; Foreyt et al. 1994; Ward et al. 1997; Green et al. 1999). Serology for antibody to somatic antigens, surface antigens, or toxins may be useful in investigation of suspected pasteurellosis epidemics (Mohler and Eichhorn 1912–13; Heddleston and Gallagher 1969; Deeb 1997; Kraabel et al. 1998).

DIFFERENTIAL DIAGNOSES. For isolated cases involving localized *Pasteurella* spp. infections, differential diagnoses should include trauma, foreign-body reactions, other common bacterial infections (actinomycosis or staphylococcosis) or mycotic infections, and parasitism. Depending on the host species affected, differential diagnoses for pneumonic or systemic pasteurellosis may include diseases caused by a variety of viruses (caliciviruses, respiratory syncytial viruses, morbilliviruses, or orbiviruses), bacteria (*Francisella tularensis, Salmonella* spp., or *Yersinia* spp.), mycoplasma, chlamydia, and parasites, as well as trauma, aspiration pneumonia, and intoxication.

IMMUNITY. The development of natural immunity to pasteurellosis (Shewen 1995) is complex and, among wild mammal hosts, poorly understood. For commensal *Pasteurella* spp. strains that survive on nasopharyngeal mucosae, nonspecific host defense mechanisms and local immune responses are probably sufficient to prevent deeper colonization under most conditions. When these defenses are temporarily overwhelmed or compromised, deeper colonization may occur. Whether such incursions are successfully defended or lead to disease appears to be affected by strain virulence and host response. In many cases, host responses are ultimately responsible for damage caused during these deeper infections (Slocombe et al. 1985). Humoral immunity appears more important than cell-mediated immunity in host responses to *Pasteurella* spp. Because mucosal exposure alone may not be sufficient to stimulate protective antibodies to soluble virulence factors, deeper colonization and subclinical disease may be beneficial or necessary in priming both local and systemic antibody production in otherwise healthy individuals (Shewen 1995). Alternatively, deeper colonization allowed via compromise of host defenses by stress or intercurrent disease may lead to pneumonic or septicemic pasteurellosis. Similar processes apparently occur in some wild mammal species. Bighorns that recovered from pneumonic pasteurellosis remained healthy during subsequent epidemics (Miller et al. 1991; M.W. Miller unpublished). High serum antibody titers to surface antigens stimulated by commensal *Pasteurella* spp. failed to protect bighorn sheep from challenge with pathogenic *P. trehalosi;* in contrast, most individuals with previous parenteral exposure to leukotoxin survived challenge (Kraabel et al. 1998).

Some *Pasteurella* spp. strains appear to be novel to, or at least rarely encountered by, some wild mammal hosts. For these strains, it seems likely that little or no immunity is present prior to initial exposure. In contrast to isolates from domestic sheep and cattle (Shewen and Wilke 1983; Sweeney et al. 1994), commensal *M. haemolytica* and *P. trehalosi* from free-ranging bighorn sheep rarely produce cytotoxin (Sweeney et al. 1994; M.W. Miller unpublished). Polymerase chain reaction-based analyses have revealed that these noncytotoxic isolates lack portions of the gene coding for leukotoxin A (Green et al. 1999). It follows that even deep colonization with such strains probably would not stimulate leukotoxin-neutralizing antibody production; limited experimental data (Foreyt and Silflow 1996) support this conclusion. Aside from domesticated water buffalo, *P. multocida* B:2 does not appear to be carried by wild ruminants. Consequently, most exposed populations would be expected to show little or no natural resistance to hemorrhagic septicemia unless other commensal *P. multocida* stimulate cross-protective immunity.

PREVENTION, CONTROL, AND TREATMENT. In captive and domesticated mammalian species, proper husbandry, nutrition, and veterinary care aid in preventing pasteurellosis. Commercial or autogenous vaccines have been used in captive and domestic settings to prevent clinical disease in bison (Mohler and Eichhorn 1912–13; Heddleston and Wessman 1973), water buffalo (De Alwis 1995), rabbits (Deeb 1997), pinnipeds (Dunn 1990), and mink and other furbearers (Lyubashenko and Dukur 1983). Field strains of *Pasteurella* and *Mannheimia* spp. isolated from wild mammals are usually susceptible to common antibiotics, including penicillins, tetracyclines, and cephalosporins. With few exceptions, attempts to treat isolated cases of pasteurellosis in free-ranging animals appear unwarranted.

Practical approaches for preventing, controlling, or treating epidemic pasteurellosis in free-ranging wild mammals remain elusive. Because most *environmental stressors* of free-ranging wildlife (e.g., severe weather) cannot be managed, strategies for preventing epidemics have focused largely on controlling densities of susceptible species and preventing introduction of novel pathogenic *Pasteurella* spp. strains; unfortunately, the efficacy of such approaches remains undetermined. Anthelmintic therapy aimed at reducing lungworm burdens has failed to prevent epidemics in free-ranging bighorn sheep populations in Colorado (M.W. Miller unpublished). At least one vaccine that stimulates protective immunity to *P. trehalosi* has been developed for bighorn sheep (Miller et al. 1997; Kraabel et al. 1998), but lack of practical delivery options may limit its use in free-ranging population management (M.W. Miller unpublished). In order for vaccination to become a practical tool for managing pasteurellosis in bighorn sheep or other free-ranging species, self-replicating, antigenic but nonpathogenic, vaccine strains of *Pasteurella* spp. will probably need to be developed. Antibiotic therapy, though potentially effective, is impractical for managing pasteurellosis epidemics in free-ranging mammals.

PUBLIC HEALTH CONCERNS. Although pasteurellosis can be regarded as a zoonotic disease (Frederiksen 1989), *Pasteurella* spp. infections of wild mammals have relatively few implications for public health. *Pasteurella* spp., particularly *P. multocida,* can infect humans via animal bites or scratches or by wound contamination. When human infections occur, they commonly localize in soft tissue but can involve bones or joints or become septicemic. It follows that reasonable caution should be exercised in handling wild mammals, particularly carnivores and rodents, as well as in examining carcasses or tissues from wild mammals where pasteurellosis is among the differential diagnoses. Animals suffering from chronic pasteurellosis may be unfit for human consumption.

DOMESTIC ANIMAL HEALTH CONCERNS. With the exception of possible *P. multocida* B:2 and E:2 infections, which are reportable in most countries, cases of pasteurellosis in wild mammals usually pose no immediate threat to domestic animal health. Pasteurellosis epidemics in domestic livestock attributable to interactions with free-ranging ruminants have not been reported, and domestic ruminant species appear relatively resistant to some *Pasteurella* spp. strains that are highly pathogenic to wild ruminants (Onderka et al. 1988; Foreyt et al. 1994). Pathogenic *Pasteurella* spp. common to both wild and farmed hares (Devriese et al. 1991) suggest a potential for sylvatic reservoirs among lagomorph species, but no supporting epidemiologic data have been reported. Carcasses or tissues from wild mammals infected with pathogenic *Pasteurella* spp. probably should not be fed to farmed or domestic canids or mustelids.

MANAGEMENT IMPLICATIONS. Isolated cases of pasteurellosis among wild mammals usually have no appreciable impacts on populations or species. In contrast, epidemic pasteurellosis can cause widespread mortality that may hamper effective population or species management, or at least public perceptions of management efficacy. In most cases, the effects of epidemics are limited and/or ephemeral and present only temporary obstacles to resource management. One notable exception is bighorn sheep, whose abundance appears to be limited by recurrent pasteurellosis epidemics (Hobbs and Miller 1992). In addition to significant mortality exacted across all age classes during these epidemics, pneumonia and septicemia in neonatal lambs may suppress recruitment for 1–15 years afterward, thereby impairing population recovery and stability. In addition to intensive population management designed to keep some bighorn herds below perceived density-dependent epidemic thresholds, livestock-grazing policies on some public lands in the western United States (Bureau of Land Management 1992) have been modified to prevent contact between bighorn and domestic sheep in an attempt to reduce the frequency and severity of these epidemics via introduction of novel pathogenic *Pasteurella* and *Mannheimia* spp. strains. Because of its status as a foreign animal disease in most countries around the world, hemorrhagic septicemia epidemics also have potentially serious management implications. A confirmed hemorrhagic septicemia epidemic in free-ranging wild ruminants could devastate affected populations, threaten sympatric livestock, and impact local and perhaps international economies. Management actions necessary to eliminate hemorrhagic septicemia carriers from free-ranging populations also could have severe impacts on wildlife resources. In light of the serious implications of either pasteurellosis or some of its differential diagnoses, suspected epidemics in wild mammal populations warrant thorough diagnostic evaluation and epidemiologic investigation.

LITERATURE CITED

Adlam, C. 1989. The structure, function, and properties of cellular and extracellular components of *Pasteurella haemolytica.* In *Pasteurella and pasteurellosis,* ed. C. Adlam and J.M. Rutter. San Diego: Academic, pp. 75–92.

Angen, O., R. Mutters, D.A. Caugant, J.E. Olson, and M. Bisgaard. 1999. Taxonomic relationships of the [*Pasteurella*] *haemolytica* complex as evaluated by DNA-DNA hybridizations and 16S rRNA sequencing with proposal of *Mannheimia haemolytica* gen. nov., comb. nov., *Mannheimia granulomatis* comb. nov., *Mannheimia glucosida* sp. nov., *Mannheimia ruminalis* sp. nov. and *Mannheimia varigena* sp. nov. *International Journal of Systematic Bacteriology* 49:67–86.

Benjamin, M.M. 1978. *Outline of veterinary clinical pathology.* Ames: Iowa State University Press, 351 pp.

Biberstein, E.L. 1981. *Haemophilus-Pasteurella-Actinobacillus:* Their significance in veterinary medicine. In *Haemophilus, Pasteurella, and Actinobacillus,* ed. M. Kilian, W. Frederiksen, and E.L. Biberstein. San Francisco: Academic, pp. 61–73.

Biberstein, E.L., M. Gills, and H. Knight. 1960. Serologic types of *Pasteurella haemolytica. Cornell Veterinarian* 50:283–300.

Bisgaard, M. 1995. Taxonomy of the family Pasteurellaceae Pohl 1981. In *Haemophilus, Actinobacillus, and Pasteurella,* ed. W. Donachie, F.A. Lainson, and J.C. Hodgson. New York: Plenum, pp. 1–8.

Bisgaard, M., and R. Mutters. 1986. Re-investigation of selected bovine and ovine strains previously classified as *Pasteurella haemolytica* and description of some new taxa within the *Pasteurella haemolytica*-complex. *Acta Pathologica Microbiologica Immunologica Scandinavica [B]* 94:185–193.

Brogden, K.A., and K.R. Rhoades. 1983. Prevalence of serologic types of *Pasteurella multocida* from 57 species of birds and mammals in the United States. *Journal of Wildlife Diseases* 19:315–320.

Buechner, H.K. 1960. The bighorn sheep in the United States: Its past, present, and future. *Wildlife Monographs* 4:1–174.

Bureau of Land Management. 1992. *Guidelines for managing domestic sheep in bighorn sheep habitats.* Washington, DC: Bureau of Land Management, 2 pp; Information Bulletin 92–212.

Carrigan, M.J., H.J.S. Dawkins, F.A. Cockram, and A.T. Hansen. 1991. *Pasteurella multocida* septicemia in fallow deer (*Dama dama*). *Australian Veterinary Journal* 68:201–203.

Carter, G.R. 1955. Studies on *Pasteurella multocida:* I. A hemagglutination test for the identification of serological types. *American Journal of Veterinary Research* 16:481–484.

Carter, G.R., and M.C.L. de Alwis. 1989. Hemorrhagic septicemia. In *Pasteurella and pasteurellosis,* ed. C. Adlam and J.M. Rutter. San Diego: Academic, pp. 131–160.

Collins, F.M. 1977. Mechanisms of acquired resistance to *Pasteurella multocida* infection: A review. *Cornell Veterinarian* 67:103–136.

Confer, A.W., R.J. Panciera, K.D. Clinkenbeard, and D.A. Mosier. 1990. Molecular aspects of virulence of *Pasteurella haemolytica. Canadian Journal of Veterinary Research* 54:S48–S52.

De Alwis, M.C.L. 1992. Haemorrhagic septicaemia: A general review. *British Veterinary Journal* 148:99–112.

———. 1995. Haemorrhagic septicaemia (*Pasteurella multocida* B:2 and E:2 infection) in cattle and buffaloes. In *Haemophilus, Actinobacillus, and Pasteurella,* ed. W. Donachie, F.A. Lainson, and J.C. Hodgson. New York: Plenum, pp. 9–24.

Deeb, B.J. 1997. Respiratory disease and the *Pasteurella* complex. In *Ferrets, rabbits, and rodents: Clinical medicine and surgery,* ed. E.V. Hillyer and K.E. Quesenberry. Philadelphia: W.B. Saunders, pp. 189–201.

Devriese, L.A., M. Bisgaard, J. Hommez, E. Uyttebroek, R. Ducatelle, and F. Haesebrouck. 1991. Taxon 20 (fam. Pasteurellaceae) infections in European brown hares (*Lepus europaeus*). *Journal of Wildlife Diseases* 27:685–687.

DiGiacomo, R.F., Y. Xu, V. Allen, M.H. Hinton, and G.R. Pearson. 1991. Naturally acquired *Pasteurella multocida* infection in rabbits: Clinicopathological aspects. *Canadian Journal of Veterinary Research* 55:234–238.

Dunn, J.L. 1990. Bacterial and mycotic diseases of cetaceans and pinnipeds. In *CRC handbook of marine mammal medicine: Health, disease, and rehabilitation,* ed. L.A. Dierauf. Boca Raton, FL: CRC, pp. 73–87.

Fenwick, B. 1995. Liposaccharides and capsules of the HAP group bacteria. In *Haemophilus, Actinobacillus, and Pasteurella,* ed. W. Donachie, F.A. Lainson, and J.C. Hodgson. New York: Plenum, pp. 75–87.

Foreyt, W.J. 1989. Fatal *Pasteurella haemolytica* pneumonia in bighorn sheep after direct contact with clinically normal domestic sheep. *American Journal of Veterinary Research* 50:341–344.

Foreyt, W.J., and D.A. Jessup. 1982. Fatal pneumonia of bighorn sheep following association with domestic sheep. *Journal of Wildlife Diseases* 18:163–168.

Foreyt, W.J., and J.E. Lagerquist. 1994. A reliable transport method for isolating *Pasteurella haemolytica* from bighorn sheep. *Journal of Wildlife Diseases* 30:263–266.

Foreyt, W.J., and R.M. Silflow. 1996. Attempted protection of bighorn sheep (*Ovis canadensis*) from pneumonia using a nonlethal cytotoxic strain of *Pasteurella haemolytica,* biotype A, serotype 11. *Journal of Wildlife Diseases* 32:315–321.

Foreyt, W.J., K.P. Snipes, and R.W. Kasten. 1994. Fatal pneumonia following inoculation of healthy bighorn sheep with *Pasteurella haemolytica* from healthy domestic sheep. *Journal of Wildlife Diseases* 30:137–145.

Foreyt, W.J., K.P. Snipes, and J.E. Lagerquist. 1996. Susceptibility of Dall sheep (*Ovis dalli dalli*) to pneumonia caused by *Pasteurella haemolytica. Journal of Wildlife Diseases* 32:586–593.

Fowler, M.E., ed. 1986. *Zoo and wild animal medicine,* 2d ed. Philadelphia: W.B. Saunders, 1127 pp.

Frank, G.H., and G.E. Wessman. 1978. Rapid plate agglutination procedure for serotyping *Pasteurella haemolytica. Journal of Clinical Microbiology* 7:142–145.

Franson, J.C., and B.L. Smith. 1988. Septicemic pasteurellosis in elk (*Cervus elaphus*) on the United States National Elk Refuge, Wyoming. *Journal of Wildlife Diseases* 24:715–717.

Frederiksen, W. 1989. Pasteurellosis of man. In *Pasteurella and pasteurellosis,* ed. C. Adlam and J.M. Rutter. San Diego: Academic, pp. 303–320.

Gilmour, N.J.L., and J.S. Gilmour. 1989. Pasteurellosis of sheep. In *Pasteurella and pasteurellosis,* ed. C. Adlam and J.M. Rutter. San Diego: Academic, pp. 223–262.

Gochenour, W.S. 1924. Hemorrhagic septicemia studies: The development of a potent immunizing agent (natural aggressin) by the use of highly virulent strains of hemorrhagic septicemia organisms. *Journal of the American Veterinary Medical Association* 65:433–441.

Green, A.L., N.M. DuTeau, M.W. Miller, J.M. Triantis, and M.D. Salman. 1999. Polymerase chain reaction techniques for differentiating cytotoxic and noncytotoxic *Pasteurella trehalosi* from Rocky Mountain bighorn sheep. *American Journal of Veterinary Research* 60:583–586.

Grey, C.L., and R.G. Thomson. 1971. *Pasteurella haemolytica* in the tracheal air of calves. *Canadian Journal of Comparative Medicine* 35:121–128.

Grinnell, G.B. 1928. Mountain sheep. *Journal of Mammalogy* 9:1–9.

Heddleston, K.L., and J.E. Gallagher. 1969. Septicemic pasteurellosis (hemorrhagic septicemia) in the American bison: A serologic survey. *Bulletin of the Wildlife Disease Association* 5:206–207.

Heddleston, K.L., and G. Wessman. 1973. Vaccination of American bison against *Pasteurella multocida* serotype 2 infection (hemorrhagic septicemia). *Journal of Wildlife Diseases* 9:306–310.

Heddleston, K.L., K.R. Rhoades, and P.A. Rebers. 1967. Experimental pasteurellosis: Comparative studies on *Pasteurella multocida* from Asia, Africa, and North America. *American Journal of Veterinary Research* 28:1003–1012.

Heddleston, K.L., J.E. Gallagher, and P.A. Rebers. 1972. Fowl cholera: Gel diffusion precipitin test for serotyping *Pasteurella multocida* from avian species. *Avian Disease* 16:925–936.

Hobbs, N.T., and M.W. Miller. 1992. Interactions between pathogens and hosts: Simulation of pasteurellosis epizootics in bighorn sheep populations. In *Wildlife 2001: Populations,* ed. D.R. McCullough and R.H. Barrett. London: Elsevier Science, pp. 997–1007.

Jakab, G.J. 1981. Interactions between Sendai virus and bacterial pathogens in the murine lung: A review. *Laboratory Animal Science* 31:170–177.

Jaworski, M.D., A.C.S. Ward, D.L. Hunter, and I.V. Wesley. 1993. Use of DNA analysis of *Pasteurella haemolytica* biotype T isolates to monitor transmission in bighorn sheep (*Ovis canadensis canadensis*). *Journal of Clinical Microbiology* 31:831–835.

Jaworski, M.D., D.L. Hunter, and A.C.S. Ward. 1997. Biovariants of isolates of *Pasteurella* from domestic and wild ruminants. *Journal of Veterinary Diagnostic Investigation* 10:49–55.

Jones, T.O., and S.N. Hussaini. 1982. Outbreak of *Pasteurella multocida* septicaemia in fallow deer (*Dama dama*). *Veterinary Record* 110:451–452.

Kelley, K.W. 1988. Cross-talk between the immune and endocrine systems. *Journal of Animal Sciences* 66:2095–2108.

Kilian, M., and W. Frederiksen. 1981. Identification tables for the *Haemophilus-Pasteurella-Actinobacillus* group. In *Haemophilus, Pasteurella, and Actinobacillus,* ed. M. Kilian, W. Frederiksen, and E.L. Biberstein. San Francisco: Academic, pp. 281–287.

Kirby, S.D., J.A. Ogunnariwo, and A.B. Schryvers. 1995. Receptor-mediated iron acquisition from transferrin in the Pasteurellaceae. In *Haemophilus, Actinobacillus, and Pasteurella,* ed. W. Donachie, F.A. Lainson, and J.C. Hodgson. New York: Plenum, pp. 115–127.

Kraabel, B.J., and M.W. Miller. 1997. Effect of simulated stress on susceptibility of bighorn sheep neutrophils to *Pasteurella haemolytica* leukotoxin. *Journal of Wildlife Diseases* 33:558–566.

Kraabel, B.J., M.W. Miller, J.A. Conlon, and H.J. McNeil. 1998. Evaluation of a multivalent *Pasteurella haemolytica* vaccine in bighorn sheep: Protection from experimental challenge. *Journal of Wildlife Diseases* 34:325–333.

Lyubashenko, S.Y., and I.I. Dukur. 1983. Pasteurellosis. In *Diseases of fur-bearing animals* [Bolezni pushnykh zverei], ed. S.Y. Lyubashenko. New Delhi: Amerind, pp. 133–138.

Miller, M.W., N.T. Hobbs, and E.S. Williams. 1991. Spontaneous pasteurellosis in captive Rocky Mountain bighorn sheep (*Ovis canadensis canadensis*): Clinical, laboratory, and epizootiological observations. *Journal of Wildlife Diseases* 27:534–542.

Miller, M.W., J.A. Conlon, H.J. McNeil, J.M. Bulgin, and A.C.S. Ward. 1997. Evaluation of a multivalent *Pasteurella haemolytica* vaccine in bighorn sheep: Safety and serologic responses. *Journal of Wildlife Diseases* 33:738–748.

Mohler, J.R., and A. Eichhorn. 1912–13. Immunization against hemorrhagic septicemia. *American Veterinary Review* 42:409–418.

Mutters, R., W. Mannheim, and M. Bisgaard. 1989. Taxonomy of the group. In *Pasteurella and pasteurellosis,* ed. C. Adlam and J.M. Rutter. San Diego: Academic, pp. 3–34.

Onderka, D.K., and W.D. Wishart. 1984. A major bighorn sheep dieoff from pneumonia in southern Alberta. *Biennial Symposium of the Northern Wild Sheep and Goat Council* 4:356–363.

———. 1988. Experimental contact transmission of *Pasteurella haemolytica* from clinically normal domestic sheep causing pneumonia in Rocky Mountain bighorn sheep. *Journal of Wildlife Diseases* 24:663–667.

Onderka, D.K., S.A. Rawluk, and W.D. Wishart. 1988. Susceptibility of Rocky Mountain bighorn sheep and domestic sheep to pneumonia induced by bighorn and domestic livestock strains of *Pasteurella haemolytica. Canadian Journal of Veterinary Research* 52:439–444.

Nordkvist, M., and K.-A. Karlsson. 1962. Epizootiskt förlöpande infektion med *Pasteurella multocida* hos ren. *Nordisk Veterinarmedicin* 14:1–15.

Post, G. 1962. Pasteurellosis of Rocky Mountain bighorn (*Ovis canadensis canadensis*). *Wildlife Disease* 23:1–14.

Potts, M.K. 1937. Hemorrhagic septicemia in the bighorn of Rocky Mountain National Park. *Journal of Mammalogy* 18:105–106.

Quan, T.J., K.R. Tsuchiya, and L.G. Carter. 1986. Recovery and identification of *Pasteurella multocida* from mammals and fleas collected during plague investigations. *Journal of Wildlife Diseases* 22:7–12.

Quortrup, E.R. 1942. Hemorrhagic septicemia in mule deer. *North American Veterinarian* 23:34–36.

Rhoades, K.R., and R.B. Rimler. 1992. Serological characterization of *Pasteurella multocida* strains isolated from wild ruminants as capsular serogroup B. *Veterinary Record* 130:331–332.

Rimler, R.B., and K.R. Rhoades. 1987. Serogroup F, a new capsular serogroup of *Pasteurella multocida. Journal of Clinical Microbiology* 25:615–618.

———. 1989. *Pasteurella multocida.* In *Pasteurella and pasteurellosis,* ed. C. Adlam and J.M. Rutter. San Diego: Academic, pp. 37–73.

Rimler, R.B., and M.A. Wilson. 1994. Re-examination of *Pasteurella multocida* serotypes that caused haemorrhagic septicaemia in North America. *Veterinary Record* 134:256.

Rimler, R.B., M.A. Wilson, and T.O. Jones. 1987. Serological and immunological study of *Pasteurella multocida* strains that produced septicaemia in fallow deer. *Veterinary Record* 121:300–301.

Rosen, M.N. 1981a. Miscellaneous bacterial and mycotic diseases. In *Diseases and parasites of white-tailed deer,* ed. W.R. Davidson, F.A. Hayes, V.F. Nettles, and F.E. Kellog. Tallahassee, FL: Tall Timbers Research Station, pp. 175–192; Miscellaneous Publications 7.

———. 1981b. Pasteurellosis. In *Infectious diseases of wild mammals,* ed. J.W. Davis, L.H. Karstad, and D.O. Trainer, 2d ed. Ames: Iowa State University Press, pp. 244–252.

Rush, W.M. 1927. Notes on diseases in wild game animals. *Journal of Mammalogy* 8:163–165.

Shewen, P.E. 1995. Host response to infection with HAP: Implications for vaccine development. In *Haemophilus, Actinobacillus, and Pasteurella,* ed. W. Donachie, F.A. Lainson, and J.C. Hodgson. New York: Plenum, pp. 165–171.

Shewen, P.E., and J.A. Conlon. 1993. *Pasteurella.* In *Pathogenesis of bacterial infections in animals,* ed. C.L. Gyles and C.O. Thoen, 2d ed. Ames: Iowa State University Press, pp. 216–225.

Shewen, P.E., and B.N. Wilke. 1983. *Pasteurella haemolytica* cytotoxin: Production by recognized serotypes and neutralization by type-specific rabbit antisera. *American Journal of Veterinary Research* 44:715–719.

Silflow, R.M., and W.J. Foreyt. 1994. Susceptibility of phagocytes from elk, deer, bighorn sheep, and domestic sheep to *Pasteurella haemolytica* cytotoxins. *Journal of Wildlife Diseases* 30:529–535.

Silflow, R.M., W.J. Foreyt, S.M. Taylor, W.W. Laegried, H.D. Liggitt, and R.W. Leid. 1989. Comparison of pulmonary defense mechanisms in Rocky Mountain bighorn sheep (*Ovis canadensis canadensis*) and domestic sheep. *Journal of Wildlife Diseases* 25:514–520.

Silflow, R.M., W.J. Foreyt, S.M. Taylor, W.W. Laegried, H.D. Liggitt. 1991. Comparison of arachidonate metabolism by alveolar macrophages from bighorn and domestic sheep. *Inflammation* 15:43–54.

Silflow, R.M., W.J. Foreyt, and R.W. Leid. 1993. *Pasteurella haemolytica* cytotoxin dependent killing of neutrophils from bighorn and domestic sheep. *Journal of Wildlife Diseases* 29:30–35.

Skinner, M. P. 1928. The elk situation. *Journal of Mammalogy* 9:309–317.

Slocombe, R., J. Malark, R. Ingersol, F. Derksen, and N. Robinson. 1985. Importance of neutrophils in the pathogenesis of acute pneumonic pasteurellosis in calves. *American Journal of Veterinary Research* 46:2253–2258.

Smith, G.R. 1961. The characteristics of two types of *Pasteurella haemolytica* associated with different pathological conditions in sheep. *Journal of Pathology and Bacteriology* 81:431–440.

Sneath, P.H.A., and M. Stevens. 1990. *Actinobacillus seminis* sp. nov., nom. rev., *Pasteurella bettii* sp. nov., *Pasteurella lymphangitidis* sp. nov., *Pasteurella mairi* sp. nov., and *Pasteurella trehalosi* sp. nov. *International Journal of Systematic Bacteriology* 40:148–153.

Spraker, T.R., and C.P. Hibler. 1982. An overview of the clinical signs, gross and histological lesions of the pneumonia complex of bighorn sheep. *Biennial Symposium of the Northern Wild sheep and Goat Council* 3:163–172.

Spraker, T.R., C.P. Hibler, G.G. Schoonveld, and W.S. Adney. 1984. Pathologic changes and microorganisms found in bighorn sheep during a stress-related die-off. *Journal of Wildlife Diseases* 20:319–327.

Spencer, C.C. 1943. Notes on the life history of Rocky Mountain bighorn sheep in the Tarryall Mountains of Colorado. *Journal of Mammalogy* 24:1–11.

Strathdee, C.A., and R.Y.C. Lo. 1987. Extensive homology between the leukotoxin of *Pasteurella haemolytica* and the alpha-hemolysin of *Escherichia coli*. *Infection and Immunity* 55:3233–3236.

Sweeney, J. 1986. Marine mammals (Cetacea, Pinnipedia, and Sirenia): Infectious diseases. In *Zoo and wild animal medicine*, ed. M.E. Fowler, 2d ed. Philadelphia: W.B. Saunders, pp. 777–781.

Sweeney, S.J., R.M. Silflow, and W.J. Foreyt. 1994. Comparative leukotoxicities of *Pasteurella haemolytica* isolates from domestic sheep and free-ranging bighorn sheep (*Ovis canadensis*). *Journal of Wildlife Diseases* 30:523–528.

Wallach, J.D., and W.J. Boever. 1983. *Diseases of exotic animals: Medical and surgical management.* Philadelphia: W.B. Saunders, 1159 pp.

Ward, A.C.S., L.R. Stevens, B.J. Winslow, R.P. Gogolewski, D.C. Schaefer, S.K. Wasson, and B.L. Williams. 1986. Isolation of *Haemophilus somnus:* A comparative study of selective media. *Proceedings of the American Association of Veterinary Laboratory Diagnosticians* 29:479–486.

Ward, A.C.S., D.L. Hunter, M.D. Jaworski, M.P. Dobel, J.B. Jeffress, and G.A. Tanner. 1997. *Pasteurella* spp. in sympatric bighorn and domestic sheep. *Journal of Wildlife Diseases* 33:544–557.

Wild, M.A., and M.W. Miller. 1991. Detecting nonhemolytic *Pasteurella haemolytica* infections in healthy Rocky Mountain bighorn sheep (*Ovis canadensis canadensis*): Influences of sample site and handling. *Journal of Wildlife Diseases* 27:53–60.

———. 1994. Effects of modified Cary and Blair medium on recovery of nonhemolytic *Pasteurella haemolytica* from Rocky Mountain bighorn sheep (*Ovis canadensis canadensis*) pharyngeal swabs. *Journal of Wildlife Diseases* 30:16–19.

Wilson, M.A., R.M. Duncan, T.J. Roffe, G.E. Nordholm, and B.M. Berlowski. 1995. Pasteurellosis in elk (*Cervus elaphus*): DNA fingerprinting of isolates. *Veterinary Record* 137:195–196.

21 MYCOBACTERIAL DISEASES

MYCOBACTERIUM BOVIS INFECTIONS

RICHARD S. CLIFTON-HADLEY, CAROLA M. SAUTER-LOUIS, IAN W. LUGTON, RONALD JACKSON, PETER A. DURR, AND JOHN W. WILESMITH

Synonyms: Tuberculosis, mycobacteriosis, scrofula.

INTRODUCTION. There are numerous reports of *Mycobacterium bovis* infection in both free-ranging and captive wildlife [Ministry of Agriculture, Fisheries and Food (MAFF) 1986; Thoen 1993; Thoen et al. 1995]. It has an exceptionally wide host range with the capacity to cause lesions in primates, most domestic animals, and a large number of wild mammals (O'Reilly and Daborn 1995).

Mycobacterium is a genus of nonsporing, nonmotile Gram-positive bacteria. Generally, they are straight or curved rods, although coccobacillary, filamentous, and branched forms may occur (McFadden 1992). They are distinguished by the large amounts of lipid present in their cell walls, mostly in the form of branched, long-chain fatty acids, the mycolic acids. Consequently, they form hydrophobic colonies on solid media and are difficult to stain by conventional stains such as Gram's. However, once stained, they are resistant to decoloration by acid-alcohol, which is the basis for the Ziehl-Neelsen method, and their identification property of acid-fastness.

The mycobacteria are divisible into a few obligate pathogens and a much larger number of free-living saprophytes living on dead and decaying matter in the environment (Grange 1996). Of the obligate pathogens, the most important include the mammalian *tubercle bacilli.* Besides *M. bovis,* this complex includes *Mycobacterium tuberculosis,* the predominant cause of tuberculosis (TB) in humans; *Mycobacterium africanum,* a rarer cause of TB in humans that was first isolated in equatorial Africa; and *Mycobacterium microti,* which occurs in voles and certain other rodents.

Considerable controversy surrounds the classification of this tubercle bacilli group. Originally, Koch, who first isolated these bacilli in 1882, did not discriminate between isolates from tuberculous humans and cattle (Collins and Grange 1983). Recognition of the small but consistent cultural differences between the two gradually led to the use of *M. tuberculosis* for the human isolates and *M. bovis* for those from cattle. More recently, the validity of this separation has again been doubted, because neither DNA homology studies nor immunologic distance between catalase proteins separate *M. bovis* and *M. tuberculosis* (McFadden 1992). Nevertheless, the fact that *M. tuberculosis* is primarily a human disease, and *M. bovis* one of animals and zoonotic, means that most veterinarians and physicians support the continued separation on pragmatic and familiar grounds.

Survival of *M. bovis* outside mammalian hosts is highly variable and dependent on ambient environmental conditions (Wray 1975). Maximum survival occurs in cold, damp conditions, and survival in cow feces in winter may be for at least 5 months (Williams and Hoy 1930). In contrast, with exposure to direct sunlight under dry conditions, survival is less than 4 weeks (Duffield and Young 1985).

Infection with *M. bovis* generally results in development of chronic granulomatous lesions, which sometimes become necrotic, caseous, and calcified. Although typically identified with lesions in cattle, there is no evidence that *M. bovis* is particularly adapted to this species.

Tuberculosis in cattle is most commonly transmitted by inhalation or ingestion, and these routes of infection determine the resulting lesions (Radostits et al. 1994). Following inhalation, *M. bovis* lodges in the terminal bronchi and, if infection is successfully established, then typically disseminates to the regional lymph node. Both components—the initial focus and the reactive lymph node—are known as the *primary complex* (Dungworth 1993). Following ingestion, a similar process occurs in the pharynx or the intestine.

The pathogenesis of mammalian TB is broadly known (Thoen and Bloom 1995). Following detection by the mononuclear macrophage system, the bacteria are taken into an intracellular phagosome. This is followed by an attempt to fuse the phagosome with a lysosome, the resulting phagolysosome leading to the destruction of most bacteria. However, virulent mycobacteria elude this process, probably by escaping directly into the cytoplasm of the macrophages. The bacilli are then free to multiply and destroy the phagocyte. Consequently, more macrophages enter the area and ingest the increasing number of tubercle bacilli. The resulting cluster of cells is then known as a granuloma and, when macroscopic, a tubercle.

Progress of the disease at this stage is not inevitable and depends on the host mounting a successful cell-mediated immunity (CMI) as mediated by T lymphocytes. Release of lymphokines from the sensitized T lymphocytes results in a cellular hypersensitivity, which through localized tissue destruction may cause the elimination of the mycobacteria or, alternatively, their successful containment within a granuloma. However, if the localized cellular hypersensitivity is inadequate due to either the virulence of the tubercle bacilli or any compromise in the host immune system, then the bacilli may spread. This in turns provokes more and more tissue destruction by the T lymphocytes, and this inappropriate reaction is largely responsible for the characteristic lesions of the disease, such as caseous necrosis. Clinical signs then depend on which organs are affected by these necrotic lesions. With progression, generalized TB develops, manifested by weakness, debility, and eventual death (Radostits et al. 1994).

Mycobacterium bovis infection was previously a major public health problem when this organism was transmitted to people in milk from infected cows (Collins and Grange 1987). The disease was one primarily of children, the typical clinical entity (scrofula) being characterized by swollen and discharging tuberculous lymph nodes in the neck. Introduction of pasteurization of milk and milk products, as well as the low prevalence of the disease in cattle, has eliminated this problem, at least in developed countries (Grange and Yates 1994). In developing countries where routine pasteurization is not undertaken, this means of transmission may still occur (Daborn et al. 1996). In addition to infection of humans, captive and free-ranging nonhuman primates may contract bovine TB by consumption of tissues from infected cattle. Slaughterhouse refuse was the likely source of infection for free-ranging Olive baboons *Papio anubis* in Kenya (Tarara et al. 1985; Sapolsky and Else 1987).

Nowadays, most *M. bovis* infections in humans occur via the respiratory route following close contact with tuberculous cattle, the resulting lesions often resembling closely those caused by *M. tuberculosis* (Grange and Yates 1994). A third, even more uncommon route of infection is via cuts and abrasions on the skin. Butchers and slaughterhouse workers are particularly at risk of contracting the resulting skin TB in circumstances where tuberculous meat is handled.

Transmission of *M. bovis* from humans to cattle is also known to occur, an example of *reverse zoonosis.* Most documented cases have involved humans with genitourinary TB and have been associated with infected farmworkers urinating in cow sheds (Huitema 1969). There is also the possibility that human-to-human transmission of *M. bovis* may occur, although it is generally agreed that it is a rare occurrence (Grange and Yates 1994).

Because of the perceived risk to eradication programs for TB in domestic cattle *Bos taurus* and *B. indicus,* some reservoir species have achieved particular prominence, attracting in the process more detailed study. Because our knowledge of *M. bovis* infection is greatest in these species, this chapter describes in detail bovine TB in the European badger *Meles meles,* the Australian brushtail possum *Trichosurus vulpecula,* ferrets *Mustela putorius furo,* and various species of cervids and bovids.

MYCOBACTERIUM BOVIS IN EUROPEAN BADGER

Introduction and History. Badgers infected with *M. bovis* were first described in Switzerland (Bouvier et al. 1957), but it was 14 years later, in 1971, that interest in the species as a possible wildlife reservoir of infection for cattle was aroused when a tuberculous badger was discovered dead on an English farm where TB had recently been confirmed (Muirhead et al. 1974). Attestation of the national herd had been completed in 1960, but eradication had not been achieved. There still remained localized areas of infection, particularly in southwest England, where the herd incidence persisted unexplained at many times the national average.

The possibility of a significant wildlife reservoir of infection resulted in the MAFF initiating detailed surveys of wildlife (Zuckerman 1980). Particular effort was directed at establishing the geographic extent of infection in badgers. Badger carcasses, and sometimes feces, were collected, initially in the vicinity of where the first tuberculous badger was discovered, then in the county of Gloucestershire, and finally throughout Great Britain.

Tuberculous badgers have also been found in the Republic of Ireland (Noonan et al. 1975) and in Northern Ireland [Her Majesty's Stationery Office (HMSO) 1978]. In both countries, but to a greater extent in the Republic of Ireland, cattle infections have been associated with the disease in badgers (Griffin and Hahesy 1992; Martin 1995).

Distribution and Epidemiology. In the initial survey in Gloucestershire, *M. bovis* was detected in 36 (22%) of 165 carcasses and 12 (11%) of 112 fecal samples. Infection was identified in 23 separate geographic locations within the county, 17 being within 0.8 km of cattle herds with persistent TB problems (Muirhead et al. 1974).

The national Great Britain badger survey, in which animals were subject to detailed postmortem and bacteriologic examination, was based principally on collection of roadside badger carcasses reported to MAFF by the general public. It started in 1972 and continued until its official suspension in August 1990. Submission of carcasses by the public continued after this time, especially in the southwest counties. During the period 1972–95, over 22,000 badgers were collected from throughout all counties in Great Britain. The sample cannot be considered random, because public awareness was no doubt influenced by the local cattle TB incidence. The sample size was also quite restricted in some areas, for example, in Scotland, where only 48 badgers were examined.

Nonetheless, *M. bovis* was isolated from badgers in 16 (27%) of 59 counties by 1987 (Cheeseman et al. 1989) and in 23 (39%) of 59 counties by 1995 (MAFF 1996). It can be surmised that if the sample size were increased from some areas, infection would be confirmed in other counties. Overall prevalence of infection within the total sample was 4.3%. However, this masked a wide regional and within-region variation. For example, a highly significant difference was seen if the southwest regional prevalence (6.5%) was compared with that in the rest of Great Britain (1.0%). At an even more geographically restricted level, as applies in MAFF badger control operations, prevalences up to 100% were found.

Where prevalences of badger infection were low, there were generally no cases of TB in cattle considered to be associated with local badger populations. It was suggested that low badger density and restricted contact with cattle accounted for the lack of transmission between species (Cheeseman et al. 1989; Nolan and Wilesmith 1994). However, the relationship between badger density and prevalence is not a simple linear one, as demonstrated by national survey and badger removal operation data (Cheeseman et al. 1989) and by results from a prospective study of infection in a population encompassing 32 social groups (Cheeseman et al. 1988; Clifton-Hadley and Cheeseman 1995). In the latter study, started in 1975, annual badger density increased progressively (Rogers et al. 1997), while prevalence of infection showed evidence of cyclicity. Infection was slow to spread into uninfected groups, and, during the period 1981–94, four social groups contained 60% of infected animals; no infection was detected in 11 social groups (Clifton-Hadley and Cheeseman 1995). Between infected social groups, there was no indication of synchronous waxing and waning of the TB epidemic (R.S. Clifton-Hadley unpublished).

It was suggested that wildlife other than badgers or perhaps an undisclosed pool of infected cattle were responsible for appearance of TB in cattle herds. Such a proposal argued against TB being endemic in badger populations, although results from both the national survey and the prospective study strongly supported such a state, with no external source of infection necessary to maintain the disease in the badger population. Other data indicated that TB did not necessarily disturb population structure or increase the mortality rate substantially, at least not in the adult population. In one study, only 19 (5.2%) of 363 badgers at autopsy had TB as the primary cause of death; individual infected animals within a social group were detected as excreting *M. bovis,* albeit intermittently, for periods of nearly 2 years. Infected adults reproduced successfully (Clifton-Hadley et al. 1993). All these are traits of an ideal maintenance host (Clifton-Hadley 1996).

In the Republic of Ireland between 1980 and 1991, a total of 7143 badgers were examined. Infected badgers were found in all 26 counties, and 14.3% had gross lesions of TB (Collins 1995). In Northern Ireland, TB was also detected in badgers; the prevalence of infection based on one sample of 450 badgers was about 10% (Dunnet et al. 1986).

Transmission. Badgers excrete *M. bovis* in sputum, urine, feces, and discharge from bite wounds and submandibular abscesses (Clifton-Hadley et al. 1993). Although badgers may be exposed to contaminated excreta in latrines or on the pasture, as is currently thought to be the principal means by which cattle become infected (Nolan and Wilesmith 1994), studies of the lesions suggest that the principal route of infection, as with most other mammalian species, is by inhalation. In one study, 44 (94%) of 47 tuberculous badgers had visible lesions in the lungs or mediastinal lymph nodes (Clifton-Hadley et al. 1993). Close contact between badgers occurs both above ground and within their underground setts, especially during grooming and sleeping (Neal and Cheeseman 1996), and could favor this route of transmission. Similarly, histopathologic evidence of silicosis, a cause of alveolar macrophage suppression, supports direct spread via the respiratory route (Higgins et al. 1985). At present, pseudovertical spread from mother to offspring is considered an important pathway by which infection is maintained in a population (Cheeseman et al. 1989).

Badgers are territorial animals, and infection through bite wounds is another route of transmission. This may help account for the male bias in a sex-based difference of prevalence indicated by several studies (Gallagher and Nelson 1979; Nolan 1991). Although not statistically significant because of restricted sample size, such a differential might be expected if territoriality increased the risk particularly to male animals or if males were more susceptible to infection. Certainly, there is some evidence to suggest that the frequency of bite wounding may be greater among males, perhaps related specifically to this role of territory defense (Gallagher and Nelson 1979; Cheeseman et al. 1988). Findings from a trial of an enzyme-linked immunosorbent assay (ELISA) may support the second hypothesis (Clifton-Hadley et al. 1995a). Infected female badgers, whether with visible lesions at postmortem examination or not, were significantly less likely to produce a positive ELISA result than male animals. This could be explained by the female cell-mediated response being more successful at restricting the number of *M. bovis* organisms, i.e., the antigenic load reaching the antibody-producing cells.

Clinical Signs and Pathology. The main clinical signs in tuberculous badgers are loss of condition and emaciation. Behavior may also change: Terminally ill animals may leave their long-term social group and wander erratically between setts. They may enter farm buildings in search of food and to some extent may lose their fear of humans (Cheeseman and Mallinson 1981). Although TB is seen in all ages of badgers, it has been suggested that in some cubs the disease may progress more rapidly than in adult animals (Nolan 1991). This may indicate that TB has an impact on cub mortality.

The gross and microscopic pathology has been described (Gallagher et al. 1976, 1998; Nolan 1991). Lesions may be seen in most sites but are particularly found in head lymph nodes, lungs and associated lymph nodes, and kidneys (Clifton-Hadley et al. 1993). A possible primary focus in the lungs was described as measuring about 0.5 mm. Unlike the progressive lesions, these foci are generally single and characterized by mineralization and fibrosis. *Mycobacterium bovis* within these foci, considered to be the equivalent of Ghon foci in humans, are often poorly stained, suggestive of successful containment (Gallagher et al. 1998). Where containment is not successful, the picture is mainly one of chronic, progressive pneumonia; but unlike in cattle, there is little evidence of fibrosis, encapsulation of organisms, cavitation, or calcification. Early tubercles, measuring 1–2 mm in diameter, are characterized by the presence of epithelioid cells surrounding an area of central necrosis. Organisms may be scarce. As lesions develop, coagulative necrosis becomes extensive, and numbers of *M. bovis* greatly increase, allowing smaller lesions to coalesce and form tubercles 3–4 mm in diameter. Hematogenous spread may occur through extension of lesions into blood vessels, whereas local spread is by extension into bronchioles. In the former, generalized miliary TB or spread to a limited number of organs, particularly the kidney, may result. Miliary lesions often develop in the lungs by the end stage of disease. The histologic picture suggests that lung lesions could be a potent source of *M. bovis* in sputum, released directly as such or, following swallowing, indirectly in the feces. Abscesses or extensive ulceration may develop at the site of infected bite wounds.

Immunity. Various studies have suggested that badgers have poor immunologic responsiveness and are particularly susceptible to infection (Higgins and Gatrill 1984; Higgins 1985; Higgins et al. 1985). However, it has been hypothesized that many badgers may be immunosuppressed by having silica in their lungs, as a result of their subterranean lifestyle. This could affect the ability of a badger's alveolar macrophages to contain *M. bovis* multiplication (Higgins et al. 1985).

Initial studies also indicated that badgers mount only limited cell-mediated responses. Tissue changes reflecting a hypersensitivity reaction, as seen in cattle tissues, were limited (Gallagher et al. 1976). Injected mycobacterial purified protein derivative elicited no skin hypersensitivity response, and lymphocytes failed to proliferate in vitro (Morris et al. 1978; Little et al. 1982a). However, later studies provided evidence for both these abilities (Higgins 1985; Mahmood et al. 1987), and it has been suggested that the concept of a spectrum of disease, as defined for human TB, incorporating primary infection with subsequent latency in most cases and reactivation and progressive disease in some of these (Barnes et al. 1994), may also be applicable to TB in badgers (Mahmood et al. 1987; Clifton-Hadley and Cheeseman 1995).

Diagnosis. Postmortem examination and bacterial culture from tissue samples is the most sensitive method of detecting *M. bovis* infection in badgers (Pritchard et al. 1986). Efforts have been made to develop nondestructive diagnostic methods. For example, bacterial isolation from clinical samples, such as sputum, urine, and feces, was investigated, but when confined to a single occasion was not considered sufficiently sensitive. The length of time before bacterial colonies could be identified also rendered the method impractical. The possibilities of using skin testing were explored using several reagents, but the results, irrespective of the problems associated with holding badgers for 72 hours prior to reading the tests, indicated both a lack of sensitivity and specificity (Pritchard et al. 1986).

Enzyme-linked immunosorbent assays were developed using various antigen preparations (Morris et al. 1979; Mahmood et al. 1987). Poor specificity, especially because of cross-reactions with *M. avium* and *M. vaccae,* precluded further refinement of these assays. However, an indirect ELISA using a 25-kDa antigen of *M. bovis* has now been developed and validated using nearly 2000 blood samples from badgers taken in MAFF control operations (Goodger et al. 1994; Clifton-Hadley et al. 1995a). Sensitivity was 40.7% and specificity 94.3%. Sensitivity was significantly greater in males and in animals with gross lesions of TB. Enhanced performance of the test achieved by grouping ELISA results according to control operation suggested that it might be practical to use this ELISA as a screening test and, in this way, form the basis of a control strategy (MAFF 1994). A subsequent study of test performance, when considering whether any animal in a sett was infected or not, provided sensitivity estimates ranging from 34.9% to 67.2% and specificity estimates from 91.0% to 91.8%, depending on the criteria used to define the infection status of individual badgers (Clifton-Hadley and Cheeseman 1997). Preliminary data indicate that antigen-specific lymphocyte proliferation assays are more sensitive than the ELISA (R.G. Hewinson personal communication).

Domestic Animal Health Concerns. Evidence to support the role played by badgers in the transmission of TB to cattle in Great Britain has accumulated over the 25 years since the first tuberculous badger was found in Gloucestershire. Under laboratory conditions, TB developed in calves housed with infected badgers

(Little et al. 1982a). The prevalence of infection in badger populations has been shown on many occasions to be much higher than in other wildlife species, such as foxes *Vulpes vulpes,* moles *Talpa europaea,* rats *Rattus norvegicus,* and deer, which have occasionally been detected with infection (MAFF 1986), even when these have been collected from the same areas as infected badgers (Little et al. 1982b). A correlation was demonstrated between increasing badger sett density and an increased risk of TB in cattle where the source of infection was either unknown or ascribed to infected badgers (Wilesmith 1983). Two large-scale interventions in Great Britain, around Steeple Leaze in Dorset and Thornbury in Avon, and one in East Offaly, Republic of Ireland, have demonstrated that sustained action to remove badgers from areas with persistent TB in the local cattle herds, even when no other species are removed, not only reduces the cattle reactor rate but can stop infection of cattle herds completely for a number of years (Little et al. 1982c; Clifton-Hadley et al. 1995b; Dolan et al. 1995).

Much of this information can be termed circumstantial, but the weight of epidemiologic evidence supports the hypothesis that badgers are an important reservoir of infection for cattle. This has been the considered view of three independent reviews (Zuckerman 1980; Dunnet et al. 1986; Krebs 1997).

The precise way that transmission between species occurs is more open to speculation. Badgers and cattle generally avoid direct contact (Benham and Broom 1989). Pasture contamination, especially with infected badger sputum, urine, and feces, presents a probable risk of exposure for grazing cattle, and, although cattle have been shown to demonstrate avoidance behavior toward pasture treated with badger feces and urine, some cattle do not appear to mind. Also, grazing pressure may counteract these avoidance tactics (Benham and Broom 1991; Hutchings and Harris 1997). Analysis of habitat type, landscape features such as fences, and badger urinary behavior suggested that places where badger runs converge to cross linear obstacles, like hedgerows and fences, would be far more heavily contaminated than others areas and may represent the most important source of infection for cattle (White et al. 1993). However, infected badgers may modify their behavior in the terminal stages of disease, entering farm buildings for shelter and in the search for food (Cheeseman and Mallinson 1981). Such occurrences may present particular risks of transmission for cattle.

The advent of DNA-fingerprinting techniques for *M. bovis* presents new opportunities for investigating transmission pathways. One study using spoligotyping has indicated a high correlation between strains in badgers and cattle in the same localities (Clifton-Hadley et al. 1998).

Control. In Great Britain, control operations directed against infected badger populations were started in 1975. Initially, this involved defining infected badger social groups and possible contacts that were associated with occurrence of TB in cattle. Setts were then gassed with hydrocyanic acid (HCN). No recolonization was allowed for 1 year. Control areas were extensive, with a mean area of 7 km^2, and costly (Power and Watts 1987). This strategy continued until 1982, when research indicated that badgers were relatively resistant to HCN and that the required concentration of the gas in setts necessary to kill badgers quickly and humanely might not always be reached.

Shooting badgers, after capture principally using cage trapping, replaced the use of HCN, allowing more information, especially about disease prevalence, to be collected about badger populations associated with TB in cattle. Badgers were removed from infected and adjacent groups until a ring of uninfected groups was removed. These operations, which were again extensive, covering land well beyond the boundaries of the index farm, were particularly expensive to perform, and the average prevalence in badgers taken in this type of operation was only about 12%.

From 1986, operations were restricted to trapping badgers within the farm boundaries of the index herd (Dunnet et al. 1986). This strategy was seen as an interim measure until a discriminatory test in the live badger was developed that could be used as the basis of a fourth strategy. The indirect ELISA described previously was the basis of a controlled trial. Using the test, the infection status of setts used by badgers considered to have caused the original transmission to cattle was established and then only animals from infected setts were destroyed (MAFF 1994). The conclusion of the Krebs Review (Krebs 1997) was that the ELISA would not form the basis for a cost-effective control strategy, with the result that the trial was discontinued.

The effects of the first three strategies have been difficult to assess, given the lack of any comparable areas where action against badgers was not instigated. There was a significant decline in incidence of TB reactor cattle herds following implementation of the gassing strategy, but the causal relationship has been disputed. Although some decrease in risk of subsequent cattle TB has been described for index herds following interim control operations (MAFF 1994), the overall number of new, confirmed affected cattle herds attributed to a badger source has increased from 36 in 1986 to 262 in 1995.

In this context, the results from modeling are of interest. The first model, a deterministic one, indicated that eradication of TB from a badger population by trapping was possible, but that the badger density could only be reduced sufficiently for this to happen if the process was repeated every 3–4 years (Anderson and Trewhella 1985). In one recolonization study, it was shown that the badger population took about 10 years to recover to the numbers prior to clearance (Cheeseman et al. 1993). During that period, *M. bovis* infection was reintroduced.

Studies incorporating data from the same extensively studied badger population suggested that, for TB to be sustained within a social group, group sizes of eight or more members were required. A spatial stochastic simulation model was used to assess and compare the effects of past and current control strategies on infected badger populations (White and Harris 1995). The potential role of badger vaccination was also considered. As might be expected, strategies incorporating extensive removal of badgers were more successful in eliminating infection, while the success of a vaccination strategy was highly dependent on frequency of application and vaccine efficacy. Fertility control was also modeled and found to be less effective at reducing prevalence of infection than were control measures based on killing badgers. However, the risk of perturbation could negate the positive effects on prevalence of removing infected badgers (Swinton et al. 1997).

The Krebs Review (Krebs 1997) recommended that a randomized block trial comprising a minimum of 30 areas of approximately 100 km^2 be undertaken to establish beyond doubt the role of badgers in causing TB in cattle, and to provide the basis for a cost-benefit analysis of different control strategies.

MYCOBACTERIUM BOVIS INFECTION IN AUSTRALIAN BRUSHTAIL POSSUM

Introduction and History. The Australian brushtail possum is a small, nocturnal, predominantly arboreal, herbivorous marsupial native to mainland Australia, Tasmania, and some offshore islands (Strahan 1983). Its introduction into New Zealand from Tasmania in the 1840s to create a fur-trade resource was followed by many more importations and widespread releases (Pracy 1974). An estimated 60–70 million possums are now firmly established as an endemic species throughout most of the North and South Islands of New Zealand (Batcheler and Cowan 1988) and are relentlessly expanding into the few remaining uncolonized areas.

Bovine TB was probably introduced into Australia and New Zealand by tuberculous cattle at the time of early European settlement, yet the disease in wild possums occurs only in New Zealand (Presidente 1984). The presence of predators in Australia and their effect on denning habits, together with lower densities of possums allowing less interaction with livestock there, offer the most likely explanation. It has also been suggested that exposure from tuberculous deer in New Zealand may have been responsible for introducing the disease into possums. The issue of which species was responsible for infecting possums will probably never be resolved. If cattle were, as suggested, responsible (Julian 1981), transmission was infrequent because the disease did not regularly establish in nearby possums despite being common in cattle.

The possum is a reservoir host for TB, and the disease is endemic throughout much of New Zealand's possum population. Tuberculous possums are responsible for many repeated episodes of TB infection in cattle herds and have seriously delayed the eradication of TB from livestock. In June 1995, 2.4% of cattle and 4% of deer herds were classified as being currently or recently infected (O'Neil and Pharo 1995). National control programs are targeted to achieve an internationally recognized *Bovine TB Free Status* that allows a maximum of 0.2% of infected herds at any time.

Bovine TB was first detected in New Zealand possums in 1967 by a trapper on a farm in Westland on the South Island (Ekdahl et al. 1970). A serious outbreak of TB soon after in a dairy herd a few miles upstream from where the tuberculous possums were found drew attention to the possible role of possums in the transmission of the disease. A series of epidemiologic studies were then conducted locally and provided evidence of a strong association between possums and TB (Davidson 1976). Nationwide surveys conducted about the same time found tuberculous possums widely spread on both North and South Islands and indicated that the disease could not have spread from the site of its first detection and must have developed independently in several different areas.

Distribution and Epidemiology. Endemic areas for TB in possums comprise 22% of New Zealand, and for control purposes adjacent areas amounting to another 17% are subject to regular intensive possum control to prevent further spread of the disease among wildlife. Postmortem surveys based on the presence of gross lesions commonly record prevalences of 1%–3% in endemic areas, but one study at a forest pasture location reported 60% prevalence of infection (Coleman et al. 1994). In a longitudinal study of an infected possum population conducted at Castlepoint in the Wairarapa on the North Island, the cumulative prevalence in 979 possums examined over 5.5 years was 11.1% (I.W. Lugton unpublished). This incidence would have underestimated the true level of occurrence of disease, since most diagnoses were made from clinical examinations of trapped animals, a method that is inferior to necropsy and cultural examination. Monthly point prevalence varied from 0.5% to 20% during the Castlepoint study. Cross-sectional survey results need therefore to be interpreted with care and with regard both to the short life expectancy of possums once disease has progressed to the clinically detectable stage and the temporal variation in onset of detectable disease states. Despite the relatively high incidence of TB in infected populations, there is little evidence to suggest that the disease significantly perturbs population size or overall population survival probabilities.

Transmission. Experimental studies (O'Hara et al. 1976; Corner and Presidente 1981; Buddle et al. 1994) and epidemiologic studies of infected possums in the wild (Jackson et al. 1995b) point to the respiratory route as the principal route of transmission of TB among possums. About 85% of affected animals have lung lesions, and *M. bovis* can be readily cultured from tracheal washings of such animals, whereas urine and fecal shedding seem unimportant and occurred in only about 2% of diseased animals (Jackson et al. 1995b). The disease exhibits spatial and temporal clustering, as expected for a slowly spreading contagious disease such as TB with a predominantly respiratory route of infection. Two major pathways and a minor one for transmission are considered to be important.

The first major pathway is pseudovertical transmission from mother to pouch young during the rearing process. It is highly likely that many offspring of tuberculous females become infected while still in the pouch during their first 6 months of life, after a gestation period of about 20 days. Although many of these animals die before maturity, they pose a danger of spreading the disease geographically through dispersal from their natal area. The second major transmission pathway is direct horizontal transmission among adult possums and is thought to occur during direct contact between possums in the vicinity of dens, especially during courting and competition during the breeding seasons (Morris et al. 1994; Morris and Pfeiffer 1995).

Indirect transmission among mature possums through sequential den sharing, sequential marking, or contamination of some other commonly shared locations, such as tracks, is considered to be of minor importance (Morris and Pfeiffer 1995). Poor environmental survival of *M. bovis* (Jackson 1995) and the nature of lesion distribution in tuberculous possums (Jackson et al. 1995a) suggest that the oral route is not important.

Clusters or foci of the disease in particular locations have been remarkably constant over extended periods and can persist despite intensive possum control (Hickling 1989; Morris and Pfeiffer 1995). The formation of clusters and their persistence is best explained by consideration of possum dispersal behavior. Most male juveniles leave their natal area and disperse into new areas. If they survive to the stage of becoming established in a new social community despite being infected prior to dispersal and then have social contact with other members, they can set up new foci of infection (Morris and Pfeiffer 1995). Dispersal distances of young possums range 2–5 km (maximum, 11.6 km) from their natal area. Dispersal occurs in two waves per year, at about 10–12 months prior to sexual maturity and again as adults at about 18–25 months (Efford 1991). In contrast to males, female offspring tend to settle close by their natal areas and, if pseudovertically infected, would be likely to be responsible for maintaining the tuberculous cluster in the same location over time.

Clinical Signs and Pathology. The typical form of the disease involves establishment of infection at one or more initial locations in the animal, followed by rapid extension to other body sites, often by the hematogenous route. Initial lymphatic spread appears to expand into hematogenous spread unusually early in the course of the disease compared with other species. In a detailed study of the lesion distribution in tuberculous possums with gross lesions, an average of 4.6 (range, 1–10) different body-organ sites were affected, with a much higher number (average, 11.6; range 1–28) if microscopic lesions were included, indicating that the degree of generalization of disease is much greater than appears grossly. Of 119 possums that had no gross lesions and were subjected to additional examinations, TB was diagnosed in 10 (8.4%) by histology or culture of pooled lymph nodes (Jackson et al. 1995a).

Lesions were found in lung in 85% of affected animals, in axillary lymphocenters in 85%, in inguinal lymphocenters in 69%, and in either axillary or inguinal lymphocenters in 95%. Terminally ill tuberculous possums had a significantly higher average number of gross lesions than possums in other stages of the disease (Jackson et al. 1995a). The high prevalence of lesions in axillary and inguinal lymphocenters may be a reflection of the patterns of lymphatic drainage that are peculiar to possums and other marsupials (Jackson and Morris 1996).

The nature and appearance of microscopic lesions indicate only a partly effective host response that is unable to wall off established lesions and allows formation of satellite lesions close to established central lesions (Cooke et al. 1995). Despite this apparently limited host response that allows progressive development of lesions in multiple organ sites, growth and behavior of the possum are unaffected until the terminal stages. Mycobacteria are not toxigenic, and dysfunction probably becomes apparent only when there is interference with function from expanding and obliterating lesions or when the toxic products of tissue breakdown from extensive lesions interfere with metabolic processes of the body.

The terminally ill group is characterized by widespread proliferative lesions that have gradually replaced normal tissue, particularly lung, accompanied by debility from associated catabolic effects and interference with normal function. This group is potentially highly infectious, but such animals are debilitated and their abnormal behavior may change the extent of exposure of other possums and domestic livestock compared with potential for exposure before they reach the terminal debilitated state. A combination of high prevalence of lung lesions and generalized disease in a highly susceptible host whose behavior is unaffected until a terminal stage is ideal for ensuring high levels of contagion.

From the limited information available on progress of the disease under natural conditions, no specific rates of disease development can be given, apart from being very variable. This applies both for the incubation period and clinical duration until death, which in the Castlepoint study averaged about 6 months, with occasional animals surviving for much longer periods (Jackson 1995). Most new cases of clinical disease occur in temporal waves, suggesting ecological initiating factors. Only limited information is available on these factors, but adverse weather conditions and nutritional stress may be involved (Morris and Pfeiffer 1995; Van den Oord et al. 1995).

From initial experimental work (O'Hara et al. 1976) it is presumed that early lesions comprise small foci of necrosis infiltrated by neutrophils, followed by invasion of large numbers of macrophages that do not organize themselves as epithelioid cells, as occurs in some other species with TB. At the center of the macrophage nodules, coagulative necrosis occurs that becomes caseous, and numbers of bacilli increase over time. Lymph nodes with gross lesions are palpably enlarged and caseous; some discharge via fistulae. Lesions are usually pale, turgid, and occasionally edematous and suppurative. Sometimes the lymph nodes show a thick fibrous capsule, containing lime-green, liquefactive-to-cream, firm caseous material; mineralization of lesions is not a feature of the disease (Corner and Presidente 1980; Cooke et al. 1995).

Lesions of TB in lung tissue of possums range in size up to about 60 mm in diameter. Discrete nodular lesions are the most common manifestation in lung, but occasionally there is firm gray consolidation of parts of a lobe or several lobes with pleural adhesions (Julian 1981; Cooke et al. 1995). Lesions in the liver and kidney are usually 1–2 mm in diameter, creamy white, firm-to-soft caseous, and in the kidney mostly involve the cortex (Cooke et al. 1995).

Large tuberculous lesions are infiltrated by neutrophils that in extensive lesions are often accompanied by areas of coagulative or caseous necrosis (Cooke et al. 1995). Acid-fast organisms are demonstrable within the cytoplasm of macrophages and sometimes within neutrophils. In necrotic foci, massive numbers of acid-fast organisms occur intracellularly at the periphery of the lesion and in much smaller numbers extracellularly. Once liquefaction in lesions occurs, the number of acid-fast organisms decreases (Julian 1981; Cooke et al. 1995). Liver lesions tend to contain very few acid-fast organisms (Corner and Presidente 1980; Cooke et al. 1995). Multinucleate giant cells are rare and occur in very few lesions randomly scattered throughout (O'Hara et al. 1976; Corner and Presidente 1980; Cooke et al. 1995).

Diagnosis. In live possums, a preliminary diagnosis of TB can be made clinically by finding enlarged superficial lymph nodes or discharging fistulae, with confirmation by culture of fluid aspirates from open or closed lesions (Pfeiffer and Morris 1991). Most infected animals can be identified by gross examination, but, in one series of studies, additional cultural and histopathologic testing of possums without gross lesions found prevalences 8.4% higher on average than those recorded from gross pathology alone (Jackson et al. 1995a). Gross lesions of TB in possums are generally easy to recognize, but it is not possible to distinguish grossly between very small tuberculous lesions in the lung and adiaspiromycosis (Johnstone et al. 1993). Other mycobacterial infections, such as *M. avium* and *M. fortuitum,* are apparently very rare in possums (Julian 1981; Jackson et al. 1995a). *Mycobacterium vaccae* was found a few times in early investigations and was considered saprophytic and opportunistic (Julian 1981).

Efforts have been made to develop a simple serologic test capable of diagnosing TB in live possums, but results have been disappointing. All of three ELISAs evaluated were insufficiently sensitive (Buddle et al. 1995b). A lymphocyte transformation assay gave reasonable predictions of infection status both in experimental settings (Buddle et al. 1994) and in the field (I.W. Lugton unpublished), but field application is not yet practical due to economic and logistic considerations.

Domestic Animal Health Concerns. At first, it was widely believed that cattle became infected principally by ingestion of pasture contaminated by tuberculous possums. The reason for this belief was the high prevalence of discharging sinuses in clinically affected possums and the high concentration of *M. bovis* organisms in exudate.

Poor survival of organisms on pasture and the inefficiency of the oral route, which requires large doses of organisms to infect cattle reliably, do not support this hypothesis (Morris et al. 1994). Transmission between possums and livestock appears to occur principally when terminally ill tuberculous possums, due to aberrant behavior, come into uncharacteristically close contact with livestock and fail to show the normal avoidance behavior. Terminally ill tuberculous possums have been observed on pasture during the day, feeding and moving very slowly, apparently unaware of their surroundings (Paterson 1993). Such behavior was simulated by sedating possums and exposing them to cattle, red deer *Cervus elaphus,* and sheep *Ovis aries* (Paterson and Morris 1995; Sauter and Morris 1995a,b). Animals within each herd, attracted by the unusual behavior, came close, and some investigated the sedated possum at length. Cattle and deer showed extensive investigative behavior, such as snorting, inhaling, licking, and lifting the possum. This type of behavior provides an ideal opportunity for transmission from terminally ill tuberculous possums, which have extensive lesions throughout all major organs and are potentially highly infectious. Sheep, in contrast, showed little interest in the possum (Sauter and Morris 1995a), a finding which agrees with field observations that this species, despite being experimentally susceptible to TB, only rarely contracts the disease in the field (Cordes et al. 1981; Davidson et al. 1981). In cattle and deer herds, a positive association was found between an animal's rank in the dominance hierarchy and indicators of infection risk, such as intensity of exploration of a sedated possum, and actual occurrence of infection when it was exposed to infected possum populations (Sauter and Morris 1995b). A disproportionate number of tuberculous possums die on pasture, but dead possums are investigated to a far lesser extent by cattle and deer than live, simulated terminally ill ones (Paterson 1993).

The pharyngeal tonsil appears to be the most common site of entry of infection during transmission of the disease from possums to deer (Lugton et al. 1997b), whereas alimentary tract infections seem to predominate in predator species, such as ferrets and cats *Felis catus* (Ragg et al. 1995b).

Control. Free-living and feral species that have been found infected with *M. bovis* in New Zealand include possums, deer, pigs *Sus scrofa,* cats, ferrets, stoats *Mustela erminea,* weasels *Mustela nivalis,* goats *Capra hircus,* rabbits *Oryctolagus cuniculus,* hares *Lepus europaeus occidentalis,* and hedgehogs *Erinaceus europaeus,* although infection has been detected only very rarely in some. Of these species, only the possum has been shown to be a true reservoir host. Deer and ferrets are certainly *spillover* hosts and are capable of some transmission of infection, but the contentious issue of whether they are reservoir hosts is as yet undecided. For effective control, the greatest attention needs to be directed toward reservoir hosts, wherein the disease is self-sustaining. Reservoir host species are central to the maintenance of the disease, whereas infection in spillover hosts species will disappear progressively once the disease is eliminated from the former species (Morris and Pfeiffer 1995).

A decrease in the number of new cases of tuberculous cattle and infected herds follows a localized possum control operation aimed at reducing possum population size (Livingstone 1991). Yet despite intensive control programs conducted over many years, the area of New Zealand classified as endemic for bovine TB has increased, and it is now accepted that there is no real prospect of eradicating the disease in possums with current methods.

National control of the disease in cattle and deer is currently achieved by removal of the wildlife hosts through intensive control of tuberculous vectors, mainly possums, at a cost of about New Zealand $31 million each year (O'Neil and Pharo 1995). The pest and disease status of possums tempers public opposition to large-scale aerial and ground poisoning programs, but it is generally accepted that large-scale kill campaigns will not be sustainable in the long term.

Alternative management options are being studied, such as control of the disease in possums, using BCG vaccine in cattle (Buddle et al. 1995a), development of population regulating processes, identification of spatial determinants of clusters to allow better-targeted controls, and development of farm-based management procedures to reduce contact between infected wildlife and farmed animals (McKenzie et al. 1997).

The current national control program has now adopted the following objectives: to reduce the prevalence of herd infection in vector areas to internationally accepted levels, to prevent the establishment of tuberculous vectors in new areas, to decrease the number and size of existing areas where tuberculous vectors exist, and to encourage landowners to take action against TB on their properties and in their herds (O'Neil and Pharo 1995).

MYCOBACTERIUM BOVIS INFECTION IN FERRETS

Introduction. Tuberculosis has been described in captive ferrets (Dunkin et al. 1929; Symmers et al. 1953) and was probably due to the ingestion of infected milk. The disease in feral ferrets has not been recorded outside of New Zealand.

Until the early 1990s, ferrets were considered unimportant in the epidemiology of bovine TB in New Zealand (Allen 1991), although TB in feral ferrets was known from the 1970s and reported in 1982 (de Lisle et al. 1993). Their role began to be queried in 1992 when tuberculous ferrets were found in an endemic area free of possums (Walker et al. 1993). Circumstantial evidence from surveys of the disease status of ferrets and field reports also suggested that ferrets were infecting cattle (Ragg et al. 1995a).

New Zealand has a large feral ferret population (Nowak 1991). They were introduced in the 1880s to control rapidly expanding rabbit populations on agricultural land but failed to achieve that aim. Ironically, their abundance seems to have become dependent on the number and distribution of rabbits in regions where rabbits are their main food source (King 1984).

Epidemiology and Transmission. Disease prevalences ranging from 2.0%–17.9% (Walker et al. 1993; Ragg et al. 1995a) based on detection of gross lesions in ferrets probably underestimated the true prevalence of infection. A more detailed study, which included mycobacterial culture and histopathologic examinations, found 32% infected, of which 28% had no gross visible lesions (Lugton et al. 1997a). Other detailed studies have shown marked between-region differences in prevalence, ranging from 26% where possum densities were low to 96% in areas where possums were abundant and known to be an important source of food for ferrets (Lugton et al. 1997b).

The most detailed information on routes of infection in ferrets was provided by Lugton et al. (1997b), who found primary lesions associated with the alimentary tract in nearly 80% of animals in which initial site of entry lesions could be identified. Ferrets eat carrion, such as possums and hedgehogs, both of which are commonly infected with *M. bovis,* and it is thought that most ferrets become infected through scavenging on dead animals (Lugton et al. 1997b). Primary lesions in the remaining animals were associated with sites draining to peripheral lymph nodes and were suggestive of entry through wounds during fighting.

Gross lung lesions in naturally infected ferrets were rare (Ragg et al. 1995a), and acid-fast organisms were seldom identified in tracheal washings (Lugton et al. 1997b). Although one of eight females had *M. bovis* cultured from the mammary glands (Lugton et al. 1997b), evidence of disease is extremely rare in very young animals.

The distribution of lesions and excretion pathways indicates that horizontal transmission may account for only about 20% of infections and therefore is a minor mode of infection compared to infection acquired via the oral route. Furthermore, prevalence is not population density dependent (Ragg et al. 1995b), which would be expected if TB were commonly transmitted horizontally. Natural disease progression is independent of age and sex, and the disease does not seem to interfere with the general well-being of ferrets until the late stages of the disease when body weights are reduced (Lugton et al. 1997a).

Clinical Signs and Pathology. The most commonly affected sites in naturally infected ferrets are mesenteric lymph nodes, followed by retropharyngeal and caudal cervical lymph nodes. The average number of gross lesion sites was 2.3 (ranging from 0 to 15), with obvious gross lesions occurring mostly in the jejunal lymph node (Lugton et al. 1997a). These lesions generally did not show the pyogranulomatous or caseous features seen in possums or deer. Occasionally, small white nodular lesions were visible in lymph node parenchyma, but often the only visible signs were enlargement and edema. Liquefaction was often a feature of large lesions in the jejunal lymph node, giving the lymph node the appearance of a watery or milky-colored fluid-filled sac up to 35 mm in diameter. Leakage of this fluid into the abdominal cavity caused tuberculous peritonitis, expressed as a thin gray film of exudate.

Early reports of TB in ferrets noted that lesions were characterized histologically by very large numbers of acid-fast organisms intracellularly and especially around the periphery of lesions. Giant cells were nonexistent or very rare and only small in size; fibrosis was seen only in capsules of lymph nodes, and the reactions were classified as histiocytic and necrotic (Dunkin et al. 1929; Iland et al. 1951).

In a recent study in New Zealand, acid-fast organisms were not found in all animals from which *M. bovis* was cultured, and the average number of organ sites containing acid-fast organisms was 5.8 (ranging from 0 to 31) (Lugton et al. 1997a). Histopathologic findings in lymph nodes included discrete aggregations of a few to extensive numbers of epithelioid cells throughout the lymph node. Complexes were often seen, where amorphous necrotic tissue was surrounded by such epithelioid cells, which in turn were surrounded by lymphocytes and a few plasma cells. Acid-fast organisms were not commonly found in necrotic tissue but rather in the epithelioid cells surrounding it. A strong positive association between the presence of necrosis and the abundance of acid-fast organisms was found. Fibrosis and neutrophils were seen infrequently, and mineralization and giant cells never occurred in affected animals, a finding in agreement with TB lesions found in other mustelids (Head 1959; Thorns et al. 1982).

Gross lesions were not detected in liver, yet this organ was the site most commonly affected with microscopic lesions. Nodular accumulations of macrophages were found throughout the parenchyma, with some containing lymphocytes and plasma cells. Ferrets in advanced stages of the disease showed more of these nodules than ferrets with early-stage disease.

Small lung lesions consisted of focal collections of foamy macrophages and epithelioid cells within alveoli, while larger lesions had epithelioid cells at the periphery. Other sites with histopathologic lesions were spleen, tonsils, adrenal glands, and bone marrow, which commonly showed small aggregations of epithelioid cells and very few acid-fast organisms. The few observed renal lesions occurred mostly in ferrets with advanced TB and were characterized by lymphocytes, plasma cells, and a few macrophages in the interstitium of the cortex.

Diagnosis and Differential Diagnoses. A tentative diagnosis of TB can be made based on the presence of gross lesions (Walker et al. 1993; Ragg et al. 1995a), although early-stage cases may be missed (Lugton et al. 1997a). Histopathology or cultural examination for mycobacteria may be used to confirm diagnoses based on gross lesions. Histopathology of livers from animals with no gross lesions has potential value as a screening test, because microscopic granulomas in the liver appear to be pathognomonic for TB in ferrets in New Zealand (Lugton et al. 1997a).

Some tuberculous lesions in lymph nodes are unspectacular and are difficult to distinguish from normal hypertrophy. Follicular hyperplasia can be confused with white nodular lesions of TB (Lugton et al. 1997a). Infection with adiaspores of the fungus *Chrysosporium parvum* var. *crescens* occurs commonly in ferrets and, if lesions are numerous, could be confused with miliary TB (Lugton et al. 1997a).

Domestic Animal Health Concerns. Contact between ferrets and livestock was investigated by sedating ferrets to simulate the behavior of sick or moribund animals with advanced TB (Sauter and Morris 1995a). Some, but not all, individual cattle and deer advanced to within 1.5 μ and sniffed the ferrets and were in direct physical contact for short periods. The degree and intensity of interest shown was considerably less than that shown to possums but may have been due to the known repellent properties of some mustelid secretions (Sullivan et al. 1990). Ferrets in an advanced stage of TB are believed to be highly infectious because they excrete bacteria via multiple routes (Lugton et al. 1997b).

Control. In regions where ferrets are thought to be responsible for transmitting the disease to livestock, control is at present mainly by removal by trapping. Indirect control of ferrets is also achieved by reducing rabbit populations, mainly by poisoning.

The role of the ferret in the epidemiology of bovine TB is not yet fully understood, and efforts are currently directed to resolving the nature of its host status. If it is a maintenance host in which infection can be transmitted without seeding from other animals, then control measures directed specifically to ferrets are needed to reduce infection rates among livestock. On the other hand, if it is a spillover host and is dependent on other species in the form of carrion for infection, control measures directed to the true source of infection are required, even though infected ferrets may be responsible for some cases of disease among livestock. Although interim control until the issues are resolved may be prudent, spillover hosts generally do not require control. Resources and effort are best directed toward controlling maintenance hosts (Morris and Pfeiffer 1995).

MYCOBACTERIUM BOVIS INFECTION IN DEER

Introduction. The occurrence of *M. bovis* infection in both free-living and captive wild deer of various species has been described on many occasions, often as case reports. In free-living and captive animals, infection was of concern because of the possible impact on eradication programs for cattle, for example, in Germany and the United States (Bouvier et al. 1957; Bouvier 1963; Stumpff 1982; Sibartie et al. 1983; Dodd 1984; Stumpff et al. 1984; Tessaro 1986), and the risk presented to other species, such as bison *Bison bison* (Essey and Vantiem 1995). Cases in captive animals were also of particular concern because of zoonotic implications (Towar et al. 1965; Basak et al. 1975). The question of risk to human health was raised, especially for those either in close contact with live deer or handling infected carcasses (Beatson et al. 1984; Dodd 1984; McKeating and Lehner 1988). Transmission to a veterinarian handling an elk *Cervus elaphus* with bovine TB was confirmed in Canada (Fanning et al. 1991).

Tuberculosis among farmed deer occurred in several countries, including Canada, Great Britain, New Zealand, Sweden, and the United States (Beatson and Hutton 1981; Stuart 1988; Miller et al. 1991; Whiting and Tessaro 1994; Bolske et al. 1995) and focused attention on wild populations as both potential reservoirs of infection and targets for spread from domesticated animals (Munro and Hunter 1985; Guerin 1989). In fact, bovine TB spread from farmed elk to free-ranging mule deer *Odocoileus hemionus,* based on epidemiology and typing of *M. bovis* isolates by restriction fragment length polymorphism analysis (Rhyan et al. 1995; Whipple et al. 1997).

Distribution. Bovine TB has been reported as a sporadic disease of free-ranging cervids, linked on several occasions to disease in other species (Clifton-Hadley and Wilesmith 1991). Cervid species reported with bovine TB, whether in free-ranging or captive deer, are summarized in Table 21.1.

Samples collected from wild deer have tended to be of an ad hoc nature and generally confined to animals where pathologic lesions were found after death. Therefore, accurate estimates of disease prevalence within defined populations have not been possible. However, estimates have tended to suggest prevalences of less than 5% in free-ranging wild deer and more, on occasion, in park deer. Recent studies of a focus of bovine TB in free-ranging white-tailed deer *Odocoileus virginianus* in Michigan demonstrated a prevalence of greater than 4% (Schmitt et al. 1997). This is at variance with a prevalence in wild red deer of 37% recorded in one study from New Zealand (Nugent and Lugton 1995), where it was thought tuberculous possums might have been responsible for the intensity of infection. Reported prevalence estimates are summarized in Table 21.2.

Transmission and Epidemiology. There have been no reported cases of congenital TB in deer, which contrasts with its occurrence in cattle (Dungworth 1993). No evidence of transmission was found in 80 6-month calves from a herd where the adult prevalence was 90% (Griffin 1988). Transmission via contaminated milk has not been confirmed, although calves from tuberculous dams have been infected, despite removal from the mother at birth (Fletcher 1990); mammary gland lesions have been described (Rhyan et al. 1992).

Lesions in infected deer are found particularly in the retropharyngeal, mediastinal, and mesenteric lymph nodes (Nikanorov 1969; Beatson and Hutton 1981; Beatson 1985; de Lisle and Havill 1985; Stuart 1988; Mackintosh and Griffin 1994). This distribution is indicative of infection by both the respiratory and alimentary routes (Quinn and Towar 1963; Beatson et al. 1984), with lesions in the retropharyngeal lymph node representing either possible route (Stamp 1944).

In challenge experiments, attempts have been made to mimic natural infection by the intravenous, intradermal, intratracheal, intranasal, and intratonsillar routes (de Lisle et al. 1983; Brooks 1984; Corrin et al. 1993; Mackintosh et al. 1993, 1995; Palmer et al. 1999). Infections subsequent to intratonsillar challenge reproduced the pattern of disease seen in the field most consistently, suggesting that oral challenge may be the most common route of infection.

In wild deer populations where the prevalence of infection is low, it is uncertain whether endemic disease would be maintained without the existence of other reservoirs of infection. In Great Britain and Ireland, infected badgers are known to be in areas where cases of TB have occurred among wild deer (Dodd 1984; MAFF 1986). Similarly, in New Zealand, the possum is a natural reservoir of infection for cattle, and natural infection of red deer within an enclosure after contact with tuberculous possums has been described (Lugton et al. 1995). The high prevalence recorded among wild deer in one New Zealand locality was considered most likely to have occurred through transmission from infected possums rather than from deer-to-deer transmission (Nugent and Lugton 1995). Elsewhere, wild pigs were considered the source of

TABLE 21.1—*Mycobacterium bovis* infections reported in free-ranging and captive deer

Common Name	Scientific Name	Country (Reference)
Fallow deer	*Dama dama*	Denmark (Jørgensen et al. 1988[a]), New Zealand (de Lisle and Havill 1985), Republic of Ireland (Wilson and Harrington 1976), United Kingdom (Fleetwood et al. 1988), USA (Quinn and Towar 1963[b]; Rhyan and Saari 1995[a]), Sweden (Bolske et al. 1995[a])
Axis deer	*Axis axis*	Hawaii (Sawa et al. 1974), India (Liston and Soparkar 1924[b]), United Kingdom (Jones et al. 1976[b])
Hog deer	*Axis porcinus*	India (Basak et al. 1975)
Sambar deer	*Cervus unicolor*	India (Liston and Soparkar 1924[b])
Sika deer	*Cervus nippon*	New Zealand (de Lisle and Havill 1985), Republic of Ireland (Dodd 1984), United Kingdom (Rose 1987), USA (Mirsky et al. 1992[a])
Red deer, wapiti, elk	*Cervus elaphus*	Canada (Hadwen 1942; Hutchings and Wilson 1995[a]), New Zealand (Beatson et al. 1984[a]), Republic of Ireland (Dodd 1984), Switzerland (Bouvier 1963), United Kingdom (Stuart et al. 1988), USA (Stumpff 1982; Thoen et al. 1992[a]; Rhyan et al. 1992[a]), USSR (Nikanorov 1969[a]), West Germany (Witte 1940)
Mule deer	*Odocoileus hemionus*	Canada (Hadwen 1942), USA (Rhyan et al. 1995)
White-tailed deer	*Odocoileus virginianus*	Canada (Belli 1962), USA (Levine 1934; Essey and Davis 1997)
Moose, elk	*Alces alces*	Canada (Hadwen 1942), Sweden (Hülphers and Lilleengen 1947)
Caribou, reindeer	*Rangifer tarandus*	United Kingdom (Lovell 1930[a])
Roe deer	*Capreolus capreolus*	Germany (Schmidt 1938), Switzerland (Bouvier 1963), United Kingdom (Rose, 1987)

[a] Farmed or park deer.
[b] Captive wild deer.
(Modified from Clifton-Hadley and Wilesmith 1991.)

TABLE 21.2—Prevalence of infection and/or disease caused by *Mycobacterium bovis* in free-living deer populations

Species	Prevalence	Origin	Comments
Fallow deer, red deer, sika deer	5/130 (3.8%)	Republic of Ireland (Dodd 1984)	Survey of deer killed by hunters in cattle breakdown areas
Sika deer, roe deer	5/117 (4.3%)	England (Rose 1987)	Investigation of cattle breakdowns
Unreported species	8/734 (1.1%)	England (Philip 1989)	Cases associated with cattle breakdowns and infected badgers
Red deer	(0.09%)	Germany (Witte 1940)	Based on gross pathology
	(2.0%)	New Zealand (Nugent and Mackereth 1996)	Associated with cattle
	24/65 (37%)	New Zealand (Nugent and Lugton 1995)	Survey of wild deer from the Hauhungaroa Range, North Island
Elk	73/1329 (5.5%)	Canada (Hadwen 1942; Tessaro 1986)	Elk Island National Park, ranging with infected bison *Bison bison*
	0/770 (<1.0%)	Wyoming (Williams et al. 1995)	Survey of elk in northwestern Wyoming
Axis deer	(<5.0%)	Hawaii (Essey et al. 1981)	No confirmed cases
Roe deer	11/892 (1.2%)	Switzerland (Bouvier 1963)	Associated with infected cattle
White-tailed deer	1/440 (0.2%)	USA (Belli 1962)	Survey
	(>4.0%)	Michigan (Schmitt et al. 1997)	Hunter survey of wild deer in area where deer are fed in winter
Mule deer	2/41 (4.9%)	USA (Rhyan et al. 1995)	Infected elk on adjacent game ranch
	2/242 (0.8%)	Canada (Hadwen 1942)	Elk Island National Park, ranging with infected bison
Moose	6/107 (5.6%)	Canada (Hadwen 1942)	Elk Island National Park, ranging with infected bison

(Modified from Clifton-Hadley and Wilesmith 1991).

infection for the native deer population on the Hawaiian island of Molokai (Essey et al. 1981).

In Switzerland and the United States, wild deer populations may have become infected from local cattle, and in Switzerland, after TB was eradicated from cattle, cases among the wild deer population were no longer reported (Levine 1934; Sawa et al. 1974; Pastoret et al. 1988). However, under circumstances of high density, such as concentration of white-tailed deer in winter yards or over supplemental feed, bovine TB may persist in free-ranging cervids (Schmitt et al. 1997).

Given the general lack of accurate prevalence data, estimates of the risk that infected wild deer pose to other wild and farmed species are likely to be inaccurate (Tessaro 1986). However, the risk will vary depending on degree of contact with infected animals and their products, the population density of the reservoir, and the prevalence of infection. It has been suggested that grazing patterns may influence prevalence within different species of wild deer: for example, a higher prevalence was found among elk, which feed in herds, compared to mule deer, which feed in small groups or in isolation (Hadwen 1942). In Canada and the United Kingdom, white-tailed deer and roe deer *Capreolus capreolus*, respectively, were considered a source of infection for cattle (Belli 1962; Gunning 1985). However, even when the prevalence is high, excretion site sampling suggests that most infected deer shed few mycobacteria, except in the terminal stages of disease (Nugent and Lugton 1995).

Affected cervids may serve as sources of infection for sympatric predators and scavengers. In Switzerland, diseased roe deer carcasses were considered a source of infection for badgers and foxes (Bouvier 1963). Coyotes *Canis latrans* were infected by consumption of infected cervids; in Montana, exposure was probably from mule deer or elk (Rhyan et al. 1995) and, in Michigan, white-tailed deer were the source of *M. bovis* (Whipple et al. 1997; Bruning-Fann et al. 1998).

Clinical Signs and Pathology. Disease in deer caused by *M. bovis* is subacute or chronic, with some animals showing clinical signs within 6 months of infection, whereas others may survive for several years without apparent evidence of infection (Fedoseev et al. 1982; Williams 1987). There may be no clinical signs throughout an animal's life if lesions are confined to internal lymph nodes or restricted areas of lung. Although generalized disease involving the lungs may result in emaciation (Basak et al. 1975; Jørgensen et al. 1988), tuberculous deer may die suddenly, without prior clinical signs (Beatson and Hutton 1981; Beatson 1985), even when little functional lung tissue remains (Quinn and Towar 1963). Nonspecific signs have been described, including retardation of antler growth, sexual indifference in stags in the rutting season, and failure of hinds to come into estrus (Friend et al. 1963; Fedoseev et al. 1982).

Coughing and respiratory rales, although sometimes present, are not typical features of the disease in deer (Krucky et al. 1982). A more common finding is enlargement of one or more superficial lymph nodes as abscesses develop (Beatson 1985; Jørgensen et al. 1988). These may form sinuses and discharge pus through the skin or mucosal surfaces (Beatson et al. 1984; Fleetwood et al. 1988).

Pathologic changes may be in the form of the classic, proliferative granuloma accompanied by central caseation and calcification as lesions develop and age, such as are found routinely in cattle, or thin-walled abscesses involving lymph nodes may predominate (Stumpff 1982; Fleetwood et al. 1988; Buchan and Griffin 1990; Rhyan et al. 1992; Rhyan and Saari 1995; Palmer et al. 1999a). This tendency toward abscess rather than granuloma formation has prompted the suggestion that deer are more susceptible to *M. bovis* infection than cattle (Towar et al. 1965). Lesions vary in shape, and little recognizable lymph node tissue may remain in some instances (Towar et al. 1965; Beatson et al. 1984). Mesenteric lymph nodes may abscess (Fleetwood et al. 1988). Apart from caseous necrosis and mineralization, epithelioid cells, fibroblasts, and Langhans' giant cells may be present. Lesions in fallow *Dama dama* and red deer appear similar, but those in sika deer *Cervus nippon* have fewer neutrophils, more fibrosis, and larger giant cells containing more nuclei than in the other species (Rhyan and Saari 1995).

Acid-fast bacilli are highly variable in number within both principle types of lesion. Fewer organisms are found in lesions characterized by fibrosis and encapsulation (Beatson and Hutton 1981; Fleetwood et al. 1988; Corrin et al. 1993).

In the head, purulent tonsillitis has been described. Lesions have also been found in submandibular and retropharyngeal lymph nodes (Beatson et al. 1984; Brooks 1984; Fleetwood et al. 1988). A variety of changes have been recorded in the thorax, apart from lesions in the bronchial and mediastinal lymph nodes. In the lungs, lesions in the caudal lobes may predominate as in cattle, although localization in the cardiac and apical lobes has also been described. Lesions may be restricted to occasional, discrete, 1–2-cm-diameter nodules or be spread throughout the lung tissue, sometimes comprising extensive consolidation. Miliary lesions, especially in deer calves, and cavitation also have been described (Basak et al. 1975; Fleetwood et al. 1988; Stuart et al. 1988; Philip 1989). Spread from the lungs may result in abscess formation on the diaphragm, subcutaneous swellings from penetration of the thoracic wall, and lesions in the pericardium and pleura, although pleural effusions are rare (Levine 1934; Basak et al. 1975; Beatson et al. 1984; Stuart et al. 1988). Purulent or serous bronchopneumonia may develop around the lesions, with erosion of vascular and bronchiolar walls. Atelectasis and emphysema may also be sequelae of infection (Towar et al. 1965; Sawa et al. 1974; Singh et al. 1986).

Abdominal lesions particularly include abscesses in mesenteric lymph nodes, but these have also been reported in ruminal and hepatic lymph nodes, on serosal surfaces of peritoneum and omentum, and in spleen and kidney (Towar et al. 1965; Gunning 1985; Fleetwood et al. 1988; Stuart et al. 1988). Orchitis and metritis may also occur (Kovalev 1980).

All ages of deer may be infected, with prevalence increasing with age (Nugent and Lugton 1995). Under some circumstances, however, it may be that young animals are more prone to infection and acutely progressive disease (HMSO 1990). It has been suggested that in stock under 6 months of age the disease is different from that in adults, both in terms of the lesions found and the immunologic response to infection. Large numbers of *M. bovis* organisms have been described in lymphatic tissue of young deer, without any obvious macroscopic change (Griffin 1988; Griffin and Buchan 1994). Recently, a sex difference was recorded in the severity of disease: males were more likely to have gross lesions and more advanced lesions than were females (I.W. Lugton unpublished).

Diagnosis. In wild deer, a presumptive diagnosis may rest solely on postmortem and, in some cases, microscopic appearance of lesions. However, neither is pathognomonic for *M. bovis* infection. Changes in certain hematologic parameters, for example, fibrinogen concentration, lymphocyte counts, and plasma viscosity, may occur but in isolation lack specificity (Griffin and Cross 1986; Cross 1987). As in other species, definitive diagnosis rests to date on bacterial culture and identification (Pritchard 1988), although problems of bacterial isolation may be encountered even from gross lesions (Hunter 1984). The use of polymerase chain reaction to detect *M. bovis* in cattle and elk tissues has shown promise when applied to histologic sections with lesions and acid-fast organisms characteristic of TB (Miller et al. 1997). This technique may help overcome some of the deficiencies of culture.

Tuberculin skin testing is more challenging in deer than in cattle, requiring better site preparation and more care in injecting intradermally. The test's sensitivity has, in consequence, been quite variable (Clifton-Hadley and Wilesmith 1991; Griffin et al. 1994). Nonspecific sensitization also presents particular problems. For these reasons, attention has focused on developing and validating blood-based diagnostic tests (Palmer et al. 2000), and a composite test, comprising a lymphocyte transformation test, ELISA, and serum haptoglobin estimation is now in use in New Zealand and the United States as an ancillary test (Essey and Vantiem 1995). Lymphocyte transformation tests alone were not reliable (Hutchings and Wilson 1995).

Control. Most reports indicate the sporadic nature of infection in free-ranging wild deer populations. This will generally make control impractical. However, limited evidence suggests that, in some populations, control of infection in sympatric reservoir species may have the effect of reducing or even eliminating infection from wild deer in the vicinity.

Tuberculosis recently was found in free-ranging white-tailed deer in Michigan (Schmitt et al. 1997), and it is considered the greatest potential threat to TB eradication in the United States, given the difficulty in removing infection from wildlife populations (Essey and Davis 1997; McCarty and Miller 1998). Control will be focused on decreasing deer density and eliminating supplemental winter feeding.

MYCOBACTERIUM BOVIS IN WILD BOVIDS

Introduction. The family Bovidae includes cattle and their near relatives of the subfamily Bovinae, the bison and buffaloes; the African and Asian antelopes; and the subfamily Caprinae, which includes goats and sheep (Nowak 1991). Infections by *M. bovis* have been reported in many nondomestic bovid species. Many of these reports are descriptive single cases, frequently where *M. bovis* was assumed to be causative on the basis of tubercle-like lesions. Some reports, such as those of TB in free-ranging lechwe *Kobus leche* and bushbuck *Tragelaphus scriptus* in Zambia (Gallagher et al. 1972; Zieger et al. 1998), suggest a focus of TB activity. In general, there has been little epidemiologic study of prevalence and distribution of the disease in the originating population. To date, there has been detailed investigative work on TB in only three wild bovid populations: bison in Wood Buffalo National Park (WBNP) in northern Canada, African buffalo in Kruger National Park (KNP) in South Africa, and feral water buffalo in the Northern Territory of Australia.

Tuberculosis in Bison in Wood Buffalo National Park, Canada. Two subspecies of bison are recognized: wood bison *Bison bison athabascae* and plains bison *Bison bison bison.* These subspecies were geographically isolated, with *B. bison athabascae* occurring to the north of a dividing line extending across west-central Canada, and the more abundant plains bison to the south (Nowak 1991). The original number of wood bison in western Canada was estimated at 168,000, but by 1891 no more than 300 survived, being confined to western Alberta and adjacent parts of the Northwest Territories. In recognition of the need for protection, legislation was introduced in 1897 and, in 1922, WBNP was established. However, shortly thereafter, some 6000 plains bison were translocated from the Buffalo National Park at Wainwright, Alberta, and released at WBNP. Consequently, most of the population of wood bison was genetically swamped (Choquette et al. 1961). In 1957, an isolated, remnant herd of wood bison was discovered in the northern part of the park, and some of these were then translocated to a new reserve some 100 km to the north: the Mackenzie Bison Sanctuary. The wood bison population in this reserve has thrived and currently comprises over 2000 animals (Tessaro et al. 1993); meanwhile, the population in WBNP has declined, from an estimated 11,000 animals in 1970 to a present population of 2300 (Joly et al. 1998).

Tuberculosis was first positively identified in the hybrid bison of WBNP in a detailed survey during 1959–60 (Choquette et al. 1961). Over 1000 animals were tested with bovine tuberculin, with 13.5% giving a positive skin reaction. These reactors were subsequently slaughtered and, on postmortem examination, most had tuberculous lesions, particularly in the lymph nodes of the head. The disease in the park population was almost certainly introduced by translocation of Wainwright plains bison; Hadwen (1942) had examined this herd and found the disease to be common.

Impetus to investigate the disease status of the WBNP bison was increased by the near eradication of

TB among the Canadian cattle population by the mid-1980s. In particular, there was concern that the infected bison might act as a reservoir of disease (Tessaro 1986). This was confirmed by a survey of 72 animals that had been hunted or found dead in 1983–85 (Tessaro et al. 1990), in which *M. bovis* was isolated or suspected in 15 animals. Lesions were most frequent in the lymph nodes, particularly involving the retropharyngeal and bronchial nodes. The survey prevalence of 21% was considered to be an underestimate, because sampling was biased for healthy, younger animals.

In 1989, the situation in the WBNP bison was examined by a Federal Environmental Assessment Review Panel, and it was recommended that the diseased bison be destroyed and replaced by healthy wood bison. The Bison Research and Containment Program was established to implement this (Lavigne personal communication). Consequently, a project was funded specifically to investigate TB and brucellosis in the herd and their effects on its population dynamics. Based on the caudal fold test and ELISA blood test, the prevalence of TB was estimated to be 51% (Joly et al. 1998). Studies on the effect of the diseases in causing the population decline are continuing, currently emphasizing the synergistic effects between disease and predation by wolves.

Other wild bison populations in North America are currently considered to be free of TB. In particular, the wood bison in the Mackenzie Sanctuary, considered the population most at risk, were examined between 1986 and 1988 and found free of disease (Tessaro et al. 1993). However, Stumpff et al. (1984) reported on the rapid spread of the disease from an infected elk to a ranched population of plains bison in South Dakota, thus demonstrating the need for vigilance in ensuring that the remaining wild populations remain free of the disease.

Tuberculosis in African Buffalo in Kruger National Park, South Africa. The African buffalo *Syncerus caffer* originally occurred in most of Africa south of the Sahara (Nowak 1991). Both subspecies of buffalo suffered population declines over the past 100 years, particularly from the combined effects of habitat loss and persecution. The national park system plays an important role in their conservation, with large populations in the Serengeti in Tanzania and Kruger National Park (KNP), the latter having an estimated buffalo population of about 30,000 (de Vos et al. 1995).

Tuberculosis was first diagnosed in a buffalo in KNP in 1990 during an opportunistic necropsy performed on an emaciated, moribund 2-year-old bull (Bengis et al. 1996). The postmortem diagnosis of TB was based on finding lesions in the lungs and thoracic lymph nodes. Subsequently, random sampling of 57 buffalo from two herds in close proximity to the index case showed nine other cases with similar or suspect lesions. Bovine TB was confirmed by isolation of *M. bovis* from lesions of two of the affected animals. It was considered that the disease had entered the buffalo population not more than 10–15 years previously, because intensive sampling of the same population during 1966 (Basson et al. 1970), as well as routine meat inspection in the 1970s, had not detected its presence. The source of the infection was considered most likely to be infected cattle on adjoining farms. However, the disease has become established in the buffalo herd.

Further survey work established that the disease is concentrated in the buffalo populations of the south section of the park (de Vos et al. 1995). Furthermore, the prevalence among the affected herds is variable, ranging from 2%–67%. The disease is spreading rapidly, and the entire buffalo population of the park is expected to become infected by 2003–2008. Nevertheless, the disease has not established itself among other wild ungulates in KNP, although it has been diagnosed in lions *Panthera leo,* cheetah *Acinonyx jubatus,* and baboon *Papio ursinus* (Keet et al. 1996).

To develop a containment strategy, investigations have been ongoing to determine the most appropriate antemortem testing procedure. Raath et al. (1995) report on a detailed trial in which skin testing, which necessitated holding the animals for 72 hours, was compared to the existing blood tests. Of the latter, the ELISA gave inconsistent results. The gamma-interferon test performed better, having a sensitivity of 76%, but was inferior to the skin test sensitivity of 84%–91%.

The current TB status in other buffalo populations of Africa is uncertain. The only other report of the disease was from Ruwenzori National Park in Uganda, where its occurrence was documented during the 1960s (Guilbride et al. 1963; Thurlbeck et al. 1965). Subsequently, Woodford (1982) estimated the prevalence to be between 10% and 38%. The source of the infection was considered to be the cattle kept by neighboring tribesmen. There are no recent reports; therefore, the current prevalence of the disease in this population is unknown.

Tuberculosis in Feral Water Buffalo in the Northern Territory, Australia. It is estimated there are at least 150 million domestic water buffalo *Bubalus bubalis* in the world, principally in India, Southeast Asia, and China (Nowak 1991). In contrast, the nondomestic source population has disappeared from most of its range due to habitat loss and hunting, and this wild species is endangered. Currently, the largest nondomesticated population occurs in northern Australia, being derived from domesticated animals introduced during the middle part of the 19th century (Letts 1964). The feral population expanded rapidly, and at its maximum during the mid 1980s numbered almost 350,000 individuals (Bayliss and Yeomans 1989). These buffalo did not extend their range much beyond the subcoastal plain between Darwin and the Arnhem Land escarpment to the east where the seasonally flooded riverine landscape provides an appropriate habitat (Tulloch 1978).

Exploitation of this population has varied (Letts 1964). In the late 19th century, an industry developed for the tanning and export of hides of animals shot in the field. By the 1950s, this industry collapsed through successful competition from Asia. In the 1960s, the focus then shifted to exploitation for meat, both for pet feed and human consumption. Animals were brought in alive to the abattoirs, and this enabled high standards of disease surveillance. As a result of this abattoir surveillance, bovine TB was recognized to be a common problem in the buffalo population. Letts (1964) reported a mean prevalence in the early 1960s of 16.4%. A subsequent survey in 1979 showed a lower prevalence, between 0.3% and 8.2%, with higher prevalence occurring among animals originating from the coastal plain (Hein and Tomasovic 1981). This decline in prevalence over the period was attributed to the selective harvesting of mature animals, those most likely to be infective. Tuberculous lesions in the slaughtered buffalo were predominately in the thoracic cavity, which was taken as evidence that the aerogenous route was the most important means of transmission in the population (McCool and Newton-Tabrett 1979; Hein and Tomasovic 1981).

By the 1980s, TB was successfully eradicated from the cattle population of southern Australia, and the emphasis shifted to its control in feral cattle and buffaloes of the Northern Territory (Lehane 1996). To achieve this eradication, intensive culling of the buffalo and cattle population was undertaken, with whole areas depopulated of animals. To detect small pockets of animals in difficult and inaccessible terrain, the "judas" animal technique was developed. Single buffalo were anesthetized with a tranquilizer dart and fitted with a radio collar. When these animals rejoined their herd, this could then be detected and the animals shot from helicopters. In areas cleared by these means, restocking was undertaken from a TB-free population in southern Arnhem Land (Radunz personal communication). Nevertheless, repopulating was only partial, and currently the total feral buffalo population numbers only about 35,000 animals. Skin testing was little used, as despite experimental evidence of its efficiency (Small and Thomson 1986), test and slaughter based on it were not found to be practical (Lehane 1996).

These operations were successful in controlling bovine TB in northern Australia, and by 1998 the disease was believed effectively eradicated from all ruminant populations of the country (Scanlan personal communication). Because it is still possible that an occasional case may present, a final surveillance is currently in operation whereby any granuloma encountered in an abattoir is submitted for examination for *M. bovis*. This program is expected to run until 2002, when, given continued negative results, complete eradication of bovine TB in the cattle and buffalo populations will have been achieved.

LITERATURE CITED

Allen, G.M. 1991. Other animals as sources of TB infection. In *Proceedings of a symposium on tuberculosis,* ed. R. Jackson. Palmerston North, New Zealand: New Zealand Veterinary Association Foundation for Continuing Education, pp. 197–201.

Anderson, R.M., and W. Trewhella. 1985. Population dynamics of the badger (*Meles meles*) and the epidemiology of bovine tuberculosis (*Mycobacterium bovis*). *Philosophical Transactions of the Royal Society, London [B]* 310:327–381.

Barnes, P.F., R.L. Modlin, and J.J. Ellner. 1994. T-cell responses and cytokines. In *Tuberculosis: Pathogenesis, protection, and control,* ed. B.R. Bloom. Washington, DC: American Society for Microbiology, pp. 417–435.

Basak, D.K., A. Chatterjee, M.K. Neogi, and D.P. Samanta. 1975. Tuberculosis in captive deer. *Indian Journal of Animal Health* 14:135–137.

Basson, P.A., R.M. McCully, S.P. Kruger, J.W. van Niekerk, E. Young, and V. de Vos. 1970. Parasitic and other diseases of African buffalo in the Kruger National Park. *Onderstepoort Journal of Veterinary Research* 37:11–28.

Batcheler, C.L., and P.E. Cowan. 1988. *Review of the status of the possum (*Trichosurus vulpecula*) in New Zealand, for the Agricultural Pest Destruction Council.* Ministry of Agriculture and Fisheries, New Zealand: Department of Conservation MAF Qual, 129 pp.

Bayliss, P., and K.M. Yeomans. 1989. Distribution and abundance of feral livestock in the "top end" of the Northern Territory (1985–86), and their relation to population control. *Australian Wildlife Research* 16:651–676.

Beatson, N.S. 1985. Tuberculosis in red deer in New Zealand: Biology of deer production. *Royal Society of New Zealand Bulletin* 22:147–150.

Beatson, N.S., and J.B. Hutton. 1981. Tuberculosis in farmed deer in N.Z. In *Proceedings of a deer seminar for veterinarians.* Queenstown: New Zealand Veterinary Association Deer Advisory Panel, pp. 143–151.

Beatson, N.S., J.B. Hutton, and G.W. de Lisle. 1984. Tuberculosis test and slaughter. In *Proceedings of a deer course for veterinarians,* vol. 1. Palmerston North: Deer Branch of the New Zealand Veterinary Association, pp. 18–27.

Belli, L.B. 1962. Bovine tuberculosis in a white-tailed deer (*Odocoileus virginianus*). *Canadian Veterinary Journal* 3:356–358.

Bengis, R.G., N.P.J. Kriek, D.F. Keet, J.P. Raath, V. de Vos, and H.F.A.K. Huchzermeyer. 1996. An outbreak of bovine tuberculosis in a free-living African buffalo (*Syncerus caffer* Sparrman) population in the Kruger National Park: A preliminary report. *Onderstepoort Journal of Veterinary Research* 63:15–18.

Benham, P.F.J., and D.M. Broom. 1989. Interactions between cattle and badgers at pasture with reference to bovine tuberculosis transmission. *British Veterinary Journal* 145:226–241.

———. 1991. Responses of dairy cows to badger urine and feces on pasture with reference to bovine tuberculosis transmission. *British Veterinary Journal* 147:517–532.

Bolske, G., L. Englund, H. Wahlstrom, G.W. de Lisle, D.M. Collins, and P.S. Croston. 1995. Bovine tuberculosis in Swedish deer farms: Epidemiological investigations and tracing using restriction fragment analysis. *Veterinary Record* 136:414–417.

Bouvier, G. 1963. Possible transmission of tuberculosis and brucellosis from game animals to man and to domestic animals. *Bulletin de l'Office International des Epizooties* 59:433–436.

Bouvier, G., H. Burgisser, and P.A. Schneider. 1957. Observations sur les maladies du gibier, des oiseaux et des poissons faites en 1955 et 1956. *Schweizer Archiv für Tierheilkunde* 99:461–477.

Brooks, H.V. 1984. Pathology of tuberculosis in red deer (*Cervus elaphus*). In *Proceedings of a deer course for veterinarians,* vol. 1. Palmerston North: Deer Branch of the New Zealand Veterinary Association, pp. 13–15.

Bruning-Fann, C.S., S.M. Schmitt, S.D. Fitzgerald, J.B. Payeur, D.L. Whipple, T.M. Cooley, T. Carlson, and P. Fredrich. 1998. *Mycobacterium bovis* in coyotes from Michigan. *Journal of Wildlife Diseases* 34:632–636.

Buchan, G.S., and J.F.T. Griffin. 1990. Tuberculosis in domesticated deer (*Cervus elaphus*): A large animal model for human tuberculosis. *Journal of Comparative Pathology* 103:11–22.

Buddle, B.M., F.E. Aldwell, A. Pfeffer, and G.W. de Lisle. 1994. Experimental *Mycobacterium bovis* infection in the brushtail possum (*Trichosurus vulpecula*): Pathology, haematology and lymphocyte stimulation responses. *Veterinary Microbiology* 38:241–254.

Buddle, B.M., G.W. de Lisle, A. Pfeffer, and F.E. Aldwell. 1995a. Immunological responses and protection against *Mycobacterium bovis* in calves vaccinated with a low dose of BCG. *Vaccine* 13:1123–1130.

Buddle, B.M., A. Nolan, A.R. McCarthy, J. Heslop, F.E. Aldwell, R. Jackson, and D.U. Pfeiffer. 1995b. Evaluation of three serological assays for the diagnosis of *Mycobacterium bovis* infection in brushtail possums. *New Zealand Veterinary Journal* 43:91–95.

Cheeseman, C.L., and P.J. Mallinson. 1981. Behavior of badgers (*Meles meles*) infected with bovine tuberculosis. *Journal of Zoology* 194:284–289.

Cheeseman, C.L., J.W. Wilesmith, F.A. Stuart, and P.J. Mallinson. 1988. Dynamics of tuberculosis in a naturally infected badger population. *Mammal Review* 18:61–72.

Cheeseman, C.L., J.W. Wilesmith, and F.A. Stuart. 1989. Tuberculosis: The disease and its epidemiology in the badger: A review. *Epidemiology and Infection* 103:113–125.

Cheeseman, C.L., P.J. Mallinson, J. Ryan, and J.W. Wilesmith. 1993. Recolonization by badgers in Gloucestershire. In *The badger,* ed. T.J. Hayden. Dublin: Royal Irish Academy, pp. 78–93.

Choquette, L.P.E., J.F. Gallivan, J.L. Byrne, and J. Pilipavicius. 1961. Parasites and diseases of bison in Canada: I. Tuberculosis and some other pathological conditions in bison at Wood Buffalo and Elk Island National Parks in the fall and winter of 1959–60. *Canadian Veterinary Journal* 2:168–174.

Clifton-Hadley, R.S. 1996. Badgers, bovine tuberculosis and the age of reason. *British Veterinary Journal* 152:243–246.

Clifton-Hadley, R.S., and C.L. Cheeseman. 1995. *Mycobacterium bovis* infection in a wild badger (*Meles meles*) population. In *Tuberculosis in wildlife and domestic animals: Proceedings of the second international conference on* Mycobacterium bovis. *Otago Conference Series,* no. 3, ed. F. Griffin and G. de Lisle. Dunedin, New Zealand: University of Otago Press, pp. 264–266.

———. 1997. Performance of an ELISA in determining the *Mycobacterium bovis* status of badger (*Meles meles*) setts. *Epidemiologie et Sante Animale* 31:1–3.

Clifton-Hadley, R.S., and J.W. Wilesmith. 1991. Tuberculosis in deer: A review. *Veterinary Record* 129:5–12.

Clifton-Hadley, R.S., J.W. Wilesmith, and F.A. Stuart. 1993. *Mycobacterium bovis* in the European badger (*Meles meles*): Epidemiological findings in tuberculous badgers from a naturally infected population. *Epidemiology and Infection* 111:9–19.

Clifton-Hadley, R.S., A.R. Sayers, and M.P. Stock. 1995a. Evaluation of an ELISA for *Mycobacterium bovis* infection in badgers (*Meles meles*). *Veterinary Record* 137:555–558.

Clifton-Hadley, R.S., J.W. Wilesmith, M.S. Richards, P. Upton, and S. Johnston. 1995b. The occurrence of *Mycobacterium bovis* infection in and around an area subject to extensive badger (*Meles meles*) control. *Epidemiology and Infection* 114:179–193.

Clifton-Hadley, R.S., J. Inwald, J. Archer, S. Hughes, N. Palmer, A.R. Sayers, K. Sweeney, J.D.A. van Embden, and R.G. Hewinson. 1998. Recent advances in DNA fingerprinting using spoligotyping: Epidemiological applications in bovine tuberculosis. *Journal of the British Cattle Veterinary Association* 6:79–82.

Coleman, J.D., R. Jackson, M.M. Cooke, and L. Grueber. 1994. Prevalence and spatial distribution of bovine tuberculosis in brushtail possums on a forest-scrub margin. *New Zealand Veterinary Journal* 42:128–132.

Collins, J.D. 1995. Regional and country status reports: Ireland. In Mycobacterium bovis *infection in animals and humans,* ed. C.O. Thoen and J.H. Steele. Ames: Iowa State University Press, pp. 224–238.

Collins, C.H., and J.M. Grange. 1983. A review: The bovine tubercle bacillus. *Journal of Applied Bacteriology* 55:13–29.

———. 1987. Zoonotic implication of *Mycobacterium bovis* infection. *Irish Veterinary Journal* 41:363–366.

Cooke, M.M., R. Jackson, J.D. Coleman, and M.R. Alley. 1995. Naturally occurring tuberculosis caused by *Mycobacterium bovis* in brushtail possums (*Trichosurus vulpecula*): II. Pathology. *New Zealand Veterinary Journal* 43:315–321.

Cordes, D.O., J.A. Bullians, D.E. Lake, and M.E. Carter. 1981. Observations on tuberculosis caused by *Mycobacterium bovis* in sheep. *New Zealand Veterinary Journal* 29:60–62.

Corner, L.A., and P.J.A. Presidente. 1980. *Mycobacterium bovis* infection in the brush-tailed possum (*Trichosurus vulpecula*): 1. Preliminary observations on experimental infection. *Veterinary Microbiology* 5:309–321.

———. 1981. *Mycobacterium bovis* infection in the brush-tailed possum (*Trichosurus vulpecula*): II. Comparison of experimental infections with an Australian cattle strain and a New Zealand possum strain. *Veterinary Microbiology* 6:351–366.

Corrin, K.C., C.E. Carter, R.C. Kissling, and G.W. de Lisle. 1993. An evaluation of the comparative tuberculin skin test for detecting tuberculosis in farmed deer. *New Zealand Veterinary Journal* 41:12–20.

Cross, J.P. 1987. Haematology-based prediction of lesion status in bovine tuberculosis of farmed red deer. In *Proceedings of a deer course for veterinarians,* vol. 4. Dunedin: Deer Branch of the New Zealand Veterinary Association, pp. 147–154.

Daborn, C.J., J.M. Grange, and R.R. Kazwala. 1996. The bovine tuberculosis cycle: An African perspective. *Journal of Applied Bacteriology* 81(Symposium Suppl.):27S–32S.

Davidson, R.M. 1976. The role of the opossum in spreading tuberculosis. *New Zealand Journal of Agriculture* 133:21–25.

Davidson, R.M., M.R. Alley, and N.S. Beatson. 1981. Tuberculosis in a flock of sheep. *New Zealand Veterinary Journal* 29:1–2.

de Lisle, G.W., and P.F. Havill. 1985. Mycobacteria isolated from deer in New Zealand from 1970–1983. *New Zealand Veterinary Journal* 33:138–140.

de Lisle, G.W., P.J. Welch, P.F. Havill, A.F. Julian, W.S.H. Poole, K.C. Corrin, and N.R. Gladden. 1983. Experimental tuberculosis in red deer (*Cervus elaphus*). *New Zealand Veterinary Journal* 31:213–216.

de Lisle, G.W., K. Crews, J. de Zwart, R. Jackson, G.J.E. Knowles, K.D. Paterson, R.W. MacKensie, K.A. Waldrup, and R. Walker. 1993. *Mycobacterium bovis* infections in wild ferrets. *New Zealand Veterinary Journal* 41:148–149.

de Vos, V., J.P. Raath, R. Bengis, N.P.J. Kriek, H. Huchzermeyer, and D.F. Keet. 1995. The epidemiology of bovine tuberculosis in the Kruger National Park, South Africa. In *Tuberculosis in wildlife and domestic animals: Proceedings of the second international conference on* Mycobacterium bovis. *Otago Conference Series,* no. 3, ed. F. Griffin and G. de Lisle. Dunedin, New Zealand: University of Otago Press, pp. 255–259.

Dodd, K. 1984. Tuberculosis in free living deer. *Veterinary Record* 115:592–593.

Dolan, L.A., J.A. Eves, and D. Bray. 1995. *East Offaly badger research project (EOP): Interim report for the period, January 1989 to December 1995.* Dublin: Tuberculosis Investigation Unit, University College, pp. 18–20.

Duffield, B.J., and D.A. Young. 1985. Survival of *Mycobacterium bovis* in defined environmental conditions. *Veterinary Microbiology* 10:193–197.

Dungworth, D.L. 1993. The respiratory system. In *Pathology of domestic animals,* vol. 2, ed. K.V.F. Jubb, P.C. Kennedy, and N. Palmer, 4th ed. San Diego: Academic, pp. 539–698.

Dunkin, G.W., P.P. Laidlaw, and A.S. Griffith. 1929. A note on tuberculosis in the ferret. *Journal of Comparative Pathology* 42:46–49.

Dunnet, G.M., D.M. Jones, and J.P. McInerney. 1986. *Badgers and bovine tuberculosis: Review of policy.* London: Her Majesty's Stationery Office, 73 pp.

Efford, M.G. 1991. *Contract report between DSIR Land Resources and Animal Health Board, 91/41.* Lower Hutt, New Zealand: DSIR Land Resources [cited by Jackson (1995)].

Ekdahl, M.O., B.L. Smith, and D.F.L. Money. 1970. Tuberculosis in some wild and feral animals in New Zealand. *New Zealand Veterinary Journal* 18:44–45.

Essey, M.A., and J.P. Davis. 1997. Status of the national cooperative state-federal bovine tuberculosis eradication program fiscal year 1997. *Proceedings of the United States Animal Health Association* 101:561–581.

Essey, M.A., and J.S. Vantiem. 1995. *Mycobacterium bovis* infection in captive Cervidae: An eradication program. In Mycobacterium bovis *infection in animals and humans,* ed. C.O. Thoen and J.H. Steele. Ames: Iowa State University Press, pp. 145–157.

Essey, M.A., R.L. Payne, E.M. Himes, and D.W. Luchsinger. 1981. Bovine tuberculosis surveys of axis deer and feral swine on the Hawaiian island of Molokai. *Proceedings of the United States Animal Health Association* 85:538–549.

Fanning, A., S. Edwards, and G. Hauer. 1991. *Mycobacterium bovis* infection in humans exposed to elk in Alberta. *Canadian Disease Weekly Report* 17:239–240 and 243.

Fedoseev, V.S., L.N. Rubtsova, E.O. Omarbekov, and N.G. Kirilenko. 1982. Anti-tuberculosis measures in maral rearing. *Veterinariya (Moscow)* 4:35–36.

Fleetwood, A.J., F.A. Stuart, R. Bode, and J.P. Sutton. 1988. Tuberculosis in deer. *Veterinary Record* 123:279–280.

Fletcher, J. 1990. Tuberculosis in British deer. In *Proceedings of a deer course for veterinarians,* vol. 7. Auckland: Deer Branch of the New Zealand Veterinary Association, pp. 49–51.

Friend, M., E.T. Kroll, and H. Gruft. 1963. Tuberculosis in a wild white-tailed deer. *New York Fish and Game Journal* 10:118–123.

Gallagher, J., and J. Nelson. 1979. Causes of ill health and natural death in badgers in Gloucestershire. *Veterinary Record* 105:546–551.

Gallagher, J., I. Macadam, J. Sayer, and L.P. van Lavieren. 1972. Pulmonary tuberculosis in free-living lechwe antelope in Zambia. *Tropical Animal Health and Production* 4:204–413.

Gallagher, J., R.H. Muirhead, and K.J. Burn. 1976. Tuberculosis in wild badgers (*Meles meles*) in Gloucestershire: Pathology. *Veterinary Record* 98:9–14.

Gallagher, J., R. Monies, M. Gavier-Widen, and B. Rule. 1998. Role of infected, non-diseased badgers in the pathogenesis of tuberculosis in the badger. *Veterinary Record* 142:710–714.

Goodger, J., A. Nolan, W.P. Russell, D.J. Dalley, C.J. Thorns, F. A. Stuart, P. Crostan, and D.G. Newell. 1994. Serodiagnosis of *Mycobacterium bovis* infection in badgers: Development of an indirect ELISA using a 25 kDa antigen. *Veterinary Record* 135:82–85.

Grange, J.M. 1996. The biology of the genus *Mycobacterium. Journal of Applied Bacteriology* 81(Symposium Suppl.):1S–9S.

Grange, J.M., and M.D. Yates. 1994. Zoonotic aspects of *Mycobacterium bovis* infection. *Veterinary Microbiology* 40:137–151.

Griffin, J.F.T. 1988. The aetiology of tuberculosis and mycobacterial diseases in farmed deer. *Irish Veterinary Journal* 42:23–26.

Griffin, J.F.T., and G.S. Buchan. 1994. Aetiology, pathogenesis and diagnosis of *Mycobacterium bovis* in deer. *Veterinary Microbiology* 40:193–205.

Griffin, J.F.T., and J.P. Cross. 1986. In vitro tests for tuberculosis in farmed deer. In *Proceedings of a deer course for veterinarians,* vol. 3. Rotorua: Deer Branch of the New Zealand Veterinary Association, pp. 71–77.

Griffin, J.F.T., J.P. Cross, D.N. Chinn, C.R. Rodgers, and G.S. Buchan. 1994. Diagnosis of tuberculosis due to *Mycobacterium bovis* in New Zealand red deer (*Cervus elaphus*) using a composite blood test and antibody assays. *New Zealand Veterinary Journal* 42:173–179.

Griffin, J.M., and T. Hahesy. 1992. *Analysis of epidemiology reports on 3957 herd breakdowns in ten DVO regions during 1987–90.* Dublin: Tuberculosis Investigation Unit, University College, p. 21.

Guerin, L. A. 1989. Deer in Ireland in 1988. *Irish Veterinary News* 11:21–23.

Guilbride, P.D.L., D.H.L. Rollinson, and E.G. McAnulty. 1963. Tuberculosis in the free living African (Cape) buffalo (*Syncerus caffer caffer* Sparrman). *Journal of Comparative Pathology* 73:337–348.

Gunning, R.F. 1985. Bovine tuberculosis in roe deer. *Veterinary Record* 116:300–301.

Hadwen, S. 1942. Tuberculosis in the buffalo. *Journal of the American Veterinary Medical Association* 100:19–22.

Head, K.W. 1959. Diseases of mink. *Veterinary Record* 71:1025–1032.

Hein, W.R., and A.A. Tomasovic. 1981. An abattoir survey of tuberculosis in feral buffaloes. *Australian Veterinary Journal* 57:543–547.

Her Majesty's Stationery Office (HMSO). 1978. *Annual report on research and technical work, 1978, of the Department of Agriculture for Northern Ireland.* London: HMSO, p. 282.

———. 1990. *Animal health 1989: Report of the Chief Veterinary Officer.* London: HMSO, pp. 14–16.

Hickling, G.H. 1989. Assessment of possum control in an area of endemic bovine tuberculosis. *Forestry Research Institute Progress Report,* pp. 16.

Higgins, D.A. 1985. The skin inflammatory response of the badger (*Meles meles*). *British Journal of Experimental Pathology* 66:643–653.

Higgins, D.A., and A.J. Gatrill. 1984. A comparison of the antibody responses of badgers (*Meles meles*) and rabbits (*Oryctolagus cuniculus*) to some common antigens. *International Archives of Allergy and Applied Immunology* 75:219–226.

Higgins, D.A., I.T.M. Kung, and R.S.B. Or. 1985. Environmental silica in badger lungs: A possible association with susceptibility to *Mycobacterium bovis* infection. *Infection and Immunity* 48:252–256.

Huitema, H. 1969. The eradication of bovine tuberculosis in cattle in the Netherlands and the significance of man as a source of infection. *Selected Papers, Royal Netherlands Tuberculosis Association* 12:62–67.

Hülphers, G., and K. Lilleengen. 1947. Tuberculosis in wild mammals and birds. *Svensk Veterinartidskrift* 52:193–208.

Hunter, J.W. 1984. Tuberculin testing in deer: A veterinary practitioners view. In *Proceedings of a deer course for veterinarians,* vol. 1. Palmerston North: Deer Branch of the New Zealand Veterinary Association, pp. 28–33.

Hutchings, D.L., and S.H. Wilson. 1995. Evaluation of lymphocyte stimulation tests for diagnosis of bovine tuberculosis in elk (*Cervus elaphus*). *American Journal of Veterinary Research* 56:27–33.

Hutchings, M.R., and S. Harris. 1997. Effects of farm management practices on cattle grazing behaviour and the potential for transmission of bovine tuberculosis from badgers to cattle. *Veterinary Journal* 153:149–162.

Iland, C.N., W.St.C. Symmers, and A.P.D. Thomson. 1951. A note on tuberculosis in the ferret (*Mustela furo* L.). *Journal of Comparative Pathology* 63:554–556.

Jackson, R. 1995. Transmission of tuberculosis (*Mycobacterium bovis*) by possums. Ph.D. thesis, Massey University, Palmerston North, New Zealand, 282 pp.

Jackson, R., and R.S. Morris. 1996. A study of the topography of the lymphatic system of the Australian brushtail possum (*Trichosurus vulpecula*). *Journal of Anatomy* 188:603–609.

Jackson, R., M.M. Cooke, J.D. Coleman, and R.S. Morris. 1995a. Naturally occurring tuberculosis caused by *Mycobacterium bovis* in brushtail possums (*Trichosurus vulpecula*): I. An epidemiological analysis of lesion distribution. *New Zealand Veterinary Journal* 43:306–314.

Jackson, R., M.M. Cooke, J.D. Coleman, R.S. Morris, G.W. de Lisle, and G.F. Yates. 1995b. Naturally occurring tuberculosis caused by *Mycobacterium bovis* in brushtail possums (*Trichosurus vulpecula*): III. Routes of infection and excretion. *New Zealand Veterinary Journal* 43:322–327.

Johnstone, A.C., H.M. Hussein, and A. Woodgyer. 1993. Adiaspiromycosis in suspected cases of pulmonary tuberculosis in the common brushtail possum (*Trichosurus vulpecula*). *New Zealand Veterinary Journal* 41:175–178.

Joly, D.O., F.A. Leighton, and F. Messier. 1998. Tuberculosis and brucellosis infection of bison in Wood Buffalo National Park, Canada: Preliminary results. In *International symposium on bison ecology and management in North America,* ed. L. Irby and J. Knight. Bozeman: Montana State University Press, pp. 23–31.

Jones, D.M., V.J.A. Manton, and P. Cavanagh. 1976. Tuberculosis in a herd of axis deer (*Axis axis*) at Whipsnade Park. *Veterinary Record* 98:525–526.

Jørgensen, J.B., P. Husum, and I. Sørensen. 1988. Bovin tuberkulose I en hjortefarm. *Dansk Veterinaertidsskrift* 71:806–808.

Julian, A.F. 1981. Tuberculosis in the possum *Trichosurus vulpecula.* In *Proceedings of the first symposium on marsupials in New Zealand,* ed. B.D. Bell. Wellington, New Zealand: Zoology Publications, Victoria University, pp. 163–175.

Keet, D.F., N.P. Kriek, M.L. Penrith, A. Michel, and H. Huchzermeyer. 1996. Tuberculosis in buffaloes (*Syncerus caffer*) in the Kruger National Park: Spread of the disease to other species. *Onderstepoort Journal of Veterinary Research* 63:239–244.

King, C. M. 1984. *Immigrant killers: Introduced predators and the conservation of birds in New Zealand.* Auckland, New Zealand: Oxford University Press, 224 pp.

Kovalev, G.K. 1980. Tuberculosis in wildlife. *Journal of Hygiene, Epidemiology, Microbiology and Immunology* 24:495–503.

Krebs, J.R. 1997. Bovine tuberculosis in cattle and badgers. London: MAFF, 191 pp.

Krucky, J., D. Zajicek, and J. Pavlista. 1982. Vyskyt tuberkulozy u jeleni zvere. *Veterinarstvi* 32:501–503.

Lehane, R. 1996. Beating the odds in a big country: The eradication of bovine brucellosis and tuberculosis in Australia. Collingwood, Victoria: CSIRO, pp. 195–226.

Letts, G.A. 1964. Feral animals in the Northern Territory. *Australian Veterinary Journal* 40:84–88.

Levine, P.P. 1934. A report on tuberculosis in wild deer (*Odocoileus virginianus*). *Cornell Veterinarian* 24:264–266.

Liston, W.G., and M.B. Soparkar. 1924. An outbreak of tuberculosis among animals in the Bombay zoological gardens. *Indian Journal of Medical Research* 2:671–680.

Little, T.W., P.F. Naylor, and J.W. Wilesmith. 1982a. Laboratory study of *Mycobacterium bovis* infection in badgers and calves. *Veterinary Record* 111:550–557.

Little, T.W., C. Swan, H.V. Thompson, and J.W. Wilesmith. 1982b. Bovine tuberculosis in domestic and wild mammals in an area of Dorset: II. The badger population, its ecology and tuberculosis status. *Journal of Hygiene* 89:211–224.

———. 1982c. Bovine tuberculosis in domestic and wild mammals in an area of Dorset: III. The prevalence of tuberculosis in mammals other than badgers and cattle. *Journal of Hygiene* 89:225–234.

Livingstone, P.G. 1991. TB in New Zealand: Where have we reached? In *Proceedings of a symposium on tuberculosis,* ed. R. Jackson. Palmerston North, New Zealand: New Zealand Veterinary Association Foundation for Continuing Education, pp. 113–124.

Lovell, R. 1930. The isolation of tubercle bacilli from captive wild animals. *Journal of Comparative Pathology* 43:205–215.

Lugton, I., R. Morris, P. Wilson, and F. Griffin. 1995. Natural infection of red deer with tuberculosis. In *Tuberculosis in wildlife and domestic animals: Proceedings of the second international conference on* Mycobacterium bovis. *Otago Conference Series,* no. 3, ed. F. Griffin and G. de Lisle. Dunedin, New Zealand: University of Otago Press, pp. 280–283.

Lugton, I., G. Wobeser, R.S. Morris, and P. Caley. 1997a. Epidemiology of *Mycobacterium bovis* infection in feral ferrets (*Mustela furo*) in New Zealand: I. Pathology and diagnosis. *New Zealand Veterinary Journal* 45:150–160.

———. 1997b. Epidemiology of *Mycobacterium bovis* infection in feral ferrets (*Mustela furo*) in New Zealand: II. Routes of infection and excretion. *New Zealand Veterinary Journal* 45:161–167.

Mackintosh, C.G., and J.F.T. Griffin. 1994. Epidemiological aspects of deer tuberculosis research. In *Proceedings of a deer course for veterinarians,* vol. 11. Queenstown: Deer Branch of the New Zealand Veterinary Association, pp. 106–113.

Mackintosh, C.G., K. Waldrup, R. Labes, F. Griffin, G. Buchan, J. Cross, and G. de Lisle. 1993. Experimental *Mycobacterium bovis* infection in red deer weaners: Preliminary findings. In *Proceedings of a deer course for veterinarians,* vol. 10. Palmerston North: Deer Branch of the New Zealand Veterinary Association, pp. 297–304.

Mackintosh, C.G., K. Waldrup, and J.F.T. Griffin. 1995. Investigations into the role of genetic resistance in the epidemiology of tuberculosis in deer. In *Proceedings of a deer course for veterinarians,* vol. 12. Rotorua: Deer

Branch of the New Zealand Veterinary Association, pp. 151–154.

Mahmood, K.H., G.A.W. Rook, J.L. Stanford, F.A. Stuart, and D.G. Pritchard. 1987. The immunological consequences of challenge with bovine tubercle bacilli in badgers (*Meles meles*). *Epidemiology and Infection* 98:155–163.

Martin, R.N. 1995. Regional and country status reports: Northern Ireland. In Mycobacterium bovis *infection in animals and humans,* ed. C.O. Thoen and J.H. Steele. Ames: Iowa State University Press, p. 257.

McCarty, C.W., and M.W. Miller. 1998. A versatile model of disease transmission applied to forecasting bovine tuberculosis dynamics in white-tailed deer populations. *Journal of Wildlife Diseases* 34:722–730.

McCool, C.J., and D.A. Newton-Tabrett. 1979. The route of infection in tuberculosis in feral buffalo. *Australian Veterinary Journal* 55:401–402.

McFadden, J. 1992. Mycobacteria. In *Encyclopedia of microbiology,* vol. 3, ed. J. Lederberg. San Diego: Academic, pp. 203–215.

McKeating, F.J., and R.P. Lehner. 1988. Tuberculosis in red deer. *Veterinary Record* 123:62–63.

McKenzie, J., R. Morris, and D. Pfeiffer. 1997. Identification of environmental predictors of disease in a wildlife population. *Epidemiologie et Sante Animale* 31:1–3.

Miller, J., A. Jenny, J. Rhyan, J., D. Saari, and D. Suarez. 1997. Detection *of Mycobacterium bovis* in formalin-fixed, paraffin-embedded tissues of cattle and elk by PCR amplification of an IS6110 sequence specific for *Mycobacterium tuberculosis* complex organisms. *Journal of Veterinary Diagnostic Investigation* 9:244–249.

Miller, M.W., J.M. Williams, T.J. Schiefer, and J.W. Seidel. 1991. Bovine tuberculosis in a captive elk herd in Colorado: Epizootiology, diagnosis, and management. *Proceedings of the United States Animal Health Association* 95:533–542.

Ministry of Agriculture, Fisheries and Food (MAFF). 1986. *Bovine tuberculosis in badgers: Tenth report by the Ministry of Agriculture, Fisheries and Food.* London: MAFF, p. 21.

———. 1994. *Bovine tuberculosis in badgers: 17th report by the Ministry of Agriculture, Fisheries and Food.* London: MAFF, pp. 37.

———. 1996. *Bovine tuberculosis in badgers. 19th report by the Ministry of Agriculture, Fisheries and Food.* London: MAFF, p. 32.

Mirsky, M.L., D. Morton, J.W. Piehl, and H. Gelberg. 1992. *Mycobacterium bovis* in infection in a captive herd of sika deer. *Journal of the American Veterinary Medical Association* 200:1540–1542.

Morris, J.A., A.E. Stevens, T.W. Little, and P. Stuart. 1978. Lymphocyte unresponsiveness to PPD tuberculin in badgers infected with *Mycobacterium bovis. Research in Veterinary Science* 25:390–392.

Morris, J.A., A.E. Stevens, P. Stuart, and T.W. Little. 1979. A pilot study to assess the usefulness of ELISA in detecting tuberculosis in badgers. *Veterinary Record* 104:14.

Morris, R.S., and D.U. Pfeiffer. 1995. Directions and issues in bovine tuberculosis epidemiology and control in New Zealand. *New Zealand Veterinary Journal* 43:256–265.

Morris, R.S., D.U. Pfeiffer, and R. Jackson. 1994. The epidemiology of *Mycobacterium bovis* infections. *Veterinary Microbiology* 40:153–177.

Muirhead, R.H., J. Gallagher, and K.J. Burn. 1974. Tuberculosis in wild badgers in Gloucestershire: Epidemiology. *Veterinary Record* 95:552–555.

Munro, R., and A.R. Hunter. 1985. Lung, heart and liver lesions in adult red deer: A histopathological survey. *British Veterinary Journal* 141:388–396.

Neal, E., and C. Cheeseman. 1996. *Badgers.* London: T. and A.D. Poyser, 271 pp.

Nikanorov, B.A. 1969. Experimental study on TB in marals. *Veterinariya (Moscow)* 1:35–36.

Nolan, A. 1991. An investigation of the development of specific antibody responses of badgers (*Meles meles*) to infection with *Mycobacterium bovis* with reference to the pathogenesis and epidemiology of disease. Ph.D. thesis, Brunel University, West London, UK, 287 pp.

Nolan, A., and J.W. Wilesmith. 1994. Tuberculosis in badgers (*Meles meles*). *Veterinary Microbiology* 40:179–191.

Noonan, N.L., W.D. Sheane, L.R. Harper, and P.J. Ryan. 1975. Wildlife as a possible reservoir of bovine tuberculosis. *Irish Veterinary Journal* 29:1.

Nowak, R.M. 1991. *Walker's mammals of the world,* vol. 2, 5th ed. Baltimore: Johns Hopkins University Press, pp. 1113 and 1407–1433.

Nugent, G., and I. Lugton. 1995. Prevalence of bovine tuberculosis in wild deer in the Hauhungaroa range, North Island. In *Proceedings of the second international conference on* Mycobacterium bovis. *Otago Conference Series,* no. 3, ed. F. Griffin and G. de Lisle. Dunedin, New Zealand: University of Otago Press, pp. 273–275.

Nugent, G., and G. Mackereth. 1996. Tuberculosis prevalence in wild deer and possums on Timahanga Station, Rangitikei. *Surveillance* 23:22–24.

O'Hara, P.J., A.F. Julian, and M.O. Ekdahl. 1976. Tuberculosis in the opossum (*Trichosurus vulpecula*): An experimental study. In *Ministry of Agriculture and Fisheries Tuberculosis Seminar, Hamilton, New Zealand* [cited by Jackson (1995)].

O'Neil, B.D., and H.J. Pharo. 1995. The control of bovine tuberculosis in New Zealand. *New Zealand Veterinary Journal* 43:249–255.

O'Reilly, L.M., and C.J. Daborn. 1995. The epidemiology of *Mycobacterium bovis* infections in animals and man: A review. *Tubercle and Lung Disease* 76(Suppl. 1):1–46.

Palmer, M.V., D.L. Whipple, and S.C. Olsen. 1999. Development of a model of natural infection with *Mycobacterium bovis* in white-tailed deer. *Journal of Wildlife Diseases* 35:450–457.

Palmer, M.V., D.L. Whipple, S.C. Olsen, and R.H. Jacobson. 2000. Cell mediated and humoral immune responses of white-tailed deer experimentally infected with *Mycobacterium bovis. Research in Veterinary Science* 68:95–98.

Pastoret, P.P., E. Thiry, B. Brochier, A. Schwers, I. Thomas, and J. Dubuisson. 1988. Diseases of wild animals transmissible to domestic animals. *Revue Scientifique et Technique O.I.E* 7:705–736.

Paterson, B.M. 1993. Behavioural patterns of possums and cattle which may facilitate the transmission of tuberculosis. M.V.Sc. thesis, Massey University, Palmerston North, New Zealand [cited by Jackson (1995)].

Paterson, B.M., and R.S. Morris. 1995. Interactions between beef-cattle and simulated tuberculous possums on pasture. *New Zealand Veterinary Journal* 43:289–293.

Pfeiffer, D.U., and R.S. Morris. 1991. A longitudinal study of bovine tuberculosis in possums and cattle. In *Proceedings of a symposium on tuberculosis,* ed. R. Jackson. Palmerston North, New Zealand: New Zealand Veterinary Association Foundation for Continuing Education, pp. 17–39.

Philip, P.M. 1989. Tuberculosis in deer in Great Britain. *State Veterinary Journal* 43:193–204.

Power, A.P., and B.G.A. Watts. 1987. *The badger control policy: An economic assessment.* Government Economic Service Working Paper 96. London: Ministry of Agriculture, Fisheries and Food, pp. 41.

Pracy, L.T. 1974. *Introduction and liberation of the opossum (*Trichosurus vulpecula *Kerr) into New Zealand.* Wellington: New Zealand Forest Service; Information Series 45 [cited by Jackson (1995)].

Presidente, P.J.A. 1984. Parasites and diseases of brushtail possums (*Trichosurus* spp.): Occurrence and significance. In *Possums and gliders,* ed. A.P. Smith and I.D. Hume. Sydney: Australian Mammal Society, pp. 171–190.

Pritchard, D.G. 1988. A century of bovine tuberculosis 1888–1988: Conquest and controversy. *Journal of Comparative Pathology* 99:357–399.

Pritchard, D.G., F.A. Stuart, J.W. Wilesmith, C.L. Cheeseman, J.I. Brewer, R. Bode, and P. Sayers. 1986. Tuberculosis in East Sussex: III. Comparison of post mortem and clinical methods for the diagnosis of tuberculosis in badgers. *Journal of Hygiene* 97:27–36.

Quinn, J.F., and D.R. Towar. 1963. Tuberculosis problems at a deer park in Michigan. In *Scientific proceedings of the 100th annual meeting of the American Veterinary Medical Association,* pp. 262–264.

Raath, J.P., R.G. Bengis, M. Bush, H. Huchzermeyer, D.F. Keet, D.J. Kernes, N.P.J. Kriek, and A. Michel. 1995. Diagnosis of tuberculosis due to *Mycobacterium bovis* in the African buffalo (*Syncerus caffer*) in the Kruger National Park. In *Tuberculosis in wildlife and domestic animals,* ed. F. Griffin and D. de Lisle. Dunedin, New Zealand: University of Otago Press, pp. 313–315.

Radostits, O.M., D.C. Blood, and C.C. Gay. 1994. *Veterinary medicine: A textbook of the diseases of cattle, sheep, pigs, goats and horses,* 8th ed. London: Ballière Tindall, pp. 830–838.

Ragg, J.R., H. Moller, and K.A. Waldrup. 1995a. The prevalence of bovine tuberculosis (*Mycobacterium bovis*) infections in feral populations of cats (*Felis catus*), ferrets (*Mustela furo*) and stoats (*Mustela erminea*) in Otago and Southland, New Zealand. *New Zealand Veterinary Journal* 43:333–337.

Ragg, J.R., K.A. Waldrup, and H. Moller. 1995b. The distribution of gross lesions of tuberculosis caused by *Mycobacterium bovis* in feral ferrets (*Mustela furo*) from Otago, New Zealand. *New Zealand Veterinary Journal* 43:338–341.

Rhyan, J.C., and D.A. Saari. 1995. A comparative study of the histopathologic features of bovine tuberculosis in cattle, fallow deer (*Dama dama*), sika deer (*Cervus nippon*), and red deer and elk (*Cervus elaphus*). *Veterinary Pathology* 32:215–220.

Rhyan, J.C., D.A. Saari, E.S. Williams, M.W. Miller, A.J. Davis, and A.J. Wilson. 1992. Gross and microscopic lesions of naturally occurring tuberculosis in a captive herd of wapiti (*Cervus elaphus nelsoni*) in Colorado. *Journal of Veterinary Diagnostic Investigation* 4:428–433.

Rhyan, J.C., K. Aune, B. Hood, R. Clarke, J. Payeur, J. Jarnagin, and L. Stackhouse. 1995. Bovine tuberculosis in a free-ranging mule deer (*Odocoileus hemionus*) from Montana. *Journal of Wildlife Diseases* 31:432–435.

Rogers, L.M., C.L. Cheeseman, P.J. Mallinson, and R. Clifton-Hadley. 1997. The demography of a high-density badger (*Meles meles*) population in the west of England. *Journal of Zoology* 242:705–728.

Rose, H.R. 1987. Bovine tuberculosis in deer. *Deer* 7:78.

Sapolsky, R.M., and J.G. Else. 1987. Bovine tuberculosis in a wild baboon population: Epidemiologic aspects. *Journal of Medical Primatology* 16:229–235.

Sauter, C.M., and R.S. Morris. 1995a. Behavioral studies on the potential for direct transmission of tuberculosis from feral ferrets (*Mustela furo*) and possums (*Trichosurus vulpecula*) to farmed livestock. *New Zealand Veterinary Journal* 43:294–300.

———. 1995b. Dominance hierarchies in cattle and red deer (*Cervus elaphus*): Their possible relationship to the transmission of bovine tuberculosis. *New Zealand Veterinary Journal* 43:301–305.

Sawa, T.R., C.O. Thoen, and W.T. Nagao. 1974. *Mycobacterium bovis* infection in wild axis deer in Hawaii. *Journal of the American Veterinary Medical Association* 165:998–999.

Schmidt, H.W. 1938. Tuberculosis in roe deer. *Deutsche Tierarztliche Wochenschrift* 46:482–485.

Schmitt, S.M., S.D. Fitzgerald, T.M. Cooley, C.S. Bruning-Fann, L. Sullivan, D. Berry, T. Carlson, R.B. Minnis, J.B. Payeur, and J. Sikarskie. 1997. Bovine tuberculosis in free-ranging white-tailed deer from Michigan. *Journal of Wildlife Diseases* 33:749–758.

Sibartie, D., L.L. Beeharry, and M.R. Jaumally. 1983. Some diseases of deer (*Cervus timorensis russa*) in Mauritius. *Tropical Veterinary Journal* 1:8–14.

Singh, C.D.N., L.N. Prasad, and H.N. Thakur. 1986. Some observations on tuberculosis in deer. *Indian Veterinary Journal* 63:867–868.

Small, K.J., and D. Thomson. 1986. The efficiency of bovine PPD tuberculin in the single caudal fold test to detect tuberculosis in water buffalo. *Buffalo Bulletin* 5:62–64.

Stamp, J.T. 1944. A review of the pathogenesis and pathology of bovine tuberculosis with special reference to practical problems. *Veterinary Record* 56:443–446.

Strahan, R. 1983. *Complete book of Australian mammals.* Sydney: Angus and Robertson, 530 pp.

Stuart, F.A. 1988. Tuberculosis in farmed red deer (*Cervus elaphus*). In *The management and health of farmed deer,* ed. H.W. Reid. London: Kluwer Academic, pp. 101–111.

Stuart, F.A., P.A. Manser, and F.G. McIntosh. 1988. Tuberculosis in imported red deer (*Cervus elaphus*). *Veterinary Record* 122:508–511.

Stumpff, C.D. 1982. Epidemiological study of an outbreak of bovine TB in confined elk herds. *Proceedings of the United States Animal Health Association* 86:524–527.

Stumpff, C.D., M.A. Essey, D.H. Person, and D. Thorpe. 1984. Epidemiologic study of *M. bovis* in American bison. *Proceedings of the United States Animal Health Association* 88:564–570.

Sullivan, T.P., D.P. Crump, H. Weiser, and E.A. Dixon. 1990. Response of pocket gophers (*Thomomys talpoides*) to an operational application of synthetic semiochemicals of stoat (*Mustela erminea*). *Journal of Chemical Ecology* 16:941–949.

Swinton, J., F. Tuyttens, D. MacDonald, D.J. Nokes, C.L. Cheeseman, and R. Clifton-Hadley. 1997. A comparison of fertility control and lethal control of bovine tuberculosis in badgers: The impact of perturbation induced transmission. *Philosophical Transactions of the Royal Society, London [B]* 352:619–631.

Symmers, W.St.C., A.P.D. Thomson, and C.N. Iland. 1953. Observations on tuberculosis in the ferret (*Mustela furo* L). *Journal of Comparative Pathology* 63:20–30.

Tarara, R., M.A. Suleman, R. Sapolsky, M.J. Wabomba, and J.G. Else. 1985. Tuberculosis in wild olive baboons, *Papio cynocephalus anubis* (Lesson), in Kenya. *Journal of Wildlife Diseases* 21:137–140.

Tessaro, S.V. 1986. The existing and potential importance of brucellosis and tuberculosis in Canadian wildlife: A review. *Canadian Veterinary Journal* 27:119–124.

Tessaro, S.V., L.B. Forbes, and C. Turcotte. 1990. A survey of brucellosis and tuberculosis in bison in and around Wood Buffalo National Park, Canada. *Canadian Veterinary Journal* 31:174–180.

Tessaro, S.V., C.C. Cormack, and L.B. Forbes. 1993. The brucellosis and tuberculosis status of wood bison in the Mackenzie Bison Sanctuary, Northwest Territories, Canada. *Canadian Journal of Veterinary Research* 57:231–235.

Thoen, C.O. 1993. Tuberculosis and other mycobacterial diseases in captive wild animals. In *Zoo and wild animal medicine,* ed. M.E. Fowler. Philadelphia: W.B. Saunders, pp. 45–50.

Thoen, C.O., and B.R. Bloom. 1995. Pathogenesis of *Mycobacterium bovis.* In Mycobacterium bovis *infection in animals and humans,* ed. C.O. Thoen and J.H. Steele. Ames: Iowa State University Press, pp. 3–14.

Thoen, C.O., W.J. Quinn, L.D. Miller, L.L. Stackhouse, B.F. Newcomb, and J.M. Ferrell. 1992. *Mycobacterium bovis* infection in North American elk (*Cervus elaphus*). *Journal of Veterinary Diagnostic Investigation* 4:423–427.

Thoen, C.O., T. Schliesser, and B. Kormendy. 1995. Tuberculosis in captive wild animals. In Mycobacterium bovis *infection in animals and humans,* ed. C.O. Thoen and J.H. Steele. Ames: Iowa State University Press, pp. 93–104.

Thorns, C.J., J.A. Morris, and T.W. Little. 1982. A spectrum of immune responses and pathological conditions between certain animal species to experimental *Mycobacterium bovis* infection. *British Journal of Experimental Pathology* 63:562–572.

Thurlbeck, W.M., C.A. Butas, E.M. Mankiewicz, and R.M. Laws. 1965. Chronic pulmonary disease in the wild buffalo (*Syncerus caffer*) in Uganda. *American Review of Respiratory Diseases* 92:801–805.

Towar, D.R., R.M. Scott, and L.S. Goyings. 1965. Tuberculosis in a captive deer herd. *American Journal of Veterinary Research* 26:339–346.

Tulloch, D.G. 1978. The water buffalo, *Bubalus bubalis,* in Australia: Grouping and home range. *Australian Wildlife Research* 5:327–354.

Van den Oord, Q.G.W., E.J.A. van Wijk, I.W. Lugton, R S. Morris, and C.W. Holmes. 1995. Effects of air temperature, air movement and artificial rain on the heat production of brushtail possums (*Trichosurus vulpecula*): An exploratory study. *New Zealand Veterinary Journal* 43:328–332.

Walker, R., B. Reid, and K. Crews. 1993. Bovine tuberculosis in predators in the Mackenzie Basin. *Surveillance* 20:11–14.

Whipple, D.L., P.R. Clarke, J.L. Jarnagin, and J.B. Payeur. 1997. Restriction fragment length polymorphism analysis of *Mycobacterium bovis* isolates from captive and free-ranging animals. *Journal of Veterinary Diagnostic Investigation* 9:381–386.

White, P.C.L., and S. Harris. 1995. Bovine tuberculosis in badger (*Meles meles*) populations in southwest England: An assessment of past, present and possible future control strategies using simulation modelling. *Philosophical Transactions of the Royal Society, London [B]* 349:415–432.

White, P.C.L., J.A. Brown, and S. Harris. 1993. Badgers (*Meles meles*), cattle and bovine tuberculosis *Mycobacterium bovis:* A hypothesis to explain the influence of habitat on the risk of disease transmission in southwest England. *Proceedings of the Royal Society of London [B]* 253:277–284.

Whiting, T.L., and S.V. Tessaro. 1994. An abattoir study of tuberculosis in a herd of farmed elk. *Canadian Veterinary Journal* 35:497–501.

Wilesmith, J.W. 1983. Epidemiological features of bovine tuberculosis in cattle herds in Great Britain. *Journal of Hygiene* 90:159–176.

Williams, D. 1987. Bovine tuberculosis in deer: What to look for and what to do. *Deer* 7:143.

Williams, E.S., S.G. Smith, R.M. Meyer, and E.T. Thorne. 1995. Three-year survey for bovine tuberculosis in hunter-killed free-ranging elk (*Cervus elaphus nelsoni*) in northwestern Wyoming. *Proceedings of the United States Animal Health Association* 99:631–637.

Williams, R.S., and W.A. Hoy. 1930. The viability of *B. tuberculosis* (Bovinus) on pasture land, in stored faeces and in liquid manure. *Journal of Hygiene* 30:413–419.

Wilson, P., and R. Harrington. 1976. A case of bovine tuberculosis in fallow deer. *Veterinary Record* 98:74.

Witte, J. 1940. Tuberculosis in red deer in the wild. *Berliner und Munchener Tierarztliche Wochenschrift* 29:349–350.

Woodford, M.H. 1982. Tuberculosis in wildlife in the Ruwenzori National Park, Uganda (Part I). *Tropical Animal Health and Production* 14:81–88.

Wray, C. 1975. Survival and spread of pathogenic bacteria of veterinary importance within the environment. *Veterinary Bulletin* 45:543–550.

Zieger, U., G.S. Pandey, N.P. Kriek, and A.E. Cauldwell. 1998. Tuberculosis in Kafue lechwe (*Kobus leche kafuensis*) and in a bushbuck (*Tragelaphus scriptus*) on a game ranch in central province, Zambia. *Journal of the South African Veterinary Association* 69:98–101.

Zuckerman, Lord. 1980. *Badgers, cattle and tuberculosis.* London: Her Majesty's Stationery Office, 107 pp.

PARATUBERCULOSIS AND OTHER MYCOBACTERIAL DISEASES

ELIZABETH S. WILLIAMS

PARATUBERCULOSIS

Synonym: Johne's disease.

Introduction. Paratuberculosis is caused by the bacterium *Mycobacterium avium* subsp. *paratuberculosis* (*M. paratuberculosis)* and is a typical mycobacterial disease with long incubation periods and prolonged clinical duration. The digestive tract is the primary target of infection, and clinical signs are generally referable to lesions in these organs. Loss of body condition and variable presence of diarrhea are hallmark clinical signs of paratuberculosis.

Paratuberculosis is a well-known disease of domestic ruminants, including cattle *Bos taurus,* sheep *Ovis aries,* and goats *Capra hircus,* but is also a significant pathogen in farmed deer. In fact it was recognized in park deer in England as early as 1907 (M'Fadyean and Sheather 1916). These deer had chronic diarrhea, and when examined postmortem, the lesions were consistent with paratuberculosis. This report indicated, however, that the fact that "deer may suffer from the disease is not one of much direct interest." But now it is.

Paratuberculosis in domestic livestock has been reviewed (Sweeney 1996a). Although it is seldom a problem in free-ranging wildlife, a few foci of paratuberculosis in free-ranging populations cause difficult management situations. Thoen and Johnson (1981) in the previous edition of this book reviewed paratuberculosis in wild mammals through the 1970s, and several recent short reviews cover more current information on paratuberculosis in free-ranging animals (Jessup and Williams 1999) and zoo animals (Manning and Collins 1999). There has been increasing interest in paratuberculosis in wildlife with the growth of the game-farming industry, particularly where cervids and bison *Bison bison* are involved, and with government efforts to control the disease in domestic livestock.

History and Distribution. Paratuberculosis was described approximately 100 years ago by Johne and Frothingham (1895) in Germany and was recognized as being distinct from tuberculosis. Since then, the disease has been the subject of considerable research because of its economic impact on livestock industries, difficulties inherent in diagnosis, and its worldwide distribution. In nondomestic species, paratuberculosis has been reported in Europe (Gilmour 1984), North America (Temple et al. 1979), and New Zealand (de Lisle et al. 1993).

Host Range. Paratuberculosis is primarily a disease of ruminants and camelids, and all species in these groups are probably susceptible to infection. Table 21.3 lists reports of clinical paratuberculosis in species other than domestic livestock. Paratuberculosis is most commonly recognized among captive wild ruminants in zoological collections or on game farms. It appears to be an increasing problem in farmed elk *cervus elaphus* and red deer *Cervus elaphus* (Gilmour 1984; Power et al. 1993; Fawcett et al. 1995; Manning et al. 1998) and bison (D. Hunter personal communication).

Paratuberculosis is rarely reported in free-ranging populations. In North America, clinical disease has been reported only in a few populations of Rocky Mountain bighorn sheep *Ovis canadensis* in Colorado and Wyoming (Williams et al. 1979) and Rocky mountain goats *Oreamnos americanus* in Colorado (Williams et al. 1979), tule elk *Cervus elaphus nannodes* in California (Jessup et al. 1981; Cook et al. 1997), and Key deer *Odocoileus virginianus* in Florida (C. Quist personal communication). *Mycobacterium avium paratuberculosis* has been cultured from feces of clinically healthy free-ranging fallow deer *Dama dama* and axis deer *Axis axis* (Riemann et al. 1979), tule elk (Cook et al. 1997), and white-tailed deer *Odocoileus virginianus* (Chiodini and Van Kruiningen 1983). Paratuberculosis has been identified in free-ranging rabbits *Oryctolagus cuniculus* in Scotland (Angus 1990; Greig et al. 1997).

Etiology. Thorel et al. (1990) proposed that *M. paratuberculosis* is a subspecies of *M. avium,* and this proposed taxonomic change has been accepted in general use in Europe but to a lesser degree in North America (Sweeney 1996b). Some of the characteristic features of the mycobacteria have been described in the previous section on *M. bovis* and are also typical of *M. avium paratuberculosis.* In addition, mycobactin, an iron-binding compound produced by most mycobacteria, is required in culture medium for growth of this

TABLE 21.3—Reports of paratuberculosis in species other than domestic livestock

Common Name	Scientific Name	Reference
Bovidae		
Mouflon	*Ovis musimon*	Boever and Peters 1977[a]
Rocky Mountain bighorn	*Ovis canadensis*	Williams et al. 1979[b]
Aoudad	*Ammotragus lervia*	Boever and Peters 1977[a]
Rocky Mountain goat	*Oreamnos americanus*	Williams et al. 1979[b]
Saiga	*Saiga tatarica*	Dukes et al. 1992[a]
Gnu	*Connochaetes taurinus*	Jarmai 1922[a]
Topi	*Damaliscus lunatus*	Steinberg 1981[a]
African buffalo	*Syncerus caffer*	Katic 1961[a]
Yak	*Bos grunniens*	Katic 1961[a]
Bison	*Bison bison*	D. Hunter, personal communication
Cervidae		
White-tailed deer	*Odocoileus virginianus*	Libke and Walton 1975[a]; Williams et al. 1983a[c],b[c]; C. Quist, personal communication[b]
Mule deer	*Odocoileus hemionus*	Williams et al. 1983a[c],b[c]
Moose	*Alces alces*	Soltys et al. 1967[a]
Sika deer	*Cervus nippon*	Bourgeois 1940[a]; Temple et al. 1979[a]
Red deer	*Cervus elaphus*	Dorofeev and Kalachev 1949[a]; Vance 1961[a]; Power et al. 1993[a]; Fawcett et al. 1995[a]
Elk	*Cervus elaphus*	Williams et al. 1983a[c],b[c]; Manning et al. 1998[a]
Tule elk	*Cervus elaphus nannodes*	Jessup et al. 1981[b]; Cook et al. 1997[b]
Axis deer	*Axis axis*	Riemann et al. 1979[b]
Fallow deer	*Dama dama*	Riemann et al. 1979[b]; Temple et al. 1979[a]
Reindeer	*Rangifer tarandus*	Poddoubski 1957[a]
Camelidae		
Llama	*Lama glama*	Belknap et al 1994[a]
Alpaca	*Lama pacos*	Ridge et al. 1995[a]
Camel	*Camelus bactrianus*	Poddoubski 1957[a]; Katic 1961[a]; Radwan et al. 1991[a]
Lagomorpha		
Rabbit	*Oryctolagus cuniculus*	Angus 1990[b]; Greig et al. 1997[b]
Primate		
Stump-tailed macaques	*Macaca arctoides*	McClure et al. 1987

[a] Captive animals in zoos or game farms.
[b] Free-ranging animals.
[c] Experimental infections.

subspecies (Wayne and Kubica 1986). Primary isolation on egg-yolk medium may require 3–4 months, but advances using radiometric culture have reduced culture times to 1–7 weeks (Cousins et al. 1995; Whittington et al. 1998). Isolates are tested for a unique insertion sequence (IS900) by polymerase chain reaction (PCR) amplification (Collins et al. 1998).

Variations exist among isolates of the bacterium from different species; strains of *M. avium paratuberculosis* from domestic sheep are notoriously difficult to isolate in culture. Most isolates from deer are similar to bovine strains (de Lisle et al. 1993), but sheep strains have been found in deer and saiga antelope *Saiga tatarica* (de Lisle et al. 1993; Dukes et al. 1992). However, transmission of various strains among ruminants should be assumed to be possible.

Mycobacteria survive in feces, soil, and water for up to a year (Mitscherlich and Marth 1984), though recent studies in New Zealand suggest that survival of mycobacteria on pasture may be considerably shorter (Jackson et al. 1995).

Transmission and Epidemiology. The epidemiology of paratuberculosis has been studied extensively in domestic species and, even though wild mammals have not be studied as intensely, the general epidemiologic features are probably similar. The primary site of lesions in animals with paratuberculosis is the intestinal tract; hence, fecal shedding by an infected animal and ingestion of the organism by susceptible hosts is the typical route of transmission. Thus, when wild mammals are artificially managed under higher than natural densities, transmission of many pathogens, including *M. avium paratuberculosis,* is enhanced, increasing the chance that disease may be maintained in these populations. Young animals, particularly neonates, appear to be most susceptible to infection, though it is possible to infect adults. Domestic lambs were experimentally infected by as few as 1000 organisms (Gilmour 1976). Factors that influence the likelihood of transmission include host characteristics such as age and immunocompetence, bacterial factors such as strain, and environmental features that influences dose such as degree of contamination, humidity, and exposure to sunlight. Transmission via colostrum and milk is also possible. In utero transmission may occur in bighorn sheep (Williams 1981); this is most likely to occur late in the dam's illness. The organism has been isolated from semen of domestic bulls, suggesting the possibility of venereal transmission (Larsen et al. 1981).

Infected but apparently healthy animals may shed organisms and serve as sources for infection of susceptible herdmates for years. The majority of infected animals never develop clinical disease but yet may shed the organism in their feces.

Maintenance of paratuberculosis within populations of free-ranging mammals reflects artificial or behavioral propensity for high densities that facilitate transmission. At Point Reyes National Seashore, California, tule elk probably maintain paratuberculosis due to high animal density as well as conditions favorable for survival of the bacteria in the environment (Cook et al. 1997; Jessup and Williams 1999). One bighorn sheep population in Colorado probably maintains the infection due to gregarious behavior and traditional use of bedding areas that facilitate fecal-oral transmission (Jessup and Williams 1999).

Clinical Signs. Paratuberculosis is a chronic disease, and both the incubation period and the duration of illness are prolonged and may last years. The clinical signs of paratuberculosis in wild ruminants have been reviewed (Stehman 1996; Jessup and Williams 1999). Affected wild ruminants may be 1 year or older; clinical disease seems to occur in wild species at younger ages than is typical in cattle. The primary signs of paratuberculosis are gradual loss of body condition and poor hair coats. Diarrhea is typical in cattle with paratuberculosis, but it may not occur until the later stages of disease in nondomestic species and then may be intermittent. Affected animals remain alert and continue to eat until late in the course of disease, when they become weak and depressed. Hypoproteinemia and intermandibular edema may occur. Growth of antlers and horns may be adversely affected. The disease is fatal once clinical signs appear.

Pathogenesis. *Mycobacterium avium paratuberculosis* infects and proliferates in the distal small intestine, spiral colon, colon, and associated lymph nodes and lymphatics. Entry is thought to be through the μ cells in the ileum of the small intestine (Momotani et al. 1988), though entry through the tonsil may occur with heavy exposure (Payne and Rankin 1961). Bacteria are ingested by macrophages and proliferate within the cytoplasm (Bendixen et al. 1981). Occasionally, bacteria and associated inflammation occur in extraintestinal tissues, especially lung and liver, late in the course of disease. As with other mycobacterial infections, granulomatous inflammation is induced by the bacteria, resulting in thickened intestinal walls, enlargement of mesenteric and ileocecal lymph nodes, and thickening and possibly blockage of afferent mesenteric lymphatics. Extensive inflammation in the gut results in diarrhea, malabsorption, and malnutrition.

Pathology. Changes in clinical chemistry of an animal with paratuberculosis are nonspecific and reflect emaciation. Hypoproteinemia may be the only abnormality detected.

Gross lesions are variable, and even in cases with severe clinical disease, lesions may be relatively mild as compared with classic paratuberculosis in cattle. Typical lesions (Williams et al. 1983a; Manning et al. 1998) include poor body condition or emaciation with serous atrophy of fat; increased pericardial, thoracic, and abdominal fluid; thickened mucosa of the distal small intestine and, rarely, cecum and spiral colon;

enlarged mesenteric lymph nodes, which may be edematous, have indistinct corticomedullary junctions, and may contain foci of mineralization; mesenteric and subserosal edema; and thickening of lymphatics in the mesentery and serosa.

Microscopic lesions are characterized by granulomatous inflammation in the small intestine and associated lymphatics and lymph nodes (Williams et al. 1983a). Epithelioid macrophages, Langhans' giant cells, and lymphocytes are present with variable numbers of intracellular acid-fast bacteria that stain with Ziehl-Neelsen or Kinyon's acid-fast stains. Necrosis and mineralization sometimes occur, especially in cervids.

Two distinct forms of inflammatory responses to *M. avium paratuberculosis* are described in domestic sheep, and similar patterns probably also occur in wild species. These categories of disease are associated with immune responsiveness (Clarke and Little 1996; Clarke et al. 1996; Burrells et al. 1998). The multibacillary, pleuribacillary, or lepromatous form is characterized by large numbers of intracellular bacteria and epithelioid macrophage infiltrates. This occurs in animals with weak cell-mediated immune (CMI) responses. Pleuribacillary-affected animals do, however, develop detectible serum antibodies. The paucibacillary or tuberculoid type is characterized by prominence of lymphocytes in the inflammatory reaction, strong CMI, and poor sensitivity of serologic tests. Bighorn sheep typically showed the tuberculoid pattern of granulomatous inflammation as compared to mountain goats that had a lepromatous appearance; however, the classifications were not clear-cut (Williams et al. 1983a).

Diagnosis. As with other mycobacterial diseases, antemortem diagnosis of paratuberculosis is problematic [reviewed by Collins (1996)]. Because the bacterium grows slowly, culture requires weeks to months, even with the recent improvements in techniques (Cousins et al. 1995; Whittington et al. 1998). Serologic tests tend to have poor sensitivity, especially early in infection prior to appearance of clinical signs. Measures of CMI have been developed, but these may not be practical in the field, are relatively difficult to perform, and are expensive.

Antemortem diagnostic tests for paratuberculosis in wild species have not been validated. Because of the characteristics of the immunologic response to the organism during the course of infection, sensitivity of serologic tests tends to be low prior to appearance of clinical signs. Commonly used serologic tests are complement fixation (Williams et al. 1985), agar gel diffusion (Manning et al. 1998), and enzyme-linked immunosorbent assays (ELISAs) (Shulaw et al. 1986); ELISA is the most useful in cattle (Collins 1996). Refinements in ELISAs should improve their usefulness in all species (Clarke et al. 1996; Ellis et al. 1998).

Measures of CMI have the potential to diagnose paratuberculosis earlier in the course of disease than serologic tests; however, these tests are less practical and more expensive than tests for antibodies. Skin tests of delayed-type hypersensitivity using johnin, an extract of *M. avium paratuberculosis,* have low sensitivity and specificity and are seldom used even in domestic species. Other measures of CMI include lymphocyte blastogenesis tests (Williams et al. 1985) and tests for gamma-interferon production by blood mononuclear cells (Stabel 1996). Although these tests hold some promise, they are unlikely to be practical for free-ranging species because of the need to collect and rapidly process viable lymphocytes for use in these assays.

Culture of feces, tissues, and environmental samples is widely used for diagnosis of paratuberculosis (Whipple et al. 1991). It is highly specific, but sensitivity is low in the early stages of infection due to intermittent fecal shedding and the need for adequate numbers of organisms to be present in the sample before shedding can be detected. Improved culture techniques employ the Bactec detection system with confirmation of the identity of isolates by a genetic probe for IS900 (Cousins et al. 1995; Whittington et al. 1998). However, even with improved techniques, weeks or months are required for culture. The genetic probe for IS900 can be used directly on feces (Sockett et al. 1992; Whipple et al. 1992) and rapidly identifies animals shedding relatively large numbers of bacteria (approximately 10^4), but cultural methods are considered more sensitive. In addition, PCR tests have been developed for detection of *M. avium paratuberculosis* in tissues and blood (Gwóżdż et al. 1997). These methods hold promise for use in free-ranging species when appropriately validated.

Gross and histologic lesions may be suggestive of paratuberculosis but are not diagnostic. Impression smears of thickened small intestine or enlarged mesenteric lymph nodes may contain acid-fast bacteria and suggest mycobacterial infection, but additional testing for confirmation and identification of the bacteria is required. Immunohistochemistry improves histopathologic diagnosis of paratuberculosis (Plante et al. 1996; Stabel et al. 1996; Coetsier et al. 1998), and PCR on formalin-fixed tissues is also possible (Plante et al. 1996).

Serologic surveys of free-ranging populations for paratuberculosis have been conducted (Shulaw et al. 1986), but problems with both sensitivity and specificity need to be recognized in evaluating results. Surveys using fecal culture (Cook et al. 1997) and collection of digestive tracts from hunter-harvested animals (Williams 1981; Rhyan et al. 1997) or road-killed animals (C. Quist personal communication) may be useful in free-ranging populations.

Differential Diagnoses. A wide variety of diseases of wild ruminants present with signs of loss of body condition. Some diseases that should be considered include malnutrition or starvation, dental attrition, musculoskeletal or traumatic lesions hindering movements and ability to forage, a variety of intoxications, parasitism, and chronic wasting disease. Most of these are readily ruled out via evaluation of the history and diagnostic testing.

Because diarrhea often does not occur until the terminal stages of disease, this may not be part of the clinical presentation. But other conditions that should be considered in animals with chronic diarrhea include intestinal parasitism and a variety of nutritional diseases of deficiency or intoxication.

Immunity. The immune responses of wild species to *M. avium paratuberculosis* have not been studied, but likely they are similar to those of domestic animals. Chiodini (1996) provided a comprehensive review of this subject in domestic animals. The immune response to mycobacteria is complicated and involves both humoral and CMI responses. One of the consequences of the long-term complex relationship between the pathogen and the animal's immune response is that there are limitations on current diagnostic tests that rely solely on measure of antibodies or CMI.

Recent studies of genetic resistance of cattle to intracellular bacteria such as *Brucella abortus* and *Mycobacterium bovis* (Qureshi et al. 1996) suggest that similar factors may play a role in resistance to *M. avium paratuberculosis.* There may be genetic control of susceptibility of bison to paratuberculosis by these same mechanisms (D. Davis and J. Templeton personal communication).

Control and Treatment. Control of paratuberculosis is difficult in captive wild species, and efforts at control have not been systematically evaluated in free-ranging species. Prevention of introduction of the disease is always preferable to attempting control. Veterinary oversight of herd health and management is very important. In captive wild mammals, techniques such as quarantine, testing (serology and bacteriology), culling of test-positive or clinically affected individuals, sanitation to reduce or prevent fecal contamination of feed and water, pasture management such as tilling, and rearing neonates in clean environments can be practiced [reviewed by Manning and Collins (1999)].

Most control techniques used in captive animals are impractical or impossible in the wild. Quarantine with limited culling and increased hunting pressure have been used in tule elk in California and bighorn sheep in Colorado, respectively (Jessup and Williams 1999). However, paratuberculosis has persisted in both populations for over 20 years. Populations of wild ruminants known to be infected with paratuberculosis should never be used as sources of stock for translocation or reintroduction.

Animals being brought into paratuberculosis-free herds should come only from herds or premises without a history of the disease. In addition, at least three negative fecal samples collected weeks to months apart should be obtained before introduction of the new animal into the herd.

Expensive, long-term antibiotic therapies have been used to treat cattle (St. Jean 1996) and could be applied to exceptionally valuable captive individuals. No treatment is available for use in free-ranging populations.

Vaccination against paratuberculosis has been attempted in cattle, sheep, and goats but has not been reported in nondomestic ruminants; results have been mixed (Körmendy 1994; Van Schaik et al. 1996). Live and heat-killed organisms in oil or adjuvant have been tested. Localized inflammatory reactions may occur at vaccination sites.

Public Health Concerns. For many years, there has been discussion about the possible relationship of *M. avium paratuberculosis* and Crohn's disease, a chronic granulomatous ileocolitis of humans (Chiodini 1989). This issue is controversial and continues to be studied (Chiodini and Rossiter 1996).

Domestic Animal Health Concerns. Paratuberculosis is primarily a disease of domestic ruminants (Sweeney 1996a), where it may cause significant economic impact. More attention has been directed at control and management of this disease in recent years. Free-ranging ruminants are unlikely to be significant sources of *M. avium paratuberculosis* for domestic species, because transmission is most efficient in neonates when contact with wildlife and their feces is not likely. In addition, only a very few populations of free-ranging ruminants maintain the disease. The role of free-ranging rabbits in the epidemiology of paratuberculosis is not yet known.

Management Implications. Because paratuberculosis is maintained in only a few populations of free-ranging wildlife, management should be directed at preventing introduction of the disease into populations where it does not exist. This might include minimizing contact between wild species and domestic or nondomestic herds with paratuberculosis, only purchasing and moving individuals from paratuberculosis-free herds, and including tests for paratuberculosis in herd health-monitoring plans. Infected wild mammals or animals from infected herds should not be used in reintroduction or translocation programs.

Some wildlife managers have expressed concern about domestic pack animals introducing paratuberculosis into bighorn sheep herds in the U.S. desert southwest. However, the potential for clinically healthy pack goats or llamas *Lama glama* to transmit paratuberculosis to bighorn sheep is exceedingly remote. Healthy animals are unlikely to excrete large numbers of bacteria into the environment, the habitat of bighorn sheep is not conducive to survival of the pathogen, and bighorn sheep would likely not ingest adequate numbers of the organism to become infected.

OTHER MYCOBACTERIAL INFECTIONS

Tuberculosis due to *Mycobacterium tuberculosis.* Disease in wildlife associated with the bacterium that typically causes human tuberculosis, *Mycobacterium tuberculosis,* is relatively uncommon. Thoen and Himes (1981) reviewed tuberculosis, including *M. tuberculosis* infections, in wild mammals. It most commonly causes infection and disease in captive nonhuman primates and occasionally hoofed stock (Thoen et al. 1977). In many cases of *M. tuberculosis* infection in wild mammals, humans were considered the source of the pathogen (Thoen and Himes 1981; Michel and Huchzermeyer 1998).

Tuberculosis due to *M. tuberculosis* has been reported in a wide variety of nonhuman primates and is an important consideration in management of these species (Bennett et al. 1998; Michel and Huchzermeyer 1998). The disease caused by this pathogen primarily affects the respiratory system with typical granulomatous lesions and fibrosis, but disseminated disease may occur (Bennett et al. 1998).

Tuberculin testing of nonhuman primates is complicated by differences in response among various species and appropriate doses, routes, and tuberculin formulations (Manning et al. 1980; Thoen and Himes 1981; Calle 1999). In general, mammalian old tuberculin formation is recommended for skin testing, and nonhuman primates require higher doses of tuberculin than humans (Calle 1999). Commonly used sites for intradermal tuberculin testing include the eyelid, arm, chest, or abdomen. Some species, such as orangutan *Pongo pygmaeus,* present special problems in tuberculin testing because they tend to respond even when not infected (Calle 1999). Long-term isoniazid treatment alone or in combination with other antituberculosis drugs has been used for valuable individuals or endangered species (Thoen 1993).

Mycobacterium tuberculosis was isolated from captive oryx *Oryx gazella beisa* (Lomme et al. 1976) and addax *Addax nasomaculatus* (Thoen and Himes 1981). In oryx, the disease affected lung and liver, with dissemination to uterus and lymph nodes in one animal (Lomme et al. 1976). The source of the bacterium for the hoofed stock was probably effluent from an area occupied by rhesus macaques *Macaca mulatta* infected with *M. tuberculosis.*

In recent years, the problem of tuberculosis in domestic, circus, and zoo elephants has been highlighted due to transmission of the bacterium from elephants to humans (Michalak et al. 1998), though Pinto et al. (1973) suggested this as a potential occupational hazard more than 25 years ago. Tuberculosis has been documented most often in Asian elephants *Elephas maximus,* but African elephants *Loxodonta africana* are also susceptible (Pinto et al. 1973; Saunders 1983; Furley 1997; Michalak et al. 1998); it has not been diagnosed in free-ranging elephants. Clinically, tuberculosis in elephants is characterized by chronic weight loss; lesions are widespread caseous necrosis in lung and mediastinal lymph nodes with considerable fibrosis and variable numbers of acid-fast bacteria. Trunk cultures are the primary antemortem procedure used to identify infected individuals shedding the organism. Ancillary tests such as intradermal tuberculin testing and blood tests such as the blood tuberculosis test, ELISA, and gamma-interferon test are being evaluated but have not yet been validated (Montali et al. 1998b).

Mycobacteria of the *M. tuberculosis* complex (which includes *M. tuberculosis* and *M. bovis*) were isolated from captive New Zealand fur seals *Arctocephalus forsteri,* Australian fur seals *Arctocephalus pusillus doriferus,* and Australian sea lions *Neophoca cinerea.* The isolates were similar to *M. bovis* but had unique features on Western blotting that suggested a closer relationship to *M. tuberculosis* (Cousins et al. 1990). In this outbreak, four seals died of mycobacteriosis (Forshaw and Phelps 1991). Lesions primarily involved the respiratory tract, with occasional dissemination to liver, lymph nodes, and meninges. Microscopically, the lesions were characterized by necrosis and fibrosis; mineralization and giant cells were not observed. The seals involved in this incident may have been infected when captured in the wild (Forshaw and Phelps 1991). Evidence that this pathogen is present in the wild was provided by isolation and characterization of mycobacterial isolates from beached Australian sea lions and New Zealand fur seals (Cousins et al. 1993). These authors also demonstrated that a seal trainer working with the infected captive seals was infected with the same mycobacterium. This agent appears to be a unique member of the *M. tuberculosis* complex (Cousins et al. 1993).

Mycobacteria of the *M. tuberculosis* complex, distinguishable from human and cattle isolates of *M. bovis,* were isolated from stranded sea lions *Otaria flavescens* and fur seals *Arctocephalus australis* on the southwest coast of Argentina (Bernardelli et al. 1996). The relationship of the pinniped isolates from Argentina and those from Australia is not known.

An unusual *M. tuberculosis* complex isolate (*Mycobacterium microti*-like) was obtained from wild-caught hyrax *Procavia capensis* imported from South Africa to Australia, highlighting the difficulty in testing for and managing tuberculosis in captive wild animals (Cousins et al. 1994). The affected hyrax were emaciated and had granulomas in lung, spleen, kidney, and liver.

***Mycobacterium avium* subsp. *avium* and *M. avium* subsp. *sylvaticum* Infections.** The taxonomy of *M. avium* was recently restructured to include three subspecies: *M. avium* subsp. *avium,* which includes the type strains of *M. avium* and *Mycobacterium intracellulare; M. avium* subsp. *paratuberculosis,* which includes mycobactin-dependent isolates previously identified as *M. paratuberculosis;* and *M. avium* subsp. *sylvaticum,* which includes wood-pigeon mycobacteria (Thorel et al. 1990). Wild mammals are commonly exposed to *M. avium* subspecies, though clinical disease due to these organisms is rare. At least some individuals developing progressive disease are likely immunosuppressed and unable to mount appropriate CMI responses. Carnivores, marsupials, ungulates, and nonhuman primates have been reported with mycobacteriosis due to these pathogens. These infections may cause cross-reactions and confusion in testing of many species for tuberculosis (*M. bovis* and *M. tuberculosis*).

Mink *Mustela vison* are susceptible to *M. avium* infection (Hall and Winkel 1957). In one reported outbreak, large numbers of mink were affected, and clinical signs included anorexia and emaciation. Granulomatous lesions without fibrosis or calcification were present in many organs, and the source may have been pork in the diet.

An unusual host-parasite relationship exists in Matschie's tree kangaroo *Dendrolagus matschiei;* many of these animals develop primary infections with *M. avium* complex organisms from environmental sources. Lymphocyte proliferation tests showed that even healthy individuals have low cellular immune reactivity, possibly predisposing them to progressive disease (Montali et al. 1998a).

A variety of mycobacteria have been isolated from cervids, often during surveillance for bovine tuberculosis or to determine the cause of skin reactions when testing for bovine tuberculosis. *Mycobacterium avium* complex bacteria are among the most common isolates in surveys of lymph nodes from red deer and fallow deer; 49 of 115 tuberculous lesions in slaughtered farmed deer in Ireland yielded *M. avium* (Quigley et al. 1997), and *M. avium* and *M. intracellulare* were isolated from resident farmed deer in New Zealand and from imported deer (Wards et al. 1991). Orr et al. (1978) experimentally infected red deer with *M. avium;* lesions were very mild. Disseminated *M. avium* infection is possible and may occur in free-ranging cervids: a calf elk with widespread mycobacteriosis was found dead in Wyoming (W. Cook and E.S. Williams unpublished), and disseminated *M. avium sylvaticum* was described in roe deer *Capreolus capreolus* (Jørgensen and Clausen 1976). Lesions in deer typically contain more acid-fast organisms than those from which *M. bovis* is cultured (Quigley et al. 1997).

Mycobacterium avium may infect and cause clinical disease in nonhuman primates. Thoen and Himes (1981) reported that a high percentage of *M. avium* isolates submitted from captive wildlife were from nonhuman primates. Involvement of the digestive tract and associated lymph nodes with persistent diarrhea and weight loss is the most common manifestation of progressive *M. avium* infection in primates. Microscopically, lesions of *M. avium* infection in primates are characterized by diffuse accumulations of epithelioid macrophages containing large numbers of acid-fast bacteria (Holmberg et al. 1982) in contrast to the typical respiratory tract tuberculous lesions caused by *M. tuberculosis* and *M. bovis* (Thoen 1993). *Mycobacterium avium* serovars 1 and 8 are most commonly isolated from nonhuman primates (Thoen 1993).

***Mycobacterium leprae* in Armadillos.** *Mycobacterium leprae,* the etiologic agent of Hansen's disease (leprosy), is found in free-ranging armadillos *Dasypus novemcinctus* in Texas and Louisiana (Smith et al. 1983; Truman et al. 1986, 1991). It was not detected in surveys of armadillos in Florida, Alabama, Arkansas, Georgia, and Mississippi (Howerth et al. 1990; Truman et al. 1990). Surveillance of free-ranging armadillo populations is conducted by histologic examination of ear tissue for evidence of disseminated disease (Smith et al. 1983; Howerth et al. 1990) and serology using ELISA (Truman et al. 1986, 1990, 1991). Prevalence of infection varied from 1%–15% in these surveys, but in one study approximately a third of adult armadillos were considered infected (Truman et al. 1991). In another study, seropositivity (13%) was found to be about five times higher than disseminated disease (3%) detected by histopathology (Truman et al. 1990). A retrospective serologic survey using banked serum indicated *M. leprae* was enzootic in Gulf Coast armadillos as early as 1961 (Truman et al. 1986).

The low body temperature of armadillos and perhaps their lack of adequate immune response appear to facilitate growth of the organism, and these animals develop severe disseminated leprosy. Lesions in armadillos tend to be lepromatous, and large numbers of acid-fast bacilli are found in many tissues (Job et al. 1985). During surveys for *M. leprae,* a variety of other mycobacteria were isolated from armadillos (Smith et al. 1983). The predilection of armadillos to develop disseminated mycobacterial disease has been exploited in the study of this organism in the laboratory due to inability to cultivate it in vitro.

Historically, humans were thought to be the only hosts of *M. leprae.* The origin of leprosy in Gulf Coast armadillos is not known. It has been hypothesized to be from environmental contamination with human materials prior to effective chemotherapy for leprosy (Meyers et al. 1980), to be from inadvertent introduction of the organism to the wild associated with studies of leprosy in armadillos beginning in 1968, or to be a natural endemic infection of these populations of armadillos. Truman et al. (1986) provide serologic evidence that leprosy was present in armadillo populations prior to studies of the disease in this species, thus eliminating the theory of inadvertent introduction of the disease into wild populations from experimental animals or materials. Some believe that armadillos play a role in the epidemiology of human leprosy (Lumpkin et al. 1983; Meyers et al. 1992), but this relationship has not been confirmed.

Other Nontuberculous Mycobacterial Infections. Important features of the nontuberculous mycobacterial species as zoonotic agents have been recently reviewed by Lamberski (1999). Mycobacterial infections are very important diseases of immunosuppressed humans; however, even healthy humans may become infected, and thus these agents and animals suspected of having mycobacterial infections should be handled so as to minimize transmission to humans. Many of these mycobacteria are found in the soil and environment and thus are not true zoonotic agents.

Many case reports of uncommon mycobacterial infections in wild mammals are in the literature. These are mostly incidental findings or, rarely, reports of significant disease in individuals. Often these infections are found in immunosuppressed animals.

LITERATURE CITED

Angus, K.W. 1990. Intestinal lesions resembling paratuberculosis in a wild rabbit (*Oryctolagus cuniculus*). *Journal of Comparative Pathology* 103:101–105.

Belknap, E.B., D.M. Getzy, L.W. Johnson, R.P. Ellis, G.L. Thompson, and W.P. Shulaw. 1994. *Mycobacterium paratuberculosis* infection in two llamas. *Journal of the American Veterinary Medical Association* 204:1805–1808.

Bendixen, P.H., B. Bloch, and J.B. Jørgensen. 1981. Lack of intracellular degradation of *Mycobacterium paratuberculosis* by bovine macrophages infected in vitro and in vivo: Light microscopic and electron microscopic observations. *American Journal of Veterinary Research* 42:109–113.

Bennett, B.T., C.R. Abee, and R. Henrickson. 1998. *Nonhuman primates in biomedical research,* vol. 2. San Diego: Academic, 512 pp.

Bernadelli, A., R. Bastida, J. Loureiro, H. Michelis, M.I. Romano, A. Cataldi, and E. Costa. 1996. *Revue Scientifique et Technique O.I.E.* 15:985–1005.

Boever, W.J., and D. Peters. 1977. Paratuberculosis in two herds of exotic sheep. *Journal of the American Veterinary Medical Association* 170:987–990.

Bourgeois, E. 1940. Paratuberkulose Darmentzundung bei eineum Sikahirsch. *Schweizer Archiv für Tierheilkunde* 86:119–120.

Burrells, C., C.J. Clarke, A. Colston, J.M. Kay, J. Porter, D. Little, and J.M. Sharpe. 1998. A study of immunological responses of sheep clinically-affected with paratuberculosis (Johne's disease): The relationship of blood, mesenteric lymph node and intestinal lymphocyte responses to gross and microscopic pathology. *Veterinary Immunology and Immunopathology* 66:343–358.

Calle, P.P. 1999. Tuberculin responses in orangutans. In *Zoo and wild animal medicine,* ed. M.E. Fowler and R.E. Miller, 4th ed. Philadelphia: W.B. Saunders, pp. 392–396.

Chiodini, R. 1989. Crohn's disease and the mycobacterioses: A review and comparison of two disease entities. *Clinical Microbiology Reviews* 2:90–117.

———. 1996. Immunology: Resistance to paratuberculosis. *Veterinary Clinics of North America: Food Animal Practice* 12:313–343.

Chiodini, R., and C.A. Rossiter. 1996. Paratuberculosis: A potential zoonosis. *Veterinary Clinics of North America: Food Animal Practice* 12:457–467.

Chiodini, R., and H.J. van Kruiningen. 1983. Eastern white-tailed deer as a reservoir of ruminant paratuberculosis. *Journal of the American Veterinary Medical Association* 182:168–169.

Clarke, C.J., and D. Little. 1996. The pathology of ovine paratuberculosis: Gross and histological changes in the intestine and other tissues. *Journal of Comparative Pathology* 114:419–437.

Clarke, C.J., I.A. Patterson, K.E. Armstrong, and J.C. Low. 1996. Comparison of the absorbed ELISA and agar gel immunodiffusion test with clinicopathological findings in ovine clinical paratuberculosis. *Veterinary Record* 139:618–121.

Coetsier, C., X. Havaux, F. Mattelard, S. Sadatte, F. Cormont, K. Buergelt, B. Limbourg, D. Latinne, H. Bazin, J.F. Denef, and C. Cocito. 1998. Detection of *Mycobacterium avium* subsp. *paratuberculosis* in infected tissues by new species-specific immunohistological procedures. *Clinical and Diagnostic Laboratory Immunology* 5:446–451.

Collins, M.T. 1996. Diagnosis of paratuberculosis. *Veterinary Clinics of North America: Food Animal Practice* 12:357–371.

Collins, M.T., S. Cavaignac, and G.W. de Lisle. 1998. Use of four DNA insertion sequences to characterize strains of the *Mycobacterium avium* complex isolated from animals. *Molecular and Cellular Probes* 11:373–380.

Cook, W., T.E. Cornish, S. Shideler, B. Lasley, and M.T. Collins. 1997. Radiometric culture *Mycobacterium avium paratuberculosis* from the feces of tule elk. *Journal of Wildlife Diseases* 33:635–637.

Cousins, D.V., B.R. Francis, B.L. Gow, D.M. Collins, C.H. McGlashan, A. Gregory, and R.M. Mackenzie. 1990. Tuberculosis in captive seals: Bacteriological studies on an isolate belonging to the *Mycobacterium tuberculosis* complex. *Research in Veterinary Science* 48:196–200.

Cousins, D.V., S.N. Williams, R. Reuter, D. Forshaw, B. Chadwick, D. Coughran, P. Collins, and N. Gales. 1993. Tuberculosis in wild seals and characterisation of the seal bacillus. *Australian Veterinary Journal* 70:92–97.

Cousins, D.V., R.L. Peet, W.T. Gaynor, S.N. Williams, and B.L. Gow. 1994. Tuberculosis in imported hyrax (*Procavia capensis*) caused by an unusual variant belonging to the *Mycobacterium tuberculosis* complex. *Veterinary Microbiology* 42:135–145.

Cousins, D.V., R.J. Evans, and B.R. Francis. 1995. Use of BACTEC radiometric culture method and polymerase chain reaction for the rapid screening of faeces and tissues for *Mycobacterium paratuberculosis. Australian Veterinary Journal* 72:458–462.

de Lisle, G.W., G.F. Yates, and D.M. Collins. 1993. Paratuberculosis in farmed deer: Case reports and DNA characterization of isolates of *Mycobacterium paratuberculosis. Journal of Veterinary Diagnostic Investigation* 5:567–571.

Dorofeev, K.A., and L.A. Kalachev. 1949. Johne's disease in sheep and wild animals. *Veterinariya (Moscow)* 26:21–24 [*Veterinary Bulletin* 22:173. Abstract 917].

Dukes, T.W., G.J. Glover, B.W. Brooks, J.R. Duncan, and M. Swendrowski. 1992. Paratuberculosis in saiga antelope (*Saiga tatarica*) and experimental transmission to domestic sheep. *Journal of Wildlife Diseases* 28:161–170.

Ellis, T.M., B.A. Carson, M.J. Kalkhoven, and P.A. Martin. 1998. Specificity of two absorbed ELISAs for surveys of *Mycobacterium paratuberculosis* in cattle. *Australian Veterinary Journal* 76:497–499.

Fawcett, A.R., P.J. Goddard, W.A.C. McKelvey, D. Buxton, H.W. Reid, A. Greig, and A.J. Macdonald. 1995. Johne's disease in a herd of farmed red deer. *Veterinary Record* 136:165–169.

Forshaw, D., and G.R. Phelps. 1991. Tuberculosis in a captive colony of pinnipeds. *Journal of Wildlife Diseases* 27:288–295.

Furley, C.W. 1997. Tuberculosis in elephants. *Lancet* 350:224.

Gilmour, N.J.L. 1976. The pathogenesis, diagnosis and control of Johne's disease. *Veterinary Record* 99:433–434.

———. 1984. Paratuberculosis. In *The management and health of farmed deer,* ed. H.W. Reid. Boston: Kluwer Academic, pp. 113–119.

Greig, A., K. Stevenson, V. Perez, A.A. Pirie, J.M. Grant, and J.M. Sharp. 1997. Paratuberculosis in wild rabbits. *Veterinary Record* 140:141–143.

Gwóźdź, J.M., M.P. Reichel, A. Murray, W. Manktelow, D.M. West, and K.G. Thompson. 1997. Detection of *Mycobacterium avium* subsp. *paratuberculosis* in ovine tissues and blood by the polymerase chain reaction. *Veterinary Microbiology* 51:233–244.

Hall, R., and F. Winkel. 1957. Avian tuberculosis in mink: A case report. *Journal of the American Veterinary Medical Association* 131:49–51.

Holmberg, C.A., R.V. Hedrickson, C. Malaga, R. Schneider, and D. Gribble. 1982. Nontuberculous mycobacterial disease in rhesus monkeys. *Veterinary Pathology Supplement* 7:9–16.

Howerth, E.W., D.E. Stallknecht, W.R. Davidson, and E.J. Wentworth. 1990. Survey for leprosy in nine-banded armadillos (*Dasypus novemcinctus*) from the southeastern United States. *Journal of Wildlife Diseases* 26:112–115.

Jackson, R., G.W. de Lisle, and R.S. Morris. 1995. A study of the environmental survival of *Mycobacterium bovis* on a farm in New Zealand. *New Zealand Veterinary Journal* 43:346–352.

Jarmai, K. 1922. *Deutsche Tierarztliche Wochenschrift* 30:257 [cited by Katic (1961)].

Jessup, D.A., and E.S. Williams. 1999. Paratuberculosis in free-ranging wildlife in North America. In *Zoo and wild animal medicine,* ed. M.E. Fowler and R.E. Miller, 4th ed. Philadelphia: W.B. Saunders, pp. 616–620.

Jessup, D.A., B. Abbas, D. Behymer, and P. Gogan. 1981. Paratuberculosis in tule elk in California. *Journal of the American Veterinary Medical Association* 179:1252–1254.

Job, C.K., R.M. Sanchez, and R.C. Hastings. 1985. Manifestations of experimental leprosy in the armadillo. *American Journal of Tropical Medicine and Hygiene* 34:151–161.

Johne, H.A., and L. Frothingham. 1895. Ein eigentumlicher Fall von Tuberberculose beim Rinde. *Deutsche Zeitschrift für Tiermedizin und Vergleich* 21:438–454 [cited by Twort and Ingram (1913)].

Jørgensen, J.B., and B. Clausen. 1976. Mycobacteriosis in a roe-deer caused by wood pigeon mycobacteria. *Nordisk Veterinaermedicin* 28:539–546.

Katic, I. 1961. Paratuberculosis (Johne's disease) with special reference to captive wild animals. *Nordisk Veterinaermedicin* 13:205–214.

Körmendy, B. 1994. The effect of vaccination on the prevalence of paratuberculosis in large dairy herds. *Veterinary Microbiology* 41:117–125.

Lamberski, N. 1999. Nontuberculous mycobacteria: Potential for zoonosis. In *Zoo and wild animal medicine,* 4th ed., ed. M.E. Fowler and R.E. Miller. Philadelphia: W.B. Saunders, pp. 146–150.

Larsen, A.B., O.H. Stalheim, D.E. Hughes, L.H. Appell, W.D. Richards, and E.M. Himes. 1981. *Mycobacterium paratuberculosis* in the semen and genital organs of a semen-donor bull. *Journal of the American Veterinary Medical Association* 179:169–171.

Libke, K.G., and A.M. Walton. 1975. Presumptive paratuberculosis in a Virginia white-tailed deer. *Journal of Wildlife Diseases* 11:552–553.

Lomme, J.R., C.O. Thoen, E.M. Himes, J.W. Vinson, and R.E. King. 1976. *Mycobacterium tuberculosis* infection in two East African oryxes. *Journal of the American Veterinary Medical Association* 169:912–914.

Lumpkin III, L.R., G.F. Cox, and J.E. Wolf Jr. 1983. Leprosy in five armadillo handlers. *Journal of the American Academy of Dermatology* 9:899–903.

Manning, E.J.B., and M.T. Collins. 1999. Paratuberculosis in zoo animals. In *Zoo and wild animal medicine,* ed. M.E. Fowler and R.E. Miller, 4th ed. Philadelphia: W.B. Saunders, pp. 612–616.

Manning, E.J.B., H. Steinberg, K. Rossow, G.R. Ruth, and M.T. Collins. 1998. Epizootic of paratuberculosis in farmed elk. *Journal of the American Veterinary Medical Association* 213:1320–1322.

Manning, P.D., F.C. Cadigan, and E.I. Goldsmith. 1980. Detection of tuberculosis. In *Laboratory animal management: Nonhuman primates,* vol. 23, no. 2–3, ed. P.D. Manning, F.C. Cadigan, and E.I. Goldsmith. Washington, DC: Institute of Laboratory Animal Resources, National Academy Press, pp. 27–28.

McClure, H.M., R.J. Chiodini, D.C. Anderson, R.B. Swenson, W.R. Thayer, and J.A. Coutu. 1987. *Mycobacterium paratuberculosis* infection in a colony of stumptail macaques (*Macaca arctoides*). *Journal of Infectious Diseases* 155:1011–1019.

Meyers, W.M., C.H. Binford, and G.P. Walsh. 1980. Leprosy in wild armadillos. In *The comparative pathology of zoo animals,* ed. R.J. Montali and G. Migaki. Washington, DC: Smithsonian Institution, pp. 247–251.

Meyers, W.M., B.J. Gormus, and G.P. Walsh. 1992. Nonhuman sources of leprosy. *International Journal of Leprosy and Other Mycobacterial Diseases* 60:477–480.

M'Fadyean, J., and A.L. Sheather. 1916. Johne's disease: The experimental transmission of the disease to cattle, sheep and goats, with notes regarding the occurrence of natural cases in sheep and goats. *Journal of Comparative Pathology and Therapeutics* 29:62–94.

Michalak, K., C. Austin, S. Diesel, J.M. Bacon, P. Zimmerman, and J.N. Maslow. 1998. *Mycobacterium tuberculosis* infection as a zoonotic disease: Transmission between humans and elephants. *Emerging Infectious Diseases* 4:283–287 [*www.cdc.gov/ncidod/eid/vol4no2/michalak.htm]*.

Michel, A.L., and H.F. Huchzermeyer. 1998. The zoonotic importance of *Mycobacterium tuberculosis:* Transmission from human to monkey. *Journal of the South African Veterinary Association* 69:64–65.

Mitscherlich, E., and E.H. Marth. 1984. *Microbial survival in the environment.* New York: Springer-Verlag, 802 pp.

Momotani, E., D.L. Whipple, A.B. Thiermann, and N.F. Cheville. 1988. Role of μ cells and macrophages in the entrance of *Mycobacterium paratuberculosis* into domes of ileal Peyer's patches in calves. *Veterinary Pathology* 25:131–137.

Montali, R.J., M. Bush, R. Cromie, S.M. Holland, J.N. Maslow, M. Worley, F.G. Witebsky, and T.M. Phillips. 1998a. Primary *Mycobacterium avium* complex infections correlate with lowered cellular immune reactivity in Matschie's tree kangaroos (*Dendrolagus matschiei*). *Journal of Infectious Disease* 178:1719–1725.

Montali, R.J., L.H. Spelman, R.C. Cambre, D. Chatterjee, and S.K. Mikota. 1998b. Factors influencing interpretation of indirect testing methods for tuberculosis in elephants. In *Proceedings of the American Association of Zoo Veterinarians and American Association of Wildlife Veterinarians joint conference,* ed. C.K. Baer. Omaha, NE: American Association of Zoo Veterinarians, pp. 109–112.

Orr, M.B., A.B. Hunter, T. Brand, and D. Owen. 1978. Experimental challenge of red deer with *Mycobacterium avium. Veterinary Record* 102:484–485.

Payne, J.M., and J.D. Rankin. 1961. The pathogenesis of experimental Johne's disease in calves. *Research in Veterinary Science* 2:167–174.

Pinto, M.R., M.R. Jainudeen, and R.G. Panabokke. 1973. Tuberculosis in a domesticated Asiatic elephant *Elaphas maximus. Veterinary Record* 93:662–664.

Plante, Y., B.W. Remenda, B.J. Chelack, and D.M. Haines. 1996. Detection of *Mycobacterium paratuberculosis* in formalin-fixed paraffin-embedded tissues by the polymerase chain reaction. *Canadian Journal of Veterinary Research* 60:115–120.

Poddoubski, I.V. 1957. La paratuberculose. *Bulletin of the Office of International Epizootics* 48:469–476.

Power, S.B., J. Haagsma, and D.P. Smyth. 1993. Paratuberculosis in red deer (*Cervus elaphus*) in Ireland. *Veterinary Record* 132:213–216.

Quigley, F.C., E. Costello, O. Flynn, A. Gogarty, J. McGuirk, A. Murphy, and J. Egan. 1997. Isolation of mycobacteria from lymph node lesions in deer. *Veterinary Record* 141:516–518.

Qureshi, T., J.W. Templeton, and L.G. Adams. 1996. Intracellular survival of *Brucella abortus, Mycobacterium bovis* BCG, *Salmonella dublin,* and *Salmonella typhimurium* in macrophages from cattle genetically resistant to *Brucella abortus. Veterinary Immunology and Immunopathology* 50:55–65.

Radwan, A.E., S. El-Magawry, A. Hawari, D. Al-Bekiri, S. Aziz, and R.M. Rebleza. 1991. Paratuberculosis enteritis (Johne's disease) in camels in Saudi Arabia. *Biological Science* 1:57–66.

Rhyan, J.C., K. Aune, D.R. Ewalt, J. Marquardt, J.W. Mertins, J.B. Payeur, D.A. Saari, P. Schladweiler, E.J. Sheehan, and D. Worley. 1997. Survey of free-ranging elk from Wyoming and Montana for selected pathogens. *Journal of Wildlife Diseases* 33:290–298.

Ridge, S.E., J.T. Harkin, R.T. Badman, A.M. Mellor, and J.W. Larsen. 1995. Johne's disease in alpacas (*Lama pacos*) in Australia. *Australian Veterinary Journal* 72:150–153.

Riemann, H., M.R. Zaman, R. Ruppaner, O. Aalund, J.B. Jorgensen, H. Worsaae, and D. Behymer. 1979. Paratuberculosis in cattle and free-living deer. *Journal of the American Veterinary Medical Association* 174:841–843.

Saunders, G. 1983. Pulmonary *Mycobacterium tuberculosis* infection in a circus elephant. *Journal of the American Veterinary Medical Association* 183:1311–1312.

Shulaw, W.P., J.C. Gordon, S. Bech-Nielsen, C.I. Pretzman, and G.F. Hoffsis. 1986. Evidence of paratuberculosis in Ohio's white-tailed deer as determined by an enzyme-linked immunosorbent assay. *American Journal of Veterinary Research* 47:2539–2542.

Smith, J.H., D.S. Folse, E.G. Long, J.D. Christie, D.T. Crouse, M.E. Tewes, A.M. Gatson, R.L. Ehrhardt, S.K. File, and M.T. Kelly. 1983. Leprosy in wild armadillos (*Dasypus novemcinctus*) of the Texas Gulf Coast: Epidemiology and mycobacteriology. *Journal of Reticuloendothelial Society* 34:75–88.

Sockett, D.C., D.J. Carr, and M.T. Collins. 1992. Evaluation of conventional and radiometric fecal culture and a commercial DNA probe for diagnosis of *Mycobacterium paratuberculosis* infections in cattle. *Canadian Journal of Veterinary Research* 56:148–153.

Soltys, M.A., C.E. Andress, and A.L. Fletch. 1967. Johne's disease in a moose (*Alces alces*). *Bulletin of the Wildlife Disease Association* 3:183–184.

Stabel, J.R. 1996. Production of gamma-interferon by peripheral blood mononuclear cells: An important diagnostic tool for detection of subclinical paratuberculosis. *Journal of Veterinary Diagnostic Investigation* 8:345–350.

Stabel, J.R., M.R. Ackermann, and J.P. Goff. 1996. Comparison of polyclonal antibodies to three different preparations of *Mycobacterium paratuberculosis* in immunohistochemical diagnosis of Johne's disease in cattle. *Journal of Veterinary Diagnostic Investigation* 8:469–473.

Stehman, S.M. 1996. Paratuberculosis in small ruminants, deer, and South American camelids. *Veterinary Clinics of North America: Food Animal Practice* 12:441–455.

Steinberg, H. 1981. Johne's disease (*Mycobacterium paratuberculosis*) in a Jemela topi (*Damaliscus lunatus jimela*). *Journal of Zoo Animal Medicine* 19:33–41.

St. Jean, G. 1996. Treatment of clinical paratuberculosis in cattle. *Veterinary Clinics of North America: Food Animal Practice* 12:417–430.

Sweeney, R.W., ed. 1996a. Paratuberculosis (Johne's disease). *Veterinary Clinics of North America: Food Animal Practice* 12:1–471.

Sweeney, R.W. 1996b. Transmission of paratuberculosis. *Veterinary Clinics of North America: Food Animal Practice* 12:305–312.

Temple, R.M., C.C. Muscoplat, C.O. Thoen, E.M. Himes, and D.W. Johnson. 1979. Observations on diagnostic tests for paratuberculosis in a deer herd. *Journal of the American Veterinary Medical Association* 175:914–915.

Thoen, C.O. 1993. Tuberculosis and other mycobacterial diseases in captive wild animals. In *Zoo and wild animal medicine: Current therapy 3,* ed. M.E. Fowler. Philadelphia: W.B. Saunders, pp. 45–49.

Thoen, C.O., and E.M. Himes. 1981. Tuberculosis. In *Infectious diseases of wild mammals,* ed. J.W. Davis, L.H. Karstad, and D.O. Trainer. Ames: Iowa State University Press, pp. 263–274.

Thoen, C.O., and D.W. Johnson. 1981. Johne's disease (Paratuberculosis). In *Infectious diseases of wild mammals,* ed. J.W. Davis, L.H. Karstad, and D.O. Trainer. Ames: Iowa State University Press, pp. 275–279.

Thoen, C.O., W.D. Richards, and J.L. Jarnagin. 1977. Mycobacteria isolated from exotic animals. *Journal of the American Veterinary Medical Association* 170:987–989.

Thorel, M.F., M. Krichevsky, and V.V. Lévy-Frébault. 1990. Numerical taxonomy of mycobactin-dependent mycobacteria, emended description of *Mycobacterium avium,* and description of *Mycobacterium avium* subsp. *avium* subsp. nov., *Mycobacterium avium* subsp. *paratuberculosis* subsp. nov., and *Mycobacterium avium* subsp. *silvaticum* subsp. nov. *International Journal of Systematic Bacteriology* 40:254–260.

Truman, R.W., E.J. Shannon, H.V. Hagstad, M.E. Hugh-Jones, A. Wolff, and R.C. Hastings. 1986. Evaluation of the origin of *Mycobacterium leprae* infections in the wild armadillo, *Dasypus novemcinctus. American Journal of Tropical Medicine and Hygiene* 35:588–593.

Truman, R.W., C.K. Job, and R.C. Hastings. 1990. Antibodies to the phenolic glycolipid-1 antigen for epidemiologic investigations of enzootic leprosy in armadillos (*Dasypus novemcinctus*). *Leprosy Review* 61:19–24.

Truman, R.W., J.A. Kumaresan, C.M. McDonough, C.K. Job, and R.C. Hastings. 1991. Seasonal and spacial trends in the detectability of leprosy in wild armadillos. *Epidemiology and Infection* 106:549–560.

Twort, F.W., and G.L.Y. Ingram. 1913. *A monograph on Johne's disease (enteritis chronica pseudotuberculosia bovis).* London: Bailliere, Tindall and Cox, 179 pp.

Vance, H.N. 1961. Johne's disease in a European red deer. *Canadian Veterinary Journal* 2:305–307.

Van Schaik, G., C.H. Kalis, G. Benedictus, A.A. Dijkhuizen, and R.B. Huirne. 1996. Cost-benefit analysis of vaccination against paratuberculosis in dairy cattle. *Veterinary Record* 139:624–627.

Wards, B.J., G.W. de Lisle, G.F. Yates, and D.J. Dawson. 1991. Characterization by restriction endonuclease analysis and seroagglutination of strains of *Mycobacterium avium* and *Mycobacterium intracellulare* obtained from farmed deer. *American Journal of Veterinary Research* 52:197–201.

Wayne, L.G. and G.P. Kubica. 1986. Genus *Mycobacterium.* In *Bergey's manual of systematic bacteriology,* vol. 2, ed. P.H.A. Sneath, N.S. Mair, M.E. Sharpe, and J.G. Holt. Baltimore: Williams and Wilkins, pp. 1436–1457.

Whipple, D.L., D.M. Callihan, and J.M. Jarnagin. 1991. Cultivation of *Mycobacterium paratuberculosis* from bovine fecal specimens and a suggested standardized procedure. *Journal of Veterinary Diagnostic Investigation* 3:368–373.

Whipple, D.L., P.A. Kapke, and P.R. Andersen. 1992. Comparison of a commercial DNA probe test and three cultivation procedures for detection of *Mycobacterium paratuberculosis* in bovine feces. *Journal of Veterinary Diagnostic Investigation* 4:23–27.

Whittington, R.J., I. Marsh, M.J. Turner, S. McAllister, E. Choy, G.J. Eamens, D.J. Marshall, and S. Ottaway. 1998. Rapid detection of *Mycobacterium paratuberculosis* in clinical samples from ruminants and in spiked environmental samples by modified BACTEC 12B radiometric culture and direct confirmation by IS900 PCR. *Journal of Clinical Microbiology* 36:701–707.

Williams, E.S. 1981. Spontaneous and experimental infection of wild ruminants with *Mycobacterium paratuberculosis.* Ph.D. diss., Colorado State University, Fort Collins, Colorado, 362 pp.

Williams, E.S., T.R. Spraker, and G.G. Schoonveld. 1979. Paratuberculosis (Johne's disease) in bighorn sheep and a mountain goat in Colorado. *Journal of Wildlife Diseases* 15:221–227.

Williams, E.S., S.P. Snyder, and K.L. Martin. 1983a. Experimental infection of some North American wild ruminants and domestic sheep with *Mycobacterium paratuberculosis:* Clinical and bacteriological findings. *Journal of Wildlife Diseases* 19:185–191.

———. 1983b. Pathology of spontaneous and experimental infection of North American wild ruminants with *Mycobacterium paratuberculosis. Veterinary Pathology* 20:274–291.

Williams, E.S., J.C. DeMartini, and S.P. Snyder. 1985. Lymphocyte blastogenesis, complement fixation, and fecal culture as diagnostic tests for paratuberculosis in North American wild ruminants and domestic sheep. *American Journal of Veterinary Research* 46:2317–2321.

22

BRUCELLOSIS

E. TOM THORNE

Synonyms: Mediterranean fever, Malta fever, undulant fever, Bang's disease, contagious abortion, besmetlike misgeboorte, fistulous withers, poll evil.

INTRODUCTION. Brucellosis is an infectious, contagious disease of animals and humans caused by bacteria of the genus *Brucella.* There are six species, each with its principal host: *B. abortus* (cattle, *Bos taurus*), *B. melitensis* (goats, *Capra hircus*), *B. suis* (swine, *Sus scrofa*), *B. neotomae* (desert wood rat, *Neotoma lepida*), *B. ovis* (sheep, *Ovis aries*), and *B. canis* (dog, *Canis familiaris*). *Brucella* spp. recently isolated from marine mammals have not yet been identified to species. *Brucella* are worldwide in distribution and of considerable economic concern because of diseases some of them cause in domestic animals and humans. Clinical manifestations typically involve the reproductive tract, causing abortion or birth of nonviable offspring and orchitis, epididymitis, and infertility. Due to public health and economic importance, many countries have implemented brucellosis control and/or eradication programs. *Brucella abortus* is responsible for the most notable *Brucella*-related wildlife disease problems. Consequently, this chapter focuses primarily on *B. abortus.* A vast amount of literature on *Brucella* and brucellosis exists, and two notable publications are by Young and Corbel (1989) and by Nielsen and Duncan (1990).

HISTORY. The cause of human Malta fever, *Micrococcus melitensis,* was isolated in 1887 by Sir David Bruce (Bruce 1887; Meyer 1990). Bang (1897) identified the organism now known as *Brucella abortus* as a cause of abortion in cattle in Denmark. The organism was first isolated from cattle in the United States in 1910 (Metcalf 1986). Because the organisms that caused Malta fever and abortion in cattle shared infectious properties distinct from other similar pathogens, Meyer and Shaw (1920) placed them in the genus *Brucella* as *B. melitensis* and *B. abortus,* respectively (Meyer 1990). Subsequently, *B. suis* was added to the genus, and several biotypes were identified (Meyer 1990). Beginning in the 1980s, *B. neotomae, B. ovis,* and *B. canis* were added to the genus. *Brucella* spp. have been isolated from marine mammals (Ross et al. 1994; Foster et al. 1996) but have not yet been classified to species (Ewalt et al. 1994; Jensen et al. 1999).

Brucellosis in wildlife has been reviewed (Witter 1981; Davis 1990). In the United States, brucellosis was first detected in wildlife in 1917, when bison *Bison bison* in Yellowstone National Park, Wyoming, were tested for antibodies against *B. abortus.* Mohler (1917) found that two bison that had recently aborted were seroagglutination positive. Although the organism was not isolated, this demonstrated susceptibility of bison to brucellosis and suggested it may induce abortion (Williams et al. 1997).

Brucellosis was first detected, based on the presence of antibodies, in free-ranging elk *Cervus elaphus* in 1930 (Murie 1951). In the early 1930s, Tunnicliff and Marsh (1935) detected antibodies in 22% of 105 elk from Yellowstone National Park that shared range with seropositive bison. *Brucella abortus* was first recovered from aborted elk fetuses from the National Elk Refuge, Wyoming, in 1969 (Thorne et al. 1978a).

Brucella antibodies were found in African ungulates in 1962 (Rollinson 1962). *Brucella abortus* was reported from a waterbuck *Kobus ellipsiprymnus* in 1969 (Condy and Vickers 1969). In the 1970s, *B. abortus* was isolated from African buffalo *Syncerus caffer* (Kaliner and Staak 1973; Gradwell et al. 1977), and *B. melitensis* was recovered from an impala *Aepyceros melampus* (Schiemann and Staak 1971).

In 1963, a *Brucella* sp. that was pathogenic for humans was isolated from caribou *Rangifer tarandus* (Huntley et al. 1963). This organism, also responsible for brucellosis in reindeer *Rangifer tarandus,* is now known as *B. suis* biovar 4, and the disease is frequently referred to as rangiferine brucellosis (Meyer 1966).

In the United States, *B. suis* biovar 1 was reported in feral swine in 1976 (Wood et al. 1976). Subsequently, brucellosis in feral swine was demonstrated to be widespread in the United States (Bigler et al. 1977; Becker et al. 1978).

DISTRIBUTION AND HOST RANGE. Historically, brucellosis was found worldwide, but with successful eradication programs in domestic livestock, its distribution is now more limited. Corbel (1997) reviews the distribution of *B. abortus* in 1994. Occurrence of *B. abortus* in free-ranging wildlife reflects their previous or current association with infected cattle, their behavior, and/or the way they are managed. *Brucella abortus* is likely exotic to North America and was probably introduced with European domestic cattle (Meagher and Meyer 1994; Cheville et al. 1998).

One of the most notable wildlife disease problems is brucellosis in the Greater Yellowstone Area (GYA) of North America. Numerous serologic and a few microbiologic surveys have demonstrated the presence of brucellosis in elk in the GYA and its absence elsewhere in North America (Lee and Turner 1937; Vaughn et al. 1973; Adrian and Keiss 1977; Thorne et al. 1978a; McCorquodale and DiGiacomo 1985; Herriges et al. 1989; Smith and Robbins 1994; Aguirre et al. 1995; Rhyan et al. 1997a; Toman et al. 1997). Brucellosis has been reported in a number of bison populations in North America (Creech 1930; Rush 1932; Corner and Connell 1958; Choquette et al. 1978), but attention is now focused on the GYA and Wood Bison National Park, Canada (Tessaro 1986; Williams et al. 1993; Rhyan et al. 1994).

Rangiferine brucellosis is circumpolar in distribution (Neiland et al. 1968; Broughton et al. 1970; Zarnke and Yuill 1981; Dieterich 1981; Zarnke 1983; Tessaro and Forbes 1986; Davis 1990).

In 1994, *Brucella* spp. were reported from a bottle-nosed dolphin *Tursiops truncatus* in California (Ewalt et al. 1994) and from common seals *Phoca vitulina,* harbor porpoises *Phocoena phocoena,* and a common dolphin *Delphinus delphis* around the Scottish coast (Ross et al. 1994). Foster et al. (1996) further reported on the isolation of *Brucella* spp. from several cetaceans, seals, and a European otter *Lutra lutra.*

Table 22.1 presents species of wild mammals in which *Brucella* susceptibility has been reported. The list is not complete, but the number and diversity of species susceptible to *Brucella* infection are striking and suggest that at least 91 species of mammals from nine orders are susceptible to *Brucella* infection. The well-known susceptibility of humans to *Brucella* infections suggests that under the right circumstances other primates are likely also susceptible. Although most mammals may be susceptible to *Brucella* infection, clinical disease seems to be rare. With few exceptions, clinical brucellosis occurs primarily among Artiodactyla. A potential exception may be infections in marine mammals.

Although a broad host range has been demonstrated for *Brucella,* virulence varies among *Brucella* spp., and infections in hosts other than the primary host often are self-limiting (Smith and Ficht 1990). It is likely that many infections in wild mammals are self-limiting and of little or no consequence to the atypical host. Except for *B. suis* biovar 4, *B. neotomae,* and possibly brucellae from marine mammals, free-ranging wild mammals have not been identified as primary hosts of the recognized species of *Brucella.*

ETIOLOGY. *Brucella melitensis* is the type strain; it has three biovars and is typically pathogenic for sheep and goats. *Brucella abortus* has biovars 1, 2, 3, 4, 5, 6, 7, and 9, and cattle are considered the natural host. *Brucella suis* has four biovars and is usually pathogenic for swine except biovar 4, which affects reindeer and caribou, and biovar 2, which affects European hares *Lepus europaeus.* No biovars are recognized in *B. neotomae,* known only from desert wood rats; *B. ovis* primarily affects sheep; and *B. canis* is pathogenic for dogs (Corbel and Morgan 1982; Holt et al. 1994).

Brucella spp. recently recovered from marine mammals have not yet been classified. Ewalt et al. (1994) suggested that an isolate from the aborted fetus of a bottle-nosed dolphin was more closely related to *B. abortus* or *B. melitensis* than other species of *Brucella.* However, Jensen et al. (1999), analyzing DNA by pulsed-field gel electrophoresis, found differences between the known species of *Brucella,* the dolphin isolates, and those from seals and porpoise.

Brucella are Gram-negative aerobic/microaerophilic cocci, coccobacilli, or short rods (Holt et al. 1994) that usually occur as intracellular clusters of cocci but may be pleomorphic (Corbel 1989b). Taxonomic differentiation of species and biovars is based on utilization of certain amino acids, urea-cycle components, and carbohydrates (Corbel 1989b). Many *Brucella,* especially *B. abortus,* use erythritol in preference to glucose as an energy source, which enhances their growth (Corbel 1989b). Most *Brucella* form raised, convex circular colonies 0.5–1.0 μm in diameter after 48-hour growth on solid media at 37° C, and at 7 days are 1.5–2.0 μm in diameter and circular, convex, slightly opaque, and butyrous in nature (Corbel 1989b). Usually, 10% CO_2 atmosphere is used for primary isolation.

Effective use of *Brucella* phages for identification of *Brucella* began in the 1950s when Russian workers isolated phages named the Tbilisi (Tb) phages (Rigby 1990), which were active against smooth strains of *B. abortus.* Some phages lyse all six identified species of *Brucella* and are specific to smooth or rough strains (Rigby 1990). Phage lysis and oxidative metabolism tests are used for species identification, and conventional tests are used to identify biovars (Corbel and Morgan 1982; Gargani and Tolari 1986; Corbel 1989b).

For regulatory and epidemiologic purposes, *B. abortus* strain 19 (S19), an important vaccine strain, must often be differentiated from field strain *B. abortus.* Strain 19 grows in normal atmosphere at 37° C, is inhibited by thionin blue dye (Thomas et al. 1981) and penicillin, and is usually inhibited by erythritol (Alton et al. 1988).

Culture of *Brucella* from tissue or fluid, including milk, is the traditional and definitive diagnosis of infection. Primary isolation of *Brucella* from tissues may be enhanced by freezing and thawing before inoculation into culture media (Corbel 1989b). *Brucella* grow slowly in the laboratory but are not highly exacting in growth requirements. Most will grow on high-quality peptone-based media. Blood or serum enhances growth (Corbel 1989b). Many selective media have been developed for isolation of *Brucella* from contaminated material and milk (Alton et al. 1988; Mayfield et al. 1990).

TABLE 22.1—Wild mammals in which *Brucella* susceptibility has been demonstrated by serology and/or culture from naturally or artificially infected hosts

Common Name	Scientific Name	Reference
Marsupialia		
Opossum	*Didelphis virginiana*	Swann et al. 1980; Moore and Schurrenberger 1981
Lagomorpha		
Nuttal's cottontail	*Sylvilagus nuttallii*	Thorpe et al. 1965
Desert cottontail	*S. audubonii*	Thorpe et al. 1965
Brown hare	*Lepus europaeus*	Jacotot and Valle 1951; Christiansen and Thomsen 1956
Mountain hare	*L. timidus*	Rementsova 1962
Black-tailed jack rabbit	*L. californicus*	Stoenner et al. 1959; Thorpe et al. 1967
Rodentia		
Antelope ground squirrel	*Ammospermophilus leucurus*	Thorpe et al. 1965, 1967
Yellow suslik	*Spermophilus fulvus*	Rementsova 1962
Lesser suslik	*S. pygmaeus*	Rementsova 1962
Ground squirrel	*S. townsendii*	Thorpe et al. 1965
Kangaroo rat	*Dipodomys ordii*	Thorpe et al. 1965, 1967
Chisel-toothed kangaroo rat	*D. microps*	Thorpe et al. 1965
Long-tailed pocket mouse	*Perognathus parvus*	Thorpe et al. 1965
Desert wood rat	*Neotoma lepida*	Stoenner and Lackman 1957; Thorpe et al. 1965, 1967
Grasshopper mouse	*Onychomys leucogaster*	Thorpe et al. 1965
Western harvest mouse	*Reithrodontomys megalotis*	Thorpe et al. 1967
Deer mouse	*Peromyscus maniculatus*	Corey et al. 1964; Thorpe et al. 1965, 1967
Pinyon mouse	*P. truei*	Thorpe et al. 1967
Bank vole	*Clethrionomys glareolus*	Redwood and Corbel 1985
Common field vole	*Microtus arvalis*	Rementsova 1962
Montane vole	*M. montanus*	Thorpe et al. 1967
Norway rat	*Rattus norvegicus*	Hagan 1922
Allied rat	*R. assimilis (fuscipes?)*	Cook et al. 1966
Large climbing rat	*Melomys cervinipes*	Cook et al. 1966
Small climbing rat	*M. lutillus*	Cook et al. 1966
House mouse	*Mus musculus*	Corey et al. 1964
Porcupine	*Erethizon dorsatum*	Thorpe et al. 1965
Capybara	*Hydrochaeris hydrochaeris*	Lord and Flores 1983
Cetacea		
Bottle-nosed dolphin	*Tursiops truncatus*	Ewalt et al. 1994
Striped dolphin	*Stenella coeruleoalba*	Foster et al. 1996
Common dolphin	*Delphinus delphis*	Ross et al. 1994
Atlantic white-sided dolphin	*Lagenorhynchus acutus*	Foster et al. 1996
Harbor porpoise	*Phocoena phocoena*	Ross et al. 1994; Foster et al. 1996
Carnivora		
Red fox	*Vulpes vulpes*	McCaughey 1969; Neiland 1975
Gray fox	*Urocyon cinereoargenteus*	Scanlan et al. 1984
Arctic fox	*Alopex lagopus*	Pinigin and Zabrodin 1970
Black-backed jackal	*Canis mesomelas*	Sachs et al. 1968
Coyote	*C. latrans*	Hoff et al. 1974; Hoq 1978; Davis et al. 1979
Gray wolf	*C. lupus*	Neiland 1970, 1975; Neiland and Miller 1981; Tessaro 1986
African hunting dog	*Lycaon pictus*	Sachs et al. 1968
Black-bear	*Ursus americanus*	Binninger et al. 1980; Zarnke and Yuill 1981
Grizzly bear	*U. arctos*	Neiland 1975
Raccoon	*Procyon lotor*	Hoq 1978; Swann et al. 1980
Pampas fox	*Pseudalopex gymnocerus*	Szyfres and Gonzalez-Tome 1966; Lord and Flores 1983
Patagonian fox	*P. griseus*	Szyfres and Gonzalez-Tome 1966
European badger	*Meles meles*	Corbel et al. 1983
American badger	*Taxidea taxus*	Hoq 1978
Spotted skunk	*Spilogale putorius*	Hoq 1978
Striped skunk	*Mephitis mephitis*	Hoq 1978
American mink	*Mustela vison*	Prichard et al. 1971
European otter	*Lutra lutra*	Foster et al. 1996
Spotted hyena	*Crocuta crocuta*	Sachs et al. 1968; de Vos and van Niekerk 1969
Bobcat	*Felis rufus*	Hoq 1978
Lion	*Panthera leo*	de Vos and van Niekerk 1969
Pinnipedia		
Hooded seal	*Cystophora cristata*	Foster et al. 1996
Gray seal	*Halichoerus grypus*	Foster et al. 1996
Common seal	*Phoca vitulina*	Ross et al. 1994; Garner et al. 1997

(continued)

TABLE 22.1—*Continued*

Common Name	Scientific Name	Reference
Tubulidentata		
Ant bear	*Orycteropus afer*	Condy and Vickers 1972
Perissodactyla		
Burchell's zebra	*Equus burchelli*	Rollinson 1962; Condy and Vickers 1972; Thimm and Wundt 1976; Bishop et al. 1994
Artiodactyla		
Feral swine	*Sus scrofa*	Wood et al. 1976; Bigler et al. 1977; Becker et al. 1978; Hubálek et al. 1993
Hippopotamus	*Hippopotamus amphibius*	de Vos and van Niekerk 1969; Condy and Vickers 1972; Thimm and Wundt 1976; Bishop et al. 1994
Dromedary camel	*Camelus dromedarius*	al-Khalaf and el-Khaladi 1989; Agab et al. 1994; Obied et al. 1996
Fallow deer	*Dama dama*	McDiarmid 1951
Sika deer	*Cervus nippon*	Daniel 1967
Elk, red deer	*C. elaphus*	Rush 1932; Tunnicliff and Marsh 1935; Lee and Turner 1937; Murie 1951; Honess and Winter 1956; Corner and Connell 1958; Rementsova 1962; Adrian and Keiss 1977; Merrell and Wright 1978; Thorne et al. 1978a, 1978b; Morton et al. 1981; Thorne 1982; Rhyan et al. 1997a; Hutching 1998
Mule deer	*Odocoileus hemionus*	Thorpe et al. 1965, 1967
White-tailed deer	*O. virginianus*	Fenstermacher et al. 1943; Shotts et al. 1958; Youatt and Fay 1959; Hayes et al. 1960; Fay 1961; Baker et al. 1962; Corey et al. 1964; Barron et al. 1985
Moose	*Alces alces*	Fenstermacher 1937; Fenstermacher and Olson 1942; Corner and Connell 1958; Jellison et al. 1953; Zarnke 1983; Dieterich et al. 1991; Honour and Hickling 1993; Forbes et al. 1996; O'Hara et al. 1998
Caribou, reindeer	*Rangifer tarandus*	Golosov and Zabrodin 1959; Huntley et al. 1963; Meyer 1966; Neiland et al. 1968; Broughton et al. 1970; Rausch and Huntley 1978; Dieterich 1981; Zarnke and Yuill 1981; Zarnke 1983; Tessaro and Forbes 1986; Ferguson 1997
Roe deer	*Capreolus capreolus*	Rementsova 1962; Bouvier 1964
Giraffe	*Giraffa camelopardalis*	Rollinson 1962
Bushbuck	*Tragelaphus scriptus*	Condy and Vickers 1969, 1972; Thimm and Wundt 1976
Greater kudu	*T. strepsiceros*	Condy and Vickers 1972
Common eland	*Taurotragus oryx*	Rollinson 1962; Condy and Vickers 1972; Thimm and Wundt 1976; Bishop et al. 1994
Asian water buffalo	*Bubalus bubalis*	Zaki 1948
African buffalo	*Syncerus caffer*	Sachs et al. 1968; de Vos and van Niekerk 1969; Condy and Vickers 1972; Kaliner and Staak 1973; Thimm and Wundt 1976; Gradwell et al. 1977; Herr and Marshall 1981; Waghela and Karstad 1986; Bishop et al. 1994
American bison	*Bison bison*	Mohler 1917; Creech 1930; Rush 1932; Tunnicliff and Marsh 1935; Moore 1947; Honess and Winter 1956; Corner and Connell 1958; Thorpe et al. 1965; Choquette et al. 1978; Tessaro 1986, 1987; Davis et al. 1990, 1991; Williams et al. 1993; Meagher and Meyer 1994; Rhyan et al. 1994; Bevins et al. 1996; Rhyan et al. 1997a,b
Common duiker	*Sylvicapra grimmia*	Condy and Vickers 1969, 1972
Waterbuck	*Kobus ellipsiprymnus*	Condy and Vickers 1969, 1972; de Vos and van Niekerk 1969
Sable antelope	*Hippotragus niger*	Condy and Vickers 1972
Arabian oryx	*Oryx leucoryx*	Greth et al. 1992
Topi	*Damaliscus lunatus*	Sachs and Staak 1966
Blue wildebeest	*Connochaetes taurinus*	Sachs et al. 1968; Thimm and Wundt 1976
Sharpe's grysbok	*Raphicerus sharpei*	Condy and Vickers 1972
Impala	*Aepyceros melampus*	Schiemann and Staak 1971; Condy and Vickers 1972, 1977; Thimm and Wundt 1976; Bishop et al. 1994
Thompson's gazelle	*Gazella thomsonii*	Sachs and Staak 1966
Grant's gazelle	*G. granti*	Sachs and Staak 1966
Saiga	*Saiga tatarica*	Rementsova 1962
Chamois	*Rupicapra rupicapra*	Bouvier 1964
Muskox	*Ovibos moschatus*	Gates et al. 1984
Dall's sheep	*Ovis dalli*	Foreyt et al. 1983

In 1968, Diaz et al. suggested important surface antigens of smooth *Brucella* were in lipopolysaccharide (LPS)-protein complexes. The O-polysaccharide side chain, core oligosaccharide, and the lipid-A backbone of *Brucella* LPS induce strong antibody response and are important to serologic identification of *Brucella* spp. and strains (Moreno et al. 1981; Alton et al. 1988; Rigby 1990). Only smooth strains have an O side chain (Moreno et al. 1979), which contains the major antigenic epitopes.

Virulence of *Brucella* is apparently determined, in part, by smooth LPS and is presumably a function of the polymer-forming O chains. Rough variants of smooth species are usually reduced in virulence. Rough *B. ovis* and *B. canis* have limited host ranges and ability to cause infection in other species (Corbel 1989b), although *B. ovis* infections have occurred in experimentally inoculated white-tailed deer *Odocoileus virginianus* (Barron et al. 1985) and farmed red deer *Cervus elaphus* (Hutching 1998).

Under suitable conditions, *Brucella* are relatively hardy in the environment, especially in the presence of protein and when protected from sunlight (Kuzdas and Morse 1954; Wray 1975). *Brucella abortus* may survive up to 6 months in shaded fetuses (Stableforth 1959).

TRANSMISSION AND EPIDEMIOLOGY. *Brucella abortus* is usually transmitted among cattle, and probably most wild ruminants, through ingestion of contaminated feed; through licking an infected fetus, calf, or placenta; or through licking the genitalia of an infected dam shortly after a birth event when large numbers of the organism are present. Transmission of *B. abortus* also may occur through conjunctival contamination and inhalation (Nicoletti 1980). High numbers of *B. abortus* were cultured from umbilicus [2.4×10^8–4.3×10^9 colony-forming units (CFU)/g], fetal fluid (9.5×10^{10} CFU/ml), and cotyledon (5.2×10^{11}–1.4×10^{13} CFU/g) of two naturally infected cows (Alexander et al. 1981). It is generally felt that cattle remain chronically infected with *B. abortus* and, after the first abortion, continue to shed the organism in milk and in reproductive tract discharges following subsequent normal births (Cunningham 1977a; Morgan 1977; Nicoletti 1980). However, persistent shedding is poorly documented and may be the exception rather than the rule (Enright 1990a).

Among cattle, susceptibility to infection with *B. abortus* is associated with sexual maturity: sexually mature heifers, especially if pregnant, are more susceptible than immature heifers and calves. This also appears to be true with elk (Thorne et al. 1978a,b) and is likely the case with bison and other artiodactyls.

Bovine calves may become infected in utero or via ingestion of *B. abortus* with colostrum and milk. Most neonatal infections are cleared within a few months (Nicoletti 1980). A small percentage of heifer calves born to infected cattle may remain latently infected until their first pregnancy, when antibodies first become detectable and abortion may occur (Wilesmith 1978; Dolan 1980; La Parik and Moffat 1982). Abortion by latent carriers can cause unexpected transmission of brucellosis in susceptible herds. Experimental infections suggested elk also may experience similar latent infections (Thorne et al. 1978b).

Although *Brucella* are capable of lengthy survival in the environment (Kuzdas and Morse 1954; Wray 1975), there is no evidence of replication outside the host. Infected hosts are the only important mode of transmission and persistence of *Brucella* (Corbel 1989a). Disinfection of facilities and removal of infected cattle from premises for 1 month generally are sufficient to prevent infection of newly introduced cattle (Ray 1977).

Transmission of brucellosis from one cattle herd to another by ticks, fleas, or mosquitoes has never been proven (Meyer 1977); but, experimentally, face flies *Musca autumnalis* that fed on *B. abortus*-contaminated material shed them in feces, suggesting a short-term mechanism of transmission (Cheville et al. 1989). Based on recovery of *Brucella* sp. from *Parafilaroides* lungworms from a Pacific harbor seal *Phoca vitulina richardsi,* Garner et al. (1997) suggested that brucellosis might be transmitted among pinnipeds by infected lungworms.

Transmission of *B. abortus* by elk (Thorne 1982; Thorne et al. 1978b, 1979) and bison (Davis et al. 1990) is similar to that by cattle; presumably transmission among other species of free-ranging artiodactyls is also similar. Epidemiology of *B. abortus* is influenced by a variety of management and environmental factors, the host susceptibility to infection, and the amount of exposure, which is largely influenced by the number of abortions that occur in a herd (Crawford et al. 1988, 1990; Cheville et al. 1998).

Multiple factors determine the level of exposure to susceptible animals following a *B. abortus*-contaminated parturition (Crawford et al. 1990). The number of organisms shed into the environment may be influenced by the affected host's genetic resistance or acquired immunity due to chronicity of infection or previous vaccination. Survival of organisms in the environment is enhanced by cool temperatures and moisture and is decreased by high temperatures, direct sunlight, and dryness (Crawford et al. 1990). Management and behavior of exposed animals play important roles in determining the probability that they encounter an infectious dose of viable *B. abortus.* Susceptibility to infection based on age, genetic resistance, and vaccination status influences the magnitude of the dose of organisms required to induce infection.

Epidemiology and transmission of *B. abortus* in free-ranging bison have been reviewed (Williams et al. 1997; Meyer and Meagher 1997; Cheville et al. 1998). Bison, like cattle, are social animals, which results in opportunities for contagious disease transmission. Bison infected with *B. abortus* abort (Davis et al. 1990; Williams et al. 1993; Rhyan et al. 1994), although the incidence of abortion among free-ranging bison is unknown. Experimentally infected bison experienced high abortion rates, and *B. abortus* may be shed from the reproductive tract 7–14 days following abortion or birth of a nonviable calf [D.S. Davis unpublished, in Williams et al. (1997)]. Cattle may shed the organism following abortion and normal birth for a few weeks (Kennedy and Miller 1993).

Because venereal transmission of brucellosis from bovine bulls to cows is not considered important in the epidemiology of brucellosis (Crawford et al. 1990; Kennedy and Miller 1993), bull bison also are not considered to be a significant source of transmission. Robison et al. (1998) successfully bred four bison cows by using an infected bull and concluded brucellosis was not readily transmitted venereally in bison. However, *B. abortus* has been cultured on a number of occasions from reproductive tracts of bull bison (Creech 1930; Tunnicliff and Marsh 1935; Corner and Connell 1958; Williams et al. 1993).

Meyer and Meagher (1995) proposed that, in the chronically infected bison of Yellowstone National Park, transmission from the bison dam to calf via contaminated milk is the most important route. Excretion of *B. abortus* in milk by cattle is well known (Kennedy and Miller 1993); *B. abortus* and *B. suis* biovar 4 have been isolated from milk of experimentally infected bison 3 weeks after parturition [D.S. Davis unpublished, in Williams et al. (1997)] and from mammary glands and supramammary lymph nodes of infected bison (Davis et al. 1990; Williams et al. 1993; Bevins et al. 1996), suggesting that transmammary transmission could occur. Transmission might occur via contact with feces of bison calves nursing infected dams (Meyer and Meagher 1995), and this is supported by recovery of *B. suis* biovar 4 from feces of calves of artificially infected bison (Bevins et al. 1996).

A presently undocumented, but likely, means of *B. abortus* transmission from female to male bison is oral-nasal contact by males with genital discharges of infected cow bison. Bison are highly social animals, with younger bulls remaining in association with cow-calf herds much of the year. Males frequently approach and smell the anogenital region of females to determine stage of estrus (Berger and Cunningham 1994). It is likely males are attracted to and investigate females that have recently aborted and calved. If the female is infected with *B. abortus,* these investigations likely serve as efficient routes of exposure to susceptible bulls.

Using experimental *B. abortus* strain 2308 (S2308) infections, Davis et al. (1990) demonstrated transmission from bison to susceptible cattle in a 1.1-ha paddock; cattle had contact with fetuses, reproductive products, and exudates from bison. Under ranch conditions, brucellosis was transmitted from a highly infected bison herd to a herd of cattle with which the bison had some contact during the winter (Armstrong 1983). The risk of brucellosis transmission from bison to cattle in the GYA is of considerable concern and debate (Thorne et al. 1991b; Keiter and Froelicher 1993; Keiter 1997b; Thorne et al. 1997a; U.S. Department of Interior 1998). Brucellosis in bison probably originated from cattle (Meagher and Meyer 1994) and, without doubt, could be transmitted from free-ranging bison to cattle. However, the continued absence of brucellosis in cattle that associate with bison in the GYA demonstrates the risk of transmission is small.

Transmission of *B. abortus* by elk is via contact with fetal fluids, vaginal exudates, and aborted fetuses from infected elk (Thorne et al. 1978b). In controlled studies, licking or consuming *Brucella*-contaminated materials, e.g., fetus and feed, were the most likely routes of exposure. Confined elk and free-ranging elk on feedgrounds become exposed when they investigate fetuses and neonates other than their own, and when they consume or attempt to consume fetuses (Thorne et al. 1978b). Vaginal exudates of infected cow elk were shown to contain *B. abortus* following abortion and normal birth up to 17 and 9 days, respectively (Thorne et al. 1978b). Unlike cattle and bison, the placenta of an infected elk is unlikely to be a source of environmental contamination or exposure, because elk almost always consume the placenta and are sufficiently agile that the placenta seldom touches the ground (Thorne et al. 1978b). Elk are very secretive during a normal parturition and thoroughly clean up reproductive products (Geist 1982); this leaves almost no opportunity for inter- or intraspecific transmission of brucellosis by an infected elk during a normal birth (Thorne et al. 1997c; Cheville et al. 1998).

Brucellosis occurs in free-ranging elk only in the GYA and is widely agreed to be associated with the management practice, primarily in Wyoming, of artificially feeding elk during winter (Thorne et al. 1978a, 1997c; Smith et al. 1997a,b). Brucellosis among elk is readily transmitted on crowded feedgrounds where normal behavior is disrupted. A single abortion on a crowded elk feedground assures exposure of many elk sharing the feedground (Thorne et al. 1997c).

Based on serologic tests, the prevalence of brucellosis in Wyoming on 23 elk feedgrounds, including the National Elk Refuge, ranges from 13% to 58% and averages about 34% (Herriges et al. 1991; Thorne et al. 1997c; Toman et al. 1997; Wyoming Game and Fish Department unpublished data). Elk are probably not capable of maintaining the disease within a population in the absence of feedgrounds. In the Wyoming portion of the GYA, seroprevalence is about 2.2% where elk winter on native range adjacent to feedground (Toman et al. 1997).

Brucellosis was experimentally transmitted from elk to susceptible cattle under conditions of close contact and the occurrence of birth events by infected elk (Thorne et al. 1979). Transmission of *B. abortus* from elk to cattle under natural conditions likely could occur only when infected pregnant elk feed during winter with cattle on a cattle feedground (Thorne et al. 1997c). Transmission to cattle during normal parturition is very unlikely to occur (Thorne et al. 1997c). A small number of outbreaks of brucellosis among cattle in Wyoming have been attributed by the U.S. Department of Agriculture, Veterinary Services, to transmission from elk or bison. Elk that winter on the National Elk Refuge are likely the source of brucellosis in the Jackson bison herd that feeds with elk during winter (Williams et al. 1993, 1997).

The fluid of *B. abortus*-infected hygromas, tendon sheaths, and joint cavities of elk may contain numerous organisms (Thorne et al. 1978b), but because the fluids are restricted to the lesion they are not likely to be important in transmission, except, perhaps, to an unwary hunter or scavenger.

Brucella abortus was transmitted to susceptible cattle following abortion by experimentally infected dogs (Crawford et al. 1990). Congenital transmission and shedding of *B. abortus* in vaginal exudates for as long as 11 days postpartum by a naturally infected coyote *Canis latrans* demonstrated a means of environmental contamination (Davis et al. 1979). Coyotes were a source of *B. abortus* to cattle only when, or shortly after, they fed on contaminated material (Davis et al. 1988). It is generally believed that wild canids and bears are susceptible to infection with *Brucella* if they scavenge on highly contaminated fetuses, placentas, and, perhaps, carcasses, but their significance in transmission and epidemiology of brucellosis is low and offset by the positive role they play in removing contaminated materials from the environment (Davis et al. 1988; Cheville et al. 1998).

Brucella melitensis causes natural infections in domestic sheep and goats. Transmission is associated with persistent shedding of large numbers of organisms following abortion or parturition (Herr 1994). Presumably, the occurrence of *B. melitensis* in free-ranging wildlife reflects transmission from infected domestic sheep and goats, although species with strong herding tendencies may experience intraspecific transmission. *Brucella ovis* infection in domestic sheep is transmitted primarily by rams, and a similar mode of transmission is suspected in red deer (Hutching 1998).

Biovars 1, 2, and 3 of *B. suis* primarily infect domestic and feral swine, and biovars 4 and 5 infect reindeer or caribou and wild rodents, respectively (Bishop and Bosman 1994). Intraspecific transmission among swine occurs mainly through coitus. Ingestion of contaminated material may also serve as a method of transmission (Bishop and Bosman 1994). The primary means of transmission among reindeer and caribou is believed to be through contact with infective uterine discharges following abortion (Dieterich 1981). Under confined conditions, *B. suis* biovar 4 was transmitted from reindeer to cattle following abortion and parturition by infected reindeer, with probable routes of exposure being oral, nasal, or conjunctival (Forbes and Tessaro 1993).

CLINICAL SIGNS. Abortion and retained placenta are the most important signs of brucellosis in bison. Corner and Connell (1958) reported numerous abortions among bison in Elk Island National Park, Canada, that presumably were due to brucellosis, and they felt low productivity may have been associated with calf loss due to brucellosis. Retained placentas were reported among bison at the National Bison Range, Montana (Creech 1930), and Yellowstone National Park (Rush 1932). More recently, abortion and retained placenta were documented in free-ranging bison in the GYA (Williams et al. 1993; Rhyan et al. 1994). High rates of abortion and retained placenta were observed in a series of studies in bison (Davis et al. 1990, 1991; D.S. Davis personal communication; Bevins et al. 1996) [summarized by Williams et al. (1997)]. Although abortions are clearly the most important sign of brucellosis in bison, the frequency of abortion in free-ranging bison, especially those of Yellowstone National Park, which have harbored *B. abortus* for 80–90 years, is unknown and disputed (Meyer and Meagher 1995, 1997).

Enlarged testicles and pendulous scrotums have been described in bison with *B. abortus* infection (Creech 1930; Corner and Connell 1958; Choquette et al. 1978). A bull described by Corner and Connell (1958) had such marked scrotal enlargement that it moved in a "crow-hop fashion" and had rubbed the hair from the medial aspects of the rear legs. Arthritis or hygroma was observed in 4% of 72 bison examined in Wood Buffalo National Park (Tessaro 1987). All of these may not have been clinically apparent, but one bull had a very large hygroma that probably inhibited mobility and contributed to emaciation.

Brucella abortus was recovered from a dead fetus in a culled African buffalo in South Africa (Gradwell et al. 1977). Pathologic changes suggested that abortion was imminent and that brucellosis will cause at least some buffalo to abort.

Effects of brucellosis in elk have been well studied; all clinical signs identified through artificial *B. abortus* infections in captive elk have also been observed in free-ranging elk in western Wyoming where *B. abortus* biovars 1 and 4 have been recovered (Murie 1951; Thorne et al. 1997c). Abortion and birth of nonviable calves were the most frequent and important signs of experimental brucellosis in captive elk; overall, loss of the first calf following infection among 95 elk was 57% (Thorne et al. 1978b, 1981; Herriges et al. 1989, 1991). One of nine cows lost a second calf. It appeared that exposure earlier in pregnancy was more likely to induce abortion than later exposure.

Retained placenta did not occur in any of the artificially infected elk. Clinical orchitis and infertility were not observed. *Brucella*-infected hygroma occurred in 18% of 65 elk 6 months of age or older but seemed to cause little or no discomfort, and the swelling often receded spontaneously. On the other hand, synovitis associated with chronic brucellosis often led to severe lameness that could have resulted in the death of free-ranging elk. Swelling was often apparent in the fetlock and sheath of a deep flexor tendon of either fore or hind limbs, and sometimes multiple limbs were affected simultaneously. Synovitis due to *B. abortus* was found in 20% of 65 elk at necropsy (Thorne et al. 1978b).

Based on high seroprevalence on winter feedgrounds in Wyoming, which suggests every female will become infected if she lives long enough, and hunting pressure, which determines turnover rate of the population, annual calf loss due to brucellosis was estimated as 12.5% prior to initiation of an elk vaccination program (Herriges et al. 1989). On the National Elk Refuge, where hunting pressure is lower and cow elk live longer, a calf loss of 7% was estimated (Smith and Robbins 1994).

Clinical signs in a moose *Alces alces* dying due to *B. abortus* infection were progressive weakness over a period of a few days (Fenstermacher and Olson 1942). Only one of four moose artificially infected with *B. abortus* biovar 1 showed clinical signs before being euthanized, even though all became infected (Forbes et al. 1996). This moose developed congested mucous membranes, fever, and weakness 83 days after inoculation. She became moribund only 2 hours before death. Forbes et al. (1996) concluded that *B. abortus* infection in moose may be fatal. Cheville et al. (1998) acknowledged that failure to detect antibodies to *Brucella* in moose in North America may contribute to the belief that *B. abortus* infection is generally fatal in moose (Thorne et al. 1978b; Hudson et al. 1980; Thorne 1982; Zarnke 1983; Dieterich et al. 1991) but questioned the veracity of that conclusion.

A coyote naturally infected with *B. abortus* aborted after being brought into captivity (Davis et al. 1979). Pups of two wolves *Canis lupus* artificially infected with *B. suis* biovar 4 died within 24 hours after birth (Neiland and Miller 1981). Although *B. suis* was recovered from the pups and their dams, it was not certain these deaths were due to brucellosis.

Brucella suis biovar 4 does not appear to be pathogenic for bison (Bevins et al. 1996). No clinical signs were observed in six inoculated bison and, at necropsy, *B. suis* was recovered from lymph nodes of only two cows.

Orchitis-epididymitis, bursitis-synovitis, and metritis-abortion syndromes are found among free-ranging Alaskan caribou infected with *B. suis* (Neiland et al. 1968). In orchitis-epididymitis syndrome, testis and epididymis become greatly enlarged. Bursitis-synovitis was characterized by animals with large hygromas called *caribou knees,* swollen hocks, and poor body condition. In the animals with metritis-abortion syndrome, retained placentas and/or excessive bleeding occurred, although it was acknowledged these signs in some animals may have been due to poor nutrition. Rausch and Huntley (1978) artificially infected pregnant reindeer with *B. suis* biovar 4. Although abortions occurred, they did not observe retained placentas and excessive bleeding.

A debilitated moose in the Northwest Territories, Canada, was infected with *B. suis* biovar 4 (Honour and Hickling 1993). Lesions were restricted to the forelegs and involved severe osteomyelitis and fractures. The emaciated cow was found kneeling in the snow and appeared to have been at that site for some time. A 9-month-old male moose artificially inoculated with *B. suis* biovar 4 experienced transient clinical signs beginning 42 days after inoculation that probably would have been fatal if it had not been captive (Dieterich et al. 1991). Fever, leukocytosis, recumbence, anorexia, and depression were the signs observed, but normal appetite and activity resumed by 58 days after inoculation. Antibodies against *B. suis* were detected in 19% of 42 cow moose in northern Alaska, suggesting that rangiferine brucellosis may not always be a serious disease in moose (O'Hara et al. 1998).

Brucella sp. was recovered from an aborted bottlenose dolphin fetus. However, guinea-pig inoculation indicated the organism was relatively avirulent (Ewalt et al. 1994), and its role in the abortion was uncertain.

PATHOGENESIS. *Brucella* are facultative intracellular parasites adapted to survival within host cells. Virulence is associated with several factors, including resistance to normal serum components, attachment and penetration into cells, and survival and multiplication within a variety of cells, including some phagocytes (Corbel 1989b). The pathogenesis and virulence of *Brucella* infections at the molecular level have been reviewed by Smith and Ficht (1990).

Brucella generally cause nonlethal infections, and *B. melitensis* is generally regarded as the most pathogenic member of the genus. Infections are usually chronic and generally persist for the life of the host. Infection in cattle with *B. abortus* may be perpetuated by establishment of an intracellular reservoir of organisms (Campbell et al. 1994). In domestic animals, and probably wild animals, the pathogenesis of infection is quite similar, regardless of species of *Brucella* involved (Enright 1990b; Samartino and Enright 1993).

Virulent strains of *B. abortus* are quite invasive through mucous membranes of the oropharynx and upper respiratory tract, conjunctiva, scarified skin, and cervix. Bacteremia follows initial localization in regional lymph nodes, resulting in bacterial seeding in phagocytes in lymph nodes and spleen. Prolonged periods of bacteremia are common, during which *Brucella* are able to survive within leukocytes and may use neutrophils and macrophages for protection against humoral and cellular bactericidal mechanisms (Enright 1990b). If the host is pregnant, the endometrium of the uterus is invaded followed by colonization of the placenta. In addition to the gravid uterus, udder, and supramammary lymph nodes, localization may occur in other lymph nodes, spleen, testes, accessory male sex glands, synovial surfaces, and carpal bursae (Enright 1990b; Bishop et al. 1994). *Brucella suis* and *B. canis,* which tend to induce prolonged bacteremias, localize in a wider variety of secondary sites than other species (Enright 1990b).

Virulent *B. abortus,* and to a lesser extent *B. melitensis* and *B. suis,* are stimulated by erythritol, a sugar that occurs in ungulate fetal placental tissues and fluids in the later months of pregnancy. It may be important in determining the site of bacterial localization (Pearce et al. 1962; Keppie et al. 1965; Enright 1990b; Smith and Ficht 1990; Samartino and Enright 1993; Bishop et al. 1994). It is likely erythritol accumulates in the placenta of wild ungulates. Although prolific growth of *B. abortus* in placental tissue may be explained by erythritol, the host's immune response undoubtedly also plays a role. Pregnancies subsequent to an initial abortion are not characterized by such extensive growth in the placenta (Corbel 1989b), and it has not been clearly established that erythritol is the primary factor that influences *Brucella* to localize and multiply in the placenta (Samartino and Enright 1993). Strain 19 is inhibited by erythritol but may colonize the bovine placenta, which suggests that factors other than erythritol may influence *Brucella* colonization (Corner and Alton 1981; Bosseray 1987; Enright 1990b).

Studies in mice *Mus musculus* did not indicate a tropism for the placenta, but rather that the placenta is a privileged site for multiplication of the few *B. abortus* carried fortuitously by blood flow to this organ (Bosseray 1987). Prolonged pregnancy of ruminants provides opportunity for extremely high growth rates of *Brucella* in placenta.

The mechanisms by which *Brucella* induce abortions are poorly understood. Enright (1990b) and Samartino and Enright (1993) reviewed proposed mechanisms of abortion. Severe placentitis was thought to prevent oxygen and nutrient delivery to the fetus and removal of waste products, resulting in fetal death. But abortions may also occur when minimal placentitis is present, and endotoxins may be an alternative explanation for abortions. Infections of the placenta and fetus may interfere with production of progesterone, which is necessary during the last half of gestation for maintenance of pregnancy. Fetal stress due to *Brucella* infection may result in elevated cortisol levels leading to decreased progesterone and increased estrogen production by the placenta; shifts toward estrogen induce the delivery process. However, S19 and a rough mutant have induced fetal infection and inflammatory reactions without inducing abortion, suggesting effects on placenta may be more important than those in the fetus (Enright 1990b).

The pathogenesis of *B. melitensis* infection is similar to the pathogenesis of *B. abortus* infection in cattle (Herr 1994). *Brucella suis* in swine has little tissue or organ predilection: following multiplication at the site of entry and bacteremia, the organism may localize in reproductive organs, skeletal system, joints, mammary glands, lymph nodes, spleen, liver, kidney, and urinary bladder (Bishop and Bosman 1994).

PATHOLOGY. Because most of the important *Brucella* infections of wild mammals are in artiodactyls, bovine brucellosis is an appropriate template for discussion of the pathology of brucellosis. Changes in the genital system merit the most attention and are reviewed by Kennedy and Miller (1993).

Following infection, bacteremia and rapid invasion of regional lymph nodes cause acute lymphadenitis (Kennedy and Miller 1993). Sinuses are infiltrated with neutrophils and eosinophils, with a gradual extensive accumulation of plasma cells.

Brucella abortus reaches the pregnant endometrium and fetal placenta during a bacteremic episode. Externally, an infected pregnant uterus appears normal. Placental lesions are not pathognomonic but are characteristic. There is a variable amount of exudate, which is odorless, slightly viscid, and contains gray-yellow flecks between the endometrium and chorion in the intercotyledonary area. Affected cotyledons are necrotic, soft, gray-yellow, and occasionally covered with sticky odorless, brown exudate (Kennedy and Miller 1993). Placental lesions are not diffuse; some cotyledons appear normal and others extensively necrotic. Fetal fluids generally appear normal. Considerable variation occurs in severity of lesions; if lesions are severe, abortion or premature birth is likely, whereas if they are mild, the calf likely is delivered at term.

Mononuclear cells and neutrophils are present in the edematous placental stroma; chorionic epithelial cells are filled with the bacteria, which also occur free in the exudate; and syncytial trophoblasts may be necrotic. Early in infection, the endometrium is relatively unaffected; later, there is severe endometritis (Kennedy and Miller 1993).

The *B. abortus*-infected fetus is usually edematous with blood-tinged fluid in the subcutis and body cavities. Abomasal content is frequently turbid, flocculent, and yellow. Pneumonia is the most important lesion and occurs to some degree in most fetuses aborted in the last half of pregnancy. Although lungs may appear grossly normal, scattered microscopic foci of bronchitis and bronchopneumonia are present. Severely affected lungs are enlarged and firm, reddened or hemorrhagic, and have fine yellow-white strands of fibrin on the pleura. Microscopic lesions range from relatively mild bronchitis to severe bronchopneumonia. Inflammatory cells are primarily mononuclear, but neutrophils are typically present. By the time abortion occurs, fetal lesions may include necrotizing arteritis, focal areas of necrosis, and granulomas, with giant cell formation in lymph nodes, liver, spleen, and kidney (Kennedy and Miller 1993).

Lesions associated with *B. abortus* infection of mammary glands are subtle. Microscopic changes usually include mononuclear inflammatory infiltrates in the interstitium and exudation of neutrophils in acini (Kennedy and Miller 1993).

Necrosis is characteristic of *B. abortus* infection of the bovine testicle (Ladds 1993). Orchitis is often acute; it may be unilateral, but infertility occurs due to mixing of inflammatory products with semen. Swelling due to inflammatory changes in the tunics and epididymis develops quickly but is limited due to toughness of the tunica albuginea, which predisposes the testis to pressure necrosis. The tunica vaginalis may become filled with fibrinopurulent exudate. Initially, the testis and epididymis may appear grossly normal, but scattered foci of necrosis soon appear that may coalesce and liquefy (Ladds 1993).

There are modest differences in the diseases and lesions caused by *B. abortus* and *B. suis,* especially in the frequency with which *B. suis* causes granulomatous lesions, its affinity for the skeleton and joints, and its tendency to remain in granulomas in the nonpregnant uterus (Kennedy and Miller 1993). Articular lesions that begin as synovitis and affect large and compound joints are common; inflammation is typically purulent or fibrinopurulent. Lumbar vertebral osteomyelitis is common. There is a tendency toward diskospondylitis and destruction of intervertebral cartilages, and lesions may extend to the meninges or fistulate to cause paravertebral abscesses (Kennedy and Miller 1993).

In spite of the large number of references to natural and artificial *Brucella* infections in wild mammals (Table 22.1), few researchers have reported the lesions of brucellosis. When described, the lesions were similar to those associated with *B. abortus* and *B. suis.* Williams et al. (1997) reviewed the pathology of *B. abortus* infection in bison that was similar to the disease in cattle. There was loss of uterine epithelium, lymphoplasmacytic infiltrates in the lamina propria, and neutrophils within uterine glands of a cow bison that aborted (Williams et al. 1993). Abdominal hemorrhages occurred in an aborted bison fetus. Microscopically, there were neutrophils, mononuclear inflammatory cells, and degenerate leukocytes in bronchioles and alveoli; meconium and amniotic debris were in the airways, and the alveolar septa were mildly thickened by increased mononuclear cells and neutrophils (Rhyan et al. 1994). Davis et al. (1990) followed progressive changes in six bison cows artificially inoculated with S2308 and concluded that fetal and placental lesions were similar to those described for cattle.

Orchitis and epididymitis in bison are similar to lesions in the bovine. Microscopic lesions in the epididymis, seminal vesicle, and ampulla of a *B. abortus* culture-positive free-ranging bison bull consisted of interstitial lymphoplasmacytic infiltrates with some neutrophils in the interstitium, in intraepithelial migration, and within glandular lumens (Williams et al. 1993). Severe unilateral necrotizing and pyogranulomatous orchitis in a young bull and mild to marked seminal vesiculitis in four bulls were found in a herd of ranched bison (Rhyan et al. 1997b).

Hygroma and arthritis occurred in *B. abortus* culture-positive bison (Tessaro 1987). One bull had a huge hygroma over the carpus and metacarpus, and others had swollen stifle joints. Grossly, arthritis consisted of lost articular cartilage with eburnation and lysis of bone, synovial villous hyperplasia, pannus formation, joint capsule thickening, and abundant viscous translucent yellow fluid containing small flecks of debris. Microscopically, there was serous arthritis with lymphoplasmacytic infiltrates and hemosiderin-laden macrophages in synovium and proliferation of synovial villi. Within the synovium and joint capsule were foci of necrosis and mineralization (Tessaro 1987). Lymphoid hyperplasia is a common but nonspecific lesion (Tessaro 1987; Davis et al. 1990; Williams et al. 1997). Histologic lesions in lymph nodes of bison inoculated with *B. abortus* vaccine strains S19 and RB51 (RB51) were similar to those reported in cattle (Olsen et al. 1997).

Testicular lesions in an African buffalo from which *B. abortus* biovar 3 was isolated were similar to those in chronically affected bovines (Kaliner and Staak 1973). The testicle was slightly enlarged, and there were large necrotic foci. Granulomatous inflammation, necrosis, and calcification in the testicular interstitium were present.

Lesions in elk with brucellosis were not dramatic nor diagnostic (Thorne 1982). Early stages of brucellosis in elk were characterized by enlarged, edematous lymph nodes; microscopically, lymphoid follicles were increased in size and number, and many neutrophils and some eosinophils were present during acute infection (Thorne 1982).

Gross changes in aborted elk fetuses were similar to those reported for cattle, except elk fetuses were frequently mutilated by elk attempting to consume them. Serosanguinous fluid often was present subcutaneously and within thoracic and abdominal cavities. Frequently, interlobular edema occurred in lungs, and meconium was in the abomasum. Fetal lungs had focal to diffuse infiltrates of lymphoreticular cells around bronchioles and associated vessels and suppurative to fibrinopurulent bronchopneumonia. A placenta was thickened and covered with purulent exudate. Necrosis and mixed inflammatory cell infiltrates occurred in the placentome and intercotyledonary epithelium. Bacteria were numerous in some necrotic chorionic and trophoblastic cells. In the infected uterus, giant cells and epithelioid macrophages were present, and uterine glands were distended by necrotic debris and neutrophils. Mild chronic endometritis may persist for many months (Thorne 1982).

Hygromas had moist, rough, red-brown internal linings and sometimes contained opaque, straw-colored fluid, and fibrinous debris. Histologically, hygromas were characterized by roughened surfaces covered with tags of fibrin in various stages of organization. Underlying tissues were edematous and infiltrated by neutrophils, plasma cells, macrophages and, occasionally, giant cells (Thorne 1982). Synovitis and arthritis were characterized by villous proliferation of synovium. Joint spaces and tendon sheaths were distended with opaque, straw-colored fluid, usually containing white fibrinous debris resembling rice kernels (Thorne 1982).

Brucella abortus-induced lesions in the reproductive tract of male elk were mild and consisted of focal to diffuse infiltrates of mononuclear cells, primarily lymphocytes and plasma cells, in the lamina propria and interstitium of the prostate, seminal vesicles, and ampullae. Mild focal epididymitis and orchitis were occasionally present (Thorne 1982).

Moose seem especially susceptible to *B. abortus* infections (Fenstermacher and Olson 1942). Pulmonary adhesions to the parietal pleura, pericarditis, fibrin on the liver, enlarged kidneys with small white foci on the cut surface, enlarged testis with thickened tunica vaginalis, and enlarged lymph nodes were observed in a moose with brucellosis (Fenstermacher and Olson 1942). *Brucella abortus* was isolated from a naturally infected moose in Montana with pericarditis, pleuritis, peritonitis, lymphadenitis, and arteritis (Jellison et al. 1953).

The most consistent lesions of experimental *B. abortus* infection in moose involved the lymph nodes, which were remarkably enlarged in some animals and had follicular hyperplasia with a few giant cells in sinusoids and subcapsular sinuses (Forbes et al. 1996). Fibrin was present on pleura, liver, other abdominal viscera, and synovium of carpal joints. Two moose had villus hypertrophy and pannus formation in joints. Forbes et al. (1996) concluded that *B. abortus* infection in moose is likely to be fatal and that moose appear to be dead-end hosts for brucellosis.

There are few reports of lesions of brucellosis in nonruminant wild mammals. Six months after inoculation with *B. abortus,* infection was detected in two of three European badgers *Meles meles,* and no lesion attributable to brucellosis were observed (Corbel et al. 1983). A few bank voles *Clethrionomys glareolus* artificially inoculated with *B. abortus* strain 544 had enlarged spleens at necropsy, but no other gross lesion of brucellosis was observed (Redwood and Corbel 1985).

Most descriptions of lesions associated with *B. suis* in wild mammals are of biovar 4 in reindeer, caribou, and moose. Bursitis, tenosynovitis, arthritis, osteomyelitis, hygroma formation, orchitis, epididymitis, metritis, mastitis, lymphadenitis, subcutaneous and visceral abscessation, and mineralized granulomas have been reported (Neiland et al. 1968; Rausch and Huntley 1978; Tessaro and Forbes 1986; Dieterich et al. 1991; Honour and Hickling 1993)

Brucella suis biovar 4 was isolated from a muskox *Ovibos moschatus* with bursitis (Gates et al. 1984). Epididymitis developed in mature white-tailed deer artificially inoculated with *B. ovis* (Barron et al. 1985). Lesions were unilateral and consisted of epididymal enlargement, adhesions, and cavities containing cream-white fluid on the cut surface. Microscopic lesions were granulomas containing some spermatozoa. Epithelial cells of many tubules were hyperplastic, and cysts or vacuoles occurred in areas of epithelial hyperplasia (Barron et al. 1985).

Porpoises, common dolphin, and striped dolphin *Stenella coeruleoalba* had subcutaneous lesions from which *Brucella* sp. was consistently isolated (Foster et al. 1996), but the lesions were not described. Lesions in a *Brucella*-infected Pacific harbor seal were associated with lungworms, which also harbored *Brucella* sp.; there was chronic inflammation and wasting (Garner et al. 1997).

DIAGNOSIS. Because of its variable incubation period, frequent absence of detectable clinical signs, lack of unique clinical or pathologic signs, and chronicity, brucellosis may be difficult to diagnose and requires demonstration of the organism or demonstration of the host's specific immune response to infection. Most attempts to diagnose brucellosis are for surveillance purposes to determine the disease's presence or absence within a population or a geographic area. Consequently, serologic tests to detect *Brucella* antibodies are most frequently used, but isolation and identification of the bacteria should be used with individuals or populations for confirmation.

Detection of *Brucella* serum antibodies is indirect evidence of infection but is quite valuable for surveillance of free-ranging mammals. Current serologic tests use *Brucella* LPS as test antigen (Cheville et al. 1998). Probably more serologic tests have been developed to diagnose *Brucella* infections than any other animal disease, which reflects the economic importance of animal brucellosis and that no single ideal test has yet been developed. Serologic tests used on free-ranging mammals were developed for domestic animals and have been validated for only a few wild mammals, notably elk (Thorne et al. 1978a; Morton 1978; Morton et al. 1981), bison (Davis et al. 1990), and coyotes (Davis et al. 1979), for which reliance on a battery rather than a single test was recommended.

Difficulty with serologic diagnosis of bovine brucellosis occurs because of problems associated with low specificity of many tests in cattle previously vaccinated with S19 and because of the occurrence of serologically nonreactive but infected animals (Alton 1978; Sutherland and Searson 1990). However, problems with serologic diagnosis of brucellosis should be rare with free-ranging wildlife because, with very rare exceptions, wildlife are not vaccinated against brucellosis; serologic tests are used with wildlife for surveillance rather than eradication programs, making sensitivity less important; and serologically nonreactive wildlife are likely sufficiently rare to be inconsequential in surveys.

Typically, antibody titers of most calves born to infected dams decline to nondetectable levels by about 6 weeks of age (Fitch et al. 1941). Thorne et al. (1978b) reported the probable occurrence of latent infection in elk born to infected dams, but latent carriers have not been reported or studied in other species of wild mammals. In addition, some infected mammals retain *Brucella* in lymphoid tissues in small numbers and in an inactive state that fails to stimulate sufficient antibodies to produce reactions to serologic tests (Cheville et al. 1998).

In cattle, and presumably in wild mammals, antibodies that cross-react with *B. abortus* antigens sometimes occur. Cross-reactions are due to exposure and/or infection with certain bacteria with antigenic similarity to *B. abortus,* usually due to similarities in the O chains of LPS antigens (Sutherland and Searson 1990). Cross-reactions have been recorded between *B. abortus* and *Yersinia enterocolitica* serogroup O9, *Escherichia coli* O116 and O157, *Salmonella* serotypes of the Kauffman-White group N, *Francisella tularensis, Vibrio cholerae,* and *Pseudomonas maltophilia* (Corbel 1989b; Sutherland and Searson 1990).

In addition, nonspecific reactions to serologic tests may occur (Sutherland and Searson 1990). Cross-reactions and nonspecific reactions, although rare, undoubtedly occur in sera of wild mammals, and their possible occurrence necessitates veterinary epidemiologic evaluation when serologic surveys of wild mammal populations for *Brucella* antibodies are conducted.

Davis et al. (1990) artificially infected bison with S2308 and evaluated serologic tests for diagnosis of brucellosis. They used the card, rivanol (Riv), standard plate test (SPT), standard tube test (STT), cold complement fixation tube (CCFT), warm complement fixation tube, buffered acidified antigen, rapid screen, bovine conjugated enzyme-linked immunosorbent assay (ELISA), bison conjugated ELISA (BisELISA), and hemolysis in gel (HIG) tests. During the 8 weeks after exposure, no single serologic test was reliable in diagnosing infection. After 8 weeks, the card test in combination with BisELISA and HIG test was a valuable battery for serologic diagnosis of brucellosis in bison. The response of the bison to common serologic tests was 2–3 weeks slower than that expected for cattle.

Serologic responses of elk naturally and experimentally infected with *B. abortus* were studied (Thorne et al. 1978a; Morton 1978; Morton et al. 1981). The SPT, complement fixation test (CFT), card, and Riv tests were used. Of the positive sera from free-ranging elk, the SPT correctly identified only 29%, the card 83%, the Riv 86%, and the CFT 85%. On 17 elk from which *B. abortus* was isolated at necropsy, the SPT correctly identified 59%, the card 94%, and the Riv 88%; the CFT correctly identified all nine culture-positive elk on which it was used. No single serologic test should be used on elk (Thorne et al. 1978a).

Nearly 3000 sera from elk infected with S2308 were evaluated by using the SPT, card, Riv, and CFT (Morton 1978; Morton et al. 1981). The serologic response of elk to infection was similar to that of cattle. Low titers to the SPT appeared within 2 weeks after inoculation; the other tests began detecting antibodies in 3 weeks. After 6 months, the percentage of elk reacting to the SPT at 1:100 rapidly declined. The CFT correctly identified more culture-positive elk than did the other tests, and CFT titers were persistent. Antibodies were detected by all four tests in some elk as long as 4 years after inoculation, but males did not seem to maintain titers as long as females. Isolation of *B. abortus* from four calves seronegative at necropsy suggested the occurrence of latent infections (Morton et al. 1981). A competitive ELISA was shown capable of distinguishing between S19-vaccinated and *B. abortus*-infected cattle (Gorrel et al. 1984; Mia et al. 1992) and has been used experimentally to identify S19-vaccinated elk in Wyoming (Smith et al. 1997b).

Davis et al. (1979) evaluated 51 sera from wild coyotes by using the card and Riv tests, STT, and CCFT, and 20%, 12%, 20%, and 19%, respectively, were positive. *Brucella abortus* was isolated from seven of the 51 coyotes but not from five serologically positive animals. Three seronegative coyotes were culture positive.

In cattle, the most frequently used method to demonstrate the presence of *Brucella* is culture. Placenta, lungs, abomasal content, liver, and spleen of fetuses or full-term calves; uterine discharge, colostrum, or milk of suspect infected cows; semen; and from carcasses the supramammary lymph node are most suitable, but retropharyngeal, mandibular, iliac, prescapular, and parotid lymph nodes and uterus, milk, udder, fluid from hygroma, male accessory sex glands, and testes may be cultured. At slaughter, 90% of infected adult cattle can be identified by culture of the supramammary lymph nodes, and recovery can approach 100% if the supramammary, parotid, mandibular, and subiliac lymph nodes are cultured (Bishop et al. 1994). Isolation of *Brucella* tends to correlate with high serologic titers; animals infected with large numbers of organisms in numerous tissues usually respond strongly to serologic tests.

In a study of experimentally infected bison, the organism was isolated from a wide variety of tissues similar to what would be expected with artificially infected cattle (Davis et al. 1990). Thorne et al. (1978a) cultured approximately 38 tissues from each of 45 elk naturally exposed to *B. abortus,* most of which were serologically positive for *Brucella* antibodies. *Brucella abortus* biovar 1 was isolated from 17 (38%). Most frequent sites of isolation were seminal vesicles, ampullae, and mandibular, suprapharyngeal, popliteal, and retropharyngeal lymph nodes. Udder and milk were not reliable sources for bacterial isolation in elk (Thorne et al. 1978b).

Culture of *Brucella* takes several days or more, and reasonably clean and fresh tissues are required. Organs considered appropriate for culture frequently harbor living organisms only intermittently. When the infection is systemic or quite active (e.g., acute infection and shortly before or after abortion in elk or bison), *Brucella* is relatively easy to isolate; but when infection is latent or chronic, *Brucella* can be very difficult to recover, especially when necropsies are conducted and tissues collected under field conditions (Cheville et al. 1998). Although culture may be quite accurate for diagnosis of acute bovine brucellosis (Bishop et al. 1994), that is not the case for free-ranging wild mammals. This has led to frustration and controversy in some situations regarding reliability of serologic tests, actual infection rates, and risk of infectivity, especially regarding bison (Meyer and Meagher 1995; Dobson and Meagher 1996; Schubert et al. 1997; Cheville et al. 1998). Discrepancies between serology and culture are most likely due to inadequate sampling or sampling during chronic infection when few bacteria are present. Inappropriate sampling, improper storage of samples, use of insufficient amounts of tissues, or overgrowth by more rapidly growing bacteria can lead to failure to culture *Brucella* from an infected animal (Cheville et al. 1998).

Diagnostic tests based on measure of cell-mediated immunity have been used in cattle (Nicoletti and Winter 1990) but are, in general, not practical or validated in wild species. Polymerase chain reaction (PCR) techniques have been developed to detect and differentiate *Brucella* DNA, even in very small amounts (Bricker and Halling 1994, 1995; Leal-Klevezas et al. 1995; Romero et al. 1995). Although PCR may have future application to the study of brucellosis in wild mammals, the technique has not yet been established for official use with domestic animals and has not been used with wildlife.

DIFFERENTIAL DIAGNOSES. Clinical and pathologic signs of *Brucella* infections are extremely variable. In most species of wild mammals known to be susceptible to *Brucella* infection, detectable signs have not been described and may not occur. These factors make differential diagnoses almost limitless. From a practical approach, differential diagnoses must be keyed on the predominant signs of brucellosis: fetal loss and retained placenta; arthritis, synovitis, osteomyelitis, and bursitis; and orchitis and epididymitis.

Numerous infectious diseases cause fetal loss, so a multidisciplinary diagnostic approach that includes pathology, microbiology, and toxicology must be used (Bishop et al. 1994). Fetal loss is rarely identified in free-ranging mammals, and with the possible exceptions of a few specific populations of elk, caribou, bison, and African buffalo, brucellosis is an unlikely cause of fetal loss. Identification of the etiologic causes of arthritis, osteomyelitis, synovitis, bursitis, and related syndromes also should be multidisciplinary.

Most diagnoses of *Brucella* infections in wild mammals are based upon serologic tests. In such cases, the possibility of cross-reaction requires that infection or previous exposure to antigenically similar bacteria (Corbel 1989b; Sutherland and Searson 1990) should be considered.

The recent recovery of *Brucella* spp. from cetaceans and pinnipeds at widespread geographic locations suggests that brucellosis should be considered when these animals are necropsied.

IMMUNITY. *Brucella* are facultative intracellular Gram-negative bacteria, and resistance to infection involves both cellular and humoral immunity. Studies of the immune response of cattle (Sutherland and Searson 1990), elk (Thorne et al. 1978a, 1981; Morton et al. 1981; Herriges et al. 1989), and bison (Davis et al. 1990, 1991) usually involved evaluation of the humoral response and/or ability to resist challenge after vaccination.

In cattle exposed to *B. abortus*, immunoglobulin M (IgM) is the initial predominant antibody but persists only a few weeks (Rice et al. 1966; Sutherland and Searson 1990). Subsequently, IgG rapidly dominates, with more IgG_1 than IgG_2 (Sutherland and Searson 1990). In some infected cattle, IgG_2 dominates; at the onset of lactation, the ratio of IgG_2 to IgG_1 temporarily increases, with selective secretion of IgG_1 into colostrum (Williams and Millar 1979). In elk artificially inoculated with S2308, nonspecific IgMs appeared first and were followed by specific IgG_1s (Morton 1978).

In cattle, there is not good correlation between serum antibodies and disease resistance. Acquired immunity attained after infection or vaccination was referred to as *relative immunity*, because it does not provide complete protection (Nelson 1977).

Cell-mediated immunity is the primary defense against *Brucella* infection [reviewed by Nicoletti and Winter (1990), Bishop et al. (1994), and Hoffman and Houle (1995)] and probably works in conjunction with antibodies specific for O polysaccharides. Protection against *B. abortus* is especially dependent on CD8+ T lymphocytes (Splitter et al. 1996). Gamma-interferon (IFN-γ) appears to play a role in resistance; cytokine interleukin 12 is also involved, probably through its induction of IFN-γ production (Zhan and Cheers 1995).

In mammals with epitheliochorial placentation, there is little or no placental transfer of maternal immunoglobulins. Instead, the neonate ingests antibodies with colostrum that are absorbed from the small intestine (Sutherland and Searson 1990). In bovine neonates, antibody levels increase dramatically following ingestion of colostrum if the dam has high antibody levels (Sutherland and Searson 1990), and colostral antibodies were shown to be protective against exposure to large doses of *B. abortus* via uterine discharges and milk in most calves (Cunningham 1977b). Similarly, elk neonates born to infected dams with elevated serologic titers also had high antibody levels after nursing (Morton 1978; Thorne et al. 1978b).

Some cattle are genetically resistant to infection with *B. abortus* (Campbell et al. 1994; Newman et al. 1996). At least two separate genes or genetic systems appear to contribute to control of antibody responses of cattle to *B. abortus* vaccination, and peripheral blood monocyte-derived macrophages of cattle bred for resistance to S2308 are more effective at controlling in vitro intracellular replication than are macrophages from susceptible cattle (Qureshi et al. 1996). Similar mechanisms of resistance may be present in wild ruminants.

PUBLIC HEALTH CONCERNS. Brucellosis is a zoonotic disease (Nicoletti 1989) and, in humans, may be caused by *B. abortus, B. melitensis, B. suis,* or *B. canis.* No clear evidence of *B. ovis* infection has been demonstrated, and *B. neotomae* has not been recorded as a human pathogen (Corbel 1989a). Although somewhat limited in geographic distribution, *B. melitensis* is the most important cause of human brucellosis (Corbel 1989a; Nicoletti 1989). *Brucella abortus* has a greater geographic distribution, but it is less pathogenic for humans and results in a greater proportion of mild, subclinical illnesses. *Brucella suis* is even more limited geographically, and most human infections are derived from swine. *Brucella suis* biovar 2 largely occurs in wild hares in continental Europe and is of low virulence in humans, but *B. suis* biovar 4 from reindeer and caribou causes disease in humans that is similar to the other more pathogenic forms of *Brucella* (Corbel 1989a; Nicoletti 1989).

Brucellosis in humans is largely a reflection of occupational hazards (e.g., abattoir workers, veterinarians, and farmers). The most important means of transmission is through contact with reproductive products from a recently aborted or parturient animal (Corbel 1989a). Exposure may be via splashing and aerosol contact with the conjunctivae, via the respiratory tract, via the oral route, or through minor skin abrasions (Corbel 1989a; Nicoletti 1989). Meat products from animals infected with *B. abortus* are not regarded as important sources of human exposure because they rarely are consumed raw and numbers of organisms contained in muscle usually are low (Corbel 1989a; Nicoletti 1989).

The prevalence of human brucellosis reflects the prevalence of brucellosis in animals with which they are associated, and human disease is best prevented by elimination of the disease in animals (Nicoletti 1989). This is illustrated by the decline in human cases in the United States that correlates with reduction of the disease in cattle (Nicoletti 1989). The annual number of human cases in the United States reported by the Centers for Disease Control and Prevention reached a high in 1947 of 6321; by 1994, the number was approximately 100 (Young and Nicoletti 1997).

Risk to humans associated with hunting *Brucella*-infected wild mammals appears to be low. Canadian reviewers concluded that risk associated with *B. abortus*-infected bison in and around Wood Buffalo National Park is not serious and is not perceived to be a problem by hunters (Environmental Assessment Panel 1990). Another example of the apparent low risk to humans occurs in the GYA. Although thousands of *B. abortus*-exposed and infected elk and, sometimes, hundreds of bison are killed and processed annually by hunters, agency personnel, and Native Americans, the known occurrence of human brucellosis associated with infected wildlife is limited to two elk hunters who harvested their animals during February north of Yellowstone National Park. The most logical explanation for lack of apparent infection of hunters has to do with the timing of hunting seasons. *Brucella abortus* localizes in the placenta of elk and bison in the later stages of pregnancy in late winter and spring. Most hunting seasons in the GYA occur in the fall, before *B. abortus* is present in large numbers in the uterus. However, hunters where *Brucella*-infected wild mammals occur should recognize that the risk of exposure increases as pregnancy progresses and should be encouraged to minimize contact with reproductive organs and fetuses, avoid opening enlarged joints and hygroma, and not consume raw tissues that might harbor *Brucella,* especially blood, bone marrow, and lymph nodes. The source of *B. suis* infections among Native Alaskans is believed to be due to handling and consumption of raw material, particularly bone marrow, of infected caribou (Toshach 1963; Brody et al. 1966; Chan et al. 1989).

The incubation period of human brucellosis is about 3 weeks, but there is a great deal of variability (Young 1989). Symptoms of human brucellosis are numerous and nonspecific, but the predominant ones include weakness, easy fatigue, malaise, anorexia, sweats (often with a curious malodor), myalgias, arthralgias, body aches, mental inattention, and depression. Fever is intermittent and, in the absence of antibiotics or antipyretics, undulant (Young 1989).

Although abortion is an important characteristic of brucellosis in domestic and wild animals, in humans it is not as important. Humans, and other *Brucella*-resistant species, have low concentrations of erythritol in the placenta (Keppie et al. 1965; Young 1989). Although human abortions occasionally occur, it appears they are due more to complications of bacteremia, such as disseminated intravascular coagulation, than to a proclivity to infection of the placenta (Corbel 1989a).

The only definitive diagnostic test for human brucellosis is bacteriologic isolation of *Brucella,* but cultures from infected individuals are not always positive, and serology frequently is relied on for diagnosis (Diaz and Moriyón 1989). The serum agglutination test (SAT) is the most frequently used serologic test for diagnosis of human brucellosis; other serologic tests include the rose bengal test, CFT, Coomb's test, and ELISA (Diaz and Moriyón 1989).

MANAGEMENT IMPLICATIONS AND DOMESTIC ANIMAL HEALTH CONCERNS. By 1985, there had been no published report of brucellosis in wild ungulates, hindering eradication of bovine brucellosis (McCorquodale and DiGiacomo 1985). However, Davis (1990) pointed out that there does not seem to be any country where brucellosis has been eradicated that also had large numbers of brucellosis-infected wildlife. In certain parts of Africa, large numbers of brucellosis-infected ungulates are reasonable sources of concern regarding health of domestic livestock (Davis 1990). The presence of brucellosis in bison in northern Canada and in both free-ranging elk and bison of the GYA has become a major socioeconomic problem that is seen by many to threaten livestock industries, goals of brucellosis eradication programs, and wildlife management (Boyce 1989; Davis 1990; Environmental Assessment Panel 1990; Thorne and Herriges 1992; Thorne et al. 1991a,b, 1995, 1996, 1997a,b; Anonymous 1992; Keiter and Froelicher 1993; Carlman 1994;

Smith and Robbins 1994; Gates et al. 1997; Keiter 1997a,b; Cheville et al. 1998).

Brucellosis in free-ranging bison is a national issue in Canada because of infected bison in and adjacent to Wood Buffalo National Park. The problem began in the 1920s when over 6000 bison known to be infected or exposed to both *B. abortus* and *Mycobacterium bovis* were translocated from Wainwright Buffalo Park in central Alberta to Wood Buffalo National Park in northern Alberta and the southeast Northwest Territories (Tessaro 1987; Environmental Assessment Panel 1990; Gates et al. 1997). Canada's national cattle herd achieved brucellosis-free status in 1985, and bovine tuberculosis was by then nearly eliminated. The desire to protect the cattle brucellosis-free status and concern about transmission of the diseases to other nearby nondiseased bison populations in the Mackenzie Delta and elsewhere raised the issue to one of national concern [reviewed by Gates et al. (1997)]. A federal Environmental Assessment Panel (1990) concluded that the diseases constituted a very small risk to cattle and other bison; fences and buffer zones to separate diseased and nondiseased populations were only a short-term solution; there was no feasible treatment, vaccination, or test and slaughter solution; the only practical solution was elimination of the diseased bison, with provisions for genetic conservation and replacement with nondiseased bison; and a multiple-stockholders group should be created to plan for disease elimination and bison reestablishment. Only the latter recommendation has been implemented. Currently, bison that enter large buffer zones separating diseased and disease-free bison populations are destroyed (Gates et al. 1997).

Brucella abortus in free-ranging elk and bison of the GYA in the United States is also an intense national issue (Thorne et al. 1996, 1997a). Conflicts between elk and cattle have their origins with the first permanent settlers in the region, probably in the 1880s. During the next several decades, elk winter range was lost to development, so elk began feeding during winter on hay stored for cattle and feeding with cattle; *B. abortus* was transmitted to elk; and, by 1912, federal legislation created the National Elk Refuge, the first of almost two dozen feedgrounds that were established by the 1980s. Artificial crowding on winter feedgrounds and elk interchange between feedgrounds resulted in high rates of infection in approximately 25,000 feedground elk and much lower infection rates among another 90,000–100,000 elk of the GYA. Conflicts between bison and cattle did not begin until much later, because bison were nearly extirpated at the time settlement began in the GYA. Early in the process of restoring bison to viable, free-ranging populations in the GYA, *B. abortus* was accidentally introduced into the region's bison (Meagher and Meyer 1994).

Cattle ranching quickly became an important part of the lifestyle and economy of the GYA, along with wildlife-related hunting and tourism. Beginning in the 1960s, animal health officials began expressing concern about brucellosis in bison and elk. Brucellosis-related conflicts began to occur in the 1970s when increasing bison numbers left Yellowstone National Park for areas where cattle were raised. Pressure on brucellosis-infected and exposed wildlife of the GYA mounted as the target date for eradication of brucellosis from all U.S. cattle approached. During the particularly severe winter of 1996–97, over 1100 bison that had left Yellowstone National Park were destroyed because they were infected or exposed to *B. abortus* (Cheville et al. 1998).

Brucella suis in feral swine will likely be a complicating factor in programs to eradicate the organism from domestic swine (Davis 1990). Forbes and Tessaro (1993) demonstrated that *B. suis* biovar 4 can be transmitted to cattle, and they cautioned against moving reindeer or caribou into areas of traditional agriculture with cattle.

According to Thimm and Wundt (1976), serologic evidence of brucellosis in a wide variety of wildlife and indigenous domestic cattle breeds suggests that brucellosis is an old disease in Africa and is self-sustained in some wildlife populations. Waghela and Karstad (1986) suggested that the use of shared watering and grazing areas between livestock and wildlife will become important in African brucellosis control programs. However, several species of African wildlife were not considered to be significant sources of *B. abortus* for domestic cattle (Bishop et al. 1994). This has been attributed to infrequent contact between wildlife and cattle, in part due to strict measures taken to prevent spread of foot-and-mouth disease from wildlife to cattle (de Vos and van Niekerk 1969; Condy and Vickers 1972; Gradwell et al. 1977; Bishop et al. 1994).

CONTROL. Although brucellosis may cause fetal wastage in some populations of free-ranging mammals, control and eradication of brucellosis among wild mammals are likely to be attempted as much because its presence is a threat to health of domestic livestock or humans or is in conflict with a brucellosis eradication program for livestock. Brucellosis in elk and bison of the GYA is an example where there are multiple justifications for its control (Thorne et al. 1997a; Cheville et al. 1998).

Without providing specifics, Davis (1990) stated there is no example of eradication of brucellosis from a free-ranging wildlife population without accompanying elimination of the wildlife population. He, therefore, predicted that wildlife will ultimately be the last reservoirs of brucellosis and that innovative prevention and control methods should be tested for efficacy in wildlife. With the exception of a few confined bison herds from which brucellosis has been eliminated, the only free-ranging populations to date for which eradication of brucellosis is being considered or seriously attempted are bison in the GYA and in Wood Buffalo National Park and elk in the GYA.

The history and procedures of the U.S. brucellosis program are reviewed by Metcalf (1986) and Frye and Hillman (1997). Similar eradication efforts were adopted elsewhere, and by 1986 at least 33 countries reported they were free of bovine brucellosis (Crawford et al. 1990).

Bison that leave Yellowstone National Park are destroyed by shooting or capture and slaughter in order to assure they do not transmit *B. abortus* to cattle (Cheville et al. 1998). Attempts to control or prevent movement of bison out of Yellowstone National Park, where they come into conflict with cattle because they harbor brucellosis, were not successful (Meagher 1989). Although destroying bison to prevent disease transmission to cattle or other bison may be regarded as a form of brucellosis control, it does nothing to reduce the prevalence of brucellosis within affected populations. Test and slaughter, in combination with vaccination, have been used to eradicate brucellosis from confined ranched and park bison herds. Logistic difficulties and general lack of public acceptance make test and slaughter and depopulation strategies unlikely to be successfully used with large free-ranging wild mammal populations.

Movements of free-ranging wild mammals cannot be controlled. Therefore, any brucellosis control or eradication effort would have to involve all susceptible species and populations simultaneously within a geographic area sufficiently large to assure no interchange with other exposed or affected populations in order to prevent reinfection.

Public education is important in brucellosis control programs for wild mammals. This is especially true where wildlife are public resources, because public acceptance, and consequently success, of a brucellosis control program require understanding of techniques used and the necessity of the program.

Davis (1990) stated that the only realistic hope for control or eradication of brucellosis in wildlife lies in prevention through vaccination. And brucellosis control or eradication programs for free-ranging wild mammals will almost certainly require massive immunization efforts as their main strategy (Cheville et al. 1998). Vaccination is the cornerstone strategy for most national programs to control or eradicate brucellosis in domestic animals (Bishop et al. 1994; Frye and Hillman 1997). Although vaccination alone may or may not eliminate brucellosis from wild mammal populations, it can bring about a certain measure of control and reduce the prevalence of disease to a level low enough that other strategies may bring about eradication (Enright and Nicoletti 1997; Cheville et al. 1998). Vaccination of free-ranging wildlife is difficult and fraught with potential problems or questions that must be addressed before implementation. Vaccines must be effective in the target species, especially at reducing *Brucella*-induced abortion or other relevant means of shedding the organism into the environment; the vaccine must be safe in the intended host and appropriate nontarget species; and there must be a cost-effective and humane delivery system.

It is likely there has been more research effort on vaccines against brucellosis than any other disease of domestic livestock, but research on vaccines for use in free-ranging mammals is limited. The most widely used *Brucella* vaccine is S19, a live attenuated bacterium (Worthington et al. 1974; Nicoletti 1990; Sutherland and Searson 1990). If vaccination levels of 80% of the cows are maintained, a gradual reduction in the prevalence of brucellosis can be expected (Bishop et al. 1994).

Brucella melitensis Rev. 1 is an attenuated strain used primarily as a vaccine in sheep and goats (Nicoletti 1990). *Brucella suis* strain 2 is a stable strain of biovar 1 with approximately the same virulence as S19 (Xin 1986). It was developed and is used in China as an oral vaccine that is reported to provide good protection against bovine brucellosis without causing persistent antibodies (Nicoletti 1990).

In the early 1980s, a rough mutant of smooth pathogenic S2308 was discovered that in culture remained stable and was highly deficient in O side-chain LPS; it was named SRB51 (Schurig et al. 1991). Strain RB51 is less virulent than S19 in mice, guinea pigs, cattle, and goats (Cheville et al. 1992; Enright and Nicoletti 1997). Probably the single most important aspect about SRB51 is that it does not cause positive reactions to standard serologic tests (Enright and Nicoletti 1997).

In a controlled study in which 92 pregnant bison were vaccinated manually or ballistically with S19, 58% aborted due to S19 at about 60 days after vaccination (Davis et al. 1991). In addition, one female aborted during the second pregnancy and S19 was recovered, indicating chronic infection. Vaccine-induced antibody titers persisted at least 10 months in 73% of the bison. Although protection against abortion after challenge with S2308 was 67% for vaccinates compared to 4% in 30 nonvaccinated controls and protection against infection was 39% in vaccinates compared to 0% in controls, evaluation of protection was confounded by the fact that many of the bison had aborted due to S19 during the previous pregnancy. Davis et al. (1991) concluded S19 at the doses evaluated was not suitable for pregnant bison. However, S19 has been used since the 1960s as a calfhood vaccine in commercial bison herds without adverse effects (Cheville et al. 1998).

Strain RB51 has been evaluated in bison and may be a more suitable vaccine. It was shown to have a tropism for the placenta and to cause placentitis, which can cause abortion (Palmer et al. 1996). However, the abortion-inducing potential of SRB51 was less than that of S19 (Davis et al. 1991; Palmer et al. 1996). Olsen et al. (1997) compared pathologic and serologic response and bacterial persistence in six 3-month-old bison vaccinated with SRB51 against three vaccinated with S19. Strain RB51-vaccinated bison did not develop antibodies detected by the SAT. Lesions in lymph nodes of both groups of vaccinates were similar to those observed in cattle, but bison did not clear SRB51 as quickly as cattle.

According to models, brucellosis in bison increases with population density, and slightly over 50% of a bison population would have to be effectively vaccinated in order to eradicate brucellosis by vaccination alone (Dobson and Meagher 1996). In addition, Dobson and Meagher (1996) examined the recent policy of removal of bison that leave Yellowstone National Park in light of historical records and relationships among population sizes, recruitment, culling intensity, and prevalence of brucellosis and predicted that the bison would be nearly eradicated before prevalence was reduced significantly.

The response of elk to vaccination with S19 followed by challenge with S2308 was evaluated in controlled studies from 1979 through 1989, and efficacy of S19 in free-ranging elk on winter feedgrounds was measured beginning in 1985 (Thorne et al. 1981; Angus 1989; Herriges et al. 1989, 1991; Enright and Nicoletti 1997; Thorne et al. 1997c; Smith et al. 1997b). Ten separate controlled studies using over 150 calf and pregnant elk vaccinated manually or ballistically, and nonvaccinated controls were used. There was no local adverse reaction to inoculation with S19, but approximately 27% of the vaccinated pregnant elk aborted, at least in part due to S19, using the standard bovine calfhood vaccination dose of S19. Vaccine-induced abortions did not occur when reduced doses of S19 were used and stress was reduced in vaccinated elk. Some trials were hampered by small numbers of elk, especially controls, and a few trials showed that little or no protection was afforded by S19. Overall, however, the vaccine trials indicated that S19 protected elk against abortion (60%–62% live calves by vaccinates compared to 31%–33% live calves by controls) and reduced infection rates when vaccinates and controls were challenged with S2308 (Thorne et al. 1981; Herriges et al. 1989; Enright and Nicoletti 1997). Serologic response of elk to S19 was similar to that of cattle, and a few elk maintained detectable serologic responses to S19 up to 2 years after vaccination (Thorne et al. 1981; Herriges et al. 1989).

Ballistic vaccination of free-ranging feedground elk in Wyoming was evaluated using air-gun-delivered hydroxypropyl methylcellulose biobullets with a hollow cavity in the rear into which a lactose pellet containing lyophilized S19 was loaded (Angus 1989; Herriges et al. 1989, 1991; Smith et al. 1997). Vaccination was conducted from a horse-drawn feed sled or tractor-drawn feed wagon. Calves of both sexes and cow elk were vaccinated the first 2–4 years, after which only calves were vaccinated. On one feedground where approximately 1000 elk winter, a retrospective evaluation during a period when other management practices did not change indicated S19 vaccination reduced the *Brucella* seroprevalence by at least 50% in mature cow elk, from 46% before vaccination to 8%–23% after vaccination (Smith et al. 1997b; Cheville et al. 1998).

Killed *B. abortus* strain 45/20 vaccine with added adjuvant was evaluated in a small number of pregnant reindeer that were challenged with *B. suis* biovar 4 (Dieterich et al. 1981). All vaccinates produced live fawns, but one was nonviable; *B. suis* was recovered from only one vaccinate. The authors concluded that this vaccine increased resistance of reindeer to brucellosis.

In Wyoming, habitat enhancements are being conducted to encourage feedground elk to spend less time on feedgrounds and use native winter ranges, especially in early spring where occurrence of a *B. abortus*-induced abortion is less likely to result in intraspecific transmission (Smith et al. 1997b; Thorne et al. 1997c).

Mechanisms of genetic resistance by cattle to infection with *B. abortus* have been studied (Campbell et al. 1994; Qureshi et al. 1996). The frequency of genetically controlled natural resistance to *B. abortus* was approximately 18% (Templeton and Adams 1990), and similar genetic resistance in other ruminants has been suggested. It is difficult at this time to imagine how increasing genetic controlled resistance, requiring controlled breeding, could be used to control brucellosis in free-ranging mammals.

Finally, because of multiagency jurisdictional nature of brucellosis in the GYA, the Greater Yellowstone Interagency Brucellosis Committee was formed in 1994–95. It consists of an executive committee and technical and information and education subcommittees and has a goal, mission, and guidelines intended to protect the integrity of the region's elk and bison populations and its livestock industries while planning for the eventual elimination of brucellosis (Keiter 1997b; Petera et al. 1997; Thorne et al. 1996, 1997a).

LITERATURE CITED

Adrian, W.J., and R.E. Keiss. 1977. Survey of Colorado's wild ruminants for serologic titers to brucellosis and leptospirosis. *Journal of Wildlife Diseases* 13:429–431.

Agab, H., B. Abbas, H. el Jack Ahmed, and I.E. Maoun. 1994. First report on the isolation of *Brucella abortus* biovar 3 from camel (*Camelus dromedarius*) in the Sudan. *Revue Elevage et de Medicine Veterinaire des Pays Tropicaux* 47:361–363.

Aguirre, A.A., D.E. Hansen, E.E. Starkey, and R.G. McLean. 1995. Serologic survey of wild cervids for potential disease agents in selected national parks in the United States. *Preventive Veterinary Medicine* 21:313–322.

Alexander, B., P.R. Schnurrenberger, and R.R. Brown. 1981. Numbers of *Brucella abortus* in the placenta, umbilicus and fetal fluid of two naturally infected cows. *Veterinary Record* 108:500–501.

Al-Khalaf, S., and A. el-Khaladi. 1989. Brucellosis of camels in Kuwait. *Comparative Immunology, Microbiology and Infectious Diseases* 12:1–4.

Alton, G.G. 1978. Recent developments in vaccination against bovine brucellosis. *Australian Veterinary Journal* 54:551–557.

Alton, G.G., L.M. Jones, R.D. Angus, and J.M. Verger. 1988. *Techniques for the brucellosis laboratory.* Paris: Institut National de la Recherche Agronomic, 190 pp.

Angus, R.D. 1989. Preparation, dosage, delivery, and stability of a *Brucella abortus* strain 19 vaccine ballistic implant. *Proceedings of the United States Animal Health Association* 93:656–666.

Anonymous. 1992. *Wildlife management: Many issues unresolved in Yellowstone bison-cattle brucellosis conflict.* Washington, DC: U.S. General Accounting Office, 35 pp.

Armstrong, J.B. 1983. Report of the committee on brucellosis. *Proceedings of the United States Animal Health Association* 87:162–177.

Baker, M.F., G.J. Dills, and F.A. Hayes. 1962. Further experimental studies on brucellosis in white-tailed deer. *Journal of Wildlife Management* 25:27–31.

Bang, B. 1897. The etiology of epizootic abortion. *Journal of Comparative Pathology and Therapy* 10:125.

Barron, S.J., A.A. Kocan, R.J. Morton, T.R. Thedford, and C.S. McCain. 1985. Susceptibility of male white-tailed deer (*Odocoileus virginianus*) to *Brucella ovis* infection. *American Journal of Veterinary Research* 46:1762–1764.

Becker, H.N., R.C. Belden, T. Breault, M.J. Burridge, W.B. Frankenberger, and P. Nicoletti. 1978. Brucellosis in feral swine in Florida. *Journal of the American Veterinary Medical Association* 173:1181–1182.

Berger, J., and C. Cunningham. 1994. *Bison: Mating and conservation in small populations.* New York: Columbia University Press, 330 pp.

Bevins, J.S., J.E. Blake, L.G. Adams, J.W. Templeton, J.K. Morton, and D.S. Davis. 1996. The pathogenicity of *Brucella suis* biovar 4 for bison. *Journal of Wildlife Diseases* 32:581–585.

Bigler, W.J., G.L. Hoff, W.H. Hemmert, J.A. Tomas, and H.T. Janowski. 1977. Trends of brucellosis in Florida: An epidemiologic review. *American Journal of Epidemiology* 105:245–251.

Binninger, C.E., J.J. Beecham, L.A. Thomas, and L.D. Winward. 1980. A serologic survey for selected infectious diseases of black bears in Idaho. *Journal of Wildlife Diseases* 16:423–430.

Bishop, G.C., and P.P. Bosman. 1994. *Brucella suis* infection. In *Infectious diseases of livestock with special reference to South Africa,* vol. 2, ed. J.A.W. Coetzer, G.R. Thomson, and R.C. Tustin. Oxford: Oxford University Press, pp. 1076–1077.

Bishop, G.C., P.P. Bosman, and S. Herr. 1994. Bovine brucellosis. In *Infectious diseases of livestock with special reference to South Africa,* vol. 2, ed. J.A.W. Coetzer, G.R. Thomson, and R.C. Tustin. Oxford: Oxford University Press, pp. 1053–1066.

Bosseray, N. 1987. *Brucella* infection and immunity in placenta. *Annals of the Institute Pasteur Microbiology* 138:110–1113.

Bouvier, G. 1964. Distribution geographique de quelques maladies du gibier et des animaux sauvages de la Suisse. *Bulletin de l'Office International des Epizooties* 61:67–89.

Boyce, M.S. 1989. *The Jackson elk herd: Intensive wildlife management in North America.* New York: Cambridge University Press, 306 pp.

Bricker, B.J., and S.M. Halling. 1994. Differentiation of *Brucella abortus* bv. 1, 2, and 4, *Brucella melitensis, Brucella ovis,* and *Brucella suis* bv. 1 by PCR. *Journal of Clinical Microbiology* 32:2660–2666.

———. 1995. Enhancement of the *Brucella* AMOS PCR assay for differentiation of *Brucella abortus* vaccine strains S19 and RB51. *Journal of Clinical Microbiology* 33:1640–1642.

Brody, J.A., B. Huntley, T.M. Overfield, and J. Maynard. 1966. Studies of human brucellosis in Alaska. *Journal of Infectious Diseases* 116:263–269.

Broughton, E., L.P.E. Choquette, J.G. Cousineau, and F.L. Miller. 1970. Brucellosis in reindeer, *Rangifer tarandus* L., and the migratory barren-ground caribou, *Rangifer tarandus groenlandicus* L. in Canada. *Canadian Journal of Zoology* 48:1023–1027.

Bruce, D. 1887. Note on the discovery of a microorganism in Malta fever. *Practitioner* 39:161–170.

Campbell, G.A., L.G. Adams, and B.A. Sowa. 1994. Mechanisms of binding of *Brucella abortus* to mononuclear phagocytes from cows naturally resistant or susceptible to brucellosis. *Veterinary Immunology and Immunopathology* 41:295–306.

Carlman, L.R. 1994. Wildlife-private property damage law. *Land and Water Law Review* 29:89–115.

Chan, J., C. Baxter, and W.M. Wenman. 1989. Brucellosis in an Inuit child, probably related to caribou meat consumption. *Scandinavian Journal of Infectious Diseases* 21:337–338.

Cheville, N.F., D.G. Rogers, W.L. Deyoe, E.S. Krafsur, and J.C. Cheville. 1989. Uptake and excretion of *Brucella abortus* in tissues of the face fly (*Musca autumnalis*). *American Journal of Veterinary Research* 50:1302–1306.

Cheville, N.F., A.E. Jensen, S.M. Halling, F.M. Tatum, D.C. Morfitt, S.G. Hennager, W.M. Freriches, and G. Schurig. 1992. Bacterial survival, lymph node changes, and immunologic responses of cattle vaccinated with standard and mutant strains of *Brucella abortus. American Journal of Veterinary Research* 53:1881–1888.

Cheville, N.F., D.R. McCullough, and L.R. Paulson. 1998. *Brucellosis in the Greater Yellowstone Area.* Washington, DC: National Research Council, National Academy Press, 186 pp.

Choquette, L.P.E., E. Broughton, A.A. Currier, J.G. Cousineau, and N.S. Novakowski. 1978. Parasites and diseases of bison in Canada: IV. Serologic survey for brucellosis in bison in northern Canada. *Journal of Wildlife Diseases* 14:329–332.

Christiansen, M., and A. Thomsen. 1956. A contribution to surveying of the spread of brucellosis in hares of Denmark. *Nordisk Veterinaermedicin* 8:841–858.

Condy, J.B., and D.B. Vickers. 1969. The isolation of *Brucella abortus* from a waterbuck (*Kobus ellipsiprymnus*). *Veterinary Record* 85:200.

———. 1972. Brucellosis in Rhodesian wildlife. *Journal of the South African Veterinary Association* 43:175–179.

Cook, I., R.W. Campbell, and G. Barrow. 1966. Brucellosis in North Queensland rodents. *Australian Veterinary Journal* 42:5–8.

Corbel, M.J. 1989a. Brucellosis: Epidemiology and prevalence worldwide. In *Brucellosis: Clinical and laboratory aspects,* ed. E.J. Young and M.J. Corbel. Boca Raton, FL: CRC, pp. 25–40.

———. 1989b. Microbiology of the genus *Brucella.* In *Brucellosis: Clinical and laboratory aspects,* ed. E.J. Young and M.J. Corbel. Boca Raton, FL: CRC, pp. 53–72.

———. 1997. Brucellosis: An overview. *Emerging Infectious Diseases* 3:213–221.

Corbel, M.J., and W.J.B. Morgan. 1982. Classification of the genus *Brucella:* The current position. *Revue Scientifique et Technique O.I.E.* 1:281–289.

Corbel, M.J., J.A. Morris, C.J. Thorns, and D.W. Redwood. 1983. Response of the badger (*Meles meles*) to infection with *Brucella abortus. Research in Veterinary Science* 34:296–300.

Corey, R.R., L.J. Paulissen, and D. Swartz. 1964. Prevalence of brucellae in the wildlife of Arkansas. *Journal of Wildlife Diseases* 36:1–9.

Corner, L.A., and G.G. Alton. 1981. Persistence of *Brucella abortus* strain 19 infection in adult cattle vaccinated with reduced doses. *Research in Veterinary Science* 31:342–344.

Corner, A.H., and R. Connell. 1958. Brucellosis in bison, elk, and moose in Elk Island National Park, Alberta, Canada. *Canadian Journal of Comparative Medicine* 22:9–21.

Crawford, R.P., L.G. Adams, and B.E. Richardson. 1988. Correlation of field strain exposure with new cases of brucellosis in six beef herds vaccinated with strain 19. *Journal of the American Veterinary Medical Association* 192:1550–1552.

Crawford, R.P., J.D. Huber, and B.S. Adams. 1990. Epidemiology and surveillance. In *Animal brucellosis,* ed. K. Nielsen and J.R. Duncan. Boca Raton, FL: CRC, pp. 131–151.

Creech, B.T. 1930. *Brucella abortus* infection in a male bison. *North American Veterinarian* 11:35–36.

Cunningham, B. 1977a. A difficult disease called brucellosis. In *Bovine brucellosis: An international symposium,* ed. R.P. Crawford and R.J. Hidalgo. College Station: Texas A&M University Press, pp. 11–20.

———. 1977b. Protective effects of colostral antibodies to *Brucella abortus* on strain 19 vaccination and field infection. *Veterinary Record* 101:521–524.

Daniel, M.J. 1967. A survey of diseases in fallow, Virginia and Japanese deer, chamois, tahr and feral goats and pigs in New Zealand. *New Zealand Journal of Science* 10:949–963.

Davis, D.S. 1990. Brucellosis in wildlife. In *Animal brucellosis,* ed. K. Nielsen and J.R. Duncan. Boca Raton, FL: CRC, pp. 321–334.

Davis, D.S., W.J. Boeer, J.P. Mims, F.C. Heck, and L.G. Adams. 1979. *Brucella abortus* in coyotes: I. A serologic and bacteriologic survey in eastern Texas. *Journal of Wildlife Diseases* 15:367–372.

Davis, D.S., F.C. Heck, J.D. Williams, T.R. Simpson, and L.G. Adams. 1988. Interspecific transmission of *Brucella abortus* from experimentally infected coyotes (*Canis latrans*) to parturient cattle. *Journal of Wildlife Diseases* 24:533–537.

Davis, D.S., J.W. Templeton, T.A. Ficht, J.D. Williams, J.D. Kopec, and L.G. Adams. 1990. *Brucella abortus* in captive bison: I. Serology, bacteriology, pathogenesis, and transmission to cattle. *Journal of Wildlife Diseases* 26:360–371.

Davis, D.S., J.W. Templeton, T.A. Ficht, J.D. Huber, R.D. Angus, and L.G. Adams. 1991. *Brucella abortus* in Bison: II. Evaluation of strain 19 vaccination of pregnant cows. *Journal of Wildlife Diseases* 27:258–264.

de Vos, V., and C.A.W.J. van Niekerk. 1969. Brucellosis in the Kruger National Park. *Journal of the South Africa Veterinary Medical Association* 40:331–334.

Diaz, R., and I. Moriyón. 1989. Laboratory techniques in the diagnosis of human brucellosis. In *Brucellosis: Clinical and laboratory aspects,* ed. E.J. Young and M.J. Corbel. Boca Raton, FL: CRC, pp. 73–83.

Diaz, R., L.M. Jones, D. Leong, and J.B. Wilson. 1968. Surface antigens of smooth brucellae. *Journal of Bacteriology* 96:893–901.

Dieterich, R.A. 1981. Brucellosis. In *Alaskan wildlife diseases,* ed. R.A. Dieterich. Fairbanks: University of Alaska, pp. 53–58.

Dieterich, R.A., B.L. Deyoe, and J.K. Morton. 1981. Effects of killed *Brucella abortus* strain 45/20 vaccine on reindeer later challenge exposed with *Brucella suis* type 4. *American Journal of Veterinary Research* 42:131–134.

Dieterich, R.A., J.K. Morton, and R.L. Zarnke. 1991. Experimental *Brucella suis* biovar 4 infection in a moose. *Journal of Wildlife Diseases* 27:470–472.

Dobson, A., and M. Meagher. 1996. The population dynamics of brucellosis in the Yellowstone National Park. *Ecology* 77:1026–1036.

Dolan, L.A. 1980. Latent carriers of brucellosis. *Veterinary Record* 106:241–243.

Enright, F.M. 1990a. Mechanisms of self cure in *Brucella abortus* infected cattle. In *Advances in brucellosis research,* ed. L.G. Adams. College Station: Texas A&M University Press, pp. 191–196.

———. 1990b. The pathogenesis and pathobiology of *Brucella abortus* infection in domestic animals. In *Animal brucellosis,* ed. K. Nielsen and J.R. Duncan. Boca Raton, FL: CRC, pp. 301–320.

Enright, F.M., and P. Nicoletti. 1997. Vaccination against brucellosis. In *Brucellosis, bison, elk, and cattle in the Greater Yellowstone Area: Defining the problem, exploring solutions,* ed. E.T. Thorne, M.S. Boyce, P. Nicoletti, and T.J. Kreeger. Cheyenne: Wyoming Game and Fish Department, pp. 86–95.

Environmental Assessment Panel. 1990. *Northern diseased bison: Report 35.* Hull, Quebec: Federal Environmental Assessment Review Office, 47 pp.

Ewalt, D.R., J.B. Payeur, B.M. Martin, D.R. Cummins, and W.G. Miller. 1994. Characteristics of a *Brucella* species from a bottlenose dolphin (*Tursiops truncatus*). *Journal of Veterinary Diagnostic Investigation* 6:448–452.

Fay, L.D. 1961. The current status of brucellosis in white-tailed and mule deer in the United States. *Transactions of the North American Wildlife and Natural Resources Conference* 26:203–211.

Fenstermacher, R. 1937. Further studies of diseases affecting moose: II. *Cornell Veterinarian* 27:25–37.

Fenstermacher, R., and O.W. Olson. 1942. Further studies of diseases affecting moose: III. *Cornell Veterinarian* 32:241–254.

Fenstermacher, R., O.W. Olson, and B.S. Pomery. 1943. Some diseases of white-tailed deer in Minnesota. *Cornell Veterinarian* 33:323–332.

Ferguson, M.A. 1997. Rangiferine brucellosis on Baffin Island. *Journal of Wildlife Diseases* 33:536–543.

Fitch, W.L. Boyd, M.D. Kelly, and L.M. Bishop. 1941. An extended study of female offspring of positive Bang's diseased cattle. *Journal of the American Veterinary Medical Association* 99:413.

Forbes, L.B., and S.V. Tessaro. 1993. Transmission of brucellosis from reindeer to cattle. *Journal of the American Veterinary Medical Association* 203:289–294.

Forbes, L.B., S.V. Tessaro, and W. Lees. 1996. Experimental studies on *Brucella abortus* in moose (*Alces alces*). *Journal of Wildlife Diseases* 32:94–104.

Foreyt, W.J., T.C. Smith, J.F. Evermann, and W.E. Heimer. 1983. Hematologic, serum chemistry and serologic values of Dall's sheep (*Ovis dalli dalli*) in Alaska. *Journal of Wildlife Diseases* 19:136–139.

Foster, G., K.L. Jahans, R.J. Reid, and H.M. Ross. 1996. Isolation of *Brucella* species from cetaceans, seals and an otter. *Veterinary Record* 138:583–586.

Frye, G.H., and B.R. Hillman. 1997. National Cooperative Brucellosis Eradication Program. In *Brucellosis, bison, elk, and cattle in the Greater Yellowstone Area: Defining the problem, exploring the solutions,* ed. E.T. Thorne, M.S. Boyce, P. Nicoletti, and T.J. Kreeger. Cheyenne: Wyoming Game and Fish Department, pp. 79–85.

Gargani, G., and F. Tolari. 1986. *Brucella* phagotypes: Their relation to the spread of infection in Italy. *European Journal of Epidemiology* 2:67–79.

Garner, M.M., D.M. Lambourn, S.J. Jeffries, P.B. Hall, J.C. Rhyan, D.R. Ewalt, L.M. Polzin, and N.F. Cheville. 1997. Evidence of *Brucella* infection in *Parafilaroides* lungworms in a Pacific harbor seal (*Phoca vitulina richardsi*). *Journal of Veterinary Diagnostic Investigation* 9:298–303.

Gates, C.C., G. Wobeser, and L.B. Forbes. 1984. Rangiferine brucellosis in a muskox, *Ovibos moschatus moschatus* (Zimmermann). *Journal of Wildlife Diseases* 20:233–234.

Gates, C.C., B.T. Elkin, and L.N. Carbyn. 1997. The diseased bison issue in northern Canada. In *Brucellosis, bison, elk, and cattle in the Greater Yellowstone Area: Defining the problem, exploring solutions,* ed. E.T. Thorne, M.S. Boyce, P. Nicoletti, and T.J. Kreeger. Cheyenne: Wyoming Game and Fish Department, pp. 120–132.

Geist, V. 1982. Adaptive behavioral strategies. In *Elk of North America: Ecology and management,* ed. J.W. Thomas and D.E. Toweill. Harrisburg, PA: Stackpole Books, pp. 219–277.

Golosov, I.M., and V.A. Zabrodin. 1959. Brucellosis in reindeer. *Veterinariya (Moscow)* 36:23–25.

Gorrell, M.D., G.L. Milliken, B.J. Anderson, and A. Pucci. 1984. An enzyme immunoassay for bovine brucellosis using a monoclonal antibody specific for field strains of *Brucella abortus. Developments in Biological Standardization* 56:491–494.

Gradwell, D.V., A.P. Schutte, C.A.W.J. van Niekerk, and D.J. Roux. 1977. The isolation of *Brucella abortus* biotype I from African buffalo in the Kruger National Park. *Journal of the South Africa Veterinary Association* 48:41–43.

Greth, A., D. Calvez, M. Vassart, and P.C. Lefevre. 1992. Serological survey for bovine bacterial and viral pathogens in captive oryx (*Oryx leucoryx* Pallas, 1776). *Revue Scientifique et Technique O.I.E.* 11:1163–1168.

Hagan, W.A. 1922. The susceptibility of mice and rats to infection with *Brucella abortus. Journal of Experimental Medicine* 36:727–731.

Hayes, F.A., W.T. Gerard, E.B. Shotts, and G.J. Dills. 1960. Brucellosis in white-tailed deer. *Journal of the American Veterinary Medical Association* 137:190–191.

Herr, S. 1994. *Brucella melitensis* infection. In *Infectious diseases of livestock with special reference to South Africa,* ed. J.A.W. Coetzer, G.R. Thomson, and R.C. Tustin. Oxford: Oxford University Press, pp. 1073–1075.

Herr, S., and C. Marshall. 1981. Brucellosis in free-living African buffalo (*Syncerus caffer*): A serological survey. *Onderstepoort Journal of Veterinary Research* 48:133–134.

Herriges, J.D., Jr., E.T. Thorne, S.L. Anderson, and H.A. Dawson. 1989. Vaccination of elk in Wyoming with reduced dose strain 19 *Brucella:* Controlled studies and ballistic implant field trials. *Proceedings of the United States Animal Health Association* 93:640–653.

Herriges, J.D., Jr., E.T. Thorne, and S.L. Anderson. 1991. Brucellosis vaccination of free-ranging elk (*Cervus elaphus*) on western Wyoming feedgrounds. In *The biology of deer,* ed. R.D. Brown. New York: Springer-Verlag, pp. 107–112.

Hoff, G.L., W.J. Bigler, D.O. Trainer, J.G. Debbie, G.M. Brown, W.G. Winkler, S.H. Richards, and M. Reardon. 1974. Survey of selected carnivore and opossum serums for agglutinins to *Brucella canis. Journal of the American Veterinary Medical Association* 165:830–831.

Hoffman, E.M., and J.J. Houle. 1995. Contradictory roles for antibody and complement in the interaction of *Brucella abortus* with its host. *Critical Reviews in Microbiology* 21:153–163.

Holt, J.G., N.R. Krieg, P.H.A. Sneath, J.T. Staley, and S.T. Williams. 1994. *Bergey's manual of determinative bacteriology.* Baltimore: Williams and Wilkins, 787 pp.

Honess, R.F., and K.B. Winter. 1956. *Diseases of wildlife in Wyoming.* Cheyenne: Wyoming Game and Fish Commission, 279 pp.

Honour, S., and K.M. Hickling. 1993. Naturally occurring *Brucella suis* biovar 4 infection in a moose (*Alces alces*). *Journal of Wildlife Diseases* 29:596–598.

Hoq, M.A. 1978. A serologic survey of *Brucella* agglutinins in wildlife and sheep. *California Veterinarian* 32:15–17.

Hubálek, Z., Z. Juricova, I. Svobodova, and J. Halouzka. 1993. A serologic survey for some bacterial and viral zoonoses in game animals in the Czech Republic. *Journal of Wildlife Diseases* 29:604–607.

Hudson, M., K.N. Childs, D.F. Halter, K.K. Fujino, and K.A. Hudson. 1980. Brucellosis in moose (*Alces alces*): A serologic survey in an open range cattle area of north central British Columbia recently infected with bovine brucellosis. *Canadian Veterinary Journal* 21:47–49.

Huntley, B.E., R.N. Phillip, and J.E. Maynard. 1963. Survey of brucellosis in Alaska. *Journal of Infectious Diseases* 112:100–106.

Hutching, B. 1998. *Brucella ovis* discovered in deer. *Deer Farmer* 150:32.

Jacotot, H., and E. Valle. 1951. Deuziéme cas d'infection brucellique du liévre en France. *Annales de l'Institut Pasteur (Paris)* 80:214–215.

Jellison, W.L., C.W. Fishel, and E.L. Cheatum. 1953. Brucellosis in a moose, *Alces americanus. Journal of Wildlife Management* 17:217–218.

Jensen, A.E., N.F. Cheville, C.O. Thoen, A.P. MacMillan, and W.G. Miller. 1999. Genomic fingerprinting and development of a dendrogram for *Brucella* spp. isolated from seals, porpoises, and dolphins. *Journal of Veterinary Diagnostic Investigation* 11:152–157.

Kaliner, G., and C. Staak. 1973. A case of orchitis caused by *Brucella abortus* in the African buffalo. *Journal of Wildlife Diseases* 9:251–253.

Keiter, R.B. 1997a. Brucellosis and law in the Greater Yellowstone Area. In *Brucellosis, bison, elk, and cattle in the Greater Yellowstone Area: Defining the problem, exploring solutions,* ed. E.T. Thorne, M.S. Boyce, P. Nicoletti, and T.J. Kreeger. Cheyenne: Wyoming Game and Fish Department, pp. 181–189.

———. 1997b. Greater Yellowstone's bison: Unraveling of an early American wildlife conservation achievement. *Journal of Wildlife Management* 61:1–11.

Keiter, R.B., and R.H. Froelicher. 1993. Bison, brucellosis, and law in the Greater Yellowstone Ecosystem. *Land and Water Law Review* 28:1–75.

Kennedy, P.C., and R.D. Miller. 1993. The female genital system. In *Pathology of domestic animals,* vol. 3, ed. K.V.F. Jubb, P.C. Kennedy, and N. Palmer, 4th ed. New York: Academic, pp. 349–470.

Keppie, J., A.E. Williams, K. Witt, and H. Smith. 1965. The role of erythritol in the tissue localization of the brucellae. *British Journal of Experimental Pathology* 46:104–108.

Kuzdas, C.D., and E.V. Morse. 1954. The survival of *Brucella abortus,* U.S.D.A. strain 2308, under controlled conditions in nature. *Cornell Veterinarian* 44:216–228.

Ladds, P.W. 1993. The male genital system. In *Pathology of domestic animals,* vol. 3, ed. K.V.F. Jubb, P.C. Kennedy, and N. Palmer, 4th ed. New York: Academic, pp. 471–529.

La Parik, R.D., and R. Moffat. 1982. Latent bovine brucellosis. *Veterinary Record* 111:578–579.

Leal-Klevezas, D.S., I.O. Martinez-Vazquez, A. Lopez-Merino, and J.P. Martinez-Soriano. 1995. Single-step PCR for detection of *Brucella* spp. from blood and milk of infected animals. *Journal of Clinical Microbiology* 33:3087–3090.

Lee, A.M., and M.E. Turner. 1937. A comparison of the tube and plate methods of testing for Bang's disease in elk. *Journal of the American Veterinary Medical Association* 90:637–640.

Lord, V.R., and R. Flores. 1983. *Brucella* spp. from the capybara (*Hydrochaeris hydrochaeris*) in Venezuela: Serologic studies and metabolic characterization of isolates. *Journal of Wildlife Diseases* 19:308–314.

Mayfield, J.E., J.A. Bantle, D.R. Ewalt, V.P. Meador, and L.B. Tabatabai. 1990. Detection of *Brucella* cells and cell components. In *Animal brucellosis,* ed. K. Nielsen and J.R. Duncan. Boca Raton, FL: CRC, pp. 97–120.

McCaughey, W.J. 1969. Brucellosis in wildlife. *Symposia of the Zoological Society of London* 24:99–105.

McCorquodale, S.M., and R.F. DiGiacomo. 1985. The role of wild North American ungulates in the epidemiology of bovine brucellosis: A review. *Journal of Wildlife Diseases* 21:351–357.

McDiarmid, A. 1951. The occurrence of agglutinins for *Brucella abortus* in the blood of wild deer in the south of England. *Veterinary Record* 63:469–470.

Meagher, M. 1989. Evaluation of boundary control for bison of Yellowstone National Park. *Wildlife Society Bulletin* 17:15–19.

Meagher, M., and M.E. Meyer. 1994. On the origin of brucellosis in bison of Yellowstone National Park: A review. *Conservation Biology* 8:645–653.

Merrell, C.L., and D.N. Wright. 1978. A serologic survey of mule deer and elk in Utah. *Journal of Wildlife Diseases* 14:471–478.

Metcalf, H.E. 1986. Control of bovine brucellosis in the United States. In *Practices in veterinary public health and preventive medicine in the United States,* ed. G.G. Woods. Ames: Iowa State University Press, pp. 97–126.

Meyer, M.E. 1966. Identification and virulence studies of *Brucella* strains isolated from Eskimos and reindeer in Alaska, Canada, and Russia. *American Journal of Veterinary Research* 27:353–358.

———. 1977. Epidemiological odds and ends. In *Bovine brucellosis: An international symposium,* ed. R.P. Crawford and R.J. Hidalgo. College Station: Texas A&M University Press, pp. 135–142.

———. 1990. Evolutionary development and taxonomy of the genus *Brucella.* In *Advances in brucellosis research,* ed. L.G. Adams. College Station: Texas A&M University Press, pp. 12–35.

Meyer, M.E., and M. Meagher. 1995. Brucellosis in free-ranging bison (*Bison bison*) in Yellowstone, Grand Teton, and Wood Bison National Parks: A review. *Journal of Wildlife Diseases* 31:579–598.

———. 1997. *Brucella abortus* infection in the free-ranging bison of Yellowstone National Park. In *Brucellosis, bison, elk, and cattle in the Greater Yellowstone Area: Defining the problem, exploring solutions,* ed. E.T. Thorne, M.S. Boyce, P. Nicoletti, and T.J. Kreeger. Cheyenne: Wyoming Game and Fish Department, pp. 20–32.

Meyer, K.F., and E.B. Shaw. 1920. A comparison of the morphological, cultural, and biochemical characteristics of *B. abortus* and *B. melitensis. Journal of Infectious Diseases* 27:173.

Mia, A.S., B. Freeze, and C.J. Frank. 1992. Extended field evaluation of "D-Ten *Brucella* A." monoclonal antibody-based competitive enzyme linked immunosorbent assay for serodiagnosis of brucellosis. *Proceedings of the United States Animal Health Association* 96:45–60.

Mohler, J.R. 1917. Abortion disease. In *Annual report of the U.S. Bureau of Animal Industry.* Washington, DC: U.S. Government Printing Office, pp. 105–106.

Moore, T. 1947. A survey of the buffalo and elk herds to determine the extent of *Brucella* infection. *Canadian Journal of Comparative Medicine* 11:131.

Moore, C.G., and P.R. Schnurrenberger. 1981. Experimental infection of opossums with *Brucella abortus. Journal of the American Veterinary Medical Association* 179:1113–1116.

Moreno, E., M.W. Pitt, L.M. Jones, G.G. Schurig, and D.T. Berman. 1979. Purification and characterization of smooth and rough lipopolysaccharides from *Brucella abortus. Journal of Bacteriology* 138:361–369.

Moreno, E., D.T. Berman, and L.A. Boettcher. 1981. Biological activities of *Brucella abortus* lipopolysaccharides. *Infection and Immunity* 31:362–370.

Morgan, W.J.B. 1977. The diagnosis of *Brucella abortus* infection in Britain. In *Bovine brucellosis: An international symposium,* ed. R.P. Crawford and R.J. Hidalgo. College Station: Texas A&M University Press, pp. 21–39.

Morton, J.K. 1978. Serologic investigations of brucellosis in elk. M.S. thesis, University of Wyoming, Laramie, Wyoming, 105 pp.

Morton, J.K., E.T. Thorne, and G.M. Thomas. 1981. Brucellosis in elk: III. Serologic evaluation. *Journal of Wildlife Diseases* 17:23–31.

Murie, O. 1951. *The elk of North America.* Washington, DC: The Stackpole Company and the Wildlife Management Institute, 376 pp.

Neiland, K.A. 1970. Rangiferine brucellosis in Alaskan canids. *Journal of Wildlife Diseases* 6:136–139.

———. 1975. Further observations on rangiferine brucellosis in Alaskan carnivores. *Journal of Wildlife Diseases* 11:45–53.

Neiland, K.A., and L.G. Miller. 1981. Experimental *Brucella suis* type 4 infections in domestic and wild Alaskan carnivores. *Journal of Wildlife Diseases* 17:183–189.

Neiland, K.A., J.A. King, B.E. Huntley, and R.O. Skoog. 1968. The diseases and parasites of Alaskan wildlife populations: Part I. Some observations on brucellosis in caribou. *Bulletin of the Wildlife Disease Association* 4:27–36.

Nelson, C.J. 1977. Immunity to *Brucella abortus.* In *Bovine brucellosis: An international symposium,* ed. R.P. Crawford and R.J. Hidalgo. College Station: Texas A&M University Press, pp. 177–188.

Newman, M.J., R.E. Traux, D.D. French, M.A. Dietrich, D. Franke, and M.J. Stear. 1996. Evidence for genetic control of vaccine-induced antibody responses in cattle. *Veterinary Immunology and Immunopathology* 50:43–54.

Nicoletti, P. 1980. The epidemiology of bovine brucellosis. *Advances in Veterinary Science and Comparative Medicine* 24:69–98.

———. 1989. Relationship between animal and human disease. In *Brucellosis: Clinical and laboratory aspects,* ed. E.J. Young and M.J. Corbel. Boca Raton, FL: CRC, pp. 41–51.

———. 1990. Vaccination. In *Animal brucellosis,* ed. K. Nielson and J.R. Duncan. Boca Raton, FL: CRC, pp. 283–299.

Nicoletti, P., and A.J. Winter. 1990. The immune response to *Brucella abortus:* The cell mediated response to infections. In *Animal brucellosis,* ed. K. Nielsen and J.R. Duncan. Boca Raton, FL: CRC, pp. 83–95.

Nicoletti, P., L.M. Jones, and D.T. Berman. 1978. Adult vaccination with standard and reduced doses of *Brucella abortus* strain 19 vaccine in a dairy herd infected with brucellosis. *Journal of the American Veterinary Medical Association* 173:1445–1449.

Nielsen, K., and J.R. Duncan, eds. 1990. *Animal brucellosis.* Boca Raton, FL: CRC, 453 pp.

Obied, A.I., H.O. Bagadi, and M.M. Mukhtar. 1996. Mastitis in *Camelus dromedarius* and the somatic cell content of camels' milk. *Research in Veterinary Science* 61:55–58.

O'Hara, T.M., J. Dau, G. Carroll, J. Bevins, and R.L. Zarnke. 1998. Evidence of exposure to *Brucella suis* biovar 4 in northern Alaska moose. *Alces* 34:31–40.

Olsen, S.C., N.F. Cheville, R.A. Kunkle, M.V. Palmer, and A.E. Jensen. 1997. Bacterial survival, lymph node pathology, and serological responses of bison (*Bison bison*) vaccinated with *Brucella abortus* strain RB51 or strain 19. *Journal of Wildlife Diseases* 33:146–151.

Palmer, M.V., S.C. Olsen, M.J. Gilsdorf, L.M. Philo, P.R. Clarke, and N.F. Cheville. 1996. Abortion and placentitis in pregnant bison (*Bison bison*) induced by the vaccine candidate, *Brucella abortus* strain RB51. *American Journal of Veterinary Research* 57:1604–1607.

Pearce, J.H., A.E. Williams, P.W. Harris-Smith, R.B. Fitzgeorge, and H. Smith. 1962. The chemical basis of the virulence of *Brucella abortus:* II. Erythritol, a constituent of bovine foetal fluids which stimulates the growth of *Brucella abortus* in bovine phagocytes. *British Journal of Experimental Pathology* 43:31–37.

Petera, F., J. Mundinger, and L.L. Kruckenberg. 1997. The Greater Yellowstone Interagency Brucellosis Committee. In *Brucellosis, bison, elk, and cattle in the Greater Yellowstone Area: Defining the problem, exploring solutions,* ed. E.T. Thorne, M.S. Boyce, P. Nicoletti, and T.J. Kreeger. Cheyenne: Wyoming Game and Fish Department, pp. 193–194.

Pinigin, A.F., and V.A. Zabrodin. 1970. On the nidality of brucellosis. *Vestnik Selskokhozyaistvennoi Nauki (Moscow)* 7:96–99.

Prichard, W.D., K.W. Hagen, J.R. Gorham, and F.C.J. Stiles. 1971. An epizootic of brucellosis in mink. *Journal of the American Veterinary Medical Association* 159:635–637.

Qureshi, T., J.W. Templeton, and L.G. Adams. 1996. Intracellular survival of *Brucella abortus, Mycobacterium bovis* BCG, *Salmonella dublin,* and *Salmonella typhimurium* in macrophages from cattle genetically resistant to *Brucella abortus. Veterinary Immunology and Immunopathology* 50:55–65.

Rausch, R.L., and B.E. Huntley. 1978. Brucellosis in reindeer, *Rangifer tarandus* L., inoculated experimentally with *Brucella suis,* type 4. *Canadian Journal of Microbiology* 24:129–135.

Ray, W.C. 1977. The epidemiology of *Brucella abortus.* In *Bovine brucellosis: An international symposium,* ed. R.P. Crawford and R.J. Hidalgo. College Station: Texas A&M University Press, pp. 103–115.

Redwood, D.W., and M.J. Corbel. 1985. *Brucella abortus* infection in the bank vole (*Clethrionomys glareolus*). *British Veterinary Journal* 141:397–400.

Rementsova, M.M. 1962. Wild vertebrates as carriers of brucellosis. In *Brucellosis in wild animals.* Alma-Ata, USSR: Academy of Sciences of the Kazakustan SSR, p. 272.

Rhyan, J.C., W.J. Quinn, L.S. Stackhouse, J.J. Henderson, D.R. Ewalt, J.B. Payeur, M. Johnson, and M. Meagher. 1994. Abortion caused by *Brucella abortus* biovar 1 in a free-ranging bison (*Bison bison*) from Yellowstone National Park. *Journal of Wildlife Diseases* 30:445–446.

Rhyan, J.C., K. Aune, D.R. Ewalt, J. Marquardt, J.W. Mertins, J.B. Payeur, D.A. Saari, P. Schladweiler, E.J. Sheehan, and D. Worley. 1997a. Survey of free-ranging elk from Wyoming and Montana for selected pathogens. *Journal of Wildlife Diseases* 33:290–298.

Rhyan, J.C., S.D. Holland, T. Gidlewski, D.A. Saari, A.E. Jensen, D.R. Ewalt, S.G. Hennager, S.C. Olsen, and N.F. Cheville. 1997b. Seminal vesiculitis and orchitis caused by *Brucella abortus* biovar 1 in young bison bulls from South Dakota. *Journal of Veterinary Diagnostic Investigation* 9:368–374.

Rice, C.E., J. Tailyuor, and D. Cochrane. 1966. Ultracentrifugal studies of sera from cattle vaccinated or naturally infected with *Brucella abortus. Canadian Journal of Comparative Medicine and Veterinary Science* 30:270–278.

Rigby, C.E. 1990. The bacteriophages. In *Animal brucellosis,* ed. K. Nielsen and J.R. Duncan. Boca Raton, FL: CRC, pp. 121–130.

Robison, C., D. Davis, J. Templeton, M. Westhusin, W. Foxworth, M. Gilsdorf, and L. Adams. 1998. Conservation of germ plasm from bison infected with *Brucella abortus. Journal of Wildlife Diseases* 34:582–589.

Rollinson, D.H.L. 1962. *Brucella* agglutinins in East African game animals. *Veterinary Record* 74:904.

Romero, C., C. Gamazo, M. Pardo, and I. Lopez-Goni. 1995. Specific detection of *Brucella* DNA by PCR. *Journal of Clinical Microbiology* 33:615–617.

Ross, H.M., G. Foster, R.J. Reid, K.L. Jahans, and A.P. MacMillan. 1994. *Brucella* species infection in seamammals. *Veterinary Record* 134:359.

Rush, W.M. 1932. Bang's disease in the Yellowstone National Park buffalo and elk herds. *Journal of Mammalogy* 13:371–372.

Sachs, R., and C. Staak. 1966. Evidence of brucellosis in antelopes of the Serengeti. *Veterinary Record* 79:857–858.

Sachs, R., C. Staak, and C.M. Groocock. 1968. Serological investigation of brucellosis in game animals in Tanzania. *Bulletin of Epizootic Diseases in Africa* 16:91–100.

Samartino, L.E., and F.M. Enright. 1993. Pathogenesis of abortion of bovine brucellosis. *Comparative Immunology, Microbiology and Infectious Diseases* 16:95–101.

Scanlan, C.M., G.L. Pidgeon, L.J. Swango, S.S. Hannon, and P.A. Galik. 1984. Experimental infection of gray foxes (*Urocyon cinereoargenteus*) with *Brucella abortus. Journal of Wildlife Diseases* 20:27–30.

Schiemann, B., and C. Staak. 1971. *Brucella melitensis* in impala (*Aepyceros melampus*). *Veterinary Record* 88:344.

Schubert, D.J., A. Rutberg, and P. Knight. 1997. *Brucella abortus* in the Greater Yellowstone Area: The animal protection perspective. In *Brucellosis, bison, elk, and cattle in the Greater Yellowstone Area: Defining the problem, exploring solutions,* ed. E.T. Thorne, M.S. Boyce, P. Nicoletti, and T.J. Kreeger. Cheyenne: Wyoming Game and Fish Department, pp. 169–177.

Schurig, G.G., R.M. Roop II, T. Bagchi, S. Boyle, D. Buhrman, and N. Sriranganathan. 1991. Biological properties of RB51: A stable rough strain of *Brucella abortus. Veterinary Microbiology* 28:171–188.

Shotts, E.B., W.E.G. Greer, and F.A. Hayes. 1958. A preliminary survey of the incidence of brucellosis and leptospirosis among white-tailed deer (*Odocoileus virginianus*) of the southeast. *Journal of the American Veterinary Medical Association* 133:359–361.

Smith, L.D., and T.A. Ficht. 1990. Pathogenesis of *Brucella. Critical Reviews of Microbiology* 17:209–230.

Smith, B.L., and R.L. Robbins. 1994. Migrations and management of the Jackson elk herd. Washington, DC: U.S. Department of the Interior, National Biological Survey, 61 pp.

Smith, S.G., S. Kilpatrick, A.D. Reese, B.L. Smith, T. Lemke, and D.L. Hunter. 1997a. Wildlife habitat, feedgrounds, and brucellosis in the Greater Yellowstone Area. In *Brucellosis, bison, elk, and cattle in the Greater Yellowstone Area: Defining the problem, exploring solutions,* ed. E.T. Thorne, M.S. Boyce, P. Nicoletti, and T.J. Kreeger. Cheyenne: Wyoming Game and Fish Department, pp. 65–78.

Smith, S.G., T.J. Kreeger, and E.T. Thorne. 1997b. Wyoming's integrated management program for brucellosis in free-ranging elk. *Proceedings of the United States Animal Health Association* 101:52–61.

Splitter, G., S. Oliveira, M. Carey, C. Miller, J. Ko, and J. Covert. 1996. T lymphocyte mediated protection against facultative intracellular bacteria. *Veterinary Immunology and Immunopathology* 54:309–319.

Stableforth, A.W. 1959. Brucellosis. In *Infectious diseases of animals.* Vol. 1, *Diseases due to bacteria,* ed. A.W. Stableforth, and I.A. Galloway. New York: Academic, pp. 53–159.

Stoenner, H.B., and D.B. Lackman. 1957. A new *Brucella* species isolated from the desert wood rat *Neotoma lepida* Thomas. *American Journal of Veterinary Research* 18:947–951.

Stoenner, H.B., R. Holdenried, D. Lackman, and J.S. Orsborn Jr. 1959. The occurrence of *Coxiella burnetti, Brucella,* and other pathogens among fauna of the Great Salt Lake Desert in Utah. *American Journal of Tropical Medicine* 8:590–596.

Sutherland, S.S., and J. Searson. 1990. The immune response to *Brucella abortus:* The humoral response. In *Animal brucellosis,* ed. K. Nielson and J.R. Duncan. Boca Raton, FL: CRC, pp. 65–95.

Swann, A.I., P.R. Schnurrenberger, R.R. Brown, and C.L. Garby. 1980. *Brucella abortus* isolations from wild animals. *Veterinary Record* 106:57.

Szyfres, B., and J. Gonzalez-Tome. 1966. Natural *Brucella* infection in Argentine wild foxes. *Bulletin of the World Health Organization* 34:919–923.

Templeton, J.W., and L.G. Adams. 1990. Natural resistance to bovine brucellosis. In *Advances in brucellosis research,* ed. L.G. Adams. College Station: Texas A&M University Press, pp. 144–150.

Tessaro, S.V. 1986. The existing and potential importance of brucellosis and tuberculosis in Canadian wildlife: A review. *Canadian Veterinary Journal* 27:119–124.

———. 1987. A descriptive and epizootiologic study of brucellosis and tuberculosis in bison in northern Canada. Ph.D. diss., University of Saskatchewan, Saskatoon, Saskatchewan, 320 pp.

Tessaro, S.V., and L.B. Forbes. 1986. *Brucella suis* biotype 4: A case of granulomatous nephritis in a barren ground caribou (*Rangifer tarandus groenlandicus* L.) with a review of the distribution of rangiferine brucellosis in Canada. *Journal of Wildlife Diseases* 22:479–483.

Thimm, B., and W. Wundt. 1976. The epidemiological situation of brucellosis in Africa. *Developments of Biological Standardization* 31:201–217.

Thomas, E.L., C.D. Bracewell, and M.J. Corbel. 1981. Characterisation of *Brucella abortus* strain 19 cultures isolated from vaccinated cattle. *Veterinary Record* 108:90–93.

Thorne, E.T. 1982. Brucellosis. In *Diseases of wildlife in Wyoming,* ed. E.T. Thorne, N. Kingston, W.R. Jolley, and R.C. Bergstrom. Cheyenne: Wyoming Game and Fish Department, pp. 54–63.

Thorne, E.T., and J.D. Herriges Jr. 1992. Brucellosis, wildlife and conflicts in the Greater Yellowstone Area. *Transactions of the North American Wildlife and Natural Resources Conference* 57:453–465.

Thorne, E.T., J.K. Morton, and G.M. Thomas. 1978a. Brucellosis in elk: I. Serologic and bacteriologic survey in Wyoming. *Journal of Wildlife Diseases* 14:74–81.

Thorne, E.T., J.K. Morton, F.M. Blunt, and H.A. Dawson. 1978b. Brucellosis in elk: II. Clinical effects and means of transmission as determined through artificial infections. *Journal of Wildlife Diseases* 14:280–291.

Thorne, E.T., J.K. Morton, and W.C. Ray. 1979. Brucellosis, its effects and impact on elk in western Wyoming. In *North American elk: Ecology, behavior and management,* ed. M.S. Boyce and L.O. Hayden-Wing. Laramie: University of Wyoming, pp. 212–220.

Thorne, E.T., T.J. Walthall, and H.A. Dawson. 1981. Vaccination of elk with strain 19 *Brucella abortus. Proceedings of the United States Animal Health Association* 85:359–374.

Thorne, E.T., J.D. Herriges, and A.D. Reese. 1991a. Bovine brucellosis in elk: Conflicts in the Greater Yellowstone Area. In *Proceedings, elk vulnerability symposium,* ed. A.G. Christensen, L.J. Lyon, and T.N. Lonner. Bozeman: Montana State University, pp. 296–303.

Thorne, E.T., M. Meagher, and R. Hillman. 1991b. Brucellosis in free-ranging bison: Three perspectives. In *The Greater Yellowstone ecosystem,* ed. R.B. Keiter and M.S. Boyce. New Haven: Yale University Press, pp. 275–287.

Thorne, E.T., S. Smith, and A.D. Reese. 1995. Cattle, elk, bison, and brucellosis in the Greater Yellowstone Area: Is there a solution? In *Proceedings of the international wildlife conference,* ed. J.A. Bissonette and P.R.K. Krausman. Bethesda, MD: Wildlife Society, pp. 386–389.

Thorne, E.T., A.D. Reese, and S.G. Smith. 1996. Brucellosis, wildlife, and cattle in the Greater Yellowstone Area of the United States: The problem, conflicts, and solutions. *Supplemento alle Ricerche di Biologia della Selvaggina* 24:557–573.

Thorne, E.T., M.S. Boyce, P. Nicoletti, and T.J. Kreeger, eds. 1997a. *Brucellosis, bison, elk, and cattle in the Greater Yellowstone Area: Defining the problem, exploring solutions.* Cheyenne: Wyoming Game and Fish Department, 219 pp.

Thorne, E.T., D. Price, J. Kopec, D. Hunter, S.G. Smith, T.J. Roffe, and K. Aune. 1997b. Efforts to control and eradicate brucellosis in wildlife of the Greater Yellowstone Area. In *Brucellosis, bison, elk, and cattle in the Greater Yellowstone Area: Defining the problem, exploring solutions,* ed. E.T. Thorne, M.S. Boyce, P. Nicoletti, and T.J. Kreeger. Cheyenne: Wyoming Game and Fish Department, pp. 101–132.

Thorne, E.T., S.G. Smith, K. Aune, D. Hunter, and T.J. Roffe. 1997c. Brucellosis: The disease in elk. In *Brucellosis, bison, elk, and cattle in the Greater Yellowstone Area: Defining the problem, exploring solutions,* ed. E.T. Thorne, M.S. Boyce, P. Nicoletti, and T.J. Kreeger. Cheyenne: Wyoming Game and Fish Department, pp. 33–44.

Thorpe, B.D., R.W. Sidwell, J.B. Bushman, K.L. Smart, and R. Moyes. 1965. Brucellosis in wildlife and livestock of west central Utah. *Journal of the American Veterinary Medical Association* 146:225–232.

Thorpe, B.D., R.W. Sidwell, and D.L. Lundgren. 1967. Experimental studies with four species of *Brucella* in selected wildlife, laboratory, and domestic animals. *American Journal of Tropical Medicine and Hygiene* 16:665–674.

Toman, T.L., T. Lemke, L. Kuck, B.L. Smith, S.G. Smith, and K. Aune. 1997. Elk in the Greater Yellowstone Area: Status and management. In *Brucellosis, bison, elk, and cattle in the Greater Yellowstone Area: Defining the problem, exploring solutions,* ed. E.T. Thorne, M.S. Boyce, P. Nicoletti, and T.J. Kreeger. Cheyenne: Wyoming Game and Fish Department, pp. 56–64.

Toshach, S. 1963. Brucellosis in the Canadian Arctic. *Canadian Journal of Public Health* 54:271–275.

Tunnicliff, E.A., and H. Marsh. 1935. Bang's disease in bison and elk in the Yellowstone National Park and on the National Bison Range. *Journal of the American Veterinary Medical Association* 86:745–752.

U.S. Department of the Interior, National Park Service. 1998. *Draft environmental impact statement for the interagency bison management plan for the state of Montana and Yellowstone National Park: Interagency bison management plan DSC-RP,* Denver, CO: U.S. Department of the Interior, National Park Service, 395 pp.

Vaughn, H.W., R.R. Knight, and F.W. Frank. 1973. A study of reproduction, disease and physiological blood and serum values in Idaho elk. *Journal of Wildlife Diseases* 9:296–301.

Waghela, S., and L. Karstad. 1986. Antibodies to *Brucella* spp. among blue wildebeest and African buffalo in Kenya. *Journal of Wildlife Diseases* 22:189–192.

Wilesmith, J.W. 1978. The persistence of *Brucella abortus* infection in calves: A retrospective study of heavily infected herds. *Veterinary Record* 103:149–153.

Williams, M.R., and P. Millar. 1979. Changes in serum immunoglobulin levels in Jerseys and Friesians near calving. *Research in Veterinary Science* 26:81–84.

Williams, E.S., E.T. Thorne, S.L. Anderson, and J.D. Herriges. 1993. Brucellosis in free-ranging bison (*Bison bison*) from Teton County, Wyoming. *Journal of Wildlife Diseases* 29:118–122.

Williams, E.S., S.L. Cain, and D.S. Davis. 1997. Brucellosis: The disease in bison. In *Brucellosis, bison, elk, and cattle in the Greater Yellowstone Area: Defining the problem, exploring solutions,* ed. E.T. Thorne, M.S. Boyce, P. Nicoletti, and T.J. Kreeger. Cheyenne: Wyoming Game and Fish Department, pp. 7–19.

Witter, J.F. 1981. Brucellosis. In *Infectious diseases of wild mammals,* ed. J.W. Davis, L.H. Karstad, and D.O. Trainer. Ames: Iowa State University Press, pp. 280–287.

Wood, G.W., J.B. Hendricks, and D.E. Goodman. 1976. Brucellosis in feral swine. *Journal of Wildlife Diseases* 12:579–582.

Worthington, R.W., F.D. Horwell, M.S. Mulders, I.S. McFarlane, and A.P. Schutte. 1974. An investigation of the efficacy of three *Brucella* vaccines in cattle. *Journal of the South African Veterinary Association* 45:87–91.

Wray, C. 1975. Survival and spread of pathogenic bacteria of veterinary importance within the environment. *Veterinary Bulletin* 45:543–550.

Xin, X. 1986. Orally administrable brucellosis vaccine: *Brucella suis* strain 2 vaccine. *Vaccine* 4:212–214.

Youatt, W.G., and L.D. Fay. 1959. Experimental brucellosis in white-tailed deer. *American Journal of Veterinary Research* 20:925–926.

Young, E.J. 1989. Clinical manifestations of human brucellosis. In *Brucellosis: Clinical and laboratory aspects,* ed. E.J. Young and M.J. Corbel. Boca Raton, FL: CRC, pp. 97–126.

Young, E.J., and M.J. Corbel, eds. 1989. *Brucellosis: Clinical and laboratory aspects.* Boca Raton, FL: CRC, 187 pp.

Young, E.J., and P. Nicoletti. 1997. Brucellosis in humans. In *Brucellosis, bison, elk, and cattle in the Greater Yellowstone Area: Defining the problem, exploring solutions,* ed. E.T. Thorne, M.S. Boyce, P. Nicoletti, and T.J. Kreeger. Cheyenne: Wyoming Game and Fish Department, pp. 147–153.

Zaki, R. 1948. *Brucella abortus* infection among buffalos in Egypt. *Journal of Comparative Pathology* 58:73.

Zarnke, R.L. 1983. Serologic survey for selected microbial pathogens in Alaskan wildlife. *Journal of Wildlife Diseases* 19:324–329.

Zarnke, R.L., and T.M. Yuill. 1981. Serologic survey for selected microbial agents in mammals from Alberta, 1976. *Journal of Wildlife Diseases* 17:453–461.

Zhan, Y., and C. Cheers. 1995. Endogenous interleukin-12 is involved in resistance to *Brucella abortus* infection. *Infection and Immunity* 63:1387–1390.

23

ANTHRAX

C. CORMACK GATES, BRETT ELKIN, AND DAN DRAGON

Synonyms: Charbon, malignant edema, woolsorter's disease, malignant pustule, ragpicker's disease, splenic fever, sang de rate, milzbrand, miltsiekte, fievre charbonneuse, bloedsiekte, malnair dal sang, murraine, Cumberland disease, Siberian plague, milzfieber, Bradford disease, black bane/bain, pustula maligna, karbunkelkrankheit.

INTRODUCTION. Anthrax is an infectious, often fatal disease of wild and domestic animals and humans that is caused by the endospore-forming bacterium *Bacillus anthracis.* Anthrax epidemics have occurred on virtually every continent and have been recorded throughout written history.

The life history of *B. anthracis* differs markedly from most other infectious bacteria: its persistence depends on extreme virulence, death of its host, and survival of highly resistant, infectious endospores during prolonged periods outside the host. The bacillus replicates rapidly in the bloodstream to high concentrations and releases toxins resulting in septicemia, which soon kills the host. Scavengers attracted to the carcass rend it apart, freeing a multitude of *B. anthracis* organisms and through ingestion dispersing them over a wide area. In the soil and vegetation, the microorganism, in its spore form, can remain viable and infectious for years until it comes into contact with and enters a new host, where it germinates and begins the cycle anew (Fig. 23.1).

Anthrax epidemics still appear regularly in many areas around the globe. Effective vaccines have reduced the economic significance of the disease in developed countries, where it now occurs sporadically in unvaccinated domestic stock and wildlife populations. It remains a serious zoonosis in areas of the Middle East, southern Russia, Africa, and Asia.

HISTORY. Anthrax is putatively the oldest infectious disease known to humankind (Fig. 23.2) (Klemm and Klemm 1959). The earliest known reference to an anthrax epidemic is described in the Book of Exodus (9:2–7) circa 1491 B.C. Moses warned the Pharaoh, "The hand of the Lord will fall on your livestock in the fields, on horses, asses, camels, herds, and flocks, with a very severe murrain." The prophesy was fulfilled, and all the Egyptian cattle died of a carbuncular disease whose description matched that of anthrax. The Egyptian cattle that were grazed in the vegetation-rich floodplains of the Nile died, but the livestock of the Israelites, which were forced to graze the dry, sparsely vegetated highlands above the river, were spared (Klemm and Klemm 1959). This provided the earliest observation of a key aspect of the epidemiology of anthrax: low-lying areas subject to periodic flooding are prime locations for anthrax epidemics.

Other ancient cultures also recognized the disease in their livestock (Klemm and Klemm 1959). By 500 B.C., the Hindu of India had classified and described a number of cattle diseases, including anthrax. In ancient Greece, Homer (1000 B.C.), Hippocrates (400 B.C.), and Galen (200 A.D.) observed the disease and wrote about its effects on livestock and people.

The Romans also encountered problems with anthrax. Virgil (70 to 19 B.C.) described an anthrax epidemic that he called the murrain of Noricum, stating there seemed to be no cure for the animals (McSherry and Kilpatrick 1992). During the outbreak, he observed that the disease could be transmitted by the consumption of flesh from dead animals (Klemm and Klemm 1959). He also described a human form of the disease resulting from wearing hides or wool taken from animals that had died in the epidemic. Based on his observations, Virgil recommended that all animals affected by the disease should be slaughtered rather than trying to treat them and that the carcasses should be destroyed. Present-day control and containment of anthrax is still largely based on these procedures.

From the end of the Dark Ages to the late 19th century, anthrax spread across much of the known world due to a growing livestock trade and colonialism that resulted in the introduction of large herds of domestic animals to new areas. Between 1613 and 1617, Black Bane, a severe epidemic of anthrax, raged throughout southern Europe, killing over 60,000 people and countless livestock (Maxy 1951). By the mid-19th century, anthrax was pandemic throughout much of Europe and had spread into Russia and the New World. In France, for example, anthrax was responsible for killing 20%–30% of the country's sheep and cattle. In some French provinces, the disease claimed almost 50% of all livestock (Klemm and Klemm 1959).

Fortunately, in the later half of the 19th century, scientists in the fledgling disciplines of bacteriology, epidemiology, and immunology used *B. anthracis* as a

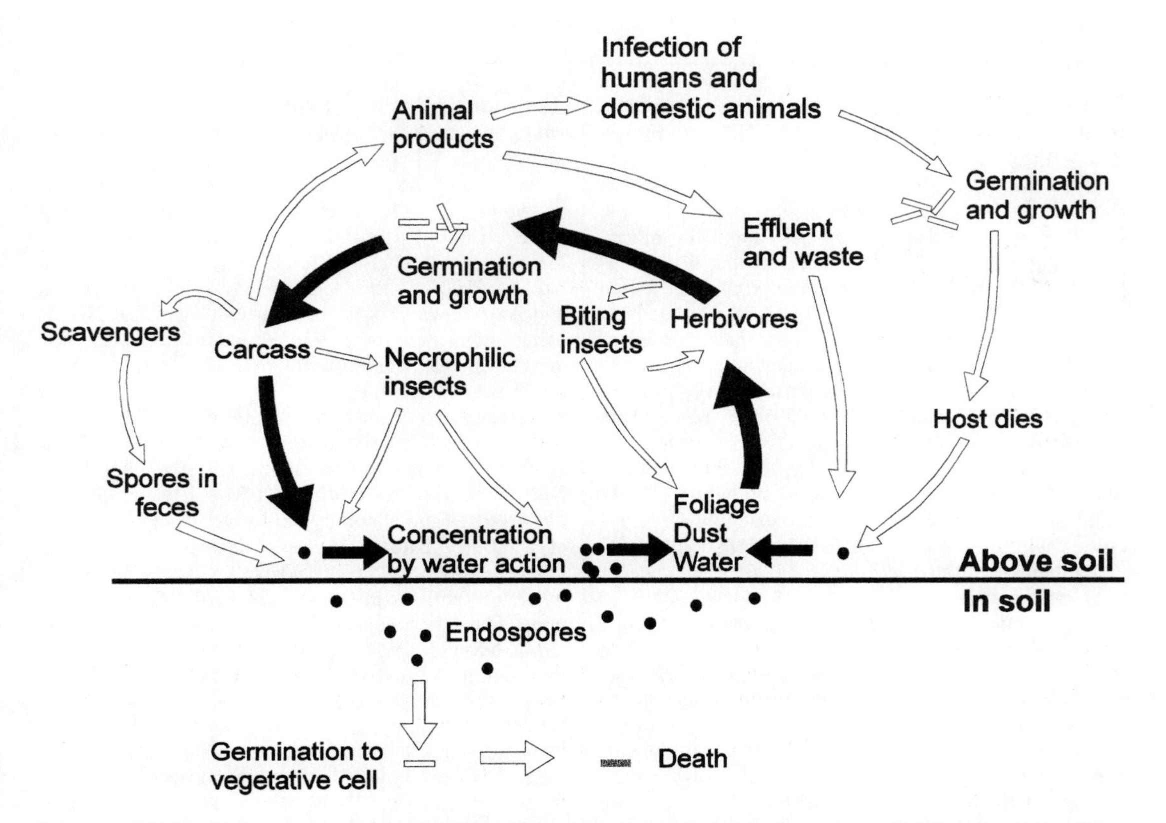

FIG. 23.1—The infectious cycle of *Bacillus anthracis* in the environment. Major routes (*thick arrows*) and minor routes (*thin arrows*) of cycling of vegetative cells (*rods*) or spores (*dots*) are depicted.

B.C.	1491	5th Plague of Egypt
	1000	Homer
	500	Hindu descriptions
	400	Hippocrates
	100	Varro
	70-19	Virgil
A.D.	200	Galen
	996	Pandemic in France
	1613-17	Black Bane pandemic in Southern Europe
	1709-12	Pandemics in Germany, Hungary, and Poland
	1752	Maret's description of Malignant Pustule
	1769	Dijon's scientific article
	1834	First human case in North America
	1838	First microscopy by Delafond
	1850's	Rayer and Davaine demonstrate the bacillus produces disease
	1875	Naming by Cohn
	1879	Southwestern Africa, Otjindimba: Year of the anthrax pox
	1877	Koch's postulates
	1880	Greenfield's vaccine
	1881	Pasteur's vaccine
	1895	Sclavo's serum therapy
	1926	Neoarsphenamine treatment
	1937	Gruinard Island anthrax bomb
	1944	Penicillin
	1979	Sverdlovsk military accident

FIG. 23.2—Significant events in the history of anthrax.

model for their hypotheses. The first scientific paper on anthrax was published in 1769 by Dijon of France (Klemm and Klemm 1959).

Delafond, a professor of veterinary medicine at the Alfort Veterinary School in France, is credited as the first person to observe *B. anthracis* under the microscope. As early as 1838, he observed that blood collected from animals with anthrax contained "little rods." By 1868, Davaine had successfully isolated the bacterium from blood and had shown that upon injection into a healthy animal they caused anthrax.

In 1877, Pasteur began experimenting with *B. anthracis.* He filtered infected blood to remove the bacilli and demonstrated that injection of cleared filtrate did not cause anthrax, but that sediment from liquid cultures caused disease (Klemm and Klemm 1959). Many authors cite Pasteur's work in isolating *M. tuberculosis* as the sole proof of his germ theory, but his experiments with *B. anthracis* actually predated his work on tuberculosis.

At the same time, Robert Koch, a German physician, had completed similar work (Klemm and Klemm 1959). Through meticulous observation by microscope, Koch worked out the complete life cycle of the anthrax bacillus. Koch was able to isolate the bacillus in pure culture from infected animals, transfer it to healthy animals, who then developed anthrax, and reisolate pure cultures of the bacillus from their blood, thus proving the postulates that now bear his name. Koch's isolation of the anthrax bacillus was of momentous importance, because it was the first time that the causative agent of an infectious disease had been isolated in pure culture and its infectious nature demonstrated beyond a reasonable doubt.

The recognition of inhalation anthrax was an important stimulus to development of industrial hygiene and microbiology. In 1837, mohair from Asia Minor and alpaca *Lama pacos* hair from Peru were introduced into the developing British textile industry. Soon after, woolsorter's disease, a mysterious and rapidly fatal illness, began to occur among the workers. At the same time, a similar disease was seen in Germany among persons who handled rags (ragpicker's disease). In the late 1870s, Bell began a study of woolsorter's disease and in short order identified the causative bacilli in the blood of patients. Upon injection of the microorganism into mice *Mus musculus* and rabbits *Oryctolagus cuniculus,* it caused a condition identical to anthrax, and he correctly concluded the disease in the textile workers was a new form of anthrax. Bell recommended that manufacturers should wash potentially hazardous material thoroughly before sorting.

In 1905, F.W. Eurich evaluated the sporicidal ability of a large number of compounds on the bacillus and found formaldehyde to be the most effective. Decontamination of imported mohair, raw wool, and alpaca from anthrax-endemic areas led to decreased rates of inhalation anthrax. Other countries quickly adopted the British policy and experienced similar decreases in industrial inhalation anthrax (Brachman 1980). Even today, formaldehyde is still the disinfectant of choice when decontaminating premises and equipment of *B. anthracis.*

Because of the extreme environmental resistance of its spores and its high virulence, *B. anthracis* has long been of interest to military powers as a potential biologic weapon. Although biologic weapons were outlawed in 1925 by the Geneva Convention, both Allied and Axis powers experimented with the feasibility of using anthrax and other human and animal pathogens as weapons of war during World War II (Humphrey 1987; Bryden 1989; Harris 1992). The most publicized of these experiments was detonation and dropping of anthrax bombs on Gruinard Island, off the northwest coast of Scotland, by the British in 1942 and 1943. During the trials, an estimated 4 4×10^{14} spores were dispersed on the island by explosive means (Manchee et al. 1990). The trials were a success, showing that anthrax spores could be explosively released and cause inhalation anthrax in sheep *Ovis aries* tethered downwind. The experiments also left the island heavily contaminated with spores and unfit for human and animal occupation. Annual tests for more than 20 years following the war showed the persistence of large numbers of fully virulent anthrax spores in the soil. In 1986, after extensive microbiologic sampling revealed that the spores were limited to 4 ha around the original detonation site, the island was disinfected with a mixture of 280 tonnes of 40% formaldehyde and 2000 tonnes of sea water (Mierzejewski and Bartoszcze 1996). Resampling studies in 1987 showed that the soil no longer harbored *B. anthracis,* and the island was returned to normal agricultural use. Sheep have been grazing unharmed on the island since 1991 (Mierzejewski and Bartoszcze 1991).

Despite the Biological Weapons Convention of 1972 forbidding research, development, and stockpiling of such weapons, the threat of their use has not been eliminated. There are several reports that the Iraqis were developing anthrax weapons during the Gulf War (Kahler 1992). In 1979, an epidemic of anthrax occurred in the city of Sverdlovsk, 1400 km east of Moscow. There were at least 96 human cases of anthrax, with 64 deaths (Meselson et al. 1994). Soviet officials later reported an anthrax outbreak among livestock south of the city. For years, the epidemic was the focus of international debate and speculation. In 1993, Russian pathologists who performed autopsies during the epidemic published pathologic evidence that the fatal cases were indeed due to inhalation anthrax (Abramova et al. 1993). Later, a joint United States-Russian research team reported that the daytime locations of 90% of the patients for which the information could be collected could be traced within a narrow band running southwest through the southern part of the city, originating from a military microbiology facility at its northeast end (Meselson et al. 1994). They also determined that the animal outbreaks south of the city occurred after the human epidemic and were on farms that fell within the same southwest vector from the military facility to a distance of over 40 km. One of the main problems enforcing the Biological Weapons Convention is that research facilities required for vaccine and biologic weapon development are essentially identical.

DISTRIBUTION. Anthrax is global in distribution and is endemic to many parts of Europe, Asia, Africa, Australia, and North, Central, and South America. The organism is endemic to several regions of North America. Recurrent outbreaks have been reported among free-ranging bison *Bison bison* in nonagricultural areas in the Northwest Territories and northeastern Alberta, Canada (Gates et al. 1995). The disease has also been reported in cattle *Bos taurus* herds in a farming district near the southwest corner of Wood Buffalo National Park, Alberta, and the Northwest Territories (Gainer and Saunders 1989). Sporadic outbreaks are widespread in the prairie regions of Alberta and Saskatchewan (MacDonald et al. 1992). In the United States, there are two main endemic areas: northwestern Mississippi and adjacent southeastern Arkansas, where outbreaks occur primarily in cattle; and western Texas and adjacent Mexico, where outbreaks are reported sporadically in sheep and white-tailed deer *Odocoileus virginianus.* In Mexico, reports of anthrax in cattle and sheep are widespread. The epidemiologic pattern in livestock suggests that contamination of feed with anthrax carcasses is primarily responsible for the continuing occurrence of outbreaks (Hugh-Jones 1996).

Anthrax is a serious problem in Central America, with the exception of Belize (Hugh-Jones 1996). The disease is epidemic in Guatemala. It is absent from the Caribbean, with the exception of Haiti, where it is a serious zoonosis. The human case rate in Haiti may exceed 2000 per year. Pierre and Valbrun (1996) attributed this high prevalence to lack of education in preventative measures, absence of long-term veterinary control programs, and lack of resources to support consistent vaccination of susceptible livestock.

Anthrax is endemic in many South American countries. Hugh-Jones (1996) considered that the status of endemism in Chile reflected an improvement in veterinary surveillance. Other countries in which anthrax has been reported include Brazil, Uruguay (Hugh-Jones 1996), and Paraguay (Harrison et al. 1989).

In Europe, the major affected regions remain Turkey, Greece, Albania, southern Italy, Romania, and central Spain. Norway, which had not recorded a case of the disease in years, recently reported several cases in cattle; the organism probably was transmitted in contaminated feedstuffs (Hugh-Jones 1996). Southern Russia experiences the greatest frequency of outbreaks in that nation, although an increase in vaccination has reduced the incidence of the disease in livestock in recent years (Cherkasskiy 1996). Anthrax is a serious endemic zoonosis in Turkey, where 1779 human cases were reported between 1990 and 1994 (Doganay 1996).

In 1994, anthrax was epidemic in sheep and goats *Capra hircus* in Syria. Anthrax outbreaks occur sporadically in Pakistan and India; the state of Tamil Nadu lies in a belt where the reported incidence of anthrax is highest in India (Lalitha et al. 1996). Anthrax is endemic in southeast Asia, where it frequently affects water buffalo *Bubalus bubalis* and pigs *Sus scrofa.* Although no cases had been reported since 1978, Korea had two confirmed cattle and 24 human cases in 1994 (Hugh-Jones 1996). In China, vast areas are contaminated with anthrax spores, and the disease is epidemic in ten provinces (Xudong et al. 1996). According to a national communicable diseases report, during 1990–93 there were 8122 human cases of anthrax in China, of which 324 were fatal. The annual incidence of anthrax in humans was 0.0195 per million, with a mortality rate of 3.99%. The main source of human infection was reported to be sheep in the north and cattle in southern China.

The disease is endemic in Africa, with human, wildlife, and livestock cases occurring especially in west Africa. South Africa suffers from sporadic epizootics with large numbers of mortalities in afflicted herds (Hugh-Jones 1996). Anthrax cases are diagnosed every year in the wild herbivores of Etosha National Park in Nambia (Lindeque et al. 1996), but the incidence has been low in recent years. In the park, the disease primarily affects Burchell's zebra *Equus burchelli,* blue wildebeest *Connochaetes taurinus,* and elephants *Loxodonta africana.* The disease occurs sporadically in the wild herbivore population of Kruger National Park in South Africa, with major epidemics occurring in a cyclical pattern with a periodicity of 10 years or multiples thereof. The principal animals affected in the park are greater kudu *Tragelaphus strepsiceros,* although dozens of species may be affected during major outbreaks (de Vos and Bryden 1996).

Australia has traditionally had a small area affected by the disease in New South Wales but has recently seen cases in Western Australia as well. With aid from the Australian government, the anthrax situation in Indonesia has improved and is now a localized problem (Hugh-Jones 1996).

HOST RANGE. Anthrax has been reported in a wide range of homeothermic species (Table 23.1). The body temperature of the host is critical in determining whether it develops the disease. In his work with anthrax, Pasteur took groups of chickens *Gallus gallus,* which are normally resistant to the disease, immersed them in cold water, and lowered the birds' body temperature from 42° C to 37° C. He inoculated half the chilled chickens with *B. anthracis,* leaving the other half as controls, and left both sets of birds in the baths. Chickens that had been inoculated with the anthrax bacillus died, and the organism was recovered from their blood. Noninoculated control birds did not die. Ostriches *Struthio camelus* are the only avian species that have been reported naturally infected with anthrax (Table 23.1).

TABLE 23.1—Wild mammals confirmed to be susceptible to anthrax

Artiodactyla		
Suidae		
Wild boar, feral swine	*Sus scrofa*	Worldwide
Bush pig	*Potamochoerus porcus*	Africa
Wart hog	*Phacochoerus aethiopicus*	Africa
Hippopotamidae		
Hippopotamus	*Hippopotamus amphibius*	Africa
Camelidae		
Llama	*Lama glama*	Central and South America
Alpaca	*Lama pacos*	South America
Camel	*Camelus dromedarius*	Africa, Middle East
Giraffidae		
Giraffe	*Giraffa camelopardalis*	Africa
Cervidae		
Fallow deer	*Dama dama*	Europe
Sambar	*Cervus unicolor*	Asia
Red deer	*Cervus elaphus*	Europe
White-tailed deer	*Odocoileus* virginianus	North America
Moose	*Alces alces*	North America, Europe
Roe deer	*Capreolus capreolus*	Europe
Bovidae		
Nyala	*Tragelaphus angasii*	Africa
Bushbuck	*Tragelaphus scriptus*	Africa
Greater kudu	*Tragelaphus strepsiceros*	Africa
Eland	*Taurotragus oryx*	Africa
Water buffalo	*Bubalus bubalis*	Southeast Asia
African buffalo	*Syncerus caffer*	Africa
American bison	*Bison bison*	North America
Common duiker	*Sylvicapra grimmia*	Africa
Waterbuck	*Kobus ellipsiprymnus*	Africa
Reedbuck	*Redunca arundinum*	Africa
Roan antelope	*Hippotragus equinus*	Africa
Gemsbok	*Oryx gazella*	Africa
Tsessebe	*Damaliscus lunatus*	Africa
Lichtenstein's hartebeest	*Sigmoceros lichtensteinii*	Africa
Black wildebeest	*Connochaetes gnou*	Africa
Blue wildebeest	*Connochaetes taurinus*	Africa
Klipspringer	*Oreotragus oreotragus*	Africa
Steenbuck	*Raphicerus campestris*	Africa
Sharpe's grysbok	*Raphicerus sharpei*	Africa
Impala	*Aepyceros melampus*	Africa
Springbok	*Antidorcas marsupialis*	Africa
Goat	*Capra* spp.	Africa
Sheep	*Ovis* spp.	Worldwide
Carnivora		
Canidae		
Black-backed jackal	*Canis mesomelas*	Africa
Wild dog	*Canis familiaris*	Australia
Dingo	*Canis familiaris*	Australia
African wild dog	*Lycaon pictus*	Africa
Felidae		
Bobcat	*Felis rufus*	North America
Cougar	*Felis concolor*	North America
Leopard	*Panthera pardus*	Africa
Lion	*Panthera leo*	Africa
Cheetah	*Acinonyx jubatus*	Africa
Procyonidae		
Raccoon	*Procyon lotor*	North America
Mustelidae		
Mink	*Mustela vison*	North America
Ferret (polecat)	*Mustela putorius*	Europe
Honey badger	*Mellivora capensis*	Africa
European badger	*Meles meles*	Europe
Viverridae		
Civet	*Civettictis civetta*	Africa
Genet cat	*Genetta genetta*	Africa
Hyaenidae		
Hyena	*Hyaena hyaena*	Africa

(continued)

TABLE 23.1—Continued

Lagomorpha		
Leporidae		
Hare	*Lepus* spp.	Europe
Perssiodactyla		
Rhinocerotidae		
Indian rhinoceros	*Rhinoceros unicornis*	Asia
Black rhinoceros	*Diceros bicornis*	Africa
Equidae		
Hartmann's zebra	*Equus zebra hartmannae*	Africa
Burchell's zebra	*Equus burchelli*	Africa
Primates		
Pongidae		
Chimpanzee	*Pan troglodytes*	Experimental
Cercopithecidae		
Rhesus monkey	*Macaca mulatta*	Experimental
Proboscidea		
Elephantidae		
Asian elephant	*Elephas maximus*	Asia
African elephant	*Loxodonta africana*	Africa

In general, herbivores are much more susceptible to anthrax than are carnivores, and the disease is rapidly fatal. Some herbivores, such as impala *Aepyceros melampus* and bison, are so susceptible that death can occur within hours of exposure. Carnivores, on the other hand, are more likely to develop chronic anthrax, which is rarely fatal. During an epidemic, carnivores may eat vast quantities of contaminated meat without developing anthrax. In an epidemic among bison in northern Canada, workers observed numerous apparently healthy wolves *Canis lupus* scavenging on carcasses, even though they had consumed so much contaminated meat that their abdomens were distended almost to the ground and they could barely run. No dead wolves have ever been found during anthrax epidemics in northern Canada. The reason for this discrepancy appears to be the development of a more effective immune response against *B. anthracis* in carnivores compared with herbivores. In Etosha National Park, naturally acquired *B. anthracis* antibodies are rare in herbivores but common in carnivores, in which they appear to reflect prevalence of the disease in the predator's range (Turnbull et al. 1992).

Though resistant to the disease, many mammalian carnivores and avian scavengers act as carriers of *B. anthracis* spores. In Etosha National Park, anthrax spores were obtained from 72% of jackal *Canis mesomelas,* 60% of hyena *Hyaena* spp., and 50% of vulture *Gyps africanus* feces collected around carcasses of animals that had died of anthrax (Lindeque and Turnbull 1994). Similarly, herring gulls *Larus argentatus* and ravens *Corvus corax* collected during anthrax epidemics in bison were found to contain spores in their gastrointestinal tracts (Choquette 1970; Gates et al. 1995). Not all carriers of anthrax are scavengers: in 1952, *B. anthracis* was isolated from the digestive tracts of two house sparrows *Passer domesticus domesticus* in Great Britain, and there were no cases of anthrax in livestock in the area at the time of isolation (Shrewsbury and Barson 1952). In Etosha National Park, the microorganism was recovered from the bill and gizzard of an Egyptian goose *Alophochen aegyptiacus* shot on a water pan contaminated with anthrax spores (Ebedes 1976).

Fatal anthrax in several species of mammalian carnivores and birds has been observed only in captivity. Meat from lame animals or carcasses of animals dying of unexplained causes, which is sold cheaply to fur farms and zoological gardens, is often the cause. Although rodents are susceptible to anthrax in the laboratory, no confirmed cases of rodent anthrax have been found, and there have been no reports of sudden declines in rodent populations following an anthrax epidemic.

ETIOLOGY. Anthrax is caused by the Gram-positive, nonmotile, bacterial rod *Bacillus anthracis.* Vegetative growth and multiplication occur in a homeothermic host where, in the circulatory system, the bacterium is exposed to aerobic, nutrient-rich conditions and is protected from competition from other microbes. Like other *Bacillus* species, *B. anthracis* can form metabolically inactive endospores (Fig. 23.3). Each vegetative cell is capable of forming one spore. Hence, sporulation is not a form of replication, but a resting stage that protects the bacillus until environmental conditions become favorable for growth.

A common misconception about *B. anthracis* is that its vegetative cells produce spores upon exposure to atmospheric oxygen (for example, when a carcass is opened by scavengers). In all *Bacillus* species, sporulation is a response to low-nutrient conditions or dehydration, which effectively limits diffusion of nutrients to the bacillus. The response is totally independent of the presence of oxygen (Gould 1977; Claus and Berkeley 1986). When the host dies, its circulatory system stops transporting oxygen to the tissues, and they become anaerobic. Within this still nutrient-rich environment, the aerobic *B. anthracis* bacilli are held in a

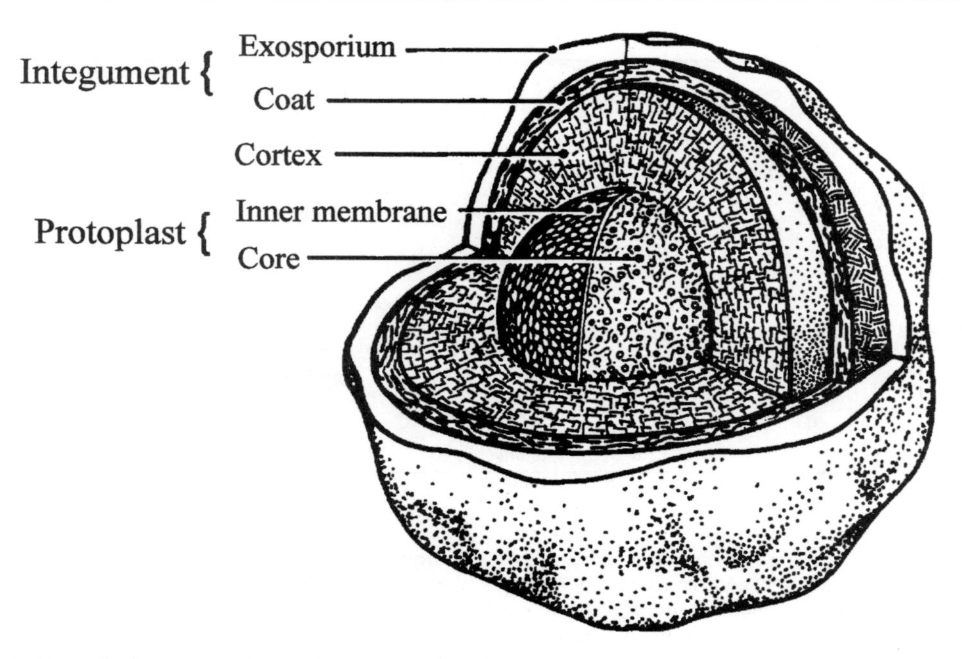

FIG. 23.3—Generalized structure of bacterial endospores. Note that not all *Bacillus* species have exosporia. (Reproduced with the permission of Dr. K. Johnstone and Blackwell Scientific Publications.)

static condition, unable to replicate or to sporulate. Anaerobic bacteria, particularly *Clostridium* spp. from the gastrointestinal tract, start to decompose the carcass rapidly. The vegetative form of *B. anthracis* is very susceptible to competition from other microbes, and if putrefactive organisms reach them before the carcass is opened, they are quickly eliminated (Sterne 1959; Tolstova 1960; Vasil'eva 1960; Turnbull et al. 1989). In nature, however, carcasses are rarely left intact long enough for putrefaction to eliminate all of the vegetative *B. anthracis*. Instead, opening the carcass helps to disperse vegetative *B. anthracis* into aerobic microenvironments where, either through metabolic activity or dehydration, nutrients become limited and sporulation can then proceed (Choquette 1970).

Vegetative cells of *B. anthracis* have very specific nutrient and physiologic requirements and survive poorly outside a host or complex artificial medium (Turnbull et al. 1989). Even when adequate nutrients are provided to trigger germination of spores, vegetative cells are very susceptible to competition from other bacterial species (Vasil'eva 1960). In contrast, *B. anthracis* spores can resist prolonged exposure to chemical disinfectants; desiccation; ultraviolet and ionizing radiation; and extremes in pH, temperature, and pressure (Roberts and Hitchins 1969; Gould 1977). Infectious viable *B. anthracis* spores can persist for decades in the environment. In Kruger National Park, spores have been recovered from bones estimated to be approximately 200 years old (de Vos 1990). A museum conservator in Poland developed cutaneous anthrax after pricking his palm with a nail while extracting it from furniture covered with goat leather dating from the 15th century (Przybyszewska et al. 1996).

The characteristics of dormancy and resistance of endospores are a function of their structure. The surface layers of the spore, known as the exosporium and spore coat (Fig. 23.3), are responsible for surface hydrophobicity (Doyle et al. 1984; Koshikawa et al. 1989) and protect the inner spore from the effects of harmful chemicals, disinfectants, and degradative enzymes (Foster and Johnstone 1990). In response to changes in pH or ionic strength, the spore cortex may undergo significant changes in volume that increase pressure on the core, squeezing water out (Gould 1977; Popham and Setlow 1993). Consequently, endospores have a high buoyant density. The low water content of the spore confers resistance to heat and ultraviolet light and is believed to contribute to spore dormancy (Gould 1984).

Calcium may be an important structural component of *Bacillus* spores that is affected by prolonged interactions with the environment (Rode and Foster 1966a). Calcium cations participate in germination and in maintaining dormancy (Rode and Foster 1966a,b; Shibata et al. 1992). In combination with dipicolinic acid (DPA), calcium forms an extensive salt lattice that immobilizes enzymes, DNA, and metabolically active components in the core. Immobilization is thought to play a role in maintaining metabolic dormancy and heat resistance of core components (Gould 1984). Dragon and Rennie (1995) speculated that exchange of core water along with diffusion of protons or other cations into the core from the environment could result, over time, in the leaching of calcium cations and could disrupt the calcium-DPA lattice. This in turn could adversely affect spore dormancy and resistance, leading to a reduction in germinability, or to premature germination resulting in death from inadequate nutrients or competition from other bacteria. They also hypothesized that high levels of calcium in the soil may act to buffer the internal supplies of calcium and greatly extend their viability.

Anthrax spores germinate optimally under specific temperature, density, and micronutrient conditions. Titball and Manchee (1987) found that in vitro germination of anthrax spores occurred optimally at 22° C. Germination was reduced considerably at 37° C and 9° C. They suggested that the low optimal germination temperature for spores may be important for establishment of cutaneous infection. Spore density adversely affects germination. The mechanism is thought to involve the release of the enzyme alanine racemase by germinating spores (Gould 1996). Density-dependent inhibition of germination is thought to be important for ensuring that if spore germination occurs outside of living hosts (for example, in soil) a reservoir of inactivated spores will be maintained (Titball and Manchee 1987). Germination is known to be triggered by l-alanine, a component of blood serum; tyrosine may also be an additional requirement (Hills 1949, 1950).

TRANSMISSION AND EPIDEMIOLOGY. There is debate about whether *B. anthracis* cycles outside of hosts. Based on observed patterns in the distribution of areas of repeated anthrax epidemics in livestock in the southern United States, Van Ness (1971) postulated the controversial "incubator area hypothesis," which states that under conditions of alkaline pH, high soil moisture, and the presence of organic matter, *B. anthracis* may undergo cycles of spore germination, vegetative cell outgrowth, and resporulation leading to an increase in spore density. There is little experimental evidence to support this hypothesis, except under experimental conditions when soil or water was artificially enriched with animal blood or viscera (Manchee et al. 1981; Turnbull et al. 1989, 1991; Lindeque and Turnbull 1994). Despite high concentrations of *B. anthracis* at the death of host animals, Lindeque and Turnbull (1994) found that extensive spore contamination of soil from blood did not always occur and, when it did, concentrations were typically low. It was evident that only a small proportion of released bacilli sporulated successfully. The evidence to date is that *B. anthracis* depends to the greatest extent on multiplication within a host, not in the environment.

Anthrax epidemics are commonly associated with low-lying depressions and rockland seep areas with high moisture content, high organic content in the soil, and an alkaline pH. They generally occur during dry summer months following periods of heavy rain or floods. These observations suggest that water plays an important role in the epidemiology of anthrax. Prolonged rain or flooding preceding outbreaks promotes runoff and eventual pooling of standing water. *Bacillus anthracis* spores have high surface hydrophobicity (Doyle et al. 1984) and as a result could be carried along in clumps of organic matter by runoff to collect in standing pools. Spores also have high buoyant density (Bakken 1985) and can float free of resting sites in soil. Drying weather, which often precedes outbreaks, may concentrate spores in depressions because such storage areas would be the last to experience evaporation. Dragon and Rennie (1995) suggested that successive cycles of runoff and evaporation may slowly concentrate anthrax spores into storage areas.

There are several mechanisms by which herbivores become infected with *B. anthracis*. Changes in grazing behavior during dry periods may lead grazers to crop vegetation closer to the soil (Christie 1987) or to increase grazing in depressions where forage may be more nutritious and where spores may also be concentrated. Other behaviors have also been implicated in transmission. In northern Canada, anthrax epidemics in bison occur during hot, dry weather between late June and early September (Broughton 1987; Gates et al. 1995). Part of this period (early August–early September) coincides with the rut (Komers 1992), and rutting behavior of bulls may greatly increase the chance of exposure to spores. At any time of the year, but typically toward the onset of the rut, bulls engage in wallowing. Using their horns, they dig up dirt and then roll and thrash in the pit, throwing up clouds of dust when the soil is dry. Wallowing pits are reused by other bison and tend to become larger over time. The depressions may become storage areas for anthrax spores. Wallowing in these sites may create clouds of aerosolized spores and expose bison in the immediate vicinity through inhalation.

The strong seasonality of anthrax epidemics in other areas may also be attributed to combinations of weather and animal behavior. In the Etosha National Park, the seasonal peak of anthrax mortalities among elephants is in November at the end of the dry season, whereas among plains ungulates it occurs in March at the end of the rainy season (Lindeque and Turnbull 1994). Perhaps the difference in seasonal mortality patterns between these species can be attributed to differences in foraging behavior or use of water holes in relation to spore concentration sites. Information on the role of animal behavior in the epidemiology of anthrax is largely unavailable but is worthy of investigation to provide a more complete understanding of the infectious cycle in different ecosystems.

Direct transmission between herbivore hosts is considered to be of minor importance in anthrax epidemics. Furniss and Hahn (1981) refer to one instance where a kudu that licked the face of a recently dead member of the social group may have ingested spores directly from the carcass. We were unable to find reference to any other cases in the literature.

Infected herbivores shed spores in their feces and urine (Choquette 1970). Carcasses of animals killed by anthrax rarely remain intact long enough for putrefaction to eliminate the bacillus. Scavenging species that are typically resistant to anthrax open the carcasses, thereby promoting sporulation. Spores ingested by carnivore scavengers may pass through the gastrointestinal tract without infecting the animal and, through defecation, contaminate new sites (Pienaar 1967; Ebedes 1976; Turnbull et al. 1989).

Other routes of transmission have also been described. Necrophilic insects may contaminate forage (Pienaar 1967; de Vos 1990). Mechanical transmission of *B. anthracis* by hematophagous insects has been documented under laboratory and field conditions. Turell and Knudson (1987) confirmed the ability of stable flies *Stomoxys calcitrans* and two species of mosquitoes *Aedes aegypti* and *Aedes taeniorhynchus* to transmit the bacillus. They determined that all three insect species could transmit lethal anthrax infections to guinea pigs *Cavia porcellus* and mice in the laboratory. Large biting flies of the genus *Tabanus* can act as vectors of anthrax. Morris (1918) demonstrated that anthrax was transmitted most efficiently by tabanid flies that bit an infected host between 4 hours before death and a few minutes after death. This coincides with the highest concentration of bacilli in the peripheral circulation of the host (Krishna Rao and Mohiyudeen 1958). Biting flies have been implicated as vectors of anthrax in epidemics involving bison in northern Canada (Broughton 1987; Gates et al. 1995), large herbivores in Etosha National Park (Ebedes 1976), and livestock in northern Russia and India (Krishna Rao and Mohiyudeen 1958). Various lines of evidence suggest that biting insects play a role in transmitting anthrax to humans during an epidemic. Cutaneous anthrax infections associated with insect bites have been reported in Zimbabwe (McKendrick 1980; Levy et al. 1981; Davies 1983, 1985) and India (Krishna Rao and Mohiyudeen 1958).

Transmission to humans typically occurs as a result of incidental contact with infected animals or contaminated animal products or as a consequence of fly bites (McKendrick 1980). Anthrax epidemics continue to occur in developing countries where rural residents may use carcasses of infected or recently dead animals for food. Infections in such cases result from skinning, butchering, and transporting infected animal parts and from ingestion of infected meat.

CLINICAL SIGNS. Anthrax in herbivores is characterized by sudden onset, fever, rapidly progressing debility, respiratory distress, disorientation, and death within a few hours to several days (Pienaar 1961, 1967). Peracute and acute forms of the disease are recognized in herbivores. Animals that die of anthrax are otherwise often in good physical condition owing to the acute nature of the disease. Clinical signs in elephants were reported to be frequent micturition, restlessness, weakness of the hindquarters, and ataxia (Okewole et al. 1993). Cutaneous or localized swelling may be evident in various parts of the body (Stein and Van Ness 1955). Gates et al. (1995) observed extensive edema in the inner thigh, testicular, and perineal areas of several moribund or recently dead bison. Signs of antemortem struggling are typically absent. In dead animals, bloating and putrefaction are rapid (Pienaar 1967). Rigor mortis is incomplete or nonexistent. Carcasses become distended with putrefactive gases, resulting in a characteristic "sawhorse" position. Externally, there may be bloody discharges from the natural body orifices, frothy nasal discharges, and subcutaneous swellings.

PATHOGENESIS. Several forms of anthrax are recognized, depending on route of spore entry into the host. Three routes of infection occur: cutaneous, pulmonary, and gastrointestinal (Laforce et al. 1969; Borts 1972; Harrison et al. 1989). In wildlife, the gastrointestinal and pulmonary routes predominate. Following ingestion of spores, infection may occur through intact mucous membranes or defects in the epithelium of the oropharynx or intestinal tract. Following inhalation of spores, infection occurs through active transport of spores across alveolar membranes. In cutaneous anthrax, spores gain entry into the body through cuts and abrasions in the skin. Shortly after entry into the host, *B. anthracis* spores germinate into an encapsulated vegetative stage. The process by which this happens remains poorly understood (Titball and Manchee 1987). Anthrax characteristically begins as a lymphangitis and lymphadenitis and then proceeds to fatal septicemia.

In pulmonary anthrax, inhaled spores are ingested by alveolar macrophages and carried by draining lymphatics to major lymph nodes of the neck (Brachman et al. 1962; Brachman 1980). Airborne spores between 2 and 5 μm in diameter are responsible for infection, because spores of 5 μm or larger are either physically trapped in the nasopharynx or are cleared by the mucociliary system (Druett et al. 1953). The number of spores required to establish pulmonary infection may be fewer than 10 (Lincoln et al. 1967; Whitford 1979). The spores germinate into vegetative cells, which escape from the macrophage's phagosomes into its nutrient-rich cytoplasm and undergo rapid multiplication and produce toxins (Shafa et al. 1966). Intracellular replication continues until the macrophage can no longer contain the replicating bacilli and ruptures. The encapsulated bacilli continue to replicate and produce toxin within the regional lymph node (Brachman 1980). A similar mechanism occurs in the gastrointestinal form, in which spores are absorbed by mast cells in the intestine and transported to the mesenteric lymph nodes. There, the spores germinate and undergo replication and toxin production (Laforce 1994).

After proliferation within regional nodes, the microorganisms pass via lymphatic vessels into the bloodstream. Bacilli that enter the blood are taken up in other parts of the reticuloendothelial system, particularly the spleen, to establish secondary centers of infection and proliferation. Infection progresses to septicemia, with massive invasion of all body tissues. Terminal blood concentrations of bacilli are generally greater than 10^7 colony-forming units/ml (Lindeque and Turnbull 1994).

The vegetative stage of *B. anthracis* produces a lethal combination of exotoxins responsible for the signs of severe anthrax (Hoover et al. 1994). The bacilli themselves and their capsular material are virtually nontoxic, although the capsular material may inhibit the activity of leukocytes. The toxin complex consists of two separate protein toxins, designated *edema factor* (EF) and *lethal factor* (LF), and a cell receptor-binding protein called *protective antigen* (PA). Protective antigen combines with EF or LF to form *edema toxin* (ET) and *lethal toxin* (LT), respectively (Iacono-Connors et al. 1990). The individual components are not toxic (Iacono-Connors et al. 1990; Hoover et al. 1994), and all three appear necessary for virulence in an anthrax infection (Turnbull et al. 1992). The toxin complex acts to reduce phagocytosis, increase capillary permeability, and damage blood-clotting mechanisms. The net effect produces edema (Hoover et al. 1994), shock (Sterne 1959), acute renal failure (Harris-Smith et al. 1958), terminal anoxia mediated by the central nervous system, and ultimately death.

PATHOLOGY. Lesions associated with anthrax have a range of typical characteristics, although considerable variation in expression occurs among domestic species (Hutyra et al. 1938; Henning 1956; Sterne 1959) and wildlife species (Edelstein et al. 1990; Kriek and de Vos 1995). The primary lesions of anthrax result from widespread damage to the reticuloendothelial system and vasculature. In carcasses, the blood is generally dark and thick, clots poorly, and flows freely from cut surfaces (Kaufmann 1993). Blood-tinged discharges may exude from the nose, mouth, and anus. Small hemorrhages are common on mucous and serous membranes and in subcutaneous tissues. Hemorrhages of various sizes are common on serosal surfaces of the abdomen and thorax, as well as the epicardium and endocardium. A small amount of serous to blood-tinged fluid is usually present in body cavities.

Edema is a common feature of the disease. Frothy nasal discharges and marked subcutaneous edema may be noted externally (Turnbull et al. 1991; Gates et al. 1995). Localized to diffuse gelatinous edema is commonly present under the serosa of various organs, between skeletal muscle groups, in loose connective tissue, and in subcutaneous tissue (Pienaar 1967). Pulmonary and mediastinal edema may be seen in inhalation anthrax, and mesenteric edema may be seen with gastrointestinal infection. Severe edema of the brain has been reported consistently in cases of peracute septicemia (Kriek and de Vos 1995).

Severe local tissue damage may occur at the site of bacterial entry and in draining lymphatics and lymph nodes (Hoover et al. 1994). Acute pulmonary congestion and edema, consolidation of portions of the lung, and regional hemorrhagic lymphadenitis may occur in inhalation anthrax. Histologically, multifocal atelectasis, intra-alveolar fibrin exudation, and the presence of bacteria and neutrophils in alveolar capillaries occur (McGee et al. 1994). Ulcerative hemorrhagic enteritis is most common in gastrointestinal infection, although there may also be acute inflammation in the abomasum and large intestine. The lesions are most severe in the intestinal lymphoid tissue. The local mesentery may be edematous and hemorrhagic as a result of acute lymphangitis. Mesenteric lymph nodes are enlarged, congested, and hemorrhagic with bacilli and leukocytes in the sinuses and pulp.

The spleen is almost always grossly enlarged, dark, and soft. The necrotic and engorged red pulp readily exudes from incisions in the splenic capsule. There is generalized congestion, multifocal lymphocyte necrosis, and fibrin exudation. Histologically, the splenic vessels and pulp are obscured and suffused with blood, and lymphoid follicles are not readily apparent. Smears and sections of the spleen reveal very large numbers of leukocytes and bacilli if the carcass is fresh, but with advanced decomposition the organisms are destroyed by putrefactive changes. Congestion, swelling, and degeneration of other parenchymatous organs also occur.

Carnivores show a degree of resistance to anthrax (Turnbull et al. 1990; Creel et al. 1995). When infection occurs through the oropharynx, hemorrhagic lymphadenitis is seen in the nodes of the throat, with edema of local connective tissues (McGee et al. 1994). Occasionally, the lips, gingiva, and jowls are also involved, with locally extensive necrotic cellulitis and severe edema (Kriek and de Vos 1995). Less frequently, carnivores may develop gastrointestinal anthrax, with clinical signs and lesions of acute gastroenteritis.

DIAGNOSIS. Isolation of *B. anthracis* provides the standard for confirming anthrax. Direct sampling of cutaneous lesions of humans suspected of having anthrax is recommended. The organism may be easily recovered from an early lesion. Fluid and scrapings from the base of a previously unopened vesicle are best for culture. Specimens for bacteriologic study should be obtained before therapy is initiated, as vegetative cells are rapidly eliminated by antibiotics. Collection of blood, sputum, cerebrospinal fluid, or fecal samples may also demonstrate the organism, depending on the route of entry and seriousness of infection.

Carcasses of animals suspected of having anthrax should not be opened. Samples from external lesions or superficial tissues should be taken to minimize exposure of personnel to the organism. Protective clothing, especially gloves, should always be worn and later disposed of or disinfected safely. An enlarged spleen is suggestive, but not diagnostic, of anthrax, and all autopsy procedures should be terminated if this is seen and there is reason to suspect anthrax. Concentrations of *B. anthracis* in the body are greatest in the last few hours before death and in the first hour after death and can be isolated with little difficulty from the blood. After the first hour of death, putrefaction becomes a problem, and it becomes more difficult to isolate the organism due to overgrowth of anaerobes present in decomposing tissue. Blood, extracellular fluid at sites of swelling, and swabbings of the mouth, nares, and anus can be taken to attempt culture of the organism. Because contaminating microorganisms originate from the intestinal tract, the extremities are the last place to experience the putrefactive process and offer the best chance of recovery of the organism. Ears and sections of tail may be collected for submission for bacteriologic culturing. In bison, *B. anthracis* was most consistently detected from intact ear samples (Gates et al. 1995). The outside of the tissue should be thoroughly soaked with 70% ethanol to decrease the chances of contamination of cultures by commensal organisms. All samples should be carefully packaged in airtight biohazard-approved packaging and transported cool. *Bacillus anthracis*-infected specimens may be frozen to −70° C; however, repeated freezing and thawing should be avoided (Doyle et al. 1985).

On microscopic examination of Gram-stained smears, anthrax bacilli appear as long rods (3–5 μm long and 1–1.5 μm wide) with flattened ends (Doyle et al. 1985). In clinical specimens, vegetative cells usually occur as short chains and spores are rarely observed. In culture, long chains are typically formed with an appearance often described as resembling a jointed bamboo shoot. Endospores are central to subterminal and nonswollen and occur frequently in a free state in culture.

The Ascoli thermoprecipitin reaction may be useful in postmortem diagnosis. Potentially infected material is ground in saline, heated for 5 minutes, and filtered. The filtrate is then layered over appropriate antianthrax serum. In positive reactions, a ring of precipitate is formed at the interface of the two fluids (Wright 1965). Great care and the use of proper containment facilities are essential while performing the assay in order to prevent aerosolization of potentially infectious agents.

On sheep blood agar after 18–24 hours at 35° C, *B. anthracis* colonies reach 2–5 μm in diameter and are nonhemolytic except for weak hemolysis by some strains in areas of high density (Doyle et al. 1985). Colonies older than 36 hours have a rough texture (ground-glass appearance) and a serrated edge. They often form curled peripheral projections, producing the classic medusa head appearance.

Specimens should be cultured on nutrient or brain-heart infusion agar supplemented with 0.5% sodium bicarbonate and incubated in a 5% CO_2 atmosphere or in a candle jar. Virulent *B. anthracis* strains become encapsulated, resulting in mucoid colonies in the presence of elevated CO_2. No other *Bacillus* species does this. For isolation of *B. anthracis* from contaminated tissue or soil samples, selective PLET medium may be useful (Knisely 1966; Carman et al. 1985).

A motility test is a useful and reliable procedure for preliminary screening of suspected *B. anthracis* isolates (Doyle et al. 1985). Any *Bacillus* isolate exhibiting motility may safely be assumed to be a species other than *B. anthracis.* Susceptibility to penicillin is also a useful preliminary test. *Bacillus anthracis* strains are generally highly susceptible to penicillin, whereas *B. cereus, B. cereus* var. *mycoides,* and *B. thuringiensis* are resistant.

Detection of the components of the anthrax toxin is also diagnostic. This can be accomplished with various monoclonal and polyclonal sera against the components in immunodiffusion or enzyme-linked immunosorbent assay (ELISA) using tissue specimens or cultured organisms. Recently, a hand-held ELISA for detection of PA has become commercially available (Burans et al. 1996). The assay detects PA in human or animal cases with a calorimetric qualitative result. The assay has a built-in positive control and can give results within 2 hours. Routine serology of carnivores by using this test could be a useful diagnostic aid to indicate the presence or prevalence of anthrax in a known or suspect area (Turnbull et al. 1992).

In 1962, the former Soviet Union began licensing anthraxin, an antigenic preparation derived from an nonencapsulated *B. anthracis* strain, as an assay to detect an antianthrax delayed hypersensitivity reaction in suspected cases (Shlyakhov 1996). Anthraxin works in the same way as tuberculin testing: it is administered intradermally and, after 18–24 hours, elicits a local erythema and skin induration in acute or recovered anthrax patients. The assay can confirm anthrax infection up to 31 years after initial infection (Shlyakhov 1996).

DIFFERENTIAL DIAGNOSES. In humans, cutaneous anthrax must be differentiated from other ulceration skin diseases involving regional lymphadenopathy, including rat-bite fever, ulceroglandular tularemia, plague, glanders, rickettsial pox, orf (contagious ecthyma) virus infection, and cowpox. Staphylococcal lymphangitis may be distinguished from anthrax by the discharge of purulent material and by the inflammatory response observed microscopically. Early clinical diagnosis of inhalation anthrax is next to impossible. Initial symptoms mimic an influenza-like infection (Laforce 1994). The fulminating course of the advanced disease may resemble cardiac failure or cerebrovascular accident. Mediastinal widening demonstrable by radiology is suggestive of inhalation anthrax (Wright 1965). Gastrointestinal anthrax must be differentiated from other causes of abdominal catastrophe and, if there is hemorrhage, from duodenal ulcer, typhoid, and intestinal tularemia. Anthrax can progress to hemorrhagic meningitis, which must be differentiated from subarachnoid hemorrhage (Haight 1952). Because of the nonspecific nature of the first stage of inhalation anthrax and the clinical presentation of the gastrointestinal form, a history of exposure and isolation of the organism are extremely important for establishing the diagnosis.

In animals, anthrax must be differentiated from other conditions that cause sudden death. In herbivores, clostridial infections such as blackleg, malignant edema, and bacillary hemoglobinuria may be confused with anthrax. Other conditions that should be considered include lightning strike, trauma, bloat, and acute poisonings induced by bracken fern or lead poisoning. Agents causing other systemic infections such as acute leptospirosis, yersiniosis, and malignant catarrhal fever should also be ruled out. In carnivores, differential diagnoses include acute systemic infections and pharyngeal swellings due to other causes.

IMMUNITY. Immunity to anthrax is not completely understood. Humoral immunity plays a role in protection against anthrax; PA is the main immunogen, although anti-PA titers alone do not reliably predict survival from spore challenges (Ivins et al. 1994). Antibodies against the bacterial cell and capsule themselves are not protective. Although antibodies play a role in specific immunity (Gladstone 1946; Belton and Strange 1954; Little et al. 1994), cell-mediated immunity may be a critical part of the response (Ivins et al. 1995).

Credit for the first recorded demonstration of protection induced by attenuated strains of *B. anthracis* belongs to W.S. Greenfield of the Brown Animal Sanatory Institution in London (Greenfield 1880a,b). Greenfield demonstrated that the microorganism progressively lost virulence when continuously subcultured in a fluid medium. Injection of suitably attenuated organisms into cattle rendered them immune to subsequent challenge with fully virulent bacilli. However, due to a combination of lack of funding and opposition by the antivivisectionist movement that was popular in Britain at the time, Greenfield was forced to confine his experiments to a small number of animals, and his findings, though conclusive, were overlooked.

Instead, it was Pasteur's vaccine, field tested less than 1 year after Greenfield described his results, that was adapted for wide-scale use in cattle and sheep, particularly in Europe and South America, over the next 50 years, with various modifications (Turnbull 1991).

In 1939, Sterne isolated an avirulent nonencapsulated variant of *B. anthracis,* $34F_2$, from which he developed a live attenuated spore vaccine (Hambleton et al. 1984). The vaccine included saponin as an adjuvant. The formulation of livestock vaccines used in most of the world today remains essentially as specified by Sterne and use derivatives of $34F_2$. Since its introduction, the Sterne vaccine has proved safe and extremely effective.

Presently, there are three primary vaccines for use in humans. The Russian vaccine is comprised of a live spore vaccine reported to be a derivative of Sterne's $34F_2$ strain (Turnbull 1991). The vaccine used in the United Kingdom consists of an alum-precipitated cell-free filtrate of Sterne strain cultures grown so as to maximize PA content. The vaccine used in the United States is an aluminum hydroxide-absorbed cell-free filtrate of a nonencapsulated, nonproteolytic derivative of an isolate from a bovine case of anthrax in Florida in 1951. This strain produces higher PA and lower EF and LF concentrations than the U.K. vaccine.

Because of its central role in the delivery and action of anthrax toxin, PA is an ideal target for the immune response. A humoral response to PA and, to a more limited extent, to LF and EF plays the dominant role in providing protection from anthrax infection. There is poor correlation, however, between elevated anti-PA titers and protection against virulent anthrax strains in experimental animals. Positive anthraxin results correlate extremely well with protective immunity elicited by vaccines in animal experiments. The delayed hypersensitivity response to anthraxin in infected or vaccinated patients indicates that the cell-mediated arm of the immune response is also involved in protection against anthrax infection. The exact nature of this protection is unknown.

Elevated levels of stress have been implicated in depressing the immune system and increasing an animal's susceptibility to anthrax and other diseases (Klein 1993). Gainer and Saunders (1989) hypothesized that conditions often associated with anthrax epidemics, including insect harassment, overgrazing, heat, breeding activity, and concentration of animals around limited resources, may modify host resistance to anthrax bacilli already present at low levels in the body and allow disease to develop. Provost and Tronette (1957) and Stein (1948) observed anthrax bacilli at low levels in the retropharyngeal lymph nodes of otherwise healthy animals from endemic areas.

Although stress can and probably does play a role in host susceptibility to anthrax, the extent of its role in the epidemiology of an outbreak is unknown. The aforementioned conditions or other factors that cause them could just as easily be responsible for altering contact patterns between the herbivore host and the pathogen. Prolonged rain or flooding followed by drought, which often precedes anthrax outbreaks, creates ideal conditions for increased populations of mosquitoes and other biting insects, yet could also act to redistribute and concentrate anthrax spores in low-lying areas and on vegetation. Feed shortages due to drought would result in animals crowding to the last supplies of feed, which are often in low-lying areas close to evaporating water sources and, if these areas were contaminated with anthrax spores, there would be an increased probability of transmission of the disease. Feeding on dried vegetation, or change of diet due to drought, could result in damage to the mucosa of the mouth and esophagus, providing easier entrance into the animal for the bacilli. More research on the location and concentration of anthrax spores in the environment and on seasonal modification of herbivore immune responses to *B. anthracis* is needed to determine how conditions associated with anthrax epidemics precipitate an outbreak.

CONTROL AND TREATMENT. Prior to the 1890s, there was no effective treatment for anthrax aside from making the patients comfortable or killing the animals. Charms, incantations, and medicines based on superstitions abounded but were not efficacious. In 1895, Sclavo of Italy produced an antiserum to *B. anthracis* by injecting horses with low concentrations of the microorganism (Klemm and Klemm 1959). This antiserum when injected into patients proved a very effective treatment, and the procedure was put into widescale use. The horse antiserum remained the most effective anthrax treatment for the next 30–40 years. In 1926, Pijer of Great Britain introduced neoarsphenamine as an effective treatment of the disease. In 1933, Eurich treated 200 human anthrax cases with a combined treatment of neoarsphenamine and horse antiserum and witnessed only 5% fatality.

During World War II, penicillin was mass produced and available for the first time as a potential treatment for a number of bacterial pathogens. The first recorded use of penicillin against anthrax was by Murphy et al. (1944), who successfully treated three female wool workers who had cutaneous anthrax. The dosage of penicillin used ranged from 150,000 to 475,000 units. One of the first reports of the use of penicillin in an animal anthrax case was by Sugg (1948). He used 10^6 units in eight affected calves. Although the animals' temperatures soared over 41° C, all eight recovered completely.

Although there have been a few reports of penicillin-resistant strains of *B. anthracis* (Doganay and Aydin 1991; Bradaric and Punda-Polic 1992), penicillin is still the treatment of choice for anthrax and is given for 5–7 days. Tetracycline, erythromycin, or chloramphenicol are also effective (Benenson 1995). Antibiotic is only effective against the vegetative form of *B. anthracis* and not the spore form. Spores can remain in the alveolar spaces of the lungs for extended periods following inhalation. Because of this, it has been found effective to use vaccination in conjunction with antibiotic treatment when dealing with inhalation anthrax (Friedlander et al. 1993). The antibiotic eliminates any *B. anthracis* that germinate in the body prior to the development of immunity, which then provides protection once the antibiotic treatment is discontinued.

Although antibiotic treatment is also effective in animal anthrax, it is often not a practical option in controlling epidemics in large animal populations. Instead, control of an epidemic is commonly limited to removing diseased carcasses as sources of contamination from the environment. This is accomplished by burning or deep burial of the carcass, preferably at the site it was found. In many developing countries, shortages of wood and other combustibles create serious problems in controlling outbreaks. With large animals, such as giraffe *Giraffa camelopardalis* and elephant, there are also logistic problems of being able to dig a big enough hole. In some African countries, the problem of carcass disposal is circumvented by covering the carcass with branches of large thorn bushes, such as *Acacia seyal,* to discourage scavengers from opening the carcass, allowing the bacilli in the carcass to be destroyed naturally by the putrefactive process (Bbalo 1996; Jiwa 1996). With animal species possessing thick hides, such as giraffe and hippopotamus *Hippopotamus amphibius,* thorn branches are only required to cover the natural body openings. Gates et al. (1995) applied dilute formaldehyde to the exterior of carcasses, thereby effectively preventing scavenging.

Vaccination may be used as a preventative measure in populations potentially at risk. Aerial vaccination of susceptible wild herbivores in the Kruger National Park has helped to bring an abrupt end to anthrax epizootics raging there (de Vos and Scheepers 1996).

PUBLIC HEALTH CONCERNS. Preventative measures against human anthrax (Benenson 1995; Turnbull 1996) include: (1) Immunization of persons in high-risk occupations, such as wildlife management personnel working in anthrax endemic areas, laboratory personnel routinely working with *B. anthracis,* and workers handling potentially contaminated industrial raw materials such as hair, wool, or hides. Annual boosters are recommended, if exposure is continual. (2) Education of persons, including hunters and trappers, who handle potentially contaminated materials about anthrax. In anthrax-endemic areas, hunters and other persons should be advised against opening carcasses of animals found dead. (3) Proper ventilation and dust control when handling potentially contaminated raw animal materials. Appropriate protective clothing and gloves should be worn. Eating facilities should be located away from animal-processing areas. There should be prompt medical care of all suspicious skin lesions. (4) Thorough washing, disinfection, or sterilization of possibly contaminated hair, wool, hides, and bonemeal or other feed of animal origin, prior to processing. (5) Avoid selling the hides of animals exposed to anthrax or using their carcasses as food or feed supplements. In anthrax-endemic areas, this should include animal products derived from animals dead of unknown causes. (6) If anthrax is suspected, do not necropsy or open the animal. If the carcass is inadvertently opened, autoclave, incinerate, or chemically disinfect all instruments and materials that come in contact with the carcass. (7) In anthrax-endemic areas, avoid drinking unboiled or untreated surface water. Water should be boiled for at least 30 minutes before ingestion. (8) Immunize animals at risk, when appropriate and practical.

DOMESTIC ANIMAL HEALTH CONCERNS. Areas where anthrax is endemic among wild animals may also constitute important reservoirs from which the disease might spread to agricultural species (Peinaar 1967) and vice versa. Transmission may occur through the dissemination of spores by birds and other carrion feeders that have fed on anthrax carcasses. Anthrax occurs very irregularly even in anthrax-endemic areas (Kaufmann 1993), but given the extreme susceptibility of some species, steps should be taken to protect livestock in these areas. Annual vaccination of susceptible livestock is recommended. Vaccinations should be done 2–4 weeks prior to the season when outbreaks are expected, and animals should not be vaccinated within 60 days of slaughter (Kaufmann 1993).

Anthrax is a reportable disease in most countries, and all outbreaks should be reported to local regulatory and public health officials as stipulated by laws enacted to control and eradicate the disease. When an outbreak occurs on a farm, animal disease control measures are implemented that include facility quarantine, prompt disposal of carcasses and other contaminated material, isolation of sick animals, vaccination of surviving animals, control of scavengers and insects, and restrictions on the movement of meat and milk from the farm during the quarantine period. Carcasses of animals that die of anthrax should be either burned completely or buried deeply (2 m of soil) to reduce environmental contamination. Bedding and other contaminated material should also be burned or buried, and chemical decontamination of facilities may be required. In the face of an outbreak, sick animals should be isolated and unaffected animals separated from the contaminated area. The administration of hyperimmune serum to in-contact animals may prevent further losses (Blood and Radostits 1989), and prophylactic treatment with penicillin may also be effective in animals at risk (Fraser et al. 1991).

MANAGEMENT IMPLICATIONS. The effect of anthrax on wild mammal populations depends on the frequency of epidemics, their severity, and the specific susceptibility of age and sex classes. The characteristics of epidemics vary between host species. In highly susceptible species such as impala (Prins and Weyerhaeuser 1987) and roan antelope *Hippotragus equinus* (de Vos and Scheepers 1996), anthrax epidemics may be catastrophic, resulting in massive mortality and affecting all sectors of the population. An outbreak could potentially result in the extirpation of a population from direct mortality or through cascading ecological effects. In other species such as bison (Gates et al. 1995), springbok, blue wildebeest, and elephants (Lindeque and Turnbull 1994), mortality may be highly sex and age specific, affecting mainly adult males, thus having little impact on the viability of populations or on population growth. The sporadic occurrence of anthrax epidemics in endemic areas reduces the potential impact on the long-term viability of demographically robust populations. Endangered or small remnant populations of highly susceptible herbivores risk the most serious impact from an anthrax outbreak, yet there is no record of a population being eliminated by this disease.

Population management prescriptions may need to take anthrax into account. For example, it may be necessary to limit hunting quotas for trophy-class animals following a heavy loss of mature males to anthrax. A complete ban on hunting may be necessary where high mortality has been caused by anthrax. The potential for infectious diseases such as anthrax to occur should be examined to minimize the probability of failure of introductions or reintroductions of endangered or previously extirpated herbivores. Vaccination of animals prior to release should provide protection in the short term. Subsequent vaccinations may be required periodically to ensure that a small reintroduced population continues to grow without undue losses during the initial years following release into a contaminated area.

Control measures may impact on human use of an endemic area. For example, public access to the Mackenzie Bison Range in the Northwest Territories was closed during an anthrax outbreak in 1993 (Gates et al. 1995). This precluded use of the area for wilderness travel, hunting, and other activities during 2 months. Control measures are expensive to implement and could detract from resources available to other management programs when budgets are limited. Prins and Weyerhaeuser (1987) reported that the budget of Tanzania National Parks did not allow for treatment of watering holes with antibiotics (Pienaar 1961 1967) or for aerial immunization (de Vos et al. 1973) as was practiced in Kruger National Park. Measures such as incineration and burials may not always be practical or affordable. Other methods to reduce environmental contamination, such as protecting carcasses from scavengers to allow natural putrefaction or letting nature take its course, may in some cases be the only practical alternatives. Local endemism of anthrax has implications for the management of the health of livestock on game or conventional ranches. Annual vaccinations of all stock would be well advised in these areas.

The response of a local human population to an outbreak may be quite fearful, depending on how effectively information is provided. The fear of infection may influence the future use of the resource unless there is clear understanding of how the risk can be managed.

There appears to be no restriction on international trade in animal products from endemic areas, although some nations may require decontamination of goods imported from areas known to experience anthrax outbreaks. Anthrax is a reportable disease in European and North American countries. In many nations, anthrax in domestic stock is controlled through required vaccinations on endemic properties. We could find no examples of legislated control measures for wild mammal populations in any country.

LITERATURE CITED

Abramova, F.A., L.M. Grinberg, O.V. Yampolskaya, and D.H. Walker. 1993. Pathology of inhalational anthrax in 42 cases from the Sverdlovsk outbreak of 1979. *Proceedings of the National Academy of Sciences U.S.A.* 90:2291–2294.

Bakken, L.B. 1985. Separation and purification of bacteria from soil. *Applied and Environmental Microbiology* 49:1482–1487.

Bbalo, G. 1996. Anthrax in Western Province, Zambia. *Salisbury Medical Bulletin* 87(Special Suppl.):11–12.

Belton, F.C., and R.E. Strange. 1954. Studies on a protective antigen produced *in vitro* from *Bacillus anthracis:* Medium and methods of production. *British Journal of Experimental Pathology* 35:144–152.

Benenson, A.S., ed. 1995. Anthrax. In *Control of communicable diseases manual.* Washington, DC: American Public Health Association, pp. 18–22.

Blood, D.C., and O.M. Radostits. 1989. *Veterinary medicine.* Toronto: Bailliere Tindall, pp. 592–596.

Borts, I.H. 1972. Anthrax. In *Communicable and infectious diseases,* ed. F.H. Top and P.F. Wehrle, 7th ed. St. Louis: C.V. Mosby, pp. 108–112.

Brachman, P.S. 1980. Inhalation anthrax. *Annals of the New York Academy of Sciences* 353:83–93.

Brachman, P.S., H. Gold, S.A. Plotkin, F.R. Fekety, M. Werrin and N.R. Ingraham. 1962. Field evaluation of a human anthrax vaccine. *American Journal of Public Health* 52:632–645.

Bradaric, N., and V. Punda-Polic. 1992. Cutaneous anthrax due to penicillin-resistant *Bacillus anthracis* transmitted by an insect bite. *Lancet* 340:306–307.

Broughton, E. 1987. Diseases affecting bison. In *Bison ecology in relation to agricultural development in the Slave River Lowlands, NWT,* ed. H.W. Reynolds and A.W.L. Hawley. Ottawa: Canadian Wildlife Service, pp. 34–38; Occasional Paper 63.

Bryden, J. 1989. Deadly allies: Canada's secret war 1937–1947. Toronto: McClelland and Stewart, 316 pp.

Burans, J., A. Keleher, T. O'Brien, J. Hager, A. Plummer, and C. Morgan. 1996. Rapid method for the diagnosis of *Bacillus anthracis* infection in clinical samples using a hand-held assay. *Salisbury Medical Bulletin* 87(Special Suppl.):36–37.

Carman, J.A., P. Hambleton, and J. Melling. 1985. *Bacillus anthracis.* In *Isolation and identification of micro-organisms of medical and veterinary importance,* ed. C.H. Collins and J.M. Grange. London: Academic, pp. 207–213.

Cherkasskiy, B. 1996. Anthrax in Russia. *Salisbury Medical Bulletin* 87(Special Suppl.):6–7.

Choquette, L.P.E. 1970. Anthrax. In *Infectious diseases of wild mammals,* ed. J.W. Davis, L.H. Karstad, and D.O. Trainer. Ames: Iowa State University Press, pp. 256–266.

Christie, A.B. 1987. *Infectious diseases: Epidemiology and clinical practice,* 4th ed. Edinburgh: Churchill Livingstone, pp. 983–1003.

Claus, D., and R.C.W. Berkeley. 1986. Genus *Bacillus.* In *Bergey's manual of systematic bacteriology,* ed. P.H.A. Sneath. Los Angeles: Williams and Wilkins, pp. 1105–1139.

Creel, S., N. Marusha Creel, J.A. Matovelo, M.M.A. Mtambo, E.K. Batamuzi, and J.E. Cooper. 1995. The effects of anthrax on endangered African wild dogs (*Lycaon pictus*). *Journal of Zoology (London)* 236:199–209.

Davies, J.C. 1983. A major epidemic of anthrax in Zimbabwe: II. Distribution of cutaneous lesions. *Central African Journal of Medicine* 29:8–12.

———. 1985. A major epidemic of anthrax in Zimbabwe: The experience at the Beatrice Road Infectious Disease Hospital, Harare. *Central African Journal of Medicine* 31:176–180.

de Vos, V. 1990. The ecology of anthrax in Kruger National Park, South Africa. *Salisbury Medical Bulletin* 87(Special Suppl.):19–23.

de Vos, V., and H.B. Bryden. 1996. Anthrax in the Kruger National Park: Temporal and spatial patterns of disease occurrence. *Salisbury Medical Bulletin* 87(Special Suppl.):26–30.

de Vos, V., and G.J. Scheepers. 1996. Remote mass vaccination of large free-ranging wild animals for anthrax using Sterne spore vaccine. *Salisbury Medical Bulletin* 87(Special Suppl.):116–121.

de Vos, V., G.L. Rooyen, and J.J. Kloppers. 1973. Anthrax immunization of free-ranging roan antelope *Hippotragus equinus* in the Kruger National park. *Koedoe* 16:11–25.

Doganay, M. 1996. Human anthrax in Turkey. *Salisbury Medical Bulletin* 87(Special Suppl.):8.

Doganay, M., and N. Aydin. 1991. Antimicrobial susceptibility of *Bacillus anthracis. Scandinavian Journal of Infectious Diseases* 23:333–335.

Doyle, R.J., F. Nedjat-Haiem, and J.S. Singh. 1984. Hydrophobic characteristics of *Bacillus* spores. *Current Microbiology* 10:329–332.

Doyle, R.J., K.F. Keller, and J.W. Ezzell. 1985. *Bacillus.* In *Manual of clinical microbiology,* ed. E.H. Lennette, A. Balows, W.J.J. Hausler, and H.J. Shadomy, 4th ed. Washington, DC: American Society for Microbiology, pp. 211–215.

Dragon, D.C., and R.P. Rennie. 1995. The ecology of anthrax spores: Tough but not invincible. *Canadian Veterinary Journal* 36:295–301.

Druett, H.A., D.W. Henderson, L. Packman, and S. Peacock. 1953. Studies on respiratory infection: I. The influence of particle size on respiratory infection with anthrax spores. *Journal of Hygiene* 51:359–371.

Ebedes, H. 1976. Anthrax epizootics in Etosha National Park. *Madoqua* 10:99–118.

Edelstein, R.M., R.N. Gourlay, G.H.K. Lawson, A.N. Morrow, and S. Ramachandran. 1990. Diseases caused by bacteria. In *Handbook on animal diseases in the tropics,* ed. M.M.H. Sewell and D.W. Brocklesby, 4th London: Bailliere Tindall, pp. 33–103.

Foster, S.J., and K. Johnstone. 1990. Pulling the trigger: The mechanism of bacterial spore germination. *Molecular Microbiology* 4:137–141.

Fraser, C.M., J.A. Bergeron, A. Mays, and S.E. Aiello, eds. 1991. Anthrax. In *The Merck veterinary manual,* 7th ed. Rahway, NJ: Merck, pp. 320–322.

Friedlander, A.M., S.L. Welkos, M.L.M. Pitt, J.W. Ezzell, P.L. Worsham, K.J. Rose, B.E. Ivins, J.R. Lowe, G.B. Howe, P. Mikesell, and W.B. Lawrence. 1993. Postexposure prophylaxis against experimental inhalation anthrax. *Journal of Infectious Diseases* 167:1239–1243.

Furniss, P.R., and B.D. Hahn. 1981. A mathematical model of an anthrax epizootic in the Kruger National Park. *Applied Mathematical Modelling* 5:130–136.

Gainer, R.S., and J.R. Saunders. 1989. Aspects of the epidemiology of anthrax in Wood Buffalo National Park and environs. *Canadian Veterinary Journal* 30:953–956.

Gates, C.C., B.T. Elkin, and D.C. Dragon. 1995. Investigation, control and epizootiology of anthrax in a geographically isolated, free-roaming bison population in northern Canada. *Canadian Journal of Veterinary Research* 59:256–264.

Gladstone, G.P. 1946. Immunity to anthrax: Protective antigen present in cell-free culture filtrates. *British Journal of Experimental Pathology* 27:394–418.

Gould, G.W. 1966. Stimulation of L-alanine induced germination of *Bacillus cereus* spores by D-cycloserine and O-carbamyl-D- serine. *Journal of Bacteriology* 92:1261–1262.

———. 1977. Recent advances in the understanding of resistance and dormancy in bacterial spores. *Journal of Applied Bacteriology* 42:297–309.

———. 1984. Mechanisms of resistance and dormancy. In *The bacterial spore,* ed. A. Hurst and G.W. Gould. Toronto: Academic, pp. 173–209.

Greenfield, W.S. 1880a. Lectures on some recent investigations into the pathology of infectious and contagious diseases: Lecture III—Part 1. *Lancet* 1:865–867.

———. 1880b. Quotation in notes, comments and answers to correspondents. *Lancet* 1:586.

Haight, T.H. 1952. Anthrax meningitis: Review of literature and report of two cases with autopsies. *American Journal of Medical Sciences* 224:57–69.

Hambleton, P., J.A. Carman, and J. Melling. 1984. Anthrax: The disease in relation to vaccines. *Vaccine* 2:125–132.

Harris, S. 1992. Japanese biological warfare research on humans: A case study of microbiology and ethics. *Annals of the New York Academy of Sciences* 666:21–52.

Harrison, L.H., J.W. Ezzell, T.G. Abshire, S. Kidd, and A.F. Kaufmann. 1989. Evaluation of serologic tests for diagnosis of anthrax after an outbreak of cutaneous anthrax in Paraguay. *Journal of Infectious Diseases* 160:706–710.

Harris-Smith, P.W., H. Smith, and J. Keppie. 1958. Production in vitro of the toxin of *Bacillus anthracis* previously recognised in vivo. *Journal of General Microbiology* 19:91–103.

Henning, M.W.H. 1956. Anthrax. In *Animal diseases in South Africa,* 3d ed. South Africa: Central News Agency, pp. 3–29.

Hills, G.M. 1949. Chemical factors in the germination of spore bearing aerobes: The effect of yeast extract on the germination of *Bacillus anthracis* and its replacement by adenosine. *Biochemical Journal* 45:363–370.

———. 1950. Chemical factors in the germination of spore bearing aerobes: Observation on the influence of species, strain and conditions of growth. *Journal of General Microbiology* 4:38–47.

Hoover, D.L., A.M. Friedlander, L.C. Rogers, I.-K. Yoon, R.L. Warren, and A.S. Cross. 1994. Anthrax edema toxin differentially regulates lipopolysaccharide-induced monocyte production of tumor necrosis factor alpha and interleukin-6 by increasing intracellular cyclic AMP. *Infection and Immunity* 62:4432–4439.

Hugh-Jones, M.E. 1996. World situation 1993/94. *Salisbury Medical Bulletin* 87(Special Suppl.):1–2.

Humphrey, J.H. 1987. Biological weapons: Banned but gone forever? *Medicine and War* 3:23–30.

Hutyra, F., J. Marek, and R. Manninger. 1938. *Special pathology and therapeutics of the diseases of domestic animals,* 4th ed. London: Bailliere, Tindall and Cox, 320 pp.

Iacono-Connors, L.C., C.S. Schmaljohn, and J.M. Dalrymple. 1990. Expression of the *Bacillus anthracis* protective antigen gene by baculovirus and vaccinia virus recombinants. *Infection and Immunity* 58:366–372.

Ivins, B.E., P.F. Fellows, and G.O. Nelson. 1994. Efficacy of a standard human anthrax vaccine against *Bacillus anthracis* spore challenge in guinea pigs. *Vaccine* 12:872–874.

Ivins, B.E., P. Fellows, L. Pitt, J. Estep, J. Farchaus, A. Friedlander, and P. Gibbs. 1995. Experimental anthrax vaccines: Efficacy of adjuvants combined with protective antigen against an aerosol *Bacillus anthracis* spore challenge in guinea pigs. *Vaccine* 13:1779–1784.

Jiwa, S.F.H. 1996. Experience with anthrax control in areas of Tanzania. *Salisbury Medical Bulletin* 87(Special Suppl.):10–11.

Kahler, S. 1992. Veterinarian leads team of UN inspectors in Iraq. *Journal of the American Veterinary Medical Association* 200:259–260.

Kaufmann, A.F. 1993. Anthrax. In *Current veterinary therapy 3: Food animal therapy,* ed. J.L. Howard. Philadelphia: W.B. Saunders, pp. 565–567.

Klein, T.W. 1993. Psychoimmunology and infection. *Clinical Microbiology Newsletter* 15:17–22.

Klemm, D.M., and W.R. Klemm. 1959. A history of anthrax. *Journal of the American Veterinary Medical Association* 135:458–462.

Knisely, R.F. 1966. Selective medium for *Bacillus anthracis. Journal of Bacteriology* 92:784–786.

Komers, P.E. 1992. Mating strategies of male wood bison. Ph.D. thesis, University of Saskatchewan, Saskatoon, Canada, 121 pp.

Koshikawa, T., M. Yamazaki, M. Yoshimi, S. Ogawa, A. Yamada, M. Watabe, and M. Torii. 1989. Surface hydrophobicity of spores of *Bacillus* spp. *Journal of General Microbiology* 135:2717–2722.

Kriek, N.P.J., and V. de Vos. 1995. Species differences in the pathology of wildlife in the Kruger National Park, South Africa. *Salisbury Medical Bulletin* 87(Special Suppl.):82.

Krishna Rao, N.S., and S. Mohiyudeen. 1958. *Tabanus* flies as transmitters of anthrax: A field experience. *Indian Veterinary Journal* 35:348–353.

Laforce, F.M. 1994. Anthrax. *Clinical Infectious Disease* 19:1009–1014.

Laforce, F.M., F.H. Bumford, J.C. Feeley, S.L. Stokes, and D.B. Snow. 1969. Epidemiologic study of a fatal case of inhalation anthrax. *Archives of Environmental Health* 18:798–805.

Lalitha, M.K., D. Mathai, K. Thomas, A. Kumar, A. Ganesh, M. Jacob, and T.J. John. 1996. Anthrax: A continuing problem in Southern India. *Salisbury Medical Bulletin* 87(Special Suppl.):14–15.

Levy, L.M., N. Baker, M.P. Meyer, P. Crosland, and J. Hampton. 1981. Anthrax meningitis in Zimbabwe. *Central African Journal of Medicine* 27:101–104.

Lincoln, R.E., J.S. Walker, F. Klein, A.J. Rosenwald, and W.I. Jones Jr. 1967. Value of field data for extrapolation in anthrax. *Federation Proceedings* 26:1558–1562.

Lindeque, P.M., and P.C.B. Turnbull. 1994. Ecology and epidemiology of anthrax in the Etosha National Park, Namibia. *Onderstepoort Journal of Veterinary Research* 61:71–83.

Lindeque, P.M., C. Brain, and P.C.B. Turnbull. 1996. A review of anthrax in the Etosha National Park. *Salisbury Medical Bulletin* 87(Special Suppl.):24–26.

Little, S., B. Ivins, P. Fellows, and A. Friedlander. 1994. Passive protection of guinea pigs with monoclonal antibodies against *Bacillus anthracis* infection. In *Abstracts of the 94th annual meeting of the American Society for Microbiology.* Las Vegas: American Society for Microbiology, p. 154.

MacDonald, D.W., S.R. Rawluk, and V.P.J. Gannon. 1992. Anthrax in cattle. *Canadian Veterinary Journal* 33:135.

Manchee, R.J., M.G. Broster, J. Melling, R.M. Henstridge, and A.J. Stagg. 1981. *Bacillus anthracis* on Gruinard Island. *Nature* 294:254–255.

Manchee, R.J., M.G. Broster, A.J. Stagg, S.E. Hibbs, and B. Patience. 1990. Out of Gruinard Island. *Salisbury Medical Bulletin* 87(Special Suppl.):17–18.

Maxy, K.F. 1951. *Preventative medicine & hygiene.* New York: Appleton-Century-Crofts, 518 pp.

McGee, E.D., D.L. Fritz, J.W. Ezzell, H.L. Newcomb, R.J. Brown, and N.K. Jaax. 1994. Anthrax in a dog. *Veterinary Pathology* 31:471–473.

McKendrick, D.R.A. 1980. Anthrax and its transmission to humans. *Central African Journal of Medicine* 26:126–129.

McSherry, J., and R. Kilpatrick. 1992. The plague of Athens. *Journal of the Royal Society of Medicine* 85:713.

Meselson, M., J. Guillemin, M. Hugh-Jones, A. Langmuir, I. Popova, A. Shelokov, and O. Yampolskaya. 1994. The Sverdlovsk anthrax outbreak of 1979. *Science* 266:1202–1208.

Mierzejewski, J., and M. Bartoszcze. 1991. Decontamination of soil after bacterial warfare experiments on Gruinard Island. *Przeglad Epidemiologiczny* 45:197–205.

———. 1996. Literature on anthrax as a biological weapon. *Salisbury Medical Bulletin* 87(Special Suppl.):67–68.

Morris, H. 1918. Blood-sucking insects as transmitters of anthrax. *Bulletin of the Louisiana Agricultural Experiment Station* 163.

Murphy, F.D., A.C. LaBocetta, and J.A. Lockwood. 1944. Treatment of human anthrax with penicillin. *Journal of the American Medical Association* 126:948–950.

Okewole, P.A., I.L. Oyetunde, E.A. Irokanulo, J.C. Chima, N. Nwankpa, Y. Laleye, and C. Bot. 1993. Anthrax and cowdriosis in an African elephant (*Loxodonta africana*). *Veterinary Record* 133:168.

Pienaar, U.d.V. 1961. A second outbreak of anthrax amongst game animals in the Kruger National Park. *Koedoe* 4:4–17.

———. 1967. Epidemiology of anthrax in wild animals and the control of epizootics in the Kruger National Park, *South Africa. Federation Proceedings* 26:1496–1502.

Pierre, J.N., and J. Valbrun. 1996. Incidence ou prevalence de la fievre charbonneuse en Haiti: Efforts de lutte. *Salisbury Medical Bulletin* 87(Special Suppl.):21.

Popham, D.L., and P. Setlow. 1993. The cortical peptidoglycan from spores of *Bacillus megaterium* and *Bacillus subtilis* is not highly cross-linked. *Journal of Bacteriology* 175:2767–2769.

Prins, H.H.T., and F.J. Weyerhaeuser. 1987. Epidemics in populations of wild ruminants: Anthrax and impala, rinderpest and buffalo in Lake Manyara National Park, Tanzania. *Oikos* 49:28–38.

Provost, A., and M. Tronette. 1957. Reflexions sur quelques cas de charbon bacterdien (cryptique) chez des bovins. *Revue d'Elevage et de Médecine Vétérinaire des Pays Tropicaux* 1:25–26.

Przybyszewska, M., J. Matras, M. Gorecka, J. Knap, M. Bartoszcze, and J. Mierzejewski. 1996. Dermal anthrax, probably originating from a XVth century goat leather "cordovan" and apparently attributable to a non-encapsulating *Bacillus anthracis. Salisbury Medical Bulletin* 87(Special Suppl.):21.

Roberts, T.A., and A.D. Hitchins. 1969. Resistance of spores. In *The bacterial spore,* ed. G.W. Gould and A. Hurst. London: Academic, pp. 611–656.

Rode, L.J., and J.W. Foster. 1966a. Quantitative aspects of exchangeable calcium in spores of *Bacillus megaterium. Journal of Bacteriology* 91:1589–1593.

———. 1966b. Influence of exchangeable ions on germinability of bacterial spores. *Journal of Bacteriology* 91:1582–1588.

Shafa, F., B.J. Moberly, and P. Gerhardt. 1966. Cytological features of anthrax spores phagocytized in vitro by rabbit alveolar macrophages. *Journal of Infectious Diseases* 116:401–413.

Shibata, H., S. Miyoshi, T. Osato, I. Tani, and T. Hashimoto. 1992. Involvement of calcium in the germination of coat-modified spores of *Bacillus cereus* T. *Microbiology and Immunology* 36:935–946.

Shlyakhov, E. 1996. Anthraxin: A skin test for early and retrospective diagnosis of anthrax and anthrax vaccination assessment. *Salisbury Medical Bulletin* 87(Special Suppl.):109–110.

Shrewsbury, J.F.D., and G.J. Barson. 1952. A bacteriological study of the house sparrow, *Passer domesticus domesticus. Journal of Pathology and Bacteriology* 64:605–618.

Stein, C.D. 1948. Incidence of anthrax in livestock during 1945, 1946 and 1947 with special reference to control measures in the various states. *Veterinary Medicine* 43:463–469.

Stein, C.D., and G.B. Van Ness. 1955. A ten year survey of anthrax in livestock with special reference to outbreaks in 1954. *Veterinary Medicine* 50:579–588.

Sterne, M. 1959. Anthrax. In *Infectious diseases of animals: Diseases due to bacteria,* ed. A.W. Stableforth and I.A. Galloway. New York: Academic, pp. 16–52.

Sugg, R.S. 1948. Penicillin on treatment of anthrax. *Journal of the American Veterinary Medical Association* 113:467.

Titball, R.W., and R.J. Manchee. 1987. Factors affecting the germination of spores of *Bacillus anthracis. Journal of Applied Bacteriology* 62:269–273.

Tolstova, A.G. 1960. Antagonism of microflora in the gastrointestinal tract of laboratory animals to *Bacillus anthracis. Veterinary Bulletin* 30:668. Abstract.

Turell, M.J., and G.B. Knudson. 1987. Mechanical transmission of *Bacillus anthracis* by stable flies (*Stomoxys calcitrans*) and mosquitoes (*Aedes aegypti* and *Aedes taeniorhynchus*). *Infection and Immunity* 55:1859–1861.

Turnbull, P.C.B. 1991. Anthrax vaccines: Past, present and future. *Vaccine* 9:533–539.

———. 1996. Guidance on environs known to be or suspected of being contaminated with anthrax spores. *Land Contamination and Reclamation* 4:37–45.

Turnbull, P.C.B., J.A. Carman, D.M. Lindeque, F. Joubert, O.J.B. Hubschle, and G.S. Snoeyenbos. 1989. Further progress in understanding anthrax in Etosha National Park. *Madoqua* 16:93–104.

Turnbull, P.C.B., C.P. Quinn, R. Hewson, M.C. Stockbridge, and J. Melling. 1990. Protection conferred by microbially-supplemented UK and purified PA vaccines. *Salisbury Medical Bulletin* 87(Special Suppl.):89–91.

Turnbull, P.C.B., R.H. Bell, K. Saigawa, F.E. Munyenyembe, C.K. Mulenga, and L.H. Makala. 1991. Anthrax in wildlife in the Luangwa Valley, Zambia. *Veterinary Record* 128:399–403.

Turnbull, P.C.B., M. Doganay, P.M. Lindeque, B. Aygen, and J. McLaughlin. 1992. Serology and anthrax in humans, livestock and Etosha National Park wildlife. *Epidemiology and Infection* 108:299–313.

Van Ness, G.B. 1971. Ecology of anthrax. *Science* 172:1303–1307.

Vasil'eva, V.M. 1960. Soil bacteria as antagonists of anthrax bacilli: II. *Veterinary Bulletin* 30:668. Abstract.

Whitford, H.W. 1979. Anthrax. In *CRC handbook series in zoonoses,* section A: *Bacterial, rickettsial and mycotic diseases,* ed. H. Stoenner, W. Kaplan, and M. Torten. Boca Raton, FL: CRC, pp. 31–66.

Wright, G.G. 1965. The anthrax bacillus. In *Bacterial and mycotic infections of man,* ed. R.J. Dubos and J.G. Hirsch, 4th ed. Philadelphia: J.B. Lippincott, pp. 530–544.

Xudong, L., M. Fenggin, and L. Aifang. 1996. Anthrax surveillance and control in China. *Salisbury Medical Bulletin* 87(Special Suppl.):16–18.

24 DISEASES DUE TO MYCOPLASMAS

KEVIN WHITHEAR

Synonyms: Pleuropneumonia-like organism (PPLO) is an obsolete term for mycoplasmas.

INTRODUCTION. Mycoplasmas are a diverse group of small bacteria that lack a cell wall. They belong in the class Mollicutes (*mollis,* soft; *cutis,* skin) of the Procaryotae. Species from the genera *Mycoplasma, Ureaplasma,* and *Acholeplasma* may be found in clinical specimens from animals; most of the animal pathogens are members of the genus *Mycoplasma.* See Rosenbusch (1994), Razin (1996), Tully (1996), Nicolet (1996), and Baseman and Tully (1997) for succinct introductions to the taxonomy and biology of mycoplasmas. Consult Rosenbusch (1994) and Tully (1996) for lists or citations of the species described in humans and in wild and domestic animals.

Assessing the pathogenic role of mycoplasmas is often problematic. Their isolation from mucosal surfaces in disease states does not necessarily imply etiologic significance, because the same species often can be isolated from the same sites in clinically normal animals. The pathogenicity of mycoplasmas varies widely, both between species and among strains within species. Invasive pathogenic species can coexist with their host in clinically inapparent infections, sometimes in unusual sites, as, for example, members of the *Mycoplasma mycoides* cluster in the ears of goats (Cottew and Yeats 1982). The dilemma of understanding the pathogenesis of mycoplasmal disease, summarized by Lewis Thomas decades ago, remains valid today: "At their pathogenic best, there was something not quite straightforward about their invasive relation to the host; sometimes they would, but more often they wouldn't and there usually seemed to be some kind of helper" (Thomas 1968).

In this chapter, a general overview of *Mycoplasma* infection and disease (mycoplasmosis) precedes a description of specific mycoplasmas that have been associated with disease in various wild mammals. Significant historic aspects are considered with descriptions of the more important species.

DISTRIBUTION. The distribution of mycoplasmas mirrors the distribution of their hosts. Some species of mycoplasmas, notably those causing contagious bovine pleuropneumonia and contagious caprine pleuropneumonia, are restricted in their distribution as a result of disease quarantine policies in domestic livestock. However, the confined geographic distribution of most mycoplasmoses in wild mammals, described below with specific entities, probably is more a reflection of sporadic occurrence and reporting within the geographic range of susceptible hosts than of restricted distribution of the etiologic agents themselves.

HOST RANGE. All mammalian species probably serve as hosts to mycoplasmas, even if infection is undescribed. Mycoplasmal diseases likely have gone unreported in wild mammals, either because the diagnostic procedures required to recognize these fastidious pathogens were not used, or because the nature of the disease, often mild and chronic, did not warrant detailed investigation. Wild mammals in which mycoplasmal diseases have been identified are listed in Table 24.1.

Mycoplasmas usually are considered to be relatively host specific. Infection may occur in more than one host, but the agent tends to persist in a primary host, to which disease is often confined. However, variations occur, as, for example, with *M. ovipneumoniae* and *M. conjunctivae,* which cause more severe disease in some wild ruminant species than in their primary host, the domestic sheep (see below). Species with little host specificity, such as *A. laidlawii* and *M. arginini,* tend to be nonpathogenic or weakly pathogenic.

ETIOLOGY. The *Mycoplasma* species reported to cause disease in captive or free-ranging wild mammals are listed in Table 24.1. Despite the aforementioned generalizations about host specificity, it is clear from Table 24.1 that many of the mycoplasmoses reported in free-ranging or captive wild mammals are associated with *Mycoplasma* species that cause similar disease in related domestic mammals. Given the wide range of animal species involved and the variety of clinical signs that may be associated with particular mycoplasmal conditions, *Mycoplasma* infections of broad taxa of hosts are described on an etiologic basis below.

TRANSMISSION AND EPIDEMIOLOGY. In general, animal-to-animal transmission is the means of

Table 24.1—Mycoplasma reported to cause natural disease in free-living or captive wild mammals

Mycoplasma Species	Primary Host	Wildlife Host	Disease[a]	Reference
M. arginini	Various	Bighorn sheep *Ovis canadensis*	Pneumonia, probably secondary Polyarthritis	Woolf et al. 1970
M. capricolum ssp. *capripneumoniae* (F-38)	Goat	West Caucasian tur *Capra ibex severtzovi*		
M. capricolum ssp. *capricolum*	Goat	Alpine ibex *Capra ibex ibex*	Septicemia	Schweighardt et al. 1989
M. conjunctivae	Sheep, goat	Alpine ibex	Keratoconjunctivitis	Mayer et al. 1996
M. conjunctivae	Sheep, goat	Alpine chamois *Rupicapra rupicapra*	Keratoconjunctivitis	Klinger et al. 1969; Nicolet and Freundt 1975
M. mycoides ssp. *mycoides* (large colony)	Goat	Wild goat *Capra aegagrus cretica*	Septicemia	Perrin et al. 1994
M. ovipneumoniae	Sheep	Dall's sheep *Ovis dalli*	Pneumonia	Black et al. 1988
M. phocacerebrale	Harbor seal	Harbor seal *Phoca vitulina*	Secondary infections	Kirchhoff et al. 1989
M. phocarhinis	Harbor seal	Harbor seal	Secondary infections	Kirchhoff et al. 1989
M. phocidae	Harbor seal	Harbor seal	Secondary pneumonia	Madoff et al. 1982
M. pneumoniae	Human	Leaf monkey *Trachypithecus (Presbytus) cristata*	Pneumonia	Stipkovits et al. 1989
Unclassified *Mycoplasma* sp.	Raccoon	Raccoon *Procyon lotor*	Epiphysitis and periostitis	Hunter et al. 1988

[a]For *Mycoplasma* spp. where a domestic animal is the primary host, a similar disease occurs in both the domestic animal and wildlife. With *M. conjunctivae* and *M. ovipneumoniae,* disease is reported to be more severe in wildlife than in domestic animals.

contagion, either through direct contact, or by aerosol transmission over short distances. Venereal transmission and the presence of mycoplasmas in semen have implications in the epidemiology of urogenital infections. Healthy carrier animals may be the basis of recrudescent disease in a population and can act as the vehicle for spread associated with animal movements.

Since *Mycoplasma* spp. from domestic mammals often are implicated, the involvement of domesticated species should always be considered in epidemiologic investigations of outbreaks of mycoplasmosis in wildlife. Segregation of wildlife from related species of domestic animals should be a tenet of programs to prevent disease caused by these agents. Details on epidemiology and transmission are presented in the discussions of specific species below.

PATHOGENESIS. Pathogenic mammalian mycoplasmas tend to be parasites of the moist mucosal surfaces of the respiratory and genital tracts, where they establish persistent and often clinically inapparent infections. High-frequency phase variation of major surface proteins, including adhesins, likely plays an important role in mucosal colonization and the pathogenesis of disease (Nicolet 1996; Baseman and Tully 1997).

Disease, when it occurs, results from a complex interaction among the virulence of the organism, the resistance of the host, and a variety of environmental factors. Synergistic interactions between the *Mycoplasma* and other infectious agents can play a role in the development and severity of disease.

Localized or systemic spread may follow mucosal colonization. Respiratory disease, including pneumonia and pleuropneumonia, is probably the most important clinical manifestation of mycoplasmosis in mammals. Ocular disease, genital disease, and mastitis also may occur. Systemic spread via the bloodstream, particularly when host defenses are compromised, may result in involvement of joints (usually multiple) and other serosal surfaces, or lead to septicemia. Clinical signs associated with each pathogen are described below.

DIAGNOSIS. Particular clinical and pathologic syndromes are characteristic of a *Mycoplasma* etiology. Mycoplasmosis should be suspected in animals with keratoconjunctivitis; bronchopneumonia; arthritis, serositis, or polyserositis, especially if fibrinous; and abortion or infertility (see Jubb et al. 1993); and as a cause of septic sequelae to carnivore bites (Walker et al. 1995). The histologic appearance of the lesion, particularly in pneumonias, often is characteristic of mycoplasmosis.

However, etiologic diagnosis requires isolation and identification of the organism. Compared with most bacterial pathogens, mycoplasmas are moderately to highly fastidious organisms. Specimens should be inoculated immediately into liquid and onto solid medium, or into a suitable transport medium. If this is not possible, then specimens should be immediately frozen on dry ice. Special media are required for successful cultivation, and not all diagnostic laboratories are equipped for *Mycoplasma* diagnosis. Immunofluorescence or immunohistochemistry may be applied to frozen or fixed tissues. Molecular genetic techniques such as DNA probes and polymerase chain reaction are now being widely adapted for the identification of *Mycoplasma* species and strains from clinical materials, supplementing or supplanting serologic identification.

Detection of antibodies by a variety of serologic tests is also commonly done, particularly in epidemiologic investigations. Serologic tests that have been used in wild mammals include complement fixation (*M. mycoides* cluster) and indirect hemagglutination (*M. ovipneumoniae*).

Handbooks for laboratory diagnosis of *Mycoplasma* infections are available (Whitford et al. 1994; Tully and Razin 1996).

IMMUNITY. Since mycoplasmas are pathogens of mucosal surfaces, local immune mechanisms are important. Serum antibody levels need not be correlated with protection, and efficacious vaccines are not available to prevent disease caused by the species of *Mycoplasma* of most significance in wild mammals. The characteristic lymphocyte infiltrate that occurs in mycoplasmal disease, especially pneumonias, suggests an immunologic component in the development of the lesions.

TREATMENT. Because they lack the ability to synthesize a cell wall, mycoplasmas are not susceptible to β-lactam antibiotics. They are intrinsically sensitive to tetracyclines, macrolides (although erythromycin is a common exception), lincomycin, tiamulin, and fluoroquinolones. However, results of treatment of clinically affected animals may be disappointing.

PUBLIC HEALTH CONCERNS. As relatively host-specific organisms, mycoplasmas are not usually implicated in zoonotic infections. *Mycoplasma arginini,* an exception to the rule of host specificity, has been isolated from a severely immunocompromised person with fatal pneumonia and septicemia, although probably not a primary pathogen (Yechouron et al. 1992). The infection may have been occupationally acquired, since the patient was an abattoir worker, and the agent is common in the respiratory and genital tracts of food-producing animals.

Madoff et al. (1991) [cited in Stadtländer and Madoff (1994)] isolated *M. phocacerebrale* from a lesion ("seal finger") on a human handler bitten by a seal and also from the teeth of the offending seal. Though mycoplasmas are not a common cause of wound infections in people, apparently septic wounds, due to bites or scratches of carnivores, that fail to respond to conventional antibiotic therapy should be cultured for *Mycoplasma,* and/or the antibiotic should be switched to one that is effective against mycoplasmas (McCabe et al. 1987).

DOMESTIC ANIMAL HEALTH CONCERNS. The major potential concerns for domestic animals are *M. mycoides* ssp. *mycoides* (small colony type) and *M. capricolum* ssp. *capripneumoniae,* the organisms associated with contagious bovine pleuropneumonia and contagious caprine pleuropneumonia, respectively; the possibility that wild mammals act as reservoirs of infection is discussed below with these species.

MANAGEMENT IMPLICATIONS. Mycoplasmoses are significant spontaneous diseases of free-ranging and captive wildlife mainly among sheep, goats, and their relatives. The *Mycoplasma* species involved are frequently those native to related domestic animals, in which infection is often subclinical and inapparent, or less severe than in the wild counterpart. Hence, consideration should be given to preventing contact between related species known to be susceptible to mycoplasmoses, in wild and captive environments.

Inapparent carriers also may exist among wildlife hosts, and adverse environmental conditions, crowding, or stress may predispose individuals to transmission or to clinical expression of infection. Hence, testing for carriers of potentially pathogenic *Mycoplasma* species should be considered if wildlife translocations are contemplated, and the conditions under which animals are held and handled should be optimized.

Although elimination of infected animals from a population has been undertaken in an attempt to truncate an outbreak of mycoplasmal keratoconjunctivitis among free-ranging chamois, it was concluded that such measures may not be warranted (Loison et al. 1996). Rather, management procedures should be taken to promote recovery of the population subsequent to the outbreak, if possible. Once infection is established in a wild population, it may well persist among inapparent carriers, and it should be noted that elimination of major mycoplasmoses from domestic animal populations has required a policy of eradication ("stamping out") of the affected population.

In captive situations, clinically affected animals might be segregated from unaffected members of the cohort as soon as they are observed, and treated aggressively. Clinically normal animals should be watched closely and removed for treatment immediately if they begin to show signs. Antibiotic therapy of clinically normal contacts might be considered in the face of an outbreak, particularly if the species involved is rare or valuable.

MYCOPLASMA INFECTIONS OF NONHUMAN PRIMATES. The literature to 1979 has been reviewed by Somerson and Cole (1979). Most species isolated from humans have also been found in nonhuman primates, a fact that prompted Del Guidice et al. (1969) to suggest that these species should be considered of primate, rather than human, origin. However, most of the information about the *Mycoplasma* flora of nonhuman primates has been obtained from captive animals, so the possibility of anthroponotic infection cannot be ruled out. A review of biologic and serologic comparisons of strains of *Mycoplasma* spp. from human and nonhuman primates led Somerson and Cole (1979) to conclude that any differences were probably not due to the host source of the isolates.

Reports of fully characterized *Mycoplasma* spp. that are unique to nonhuman primates are few. *Mycoplasma indiense* has thus far been isolated only from the throats of rhesus monkeys *Macaca mulatta* and a baboon *Papio anubis* (Hill 1993). *Mycoplasma moatsii* was initially isolated from recently captured grivet monkeys *Cercopithecus aethiops* (Madden et al. 1974) but was subsequently found in the intestines of wild Norway rats *Rattus norvegicus,* suggesting that rats may have been the source of infection for the monkeys (Giebel et al. 1990).

As in humans, most mycoplasmas in nonhuman primates occur as clinically inapparent infections. Occasionally, they may cause opportunistic disease or contribute to disease as secondary pathogens. Nonhuman primates have been used as animal models for some human mycoplasmal diseases, including respiratory disease [*M. pneumoniae* (Barile et al. 1994)]; urogenital tract disease [*M. genitalium* (Møller et al. 1985; Tully et al. 1986) and *U. urealyticum* (Kirchoff and Hill 1987)]; and systemic disease [*M. fermentans* (Lo et al. 1993)]. However, with the exception of *M. pneumoniae,* naturally occurring disease involving human pathogens has not been reported in nonhuman primates.

Mycoplasma Pneumoniae. *Mycoplasma pneumoniae* is an important cause of primary atypical pneumonia in people. It was considered an obligate human pathogen by Somerson and Cole (1979), although subsequently an outbreak was reported in silvered leaf monkeys *Trachypithecus (Presbytis) cristatus* imported from Indonesia to the former Soviet Union (Stipkovits et al. 1989). The source of infection for the monkeys was not described, but since people are the natural host of *M. pneumoniae,* a human source may have been involved and airborne transmission the likely route of infection. Stress of transportation may have been an important contributing factor in this case (Stipkovits et al. 1989). All of 13 monkeys developed severe respiratory disease and died 20–60 days after transportation. They had severe alveolar and interstitial pneumonia characterized by intensive perivascular and interalveolar lymphocyte infiltration; *Mycoplasma pneumoniae* was isolated from a variety of tissues (Stipkovits et al. 1989).

Other reports possibly incriminating *M. pneumoniae* in naturally occurring infections in nonhuman primates have been based on serologic evidence (Hutchison et al. 1970; Obeck et al. 1976). A rhesus monkey with severe polyarthritis had an increasing and persistent antibody titer to *M. pneumoniae.* However, no *Mycoplasma* or other bacteria could be isolated from joint fluid (Obeck et al. 1976).

On balance, it seems that despite an apparent susceptibility of some nonhuman primate species, *M. pneumoniae* does not commonly naturally infect nor cause disease in species other than humans.

MYCOPLASMAL DISEASES OF WILD RUMINANTS

The *Mycoplasma mycoides* Cluster. It is a century since the isolation of the first *Mycoplasma,* the etiologic agent of contagious bovine pleuropneumonia (Nocard and Roux 1898). The so-called "mycoides cluster" includes the following related pathogens of cattle or goats or sheep: *M. mycoides* ssp. *mycoides* small colony (SC) type, *M. mycoides* ssp. *mycoides* large colony (LC) type, *M. mycoides* ssp. *capri, M. capricolum* ssp. *capricolum,* and *M. capricolum* ssp. *capripneumoniae,* as well as several unnamed taxa (Damassa et al. 1992; Leach et al. 1993).

The members of the mycoides cluster cause respiratory disease, and infection by respiratory aerosol is a major mode of transmission. However, several species and subspecies have been found in the ears of ruminants and in ear mites, suggesting that mites may act as mechanical vectors (Cottew and Yeats 1982).

MYCOPLASMA MYCOIDES SSP. *MYCOIDES* SC TYPE. The SC type of *M. mycoides* ssp. *mycoides* is the cause of contagious bovine pleuropneumonia (CBPP), the only mycoplasmal disease classified as an Office International des Epizooties (O.I.E.) List A disease. CBPP has been eradicated from much of the globe but remains a serious problem in parts of Africa and Asia and recently has occurred in Southern Europe (Martel et al. 1991).

Because of the international effort to eradicate CBPP from cattle, it has been important to learn whether wild ruminants might act as reservoirs of infection. Water buffalo *Bubalus bubalus* (Letts 1964), African buffalo *Syncerus caffer* (Leach 1957; Shifrine et al. 1970), bison *Bison bison* (Leach 1957), and yak *Bos grunniens mutus* (Leach 1957) were susceptible to experimental infection, but these species do not appear to be important in transmission. Several other species, including camels *Camelus dromedarius* (Paling et al. 1988), have been tested for susceptibility to infection with *M. mycoides,* with negative results. It can be reasonably concluded that CBPP is confined to cattle and that wild animals are unimportant as reservoirs of infection.

MYCOPLASMA MYCOIDES SSP. *MYCOIDES* LC TYPE. This agent originally was isolated from a goat with fibrinous peritonitis (Laws 1956), and subsequently it has been isolated from young goats with septicemia and polyarthritis and from adults with mastitis, arthritis, and keratoconjunctivitis. It also was isolated from a condition resembling contagious caprine pleuropneumonia.

Mycoplasma mycoides ssp. *mycoides* LC was isolated at necropsy from 2- to 6-week-old wild goat kids *Capra aegagrus cretica* that had died of septicemia in a Swiss zoo. The kids showed signs of peritonitis, pneumonia, and enteritis. The high density of the animals, the presence of concomitant diseases, and the high rate of carriage in ear canals of healthy animals were suspected of being significant predisposing factors in the Swiss outbreak (Perrin et al. 1994).

MYCOPLASMA CAPRICOLUM SSP. *CAPRIPNEUMONIAE.* (**Synonym:** Mycoplasma sp. Taxon F-38.) Taxon F-38, the etiologic agent now considered to be responsible for classic contagious caprine pleuropneumonia (CCCP), was first isolated by MacOwan (1976) and has been named *M. capricolum* ssp. *capripneumoniae* (Leach et al. 1993). Classical CCCP naturally occurs in goats.

Disease in wildlife due to this agent has not been formally reported, but an agent identified as *Mycoplasma capricolum*/F-38 was repeatedly isolated from joint fluid and lungs of neonatal West Caucasian tur *Capra ibex severtzovi* during recurrent outbreaks of severe and sometimes fatal polyarthritis at the Toronto Zoo (I.K. Barker personal communication).

Antibody titers were found in a relatively high percentage of African buffalo and camels but not in eland *Taurotragus oryx* or oryx *Oryx beisa* in a farming enterprise in Kenya where domesticated wild herbivores and farm livestock commingled (Paling et al. 1978, 1988). While African buffalo and camel may be susceptible to *M. capricolum* ssp. *capripneumoniae,* there was no clinical evidence of disease, attempts to isolate the organism from nasal secretions were unsuccessful, and there was no evidence of transmission to sheep and goats (Paling et al. 1988).

MYCOPLASMA CAPRICOLUM SSP. *CAPRICOLUM.* (**Synonym:** Mycoplasma capricolum.) This agent is an invasive pathogen of goats and sheep, causing outbreaks similar to the septicemic disease caused by *M. mycoides* ssp. *mycoides* LC. The first report of disease involving *M. capricolum* ssp. *capricolum* probably was that by Cordy et al. (1955), who described an acute septicemic syndrome in goats.

Mycoplasma capricolum ssp. *capricolum* caused an outbreak of acute septicemia with diarrhea, 50% mortality, and focal hepatic necrosis among young Alpine ibex *Capra ibex* kept in an animal park in western Austria. Adult ibex were unaffected (Schweighardt et al. 1989).

Mycoplasma ovipneumoniae. This agent was first isolated in Scotland from lungs of sheep with pulmonary adenomatosis (MacKay et al. 1963). It causes a mild proliferative pneumonia in sheep, but predisposes individuals to secondary infection with pathogens, such as *Pasteurella haemolytica,* that result in a more severe exudative pneumonia. Environmental stressors also are important in exacerbating disease.

Sheep are the primary host. Although *M. ovipneumoniae* has been isolated from goats, its role as a cause of pneumonia in goats is unclear.

Free-ranging populations of Dall's sheep *Ovis dalli* in Alaska showed no evidence of natural infection with this agent (Zarnke and Rosendal 1989), and contact with subclinically infected domestic sheep was the likely source of a serious outbreak of *M. ovipneumoniae* pneumonia in a group of previously unexposed captive Dall's sheep (Black et al. 1988). The environmental and social stress associated with relocation to a new exhibit area at the Toronto Zoo, in the heat of midsummer, may have contributed to induction of severe disease (Black et al. 1988).

Clinical signs observed in the Dall's sheep were more severe than those seen in domestic sheep. The disease ran a chronic course, and signs (including mucopurulent nasal discharge, coughing, and rales) were seen in all animals. It did not respond to antimicrobial or anthelminthic treatment, and over a 6-month period, two sheep died and one was euthanatized. Gross and microscopic pulmonary lesions in all three were consistent with a *Mycoplasma* infection, particularly the evidence of lymphocytic cuffing of airways (Black et al. 1988). Increasing levels of antibody to *M. ovipneumoniae* were detected in paired serum samples, and *M. ovipneumoniae* was isolated from lungs at necropsy and from the nasal cavities of survivors. No other viral or bacterial pathogen was consistently isolated, although both diseased and a separate group of uninfected Dall's sheep had high antibody levels to bovine respiratory syncytial virus (BRSV).

Black et al. (1988) cited unpublished data which suggested that Dall's sheep may have reduced immunologic competence when compared with domestic sheep and mouflon *Ovis musimon.* This, together with the fact that the flock had never previously been exposed to *M. ovipneumoniae,* may help account for the apparent high susceptibility of the Dall's sheep to this agent. Synergistic interactions between respiratory viruses and mycoplasmas are well recognized, and BRSV also may have contributed to the severity of the disease.

Although interspecific contact in the wild is unlikely, the *M. ovipneumoniae*-free status of captive and free-ranging populations of Dall's sheep should be maintained by avoiding interaction with domestic sheep.

Mycoplasma arginini. *Mycoplasma arginini* has been isolated from a wide variety of species, including sheep, goats, and cattle. It generally is regarded as poorly pathogenic. A *Mycoplasma* (Woolf et al. 1970) subsequently identified as *M. arginini* (Al-Aubaidi et al. 1972) was isolated from lung lesions of captive bighorn sheep *Ovis canadensis* that died of a chronic active pneumonia, and it also was isolated from the nasal cavities of clinically affected sheep. Similarities between the lesions in this condition and those in an outbreak of *M. ovipneumoniae* pneumonia in Dall's sheep have been noted (Black et al. 1988). Since *M. ovipneumoniae* is a more fastidious species than *M. arginini* and does not produce typical "fried egg" colonies, it might have been involved but not isolated or recognized in cultures from lung lesions in the bighorn sheep.

Mycoplasma conjunctivae. *Mycoplasma conjunctivae* was isolated from sheep with infectious keratoconjunctivitis and implicated as a cause of this condition by Surman (1968). It commonly is isolated from keratoconjunctivitis in sheep and goats, and the disease has been reproduced with pure cultures (Dagnall 1993).

Natural disease has also been reported in Alpine chamois *Rupicapra rupicapra* (Klinger et al. 1969), in which it is common in the alpine regions of Germany, Austria, Switzerland, and France (Nicolet and Freundt 1975; Loison et al. 1996), and in Alpine ibex *Capra ibex* in Switzerland (Mayer et al. 1996, 1997). *Mycoplasma conjunctivae* is the pathogen most consistently isolated from keratoconjunctivitis in chamois, and it has been isolated from cases of pneumonia (Nicolet and Freundt 1975). It also was isolated from clinically affected ibex in Switzerland, in which it was considered to be etiologically significant (Mayer et al. 1997).

Keratoconjunctivitis was reproduced by transfer of conjunctival material from the eye of a diseased chamois to the eye of a susceptible animal (Klinger et al. 1969), though apparently the disease could not be reproduced in healthy chamois by inoculation of *M. conjunctivae* cultures (Loison et al. 1996). The commingling of domestic sheep and chamois in regions where infectious keratoconjunctivitis is endemic was considered significant epidemiologically (Nicolet and Freundt 1975). Contact, and mechanical vectors such as flies, may be involved in transmission (see Loison et al. 1996). The threshold population density for transmission of keratoconjunctivitis in chamois is likely quite low, and transmission seems unrelated to physical condition of the host (see Loison et al. 1996).

Chamois and Alpine ibex have more severe clinical signs than domestic ruminants (Nicolet and Freundt 1975; Mayer et al. 1997), including blindness (Nicolet and Freundt 1975). In outbreaks of keratoconjunctivitis in ibex in the Alps, affected animals had inflamed conjunctivae, ocular discharge, impaired vision, progressing to corneal opacity, ulceration, and perforation (Mayer et al. 1997). Similar signs, also sometimes progressing to corneal ulceration and perforation, were described in chamois; up to 90% of the population was affected, and deaths occurred mainly due to accidents caused by blindness (see Loison et al. 1996).

Prevention of the disease through management of chamois populations was not considered practical, but to promote rehabilitation of the population, a coping strategy should be developed setting out steps to be taken subsequent to significant outbreaks. These might include restricting hunting and removal of animals (Loison et al. 1996).

In an outbreak of infectious keratoconjunctivitis causing 80% morbidity among mouflon *Ovis ammon musimon* in the southern French Alps, *M. conjunctivae* could not be isolated, although domestic sheep running on the same pasture were a possible source of infection, and the disease could be transferred experimentally via conjunctival material from mouflon to sheep (Sarrazin et al. 1990).

MYCOPLASMA INFECTIONS OF ELEPHANTS. Uncharacterized mycoplasmas, possibly including *Ureaplasma* sp., were isolated from the lower urogenital tracts of a high proportion of captive Asian elephants *Elephas maximus* and African elephants *Loxodonta africana* that were cultured. It was speculated that infection with mycoplasmas might be associated with the pathogenesis of arthritis in elephants (Clark et al. 1980, 1981), but this is unsubstantiated.

MYCOPLASMA INFECTIONS OF WILD RODENTS. *Mycoplasma pulmonis* is a well-characterized pathogen of laboratory-bred mice and rats, causing murine respiratory mycoplasmosis. *Mycoplasma arthritidis* causes arthritis in laboratory rats but more commonly is subclinical. *Mycoplasma pulmonis* was isolated from several species of wild rats, and *M. arthritidis* was isolated from one rat species in a Japanese survey. No mycoplasmas were isolated from several other wild rodents, including the house mouse *Mus musculus* (Koshimizu et al. 1993). In a serologic field survey in the Darling Downs region of Queensland, Australia, Smith et al. (1993) found antibodies to *M. pulmonis* in only 1 of 81 wild house mice collected during and after mouse "plagues." This is surprising, given that plague conditions should favor the dissemination of *M. pulmonis.* There appear to be no published reports of clinical disease caused by mycoplasmas in wild mice and rats or other rodents.

MYCOPLASMA INFECTIONS OF CARNIVORES. Although mycoplasmas are widespread among domestic carnivores, few reports implicate them in disease. With the notable exception of raccoons, a similar situation appears to apply to wild carnivores.

Mycoplasma arginini. Heyward et al. (1969) isolated a *Mycoplasma* from the brain and lungs of a lion that had died of an encephalitis-like illness. The isolate, named *M. leonis,* was subsequently identified as *M. arginini* (Tully et al. 1972). While the principal host range of *M. arginini* includes herbivores (cattle, sheep, goats, horses, and pigs) and domestic carnivores (dogs and cats), it also has been isolated from the throats of captive wild carnivores, including lion *Panthera leo,* lynx *Felis lynx,* tiger *Panthera tigris,* cheetah *Acinonyx jubatus,* puma *Felis concolor,* and leopard *Panthera pardus* (Hill 1972, 1975). Carnivores may be exposed to *M. arginini* through their food source.

Mycoplasma felis. Mycoplasma felis has been isolated from the lower respiratory tract of a pneumonic juvenile serval *Felis serval* (Johnsrude et al. 1996), but its causal association with the pneumonia is uncertain.

Acholeplasma laidlawii. Typically considered a commensal organism, *A. laidlawii* was isolated in association with *Pasteurella multocida* from the pneumonic lung of a lynx found moribund in the wild in Alberta, Canada (Langford 1974). The author correctly equivocated on the possible implication of *A. laidlawii* in the pathogenesis of the pneumonia in this case.

***Mycoplasma*-associated Epiphysitis and Periostitis in Raccoons.** This condition was first identified in 1981 in orphaned wild raccoons *Procyon lotor* kept in a rehabilitation facility in Toronto, Canada (Hunter et al. 1988). A *Mycoplasma* sp. was isolated from purulent joint fluid; apparently it is unrelated to currently described species. No other bacteria, including *Chlamydia* sp., or fungi were isolated from joint fluid (Hunter et al. 1988). The condition has been reproduced experimentally by intravenous inoculation of the *Mycoplasma* sp. isolate. The mechanism of natural transmission is unknown. Since 1981, sporadic cases have been identified in free-ranging raccoons from various regions of Ontario (D.B. Hunter personal communication).

Affected animals developed a progressive lameness associated with one or more persistently swollen, pus-filled joints. Metacarpal and metatarsal joints were most commonly involved, although other joints, including those between spinal vertebrae, also may be affected (D.B. Hunter personal communication). Joint lesions included epiphysitis, osteomyelitis, and periarthritis.

There appeared to be an age susceptibility to experimental infection, with raccoons less than 6 weeks old developing lesions while those more than 12 weeks old were resistant (Hunter et al. 1988). The organism was reisolated from joint lesions, lungs, and lymph nodes of experimentally infected animals (D.B. Hunter personal communication).

MYCOPLASMA INFECTION OF PINNIPEDS

Mycoplasma phocidae. A new species, *M. phocidae* (Ruhnke and Madoff 1992), was isolated from the respiratory tracts and other organs (Madoff et al. 1982) of harbor seals *Phoca vitulina* during an epidemic of pneumonia along the New England coast of the United States in 1979 and 1980. Lung lesions were characterized by interstitial pneumonia with some microabscesses (Madoff et al. 1982). An influenza-A virus was isolated from the lungs, with *M. phocidae* probably contributing to the disease as a secondary pathogen. An isolate from this outbreak, designated strain M 4359, was shown to exert a cytotoxic effect in rat tracheal explants (Stadtländer et al. 1989).

***Mycoplasma phocarhinis* and *Mycoplasma phocacerebrale*.** Mycoplasmas were isolated from the nasal cavity and various internal organs from moribund, dead, or orphaned seals during a phocine morbillivirus epidemic in harbor seals in the North and Baltic Seas in 1988 and 1989 (Kirchoff et al. 1989). Two subsequently were named *M. phocarhinis* and *M. phocacerebrale* (Giebel et al. 1991). Recovered from the respiratory tracts (including lungs), hearts, brains, and eyes of the seals, they were considered to be secondary invaders complicating the morbillivirus infection (Kirchoff et al. 1989). Strains representative of both species caused a cytotoxic effect in rat tracheal explants (Stadtländer et al. 1989).

As noted earlier, *M. phocacerebrale* has been implicated as a cause of "seal finger" in the case of an aquarium worker bitten by a harbor seal (see Stadtländer and Madoff 1994).

LITERATURE CITED

Al-Aubaidi, J.M., W.D. Taylor, G.R. Bubash, and A.H. Dardiri. 1972. Identification and characterization of *Mycoplasma arginini* from bighorn sheep (*Ovis canadensis*) and goats. *American Journal of Veterinary Research* 33:87–90.

Barile, M.F., M.W. Grabowski, K. Kapatais-Zoumbois, B. Brown, P.C. Hu, and D.K.F. Chandler. 1994. Protection of immunized and previously infected chimpanzees challenged with *Mycoplasma pneumoniae*. *Vaccine* 12:707–714.

Baseman, J.B., and J.G. Tully. 1997. Mycoplasmas: Sophisticated, reemerging, and burdened by their notoriety. *Emerging Infectious Diseases* 3:21–32.

Black, S.R., I.K. Barker, K.G. Mehren, G.J. Crawshaw, S. Rosendal, L. Ruhnke, J. Thorsen, and P.S. Carman. 1988. An epizootic of *Mycoplasma ovipneumoniae* infection in captive Dall's sheep (*Ovis dalli dalli*). *Journal of Wildlife Diseases* 24:627–635.

Clark, H.W., D.C. Laughlin, J.S. Bailey, and T.M. Brown. 1980. *Mycoplasma* species and arthritis in captive elephants. *Journal of Zoo Animal Medicine* 11:3–15.

Clark, H.W., D.C. Laughlin, and T.M. Brown. 1981. Rheumatoid arthritis in elephants: A review to date. In *Proceedings of the 1981 conference of the American Association of Zoo Veterinarians.* Seattle: American Association of Zoo Veterinarians, pp. 95–100.

Cordy, D.R., H.E. Adler, and R. Yamamoto. 1955. A pathogenic pleuropneumonia-like organism from goats. *Cornell Veterinarian* 45:50–68.

Cottew, G.S., and F.R. Yeats. 1982. Mycoplasmas and mites in the ears of clinically normal goats. *Australian Veterinary Journal* 59:77–81.

Dagnall, G.J. 1993. Experimental infection of the conjunctival sac of lambs with *Mycoplasma conjunctivae*. *British Veterinary Journal* 149:429–435.

Damassa, A.J., P.S. Wakenell, and D.L. Brooks. 1992. Mycoplasmas of goats and sheep: Review article. *Journal of Veterinary Diagnostic Investigation* 4:101–113.

Del Guidice, R.A., T.R. Carski, M.F. Barile, H.M. Yamashiroya, and J. E. Verna. 1969. Recovery of human mycoplasmas from simian tissues. *Nature* 222:1088–1089.

Giebel, J., A. Binder, and H. Kirchhoff. 1990. Isolation of *Mycoplasma moatsii* from the intestine of wild Norway rats (*Rattus norvegicus*). *Veterinary Microbiology* 22:23–29.

Giebel, J., J. Meier, A. Binder, J. Flossdorf, J.B. Poveda, R. Schmidt, and H. Kirchhoff. 1991. *Mycoplasma phocarhinis,* new species and *Mycoplasma phocacerebrale,* new species, two new species from harbor seals (*Phoca vitulina* L.). *International Journal of Systematic Bacteriology* 41:39–44.

Heyward, J.T., M.Z. Sabry, and W.R. Dowdle. 1969. Characterization of *Mycoplasma* species of feline origin. *American Journal of Veterinary Research* 30:615–622.

Hill, A. 1972. The isolation of *Mycoplasma arginini* from captive wild cats. *Veterinary Record* 91:224–225.

———. 1975. Comparison of mycoplasmas isolated from captive wild felines. *Research in Veterinary Science* 18:139–143.

Hill, A.C. 1993. *Mycoplasma indiense* sp. nov., isolated from the throats of nonhuman primates. *International Journal of Systematic Bacteriology* 43:36–40.

Hunter, D.B., I.K. Barker, S. Rosendal, and P.Y. Daoust. 1988. Mycoplasma epiphysitis/periostitis in raccoons (*Procyon lotor*) in Ontario. In *Proceedings of the 1988 joint conference, American Association of Zoo Veterinarians and American Association of Wildlife Veterinarians.* Toronto: American Association of Zoo Veterinarians and American Association of Wildlife Veterinarians, p. 105.

Hutchison, V.E., M.E. Pinkerton, and S.S. Kalter. 1970. Incidence of *Mycoplasma* in nonhuman primates. *Laboratory Animal Care* 20:914–922.

Johnsrude, J.D., M.M. Christopher, N.P. Lung, and M.B. Brown. 1996. Isolation of *Mycoplasma felis* from a serval (*Felis serval*) with severe respiratory disease. *Journal of Wildlife Diseases* 32:691–694.

Jubb, K.V.F., P.C. Kennedy, and N. Palmer, eds. 1993. *Pathology of domestic animals,* vols. 1–3, 4th ed. San Diego: Academic, 780 pp., 747 pp., 652 pp.

Kirchoff, H., and A. Hill. 1987. Mycoplasmal infections of laboratory animals: Current status. *Israel Journal of Medical Sciences* 23:775–777.

Kirchoff, H., A. Binder, B. Liess, K.T. Friedhoff, J. Pohlenz, M. Stede, and T. Willhaus. 1989. Isolation of mycoplasmas from diseased seals. *Veterinary Record* 124:513–514.

Klinger, K., J. Nicolet, and E. Schipper. 1969. Neue Befunde über die Gemsblindheit. *Schweizer Archiv für Tierheilkunde* 111:587–602.

Koshimizu, K., T. Saito, Y. Shinozuka, K. Tsuchiya, and R. O. Cerda. 1993. Isolation and identification of *Mycoplasma* strains from various species of wild rodents. *Journal of Veterinary Medical Science* 55:323–324.

Langford, E.V. 1974. *Acholeplasma laidlawii* and *Pasteurella multocida* isolated from the pneumonic lung of a lynx. *Journal of Wildlife Diseases* 10:420–422.

Laws, L. 1956. A pleuropneumonia-like organism causing peritonitis in goats. *Australian Veterinary Journal* 32:326–329.

Leach, T.M. 1957. The occurrence of contagious bovine pleuropneumonia in species other than domesticated cattle. *Bulletin of Epizootic Diseases of Africa* 5:325–328.

Leach, R.H., H. Erno, and K.J. MacOwan. 1993. Proposal for designation of F38-type caprine mycoplasmas as *Mycoplasma capricolum* subsp. *capripneumoniae* subsp. nov. and consequent obligatory relegation of strains currently classified as *M. capricolum* (Tully, Barile, Edward, Theodore, and Erno 1974) to an additional new subspecies, *M. capricolum* subsp. *capricolum* subsp. nov. *International Journal of Systematic Bacteriology* 43:603–605.

Letts, G.A. 1964. Feral animals in the Northern Territory. *Australian Veterinary Journal* 40:84–88.

Lo, S.C., D.J. Wear, J.W. Shih, R.Y. Wang, P.B. Newton, and J.F. Rodriguez. 1993. Fatal systemic infections of nonhuman primates by *Mycoplasma fermentans* (incognitus strain). *Clinical Infectious Diseases* 17(Suppl.): S283–S288.

Loison, A., J.-M. Gaillard, and J.-M. Jullien. 1996. Demographic patterns after an epizootic of keratoconjunctivitis in a chamois population. *Journal of Wildlife Management* 60:517–527.

MacKay, J.M.K., D.K. Nisbet, and A. Foggie. 1963. Isolation of pleuropneumonia-like organisms (genus *Mycoplasma*) from cases of sheep pulmonary adenomatosis (S.P.A.). *Veterinary Record* 75:550–551.

MacOwan, K.J. 1976. A mycoplasma from chronic caprine pleuropneumonia in Kenya. *Tropical Animal Health and Production* 8:28–36.

Madden, D.L., K.E. Moats, W.T. London, E.B. Mathew, and J.L. Sever. 1974. *Mycoplasma moatsii,* a new species isolated from recently imported grivit monkeys (*Cercopithecus aethiops*). *International Journal of Systematic Bacteriology* 24:459–464.

Madoff, S., R.T. Schooley, H.L. Ruhnke, R.A. del Guidice, I.K. Barker, J. Geraci and A.S. Baker. 1982. Mycoplasmal pneumonia in phocid (harbor) seals. *Review of Infectious Diseases* S4:241.

Martel, J.L., P. Belli, M. Perrin, F. Poumarat, G. Dannacher, and M. Savey. 1991. Contagious bovine pleuropneumonia in 1991 in southern Europe. *Point Vétérinaire* 23:355–360.

Mayer, D., J. Nicolet, M. Giacometti, M. Schmitt, T. Wahli, and W. Meier. 1996. Isolation of *Mycoplasma conjunctivae* from conjunctival swabs of alpine ibex (*Capra ibex ibex*) affected with infectious keratoconjunctivitis. *Journal of Veterinary Medicine [B]* 43:155–161.

Mayer, D., M.-P. Degiorgis, W. Meier, J. Nicolet, and Marco Giacometti. 1997. Lesions associated with infectious keratoconjunctivitis in Alpine ibex. *Journal of Wildlife Diseases* 33:413–419.

McCabe, S.J., J.F. Murray, H.L. Ruhnke, and A. Rachlis. 1987. *Mycoplasma* infection of the hand acquired from a cat. *Journal of Hand Surgery [Am]* 12:1085–1088.

Møller, B.R., D. Taylor-Robinson, P.M. Furr, and E.A. Freundt. 1985. Acute upper genital-tract disease in female monkeys provoked experimentally by *Mycoplasma genitalium. British Journal of Experimental Pathology* 66:417–426.

Nicolet, J. 1996. Animal mycoplasmoses: A general introduction. *Revue Scientifique et Technique O.I.E.* 15:1233–1240.

Nicolet, J., and E.A. Freundt. 1975. Isolation of *Mycoplasma conjunctivae* from chamois and sheep affected with kerato-conjunctivitis. *Zentralblatt für Veterinärmedizin [B]* 22:302–307.

Nocard, E., and E.R. Roux. 1898. Le microbe de la peripneumoniae. *Annals de l'Institute Pasteur (Paris)* 12:240–262.

Obeck, D.K., J.D. Toft, and H.J. Dupuy. 1976. Severe polyarthritis in a rhesus monkey: Suggested *Mycoplasma* etiology. *Laboratory Animal Science* 26:613–618.

Paling, R.W., K.J. MacOwan, and L. Karstad. 1978. The prevalence of antibody to contagious caprine pleuropneumonia (*Mycoplasma* strain F38) in some wild herbivores and camels in Kenya. *Journal of Wildlife Diseases* 14:305–308.

Paling, R.W., S. Waghela, K.J. MacOwan, and B.R. Heath. 1988. The occurrence of infectious diseases in mixed farming of domesticated wild herbivores and livestock in Kenya: II. Bacterial diseases. *Journal of Wildlife Diseases* 24:308–316.

Perrin, J., M. Muller, N. Zangger, and J. Nicolet. 1994. *Mycoplasma mycoides* subsp. *mycoides* LC (large colony type) infection in wild goat kids (*Capra aegagrus cretica*) in Bern Zoo (Switzerland). *Schweizer Archiv für Tierheilkunde* 136:270–274.

Razin, S. 1996. Molecular properties of mollicutes: A synopsis. In *Molecular diagnostic procedures in mycoplasmology,* vol. 1, ed. S. Razin and J. G. Tully. San Diego: Academic, pp. 1–25.

Rosenbusch, R.F. 1994. Biology and taxonomy of the mycoplasmas. In *Mycoplasmosis in animals: Laboratory diagnosis,* ed. H.W. Whitford, R.F. Rosenbusch, and L.H. Lauerman. Ames: Iowa State University Press, pp. 3–11.

Ruhnke, H.L., and S. Madoff. 1992. *Mycoplasma phocidae* sp. nov., isolated from harbor seals (*Phoca vitulina* L.). *International Journal of Systematic Bacteriology* 42:211–214.

Sarrazin, C., J. Oudar, M. Prave, Y. Richard, and J. Borel. 1990. First description in mouflon (*Ovis ammon musimon*) of an outbreak of contagious infectious keratoconjunctivitis in the southern French Alps. *Gibier Faune Sauvage* 7:389–399.

Schweighardt, H., P. Pechan, E. Lauermann, and G. Krassnig. 1989. *Mycoplasma capricolum* infection in an alpine ibex (*Capra ibex ibex*). *Kleintierpraxis* 34:297–299.

Shifrine, M., S.S. Stone, and C. Staak. 1970. Contagious bovine pleuropneumonia in the African buffalo (*Syncerus caffer*). *Bulletin of Epizootic Diseases of Africa* 18:201–205.

Smith, A.L., G.R. Singleton, G.M. Hansen, and G. Shellam. 1993. A serologic survey for viruses and *Mycoplasma pulmonis* among wild house mice (*Mus domesticus*) in southeastern Australia. *Journal of Wildlife Diseases* 29:219–229.

Somerson, N.L., and B.C. Cole. 1979. The mycoplasma flora of human and nonhuman primates. In *The mycoplasmas,* ed. J.G. Tully and R.F. Whitcomb. New York: Academic, pp. 191–216.

Stadtländer, C.T.K.-H., and S. Madoff. 1994. Characterization of cytopathogenicity of aquarium seal mycoplasmas and seal finger mycoplasmas by light and scanning electron microscopy. *Zentralblatt für Bacteriologie* 280:458–467.

Stadtländer, C., D. Harman, A. Binder, and H. Kirchhoff. 1989. Investigation of seal mycoplasmas for their cytotoxic potential on tracheal organ cultures of SPF and gnotobiotic rats. *International Journal of Medical Microbiology* 272:216–224.

Stipkovits, L., A.N. Marantidi, E.K. Dzikidze, R.I. Krylova, and J.V. Vulvovich. 1989. Isolation of *Mycoplasma pneumoniae* from monkeys (*Presbitus cristata*). *Journal of Veterinary Medicine [B]* 36:134–138.

Surman, P.G. 1968. Cytology of "pink-eye" of sheep, including a reference to trachoma in man, by employing acridine orange and iodine stains and isolation of mycoplasma agents. *Australian Journal of Experimental Biology and Medical Science* 21:447–467.

Thomas, L. 1968. Mechanisms of pathogenesis in *Mycoplasma* infection. *Harvey Lectures* 63:73–98.

Tully, J.G. 1996. Mollicute-host interrelationships: Current concepts and diagnostic implications. In *Molecular diagnostic procedures in mycoplasmology,* vol. 2, ed. J.G. Tully and S. Razin. San Diego: Academic, pp. 1–21.

Tully, J.G., and S. Razin, eds. 1996. *Molecular diagnostic procedures in mycoplasmology,* vol. 2. San Diego: Academic.

Tully, J.G., R.A. del Guidice, and M.F. Barile. 1972. Synonymy of *Mycoplasma arginini* and *Mycoplasma leonis.*

International Journal of Systematic Bacteriology 22:47–49.

Tully, J.G., R.D. Taylor, D.L. Rose, P.M. Furr, C.E. Graham, and M.F. Barile. 1986. Urogenital challenge of primate species with *Mycoplasma genitalium* and characteristics of infection induced in chimpanzees. *Journal of Infectious Diseases* 153:1046–1054.

Walker, R.D., R. Walshaw, C.M. Riggs, and T. Mosser. 1995. Recovery of two mycoplasma species from abscesses in a cat following bite wounds from a dog. *Journal of Veterinary Diagnostic Investigation* 7:154–156.

Whitford, H.W., R.F. Rosenbusch, and L.H. Lauerman, eds. 1994. *Mycoplasmosis in animals: Laboratory diagnosis.* Ames: Iowa State University Press, 173 pp.

Woolf, A., D.C. Kradel, and G.R. Bubash. 1970. Mycoplasma isolates from pneumonia in captive Rocky Mountain bighorn sheep. *Journal of Wildlife Diseases* 6:169–170.

Yechouron, A., J. Lefebvre, H.G. Robson, D.L. Rose, and J.G. Tully. 1992. Fatal septicemia due to *Mycoplasma arginini:* A new human zoonosis. *Clinical Infectious Diseases* 15:434–438.

Zarnke, R.L., and S. Rosendal. 1989. Serologic survey for *Mycoplasma ovipneumoniae* in free-ranging Dall sheep (*Ovis dalli*) in Alaska. *Journal of Wildlife Diseases* 25:612–613.

25 CHLAMYDIOSIS OF KOALAS

RICHARD WHITTINGTON

Synonyms: Dirty tail, wet bottom, pink eye, koala infertility.

INTRODUCTION. The chlamydiae are cosmopolitan parasites and pathogens of a wide range of vertebrates. Diseases caused by *Chlamydia* occur in mammals, birds, reptiles, amphibians, and fish (Vanrompay et al. 1995). Mammalian infections typically involve mucosae, including the conjunctiva, urogenital tract, respiratory tract, and intestine, but systemic conditions such as encephalomyelitis and polyarthritis occur (Storz 1971; Vanrompay et al. 1995). The most-studied syndromes are keratoconjunctivitis and infertility/abortion in humans and in domestic ruminants.

Little is known about chlamydial infections in wild mammals. However, several epidemics have been reported in ungulates in the United States. These include polyarthritis in bighorn sheep *Ovis canadensis* in Wyoming (E.S. Williams personal communication 1996), keratoconjunctivitis in bighorn sheep in Montana (Meagher et al. 1992) and in mule deer *Odocoileus hemionus* in Utah (Taylor et al. 1996), and systemic infection in captive blackbuck *Antilope cervicapra* in Georgia (Mansell et al. 1995). *Chlamydia* was not necessarily solely responsible for the lesions in these outbreaks (Taylor et al. 1996). Chlamydiosis has been reported also in springbok *Antidorcas marsupialis* on a farm in South Africa (Van der Lugt and Kriek 1988).

Inapparent chlamydial infection may be more common than is recognized in wild ruminants. For example, a high proportion of fallow deer *Dama dama* in Tuscany, Italy (Giovannini et al. 1988), and Alpine ibex *Capra ibex ibex* in Switzerland (Giacometti et al. 1995) were found to be seropositive to *C. psittaci.* Red deer *Cervus elaphus* are susceptible to pathogenic *Chlamydia* of domestic ruminants, at least experimentally (McMartin et al. 1979). The prevalence of chlamydial infection in wild mammals in part may be related to contact with domestic animals and could rise as wildlife increasingly become restricted to habitat reserves adjacent to farmland.

The most intensively studied chlamydial infection of wild mammals affects the koala *Phascolarctos cinereus.* Chlamydiosis of koalas is a multisystemic mucosal infection with significant implications for management of the host. The syndromes recognized in koalas are similar in many respects to those known from other mammals, including humans (Storz 1971; Vanrompay et al. 1995).

HISTORY OF CHLAMYDIOSIS IN KOALAS. Lesions of the female genitalia consistent with chlamydial infection have been recognized in koalas since 1919 (MacKenzie 1919), and soon after there were accounts of other putative chlamydial syndromes: cystitis, nephritis, ophthalmia and conjunctivitis, and pulmonary disease (Pratt 1937). The association between keratoconjunctivitis and *Chlamydia* was demonstrated 30 years later in New South Wales (Cockram and Jackson 1974, 1981), while the link between infertility in older animals, reproductive tract lesions, and chlamydial infection was established by researchers in Victoria and Queensland (Martin 1981; Obendorf 1981; Brown et al. 1984; McColl et al. 1984). Concurrently, *Chlamydia* was isolated from koalas with these syndromes (Brown and Grice 1984), and transmission trials were reported briefly (Brown and Grice 1986).

Pratt (1937) expressed concern that ocular and urogenital diseases would be responsible for the extinction of the koala in Queensland and New South Wales. Conjecture about the impact of chlamydiosis has continued (Troughton 1941; Dayton 1990), because chlamydiosis can be a spectacular and emotive condition, the koala is a popular national icon in Australia, and disease may be observed coincident with population declines.

DISTRIBUTION. Chlamydial infection is widespread in koala populations in Australia. Disease occurs throughout the range of the koala, but some infected free-living populations seem to be free of clinical signs (Close 1993). The disease also occurs commonly in captive koalas in Australia and elsewhere (P.T. Robinson 1978; Canfield et al. 1991a).

The historical distribution of chlamydiosis was different. In Victoria, reproductive tract cysts were rarely seen at necropsy, and no ocular lesions were observed over a 15-year period (Pratt 1937). Meanwhile, koalas in New South Wales and Queensland were "disease-ridden"; reproductive tract cysts were "affected 50%" and ophthalmia was "a very prevalent disorder" (Pratt 1937). Since that time, the ocular form of chlamydiosis has become endemic in Victoria, and the urogenital form has probably increased in prevalence (Obendorf 1983; Brown et al. 1984), for reasons that are uncertain.

ETIOLOGY. There is a single genus, *Chlamydia,* in the family Chlamydiaceae, order Chlamydiales. Chlamydiae are coccoid, nonmotile, obligatory intracellular bacteria. The cell wall resembles that of Gram-negative bacteria, in that there is a double-unit membrane, but *Chlamydia* lacks peptidoglycan. Much of the cell wall is lipopolysaccharide (LPS) and a 40-kDa major outer-membrane protein (MOMP). Infection is by small elementary bodies (0.2- to 0.4-μm diameter, condensed cores), which survive in the external environment. Within membrane-bound vacuoles in the host cell cytoplasm, they enlarge through an intermediate stage to form reticulate bodies (0.5- to 1.5-μm diameter) that undergo binary fission to produce the infective elementary bodies (Moulder et al. 1984; Obendorf and Handasyde 1990).

Chlamydia spp. have been grouped according to phenotypic characteristics, including host origin, morphology of intracellular inclusions and elementary bodies, susceptibility to growth inhibitors, and antigenic analysis. Until recently, they were classified into two species: *C. trachomatis,* a pathogen of man, and *C. psittaci,* which infects many species (Moulder et al. 1984). As a result of recent molecular studies, isolates classified within *C. psittaci,* some of which had less than 10% DNA homology, were separated. Two additional species are now recognized: *C. pneumoniae,* comprising isolates from the human and equine respiratory tracts, and *C. pecorum,* which was recovered from cattle, sheep, and pigs (Fukushi and Hirai 1992, 1993; Storey et al. 1993). The nature of chlamydial isolates from koalas has been progressively elucidated.

A koala conjunctivitis isolate was shown to be distinct from avian, ovine, and bovine *C. psittaci* isolates, using restriction endonuclease analysis (REA), and Southern analysis with DNA probes (Timms et al. 1988a). Using similar methods, koalas in southeast Queensland were then shown to be infected with two strains of *Chlamydia,* named type I (conjunctival) and type II (urogenital) (Girjes et al. 1988). Only the type-II isolates contained a 7.4-kb plasmid (Girjes et al. 1988; Timms et al. 1988b). It was inferred later that type-I and type-II isolates differed also in the sequence of their MOMP gene (White and Timms 1994).

Remarkably, the REA and Southern hybridization patterns of the type-II isolates were identical to those of a sporadic bovine encephalomyelitis (SBE) isolate, which also contained the 7.4-kb plasmid (Girjes et al. 1988). However, the type-II isolates were later shown to be distinct from SBE and other ruminant isolates by using Southern analysis with a koala type-II strain DNA probe (Girjes et al. 1993c).

The phenotypic and genotypic characteristics of type-I and type-II koala isolates subsequently were compared with those of other chlamydiae (Girjes et al. 1993c). The type-I strain was more infective in culture systems than was the type-II strain. There was only 10% DNA homology between the two koala-type strains and a similar level of homology between these and avian *C. psittaci.* The type-II strains were found to be related to bovine strains of *C. pecorum* proposed by Fukushi and Hirai (1992).

The sequences of the MOMP gene of koala type-I isolates had greater than 97% homology with human *C. pneumoniae* (Kaltenboeck et al. 1993; Girjes et al. 1994). However, the results of comparative REA of genomic DNA, and the failure of monoclonal antibodies against human *C. pneumoniae* to react with the koala strain, confirmed that the koala strain was distinct from human *C. pneumoniae* (Girjes et al. 1994).

The distinction between *Chlamydia* from koalas and *C. psittaci* has been further reinforced (Glassick et al. 1996), using DNA sequence data from the *omp*2 gene. A cluster of isolates termed group-A *omp*2 corresponded to koala type-I and was 99% similar to a *C. pneumoniae*-type strain. A cluster termed group-B *omp*2 corresponded to koala type II and was 99% similar to a *C. pecorum*-type strain from a calf. Group B had only 71% DNA sequence similarity to group A and 77% similarity to *C. psittaci.* The relationship between koala and ruminant isolates of *C. pecorum* is the subject of further study, but to account for the genetic diversity of chlamydial species now endemic in koalas, Timms et al. (1996) proposed that koalas may have become infected with *Chlamydia* from sheep or cattle sources, but not from avian sources, on more than one occasion in the past. There is a proposal to name koala chlamydial type-I isolates koala *C. pneumoniae,* and koala type-II isolates *C. pecorum* (Glassick et al. 1996).

Both type-I and type-II isolates occur across the range of the koala in eastern Australia, but type-II infection may be more common (Girjes et al. 1993c). Type-I strains have now been recovered from conjunctiva, trachea, penis, and kidney, whereas type-II strains have been found in conjunctiva, trachea, vagina, uterus, cervix, penis, rectum, and bladder. Thus, type-I strains cannot be regarded as "conjunctival," and type-II strains cannot be regarded as "urogenital." Both types may coexist in the same tissue of the same animal (Girjes et al. 1993c). Furthermore, two type-I isolates from one koala, one of which came from the urogenital tract, contained a plasmid. While of similar size to the plasmid found in all type-II isolates, it differed in DNA sequence. Clearly, further research is required.

SYNDROMES ASSOCIATED WITH CHLAMYDIAL INFECTION IN THE KOALA

Keratoconjunctivitis

CLINICAL SIGNS AND PATHOLOGY. Serous ocular discharge, mild blepharospasm, and reddening of the conjunctiva are followed by a mucopurulent discharge within 2 weeks and by swelling of the conjunctiva and nictitating membrane (Cockram and Jackson 1974,

1976, 1981). The hair around the eyes becomes matted and may be lost, and the eyelids become gummed together. Peripheral corneal edema with superficial neovascularization moves centrally from week 3; chemosis may be severe and protrude through closed eyelids. Pannus and marked corneal opacity develop later.

Some animals recover spontaneously, but lesions may persist for years. Traumatic lesions due to rubbing, secondary bacterial infections, and anterior synechia secondary to corneal ulceration also occur (Obendorf 1983). Koalas may be blind from the early stages, severely affecting mobility and prehension, leading to mortality. Keratoconjunctivitis probably is more common in summer than winter (Cockram and Jackson 1976; Obendorf 1983). It occurs in animals of all ages but is more frequent in mature koalas.

Microscopically, in advanced cases there is villous hyperplasia of conjunctiva, with plasma cell and neutrophil infiltration; suppurative conjunctivitis; episcleral congestion; and corneal edema and ulceration, neovascularization, fibrosis, and pigmentation (Obendorf 1983).

CLINICAL PATHOLOGY. There were no consistent trends in hematologic or serum chemical parameters in eight koalas with keratoconjunctivitis (Canfield et al. 1989).

PATHOGENESIS. The incubation period after experimental inoculation was 7–19 days (Brown and Grice 1986). Concurrent ocular and urogenital infections occur in both female and male koalas (Canfield et al. 1991a; White and Timms 1994), but the relation to chlamydial type and route of infection requires further study. Both type-I and type-II koala *Chlamydia* have been isolated from the eye (Girjes et al. 1993c), but the influence of chlamydial type on disease outcome is unknown.

Urinary Tract Infection

CLINICAL SIGNS. There was apparent urinary incontinence with wetting of the fur on the rump, increased frequency of urination, dysuria, and tenesmus with cloacal eversion (Obendorf 1983; Brown et al. 1987). These signs are due to cystitis, from which the infection may ascend to involve the kidneys (Canfield 1989). Acute cases remain bright and continue to eat, but chronic cases lose weight and develop a coarse coat; secondary myiasis of the perineum may occur (Obendorf 1983). Both sexes are affected. Disease is more common in mature than immature koalas and in free-living compared to captive animals (Canfield 1989). Captives may recover spontaneously.

CLINICAL PATHOLOGY. There were consistent abnormalities in urine that contained erythrocytes, leukocytes, epithelial cells, occasional renal tubular casts, bacteria (coagulase-positive *Staphylococcus,* hemolytic *Streptococcus, Escherichia coli, Proteus,* and diphtheroids), and yeasts (Obendorf 1983). In addition, anemia and hypoproteinemia occurred. However, no consistent hematologic or serum chemical abnormalities were observed in koalas with uncomplicated urinary tract infections (Canfield 1989).

PATHOLOGY. Mucosal ulceration, luminal and mural hemorrhage and edema of the bladder, urethra, and sometimes also the ureters; prominence of ureteric papillae; dilated thickened ureters; pyelonephritis; and hydronephrosis and renal fibrosis were observed (Obendorf 1983; Canfield 1989).

Microscopically, erosion, deep ulceration, pseudomembrane formation, infiltration of plasma cells, and lymphoid cell aggregates were present in the urethra, bladder, and ureters. Tubular degeneration, protein casts, focal to generalized pyelonephritis, hydronephrosis, glomerular atrophy, fibrosis, and subacute to chronic interstitial nephritis were noted in the kidney (Obendorf 1983; Canfield 1989).

PATHOGENESIS. The incubation period following experimental infection was 25–27 days (Brown and Grice 1986). Ascending infection is a sequel to cystitis, due to interference with urinary outflow at the ureteric papillae, causing dilation and thickening of the ureters, hydronephrosis, and pyelonephritis (Obendorf 1983; Canfield 1989). Females frequently have concurrent ascending genital tract infection, probably due to the proximity of the openings of the urinary and genital systems in a common urogenital sinus (Obendorf 1981).

Genital Tract Infection

CLINICAL SIGNS. Infertility may be the only sign (Brown et al. 1987). However, a persistent purulent cloacal discharge may be detected, if secondary bacterial infections have resulted in metritis or pyometra, and there also may be signs of cystitis (Canfield et al. 1983).

CLINICAL PATHOLOGY. Koalas with pyometra or pyovagina had low-grade anemia, leukocytosis with neutrophilia or leukopenia, azotemia, and hypoproteinemia, reflecting the severity of the disease (Canfield et al. 1989). A mixed bacterial flora was present in uteri with pyometra (Obendorf 1981; Canfield et al. 1983). Nonhemolytic *E. coli* were recovered from cases of salpingitis with bursal adhesions (Obendorf 1981).

PATHOLOGY. Lesions in the genital tract varied from mild urogenital sinusitis and vaginitis without abnormalities in the upper tract, to pyometra, salpingitis, and cystic changes of the bursa (Obendorf and Handasyde 1990), described definitively by Obendorf (1981). There was enlargement of one or both uterine horns, with thickening of the walls and accumulation of purulent material in the lumen. The uterine (fallopian) tubes were sometimes also enlarged, associated with closed cervices. Unilateral or, rarely, bilateral fluid-filled cysts, 1.5–8.0 cm in diameter, were present in the paralumbar region. Each was a dilation of the ovarian bursa and the fimbrionated end of the uterine tube. The peritoneal ostium of the bursa was sometimes inapparent. Adhesions sometimes were present between the cyst and adjacent organs. Fibrous strands and clear straw-colored fluid, or red-brown turbid fluid, were present within the cysts. The ovaries were normal in all cases. Lesions have not been reported in the male genital tract (Obendorf and Handasyde 1990).

Microscopic lesions, which vary greatly in severity and extent (Obendorf and Handasyde 1990), were described by Obendorf (1981). There was mild acute erosive vaginitis, associated with focal chronic active erosive metritis and cystic dilation of uterine glands. In pyometra, there was generalized necrotizing ulcerative metritis. Chronic active salpingitis, sometimes obliterating the lumen of the uterine tube, and hydrosalpinx were noted. The wall of cystic bursae was normal and lined by flattened serosal cells, but inflammation and fibrous adhesions occurred in nondistended bursae. Ovaries were normal. Hyperplasia of the ovarian lining epithelium was noted both in normal koalas and in those with genital tract lesions. A previous description of serous cystadenomata in koalas (Finckh and Bolliger 1963) was probably a misnomer.

PATHOGENESIS. The lesions are the chronic sequelae of ascending inflammation of the genital tract. The syndrome resembles human genital chlamydiosis (Obendorf 1981, 1988; Obendorf and Handasyde 1990) in which ascending infection with acute salpingitis results in infertility (Mardh 1986) and in which urethral infection is common in men (McCormack 1986). *Chlamydia* has been cultured from the bursa and elsewhere in the genital tract of affected female koalas (Brown and Grice 1984), but chlamydial inclusions were not seen in sections of the bursal lesions (Obendorf 1988), which may be end-stage lesions. The mixed bacterial flora recovered from affected genital tracts probably represents secondary invasion (Obendorf 1988).

Obliteration of the peritoneal ostium of the ovarian bursa associated with inflammation may lead to fluid accumulation in the bursa (Obendorf 1981). This is predisposed by the unusual anatomy of the koala ovary and bursa, in which the ovary is closely enclosed within a fold of peritoneum (the ovarian marsupium) and is not exposed in the peritoneal cavity as in most mammals (Finckh and Bolliger 1963). Fluid accumulation in the uterus was thought to be due to the narrow lumen of the lateral vaginae through which fluid drains into the urogenital sinus. Drainage might be inadequate during estrus when there is increased keratinization, and cellular debris could cause a blockage (Backhouse and Bolliger 1961), for example, during acute chlamydial vaginitis.

Ingestion of phytoestrogens present in the diet of eucalypt leaves was implicated in the development of reproductive tract lesions (Martin 1981). However, the changes in the female genital tract associated experimentally with low doses of estrogen in *Chlamydia*-free koalas differed from those due to chlamydiosis (Handasyde et al. 1990). Phytoestrogen exposure may predispose to ascending infection of the urogenital tract, but the link remains uncertain (Obendorf 1988).

Reproductive tract lesions tend to develop after sexual maturity and are associated with reproductive failure (Brown et al. 1984; White and Timms 1994). Infertility is the result of irreversible lesions of the upper genital tract that interfere with fertilization and implantation. Seropositive females may continue to breed for several seasons, and fertility may be unaffected if only the urogenital sinus, vaginae, and cervices are involved (Obendorf and Handasyde 1990). Uterine and bursal enlargement may develop over a period of only 4 months (Brown et al. 1984); hence, females with advanced lesions may carry older young on their back, evidence of reproduction within the previous year. This can confound assessment of fertility in ecological and epidemiologic studies.

Proctitis. Proctitis, in which *Chlamydia* was demonstrated, has been reported in an old male koala with bilateral conjunctivitis and urogenital tract disease (Hemsley and Canfield 1996). No clinical signs were associated with the proctitis, but chlamydiae would have been shed in feces.

Respiratory Tract Infection. This apparently uncommon syndrome is poorly documented. *Chlamydia* has been isolated from the respiratory tract of some koalas with signs of respiratory disease (Brown and Grice 1984; Brown et al. 1987), but lesions are not described. Pneumonia was noted in koalas early this century (Pratt 1937) and has been mentioned in surveys of mortality (Backhouse and Bolliger 1961; McKenzie 1981; Canfield 1987; Weigler et al. 1987). Bacteria (*Bordetella bronchiseptica* and *Pseudomonas aeruginosa*) or fungi (*Cryptococcus neoformans*) have been isolated, and there is one report of idiopathic interstitial pneumonia (McKenzie 1981), but an etiology often was not sought, or *Chlamydia* was not excluded in these studies.

TRANSMISSION AND EPIDEMIOLOGY.

Chlamydia are maintained in a population by chronically infected individuals. The condition of free-living koalas with clinical urogenital infection has been followed for over 22 weeks (Weigler et al. 1988b), and chronic ocular disease is common (Cockram and Jackson 1976, 1981). Subclinical carriers also occur, many being recovered clinical cases (Canfield et al. 1991a). The prevalence of clinical disease is far less than the prevalence of infection in both free-living and captive populations (Mitchell et al. 1988; Weigler et al. 1988b; Canfield et al. 1991a; White and Timms 1994).

Chlamydia may be present in the conjunctiva and upper respiratory, urogenital, and intestinal tracts, suggesting that transmission may be direct or indirect by aerosol, by contact with excretions, and by the venereal route (Brown and Grice 1984; Brown and Woolcock 1988; Hemsley and Canfield 1996).

Survival of elementary bodies of koala *Chlamydia* in nature is permissive of indirect transmission. They were tolerant of a wide range of pH in vitro. Though they were inactivated by heating to 56° C for 5 minutes, 50% survived at 4° C or 20° C for 3 days, and 20% survived at these temperatures for 6 days. At 35° C, however, few elementary bodies survived to 6 days (Rush and Timms 1996). When dried on the leaves of young *Eucalyptus* plants in a laboratory, 1% of elementary bodies survived for 3 days (Rush and Timms 1996). The effect of ultraviolet radiation, an important consideration in the Australian environment, is unknown.

Direct transmission of the ocular infection during contact between mature males is likely (White and Timms 1994). The higher prevalence of ocular infection in the warmer months (Cockram and Jackson 1976; Obendorf 1983) supports a role for arthropod vectors in spread (Brown and Woolcock 1988; Handasyde et al. 1988). The sheep blowfly (*Lucilia cuprina*) was capable of carrying and transmitting koala *Chlamydia* to uninfected laboratory cultures (Wati 1991). Close contact between dam and offspring, and coprophagy in weaners, may contribute to the spread of the disease between generations. Infection of the offspring by contact with the genital epithelium of the dam during parturition is also possible. These factors could account for the occurrence of ocular or respiratory infection in koalas at any stage after leaving the pouch.

Venereal transmission may occur (Obendorf 1983) and is supported by the demonstration of *Chlamydia* in both male and female genitalia (Brown and Grice 1984; Hemsley and Canfield 1996), by the association of seroconversion with breeding, and by apparent freedom from infection among juvenile females in some Victorian populations (Handasyde et al. 1988). While it has yet to be proven, venereal transmission may account for most cases of urogenital infection in some populations. In other populations, however, urogenital infection was common in subadult females, suggesting nonvenereal spread (Weigler et al. 1988b).

All syndromes associated with chlamydial infection probably occur throughout the range of the koala, but there may be regional differences in prevalence. For example, urinary tract infection seems to be more common in northern New South Wales and Queensland than in Victoria (Canfield 1987).

Field surveys based on clinical examination are not sufficient to rule out the presence of chlamydial disease, because it is difficult to locate and examine every individual, and there may be long periods when affected individuals cannot be observed in populations known to be infected (Cockram and Jackson 1981). Marked variations in prevalence also occur among subgroups of a regional population (Cockram and Jackson 1976).

Predisposing factors are likely to be important in disease expression. The uneven prevalence of disease among populations of koalas probably reflects these factors. Overt chlamydiosis in captive koalas resulted from managerial procedures or concomitant diseases that acted as stressors (Canfield et al. 1991a). Other stressors probably exist in the wild, including the availability and quality of habitat (Ellis et al. 1993), particularly deforestation, crowding, competition, and urban encroachment (Weigler et al. 1988b; Close 1993); in some relatively undisturbed environments, overt disease is unknown despite chlamydial infection (Close 1993). Koalas are subject to a number of infectious and noninfectious diseases (Blanshard 1994), and some agents, such as retroviruses, might cause immunosuppression, triggering expression of chlamydial disease and enhanced transmission.

DIAGNOSIS. Chlamydiosis is diagnosed on the basis of characteristic clinical signs and pathologic lesions, supported by ancillary tests described below, particularly culture and antigen or DNA detection.

Ovarian bursal cysts and uterine enlargement may be detected radiographically in anesthetized koalas after inducing pneumoperitoneum (Brown et al. 1984). Large cysts are also palpable.

Hematologic and serum chemical analyses are unlikely to be useful aids to diagnosis, especially in individuals with uncomplicated chlamydial infections affecting single body systems. Urinalysis results compatible with chlamydiosis are discussed above with the clinical syndromes.

Laboratory tests have been used to demonstrate *Chlamydia,* chlamydial antigen, or chlamydial DNA in specimens such as ocular and urogenital swabs. Tests for antichlamydial antibody in serum also have been used. In general, data for sensitivity and specificity of these tests are scant and difficult to evaluate. Standard procedures for detection of chlamydial diseases in animals are given in Timms (1993) and Kennedy (1985).

Chlamydia can be cultured from swabs of exudate or from tissues of any of the affected body systems (Cockram and Jackson 1974, 1981; McColl et al. 1984; Brown and Grice 1984; Brown et al. 1984; Girjes et al. 1993c).

Chick embryo yolk sacs (CEYSs) were used initially, with good success, to isolate *Chlamydia* from clinical cases (Cockram and Jackson 1974, 1981); false-negative results were associated with early clinical cases (Cockram and Jackson 1981). More recently, buffalo-green monkey (BGM) cells have been used; lower rates of isolation generally have been reported than with CEYSs (Brown and Grice 1984; Grice and Brown 1985; Girjes et al. 1993b; Timms 1993), though some have described very good success (Canfield et al. 1991a).

Attention must be paid to specimen collection, transport, and storage if isolation is to be successful (Canfield et al. 1991a,b; Wood and Timms 1992; Timms

1993). Sucrose phosphate glucose (SPG) transport medium with antibiotics is beneficial for isolation of nonkoala strains of *C. psittaci* (Spencer and Johnson 1983) and has been used in several koala studies (Girjes et al. 1993c; Rush and Timms 1996), but without critical evaluation. If there will be a delay in culture, swabs in SPG medium should be placed at –70° C (Timms 1993).

Despite its limitations, culture is specific and sensitive, and remains the test against which other tests are compared.

Elementary bodies may be demonstrated in direct smears of lesion material, histologic sections, and tissue cultures by using stains such as Giemsa or Macchiavello (Timms 1993). However, chlamydial inclusions are quite difficult to identify with these stains, and immunologic stains are more sensitive (Canfield et al. 1991a; Hemsley and Canfield 1996).

Commercial chlamydial antigen detection kits are available that incorporate antibodies against LPS or MOMP antigens. The formats include enzyme-linked immunosorbent assay (ELISA) in microtiter plates or other solid-phase supports, and direct labeled-antibody stains for tissue smears on glass slides. Some ELISAs are particularly simple and quick to use, and immunologic methods also are advantageous because of less stringent requirements than culture for sample collection and handling in the field (Wood and Timms 1992). Antigen detection tests require the presence of epithelial cells, and the sample collection protocol must be followed carefully (Wood and Timms 1992).

The sensitivity of antigen detection kits using swabs from lesions may be as high as 70%–90% (Canfield et al. 1991b; Wood and Timms 1992), although a false-negative rate of 89% was obtained with one of the ELISAs (Weigler et al. 1988a). The specificity of these tests relative to cell culture was 80%–90%. An immunoblot test has also been described (Girjes et al. 1989).

Retrospective diagnosis may be achieved by immunologic staining of histologic sections using fluorescein- or peroxidase-labeled antichlamydial monoclonal antibody (Palmer et al. 1988; Hemsley and Canfield 1996).

Gene probes and the polymerase chain reaction (PCR) have been used to detect the DNA of koala *Chlamydia.* PCR amplification using primers for the conserved MOMP gene detected fewer than 10 koala chlamydial elementary bodies, but an analytical sensitivity of about 300 elementary bodies was suggested for clinical samples (Rasmussen and Timms 1991). DNA tests may be more sensitive than culture or antigen detection tests (Girjes et al. 1989, 1993b; Ellis et al. 1993) but require further evaluation.

To detect serologic evidence of exposure, the complement fixation test (CFT), using chlamydial group antigen ("group-CFT") derived from ruminant isolates or a more specific antigen derived from koala type-I keratoconjunctivitis strain ("koala-CFT"), and several formats of ELISA, have been employed. Serum, or eluate from blood dried on absorbent paper, may be used as the sample for these tests (Osawa et al. 1990; Ueno et al. 1991).

Antibody against *Chlamydia* may be detected in a high proportion of diseased koalas (Obendorf 1983; Mizuno 1990; Ueno et al. 1991; Girjes et al. 1993b), but the group-CFT was found to be relatively insensitive in some studies (Brown et al. 1984; Ueno et al. 1991; Girjes et al. 1993b). This might relate more to the stage of the disease than to the antigen used, since in one study only acute cases were seronegative (Cockram and Jackson 1981). High group-CFT titers may be found in clinically normal koalas (Cockram and Jackson 1981; Obendorf 1983), so such a titer is not necessarily indicative of the etiology of a clinical problem.

Serology may be inefficient in detecting chlamydial infection in a population when there is a low prevalence of clinical signs, and the low sensitivity of the group-CFT in such populations has been recognized (Girjes et al. 1989; White and Timms 1994). The 3- to 4-month interval between infection and seroconversion by the group-CFT (Brown and Grice 1986; Brown et al. 1987) is a serious problem if this test is used in epidemiologic studies or to assess spread of chlamydiosis.

A microtiter-plate ELISA based on a group-specific (nonkoala) recombinant chlamydial antigen had an analytical sensitivity 16 times greater than a CFT, and antibodies were detected on many occasions in koalas that did not yield *Chlamydia* in culture (Emmins 1996).

There are still insufficient data from the koala-CFT and the various ELISAs to assess accurately their sensitivities during various stages of infection, but preliminary findings suggest a useful role for serology in population surveys (Mizuno 1990; Ueno et al. 1991; Emmins 1996).

Caution is required in the interpretation of laboratory tests. As *Chlamydia* and antichlamydial antibody are often present in apparently healthy koalas (Weigler et al. 1988a; Girjes et al. 1993b), the positive predictive value of these tests for confirmation of clinical chlamydiosis is low. Demonstration of *Chlamydia,* antigen, or antibody merely supports a diagnosis of chlamydiosis, while the absence of viable *Chlamydia,* antigen, or antibody suggests an alternative etiology. The demonstration of a rising antibody titer in paired serum samples would be useful in confirming a diagnosis in acute cases.

Serology is useful for surveys in koala populations because samples are simple to collect, and it provides retrospective information. However, since none of the tests for *Chlamydia* or antibody against it are sufficiently sensitive, it is not possible to determine whether a population is free of infection based on point-in-time field surveys. For that purpose, it is necessary to utilize appropriate sampling strategies to detect an assumed prevalence of infection (usually 2%) with an adequate level of confidence [usually 95%—see Cannon and Roe (1982)]. Using current serologic and culture or antigen/DNA detection tests, it would be necessary to sample every member of a population on a number of occasions over several years in order to be confident that *Chlamydia* was absent from that population.

DIFFERENTIAL DIAGNOSIS. The numerous causes of blepherospasm, conjunctivitis, and keratitis in koalas, including chlamydiosis, were reviewed by Blanshard (1994). Similarly, the signs of urogenital tract infection and respiratory tract infection could be due to a wide range of infectious and noninfectious causes. A thorough evaluation of history, a physical examination, and appropriate diagnostic tests need to be undertaken in all cases presented for veterinary treatment. Koalas with chlamydiosis may have concurrent diseases or secondary bacterial infections, and these may be more significant than obvious signs of chlamydiosis.

IMMUNITY. The immune response of the koala is poorly understood. Humoral responses detected using CFT or ELISA are associated with antibodies against LPS. There has been insufficient investigation of antibody responses after infection to reach any conclusion about the role of humoral immunity in protection of koalas soon after infection, though the antibody response seems poor (Brown and Grice 1986). CFT titers seem not to contribute to protective immunity, given the high rate of both infection and seropositivity in infected populations.

Antibodies were produced against multiple chlamydial antigens and, while some partially neutralized the infectivity of elementary bodies, there was strain restriction; type-I koala *Chlamydia* was not inhibited (Girjes et al. 1993a). T cells predominate in the lymphocytic infiltrates in both conjunctival and urogenital lesions of koalas as is the case in other species, but reagents are not yet available to establish T-cell subtypes and cell-mediated immune mechanisms in the koala (Hemsley and Canfield 1997). There are no data on innate resistance mechanisms in the koala.

CONTROL AND TREATMENT. Captive koalas with chlamydiosis should be handled in such a way that infection is not inadvertently transmitted to healthy animals, and a high level of sanitation should be observed. A glutaraldehyde-based disinfectant was effective in killing koala type-I and II organisms in 1 minute in cell cultures in the presence of 5% yeast extract or 5% koala feces, while a chloramine-based disinfectant was less effective in the presence of organic matter (Wati 1991). On the basis of that study, a disinfection protocol for enclosures has been recommended: (1) physically remove organic matter, including feces, soil, and plant material; (2) clean the enclosure to remove organic matter adherent to walls, floor, and climbing poles; (3) apply an effective disinfectant (glutaraldehyde- or chloramine-based) for at least 1–10 minutes; and (4) thoroughly rinse the enclosure to remove residual disinfectant.

Attempts to vaccinate koalas against chlamydiosis have been unsuccessful (Brown and Woolcock 1988).

Prevention of chlamydiosis in captive displays should be based on quarantine of newly introduced animals. These should be subjected to repeated serologic examination or other tests for detection of *Chlamydia* (Drake et al. 1990; Blanshard 1994).

Koala translocation programs have been undertaken in southern Australia to alleviate overpopulation or to encourage colonization of other habitats (A.C. Robinson 1978; Warneke 1978). Introduction of *Chlamydia* to a naive population could cause a substantial decline in fertility and cause clinical disease and some mortalities.

Chlamydial infection status is an important consideration in translocation programs. Fecundity in *Chlamydia*-free females decreased from 71% to 23% after introduction to a *Chlamydia*-infected population (Lee et al. 1990); two *Chlamydia*-free females transferred to an endemic area developed clinical disease within 14 months (McColl et al. 1984); and 14% of a group of *Chlamydia*-free koalas placed in an endemic area developed keratoconjunctivitis, compared to a rate of 3% in the local population (Handasyde et al. 1988). The ability of koalas to disperse quickly and to cross unsuitable habitat (Martin 1985) means that it would be difficult to contain infection after it had been introduced.

Wildlife rehabilitation programs also create the potential for introduction of *Chlamydia.* Individuals rescued from the wild and placed in a rehabilitation facility may mix with animals from other localities, exchanging infectious agents. To obviate the need to assess each animal for *Chlamydia* infection, and in light of the insensitivity of diagnostic tests, Blanshard (1994) recommended that animals be released only at the site from which they were originally collected. This presupposes that animals are isolated during captivity and that there is a high level of sanitation to prevent cross-infection.

The tetracyclines, erythromycin, and rifampicin have high activity against *Chlamydia* in vitro (Oriel 1986). Early attempts to treat koalas suffering from clinical chlamydiosis by using systemic oxytetracycline or erythromycin were unsuccessful. Despite improvement in the clinical signs of chlamydiosis, most animals died, with weight loss of up to 30% over several weeks (Brown and Woolcock 1988, 1990). This probably was due to interference by antibiotics with the microbial flora of the hindgut (Brown and Woolcock 1988), which plays a role in digestion of the koala's diet of eucalypt leaves (Cork and Sanson 1990).

Dietary supplementation with soy-based human infant formula throughout the period of systemic treatment with oxytetracycline, and for the subsequent 2–3 weeks, enabled koalas to maintain body weight (Osawa and Carrick 1990) and became routine in medical management of koalas with chlamydiosis (Blanshard 1994). Recently, the fluoroquinolones ciprofloxacin and enrofloxacin have been used successfully to treat both cystitis and conjunctivitis in koalas (Booth and Blanshard 1997).

Topical ophthalmic treatments for chlamydial keratoconjunctivitis have been used with variable success, prolonged courses of tetracycline-based ointments being the most useful; chronic cases were refractory (Cockram and Jackson 1976; Obendorf 1983). Systemic treatments, including weekly injections of long-acting tetracycline, have been somewhat successful (Osawa and Carrick 1990), and systemic doxycycline hydrochloride and chloramphenicol have been used to treat ocular or concurrent ocular and urinary tract infections (Blanshard 1994). Elimination of *Chlamydia* by antibiotic therapy may be an unrealistic goal, based on studies of intensive treatment of human neonatal *C. trachomatis* conjunctivitis (Oriel 1986). Topical treatment of ocular lesions in koalas would not eliminate *Chlamydia,* as many animals have concurrent urogenital infection (Brown and Woolcock 1988).

Chlamydial cystitis has been treated with some success by using systemic oxytetracycline, and the bladder of refractory cases may be flushed with oxytetracycline solution (Blanshard 1994). Chloramphenicol by subcutaneous injection also has been used, but prolonged treatment was required (Blanshard 1994). Relapses are common after apparently successful treatment (Brown et al. 1987). Other treatments are reviewed by Booth and Blanshard (1997).

There are no reports of successful treatment of genital or respiratory chlamydiosis.

PUBLIC HEALTH CONCERNS. *Chlamydia psittaci* is an important zoonotic agent in birds, and human infection is common (Wills 1986; Vanrompay et al. 1995). Although mammalian *Chlamydia* are considered to be poorly infective for humans (Wills 1986), there are accounts of infection with *C. psittaci* of feline origin causing conjunctivitis, infection of ovine origin causing abortion, and infection of bovine origin causing death (Dawson 1986; Schachter 1986; Wills 1986).

There are no reports of human infection derived from koalas. Circumstantial evidence suggests that chlamydial strains from koalas are poorly infective to humans. People who have worked intensively with affected koalas for many years have experienced no ill-effects (A.R.B. Jackson, cited in Cockram 1978; Blanshard 1994; Emmins 1996). These observations are consistent with koala strains being distinct from *C. psittaci.*

Nevertheless, basic sanitary precautions should be observed when handling infected koalas, because koala *Chlamydia* might prove pathogenic in unusual circumstances (for example, in immunocompromised individuals). During necropsies, surface disinfectants should be used and aerosolation of fluids should be avoided. *Chlamydia* cultures should be handled in a class II laminar flow biosafety cabinet, and an approved dangerous-goods container must be used to transport cultures. Contaminated materials should be autoclaved before disposal.

DOMESTIC ANIMAL HEALTH CONCERNS. The susceptibility of laboratory or domestic animals to *Chlamydia* of koala origin has not been studied. The possibility of ruminant-koala cross-infection, based on the close genetic relationship of koala isolates to ruminant *C. pecorum,* warrants investigation.

MANAGEMENT IMPLICATIONS. Perceptions that chlamydiosis is an important issue in management of wild koala populations require close examination.

When records commenced in the 1800s, the koala was not abundant, could be located only by "diligent search" even in favorable habitat (Warneke 1978), and in fact was not noticed until 10 years after European settlement of Australia (Troughton 1941). A rapid increase in the number of koalas occurred in the late 1800s, as indigenous subsistence hunters were displaced by European settlers. The commercial fur trade then was associated with a dramatic decline in population: one million koala skins were sold in Queensland in 1919 (Gordon and McGreevy 1978). Local events such as bushfires accounted for declines of some populations. The koala became extinct in South Australia and was reduced to 500–1000 individuals in Victoria by 1920 (Warneke 1978).

A hunting ban, protection of habitat, and restocking programs resulted in recovery of the species, and it is now overabundant in some areas. In isolated and restricted habitats, overbrowsing may result in dramatic population decline, as may drought, heat wave, and interactions with the human environment, particularly motor vehicles (Martin 1985; Backhouse and Crouch 1990; Gordon et al. 1990; Martin and Handasyde 1990). Thus, any effects of chlamydiosis must be superimposed on major environmental influences and must be assessed in the light of the substantial historical fluctuations in koala numbers.

Lesions consistent with chlamydiosis were thought severe enough to account for death in 25%–50% of koalas in mortality surveys (Weigler et al. 1987; Backhouse and Bolliger 1961; White and Kunst 1990). However, the validity of such assessments is uncertain, because little is known about the physiology of the koala, its response to stress, and the significance of some of the lesions observed in this species (Canfield 1990).

The potential virulence of *Chlamydia* in koalas has been demonstrated experimentally: three of four animals inoculated with *Chlamydia* died within 4 months (Brown and Grice 1986). In a prospective case-control study in Queensland, wild koalas with cystitis had significantly reduced survival, while survival of animals with keratoconjunctivitis did not differ from controls (Gordon et al. 1990). Thus, chlamydiosis can be assumed to cause mortality in a wild population. Reduced survival might limit the number of koalas available for emigration and colonization of peripheral areas (Gordon et al. 1990).

Lower fertility rates (0–56%) were observed in *Chlamydia*-infected populations compared with

Chlamydia-free populations (56%–70%) in Victoria (Martin and Handasyde 1990), but in Queensland and northern New South Wales fertility in infected populations was high (67%–78%) (Gall 1978; White and Kunst 1990). There also were inconsistencies within Victoria: fertility in a population introduced to the Brisbane Ranges was 61%, compared with 20% in the parent population on Phillip Island (Martin 1981; Mitchell et al. 1988). As infertility occurs late in the disease process, the older age classes are more likely to fail to breed; it was the young females that bred on Phillip Island (Martin 1981). Fertility also is less likely to be impaired where lesions are mild. In the Queensland population, the infection rate was 71%, but only 7% had clinical signs (White and Kunst 1990). Thus, although fertility may be suppressed in infected populations, the effect is very variable. Future assessment of the effects of *Chlamydia* on fertility must involve adjustments for differences in age structure and stage of disease among infected populations.

Despite a high rate of *Chlamydia* infection and a relatively low level of fertility, the cycle of population growth, defoliation of food trees, and population decline has persisted on Raymond Island, Victoria, and culling of the infected population of koalas has been required on Phillip Island, Victoria, to control habitat degradation (McColl et al. 1984; Mitchell et al. 1988; Lee et al. 1990; Martin and Handasyde 1990).

CONCLUSION. While some have stated that chlamydiosis is not a significant threat to the continued survival of the koala (White and Kunst 1990), it must remain an important consideration in the management of free-living and captive populations. The aim of management is to maintain a disease-free status, where this exists, and to avoid introduction of clinically diseased animals to any population, as these may become the source of an outbreak. Translocations of koalas should not be undertaken without first ascertaining the infection status of populations at both the origin and destination, including genetic typing of the strains of *Chlamydia* that are present. Issues associated with wildlife rehabilitation programs are discussed above. Translocation of agents of wildlife disease is an emerging issue internationally (Woodford 1993).

Chlamydiosis is an important cause of morbidity and mortality among free-living and captive koalas. However, its significance in the population dynamics of the koala is unclear, because most studies have been cross-sectional. The effects of chlamydiosis over time, and changes associated with environmental cycles, are unknown. The factors that lead to increased prevalence and severity of disease are unknown but are clearly important determinants of the impact of chlamydial infection in a population. Long-term (at least 10-year) prospective studies in a number of populations are required to enable prediction of the impact of chlamydial infection.

There is consensus that, although environmental and human factors have major impacts on populations, chlamydiosis must be considered in koala management (Martin and Handasyde 1990; Reed et al. 1990; White and Kunst 1990). Further investigation of the molecular epidemiology of *Chlamydia* in the koala over its entire range is required. Correlations among microbiologic findings, and clinical (virulence, organ system, circumstances of expression) and epidemiologic patterns of disease should be sought and related to the population ecology of the koala. Better understanding of the pathogen and its interaction with the host will assist long-term management of the koala.

NOTE ADDED IN PROOF. While this chapter was in press, the taxonomy of the order *Chlamydiales* was revised based on extensive molecular investigations. The species infecting koalas are now found in the new genus *Chlamydophila,* as *C. pecorum,* and as *C. pneumoniae,* biovar Koala (K. D. E. Everett, R. M. Bush, and A. A. Andersen, 1999. Emended description of the order *Chlamydiales,* proposal of *Parachlamydiaceae* fam. nov. and *Simkaniaceae* fam. nov., each containing one monotypic genus, revised taxonomy of the family *Chlamydiaceae,* including a new genus and five new species, and standards for the identification of organisms. *International Journal of Systematic Bacteriology* 49:415-440).

While *C. pneumoniae* seems less pathogenic than *C. pecorum,* it has been associated with respiratory disease (S. Wardrop, A. Fowler, P. Giffard, and P. Timms, 1999. Characterization of the koala biovar of *Chlamydia pneumoniae* at four gene loci-ompAVD4, ompB, 16S RNA, groESL, spacer region. *Systematic and Applied Microbiology* 22:22-27).

A simulation model suggested that *Chlamydophila* infection was unlikely to pose a significant threat to persistence of koala populations under the conditions evaluated (D. J. Augustine, 1998. Modelling *Chlamydia*-koala interactions: coexistence, population dynamics and conservation implications. *Journal of Applied Ecology* 35:261-272). *The Editors*

LITERATURE CITED

Backhouse, G., and A. Crouch. 1990. Koala management in the Western Port region, Victoria. In *Biology of the koala,* ed. A.K. Lee, K.A. Handasyde, and G.D. Sanson. Chipping Norton, Australia: Surrey Beatty and Sons, pp. 313–317.

Backhouse, T.C., and A. Bolliger. 1961. Morbidity and mortality in the koala (*Phascolarctos cinereus*). *Australian Journal of Zoology* 9:24–37.

Blanshard, W.H. 1994. Medicine and husbandry of koalas. In *Wildlife.* Proceedings 233. Sydney: Post Graduate Committee in Veterinary Science, University of Sydney, pp. 547–629.

Booth, R., and Blanshard W. 1998. Diseases of koalas. In *Zoo and wild animal medicine: Current therapy 4,* ed. M.E. Fowler, and R.E. Miller. Philadelphia: W.B. Saunders. In press.

Brown, A.S., and R.G. Grice. 1984. Isolation of *Chlamydia psittaci* from koalas (*Phascolarctos cinereus*). *Australian Veterinary Journal* 61:413.

———. 1986. Experimental transmission of *Chlamydia psittaci* in the koala. In *Chlamydial infections: Proceedings of the sixth international symposium on human chlamydial infections, Sanderstead, Surrey, 15–21 June 1986,* ed. D. Oriel, G. Ridgway, J. Schachter, D. Taylor-Robinson, and M. Ward. Cambridge: Cambridge University Press, pp. 349–352.

Brown, A.S., F.N. Carrick, G. Gordon, and K. Reynolds. 1984. The diagnosis and epidemiology of an infertility disease in the female koala *Phascolarctos cinereus* (Marsupialia). *Veterinary Radiology* 25:242–248.

Brown, A.S., A.A. Girjes, M.F. Lavin, P. Timms, and J.B. Woolcock. 1987. Chlamydial disease in koalas. *Australian Veterinary Journal* 64:346–350.

Brown, S., and J. Woolcock. 1988. Epidemiology and control of chlamydial disease in koalas. In *Australian wildlife.* Proceedings 104. Sydney: Post Graduate Committee in Veterinary Science, University of Sydney, pp. 495–502.

———. 1990. Strategies for control and prevention of chlamydial diseases in captive koalas. In *Biology of the koala,* ed. A.K. Lee, K.A. Handasyde, and G.D. Sanson. Chipping Norton, Australia: Surrey Beatty and Sons, pp. 295–298.

Canfield, P.J. 1987. A mortality survey of free range koalas from the north coast of New South Wales. *Australian Veterinary Journal* 64:325–328.

———. 1989. A survey of urinary tract disease in New South Wales koalas. *Australian Veterinary Journal* 66:103–106.

———. 1990. Disease studies on New South Wales koalas. In *Biology of the koala,* ed. A.K. Lee, K.A. Handasyde, and G.D. Sanson. Chipping Norton, Australia: Surrey Beatty and Sons, pp. 249–254.

Canfield, P.J., C.J. Oxenford, D.N. Love, and R.K. Dickens. 1983. Pyometra and pyovagina in koalas. *Australian Veterinary Journal* 60:337–338.

Canfield, P.J., M.E. O'Neill, and E.F. Smith. 1989. Haematological and biochemical investigations of diseased koalas (*Phascolarctos cinereus*). *Australian Veterinary Journal* 66:269–272.

Canfield, P.J., D.N. Love, G. Mearns, and E. Farram. 1991a. Chlamydial infection in a colony of captive koalas. *Australian Veterinary Journal* 68:167–169.

———. 1991b. Evaluation of an immunofluorescence test on direct smears of conjunctival and urogenital swabs taken from koalas for the detection of *Chlamydia psittaci. Australian Veterinary Journal* 68:165–167.

Cannon, R.M., and R.T. Roe. 1982. Livestock disease surveys: A field manual for veterinarians. Canberra: Australian Government Publishing Service, 35 pp.

Close, R. 1993. Campbelltown's koalas: What is their future? *National Parks Journal* 37:22–25.

Cockram, F.A. 1978. Investigations into kerato-conjunctivitis in koalas. In *The koala,* ed. T.J. Bergin. Sydney: Zoological Parks Board of New South Wales, pp. 177–182.

Cockram, F.A., and A.R.B. Jackson. 1974. Isolation of a *Chlamydia* from cases of keratoconjunctivitis in koalas. *Australian Veterinary Journal* 50:82–83.

———. 1976. Chlamydial kerato-conjunctivitis in koalas. *Australian Veterinary Practitioner* March:36–38.

———. 1981. Keratoconjunctivitis of the koala, *Phascolarctos cinereus,* caused by *Chlamydia psittaci. Journal of Wildlife Diseases* 17:497–504.

Cork, S.J., and G.D. Sanson. 1990. Digestion and nutrition in the koala: A review. In *Biology of the koala,* ed. A.K. Lee, K.A. Handasyde, and G.D. Sanson. Chipping Norton, Australia: Surrey Beatty and Sons, pp. 129–144.

Dawson, C.R. 1986. Eye disease with chlamydial infection. In *Chlamydial infections,* ed. D. Oriel, G. Ridgway, J. Schachter, and D. Taylor-Robinson. Cambridge: Cambridge University Press, pp. 135–144.

Dayton, L. 1990. Can koalas bear the twentieth century. *New Scientist* 127:27–29.

Drake, B., M. Miller, and N.W. Morley. 1990. Management of koalas in captivity. In *Biology of the koala,* ed. A.K. Lee, K.A. Handasyde, and G.D. Sanson. *Biology of the koala,* ed. A.K. Lee, K.A. Handasyde, and G.D. Sanson. Chipping Norton, Australia: Surrey Beatty and Sons, pp. 323–329.

Ellis, W., A.A. Girjes, F.N. Carrick, and A. Melzer. 1993. Chlamydial infection in koalas under relatively little alienation pressure. *Australian Veterinary Journal* 70:427–428.

Emmins, J.J. 1996. The Victorian koala: Genetic heterogeneity, immune responsiveness and epizootiology of chlamydiosis. Ph.D. diss., Monash University, Melbourne, Australia, 345 pp.

Finckh, E.S., and A. Bolliger. 1963. Serous cystadenomata of the ovary in the koala. *Journal of Pathology and Bacteriology* 85:526–528.

Fukushi, H., and K. Hirai. 1992. Proposal of *Chlamydia pecorum* sp. nov. for *Chlamydia* strains derived from ruminants. *International Journal of Systematic Bacteriology* 42:306–308.

———. 1993. *Chlamydia pecorum:* The fourth species of genus *Chlamydia. Microbiology and Immunology* 37:515–522.

Gall, B. 1978. Koala Research, Tucki Nature Reserve. In *The koala,* ed. T.J. Bergin. Sydney: Zoological Parks Board of New South Wales, pp. 116–124.

Giacometti, M., F. Tolari, A. Mannelli, and P. Lanfranchi. 1995. Seroepidemiologic investigations in the Alpine ibex (*Capra i. ibex*) of Piz Albris in the canton of Grigioni [in Italian]. *Schweizer Archiv Tierheilkunde* 137:537–542. English abstract.

Giovannini, A., M. Cancellotti, C. Turilli, and E. Randi. 1988. Serological investigations for some bacterial and viral pathogens in fallow deer (*Cervus dama*) and wild boar (*Sus scrofa*) of the San Rossore Preserve, Tuscany, Italy. *Journal of Wildlife Diseases* 24:127–132.

Girjes, A.A., A.F. Hugall, P. Timms, and M.F. Lavin. 1988. Two distinct forms of *Chlamydia psittaci* associated with disease and infertility in *Phascolarctos cinereus* (koala). *Infection and Immunity* 56:1897–1900.

Girjes, A.A., B.J. Weigler, A.F. Hugall, F.N. Carrick, and M.F. Lavin. 1989. Detection of *Chlamydia psittaci* in free-ranging koalas (*Phascolarctos cinereus*): DNA hybridization and immuno-slot blot analyses. *Veterinary Microbiology* 21:21–30.

Girjes, A.A., W.A.H. Ellis, F.N. Carrick, and M. F. Lavin. 1993a. Some aspects of the immune response of koalas (*Phascolarctos cinereus*) and in vitro neutralization of *Chlamydia psittaci* (koala strains). *FEMS Immunology and Medical Microbiology* 6:21–30.

Girjes, A.A., W.A.H. Ellis, M.F. Lavin, and F.N. Carrick. 1993b. Immuno-dot blot as a rapid diagnostic method for detection of chlamydial infection in koalas (*Phascolarctos cinereus*). *Veterinary Record* 133:136–141.

Girjes, A.A., A.F. Hugall, D.M. Graham, T.F. McCaul, and M.F. Lavin. 1993c. Comparison of type I and type II *Chlamydia psittaci* strains infecting koalas *Phascolarctos cinereus. Veterinary Microbiology* 37:65–83.

Girjes, A.A., F.N. Carrick, and M.F. Lavin. 1994. Remarkable sequence relatedness in the DNA encoding the major outer membrane protein of *Chlamydia psittaci* (koala type I) and *Chlamydia pneumoniae. Gene* 138:139–142.

Glassick, T., P. Giffard, and P. Timms. 1996. Outer membrane protein 2 gene sequences indicate that *Chlamydia pecorum* and *Chlamydia pneumoniae* cause infections in koalas. *Systematic and Applied Microbiology* 19:457–464.

Gordon, G., and D.G. McGreevy. 1978. The status of the koala in Queensland. In *The koala,* ed. T.J. Bergin. Sydney: Zoological Parks Board of New South Wales, pp. 125–131.

Gordon, G., D.G. McGreevy, and B.C. Lawrie. 1990. Koala populations in Queensland: Major limiting factors. In *Biology of the koala,* ed. A.K. Lee, K.A. Handasyde, and G.D. Sanson. Chipping Norton, Australia: Surrey Beatty and Sons, pp. 85–95.

Grice, R.G., and A.S. Brown. 1985. A tissue culture procedure for the isolation of *Chlamydia psittaci* from koalas (*Phascolarctos cinereus*). *Australian Journal of Experimental Biology and Medical Science* 63:283–286.

Handasyde, K.A., R.W. Martin, and A.K. Lee. 1988. Field investigations into chlamydial disease and infertility in koalas in Victoria. In *Australian wildlife.* Proceedings 104. Sydney: Post Graduate Committee in Veterinary Science, University of Sydney, pp. 505–515.

Handasyde, K.A., D.L. Obendorf, R.W. Martin, and I.A. McDonald. 1990. The effect of exogenous oestrogen on fertility and the reproductive tract of the female koala. In *Biology of the koala,* ed. A.K. Lee, K.A. Handasyde, and G.D. Sanson. Chipping Norton, Australia: Surrey Beatty and Sons, pp. 243–248.

Hemsley, S., and P.J. Canfield. 1996. Proctitis associated with chlamydial infection in a koala. *Australian Veterinary Journal* 74:148–150.

———. 1997. Histopathological and immunohistochemical investigation of naturally occurring chlamydial conjunctivitis and urogenital inflammation in koalas (*Phascolarctos cinereus*). *Journal of Comparative Pathology* 116:273–290.

Kaltenboeck, B., K.G. Kousoulas, and J. Storz. 1993. Structures of and allelic diversity and relationships among the major outer membrane protein (*ompA*) genes of the four chlamydial species. *Journal of Bacteriology* 175:487–502.

Kennedy, G.A. 1985. Laboratory diagnosis of chlamydial diseases. *Proceedings of the American Association of Veterinary Laboratory Diagnosticians* 28:421–436.

Lee, A.K., R.W. Martin, and K.A. Handasyde. 1990. Experimental translocation of koalas to new habitat. In *Biology of the koala,* ed. A.K. Lee, K.A. Handasyde, and G.D. Sanson. Chipping Norton, Australia: Surrey Beatty and Sons, pp. 299–312.

MacKenzie, W.C. 1919. *The genito-urinary system in monotremes and marsupials.* Melbourne: Jenkin, Buxton, 106 pp.

Mansell, J.L., K.N. Tang, C.A. Baldwin, E.L. Styer, and A.D. Liggett. 1995. Disseminated chlamydial infection in antelope. *Journal of Veterinary Diagnostic Investigation* 7:397–399.

Mardh, P.A. 1986. Ascending chlamydial infection in the female genital tract. In *Chlamydial infections,* ed. D. Oriel, G. Ridgway, J. Schachter, and D. Taylor-Robinson. Cambridge: Cambridge University Press, pp. 173–184.

Martin, R., and K. Handasyde. 1990. Population dynamics of the koala (*Phascolarctos cinereus*) in southeastern Australia. In *Biology of the koala,* ed. A.K. Lee, K.A. Handasyde, and G.D. Sanson. Chipping Norton, Australia: Surrey Beatty and Sons, pp. 75–84.

Martin, R.W. 1981. Age specific fertility in three populations of the koala, *Phascolarctos cinereus* Goldfuss, in Victoria. *Australian Wildlife Research* 8:275–283.

———. 1985. Overbrowsing, and decline of a population of the koala, *Phascolarctos cinereus,* in Victoria: III. Population dynamics. *Australian Wildlife Research* 12:377–385.

McColl, K.A., R.W. Martin, L.J. Gleeson, K.A. Handasyde, and A.K. Lee. 1984. Chlamydia infection and infertility in the female koala *(Phascolarctos cinereus). Veterinary Record* 115:655.

McCormack, W.M. 1986. Chlamydial infections in men. In *Chlamydial infections,* ed. D. Oriel, G. Ridgway, J. Schachter, and D. Taylor-Robinson. Cambridge: Cambridge University Press, pp. 173–184.

McKenzie, R.A. 1981. Observations on diseases of free-living and captive koalas (*Phascolarctos cinereus*). *Australian Veterinary Journal* 57:243–246.

McMartin, D.A., A.R. Hunter, and J.W. Harris. 1979. Experimental pneumonia in red deer (*Cervus elaphus* L) produced by an ovine chlamydia. *Veterinary Record* 105:574–576.

Meagher, M., W.J. Quinn, and L. Stackhouse. 1992. Chlamydial-caused infectious keratoconjunctivitis in bighorn sheep of Yellowstone National Park. *Journal of Wildlife Diseases* 28:171–176.

Mitchell, P.J., R. Bilney, and R.W. Martin. 1988. Population structure and reproductive status of koalas on Raymond Island, Victoria. *Australian Wildlife Research* 15:511–514.

Mizuno, S. 1990. Development of serological diagnosis of chlamydial infection in the koala (*Phascolarctos cinereus*) by avidin-biotin enzyme-linked immunosorbent assay. *Japanese Journal of Veterinary Research* 38:66.

Moulder, J.W., T.P. Hatch, C. Kuo, J. Schachter, and J. Stortz. 1984. Genus I. *Chlamydia* Jones, Rake and Stearns 1945, ch. 55. In *Bergey's manual of systematic bacteriology,* ed. N.R. Kreig and J.G. Holt. Baltimore: Williams and Wilkins, pp. 729–739.

Obendorf, D.L. 1981. Pathology of the female reproductive tract in the koala, *Phascolarctos cinereus* (Goldfuss), from Victoria, Australia. *Journal of Wildlife Diseases* 17:587–592.

———. 1983. Causes of mortality and morbidity of wild koalas, *Phascolarctos cinereus* (Goldfuss), in Victoria, Australia. *Journal of Wildlife Diseases* 19:123–131.

———. 1988. The pathogenesis of urogenital tract disease in the koala. In *Australian wildlife.* Proceedings 104. Sydney: Post Graduate Committee in Veterinary Science, University of Sydney, pp. 649–655.

Obendorf, D.L., and K.A. Handasyde. 1990. Pathology of chlamydial infection in the reproductive tract of the female koala (*Phascolarctos cinereus*). In *Biology of the koala,* ed. A.K. Lee, K.A. Handasyde, and G.D. Sanson. Chipping-Norton: Australia: Surrey Beatty and Sons, pp. 255–259.

Oriel, J.D. 1986. Chemotherapy. In *Chlamydial infections,* ed. D. Oriel, G. Ridgway, J. Schachter, D. Taylor-Robinson, and M. Ward. Cambridge: Cambridge University Press, pp. 513–523.

Osawa, R., and F.N. Carrick. 1990. Use of a dietary supplement in koalas during systemic antibiotic treatment of chlamydial infection. *Australian Veterinary Journal* 67:305–307.

Osawa, R., F.N. Carrick, N. Hashimoto, I. Takashima, and T. Takahasi. 1990. Application of a blood sampling paper method for complement fixation test detection of antichlamydial antibody in koalas (*Phascolarctos cinereus*). In *Biology of the koala,* ed. A.K. Lee, K.A. Handasyde, and G.D. Sanson. Chipping Norton, Australia: Surrey Beatty and Sons, pp. 277–279.

Palmer, D.G., D. Forshaw, and S.L. Wylie. 1988. Demonstration of *Chlamydia psittaci* antigen in smears and paraffin tissue sections using a fluorescein isothiocyanate labelled monoclonal antibody. *Australian Veterinary Journal* 65:98–99.

Pratt, A. 1937. *The call of the koala.* Melbourne: Robertson and Mullens, 120 pp.

Rasmussen, S., and P. Timms. 1991. Detection of *Chlamydia psittaci* using DNA probes and the polymerase chain reaction. *FEMS Microbiology Letters* 77:169–174.

Reed, P.C., D. Lunney, and P. Walker. 1990. A 1986–1987 survey of the koala *Phascolarctos cinereus* (Goldfuss) in New south Wales and an ecological interpretation of its distribution. In *Biology of the koala,* ed. A.K. Lee, K.A. Handasyde, and G.D. Sanson. Chipping Norton, Australia: Surrey Beatty and Sons, pp. 55–74.

Robinson, A.C. 1978. The koala in South Australia. In *The koala,* ed. T.J. Bergin. Sydney: Zoological Parks Board of New South Wales, pp. 132–143.

Robinson, P.T. 1978. Koala management and medicine at the San Diego zoo. In *The koala,* ed. T.J. Bergin. Sydney: Zoological Parks Board of New South Wales, pp. 166–173.

Rush, C.M., and P. Timms. 1996. In vitro survival characteristics of koala chlamydiae. *Wildlife Research* 23:213–219.

Schachter, J. 1986. Human *Chlamydia psittaci* infection. In *Chlamydial infections,* ed. D. Oriel, G. Ridgway, J. Schachter, and D. Taylor-Robinson. Cambridge: Cambridge University Press, pp. 311–320.

Spencer, W.N., and F.W.A. Johnson. 1983. Simple transport medium for the isolation of *Chlamydia psittaci* from clinical material. *Veterinary Record* 113:535–536.

Storey, C., M. Lusher, P. Yates, and S. Richmond. 1993. Evidence for *Chlamydia pneumoniae* of non-human origin. *Journal of General Microbiology* 139:2621–2626.

Storz, J. 1971. *Chlamydia and chlamydia-induced diseases.* Springfield, IL: Charles C. Thomas, 358 pp.

Taylor, S.K., V.G. Vieira, E.S. Williams, R. Pilkington, S.L. Fedorchak, K.W. Mills, J.L. Cavender, A.M. Boerger-Fields, and R.E. Moore. 1996. Infectious keratoconjunctivitis in free-ranging mule deer (*Odocoileus hemionus*) from Zion national park, Utah. *Journal of Wildlife Diseases* 32:326–330.

Timms, P. 1993. Chlamydiosis in birds, wild and domestic animals. In *Australian standard diagnostic techniques for animal diseases,* ed. L.A. Corner and T.J. Bagust. East Melbourne: Commonwealth of Australia, CSIRO, pp. 1–8.

Timms, P., F.W. Eaves, A.A. Girjes, and M.F. Lavin. 1988a. Comparison of *Chlamydia psittaci* isolates by restriction endonuclease and DNA probe analyses. *Infection and Immunity* 56:287–290.

Timms, P., F.W. Eaves, A.F. Hugall, and M.F. Lavin. 1988b. Plasmids of *Chlamydia psittaci:* Cloning and comparison of isolates by Southern hybridisation. *FEMS Microbiology Letters* 51:119–124.

Timms P., M. Jackson, T. Glassick, and P. Giffard. 1996. Phylogenetic diversity of chlamydial strains (*C. pecorum* and *C. pneumoniae*) infecting koalas. In *Proceedings of the third meeting of the European Society for Chlamydia Research, Vienna, Austria, September 11–14, 1996,* ed. A. Stary. Vienna: Study Group for STD and Dermatological Microbiology of the Austrian Society for Dermatology and Venereology, p. 119.

Troughton E. 1941. *Furred animals of Australia.* Sydney: Angus and Robertson, 376 pp.

Ueno H., S. Mizuno, I. Takashima, R. Osawa, W. Blanshard, P. Timms, N. White, and N. Hashimoto. 1991. Serological assessment of chlamydial infection in the koala by a slide EIA technique. *Australian Veterinary Journal* 68:393–396.

Van der Lugt, J.J., and J.C. Kriek. 1988. Chlamydiosis in a springbok (*Antidorcas marsupialis*). *Journal of the South African Veterinary Association* 59:33–37.

Vanrompay D., R. Ducatelle, and F. Haesebrouck. 1995. *Chlamydia psittaci* infections: A review with emphasis on avian chlamydiosis. *Veterinary Microbiology* 45:93–119.

Warneke R.M. 1978. The status of the koala in Victoria. In *The koala,* ed. T.J. Bergin. Sydney: Zoological Parks Board of New South Wales, pp. 109–114.

Wati S. 1991. Transmission and control of *Chlamydia psittaci* in koalas. B.Sc. Honours diss., Queensland University of Technology, Brisbane, Australia, 88 pp.

Weigler B.J., R.J. Booth, R. Osawa, and F.N. Carrick. 1987. Causes of morbidity and mortality in 75 free-ranging and captive koalas in south east Queensland, Australia. *Veterinary Record* 121:571–572.

Weigler B.J., F.C. Baldock, A.A. Girjes, F.N. Carrick, and M.F. Lavin. 1988a. Evaluation of an enzyme immunoassay test for the diagnosis of *Chlamydia psittaci* infection in free-ranging koalas (*Phascolarctos cinereus*) in southeastern Queensland, Australia. *Journal of Wildlife Diseases* 24:259–263.

Weigler B.J., A.A. Girjes, N.A. White, N.D. Kunst, F.N. Carrick, and M.F. Lavin. 1988b. Aspects of the epidemiology of *Chlamydia psittaci* infection in a population of koalas (*Phascolarctos cinereus*) in southeastern Queensland, Australia. *Journal of Wildlife Diseases* 24:282–291.

White N.A., and N.D. Kunst. 1990. Aspects of the ecology of the koala in southeastern Queensland. In *Biology of the koala,* ed. A.K. Lee, K.A. Handasyde, and G.D. Sanson. Chipping Norton, Australia: Surrey Beatty and Sons, pp. 109–116.

White N.A., and P. Timms. 1994. *Chlamydia psittaci* in a koala (*Phascolarctos cinereus*) population in south-east Queensland. *Wildlife Research* 21:41–47.

Wills J.M. 1986. *Chlamydia* zoonosis. *Journal of Small Animal Practice* 27:717–731.

Wood M.N., and P. Timms. 1992. Comparison of nine antigen detection kits for diagnosis of urogenital infections due to *Chlamydia psittaci* in koalas. *Journal of Clinical Microbiology* 30:3200–3205.

Woodford M.H. 1993. International disease implications for wildlife translocation. *Journal of Zoo and Wildlife Medicine* 24:265–270.

26 LYME BORRELIOSIS

RICHARD N. BROWN AND ELIZABETH C. BURGESS

Synonyms: Lyme disease, erythema chronicum migrans, erythema migrans, acrodermatitis chronica atrophicans, lymphadenosis benigna cutis, Bannwarth's syndrome, Garin-Bujadoux syndrome, chronic lymphocytic meningoradiculoneuritis, tick-borne meningoradiculoneuritis.

INTRODUCTION. Lyme borreliosis is a tick-borne disease of people and some domestic animals, most commonly dogs, caused by spirochetal bacteria in the species group referred to as *Borrelia burgdorferi* sensu lato (s. l., meaning "in the broad sense"). Certain species of wild mammals and birds act as reservoir hosts for Lyme borreliosis spirochetes that are transmitted among wildlife and incidentally to people and domestic animals by some species of ticks. Disease in wildlife infected with *B. burgdorferi* appears rare. Lyme borreliosis is distinct from the relapsing fevers, which are caused by other species of *Borrelia* (Ras et al. 1996).

In people, Lyme borreliosis may mimic other conditions, including rheumatoid arthritis, Bell's palsy, and other disorders of the central and peripheral nervous systems. It is the debilitating effect of chronic infection, rather than mortality, that makes this disease so significant. Lyme borreliosis is the most important tick-borne infection in the world, occurring throughout temperate regions of the Northern Hemisphere, and it is the most prevalent vector-borne human disease in both Europe and the United States [Lane 1994; Piesman and Gray 1994; Centers for Disease Control and Prevention (CDC) 1997a].

The literature regarding this disease, its etiology, epidemiology, and the ecology of its tick vectors is voluminous. Here, we briefly review the history of Lyme borreliosis, its etiology, geographic distribution, ecology and epidemiology, clinical signs in people and domestic animals, pathology, immunology, diagnosis, management, and control. The role of wildlife in the ecology and epidemiology of Lyme borreliosis, and measures that may be taken to reduce the risk of disease, are emphasized.

HISTORY. The early history of Lyme borreliosis has been reviewed by Steere (1989). An expanding annular skin rash, "erythema chronicum migrans" (ECM; now preferably termed *erythema migrans,* EM) described by the Swedish dermatologist Afzelius in 1909 as a sequel to a tick bite, was likely the first reference to this disease. An Austrian dermatologist, Lipshutz, also described an expanding skin rash following the bite of a sheep tick *Ixodes ricinus* in 1913 and speculated that the cause was tick transmitted. Erythema migrans was associated with meningitis by Hollstrom in Sweden in 1930, and in 1949 he reported that penicillin had a beneficial action in the treatment of EM. In 1948, Lennhoff speculated on the spirochetal etiology of the disease based on the appearance of organisms in skin sections from EM patients in Sweden. In the 1940s, Bannwarth defined a syndrome of radicular (nerve root) pain followed by meningitis and sometimes cranial or peripheral neuritis, which subsequently has been associated with Lyme borreliosis.

The first case of Lyme borreliosis described in the United States was observed in a grouse hunter in Wisconsin in 1969 by Scrimenti (Burgdorfer 1986). In 1975, two mothers in rural Connecticut became concerned when an unusual outbreak of arthritis occurred in children in the Connecticut communities of Old Lyme, Lyme, and East Haddam. The syndrome included a characteristic skin rash following tick bite, leading to the description of a tick-borne disease complex of unknown etiology, named Lyme arthritis (Steere et al. 1977). The tick suspected of being the vector of this syndrome was described as *Ixodes dammini,* since synonymized with *Ixodes scapularis,* the blacklegged tick (Oliver et al. 1993).

The etiologic agent of Lyme arthritis was isolated from the midgut of an *I. scapularis* and named *Borrelia burgdorferi* in 1982 (Burgdorfer et al. 1982). This agent subsequently was isolated from the skin, blood, and cerebrospinal fluid of people with Lyme disease (Steere et al. 1983). *Ixodes pacificus,* the western blacklegged tick, was incriminated as the vector of *B. burgdorferi* in California by demonstration of spirochetes in the midgut, and *Ixodes ricinus* collected in Switzerland in 1982 was similarly implicated as a vector in Europe (Burgdorfer 1986).

Since 1982, knowledge of the pathogenesis, epidemiology, diagnosis, treatment, and prevention of this disease has grown exponentially, abetted perhaps as much by the capacity to culture the causative spirochetes in vitro as by any other technical advance (Anderson and Magnarelli 1992). However, much still remains to be learned about the complex interactions of the agent with mammalian hosts, and about the cycles of infection in nature.

ETIOLOGY. *Borrelia burgdorferi* s. l. is comprised of many strains (Postic et al. 1994), at least eight of which are sufficiently distinct genetically to be named genospecies: *B. burgdorferi* sensu stricto (s. s.), DN127, *Borrelia andersonii, Borrelia garinii, Borrelia afzelii, Borrelia valaisiana, Borrelia lusitaniae* (Saint-Girons et al. 1998), and *Borrelia japonica* (Kawabata et al. 1993). Isolates from a single host or vector containing two or more strains or genotypes of *B. burgdorferi* s. l. have been reported (Casjens et al. 1995; Marconi et al. 1995).

These spirochetes are spiral-shaped bacteria averaging approximately 0.2 μm in diameter and up to 30 μm in length. They have 7–11 periplasmic flagella or axial fibrils at one end (Johnson et al. 1984), which provide motility, particularly in high-viscosity fluids such as occur in dermal connective tissues (Kimsey and Spielman 1990).

A 41-kDa antigen associated with the flagella, one of the major antigens stimulating an antibody response in infected individuals, is shared with other spirochetes (Barbour et al. 1986). Two outer-surface membrane proteins (OspA and OspB) are specific to *B. burgdorferi* s. l., but they have not been associated with all strains or isolates (Barbour et al. 1985; Anderson et al. 1989). Plasmid DNA contains the genetic codes for OspA (30–32 kDa) and OspB (34–36 kDa), and possibly for some virulence attributes (Barbour 1989), but it is not possible to correlate loss of mammalian infectivity with presence or absence of these specific plasmids. A 39-kDa protein thought to be specific for *B. burgdorferi* s. l. may be related to recently acquired infection (Simpson et al. 1991).

These organisms are micro-aerophilic and difficult to culture, requiring incubation for weeks at 33° C–35° C in complex liquid media, such as Barbour-Stoenner-Kelly (BSK) medium (Barbour 1984). They may be specifically detected in various substrates by the polymerase chain reaction (PCR) (Rosa and Schwan 1989). Borreliae may be visualized by dark-field or phase microscopy but typically not by bright-field microscopy. They may be demonstrated in tissue nonspecifically by Giemsa staining and by Warthin-Starry or Steiner silver staining, and with greater specificity by immunofluorescent or immunohistochemical techniques (Abele and Anders 1990).

Classification and identification of spirochetes within *B. burgdorferi* s. l. are based on the reactivity to specific monoclonal antibodies, sodium dodecyl sulfate-polyacrylamide gel electrophoresis, immunoblotting, multilocus enzyme electrophoresis, DNA homology, restriction enzyme analysis, rRNA gene restriction, and plasmid profile analysis (Zingg et al. 1993; Saint-Girons et al. 1998).

Genospecies of *B. burgdorferi* s. l. vary regionally, as well as among vectors and hosts (Barthold 1996b). *Borrelia burgdorferi* s. s. occurs most commonly in North America and in Europe but is absent from most of Asia, whereas *B. garinii* and *B. afzelii* are relatively common in Asia and the Far East (Hubalek and Halouzka 1997; Li et al. 1998). *Borrelia afzelii* infections have been associated with rodents, whereas *B. garinii* is most often associated with birds (Nakao et al. 1994; Humair et al. 1995; Kurtenbach et al. 1998b). *Ixodes uriae* and seabirds maintain a cycle of *B. garinii* (Bunikis et al. 1996). *Ixodes ovatus* transmits *B. japonica* in Japan (Nakao et al. 1996), and *Ixodes dentatus* and eastern cottontail rabbits *Sylvilagus floridanus* maintain a cycle of *B. andersonii* in the northeastern United States (Anderson et al. 1989). While recognizing the necessity for complete characterization of the genospecies involved in particular ecologic or epidemiologic circumstances, we subsequently use the term *B. burgdorferi* when referring to the agents comprising *B. burgdorferi* s. l.

DISTRIBUTION OF LYME BORRELIOSIS. Lyme borreliosis occurs in most temperate areas of the Holarctic region. It is now regularly reported from Japan (Nakao et al. 1994, 1996), China (Ai et al. 1994; Li et al. 1998), and countries of the former Soviet Union (Korenberg et al. 1993), as well from as Western Europe (O'Connell et al. 1998) and much of the United States (CDC 1997a). Cases also have been reported from southern Canada and Australia (Dos Santos and Kain 1998; Nash 1998) and occasionally from Africa and South America (Azulay et al. 1991; Hammouda et al. 1995; Strijdom and Berk 1996).

The number of cases reported annually in the United States has risen since 1982 to over 16,000 (CDC 1997b). Initially, most cases were reported from the northeast/mid-Atlantic region, the upper midwest, and California. By 1996, however, cases had been reported from almost all states (CDC 1997a), although the original regions of high incidence are still evident.

The states with the most reported cases are (in order of decreasing number) New York, Connecticut, Pennsylvania, New Jersey, Rhode Island, Maryland, and Massachusetts (CDC 1997a). In 1996, the annual incidence of Lyme disease was greatest in Nantucket County, Massachusetts (more than 1200 cases per 100,000 population), and on a statewide basis in Connecticut and Rhode Island, with 95 and 53.5 new cases reported (respectively) per 100,000 population (CDC 1997a).

Wisconsin and Minnesota in the upper midwest have reported more cases than the other remaining states. Although California was initially considered to have a relatively high incidence, this impression has changed as the incidence has decreased in California and increased in many other states (CDC 1997a).

In Europe, the prevalence of Lyme borreliosis increases generally from west to east, and annual incidence ranges nationally from 0.3 per 100,000 in the United Kingdom to 130 per 100,000 in Austria. The highest estimated incidence occurs in eastern and southern Austria, where 300–350 people per 100,000 population develop disease annually (O'Connell et al. 1998).

HOST RANGE. *Borrelia burgdorferi* has been demonstrated in many species of mammals and birds. Any summary of host range is bound to be incomplete due to the addition of new host records. At least 53 species of wild mammals have been reported as naturally infected with *B. burgdorferi* (Table 26.1). Additionally, *B. burgdorferi* has been demonstrated in naturally infected domestic mammals, including dogs, horses, and cattle (Burgess 1986; Burgess and Mattison 1987; Burgess et al. 1987).

TABLE 26.1—Wild mammals naturally infected with *Borrelia burgdorferi* sensu lato as determined by xenodiagnosis, isolation, polymerase chain reaction (PCR), or demonstration in tissue using monoclonal antibodies

Scientific Name	Geographic Location	Selected References
Insectivora		
Erinaceus europaeus	Europe	Gray et al. 1994
Blarina brevicauda	North America	Telford et al. 1990
Cryptotis parva	North America	Sonenshine et al. 1995
Neomys fodiens	Europe	Talleklint and Jaenson 1994
Sorex araneus	Europe	Talleklint and Jaenson 1994
Sorex minutus	Europe	Talleklint and Jaenson 1994
Carnivora		
Canis latrans	North America	Burgess and Windberg 1989
Vulpes vulpes	Asia	Isogai et al. 1994a
Meles meles	Europe	Gern et al. 1998
Procyon lotor	North America	Ouellette et al. 1997
Artiodactyla		
Cervus nippon	Asia	Kimura et al. 1995
Rodentia		
Sciurus vulgaris	Europe	Humair and Gern 1998
Sciurus carolinensis	Europe	Craine et al. 1997
Tamias striatus	North America	Anderson et al. 1985
Dipodomys californicus	North America	Lane and Brown 1991
Oryzomys plaustris	North America	Sonenshine et al. 1995
Peromyscus maniculatus	North America	Rand et al. 1993
Peromyscus gossypinus	North America	Oliver et al. 1995
Peromyscus boylii	North America	Brown and Lane 1996
Peromyscus truei	North America	Brown and Lane 1996
Peromyscus leucopus	North America	Anderson et al. 1985
Peromyscus difficilis	North America	Maupin et al. 1994
Sigmodon hispidus	North America	Oliver et al. 1995
Neotoma mexicana	North America	Maupin et al. 1994
Neotoma fuscipes	North America	Lane and Brown 1991
Neotoma cinerea	North America	Gordus and Theis 1993
Eothenomys andersoni	Asia	Nakao et al. 1996
Eothenomys smithii	Asia	Masuzawa et al. 1996
Clethrionomys glareolus	Europe	Kurtenbach 1996
Clethrionomys rufocanus	Europe, Asia	Gorelova et al. 1995
Clethrionomys rutilus	Europe	Gern et al. 1998
Microtus agrestis	Europe	Talleklint and Jaenson 1994
Microtis californicus	Western North America	Peavey et al. 1997
Microtus montebelli	Asia	Nakao et al. 1996
Microtus oeconomus	Europe	Gorelova et al. 1995
Microtus pennsylvanicus	North America	Anderson et al. 1986
Apodemus agrarius	Asia	Park et al. 1993
Apodemus argenteus	Asia	Nakao et al. 1996
Apodemus speciosus	Asia	Masuzawa et al. 1996
Apodemus sylvaticus	Europe	De Boer et al. 1993
Apodemus flavicollis	Europe	Kurtenbach 1996
Apodemus peninsulae	Asia	Takada et al. 1998
Rattus confucianus	Asia	Pan 1992
Rattus norvegicus	North America and Madeira	Matuschka et al. 1994a; Piesman and Gray 1994
Rattus rattus	North America and Madeira	Peavey et al. 1997; Matuschka et al. 1994a
Mus musculus	North America	Sonenshine et al. 1995
Glis glis	Europe	Matuschka et al. 1994b
Napaeozapus insignis	North America	Anderson and Magnarelli 1984
Lagomorpha		
Sylvilagus floridanus	North America	Anderson et al. 1989
Sylvilagus bachmanni	North America	Peavey et al. 1997
Lepus timidus	Europe	Talleklint and Jaenson 1994
Lepus californicus	North America	Burgess and Windberg 1989
Lepus europaeus	Europe	Talleklint and Jaenson 1994

Birds in which *B. burgdorferi* has been detected include the mallard *Anas platyrhynchos,* ring-necked pheasant *Phasianus colchicus,* wild turkey *Meleagris gallopavo,* common murre *Uria aalge,* razorbill *Alca torda,* black guillemot *Cepphus grylle,* Atlantic puffin *Fratercula arctica,* house wren *Troglydytes aedon,* veery *Catharus fuscescens,* robin *Erithacus rubecula,* Eurasian blackbird *Turdus merula,* song thrush *Turdus philomelos,* American robin *Turdus migratorius,* gray catbird *Dumetella carolinensis,* blackcap *Sylvia atricapilla,* prairie warbler *Dendroica discolor,* common yellowthroat *Geothlypis trichas,* song sparrow *Melospiza melodia,* orchard oriole *Icterus spurius,* and the house sparrow *Passer domesticus* (Anderson and Magnarelli 1984; Anderson et al. 1985; Schulze et al. 1986; Burgess 1989; Burgess et al. 1993; McLean et al. 1993; Olsen et al. 1993; Craine et al. 1997; Gern et al. 1998; Humair et al. 1998).

EPIDEMIOLOGY. The maintenance of a vector-borne disease involves the interaction of agent, vectors, and hosts in habitats that support all facets of the system. The ecology and epidemiology of Lyme borreliosis have been reviewed (Lane et al. 1991; Anderson and Magnarelli 1993; Lane 1994; Piesman and Gray 1994; O'Connell et al. 1998), and readers are directed to these sources for detailed treatments of these topics.

Basic Vector Biology. The epidemiology of Lyme borreliosis is tied to the life history strategies of its vectors. Vectors of *B. burgdorferi* are all "three-host" ticks, in that they typically obtain blood meals from three different vertebrate hosts: once as a larva, again as a nymph, and lastly as an adult female. Adult males may take several small blood meals while searching for females with which to mate. Fully fed (replete) females detach from the host and oviposit a mass of eggs from which larvae eventually hatch.

Amplification of the population of infected vectors via transovarial transmission (transmission of the agent from the female to her eggs and subsequently to the larvae that hatch from those eggs) occurs only rarely in vectors of *B. burgdorferi* (Lane 1994); hence, unfed larvae are typically considered to be uninfected. Transstadial transmission (maintenance of the agent in the vector through the molt from one stage to the next) is therefore essential, because each stage feeds only once; nymphs or adults may infect a host with bacteria acquired while feeding at an earlier stage.

Vectorial capacity is a broad term that describes the relative ability of an arthropod to transmit an agent effectively among vertebrates in nature. This involves a complex of interacting variables, including intrinsic attributes of the vector that determine its vector competence (including the capacity for transstadial transmission of the agent and its ability to transmit the organism to subsequent hosts); vector abundance and distribution (geographically and among species of hosts); host abundance, distribution, and capacity to sustain *B. burgdorferi* infections; interaction of the agent with other microbes (e.g., within a tick's gut); and abiotic factors that alter temporal and spatial distribution of hosts and vectors. Vectorial capacity is a dynamic phenomenon that varies seasonally, among years, among communities of hosts and vectors, and across landscapes.

Wildlife Host Factors. Although *B. burgdorferi* may infect many vertebrate species, the tendency for vertebrates to serve as hosts of immature vector ticks, to become infected, and to remain infectious to ticks over time varies considerably. Vertebrates in which *B. burgdorferi* infections are relatively common, and which serve as hosts for tick vectors to which they transmit the agent, are said to comprise the local wildlife reservoir of infection. Different stages of vector ticks may prefer to feed on different host species. However, only hosts that become infected when fed upon by the immature stages can act as reservoirs of *B. burgdorferi* in nature, since transovarial transmission is not significant. Hence, hosts upon which primarily adult ticks feed do not contribute significantly to the combined reservoir. Thus, immature ticks provide the key to maintenance of transmission cycles.

Reservoir competence incorporates several variables, including the proportion of the vector population infected by feeding on a specific host, which involves host preference by the vector; patterns of abundance and distribution of the hosts; and the intrinsic elements of the host-parasite relationship that enable a host to remain infectious for vectors.

Host populations contribute to maintenance of endemic cycles of *B. burgdorferi* in two ways. Some, such as some lizards and deer, are important hosts for one or more stages of vector ticks (Lane 1994). However, they do not become persistently infected with *B. burgdorferi* and therefore do not serve as significant reservoirs of the agent. Other vertebrates, typically certain rodents, and some birds, become persistently infected, remain infective to feeding ticks, and are fed upon by a large proportion of the immature vector population; these hosts have a high degree of reservoir competence.

Although *B. burgdorferi* has been associated with many species of mammals, few are known to be important reservoirs. Mammalian reservoirs in North America include dusky-footed wood rats *Neotoma fuscipes,* Mexican wood rats *Neotoma mexicana,* and California kangaroo rats *Dipodomys californicus* in the western United States (Brown and Lane 1992, 1996; Maupin et al. 1994), white-footed mice *Peromyscus leucopus* and eastern chipmunks *Tamias striatus* in the eastern and upper midwestern United States (Piesman and Gray 1994; Slajchert et al. 1997), and black rats *Rattus rattus* or Norway rats *Rattus norvegicus* in specific environments in California and on islands along the east coast of the United States (Smith et al. 1993; Matuschka et al. 1994a; Peavey et al. 1997).

In Europe, mammals with reservoir potential include gray squirrels *Sciurus carolinensis,* European tree squirrels *Sciurus vulgaris,* yellow-necked mice *Apodemus flavicollis,* wood mice *Apodemus sylvaticus,* and bank voles *Clethrionomys glareolus* (Piesman and Gray 1994; Craine et al. 1997; Gern et al. 1998; Humair and Gern 1998).

Nonrodent reservoirs include several insectivores, most notably the Eurasian hedgehog *Erinaceus europaeus* (Gern et al. 1997), and two lagomorphs, the eastern cottontail in North America (Telford and Spielman 1989) and the mountain hare *Lepus timidus* in Europe (Jaenson and Talleklint 1996). Additional endemic cycles, with additional species of reservoir hosts, likely will continue to be reported.

Other species of mammals are fed on commonly by vectors and may remain infected for an indeterminate period. Although some immature ticks may acquire infection while feeding on such hosts, the proportion of the vector population infected may be too small (e.g., due to low host population density) for these hosts to be considered significant reservoirs in themselves. Cumulatively, however, multiple minor reservoir species, such as Virginia opossums *Didelphis virginianus,* raccoons *Procyon lotor,* and striped skunks *Mephitis mephitis* in parts of North America, may contribute to the overall reservoir of the agent (Fish and Daniels 1990; Ouellette et al. 1997).

One or two vertebrate species typically act as the major reservoir for *B. burgdorferi* over a broad geographic region, but other species may serve as the principal reservoir in local areas within these regions. For example, eastern chipmunks appear to be important reservoirs in habitats where they are common in *B. burgdorferi*-endemic regions of North America (Slajchert et al. 1997), and Norway rats serve as a reservoir in the absence of white-footed mice on Monhegan Island, Maine (Smith et al. 1993).

Cervids, including white-tailed deer *Odocoileus virginianus* (Wilson et al. 1990; Luttrell et al. 1994), black-tailed deer *Odocoileus hemionus columbianus* (Lane et al. 1994a), roe deer *Capreolus capreolus* (Matuschka et al. 1993), fallow deer *Dama dama* (Gray et al. 1992; Matuschka et al. 1993), and red deer *Cervus elaphus* (Gray et al. 1992; Matuschka et al. 1993), are major hosts for adults of important tick vectors. Abundance of the primary vectors may be greater where deer are common (Wilson et al. 1990; Gray et al. 1992), but deer do not serve as reservoirs of the spirochetes. This is due, in part, to the fact that a relatively small proportion of the immature vector population feeds on deer, as well as the absence of transovarial transmission from adult ticks to the next generation.

Deer are also innately poor hosts for *B. burgdorferi.* Matuschka et al. (1993) considered the role of European ungulates (including roe deer, red deer, fallow deer, and wild sheep *Ovis ammon*) to be *zooprophylactic* because the feeding of a large proportion of a vector population on nonreservoir hosts dilutes the force of transmission by that vector. *Ixodes ricinus* not only failed to become infected while feeding on deer, but *B. burgdorferi* was cleared from infected ticks that fed on deer (Matuschka et al. 1993), and serum from wild sika deer contained borreliacidal activity (Isogai et al. 1994b).

However, deer are epidemiologically important because they serve as major hosts to adult stages of the primary tick vectors. The recent expansion of the distribution of Lyme borreliosis in the eastern United States is related, in part, to expansion of once overexploited and depleted white-tailed deer populations, associated with changes in land use and harvest regulations during the 20th century. As deer herds have moved back onto former range, the ticks that depend on them (and the agents of disease that they carry) have moved with them (Barbour and Fish 1993; Spielman et al. 1993).

Dissemination of vector ticks northward by migrating birds also contributes to extension of endemic areas (Klich et al. 1996), and large areas of North America may be climatologically susceptible to further range expansion by *I. scapularis* into areas of suitable habitat (Mount et al. 1997a).

Other explanations proposed for the current apparent epidemic of Lyme borreliosis in the United States include increased abundance and quality of tick habitat with changes in land use, a proportional increase in the human population in suburban and rural areas with favorable habitat, and increased opportunity for human exposure related to recreational use of natural areas (Ginsberg 1994).

Vectors and Endemic Cycles. The geographic distribution of human Lyme borreliosis roughly coincides with the distribution of the anthropophilic species (those that readily feed on people) of the *Ixodes ricinus* species complex. This group of related ticks includes four species (*I. ricinus, I. persulcatus, I. scapularis,* and *I. pacificus*) that are generalists, feeding on a variety of hosts, including wild and domestic animals and people. All are established vectors of several human diseases and are the primary vectors of *B. burgdorferi.*

Ixodes ricinus is found throughout Europe from Ireland eastward to the Caspian Sea and western Russia. *Ixodes persulcatus* occurs throughout Eastern Europe and northern Asia from Poland to Japan and south to Korea. Within North America, *I. scapularis* is found in most of the United States east of a line drawn approximately from Oklahoma to Minnesota. *Ixodes pacificus* occurs in western North America from northern Baja California to southwestern British Columbia (Keirans et al. 1997).

Other members of the *I. ricinus* complex may be capable of transmitting *B. burgdorferi* among wildlife. Isolates have been obtained from *Ixodes jellisoni* (found mainly in California) and its primary host, the California kangaroo rat (Brown and Lane 1996), but vector competence of *I. jellisoni* remains to be shown. The remaining *I. ricinus* complex species—*I. affinis* (distributed from the southeastern United States through Central America and into South America), *I. pararicinus* (from South America), *I. gibbosus* (from

TABLE 26.2—Examples of endemic cycles maintaining *Borrelia burgdorferi* sensu lato

Endemic Maintenance Vectors	Suggested Reservoirs	Geographic Region	Selected References
***Ixodes ricinus* complex**			
I. pacificus[1]	Unknown	W. North America	
I. persulcatus	*Apodemus* sp. *and Clethrionomys* sp.	Asia	Nakao et al. 1994
I. ricinus	*Apodemus sylvaticus*	Europe	De Boer et al. 1993
I. ricinus	*Clethrionomys glareolus* and *Apodemus flavicollis*	Europe	Hovmark et al. 1988; Kurtenbach 1996
I. ricinus	*Lepus timidus*	Baltic Sea Islands	Jaenson and Talleklint 1996
I. ricinus	*Rattus rattus* and *R. norvegicus*	Madeira Island	Matuschka et al. 1994a
I. ricinus	*Ovis aries*	Europe	Ogden et al. 1997
I. ricinus	*Phasianus colchicus*	Europe	Kurtenbach et al. 1998a
I. scapularis	*Peromyscus leucopus*	E. North America	Spielman et al. 1985
I. scapularis	*Peromyscus maniculatus*	E. North America	Rand et al. 1993
I. scapularis	*Rattus norvegicus*	E. North America	Smith et al. 1993
I. scapularis	*Tamias striatus*	E. North America	Slajchert et al. 1997
Non-*Ixodes ricinus* group			
I. dentatus	*Sylvilagus floridanus*	E. North America	Telford and Spielman 1989
I. hexagonus	*Erinaceus europaeus*	Europe	Gern et al. 1997
I. neotomae[2]	*Neotoma fuscipes*	W. North America	Brown and Lane 1992
I. spinipalpis	*Neotoma mexicana*	W. North America	Maupin et al. 1994
I. spinipalpis	*Neotoma fuscipes*	W. North America	Peavey et al. 1997
I. spinipalpis	*Rattus rattus*	W. North America	Peavey et al. 1997
I. uriae	Seabirds incl. *Alca torda*	N. Coastal Holarctic	Olsen et al. 1993, 1995

[1] *Ixodes pacificus* is a competent laboratory vector; it is found infected commonly in nature, and it serves as the primary vector to humans. Although specific maintenance cycles have not been substantiated, *I. pacificus* is expected to serve as a maintenance vector where it occurs with competent reservoir hosts but where zooprophylactic hosts are rare. [2] *Ixodes neotomae* considered a junior synonym of *I. spinipalpis* (Norris et al. 1997).

Europe), and *I. hyatti, I. kashmiricus, I. kazakstani, I. nipponensis, I. nuttallianus,* and *I. pavlovskyi* (from Asia) (Keirans et al. 1997)—have yet to be associated with transmission of *B. burgdorferi.*

Ticks unrelated to the *I. ricinus* complex, including *I. neotomae, I. spinipalpis, I. dentatus,* and *I. hexagonus,* also have been established as vectors of *B. burgdorferi* in wildlife cycles (Table 26.2).

Alternative Vectors. Many other species of blood-sucking arthropods have been associated with *B. burgdorferi* by isolation or by PCR amplification of DNA from tissue (Table 26.3). These include at least 30 species of hard ticks (Ixodidae), 2 species of soft ticks (Argasidae), and 18 species of insects, mainly mosquitoes and biting flies. The potential for vector competence, or even occasional transmission, of *B. burgdorferi* has yet to be demonstrated for these species.

Implication of vector competence requires experimental demonstration that the putative vector can acquire, maintain, and transmit the infectious agent. Among putative vectors of Lyme borreliosis, vector competence has been proven for only six tick species (*I. dentatus, I. hexagonus, I. neotomae, I. pacificus, I. ricinus,* and *I. scapularis*).

Of the potential vectors listed in Table 26.3, only *Amblyomma americanum, Dermacentor occidentalis, Dermacentor variabilis, Ixodes cookei,* and *Ixodes holocyclus* have been tested experimentally; none proved to be competent vectors in laboratory studies (Barker et al. 1993; Lane 1994; Lane et al. 1994b; Piesman and Gray 1994). Other blood-sucking arthropods, including those listed in Table 26.3, may have the potential to serve as local vectors. However, it should be recognized that many acquire the organism accidentally while feeding on an infected host, without possessing the capacity to transmit it with epidemiologic significance. Without knowledge of vector competence, associations of an agent and a vector are merely correlative (Lane 1994).

Vectorial Capacity. The host range, patterns of abundance and distribution on different species of hosts, and habitat associations, all may affect the epidemiologic importance of a vector. The feeding by different stages on separate hosts, and often multiple host species, increases the potential of a vector species to amplify an agent throughout a host community. *Ixodes neotomae* and *I. hexagonus* are relative specialists that feed on only a few species of hosts, whereas *I. pacificus, I. scapularis, I. ricinus,* and *I. persulcatus* are generalists, feeding on a wide variety of vertebrates.

Ixodes pacificus and *I. scapularis* are both competent vectors of *B. burgdorferi,* although *I. scapularis* is somewhat more efficient than *I. pacificus* (Lane et al. 1994b; Piesman and Gray 1994). This relative difference in vector competence may explain part of the disparity in the prevalence of *B. burgdorferi* infections in ticks, and in the number of people infected, in the northeast and upper midwest versus the western USA. However, the primary difference in the vectorial capacity of these ticks may lie in the host preference of *I. pacificus* larvae and nymphs (Lane 1994).

TABLE 26.3—Arthropods with which *Borrelia burgdorferi* sensu lato has been directly associated, but for which vector competence or vectorial capacity appears limited or remains unknown

Vector Species	Region	Selected References
Ixodidae		
Ixodes acuminatus	Europe	Lane 1994
Ixodes angustus	W. North America	Damrow et al. 1989
Ixodes canisuga	Europe	Piesman and Gray 1994
Ixodes columnae	Asia	Fukunaga et al. 1996
Ixodes cookei	E. North America	Barker et al. 1993
Ixodes frontalis	Europe	Piesman and Gray 1994
Ixodes granulatus	Asia	Tian et al. 1998
Ixodes hexoganus	Europe	Gern et al. 1997
Ixodes jellisoni	W. North America	Brown and Lane 1996
Ixodes ovatus	Asia	Nakao et al. 1996
Ixodes rangtangensis	Asia	Tian et al. 1998
Ixodes tanuki	Asia	Masuzawa et al. 1996
Ixodes triangulaceps	Europe	Lane 1994
Ixodes turdus	Asia	Fukunaga et al. 1996
Amblyomma americanum	North America	Schulze et al. 1984; Ouellette et al. 1997
Amblyomma maculatum	W. North America	Teltow et al. 1991
Boophilus microplus	Asia	Tian et al. 1998
Dermacentor albipictus	E. North America	Lane 1994
Dermacentor occidentalis	W. North America	Lane 1994
Dermacentor parumapertus	W. North America	Rawlings 1986
Dermacentor reticulatus	Europe	Piesman and Gray 1994
Dermacentor silvarum	Asia	Tian et al. 1998
Dermacentor variabilis	E. North America	Anderson et al. 1985
Haemaphysalis bispinosa	Asia	Tian et al. 1998
Haemaphysalis concinna	Asia	Tian et al. 1998
Haemaphysalis flava	Asia	Zhang 1992
Haemaphysalis japonicum	Asia	Tian et al. 1998
Haemaphysalis leporuspalustris	North America	Anderson and Magnarelli 1984; Lane and Burgdorfer 1988
Haemaphysalis longicornis	Asia	Tian et al. 1998
Rhipicephalus sanguineus	W. North America	Hubbard et al. 1998
Argasidae		
Argas reflexus	Europe	Stanek and Simeoni 1989
Ornithodoros coriaceus	W. North America	Lane 1994
Insecta		
Chrysops callidus	E. North America	Magnarelli et al. 1986
Chrysops macquarti	E. North America	Magnarelli et al. 1986
Chrysops univittatus	E. North America	Magnarelli et al. 1986
Chrysops vittatus	E. North America	Magnarelli et al. 1986
Hybomitra epistates	E. North America	Magnarelli et al. 1986
Hybomitra hinei	E. North America	Magnarelli et al. 1986
Hybomitra lasiophthalma	E. North America	Magnarelli et al. 1986
Hybomitra sodalis	E. North America	Magnarelli et al. 1986
Tabanus pumilus	E. North America	Magnarelli et al. 1986
Tabanus quinquevittatus	E. North America	Magnarelli et al. 1986
Tabanus lineola	E. North America	Magnarelli et al. 1986
Tabanus pumilus	E. North America	Magnarelli et al. 1986
Aedes canadensis	E. North America	Magnarelli et al. 1986
Aedes stimulans	E. North America	Magnarelli et al. 1986
Aedes vexans	E. North America	Magnarelli et al. 1986
Cuterebra fontinella	E. North America	Anderson and Magnarelli 1984
Orchopeas leucopus	E. North America	Anderson and Magnarelli 1984
Ctenocephalides felis	W. North America	Teltow et al. 1991

Immature *I. pacificus* feed heavily on the western fence lizard *Scleroporus occidentalis,* which appears to be unable to support infections of *B. burgdorferi* (Lane 1994). Moreover, *B. burgdorferi* have been shown to be killed by constituents of the blood from some species of lizards, but not others (Lane and Quistad 1998). The reservoir incompetence of lizards impedes the vectorial capacity of *I. pacificus* and reduces the risk of transmission of *B. burgdorferi* to people in western North America (Lane 1994).

A similar phenomenon may account, in part, for the geographic variation in prevalence of *B. burgdorferi* infection in *I. scapularis.* The prevalence is high (often 25% or more of nymphs and more than 50% of adults) in *I. scapularis* in the northeastern and upper midwestern regions of the United States. In the southeast, where immature *I. scapularis* feed heavily on lizards, prevalence is much lower (Lane 1994; Piesman and Gray 1994), although the southeastern five-lined skink *Eumeces inexpectatus* and the green anole *Anolis carolinensis* appear to be competent reservoirs of *B. burgdorferi* in the laboratory (Levine et al. 1996).

The seasonal activity of life stages of ticks may influence epidemiologic patterns. Immature *I. scapularis* feed heavily on white-footed mice, important reservoirs of *B. burgdorferi* in the northeastern and upper midwestern United States (Lane 1994; Piesman and Gray 1994). Nymphs, infected with *B. burgdorferi* when they fed as larvae in the summer of the preceding year, feed on mice in the spring and ensure that reservoir hosts are available to infect the cohort of larvae that emerges from eggs during the summer. This reversal of the feeding phenology (nymphs before larvae) amplifies the prevalence of infection in both the reservoir hosts and the vectors, therefore increasing the risk of transmission to people in the northeastern United States (Spielman et al. 1985; Lane 1994; Piesman and Gray 1994).

Reservoir hosts may remain infective for ticks for long periods (Brown and Lane 1992). However, in some cycles involving both *I. ricinus* and *I. scapularis,* the duration of infectivity of vertebrate hosts for ticks is brief, and transmission of *B. burgdorferi* between generations of ticks requires concurrent feeding of larvae with nymphs. Larvae must be exposed during the relatively short period of nonsystemic infection with *B. burgdorferi* following inoculation of organisms by nymphal tick bite (Randolph et al. 1996; Lindsay et al. 1997; Ogden et al. 1997). In such circumstances, infected nymphs, rather than vertebrates, may be the effective reservoir of *B. burgdorferi* over the season when immature ticks are inactive (Lindsay et al. 1997; Ogden et al. 1997).

Habitat Effects. There are similarities among maintenance systems of *B. burgdorferi* throughout the world. In general, they all require habitats with microclimates that provide high humidity, as well as appropriate hosts. Vector ticks spend most of their lives sequestered in equable humid microenvironments; a relatively small amount of time is spent questing for, or feeding on, hosts.

Typically, these species are found in deciduous woodlands, dense brush, along ecotones, or in areas where habitat heterogeneity allows a sufficient number of sites with adequate shade and moisture. Leaf-litter accumulation and thick vegetation appear to be important variables in most regions. Open areas with short grass or agricultural fields typically do not provide suitable habitat.

In Europe, heterogeneous deciduous woodlands are thought to provide food and cover for hosts and equable microhabitats that support large populations of *B. burgdorferi*-infected *I. ricinus* ticks (Gray et al. 1998). Habitat heterogeneity promotes a diverse host community and high density of some species of rodents. However, host diversity may diminish the prevalence of *B. burgdorferi*-infected ticks unless reservoir hosts are relatively numerous. Prevalence of *B. burgdorferi* may be relatively low in ticks found questing in open areas where ungulates serve as the major hosts. However, diversity of hosts, including ungulates, may be a prerequisite for the development of areas of high risk (Gray et al. 1998).

In southern New York State, the risk of human infection was linked to interactions among populations of deer, white-footed mice, ticks, gypsy moths, and oak trees (Jones et al. 1998). In this system, acorn production varies seasonally and among years. White-tailed deer migrate in the autumn into areas where acorn production is high. Adult female *I. scapularis* feeding on migrating deer are carried into the area, where they drop off and lay eggs. High acorn yields also provide rodent populations with adequate food to support greater overwinter survival. Consequently, densities of mice and larval ticks are greatest in years following high acorn production. Gypsy moth populations, not otherwise linked directly to cycles of borreliosis, erupt when mouse populations are low in years following poor oak mast production. Although probably not applicable across landscapes, the level of detail described for this system provides the potential for the development of sound strategies for local management of borreliosis.

Piesman and Gray (1994) noted plasticity in the habitat associations of *I. ricinus, I. scapularis,* and *I. pacificus,* and there are a number of exceptions to the generalizations about tick-habitat associations. Populations of *I. scapularis* have been associated with hemlock forest in Pennsylvania (Lord et al. 1994), and *I. pacificus* can be found in xeric scrub oak habitats at high elevations in Arizona (Piesman and Gray 1994). Further analysis of habitat associations is necessary, since the details of the maintenance systems that influence the risk of human exposure to *B. burgdorferi* may vary from one nidus to the next.

Patterns of Maintenance and Transmission. Simplistically, the risk of human exposure to *B. burgdorferi* in a given locality is a function of whether there are competent tick vectors that feed commonly on reservoir hosts and also, even if only occasionally, on people.

Several general patterns of maintenance and transmission of *B. burgdorferi* can be identified. Throughout much of the Northern Hemisphere, *I. scapularis, I. ricinus,* or *I. persulcatus* serve as the primary vectors among wildlife. These ticks feed heavily as immatures on reservoir hosts; nymphs or adults are anthropophilic and hence can transmit spirochetes to people. Therefore, human Lyme borreliosis is most prevalent in areas where these ticks are common.

Endemic maintenance cycles involving nonanthropophilic ticks may assume epidemiologic significance in several ways. Cycles such as those involving eastern cottontail rabbits and *I. dentatus* in the northeastern United States (Telford and Spielman 1989), or *I. hexagonus* and the European hedgehog *Erinaceus europaeus* in Eurasia (Gern et al. 1997) may "overlap" those vectored by a member of the *I. ricinus* complex and thereby contribute to the force of *B. burgdorferi* transmission locally. This occurs because the primary vectors (*I. scapularis* in the northeastern United States and *I. ricinus* in Europe) also feed heavily on the same reservoir hosts, effectively merging the cycles. In the Colorado cycle involving *I. spinipalpis* and wood rats, the vector rarely bites people, and, since no anthropophilic vector is present, the cycle remains epidemiologically "silent" (Maupin et al. 1994).

In California, ticks other than the primary vectors to humans maintain cycles involving rodent reservoir hosts. *Ixodes pacificus* nymphs, although anthropophilic, have a relatively low probability of exposing people to *B. burgdorferi,* because most immature *I. pacificus* feed on reservoir-incompetent lizards (Lane 1994). Transmission of *B. burgdorferi* to humans occurs when nymphal *I. pacificus* bite people after feeding as larvae on rodent reservoirs involved in maintenance cycles vectored by nonanthropophilic ticks (Brown and Lane 1992; Peavey et al. 1997).

In most of Asia, the relative importance of cycles maintained by ticks other than *I. persulcatus* is not known. However, the number of potential vectors (Table 26.2) suggests that parallel cycles of *B. burgdorferi* transmission may be present in areas where *I. persulcatus* is the primary vector.

The cycle involving *I. uriae* and seabirds occurs throughout island and coastal regions worldwide (Olsen et al. 1993, 1995), has the potential to transport the agent globally, and may overlap with other cycles involving anthropophilic vectors.

Direct and Transplacental Transmission. Direct transmission of *B. burgdorferi* was reported among white-footed mice (Burgess 1991) and was suggested by Lord et al. (1994). Likewise, Ouellette et al. (1997) speculated that raccoon-to-raccoon transmission might occur via infected urine. Burgess and coworkers demonstrated the occurrence of transplacental transmission by isolating *B. burgdorferi* from fetal tissues of white-footed mice, house mice *Mus musculus,* and coyotes *Canis latrans* (Burgess and Windberg 1989; Burgess et al. 1993). Although direct transmission may occur in some situations, it has not been easily substantiated (Mather et al. 1991) and has yet to be shown to be epidemiologically important.

PATHOGENESIS AND IMMUNOLOGY. The pathogenesis of Lyme borreliosis is complicated and varies with host and vector species, as well as with the strain or genospecies of *B. burgdorferi.* It is intimately linked with the host immune response, the antibody-mediated arm of which seems more important than the cell-mediated arm in clearing *B. burgdorferi* infections (Hu and Klempner 1997; Sigal 1997). However, the organism may persist in wild rodents in the presence of circulating antibody for many months or years (Schwan et al. 1989; Wright and Nielsen 1990; Brown and Lane 1994; Campbell et al. 1994).

Disease in adult animals is often difficult to induce experimentally by inoculation of *B. burgdorferi* and, with the exception of dogs *Canis familiaris,* disease seems uncommon in naturally exposed domestic animals. Animal models that have proven useful in research on the pathogenesis of borreliosis have been summarized by Barthold (1995). They include European rabbits *Oryctolagus cuniculus* (Miller et al. 1996), gerbils *Meriones* sp. (Preac-Mursic et al. 1990), golden hamsters *Mesocricetus auratus* (Munson et al. 1996), laboratory mice (Barthold 1996a), rhesus monkeys *Macaca mulatta* (Philipp and Johnson 1994), and dogs (Appel et al. 1993; Levy et al. 1993).

The skin is typically the primary site of infection because inoculation usually occurs via the bite of an infected tick. The initial cellular inflammatory response at the site of the tick bite is somewhat diminished due to constituents in tick saliva (Hu and Klempner 1997). Thus, polymorphonuclear leukocytes migrate into the primary site immediately following infection but may fail to eradicate the organisms.

Following a period of local replication in the skin, which incites the typical EM rash, systemic infection ensues as organisms disseminate to other parts of the body. Within days to weeks of infection, spirochetes spread to the lymph and blood, and then to the heart, eye, muscle, bone, synovium of joints, spleen, liver, meninges, or brain (Steere 1989; Barthold et al. 1991; Hu and Klempner 1997). The ability of *B. burgdorferi* to survive in the face of a specific host immune response, as well as the ability to survive antibiotic treatment, have led many to speculate on the importance of intracellular survival and/or localization within protected sites, including the synovium and central nervous system (Hu and Klempner 1997).

The mechanisms mediating the production of disease are poorly understood. Both direct and indirect effects of the spirochete probably are involved in people. Multiplication of the spirochetes, virulence factors associated with the organism, and the inflammatory response contribute to the development of chronic lesions in many tissues in those species in which they occur (Miller et al. 1996). Virulence factors include a lipopolysaccharide that produces fever, and cell wall peptidoglycans that may induce joint inflammation. Although the inflammatory response is relatively ineffective at clearing infections, the dramatic cellular infiltration of tissues associated with arthritis and meningitis indicate a potent cellular activation by spirochetal antigens (Hu and Klempner 1997).

Indirect effects of infection include immune complexes in the plasma and synovial fluid, producing an immune-mediated inflammatory reaction especially in joint disease; interleukin-1 production by macrophages, which stimulates proteolysis, collagenase, and prostaglandin E_2 secretion by synovial cells, damaging articular cartilage; and mast cell degranulation and mediator release (Hu and Klempner 1997). Treatment-resistant Lyme borreliosis in people with certain histocompatibility types is associated with autoimmunity to host antigens that cross-react with OspA produced by *B. burgdorferi* (Gross et al. 1998).

Infected humans and most animals produce immunoglobulin M (IgM), IgG, and IgE antibodies to *B. burgdorferi* antigens. IgM antibodies to the 41-kDa flagellar antigen develop within a few weeks of infection, followed by IgG and IgM antibodies to 15-, 27-, 55-, 60-, 66-, and 83-kDa proteins and the major 31- and 34-kDa surface proteins (Barbour 1988). Other spirochetal infections may induce cross-reactive antibodies to the 41- and 60-kDa proteins. Antibodies to the 31- and 34-kDa proteins are more specific for infection with *B. burgdorferi* (Barbour 1988).

Although not as clearly defined as the role of the humoral response in providing protection against chronic infections, cell-mediated immunity interacts in *B. burgdorferi* infections in humans. Along with the influx of phagocytic cells into infected tissues, antigen-specific T-cell responses occur in the synovial fluid, cerebrospinal fluid, and serum (Barbour 1988). Whereas cytokines released by subsets of CD4+ cells appear to confer some protection against the development of severe arthritis, CD8+ cells appear to aggravate arthritis and perpetuate local spirochete populations (Hu and Klempner 1997).

CLINICAL SIGNS. Clinical signs of infection with *B. burgdorferi* have not been reported in naturally infected wild mammals or birds, with the exception of *P. leucopus.* Some white-footed mice had erythema (reddening) of the ear pinnae (Anderson and Magnarelli 1984); others developed neurologic signs, including trembling, incoordination, circling, head tilt, and weakness of the hind limbs (Burgess et al. 1990). Speculatively, such problems might decrease fitness either by making them more prone to predation or by decreasing reproductive success.

Some domestic animals, including dogs, cats, cattle, and horses, have developed clinical signs when infected with *B. burgdorferi.* Fever, stiffness, lameness, and arthritis are most commonly reported, with occasional renal, neurologic, ocular, and cardiac signs (Burgess and Mattison 1987; Burgess et al. 1987; Levy et al. 1993). However, domestic animal species typically develop neither the range nor severity of disease seen in people.

In some people, infection with *B. burgdorferi* remains asymptomatic. In others, *B. burgdorferi* causes multisystemic disease, primarily affecting the skin, heart, joints, and nervous system (Pachner et al. 1989; Dattwyler 1990). Lyme borreliosis in people may become severe and debilitating but it is rarely fatal (Steere 1989; O'Connell et al. 1998).

Lyme borreliosis syndromes differ between Europe and North America, presumably due to the greater variation in *B. burgdorferi* s. l. in Europe (Evans 1995; Cimmino et al. 1998). Clinical manifestations involving the skin and neurologic system were recognized in Europe in the early 1900s (Asbrink and Hovmark 1988). Although skin rash is the most common sign of disease in both North America and Europe (Cimmino et al. 1998), arthritis and cardiac disease are much more prevalent in North America (Evans 1995; Cimmino et al. 1998).

Lyme borreliosis can cause a plethora of symptoms that vary among infected individuals. The earliest manifestation typically is an expanding skin rash (EM) that occurs in 60%–80% of people who develop signs (Asbrink and Hovmark 1988). Such rashes are usually 3–20 cm or more in diameter, have a flat to raised red periphery, and tend to clear centrally. The rash first appears 2–12 weeks after the tick bite, the site of which may be visible centrally. An EM rash should be differentiated from the smaller focus of inflammation that develops almost immediately around some tick-bite sites. One or more metastatic EM lesions may occur elsewhere on the body in about half of borreliosis patients. EM may be accompanied by flulike symptoms, including fever, vomiting, joint pain, and occasionally photophobia and conjunctivitis.

Symptoms associated with neurologic, cardiac, ocular, and musculoskeletal disease may develop as early as 4–9 weeks following the tick bite. Neurologic signs occur in approximately 15% of untreated human patients, most of whom have chronic second-stage (6 months or less) disease. The most common presentation of neuroborreliosis is a painful meningopolyradiculitis that may occur alone or with other neurologic signs (Oschmann et al. 1998). Additional neurologic signs include Bell's palsy (paralysis of facial nerve), cranial neuritis, encephalitis, myelitis, cerebellar ataxia, meningitis, ocular signs, pain in the extremities, headache, and/or stiff neck (Steere 1989; Evans 1995; Oschmann et al. 1998).

Cardiac disease is characterized by heart block or tachycardia, which usually resolves (Duray and Steere 1988). Musculoskeletal symptoms may consist of painful swollen muscles, myositis, synovitis, panniculitis, and migratory pain in the joints, tendons, and bursae (Steere 1989; Reimers et al. 1989; Evans 1995). In the United States, joint disease is reported in approximately 70% of untreated cases (Evans 1995).

Some people develop chronic, third-stage Lyme borreliosis, involving the joints and peripheral or central nervous system. Arthritis tends to affect the large joints, especially the knees. Chronic neuropsychological deficits may develop years after the initial infection and may be difficult to diagnose (Benke et al. 1995; Ravdin et al. 1996).

Congenital infections have been reported in human fetuses (MacDonald 1989). However, congenital transmission is rarely reported, and convincing evidence of substantial risk of transplacental transmission is lacking (Williams et al. 1995).

PATHOLOGY. Microscopic lesions have been found only rarely in wild mammals naturally infected with *B. burgdorferi.* Those described in white-footed mice include perivascular lymphoplasmacytic infiltrates in the kidneys (nephritis), liver (hepatitis), cerebrum and meninges (encephalitis), and lungs (pneumonitis) (Burgess et al. 1990). Dusky-footed wood rats infected with *B. burgdorferi* had mild to moderate synovitis, myositis, and myocarditis (Brown and Lane 1994), similar to those seen in naturally infected humans, horses, and dogs (Burgess 1986, 1991). Others have reported few or no lesions in experimentally infected wild rodents (Wright and Nielsen 1990; Campbell et al. 1994). The mild lesions observed in infected reservoir hosts probably are not significant.

Although they are commonly exposed to infected vectors, no disease attributable to *B. burgdorferi* has been reported in deer (Levine et al. 1987; Lane et al. 1994a; Luttrell et al. 1994), and deer seem to have a poor antibody response to infection (Gallivan et al. 1998).

In humans, the microscopic appearance of the EM rash varies, but a pattern of concurrent deep and superficial perivascular and interstitial lymphohistiocytic infiltrates with plasma cells is considered diagnostic for borreliosis (Berger 1989). During the acute phase of arthritis, joints may contain exudates with polymorphonuclear leukocytes and immune complexes. Chronic arthritis involves destructive erosion of the cartilage and bone and a proliferative synovium with infiltrates of lymphocytes, plasma cells, mast cells, and fibrin deposits (Duray 1987). Histologic changes in other tissues (meninges, liver, urogenital system, and spleen) consist of lymphoplasmacytic infiltrates (Duray 1987).

DIAGNOSIS. Infection with *B. burgdorferi* has little impact on individuals or populations of wildlife reservoir species. In wildlife, diagnosis is therefore limited to recognition of infection based on laboratory tests, discussed below.

Signs and symptoms may contribute to diagnosis in people, and in domestic and laboratory animals, in which disease occurs. Nevertheless, diagnosis is difficult, since clinical signs mimic so many other conditions or are nonspecific. Ultimately, diagnosis of Lyme borreliosis is based on a combination of compatible clinical signs, laboratory evidence of infection with *B. burgdorferi* (isolation, PCR, or more commonly serology), history of exposure, and elimination of appropriate alternative diagnostic hypotheses.

Though somewhat tedious and time-consuming, culture of *B. burgdorferi* has become routine, and culture media are available commercially. Isolation directly confirms current infection and yields living spirochetes that may be characterized or used in further testing. Culture remains the gold standard for detecting infection but is labor intensive, technically demanding, relatively expensive, and requires several weeks for spirochetes to grow. As well, culture may select for easily isolated strains of *B. burgdorferi* s. l. over more fastidious strains, leading to erroneous interpretations of strain prevalence among populations of hosts in which several strains are cycling (Norris et al. 1997; Saint-Girons et al. 1998).

The spirochete can be isolated from tissues, blood, or urine of animals and people (Barbour 1984; Bosler and Schultz 1986), but culture is not routinely used for diagnosis of human borreliosis due to the inherent difficulties. Skin biopsies of EM rashes may yield the spirochete in humans, and ear biopsies have proven a source of the spirochete in infected rodents (Sinsky and Piesman 1989). Demonstration of *B. burgdorferi* in biopsies by immunofluorescent or other immunohistochemical methods is rarely reported.

PCR has many advantages over other techniques for detecting *B. burgdorferi* infections and has been used widely in humans, as well as other animals (Olsen et al. 1993; Oliver et al. 1995; Schmidt 1997; O'Connell et al. 1998). A true positive PCR result indicates current infection and generates DNA that may be characterized for comparison. Additionally, PCR has the advantage of being relatively quick and inexpensive, while having the potential to maintain high levels of sensitivity and specificity. Several optimized protocols have been published (Rosa and Schwan 1989; Persing et al. 1990; Kaufman et al. 1993). However, PCR requires fastidious technique to prevent contamination that causes false-positive results, and hence is generally available only at research institutions or sophisticated diagnostic facilities.

Serology is currently the most practical laboratory procedure used in the diagnosis of Lyme borreliosis in people and domestic animals. The enzyme-linked immunosorbent assay (ELISA) performed on human sera is more sensitive and efficient than the indirect fluorescent antibody (IFA) test (Dattwyler et al. 1989). Immunoblotting (using modified Western blots) is used to determine the pattern of the immune response to the various protein antigens of the spirochete (Grodzicki and Steere 1988) and to confirm the outcome of the less specific ELISA and IFA test. An immunoblot assay using recombinant antigens distinguished between different genospecies of *B. burgdorferi* s. l. (Wilske et al. 1994).

Diagnosis in animals is problematic for several reasons. They may have nonspecific clinical signs, including lameness, stiffness, or arthritis (EM is not generally recognized); there may be a high prevalence of asymptomatic seroreactors, and there are problems in interpretation and interlaboratory comparison of serologic results (Greene 1990). Of the serologic tests available, ELISA systems are probably the most sensitive, though ELISA and IFA seem equally specific in dogs (Lindenmayer et al. 1990). As in people, immunoblotting may be used as a more specific confirmatory test.

DETECTION OF INFECTION IN WILDLIFE POPULATIONS. Several techniques have been reported for detecting evidence of *B. burgdorferi* infection in wildlife. Various serologic tests (IFA, ELISA, and Western blots) have been used in surveys for antibody in many species. Samples usually are collected under cross-sectional study designs, and paired serum samples are rarely available to enable evaluation of changes in specific IgG.

Importantly, the interpretation of serologic results requires that the test be validated for the mammal species involved by using an appropriate number of positive and negative controls (Campbell et al. 1994). Unfortunately, individuals of known infection status are not readily available for most species of wildlife, and results of nonvalidated tests should be interpreted with caution.

In general, positive serologic results suggest previous exposure but may yield little information about current infections. The possibility of cross-reactions with other agents also must be recognized. These techniques are employed because they are inexpensive, relatively easily performed, and can be used to evaluate exposure by using serum samples collected for other purposes. Serology is most appropriately used to evaluate exposure at the population level, using sample sizes adequate for the sensitivity required, but the results should not be overinterpreted at the level of the individual animal.

Culture and PCR are employed in the direct detection of *B. burgdorferi* infections in wildlife. Sensitivity of detecting infected individuals is, in part, a function of the number and type of tissues sampled. Many have reported successful isolation from urinary bladder, kidney, spleen, and other internal organs (Anderson et al. 1985; Schwan et al. 1988). Others successfully cultured ear-punch biopsy samples from wild rodents (Sinsky and Piesman 1989; Lane and Brown 1991). However, care must be taken in comparing studies of prevalence based on culture involving different host species or strains of *B. burgdorferi.* Brown and Lane (1994) cultured *B. burgdorferi* from ear-punch biopsies of naturally infected reservoir hosts, but not from other organs, and variation in the tissue tropisms of different genospecies of *B. burgdorferi* has been recognized (Saint-Girons et al. 1998).

Ticks may be assayed for *B. burgdorferi* to obtain information about infections in hosts. The proportion of a vector population that becomes infected while feeding on a host is an index of reservoir competence and may be estimated by evaluating prevalence of infection in ticks removed from naturally infested hosts or collected from the environment.

Alternatively, xenodiagnosis (feeding uninfected tick larvae on a host and subsequently examining them for infection) has been used to directly determine prevalence of infection in hosts (Sinsky and Piesman 1989; De Boer et al. 1993; Brown and Lane 1996; Gern et al. 1998). Xenodiagnosis is uncommonly used, since it is labor intensive and expensive, and because more convenient techniques exist, but it is arguably the most valid measure of the effective reservoir population of hosts (Lindsay et al. 1997).

Borrelia burgdorferi is detected in ticks most commonly by darkfield microscopy, immunofluorescence (typically using monoclonal antibodies), culture, and PCR.

PUBLIC HEALTH CONCERNS AND MANAGEMENT

Public Awareness and Risk Reduction. An important goal for borreliosis risk management is adequate public education concerning the local probability of exposure to ticks and the diseases that they transmit. Knowledge about potential exposure may enable individuals to avoid tick-infested areas or to take other measures to reduce personal risk of infection. Adequate education relies upon regular surveillance for vector ticks and human disease by public health authorities; White (1993) discussed the importance of standardized surveillance protocols that could be adopted by different agencies.

Public awareness generally is high in areas where Lyme borreliosis is common, due, in part, to the efforts of public health agencies and others in the medical community (Barbour and Fish 1993; White 1993). However, public "awareness" also results from panic and fear about a relatively new, potentially debilitating disease, and unjustified fears have been associated with the dissemination of misinformation. To achieve maximal public education, agencies need to take an active role and target an audience as wide as resources allow. Public education should remain a primary strategy for mitigating the risk of Lyme borreliosis (Barbour and Fish 1993; White 1993).

Personal protection is the key to avoiding tick-borne pathogens (White 1993; O'Connell et al. 1998). Exposure can be minimized by walking down the center of trails and decreasing the number of off-trail forays in tick-infested areas. Furthermore, people can wear light-colored clothing (to enhance detection and removal of ticks) with long pant legs and long sleeves to minimize skin exposure. Pant legs can be tucked into boots and socks, or taped at the cuffs, to help keep the ticks on the outside of clothing. Repellents containing permethrin or *N,N,*-diethyl toluamide (DEET), applied to clothing or skin, decrease the risk of being bitten.

Vigilance for, and early removal of, ticks decreases the risk of infection with *B. burgdorferi.* Of concern is the ability of people to locate and identify ticks that have embedded in their skin. Adult *Ixodes* sp. ticks are relatively easy to find, but nymphs, which are responsible for many cases of Lyme borreliosis, often go unnoticed because of their small size (1–2 mm).

In areas of high risk, embedded ticks should be sought visually and removed regularly with particularly careful attention given to the groin, axilla, areas above the hairline, and other parts of the body where ticks may be hard to detect. Using a set of fine forceps or tweezers, an embedded tick can be removed by grasping it by the basis capitulum (just behind the mouthparts), as close to the skin as possible, and withdrawing it using steady, firm traction, perhaps with some rotation (De Boer and Van den Bogaard 1993; White 1993). The tick should be saved (in 70% alcohol, if available) for specific identification should symptoms develop. A topical antiseptic may be applied to the site of attachment to decrease the likelihood of secondary infections, and the person bitten should continue to observe the site regularly for an extended period. A physician should be consulted about flulike symptoms, unexplained arthralgias, or if an expanding rash develops several days to months following the incident (White 1993).

Once embedded, ticks feed for up to 5–7 days, but the risk of transmission of *B. burgdorferi* is lower during the first day or so (Piesman and Gray 1994). Piesman et al. (1987) reported that little transmission occurred in the first 24 hours of nymphal *I. scapularis* attachment to rodents, and their findings have been substantiated (Peavey and Lane 1995). However, Kahl et al. (1998) reported up to 50% of rodents infected with *B. burgdorferi* after 16 hours of feeding by *I. ricinus* nymphs, and some patients developed clinical signs of EM after only 2 hours of attachment by nymphal *I. ricinus* (O'Connell et al. 1998), emphasizing that ticks should be detected and removed as soon as possible.

A vaccine for human Lyme borreliosis has been developed (Golde 1998). This first vaccine, in what may be a series, uses recombinant protein from *B. burgdorferi* s. s. to initiate an immune response against OspA. Although it may decrease the risk of acquiring disease for people who live or work in some highly endemic areas of the United States, the potential specificity of the resulting immune response to *B. burgdorferi* s. s. proteins may diminish its efficacy for use in California, Europe, or Asia where *B. burgdorferi* is more variable. Ticks also transmit several other potentially serious human diseases (including human granulocytic ehrlichiosis, human monocytic ehrlichiosis, Rocky Mountain spotted fever, and human babesiosis), and vaccination for Lyme borreliosis should only append continued vigilance and caution against tick bite (Golde 1998).

Tick Control. Public awareness and application of personal protective measures alone are insufficient to prevent borreliosis in highly endemic areas. Additional means of reducing human exposure are desirable, and tick control is an obvious complementary public health strategy to consider.

Tick control and management of tick-borne zoonoses have been discussed at length elsewhere (Wilson and Deblinger 1993; Ginsberg 1994; Piesman and Gray 1994; Schmidtmann 1994), and readers should consult those sources for more detail. However, a feasible, environmentally acceptable means to control vectors of borreliosis applicable on a broad scale in many epidemiologic circumstances has yet to be implemented.

Piesman and Gray (1994) identified five categories of management aimed at tick control in regions where *I. ricinus*-complex ticks serve as important vectors: (1) vegetation management, (2) host-targeted acaricides, (3) area-wide acaricide treatment, (4) host exclusion/eradication, and (5) integrated pest management (which includes aspects of education and self-protection).

The potential for vector population management will depend on knowledge of local tick ecology and the pattern of endemic maintenance. Eradication of the common primary vectors of borreliosis, which feed on a wide range of animal species and occur over a broad geographic area, is an unfeasible, if not an ecologically unsound, goal. Managers should focus on mitigation, rather than potential elimination, of human risk.

Habitat alteration, including brush clearing and burning or mowing, has been used to control a variety of ticks in many areas of the world, usually with initial success followed by rapid vector recolonization (Wilson 1986). Moreover, the initial successes of burning or mowing may be compromised by differential survival of subsets of the vector population. In one study, the prevalence of *B. burgdorferi* in ticks was considerably higher on burned areas than in adjacent nonburned areas even though the tick population was reduced, probably due to a disproportionate killing of ticks that had fed previously on deer rather than on reservoir-competent mice (Mather et al. 1993).

Landscape modifications around homes may decrease local populations of *I. scapularis* and therefore reduce the risk of transmission of *B. burgdorferi* to people (Maupin et al. 1991; Piesman and Gray 1994). Such modifications include the removal of brush as well as organic litter and leaves from near houses and areas of intense human activity. Although removal of leaf litter is successful for reducing local numbers of immature *I. scapularis,* it may increase erosion, diminish populations of sensitive nontarget plants and animals, and attract hosts that may be harboring large numbers of adult ticks (Wilson and Deblinger 1993; Schulze et al. 1995).

Targeted delivery of acaricides to rodents has been used for local control of *I. scapularis* in North America. Mather et al. (1987) placed tubes containing cotton balls impregnated with permethrin in the environment. Cotton impregnated with acaricide may be used by rodent hosts as bedding, and the acaricide is then directed against ticks that infest the nests of such hosts. However, this approach has not been universally successful (Ginsberg 1994; Piesman and Gray 1994; Leprince and Lane 1996), perhaps due to differences in the species composition of the community of hosts, habitat differences that made cotton balls more or less attractive to the rodents, or differences in study design.

Using another approach, rodents are treated as they crawl toward bait suspended in the middle of a tube (Sonenshine and Haines 1985). Good control of ticks on wood rats has been achieved by baiting them into short pieces of polyvinylchloride pipe lined with liquid permethrin-impregnated carpet (Gage et al. 1997).

Targeted acaricide delivery by such means offers the potential for local tick control, especially where endemic cycles depend on specialist feeding ticks. However, regional control of ticks should not be expected using only this means. Attendant risks include exposure to permethrin of nontarget species, most seriously fish. To minimize environmental impact (especially to aquatic habitats), treated tubes should not be used near wetlands or streams.

Control of tick populations by using acaricides, such as ivermectin and permethrin, to reduce the numbers of females that successfully engorge on deer also has been proposed (Schmidtmann 1994; Piesman and Gray 1994; Sonenshine et al. 1996). This requires repeated delivery, at appropriate times and intervals, of an adequate dose of acaricide to a significant proportion of a local population of deer. Pound et al. (1996) achieved greater than 92% reduction of immature *Amblyomma americanum* populations by feeding corn medicated with ivermectin to captive deer, and Sonenshine et al. (1996) demonstrated a reduction in *I. scapularis* burdens on deer by using a self-medicating permethrin applicator. However, in considering the use of broad-spectrum compounds such as these in wildlife, one must bear in mind possible community-level ecological effects and the question of tissue residues in animals taken by hunters.

Area-wide acaricide treatment using various formulations of diazinon, carbaryl, chlorpyrifos, cyfluthrin, and fluvalinate has been used to control ticks (Piesman and Gray 1994; Schmidtmann 1994; Stafford 1997). The consequences, in terms of nonspecific damage to the ecosystem through toxicity to other organisms, direct human toxicity, and liability, should direct managers and public health workers to minimize use of such chemicals and toward alternative avenues of risk management. However, the best management may require a balance of local costs and benefits. Barbour and Fish (1993) noted that the environmental costs associated with area-wide use of toxic pesticides may, in some highly endemic areas of the northeastern United States, be outweighed by the benefits of reduced tick abundance.

Likewise, broad-scale host exclusion and eradication is no longer acceptable. Moderate culling of white-tailed deer has an unpredictable effect on *I. scapularis* density, and nearly complete local elimination of deer is necessary to lower significantly the tick population in a habitat (Wilson and Deblinger 1993; Ginsberg 1994). On a small scale, however, deer exclusion can result in local reductions in the number of immature ticks on rodents (Daniels and Fish 1995). Although risk of exposure to nymphal ticks was less inside exclosures where deer were absent, adult ticks were as common inside the exclosures as outside, presumably because of the transport of immature ticks across the fence by mice, opossums, and raccoons.

Theoretically, deer repellents may be used during periods of adult tick activity to reduce local tick abundance (Wilson and Deblinger 1993). However, seasonal activity of adult *I. scapularis* is bimodal (peaking between September and December and again in early spring), so effective use of repellents would require diligent application at appropriate times. Deer repellents, as with deer exclusion (but with less expected efficacy), would have the greatest impact on local larval populations, less impact on nymphs questing locally, and probably little effect on populations of questing adult ticks.

Integrated pest management (IPM) holds the greatest promise for reduction of tick populations and of the attendant risk of transmission of *B. burgdorferi* to people. IPM has the potential to weaken multiple links in the chain of transmission simultaneously, while maintaining an appropriate degree of environmental sensitivity. IPM may incorporate technologies as diverse as area-wide chemical application, vegetation management, and host population exclusion or reduction. Bloemer et al. (1990) found that none of these strategies worked adequately alone with *Amblyomma americanum.* However, when the three were used simultaneously, synergistic effects reduced tick populations by greater than 92%.

Mount et al. (1997b), using a simulation model, considered that area-wide acaricide application and vegetation reduction, or a combination of these approaches, would offer seasonal management of tick-associated risk in small areas, such as recreational or residential sites, but that acaricide self-treatment of deer was the most cost-effective approach to long-term tick management over large areas. They also demonstrated the utility of simulation models in development of IPM strategies.

The use of IPM implies an understanding that the goal is one of reduction, rather than elimination, of tick populations and human health risks. Teams of people, including tick biologists, epidemiologists, public health officials, and wildlife managers must be involved in developing such strategies if they are to be successfully implemented (Stafford 1993).

LITERATURE CITED

Abele, D.C., and K.H. Anders. 1990. The many faces and phases of borreliosis: I. Lyme disease. *Journal of the American Academy of Dermatology* 23:167–186.

Ai, C.X., W.F. Zhang, and J.H. Zhao. 1994. Sero-epidemiology of Lyme disease in an endemic area in China. *Microbiology and Immunology* 38:505–509.

Anderson, J.F., and L.A. Magnarelli. 1984. Avian and mammalian hosts for spirochete-infected ticks and insects in a Lyme disease focus in Connecticut. *Yale Journal of Biology and Medicine* 57:627–641.

———. 1992. Epizootiology of Lyme disease and methods of cultivating *Borrelia burgdorferi. Annals of the New York Academy of Sciences* 653:52–63.

———. 1993. Natural history of *Borrelia burgdorferi* in vectors and vertebrate hosts. In *Ecology and environmental management of Lyme disease,* ed. H.S. Ginsberg. New Brunswick, NJ: Rutgers University Press, pp. 11–24.

Anderson, J.F., R.C. Johnson, L.A. Magnarelli, and F.W. Hyde. 1985. Identification of endemic foci of Lyme disease: Isolation of *Borrelia burgdorferi* from feral rodents and ticks (*Dermacentor variabilis*). *Journal of Clinical Microbiology* 22:36–38.

Anderson, J.F., R.C. Johnson, L.A. Magnarelli, F.W. Hyde, and J.E. Myers. 1986. *Peromyscus leucopus* and *Microtus pennsylvanicus* simultaneously infected with *Borrelia burgdorferi* and *Babesia microti. Journal of Clinical Microbiology* 23:135–137.

Anderson, J.F., L.A. Magnarelli, R.B. LeFebvre, T.G. Andreolis, J.B. McAninch, G.C. Perng, and R.C. Johnson. 1989. Antigenically variable *Borrelia burgdorferi* isolated from cottontail rabbits and *I. dentatus* in rural and urban areas. *Journal of Clinical Microbiology* 27:13–20.

Appel, M.J.G., S. Allen, R.H. Jacobson, T.L. Lauderdale, Y.F. Chang, S.J. Shin, J.W. Thomford, R.J. Todhunter, and B.A. Summers. 1993. Experimental Lyme disease in dogs produces arthritis and persistent infection. *Journal of Infectious Diseases* 167:651–664.

Asbrink, E., and A. Hovmark. 1988. Early and late cutaneous manifestations in *Ixodes*-borne borreliosis (erythema migrans borreliosis, Lyme borreliosis). *Annals of the New York Academy of Sciences* 539:4–15.

Azulay, R.D., L. Azulay-Abulafia, C.T. Sodre, D.R. Azulay, and M.M. Azulay. 1991. Lyme disease in Rio de Janeiro, Brazil. *International Journal of Dermatology* 30:569–571.

Barbour, A.G. 1984. Isolation and cultivation of Lyme disease spirochetes. *Yale Journal of Biology and Medicine* 57:521–525.

———. 1988. Laboratory aspects of Lyme borreliosis. *Clinical Microbiology Reviews* 1:399–414.

———. 1989. The molecular biology *of Borrelia burgdorferi. Reviews of Infectious Diseases* 11:S1470–S1474.

Barbour, A.G., and D. Fish. 1993. The biological and social phenomenon of Lyme disease. *Science* 260:1610–1616.

Barbour, A.G., T.R. Helland, and T.R. Howe. 1985. Heterogeneity of major proteins of Lyme disease borreliae: A molecular analysis of North American and European isolates. *Journal of Infectious Disease* 152:478–484.

Barbour, A.G., S.F. Hayes, R.A. Helland, M. E. Schrampf, and S.L. Tessier. 1986. A *Borrelia*-specific monoclonal antibody binds to a flagellar epitope. *Infection and Immunity* 52:549–554.

Barker, I.K., L.R. Lindsay, G.D. Campbell, G.A. Surgeoner, and S.A. McEwen. 1993. The groundhog tick *Ixodes cookei* (Acari: Ixodidae): A poor potential vector of Lyme borreliosis. *Journal of Wildlife Diseases* 29:416–422.

Barthold, S.W. 1995. Animal models for Lyme disease. *Laboratory Investigation* 72:127–130.

———. 1996a. Lyme borreliosis in the laboratory mouse. *Journal of Spirochetal and Tick-Borne Diseases* 3:22–32.

———. 1996b. Globalisation of Lyme borreliosis. *Lancet* 348:1603–1604.

Barthold, S.W., D.H. Persing, A.L. Armstrong, and R.A. Peeples. 1991. Kinetics of *Borrelia burgdorferi* dissemination and evolution of disease after intradermal inoculation of mice. *American Journal of Pathology* 139:263–273.

Benke, T., T. Gasse, M. Hittmair-Delazer, and E. Schmutzhard. 1995. Lyme encephalopathy: Long-term neuropsychological deficits years after acute neuroborreliosis. *Acta Neurologica Scandinavica* 91:353–357.

Berger, B.W. 1989. Dermatologic manifestations of Lyme disease. *Reviews of Infectious Diseases* 11(Suppl. 6):S1475–S1481.

Bloemer, S.R., G.A. Mount, T.A. Morris, R.H. Zimmerman, D.R. Barnard, and E.L. Snoddy. 1990. Management of lone star ticks (Acari: Ixodidae) in recreational areas with acaricide applications, vegetative management, and exclusion of white-tailed deer. *Journal of Medical Entomology* 27:543–550.

Bosler, E.M., and T.L. Schultz. 1986. The prevalence and significance of *Borrelia burgdorferi* in the urine of feral reservoir hosts. *Zentralblatt für Bakteriologie, Mikrobiologie und Hygiene [A]* 263:40–44.

Brown, R.N., and R.S. Lane. 1992. Lyme disease in California: A novel enzootic transmission cycle of *Borrelia burgdorferi. Science* 256:1439–1442.

———. 1994. Natural and experimental infections of *Borrelia burgdorferi* infections in woodrats and deer mice from California. *Journal of Wildlife Diseases* 30:389–398.

———. 1996. Reservoir competence of four chaparral-dwelling rodents for *Borrelia burgdorferi* in California. *American Journal of Tropical Medicine and Hygiene* 54:84–91.

Bunikis J., B. Olsen, V. Fingerle, J. Bonnedahl, B. Wilske, and S. Bergstrom. 1996. Molecular polymorphism of the Lyme disease agent *Borrelia garinii* in Northern Europe is influenced by a novel enzootic *Borrelia* focus in the North Atlantic. *Journal of Clinical Microbiology* 34:364–368.

Burgdorfer, W. 1986. Discovery of the Lyme disease spirochete: A historical review. *Zentralblatt für Bakteriologie, Mikrobiologie und Hygiene [A]* 263:7–10.

Burgdorfer, W., A.G. Barbour, S.F. Hayes, J.L. Benach, E. Grunwaldt, and J.P. Davis. 1982. Lyme disease: A tick-borne spirochetosis? *Science* 216:1317–1319.

Burgess, E.C. 1986. Natural exposure of Wisconsin dogs to the Lyme disease spirochete. *Laboratory Animal Science* 36:288–290.

———. 1989. Experimental inoculation of mallard ducks (*Anas platyrhynchos platyrhynchos*) with *Borrelia burgdorferi. Journal of Wildlife Diseases* 25:99–102.

———. 1991. The role of wild mammals in the transmission of *Borrelia burgdorferi. Bulletin of the Society for Vector Ecology* 16:50–58.

Burgess, E.C., and M. Mattison. 1987. Encephalitis associated with *Borrelia burgdorferi* infection in a horse. *Journal of the American Veterinary Medical Association* 191:1457–1458.

Burgess, E.C., and L.A. Windberg. 1989. *Borrelia* sp. infection in coyotes, black-tailed jack rabbits and desert cottontails in southern Texas. *Journal of Wildlife Diseases* 25:47–51.

Burgess, E.C., A. Gendron-Fitzpatrick, and W.O. Wright. 1987. Arthritis and systemic disease caused by *Borrelia burgdorferi* infection in a cow. *Journal of the American Veterinary Medical Association* 191:1468–1470.

Burgess, E.C., J.B. Frank, and A. Gendron-Fitzpatrick. 1990. Systemic disease in *Peromyscus leucopus* associated with *Borrelia burgdorferi* infection. *American Journal of Tropical Medicine and Hygiene* 42:254–259.

Burgess, E.C., M.D. Wachal, and T.D. Cleven. 1993. *Borrelia burgdorferi* infection in dairy cows, rodents, and birds from four Wisconsin dairy farms. *Veterinary Microbiology* 35:61–77.

Campbell, G.D., I.K. Barker, R.P. Johnson, P.E. Shewen, S.A. McEwen, and G.A. Surgeoner. 1994. Response of the meadow vole (*Microtus pennsylvanicus*) to experimental inoculation with *Borrelia burgdorferi. Journal of Wildlife Diseases* 30:408–416.

Casjens, S., M. Delange, H.L. Ley, P. Rosa, and W.M. Huang. 1995. Linear chromosomes of Lyme disease agent spirochetes: Genetic diversity and conservation of gene order. *Journal of Bacteriology* 177:2769–2780.

Centers for Disease Control and Prevention (CDC). 1997a. Lyme disease: United States, 1996. *Morbidity and Mortality Weekly Report* 46:531–535.

———. 1997b. Summary of notifiable diseases: United States 1996. *Morbidity and Mortality Weekly Report* 45(Suppl.):1–87.

Cimmino, M., M. Granstrom, J.S. Gray, E.C. Guy, S. O'Connell, and G. Stanek. 1998. European Lyme borreliosis clinical spectrum. *Zentralblatt für Bakteriologie* 287:248–252.

Craine, N.G., P.A. Nuttall, A.C. Marriott, and S.E. Randolph. 1997. Role of grey squirrels and pheasants in the transmission of *Borrelia burgdorferi* sensu lato, the Lyme disease spirochaete, in the U.K. *Folia Parasitologica* 44:155–160.

Daniels, T.J., and D. Fish. 1995. Effect of deer exclusion on the abundance of immature *Ixodes scapularis* (Acari: Ixodidae) parasitizing small and medium sized mammals. *Journal of Medical Entomology* 32:5–11.

Damrow, T., H. Freedman, R.S. Lane, and K.L. Preston. 1989. Is *Ixodes (Ixodiopsis) angustus* a vector of Lyme disease in Washington State? *Western Journal of Medicine* 150:580–582.

Dattwyler, R.L. 1990. Lyme borreliosis: An overview of the clinical manifestations. *Laboratory Medicine* 21(Suppl. 6):290–292.

Dattwyler, R.J., D.J. Volkman, and B.J. Luft. 1989. Immunologic aspects of Lyme borreliosis. *Reviews of Infectious Diseases* 11:S1494–S1498.

De Boer, R., and A.E.J.M. van den Bogaard. 1993. Removal of attached nymphs and adults of *Ixodes ricinus* (Acari: Ixodidae). *Journal of Medical Entomology* 30:748–752.

De Boer, R., K.E. Hovius, M.K. Nohlmans, and J.S. Gray. 1993. The woodmouse (*Apodemus sylvaticus*) as a reservoir of tick-transmitted spirochetes (*Borrelia burgdorferi*) in the Netherlands. *International Journal of Medical Microbiology, Virology, Parasitology, and Infectious Disease* 279:404–416.

Dos Santos, C., and K. Kain. 1998. Concurrent babesiosis and Lyme disease diagnosed in Ontario. *Canada Communicable Disease Report* 24:97–101.

Duray, P.H. 1987. The surgical pathology of human Lyme disease. *American Journal of Surgical Pathology* 1:47–60.

Duray, P.H., and A.C. Steere. 1988. Clinical pathologic correlations of Lyme disease by stage. *Annals of the New York Academy of Sciences* 539:65–79.

Evans, J. 1995. Lyme disease. *Current Opinion in Rheumatology* 7:322–328.

Fish, D., and T.J. Daniels. 1990. The role of medium-sized mammals as reservoirs of *Borrelia burgdorferi* in southern New York. *Journal of Wildlife Diseases* 26:339–345.

Fukunaga, M., A. Hamase, K. Okada, H. Inoue, Y. Tsuruta, K. Miyamoto, and M. Nakao. 1996. Characterization of spirochetes isolated from ticks (*Ixodes tanuki, Ixodes turdus,* and *Ixodes columnae*) and comparison of the sequences with those of *Borrelia burgdorferi* sensu lato strains. *Applied Environmental Microbiology* 62:2338–2344.

Gage, K.L., G.O. Maupin, J. Montenieri, J. Piesman, M. Dolan, and N.A. Panella. 1997. Flea (Siphonaptera: Ceratophyllidae, Hystrichopsyllidae) and tick (Acari: Ixodidae) control on wood rats using host-targeted liquid permethrin in bait tubes. *Journal of Medical Entomology* 34:46–51.

Gallivan, G.J., I.K. Barker, H. Artsob, L.A. Magnarelli, J.T. Robinson, and D.R. Voigt. 1998. Serologic survey for antibodies to *Borrelia burgdorferi* in white-tailed deer in Ontario, Canada. *Journal of Wildlife Diseases* 34:411–414.

Gern, L., E. Rouvinez, L.N. Toutoungi, and E. Godfroid. 1997. Transmission cycles of *Borrelia burgdorferi* sensu lato involving *Ixodes ricinus* and/or *Ixodes hexagonus* and the European hedgehog, *Erinaceus europaeus,* in suburban and urban areas in Switzerland. *Folia Parasitologica* 44:309–314.

Gern, L., A. Estrada-Pena, F. Frandsen, J.S. Gray, T.G.T. Jaenson, F. Jongejan, O. Kahl, E. Korenberg, R. Mehl, and P.A. Nuttall. 1998. European reservoir hosts of *Borrelia burgdorferi* sensu lato. *Zentralblatt für Bakteriologie* 287:196–204.

Ginsberg, H.S. 1994. Lyme disease and conservation. *Conservation Biology* 8:343–353.

Golde, W.T. 1998. A vaccine for Lyme disease: Current progress. *Infections in Medicine* 15:38 and 40–42.

Gordus, A.G., and J.H. Theis. 1993. Isolation of *Borrelia burgdorferi* from the blood of a bushy-tailed wood rat in California. *Journal of Wildlife Diseases* 29:478–480.

Gorelova, N.B., E.I. Korenberg, Y.V. Kovalevskii, and S.V. Shcherbakov. 1995. Small mammals as reservoir hosts for *Borrelia* in Russia. *International Journal of Medical Microbiology, Virology, Parasitology, and Infectious Diseases* 282:315–322.

Gray, J.S., O. Kahl, C. Janetzki, and J. Stein. 1992. Studies on the ecology of Lyme disease in a deer forest in Co. Galway, Ireland. *Journal of Medical Entomology* 29:915–920.

Gray, J.S., O. Kahl, C. Janetzki, J. Stein, and E. Guy. 1994. Acquisition of *Borrelia burgdorferi* by *Ixodes ricinus* ticks fed on the European hedgehog, *Erinaceus europaeus* L. *Experimental and Applied Acarology* 18:485–491.

Gray, J.S., O. Kahl, J.N. Robertson, M. Daniel, A. Estrada-Pena, G. Gettinby, T.G.T. Jaenson, P. Jensen, F. Jongejan, E. Korenberg, K. Kurtenbach, and P. Zeman. 1998. Lyme borreliosis habitat assessment. *Zentralblatt für Bakteriologie* 287:211–228.

Greene, R.T. 1990. An update on serodiagnosis of canine Lyme borreliosis. *Journal of Veterinary Internal Medicine* 4:167–171.

Grodzicki, R.L., and A.C. Steere. 1988. Comparison of immunoblotting and indirect enzyme-linked immunosorbent assay using different antigen preparations for diagnosing early Lyme disease. *Journal of Infectious Diseases* 157:790–797.

Gross, D.M., T. Forsthuber, M. Tary-Lehmann, C. Etling, K. Ito, Z.A. Nagy, J.A. Field, A.C. Steere, and B.T. Huber. 1998. Identification of LFA-1 as a candidate autoantigen in treatment-resistant Lyme arthritis. *Science* 281:703–706.

Hammouda, N.A., I.H. Hegazy, E.H. el-Sawy. 1995. ELISA screening for Lyme disease in children with chronic arthritis. *Journal of the Egyptian Society of Parasitology* 25:525–533.

Hovmark, A., T.G. Jaenson, E. Asbrink, A. Forsman, and E. Jansson. 1988. First isolations of *Borrelia burgdorferi* from rodents collected in Northern Europe. *APMIS* 96:917–920.

Hu, L.T., and M.S. Klempner. 1997. Host-pathogen interactions in the immunopathogenesis of Lyme disease. *Journal of Clinical Immunology* 17:354–365.

Hubalek, Z., and J. Halouzka. 1997. Distribution of *Borrelia burgdorferi* sensu lato genomic groups in Europe: A review. *European Journal of Epidemiology* 13:951–957.

Hubbard, M.J., A.S. Baker, and K.J. Cann. 1998. Distribution of *Borrelia burgdorferi* s. l. spirochaete DNA in British ticks (Argasidae and Ixodidae) since the 19th century, assessed by PCR. *Medical and Veterinary Entomology* 12:89–97.

Humair, P.-F., and L. Gern. 1998. Relationship between *Borrelia burgdorferi* sensu lato species, red squirrels (*Sciurus vulgaris*) and *Ixodes ricinus* in enzootic areas in Switzerland. *Acta Tropica (Basel)* 69:213–227.

Humair, P.-F., O. Peter, R. Wallich, and L. Gern. 1995. Strain variation of Lyme disease spirochetes isolated from *Ixodes ricinus* ticks and rodents collected in two endemic areas in Switzerland. *Journal of Medical Entomology* 32:433–438.

Humair, P.-F., D. Postic, R. Wallich, and L. Gern. 1998. An avian reservoir (*Turdus merula*) of the Lyme borreliosis spirochetes. *Zentralblatt für Bakteriologie* 287:521–538.

Isogai, E., H. Isogai, H. Kawabata, T. Masuzawa, Y. Yanagihara, K. Kimura, T. Sakai, Y. Azuna, N. Fujii, and S. Ohno. 1994a. Lyme disease spirochetes in wild fox (*Vulpes vulpes schrencki*) and in ticks. *Journal of Wildlife Diseases* 30:439–444.

Isogai, E., Y. Kamewaka, H. Isogai, K. Kimura, N. Fujii, and T. Nishikawa. 1994b. Complement-mediated killing of *Borrelia garinii:* Bactericidal activity of wild deer serum. *Microbiology and Immunology* 38:753–756.

Jaenson, T.G., and L. Talleklint. 1996. Lyme borreliosis spirochetes in *Ixodes ricinus* (Acari: Ixodidae) and the varying hare on isolated islands in the Baltic Sea. *Journal of Medical Entomology* 33:339–343.

Johnson, R.C., G.P. Schmid, F.W. Hyde, A.G. Steigerwalt, and D.J. Brenner. 1984. *Borrelia burgdorferi* sp. nov.: Etiologic agent of Lyme disease. *International Journal of Systemic Bacteriology* 34:496–497.

Jones, C.G., R.S. Ostfeld, M.P. Richard, E.M. Schauber, and J.O. Wolff. 1998. Chain reactions linking acorns to gypsy moth outbreaks and Lyme disease risk. *Science* 279:1023–1026.

Kahl, O., C. Janetzki-Mittman, J.S. Gray, R. Jonas, J. Stein, and R. de Boer. 1998. Risk of infection with *Borrelia burgdorferi* sensu lato for a host in relation to the duration of nymphal *Ixodes ricinus* feeding and the method of tick removal. *Zentralblatt für Bakteriologie* 287:41–52.

Kaufman, A.C., C.E. Greene, and R.A. McGraw. 1993. Optimization of polymerase chain reaction for the detection of *Borrelia burgdorferi* in biologic specimens. *Journal of Veterinary Diagnostic Investigation* 5:548–554.

Kawabata, H., T. Masuzawa, and Y. Yanagihara. 1993. Genomic analysis of *Borrelia japonica* sp. nov. isolated *from Ixodes ovatus* in Japan. *Microbiology and Immunology* 37:843–848.

Keirans, J.E., G.R. Needham, and J.H. Oliver Jr. 1997. The *Ixodes (Ixodes) ricinus* complex worldwide: Diagnosis of the species in the complex, hosts and distribution. *Acarology* 9:Symposia 3a.2.

Kimsey, R.B., and A. Spielman. 1990. Motility of Lyme disease spirochetes in fluids as viscous as the extracellular matrix. *Journal of Infectious Disease* 162:1205–1208.

Kimura, K., E. Isogai, H. Isogai, Y. Kamewaka, T. Nishikawa, N. Ishii, and N. Fujii. 1995. Detection of Lyme disease spirochetes in the skin of naturally infected wild sika deer (*Cervus nippon yesoensis*) by PCR. *Applied Environmental Microbiology* 61:1641–1642.

Klich, M., M.W. Lankester, and K.W. Wu. 1996. Spring migratory birds (Aves) extend the northern occurrence of blacklegged tick (Acari: Ixodidae). *Journal of Medical Entomology* 33:581–585.

Korenberg, E.I., V.N. Kryuchechnikov, and Y.V. Kovalevsky. 1993. Advances in investigations of Lyme borreliosis in the territory of the former USSR. *European Journal of Epidemiology* 9:86–91.

Kurtenbach, K. 1996. Transmission of *Borrelia burgdorferi* sensu lato by reservoir hosts. *Journal of Spirochetal and Tick-Borne Diseases* 3:53–61.

Kurtenbach, K., D. Carey, A.N. Hoodless, P.A. Nuttall, and S.E. Randolph. 1998a. Competence of pheasants as reservoirs for Lyme disease spirochetes. *Journal of Medical Entomology* 35:77–81.

Kurtenbach, K., M. Peacey, S.G. Rijpkema, A.N. Hoodless, P.A. Nuttall, and S.E. Randolph. 1998b. Differential transmission of the genospecies of *Borrelia burgdorferi* sensu lato by game birds and small rodents in England. *Applied Environmental Microbiology* 64:1169–1174.

Lane, R.S. 1994. Competence of ticks as vectors of microbial agents with an emphasis on *Borrelia burgdorferi*. In *Ecological dynamics of tick-borne zoonoses,* ed. D.E. Sonenshine and T.N. Mather. New York: Oxford University Press, pp. 45–67.

Lane, R.S., and R.N. Brown. 1991. Wood rats and kangaroo rats: Potential reservoirs of the Lyme disease spirochete in California. *Journal of Medical Entomology* 28:299–302.

Lane, R.S., and W. Burgdorfer. 1988. Spirochetes in mammals and ticks (Acari: Ixodidae) from a focus of Lyme borreliosis in California. *Journal of Wildlife Diseases* 24:1–9.

Lane, R.S., and G.B. Quistad. 1998. Borreliacidal factor in the blood of the western fence lizard (*Sceloporus occidentalis*). *Journal of Parasitology* 84:29–34.

Lane, R.S., J. Piesman, and W. Burgdorfer. 1991. Lyme borreliosis: Relation of its causative agent to its vectors and hosts in North America and Europe. *Annual Reviews of Entomology* 36:587–609.

Lane, R.S., D.M. Berger, L.E. Casher, and W. Burgdorfer. 1994a. Experimental infection of Columbia black-tailed deer with the Lyme disease spirochete. *Journal of Wildlife Diseases* 30:20–28.

Lane, R.S., R.N. Brown, J. Piesman, and C.A. Peavey. 1994b. Vector competence of *Ixodes pacificus and Dermacentor occidentalis* (Acari: Ixodidae) for various isolates of Lyme disease spirochetes. *Journal of Medical Entomology* 31:417–424.

Leprince, D.J., and R.S. Lane. 1996. Evaluation of permethrin-impregnated cotton balls as potential nesting material to control ectoparasites of woodrats in California. *Journal of Medical Entomology* 33:355–360.

Levine, S., D. Fish, L.A. Magnarellli, and J.F. Anderson. 1987. Choroid plexitis in white-tailed deer (*Odocoileus virginianus*) in southern New York state. *Veterinary Pathology* 24:207–210.

Levine, M., J.F. Levine, S. Yang, P. Howard, and C.S. Apperson. 1996. Reservoir competence of the southeastern five-lined skink (*Eumeces inexpectatus*) and the green anole (*Anolis carolinensis*) for *Borrelia burgdorferi. American Journal of Tropical Medicine and Hygiene* 54:92–97.

Levy, S.A., S.W. Barthold, D.M. Dombach, and T.L. Wasmoen. 1993. Canine Lyme borreliosis. *Compendium on Continuing Education for the Practicing Veterinarian* 15:883–846.

Li, M., T. Masuzawa, N. Takada, F. Ishiguro, H. Fujita, A. Iwaki, H. Wang, J. Wang, M. Kawabata, and Y. Yanagihara. 1998. Lyme disease *Borrelia* species in northeastern China resemble those isolated from far eastern Russia and Japan. *Applied Environmental Microbiology* 64:2705–2709.

Lindenmayer, J., M. Weber, J. Bryant, E. Marquez, and A. Onerdonk. 1990. Comparison of indirect immunofluorescent-antibody assay, enzyme-linked immunosorbent assay, and Western immunoblot for the diagnosis of Lyme disease in dogs. *Journal of Clinical Microbiology* 28:92–96.

Lindsay, L.R., I.K. Barker, G.A. Surgeoner, S.A. McEwen, and G.D. Campbell. 1997. Duration of *Borrelia burgdorferi* infectivity in white-footed mice (*Peromyscus leucopus*) for the tick vector *Ixodes scapularis* under laboratory and field conditions in Ontario. *Journal of Wildlife Diseases* 33:766–775.

Lord, R.D., V.R. Lord, J.G. Humphreys, and R.G. McLean. 1994. Distribution of *Borrelia burgdorferi* in host mice in Pennsylvania. *Journal of Clinical Microbiology* 32:2501–2504.

Luttrell, M.P., K. Nakagaki, E.W. Howerth, D.E. Stallknecht, and K.A. Lee. 1994. Experimental infection of *Borrelia burgdorferi* in white-tailed deer. *Journal of Wildlife Diseases* 30:146–154.

MacDonald, A.B. 1989. Gestational Lyme borreliosis: Implications for the fetus. *Rheumatic Disease Clinics of North America* 15:657–677.

Magnarelli, L.A., J.F. Anderson, and A.G. Barbour. 1986. The etiological agent of Lyme disease in deer flies, horse flies, and mosquitoes. *Journal of Infectious Diseases* 154:355–358.

Marconi, R.T., D. Liveris, and I. Schwartz. 1995. Identification of novel insertion elements, restriction fragment length polymorphism patterns, and discontinuous 23S rRNA in Lyme disease spirochetes: Phylogenetic analysis of rRNA genes and their intergenic spacers in *Borrelia japonica* sp. nov. and genomic group 21038 (*Borrelia andersonii* sp. nov.) isolates. *Journal of Clinical Microbiology* 33:2427–2434.

Masuzawa, T., H. Suzuki, H. Kawabata, F. Ishiguro, N. Takada, and Y. Yanagihara. 1996. Characterization of *Borrelia* spp. isolated from the tick, *Ixodes tanuki* and small rodents in Japan. *Journal of Wildlife Diseases* 32:565–571.

Mather, T.N., J.M. Ribeiro, and A. Spielman. 1987. Lyme disease and babesiosis: Acaricide focused on potentially infected ticks. *American Journal of Tropical Medicine and Hygiene* 36:609–614.

Mather, T.N., S.R. Telford III, G.H. Adler. 1991. Absence of transplacental transmission of Lyme disease spirochetes from reservoir mice (*Peromyscus leucopus*) to their offspring. *Journal of Infectious Diseases* 164:564–567.

Mather, T.N., D.C. Duffy, and S.R. Campbell. 1993. An unexpected result from burning vegetation to reduce Lyme disease transmission risks. *Journal of Medical Entomology* 30:642–645.

Matuschka, F.-R., M. Heiler, H. Eiffert, P. Fischer, H. Lotter, and A. Spielman. 1993. Diversionary role of hoofed game in the transmission of Lyme disease spirochetes. *American Journal of Tropical Medicine and Hygiene* 48:693–699.

Matuschka, F.-R., H. Eiffert, A. Ohlenbusch, D. Richter, E. Schein, and A. Spielman. 1994a. Transmission of the agent of Lyme disease on a subtropical island. *Tropical Medicine and Parasitology* 45:39–44.

Matuschka, F.-R., H. Eiffert, A. Ohlenbusch, and A. Spielman 1994b. Amplifying role of edible dormice in Lyme disease transmission in Central Europe. *Journal of Infectious Diseases* 170:122–127.

Maupin, G.O., D. Fish, J. Zultowsky, E.G. Campos, and J. Piesman. 1991. Landscape ecology of Lyme disease in a residential area of Westchester County, New York. *American Journal of Epidemiology* 133:1105–1113.

Maupin, G.O., K.L. Gage, J. Piesman, J. Montenieri, S.L. Sviat, L. VanderZanden, C.M. Happ, M. Dolan, and B.J. Johnson. 1994. Discovery of an enzootic cycle *of Borrelia burgdorferi* in *Neotoma mexicana* and *Ixodes spinipalpis* from northern Colorado, an area where Lyme disease is nonendemic. *Journal of Infectious Diseases* 170:636–643.

McLean, R.G., S.R. Ubico, C.A. Norton Hughes, S.M. Engstrom, and R.C. Johnson. 1993. Isolation and characterization of *Borrelia burgdorferi* from blood of a bird captured in the Saint Croix river valley. *Journal of Clinical Microbiology* 31:2038–2043.

Miller, J.N., D.M. Foley, J.T. Skare, C.I. Champion, E.S. Shang, D.R. Blanco, and M.A. Lovett. 1996. The rabbit as a model for the study of Lyme disease pathogenesis and immunity: A review. *Journal of Spirochetal and Tick-Borne Diseases* 3:6–14.

Mount, G.A., D.G. Haile, and E. Daniels. 1997a. Simulation of blacklegged tick (Acari: Ixodidae) population dynamics and transmission of *Borrelia burgdorferi. Journal of Medical Entomology* 34:461–484.

———. 1997b. Simulation of management strategies for the blacklegged tick (Acari: Ixodidae) and the Lyme disease spirochete, *Borrelia burgdorferi. Journal of Medical Entomology* 34:672–683.

Munson, E.L., B.D. DuChateau, D.A. Jobe, M.L. Padilla, S.D. Lovich, J.R. Jensen, L.C. Lim, J.L. Schmitz, S.M. Callister, and R.F. Schell. 1996. Hamster model of Lyme borreliosis. *Journal of Spirochetal and Tick-Borne Diseases* 3:15–21.

Nakao, M., K. Miyamoto, and M. Fukunaga. 1994. Lyme disease spirochetes in Japan: Enzootic transmission cycles in birds, rodents, and *Ixodes persulcatus* ticks. *Journal of Infectious Diseases* 170:878–882.

Nakao, M., K. Uchikawa, and H. Dewa. 1996. Distribution of *Borrelia* species associated with Lyme disease in the subalpine forests of Nagano prefecture, Japan. *Microbiology and Immunology* 40:307–311.

Nash, P.T. 1998. Does Lyme disease exist in Australia? *Medical Journal of Australia* 168:479–480.

Norris, D.E., J.S. Klompen, J.E. Keirans, R.S. Lane, J. Piesman, and W.C. Black IV. 1997. Taxonomic status of *Ixodes neotomae* and *I. spinipalpis* (Acari: Ixodidae) based on mitochondrial DNA evidence. *Journal of Medical Entomology* 34:696–703.

O'Connell, S., M. Granstrom, J.S. Gray, and G. Stanek. 1998. Epidemiology of European Lyme borreliosis. *Zentralblatt für Bakteriologie* 287:229–240.

Ogden, N.H., P.A. Nuttall, and S.E. Randolph. 1997. Natural Lyme disease cycles maintained via sheep by co-feeding ticks. *Parasitology* 115:591–599.

Oliver, J.H., Jr., M.R. Owsley, H.J. Hutcheson, A.M. James, C. Chen, W.S. Irby, E.M. Dotson, and D.K. McLain. 1993. Conspecificity of the ticks *Ixodes scapularis* and *I. dammini* (Acari: Ixodidae). *Journal of Medical Entomology* 30:54–63.

Oliver, J.H., Jr., F.W. Chandler Jr., A.M. James, F.H. Sanders Jr., H.J. Hutcheson, L.O. Huey, B.J. McGuire, and R.S. Lane. 1995. Natural occurrence and characterization of the Lyme disease spirochete, *Borrelia burgdorferi,* in cotton rats (*Sigmondon hispidus*) from Georgia and Florida. *Journal of Parasitology* 81:30–36.

Olsen, B., T.G. Jaenson, L. Noppa, J. Bunikis, and S. Bergstrom. 1993. A Lyme borreliosis cycle in seabirds and *Ixodes uriae* ticks. *Nature* 362:340–342.

Olsen, B., D.C. Duffy, T.G. Jaenson, A. Gylfe, J. Bonnedahl, and S. Bergstrom. 1995. Transhemispheric exchange of Lyme disease spirochetes by seabirds. *Journal of Clinical Microbiology* 33:3270–3274.

Oschmann, P., W. Dorndorf, C. Hornig, C. Schafer, H.J. Wellensiek, and K.W. Pflughaupt. 1998. Stages and syndromes of neuroborreliosis. *Journal of Neurology* 245:262–272.

Ouellette, J., C.S. Apperson, P. Howard, T.L. Evans, and J.F. Levine. 1997. Tick-raccoon associations and the potential for Lyme disease spirochete transmission in the coastal plain of North Carolina. *Journal of Wildlife Diseases* 33:28–39.

Pachner, A.R., P. Duray, and A.C. Steere. 1989. Central nervous system manifestations of Lyme disease. *Archives of Neurology* 46:790–795.

Pan, L. 1992. Investigation of rodents and *Ixodes* for Lyme disease and four strains of *Borrelia burgdorferi* first isolated *from Ixodes granulatus* Supino, *Rattus confucianus* and *R. norvegicus* in Fujian province [in Chinese]. *Chung Hua Liu Hsing Ping Hsueh Tsa Chih* 13:226–228.

Park, K.H., W.H. Chang, and T.G. Schwan. 1993. Identification and characterization of Lyme disease spirochetes, *Borrelia burgdorferi* sensu lato, isolated in Korea. *Journal of Clinical Microbiology* 31:1831–1837.

Peavey, C.A., and R.S. Lane. 1995. Transmission of *Borrelia burgdorferi* by *Ixodes pacificus* nymphs and reservoir competence of deer mice (*Peromyscus maniculatus*) infected by tick-bite. *Journal of Parasitology* 81:175–178.

Peavey, C.A., R.S. Lane, and J.E. Kleinjan. 1997. Role of small mammals in the ecology of *Borrelia burgdorferi* in a periurban park in North Coastal California. *Experimental and Applied Acarology* 21:569–584.

Persing, D.H., S.R.D. Telford, A. Spielman, and S.W. Barthold. 1990. Detection of *Borrelia burgdorferi* infection in *Ixodes dammini* ticks with the polymerase chain reaction. *Journal of Clinical Microbiology* 28:566–572.

Philipp, M. T., and B.J. Johnson. 1994. Animal models of Lyme disease: Pathogenesis and immunoprophylaxis. *Trends in Microbiology* 2:431–437.

Piesman, J., and J.S. Gray. 1994. Lyme disease/Lyme borreliosis. In *Ecological dynamics of tick-borne zoonoses,* ed. D.E. Sonenshine and T.N. Mather. New York: Oxford University Press, pp. 327–350.

Piesman, J., T.N. Mather, R.J. Sinsky, and A. Spielman. 1987. Duration of tick attachment and *Borrelia burgdorferi* transmission. *Journal of Clinical Microbiology* 25:557–558.

Postic, D., M. Assous, P.A.D. Grimont, and G. Baranton. 1994. Diversity of *Borrelia burgdorferi* sensu lato evidenced by restriction fragment length polymorphism of rrf(5S)-rrl(23S) intergenic spacer amplicons. *International Journal of Systemic Bacteriology* 44:743–752.

Pound, M.J., J.A. Miller, J.E. George, D.D. Oehler, and D.E. Harmel. 1996. Systemic treatment of white-tailed deer with ivermectin-medicated bait to control free-living populations of lone star ticks (Acari: Ixodidae). *Journal of Medical Entomology* 33:385–394.

Preac-Mursic, V., E. Patsouris, B. Wilske, S. Reinhardt, B. Groβ, and P. Mehraein. 1990. Persistence of *Borrelia burgdorferi* and histopathological alterations in experimentally infected animals: A comparison with histopathological findings in human Lyme disease. *Infection* 18:332–341.

Rand, P.W., E.H. Lacombe, R.P. Smith Jr., S.M. Rich, C.W. Kilpatrick, C.A. Dragoni, and D. Caporale. 1993. Competence of *Peromyscus maniculatus* (Rodentia: Cricetidae) as a reservoir host for *Borrelia burgdorferi* (Spirochaetales: Spirochaetaceae) in the wild. *Journal of Medical Entomology* 30:614–618.

Randolph, S.E., L. Gern, and P.A. Nuttall. 1996. Co-feeding ticks: Epidemiological significance for tick-borne pathogen transmission. *Parasitology Today* 12:472–479.

Ras, N.M., B. Lascola, D. Postic, S.J. Cutler, F. Rodhain, G. Baranton, and D. Raoult. 1996. Phylogenesis of relapsing fever *Borrelia* spp. *International Journal of Systematic Bacteriology* 46:859–865.

Ravdin, L.D., E. Hilton, M. Primeau, C. Clements, and W.B. Barr. 1996. Memory functioning in Lyme borreliosis. *Journal of Clinical Psychiatry* 57:282–286.

Rawlings, J. 1986. Lyme disease in Texas. *Zentralblatt für Bakteriologie, Mikrobiologie und Hygiene [A]* 263:483–487.

Reimers, C.D., D.E. Pongratz, V. Neubert, A. Pilz, G. Hubner, M. Naegele, B. Wilske, P. Duray, and J. de Koning. 1989. Myositis caused by *Borrelia burgdorferi:* Report of four cases. *Journal of the Neurological Sciences* 91:215–226.

Rosa, P.A., and T.G. Schwan. 1989. A specific and sensitive assay for the Lyme disease spirochete *Borrelia burgdorferi* using the polymerase chain reaction. *Journal of Infectious Disease* 160:1018–1029.

Saint-Girons, I., L. Gern, J.S. Gray, E.C. Guy, E. Korenberg, P.A. Nuttall, S.G.T. Rijpkema, A. Schonberg, G. Stanek, and D. Postic. 1998. Identification of *Borrelia burgdorferi* sensu lato species in Europe. *Zentralblatt für Bakteriologie* 287:190–195.

Schmidt, B.L. 1997. PCR in laboratory diagnosis of human *Borrelia burgdorferi* infections. *Clinical Microbiology Reviews* 10:185–201.

Schmidtmann, E.T. 1994. Ecologically based strategies for controlling ticks. In *Ecological dynamics of tick-borne zoonoses,* ed. D.E. Sonenshine and T.N. Mather. New York: Oxford University Press, pp. 240–279.

Schulze, T.L., G.S. Bowlen, E.M. Bosler, M.F. Lakat, W.E. Parkin, R. Altman, B.G. Ormiston, and J.K. Schisler. 1984. *Amblyomma americanum:* A potential vector of Lyme disease in New Jersey. *Science* 224:601–603.

Schulze, T.L., J.K. Shisler, E.M. Bouler, M.F. Laxat, and W.E. Parkin. 1986. Evolution of a focus of Lyme disease. *Zentralblatt für Bakteriologie, Mikrobiologie und Hygiene [A]* 263:65–71.

Schulze, T.L., R.A. Jordan, and R.W. Hung. 1995. Suppression of subadult *Ixodes scapularis* (Acari: Ixodidae) following removal of leaf litter. *Journal of Medical Entomology* 32:730–733.

Schwan, T.G., W. Burgdorfer, M.E. Schrumpf, and R.H. Karstens. 1988. The urinary bladder, a consistent source of *Borrelia burgdorferi* in experimentally infected white-footed mice *(Peromyscus leucopus)*. *Journal of Clinical Microbiology* 26:893–895.

Schwan, T.G., K.K. Kime, M.E. Schrumpf, J.E. Coe, and W.J. Simpson. 1989. Antibody response in white-footed mice (*Peromyscus leucopus*) experimentally infected with the Lyme disease spirochete (*Borrelia burgdorferi*). *Infection and Immunity* 57:3445–3451.

Sigal, L.H. 1997. Lyme disease: A review of aspects of its immunology and immunopathogenesis. *Annual Reviews in Immunology* 15:63–92.

Simpson, W.J., W. Burgdorfer, M.E. Schrumpf, R.H. Karstens, and T.G. Schwan. 1991. Antibody to a 39 kilodalton *Borrelia burgdorferi* antigen (P39) as a marker for infection in experimentally and naturally inoculated animals. *Journal of Clinical Microbiology* 29:236–243.

Sinsky, R., and J. Piesman. 1989. Ear punch biopsy method for detection and isolation of *Borrelia burgdorferi* in rodents. *Journal of Clinical Microbiology* 27:1723–1727.

Slajchert, T.L., U. Kitron, C.J. Jones, and A. Mannelli. 1997. Role of the eastern chipmunk (*Tamias striatus*) in the epizootiology of Lyme borreliosis in northwestern Illinois, USA. *Journal of Wildlife Diseases* 33:40–46.

Smith, R.P., Jr., P.W. Rand, E.H. Lacombe, S.R. Telford III, S.M. Rich, J. Piesman, and A. Spielman. 1993. Norway rats as reservoir hosts for Lyme disease spirochetes on Monhegan Island, Maine. *Journal of Infectious Diseases* 168:687–691.

Sonenshine, D.E., and G. Haines. 1985. A convenient method for controlling populations of the American dog tick, *Dermacentor variabilis* (Acari: Ixodidae), in the natural environment. *Journal of Medical Entomology* 22:577–583.

Sonenshine, D.E., R.E Ratzlaff, J. Troyer, S. Demmerle, W.E. Austin, S. Tan, A. Annis, and S. Jenkins. 1995. *Borrelia burgdorferi* in eastern Virginia: Comparison between a coastal and inland locality. *American Journal of Tropical Medicine and Hygiene* 53:123–133.

Sonenshine, D.E., S.A. Allan, R.A. Norval, and M.J. Burridge. 1996. A self-medicating applicator for control of ticks on deer. *Medical and Veterinary Entomology* 10:149–154.

Spielman, A., M.L. Wilson, J.F. Levine, and J. Piesman. 1985. Ecology of *Ixodes dammini*-borne human babesiosis and Lyme disease. *Annual Reviews of Entomology* 30:439–460.

Spielman, A., S.R. Telford III, and R.J. Pollack. 1993. The origins and course of the present outbreak of Lyme disease. In *Ecology and environmental management of Lyme disease,* ed. H.S. Ginsberg. New Brunswick, NJ: Rutgers University Press, pp. 83–96.

Stafford III, K.C. 1993. Forum. Perspectives on the environmental management of ticks and Lyme disease. In *Ecology and environmental management of Lyme disease,* ed. H.S. Ginsberg. New Brunswick, NJ: Rutgers University Press, pp. 178–181.

———. 1997. Pesticide use by licensed applicators for the control of *Ixodes scapularis* (Acari: Ixodidae) in Connecticut. *Journal of Medical Entomology* 34:552–558.

Stanek, G., and J. Simeoni. 1989. Are pigeons' ticks transmitters of *Borrelia burgdorferi* to humans? A preliminary report. *Zentralblatt für Bakteriologie* 18:42–43.

Steere, A.C. 1989. Lyme disease. *New England Journal of Medicine* 321:586–596.

Steere, A.C., S.E. Malawista, and D.R. Sandman. 1977. Lyme arthritis: An epidemic of oligoarticular arthritis in children and adults in three Connecticut communities. *Arthritis and Rheumatism* 20:7–17.

Steere, A.C., M.S. Grodzicki, A.N. Kornblatt, J.E. Craft, A.G. Barbour, W. Burgdorfer, G.P. Schmid, E. Johnson, and S.E. Malawista. 1983. The spirochetal etiology of Lyme disease. *New England Journal of Medicine* 308:733–739.

Strijdom, S.C., and M. Berk. 1996. Lyme disease in South Africa. *South African Medical Journal* 86 (Suppl. 6):741–744.

Takada, N., F. Ishiguro, H. Fujita, H.P. Wang, J.C. Wang, and T. Masuzawa. 1998. Lyme disease spirochetes in ticks from northeastern China. *Journal of Parasitology* 84:499–504.

Talleklint, L., and T.G. Jaenson. 1994. Transmission of *Borrelia burgdorferi* s. l. from mammal reservoirs to the primary vector of Lyme borreliosis, *Ixodes ricinus* (Acari: Ixodidae) in Sweden. *Journal of Medical Entomology* 31:880–886.

Telford III, S.R., and A. Spielman. 1989. Enzootic transmission of the agent of Lyme disease in rabbits. *American Journal of Tropical Medicine and Hygiene* 41:482–490.

Telford III, S.R., T.N. Mather, G.H. Adler, and A. Spielman. 1990. Short-tailed shrews as reservoirs of the agents of Lyme disease and human babesiosis. *Journal of Parasitology* 76:681–683.

Teltow, G.J., P.V. Fournier, and J.A. Rawlings. 1991. Isolation of *Borrelia burgdorferi* from arthropods collected in Texas. *American Journal of Tropical Medicine and Hygiene* 44:469–474.

Tian, W., Z. Zhang, S. Moldenhauer, Y. Guo, Q. Yu, L. Wang L, and M. Chen. 1998. Detection of *Borrelia burgdorferi* from ticks (Acari) in Hebei Province, China. *Journal of Medical Entomology* 35:95–98.

White, D.J. 1993. Lyme disease surveillance and personal protection against ticks. In *Ecology and environmental management of Lyme disease,* ed. H.S. Ginsberg. New Brunswick, NJ: Rutgers University Press, pp. 99–125.

Williams, C.L., B. Strobino, A. Weinstein, P. Spierling, and F. Medici. 1995. Maternal Lyme disease and congenital malformations: A cord blood serosurvey in endemic and control areas. *Paediatric and Perinatal Epidemiology* 9:320–330.

Wilske, B., V. Fingerle, V. Preac-Mursic, S. Jauris-Heipke, A. Hofmann, H. Loy, H.W. Pfister, D. Rossler, and E. Soutschek. 1994. Immunoblot using recombinant antigens derived from different genospecies of *Borrelia burgdorferi* sensu lato. *Medical Microbiology and Immunology* 183:43–59.

Wilson, M.L. 1986. Reduced abundance of adult *Ixodes dammini* (Acari: Ixodidae) following destruction of vegetation. *Journal of Economic Entomology* 79:693–696.

Wilson, M.L., and R.D. Deblinger. 1993. Vector management to reduce the risk of Lyme disease. In *Ecology and environmental management of Lyme disease,* ed. H.S. Ginsberg. New Brunswick, NJ: Rutgers University Press, pp. 126–156.

Wilson, M.L., A.M. Ducey, T.S. Litwin, T.A. Gavin, and A. Spielman. 1990. Microgeographic distribution of immature *Ixodes dammini* ticks correlated with that of deer. *Medical and Veterinary Entomology* 4:151–159.

Wright, S.D., and S.W. Nielsen. 1990. Experimental infection of the white-footed mouse with *Borrelia burgdorferi. American Journal of Veterinary Research* 51:1980–1987.

Zhang, Z. 1992. Survey on tick vectors of Lyme disease spirochetes in China [in Chinese]. *Chung Hua Liu Hsing Ping Hsueh Tsa Chih* 13:271–274.

Zingg, B.C., R.N. Brown, R.S. Lane, and R.B. LeFebvre. 1993. Genetic diversity among *Borrelia burgdorferi* isolates from wood rats and kangaroo rats in California. *Journal of Clinical Microbiology* 31:3109–3114.

27 ORDER RICKETTSIALES

ANAPLASMOSIS

WILLIAM R. DAVIDSON AND WILL L. GOFF

INTRODUCTION. Anaplasmosis is an infectious, noncontagious disease of ruminants caused by rickettsiae in the genus *Anaplasma.* Anaplasmosis is best known as a disease of domestic cattle *Bos taurus,* sheep *Ovis aries,* and goats *Capra hircus,* but a variety of wild ruminants also are susceptible to infection. *Anaplasma* organisms (initial bodies) infect erythrocytes and can cause anemia and icterus, although most infections are subclinical. Disease generally is less severe in younger than in older animals.

HISTORY. The genus *Anaplasma* was erected by Theiler (1910), who described *A. marginale* in the erythrocytes of cattle, although he believed the organisms were protozoan parasites. Smith and Kilborne (1893) had originally detected *Anaplasma* organisms in cattle several years earlier but had concluded they were a developmental stage of *Babesia bigemina,* the causative agent of Texas cattle fever. By 1930, bovine anaplasmosis was recognized to occur in southern and western regions of the United States, as well as in many other parts of the world (Giltner 1930). Theiler (1911) also described a second species, *Anaplasma centrale,* which was associated with a milder form of disease in cattle. Later, a third species, *Anaplasma ovis,* which also produced a mild disease, was identified from domestic sheep (Bevan 1912). Two additional species of *Anaplasma* have been described: *Anaplasma buffeli* from water buffalo *Bubalus bubalis* (Carpano 1934) and *Anaplasma mesaeterum* from sheep (Uilenberg et al. 1979). Other unnamed anaplasms have also been reported from several species of wild ruminants (Kuttler 1984).

During the 1930s, researchers in Africa (Lestoquard 1931; Neitz and DuToit 1932; Neitz 1935) and North America (Boynton and Woods 1933) confirmed that certain wild ruminants were susceptible to infection, which eventually led to recognition that wild ruminants could serve as reservoir hosts. The importance of ticks as vectors of anaplasmosis also was discovered during the 1930s (Boynton et al. 1936; Herms and Howell 1936).

DISTRIBUTION. *Anaplasma marginale* occurs throughout the tropical and subtropical regions of Africa, Asia, Australia, Europe, North America, and South America, as well as several island nations within tropical, subtropical, and temperate regions of the world. *Anaplasma centrale,* which is restricted to Africa, and *A. ovis,* which has a worldwide distribution, are less well studied than *A. marginale.*

ETIOLOGY. Anaplasmosis is caused by obligate intraerythrocytic members of the genus *Anaplasma,* within the order Rickettsiales, family Anaplasmataceae. Three well-recognized species of *Anaplasma* infect domestic livestock: *A. marginale,* the causative agent of bovine anaplasmosis; *A. centrale,* a species causing a milder disease among cattle; and *A. ovis,* the causative agent of ovine anaplasmosis. The taxonomic validity of *A. buffeli* of water buffalo and *A. mesaeterum* of sheep is uncertain. The identity of *Anaplasma* organisms present in many wild ruminant species requires elucidation (Kuttler 1984). Recent phylogenetic evaluation based on sequence analysis of the 16S rRNA gene disclosed that *A. marginale* is related to organisms in the genera *Cowdria* and *Ehrlichia,* particularly those in the *Ehrlichia phagocytophila* genogroup (Barbet 1995; Walker and Dumler 1996).

Anaplasms appear as dense, rounded, dark-blue structures about 0.3–1.0 μm in diameter in Giemsa- or Wright-stained blood films (Fig. 27.1). Organisms are non-acid-fast. As implied by the specific epithet, *A. marginale* usually occurs near the margin of the cell, and *A. centrale* occurs near the center; *A. ovis* usually occupies a marginal position. Ultrastructurally, the classic *marginale bodies* of *Anaplasma* organisms are actually an aggregate of several individual organisms (initial bodies) within a double-membrane-bound vacuole. Following invasion of an erythrocyte, initial bodies reproduce by binary fission. The genome size is approximately 1230 kb, with a G+C content of 56 mol% (Alleman et al. 1993).

HOST RANGE. Naturally occurring *Anaplasma* infections are common among domestic cattle, sheep, and goats and occur in various wild ruminants within the families Antilocapridae, Bovidae, Cervidae, and

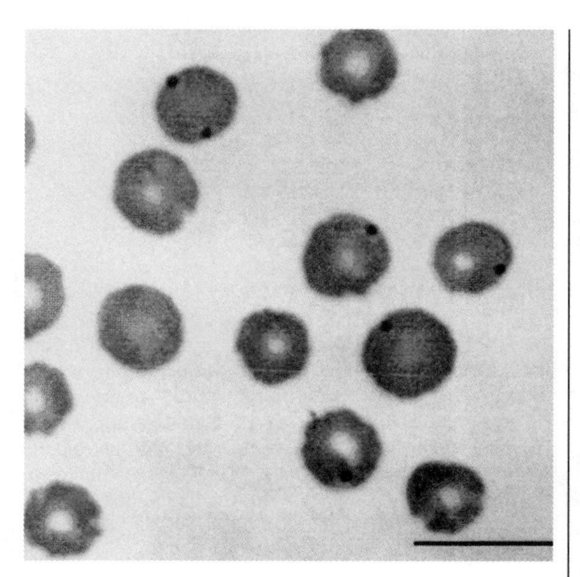

FIG. 27.1—*Anaplasma marginale* in erythrocytes of a cow (Giemsa stain). Photomicrograph courtesy of K.M. Kocan.

Giraffidae (Table 27.1). The susceptibility of many wild ruminants to infection by different species of *Anaplasma* is incompletely known, but *A. marginale, A. ovis,* and *A. centrale* are all reported to infect wild ruminants (Table 27.1). *Anaplasma marginale* infection in cattle can result in serious disease, but *A. ovis* infection in domestic sheep and goats is usually subclinical or causes only mild disease (Splitter et al. 1956). Except for two reports of acute anaplasmosis in giraffes *Giraffa camelopardalis* (Lohr and Meyer 1973; Agustyn and Bigalke 1974), all naturally occurring *Anaplasma* infections among wild ruminants have been subclinical (Kuttler 1984). Experimental infections of bighorn sheep *Ovis canadensis* with *A. ovis* have resulted in severe disease (Tibbitts et al. 1992; Goff et al. 1993). Attempts to infect numerous nonruminant species with *Anaplasma* organisms experimentally have been unsuccessful (Lignieres 1919; Dykstra et al. 1938; Summers and Gonzalez 1965; Zaugg and Newman 1985), suggesting that ruminant hosts are the only vertebrate reservoirs.

TRANSMISSION AND EPIDEMIOLOGY. Anaplasmosis is transmitted naturally by numerous blood-feeding arthropods. Certain species of ticks serve as biologic vectors (Table 27.2), whereas biting insects, particularly tabanid flies, serve as less efficient mechanical vectors (Table 27.3). Mechanical transmission can also be effected by fomites, such as vaccination needles or dehorning and castration equipment. Under natural conditions, mechanical transmission by biting insects must occur quickly because the organisms remain viable on insect mouthparts for only a brief period due to flight-associated desiccation (Howell et al. 1941). The level of rickettsemia may also be critical for mechanical transmission by biting insects because of the minuscule quantities of blood that are transferred (Piercy 1956; Foil et al. 1987). Successful transmission has not been reported by biting flies fed on carrier animals with low rickettsemias.

In contrast to mechanical transmission by biting flies, biologic transmission by ticks is highly efficient because of replication and persistence of organisms within ticks (Kocan et al. 1992a,b, 1993; Eriks et al. 1993). Intrastadial transmission through direct interhost transfer of adult *Dermacentor* spp. ticks, particularly males, may be an important means of transmission among domestic livestock (Stiller et al. 1983, 1989a; Zaugg et al. 1986; Stiller and Coan 1995). Development within ticks (Fig. 27.2) involves a complex cycle involving several morphologically distinct stages and several tick tissues, culminating with colonization of the salivary glands (Kocan 1986; Kocan et al. 1980, 1988, 1990; Stiller et al. 1989b).

The interplay of susceptible wild ruminant hosts and arthropod vectors is critical in the epizootiology of anaplasmosis. For example, black-tailed deer *Odocoileus hemionus columbianus,* mule deer *Odocoileus hemionus hemionus,* and white-tailed deer *Odocoileus virginianus* are all susceptible to infection by *A. marginale,* and all develop rickettsemias of many weeks duration without becoming clinically ill (Osebold et al. 1959; Howe and Hepworth 1965; Kuttler et al. 1967; Kuttler 1984; Zaugg 1988; Keel et al. 1995). Although each of these species of deer has an approximately equivalent biologic potential to serve as a reservoir host for *A. marginale,* the actual role that each plays may differ depending on several factors, including vertebrate and vector population densities, intensity of rickettsemia in the vertebrate hosts, and the relative efficiency of different vector species (Kuttler 1981b; Keel et al. 1995; Kollars 1996). Much of the dynamics among these factors has yet to be investigated.

Black-tailed deer along the Pacific Coast are parasitized by the three-host tick *Dermacentor occidentalis,* which is an efficient transstadial and intrastadial vector of *A. marginale* (Herms and Howell 1936; Osebold et al. 1962; Stiller et al. 1983). The presence of this efficient biologic vector can result in a high prevalence of infection, as illustrated in a black-tailed deer population in California where 31% of the fawns, 47% of the yearlings, and 92% of the adults harbored *A. marginale* (Howarth et al. 1969). *Dermacentor andersoni,* which is also capable of transstadial and intrastadial transmission (Kocan 1986; Zaugg et al. 1986), and mule deer comprise a similar but perhaps somewhat less efficient natural reservoir system in intermountain regions of the western United States (Kuttler 1981b). These combinations of efficient biologic vectors and susceptible wild ruminant hosts comprise natural reservoir systems that may serve as a source of infection for domestic livestock (Howarth et al. 1969; Kuttler 1984).

TABLE 27.1—Wild ruminants with evidence of susceptibility to infection with *Anaplasma* spp.

Animal Species	Species of *Anaplasma* and Severity of Infection[a]				References
	A. marginale	*A. ovis*	*A. centrale*	Other *Anaplasma*	
Cervidae					
Black-tailed deer *Odocoileus hemionus columbianus*	++				Boynton and Woods 1933, 1940; Christensen and McNeal 1967; Christensen et al. 1958; Osebold et al. 1959; Howarth et al. 1969, 1976; Kuttler 1981b, 1984
Mule deer *Odocoileus hemionus hemionus*	+	+			Howe et al. 1964; Howe and Hepworth 1965; Renshaw et al. 1977; Kuttler 1981b, 1984; Zaugg 1988; Waldrup et al. 1989
White-tailed deer *Odocoileus virginianus*	+				Ristic and Watrach 1961; Kuttler et al. 1967, 1968; Lancaster et al. 1968; Maas et al. 1981; Smith et al. 1982; Morley and Hugh-Jones 1989; Hungerford et al. 1989; Waldrup et al. 1989; Keel et al. 1995
Elk *Cervus elaphus*	+	+[b]			Howe et al. 1964; Magonigle and Eckblad 1979; Renshaw et al. 1979; Zaugg et al. 1996
Giraffidae					
Giraffe *Giraffa camelopardalis*				+++	Brocklesby and Vidler 1966; Lohr and Meyer 1973; Agustyn and Bigalke 1974; Lohr et al. 1974
Antilocapridae					
Pronghorn *Antilocapra americana*	+[b]	+[b]			Howe et al. 1964; Howe and Hepworth 1965; Jacobson et al. 1977; Zaugg 1987
Bovidae					
Bison *Bison bison*	++	-			Peterson and Roby 1975; Zaugg and Kuttler 1985; Zaugg 1986
Cape buffalo *Syncerus caffer*	+		+	+	Brocklesby and Vidler 1966; Potgieter 1979
Asian water buffalo *Bubalus bubalis*	+				Sharma 1987
Eland *Taurotragus oryx*				+	Lohr et al. 1974
Duiker *Sylvicapra grimmia*	+				Neitz and DuToit 1932
Blue wildebeest *Connochaetes taurinus*	+			+	Kuttler 1965; Lohr and Meyer 1973; Smith et al. 1974; Burridge 1975
Black wildebeest *Connochaetes gnou*	+				Neitz 1935
Coke's hartebeest *Alcelaphus buselaphus cokii*	+			+	Lohr and Meyer 1973
Blesbok *Damaliscus pygargus*	+	+	+		Neitz and DuToit 1932; Neitz 1939
Sable antelope *Hippotragus niger*		+		++	Thomas et al. 1982
Waterbuck *Kobus ellipsiprymnus*				+	Kuttler 1965; Lohr et al. 1974
Impala *Aepyceros melampus*				+	Kuttler 1965; Lohr et al. 1974
Thompson's gazelle *Gazella thomsonii*	+			+	Lohr and Meyer 1973
Grant gazelle *Gazella granti*				+	Lohr et al. 1974
Bighorn sheep *Ovis canadensis*	+[b]	+++			Howe et al. 1964; Tibbitts et al. 1992; Goff et al. 1993; Jessup et al. 1993
Wild goat *Capra aegagrus*				+++	Matthews 1978

[a]Severity of infection: -, refractory to infection; +, serologic evidence or subclinical infection; ++, mild clinical response; +++, severe clinical disease.[b]Experimental infections only.

TABLE 27.2—Species of ticks experimentally confirmed to be capable of biological transmission of *Anaplasma* organisms

Tick Species	Species of *Anaplasma*	Type of Transmission[a]	References
Boophilus annulatus	*A. marginale*	TS, TO[b]	Kuttler et al. 1971
B. calcaratus	*A. marginale*	TS	Howe 1981
B. decoloratus	*A. marginale*	TS	Neitz 1956; Howe 1981
	A. centrale	TS	Neitz 1956
B. microplus	*A. marginale*	TS	Connell and Hall 1972; Connell 1974
Dermacentor albipictus	*A. marginale*	TS	Herms and Howell 1936; Boynton et al. 1936
		IS	Stiller et al. 1981, 1983
	A. ovis	IS	Stiller et al. 1983; unpublished
D. andersoni	*A. marginale*	TS, TO[b]	Howell et al. 1941; Anthony and Roby 1966; Kocan 1986; Kocan et al. 1985, 1988
		IS	Anthony and Roby 1966; Zaugg et al. 1986
	A. ovis	IS	Kocan and Stiller 1992
D. hunteri	*A. marginale*	IS	Stiller et al. 1999
	A. ovis	IS	Stiller et al. 1999
D. occidentalis	*A. marginale*	TS, TO[b]	Herms and Howell 1936; Osebold et al. 1962
		IS	Stiller et al. 1983
	A. ovis	IS	Stiller unpublished
D. silvarum	*A. ovis*	TS	Neitz 1956
D. variabilis	*A. marginale*	TS	Anthony and Roby 1966; Stich et al. 1989
		IS	Anthony and Roby 1966
	A. ovis	IS	Stiller et al. 1987
Haemaphysalis cinnabarina	*A. marginale*	TS	Neitz 1956; Howe 1981
	A. centrale	TS	Neitz 1956; Howe 1981
Hyalomma excavatum	*A. marginale*	TS	Neitz 1956
H. rufipes	*A. marginale*	IS	Potgieter 1979
Ixodes ricinus	*A. marginale*	TS	Howe 1981
I. scapularis	*A. marginale*	TS	Dikmans 1950
Ornithodoros lahorensis	*A. ovis*	TS	Neitz 1956
Rhipicephalus bursa	*A. marginale*	TS	Dikmans 1950; Neitz 1956
	A. ovis	TS	Dikmans 1950; Neitz 1956
R. sanguineus	*A. marginale*	TS	Dikmans 1950
R. simus	*A. marginale*	TS	Neitz 1956; Howe 1981
		IS	Potgieter 1979
	A. centrale	IS	Potgieter 1979
R. evertsi	*A. marginale*	IS	Potgieter 1979

[a]TS, transstadial; IS, intrastadial; TO, transovarial.[b]These reports of transovarial transmission have not been confirmed and thus are in question.

TABLE 27.3—Species of biting flies experimentally confirmed to be capable of mechanical transmission of *Anaplasma marginale*

Fly Species	References
Tabanus abactor	Howell et al. 1941
T. americanus	Howell et al. 1941
T. atratus	Morris et al. 1936
T. equalis	Howell et al. 1941
T. erythraeus	Howell et al. 1941
T. fumipennis	Sanders 1933
T. fuscicostatus	Wilson and Meyer 1966; Hawkins et al. 1981
T. gracilis	Sanborne et al. 1932
T. lineola	Hawkins et al. 1981
T. mularis	Hawkins et al. 1981
T. oklahomensis	Howell et al. 1941; Piercy 1956
T. pallidescens	Hawkins et al. 1981
T. sulcifrons	Sanborne et al. 1932; Lotze and Yiengst 1941; Howell et al. 1941; Lotze 1944
T. venustus	Sanborne et al. 1932; Howell et al. 1941
Stomoxys calcitrans	Sanders 1933
Sylvius pollinosa	Howell et al. 1941
Chrysops sequax	Howell et al. 1941

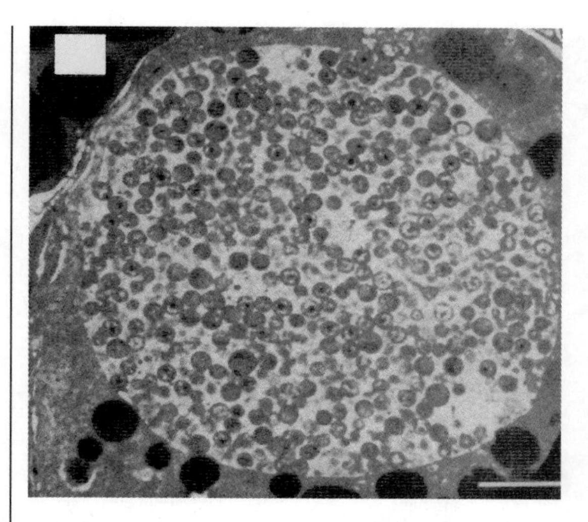

FIG. 27.2—Election photomicrograph of developing colony of *Anaplasma marginale* within the cytoplasm of an acinar cell in the salivary gland of *Dermacentor andersoni*. Bar = 10 μm). Adapted from Kocan et al. (1992a).

In contrast, numerous surveys have indicated that white-tailed deer populations in the eastern United States are not involved in the epidemiology of *A. marginale* despite being susceptible to infection (Bedell and Miller 1966; Kuttler et al. 1967, 1968; Morley and Hugh-Jones 1989; Waldrup et al. 1989; Keel et al. 1995). The principal factor preventing natural infection among white-tailed deer within the anaplasmosis-endemic regions of the eastern United States may be that transmission is by biting flies rather than by ticks (Keel et al. 1995). The numerous constraints that reduce the efficiency of mechanical transmission by biting flies apparently preclude meaningful involvement of white-tailed deer in the epidemiology of *A. marginale* in the southeastern United States (Keel et al. 1995).

Data on two species of wild ruminants from the western United States provide examples of the need for validation of diagnostic assays used in epizootiologic studies. In the first case, interest in the health status of American bison *Bison bison* from Yellowstone National Park (Wyoming-Idaho-Montana) led to a serosurvey that first detected *Anaplasma* seroreactors in wild bison (Taylor et al. 1997). *Anaplasma*-specific antibodies were detected in 8 (11%) of 76 bison by using the complement fixation (CF) test. Bison had been known to be susceptible to experimental infection (Zaugg and Kuttler 1985) and to remain carriers for as long as 496 days (Zaugg 1986). Earlier, the CF test had failed to detect antibodies in 62 Yellowstone bison collected in 1984 and 1985; however, retesting of these samples with a competitive enzyme-linked immunosorbent assay (cELISA) (Ndung'u et al. 1995) that has improved sensitivity (Torioni de Echaide et al. 1998) disclosed antibody in 11% of sera (W.L. Goff unpublished). In the second case, sera collected from about 200 elk *Cervus elaphus* were evaluated by the CF test, an indirect immunofluorescence (IIF) assay (Tibbitts et al. 1992), and the cELISA (W.L. Goff unpublished). The IIF and cELISA compared favorably and detected antibody in over 50% of the elk. The CF test was inadequate and was particularly problematic with elk sera. Elk were susceptible to experimental infection with both *A. marginale* and *A. ovis* and became chronically infected (at least 172 days) without developing clinical disease (Zaugg et al. 1996). The absence of clinical disease in experimentally infected elk, their ability to maintain a persistent infection, and serologic evidence of a rather high prevalence of infection in some locations suggest they may serve as an important reservoir.

An epidemiologic association involving *A. ovis, Dermacentor hunteri,* and desert bighorn sheep *Ovis canadensis nelsoni* has been described. Populations of bighorn sheep throughout the southwest United States demonstrated variable seroprevalence (Jessup et al. 1993; Crosbie et al. 1997), and all populations demonstrating evidence of *A. ovis* infection were infested with *D. hunteri.* There were, however, some *D. hunteri*-infested but *Anaplasma*-seronegative populations (Crosbie et al. 1997). This tick has been shown to be an efficient experimental vector of both *A. ovis* and *A. marginale* (Stiller et al. 1999). Another efficient, natural vector of *A. ovis* is *D. andersoni* (Stiller et al. 1989b; Kocan and Stiller 1992), and this tick has been demonstrated experimentally to transmit *A. ovis* between domestic sheep and bighorn sheep (Stiller et al. 1990).

CLINICAL SIGNS. Among domestic ruminants, anaplasmosis can occur as either an acute, subacute, or chronic infection, and chronically infected animals may experience recrudescence accompanied by clinical illness. Acute anaplasmosis typically occurs among previously unexposed adult animals. Chronically infected animals maintain circulating antibodies, thereby exhibiting a state of premunition. Clinical anaplasmosis has not been described among naturally infected wild ruminants except for two reports of acute fatal infections in giraffe by organisms similar to *A. marginale* (Lohr and Meyer 1973; Agustyn and Bigalke 1974). Lack of clinical disease among infected wild ruminants has been interpreted as representing either an evolutionary adaptation between the host and parasite or the product of immunity through natural exposure (Howe 1981). Despite the absence of reports of clinical disease, there is evidence that anaplasmosis may constitute a health risk for bighorn sheep populations. Jessup et al. (1993) reported a high prevalence of presumptive *A. ovis* antibodies among some bighorn sheep populations in California, and an isolate of *A. ovis* from one of these populations as well as an *A. ovis* isolate from domestic sheep in Idaho produced severe anemia and icterus in experimentally infected bighorn sheep (Tibbitts et al. 1992; Goff et al. 1993).

PATHOGENESIS. Members of the genus *Anaplasma* exhibit a strict tropism for erythrocytes and occur exclusively within this cell type. Following exposure of a susceptible host, *Anaplasma* organisms undergo an incubation period of 3–6 weeks during which clinical signs do not occur. Thereafter, *Anaplasma* organisms become detectable within erythrocytes as marginal bodies containing 2–12 individual rickettsiae. The organisms produced within marginal bodies escape and invade other erythrocytes, leading to a logarithmic increase in rickettsemia. During the exponential period of growth, the rickettsemia doubles about every 24 hours. As the rickettsemia increases and peaks, an anemia develops when erythrocytes are removed by the reticuloendothelial system (Ristic 1968). The intensity of the rickettsemia and severity of anemia vary markedly depending on the susceptibility of the host and the species of *Anaplasma.* Adult cattle that are fully susceptible to *A. marginale* may develop a rickettsemia of 50% or more, accompanied by a severe anemia with erythrocyte counts, packed cell volumes, and hemoglobin concentrations less than 25% of normal

(Ristic 1968). In contrast, in many wild ruminants, *A. marginale* rickettsemias can be 2% or less, without a detectable anemia. Animals that survive an episode of acute anaplasmosis undergo a period of convalescence during which hematologic parameters return to normal; however, episodes of recrudescent rickettsemia and anemia have been documented among cattle chronically infected with *A. marginale* (Kieser et al. 1990).

PATHOLOGY. The lesions of anaplasmosis are directly or indirectly attributable to anemia. Gross lesions associated with acute anaplasmosis include pale mucous membranes, splenomegaly, hepatomegaly, enlargement of the gallbladder, icterus, generalized lymphadenopathy, and occasional petechial hemorrhages in the endocardium, epicardium, or the serosal surfaces of other visceral organs (Ristic 1968).

DIAGNOSIS. Methods for diagnosis of *Anaplasma* infections include light microscopy of stained blood films, subinoculation of blood into susceptible animals (usually splenectomized), splenectomy of suspected carrier hosts to induce recrudescence of rickettsemia, serology, and molecular methods. The diagnostic approach typically differs between animals with acute disease and those suspected of having chronic or latent infections. Light microscopy accompanied by either serologic or molecular confirmation are the usual tools for diagnosing of acute infections. Chronic infection, which is the typical case for wild ruminants, usually is diagnosed by serologic demonstration of antibodies, with confirmation by either animal subinoculation or molecular methods.

Serodiagnosis of *Anaplasma* infections among domestic animals historically has relied on complement fixation (Hidalgo and Dimopoullos 1967), capillary agglutination (Ristic 1962), and card agglutination assays (Amerault and Roby 1968). However, use of these assays on sera from numerous species of wild ruminants, especially Cervidae, has proven to be largely unreliable, generating either false-positive or false-negative results or both (Howe et al. 1964; Howe and Hepworth 1965; Kuttler et al. 1968; Howarth et al. 1969, 1976; Peterson et al. 1973; Jacobson et al. 1977; Magonigle and Eckblad 1979; Renshaw et al. 1979; Kuttler 1984). Later, IIF was developed (Gonzalez et al. 1978; Goff et al. 1985, 1990) and modified for use with wild ruminants (Tibbitts et al. 1992). The modification consisted of using fluorescein-conjugated protein G, which binds to the Fc region of immunoglobulin G regardless of ruminant species of origin. This modified IIF assay eliminated the cross-species problems of applying the conventional livestock serologic tests to wild ruminants and has been found to detect *Anaplasma* antibodies reliably among wild ruminants (Tibbitts et al. 1992; Jessup et al. 1993; Keel et al. 1995; Zaugg et al. 1996; Crosbie et al. 1997). Although the IIF assay has proven useful, it is less practical for large numbers of samples, an important feature for routine diagnosis or extensive epidemiologic studies. A cELISA has been developed based on an epitope associated with an *Anaplasma* spp. major surface protein 5 (MSP-5) (Visser et al. 1992). This epitope appears to be expressed on all species and isolates of *Anaplasma,* and in the cELISA format (recombinant antigen and a specific monoclonal antibody) has allowed for detection of antibodies to *A. marginale* (Knowles et al. 1996) and *A. ovis* (Ndung'u et al. 1995). The cELISA also has proven useful in detecting antibody from wild ruminants (W.L. Goff et al. unpublished). While the conserved nature of several *Anaplasma* antigens results in the ability of a single assay (IIF or cELISA) to be used to assess the prevalence of *Anaplasma* exposure in wild ruminant populations, the utility of the assay to confirm infection with a particular species is limited (De Kroon et al. 1990; Tibbitts et al. 1992; Knowles et al. 1996). Although the IIF and cELISA enable relatively sensitive detection of carrier animals, when both exquisite sensitivity and specificity are required, newer molecular techniques (see below) can be used to differentiate species.

Subinoculation of blood from suspected reservoir species into susceptible hosts, usually splenectomized calves or sheep, has been used extensively to detect a carrier state among chronically infected hosts and has proven especially useful in evaluating populations of wild ruminants (Osebold et al. 1959; Howe and Hepworth 1965; Bedell and Miller 1966; Peterson et al. 1973; Peterson and Roby 1975; Howarth et al. 1976; Jacobson et al. 1977; Renshaw et al. 1977; Thomas et al. 1982). Although animal subinoculation is a proven technique and has the advantage of providing information on the susceptibility of recipient hosts to the organism, it is expensive because of the costs associated with animal care and housing.

The advent of molecular diagnostic procedures has provided an excellent alternative method for the detection of *Anaplasma* infections, which are characteristically latent among wild ruminants (Stiller 1990). Nucleic acid probes have been developed that will detect *Anaplasma* DNA in ticks and vertebrate hosts (Goff et al. 1988, 1990, 1993; Eriks et al. 1989; Shompole et al. 1989; Kieser et al. 1990; Kocan et al. 1992a,b, 1993). Nucleic acid probes can have extremely high sensitivities, enabling the detection of rickettsemias as low as 0.000025% (Eriks et al. 1989; Kieser et al. 1990). Probes for DNA have been used to identify different species of *Anaplasma,* as well as to assess the genetic similarities of different *Anaplasma* isolates (Krueger and Buening 1988; Goff et al. 1993; Eriks et al. 1994).

Polymerase chain reaction (PCR) has also been developed using primers associated with the same gene coding for MSP-1B (Barbet and Allred 1991) that was used for the design of the first DNA probe. Polymerase chain reaction has been a successful method for detecting low levels of infection in both blood and tick tissues (Figueroa et al. 1993; Stich et al. 1993a,b; Ge et al. 1995), and its levels of sensitivity appear to exceed that of strict DNA probe hybridization. More recently, development of a nested PCR produced a several-fold increase in sensitivity (Torioni de Echaide et al. 1998).

IMMUNITY. After exposure, ruminants produce circulating antibodies but remain chronically infected. This form of immune response is termed *premunition,* and premunized animals normally retain their circulating antibodies only so long as the causative organism persists as a latent infection. However, recrudescence can occur among such hosts, and premunition will be lost if latent infections are eliminated by treatment or for some other reason.

Infection of cattle by either *A. marginale* or *A. centrale* results in an immune response that is relatively protective against infection by the heterologous organism. In contrast, protective cross-immunity does not occur between *A. marginale* and *A. ovis* in experimentally infected cattle and sheep (Kuttler 1981a, 1984). Thus, regardless of antigenic similarities among species, there may remain insufficient stimulation to provide cross-immunity. Furthermore, some MSPs display polymorphism among species and isolates of the same species and are encoded by multigene families (Eid et al. 1996). These attributes may contribute to the failure of heterologous protection when much of the immune response is directed at such peptides. Little is known about this aspect in wildlife. Cross-challenge studies with known species of *Anaplasma* have not been reported for wild ruminants; however, there is evidence that blue wildebeest *Connochaetes taurinus,* Coke's hartebeest *Alcelaphus buselaphus cokii,* and Thompson's gazelle *Gazella thomsonii* harbor either strains of *A. marginale* or a closely related undescribed species that are antigenically distinct enough so that they do not protect against *A. marginale* challenge infection in cattle (Lohr and Meyer 1973).

Although protection against homologous challenge is stronger than heterologous challenge, even this protection is often incomplete. Although there is a correlation between titer of antibody to some MSPs and degree of protection, those MSPs that are encoded by multigene families may actually contribute to immune evasion. Immunity is probably not restricted to the humoral response, but little is understood concerning what cell-mediated mechanisms are involved in a successful immune response. Evidence from cattle studies suggests that cell-mediated responses are correlated with protection (Carson et al. 1977a,b; Francis et al. 1980; Wyatt et al. 1996), but nothing is known about this aspect with wild ruminants.

PREVENTION, CONTROL, AND TREATMENT. Programs for dealing with anaplasmosis among domestic livestock most frequently have included one or more of the following methods: vector control, vaccination, and antibiotic (primarily tetracycline) therapy (Kuttler 1979; Potgieter 1979; Rogers and Shiels 1979; Kuttler and Kreier 1986; Lawrence and de Vos 1990). Each of these methods has proven merit for use with domestic livestock. Conversely, others have presented the concept that, within highly endemic regions, domestic animal health programs designed to produce enzootic stability, which contain some of the aforementioned components, may be a more economical method for dealing with anaplasmosis and other tick-borne diseases (Lawrence and de Vos 1990; Meltzer et al. 1995).

The measures commonly used to combat anaplasmosis among domestic livestock are not logistically feasible for use among free-ranging wild ruminants, nor in most cases are they needed, since the health of wild ruminants is rarely compromised by *Anaplasma* infection. Among captive wild ruminants, however, vector control and antibiotic therapy can be successfully used to prevent exposure or to treat infected animals.

PUBLIC HEALTH CONCERNS. Members of the genus *Anaplasma* are not infectious to humans.

DOMESTIC ANIMAL HEALTH CONCERNS. Bovine and, to a lesser extent, ovine anaplasmosis are important diseases of domestic cattle and sheep in many regions of the world. Infection in these animals can be costly not only in terms of mortality and productivity losses but also in the cost of treatment and vaccination programs. Where anaplasmosis is enzootic, the role that wild ruminants play in its maintenance and spread can be a controversial issue between the domestic livestock and wildlife conservation interests. In situations where efficient tick vectors provide a means of transferring *Anaplasma* organisms from wild ruminant reservoir hosts to domestic livestock on common range, prevention and control of anaplasmosis among domestic animals can be extremely difficult or impossible. On the other hand, in a setting where efficient tick vectors are absent, the presence of susceptible wild ruminant populations may have little or no relationship to the occurrence of anaplasmosis in domestic livestock. Consequently, an understanding of the epizootiology of anaplasmosis in various ecologic settings is essential before accurate conclusions can be drawn regarding the risk that wild ruminants pose for anaplasmosis among domestic livestock.

MANAGEMENT IMPLICATIONS. Although a thorough understanding of anaplasmosis among wild ruminant populations is lacking, certain management implications seem clear. One issue is the potential for introduction of anaplasmosis into wild populations by the translocation of infected wildlife. Jessup et al. (1993) noted that relocation of infected bighorn sheep could pose a risk of introducing into uninfected populations strains of *A. ovis* that are pathogenic to this host under experimental conditions. Similarly, Keel et al. (1995) pointed out that continued vigilance to prevent the introduction of efficient tick vectors was an important disease management objective for white-tailed deer in the southeastern United States.

LITERATURE CITED

Agustyn, N.J., and R.D. Bigalke. 1974. *Anaplasma* infection in a giraffe. *Journal of the South African Veterinary Association* 45:229.

Alleman, A.R., S.M. Kamper, N. Viseshakul, and A.F. Barbet. 1993. Analysis of the *Anaplasma marginale* genome by pulse-field electrophoresis. *Journal of General Microbiology* 139:2439–2444.

Amerault, T.E., and T.O. Roby. 1968. A rapid card agglutination test for bovine anaplasmosis. *Journal of the American Veterinary Medical Association* 153:1828–1834.

Anthony, D.W., and T.O. Roby. 1966. The experimental transmission of bovine anaplasmosis by 3 species of North American ticks. *American Journal of Veterinary Research* 27:191–198.

Barbet, A.F. 1995. Recent developments in the molecular biology of anaplasmosis. *Veterinary Parasitology* 57:43–49.

Barbet, A.F., and D.R. Allred. 1991. The MSP1B multigene family of *Anaplasma marginale:* Nucleotide sequence analysis of an expressed copy. *Infection and Immunity* 59:971–976.

Bedell, D.M., and J.G. Miller. 1966. *A report of the examination of 270 white-tailed deer (*Odocoileus virginianus*) from anaplasmosis enzootic areas of the southeastern United States for evidence of anaplasmosis.* Tifton: University of Georgia, College of Agricultural Science, Coastal Plain Experiment Station, 158 pp.

Bevan, L.E.W. 1912. Anaplasmosis in sheep. *Veterinary Journal* 68:400–401.

Boynton, W.H., and G.M. Woods. 1933. Deer as carriers of anaplasmosis. *Science* 78:559–560.

———. 1940. Anaplasmosis among deer in the natural state. *Science* 91:168.

Boynton, W.H., W.B. Herms, D.E. Howell, and G.M. Woods. 1936. Anaplasmosis transmission by three species of ticks in California. *Journal of the American Veterinary Medical Association* 88:500–502.

Brocklesby, D.W., and B.O. Vidler. 1966. Haematozoa found in wild members of the order Artiodactyla in East Africa. *Bulletin of Epizootic Diseases in Africa* 14:285–299.

Burridge, M.J. 1975. The role of wild mammals in the epidemiology of bovine theileriosis in East Africa. *Journal of Wildlife Diseases* 11:68–74.

Carpano, M. 1934. l'Infezione da *Anaplasma* del tipo marginale dei bufali in Egitto. *Clinical Veterinarian (Milan)* 57:589–592.

Carson, C.A., D.M. Sells, and M. Ristic. 1977a. Cell-mediated immune response to virulent and attenuated *Anaplasma marginale* administered to cattle in live and inactivated forms. *American Journal of Veterinary Research* 38:173–179.

Carson, C.A., D.M. Sells, and M. Ristic. 1977b. Cell-mediated immunity to challenge exposure of cattle inoculated with virulent and attenuated strains of *Anaplasma marginale. American Journal of Veterinary Research* 38:1167–1172.

Christensen, J.F., and D.W. McNeal. 1967. *Anaplasma marginale* infection in deer in the Sierra Nevada foothill area of California. *American Journal of Veterinary Research* 28:599–601.

Christensen, J.F., J.W. Osebold, and M.N. Rosen. 1958. Infection and antibody response in deer experimentally infected with *Anaplasma marginale* from bovine carriers. *Journal of the American Veterinary Medical Association* 132:289–292.

Connell, M.L. 1974. Transmission of *Anaplasma marginale* by the cattle tick *Boophilus microplus. Queensland Journal of Agricultural Science* 31:185–193.

Connell, M.L., and W.T.K. Hall. 1972. Transmission of *Anaplasma marginale* by the cattle tick *Boophilus microplus. Australian Veterinary Journal* 48:477.

Crosbie, P.R., W.L. Goff, D. Stiller, D.A. Jessup, and W.M. Boyce. 1997. The distribution of *Dermacentor hunteri* and *Anaplasma* sp. in desert bighorn sheep (*Ovis canadensis*). *Journal of Parasitology* 83:31–37.

De Kroon, J.F., N.M. Perie, F.F.J. Franssen, and G. Uilenberg. 1990. The indirect fluorescent antibody test for bovine anaplasmosis. *Veterinary Quarterly* 12:124–128.

Dikmans, G. 1950. The transmission of anaplasmosis. *American Journal of Veterinary Research* 11:5–16.

Dykstra, R.R., H.F. Lienhardt, C.A. Pyle, and H. Farley. 1938. Studies in anaplasmosis. *Kansas Agricultural Experiment Station Report* 1:1–32.

Eid, G., D.M. French, A.M. Lundgren, A.F. Barbet, T.F. McElwain, and G.H. Palmer. 1996. Expression of major surface protein-2 antigenic variants during acute *Anaplasma marginale* rickettsemia. *Infection and Immunity* 64:836–841.

Eriks, I.S., G.H. Palmer, T.C. McGuire, D.R. Allred, and A.F. Barbet. 1989. Detection and quantification of *Anaplasma marginale* in carrier cattle by using a nucleic acid probe. *Journal of Clinical Microbiology* 27:279–284.

Eriks, I.S., D. Stiller, and G.H. Palmer. 1993. Impact of persistent *Anaplasma marginale* rickettsemia on tick infection and transmission. *Journal of Clinical Microbiology* 31:2091–2096.

Eriks, I.S., D. Stiller, W.L. Goff, M. Panton, S.M. Parish, T.F. McElwain, and G.H. Palmer. 1994. Molecular and biological characterization of a newly isolated *Anaplasma marginale* strain. *Journal of Veterinary Diagnostic Investigation* 6:435–441.

Figueroa, J.V., L.P. Chieves, G.S. Johnson, and G.M. Buening. 1993. Multiplex polymerase chain reaction-based assay for the detection of *Babesia bigemina, Babesia bovis* and *Anaplasma marginale* in bovine blood. *Veterinary Parasitology* 50:69–81.

Foil, L.D., W.V. Adams, J.M. McManus, and C.J. Issel. 1987. Bloodmeal residues on mouthparts of *Tabanus fuscicostatus* (Diptera: Tabanidae) and the potential for mechanical transmission of pathogens. *Journal of Medical Entomology* 24:613–616.

Francis, D.H., G.M. Buening, and T.E. Amerault. 1980. Characterization of immune responses of cattle to erythrocyte stroma, *Anaplasma* antigen, and dodecanoic acid-conjugated *Anaplasma* antigen: Cell-mediated immunity. *American Journal of Veterinary Research* 41:368–371.

Ge, N.L., K.M. Kocan, G.L. Murphy, and E.F. Blouin. 1995. Detection of *Anaplasma marginale* DNA in bovine erythrocytes by slot-blot and in situ hybridization with a PCR-mediated digoxigen-labeled DNA probe. *Journal of Veterinary Diagnostic Investigation* 7:465–472.

Giltner, L.T. 1930. Cattle disease called anaplasmosis. *Yearbook of the United States Department of Agriculture,* p. 156.

Goff, W.L., W.C. Johnson, and K.L. Kuttler. 1985. Development of an indirect fluorescent antibody test, using microfluorometry as a diagnostic test for bovine anaplasmosis. *American Journal of Veterinary Research* 46:1080–1084.

Goff, W.L., A.F. Barbet, D. Stiller, G.H. Palmer, D.P. Knowles, K.M. Kocan, J.R. Gorham, and T.C. McGuire. 1988. Detection of *Anaplasma marginale*-infected tick vectors by using a cloned DNA probe. *Proceedings of the National Academy of Sciences U.S.A.* 85:919–923.

Goff, W.L., D. Stiller, R.A. Roeder, L.W. Johnson, D. Falk, J.R. Gorham, and T.C. McGuire. 1990. Comparison of a DNA probe, complement-fixation and indirect immunofluorescence tests for diagnosing *Anaplasma marginale* in suspected carrier cattle. *Veterinary Microbiology* 24:381–390.

Goff, W.L., D. Stiller, D.A. Jessup, P. Msolla, W.M. Boyce, and W. Foreyt. 1993. Characterization of an *Anaplasma ovis* isolate from desert bighorn sheep. *Journal of Wildlife Diseases* 29:540–546.

Gonzalez, E.F., R.F. Long, and R.A. Todorovic. 1978. Comparisons of the complement-fixation, indirect fluorescent antibody, and card agglutination tests for the diagnosis of bovine anaplasmosis. *American Journal of Veterinary Research* 39:1538–1541.

Hawkins, J.A., J.N. Love, and R.J. Hidalgo. 1981. Mechanical transmission of anaplasmosis by tabanids (Diptera: Tabanidae). In *Proceedings of the seventh national anaplasmosis conference,* ed. R.J. Hildago and E.W. Jones. Starkville: College of Veterinary Medicine, Mississippi State University, pp. 453–462.

Herms, W.B., and D.E. Howell. 1936. The western dog tick, *Dermacentor occidentalis* Neuman, a vector of bovine anaplasmosis in California. *Journal of Parasitology* 22:283–288.

Hidalgo, R.J., and G.T. Dimopoullos. 1967. Complement-fixation microprocedures for anaplasmosis. *American Journal of Veterinary Research* 28:245–251.

Howarth, J.A., T.O. Roby, T.E. Amerault, and D.W. McNeal. 1969. Prevalence of *Anaplasma marginale* infection in California deer as measured by calf inoculation and serologic techniques. *Proceedings of the United States Animal Health Association* 73:136–147.

Howarth, J.A., Y. Hokama, and T.E. Amerault. 1976. The modified card agglutination test: An accurate tool for detecting anaplasmosis in Columbian black-tailed deer. *Journal of Wildlife Diseases* 12:427–434.

Howe, D.L. 1981. Anaplasmosis. In *Infectious diseases of wild mammals,* ed. J.W. Davis, L.H. Karstad, and D.O. Trainer, 2d ed. Ames: Iowa State University Press, pp. 407–414.

Howe, D.L., and W.G. Hepworth. 1965. Anaplasmosis in big game animals: Tests on wild populations in Wyoming. *American Journal of Veterinary Research* 26:1114–1120.

Howe, D.L., W.G. Hepworth, F.M. Blunt, and G.M. Thomas. 1964. Anaplasmosis in big game animals: Experimental infection and evaluation of serologic tests. *American Journal of Veterinary Research* 25:1271–1275.

Howell, D.E., C.E. Sanborn, L.E. Rozeboom, G.W. Stiles, and L.H. Moe. 1941. The transmission of anaplasmosis by horseflies (Tabanidae). *Oklahoma Agricultural and Mechanical College, Agricultural Experiment Station Technical Bulletin* T-11, 23 pp.

Hungerford, L.L., R.D. Smith, and A. Woolf. 1989. The role of white-tailed deer in the epidemiology of bovine anaplasmosis in Illinois. In *Proceedings of the eighth national veterinary hemoparasite conference,* ed. R.J. Hidalgo. St. Louis: C.V. Mosby, pp. 425–438.

Jacobson, R.H., D.E. Worley, and W.W. Hawkins. 1977. Studies on pronghorn antelope (*Antilocapra americana*) as reservoirs of anaplasmosis in Montana. *Journal of Wildlife Diseases* 13:323–326.

Jessup, D.A., W.L. Goff, D. Stiller, M.N. Oliver, V.C. Bleich, and W.M. Boyce. 1993. A retrospective serologic survey for *Anaplasma* spp. infection in three bighorn sheep (*Ovis canadensis*) populations in California. *Journal of Wildlife Diseases* 29:547–554.

Keel, M.K., W.L. Goff, and W.R. Davidson. 1995. An assessment of the role of white-tailed deer in the epizootiology of anaplasmosis in the southeastern United States. *Journal of Wildlife Diseases* 31:378–385.

Kieser, S.R., I.S. Eriks, and G.H. Palmer. 1990. Cyclic rickettsemia during persistent *Anaplasma marginale* infection in cattle. *Infection and Immunity* 58:1117–1119.

Knowles, D.P., S. Torioni de Echaide, G.H. Palmer, T.C. McGuire, D. Stiller, and T.F. McElwain. 1996. Antibody against an *Anaplasma marginale* MSP-5 epitope common to tick and erythrocyte stages identifies persistently infected cattle. *Journal of Clinical Microbiology* 34:2225–2230.

Kocan, K.M. 1986. Development of *Anaplasma marginale* in ixodid ticks: Coordinated development of a rickettsial organism and its tick host. In *Morphology, physiology and behavioral biology of ticks,* ed. J. Sauer and J.A. Hair. Chichester, England: Ellis Horwood, pp. 472–505.

Kocan, K.M., and D. Stiller. 1992. Development of *Anaplasma ovis* (Rickettsiales: Anaplasmataceae) in male *Dermacentor andersoni* (Acari: Ixodidae) transferred from infected to susceptible sheep. *Journal of Medical Entomology* 29:98–107.

Kocan, K.M., J.A. Hair, and S.A. Ewing. 1980. Ultrastructure of *Anaplasma marginale* Theiler in *Dermacentor andersoni* Stiles and *Dermacentor variabilis* (Say). *American Journal of Veterinary Research* 41:1966–1976.

Kocan, K.M., S.J. Barron, S.A. Ewing, and J.A Hair. 1985. Transmission of *Anaplasma marginale* by adult *Dermacentor andersoni* during feeding on calves. *American Journal of Veterinary Research* 46:1565–1567.

Kocan, K.M., K.B. Wickwire, S.A. Ewing, J.A. Hair, and S.J. Barron. 1988. Preliminary studies on the development of *Anaplasma marginale* in salivary glands of adult, feeding *Dermacentor andersoni* ticks. *American Journal of Veterinary Research* 49:1010–1014.

Kocan, K.M., R.W. Stich, P.L. Claypool, S.A. Ewing, J.A. Hair, and S.J. Barron. 1990. Intermediate site of development of *Anaplasma marginale* in feeding adult *Dermacentor andersoni* ticks that were infected as nymphs. *American Journal of Veterinary Research* 51:128–132.

Kocan, K.M., W.L. Goff, D. Stiller, P.L. Claypool, W. Edwards, S.A. Ewing, J.A. Hair, and S.J. Barron. 1992a. Persistence of *Anaplasma marginale* (Rickettsiales: Anaplasmataceae) in male *Dermacentor andersoni* (Acari: Ixodidae) transferred successively from infected to susceptible calves. *Journal of Medical Entomology* 29:657–668.

Kocan, K.M., D. Stiller, W.L. Goff, P.L. Claypool, W. Edwards, S.A. Ewing, T.C. McGuire, J.A. Hair, and S.J. Barron. 1992b. Development of *Anaplasma marginale* in male *Dermacentor andersoni* transferred from parasitemic to susceptible cattle. *American Journal of Veterinary Research* 53:499–507.

Kocan, K.M., W.L. Goff, D. Stiller, W. Edwards, S.A. Ewing, P.L. Claypool, T.C. McGuire, J.A. Hair, and S.J. Barron. 1993. Development of *Anaplasma marginale* in salivary glands of male *Dermacentor andersoni. American Journal of Veterinary Research* 54:107–112.

Kollars, T.M. 1996. Variation in infestation by *Ixodes scapularis* (Acari: Ixodidae) between adjacent upland and lowland populations of *Odocoileus virginianus* (Mammalia: Cervidae) in western Tennessee. *Journal of Entomological Science* 31:286–288.

Krueger, C.M., and G.M. Buening. 1988. Isolation and restriction endonuclease cleavage of *Anaplasma marginale* DNA *in situ* in agarose. *Journal of Clinical Microbiology* 26:906–910.

Kuttler, K.L. 1965. Serological survey of anaplasmosis incidence in East Africa, using the complement-fixation test. *Bulletin of Epizootic Diseases in Africa* 13:2557–262.

———. 1979. Current anaplasmosis control techniques in the United States. *Journal of the South African Veterinary Association* 50:314–320.

———. 1981a. Infection of splenectomized calves with *Anaplasma ovis*. *American Journal of Veterinary Research* 42:2094–2096.

———. 1981b. Anaplasmosis. In *Diseases and parasites of white-tailed deer,* ed. W.R. Davidson, F.A. Hayes, V.F. Nettles, and F.E. Kellogg. Tallahassee, FL: Tall Timbers Research Station, pp. 126–137; Miscellaneous Publications 7.

———. 1984. *Anaplasma* infections in wild and domestic ruminants: A review. *Journal of Wildlife Diseases* 20:12–20.

Kuttler, K.L., and J.P. Kreier. 1986. Trypanosomiasis, babesiosis, theileriosis, and anaplasmosis. In *Chemotherapy of parasitic diseases,* ed. W.C. Campbell and R.S. Rew. New York: Plenum, pp. 171–191.

Kuttler, K.L., R.M. Robinson, and W.P. Rogers. 1967. Exacerbation of latent erythrocytic infections in deer following splenectomy. *Canadian Journal of Comparative Medical and Veterinary Science* 31:317–319.

Kuttler, K.L., R.M. Robinson, and T.E. Franklin. 1968. Serological response to *Anaplasma marginale* infection in splenectomized deer (*Odocoileus virginianus*) as measured by complement-fixation and capillary-tube agglutination tests. In *Proceedings of the fifth national anaplasmosis conference, Stillwater, Oklahoma,* pp. 82–88.

Kuttler, K.L., O.H. Graham, and S.R. Johnson. 1971. Apparent failure of *Boophilus annulatus* to transmit anaplasmosis to white-tailed deer (*Odocoileus virginianus*). *Journal of Parasitology* 57:657–659.

Lancaster, J.L., Jr., H. Roberts, L. Lewis, L. Dinkins, and J. DeVaney. 1968. Review of anaplasmosis transmission trials with the white-tailed deer. *In Proceedings of the fifth national anaplasmosis conference, Stillwater, Oklahoma,* pp. 197–215.

Lawrence, J.A., and A.J. de Vos. 1990. Methods currently used for the control of anaplasmosis and babesiosis: Their validity and proposals for future control strategies. *Parasitologia* 32:63–71.

Lestoquard, F. 1931. Transmission au buffle des piroplasmes du boeuf: Modifications subies par *Anaplasma marginale. Bulletin de la Societe de Pathologie Exotique* 24:820–822.

Lignieres, J. 1919. La vaccination des bovides centre l'anaplasmose. *Bulletin de la Societe de Pathologie Exotique* 12:765–774.

Lohr, K.F., and H. Meyer. 1973. Game anaplasmosis: The isolation of *Anaplasma* organisms from antelope. *Zeitschrift fur Tropenmedizin und Parasitolgie* 24:192–197.

Lohr, K.F., J.P.J. Ross, and H. Meyer. 1974. Detection in game of fluorescent and agglutination antibodies to intraerythrocytic organisms. *Tropenmedizin und Parasitologie* 25:217–226.

Lotze, J.C. 1944. Carrier cattle as a source of infective material for horsefly transmission of anaplasmosis. *American Journal of Veterinary Research* 5:164–165.

Lotze, J.C., and M.J. Yiengst. 1941. Mechanical transmission of bovine anaplasmosis by the horsefly *Tabanus sulcifrons. American Journal of Veterinary Research* 2:323–326.

Maas, J., G.M. Buening, and W. Porath. 1981. Serologic evidence of *Anaplasma marginale* infection in white-tailed deer (*Odocoileus virginianus*) in Missouri. *Journal of Wildlife Diseases* 17:45–47.

Magonigle, R.A., and W.P. Eckblad. 1979. Evaluation of the anaplasmosis rapid card agglutination test for detecting experimentally infected elk. *Cornell Veterinarian* 69:402–410.

Matthews, M. 1978. Anaplasmosis in a Siberian ibex. *Journal of Zoological Animal Medicine* 9:148.

Meltzer, M.I., R.A. Norval, and P.L. Donachie. 1995. Effects of tick infestation and tick-borne disease infections (heartwater, anaplasmosis and babesiosis) on the lactation and weight gain of Mashona cattle in south-eastern Zimbabwe. *Tropical Animal Health and Production* 27:129–144.

Morley, R.S., and M.E. Hugh-Jones. 1989. Seroepidemiology of *Anaplasma marginale* in white-tailed deer (*Odocoileus virginianus*) from Louisiana. *Journal of Wildlife Diseases* 25:342–346.

Morris, H.J., A. Martin, and W.T. Oglesby. 1936. An attempt to transmit anaplasmosis by biting flies. *Journal of the American Veterinary Medical Association* 89:169–175.

Ndung'u, L.W., C. Aguirre, F.R. Rurangirwa, T.F. McElwain, T.C. McGuire, D.P. Knowles, and G.H. Palmer. 1995. Detection of *Anaplasma ovis* infection in goats by major surface protein-5 competitive inhibition enzyme-linked immunosorbent assay. *Journal of Clinical Microbiology* 33:675–679.

Neitz, W.O. 1935. Bovine anaplasmosis: The transmission of *Anaplasma marginale* to a black wildebeest (*Connochaetes gnu*). *Onderstepoort Journal of Veterinary Science* 5:8–11.

———. 1939. Ovine anaplasmosis: The transmission of *Anaplasma ovis* and *Eperythrozoon ovis* to blesbuck (*Damaliscus albifrons*). *Onderstepoort Journal of Veterinary Science and Animal Industry* 13:9–16.

———. 1956. A consolidation of our knowledge of the transmission of tick-borne diseases. *Onderstepoort Journal of Veterinary Research* 27:115–163.

Neitz, W.O., and P.J. DuToit. 1932. Bovine anaplasmosis: A method of obtaining pure strains of *Anaplasma marginale* and *Anaplasma centrale* by transmission through antelopes. In *18th Report of the Division of Veterinary Services and Animal Industry, Onderstepoort, Union of South Africa.* Onderstepoort, pp. 3–20.

Osebold, J.W., J.F. Christensen, W.M. Longhurst, and M.N. Rosen. 1959. Latent *Anaplasma marginale* infection in wild deer demonstrated by calf inoculation. *Cornell Veterinarian* 49:97–115.

Osebold, J.W., J.R. Douglas, and J.F. Christensen. 1962. Transmission of anaplasmosis to cattle by ticks obtained from deer. *American Journal of Veterinary Research* 23:21–23.

Peterson, K.J., and T.O. Roby. 1975. Absence of *Anaplasma marginale* infection in American bison raised in an anaplasmosis endemic area. *Journal of Wildlife Diseases* 11:395–397.

Peterson, K.J., T.P. Kistner, and H.E. Davis. 1973. Epizootiologic studies of anaplasmosis in Oregon mule deer. *Journal of Wildlife Diseases* 9:314–319.

Piercy, P.L. 1956. Transmission of anaplasmosis. *Annals of the New York Academy of Sciences* 64:40–48.

Potgieter, F.T. 1979. Epizootiology and control of anaplasmosis in South Africa. *Journal of South African Veterinary Association* 50:367–372.

Renshaw, H.W., H.W. Vaughn, R.A. Magonigle, W.C. Davis, E.H. Stauber, and F.W. Frank. 1977. Evaluation of free-roaming mule deer as carriers of anaplasmosis in an area of Idaho where bovine anaplasmosis is enzootic. *Journal of the American Veterinary Medical Association* 170:334–339.

Renshaw, H.W., R.A. Magonigle, and H.W. Vaughn. 1979. Evaluation of the anaplasmosis rapid card agglutination test for detecting experimentally-infected elk. *Journal of Wildlife Diseases* 15:379–386.

Ristic, M. 1962. A capillary tube-agglutination test for anaplasmosis: A preliminary report. *Journal of the American Veterinary Medical Association* 141:588–594.

———. 1968. Anaplasmosis. In *Infectious blood diseases of man and animals,* vol. 2, ed. D. Weinma, and M. Ristic. New York: Academic, pp. 473–542.

Ristic, M., and A.M. Watrach. 1961. Studies in anaplasmosis: II. Electron microscopy of *Anaplasma marginale* in deer. *American Journal of Veterinary Research* 22:109–116.

Rogers, R.J., and I.A. Shiels. 1979. Epidemiology and control of anaplasmosis in Australia. *Journal of the South African Veterinary Association* 50:363–366.

Sanborne, C.E., G.W. Stiles, and L.H. Moe. 1932. Preliminary experiments in the transmission of anaplasmosis by horseflies. *Oklahoma Agricultural and Mechanical College, Agricultural Experiment Station Bulletin* 204, 15 pp.

Sanders, D.A. 1933. Notes on the experimental transmission of bovine anaplasmosis in Florida. *Journal of the American Veterinary Medical Association* 83:799–805.

Sharma, S.P. 1987. Characterization of *Anaplasma marginale* infection in buffaloes. *Indian Journal of Animal Sciences* 57:76–78.

Shompole, S., O. Suryakant, R. Waghela, F. Rurangirwa, and T.C. McGuire. 1989. Cloned DNA probes identify *Anaplasma ovis* in goats and reveal a high prevalence of infection. *Journal of Clinical Microbiology* 27:2730–2735.

Smith, T., and F.L. Kilborne. 1893. Investigations into the nature, causation, and prevention of Texas or southern cattle fever. *United States Department of Agriculture, Bureau of Animal Industry Bulletin* 1:1–301.

Smith, K., D.W. Brocklesy, P. Bland, R.E. Purnell, C.G.D. Brown, and R.C. Payne. 1974. The fine structure of intraerythrocytic stages of *Theileria gorgonis* and a strain of *Anaplasma marginale* isolated from wildebeest (*Connochaetes taurinus*). *Tropenmedizin und Parasitologie* 25:293–300.

Smith, R.D., A. Woolf, L.L. Hungerford, and J.P. Sundberg. 1982. Serologic evidence of *Anaplasma marginale* infection in Illinois white-tailed deer. *Journal of the American Veterinary Medical Association* 181:1254–1256.

Splitter, E.J., H.D. Anthony, and M.J. Twiehaus. 1956. *Anaplasma ovis* in the United States: Experimental studies with sheep and goats. *American Journal of Veterinary Research* 17:487–491.

Stich, R.W., K.M. Kocan, G.H. Palmer, S.A. Ewing, J.A. Hair, and S.J. Barron. 1989. Transstadial and attempted transovarial transmission of *Anaplasma marginale* by *Dermacentor variabilis*. *American Journal of Veterinary Research* 50:1377–1380.

Stich, R.W., J.A. Bantle, K.M. Kocan, and A. Fekete. 1993a. Detection of *Anaplasma marginale* (Rickettsiales: Anaplasmataceae) in hemolymph of *Dermacentor andersoni* (Acari: Ixodidae) with the polymerase chain reaction. *Journal of Medical Entomology* 30:781–788.

Stich, R.W., J.R. Sauer, J.A. Bantle, and K.M. Kocan. 1993b. Detection of *Anaplasma marginale* (Rickettsiales: Anaplasmataceae) in secretagogue-induced oral secretions of *Dermacentor andersoni* (Acari: Ixodidae) with the polymerase chain reaction. *Journal of Medical Entomology* 30:789–794.

Stiller, D. 1990. Application of biotechnology for the diagnosis and control of ticks and tick-borne diseases. *Parasitologia* 32:87–111.

Stiller, D., and M.E. Coan. 1995. Recent developments in elucidating tick vector relationships for anaplasmosis and equine piroplasmosis. *Veterinary Parasitology* 57:97–108.

Stiller, D., G. Leatch, and K.L. Kuttler. 1981. *Dermacentor albipictus* (Packard): An experimental vector of bovine anaplasmosis. *Proceedings of the United States Animal Health Association* 85:65–73.

Stiller, D., L.W. Johnson, and K.L. Kuttler. 1983. Experimental transmission of *Anaplasma marginale* Theiler by males of *Dermacentor albipictus* Packard and *Dermacentor occidentalis* Marx (Acari: Ixodidae). *Proceedings of the United States Animal Health Association* 87:59–65.

———. 1987. Experimental transmission of *Anaplasma ovis* to sheep by *Dermacentor albipictus* and *D. variabilis* male ticks. In *Proceedings of the eighth annual western conference of food animal veterinary medicine, Boise, Idaho.* Boise: University of Idaho. Abstract.

Stiller, D., W.L. Goff, S.P. Shompole, L.W. Johnson, H. Glimp, F.R. Rurangirwa, J.R. Gorham, and T.C. McGuire. 1989a. *Dermacentor andersoni* Stiles: A natural vector of *Anaplasma ovis* Lestoquard on sheep in Idaho. In *Proceedings of the eighth national veterinary hemoparasite disease conference, St. Louis, Missouri,* p. 183.

Stiller, D., K.M. Kocan, W. Edwards, S.A. Ewing, J.A. Hair, and S.J. Barron. 1989b. Detection of colonies of *Anaplasma marginale* in salivary glands of three *Dermacentor* spp infected as nymphs or adults. *American Journal of Veterinary Research* 50:1381–1385.

Stiller, D., P. Msolla, W. Foreyt, W.L. Goff, S.P. Shompole, W.C. Johnson, and G. Sun. 1990. Experimental transmission of *Anaplasma ovis* Lestoquard between bighorn sheep and domestic sheep by males of the tick *Dermacentor andersoni* Stiles. In *Proceedings of the 71st conference of research workers in animal disease, Chicago, Illinois,* ed. R.A. Packer, p. 27. Abstract 156.

Stiller, D., P.R. Crosbie, W.M. Boyce, and W.L. Goff. 1999. *Dermacentor hunteri* (Acari: Ixodidae): An experimental vector of *Anaplasma marginale* and *A. ovis* (Rickettsiales: Anaplasmataceae) for calves and sheep. *Journal of Medical Entomology* 36:321–324.

Summers, W.A., and L.L. Gonzalez. 1965. Attempts to transmit bovine anaplasmosis to small laboratory animals. *Experimental Parasitology* 16:57–63.

Taylor, S.K., V.M. Lane, D.L. Hunter, K.G. Eyre, S. Frye, and M.R. Johnson. 1997. Serologic survey for infectious pathogens in free-ranging American bison. *Journal of Wildlife Diseases* 33:308–311.

Theiler, A. 1910. Anaplasma marginale *(gen. and spec. nov.): The marginal points in the blood of cattle suffering from a specific disease—Governmental Report on Veterinary Bacteriology, 1908–1909.* Transvaal, South Africa: Department of Agriculture, pp. 7–64.

———. 1911. Further investigations into anaplasmosis of South African cattle. Union of South Africa, Department of Agriculture, pp. 7–46; First Report of the Director of Veterinary Research.

Thomas, S.E., D.E. Wilson, and T.E. Mason. 1982. *Babesia, Theileria* and *Anaplasma* spp. infecting sable antelope, *Hippotragus niger* (Harris, 1838), in southern Africa. *Onderstepoort Journal of Veterinary Research* 49:163–166.

Tibbitts, T., W.L. Goff, W. Foreyt, and D. Stiller. 1992. Susceptibility of two Rocky Mountain bighorn sheep to experimental infection with *Anaplasma ovis. Journal of Wildlife Diseases* 28:125–129.

Torioni de Echaide, S., D.P. Knowles, T.C. McGuire, G.H. Palmer, C.E. Suarez, and T.F. McElwain. 1998. Detection of cattle naturally infected with *Anaplasma marginale* in a region of endemicity by nested PCR and a competitive enzyme-linked immunosorbent assay using recombinant major surface protein 5. *Journal of Clinical Microbiology* 36:777–782.

Uilenberg, G., C.J. van Vorstenbosch, and N.M. Perie. 1979. Blood parasites of sheep in the Netherlands: I. *Anaplasma mesaeterum* sp. n. (Rickettsiales: Anaplasmataceae). *Tijdschrift voor Diergeneeskunde* 104:14–22.

Visser, E.S., T.C. McGuire, G.H. Palmer, W.C. Davis, V. Shkap, E. Pipano, and D.P. Knowles. 1992. The *Anaplasma marginale* MSP-5 gene encodes a 19-kilodalton protein conserved in all recognized *Anaplasma* species. *Infection and Immunity* 60:5139–5144.

Waldrup, K.A., E. Collisson, S.E. Bentsen, C.K. Winkler, and G.G. Wagner. 1989. Prevalence of erythrocytic protozoa and serologic reactivity to selected pathogens in deer in Texas. *Preventative Veterinary Medicine* 7:49–58.

Walker, D.H., and J.S. Dumler. 1996. Emergence of the ehrlichioses as human health problems. *Emerging Infectious Diseases* 2:18–29.
Wilson, B.H., and R.B. Meyer. 1966. Transmission studies of bovine anaplasmosis with the horseflies, *Tabanus fuscicostatus* and *Tabanus nigrovittatus*. *American Journal of Veterinary Research* 27:367–369.
Wyatt, C.R., W.C. Davis, D.P. Knowles, W.L. Goff, G.H. Palmer, and T.C. McGuire. 1996. Effect on intraerythrocytic *Anaplasma marginale* of soluble factors from infected calf blood mononuclear cells. *Infection and Immunity* 64:4846–4849.
Zaugg, J.L. 1986. Experimental anaplasmosis in American bison: Persistence of infections of *Anaplasma marginale* and non-susceptibility to *A. ovis*. *Journal of Wildlife Diseases* 22:169–172.
———. 1987. Experimental infections of *Anaplasma ovis* in pronghorn antelope. *Journal of Wildlife Diseases* 23:205–210.
———. 1988. Experimental anaplasmosis in mule deer: Persistence of infection of *Anaplasma marginale* and susceptibility to *A. ovis*. *Journal of Wildlife Diseases* 24:120–126.
Zaugg, J.L., and K.L. Kuttler. 1985. *Anaplasma marginale* infections in American bison: Experimental infection and serologic study. *American Journal of Veterinary Research* 46:438–441.
Zaugg, J.L., and B.A. Newman. 1985. Evaluation of jackrabbits as nonruminant hosts for *Anaplasma marginale*. *American Journal of Veterinary Research* 46:669–670.
Zaugg, J.L., D. Stiller, M.E. Coan, and S.D. Lincoln. 1986. Transmission of *Anaplasma marginale* Theiler by males of *Dermacentor andersoni* Stiles fed on an Idaho field-infected, chronic carrier cow. *American Journal of Veterinary Research* 47:2269–2271.
Zaugg, J.L., W.L. Goff, W. Foreyt, and D.L. Hunter. 1996. Susceptibility of elk (*Cervus elaphus*) to experimental infection with *Anaplasma marginale* and *A. ovis*. *Journal of Wildlife Diseases* 32:62–66.

EHRLICHIOSES

WILLIAM R. DAVIDSON, JACQUELINE E. DAWSON, AND SIDNEY A. EWING

Synonyms: Canine ehrlichiosis (tropical canine pancytopenia), bovine petechial fever (Ondiri disease), equine granulocytotropic ehrlichiosis (equine granulocytic ehrlichiosis), equine monocytotropic ehrlichiosis (equine monocytic ehrlichiosis, Potomac horse fever), sennetsu ehrlichiosis (sennetsu rickettsiosis, sennetsu fever), human monocytotropic ehrlichiosis (human monocytic ehrlichiosis), human granulocytotropic ehrlichiosis (human granulocytic ehrlichiosis).

INTRODUCTION. Ehrlichioses are infectious, noncontagious diseases of mammals and are caused by bacteria assigned to the genus *Ehrlichia.* Ehrlichioses of veterinary importance, most notably among dogs *Canis familiaris,* cattle *Bos taurus,* sheep *Ovis aries,* goats *Capra hircus,* and horses *Equus caballus,* have been recognized for decades. Ehrlichial diseases also occur in humans, but, in contrast to domestic animals, two of the three human ehrlichioses have been recognized only recently. Compared with domestic animals and humans, relatively little is known about ehrlichial infections among wild mammals, and nearly all research on wild mammals has involved assessments of their roles in the epidemiology of the ehrlichioses of veterinary or human health importance. Thorough discussion of each ehrlichiosis affecting domestic animals or humans is beyond the scope of this chapter. Instead, because most ehrlichial diseases have many pathologic similarities and are diagnosed by using similar methods, only a general discussion of clinical signs, pathogenesis, pathology, immunity, diagnosis, treatment, and control is presented. Where available, specific information on ehrlichial infections in wild mammals is presented, particularly in regard to what is known of their role in the epidemiology of these diseases.

HISTORY. Infection by and clinical disease resulting from infection by organisms currently classified as *Ehrlichia* spp. were first recognized among dogs and ruminants during the 1930s, although the organisms were classified as members of the genus *Rickettsia* at that time. The genus *Ehrlichia* was established in 1945, in honor of the German bacteriologist Paul Ehrlich, with *Ehrlichia canis* as the type species. Subsequently, several additional species were described from hosts representing diverse taxa. The first human ehrlichial disease reported was sennetsu fever, which was described in Japan in 1954; like the type species, the etiologic agent was first considered to be a *Rickettsia* but was reclassified as an *Ehrlichia* in 1984. Although the ehrlichial diseases known as tick-borne fever, bovine ehrlichiosis, and bovine petechial fever were recognized in Europe, Africa, and the Middle East by the early 1960s, ehrlichioses of veterinary importance generally received little attention until the late 1960s, when canine ehrlichiosis wrought havoc among U.S. military dogs working in Vietnam. The etiologic agent was subsequently identified as *E. canis.*

Recognition of equine monocytic ehrlichiosis (Potomac horse fever) in the United States during the late 1970s and early 1980s provided the next major impetus for research on ehrlichial diseases. During the 1980s and 1990s, research on ehrlichioses increased dramatically as a result of the detection in the United States of human monocytotropic ehrlichiosis [HME (Maeda et al. 1987)] and human granulocytotropic ehrlichiosis (HGE). The ehrlichioses are included among those diseases described as *emerging infectious diseases* (Walker and Dumler 1996).

ETIOLOGY. Members of the genus *Ehrlichia* are obligate, intracellular bacteria that for the most part exhibit strong tropism for leukocytes, although some species infect other cell types. Ehrlichieae are classified in the order Rickettsiales, the family Rickettsiaceae, within the tribe Ehrlichieae (Ristic and Huxsoll 1984). With light microscopy, ehrlichieae appear as small, Gram-negative cocci contained within intracytoplasmic phagosomes of host leukocytes (Fig. 27.3). The organisms reproduce by binary fission within these phagosomes. Because of their "berrylike" appearance these organismal aggregates are termed morulae.

Eight species of *Ehrlichia* have been formally described. Many others, including those listed in *Bergey's Manual of Determinative Bacteriology* (Ristic and Huxsoll 1984), are considered to be *species incertae sedis.* More recently described organisms such as the HGE agent and an *Ehrlichia*-like agent from white-tailed deer *Odocoileus virginianus* also are thought to be members of the *Ehrlichieae* (Chen et al. 1994; Dawson et al. 1996a). Based on phylogenetic analyses of 16S rRNA genes, members of the genus *Ehrlichia* examined to date have been separated into three clusters of species termed *genogroups* (Rikihisa 1991; Pretzman et al. 1995; Wen et al. 1995; Walker and Dumler 1996). However, each genogroup also contains species presently assigned to other genera, viz., *Cowdria ruminantium,* causative agent of heartwater; *Anaplasma marginale,* causative agent of bovine anaplasmosis; and *Neorickettsia helminthoeca,* causative agent of salmon poisoning disease. Consequently, Pretzman et al. (1995) characterized the genus *Ehrlichia* as phylogenetically incoherent and recommended taxonomic reevaluation of members of the tribe Ehrlichieae.

HOST RANGE. Originally, most species of *Ehrlichia* were believed to occur in one species of vertebrate host or at least in closely related hosts. Although the concept of host specificity has held for some species of *Ehrlichia,* others are now known to infect hosts that are taxonomically distant (Table 27.4).

The etiologic agents of two of the three human ehrlichioses, *Ehrlichia chaffeensis* and the HGE agent, are zoonotic organisms that are maintained in nature among other mammals; the epidemiology of sennetsu ehrlichiosis in humans remains unknown. Nine of the named or proposed species of *Ehrlichia* are known to infect wild mammals under natural or experimental conditions.

EPIDEMIOLOGY. Each of the *Ehrlichia* spp. for which the epidemiology has been documented are transmitted transstadially by ixodid ticks. The white-tailed deer *Ehrlichia*-like agent also is suspected to use a tick vector (Table 27.4). Transovarial transmission by tick vectors has not been demonstrated.

Interestingly, *Ehrlichia sennetsu* and *Ehrlichia risticii,* species for which the method of natural transmission is not known, are phylogenetically distinct from other ehrlichieae and are more closely related to *Neorickettsia helminthoeca, N. elokominica,* causative agent of Elokomin fluke fever, and the *Stellantchasmus*

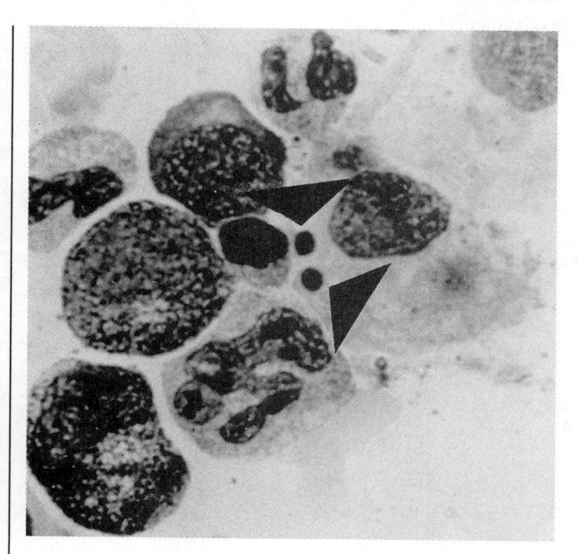

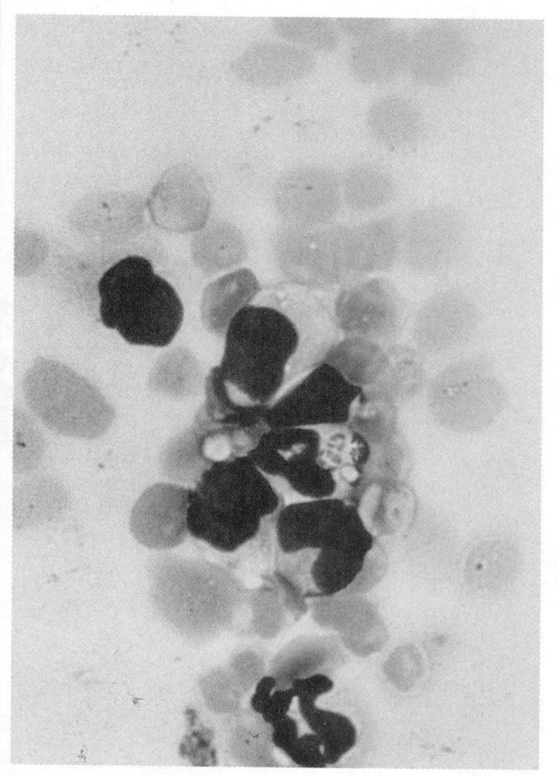

FIG. 27.3—Morulae of *Ehrlichia* organisms within circulating leukocytes. (A) *Ehrlichia chaffeensis* within a human monocyte. (B) Human granulocytotropic ehrlichiosis agent within a mouse neutrophil.

TABLE 27.4—Characteristics associated with various *Ehrlichia* spp.

Disease	Etiologic Agent	Vertebrate Hosts[a]	Cell Tropism	Vectors	Geographic Distribution
Canine ehrlichiosis (tropical canine pancytopenia)	*E. canis*	CH: dogs NH: dogs, silver-backed jackal *Canis mesomelas,* gray wolf *C. lupus* EH: African wild dog *Lycaon pictus,* coyote *C. latrans,* red fox *Vulpes vulpes,* gray fox *Urocyon cinereoargenteus*	Monocytes,macrophages	*Rhipicephalus sanguineus, Dermacentor variabilis*[b]	Worldwide
Bovine petechial fever (Ondiri disease)	*Cytoecetes ondiri*[c]	CH: cattle, sheep NH: bushbuck *Tragelaphus scriptus* EH: Cape buffalo *Syncerus caffer,* wildebeest *Connochaetes gnou*	Granulocytes	Unknown	Kenya
Equine granulocytotropic ehrlichiosis	*E. equi*	CH: horses NH: unknown EH: cats, dogs, sheep, goats, nonhuman primates	Neutrophils	*Ixodes scapularis, I. pacificus*	United States
Equine monocytotropic ehrlichiosis (Potomac horse fever)	*E. risticii*	CH: horses NH: unknown EH: laboratory mice, cats, dogs, nonhuman primates	Monocytes, macrophages, enterocytes, mast cells	Uncertain, possibly snails (*Juga*) or trematodes	North America
None	*E. muris*	CH: none known NH: vole *Eothenomys kageus* EH: laboratory mice	Monocytes, macrophages	Unknown	Japan

(continued)

TABLE 27.4—*Continued*

Disease	Etiologic Agent	Vertebrate Hosts[a]	Cell Tropism	Vectors	Geographic Distribution
Sennetsu ehrlichiosis (sennetsu rickettsiosis, sennetsu fever)	*E. sennetsu*	CH: humans NH: unknown EH: laboratory mice, nonhuman primates	Monocytes	Unknown	Japan, Malaysia
Human monocytotropic ehrlichiosis	*E. chaffeensis*	CH: humans NH: white-tailed deer *Odocoileus virginianus,* dogs EH: red foxes *Vulpes vulpes,* laboratory mice	Monocytes, macrophages, rarely lymphocytes, rarely neutrophils	*Amblyomma americanum*	United States, Africa?, Europe?
Human granulocytotropic ehrlichiosis	Unnamed *Ehrlichia* sp. (HGE agent)[d]	CH: humans NH: white-footed mice *Peromyscus leucopus,* white-tailed deer? EH: horses, dogs, laboratory mice	Granulocytes	*Ixodes scapularis*	United States, Europe
None	Unnamed *Ehrlichia*-like agent[c]	CH: none known NH: white-tailed deer EH: none reported	Unknown	*Amblyomma americanum(?)*	United States

[a]CH, host developing clinical disease; NH, natural host with inapparent infection; EH, experimentally susceptible host.
[b]Experimentally infected.
[c]Species *incertae sedis.*
[d] Telford et al. (1996) suggest that the human granulocytotropic ehrlichiosis (HGE) agent may be the same as *Cytoecetes microti* as described by Tyzzer in 1938.

Adapted from Grootenhuis (1981), Rikihisa (1991), Wen et al. (1995), Dawson et al. (1996a), Telford et al. (1996), Greig et al. (1996), Walker and Dumler (1996), Little et al. (1997), and Reubel et al. (1998).

falcatus (SF) agent (Pretzman et al. 1995). Each of the latter three organisms is associated with or transmitted by ingestion of metacercarial stage of digenetic trematodes (flukes) that use fish as a second intermediate host (Walker and Dumler 1996). Thus, the possible transmission of *E. risticii* by trematodes has been examined with encouraging results (Reubel et al. 1998).

Canine Ehrlichiosis. *Ehrlichia canis,* which is the type species of the genus and causative agent of classic canine ehrlichiosis, occurs in domestic dogs in many regions of the world and is transmitted by *Rhipicephalus sanguineus,* the brown dog tick (Ewing and Buckner 1965; Ewing 1969). Dogs are the principal vertebrate reservoir (Ewing 1969); however, the contribution of wild canids in perpetuation of *E. canis* has not been defined. Naturally acquired *E. canis* infection has been confirmed in a single captive gray wolf *Canis lupus* in the United States (Harvey et al. 1979). In Kenya, *E. canis* was demonstrated in 8 of 16 free-ranging wild black-backed jackals *Canis mesomelas,* and at least 5 of these jackals also were parasitized by *R. sanguineus* (Price and Karstad 1980). In a later study in Kenya, only 1 of 36 jackals had detectable *E. canis*-reactive antibodies (Alexander et al. 1994). Experimental infections of *E. canis* have been established in black-backed jackals, African wild dogs *Lycaon pictus* (Van Heerden 1979), coyotes *Canis latrans* (Ewing et al. 1964), red foxes *Vulpes vulpes,* and gray foxes *Urocyon cinereoargenteus* (Amyx and Huxsoll 1973).

Tick-Borne Fever. Tick-borne fever, caused by *Ehrlichia phagocytophila,* occurs among sheep, goats, and cattle in Europe (MacLeod and Gordon 1933; Foggie 1951; Woldehiwet 1983) and possibly horses and dogs in Sweden (Johansson et al. 1995; Engvall et al. 1996). *Ixodes ricinus* has been confirmed as a vector in England and on the European continent (MacLeod and Gordon 1933; Woldehiwet 1983; Webster and Mitchell 1989). Granulocytic ehrlichieae, similar to *E. phagocytophila,* have been reported in red deer *Cervus elaphus,* fallow deer *Dama dama,* and roe deer *Capreolus capreolus* (MacDiarmid 1965; Bell 1981); however, in light of data from recently developed molecular and serologic techniques, the identity of the granulocytic ehrlichieae reported from these cervids should be considered questionable.

Bovine Petechial Fever. This ehrlichial infection of cattle, also known as Ondiri disease, has been reported only from Kenya. The etiologic agent, *Ehrlichia ondiri,* was detected in the blood from 43% of healthy bushbuck *Tragelaphus scriptus* examined from an enzootic site. Eight other species of wild ruminants from the same location showed no evidence of being infected [Sileo et al. (1976 unpublished), cited by Grootenhuis (1981)]. Persistently infected bushbuck are considered to be the reservoir hosts. The means of transmission is not known, and attempts to transmit *E. ondiri* experimentally using ticks, trombiculid mites, and blood-feeding diptera have been unsuccessful (Piercy 1953; Walker et al. 1974). The disease is seasonal among cattle and occurs in high-altitude habitats with dense understory vegetation (Plowright 1962; Walker et al. 1974).

Equine Granulocytotropic Ehrlichiosis. The epidemiology of *Ehrlichia equi,* the causative agent of equine granulocytotropic ehrlichiosis, clearly involves at least some *Ixodes* spp. as vectors; however, the epidemiology of this disease is unclear because of the uncertain taxonomic relationship between *E. equi* and other ehrlichieae within the *E. phagocytophila* genogroup. *Ixodes pacificus* is the vector of *E. equi* in California (Richter et al. 1996). Vectors of *E. equi* in the eastern half of the United States and in Europe, although unconfirmed, are suspected to be *Ixodes* sp. (Richter et al. 1996; Walker and Dumler 1996). Telford et al. (1996) confirmed that *I. scapularis* (*I. dammini*) was a vector of, and white-footed mice *Peromyscus leucopus* were a reservoir for, the closely related HGE agent in Massachusetts.

Equine Monocytotropic Ehrlichiosis (Potomac Horse Fever). The natural history of *E. risticii,* the etiologic agent of this disease, is largely unknown but may involve trematodes (Reubel et al. 1998). To date, wild mammals have not been implicated in the epidemiology of this disease. Surveys of wild rodents from farms where equine monocytotropic ehrlichiosis was endemic have failed to detect *E. risticii*-reactive antibodies in white-footed mice, Norway rats *Rattus norvegicus,* house mice *Mus musculus,* or meadow voles *Microtus pennsylvanicus* (Carroll et al. 1989; Perry et al. 1989).

***Ehrlichia muris* Infection.** Wen et al. (1995) described *E. muris* from a naturally infected free-ranging vole (*Eothenomys kageus*) from Japan. The organism was originally isolated in 1993 and designated as strain AS145^T by Kawahara et al. (1993). The epidemiology of *E. muris* has not been investigated beyond its presence in this single animal.

Human Monocytotropic Ehrlichiosis. There is convincing evidence that *E. chaffeensis* (Dawson et al. 1991), the causative agent of human monocytotropic ehrlichiosis, is maintained naturally in an epidemiologic cycle involving white-tailed deer as a primary reservoir host and the lone star tick *Amblyomma americanum* as a principal vector. High levels of *E. chaffeensis*-reactive antibodies have been found in free-ranging white-tailed deer at numerous locations in the eastern United States (Dawson et al. 1994a, 1996a; Lockhart et al. 1995, 1996, 1997b; Little et al. 1997). Dawson et al. (1994b) experimentally infected deer with a human isolate of *E. chaffeensis* (Arkansas strain) and documented a rickettsemia of at least 2 weeks' duration and production of circulating antibodies, but the deer remained clinically normal. Field surveys have detected *E. chaffeensis* 16S rDNA sequences in blood, spleen, or lymph node from 20%–53% of wild deer from seropositive populations in Florida, Georgia, Missouri, and South Carolina (Little et al. 1997, 1998; Lockhart et al. 1997a,b; W.R. Davidson unpublished). Conclusive proof that deer are a reservoir host was obtained when *E. chaffeensis* was isolated from the blood of 14% of 35 deer from two of three deer populations in Georgia (Lockhart et al. 1997b). Further studies resulted in the isolation of *E. chaffeensis* from lymph node of one of five deer from another population in Georgia (Little et al. 1998).

White-tailed deer serve as a major host for all mobile life stages of *A. americanum* (Patrick and Hair 1978; Bloemer et al. 1986, 1988; Haile and Mount 1987), and *E. chaffeensis* DNA has been detected in host-seeking adults at numerous locations (Anderson et al. 1992a,b; Lockhart et al. 1997a). Ewing et al. (1995) demonstrated that *A. americanum* was capable of transstadial transmission of *E. chaffeensis* among deer. Most cases of human *E. chaffeensis* infections occur within the geographic range of *A. americanum* (Eng et al. 1990; Walker and Dumler 1996), and *A. americanum* has been identified as the source of tick bite among several human cases (Eng et al. 1990; Standaert et al. 1995).

Although lone star ticks and white-tailed deer clearly are critical components in the epidemiology of *E. chaffeensis,* the possibility that other ticks or other vertebrates also may play a role cannot be excluded. Lone star ticks exhibit minimal host preference, and as nymphs and adults feed on a wide variety of medium- to large-size mammals, which could result in transmission of infection to these hosts. Serologic surveys have disclosed *E. chaffeensis*-reactive antibodies among dogs (Dawson 1996b), raccoons *Procyon lotor,* opossums *Didelphis virginiana* (Lockhart et al. 1997a), and foxes (J.E. Dawson unpublished), all of which are common hosts for either nymphal or adult lone star ticks. In contrast, serologic testing of several species of wild rodents from enzootic sites has failed to detect *E. chaffeensis*-reactive antibodies (Lockhart et al. 1997a, 1998). There is polymerase chain reaction (PCR) evidence of naturally occurring *E. chaffeensis* infection among domestic dogs (Dawson et al. 1996b). Dogs (Dawson and Ewing 1992) and red foxes, but not gray foxes (W.R. Davidson unpublished), have been shown to develop rickettsemia when experimentally infected. Infection among mammals other than deer could expose species of ticks that do not commonly parasitize deer. Such an event may explain an earlier detection of *E. chaffeensis* DNA in a single adult *Dermacentor variabilis* removed from an opossum (Anderson et al. 1992a).

Human Granulocytotropic Ehrlichiosis. The epidemiology of HGE is incompletely known, partly because of uncertainty as to whether the agent is distinct from *E. equi* and *E. phagocytophila* (Dumler et al. 1995) and partly due to its recent discovery. However, there is mounting evidence that it is tick-borne and that wild mammals serve as reservoirs. *Ixodes scapularis* transmits the HGE agent transstadially under experimental conditions (Telford et al. 1996). Infection among field-collected deer ticks from Massachusetts has been confirmed by animal-feeding studies and PCR analyses (Telford et al. 1996). An engorged female *I. scapularis* taken from an HGE patient in Wisconsin, and field-collected *I. scapularis* from Connecticut and Wisconsin also have been found to contain 16S rDNA compatible with that of the HGE agent (Magnarelli et al. 1995; Pancholi et al. 1995).

White-footed mice were susceptible to experimental inoculation with an HGE agent and developed low-level rickettsemia of 10 days' duration. Subpassage of blood from white-footed mice that no longer had microscopically detectable rickettsemia into laboratory mice demonstrated that infection persisted (Telford et al. 1996). In 1938, Tyzzer described organisms in granulocytes, which he named *Cytoecetes microti,* in the blood of splenectomized white-footed mice and voles from Massachusetts. Telford et al. (1996) suggested that the organisms described by Tyzzer may have been the HGE agent. Additional indirect evidence of an *I. scapularis*-rodent cycle has come from the fact that HGE patients also are at increased risk of infection with both Lyme disease and babesiosis. Both of these diseases are known to have an epizootiologic cycle principally involving *I. scapularis* and white-footed mice (Pancholi et al. 1995; Telford et al. 1996). White-tailed deer also appear to be naturally infected based on detection of 16S rDNA identical to that of the HGE organism in wild deer from Georgia (Little et al. 1998) and on the development of HGE in three men who processed hunter-killed deer in Wisconsin and Minnesota. Bakken et al. (1996) speculated that infection may have occurred through cuts in the men's hands after direct contact with blood and other tissues from the wild deer; however, tick exposure could not be ruled out.

White-Tailed Deer *Ehrlichia*-like Agent. Dawson et al. (1996a) amplified *Ehrlichia*-like 16S rRNA gene fragments from the blood of ten white-tailed deer, five each from Georgia and Oklahoma. Nucleotide sequencing disclosed that gene fragments from these deer were identical but that they differed from other described members of the tribe Ehrlichieae. This gene fragment was most similar to members of the *E. phagocytophila* genogroup (Dawson et al. 1996a) and was highly prevalent among white-tailed deer populations in Arkansas, Georgia, and South Carolina but not West Virginia (Little et al. 1997, 1998; Lockhart et al. 1997a). The presence of this gene fragment was associated with *A. americanum* infestations (Little et al. 1997), and an identical 16S rRNA sequence was identified in adult ticks collected from vegetation at one site inhabited by PCR-positive deer in Georgia (Lockhart et al. 1997a). These findings have been interpreted as indicating that a previously undescribed species of *Ehrlichia* occurs commonly among white-tailed deer in the southeastern United States and that, like *E. chaffeensis,* it may be transmitted by *A. americanum* (Dawson et al. 1996a; Little et al. 1997, 1998; Lockhart et al. 1997a).

CLINICAL SIGNS, PATHOGENESIS, AND PATHOLOGY. Most species of *Ehrlichia* have been associated with clinical disease in domestic animals or in humans. The incubation period for most ehrlichial diseases is 7–21 days, and rickettsemia usually occurs during the acute phase of infection. The ehrlichioses are clinically similar and are characterized by fever, depression, anorexia, leukopenia, and usually thrombocytopenia (Rikihisa 1991). Lymphadenopathy is common in many ehrlichioses, and other lesions include varying degrees of hemorrhage, edema, and splenomegaly; however, ehrlichioses generally are not characterized by marked tissue necrosis or inflammation (Rikihisa 1991). Some ehrlichial diseases often have distinguishing pathologic features, such as epistaxis in chronic canine ehrlichiosis; multiorgan petechiation in bovine petechial fever; and watery diarrhea, intestinal hyperemia, and laminitis in Potomac horse fever (Rikihisa 1991).

The pathologic consequences of ehrlichial infections are largely unstudied among wild mammals, and only limited information is available on *E. canis, E. ondiri, E. chaffeensis,* and the white-tailed deer *Ehrlichia*-like agent. Experimental infection of red foxes and gray foxes with *E. canis* produced mild anemia, thrombocytopenia, and leukopenia (Amyx and Huxsoll 1973). African wild dogs experimentally infected with *E. canis* became anorexic and depressed and developed anemia, leukopenia, and thrombocytopenia, whereas similarly exposed black-backed jackals did not develop signs of illness (Van Heerden 1979). A captive gray wolf with naturally acquired canine ehrlichiosis had epistaxis, anorexia, and weight loss, which are characteristic of recrudescent events with chronic canine ehrlichiosis (Harvey et al. 1979). *Ehrlichia ondiri* produced persistent infection without clinical signs in naturally infected bushbuck and a transient subclinical infection in cape buffalo *Syncerus caffer* and wildebeest *Connochaetes gnou* (Grootenhuis 1981). Experimental infections of white-tailed deer with *E. chaffeensis* were subclinical (Dawson et al. 1994b; Ewing et al. 1995), and pathologic changes were not reported among naturally infected white-tailed deer (Little et al. 1997; Lockhart et al. 1997a,b). White-tailed deer were refractory to infection by *E. canis* (Dawson et al. 1994b). Pathologic changes were not observed in natural infections by the unnamed *Ehrlichia*-like agent among white-tailed deer (Little et al. 1997; Lockhart et al. 1997a,b).

DIAGNOSIS. Evidence of ehrlichial infection can be obtained by demonstration of antibodies by serologic testing, detection of organisms by light or electron microscopy, molecular detection of ehrlichial DNA, xenodiagnosis, and isolation of organisms in cell culture. Some combination of methods is desirable, and the most definitive diagnosis is confirmation by isolation or molecular detection.

Indirect fluorescent antibody (IFA) testing has been used extensively to detect circulating antibodies to ehrlichial organisms. Serologic cross-reactions are common when assaying antibodies to *Ehrlichia* with the IFA test, especially among species within the same genogroup (Rikihisa 1991; Dumler et al. 1995; Walker and Dumler 1996). Furthermore, serologic cross-reactivity has been documented between members of the genus *Ehrlichia* and other genera, such as *Cowdria* and *Neorickettsia,* which are phylogenetically closely related to the *E. canis* and *E. sennetsu* genogroups, respectively (Logan et al. 1986; Rikihisa 1991; Katz et al. 1996). Consequently, IFA seroreactions must be interpreted cautiously. In addition to IFA, enzyme-linked immunosorbent assays (ELISAs) and Western immunoblotting also have been used to detect antibodies to some species of *Ehrlichia* (Rikihisa 1991; Rikihisa et al. 1994; Dumler et al. 1995).

Ehrlichial organisms can be demonstrated microscopically with Romanowsky-, Giemsa-, or Wright-stained preparations of circulating leukocytes or platelets or in aspirates of lymph node, lung, spleen, or bone marrow. Typically, morulae are visible as clusters of purple-stained organisms within the cytoplasm of cells (Fig. 27.3). However, infected cells are infrequent, and specific identification cannot be made based on microscopic morphology alone. Electron microscopy also has been used to demonstrate ehrlichieae in blood and other tissues (Fig. 27.4).

Because of the difficulty encountered in culturing or visualizing *Ehrlichia,* recently developed molecular techniques, such as PCR and nucleotide sequencing, have been used extensively in diagnosis of ehrlichioses. Polymerase chain reaction detection of ehrlichial 16S rDNA has been used to great advantage to verify the presence of various *Ehrlichia* when other methods of confirmation have not been possible (Anderson et al. 1991, 1992a,b; Magnarelli et al. 1995; Pancholi et al. 1995; Wen et al. 1995; Dawson et al. 1996a).

Animals have been used as a xenodiagnostic aid with many species of *Ehrlichia* (Rikihisa 1991). Several species of *Ehrlichia* have been cultivated by using various cell culture systems (Fig. 27.5) (Dawson et al. 1991; Rikihisa 1991; Chen et al. 1995; Wen et al. 1995; Goodman et al. 1996). In addition to culture in mammalian cell lines, Munderloh et al. (1996a,b) have grown *E. canis* and *E. equi* in embryonic tick cell lines. Some granulocytotropic and platelet-infecting ehrlichieae have not been cultured.

IMMUNITY. In their natural hosts, several *Ehrlichia* spp. maintain persistent infections in the face of circulating antibodies (Rikihisa 1991; Telford et al. 1996).

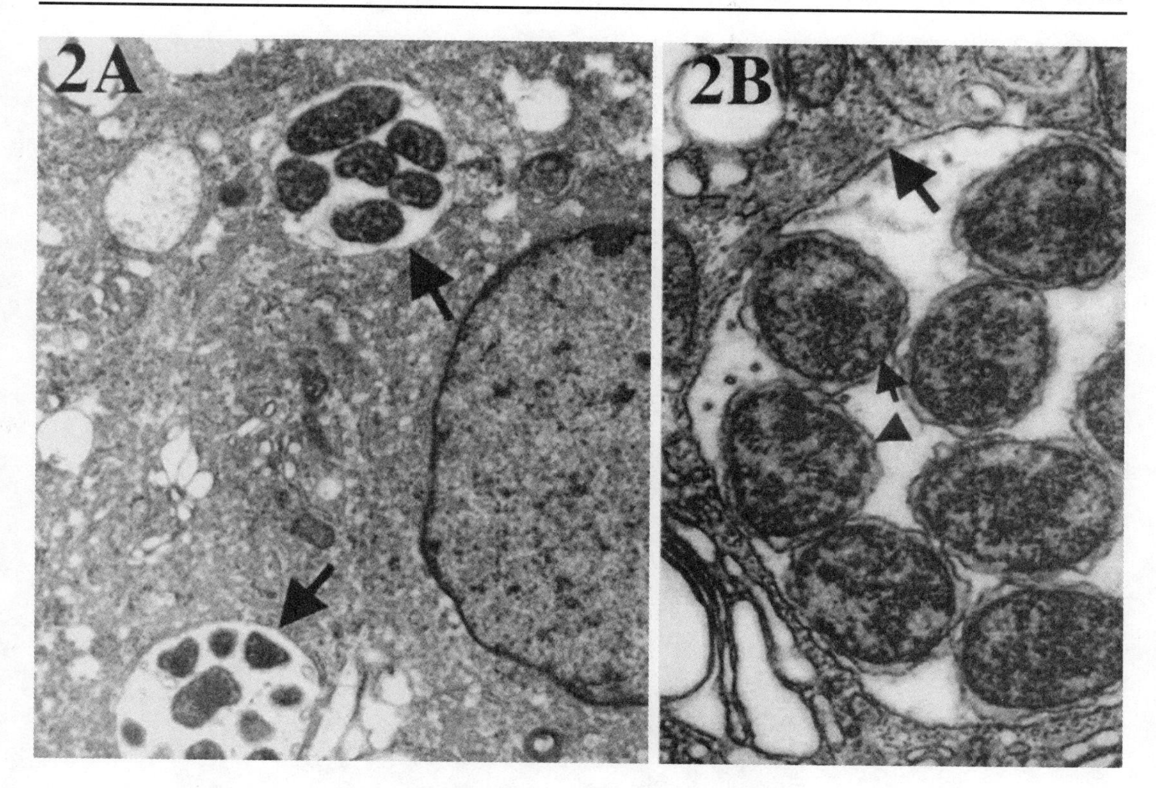

FIG. 27.4—Transmission electron photomicrographs of *Ehrlichia chaffeensis* isolated from white-tailed deer blood in the cytoplasm of DH82 canine macrophage cells. A: Organisms are seen in two membrane-bound morulae (*arrows*). Bar = 1.5 µm. B: Higher magnification of morula discloses a distinct membrane (*large arrow*) enclosing multiple organisms that have an inner plasma membrane (*small arrow*) and an outer cell wall (*arrowhead*). Bar = 0.5 µm. Adapted from Lockhart et al. (1997b). We thank Cynthia Goldsmith for the electron micrograph.

Not surprisingly, these obligate intracellular pathogens do not appear to be greatly affected by the humoral immune response.

Infections by *E. phagocytophila* and *E. risticii* have been reported to have immunosuppressive effects (Rikihisa et al. 1987; Woldehiwet 1987; Larsen et al. 1994), and there is circumstantial evidence that human infections with *E. chaffeensis* and the HGE agent also may cause immunosuppression (Walker and Dumler 1996).

CONTROL AND TREATMENT. Prevention and control of most ehrlichial infections among domestic animals and humans is predicated mainly on reducing or avoiding exposure to tick vectors. Prevention of exposure to ehrlichieae within free-ranging wildlife populations is not feasible because of the difficulties of tick control over large expanses of habitat. Furthermore, because infections among wildlife apparently rarely result in clinically detectable illness, prevention, control, and treatment of free-ranging wildlife are neither feasible nor warranted. Treatment of captive wild animals, when required, could be approached in a manner similar to that used for domestic animals. Oxytetracycline and doxycycline are efficacious against many species of *Ehrlichia* (Rikihisa 1991; Walker and Dumler 1996).

PUBLIC HEALTH CONCERNS. Two human ehrlichioses, caused by *E. chaffeensis* and the HGE agent, are known to be tick-borne zoonoses with wild mammal reservoirs. These two diseases have been categorized as emerging infections, which indicates an increase in their occurrence, presumably associated with environmental alterations that impact the vectors, reservoir hosts, or both. A recent report describes three HGE cases that were acquired after butchering white-tailed deer (Bakken et al. 1996). Use of gloves and masks may reduce the risk of exposure. The full implications of these tick-borne ehrlichioses for public health remain to be determined.

The epidemiology of the third human ehrlichiosis, sennetsu fever, is unknown. Because other members of the *E. sennetsu* genogroup are known to be transmitted by, or to occur in, digenetic trematodes that parasitize

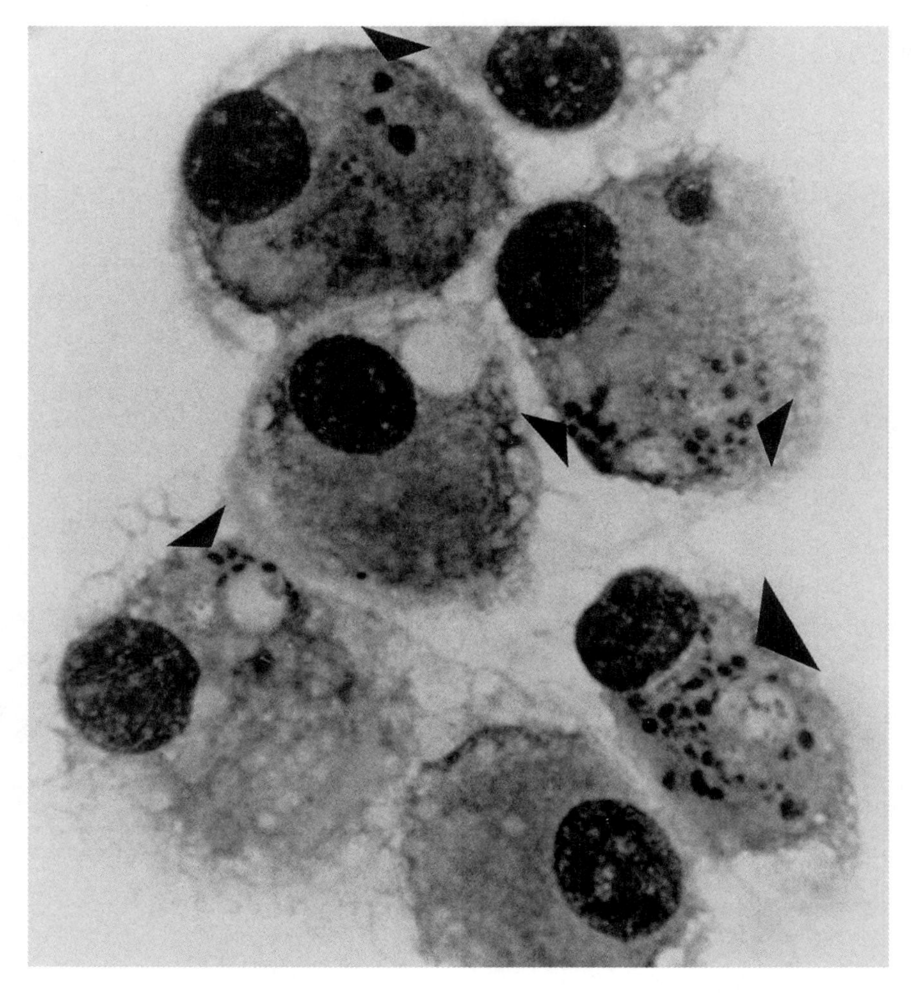

FIG. 27.5—Morulae (*arrows*) of *Ehrlichia chaffeensis* isolated from the blood of white-tailed deer in the cytoplasm of DH82 canine macrophage cells.

fish, this human ehrlichiosis has been suggested to have a similar zoonotic cycle (Rikihisa 1991; Walker and Dumler 1996; Wen et al. 1996).

DOMESTIC ANIMAL CONCERNS. Ehrlichioses have important implications for the health of domestic animals. Bovine petechial fever of cattle and sheep in Kenya is the only ehrlichiosis of veterinary medical importance that has been verified to have a wild mammal reservoir. The relative importance of wild and domestic animals as reservoirs of species such as *E. canis, E. equi,* and *E. phagocytophila,* which infect both categories of hosts, is unclear, but such infection represents a potential source of concern between wildlife and livestock or companion animal interests.

MANAGEMENT IMPLICATIONS. Although none of the ehrlichioses are known to impact the health of wild mammal populations, the known or potential roles of wild mammals in the epidemiology of ehrlichial diseases that are important to human or domestic animal health make them of concern to wildlife managers.

LITERATURE CITED

Alexander, K.A., P.W. Kat, R.K. Wayne, and T.K. Fuller. 1994. Serologic survey of selected canine pathogens among free-ranging jackals in Kenya. *Journal of Wildlife Diseases* 30:486–491.

Amyx, H.L., and D.L. Huxsoll. 1973. Red and gray foxes: Potential reservoir hosts for *Ehrlichia canis. Journal of Wildlife Diseases* 9:47–50.

Anderson, B.E., J.E. Dawson, D.C. Jones, and K.H. Wilson. 1991. *Ehrlichia chaffeensis,* a new species associated with human ehrlichiosis. *Journal of Clinical Microbiology* 29:2838–2842.

Anderson, B.E., J.W. Sumner, J.E. Dawson, T. Tzianabos, C.R. Greene, J.G. Olson, D.R. Fishbein, M. Olsen-Rasmussen, B.P. Holloway, E.H. George, and A.F. Azad. 1992a. Detection of the etiologic agent of human ehrlichiosis by polymerase chain reaction. *Journal of Clinical Microbiology* 30:775–780.

Anderson, B.E., C.E. Greene, D.C. Jones, and J.E. Dawson. 1992b. *Ehrlichia ewingii* sp. nov., the etiologic agent of canine granulocytic ehrlichiosis. *International Journal of Systematic Bacteriology* 42:299–302.

Bakken, J.S., J.K. Krueth, T. Lund, D. Malkovitch, K. Asanovich, and J.S. Dumler. 1996. Exposure to deer blood may be a cause of human granulocytic ehrlichiosis. *Clinical Infectious Diseases* 23:198.

Bell, J.F. 1981. Tick-borne fever and rickettsial pox. In *Infectious diseases of wild mammals,* ed. J.W. Davis, L.H. Karstad, and D.O. Trainer, 2d ed. Ames: Iowa State University Press, pp. 398–402.

Bloemer, S.R., E.L. Snoddy, J.C. Cooney, and K. Fairbanks. 1986. Influence of deer exclusion on populations of lone star ticks and American dog ticks (Acari: Ixodidae). *Journal of Economic Entomology* 79:679–683.

Bloemer, S.R., R.H. Zimmerman, and K. Fairbanks. 1988. Abundance, attachment sites, and density estimators for lone star ticks (Acari: Ixodidae) infesting white-tailed deer. *Journal of Medical Entomology* 25:295–300.

Carroll, J.F., E.T. Schmidtmann, and R.M. Rice. 1989. White-footed mice: Tick burdens and role in the epizootiology of Potomac horse fever in Maryland. *Journal of Wildlife Diseases* 25:397–400.

Chen, S.-M., J.S. Dumler, H.M. Feng, and D.H. Walker. 1994. Identification of the antigenic constituents of *Ehrlichia chaffeensis. American Journal of Tropical Medicine and Hygiene* 50:52–58.

Chen, S.-M., V.L. Popov, H.M. Feng, J. Wen, and D.H. Walker. 1995. Cultivation of *Ehrlichia chaffeensis* in mouse embryo, Vero, BGM, and L929 cells and study of *Ehrlichia*-induced cytopathic effect and plaque formation. *Infection and Immunity* 63:647–655.

Dawson, J.E., and S.A. Ewing. 1992. Susceptibility of dogs to infection with *Ehrlichia chaffeensis,* causative agent of human ehrlichiosis. *American Journal of Veterinary Research* 53:1322–1327.

Dawson, J.E., B.E. Anderson, D.B. Fishbein, J.L. Sanchez, C.S. Goldsmith, K.H. Wilson, and C.W. Duntley. 1991. Isolation and characterization of an *Ehrlichia* sp. from a patient diagnosed with human ehrlichiosis. *Journal of Clinical Microbiology* 29:2741–2745.

Dawson, J.E., J.E. Childs, K.L. Biggie, C. Moore, D. Stallknecht, J. Shaddock, J. Bouseman, E. Hofmeister, and J.G. Olson. 1994a. White-tailed deer as a potential reservoir of *Ehrlichia* spp. *Journal of Wildlife Diseases* 30:162–168.

Dawson, J.E., D.E. Stallknecht, E.W. Howerth, C. Warner, K. Biggie, W.R. Davidson, J.M. Lockhart, V.F. Nettles, J.G. Olson, and J.E. Childs. 1994b. Susceptibility of white-tailed deer (*Odocoileus virginianus*) to infection with *Ehrlichia chaffeensis,* the etiologic agent of human ehrlichiosis. *Journal of Clinical Microbiology* 32:2725–2728.

Dawson, J.E., C.K. Warner, V. Baker, S.A. Ewing, D.E. Stallknecht, W.R. Davidson, A.A. Kocan, J.M. Lockhart, and J.G. Olson. 1996a. *Ehrlichia*-like 16S rDNA sequence from wild white-tailed deer (*Odocoileus virginianus*). *Journal of Parasitology* 82:52–58.

Dawson, J.E., K.L. Biggie, C.K. Warner, K. Cookson, S. Jenkins, J.F. Levine, and J.G. Olson. 1996b. Polymerase chain reaction evidence of *Ehrlichia chaffeensis,* an etiologic agent of human ehrlichiosis, in dogs from southeast Virginia. *American Journal of Veterinary Research* 57:1175–1179.

Dumler, J.S., and J.S. Bakken. 1995. Ehrlichial diseases of humans: Emerging tick-borne infections. *Clinical Infectious Diseases* 20:1102–1110.

Dumler, J.S., J.R. Harkess, D.B. Fishbein, J.E. Dawson, C.N. Greene, M. A. Redus, and F.T. Satalowich. 1990. Epidemiologic, clinical, and laboratory findings of human ehrlichiosis in the United States, 1988. *Journal of the American Medical Association* 264:2251–2258.

Dumler, J.S., K.M. Asanovich, J.S. Bakken, P. Richter, R. Kimsey, and J.E. Modigan. 1995. Serologic cross-reactions among *Ehrlichia equi, Ehrlichia phagocytophila,* and human granulocytic *Ehrlichia. Journal of Clinical Microbiology* 33:1098–1103.

Eng, T.R., J.R. Harkess, D.E. Fishbein, J.E. Dawson, C.N. Greene, M.A. Redus, and F.T. Satalowich. 1990. Epidemiologic, clinical, and laboratory findings of human ehrlichiosis in the United States, 1988. *Journal of the American Medical Association* 264:2251–2258.

Engvall E.O., B. Pettersson, M. Persson, K. Artursson, and K.E. Johansson. 1996. A 16S rRNA-based PCR assay for detection and identification of granulocytic *Ehrlichia* species in dogs, horses, and cattle. *Journal of Clinical Microbiology* 34:2170–2174.

Ewing, S.A. 1969. Canine ehrlichiosis. *Advances in Veterinary Science and Comparative Medicine* 13:331–353.

Ewing, S.A., and R.G. Buckner. 1965. Manifestation of babesiosis, ehrlichiosis, and combined infections in the dog. *American Journal of Veterinary Research* 26:815–828.

Ewing, S.A., R.G. Buckner, and B.G. Stringer. 1964. The coyote, a potential host for *Babesia canis* and *Ehrlichia* sp. *Journal of Parasitology* 50:704.

Ewing, S.A., J.E. Dawson, A.A. Kocan, R.W. Barker, C.K. Warner, R.J. Panciera, J.C. Fox, K.M. Kocan, and E.F. Blouin. 1995. Experimental transmission of *Ehrlichia chaffeensis* (Rickettsiales: Ehrlichieae) among white-tailed deer by *Amblyomma americanum* (Acari: Ixodidae). *Journal of Medical Entomology* 32:368–374.

Foggie, A. 1951. Studies on the infectious agents of tick-borne fever in sheep. *Journal of Pathology and Bacteriology* 63:1–15.

Goodman, J.L., C. Nelson, B. Vitale, J.E. Madigan, J.S. Dumler, T.J. Kurtii, and U.G. Munderloh. 1996. Direct cultivation of the causative agent of human granulocytic ehrlichiosis. *New England Journal of Medicine* 334:209–215.

Greig, B., K.M. Asanovich, P.J. Armstrong, and J.S. Dumler. 1996. Geographic, clinical, serologic, and molecular evidence of granulocytic ehrlichiosis, a likely zoonotic disease, in Minnesota and Wisconsin. *Journal of Clinical Microbiology* 34:44–48.

Grootenhuis, J.G. 1981. Bovine petechial fever. In *Infectious diseases of wild mammals,* ed. J.W. Davis, L.H. Karstad, and D.O. Trainer, 2d ed. Ames: Iowa State University Press, pp. 403–404.

Haile, D.G., and G.A. Mount. 1987. Computer simulation of population dynamics of the lone star tick, *Ambylomma americanum* (Acari: Ixodidae). *Journal of Medical Entomology* 24:356–369.

Harvey, J.W., C.F. Simpson, J.M. Gaskin, and J.H. Sameck. 1979. Ehrlichiosis in wolves, dogs, and wolf-dog crosses. *Journal of the American Veterinary Medical Association* 175:901–905.

Johansson, K.E., B. Pettersson, M. Uhlen, A. Gunnarsson, M. Malmqvist, and E. Olsson. 1995. Identification of the causative agent of granulocytic ehrlichiosis in Swedish dogs and horses by direct solid phase sequencing of PCR products from the 16S rRNA gene. *Research in Veterinary Science* 58:109–112.

Katz, J.B., A.F. Barbet, S.M. Mahan, D. Kumbula, J.M. Lockhart, M.K. Keel, J.E. Dawson, J.G. Olson, and S.A. Ewing. 1996. A recombinant antigen from the heartwater agent (*Cowdria ruminantium*) reactive with antibodies in some southeastern United States white-tailed deer (*Odocoileus virginianus*), but not cattle, sera. *Journal of Wildlife Diseases* 32:424–430.

Kawahara, M., C. Suto, Y. Rikihisa, S. Yamamoto, and Y. Tsuboi. 1993. Characterization of ehrlichial organisms isolated from a wild mouse. *Journal of Clinical Microbiology* 31:89–96.

Larsen, H.J.S., G. Overnes, H. Waldeland, and G.M. Johansen. 1994. Immunosuppression in sheep experimentally infected with *Ehrlichia phagocytophila. Research in Veterinary Science* 56:216–224.

Little, S.E., J.E. Dawson, J.M. Lockhart, D.E. Stallknecht, C.K. Warner, and W.R. Davidson. 1997. Development and use of specific polymerase reaction for the detection of an organism resembling *Ehrlichia* sp. in white-tailed deer. *Journal of Wildlife Diseases* 33:246–253.

Little, S.E., D.E. Stallknecht, J.M. Lockhart, J.E. Dawson, and W.R. Davidson. 1998. Natural coinfection of a white-tailed deer (*Odocoileus virginianus*) population with three *Ehrlichia. Journal of Parasitology* 84:897–901.

Lockhart, J.M., W.R. Davidson, J.E. Dawson, and D.E. Stallknecht. 1995. Temporal association of *Amblyomma americanum* with the presence of *Ehrlichia chaffeensis* reactive antibodies in white-tailed deer. *Journal of Wildlife Diseases* 31:119–124.

Lockhart, J.M., W.R. Davidson, D.E. Stallknecht, and J.E. Dawson. 1996. Site-specific geographic association between *Amblyomma americanum* (Acari: Ixodidae) infestations and *Ehrlichia chaffeensis*-reactive (Rickettsiales: Ehrlichieae) antibodies in white-tailed deer. *Journal of Medical Entomology* 33:153–158.

Lockhart, J.M., W.R. Davidson, D.E. Stallknecht, J.E. Dawson, and S.E. Little. 1997a. Natural history of *Ehrlichia chaffeensis* (Rickettsiales: Ehrlichieae) in the Piedmont physiographic province of Georgia. *Journal of Parasitology* 83:887–894.

Lockhart, J.M., W.R. Davidson, D.E. Stallknecht, J.E. Dawson, and E.W. Howerth. 1997b. Isolation of *Ehrlichia chaffeensis* from wild white-tailed deer (*Odocoileus virginianus*) confirms their role as natural reservoir hosts. *Journal of Clinical Microbiology* 35:1681–1686.

Lockhart, J.M., W.R. Davidson, D.E. Stallknecht, and J.E. Dawson. 1998. Lack of seroreactivity to *Ehrlichia chaffeensis* in Georgia and South Carolina rodent populations. *Journal of Wildlife Diseases* 34:392–396.

Logan, L.L., C.J. Holland, C.A. Mebus, and M. Ristic. 1986. Serologic relationship between *Cowdria ruminantium* and certain ehrlichia. *Veterinary Record* 119:458–459.

MacDiarmid, A. 1965. Modern trends in animal health and husbandry: Some infectious disease of free-living wildlife. *British Veterinary Journal* 121:245–257.

MacLeod, J., and W.S. Gordon. 1933. Studies in tick-borne fever of sheep. *Parasitology* 25:273–284.

Maeda, K., N. Markowitz, R.C. Hawley, M. Ristic, D. Cox, and J.E. McDade. 1987. Human infection with *Ehrlichia canis,* a leukocytic rickettsia. *New England Journal of Medicine* 316:853–856.

Magnarelli, L.A., K.C. Stafford III, T.N. Mather, M.-T. Yeh, K.D. Horn, and J.S. Dumler. 1995. Hemocytic *Rickettsia*-like organisms in ticks: Serologic reactivity with antisera to ehrlichieae and detection of DNA of agent of human granulocytic ehrlichiosis by PCR. *Journal of Clinical Microbiology* 33:2710–2714.

Munderloh, U.G., Madigan, J.E., J.S. Dumler, J.L. Goodman, S.F. Hayes, and J.E. Barlough. 1996a. Isolation of the equine granulocytic ehrlichiosis agent, *Ehrlichia equi,* in tick cell culture. *Journal of Clinical Microbiology* 34:664–670.

Munderloh, U.G., S.F. Hayes, S.A. Ewing, K.M. Kocan, G.G. Ahlstrand, and T.J. Kurtii. 1996b. Development of ehrlichiae in tick cell culture. In *Proceedings of the 12th sesqui-annual meeting of the American Society of Rickettsiology and Rickettsial Diseases, Asilomar, California.* Abstract 56.

Pancholi, P., C.P. Kolbert, P.D. Mitchell, K.D. Reed, J.S. Dumler, J.S. Bakken, S.R. Telford III, and D.H. Persing. 1995. *Ixodes dammini* as a potential vector of human granulocytic ehrlichiosis. *Journal of Infectious Diseases* 172:1007–1012.

Patrick, C.D., and J.A. Hair. 1978. White-tailed deer utilization of three different habitats and its influence on lone star tick populations. *Journal of Parasitology* 64:263–269.

Perry, B.D., E.T. Schmidtmann, R.M. Rice, J.W. Hansen, M. Fletcher, E.C. Turner, M.G. Robl, and N.E. Hahn. 1989. Epidemiology of Potomac horse fever: An investigation into the possible role of non-equine mammals. *Veterinary Record* 125:83–86.

Piercy, S.E. 1953. Bovine infectious petechial fever. *East African Agricultural Journal* 18:65–68.

Plowright, W. 1962. Some notes on bovine petechial fever (Ondiri disease) at Muguga, Kenya. *Bulletin of Epizootic Diseases in Africa* 10:499–505.

Pretzman, C., D. Ralph, D.R. Stoddard, P.A. Fuerst, and Y. Rikihisa. 1995. 16S rRNA gene sequence of *Neorickettsia helminthoeca* and its phylogenetic alignment with members of the genus *Ehrlichia. International Journal of Systematic Bacteriology* 45:207–211.

Price, J.E., and L.H. Karstad. 1980. Free-living jackals (*Canis mesomelas*): Potential reservoir hosts for *Ehrlichia canis* in Kenya. *Journal of Wildlife Diseases* 16:469–473.

Reubel, G.H., J.E. Barlough, and J.E. Madigan. 1998. Production and characterization of *Ehrlichia risticii,* the agent of Potomac horse fever, from snails (Pleuroceridae: *Juga* spp.) in aquarium culture and genetic comparison to equine strains. *Journal of Clinical Microbiology* 36:1501–1511.

Richter, P.J., R.B. Kimsey, J.E. Madigan, J.E. Barlough, J.S. Dumler, and D.L. Brooks. 1996. *Ixodes pacificus* (Acari: Ixodidae) as a vector of *Ehrlichia equi* (Rickettsiales: Ehrlichieae). *Journal of Medical Entomology* 33:1–5.

Rikihisa, Y. 1991. The tribe *Ehrlichieae* and ehrlichial diseases. *Clinical Microbiology Reviews* 4:286–308.

Rikihisa, Y., G.J. Johnson, and C.J. Burger. 1987. Reduced immune responsiveness and lymphoid depletion in mice infected with *Ehrlichia risticii. Infection and Immunity* 55:2215–2222.

Rikihisa, Y., S.A. Ewing, and J.C. Fox. 1994. Western immunoblot analysis of *Ehrlichia chaffeensis, E. canis,* or *E. ewingii* infections in dogs and humans. *Journal of Clinical Microbiology* 32:2107–2112.

Ristic, M., and D. Huxsoll. 1984. Tribe II: *Ehrlichiae.* In *Bergey's manual of systematic bacteriology,* vol. 1, ed. N.R. Krieg and J.G. Holt. Baltimore: Williams and Wilkins, pp. 704–711.

Standaert, S.M., J.E. Dawson, W. Schaffner, J.E. Childs, K.L. Biggie, J. Singleton Jr., R.R. Gerhardt, M.L. Knight, and R.H. Hutcheson. 1995. Ehrlichiosis in a golf-oriented retirement community. *New England Journal of Medicine* 333:420–425.

Telford, S.R., J.E. Dawson, P. Katavalos, C.K. Warner, C.P. Kolbert, and D.H. Persing. 1996. Perpetuation of the agent of human granulocytic ehrlichiosis in a deer tick-rodent cycle. *Proceedings of the National Academy of Sciences U.S.A.* 93:6209–6214.

Tyzzer, E.E. 1938. *Cytoecetes microti,* N.G.N. sp.: A parasite developing in granulocytes and infective for small rodents. *Parasitology* 30:242–257.

Van Heerden, J. 1979. The transmission of canine ehrlichiosis to the wild dog *Lycaon pictus* (Temminck) and black-backed jackal *Canis mesomelas* Schreber. *Journal of the South African Veterinary Association* 50:245–248.
Walker, D.H., and J.S. Dumler. 1996. Emergence of the ehrlichioses as human health problems. *Emerging Infectious Diseases* 2:18–29.
Walker, A.R., J.E. Cooper, and D.R. Snodgrass. 1974. Investigations into the epidemiology of bovine petechial fever in Kenya and the potential of trombiculid mites as vectors. *Tropical Animal Health Production* 6:193–198.
Webster, K.A., and G.B. Mitchell. 1989. An electron microscopic study of *Cytoecetes phagocytophila* infection in *Ixodes ricinus*. *Research in Veterinary Science* 47:30–33.
Wen, B., Y. Rikihisa, J. Mott, P.A. Fuerst, M. Kawahara, and C. Suto. 1995. *Ehrlichia muris* sp. nov., identified on the basis of 16S rRNA base sequences and serological, morphological, and biological characteristics. *International Journal of Systematic Bacteriology* 45:250–254.
Wen, B., Y. Rikihisa, S. Yamamoto, N. Kawabata, and P.A. Fuerst. 1996. Characterization of the SF agent, and *Ehrlichia* sp. isolated from the fluke *Stellantchasmus falcatus,* by 16S rRNA base sequence, serological, and morphological analyses. *International Journal of Systematic Bacteriology* 46:149–154.
Woldehiwet, Z. 1983. Tick-borne fever: A review. *Veterinary Research Communications* 6:163–175.
———. 1987. The effects of tick-borne fever on some functions of polymorphonuclear cells of sheep. *Journal of Comparative Pathology* 97:481–485.

HEARTWATER

NANCY D. KOCK

Synonyms: Heartwater, cowdriosis.

INTRODUCTION. Heartwater, caused by the obligate intracellular rickettsial organism *Cowdria ruminantium,* is a complex disease that likely evolved in wild ruminants on the African continent, where it remains established without appreciable effects in wildlife. It has emerged as an important disease of domestic ruminants only relatively recently, being first described in the diary of a Voortrekker pioneer in 1838 in South Africa when his sheep died within 3 weeks of tick infestation (Provost and Bezuidenhout 1987). Since then, the disease has been reported in most African countries south of the Sahara, on islands off the African mainland, and in the Caribbean Sea (Provost and Bezuidenhout 1987). Heartwater is an important cause of death in domestic cattle *Bos taurus,* sheep *Ovis aries,* and goats *Capra hircus* in regions where tick vectors are present, making it a major threat to food production in countries struggling with overpopulation. It has posed a major impediment to introduction of exotic domestic stock into Africa for genetic improvement (Uilenberg and Camus 1993). Heartwater has also been considered a potential threat to nonendemic countries, particularly those in the Western Hemisphere, given its vector transmission and wide host range (Uilenberg 1982; Barre et al. 1987).

HOST RANGE. Investigators have long suspected a reservoir for *C. ruminantium* in African wildlife, given its extreme pathogenicity in imported domestic ruminant stock, intermittent pathogenicity in indigenous stock, and apparent nonpathogenicity in most indigenous wildlife species (Uilenberg 1983) [see Oberem and Bezuidenhout (1987)]. Fatal disease has been reported in wild bovids exotic to the sub-Saharan continent and in cervids (Young and Basson 1973; Poudelet et al. 1982), including white-tailed deer *Odocoileus virginianus* (Dardiri et al. 1987). Disease has also occurred in zoological gardens, in species not normally found in endemic areas [see Oberem and Bezuidenhout (1987)]. Natural or experimental infections with *C. ruminantium* have been reported in eland *Taurotragus* sp. (Young and Basson 1973), springbuck *Antidorcas marsupialis* (Neitz 1944), lechwe *Kobus leche kafuensis* (Pandey et al. 1992), and sitatunga *Tragelaphus spekeii* (Uilenberg and Camus 1993). Subclinical infections have been demonstrated in African buffalo *Syncerus caffer* and giraffe *Giraffa camelopardalis* (Oberem and Bezuidenhout 1987; Uilenberg and Camus 1993) and in some nonmammalian wildlife species, including guinea fowl *Numidia meleagris* and leopard tortoises *Geochelone pardalis* (Oberem and Bezuidenhout 1987). African buffalo, cattle, sheep (Andrew and Norval 1989), goats (Barre and Camus 1987), guinea fowl, leopard tortoises (Oberem and Bezuidenhout 1987), and BALB/c mice *Mus musculus* (Wassink et al. 1990) can carry *C. ruminantium* based on their ability to transmit heartwater to susceptible hosts. *Cowdria ruminantium* DNA can be amplified by polymerase chain reaction (PCR) from the bone marrow of domestic sheep months after infection, treatment, and recovery (Kock 1995), further indicating carrier status. Sera from free-ranging black rhinoceroses *Diceros bicornis* and white rhinoceroses *Ceratotherium simum* (Kock et al. 1992), elephants *Loxodonta africana,* and impala *Aepyceros melampus* (Kock 1995) from Zimbabwe tested positively for *C. ruminantium* by monoclonal antibody-mediated competitive enzyme-linked immunosorbent assay. Blood and bone marrow from free-ranging waterbuck *Kobus ellipsiprymnus,* tsessbe *Damaliscus lunatus,* and impala tested positive for *C. ruminantium* DNA by PCR (Kock et al. 1995).

ETIOLOGY AND TRANSMISSION. *Cowdria ruminantium* is classified with the Rickettsiales, family Rickettsieae, and tribe Ehrlichieae, which contains *Ehrlichia, Neorickettsia,* and *Cowdria* (Ristic and Huxsoll 1984). These three genera are grouped together based on their need for arthropod vectors and their site of multiplication within the host (Uilenberg 1983; Weiss and Moulder 1984; Scott 1987). However, the taxonomic position of *C. ruminantium* has remained controversial, and cross-reactions with *Ehrlichia* spp. continue to complicate the interpretation of serologic tests (Logan et al. 1986; Jongejan et al. 1989, 1993; Du Plessis 1993). *Cowdria ruminantium* is small (0.2–0.5 μm) and round. It stains negatively with Gram stain and from reddish-purple to pale blue with Giemsa stain. Ultrastructural studies have consistently shown the organism surrounded by two membranes within infected host cells (Prozesky 1987a). Different forms of the organism have been demonstrated within the gut, salivary glands, hemocytes, and Malpighian tubules in two of its tick vectors, *Amblyomma variegatum* and *Amblyomma hebraeum,* indicating development within the vector (Kocan and Bezuidenhout 1987; Yunker et al. 1987). These two ticks are considered the most important vectors of heartwater for domestic ruminants in Africa (Norval 1983). Twelve species of *Amblyomma* ticks are able to transmit *C. ruminantium,* either naturally or experimentally (Walker and Olwage 1987).

CLINICAL SIGNS, PATHOLOGY, AND PATHOGENESIS. The incubation period of heartwater varies with the species affected, route of infection, strain of *C. ruminantium,* and type and amount of inoculum (Van de Pypekamp and Prozesky 1987). Fever develops in cattle about 12 days after intravenous injection of the currently used blood vaccine. If infected nymph suspensions are used, fever may develop sooner, and, in naturally occurring infections, fever develops later (Uilenberg 1981). Severity of disease varies from peracute and fatal to mild. The clinical signs of heartwater in domestic livestock are somewhat vague and include lethargy, dyspnea, fever, and central nervous system (CNS) signs. Acute disease with CNS signs is common in all three domestic ruminant species (Van de Pypekamp and Prozesky 1987), and terminal cases are often found recumbent with heads twisted back. Similar signs are seen when wildlife species are affected.

Gross postmortem findings in heartwater include pulmonary edema, which is often severe enough to cause death, and cerebral edema. Histopathologic findings support the gross lesions but give no clear indication of pathogenesis (Prozesky 1987b). Smears of brain tissue often disclose the presence of characteristic round structures within endothelial cells that stain positive with Giemsa stain (Purchase 1945). In natural infections, the organism can be seen histologically in brain, lung, and heart by light microscopy (Prozesky 1987b). Further confirmation of infection can be achieved by the injection of fresh blood from affected animals into susceptible ones.

Cowdria ruminantium infects endothelium and white blood cells (Logan et al. 1987; Williams and Vodkin 1987), although the mode of infection is unclear (Williams and Vodkin 1987). Replication is thought to occur within cytoplasmic phagosomes, as occurs with *Chlamydia* spp. and *Ehrlichia* spp. (Williams and Vodkin 1987), although mechanisms underlying host cell damage and the resultant pulmonary and/or cerebral edema are unknown. Edema presumably results because of endothelial damage, although histologic and ultrastructural studies indicate only minimal endothelial damage compared with the degree of vascular leakage (Prozesky 1987b). Peracute toxemia has been suggested as a cause of death, but toxins have not been identified. Type-I hypersensitivity reactions in natural hosts have been ruled out because pretreatment with vasoactive amine antagonists does not alter the course of disease (Du Plessis et al. 1987) and because mice lacking prior sensitization develop disease.

Reservoir sites for the organism in animals capable of carrying *C. ruminantium* are under investigation. Neutrophils, macrophages, and occasionally eosinophils become infected during acute heartwater infection in goats (Logan et al. 1987), and since white blood cell circulation time is relatively brief (Valli 1985), *C. ruminantium* might be maintained in sites of hematopoiesis, the bone marrow in most species. Structures consistent with *C. ruminantium* were seen in eosinophilic metamyelocytes in bone marrow, and *C. ruminantium* DNA was recovered by PCR from bone marrow but not blood from sheep months after infection and treatment (Kock 1995). Further support for bone marrow acting as a reservoir for the organism comes from the recovery of *C. ruminantium* DNA from the bone marrow from clinically normal wild ungulates (Kock et al. 1995). Eosinophils, in particular, may play an important role in the host-parasite relationship. If the organism is harbored in polymorphonuclear white blood cell precursors, particularly eosinophil precursors, without causing cellular damage or specific host response, it is afforded a safe niche within the animal. Since eosinophils are nonspecifically recruited to sites of parasitic infestation, they could serve as vehicles for the organism to reach tick-bite sites, providing a mechanism for the transmission of disease. Because eosinophils tend to be relatively few in circulation, it is conceivable that infected eosinophils are missed in routine examination of blood smears. Further investigation into the role of eosinophils might provide answers into important aspects of the pathogenesis of heartwater. In addition, an intracellular location for *C. ruminantium* that causes neither cell damage (i.e., within polymorphonuclear cells and their precursors) nor host response might explain the apparent inconsistencies regarding antibody levels and immune status/susceptibility (Du Plessis 1970).

IMMUNITY. Humoral immune responses develop to *C. ruminantium* as expected (Du Plessis 1970). However, antibodies are not protective, and antibody levels cannot be used reliably to differentiate between susceptible and immune animals (Du Plessis 1970). Cell-mediated immunity appears to be crucial to disease resistance (Stewart 1987a,b), particularly CD8+ T lymphocytes in nude mice (Du Plessis et al. 1991), although histologic evidence for this in the form of typical cellular infiltrations is not seen (Prozesky 1987b). Solid resistance to disease is often seen in calves up to 4 weeks of age, irrespective of the immune status of the dam (Uilenberg 1981; Du Plessis 1985; Du Plessis and Malan 1987, 1988).

Indigenous sheep and goats in South Africa have greater resistance to the disease than imported breeds (Du Plessis et al. 1983). Apparent innate resistance has been attributed to latent infections in South Africa (Du Plessis et al. 1984) and, in Guadeloupe, to genetic resistance (Matheron et al. 1987).

CONTROL, TREATMENT, AND MANAGEMENT IMPLICATIONS. Heartwater remains an important disease of domestic ruminants in endemic countries. Annual mortality figures due to heartwater in Zimbabwe are thought to be greatly underestimated, in part because it is endemic in some areas and not a notifiable disease. In addition, embarrassment is associated with outbreaks, because of the implication of inadequate dipping and/or quarantine of stock. One million cattle likely died of tick-borne diseases, including heartwater, during the last few years before Zimbabwe's independence in 1980, when dipping regimens were relaxed (Hargreaves 1987). At present, vaccines against *C. ruminantium* are unavailable, and the disease is controlled through dipping, when tick burdens are high, in combination with infection and treatment. The use of infective blood vaccine, followed by treatment, confers immunity but may be impractical and is unsuited for heartwater-free areas. Both presently used control measures are expensive, often prohibitively so, for developing countries.

Wildlife species are being translocated with increasing frequency throughout Africa because of efforts aimed at preservation of species, increasing genetic variability, and stocking the increasing numbers of game farms. Entire family groups of elephants were successfully translocated from Gonerazhou National Park in Zimbabwe to locations within Zimbabwe and to South Africa in 1994. Many black rhinoceroses have been translocated from Zimbabwe to heartwater-free countries, some of which harbor *Amblyomma* spp. ticks. The identification of wildlife carriers of *C. ruminantium* has particular significance in preventing the spread of the disease into heartwater-free areas, given its potential risks to food production. If carrier animals are introduced, disease could become endemic in heartwater-free areas where tick vectors exist. Alternately, *Amblyomma* sp. ticks carried on wildlife species may be able to survive and transmit disease in otherwise tick-free regions.

LITERATURE CITED

Andrew, H.R., and R.A.I. Norval. 1989. The carrier status of sheep, cattle and African buffalo recovered from heartwater. *Veterinary Parasitology* 34:261–266.

Barre, N., and E. Camus. 1987. The reservoir status of goats recovered from heartwater. *Onderstepoort Journal of Veterinary Research* 54:193–196.

Barre, N., G. Uilenberg, P.C. Morel, and E. Camus. 1987. Danger of introducing heartwater onto the American mainland: Potential role of indigenous and exotic *Amblyomma* ticks. *Onderstepoort Journal of Veterinary Research* 54:405–417.

Dardiri, A.H., L.L. Logan, and C.A. Mebus. 1987. Susceptibility of white-tailed deer to experimental heartwater infections. *Journal of Wildlife Diseases* 23:215–219.

Du Plessis, J.L. 1970. Immunity in heartwater: I. A preliminary note on the role of serum antibodies. *Onderstepoort Journal of Veterinary Research* 51:147–150.

———. 1985. The natural resistance of cattle to artificial infection with *Cowdria ruminantium:* The role played by conglutinin. *Onderstepoort Journal of Veterinary Research* 52:273–277.

———. 1993. The relationship between *Cowdria* and *Ehrlichia:* Change in the behaviour of ehrlichial agents passaged through *Amblyomma hebraeum.* In *Proceedings of the second biennial meeting of the Society for Tropical Veterinary Medicine, Saint Francois, Guadeloupe, French West Indies,* pp. 131–143.

Du Plessis, J.L., and L. Malan. 1987. The non-specific resistance of cattle to heartwater. *Onderstepoort Journal of Veterinary Research* 54:333–336.

———. 1988. Susceptibility to heartwater of calves born to non-immune cows. *Onderstepoort Journal of Veterinary Research* 55:235–237.

Du Plessis, J.L., B.C. Jansen, and L. Prozesky. 1983. Heartwater in angora goats: I. Immunity subsequent to artificial infection and treatment. *Onderstepoort Journal of Veterinary Research* 50:137–143.

Du Plessis, J.L., J.D. Bezuidenhout, and C.J.F. Ludemann. 1984. The immunization of calves against heartwater: Subsequent immunity both in the absence and presence of natural tick challenge. *Onderstepoort Journal of Veterinary Research* 51:193–196.

Du Plessis, J.L., L. Malan, and Z. E. Kowalski. 1987. The pathogenesis of heartwater. *Onderstepoort Journal of Veterinary Research* 54:313–318.

Du Plessis, J.L., P. Berche, and L. van Gas. 1991. T cell-mediated immunity to *Cowdria ruminantium* in mice: The protective role of Lyt-2+ cells. *Onderstepoort Journal of Veterinary Research* 58:171–179.

Hargreaves, S. 1987. *Annual report.* Harare, Zimbabwe: Veterinary Research Laboratory.

Jongejan, F., L.A. Wassink, M.J.C. Thielemans, N.M. Perie, and G. Uilenberg. 1989. Serotypes in *Cowdria ruminantium* and their relationship with *Ehrlichia phagocytophila* determined by immunofluorescence. *Veterinary Microbiology* 21:31–40.

Jongejan, F., N. de Vries, J. Nieuwenhuijs, A.H.M. van Vliet, and L.A. Wassink. 1993. The immunodominant 32-kilodalton protein of *Cowdria ruminantium is* conserved within the genus *Ehrlichia. Revue d'Elevage et de Médecine Vétérinaire des Pays Tropicaux* 46:145–152.

Kocan, K.M., and J.D. Bezuidenhout. 1987. Morphology and development of *Cowdria ruminantium* in *Amblyomma* ticks. *Onderstepoort Journal of Veterinary Research* 54:177–182.

Kock, N. 1995. Studies on the pathogenesis and epidemiology of heartwater (*Cowdria ruminantium* infection). Ph.D. diss., University of California, Davis, California, 155 pp.

Kock, N., F. Jongejan, M. Kock, R. Kock, and P. Morkel. 1992. Serological evidence for exposure to *Cowdria ruminantium* in free-ranging black (*Diceros bicornis*) and white (*Ceratotherium simum*) rhinoceroses in Zimbabwe. *Journal of Zoo and Wildlife Medicine* 23:409–413.

Kock, N., A.H.M. van Vliet, K. Charlton, and F. Jongejan. 1995. Detection of *Cowdria ruminantium* in blood and bone marrow from clinically normal, free-ranging Zimbabwean ungulates. *Journal of Clinical Microbiology* 33:2501–2504.

Logan, L.L., C.J. Holland, C.A. Mebus, and M. Ristic. 1986. Serological relationship between *Cowdria ruminantium* and certain *Ehrlichia. Veterinary Record* 119:458–459.

Logan, L.L., J.C. Whyard, J.C. Quintero, and C.A. Mebus. 1987. The development of *Cowdria ruminantium* in neutrophils. *Onderstepoort Journal of Veterinary Research* 543:197–204.
Matheron, G., N. Barre, E. Camus, and J. Gogue. 1987. Genetic resistance of Guadeloupe native goats to heartwater. *Onderstepoort Journal of Veterinary Research* 54:337–340.
Neitz, 0. 1944. The susceptibility of the springbuck (*Antidorcas marsupialis*) to heartwater. *Onderstepoort Journal of Veterinary Research* 20:25–27.
Norval, R.A.I. 1983. The ticks of Zimbabwe: VII. The genus *Amblyomma. Zimbabwe Veterinary Journal* 14:3–18.
Oberem, P.T., and J.D. Bezuidenhout. 1987. Heartwater in hosts other than domestic ruminants. *Onderstepoort Journal of Veterinary Research* 54:271–276.
Pandey, G.S., D. Minyoi, F. Hasebe, and E.T. Mwase. 1992. First report of heartwater (cowdriosis) in Kafue lechwe (*Kobus leche kafuensis*) in Zambia. *Revue d'Elevage et de Médecine Vétérinaire des Pays Tropicaux* 45:23–25.
Poudelet, M., E. Poudelet, and N. Barre. 1982. Sensibilite d'un Cervide: *Cervus timorensis russa* a la cowdriose (heartwater). *Revue d'Elevage et de Médecine Vétérinaire des Pays Tropicaux* 35:23–26.
Provost, A., and J.D. Bezuidenhout. 1987. The historical background and global importance of heartwater. *Onderstepoort Journal of Veterinary Research* 54:165–169.
Prozesky, L. 1987a. Heartwater. The morphology of *Cowdria ruminantium* and its staining characteristics in the vertebrate host and in vitro. *Onderstepoort Journal of Veterinary Research* 54:173–176.
———. 1987b. The pathology of heartwater: III. A review. *Onderstepoort Journal of Veterinary Research* 54:281–286.
Purchase, H.S. 1945. A simple and rapid method for demonstrating *Rickettsia ruminantium* (Cowdry) in heartwater brains. *Veterinary Record* 57:413–414.
Ristic, M., and D.L. Huxsoll. 1984. Tribe II: Ehrlichieae. In *Bergey's manual of systematic bacteriology,* vol. 1, ed. N.R. Krieg and J.G. Hold. London: Williams and Wilkins, pp. 704–711.
Scott, G.R. 1987. The taxonomic status of the causative agent of heartwater. *Onderstepoort Journal of Veterinary Research* 54:257–260.
Stewart, C.G. 1987a. Specific immunity in farm animals to heartwater. *Onderstepoort Journal of Veterinary Research* 54:341–342.
———. 1987b. Specific immunity in mice to heartwater. *Onderstepoort Journal of Veterinary Research* 54:343–344.
Uilenberg, G. 1981. Heartwater disease. In *Diseases of cattle in the tropics: Economics and zoonotic relevance,* ed. M. Ristic and I. McIntyre. London: Martinus Nijhoff, pp. 345–360.
———. 1982. Experimental transmission of *Cowdria ruminantium* by the Gulf Coast tick, *Amblyomma maculatum:* Danger of introducing heartwater and benign African theileriosis onto the American mainland. *American Journal of Veterinary Research* 43:1279–1282.
———. 1983. Heartwater (*Cowdria ruminantium* infection): Current status. *Advances in Veterinary Science and Comparative Medicine* 27:427–480.
Uilenberg, G., and E. Camus. 1993. Heartwater (cowdriosis). In *Rickettsial and chlamydial diseases of domestic animals,* ed. Z. Woldehiwet and M. Ristic. New York: Pergamon, pp. 293–332.
Valli, V.E.O. 1985. The hematopoietic system. In *Pathology of domestic animals,* vol. 3, ed. K.V.F. Jubb, P.C. Kennedy, and N. Palmer, 3d ed. London: Academic, p. 88.
Van de Pypekamp, H.E., and L. Prozesky. 1987. Heartwater: An overview of the clinical signs, susceptibility and differential diagnosis of the disease in domestic ruminants. *Onderstepoort Journal of Veterinary Research* 54:23–266.
Walker, J.B., and A. Olwage. 1987. The tick vectors of *Cowdria ruminantium* (Ixodoidae, Ixodidae, genus *Amblyomma*) and their distribution. *Onderstepoort Journal of Veterinary Research* 54:353–359.
Wassink, L.A., F. Jongejan, E. Gruys, and G. Uilenberg. 1990. Observations on mouse-infective stocks of *Cowdria ruminantium:* Microscopical demonstration of the Kwanyanga stock in mouse tissue and the carrier-status of the Senegal stock in mice. *Research in Veterinary Science* 48:389–390.
Weiss, E., and J.W. Moulder. 1984. Order I: Rickettsiales. In *Bergey's manual of systematic bacteriology,* vol. 1, ed. N.R. Krieg and J.G. Hold. London: Williams and Wilkins, pp. 687–698.
Williams, J.C., and M.H. Vodkin. 1987. Metabolism and genetics of chlamydias and rickettsias. *Onderstepoort Journal of Veterinary Research* 54:211–222.
Young, E., and P.A. Basson. 1973. Heartwater in the eland. *Journal of the South African Veterinary Association* 44:185–186.
Yunker, C.E., K.M. Kocan, R.A.I. Norval, and M.J. Burridge. 1987. Distinctive staining of colonies of *Cowdria ruminantium* in midguts of *Amblyomma hebraeum. Onderstepoort Journal of Veterinary Research* 54:183–185.

SALMON POISONING DISEASE

WILLIAM J. FOREYT

Synonyms: Salmon disease, salmon poisoning.

INTRODUCTION. Salmon poisoning disease (SPD) is a highly fatal helminth-transmitted rickettsial disease of wild and domestic Canidae and occurs on the western slopes of the Cascade Mountains, from northern California to central Washington (Fig. 27.6). The disease is caused by *Neorickettsia helminthoeca* and is antigenically distinct from other pathogenic rickettsial species infecting canids. Most cases of SPD occur in areas coincident with the range of the snail intermediate host, *Oxytrema silicula,* but occasionally cases of SPD occur outside the indigenous range of the disease in areas where infected fish migrate or are transported. Cases of SPD in domestic dogs *Canis familiaris* in Vancouver, British Columbia, may indicate that the indigenous range of the disease is greater than previously reported. The fatal effects of the disease in domestic dogs is well documented, but the impact of SPD on free-ranging canid populations is not well known.

HISTORY. The disease was first described in 1814 in Henry's *Astoria Journal,* when dogs died after eating raw salmon [see Millemann and Knapp (1970)]. In 1849, Thornton reported that Oregon dogs that ate raw salmon became ill as early as day 2 after ingesting the fish and died on day 10. Cooper and Stukley (1859)

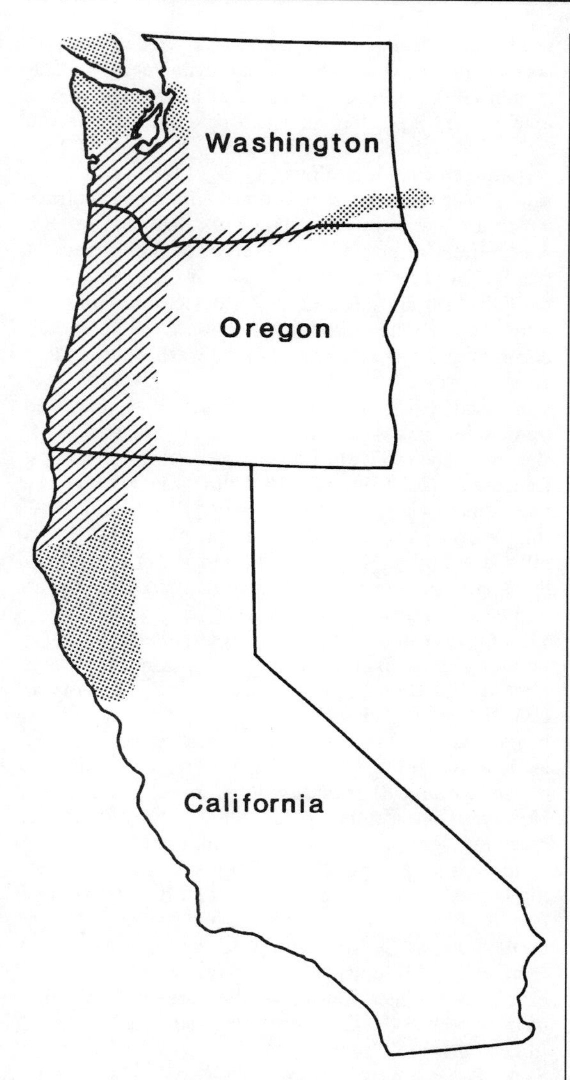

FIG. 27.6—Distribution of indigenous salmon poisoning disease (SPD). Areas indicated by *slashed lines* represent the distribution of the snail *Oxytrema silicula* and the usual distribution of SPD. Areas indicted by *dots* represent occasional cases of SPD usually resulting from infected migrating fish.

noted that very few dogs survived this "salmon sickness" in Oregon, and concluded the disease was probably distemper. They also observed that dogs that survived infection were immune to reinfection. Pernot (1911) surmised that the white, amoeba-like cysts in fish kidneys were the cause of the disease. He was the first to reproduce the disease experimentally by feeding fish kidneys to dogs and also by injecting blood from an infected dog subcutaneously into another dog. Donham et al. (1926) demonstrated that adult trematodes developed in dogs after ingestion of the cysts in fish and indicated the trematodes caused SPD. They also demonstrated that dogs surviving infection were immune to subsequent infection. Sims et al. (1931a,b) confirmed Donham and coworkers findings by inducing SPD by intraperitoneal injection of mature trematodes. They also identified the snail *Goniobasis plicifer* var. *silicula* as the intermediate host of the trematode *Nanophyetus salmincola* and indicated that the distribution of SPD was determined by the distribution of the intermediate snail host. The name of the snail was changed later by Morrison (1954) to *O. silicula.* Evidence that *N. salmincola* was the disease vector rather than the causative agent was reported by Simms and Muth (1934), who suggested that the actual agent was a rickettsia or a hemosporidian. In 1950, Cordy and Gorham were the first to discover the infectious agent, suggesting it be placed in the order Rickettsiales. Philip et al. (1954) later proposed the current name of the organism, *N. helminthoeca.*

DISTRIBUTION. Natural transmission of SPD occurs primarily within the range of the snail *O. silicula* west of the Cascade Mountains from northern California to west-central Washington (Fig. 27.6). Additional cases of SPD have been reported outside the range of *O. silicula* and are likely the result of infected fish migrating to new areas. Wilson and Foreyt (1985) reported that 51% of 251 migrating steelhead trout *Oncorhynchus mykiss* contained metacercariae of *N. salmincola* when they reached the Dworshak National Fish Hatchery at Ahsahka, Idaho. Although SPD was not demonstrated in that study, the high percentage of fish returning approximately 800 km to Idaho from the Pacific Ocean with *N. salmincola* indicated that anadromous salmonids in the Pacific Northwest have the potential of carrying SPD to all major rivers that support anadromous salmonids. Reports of SPD in dogs in Vancouver, British Columbia (Booth et al. 1984), may represent cases from migrating fish or may indicate that the indigenous range of the disease is greater than reported previously.

HOST RANGE. Only some members of the Canidae are known to be susceptible to the fatal effects of SPD. These include the domestic dog, coyote *Canis latrans,* and red fox *Vulpes vulpes.* Although the susceptibility of wolves *Canis lupus* and other canids is unknown, they likely also are susceptible to the fatal effects of SPD. The snail, *O. silicula,* is the only documented first intermediate host of *N. salmincola,* and many fish may serve as second intermediate hosts.

ETIOLOGY. The etiologic agent of SPD is *N. helminthoeca,* a coccoid or coccobacillary rickettsia that is approximately 0.3 μm in size. Pleomorphic rods, up to 2 μm long, sometimes bent in rings or crescents, have been observed. The Gram-negative rickettsial organisms appear purple with Giemsa stain, red with

Macchiavello's stain, black or dark brown with Levaditi's method, and pale blue with hematoxylin-eosin stain. The rickettsiae almost fill the cytoplasm of cells of the mononuclear phagocyte system that they primarily infect (Fig. 27.7). Initial attempts to grow the organism in embryonated chicken eggs, bacteriologic media, and duck embryo tissue cultures have been unsuccessful, but rickettsiae have been grown in canine monocytes (Frank et al. 1974a), in canine leukocytes and sarcoma cells, in mouse lymphoblasts, and in a canine macrophage cell line (Noonan 1973; Rikihisa et al. 1991). Antigenically and genetically, *N. helminthoeca* is closely related to *Ehrlichia* spp. and, based on Western blot analysis and indirect fluorescent antibody-labeling results, *N. helminthoeca* is most closely related to *Ehrlichia risticii,* the agent of Potomac horse fever, and *Ehrlichia sennetsu,* the agent of human sennetsu fever in Japan (Rikihisa 1991). All three agents are likely in the same genus.

The Elokomin fluke fever (EFF) agent (Farrell et al. 1973) probably is another strain of *N. helminthoeca* (Frank et al. 1974a,b). The disease in dogs associated with the EFF agent results in high morbidity but a lower mortality than with SPD. It appears that metacercariae can harbor both EFF and SPD agents simultaneously.

TRANSMISSION AND EPIDEMIOLOGY. The vector of SPD is a trematode, *N. salmincola,* which harbors the rickettsiae throughout its life-cycle stages from egg to adult (Millemann and Knapp 1970; Knapp and Millemann 1981). Three different hosts are required for the completion of the trematode life cycle: snails, fish, and mammals or birds (Fig. 27.8). The snail intermediate host, *O. silicula,* is a pleurocerid snail that inhabits fresh or brackish stream water in coastal areas of Washington, Oregon, and northern California (Fig. 27.6). Cercariae (free-swimming trematode larvae) leave the snail and penetrate the second intermediate host, which is usually a salmonid fish, certain species of nonsalmonid fish, or the Pacific giant salamander *Dicamptodon ensatus.* The metacercariae, which are approximately 200 μm in diameter, usually concentrate in the posterior portion of the kidneys of fish but can be found in any tissue or in the mucus on the fish. Fish are infected in freshwater and retain the trematode and the rickettsial infection throughout their ocean migration before returning to fresh water up to 3 years later (Weiseth et al. 1974).

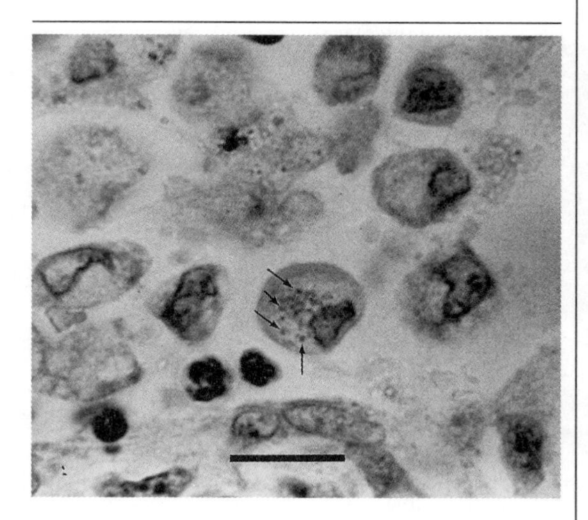

FIG. 27.7—*Neorickettsia helminthoeca* in a lymph node. Note the enlarged histiocyte with cytoplasmic *Neorickettsia* (*arrows*). Giemsa stain; bar = 20 μm.

In dead fish, rickettsiae in metacercariae (encysted trematode larvae) do not survive 30 days at 4° C (Foreyt et al. 1987). In lymph nodes, organisms resist freezing at –20° C for 31–158 days (Philip et al. 1954); they remain viable in leukocytes at 4.5° C and 52.5° C for 48 hours and 2 minutes, respectively, but not at 60° C for 5 minutes (Sims and Muth 1934). At –80° C, the agent can be maintained in cell culture fluid for up to 3 months (Brown et al. 1972).

Adult trematodes, which are approximately 1.0–2.5 mm long, develop in the intestine approximately 6 days after the ingestion of metacercariae-infected fish in certain fish-eating mammals, such as coyotes, foxes, bears, and raccoons *Procyon lotor,* and birds, that serve as definitive hosts. Clinical signs of rickettsial disease occur in Canidae, primarily dogs, foxes, and coyotes. However, two captive polar bears *Ursus maritimus* receiving long-term glucocorticoid therapy for skin conditions succumbed to an SPD-like disease after eating inadequately frozen salmon [cited by Gorham and Foreyt (1998)]. Cats are not susceptible to SPD, but trematodes will develop when infected fish are ingested (Knapp and Millemann 1981). Salmon poisoning disease also has been transmitted by parenteral injection of infected blood, spleen and lymph suspensions, adult flukes, helminth-infected snail livers, and helminth eggs. Partial transmission success was obtained by allowing ticks *Haemaphysalis leachi* and *Rhipicephalus sanguineus* that had fed on infected dogs to subsequently feed on susceptible dogs and by parenteral injection of suspensions of *R. sanguineus* into dogs (Philip 1955). Susceptible dogs also have been experimentally infected with aerosolized lymph node suspensions from infected dogs and on rare occasions, direct transmission of infection between dogs has been suspected (Bosman et al. 1970).

PATHOGENESIS. After ingestion of raw, metacercariae-infected salmonid fish by a susceptible animal, the trematode matures, attaches to the mucosa of the intestine and, by some unknown mechanism, inoculates the rickettsiae. Initial replication of rickettsiae probably takes place in the epithelial cells of the villi or in the intestinal lymphoid tissue. Inflammation of solitary lymphoid follicles and Peyer's patches along the

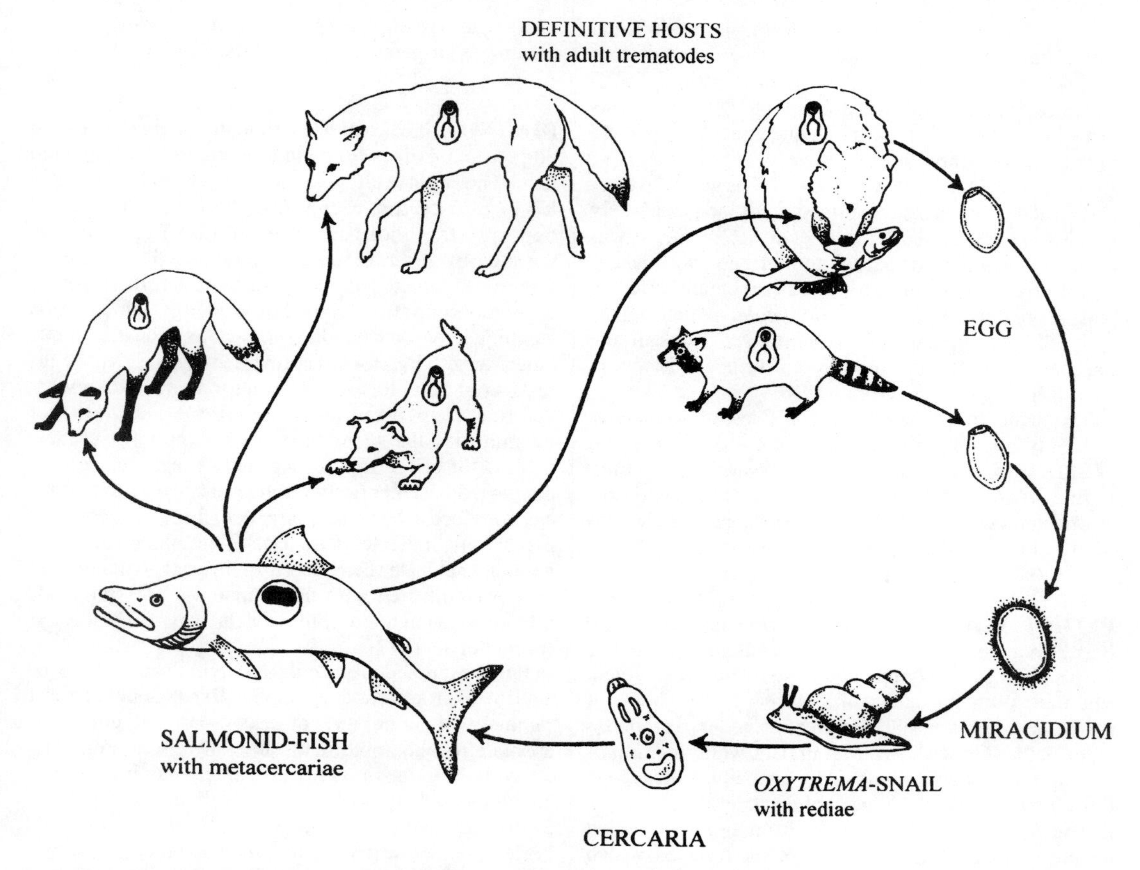

FIG. 27.8—Life cycle of the trematode *Nanophyetus salmincola. Neorickettsia helminthoeca* is carried throughout the entire life cycle of the trematode.

intestinal tract contributes to enteritis. Mild enteritis may be observed in dogs infected only with the flukes, without rickettsiae. Rickettsiae enter the blood early in the course of the disease and spread to the lymph nodes, spleen, tonsils, thymus, liver, lungs, and brain (Frank et al. 1974b). Although secondary bacterial infections often occur, the exact cause of death in SPD is unknown. Investigations to demonstrate a toxin have been limited. The trematode *N. salmincola* is generally not considered pathogenic.

CLINICAL FINDINGS. The signs of SPD infection are consistent in all Canidae. The usual incubation period following the ingestion of parasitized fish is 5–7 days, although some dogs have incubation periods as long as 19–33 days. The first sign usually is a sudden febrile response, which typically reaches a peak of 40° C–42° C. The temperature gradually decreases to normal or below normal over the next 4–8 days. Dogs are frequently hypothermic when death occurs 7–10 days after the initial clinical evidence of infection. Some animals show only a slight increase in temperature or a shortened febrile period; however, they may still die if left untreated.

Anorexia frequently accompanies or follows the onset of fever and may be marked and complete. Affected animals often continue to have inappetence throughout the course of the disease. Marked weight loss, weakness, and depression usually follow. Within 14 days of eating infected fish, experimentally infected coyotes lost approximately 58% of their body weight when compared with uninfected coyotes (Foreyt et al. 1982, 1987). Diarrhea and vomiting may occur; the diarrhea becomes progressively worse and often consists primarily of blood at the time of death. Affected animals will occasionally exhibit extreme thirst and will drink copious quantities of water. A serous nasal discharge may be observed early in the febrile period. Later, a mucopurulent conjunctival exudate may be seen. Enlarged cervical and prescapular lymph nodes can be palpated as early as 5 days after infection.

Infected animals may show severe gastrointestinal signs that are often clinically indistinguishable from canine parvovirus enteritis. Canine distemper and SPD can occur concurrently, but appropriate laboratory tests can be conducted to determine which agent is involved in a particular animal.

A free-ranging cougar kitten *Felis concolor* apparently died from massive infection of approximately 650,000 *N. salmincola* (Kistner et al. 1979). Death was attributed to malnutrition due to diarrhea and thickening of the intestine. In coyotes experimentally infected with more than 15,000 *N. salmincola,* which apparently did not contain *N. helminthoeca,* intermittent diarrhea occurred, but the coyotes survived (Foreyt et al. 1987).

The incubation period for the EFF agent is generally 5–12 days. The febrile period, which differs from that of SPD, is marked by a plateau of elevated temperature lasting 4–7 days, followed by a decline, usually to subnormal temperature (Farrell 1974). Other signs are similar to those of SPD.

PATHOLOGIC FINDINGS. The principal gross findings at necropsy are changes in lymphoid tissues. The tonsils, thymus, and lymph nodes are markedly enlarged. The most pronounced swelling occurs in the ileocecal, colic, mesenteric, portal, and lumbar nodes. The nodes are usually yellowish, with prominent white foci representing the cortical follicles. Occasionally, nodes show diffuse petechia, and edema is often observed.

The spleen frequently ranges from slightly swollen to nearly twice the normal size. Splenic follicles, which often appear as grayish-white nodules in foxes, are unaffected in dogs. The spleens of animals that die of SPD are typically a dark-bluish red, smooth, soft, and blood filled. Livers of dogs are usually normal, although those of foxes are usually soft and a pale yellow-brown. Hemorrhages may appear in the gallbladder wall. Petechia may be the only change in the pancreas. The kidneys of dogs are grossly normal, whereas those of foxes may have a slight color change toward a pale yellowish brown. The mucosa of the urinary bladder may show petechia.

Along the intestinal tract, petechia may be apparent in the mucosa of the lower esophagus, large intestine, ileocolic valve, distal colon, rectum, and gastric serosa. There may be bleeding ulcers in the pylorus. Intussusception of the ileum in the colon may also occur. The intestinal contents frequently contain free blood. Some blood may also appear in the colon and rectum. The intestines are typically empty except for some bile-stained mucus. Flukes in the intestinal tissue, primarily found in the duodenum, cause some tissue damage.

Microscopically, a characteristic pattern is observed in lymphocytic tissues. The lymph nodes show marked and consistent depletion in the number of mature lymphocytes, with hyperplasia of the mononuclear phagocytes in the cortex and medulla. In most foxes and dogs, there are foci of necrosis in the mononuclear phagocyte system. There is usually nonsuppurative meningitis or meningoencephalitis (Hadlow 1957).

DIAGNOSIS. Operculated trematode eggs appear in dog feces 5–8 days following the ingestion of infected fish. The light yellow-brown egg is approximately 87–97 × 35–55 μm, with a small, blunt point on the end opposite the indistinct operculum. Eggs can be detected by several methods, including direct smears, trichrome-stained fecal preparations, a formalin-ethyl acetic concentration method, a washing-sedimentation technique, or the standard sugar flotation technique (specific gravity, 1.27) (Foreyt 1994). Although eggs recovered by this latter method are somewhat deformed, eggs are clearly recognizable, and the author recommends the sugar floatation as the most reliable method of diagnosis. Rickettsial disease cannot be diagnosed based entirely on the presence of trematode eggs in feces, because trematode infection does not necessarily indicate rickettsial infection. In addition, animals that have recovered from the rickettsial disease may be reinfected with the trematode. The trematode infection can remain patent for at least 60–250 days (Foreyt et al. 1987).

Fluid aspirated from enlarged lymph nodes can be air dried on a microscope slide, fixed, defatted for 1 minute with a mixture of equal parts of ether and absolute alcohol, and then stained by Giemsa or Macchiavello's stain. In addition, a rapid-staining Giemsa technique, which involves staining the fixed smears for 2 minutes with equal parts of stock Giemsa and buffered water at pH 7.2 and then washing the slides, can be used (Farrell et al. 1955). Typical intracytoplasmic rickettsial bodies are characteristically seen in mononuclear phagocytes (Fig. 27.7). Extracellular organisms are not easily separated from artifacts and should not be considered diagnostic for SPD.

Hematologic and biochemical findings in SPD-infected domestic dogs are often nonspecific. Total leukocyte counts vary and range from being classified as leukopenia to leukocytosis (Schalm 1978; Mack et al. 1990). In experimentally infected coyotes, significantly higher numbers of band cells, lower numbers of eosinophils, and lower concentrations of creatinine, glucose, calcium, inorganic phosphorus, albumin, and alkaline phosphatase were detected when compared with uninfected coyotes (Foreyt et al. 1987).

CONTROL AND TREATMENT. In captive animals, *N. helminthoeca* can be controlled with oral or parenteral sulfonamides, penicillin, chlortetracycline, chloramphenicol, or oxytetracycline (Gorham and Foreyt 1998). Relief of dehydration, emesis, and diarrhea is also important. In addition, the most supportive treatment consists of keeping the animal dry, clean, and warm.

Praziquantel, when given orally or subcutaneously as one dose at 10–30 mg/kg of body weight, is highly effective against *N. salmincola* in coyotes and dogs (Foreyt and Gorham 1987). Elimination of *N. salmincola* from infected animals will minimize the diarrhea associated with fluke infections alone.

Vaccines have not been developed against SPD; therefore, keeping susceptible canids from feeding on infected fish is the best means of preventing infection. Because metacercariae can remain viable for months in rotting fish carcasses, canids will become infected with the fluke if they eat decomposed fish; however, they may not develop SPD. Also, in many areas, metacercariae apparently contain nonpathogenic rickettsiae (Green et al. 1986). Freezing fish at −20° C for 24 hours or thoroughly cooking infected fish will kill both the metacercariae and rickettsiae (Farrell et al. 1974). Supplemental prevention methods include isolation of animals with SPD and sterilization of equipment used around infected animals. Methods to prevent infection of fish with *N. salmincola,* such as elimination of snails, are impractical.

PUBLIC HEALTH CONCERNS. Salmon poisoning disease does not affect humans. However, *N. salmincola* can infect humans who eat inadequately cooked metacercaria-infected fish. Most human infections are likely asymptomatic, but some infected humans develop severe abdominal pain, associated with diarrhea, gas and bloating, fatigue, nausea, vomiting, weight loss, fever, decreased appetite, and peripheral blood eosinophilia (Fritsche et al. 1989).

DOMESTIC ANIMAL CONCERNS. Domestic dogs are highly susceptible to SPD and are likely to die without treatment. Veterinarians in the enzootic SPD areas of Washington, Oregon, and California (Fig. 27.6) routinely diagnose and treat SPD-affected dogs.

MANAGEMENT IMPLICATIONS. Because coyotes and foxes are susceptible to SPD, it is likely that these free-ranging animals usually die when infected with it. Although free-ranging coyotes and red foxes are present within the enzootic areas of SPD, its effect on their populations cannot be estimated. It is likely that SPD will always cause some mortality among susceptible canids in SPD-enzootic areas. Salmon poisoning disease has been proposed as a potential candidate for biologic control of coyotes in areas where it is necessary for economic domestic livestock production (Green et al. 1986; Foreyt et al. 1987). This could be accomplished by transporting infected fish to areas where coyote control is needed and using the infected fish as coyote bait. Because of the severe illness caused by SPD before death, humane concerns will likely preclude the use of SPD as a humane method of coyote, fox, or wolf control. The effect of SPD on wolves and other canids is unknown, but they also are likely susceptible to the fatal effects of SPD. Because wolves are being translocated into areas of the western United States where they have been absent for many years, SPD may have an effect on some wolf populations in areas where translocated wolves have access to infected fish.

LITERATURE CITED

Booth, A.J., L. Stogdale, and J.A. Grigor. 1984. Salmon poisoning disease in dogs on southern Vancouver Island. *Canine Veterinary Journal* 25:2–6.

Bosman, D.D., R.K. Farrell, and J.R. Gorham. 1970. Nonendoparasite transmission of salmon poisoning disease of dogs. *Journal of the American Veterinary Medical Association* 156:1907–1910.

Brown, J.L., D.L. Huxsoll, M. Ristic, and P.K. Hildebrandt. 1972. In vitro cultivation of *Neorickettsia helminthoeca,* the causative agent of salmon poisoning disease. *American Journal of Veterinary Research* 33:1695–1700.

Cooper, J.G., and G. Stukley. 1859. *The natural history of Washington Territory with much relating to Minnesota, Nebraska, Kansas, Oregon, and California: Between the thirty-sixth and forty-ninth parallels of latitude, being those parts of the final reports on the survey of the North Pacific Railroad route, containing the climate and physical geography, with full catalogues and descriptions of the plants and animals collected from 1853 to 1857, pt 3, Zoological Report.* New York: Ballier Brothers, p. 112.

Cordy, D.R., and J.R. Gorham. 1950. The pathology and etiology of salmon poisoning in the dog and fox. *American Journal of Pathology* 26:617–637.

Donham, C.R., B.T. Simms, and F.W. Miller. 1926. So-called salmon poisoning in dogs (progress report). *Journal of the American Veterinary Medical Association* 68:701.

Farrell, R.K. 1974. Canine rickettsiosis. In *Current veterinary therapy,* ed. R.W. Kirk, 5th ed. Philadelphia: W.B. Saunders, pp. 985–987.

Farrell, R.K., R.L. Ott, and J.R. Gorham. 1955. The clinical laboratory diagnosis of salmon poisoning. *Journal of the American Veterinary Medical Association* 127:241–244.

Farrell, R.K., R.W. Leader, and S.D. Johnson. 1973. Differentiation of salmon poisoning disease and Elokomin fluke fever: Studies with the black bear (*Ursus americana*). *American Journal of Veterinary Research* 34:919–922.

Farrell, R.K., O.A. Soave, and S.D. Johnston. 1974. *Nanophyetus salmincola* infections in kippered salmon. *American Journal of Public Health* 64:808–809.

Foreyt, W.J. 1994. *Veterinary parasitology reference manual.* Pullman: Washington State University, 178 pp.

Foreyt, W.J., and J.R. Gorham. 1987. Evaluation of praziquantel against induced *Nanophyetus salmincola* infections in coyotes and dogs. *American Journal of Veterinary Research* 48:563–565.

Foreyt, W.J., S. Thorsen, and G.R. Gorham. 1982. Experimental salmon poisoning in juvenile coyote (*Canis latrans*). *Journal of Wildlife Diseases* 18:159–162.

Foreyt, W.J., S. Thorsen, J.S. Green, C.W. Leathers, and B.R. LeaMaster. 1987. Salmon poisoning disease in juvenile coyotes: Clinical evaluation and infectivity of metacercariae and rickettsiae. *Journal of Wildlife Diseases* 23:412–417.

Frank, D.W., T.C. McGuire, J.R. Gorham, and W.C. Davis. 1974a. Cultivation of two species of *Neorickettsia* in canine monocytes. *Journal of Infectious Diseases* 129:257–262.

Frank, D.W., T.C. McGuire, J.R. Gorham, and R.K. Farrell. 1974b. Lymphoreticular lesions of canine rickettsiosis. *Journal of Infectious Diseases* 129:163–171.

Fritsche, T.R., R.L. Eastburn, L.H. Wiggins, and C.A. Terhune. 1989. Praziquantel for treatment of human *Nanophyetus salmincola* (*Troglotrema salmincola*) infection. *Journal of Infectious Diseases* 160:896–899.

Gorham, J.R., and W.J. Foreyt. 1998. Salmon poisoning disease. In *Infectious diseases of the dog and cat,* ed. C.E. Greene. Philadelphia: W.B. Saunders, pp. 135–139.

Green, J.S., B.R. LeaMaster, W.J. Foreyt, and R.A. Woodruff. 1986. Salmon poisoning disease: Research on a potential method of lethal control for coyotes. In *Proceedings, 12th vertebrate pest conference,* ed. T.P. Salmon. Davis: University of California, pp. 312–317.

Hadlow, W.J. 1957. Neuropathology of experimental salmon poisoning in dogs. *American Journal of Veterinary Research* 18:898–908.

Kistner, T.P., D. Wyse, and J.A. Schmitz. 1979. Pathogenicity attributed to massive infection of *Nanophyetus salmincola* in a cougar. *Journal of Wildlife Diseases* 15:419–420.

Knapp, S.E., and R.E. Millemann. 1981. Salmon poisoning disease. In *Infectious diseases of wild mammals,* ed. J.W. Davis, L.H. Karstad, and D.O. Trainer. Ames: Iowa State University Press, pp. 332–342.

Mack, R.E., M.G. Becovitch, G.V. Ling, R.T. Tobinger, and A.C. Belli. 1990. Salmon disease complex in dogs: A review of 45 cases. *California Veterinarian* 44:42–45.

Millemann, R.E., and S.E. Knapp. 1970. Biology of *Nanophyetus salmincola* and "salmon poisoning" disease. *Advances in Parasitology* 8:1–14.

Morrison, J.P.E. 1954. The relationships of Old and New World melanians. *Procedures of the United States Natural Museum* 103:357–394.

Noonan, W.E. 1973. *Neorickettsia helminthoeca* in cell culture. Ph.D. diss., Oregon State University, Corvallis, Oregon.

Pernot, E.F. 1911. "Salmoning" of dogs. *Oregon State Board Health Bulletin* 5:1.

Philip, C.B. 1955. There's always something new under the "parasitological sun" (the unique story of helminth-borne salmon poisoning disease). *Journal of Parasitology* 41:125–148.

Philip, C.B., W.J. Hadlow, and L.E. Hughes. 1954. Studies on salmon poisoning disease of canines: 1. The rickettsial relationships and pathogenicity of *Neorickettsia helminthoeca. Experimental Parasitology* 3:336–350.

Rikihisa, Y. 1991. Cross-reacting antigens between *Neorickettsia helminthoeca* and *Ehrlichia* spp. shown by immunofluorescence and Western immunoblotting. *Journal of Clinical Microbiology* 29:2024–2029.

Rikihisa, Y., J. Stills, and G. Zimmerman. 1991. Isolation and continuous culture of *Neorickettsia helminthoeca* in a macrophage cell line. *Journal of Clinical Microbiology* 29:1928–1933.

Schalm, O. W. 1978. Leukocyte counts and lymph node cytology in salmon poisoning of dogs. *Canine Practice* 5:59–63.

Simms B.T., and O.H. Muth. 1934. Salmon poisoning: Transmission and immunization studies. In *Proceedings of the 5th Pacific science congress,* pp. 2949–2960.

Simms B.T., C.R. Donham, and J.N. Shaw. 1931a. Salmon poisoning. *American Journal of Hygiene* 13:363–391.

Simms B.T., C.R. Donham, J.N. Shaw, and A.M. McApes. 1931b. Salmon poisoning. *Journal of American Veterinary Medical Association* 78:181–195.

Thornton, J.Q. 1849. Oregon and California in 1948, vol. 1. New York: Harper & Bros.

Weiseth, P.R., R.K. Farrell, and S.D. Johnston. 1974. Prevalence of *Nanophyetus salmincola* in ocean-caught salmon. *Journal of American Veterinary Medical Association* 165:849–850.

Wilson, R.L., and W.J. Foreyt. 1985. Prevalence of *Nanophyetus salmincola,* the vector of salmon poisoning disease in steelhead trout (*Salmo gairdneri*) in Idaho. *Proceedings of the Helminthological Society of Washington* 52:136–137.

28 MISCELLANEOUS BACTERIAL INFECTIONS

ACTINOMYCES AND *ARCANOBACTERIUM* INFECTIONS

GARY WOBESER

Synonyms: *Arcanobacterium pyogenes* was formerly known as *Corynebacterium pyogenes* and *Actinomyces pyogenes* (Ramos et al. 1997).

ETIOLOGY, TRANSMISSION, AND EPIDEMIOLOGY. The genera *Actinomyces* and *Arcanobacterium* contain a number of diverse non-spore-forming Gram-positive bacteria, at least two of which are of potential importance in wild animals. *Actinomyces bovis* is an anaerobic, filamentous organism that causes chronic mandibular osteomyelitis (actinomycosis or "lumpy jaw"). *Arcanobacterium pyogenes* is a facultatively anaerobic organism that causes a variety of suppurative lesions or abscesses, particularly in ungulates. Infections with both bacteria are sporadic.

Mandibular osteomyelitis caused by *A. bovis* has been reported in various species of North American deer *Odocoileus* spp., as well as moose *Alces alces,* pronghorn *Antilocapra americana,* caribou *Rangifer tarandus,* and mountain sheep *Ovis* spp. (Howe 1981). However, many of the descriptions have been based on gross examination of tissues without bacteriologic confirmation, and it is unknown how many of these cases in wild animals actually represent actinomycosis rather than other conditions (Drake 1951; Glaze et al. 1982). *Actinomyces* spp. occasionally cause granulomatous lesions in the peritoneum or internal organs (Gyles 1993).

Arcanobacterium pyogenes should be expected in suppurative lesions in ungulates. It was the organism most commonly isolated from cerebral abscesses in deer (Davidson et al. 1990) and from a variety of purulent processes in wild deer, moose, pronghorn, and wild sheep (Rosen and Holden 1961; Thorne et al. 1982; Hoefs and Bunch 1992), as well as captive deer and antelope (Zulty and Montali 1988). This bacterium has a synergistic relationship with *Fusobacterium necrophorum* (Smith et al. 1989), and the two bacteria often are found in mixed infections.

CLINICAL SIGNS. Infection with *A. bovis* in ungulates usually results in noticeable swelling in the mandibular area that may be associated with difficulty in mastication.

No specific clinical signs are associated with most infections with *A. pyogenes,* except in animals with intracranial abscesses, which may have obvious signs of nervous derangement (Davidson et al. 1990). Localized abscesses due to *A. pyogenes* often are found when clinically normal animals are butchered by hunters.

PATHOGENESIS AND PATHOLOGY. *Actinomyces bovis* typically produces destructive osteomyelitis of the mandible, although the infection may be limited to granulomatous inflammation of the soft tissues of the oral submucosa. If the bone is invaded, there is progressive destruction of trabeculae with associated proliferation of periosteal new bone. The affected area is greatly expanded and consists of dense connective tissue and porous bone containing many tracts filled with pus that may contain 1- to 2-mm pale granules (*sulfur granules*). Teeth in the affected areas often are lost. The inflammation usually does not spread widely and often does not involve regional lymph nodes. In the rare cases where other tissues are involved, there usually is a localized granulomatous response.

Arcanobacterium pyogenes can induce purulent inflammation almost anywhere in the body. Usually this takes the form of localized abscesses, but the inflammation may be more diffuse within body cavities, joints, or tendon sheaths.

DIAGNOSIS. Actinomycosis should be suspected when there are typical proliferative and lytic granulomatous lesions involving the mandibles of ungulates, particularly if granules containing tangled filamentous Gram-positive organisms are present in the exudate. Because similar lesions can be produced by other agents, isolation of the organism under anaerobic conditions is required for confirmation.

Arcanobacterium pyogenes infection usually is diagnosed by isolation and identification of the bacterium.

TREATMENT AND CONTROL. Because these diseases are sporadic infections by opportunistic bacteria, no control is possible or necessary in wild animals. Treatment of both infections can be difficult; those requiring such information are referred to standard texts in veterinary medicine.

LITERATURE CITED

Davidson, W.R., V.F. Nettles, L.E. Hayes, E.W. Howerth, and C.E. Couvillion. 1990. Epidemiologic features of an intracranial abscessation/suppurative meningitis complex in white-tailed deer. *Journal of Wildlife Diseases* 26:460–467.

Drake, C.H. 1951. Mistaken diagnosis of actinomycosis in an American elk. *Journal of Wildlife Management* 15:284–287.

Glaze, R.L., M. Hoefs, and T.D. Bunch. 1982. Aberrations of the tooth arcade and mandible in Dall's sheep from southeastern Yukon. *Journal of Wildlife Diseases* 18:305–309.

Gyles, C.L. 1993. *Nocardia; Actinomyces; Dermatophilus.* In *Pathogenesis of bacterial infections in animals,* ed. C.L. Gyles and C.O. Thoen, 2d ed. Ames: Iowa State University Press, pp. 124–132.

Hoefs, M., and T. D. Bunch. 1992. Cranial asymmetry in a Dall sheep ram (*Ovis dalli dalli*). *Journal of Wildlife Diseases* 28:330–332.

Howe, D.L. 1981. Miscellaneous bacterial diseases. In *Infectious disease of wild mammals,* ed. J.W. Davis, L.H. Karstad, and D.O. Trainer, 2d ed. Ames: Iowa State Press, pp. 418–422.

Ramos, C.P., G. Foster, and M.D. Collins. 1997. Phylogenetic analysis of the genus *Actinomyces* based on 16S rRNA gene sequences: Description of *Arcanobacterium phocae* sp. nov., *Arcanobacterium bernardiae* comb. nov., and *Arcanobacterium pyogenes* comb. nov. *International Journal of Systematic Bacteriology* 47:46–53.

Rosen, M.N., and F.F. Holden. 1961. Multiple purulent abscesses (*Corynebacterium pyogenes*) of deer. *California Fish and Game* 47:293–299.

Smith, G.R., D. Till, M. Wallace, and D.E. Noakes. 1989. Enhancement of the infectivity of *Fusobacterium necrophorum* by other bacteria. *Epidemiology and Infection* 102:447–458.

Thorne, E.T., N. Kingston, W.R. Jolley, and R.C. Bergstrom. 1982. *Diseases of wildlife in Wyoming,* 2d ed. Cheyenne: Wyoming Game and Fish Department, 353 pp.

Zulty, J.C., and R.J. Montali. 1988. *Actinomyces pyogenes* infection in exotic Bovidae and Cervidae: 17 cases (1978–1986). *Journal of Zoo Animal Medicine* 19:30–32.

CAMPYLOBACTER INFECTION

TORSTEN MÖRNER

Synonyms: Vibriosis, vibrionic enteritis, vibrionic abortion.

INTRODUCTION. The genus *Campylobacter* contains 13 species (Skirrow 1994). These bacteria are widespread among humans, mammals, and birds and are the most common bacterial agents isolated from primates with diarrhea. Relatively little is known about *Campylobacter* in wild mammals, and extrapolations must be made from what is known about infections in people, domestic animals, and captive wild species.

HISTORY, DISTRIBUTION, AND HOST RANGE. The first isolation of *Campylobacter* was made from the fetus of an aborted sheep in 1913 (McFadyean and Stockman 1913). The bacterium was named *Vibrio fetus* but later changed to *Campylobacter fetus.* Among the pathogenic *Campylobacter* spp. recognized today, some of the most commonly reported are *C. fetus,* which causes abortion, and *C. jejuni* and *C. coli,* which cause enteritis (Acha and Szyfres 1989).

Campylobacter spp. are found worldwide. *Campylobacter fetus* has been reported wherever sheep and cattle are kept (Skirrow 1994) and also has caused abortions in mink *Mustela vison,* domestic goats, and horses (Hunter et al. 1983; Skirrow 1994). A bacterium resembling *C. fetus* was isolated from wild bank voles *Microtus agrestis* (Fernie and Park 1977).

Enteric *Campylobacter,* such as *C. jejuni* and *C. coli,* occur naturally among wild and domesticated birds and mammals, and sometimes cause disease (Acha and Szyfres 1989). They are common causes of diarrhea in marmosets and tamarins (Callitrichidae) and macaques *Macaca* spp.(Paul-Murphy 1993). *Campylobacter* spp. also have been isolated from mink with enteritis (Hunter et al. 1986), from hares (Rosef et al. 1983), from small rodents, from the feces of a bear (Pacha et al. 1987), and from rusa deer *Cervus timorensis* (Hill et al. 1987).

ETIOLOGY. *Campylobacter* spp. are Gram-negative, curved, motile bacteria with flagella. They are microaerophilic, and many species are thermophilic (Skirrow 1994).

TRANSMISSION AND EPIDEMIOLOGY. The reservoir for *C. fetus* in cattle is infected cows or bulls, and transmission is through sexual contact (Acha and Szyfres 1989). Aborted fetuses and vaginal discharges contain high numbers of *C. fetus* and also can be sources of infection. The principal reservoirs of enteric *Campylobacter,* like *C. jejuni* and *C. coli,* are domestic and wild mammals and birds, which often are clinically healthy carriers. The enteric species are probably transmitted to people and animals most commonly through water (Skirrow 1994). Infection also can be acquired from feces or from infected carcasses. Direct transmission between infected individuals may occur but is unusual (Acha and Szyfres 1989).

Wild and domestic birds are regarded as major reservoirs of *Campylobacter* (Acha and Szyfres 1989). These bacteria have been found in 35% of migratory birds, 50% of town-dwelling pigeons, and 20%–70% of gulls. Kapperud and Rosef (1983) examined 540 wild birds and found *Campylobacter* in 28.4%, representing 11 of the 40 bird species examined.

CLINICAL SIGNS. *Campylobacter fetus* causes embryonic death and abortion. In sheep, abortion usually occurs in the final stage of gestation, whereas in cattle, it more often occurs very early and is manifest clinically as infertility (Acha and Szyfres 1989). Thus, in wild mammals, the full range from early embryonic death to late-term abortion might occur.

Enteric campylobacteriosis normally affects younger animals and causes acute diarrhea and enterocolitis that lasts about 14 days (Acha and Szyfres 1989; Paul-Murphy 1993).

PATHOLOGY. In abortions, there is acute placentitis and metritis, often with abundant neutrophils (Skirrow 1994). In the aborted fetus, there may be subcutaneous edema, serosanguineous fluid in the abdomen and thorax, and liver necrosis, but these lesions are not always present.

Ingestion of only a few hundred enteric *Campylobacter* spp. can be sufficient to cause enteritis. There is infiltration of the intestinal lamina propria with macrophages and lymphocytes, necrosis of enterocytes, and sometimes formation of crypt abscesses. Mesenteric lymphadenitis is common, and bacteria can be isolated from the inflamed lymph nodes (Paul-Murphy 1993; Skirrow 1994).

DIAGNOSIS. Diagnosis is based on occurrence of clinical disease or lesions, together with isolation of the organism from affected tissue. In abortion, diagnosis is based on culture of fetal tissue, placenta, or vaginal fluid (Acha and Szyfres 1989). In enteric campylobacteriosis, the agent can be isolated from the blood during the acute phase and from the feces in later stages (Acha and Szyfres 1989). A serologic diagnosis may be made on the basis of a rise in antibody titer in paired serum samples. Because healthy carrier animals occur, isolation of the agent alone is not proof of disease.

PUBLIC HEALTH CONCERNS. All known pathogenic species of *Campylobacter* can infect humans and commonly are isolated from patients with diarrhea. The importance of wild mammals as sources of human infection is not known.

LITERATURE CITED

Acha, A., and B. Szyfres. 1989. Campylobacteriosis. In *Zoonoses and communicable diseases common to man and animals,* 2nd ed. Washington, DC: Pan American Health Organization, pp. 45–54.

Fernie, D.S., and R.W. Park. 1977. The isolation and nature of campylobacters (microaerophilic vibrios) from laboratory and wild rodents. *Journal of Medical Microbiology* 10:325–329.

Hill, B.D., R.J. Thomas, and A.R. Mackenzie. 1987. *Campylobacter hyointestinalis*-associated enteritis in Moluccan rusa deer (*Cervus timorensis* subsp. *moluccensis*). *Journal of Comparative Pathology* 97:687–694.

Hunter, D.B., J.F. Prescott, J.R. Pettit, and W.E. Snow. 1983. *Campylobacter jejuni* as a cause of abortion in mink. *Canadian Veterinary Journal* 24:398–399.

Hunter, D.B., J.F. Prescott, J.R. Hoover, D.M. Hlywka, and J.A. Kerr. 1986. *Campylobacter* colitis in ranch mink in Ontario. *Canadian Journal of Veterinary Research* 50:47–53.

Kapperud, G., and O. Rosef. 1983. Avian wildlife reservoir of *Campylobacter fetus* subsp. *jejuni, Yersinia* spp., and *Salmonella* spp. in Norway. *Applied and Environmental Microbiology* 45:375–380.

McFadyean, J., and S. Stockman. 1913. *Report of the Department Committee appointed by the Board of Agriculture and Fisheries to inquire into epizootic abortion: Part III. Abortion in sheep.* London: HMSO.

Pacha, R.E., G.W. Clark, E.A. Williams, A.M. Carter, J.J. Scheffenmaier, and P. Debusschere. 1987. Small rodents and other mammals associated with mountain meadows as reservoirs *Giardia* spp. and *Campylobacter* spp. *Applied and Environmental Microbiology* 53:1574–1579.

Paul-Murphy, J. 1993. Bacterial enterocolitis in nonhuman primates. In *Zoo & wild animal medicine: Current therapy,* ed. M. Fowler. Philadelphia: W.B. Saunders, pp. 344–346.

Rosef, O., B. Gondrosen, G. Kapperud, and B. Underdal. 1983. Isolation and characterization of *Campylobacter jejuni* and *Campylobacter coli* from domestic and wild mammals in Norway. *Applied and Environmental Microbiology* 46:855–859.

Skirrow, M.B. 1994. Diseases due to *Campylobacter, Helicobacter* and related bacteria. *Journal of Comparative Pathology* 111:113–149.

DERMATOPHILOSIS

FREDRICK A. LEIGHTON

Synonyms: Contagious dermatitis, streptothricosis, lumpy wool disease, strawberry foot rot, proliferative dermatitis, mycotic dermatitis, Kirchi, Gasin-Gishu, Senkobo disease, Drodro-Boka, Savi, Ambarr-madow.

INTRODUCTION. Dermatophilosis is the preferred name for infections with the bacterium *Dermatophilus congolensis.* This organism infects the surface of the skin anywhere on the body, inducing marked epidermal hyperplasia and an inflammatory exudate that forms thick crusts. The disease occurs in a wide range of wild and domestic mammals and in people.

HISTORY, DISTRIBUTION, AND HOST RANGE. Dermatophilosis was first described by Van Saceghem in 1910 in Zaire as a disease of domestic cattle (Van Saceghem 1916; Zaria 1993). It has since been reported from all continents except Antarctica (Zaria 1993).

Its occurrence in wild mammals has been reviewed by Richard and Shotts (1976) and by Richard (1981), who listed reports in 21 species of wild mammals, including wild ungulates in Africa, North America, and Europe, South American primates, rabbits and rodents in North America, and various carnivores, including captive polar bears *Ursus maritimus.* More recently, dermatophilosis has been reported in woodchucks *Marmota monax* and striped skunk *Mephitis mephitis* (Salkin et al. 1981), mule deer *Odocoileus hemionus* (Williams et al. 1984), camels *Camelus dromedarius* (Gitao 1992), Kafue lechwe *Kobus leche* (Pandey et al. 1994), and orangutan *Pongo pygmaeus* (Brack et al. 1997). All mammals should be considered susceptible. Spontaneous infections also have been reported in reptiles, including crocodiles (Buenviaje et al. 1998), but not in birds (Richard 1981).

ETIOLOGY. *Dermatophilus congolensis* is a Gram-positive, acid-fast-negative bacterium with an unusual fungus-like life cycle and morphology. The form generally transferred among hosts is a dormant zoospore that is resistant to desiccation and high temperature. In a warm, moist environment, these spores develop multiple flagella and become motile in fluid. They then shed the flagella and sprout a long hypha-like germinal tube, often with multiple branches. The germinal tube eventually subdivides into discrete bacterial cells by formation of cell walls parallel to and across the axis of the germinal tube. The result is linear arrays of multiple rows of Gram-positive bacterial cells in an overall form resembling fungal hyphae (Richard et al. 1967). This morphology is unique and is diagnostic of *Dermatophilus* infection. The life cycle is completed when the individual cells become zoospores.

The organism is readily cultured on standard laboratory media such as blood agar at 36° C–37° C under aerobic or micro-aerophilic conditions (Gordon 1964; Zaria 1993).

TRANSMISSION AND EPIDEMIOLOGY. Transmission usually results from transfer of dormant zoospores from infected animals by direct contact (Zaria 1993). Zoospores also may persist for months on the skin, hair, or dried crusts of animals with resolved lesions, serving as sources of infection for new hosts and of reinfection for the original host. Mechanical transmission by arthropods such as ticks *Amblyomma variegatum* and flies *Stomoxys calcitrans* and *Musca domestica* also has been demonstrated (Macadam 1962; Richard and Pier 1966). There is no convincing evidence that the organism can persist for long periods or multiply outside a suitable vertebrate host.

The epidemiology of dermatophilosis is uncertain. Among cattle in tropical climates, the disease has a pronounced seasonal pattern, occurring during wet periods (Zaria 1993). Elsewhere and in other species, such patterns are less evident. Conditions leading to persistently wet skin are thought to favor infection by providing a medium for zoospore motility and by macerating the epidermis. Conditions leading to cuts and abrasions of the skin also are generally considered to favor infection. Ticks appear to be important factors in dermatophilosis in African cattle, and tick control has caused parallel reductions in dermatophilosis (Zaria 1993).

CLINICAL SIGNS. The disease is recognized by the typical appearance of the skin lesions. Mild infections may be apparent only on very close inspection. Lesions initially are discrete areas over which the hair is held stiffly by an enveloping crust of layers of shed epidermis and exudate. Such encrustations can be detached, usually revealing an underlying red, inflamed dermis that bleeds. Lesions can vary in severity, involving a few discrete areas, large confluent patches, or virtually the entire skin. Zones of hair loss may be the first feature noted. These areas usually are covered with thick scabs. Severely affected animals become emaciated. Secondary blowfly infestation of skin lesions has been a serious complication (Van Tonder and Horner 1994).

PATHOGENESIS AND PATHOLOGY. *Dermatophilus congolensis* inhabits the epidermis, the outer avascular cornifying layer of the skin. Following entry to the epidermis, growth of the bacterium induces an intense hyperplasia of epidermal cells and local inflammation, causing exudation of serum and neutrophils from vessels in the underlying dermis. This results in a layer of exudate and shed epidermis on the skin surface. The germinal layer of the epidermis may be partially destroyed but regenerates, and the cycle of proliferation and exudation is repeated.

Further iterations lead to progressively thicker masses of exudate and shed epidermis that entrap the hair coat. The typical gross lesions described under clinical signs (above) are the result of these cycles of proliferation, exudation, and regeneration.

Histologically, the epidermal lesion consists of layers of keratin debris and degenerate neutrophils among which the bacterium usually can be seen with its pathognomonic morphology. The stratum corneum usually is markedly thickened. The underlying dermis is infiltrated superficially with neutrophils and mononuclear inflammatory cells.

DIAGNOSIS. Diagnosis is based on identification of the bacterium in lesions, either via microscopy or by bacterial culture. When the classic morphology of parallel rows of Gram-positive bacteria in a hypha-like array is evident, a definitive diagnosis can be made without culture. However, this form is not always evident, and isolation and identification of the organism in culture then are required. Microscopic examination can be made on portions of detached crusts by direct smear if the material is moist or after moistening with sterile water or saline solution. The slides are dried and stained with a Wright's or Giemsa-type stain or Gram stain. Crusts, biopsy specimens, or portions of lesions removed at necropsy can be processed routinely for histology, stained similarly, and examined (Van Tonder and Horner 1994).

DIFFERENTIAL DIAGNOSIS. Several other infectious organisms can cause lesions that resemble those of dermatophilosis. These include mange (sarcoptic, chorioptic, psoroptic), infections with dermatophytic fungi (ringworm), certain staphylococcal infections, contagious ecthyma (orf), papillomatosis (warts), and photosensitivity dermatitis. Routine diagnostic techniques required to distinguish among these conditions include microscopic examination of skin scrapings, histopathology, and bacterial culture.

IMMUNITY. There is some experimental evidence of increased resistance after a primary infection, but repeated infections of individual animals are common, suggesting that immunity to the organism is generally weak and partial. Circulating antibodies to a variety of specific antigens of the bacterium can be detected, but none appear to afford protection. No effective vaccine has been produced. In addition to being mechanical vectors and agents of skin trauma, certain ticks may reduce resistance to *Dermatophilus* infection in cattle (Zaria 1993; Ambrose 1996).

CONTROL AND TREATMENT. Treatment of dermatophilosis is problematic. The location of the organism in the avascular epidermis beneath thick crusts of exudate affords it protection from antibiotics given either topically or parenterally. Nonetheless, successful treatment with parenteral antibiotics, including the combination of streptomycin and penicillin or oxytetracycline alone, has been reported (van Tonder and Horner 1994). Overall, control of ticks has had the most pronounced effect in reducing dermatophilosis in cattle populations. The disease has not been recognized as of sufficient importance in wild animal populations to warrant consideration of its control.

PUBLIC HEALTH CONCERNS. In people, dermatophilosis usually is a self-limiting, focal infection of the skin, without major significance. Immunocompromised individuals may suffer more serious infections (Zaria 1993).

LITERATURE CITED

Ambrose, N.C. 1996. The pathogenesis of dermatophilosis. *Tropical Animal Health and Production* 28(Suppl. 2):29S–37S.

Brack, M., C. Hochleithner, M. Hochleithner, and W. Zenker. 1997. Suspected dermatophilosis in an adult orangutan (*Pongo pygmaeus pygmaeus*). *Journal of Zoo and Wildlife Medicine* 28:336–341.

Buenviaje, G.N., P.W. Ladds, and Y. Martin. 1998. Pathology of skin diseases in crocodiles. *Australian Veterinary Journal* 76:357–363.

Gitao, C.G. 1992. Dermatophilosis in camels (*Camelus dromedarius* Linnaeus, 1758) in Kenya. *Review Scientifique et Technique de l'Office Internationale des Epizooties* 11:1079–1086.

Gordon, M.A. 1964. The genus *Dermatophilus. Journal of Bacteriology* 88:509–522.

Macadam, I. 1962. Bovine streptothricosis: Production of lesions by the bites of the tick (*Amblyomma variegatum*). *Veterinary Record* 74:634–646.

Pandey, G.S., A. Mweene, A.K. Suzuki, A. Nambota, and T. Kaji. 1994. Dermatophilosis (cutaneous streptothricosis) in Kafue lechwe (*Kobus leche kafuensis*). *Journal of Wildlife Diseases* 30:586–588.

Richard, J.L. 1981. Dermatophilosis. In *Infectious diseases of wild mammals,* ed. J.W. Davis, L.H. Karstad, and D.O. Trainer, 2d ed. Ames: Iowa State University Press, pp. 339–346.

Richard, J.L., and A.C. Pier. 1966. Transmission of *Dermatophilus congolensis* by *Stomoxys calcitrans* and *Musca domestica. American Journal of Veterinary Research* 27:419–423.

Richard, J.L., and E.B. Shotts. 1976. Wildlife reservoirs of dermatophilosis. In *Wildlife diseases,* ed. L.A. Page. New York: Plenum, pp. 205–214.

Richard, J.L., A.E. Ritchie, and A.C. Pier. 1967. Electron microscopic anatomy of motile-phase and germinating cells of *Dermatophilus congolensis. Journal of General Microbiology* 49:23–29.

Salkin, I.F., W.B. Stone, and M.A. Gordon. 1981. *Dermatophilus congolensis* infections in wildlife in New York State. *Journal of Clinical Microbiology* 14:604–606.

Van Saceghem, R. 1916. Étude complimentaire sur la dermatose contagieuse (impetigo contagieux). *Bulletin de la Société de Pathologie Exotique* 10:290–293.

Van Tonder, E.M. and R.F. Horner. 1994. Dermatophilosis. In *Infectious diseases of livestock with special reference to South Africa,* vol. 2, ed. J.A.W. Coetzer, G.R. Thomson, and R.C. Tustin. Cape Town: Oxford University Press, pp. 1472–1481.

Williams, E.S., A.C. Pier, and R.W. Wilson. 1984. Dermatophilosis in a mule deer, *Odocoileus hemionus* (Rafinesque), from Wyoming. *Journal of Wildlife Diseases* 20:236–238.

Zaria, L.T. 1993. *Dermatophilus congolensis* infection (dermatophilosis) in animals and man. An update. *Comparative Immunology, Microbiology and Infectious Diseases* 16:179–222.

ERYSIPELOTHRIX INFECTION

FREDRICK A. LEIGHTON

INTRODUCTION. Erysipelas is the disease in animals caused by the bacterium *Erysipelothrix rhusiopathiae.* In people, localized infection of the skin caused by this organism is called *erysipeloid,* whereas human *erysipelas* refers to infection with hemolytic strains of *Streptococcus* sp.

HISTORY, DISTRIBUTION, AND HOST RANGE. *Erysipelothrix* was among the bacteria first recognized by Robert Koch in the 1870s as causes of disease, in that instance a septicemia in mice. This agent now is recognized to occur worldwide and to infect an apparently unlimited range of mammalian and avian species, as well as some reptiles and, perhaps, fish (Wood and Shuman 1981; Timoney et al. 1988). Erysipelas is an important disease of domestic pigs (Wood 1984) and is significant also in cattle, sheep, and poultry, especially turkeys, as well as wild birds. Infection does not always result in disease; the bacterium has been isolated from the tonsils of numerous healthy swine and wild rodents (Wood and Shuman 1981).

Erysipelas may occur as epidemics among small mammal populations [Wayson (1927) and Van Dorssen and Jaartsveld (1959), both cited in Wood and Shuman (1981)] but also occurs sporadically in free-ranging and captive wild species. Wood and Shuman (1981) tabulated reports of *Erysipelothrix* infection (but not necessarily disease) in wild mammals, including three species of shrew, four lagomorphs, 20 species of rodents, eight species of carnivores, five species of artiodactyles, three pinnipeds, four cetaceans (all dolphins), and two species of primates.

Subsequent reports of *Erysipelothrix* infection (in most cases associated with disease) include those in marsupials: brush-tailed phascogales *Phascogale tapoatafa* (Barker et al. 1981), bandicoots *Isoodon macrouris,* bilby *Macrotis [Thylacomys] lagotis* (Eamens et al. 1988) and opossum *Didelphis virginiana* (Lonigro and LaRegina 1988); swamp beavers or coypu *Myocaster coypus* (Kohler et al. 1987); white-tailed deer *Odocoileus virginianus* (Bruner et al. 1984); moose *Alces alces* (Campbell et al. 1994); roe deer *Capreolus capreolus* (Eskens and Zschock 1991); and dolphins (Medway 1980; Thurman et al. 1983; Kinsel et al. 1997).

ETIOLOGY. *Erysipelothrix rhusiopathiae* is a Gram-positive bacillus that usually is short and narrow but occasionally grows as a long nonseptate and unbranched filament. No spores are formed (Timoney et al. 1988). There is one other species in the same genus: *E. tonsillarum,* to which 7 of the 23 serotypes formerly assigned to *E. rhusiopathiae* have been transferred. Some of these serotypes also are pathogenic to some mammalian species (Gyles 1993). Previous taxonomies have classified *E. rhusiopathiae* as *Bacillus rhusiopathiae-suis, Bacterium rhusiopathiae,* and *Erysipelothrix insidiosa,* among other names.

TRANSMISSION AND EPIDEMIOLOGY. The source of infection in erysipelas seldom is known. Healthy animals that carry subclinical infections may become diseased when stress of various kinds weakens host defenses (Barker et al. 1981). It is likely that the bacterium persists in nature primarily in healthy carrier animals, with a more transient existence in the external environment.

The common routes of infection are by ingestion and wound contamination. Exposure by ingestion is facilitated by the bacterium's persistence in soil; water; dried, smoked, or salted meat products; feces; and putrefying animal remains. There is an association of *Erysipelothrix* with meat and stored fish, and isolates can be made readily from fish-eating animals (Bauwens et al. 1992), which may explain the propensity of dolphins to contract erysipelas (Geraci et al. 1966; Medway 1980). *Erysipelothrix* also is considered among agents of human disease that are water-borne or associated with ingestion of fish or seafood (Auerbach 1987; Czachor 1992). The potential for ticks, lice, mites, and both biting and nonbiting flies to transmit the bacterium among susceptible hosts has been demonstrated (Wood and Shuman 1981).

PATHOGENESIS, PATHOLOGY, AND CLINICAL SIGNS. Little is known of the factors that determine the pathogenicity of different serotypes for different host species, or which bacterial virulence factors are of particular importance (Gyles 1993). Erysipelas is a septicemia that may be fatal or may resolve without further complication. The lesions in septicemic animals are nonspecific and often minimal. They may include patches of hemorrhage and necrosis of the skin, small hemorrhages on internal organ surfaces, congestion and edema of the lungs, and small pinpoint pale areas of necrosis in organs such as the liver. Animals with acute septicemia may be found dead with few or no premonitory signs, other than perhaps depression and skin lesions.

Acute disease also may progress to two forms of chronic, persistent disease: infection and inflammation of the heart valves (endocarditis) or of the joints (arthritis). In the former, there is failure of the right and/or left heart due to dysfunction of one or more valves, usually the atrioventricular valves, although in people the aortic valve is preferentially involved (Reboli and Farrar 1989). In addition, masses of thrombus can detach from the valve, lodge in downstream arteries, and cause ischemic necrosis (infarcts) in multiple organs and tissues. Arthritis progresses from an acute fibrinous or fibrinopurulent phase to a chronic, proliferative inflammation that can lead to destruction of cartilage and progressive lameness, with joint swelling.

DIAGNOSIS. None of the lesions evident in animals with erysipelas is pathognomonic; all can be caused by other pathogens, including Gram-negative and some other Gram-positive bacteria, and some viruses. However, the organism is readily isolated and identified by standard methods for culture of aerobic mammalian bacterial pathogens (Timoney et al. 1988), and this is the basis for diagnosis of erysipelas. Culture from at least two internal organs (spleen, red bone marrow, kidney, liver, lung) should be positive. The organism can be isolated from affected heart valves and from arthritic joints early in the disease. Chronically arthritic joints may no longer harbor the organism.

IMMUNITY, TREATMENT, AND CONTROL. There are effective vaccines in current use for turkeys and swine, but none have been developed specifically for use in wildlife. Acute disease responds well to therapy with penicillins and some other antibiotics, if it is diagnosed in time.

Control of the disease in wild populations has not been contemplated. In captive settings, sanitation, fly control, stress reduction, and quarantine, together with vaccination and treatment in certain situations, are the tools available to reduce the incidence of erysipelas.

PUBLIC HEALTH CONCERNS. Human erysipeloid is a localized infection of the skin from contamination of minor cuts and scratches with *E. rhusiopathiae.* It occurs most commonly on the hands and fingers. The infection produces localized inflammation that may be both painful and pruritic but usually is eliminated without therapy in 2–4 weeks. Very rarely, such infections progress to septicemia and endocarditis, either of which can be fatal (Reboli and Farrar 1989). The usual source of infection is infected or contaminated animal parts, and thus the disease is most common among persons who handle meat, fish, and marine animal products.

LITERATURE CITED

Auerbach, P.S. 1987. Natural microbiologic hazards of the aquatic environment. *Clinics in Dermatology* 5:52–61.

Barker, I.K., P.L. Carbonell, and A.J. Bradley. 1981. Cytomegalovirus infection of the prostate in the dasyurid marsupials *Phascogale tapoatafa* and *Antechinus stuartii. Journal of Wildlife Diseases* 17:433–441.

Bauwens, L., S. Cnops, and W. de Meurichy. 1992. Isolation of *Erysipelothrix rhusiopathiae* from frozen fish and fish-eating animals at Antwerp zoo. *Acta Zoologica et Pathologica Antverpiensia* 82:41–49.

Bruner, J.A., R.W. Griffith, J.H. Greve, and R.L. Wood. 1984. *Erysipelothrix rhusiopathiae* serotype 5 isolated from a white-tailed deer in Iowa. *Journal of Wildlife Diseases* 20:235–236.

Campbell, G.D., E.M. Addison, I.K. Barker, and S. Rosendal. 1994. *Erysipelothrix rhusiopathiae,* serotype 17, septicemia in a moose (*Alces alces*) from Algonquin Park, Ontario. *Journal of Wildlife Diseases* 30:436–438.

Czachor, J.S. 1992. Unusual aspects of water-borne illnesses. *American Family Physician* 46:797–804.

Eamens, G.J., M.J. Turner, and R.E. Catt. 1988. Serotypes of *Erysipelothrix rhusiopathiae* in Australian pigs, small ruminants, poultry, and captive wild birds and mammals. *Australian Veterinary Journal* 65:249–252.

Eskens, U., and M. Zschock. 1991. Rotlaufinfektion beim Reh: Ein Fallbericht. *Tierärztliche Praxis* 19:52–53.

Geraci, J.R., R.M. Sauer, and W. Medway. 1966. Erysipelas in dolphins. *American Journal of Veterinary Research* 27:597–606.

Gyles, C.L. 1993. *Erysipelothrix rhusiopathiae.* In *Pathogenesis of bacterial infections in animals,* ed. C.L. Gyles and C.O. Thoen, 2d ed. Ames: Iowa State University Press, pp. 80–85.

Kinsel, M.J., J.R. Boehm, B. Harris, and R.D. Murnane. 1997. Fatal *Erysipelothrix rhusiopathiae* septicemia in a captive Pacific white-sided dolphin (*Lagenorhynchus obliquidens*). *Journal of Zoo and Wildlife Medicine* 28:494–497.

Kohler, B., B. Wendland, U. Tornow, and M. Michael. 1987. The occurrence of bacterial infections in swamp beavers (*Myocaster coypus* Molina, 1782): 2. *Erysipelothrix rhusiopathiae* infections [in German]. *Archiv für experimentalle Veterinärmedizin* 41:442–446.

Lonigro, J.G., and M.C. LaRegina. 1988. Characterization of *Erysipelothrix rhusiopathiae* isolated from an opossum (*Didelphis virginiana*) with septicemia. *Journal of Wildlife Diseases* 24:557–559.

Medway, W. 1980. Some bacterial and mycotic diseases of marine mammals. *Journal of the American Veterinary Medical Association* 177:831–834.

Reboli, A.C., and W.E. Farrar. 1989. *Erysipelothrix rhusiopathiae:* An occupational pathogen. *Clinical Microbiology Reviews* 2:354–359.

Thurman, G.D., S.J. Downes, M.B. Fothergill, N.M. Goodwin, and M.M. Hegarty. 1983. Diagnosis and successful treatment of subacute erysipelas in a captive dolphin. *Journal of the South African Veterinary Association* 54:193–200.

Timoney, J.F., J.H. Gillespie, F.W. Scott, and J.E. Barlough. 1988. The genus *Erysipelothrix.* In *Hagan and Bruner's microbiology and infectious diseases of domestic animals,* 8th ed. Ithaca, NY: Comstock, pp. 197–205.

Wood, R.L. 1984. Swine erysipelas: A review of prevalence and research. *Journal of the American Veterinary Medical Association* 184:944–949.

Wood, R.L., and R.D. Shuman. 1981. *Erysipelothrix* infection. In *Infectious diseases of wild mammals,* ed. J.W. Davis, L.H. Karstad, and D.O. Trainer, 2d ed. Ames: Iowa State University Press, pp. 297–305.

FUSOBACTERIUM NECROPHORUM INFECTION

FREDRICK A. LEIGHTON

Synonyms: Necrobacillosis, foot rot, foot abscess, pododermatitis, calf diphtheria, liver abscess, Lemierre's syndrome.

INTRODUCTION AND HISTORY. Some 22 different names have been used for *Fusobacterium necrophorum* since its was first recognized in the late 1800s, the most important synonym being *Spherophorus necrophorus* (Langworth 1977). Diseases caused by this organism in people and domestic animals have many names, of which *necrobacillosis* is the most inclusive and useful.

The most common forms of necrobacillosis are infections of the feet and the mouth, either of which can progress to infection of internal organs, particularly the liver. Sporadic cases in individual animals probably are more common than large outbreaks, but the latter are more often recorded.

Rosen et al. (1951) reviewed outbreaks of necrobacillosis in California from the 1920s to 1950. Mortality among mule deer *Odocoileus hemionus* congregated at water holes under drought conditions during summer recurred throughout this period. In many cases, the habitat was shared with cattle or sheep. The disease also occurred in elk *Cervus elaphus* congregated at artificial feeding stations during winter in Wyoming. Mortality was estimated at 10% of the herd (Allred et al. 1944). A winter outbreak occurred in white-tailed deer *Odocoileus virginianus,* mule deer, and pronghorn *Antilocapra americana* in Saskatchewan, Canada (Wobeser et al. 1975). The deer were concentrated around hay bales during a time of cold stress and food shortage and were associated with both cattle and sheep. Leader-Williams (1982) determined that the prevalence of mandibular lesions, probably caused by necrobacillosis, was correlated with density of reindeer *Rangifer tarandus* on the island of South Georgia and that the disease contributed to limiting population size in a density-dependent fashion.

DISTRIBUTION AND HOST RANGE. *Fusobacterium necrophorum* occurs worldwide, with no known limit to the range of mammalian hosts in which it can be found, either as a member of the normal intestinal flora or as a cause of disease. It is an important pathogen in humans and domestic animals, particularly ruminants. Necrobacillosis has been reported in many species of free-ranging and captive wild mammals, predominantly macropodid marsupials (Munday 1988) and members of the Artyodactyla, in which major outbreaks have occurred (Table 28.1).

ETIOLOGY. *Fusobacterium necrophorum* is strictly anaerobic, Gram negative, pleomorphic but generally filamentous or rod shaped. It does not form spores but nonetheless appears to remain viable in the external environment for periods of months (Marsh et al. 1934). The species was formerly subdivided into three strains or biovars, referred to as A, B, and C. Biovars A and B now are proposed as the subspecies *F. n. necrophorum* and *F. n. funduliforme,* respectively. Biovar C has been assigned to a separate species: *F. pseudonecrophorum* (Tan et al. 1996). *Fusobacterium n. necrophorum* is highly pathogenic and is isolated from severe lesions, whereas *F. n. funduliforme,* the predominant subspecies present in the normal bovine alimentary tract, is a mild pathogen (Smith and Thornton 1993). The taxonomy of the genus *Fusobacterium* has been reviewed by Bennett and Eley (1993). Reviews by Langworth (1977), Smith (1988), and Tan et al. (1996) should be consulted for detailed information about the bacterium as an animal pathogen, whereas Eykyn (1989) has reviewed necrobacillosis in people.

TRANSMISSION AND EPIDEMIOLOGY. *Fusobacterium necrophorum* ssp. commonly are present among the intestinal flora of mammals and in their feces. Thus, their presence alone is not sufficient to cause disease.

Outbreaks of disease among wild ruminants have been associated with concentrations of animals around water holes under drought conditions, or around food sources at times of food shortage. Risk factors identified in these situations include the stress of crowding and inadequate nutrition, and heavy contamination of the local environment with feces. The presence of domestic ruminants also may be a risk factor. Outbreaks often appear to end when the causes of the abnormal concentrations of animals are relieved.

Heavy fecal contamination of the ground may also contribute to fusobacterial omphalitis and hepatitis in neonatal ruminants, and environmental contamination may play a role in oral necrobacillosis in macropodid marsupials.

An additional risk factor may be highly abrasive food material that traumatizes the oral mucosa, or other factors, such as gingival trauma or dental eruption, which permit penetration of the organism through the protective epithelium into the underlying connective tissue.

A risk factor for cattle, which may pertain at times to wild or captive ruminants, is a relatively rapid change in diet to a higher carbohydrate content. The high lev-

TABLE 28.1—Some records of necrobacillosis in wild and captive mammals

Common Name	Genus/Species	Location	Reference
Bennett's wallaby	*Macropus rufogriseus*	Zoos	Smith 1988
Red kangaroo	*Macropus rufus*	Australia	Tomlinson and Gooding 1954
Mule deer	*Odocoileus hemionus*	U.S.A.	Rosen et al. 1951
White-tailed deer	*Odocoileus virginianus*	U.S.A., Canada	Cass 1947; Wobeser et al. 1975
Elk	*Cervus elaphus*	U.S.A.	Allred et al. 1944
Red deer	*Cervus elaphus*	New Zealand	Griffin 1987
Sambar	*Cervus unicolor*	India	Chakraborty and Chaudhury 1993
Caribou and reindeer	*Rangifer tarandus*	U.S.A., Russia Sweden	Rausch 1953; Nikolaevskii 1944 Borg 1958
Pronghorn	*Antilocapra americana*	U.S.A., Canada	Thorne 1982; Wobeser et al. 1975
Gemsbok	*Oryx gazella*	Botswana	Dräger 1975
Springbok	*Antidorcas marsupialis*	Botswana	Dräger 1975
Impala	*Acepyceros melampus*	Zoo	Loomis and Wright 1989
Blue duiker	*Cephalophus monticola*	Captive	Roeder et al. 1989
Wildebeest	*Connochaetes taurinus*	Tanzania	Gainer 1983

els of lactic acid that develop in the rumen under these conditions favor proliferation of *F. necrophorum* and also can damage the mucosa of the rumen, facilitating entry of the bacteria into the submucosa and the hepatic portal bloodstream (Tan et al. 1996). Infection of internal organs often is the result.

CLINICAL SIGNS. Lameness is the important clinical sign when infection involves the feet. Its severity ranges from virtual immobility, when multiple limbs are affected acutely, to slight interference with locomotion in mild, resolving infections. Affected limbs often are greatly swollen in the infected area.

Oral infections result in excess salivation, problems in manipulating, chewing, or swallowing food, and loss of body condition due to reduced food intake. Animals with oral disease may swallow large numbers of fusobacteria, predisposing them to the development of necrobacillary lesions lower in the gastrointestinal system, especially the forestomachs. Fusobacterial stomatitis can lead to aspiration of infected exudate or necrotic tissue and subsequent pneumonia, accompanied by signs of labored breathing and general debility. Spread of infection from any location to the bloodstream is likely to result in severe lesions in highly perfused organs, including liver and lung, often followed rapidly by death. Death may be the only sign of necrobacillosis involving only internal organs.

PATHOGENESIS. The general pattern of pathogenesis involves entry of the bacterium through a physical defect in the protective epithelium of the skin or alimentary tract, and multiplication and toxin production in subepithelial connective tissue. The toxins kill local host tissues. Bacteria may gain access to local blood vessels and spread to distant internal organs via the bloodstream. At all sites, the principal disease process is necrosis of tissue.

Infection often is fatal in the acute stage due to a combination of dysfunction of infected tissues and the elaboration of toxins that have a wide range of detrimental systemic effects. Infections in subcutaneous tissues often penetrate into adjacent bones and joints, causing acute and chronic osteomyelitis and arthritis, respectively. The ability of *F. necrophorum* to cause disease is considerably enhanced by concurrent infection with certain other bacteria, particularly *Arcanobacterium pyogenes, Escherichia coli, Citrobacter freundi, Bacillus cereus, Klebsiella oxytoca,* and *Staphylococcus aureus* (Smith et al. 1989, 1991).

PATHOLOGY. The typical gross lesion in acute necrobacillosis is a zone of necrotic tissue sharply demarcated from adjacent living tissue. Such areas may be single or multiple. Oral lesions may involve zones of necrosis on buccal surfaces, tongue, pharynx, gingiva, or periodontal tissues. Patches of necrosis of the rumen wall and in the liver also are common. In subcutaneous sites of infection, such as the feet of ungulates or the faces of macropods, there often is severe swelling.

In subacute and chronic lesions, inflammation leads to liquefaction, debridement of dead tissue from affected cutaneous and mucosal surfaces, formation of abscesses in internal organs, and healing by fibrosis.

Infected bone undergoes both lysis and reactive proliferation, such that there is deformity and enlargement. Large defects in bone and loss of teeth are typical of chronic oral necrobacillosis in ruminants and macropods. Infection of joints results in acute inflammation followed by loss of cartilage and formation of periarticular new bone. A necrotizing pneumonia can arise from aspiration or from hematogenous spread. Lesions of necrobacillosis are particularly well illustrated in Wobeser et al. (1975).

Microscopically, the acute lesion is coagulation necrosis, with virtually complete absence of inflammatory cells except at the periphery of the necrotic tissue. Filamentous, Gram-negative bacteria frequently can be seen within the necrotic tissue, even with standard hematoxylin-eosin stains, often accompanied by a variety of other bacteria. Suppuration, liquefaction, and fibrosis develop in subacute and chronic lesions.

DIAGNOSIS. A diagnosis of necrobacillosis is confirmed by isolation of the bacterium from affected tissues, which requires anaerobic culture techniques (Langworth 1977). Although none of the lesions or clinical signs of necrobacillosis is pathognomonic, they may suggest a presumptive diagnosis in the absence of the capacity to isolate the agent.

IMMUNITY, CONTROL, AND TREATMENT. There is no evidence that primary infection confers resistance against future infections, and no effective vaccine against the bacterium has been developed (Smith 1988; Tan et al. 1996). Rather, control must be based on avoiding concentrations of animals in environments heavily contaminated with feces; preventing or minimizing oral or cutaneous abrasion or laceration; avoiding feeding practices that may promote acidosis of the rumen; and reducing sources of stress that often accompany suboptimal environments (Munday 1988).

Antibiotics can be therapeutic, if administered early. The tetracyclines, penicillins, and macrolide antibiotics generally are effective against *F. necrophorum,* while aminoglycosides are not (Tan et al. 1996). However, even moderately advanced lesions are very difficult to treat, requiring debridement of dead superficial tissue and drainage of affected deeper tissues, which may not be recognized. In captive situations, especially where the risk of necrobacillosis is recognized to be high, vigilance for early clinical signs, and aggressive therapy, are required for success.

LITERATURE CITED

Allred, W.J., R.C. Brown, and O.J. Murie. 1944. Disease kills feedground elk: Necrotic stomatitis takes toll of Jackson herd. *Wyoming Wild Life* 9:1–8 and 27.

Bennett, K.W., and A. Eley. 1993. Fusobacteria: New taxonomy and related diseases. *Journal of Medical Microbiology* 39:246–254.

Borg, K. 1958. Untersuchungen an 460 zugrundegegangenen Rehen in Schweden. *Zeitschrift für Jagdwissenschaft* 4:203–208.

Cass, J.S. 1947. Buccal food impactions in white-tailed deer and *Actinomyces necrophorus* in big game. *Journal of Wildlife Management* 11:91–94.

Chakraborty, A., and B. Chaudhury. 1993. Necrobacillosis in a sambar (*Cervus unicolor*). *Indian Journal of Veterinary Pathology* 17:142–143.

Dräger, N. 1975. A severe outbreak of interdigital necrobacillosis in gemsbok (*Oryx gazella*) in the northern Kalahari (Botswana). *Tropical Animal Health and Production* 7:200.

Eykyn, S.J. 1989. Necrobacillosis. *Scandinavian Journal of Infectious Diseases Supplement* 62:41–46.

Gainer, R.S. 1983. Necrobacillosis in wildebeest calves. *Journal of Wildlife Diseases* 19:155–156.

Griffin, J.F.T. 1987. Acute bacterial infections in farmed deer. *Irish Veterinary Journal* 41:328–331.

Langworth, B.F. 1977. *Fusobacterium necrophorum:* Its characteristics and role as an animal pathogen. *Bacteriological Reviews* 41:373–390.

Leader-Williams, N. 1982. Relationship between a disease, host density and mortality in a free-living deer population. *Journal of Animal Ecology* 51:235–240.

Loomis, M.R., and J.F. Wright. 1989. Systemic infections in impala (*Acepyceros melampus*) calves following ear tagging. *Journal of Zoo and Wildlife Medicine* 20:370–372.

Marsh, H., T. Hadleigh, and E.A. Tunnicliff. 1934. Experimental studies of foot rot in sheep. *Montana Agriculture Experimental Station Bulletin* 285 [cited in Rosen et al. (1951)].

Munday, B.L. 1988. Marsupial diseases. In *Australian wildlife: Proceedings 104.* Sydney: Post Graduate Committee in Veterinary Science, University of Sydney, pp. 299–365.

Nikolaevskii, L.D. 1944. Causes of necrobacillosis epizootics in reindeer. *Veterinaryia (Moscow)* 10:8–13 [cited in Rosen et al. (1951)].

Rausch, R. 1953. On the status of some arctic mammals. *Arctic* 6:91–148.

Roeder, B.L., M.M. Chengappa, K.F. Lechtenberg, T.G. Nagaraja, and G.A. Varga. 1989. *Fusobacterium necrophorum* and *Actinomyces pyogenes* associated facial and mandibular abscesses in blue duiker. *Journal of Wildlife Diseases* 25:370–377.

Rosen, M.N., O.A. Brunetti, A.I. Bischoff, and J.A. Azevedo Jr. 1951. An epizootic of foot rot in California deer. In *Transactions of the 16th North American Wildlife Conference,* pp. 164–179.

Smith, G.R. 1988. Anaerobic bacteria as pathogens in wild and captive animals. *Symposia of the Zoological Society of London* 60:159–173.

Smith, G.R., and E.A. Thornton. 1993. Pathogenicity of *Fusobacterium necrophorum* strains from man and animals. *Epidemiology and Infection* 110:499–506.

Smith, G.R., D. Till, L.M. Wallace, and D.E. Noakes. 1989. Enhancement of the infectivity of *Fusobacterium necrophorum* by other bacteria. *Epidemiology and Infection* 102:447–458.

Smith, G.R., S.A. Barton, and L.M. Wallace. 1991. Further observations on enhancement of the infectivity of *Fusobacterium necrophorum* by other bacteria. *Epidemiology and Infection* 100:305–310.

Tan, Z.L., T.G. Nagaraja, and M.M. Chengappa. 1996. *Fusobacterium necrophorum* infections: Virulence factors, pathogenic mechanism and control measures. *Veterinary Research Communications* 20:113–140.

Thorne, E.T. 1982. Bacteria. In *Diseases of wildlife in Wyoming,* ed. E.T. Thorne, N. Kingston, W.R. Jolley, and R.C. Bergstrom, 2d ed. Cheyenne: Wyoming Game and Fish Department, p. 38.

Tomlinson, A.R., and C.G. Gooding. 1954. A kangaroo disease: Investigations into "lumpy jaw" on the Murchison, 1954. *Journal of the Department of Agriculture of Western Australia* 3:715–718.

Wobeser, G., W. Runge, and D. Noble. 1975. Necrobacillosis in deer and pronghorn antelope in Saskatchewan. *Canadian Veterinary Journal* 16:3–9.

HELICOBACTER INFECTION

TORSTEN MÖRNER

INTRODUCTION. *Helicobacter pylori* is the most important cause of gastritis in people. This or other species of *Helicobacter* also are pathogens of domestic, laboratory, and zoo animals (Graham 1989; Munson 1996; Fox 1997).

DISTRIBUTION AND HOST RANGE. *Helicobacter* spp. have been reported from most parts of the world. Some specific associations include *H. acinonyx* in the cheetah *Acinonyx jubatus* (Eaton et al. 1993; La Perle et al. 1998), *H. mustelae* in ferrets *Mustela putorius* and mink *Mustela vison* (Fox et al. 1992), *H. felis* in cats (Lee et al. 1988), and the zoonotic *H. heilmanni* in dogs (Solnick et al. 1993). Isolations also have been made from pigs *Sus scrofa,* baboons *Papio anubis,* macaques *Macaca nemestrina* and *M. rhesus,* hamsters, rats, mice, terns, gulls, and house sparrows (Skirrow 1994), and from wild foxes (Valentin et al. 1997). *Helicobacter*-like organisms have been observed in section in the stomachs of a bobcat *Felis rufus,* a Pallas cat *Felis manul,* a Canada lynx *Felis lynx,* fishing cats *Felis viverrina,* margays *Felis weidii,* and sand cats *Felis margarita* (Kinsel et al. 1998)

ETIOLOGY. The type species of the genus is *H. pylori* (Holt et al. 1994), and about 18 species of *Helicobacter* have been described (Owen 1998). *Helicobacter* are helical, curved, or straight Gram-negative bacteria, 0.5–1.0 μm wide and 2.5–5.0 μm long (Holt et al. 1994). They often are difficult to culture and require special media, such as brain-heart infusion broth or agar, or chocolate agar (Holt et al. 1994).

TRANSMISSION AND EPIDEMIOLOGY. *Helicobacter* have been isolated from the stomach, intestines, liver, and gallbladder (Franklin et al. 1996; Fox 1997). Several reports have indicated that *Helicobacter* is transmitted between individuals (Skirrow 1994), but it is not clear whether transmission is by the fecal-oral or oral-oral route. Epidemiologic studies suggest that waterborne infections also can occur (Klein et al. 1991).

CLINICAL SIGNS. In most cases, infection with *Helicobacter* in animals and people is subclinical and silent (Skirrow 1994; Norris et al. 1999). Clinical signs in gastric infections include intermittent vomiting and weight loss, whereas intestinal infection can cause enteritis and diarrhea (Skirrow 1994; Munson 1996).

PATHOGENESIS. *Helicobacter* are motile bacteria that have the unusual ability to colonize the mucus overlying the gastric mucosa. Infection stimulates inflammation and ulcers (Roesch 1987; Marshall et al. 1990; Lee et al. 1993; Fox 1997).

PATHOLOGY. In most cases, the gastritis caused by *Helicobacter* is a mild infiltration of the mucosa by mononuclear cells and neutrophils (Skirrow 1994; Fox 1997). Hepatic *Helicobacter* infection in hamsters causes periportal inflammation with neutrophils, lymphocytes, and plasma cells, and fibrosis around bile ducts (Franklin et al. 1996). Hepatic infection in mice is reported to cause hepatitis and hepatic neoplasia (Ward et al. 1994).

DIAGNOSIS. The current standard diagnostic technique to demonstrate *Helicobacter* in clinically affected animals is histologic examination of biopsy samples acquired endoscopically (Skirrow 1994). Culture is necessary to identify the species or strain.

Culture of *Helicobacter* is difficult and not always successful. The different species are distinguished by culture and genetic sequence analysis. Serologic tests have been used in epidemiologic studies (Skirrow 1994; Jalava et al. 1998; Owen 1998).

CONTROL AND TREATMENT. The sources of infection and modes of transmission are not fully understood. Overcrowding and poor hygiene are predisposing factors for human infection (Vincent et al. 1994). The treatment strategy for clinically affected people is a combination of antibiotics and drugs that suppress acid secretion (Feldman 1995). *Helicobacter* infection is not recognized as a problem needing mitigation in free-ranging wildlife and only occasionally has it been recognized as such in captive collections.

PUBLIC HEALTH SIGNIFICANCE. Several *Helicobacter* species seem to infect a broad range of mammalian hosts. Thus, *Helicobacter* in animals must be regarded as a potential zoonotic risk. The dimensions of this risk remain to be defined, and generally, epidemiologic studies have not supported the contention that *Helicobacter* infections are zoonotic (Webb et al. 1996).

LITERATURE CITED

Eaton, A.K., F.E. Dewhirst, M.J. Radin, J.G. Fox, B.J. Paster, S. Krakowka, and D.R. Morgan. 1993. *Helicobacter acinonyx* sp. nov., isolated from cheetahs with gastritis. *International Journal of Systematic Bacteriology* 43:99–106.

Feldman, M. 1995. Suppression of acid secretion in peptic ulcer disease. *Journal of Clinical Gastroenterology* 20(Suppl. 1):S1–S6.

Fox, J.G. 1997. The expanding genus of *Helicobacter:* Pathogenic and zoonotic potential. *Seminars in Gastroenterology* 8:124–141.

Fox, J.G., B.J. Paster, F.E. Dewhirst, N.S. Taylor, L.-L. Yan, P.J. Macuch, and L.M. Chmura. 1992. *Helicobacter mustelae* isolated from feces of ferrets: Evidence to support fecal-oral transmission of gastric *Helicobacter. Infection and Immunity* 60:606–611.

Franklin, C.L., C.S. Beckwith, R.L. Livingston, L.K. Riley, S.V. Gibson, C.L. Besch-Wilford, and R.R. Hook Jr. 1996. Isolation of a novel *Helicobacter* species, *Helicobacter cholecystus* sp. nov. from the gallbladders of Syrian hamsters with cholangiofibrosis and centrilobular pancreatitis. *Journal of Clinical Microbiology* 34:2952–2958.

Graham, D.Y. 1989. *Campylobacter pylori* and peptic ulcer disease. *Gastroenterology* 96:615–625.

Holt, J., N.R. Krieg, P.H.A. Sneath, J.T. Staley, and S.T. Williams. 1994. *Helicobacter.* In *Bergey's manual of determinative bacteriology,* 9th ed. Baltimore: Williams and Wilkins, pp. 42–43.

Jalava, K., S.L.W. On, P.A. Vandamme, I. Happonen, A. Sukara, and M.L. Hanninen. 1998. Isolation and identification of *Helicobacter* spp. from canine and feline gastric mucosa. *Applied and Environmental Microbiology* 64:3998–4006.

Kinsel, M.J., P. Kovarik, and R.D. Murnane. 1998. Gastric spiral bacteria in small felids. *Journal of Zoo and Wildlife Medicine* 29:214–220.

Klein, P.D., D.Y. Graham, A. Gaillour, A.R. Opekun, and E. O'Brien-Smith. 1991. Water source as risk factor for *Helicobacter pylori* infection in Peruvian children. *Lancet* 337:1503–1505.

La Perle, K.M., R. Wack, L. Kaufman, and E.A. Blomme. 1998. Systemic candidiasis in a cheetah. *Journal of Zoo and Wildlife Medicine* 29:479–483.

Lee, A., S. Hazell, J. O'Rourke, and S. Kouprach. 1988. Isolation of a spiral-shaped bacterium from the cat stomach. *Infection and Immunity* 56:2843–2850.

Lee, A., J. Fox, and S. Hazell. 1993. Pathogenicity of *Helicobacter pylori:* A perspective. *Infection and Immunity* 61:1601–1610.

Marshall, B.J., L.J. Barret, C. Prakash, R.W. McCallum, and R.L. Guerrant. 1990. Urea protects *Helicobacter (Campylobacter) pylori* from the bactericidal effect of acid. *Gastroenterology* 99:697–702.

Munson, L. 1996. Emerging *Helicobacter* diseases. In *Proceedings of the American Association of Zoo Veterinarians annual conference, Puerto Vallarta, Mexico,* pp. 490–495.

Norris, C.R., S.L. Marks, K.A. Eaton, S.Z. Torabian, and J.V. Solnick. 1999. Healthy cats are commonly colonized with "*Helicobacter heilmannii*" that is associated with

minimal gastritis. *Journal of Clinical Microbiology* 37:189–194.
Owen, R.J. 1998. *Helicobacter* B species classification and identification. *British Medical Bulletin* 54:17–30.
Roesch, W. 1987. Current theories on pathogenesis of peptic ulcer disease: The pathogenic mechanism of *H. pylori*—A comment. In *Campylobacter pylori,* ed. H. Menge, M. Gregor, G.N.J. Tytgat, and B.J. Marshall. Heidelberg: Springer-Verlag, pp. 59–65.
Skirrow, M.B. 1994. Diseases due to *Campylobacter, Helicobacter* and related bacteria. *Journal of Comparative Pathology* 111:113–149.
Solnick, J.V., J. O'Rourke, A. Lee, J. Paster, F.E. Dewhirst, and L.S. Tompkins. 1993. An uncultured gastric spiral organism is a newly identified *Helicobacter* in humans. *Journal of Infectious Diseases* 168:379–385.
Valentin A., W. Jakob, M. Stolte, K. Seidel, K. Tackmann, and U.D. Wenzel. 1997. Nachweis von *Helicobacter*-ähnlichen Bakterien in der Magenschleimhaut von Fuchsen. *Verhandlungsbericht des Internationalen Symposiums über die Erkrankungen der Zoo- und Wildtiere* 38:327–332.
Vincent, P., F. Gotthard, and P. Pernes. 1994. High prevalence of *Helicobacter pylori* infection in cohabiting children: Epidemiology of a cluster, with special emphasis on molecular typing. *Gut* 35:313–316.
Ward J.M., M.R. Anver, D.C. Haines, and R.E. Benveniste. 1994. Chronic active hepatitis in mice caused by *Helicobacter hepaticus. American Journal of Pathology* 145:959–968.
Webb, P.M., T. Knight, J.B. Elder, D.G. Newell, and D. Forman. 1996. Is *Helicobacter pylori* transmitted from cats to humans? *Helicobacter* 1:79–81.

LEPTOSPIROSIS

FREDRICK A. LEIGHTON AND THIJS KUIKEN

Synonyms: Stuttgart disease (dogs); Weil's disease, canicola fever, canecutter's fever and harvest fever (among many others) in people.

INTRODUCTION. The bacterium that causes leptospirosis was first named *Spirochaeta icterohaemorrhagiae* but was renamed *Leptospira* in 1917. For a review of all aspects of this organism and the diseases it causes, consult Faine (1994).

Leptospirosis causes a variety of syndromes, most notably acute septicemia, often with hemorrhage or hemolytic jaundice; hepatitis; interstitial nephritis; and abortion or stillbirth.

Leptospirosis occurs regularly in domestic mammals and in people. Cattle, pigs, horses, and dogs are the domestic species most commonly involved, but others, such as farmed deer, also may be affected (Faine 1994). Wildlife may act as the source of infection for people and domestic animals, and the reverse is also true.

Although infection by *Leptospira* is common in a wide range of wild mammals, disease caused by such infection, i.e., leptospirosis, rarely has been reported in free-ranging wildlife. For example, despite a relatively high prevalence of serologically positive white-tailed deer *Odocoileus virginianus* in North America (about 15%–25%), and experimental proof that the species is susceptible to fatal disease, there are no reports of clinical leptospirosis in wild white-tailed deer and only one of abortion [see Shotts (1981)].

However, leptospirosis has occurred regularly in California sea lions *Zalophus californianus* since it was first recognized in 1970 (Vedros et al. 1971; Gulland et al. 1996). Between 1981 and 1994, one-third of several thousand stranded sea lions examined in California suffered from the disease, and 71% of the cases were fatal. Epidemic mortality from leptospirosis occurred at 3- to 4-year intervals during this period.

DISTRIBUTION AND HOST RANGE. Pathogenic serovars of *L. interrogans* occur throughout the world, and no species of mammal is known specifically to be refractory to infection. Every mammalian species is potentially a maintenance host or an accidental host for one or more serovars. Turner (1970), Torten (1979), and Shotts (1981) list mammals found to be infected with various serovars in different regions of the world.

Distribution and density of suitable maintenance hosts affect local or regional prevalence of infection. A requirement for warmth and moisture to survive in the external environment favors a higher incidence in the tropics and subtropics, but, seasonally, leptospirosis can be a problem in cooler temperate regions (Prescott and Zuerner 1993).

ETIOLOGY. Leptospires are helically spiraled, elongate bacteria 0.1 µm in diameter and 6–30 µm long. They are motile via a writhing motion effected by a pair of longitudinal flagella enveloped in a sheath that also surrounds the body of the bacterium. They are readily seen with darkfield microscopy and stain with silver-impregnation techniques, typically having a hooked end, or "shepherd's crook" shape. Isolation in culture is possible with special media but is technically complex, expensive, and slow, and therefore is not routine in most diagnostic laboratories.

The genus *Leptospira* has been divided into three species based on antigenic and biochemical characteristics: the pathogenic *L. interrogans* and the nonpathogenic free-living species, *L. biflexa* and *L. (Turneria) parva.* In turn, *L. interrogans* is divided into about 200 serovars, grouped into some 30 serogroups according to shared surface antigens (Faine 1994). The many serovars are distinguished by antigenic differences on the outer sheath by the microscopic agglutination test (MAT) and specific antisera. Special techniques are required to distinguish among serovars within the same serogroup.

More recently, a taxonomic classification based on genetic analysis has recognized at least seven taxa referred to as "genospecies" among what are classified by previously established criteria as *L. interrogans: L.*

interrogans, L. borgpetersenii, L. inadai, L. kirschneri, L. santarosai, L. noguchii, and *L. weilii* (Yasuda et al. 1987; Ramadass et al. 1992); further genospecies may be defined in the future. *Leptospira biflexa* now is represented similarly by the genospecies *L. biflexa, L. wolbachii,* and *L. meyeri* (Yasuda et al. 1987), although *L. meyeri* contains one pathogenic serovar, *ranarum.* Genospecies have been subdivided further into "types" on genetic grounds.

There is poor correlation of the serovars recognized on the basis of antigenicity with the types, or even the genospecies, newly defined based on genetic techniques (Prescott and Zuerner 1993; Faine 1994). Members of two genospecies may be indistinguishable antigenically and therefore fall into the same serovar (Faine 1994). A period of adaptation will be required to reassort our understanding of the epidemiology of leptospirosis on the basis of newer genetic information and to rationalize it with our understanding based on serologic classification. But there do appear to be clear relationships among some of the genetic types, geographic distribution, and clinical syndromes (Prescott and Zuerner 1993).

The inconsistency between antigenic and genetic classifications of leptospires will make correlation of seroepidemiologic data and molecular epidemiologic data difficult, and the latter may also suffer from limitations based on their dependence on bacterial DNA. The former, based on serum samples, are readily obtained and reflect the history of exposure of the host. The latter, based on detection of DNA from organisms in tissues or excreta, are technically more difficult to obtain and reflect only the infection status of the animal at a point in time.

Sensitive and specific immunologic and molecular techniques are available to detect and identify many of the pathogenic serovars or genetic types (Timoney et al. 1988; Smith et al. 1994; Corney et al. 1997), and molecular diagnosis has the advantage of speed and specificity over bacterial culture.

TRANSMISSION AND EPIDEMIOLOGY. The primary habitat of pathogenic leptospires is the mammalian renal tubule. Here, they persist for long periods, reproduce, and can be carried to new hosts via the urine, which contaminates the environment or surface water. Transmission via sexual or social contact, transplacental invasion, and ingestion of, or contact with, infected milk and tissues also can occur. The organism invades through the mucous membranes of the gastrointestinal tract, urogenital system, upper respiratory tract, and eye, or through wet or abraded skin (Faine 1994; Day et al. 1998).

The pathogenic leptospires usually do not survive for long periods in the external environment (Hanson 1982; Timoney et al. 1988), although under optimal conditions of warmth, moisture, and neutral or slightly basic groundwater pH, they may persist for up to 6 weeks (Shotts 1981; Prescott and Zuerner 1993). Leptospires have been isolated from a variety of arthropods, amphibians, reptiles, and birds (Thiermann 1984), including serovars pathogenic for mammals from amphibians (Gravekamp et al. 1991).

The epidemiology of most serovars or genetic types of *Leptospira* in most host species is not known. However, there is a consistent pattern in the epidemiology of serovars that have been studied, and it is reasonable to consider this pattern as typical of pathogenic leptospires in general.

There are two classes of mammalian hosts for pathogenic leptospires (Hathaway 1981). For any given serovar in an ecosystem in which it is endemic, there will be one or more "maintenance hosts" with which the serovar has a stable host-parasite relationship. These host species are highly susceptible to infection but, in general, suffer little or no clinical disease as a result. Leptospires will persist for long periods (months to years) in the renal tubules of an individual animal of the maintenance host species and is transmitted among individuals readily and frequently. This relationship has made *Leptospira balcanica* a candidate vector for a biologic control agent of its maintenance host, the brushtail possum *Trichosurus vulpecula,* in New Zealand (Day et al. 1998).

Other mammals in the same ecosystem are potential "accidental hosts." They may be less susceptible to infection because they require a higher infective dose or because they are ecologically separated from the main cycles of transmission. However, if infected, they may suffer clinical disease, although the organism does not persist for long periods in the kidneys of such hosts.

In New Zealand, Hathaway (1981) identified the following serovar-maintenance host associations: *pomona* with pigs; *hardjo* with cattle; *balcanica* with brushtail possums; *ballum* with the house mouse *Mus musculus,* black rat *Rattus rattus,* and hedgehog *Erinaceus europaeus;* and *copenhageni* with the Norway rat *Rattus norvegicus.* Disease generally did not occur in these host-parasite associations. In contrast, sporadic clinical leptospirosis occurred in cattle and people from infections by *pomona* and *copenhageni.* In Malaysia, rats carrying many serovars were the major wildlife maintenance host for leptospires affecting people, while cattle and pigs were suspected of being the maintenance hosts for serovars *hardjo* and *pomona* (Bahaman and Ibrahim 1988).

The serovar-maintenance host relationship appears to become established due to appropriate characteristics of the host and the serovar, and after a period of adaptation, which includes establishment of a high prevalence of antibody within the maintenance host population.

Major epidemiologic associations involving wildlife as maintenance hosts include serovar *grippotyphosa* (genospecies *kirschneri*) and raccoons *Procyon lotor* and skunks *Mephitis mephitis* in North America, and rodents worldwide; serovar *icterohaemorrhagiae* (genospecies *interrogans*) and rodents worldwide; and serovar *pomona* (genospecies *interrogans*) and various species of wildlife, including ungulates, worldwide (Prescott and Zuerner 1993).

Leptospirosis can be presumed to occur sporadically in many species of wild mammals. However, a high prevalence of antibodies or of isolation of the organism should not be interpreted as representing a high prevalence of disease. The majority of infections occur within stable host-parasite relationships, without clinical disease. However, maintenance hosts for the various serovars present can shift in importance over time within a particular ecological setting (Torten 1979). Thus, reassessment is required to maintain a current understanding of the epidemiology of *Leptospira* serovars at any particular location.

Serology, alone, does not provide reliable information about the prevalence of *Leptospira* infections in a population of animals, since culture-positive animals can be serologically negative and seropositive animals often are culture negative. There is extensive immunologic cross-reactivity among serovars, such that serology has no serovar specificity but does reflect infection by an antigenically related member of a serogroup. Hence, serovar-specific serology is possible only when the range of serovars that occur in an area is known and an appropriate range of serogroup antigens is employed to detect them (Faine 1994). Thus, to be interpretable beyond a weak qualitative index of exposure to some form of *Leptospira* sp., serologic surveys must be accompanied by isolation and identification of the bacteria (Torten 1979; Ellis 1986; Prescott 1993).

CLINICAL SIGNS. Leptospirosis can range from practically inapparent to severe and fatal. Acute disease typically begins with fever and inappetence, followed by varying degrees of hemorrhage on mucous membranes, jaundice and red urine due to lysis of red blood cells, depression, thirst, dehydration, vomition, and abdominal pain. There can be meningitis, pneumonia, abortion or stillbirth, cessation of milk production, and blood in the milk. Acute disease often is accompanied by signs of renal failure, including high concentrations of urea nitrogen, phosphorus, and creatinine in the blood (Timoney et al. 1988). Renal disease was the principal cause of death among infected sea lions (Gulland et al. 1996).

Transplacental infection can result in infertility, abortion, and infected neonates (Ellis 1986).

PATHOGENESIS AND PATHOLOGY. Following entry into the body, leptospires multiply in and spread via the bloodstream. This septicemic phase is associated with damage to vascular lining cells and with fever. Bacterial toxins and inflammatory responses to the bacteria account for clinical signs and lesions of hemorrhage, hemolysis, nephritis, meningitis, and pneumonia (Shotts 1981; Timoney et al. 1988; Faine 1994). In animals with severe hemolytic anemia, jaundice is marked, there may be deep red-black discoloration of the kidneys and urine in the bladder due to hemoglobinuria, and there may be periacinar necrosis in the liver, due to hypoxia.

Acute infection of the kidney causes mild necrosis of renal tubules as organisms leave the interstitial connective tissue and penetrate the tubular epithelium, and there is an acute inflammatory response dominated by neutrophils. This transient acute lesion is replaced by an infiltrate of lymphocytes, plasma cells, and macrophages, and in fatal cases a mixture of inflammatory cell types is often present.

Grossly, in more chronic cases, the renal parenchyma is diffusely or irregularly pale and somewhat swollen. The organisms can persist within the renal tubules, where they are protected from antibody and other host defenses. In maintenance hosts, they may persist indefinitely, eliciting chronic interstitial nephritis but negligible tubule damage (Marler et al. 1979; Prescott 1993; Gulland et al. 1996).

DIAGNOSIS. Clinical diagnosis of leptospirosis can be difficult, though a syndrome of hemolytic jaundice and pyrexia in any species is suggestive. Antibody titers, whether single or paired, may defy interpretation, and culture of blood or urine can yield false-negative results (Smith et al. 1994; Gulland et al. 1996).

Diagnosis at autopsy generally is possible due to the characteristic microscopic renal lesions. Leptospires often can be seen histologically within renal tubules or interstitium by using silver-impregnation stains, and they can be identified by fluorescent antibody or immunoperoxidase techniques. Molecular techniques such as polymerase chain reaction hold the greatest promise for fast, sensitive, and specific diagnostic tests, but primers specific for the many genetic types first must be developed (Smith et al. 1994).

IMMUNITY, CONTROL, AND TREATMENT. The antibodies detected in the MAT, reflecting exposure, are not protective and disappear within a few months, but protective, serovar-specific neutralizing antibodies can persist for several years. However, renal infection and shedding of bacteria can occur in the face of high titers of neutralizing antibody. Neutralizing antibody responses following vaccination with bacterins are weaker and less persistent than those that follow infection, but they can provide serovar-specific protection against disease for up to 1 year.

Most cases of leptospirosis appear to result from infection of accidental hosts and occur sporadically, although occasional outbreaks may occur in captive collections [for example, see Rapley et al. (1981) and Perolat et al. (1992)] or in free-ranging wildlife (Gulland et al. 1996). Studies in New Zealand revealed surprisingly little movement of serovars from their stable ecological associations with maintenance hosts into sympatric accidental hosts (Hathaway 1981).

In situations where clinical leptospirosis occurs regularly and control is warranted, the logical approach is to identify the maintenance host(s) for the causal serovar and to attempt to separate the host of concern from contact with the maintenance cycle, if that is possible or feasible. Drainage of standing water, provision of water from uncontaminated sources, exclusion of maintenance hosts from premises housing species of concern, and general rodent control, combined with serovar-specific vaccination, are examples of control measures that may be used to mitigate the risk of leptospirosis in domestic animals or captive wildlife.

Leptospires are sensitive to a variety of antimicrobial drugs. Penicillins, aminoglycosides, erythromycin, or tetracyclines may be used in animals. Treatment may not eliminate infection completely, so treated animals that survive disease may continue to shed bacteria in urine (Torten 1979; Timoney et al. 1988).

PUBLIC HEALTH CONCERNS. Humans appear to be accidental hosts and can develop a disease ranging from minimal to severe and fatal. Domestic or wild mammals are the usual sources of infection for people (Timoney et al. 1988). Outdoor recreationists, trappers, hunters, rural dwellers, farmers, wildlife biologists, wildlife rehabilitators, and zookeepers may be exposed to leptospires shed in urine by free-ranging or captive wildlife [for example, see Anderson et al. (1978), Falk (1985), Brem et al. (1995), Mikaelian et al. (1997), and Stamper et al. (1998)]. These people should be made aware of the risk, and encouraged to wear gloves and other protective clothing when appropriate and feasible; to minimize contact, especially with urine; and be encouraged to practice good personal hygiene, to minimize the probability of ingestion of the organism.

LITERATURE CITED

Anderson, D.C., J.G. Geistfeld, H.M. Maetz, C.M. Patton, and A.F. Kauffmann. 1978. Leptospirosis in zoo workers associated with bears. *American Journal of Tropical Medicine and Hygiene* 27:210–211.

Bahaman, A.R., and A.L. Ibrahim. 1988. A review of leptospirosis in Malaysia. *Veterinary Research Communications* 12:179–189.

Brem, S., O. Radu, T. Bauer, A. Schonberg, K. Reisshauer, R. Waidmann, H. Kopp, and P. Meyer. 1995. *Leptospira* infected rat population as probable cause of a fatal case of Weil's disease [in German]. *Berliner und Münchener Tierärztliche Wochenschrift* 108:405–407.

Corney, B.G., J. Colley, and G.C. Graham. 1997. Simplified analysis of pathogenic leptospiral serovars by random amplified polymorphic DNA fingerprinting. *Journal of Medical Microbiology* 46:927–932.

Day, T.D., C.E. O'Connor, J.R. Waas, A.J. Pearson, and L.R. Matthews. 1998. Transmission of *Leptospira interrogans* serovar *balcanica* infection among socially housed brushtail possums in New Zealand. *Journal of Wildlife Diseases* 34:576–581.

Ellis, W.B. 1986. The diagnosis of leptospirosis in farm animals. In *The present state of leptospirosis diagnosis and control,* ed. W.A. Ellis and T.W.A. Little. Dordrecht, The Netherlands: Commission of the European Communities and Martinus Nijhoff, pp. 13–30.

Faine, S. 1994. *Leptospira* and leptospirosis. Boca Raton, FL: CRC, 353 pp.

Falk, V.S. 1985. Leptospirosis in Wisconsin: Report of a case associated with direct contact with raccoon urine. *Wisconsin Medical Journal* 84:14–15.

Gravekamp, C., H. Korver, J. Montgomery, C.O. Everard, D. Carrington, W.A. Ellis, and W.J. Terpstra. 1991. Leptospires isolated from toads and frogs on the island of Barbados. *Zentralblatt für Bakteriologie* 275:403–411.

Gulland, F.M.D., M. Koski, L.J. Lowenstine, A. Colagross, L. Morgan, and T. Spraker. 1996. Leptospirosis in California sea lions (*Zalophus californianus*) stranded along the central California coast, 1981–1994. *Journal of Wildlife Diseases* 32:572–580.

Hanson, L.E. 1982. Leptospirosis in domestic animals: The public health perspective. *Journal of the American Veterinary Medical Association* 181:1505–1509.

Hathaway, S.C. 1981. Leptospirosis in New Zealand: An ecological view. *New Zealand Veterinary Journal* 29:109–112.

Marler, R.J., J.E. Cook, and A.I. Kerr. 1979. Experimentally induced leptospirosis in coyotes (*Canis latrans*). *American Journal of Veterinary Research* 40:1115–1119.

Mikaelian, I., R. Higgins, M. Lequient, M. Major, F. Lefebvre, and D. Martineau. 1997. Leptospirosis in raccoons in Quebec: 2 case reports and seroprevalence in a recreational area. *Canadian Veterinary Journal* 38:440–442.

Perolat, P., J.P. Poingt, J.C. Vie, C. Jouaneau, G. Baranton, and J. Gysin. 1992. Occurrence of severe leptospirosis in a breeding colony of squirrel monkeys. *American Journal of Tropical Medicine and Hygiene* 46:538–545.

Prescott, J.F. 1993. Leptospirosis. In *Pathology of domestic animals,* vol. 2, ed. K.V.F. Jubb, P.C. Kennedy, and N. Palmer, 4th ed. San Diego: Academic, pp. 503–511.

Prescott, J.F., and R.L. Zuerner. 1993. *Leptospira.* In *Pathogenesis of bacterial infections in animals,* ed. C.L. Gyles and C.O. Thoen, 2d ed. Ames: Iowa State University Press, pp. 287–296.

Ramadass, P., B.W.D. Jarvis, R.J. Corner, D. Penny, and R.B. Marshall. 1992. Genetic characterization of pathogenic *Leptospira* species by DNA hybridization. *International Journal of Systematic Bacteriology* 42:215–219.

Rapley, W.A., M.R. Cranfield, K.G. Mehren, S.I. Vas, I.K. Barker and F. Lathe. 1981. A natural outbreak of leptospirosis in a captive black-tailed deer (*Odocoileus hemionus columbianus*) herd and in Dall sheep (*Ovis dalli*) at the Metropolitan Toronto Zoo. *Proceedings of the American Association of Zoo Veterinarians, Seattle,* pp. 115–120.

Shotts, E.B., Jr. 1981. Leptospirosis. In *Infectious diseases of wild mammals,* ed. J.W. Davis, L.H. Karstad, and D.O. Trainer, 2d ed. Ames: Iowa State University Press, pp. 323–331.

Smith, C.R., P.J. Ketterer, M.R. McGowan, and B.G. Corney. 1994. A review of laboratory techniques and their use in the diagnosis of *Leptospira interrogans* serovar *hardjo* infection in cattle. *Australian Veterinary Journal* 71:290–294.

Stamper, M.A., F.M. Gulland, and T. Spraker. 1998. Leptospirosis in rehabilitated Pacific harbor seals from California. *Journal of Wildlife Diseases* 34:407–410.

Thiermann, A.B. 1984. Leptospirosis: Current developments and trends. *Journal of the American Veterinary Medical Association* 184:722–725.

Timoney, J.F., J.H. Gillespie, F.W. Scott, and J.E. Barlough. 1988. The spirochetes. In *Hagan and Bruner's microbiology and infectious diseases of domestic animals,* 8th ed. Ithaca, NY: Comstock, pp. 45–60.

Torten, M. 1979. Leptospirosis. In *CRC handbook series in zoonoses,* vol. 1, ed. H. Stoenner, W. Kaplan, and M. Torten. Boca Raton, FL: CRC, pp. 363–421.

Turner, J.C. 1970. Leptospirosis: III. Maintenance, isolation and demonstration of leptospires. *Transactions of the Royal Society of Tropical Medicine and Hygiene* 64:623–646.

Vedros, N.A., A.W. Smith, J. Schonewald, G. Migaki, and R. Hubbard. 1971. Leptospirosis epizootic among California sea lions. *Science* 172:1250–1251.

Yasuda, P.A., A.G. Steigerwalt, K.R. Sulzer, A.F. Kaufman, F. Rogers, and D.J. Brenner. 1987. Deoxyribonucleic acid relatedness between serogroups and serovars in the family Leptospiracea with proposals for seven new *Leptospira* species. *International Journal of Systematic Bacteriology* 37:407–415.

LISTERIOSIS

TORSTEN MÖRNER

Synonyms: Listerellosis, circling disease.

INTRODUCTION. Listeriosis, the disease caused by bacteria of the genus *Listeria,* occurs worldwide in domestic animals and is occasionally described in wild species. Infection may be inapparent or latent and subclinical, but it can result in encephalitis, septicemia, or metritis and abortion.

Listeriosis was first described by Murray et al. (1926) in laboratory rabbits and guinea pigs, but Hulphers's (1911) report of disease in rabbits probably also was listeriosis. The first report from a wild animal is that of Pirie (1927) in gerbils (*Tatera* sp.) during an outbreak of "Tiger River disease" in South Africa. Consult Gray and Killinger (1966) and Low and Donachie (1997) for more detailed historical reviews.

DISTRIBUTION AND HOST RANGE. Listeria monocytogenes has been isolated from over 50 species of captive or free-ranging mammals worldwide, including insectivores, rodents, lagomorphs, ruminants, carnivores, and primates [see Gray and Killinger (1966) and Dijkstra (1981)], as well as from many species of birds (Nilsson and Söderlind 1974; Fenlon 1985; Hatkin et al. 1986), from aquatic animals (Botzler et al. 1973), and from biting arthropods [see Gray and Killinger (1966) and Dijkstra (1981)].

In wild ruminants, septicemia may occur (Nilsson and Karlsson 1959; Evans and Watson 1987; Webb and Rebar 1987). However, encephalitis has been reported in moose *Alces alces* (Archibald 1960), roe deer *Capreolus capreolus* (Kemenes et al. 1983), fallow deer *Dama dama* (Eriksen et al. 1988), white-tailed deer *Odocoileus virginianus* (Trainer and Hale 1964), giraffes *Giraffa camelopardalis* (Cranfield et al. 1985), and llamas *Lama glama* (Butt et al. 1991).

Carnivores may have encephalitic or septicemic listeriosis (Avery and Byrne 1959; Black et al. 1996).

Among primates, abortion and perinatal infection with stillbirth or septicemia have been reported in *Cercopithecus mona* and *Macaca niger,* septicemia in a marmoset *Callithrix (Hapale) jaccus,* and encephalitis in a chimpanzee *Pan troglodytes* [see Heldstab and Rüedi (1982)].

In hares *Lepus timidus* and *Lepus europaeus,* septicemia and uterine infections have been described (Nilsson and Söderlind 1974) [see Dijkstra (1981)]. In rodents and other small mammals, septicemia predominates (Pirie 1927; Nordland 1959; Barker et al. 1978; Kataev et al. 1983; Tappe et al. 1984; Wilkerson et al. 1997).

ETIOLOGY. The genus *Listeria* contains a number of species, only two of which, *L. monocytogenes* and *L. ivanovii,* are pathogenic (Low and Donachie 1997). With few exceptions, listeriosis is due to *L. monocytogenes,* a short Gram-positive, aerobic to micro-aerophilic, nonspore-forming rod. Many serotypes of *L. monocytogenes* have been described, based on differences in O and H antigens [see Low and Donachie (1997)]. Phage typing, ribotyping, and molecular genetic techniques are also used in epidemiologic investigations (Low and Donachie 1997; Wiedmann et al. 1997).

TRANSMISSION. Listeriosis is not transmissible from animal to animal; rather, it is acquired from the natural habitat of *Listeria,* the external environment. *Listeria monocytogenes* has been isolated from soil, sewage sludge, effluents, stream water, mud, and dust [see Dijkstra (1981) and Rocourt and Seeliger (1985)]. It sometimes contaminates grass intended for use as feed, often is present in silage (Low and Donachie 1997), and has been found in fodder and plants from wildlife feeding grounds (Weis and Seeliger 1975). Although *L. monocytogenes* is an environmental bacterium, it is also commonly isolated from feces of healthy animals and people [see Low and Donachie (1997)].

Infection in animals occurs by two routes. The first is via transepithelial invasion in the intestine and thence to local lymphoid tissue and systemic sites, causing septicemia, perhaps with subsequent fetoplacental localization and abortion in pregnant animals. The second is possibly via damaged upper alimentary mucosa, teeth, or conjunctival mucosa, and thence along the sheath of the trigeminal nerve to the brainstem, though this mode of infection has yet to be conclusively proven (Low and Donachie 1997).

Wild animals are probably being exposed to *Listeria* by the enteric route constantly, with infection via the trigeminal nerve explaining most cases of encephalitis. Arthropod vectors may be capable of transmission [see Dijkstra (1981)], but this is likely of minor significance.

PATHOGENESIS. *Listeria* invades epithelial cells and phagocytes, probably mediated by a protein called *internalin.* Within cells, it escapes the membrane-bound phagosome by the action of the hemolysin (listeriolysin O) and perhaps phospholipases. Once free in the cytoplasm, the organism replicates and is moved toward the cell membrane by host cell actin filaments. It then invades adjacent cells directly, probably aided by a lecithinase that lyses the cell membrane (Low and Donachie 1997).

The factors that determine whether infection produces septicemia, encephalitis, or metritis are unclear. Uterine infection and abortion probably mainly follow septicemia in the dam, whereas septicemia in neonates usually occurs as an extension of intrauterine infection.

It has been difficult to incriminate intercurrent disease and immunosuppression in the pathogenesis of listeriosis in ruminants (Low and Donachie 1997). However, since *Listeria* is an intracellular pathogen, infection may remain latent within macrophages, only to recrudesce when cell-mediated immunity is compromised by disease or by elevated steroid levels associated with reproduction or stress (Barker et al. 1978; Schaffner et al. 1983; Hennebold et al. 1997). The growth of *Listeria* may be favored by the hyperglycemia associated with elevated glucocorticoid levels (Nordland 1959), and circumstances such as hemolysis that increase the availability of iron may also promote the virulence of *Listeria* and influence its survival in macrophages [see Barker et al. (1978), Coulanges et al. (1997), and Fleming and Campbell (1997)]. Most people with cases of invasive listeriosis are immunosuppressed, elderly, or pregnant (Cooper and Walker 1998), and circumstantial evidence in small mammals [see Barker et al. (1978)] and in occasional carnivores infected with canine distemper (Jakowski and Wyand 1971; Black et al. 1996) suggests that immunocompromise may play a significant role in the pathogenesis of listeriosis in animals.

EPIDEMIOLOGY. Listeriosis usually is sporadic; however, outbreaks may occur. In the case of listerial meningoencephalitis, outbreaks are likely related to exposure of many animals to high doses of *Listeria,* especially in silage (Low and Donachie 1997). Outbreaks of the septicemic form are more likely the product of immunocompromise due to common socioenvironmental stressors acting on members of a population that are heavily exposed to or latently infected with *Listeria.* Listeriosis is a major mediator of the synchronous total mortality of breeding males that occurs annually in *Antechinus,* a small dasyurid marsupial (Barker et al. 1978).

CLINICAL SIGNS. Animals with encephalitis may be depressed, ataxic, and paralyzed. Commonly, in ruminants, signs localize to one side, causing drooping of one ear and deviation of the head and neck, or circling, in one direction. Infection of the eye, with conjunctivitis and endophthalmitis, can accompany encephalitis (Low and Donachie 1997; Cooper and Walker 1998).

Animals with septicemic listeriosis rarely show clinical signs, and most are simply found dead (Barker et al. 1978; Tappe et al. 1984).

Uterine infection usually results in abortion without other signs of ill health in the dam (Low and Donachie 1997; Cooper and Walker 1998).

PATHOLOGY. Gross lesions normally are not seen in animals with encephalitis, though meninges may be edematous and cloudy. Microscopic lesions (suppurative meningoencephalitis, microabscesses, and foci of microgliosis) are most prominent in the brainstem but also can be found in other parts of the brain.

In the septicemic form, miliary hepatic foci of necrosis and acute inflammation are the most common finding, but focal necrosis and inflammation also can be seen in spleen, lung, kidney, and enteric lymphoid tissue. Abortion is characterized by pinpoint foci of acute necrosis in placental cotyledons, with intercotyledonary suppurative placentitis; lesions reflecting septicemia may be evident in the fetus (Evans and Watson 1987; Jubb and Huxtable 1993; Low and Donachie 1997).

DIAGNOSIS. Diagnosis of listeriosis usually requires isolation of *L. monocytogenes,* together with lesions compatible with the disease. The organism grows well on blood or nutrient agar and in conventional blood culture broth, although cold enrichment or selective media may be needed to promote isolation from brain or from contaminated sources such as feces or environmental samples. Use of techniques such as polymerase chain reaction, while not yet routine, can be expected in the future.

Listeria is often visible as Gram-positive coccobacilli in sections or smears of affected tissues. Immunohistochemistry and immunofluorescence are more sensitive and specific means of demonstrating the agent in tissue and are especially useful if isolation cannot be achieved (Butt et al. 1991; Johnson et al. 1995; Wilkerson et al. 1997).

Listeriosis must be differentiated from other bacterial and viral causes of meningoencephalitis, including brain abscesses, and rabies, with which it may coinfect (Avery and Byrne 1959); from encephalitic, disseminated, and placental toxoplasmosis; from Gram-negative septicemias producing focal hepatitis and splenitis, such as salmonellosis and yersiniosis; from Tyzzer's disease in small mammals; from disseminated herpesvirus infections; and from chlamydial and *Coxiella* abortion.

Serologic studies are hampered by cross-reactivity and nonspecificity, so they are not routinely employed (Low and Donachie 1997).

IMMUNITY. Antibodies play little role in immunity to *Listeria.* The most important initial defense against

Listeria resides in neutrophils, macrophages, and natural killer cells mobilized early in infection, while antigen-specific T cells, through cytokine mediators, function to potentiate intracellular killing of *Listeria* in macrophages later in infection (North and Conlan 1998).

The intracellular nature of *Listeria* infection has mitigated against development of an effective vaccine (Low and Donachie 1997).

CONTROL AND TREATMENT. Control of *Listeria* in the environment is not feasible. Feeding silage to ruminants, including wildlife, always poses some risk, which is difficult to reduce (Low and Donachie 1997; Cooper and Walker 1998). Good hygiene at sites where captive or free-ranging ruminants are fed may reduce exposure from feces and contaminated soil.

Most common antibiotics except cephalosporins are active against *Listeria.* However, treatment of clinical cases often is relatively unsuccessful, probably due to the intracellular location of the organism, or its intracerebral site in encephalitis. Antibiotic treatment in the early stages of septicemic listeriosis can be effective, but animals with this syndrome are rarely encountered alive. Ampicillin or amoxicillin, together with gentamicin, are recommended in people (Low and Donachie 1997; Cooper and Walker 1998).

PUBLIC HEALTH CONCERNS. Human listeriosis, while relatively uncommon, is increasingly recognized, and it can cause potentially fatal disease in the immunocompromised, elderly, pregnant, or neonate (Cooper and Walker 1998).

Occupational exposure to contaminated tissues such as placentas may result in minor self-resolving skin infections in immunocompetent individuals (McLaughlin and Low 1994), but animal contact is rarely incriminated as a source of human infection (Low and Donachie 1997).

The most common sources of infection in people are food products such as milk, soft cheeses, pâtés, or food items contaminated with *Listeria* after processing. Its environmental resistance and capacity to grow at low temperatures increase the risk of infection in stored products (Low and Donachie 1997; Cooper and Walker 1998). Foods of wildlife origin seem not to have been incriminated as sources of human listeriosis.

LITERATURE CITED

Archibald, R.McG. 1960. *Listeria monocytogenes* from a Nova Scotia moose. *Canadian Veterinary Journal* 1:225–226.

Avery, R.J., and J.L. Byrne. 1959. An attempt to determine the incidence of *Listeria monocytogenes* in the brain of mammals. *Canadian Journal of Comparative Medicine* 23:296–300.

Barker, I.K., I. Beveridge, A.J. Bradley, and A.K. Lee. 1978. Observations on spontaneous stress-related mortality among males of the dasyurid marsupial *Antechinus stuartii* Macleay. *Australian Journal of* Zoology 26:435–447.

Black, S.S., F.W. Austin, and E. McKinley. 1996. Isolation of *Yersinia pseudotuberculosis* and *Listeria monocytogenes* serotype 4 from a gray fox (*Urocyon cinereoargenteus*) with canine distemper. *Journal of Wildlife Diseases* 32:362–366.

Botzler, R.G., T.F. Wetzler, and A.B. Cowan. 1973. *Listeria* in aquatic animals. *Journal of Wildlife Diseases* 9:163–170.

Butt, M.T., A. Weldon, D. Step, A. DeLahunta, and C.R. Huxtable. 1991. Encephalitic listeriosis in two adult llamas (*Lama glama*): Clinical presentations, lesions and immunofluorescence of *Listeria monocytogenes* in brainstem lesions. *Cornell Veterinarian* 81:251–258.

Cooper, J., and R.D. Walker. 1998. Listeriosis. *Veterinary Clinics of North America Food Animal Practice* 14:113–125.

Coulanges, V., P. Andre, O. Ziegler, L. Buchheit, and D.J. Vidon. 1997. Utilization of iron-catecholamine complexes involving ferric reductase activity in *Listeria monocytogenes. Infection and Immunity* 65:2778–2785.

Cranfield, M., M.A. Eckhaus, B.A. Valentine, and J.D. Strandberg. 1985. Listeriosis in Angolan giraffes. *Journal of the American Veterinary Medical Association* 187:1238–1240.

Dijkstra, R.G. 1981. Listeriosis. In *Infectious diseases of wild mammals,* ed. J.W. Davis, L.H. Karstad, and D.O. Trainer, 2d ed. Ames: Iowa State University Press, pp. 306–316.

Eriksen, L., H.E. Larsen, T. Christiansen, M.M. Jensen, and E. Eriksen. 1988. An outbreak of meningo-encephalitis in fallow deer caused by *Listeria monocytogenes. Veterinary Record* 122:274–276.

Evans, M.G., and G.L. Watson. 1987. Septicemic listeriosis in a reindeer calf. *Journal of Wildlife Diseases* 23:314–317.

Fenlon, D.R. 1985. Wild birds and silage as reservoirs of *Listeria* in the agricultural environment. *Journal of Applied Bacteriology* 59:537–543.

Fleming, S.D., and P.A. Campbell. 1997. Some macrophages kill *Listeria monocytogenes* while others do not. *Immunological Reviews* 158:69–77.

Gray, M.L., and A.H. Killinger. 1966. *Listeria monocytogenes* and listeric infections. *Bacteriological* Reviews 30:309–382.

Hatkin, J.M., W.E. Phillips Jr., and G.A. Hurst. 1986. Isolation of *Listeria monocytogenes* from an eastern wild turkey. *Journal of Wildlife Diseases* 22:110–112.

Heldstab, A., and D. Rüedi. 1982. Listeriosis in an adult female chimpanzee (*Pan troglodytes*). *Journal of Comparative Pathology* 92:609–612.

Hennebold, J.D., H.H. Mu, M.F. Poynter, X.P. Chen, and R.A. Daynes. 1997. Active catabolism of glucocorticoids by 11 beta-hydroxysteroid dehydrogenase in vivo is a necessary requirement for natural resistance to infection with *Listeria monocytogenes. International Immunology* 9:105–115.

Hulphers, G. 1911. Liver necrosis in rabbit caused by a hereto not described microorganism. *Svensk Veterinär Tidskrift* 16:265–273.

Jakowski, R.M., and D.S. Wyand. 1971. Listeriosis associated with canine distemper in a gray fox (*Urocyon cinereoargenteus*). *Journal of the American Veterinary Medical Association* 159:626–628.

Johnson, G.C., W.H. Fales, C.W. Maddox, and J.A. Ramos-Vara. 1995. Evaluation of laboratory tests for confirming the diagnosis of encephalitic listeriosis in ruminants. *Journal of Veterinary Diagnostic Investigation* 7:223–228.

Jubb, K.V.F., and C.R. Huxtable. 1993. Listeriosis. In *Pathology of domestic animals,* ed. K.V.F. Jubb, P.C. Kennedy,

and N. Palmer, 4th ed. San Diego: Academic, pp. 393–397.
Kataev, G.D., K.N. Shlygina, and I.S. Meshcheriakova. 1983. Listeriosis in Norwegian lemmings [in Russian]. *Zhurnal Mikrobiologii, Epidemiologii i Immunologii* 12:99.
Kemenes, F., R. Glavits, E. Ivanics, G. Kovacs, and A. Vanyi. 1983. Listeriosis in roe-deer in Hungary. *Zentralblatt für Veterinärmedicin [B]* 30:258–265.
Low, J.C., and W. Donachie. 1997. A review of *Listeria monocytogenes* and listeriosis. *Veterinary Journal* 153:9–29.
McLaughlin, J., and J.C. Low. 1994. Primary cutaneous listeriosis in adults: An occupational disease of veterinarians and farmers. *Veterinary Record* 135:615–617.
Murray, E.G.D., R.A. Webb, and M.B.R. Swann. 1926. A disease of rabbits characterized by a large mononuclear leucocytosis, caused by a hereto undescribed bacillus *Bacterium monocytogenes* (n. sp.). *Journal of Pathology and Bacteriology* 29:407–439.
Nilsson, A., and K.A. Karlsson. 1959. *Listeria monocytogenes* isolations from animals in Sweden during 1948 to 1957. *Nordisk Veterinärmedicin* 11:305–315.
Nilsson, O., and O. Söderlind. 1974. *Listeria monocytogenes* isolerad fran djur i Sverige under Üren 1958–1972. *Nordisk Veterinärmedicin* 26:248–255.
Nordland, O.S. 1959. Host-parasite relations in initiation of infection: I. Occurrence of listeriosis in arctic mammals with a note on its possible pathogenesis. *Canadian Journal of Comparative Medicine and Veterinary Science* 23:393–400.
North, R.J., and J.W. Conlan. 1998. Immunity to *Listeria monocytogenes. Chemical Immunology* 70:1–20.
Pirie, J.H.H. 1927. A new disease of wild rodents "Tiger River disease." *Publications of the South African Institute for Medical Research* 3:163–186.
Rocourt, J., and H.P.R. Seeliger. 1985. Distribution des espèces du genre *Listeria. Zentralblatt für Bakteriologie, Mikrobiologie und Hygiene [A]* 259:317–330.
Schaffner, A., H. Douglas, and C.E. Davis. 1983. Models of T cell deficiency in listeriosis: The effects of cortisone and cyclosporin A on normal and nude BALB/c mice. *Journal of Immunology* 131:450–453.
Tappe, J.P., F.W. Chandler, W.K. Westrom, S.K. Liu, and E.P. Dolensek. 1984. Listeriosis in seven bushy-tailed jirds. *Journal of the American Veterinary Medical Association* 185:1367–1370.
Trainer, D.O., and J.B. Hale. 1964. Wildlife disease report from Wisconsin. *Bulletin of the Wildlife Disease Association* 1:4–5.
Webb, D.M., and A.H. Rebar. 1987. Listeriosis in an immature black buck antelope (*Antilope cervicapra*). *Journal of Wildlife Diseases* 23:318–320.
Weis, J., and H.P.R. Seeliger. 1975. Incidence of *Listeria monocytogenes* in nature. *Applied Microbiology* 30:29–32.
Wiedmann, M., T. Arvik, J.L. Bruce, J. Neubauer, F. del Piero, M.C. Smith, J. Hurley, H.O. Muhammed, and C.A. Batt. 1997. Investigation of a listeriosis epizootic in sheep in New York state. *American Journal of Veterinary Research* 58:733–737.
Wilkerson, M.J., A. Melendy, and E. Stauber. 1997. An outbreak of listeriosis in a breeding colony of chinchillas. *Journal of Veterinary Diagnostic Investigation* 9:320–323.

SALMONELLOSIS

TORSTEN MÖRNER

Synonyms: Enteric epizootic typhoid, paratyphoid.

INTRODUCTION. *Salmonella* species are well-recognized pathogens in people, livestock, wild mammals, birds, reptiles, and even insects.

Among wild animals, salmonellae are most frequently isolated from birds, but they also occur in mammals and reptiles. Many serotypes of *Salmonella* are human pathogens, and salmonellosis is probably the most widespread zoonosis in the world (Acha and Szyfres 1989).

The genus *Salmonella* was named for the American veterinarian and pathologist Daniel Elmer Salmon (1850–1914) who first described infection with *S. choleraesuis* in pigs.

ETIOLOGY. Salmonellae are small, straight, Gram-negative rods, $0.7–1.5 \times 2.0–5.0$ µm. Most are motile, but nonmotile forms also occur. They are facultatively anaerobic and grow well on ordinary media, although enriched media can improve the sensitivity of isolation procedures. Differentiation of members of the genus *Salmonella* into species or serotypes is based on biochemical and serologic reactions.

There are two species of *Salmonella: S. enterica,* very common and comprised of well over 2000 serotypes; and *S. bongori,* comprised of 10 serotypes, all of which are rare (Murray 1991; Clarke and Gyles 1993). The serotype is the taxon of importance. The Kauffmann-White classification system for serotypes is based on differences among somatic (O), capsular (Vi), and flagellar (H) antigens. Each is named on the basis of the place where it was first isolated (e.g., *dublin*), or the clinical syndrome it produces in a particular host (e.g., *choleraesuis*) (Clarke and Gyles 1993).

Based on host specificity, serotypes fall into two categories: those highly adapted to specific host species, like *typhi* and *paratyphi* in humans, *pullorum* in birds, *dublin* in cattle and *choleraesuis* in swine; and those with broad host ranges. Most serotypes fall into the latter category. Serotypes may vary markedly in virulence in various hosts, and only about 50 are regularly incriminated in disease (Clarke and Gyles 1993).

DISTRIBUTION. Salmonellae are ubiquitous, and *Salmonella typhimurium* is one of the most widespread bacterial pathogens in the world (Acha and Szyfres 1989).

HOST RANGE. Many *Salmonella* serotypes have been isolated from a wide variety of wild mammals. A distinction should be made between a healthy

Salmonella carrier, and an animal with salmonellosis, the disease. Ippen et al. (1987) reported *Salmonella* in free-living wild boar *Sus scrofa,* roe deer *Capreolus capreolus,* and red deer *Cervus elaphus* in Central Europe. Salmonellosis has been reported in red deer calves in New Zealand (McAllum et al. 1978) and in roe deer in Denmark (Nielsen et al. 1981). Serotypes *typhimurium, derby, muenchen, anatum, newport, orienburg, missisippi,* and *meleagridis* have been isolated from white-tailed deer *Odocoileus virginianus,* in which salmonellosis occurs in fawns (Robinson 1981).

Salmonellae have been isolated from European brown hares *Lepus europaeus* and varying hares *L. timidus* (Ippen et al. 1987; Mörner 1992), though Rattenborg (1994) found less than 1% of 975 European brown hares in Denmark to be infected. Seasonal *Salmonella* prevalence up to 100% was reported in the marsupial quokka *Setonix brachyuris* (Hart et al. 1985). Salmonellae have been isolated from wild house mice *Mus musculus* (Jones and Twigg 1976), and salmonellosis (mainly due to serotype *typhimurium)* is a significant disease in European hedgehogs *Erinaceus europaeus* (Schicht-Tinbergen 1989).

Salmonellae have been isolated from carnivores like the red fox *Vulpes vulpes* (Nielsen et al. 1981; Euden 1990; Mörner 1992), raccoon *Procyon lotor* (Bigler et al. 1974), and badger *Meles meles* (Wray et al. 1977; Euden 1990). Euden (1990) found 7% of 4881 badgers infected with 26 different serotypes.

Salmonella serotypes have been isolated from many different species of captive animals recently brought in from the wild, including macropodid marsupials (Samuel 1981; Speare and Thomas 1988), baboons (Steyn et al. 1976), and rhinoceros (Ramsay and Zainuddin 1993). In these cases, it is difficult to determine whether the animals were infected in the wild or in captivity.

TRANSMISSION AND EPIDEMIOLOGY. The worldwide distribution of *Salmonella* reflects the capacity of these bacteria to survive in the environment and to infect many different hosts, some of which act as healthy, long-term carriers. Salmonellae generally are tolerant to a wide range of temperatures, may survive for months in soil, and will multiply in water. Infections normally are acquired by ingestion of contaminated water or food (Murray 1991). A minimum number of bacteria is required to cause disease; in the presence of a conventional enteric flora, this is in excess of about 10^6 organisms (Clarke and Gyles 1993).

Depending on the serotype, species, and age of host, carrier states can occur, in which healthy animals may shed *Salmonella* for weeks or months. Latent infection, characterized by organisms in gut-associated lymphoid tissues and mesenteric lymph nodes, and intermittent shedding, also occurs. Stressors, such as overcrowding, transit, parturition, or other intercurrent diseases, are risk factors for outbreaks of salmonellosis (Clarke and Gyles 1993).

The presence of *Salmonella* in wildlife has been studied most extensively in birds like gulls and passerines [see Murray (1991)], but birds probably play a minor role in the epidemiology of salmonellosis in wild mammals. *Salmonella* infection in wildlife often may be due to transmission from domestic animals or humans and reflects in part the opportunity for exposure and the local serotypes (Murray 1991). For example, a relationship has been demonstrated between the occurrence of certain serotypes of *Salmonella* in domestic animals and in sympatric wild rodents [see Murray (1991) and Quessy and Messier (1992)].

CLINICAL SIGNS. Clinical salmonellosis may be manifest as primary enteritis and colitis, generalized infection (septicemia), or abortion (Barker et al. 1993; Clarke and Gyles 1993). The disease occurs in all forms from peracute to chronic. The clinical picture varies greatly, depending on the serotype and host species. In peracute disease in deer, clinical signs have included anorexia, listlessness, dehydration and, occasionally, diarrhea, preceding death (Robinson 1981). Diseased foxes had hemorrhagic enteritis (Nielsen 1981).

PATHOGENESIS AND PATHOLOGY. The primary site of infection is the intestines. After ingestion, the bacteria attach to and invade the enterocytes on the surface of the gut. If sufficient damage is done to the enteric lining, an enteritis ensues. To persist, the organism must invade and survive in macrophages and reach regional lymph nodes, from which they attain the bloodstream via lymphatic drainage. Salmonellae in circulation are engulfed by fixed phagocytes in the liver and spleen, but many disseminate to other sites: lung, joints, meninges, placenta, or fetus. Endotoxin released from organisms dying in circulation is an important mediator of systemic disease in salmonellosis (Barker et al. 1993; Clarke and Gyles 1993).

Gross findings in enteric salmonellosis span a spectrum from catarrhal to fibrinohemorrhagic enteritis. Septicemic salmonellosis may produce widespread hemorrhage on serous membranes, enlargement of the spleen and lymph nodes, and edema and congestion of organs such as the lung, and sometimes multifocal hepatic necrosis. Microscopic lesions in septicemia can include microvascular thrombosis in any tissue, necrosis and inflammation in liver, spleen, and lymph nodes, and focal granulomas in various organs (Barker et al. 1993).

DIAGNOSIS. A diagnosis of salmonellosis is based on culture of the bacteria together with compatible clinical disease or lesions. In living animals, fecal cultures are used, although negative cultures should not be considered definitive. In dead animals, specimens for culture should be taken from small intestine, colon, mesenteric lymph nodes, spleen, or liver. Isolation of *Salmonella* from the intestines is not sufficient, by itself, for a diagnosis of salmonellosis, since many animals are carriers; other evidence of disease compatible with salmonellosis also must be present.

CONTROL AND HUMAN HEALTH CONCERNS. Control of *Salmonella* infection in the wild is not feasible. To the extent that environmental contamination with sewage sludge, manure, or effluent from abattoirs contributes to the occurrence of *Salmonella* in wildlife (Murray 1991), improved sanitation is probably the best way to reduce the prevalence among wild mammals of serotypes infective for domestic animals and humans.

In captive situations, a high standard of husbandry and hygiene, with attention to rodent control and fecal contamination by birds, should minimize the risk of disease. Vaccines are not practical for nondomestic species. Clinically ill animals should be treated with an antibiotic to which the serotype involved is susceptible and receive supportive care such as fluid therapy if necessary. In severe cases, the prognosis is guarded (Clarke and Gyles 1993).

In general, the probability of humans or domestic animals contracting infection from wild mammals is low (Acha and Szyfres 1989; Murray 1991).

Salmonella spp. were used in baits as rodenticides in Western Europe and the United States until they were banned between the 1920s and 1960s. They have been produced more recently particularly in the former Soviet Union and Cuba and are being sold for rodent control in a number of developing countries. These baits are not effective for rodent control and, since they can cause disease in people, pose a clear public health hazard (Friedman et al. 1996; Threlfall et al. 1996).

LITERATURE CITED

Acha, A., and B. Szyfres. 1989. Salmonellosis. In *Zoonoses and communicable diseases common to man and animals,* 2d ed. Washington, DC: Pan American Health Organization, pp. 147–154.

Barker, I.K., A.A. van Dreumel, and N. Palmer. 1993. The alimentary system: Salmonellosis. In *Pathology of domestic animals,* ed. K.V.F. Jubb, P.C. Kennedy, and N. Palmer, 4th ed. San Diego: Academic, pp. 213–227.

Bigler, W.J., G.L. Hoff, A.M. Jasmon, and F.H. White. 1974. *Salmonella* infections in Florida raccoons. *Archives of Environmental Health* 28:261–162.

Clarke, R.C., and C.L. Gyles. 1993. Salmonellosis. In *Pathogenesis of bacterial infections in animals,* ed. C.L. Gyles and C.O. Then, 2d ed. Ames: Iowa State University Press, pp. 133–153.

Euden, P.R. 1990. *Salmonella* isolates from wild animals in Cornwall. *British Veterinary Journal* 146:228–232.

Friedman, C.R., G. Malcolm, J.G. Rigau-Pérez, P. Arámbulo III, and R.V. Tauxe. 1996. Public health risk from *Salmonella*-based rodenticides. *Lancet* 347:1705–1706.

Hart, R.P., S.D. Bradshaw, and J.B. Iveson. 1985. *Salmonella* infections in a marsupial, the quokka (*Setonix brachyurus*), in relation to seasonal changes in condition and environmental stress. *Applied and Environmental Microbiology* 49:1276–1281.

Ippen, R., S. Nickel, and D. Schröder. 1987. *Krankheiten des jagdbaren Wildes.* Berlin: VEB Deutscher Landwirtschverlag, 233 pp.

Jones, P.W., and G.I. Twigg. 1976. Salmonellosis in wild mammals. *Journal of Hygiene* 77:51–54.

McAllum, H.J.F., A.S. Familton, R.A. Brown, and P. Hemmingsen. 1978. Salmonellosis of red deer calves (*Cervus elaphus*). *New Zealand Veterinary Journal* 26:130–131.

Mörner, T. 1992. *Liv och Död bland vilda djur.* Stockholm: Sellin and Partner, 167 pp.

Murray, C.J. 1991. Salmonellae in the environment. *Review Scientifique et Technique de l'Office International des Epizooties* 10:765–785.

Nielsen, B.B., B. Clausen, and K. Elvestad. 1981. The incidence of *Salmonella* bacteria in wild-living animals from Denmark and in imported animals. *Nordisk Veterinärmedicin* 33:427–433.

Quessy, S., and S. Messier. 1992. Prevalence of *Salmonella* spp. *Campylobacter* spp. and *Listeria* spp. in ringed-billed gulls (*Larus delawarensis*). *Journal of Wildlife Diseases* 28:526–531.

Ramsay, E.C., and Z.-Z. Zainuddin. 1993. Infectious diseases of the rhinoceros and tapir. In *Zoo and wild animal medicine: Current therapy,* ed. M.E. Fowler. Philadelphia: W.B. Saunders, pp. 459–466.

Rattenborg, E. 1994. Diseases in the hare (*Lepus europaeus*) population in Denmark. Ph.D. diss., Royal Veterinary and Agricultural University, Copenhagen, Denmark, 115 pp.

Robinson, R.M. 1981. Salmonellosis. In *Diseases and parasites of white-tailed deer,* ed. W.R. Davidson, F.A. Hayes, V.F. Nettles, and F.E. Kellogg. Tallahassee, FL: Tall Timbers Research Station, pp. 155–160; Miscellaneous Publication 7.

Samuel, J.L. 1981. Salmonella in macropods. In *Wildlife diseases of the Pacific Basin and other countries: Proceedings of the fourth international conference of the Wildlife Disease Association, Sydney,* ed. M.E. Fowler, pp. 61–63.

Schicht-Tinbergen, M. 1989. *Der Igel.* Jena: VEB Gustav Fischer Verlag, 172 pp.

Speare, R., and A.D. Thomas. 1988. Orphaned kangaroo and wallaby joeys as a potential zoonotic source of *Salmonella* spp. *Medical Journal of Australia* 148:619, 622–623.

Steyn, D.G., M.H. Finlayson, and H.D. Brede. 1976. The Enterobacteriaceae of South African baboons. *South African Medical Journal* 50:994–996.

Threlfall, E.J., A.M. Ridley, L.R. Ward, and B. Rowe. 1996. Assessment of health risk from *Salmonella*-based rodenticides. *Lancet* 348:616–617.

Wray, C., K. Baker, J. Gallagher, and P. Naylor. 1977. *Salmonella* infection in badgers in the south west of England. *British Veterinary Journal* 133:526–529.

SHIGELLOSIS

TORSTEN MÖRNER

Synonym: Bacillary dysentery.

INTRODUCTION. Dysentery caused by *Shigella dysenteriae* can cause high mortality among people, particularly where sanitary conditions are poor, such as during natural catastrophes and times of war (Shears 1996). *Shigella* species, including *S. dysenteriae* and *S. sonnei,* but especially *S. flexneri,* are also important pathogens of nonhuman primates (Paul-Murphy 1993). In addition, *S. flexneri* has been associated with periodontal disease in rhesus monkeys *Macaca mulatta* (Armitage et al. 1982).

DISTRIBUTION AND HOST RANGE. *Shigella* has a worldwide distribution. In people, it is most common in tropical and subtropical regions (Shears 1996). Humans and nonhuman primates are the reservoirs of *Shigella,* which have been reported from many primate species, including great apes, macaques *Macaca* spp., spider monkeys *Ateles* sp., tamarins, and prosimians (Paul-Murphy 1993).

ETIOLOGY. The genus *Shigella* contains four species: *S. dysenteriae, S. flexneri, S. boydii,* and *S. sonnei.* These species often are referred to as *Shigella* subgroups A, B, C, and D, respectively. They are Gram-negative, nonmotile, straight rods and are facultative anaerobes (Shears 1996).

TRANSMISSION AND EPIDEMIOLOGY. Shigella can be a serious problem, causing outbreaks of acute diarrhea in primate collections and colonies (Paul-Murphy 1993). It has been isolated from primates newly caught in the wild (Steyn et al. 1976) and from newly imported animals (Hartley 1975). Such infections may be acquired as a result of human contact under conditions of poor sanitation after capture (Paul-Murphy 1993).

Susceptibility to infection varies among different host species. The incidence among macaques and gibbons is reported to be high, whereas infection seems to be rare among New World monkeys and prosimians (Banish et al. 1993; Paul-Murphy 1993). Chronic carriers also occur; the prevalence of carriers is reported to range from 5% to 67% in different colonies of primates (McClure et al. 1985; Banish et al. 1993). Transmission of *Shigella* is primarily fecal-oral, and it can spread rapidly in primate colonies (Paul-Murphy 1993).

CLINICAL SIGNS. Disease is most often seen in young animals, in which it occurs as acute diarrhea. The incubation period is 4 days or less. Clinical signs include abdominal discomfort, tenesmus, and frequent passage of small volumes of stool containing mucus or blood (Paul-Murphy 1993).

PATHOGENESIS AND PATHOLOGY. The primary site of infection with *S. dysenteriae* is the colon, where *Shigella* bacteria invade the epithelium and superficial lamina propria, causing epithelial necrosis, with erosions and ulceration (McClure et al. 1985; Karnell et al. 1991). The bacterium causes acute inflammation in the intestinal mucosa, beyond which it rarely penetrates, although septicemia can occur (Enurah et al. 1988). The capacity to invade epithelium is plasmid mediated, and in *S. dysenteriae* infections, a "shigatoxin" or verotoxin may promote local secretion. With endotoxin, it may be responsible for systemic lesions (Shears 1996), which may be observed in the small intestines, stomach, gingiva, skin, and submandibular air sacs (McClure et al. 1985).

At autopsy, petechial hemorrhages may be present on the serosa of the colon, and the colonic mucosa may appear lusterless and congested or hemorrhagic, with erosions. Colonic and mesenteric lymph nodes usually are edematous and swollen. Microscopically, there is necrosis and erosion or ulceration of the colonic epithelium, with numerous neutrophils in the lamina propria, and in the crypts (Kollias 1989; Karnell et al. 1991).

DIAGNOSIS. Diagnosis is based on isolation of the bacterium from feces in clinical cases or from enteric lesions at necropsy. Cultures are most successful when taken immediately from a freshly passed stool or from a rectal swab (Paul-Murphy 1993). If that is not possible, transport medium should be used, since the organism has limited viability (Shears 1996). Typing of isolates is important for epidemiologic assessment.

CONTROL AND TREATMENT. Strict quarantine and screening for *Shigella* in newly imported nonhuman primates is essential to keep primate colonies free of this organism. Diseased animals should be isolated and treated, if necessary. The correction of fluid and electrolyte imbalance is the primary therapeutic concern, and the use of antimicrobials is not always recommended. However, antimicrobials are recommended in large colonies where reinfection occurs frequently (Paul-Murphy 1993). Multiple antibiotic resistance is common in *Shigella,* and antibiotic sensitivity must be evaluated. The treatment recommended is trimethoprim-sulfamethoxazole by the oral or parenteral route (Paul-Murphy 1993), although enrofloxacin has been used with success to control extensive epidemics in captive primate collections (Line et al. 1992; Banish et al. 1993).

PUBLIC HEALTH AND CONTROL. The principal reservoir of *Shigella* infection for people is other people. However, *Shigella* is zoonotic, and occupational exposure may result in disease in primate handlers (Kennedy et al. 1993).

LITERATURE CITED

Armitage, G.C., E. Newbrun, C.I. Hoover, and J.H. Anderson. 1982. Periodontal disease associated with *Shigella flexneri* in rhesus monkeys: Clinical, microbiologic and

histopathologic findings. *Journal of Periodontal Research* 17:131–144.
Banish, L.D., R. Sims, D. Sack, R.J. Montali, L. Phillips Jr., and M. Bush. 1993a. Prevalence of shigellosis and other enteric pathogens in a zoologic collection of primates. *Journal of the American Veterinary Medical Association* 203:126–132.
Banish, L.D., R. Sims, M. Bush, D. Sack, and R.J. Montali. 1993b. Clearance of *Shigella flexneri* carriers in a zoologic collection of primates. *Journal of the American Veterinary Medical Association* 203:133–136.
Enurah, L.U., E.M. Uche, and D.R. Nawathe. 1988. Fatal shigellosis in a chimpanzee (*Pan troglodytes*) in the Jos Zoo, Nigeria. *Journal of Wildlife Diseases* 24:178–179.
Hartley, E.G. 1975. The incidence and antibiotic sensitivity of *Shigella* bacteria isolated from newly imported macaque monkeys. *British Veterinary Journal* 131:205–212.
Karnell, A., F.P. Reinholt, S. Katakura, and A.A. Lindberg. 1991. *Shigella flexneri* infection: A histopathologic study of colonic biopsies in monkeys infected with virulent and attenuated bacterial strains. *APMIS* 99:787–796.
Kennedy, F.M., J. Astbury, J.R. Needham, and T. Cheasty. 1993. Shigellosis due to occupational contact with non-human primates. *Epidemiology and Infection* 110:247–251.
Kollias, G.V. 1989. Diagnosis and management of salmonellosis and shigellosis in nonhuman primates. In *Current veterinary therapy IX: Small animal practice.* ed. R.W. Kirk. Philadelphia: W.B. Saunders, pp. 666–669.
Line, A.S., J. Paul-Murphy, D.P. Aucoin, and D.C. Hirsh. 1992. Enrofloxacin treatment of long-tailed macaques with acute bacillary dysentery due to multiresistant *Shigella flexneri* IV. *Laboratory Animal Science* 42:240–244.
McClure, H.M., A.R. Brodie, D.C. Anderson, and R.B. Swenson. 1985. Bacterial infections of nonhuman primates. In *Primates: The road to self-sustaining populations,* ed. K. Benirschke. New York: Springer-Verlag, pp. 531–556.
Paul-Murphy, J. 1993. Bacterial enterocolitis in nonhuman primates. In *Zoo and wild animal medicine: Current therapy 3,* ed. M.E. Fowler. Philadelphia: W.B. Saunders, pp. 344–351.
Shears, P. 1996. *Shigella* infections. *Annals of Tropical Medicine and Parasitology* 90:105–114.
Steyn, D.G., M.H. Finlayson, and H.D. Brede. 1976. The Enterobacteriaceae of South African baboons. *South African Medical Journal* 50:994–996.

STAPHYLOCOCCUS INFECTION

GARY WOBESER

INTRODUCTION. Members of the genus *Staphylococcus* are among the most common pyogenic or pus-inducing bacteria, causing local abscesses and, less commonly, generalized infections in a wide variety of species. Among wild mammals, staphylococcal infection is of greatest significance in rabbits *Sylvilagus* spp., *Oryctolagus* sp., and hares *Lepus* spp., in which infection may result in severe and sometimes fatal disease (Davis 1981). The bacteria occur worldwide.

ETIOLOGY AND PATHOGENESIS. Although there are many species in the genus *Staphylococcus,* most disease in wildlife is attributed to *S. aureus.* This organism is a large Gram-positive coccus that typically occurs in grapelike clusters and inhabits the skin and mucous membranes of warm-blooded vertebrates. Various strains, biotypes, or ecovars of *S. aureus* can be differentiated on the basis of phenotypic characteristics and are associated with different host species (Jonsson and Wadstrom 1993), for example, biotype D in hares (Hajek 1976; Kloos and Schleifer 1986).

The organisms are opportunistic pathogens that require some damage to the skin or mucous membrane to become established in underlying tissues. The predisposing injury can be of many types, including bites from arthropods (Kloos and Schleifer 1986) and wounds from fighting (Bell and Chalgren 1943; Campbell et al. 1981).

In most instances, infection results in a localized lesion containing pus (an abscess). Less commonly, the bacteria may spread within the fascia, causing subcutaneous inflammation, or spread to internal organs, including the heart valves. Rarely, they cause localized granulomatous inflammation, termed *botryomycosis. Staphylococcus intermedius* has been reported to cause inflammation of adnexal glands in young ranch mink *Mustela vison* (Hunter and Prescott 1991).

CLINICAL SIGNS, PATHOLOGY, AND DIAGNOSIS. Clinical signs of infection are nonspecific. Infected lagomorphs may be listless, emaciated, and lame if joints or tendons are involved. Large subcutaneous abscesses may be visible externally as swellings, and some of these may have draining tracts, resulting in crusting of the hair.

Infected areas of skin usually are encrusted with exudate. The most commonly involved internal sites are superficial lymph nodes (cervical, axillary, mandibular, and inguinal), which are enlarged and contain purulent or caseous exudate. Less commonly, there is involvement of visceral organs, with abscesses in the spleen, lungs, liver, kidneys, muscle, and heart. Other lesions may include splenomegaly without abscesses, peritonitis, metritis, mastitis, encephalitis, and valvular endocarditis. In botryomycosis, foci of pyogranulomatous or granulomatous inflammation, which may include giant cells, occur in the dermis or at other sites.

Large Gram-positive cocci, arranged in clusters, are readily found in smears of exudate from lesions or in tissue sections of suppurative or botryomycotic lesions, stained with Gram stain. The organisms grow well on 5% blood agar media.

IMMUNITY, TREATMENT, AND CONTROL. No information is available on immunity to staphylococci in wild animals. Treatment and control are unnecessary.

LITERATURE CITED

Bell, J.F., and W.S. Chalgren. 1943. Some wildlife diseases in the eastern United States. *Journal of Wildlife Management* 7:270–278.

Campbell, G.A., S.D. Kosanke, D.M. Toth, and G.L. White. 1981. Disseminated staphylococcal infection in a colony of captive ground squirrels (*Citellus lateralis*). *Journal of Wildlife Diseases* 17:177–181.
Davis, J.W. 1981. Staphylococcosis in rabbits and hares. In *Infectious diseases of wild mammals,* ed. J.W. Davis, L.H. Karstad, and D.O. Trainer, 2d ed. Ames: Iowa State Press, pp. 317–319.
Hajek, V. 1976. *Staphylococcus intermedius,* a new species isolated from animals. *International Journal of Systematic Bacteriology* 26:401–408.
Hunter, D.B., and J.G. Prescott. 1991. Staphylococcal adenitis in ranch mink in Ontario. *Canadian Veterinary Journal* 32:354–356.
Jonsson, P., and T. Wadstrom. 1993. *Staphylococcus.* In *Pathogenesis of bacterial infections in animals,* ed. C.L. Gyles and C.O. Thoen, 2d ed. Ames: Iowa State University Press, pp. 21–35.
Kloos, W.E., and K.H. Schleifer. 1986. Genus IV: *Staphylococcus* Rosenbach 1884, 18^{AL}, Nom. Cons. Opin. 17 Jud. Comm. 1958, 153. In *Bergey's manual of systematic bacteriology,* vol. 2, ed. P.H.A. Sneath, N.S. Mair, M.E. Sharpe, and J.G. Holt. Baltimore: Williams and Wilkins, pp. 1015–1035.

TYZZER'S DISEASE

GARY WOBESER

Synonyms: Hemorrhagic disease of muskrats, Errington's disease.

HISTORY, DISTRIBUTION, AND HOST RANGE. Tyzzer (1917) described a fatal disease of mice *Mus musculus* characterized by intracellular bacteria in hepatocytes surrounding foci of necrosis in the liver, and in intestinal epithelial cells. The disease, now called Tyzzer's disease (TD), occurs in a wide range of domestic and laboratory mammals (Harkness and Wagner 1995).

Errington (1946) described a disease characterized by hemorrhagic enteritis and focal hepatic necrosis in muskrats *Ondatra zibethicus* in Iowa. Epidemics attributed to this disease have occurred widely in muskrats in western and northern North America (Errington 1963). Karstad et al. (1971) proposed that Tyzzer's and Errington's diseases were synonymous; subsequently, lesions typical of TD were found when tissues from muskrats found dead in Iowa in 1947 by Errington were reexamined (Wobeser et al. 1979). The disease has been reported in several other species of wild mammal (Table 28.2), and in at least one bird (Saunders et al. 1993).

ETIOLOGY. The agent seen in mice by Tyzzer (1917) was named *Bacillus piliformis* on the basis of its morphology. Duncan et al. (1993) assigned the agent to the genus *Clostridium,* as *C. piliforme,* on the basis of 16S rRNA analysis. The relationship among organisms in different host species is unclear; however, there is antigenic heterogeneity among isolates. Boivin et al. (1993) identified six serologic groups. A single host species may be infected with more than one group (Hook et al. 1995). Cross-transmission with isolates from laboratory rodents has yielded variable results (Franklin et al. 1994).

The organism has not been cultured in cell-free media but can be cultivated in embryonated eggs and a variety of mammalian cells and can be transmitted by serial animal inoculation. The vegetative form is a slender filament approximately 0.5 × 8–10 μm, with occasional forms up to 40 μm long, arranged in parallel or crisscross bundles within host cells. Organisms stain faintly basophilic, if at all, with hematoxylin-eosin and are either nonreactive or faintly negative with Gram stain. They are stained by strongly basic aniline dyes (Giemsa, thionine, and methylene blue) and are periodic acid-Schiff positive. Silver-impregnation techniques (Warthin Starry, Warthin-Faulkner, Gomori methenamine silver, and Levaditi) are best for demonstrating organisms in tissue.

TRANSMISSION AND EPIDEMIOLOGY. In mice, TD was described originally as a highly fatal epidemic disease. Epidemics have been reported in other laboratory species and muskrats; however, subclinical infection is common in laboratory animal colonies without detectable disease. Infected animals without clinical signs can transmit the organism (Motzel and Riley 1992).

Clinical disease occurs among animals stressed by poor environmental conditions or by the administration of corticosteroids. Errington (1963) suggested that changes in the prevalence of the disease resulted from changes in resistance of the muskrat population. Occurrence of TD in muskrats shortly after capture suggests activation of a latent infection by stress (Karstad et al. 1971; Chalmers and MacNeil 1977). Cases in dogs and raccoons have occurred in animals that may have been immunocompromised by concurrent canine distemper (Wojcinski and Barker 1986; Iwanaka et al. 1993).

Transmission is by ingestion of spores from an environment contaminated by feces of infected animals. Bedding contaminated during an epidemic in mice remained infectious for 1 year at room temperature (Tyzzer 1917). Bedding from a muskrat house in which muskrats died of TD (Wobeser et al. 1978) was infective after 16 months at –10° C. Errington (1954) reported that environments remained infectious for at least 5 years in the total absence of muskrats, and believed the incubation period in muskrats to be 7–8 days (Errington 1946). Muskrats died 5–10 days after intragastric inoculation of contaminated bedding material (G. Wobeser unpublished). Transmission by cannibalism has been suggested (Fujiwara et al. 1973), and transplacental infection occurred in mice (Fries 1978).

TABLE 28.2—Reported occurrence of Tyzzer's disease in wild mammals

Common Name	Genus/Species	Location	Citation
	Free-living Animals		
Muskrat	*Ondatra zibethicus*	Canada	Wobeser et al. 1978
Cottontail rabbit	*Sylvilagus floridianus*	U.S.A.	Ganaway et al. 1976
Raccoon	*Procyon lotor*	Canada	Wojcinski and Barker 1986
Brushtail possum	*Trichosurus vulpecula*	Australia	Canfield and Hartley 1991
Common ring-tail possum	*Pseudocheirus peregrinus*	Australia	Canfield and Hartley 1991
	Captive Animals		
Coyote	*Canis latrans*	U.S.A.	Marler and Cook 1976
Gray fox	*Urocyon cinereoargenteus*	U.S.A.	Stanley et al. 1978
Snow leopard	*Panthera uncia*	U.S.A.	Schmidt et al. 1984
Herbert River ring-tail possum	*Pseudocheirus herbertensis*	Australia	Canfield and Hartley 1991
Koala	*Phascolarctos cinereus*	Australia	Canfield and Hartley 1991
Wombat	Unidentified	Australia	Canfield and Hartley 1991
Antechinus	*Antechinus* sp.	Australia	Canfield and Hartley 1991

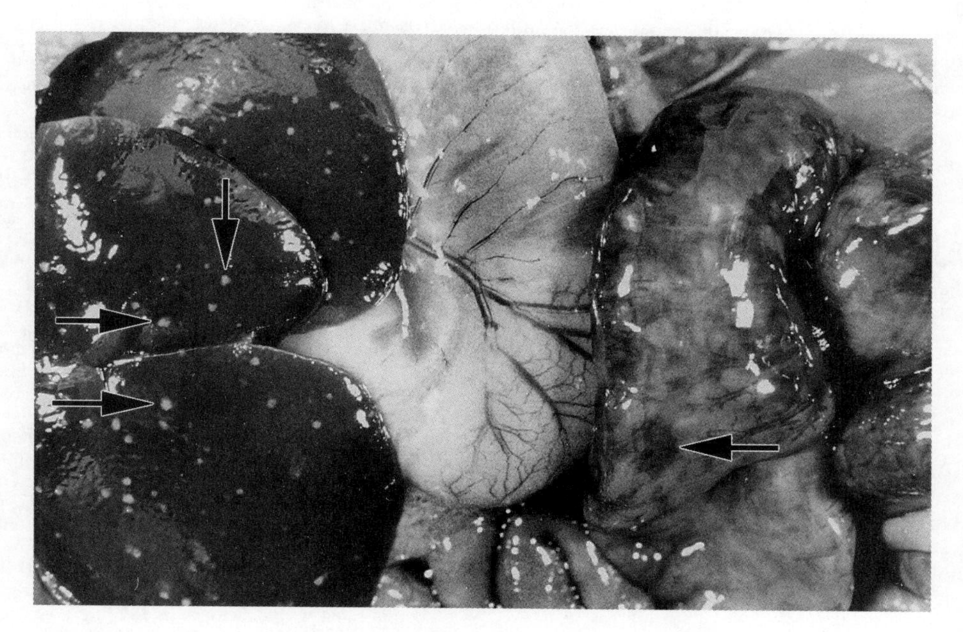

FIG. 28.1—Gross lesions of Tyzzer's disease in a muskrat. There are multiple pinhead-sized pale zones of necrosis in the liver (*arrowhead*). Dark areas of hemorrhage in the inner wall (mucosa) of the colon are visible through the outer (serosal) surface (*arrow*).

CLINICAL SIGNS. In most species, clinical TD is an acute disease with no premonitory signs. If, during the short clinical course, signs are seen, they include diarrhea, depression, anorexia, and a rough hair coat. Dysentery occurs in cottontail rabbits and muskrats.

PATHOGENESIS AND PATHOLOGY. Tyzzer's disease is an intestinal infection that spreads to the liver via the portal vein. The combination of intestinal and multifocal hepatic lesions is a feature of the disease (Fig. 28.1). Gross lesions are variable among species, but enteritis usually is present, and intestinal lesions may occur without hepatic lesions. The cecum, colon, and terminal ileum are affected most commonly, and muskrats typically have ulcerative hemorrhagic colitis and typhlitis. The intestinal contents may be fluid and in muskrats usually contain fresh blood. Liver lesions are similar in all species and consist of discrete white foci, less than 1–3 mm in diameter, within the hepatic parenchyma (Fig. 28.1). In contrast to tularemia and other Gram-negative septicemias, splenomegaly and focal splenic necrosis are never seen in TD. White streaks or foci in the myocardium occur in some species.

Histologically, there may be epithelial necrosis and ulceration in the intestine, with necrosis, hemorrhage, and edema of the underlying layers. Organisms may be found in epithelial and muscle cells. Hepatic lesions vary from foci of acute coagulation necrosis with only a few attendant granulocytes to fibrotic scars surrounded by macrophages and occasional multinucleated giant cells. Organisms are found in intact hepatocytes at the margin of acute lesions, but may be absent in chronic lesions.

DIAGNOSIS. Diagnosis of TD is usually by microscopic demonstration of the characteristic intracellular bacteria, usually by silver stains, in association with compatible lesions. Location, morphology, and staining of the organism all are important for identification; ultrastructure, particularly the presence of peritrichous flagella, also is useful. Liver and intestinal tissue should be examined. A variety of other diagnostic techniques have been described but are not widely available for routine use (Okado et al. 1986; Iwanaka et al. 1993; Hansen et al. 1994). Antibodies may be detected by immunofluorescence (Hansen et al. 1994) or by enzyme-linked immunosorbent assay using flagellar antigens (Hook et al. 1995), but interpretation is confounded by strain heterogeneity (Franklin et al. 1994).

The organism can be cultivated in embryonated hen's eggs and in several mammalian cell lines (Riley et al. 1990). Gerbils are particularly susceptible to TD and have been used as sentinels to detect subclinical infection in laboratory rodent colonies (Gibson et al. 1987).

TREATMENT AND CONTROL. No suitable measures exist for the control of TD in wild animals. Tetracyclines have been used to treat laboratory animals, and the causative organism is sensitive to a variety of other antibiotics (Harkness and Wagner 1995).

PUBLIC HEALTH CONCERNS. Antibodies have been found in people (Fries 1980), and a clinical case was reported in a man infected with HIV-1 (Smith et al. 1996). However, there is no evidence that this is an important human disease.

LITERATURE CITED

Boivin, G.P., R.R. Hook, and L.K. Riley. 1993. Antigenic diversity in flagellar epitopes among *Bacillus piliformis* isolates. *Journal of Medical Microbiology* 38:177–182.

Canfield, P.J., and W.J. Hartley. 1991. Tyzzer's disease (*Bacillus piliformis*) in Australian marsupials. *Journal of Comparative Pathology* 105:167–173.

Chalmers, G.A., and A.C. MacNeil. 1977. Tyzzer's disease in wild-trapped muskrats in British Columbia. *Journal of Wildlife Diseases* 13:114–116.

Duncan, A.J., R.J. Carman, G.J. Olsen, and K.H. Wilson. 1993. Assignment of the agent of Tyzzer's disease to *Clostridium piliforme* comb. nov. on the basis of 16S rRNA sequence analysis. *International Journal of Systematic Bacteriology* 43:314–318.

Errington, P.L. 1946. Special report on muskrat diseases. *Iowa Cooperative Wildlife Research Unit Quarterly Report* July, August, September:34–51.

———. 1954. The special responsiveness of minks to epizootics in muskrat populations. *Ecological Monographs* 24:377–393.

———. 1963. *Muskrat populations.* Ames: Iowa State University Press, 665pp.

Franklin, C.L., S.L. Motzel, C.L. Besch-Williford, R.R. Hook Jr., and L.K. Riley. 1994. Tyzzer's infection: Host specificity of *Clostridium piliforme* isolates. *Laboratory Animal Science* 44:568–572.

Fries, A.S. 1978. Demonstration of antibodies to *Bacillus piliformis* in SPF colonies and experimental transplacental infection by *Bacillus piliformis* in mice. *Laboratory Animals* 12:23–25.

———. 1980. Antibodies to *Bacillus piliformis* (Tyzzer's disease) in sera from man and other species. In *Animal quality and models of biomedical research: 7th ICLAS symposium, Utrecht.* Stuttgart, Germany: Gustav Fischer Verlag, pp. 249–252.

Fujiwara, K., N. Hirano, S. Takenaka, and K. Sato. 1973. Peroral infection in Tyzzer's disease of mice. *Japanese Journal of Experimental Medicine* 43:33–42.

Ganaway, J.R., R.S. McReynolds, and A.M. Allen. 1976. Tyzzer's disease in free-living cottontail rabbits (*Sylvilagus floridanus*) in Maryland. *Journal of Wildlife Diseases* 12:545–549.

Gibson, S.V., K.S. Waggie, J.E. Wagner, and J.R. Ganaway. 1987. Diagnosis of subclinical *Bacillus piliformis* infection in a barrier-maintained mouse production colony. *Laboratory Animal Science* 37:786–788.

Hansen, A.K., H.V. Andersen, and O. Svendsen. 1994. Studies on the diagnosis of Tyzzer's disease in laboratory rat colonies with antibodies against *Bacillus piliformis* (*Clostridium piliforme*). *Laboratory Animal Science* 44:424–429.

Harkness, J.E., and J.E. Wagner. 1995. *The biology and medicine of rabbits and rodents.* Baltimore: Williams and Wilkins, pp. 308–312.

Hook, R.R., L.K. Riley, C.L. Franklin, and C.L. Besch-Williford. 1995. Seroanalysis of Tyzzer's disease in horses: Implications that multiple strains can infect Equidae. *Equine Veterinary Journal* 27:8–12.

Iwanaka, M., S. Orita, Y. Mokuono, K. Akiyama, A. Nii, T. Yanai, T. Masegi, and K. Ueda. 1993. Tyzzer's disease complicated with distemper in a puppy. *Journal of Veterinary Medical Science* 55:337–339.

Karstad, L., P. Lusis, and D. Wright. 1971. Tyzzer's disease in muskrats. *Journal of Wildlife Diseases* 7:96–99.

Marler, R.J., and J.E. Cook. 1976. Tyzzer's disease in two coyotes. *Journal of the American Veterinary Medical Association* 169:940–941.

Motzel, S.L., and L.K. Riley. 1992. Subclinical infection and transmission of Tyzzer's disease in rats. *Laboratory Animal Science* 42:439–443.

Okada, N., W. Toriumi, K. Sakamoto, and K. Fujiwara. 1986. Pathologic observations of mouse liver infected with Tyzzer's organisms using immunoperoxidase method. *Japanese Journal of Veterinary Science* 48:89–96.

Riley, L.K., C. Besch-Williford, and K.S. Waggie. 1990. Protein and antigenic heterogeneity among isolates of *Bacillus piliformis. Infection and Immunity* 58:1010–1016.

Saunders, G.K., D.P. Sponenberg, and K.L. Marx. 1993. Tyzzer's disease in a neonatal cockatiel. *Avian Diseases* 37:891–894.

Schmidt, R.E., D.L. Eisenbrandt, and G.B. Hubbard. 1984. Tyzzer's disease in snow leopards (*Panthera uncia*). *Journal of Comparative Pathology* 94:165–167.

Smith, K.J., H.G. Skelton, E.J. Hilyard, T. Hadfield, R.S. Moeller, S. Tuur, C. Decker, K.F. Wagner, and P. Angritt. 1996. *Bacillus piliformis* infection (Tyzzer's disease) in a patient infected with HIV-1: Confirmation with 16S ribosomal RNA sequence analysis. *Journal of the American Academy for Dermatology* 34:343–348.

Stanley, S.M., R.E. Flatt, and G.N. Daniels. 1978. Naturally occurring Tyzzer's disease in gray fox. *Journal of the American Veterinary Medical Association* 173:1173–1174.

Tyzzer, E.E. 1917. A fatal disease of the Japanese waltzing mouse caused by a spore-bearing bacillus (*Bacillus piliformis,* n. sp.). *Journal of Medical Research* 37:307–338.

Wobeser, G., D.B. Hunter, and P.-Y. Daoust. 1978. Tyzzer's disease in muskrats: Occurrence in free-living animals. *Journal of Wildlife Diseases* 14:325–328.

Wobeser, G., H.J. Barnes, and K. Pierce. 1979. Tyzzer's disease in muskrats: Re-examination of specimens of hemorrhagic disease collected by Paul Errington. *Journal of Wildlife Diseases* 15:525–527.

Wojcinski, Z.W., and I.K. Barker. 1986. Tyzzer's disease as a complication of canine distemper in a raccoon. *Journal of Wildlife Diseases* 22:55–59.

29 MYCOTIC DISEASES

KATHY BUREK

Humans and animals generally contract mycotic infections by exposure to fungi growing as saprophytes in the environment. Some fungi are primary pathogens, but most pathogenic fungi are opportunists of low virulence that only rarely cause disease in healthy individuals, affecting instead those with some degree of immunocompromise. Mycotic diseases are categorized as systemic, subcutaneous, and superficial. The course of infection depends on the fungus involved, virulence, intensity of exposure, route of infection, and the presence of other underlying predisposing factors such as immunocompromise or previous treatment with antibiotics or immunosuppressive drugs. With the exception of the dermatophytoses (ringworm), sporotrichosis, and candidiasis, most pathogenic fungi are not usually contagious or zoonotic. Occurrences of multiple cases are due to simultaneous exposure to a point source.

Because most mycotic diseases occur sporadically or in small outbreaks, they are more important in captive wildlife or highly managed populations of endangered species where individuals are of particular value. Captivity may increase susceptibility to these diseases, due to use of antibiotics or corticosteroids and possible immunocompromise associated with the stress of captivity.

Diagnosis is usually based on a combination of several modalities, including culture, histopathology, serology, immunohistochemistry, immunofluorescence, and polymerase chain reaction (PCR). Most fungi can be detected in histopathologic sections using hematoxylin-eosin (H&E) stain. Special histologic stains, such as the Gomori's methenamine silver (GMS), Gridley fungus, or periodic acid-Schiff (PAS) reaction, and mucin stains, such as Mayer's mucicarmine, can be used to aid in detection and identification of the fungi.

SYSTEMIC FUNGAL INFECTIONS

Coccidioidomycosis

Synonyms: San Joaquin fever, coccidioidal granuloma, desert rheumatism, valley fever, Posada-Wernick's disease, Posada's disease, California disease.

INTRODUCTION. Coccidioidomycosis is primarily a respiratory infection with rare systemic dissemination caused by *Coccidioides immitis.* It is the most infectious cause of systemic mycoses (Davis 1981) and affects a wide variety of domestic animals and free-living and captive wild species (Table 29.1), as well as humans. It is a primary pathogen capable of affecting normal individuals but may also be an opportunistic pathogen and is recognized as an indicator of acquired immune deficiency syndrome (AIDS) in people (Pappagianis 1993).

HISTORY AND HOST RANGE. Coccidioidomycosis has been recognized in humans since 1892 with the first case in a monkey reported by Posadas in Argentina (Pappagianis et al. 1973). The etiologic agent was first classified as a protozoan, *C. immitis,* in 1896 (Blundell et al. 1961) and was correctly recognized as a fungus in 1900 (Stevens 1995).

ETIOLOGY. *Coccidioides immitis* is a dimorphic fungus, growing in a different form in tissue than in culture or in the environment. It is a free-living, soil-dwelling fungus that survives in saline soil and in seawater (Fauquier et al. 1996). It grows rapidly on Sabouraud's medium or selective media containing cyclohexamide and chloramphenicol, developing into a typical mold in 3–4 days, with moist, flat, gray colonies. The fungus then develops cottony aerial mycelium with barrel-shaped, 2–4 × 5–6 μm arthrospores (Pappagianis 1993), which are highly infectious and are aerosolized easily.

In tissue, the arthrospores develop into spherules. Immature spherules are 5–30 μm in diameter (Fig. 29.1). These develop into 30–200 μm-diameter mature, endosporulating spherules or sporangia (Chandler and Watts 1987). Spherules release endospores that may develop into more spherules. Except in very rare cases, the tissue form is not thought to be infectious.

TRANSMISSION. Inhalation of *C. immitis* arthrospores carried in dust or soil particles is the principal method of infection (Davis 1981). In mouse *Mus musculus* studies, only a few arthrospores administered intranasally are required for infection (Galgiani 1993). Traumatic inoculation can occur, resulting in a fluctuant abscess, but dissemination from such a lesion is unusual (Dungworth 1993; Pappagianis 1993).

DISTRIBUTION AND EPIDEMIOLOGY. Coccidioides immitis is endemic in the Western Hemisphere almost

TABLE 29.1—Coccidioidomycosis reported in free-ranging and captive wild mammals

Common Name	Scientific Name	Reference
Gorilla[a]	*Gorilla gorilla*	McKenney et al. 1944; Ingram 1975
Chimpanzee[a]	*Pan troglodytes*	Ingram 1975
White-faced sapajou[a]	*Cebus capucinus*	McKenney et al. 1944
Sooty mangabey[a]	*Cercocebus torquatus*	Pappagianis et al. 1973
De Brazza's guenon[a]	*Cercopithecus neglectus*	Ingram 1975
Gelada baboon[a]	*Theropithecus gelada*	Rapley and Long 1974
Yellow baboon[a]	*Papio cynocephalus*	Rosenberg et al. 1984
Mandrill baboon[a]	*Mandrillus sphinx*	Johnson et al. 1998
Rhesus monkey[a]	*Macaca mulatta*	Breznock et al. 1975
Ring-tailed lemur[a]	*Lemur catta*	Burton et al. 1986
Sea otter[b]	*Enhydra lutris*	Cornell et al. 1979
River otter[a]	*Lontra canadensis*	Harwell 1985
California sea lion[a,b]	*Zalophus californianus*	Reed et al. 1976; Osborn et al. 1985; Fauquier et al. 1996
Cougar[b]	*Felis concolor*	Clyde et al. 1990
Bengal tiger[a]	*Panthera tigris*	Henrickson and Biberstein 1972
Coyote[b]	*Canis latrans*	Straub et al. 1961
Giant red kangaroo[a]	*Macropus rufus*	Hutchinson et al. 1973
Tapir[a]	*Tapirus terrestris*	Dillehay et al. 1985.
Desert bighorn sheep[b]	*Ovis canadensis nelsoni*	Jessup et al. 1989
Collared peccary[a]	*Pecari tajacu*	Lochmiller et al. 1985
Bottle-nosed dolphin[b]	*Tursiops truncatus gilli*	Reidarson et al. 1998

[a]Captive animals.
[b]Free-ranging animals.

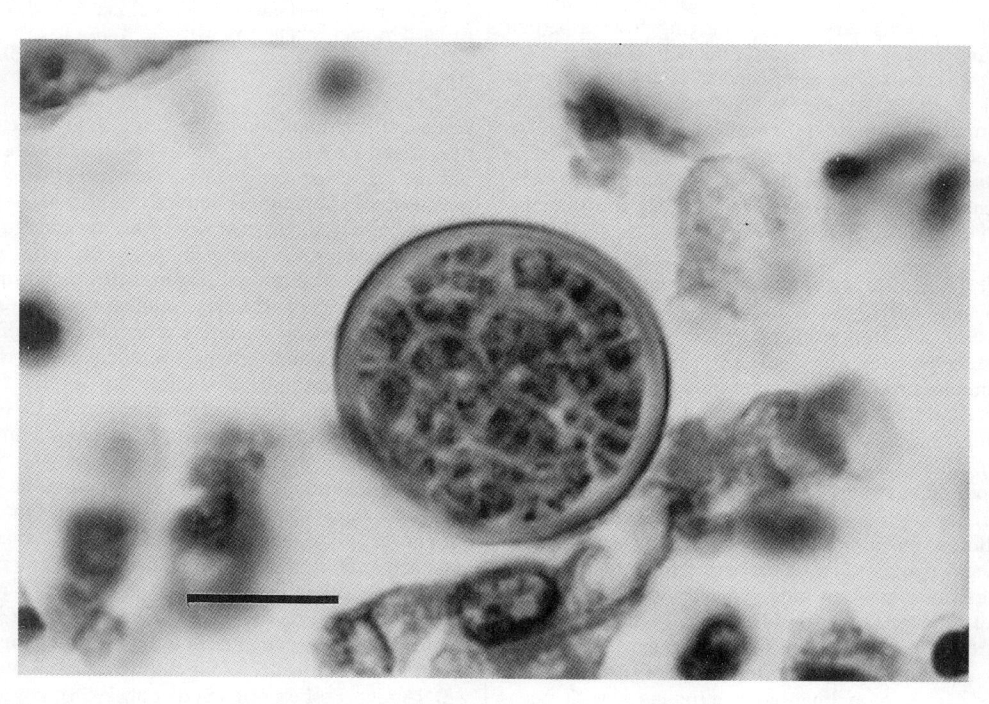

FIG. 29.1—Immature spherule of *Coccidioides immitis* from the spleen of a sea lion. Endospores have begun to form. H&E, bar = 10 μm.

exclusively between latitudes 40° N and 40° S. It occurs in the southwestern United States, mainly in California, Arizona, and Texas; in focal areas of Mexico; in Central America, including Guatemala, Honduras, and Nicaragua; and in South America, in Argentina, Paraguay, Venezuela, Colombia, and Brazil.

The endemic regions are designated as the Lower Sonoran Life Zone, where the soil is alkaline and the climate is arid, with yearly rainfall ranging from 11 to 44 cm, with hot summers, and with winters with few freezes (Galgiani 1993). Cycles of drought and rain (Campins 1970; Galgiani 1993), dust storms, and soil disturbances, including site construction and possibly earthquakes (Stevens 1995), may precede outbreaks. Occurrence of coccidioidomycosis in marine mammals distant from endemic zones also suggests that wind-blown spores extend over a very wide range (Osborn et al. 1985). High concentrations of the fungus are also associated with desert rodent burrows and their feces (Emmons and Ashburn 1942; Davis 1981).

Coccidioidomycosis can occur in nonendemic areas, usually due to a previous visit to an endemic area, reactivation of an infection acquired in an endemic area, or fomites. The most common reason for disease in a nonendemic area is reactivation in an animal transported from an endemic area, which can occur long after the initial infection.

CLINICAL SIGNS. It is estimated that most people and animals that live in endemic areas eventually become infected, but relatively few become symptomatic (Dungworth 1993). Some people develop mild, nonprogressive, flulike respiratory syndrome or pneumonia called San Joaquin Valley fever that resolves on its own. This stage is not usually recognized in animals.

Clinical signs vary greatly, ranging from a benign self-limited upper respiratory infection to chronic pulmonary disease to disseminated fatal disease. Clinical signs depend on the organs involved and often include dyspnea and anorexia (Rapley and Long 1974; Breznock et al. 1975; Dillehay et al. 1985). Bone involvement also is possible (Pappagianis et al. 1973; Rosenberg et al. 1984). Mortality is high in disseminated disease, even with treatment.

PATHOLOGY. In asymptomatic *C. immitis* infections, solid masses may occur in the lung. In disseminated infection, solid or purulent nodules may occur anywhere in the body. Gross lesions of chronic pulmonary coccidioidomycosis resemble those of tuberculosis; distinct or confluent grayish white nodules may occur in lungs and hilar and mediastinal lymph nodes (Fig. 29.2). There may be central caseous necrosis, with or without suppuration or calcification (Jones and Hunt 1983).

Microscopically, masses seen grossly are granulomas and pyogranulomas. In active infections, typical mature and immature fungal spheroids can usually be found in large numbers surrounded by the inflammatory reaction. In inactive lesions, organisms may be difficult to find. Larger spherules are often surrounded by a wide zone of epithelioid macrophages with occasional multinucleated giant cells, neutrophils, and lymphocytes. Both arthroconidia and endospores elicit a predominantly neutrophilic response.

DIAGNOSIS AND DIFFERENTIAL DIAGNOSES. Laboratory confirmation can be obtained by histopathologic or cytologic methods, culture and identification, and serologic testing (Pappagianis 1993). Histologic demonstration of typical mature spherules with endospores of *C. immitis* in characteristic lesions is considered pathognomonic (Stevens 1995). Other fungi that produce spherule-like structures that must be ruled out include *Chrysosporium parvum* and *Rhinosporidium seeberi;* immature spherules can also be confused with *Blastomyces dermatitidis*. Wet, unstained microscopic preparations of clinical materials can be examined for the typical fungal elements. If a diagnosis cannot be obtained in this way, clinical material can be cultured. Organisms in culture are highly contagious to humans, and isolation attempts should be performed by experienced personnel at a biosafety level-2 facility (Fauquier et al. 1996). The laboratory should be warned of the clinical suspicion of coccidioidomycosis in order to take proper precautions (Stevens 1995). For definitive identification, samples must either be inoculated into animals or cultured under specific conditions for conversion to the tissue form. There are exoantigen tests and a molecular probe that can be used on culture extracts for identification, but these are not readily available (Pappagianis 1993). In one case in a bottlenosed dolphin *Tursiops truncatus,* identification of *C. immitis* was confirmed from culture material by using a DNA Gen-Probe (Reidarson et al. 1998).

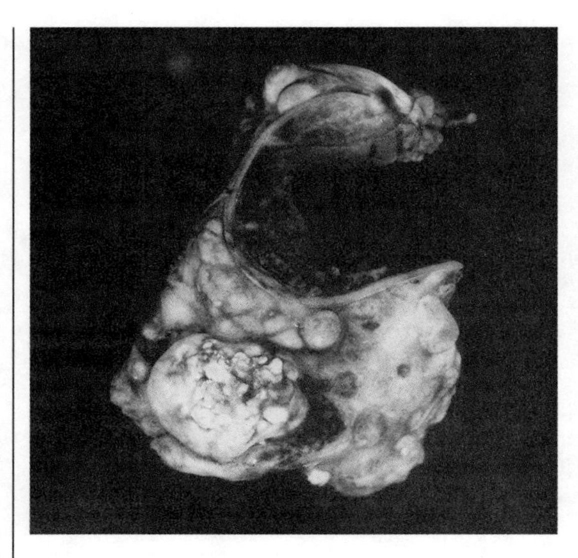

FIG. 29.2—Trachea and tracheal lymph node from a llama with coccidioidomycosis. There are multifocal to coalescing, firm, cream nodules in the lymph node, and the node is surrounded by dense connective tissue.

Standard serologic assays are the precipitin tube test (PTT) and the complement fixation (CF) test. The PTT, a nonquantitative test, detects immunoglobulin M (IgM) that is produced in the acute phase of infection and indicates early exposure, subclinical infection, or early disease (Legendre 1995). The CF test primarily detects IgG and can be used for prognosis because persistently high or rising CF antibody titer is correlated with disseminated extrapulmonary disease or refractory disease (Stevens 1995).

The PTT and the CF test are not available as kits and are difficult to adapt to general purpose clinical serologic laboratories. Dual immunodiffusion procedures are available that are highly specific and at least as sensitive. These are the immunodiffusion tube precipitin (IDTP) and the immunodiffusion complement fixing (IDCF) tests. Like the PTT, the IDTP test is qualitative. The IDCF test is generally reported qualitatively and needs to be followed by the standard CF test to obtain quantitative results (titers). Serum latex agglutination tests are also available, and an enzyme-linked immunosorbent assay (ELISA) is being developed that can be used as a rapid method for screening (Galgiani 1993). A coccidioidin skin test can also be used to detect infection and could be considered when moving an animal from an endemic area (Breznock et al. 1975).

TREATMENT. Amphotericin B and ketoconazole are recommended for animal treatment (Harwell 1985; Legendre 1995). Long-term therapy is usually required, and serial antibody titer determinations may be helpful in monitoring the response (Legendre 1995). Treatment of free-ranging species is not possible or warranted.

PUBLIC HEALTH CONCERNS. Coccidioidomycosis is a significant disease in people and is almost always contracted through the environment rather than directly from infected animals. However, even though the risk is low, caution is warranted during necropsy of an animal suspected to have coccidioidomycosis. There is one case in the literature in which an individual with no other convincing source of exposure developed systemic coccidioidomycosis following necropsy of an infected horse *Equus caballus* (Kohn et al. 1992). Carcasses should be disposed of within 5 days, since this is the time required for arthroconidiation after exposure to air (Pappagianis 1988).

DOMESTIC ANIMAL HEALTH CONCERNS. Disseminated disease has been observed mostly in dogs *Canis familiaris* and occasionally in horses, sheep *Ovis aries,* cattle *Bos taurus,* cats *Felis catus* (Dungworth 1993), burros *Equus asinus* (Cornell et al. 1979), and llamas *Lama glama* (Muir and Pappagianis 1982). As with other species, the fungus is contracted from the environment and not from other animals.

Blastomycosis

Synonyms: North American blastomycosis, Gilchrist's disease, Chicago disease.

INTRODUCTION. Blastomycosis is a disease primarily of people and dogs in North America, Africa, Europe, and Asia, and is caused by *Blastomyces dermatitidis.* It is sometimes called North American blastomycosis to differentiate it from South American blastomycosis, which is caused by *B. brasiliensis,* and European blastomycosis caused by *Cryptococcus neoformans* (Dungworth 1993). The lung is the most frequent site of primary involvement, but primary cutaneous infections also occur. Systemic dissemination is common and likely to cause clinical signs associated with lesions in lymph nodes, eyes, skin, subcutaneous tissue, bones, joints, and the urogenital tract (Dungworth 1993).

ETIOLOGY. *Blastomyces dermatitidis* is a dimorphic fungus producing mycelial growth in cultures at room temperature and yeastlike structures in tissue and in cultures at 37° C (Dungworth 1993). Colonies are white to tan and downy to fluffy, and mycelia have spherical-to-oval conidia, 3–5 μm in diameter, attached to the sides of hyphae and to the ends of simple conidiophores. The yeast form is 8–20 μm in diameter, has a thick double-contoured wall, and reproduces by single broad-based buds.

DISTRIBUTION. In North America, most cases of blastomycosis occur in the Mississippi-Ohio River Basin, the central Atlantic states of the United States, and near the northern border of Ontario and Manitoba in Canada (Dungworth 1993). It was long thought to be restricted to North America, but cases have now been reported in the Middle East and in several African countries (Chandler and Watts 1987).

HOST RANGE. Blastomycosis has been reported in a variety of marine mammals, felids, canids, bats, and ursids [Williamson et al. 1959; Medway 1980; Stroud and Coles 1980; Kennedy-Stoskopf and Russell 1983 (cited by Dunn 1990); Cates et al. 1986; Thiel et al. 1987; Morris et al. 1989; Clyde et al. 1996; Raymond et al. 1997].

TRANSMISSION AND EPIDEMIOLOGY. Most infections are due to inhalation of airborne conidia of the mycelial form of the fungus growing as a saprophyte in soil. Rarely, primary cutaneous infection results from inoculation of the fungus. Proximity of within 400 m to a waterway and exposure to excavation sites are significant risk factors for canine blastomycosis. Factors such as acidic pH, high organic content, and abundant moisture in the soil are also considered important for the growth of *B. dermatitidis* (Baumgardner et al. 1995).

CLINICAL SIGNS. Blastomycosis in animals is usually systemic and has a prolonged course with chronic debilitation, anorexia, coughing, exercise intolerance, weight loss, eye and nasal discharge, lameness, cutaneous lesions, and terminal respiratory distress. Other signs depend on the distribution of lesions in disseminated disease. Ocular blastomycosis is relatively common.

PATHOLOGY. In the pulmonary form of disease, multifocal gray-white nodules of various sizes appear throughout all lung lobes. Most of these nodules are solid granulomas, but some have central abscessation

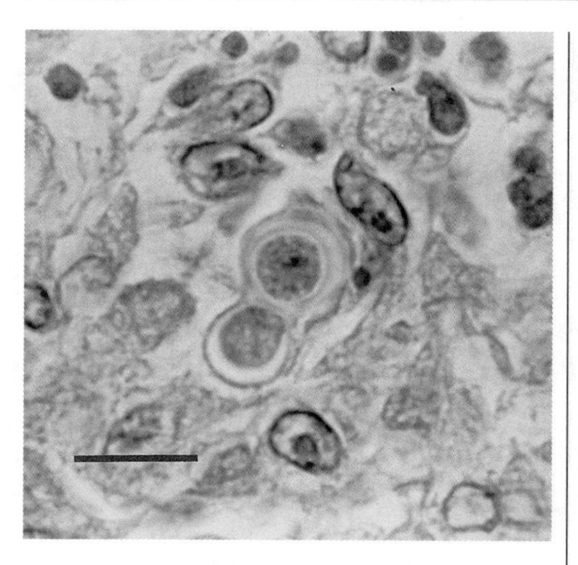

FIG 29.3—*Blastomyces dermatitidis* organism in meninges of a tiger with pyogranulomatous meningoencephalitis. Broad-based budding is demonstrated here. H&E, bar = 10 µm. Glass slide from an Armed Forces Institute of Pathology set.

or caseation. Calcification is minimal or absent. Regional lymph nodes are consistently involved and contain granulomas, abscesses, or caseous foci. In disseminated forms, there are similar granulomatous lesions in peripheral lymph nodes, eyes, skin, subcutaneous tissues, bone, and joints. Ocular lesions may involve anterior or posterior segments (Dungworth 1993).

Microscopically, inflammation is pyogranulomatous. Yeast-like fungi, both free and within the cytoplasm of macrophages and giant cells, are present. Organisms are 8–15 µm in diameter, with broad-based single buds and thick, refractile, double-contoured walls (Fig. 29.3). In early lesions, large numbers of organisms are present, whereas, in more chronic cases, organisms may be sparse (Chandler and Watts 1987).

DIAGNOSIS. Finding the persistent broad-based budding of *B. dermatitidis* in characteristic lesions is diagnostic (Chandler and Watts 1987). In live animals, thoracic radiography often reveals a reticulonodular interstitial pneumonia, sometimes with consolidation and pleural fluid. The agar gel immunodiffusion (AGID) test or ELISA for serum antibody can be helpful in establishing a diagnosis (Ward 1995). Cytologic examination is often successful in identifying organisms in samples from vitreous, skin, and lymph nodes in dogs (Arceneaux et al. 1998).

The fungus must be differentiated from similar-sized yeasts, especially *Histoplasma capsulatum* (Chandler and Watt 1987) and the dry (nonencapsulated) variant of *Cryptococcus neoformans.* Diagnosis can be confirmed by direct immunofluorescence or culture. Laboratory personnel should be warned of the clinical suspicion so proper precautions can be taken when handling specimens (Scott et al. 1995).

CONTROL AND TREATMENT. Treatment of free-ranging wild species is not possible or necessary. For domestic and captive wild mammals, recommended treatment protocols include amphotericin B alone, ketoconazole alone, and amphotericin B combined with ketoconazole and itraconazole (Dunbar et al. 1983). Itraconazole is now the treatment of choice in dogs and has been used in a captive polar bear *Ursus maritimus* (Morris et al. 1989; Legendre 1995).

PUBLIC HEALTH CONCERNS. Occurrence of blastomycosis reflects common source environmental exposure rather than transmission between mammalian hosts (Baumgardner et al. 1995). Penetrating wounds contaminated by the organisms have produced infections in humans, and care should be taken to avoid being inoculated with contaminated needles or knives or being bitten when handling infected animals (Scott et al. 1995).

DOMESTIC ANIMAL HEALTH CONCERNS. Dogs are most commonly infected with *B. dermatitidis,* though it occurs occasionally in cats and rarely in horses and other species. Blastomycosis has also been reported in a ferret *Mustela putorius furo* (Lenhard 1985).

Histoplasmosis

Synonyms: Darling's disease, reticuloendotheliosis, reticuloendothelial cytomycosis, cave sickness.

INTRODUCTION. Histoplasmosis refers to a systemic infection by the fungus *Histoplasma capsulatum.* There are two subtypes: (1) classic histoplasmosis caused by the ubiquitous small-celled variety, *H. capsulatum* var. *capsulatum,* and (2) African histoplasmosis caused by the large-celled variety, *H. capsulatum* var. *duboisii. Histoplasma capsulatum* var. *duboisii* is confined to Africa between the Sahara and the Kalahari deserts and has been reported in only humans and baboons *Papio cynocephalus papio* (Butler et al. 1988).

Classic histoplasmosis is the most prevalent of the systemic mycoses caused by exogenous fungi in people, and most infections are subclinical (Kaplan 1973).

ETIOLOGY. *Histoplasma capsulatum* is a dimorphic fungus, occurring in mycelial and yeast phases. In tissue and on enriched media incubated at 37° C, it is a uniform, budding, yeast-like cell 2–4 µm in diameter. On standard media at 25° C, the fungus grows as a mold (Chandler and Watts 1987).

HISTORY. Histoplasmosis was first reported in a man from the Panama Canal Zone by Darling in 1906. In 1934, De Monbreun isolated the causative organism, cultivated it on laboratory media, and reproduced the disease in two monkeys (Sanger 1981). The first reports of its isolation from wild animals were by Emmons et al. (1947) and Olson et al. (1947) from house mice and rats *Rattus norvegicus.*

DISTRIBUTION. *Histoplasma capsulatum* var. *capsulatum* (hereafter referred to as *H. capsulatum*) is found worldwide in temperate and in tropical zones. Hyperendemic areas include Guatemala, Mexico, Peru, Venezuela, and, in the United States, in the Ohio and Mississippi River valleys and areas along the Appalachian Mountains. It has been reported from more than 30 countries (Sanger 1981).

HOST RANGE. *Histoplasma capsulatum* has been isolated or infection diagnosed in a wide variety of wild and captive animals [Wilson et al. 1974; Lainson and Shaw 1975; Owens et al. 1975; Greer and McMurray 1981; Sanger 1981; Arias et al. 1982 (cited by Naiff et al. 1985); Naiff et al. 1985; Woolf et al. 1985; Raju et al. 1986; Weller et al. 1990; Jensen et al. 1992; Quandt and Nesbit 1992; Costa et al. 1994].

TRANSMISSION AND EPIDEMIOLOGY. Infection occurs most commonly by inhalation and rarely by ingestion or by skin contact with infective spores of *H. capsulatum.* The infection is not contagious, and clusters of disease are due to exposure to the same environmental source.

Histoplasma capsulatum generally exists in environments with high moisture and organic content, particularly environments enriched by bat or bird fecal material. Bats are the only known mammal to excrete the organisms in a medium in which it can propagate itself (Sanger 1981) and are, therefore, the only mammals believed to play a significant role in the epidemiology of the disease. A history of visiting caves and presumably being in contact with bat guano is common for people with active histoplasmosis, as reflected by the synonym *cave sickness* (Hoff and Bigler 1981).

There is controversy over whether bats are only incidentally infected or whether they play an active role in transmission of the fungus. The fungus has been isolated from lungs of 57% of naturally infected bats, and liver, spleen, and intestinal contents also readily yield the fungus [Di Salvo 1971; Tesh and Schneidau 1966 (cited by Greer and McMurray 1981)]. Greer and McMurray (1981) infected the neotropical fruit bat *Artibeus lituratus* intraperitoneally and intranasally. Rapid dissemination of fungus and death of some bats occurred with high-dose intraperitoneal inoculation. On intranasal exposure, pulmonary infection disseminated to spleen, liver, and intestine in 2 weeks, with rare gross lesions and mortality. This indicated that the bats can acquire and harbor the fungus, and the frequent involvement of the gastrointestinal tract provides a mechanism for seeding the environment with the fungus (Greer and McMurray 1981). The frequent finding of *H. capsulatum* in association with bird droppings is likely due to environmental enrichment by the bird feces rather than propagation and transmission by birds.

CLINICAL SIGNS. In animals and people, the prevalence of clinically inapparent *H. capsulatum* infection is high. Clinically apparent histoplasmosis can be a localized, benign, pulmonary disease or a disseminated, progressive, potentially fatal disease (Gatchel et al. 1986). Common clinical signs in disseminated disease include weight loss, diarrhea, fever, anemia, hepatosplenomegaly, lymphadenopathy, skin lesions, and anorexia (Woolf et al. 1985; Clinkenbeard et al. 1988; Jensen et al. 1992).

PATHOGENESIS. Primary and secondary foci of infection are established in the lungs and regional lymph nodes, followed by lymphohematogenous dissemination via the monocyte-macrophage system. Lymph nodes, spleen, bone marrow, liver, and adrenals are affected in particular. Infection is arrested by specific cellular immunity. Additional factors such as immunosuppression may be required for overt disease. In wild animals, stress of capture, transportation, confinement, and change of habitat have been implicated in induction of immunosuppression (Quandt and Nesbit 1992).

PATHOLOGY. In dogs, infection elicits granulomatous bronchopneumonia and regional lymphadenitis. Patches of bronchopneumonia may become encapsulated with possible caseation and/or calcification resulting in a *histoplasmoma. Histoplasma capsulatum* may parasitize reticuloendothelial cells and be transported to extrapulmonary organs, with resultant granulomatous inflammation. Massive proliferation of the reticuloendothelial cells, many of which contain the yeast forms, causes the displacement of normal tissues and gross enlargement of the organs (Jones and Hunt 1983). Ulceration of the gastrointestinal tract is also relatively common (Quandt and Nesbit 1992).

Large numbers of intralesional organisms can often be found within macrophages and other reticuloendothelial cells. The organisms are intracytoplasmic, irregularly egg shaped, and 2×3 to 3×4 μm. In H&E-stained sections, yeasts appear as central, spherical, usually basophilic bodies surrounded by an unstained zone or halo (Jones and Hunt 1983).

DIAGNOSIS. In live animals, organisms can be identified in monocytes and macrophages in blood smears, aspirates of bone marrow, and biopsy samples of lymph node, liver, intestine, or rectum. Thoracic radiographs, histoplasma skin test, and serologic methods such as AGID or CF can be used (Quandt and Nesbit 1992). Postmortem, diagnosis can be confirmed by histopathology, immunohistochemistry, isolation, or rodent inoculation (Quandt and Nesbit 1992).

CONTROL AND TREATMENT. It is not possible or necessary to treat free-ranging species for histoplasmosis.

Dogs and cats have been treated long-term with ketoconazole (Legendre 1995). Decreasing exposure to substrates enriched by bat and bird droppings is an important step for controlling histoplasmosis in captive animals.

PUBLIC AND DOMESTIC ANIMAL HEALTH CONCERNS. Humans and domestic species are exposed to histoplasmosis from environmental sources but not via direct transmission from wild species.

Cryptococcosis

Synonym: European blastomycosis.

INTRODUCTION. Cryptococcosis is a localized or disseminated disease with a marked propensity for the respiratory system, especially the nasal cavity and the central nervous system (CNS). It is caused by *Cryptococcus neoformans,* a soil-inhabiting, yeast-like fungus abundant in avian habitats, particularly those heavily contaminated with pigeon *Columba livia* or other bird droppings. It can be a primary pathogen, causing sporadic infections in apparently healthy individuals (Chandler and Watts 1987) but more often is an opportunist, causing disease in immunocompromised hosts. All species of mammals appear to be susceptible, though cats are affected more often than other species.

ETIOLOGY. *Cryptococcus neoformans* is a monomorphic, 4–8-μm yeast-like organism that reproduces by single buds (Dungworth 1993). The organism is surrounded by a thick mucopolysaccharide capsule that stains well with mucicarmine, PAS reaction, or Alcian blue.

Although culture is needed for definitive identification, a confident diagnosis can be made by identification of this capsule microscopically. On wet mounts, the organisms may be confused with red blood cells or lymphocytes, but, by negative staining with India ink, the characteristic capsule is easily seen. It grows rapidly on standard culture media at 25° C or 37° C into moist, mucoid, convex, white-to-pale yellow colonies. Colonies are composed of yeast-like cells 2–20 μm in diameter that replicate by budding (Chandler and Watts 1987).

There are two biovariants of *C. neoformans:* var. *neoformans,* the major environmental source of which is soil enriched with bird droppings; and var. *gattii,* which has a specific ecological association with river red gum *Eucalyptus camaldulensis* and forest red gum *Eucalyptus tereticornis* trees (Spencer et al. 1993).

DISTRIBUTION AND HOST RANGE. Cryptococcosis occurs worldwide in temperate and tropical climates. *Cryptococcus neoformans* has a very broad host range, and disease has been reported in a variety of nondomestic animal species (Table 29.2).

TRANSMISSION AND EPIDEMIOLOGY. Cryptococcosis occurs sporadically and is not contagious. Infection is acquired by inhalation of contaminated soil, particularly soil enriched with pigeon or other bird droppings. Encapsulated strains are common in pigeon feces, but cryptococcosis does not occur in birds, probably because of their high body temperature [Emmons et al. 1977 (cited by Scrimgeour and Purohit 1984)]. The association with pigeon excreta has been attributed to high concentrations of creatinine that are assimilated by *C. neoformans* but not by other species of yeasts. This gives it a competitive advantage, allowing it to

TABLE 29.2—Cryptococcosis in free-ranging or captive mammals

Common Name	Scientific Name	References
Rhesus monkey[a]	*Macaca mulatta*	Pal et al. 1984
Celebes ape[a]	*Macaca nigra*	Miller and Boever 1983
Formosan macaque[a]	*Macaca cyclopis*	Miller and Boever 1983
Lion-tailed macaque[a]	*Macaca silenus*	Miller and Boever 1983
Allen's monkey[a]	*Allenopithecus nigroviridis*	Barrie and Stadler 1995
Proboscis monkey[a]	*Nasalis larvatus*	Miller and Boever 1983
Marmosets[a]	*Saguinus geoffroyi; S. oedipus*	Miller and Boever 1983; Potkay 1992
Black lemur[a]	*Eulemur macaco*	Miller and Boever 1983
Striped dolphin[b]	*Stenella coeruleoalba*	Gales et al. 1985
Bottlenose dolphin[a]	*Tursiops truncatus*	Migaki et al. 1978b
Cheetah[a]	*Acinonyx jubatus*	Beehler 1982; Berry et al. 1997
Palm civet[a]	Not reported	Miller and Boever 1983
Mouflon[a]	*Ovis musimon*	Miller and Boever 1983
Saiga[a]	*Saiga tatarica*	Miller and Boever 1983
Vicuña[a]	*Vicugna vicugna*	Miller and Boever 1983
Tree shrews[a]	*Tupaia tana, T. minor*	Tell et al. 1997
Elephant shrew[a]	*Macroselides proboscideus*	Tell et al. 1997
Black rat[b]	*Rattus rattus*	Scrimgeour and Purohit 1984
Koala[a]	*Phascolarctos cinereus*	Spencer et al. 1993

[a]Captive animals.
[b]Free-ranging animals.

flourish and outcompete other yeasts (Kaplan 1973). The fungus may survive in pigeon excrement for more than 2 years (Miller and Boever 1983).

CLINICAL SIGNS. In chronic progressive respiratory cryptococcosis, chronic cough, low-grade fever, chest pain, nasal discharge, and weight loss may occur (Pal et al. 1984). Signs of meningoencephalitis may include headache, depression, fever, anorexia, nausea, vomiting, and a variety of CNS signs (Beehler 1982; Miller and Boever 1983; Spencer et al. 1993).

PATHOLOGY. The respiratory tract is the usual site of primary infection, with the nasal cavity more often affected than the lungs. Systemic dissemination can proceed either from the nasal region or from the lungs, with hematogenous dissemination to the CNS, eyes, lymph nodes, skin, and other organs. There may also be local spread from the nasal cavity to the CNS.

Gross lesions vary depending on the organ system involved. In parenchymal organs, lesions are discrete, whitish, gelatinous foci and may not be more than a few millimeters in diameter. The gelatinous character of lesions is due to large amounts of mucopolysaccharide capsule associated with masses of organisms. Lesions in meninges, brain, and spinal cord also have a gelatinous character when visible, but often no gross lesions are appreciated. Skin lesions are firm, small nodules that tend to ulcerate and discharge a small amount of serous exudate. Nasal lesions are often very gelatinous, polypoid nodules or space-occupying and slowly destructive masses. There may be facial swelling and extension through the bony boundaries of the nasal cavity, with involvement of the skin and occasionally the oral mucosa.

In H&E sections, *C. neoformans* is eosinophilic or lightly basophilic, uninucleate, thin-walled, spherical, oval, and elliptical yeastlike cells that vary from 2–20 μm, with an average size of 4–10 μm (Fig. 29.4) (Chandler and Watts 1987). Budding is single and narrow based. The fungal cells are surrounded by a wide, clear-to-faintly stained zone or "halo" that represents the mucinous capsule. This capsule usually stains with Alcian blue, PAS, colloidal iron, and Mayer's mucicarmine. The capsule increases the overall diameter to a maximum of 30 μm (Jones and Hunt 1983). In most animals, cellular reaction usually consists only of a few macrophages, lymphocytes, and plasma cells (Dungworth 1993); granulomatous reaction is more typical in the lungs than in other organs.

DIAGNOSIS. Diagnosis is based on identification of the organisms by cytology or tissues, antigen detection assays, and isolation of fungus in culture. Cytology preparations stained with India ink demonstrate the organisms by highlighting the capsule. In neurologic cases, the organisms can sometimes be seen in the cerebrospinal fluid (CSF). A nested PCR has also been

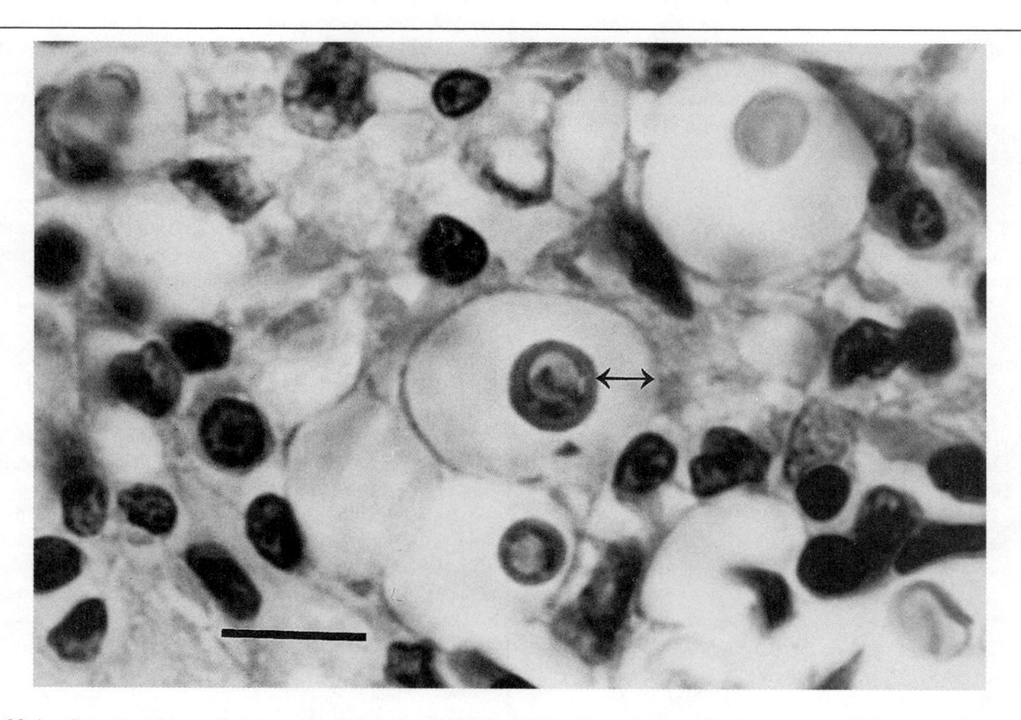

FIG 29.4—*Cryptococcus neoformans* organisms surrounded by macrophages in a domestic cat with meningoencephalitis, retinitis, and rhinitis. The *arrow* indicates the thickness of the mucopolysaccharide capsule. H&E, bar = 10 μm.

developed that has been used to detect *C. neoformans* in CSF (Rappelli et al. 1998). Because *C. neoformans* can vary greatly in size and shape and its encapsulated forms are not always evident, it should be considered in the differential diagnosis in any yeast-form mycosis. Intense staining with mucin stains is good histologic evidence because it is the only pathogenic fungus that has a mucinous capsule. Direct immunofluorescence can be used on tissue sections. In serum or CSF, antigen can be detected by a latex-cryptococcal antigen test, an agglutination test that is available commercially (Jacobs et al. 1997).

CONTROL AND TREATMENT. It is not possible or necessary to treat free-ranging species for cryptococcosis. Fluconazole and itraconazole have been used to treat domestic and captive wild species (Beehler 1982; Legendre 1995; Kano et al. 1997; Tell et al. 1997).

HUMAN AND DOMESTIC ANIMAL HEALTH CONCERNS. Many species are susceptible to cryptococcosis, but cats are affected most frequently. Transmission does not occur between domestic and wild species.

Adiaspiromycosis

Synonyms: Haplomycosis, adiasporomycosis.

INTRODUCTION. Adiaspiromycosis is a self-limited pulmonary mycotic infection found in many small animals that is caused by *Chrysosporium parvum* var. *parvum* and *C. parvum* var. *crescens*. There is no multiplication of the fungus in the host species, yet massive infections have been observed in small mammals. The variety *crescens* has the widest host range and widest geographic distribution of any mycotic pathogen and causes sporadic human infections (Chandler and Watts 1987; Nuorva et al. 1997).

ETIOLOGY. Adiaspiromycosis is caused by the ubiquitous soil-dwelling dimorphic fungi of the genus *Chrysosporium*, often referred to as *Emmonsia*. In tissue, spores grow into adiaspores, also called chlamydospores or spherules. *Chrysosporium parvum* var. *parvum* adiaspores are 20–40 µm and *C. parvum* var. *crescens* are 200–700 µm in diameter. In culture, the two fungi are indistinguishable, both producing white cottony colonies at room temperature that age to a more granular light-brown colony (Mason and Gauhwin 1982).

HISTORY. In 1920, Splendore described organisms in lungs of voles *Microtus savii* in Italy. These were identified as the eggs of a worm *vermi sconoscutti* but were almost certainly *C. parvum* var. *crescens* (*E. crescens*). In 1942, Emmons and Ashburn described fungus in lungs of desert rodents and called it *Haplosporangium parvum* (Mason and Gauhwin 1982). It was called *Emmonsia crescens* by Jellison, but in 1962 the agents of adiaspiromycosis were reclassified under the genus *Chrysosporium* by Carmichael (1962) as two varieties: *C. parvum* var. *crescens* and *C. parvum* var. *parvum*. Many authors still use the genus *Emmonsia*.

HOST RANGE. Sharapov (1969) examined 5647 members of Carnivora, Rodentia, and Insectivora of more than 45 species and found 407 infected. Dvorak et al. [1973 (cited by Jellison 1981)] prepared a lengthy list of animals examined for infection. Vole rats *Arvicola terrestris*, muskrats *Ondatra zibethica*, stoats *Mustela erminea*, sables *Mustela zibellina*, steppe polecats *Mustela eversmanni*, and European polecats *Mustela putorius* had high rates and severity of infections (Krivanec et al. 1975; Jellison 1981).

Additional reports since Dvorak's (1973) list was compiled include *C. parvum* var. *crescens* in striped skunks *Mephitis mephitis* (Albassam et al. 1986), common wombats *Vombatus ursinus* (Finnie 1986), three species of ground squirrels (Leighton and Wobeser 1978), a variety of rodents in Czechoslovakia (Otcenasek et al. 1974; Kodousek and Hejtmanek 1982; Hubalek et al. 1991), common brush-tailed possums *Trichosurus vulpecula* (Johnstone et al. 1993), Japanese pika *Ochotona alpina* (Taniyama et al. 1985); *C. parvum* var. *parvum* in a hairy-nosed wombat *Lasiorhinus latifrons* (Mason and Gauhwin 1982); and *Chrysosporium* sp. in least weasels *Mustela nivalis* (Laakkonen et al. 1998).

TRANSMISSION AND EPIDEMIOLOGY. Animals most affected are those that live in burrows and ground nests where they come into contact with the saprophytic phase of the fungus in the soil. Also affected are those living in tree cavities, heaps of stones, and straw stacks. Transmission is primarily through inhalation of the conidia or aleurospores of the saprophytic stage.

The predator-prey relationship seems to play an important role in dissemination of the fungus, since adiaspores passed in feces become viable and germinate (Krivanec et al. 1975; Krivanec and Otcenasek 1977). Predators may also disperse fungal spores by partial consumption of carcasses. Adiaspores will not germinate under anaerobic conditions; therefore, dispersal from animals left intact to decompose is unlikely. However, viable fungal spores can be released and will germinate with maceration of infected tissue (Krivanec et al. 1975).

Sharapov (1969) found the optimal habitat for *C. parvum* var. *crescens* was the forest-steppe interface. This observation was supported by high infection rates in rodents in deciduous wooded habitats at the prairie margin in Canada (Leighton and Wobeser 1978) and in rodents at windbreaks in Czechoslovakia (Hubalek et al. 1991). The peak of fungal proliferation in the soil occurs when the mean monthly soil temperature 5 cm below the surface is 3.4° C–12.2° C (Hubalek et al. 1993).

CLINICAL SIGNS. Low-level infections are insignificant. In heavy infections, there may be varying degrees of pulmonary compromise and progressive dyspnea (Chandler and Watts 1987)

PATHOLOGY. Severely infected lungs have multiple grayish white nodules, 0.5–1.0 mm in diameter, distributed throughout the lungs. Histologically, these nodules are fungal spherules (adiaspores) surrounded by granulomatous infiltrates. Adiaspores are round, thick walled, and 200–700 μm in diameter for *C. parvum* var. *crescens* and 20–40 μm for *C. parvum* var. *parvum* (Fig. 29.5) (Mason and Gauhwin 1982). Most fungal cells have a central mass of small globules and fine granular pale basophilic material; other spherules are empty. No budding or endosporulation is present. The wall and inner globules stain purple by PAS and black by silver stains. Intensity of inflammation varies in proportion to the degree of degeneration exhibited by the organisms (Johnstone et al. 1993), with more intense reactions associated with ruptured spherules. Lesions are usually restricted to the lungs; there is rare involvement of the draining lymph nodes, with a resultant granulomatous lymphadenitis (Albassam et al. 1986). For an ultrastructural description, see Albassam et al. (1986).

DIAGNOSIS. The gross and histologic features are typical enough for a presumptive diagnosis, followed by culture for a definitive diagnosis. Digestion techniques can be used for estimating severity of infection and detecting low-level infection (Leighton and Wobeser 1978).

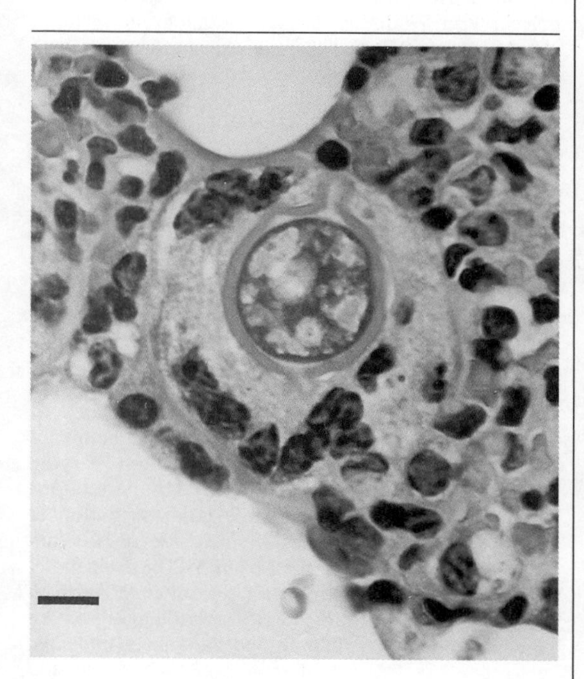

FIG 29.5—Adiaspore of *Chrysosporium parvum* var. *parvum* in the lung of a ground squirrel. The adiaspore is surrounded by a multinucleated giant cell. H&E, bar = 10 μm.

DOMESTIC ANIMAL HEALTH CONCERNS. Adiaspiromycosis is a rare sporadic disease in domestic animals.

Aspergillosis

INTRODUCTION. Aspergillosis encompasses a wide variety of disease syndromes caused by several *Aspergillus* species. These disease syndromes include mycotoxicoses; allergic reactions of the upper and lower respiratory tracts; colonization of pulmonary cavities; superficial infection of the skin, sinuses, guttural pouch, and external ear canal; invasive infection of the lower airways; and opportunistic pulmonary infections with hematogenous dissemination (Chandler and Watts 1987). Pier and Richard (1992) and O'Hara (1996) provide further discussion of mycotoxicoses. In animals, aspergillosis is usually a primary respiratory infection with rare instances of dissemination (Severo et al. 1989). Disease is largely related to the immunologic status of the host; disseminated aspergillosis is associated with debilitation, immunologic suppression, or prolonged antibiotic or corticosteroid therapy (Pickett et al. 1985).

ETIOLOGY. *Aspergillus* spp. are ubiquitous saprophytic molds of worldwide distribution. Pathogens include *A. fumigatus, A. terreus, A. flavus, A. nidulans,* and *A. niger; A. fumigatus* is the most frequent cause of aspergillosis in animals. These fungi are easily cultivated and identified by colony morphology and microscopic morphology of their conidial heads.

HOST RANGE. Sporadic cases of aspergillosis have been reported in a variety of wild species (Peden et al. 1985; Pickett et al. 1985; Jensen et al. 1989; Severo et al. 1989; Weber and Miller 1996). An unusual syndrome in camels *Camelus dromedarius* is characterized by a spectrum of respiratory and enteric signs and multisystem hemorrhages (El-Khouly et al. 1992).

TRANSMISSION AND EPIDEMIOLOGY. Growth and sporulation of *Aspergillus* spp. are encouraged by moist conditions and decaying vegetation. Disturbance of these materials results in aerosols, allowing for aspiration and deposition in bronchioles and pulmonary alveoli. Organisms may also colonize enteric structures after ingestion. This may progress by erosion and vascular dissemination to other organ systems, including brain, abdominal viscera, and the gravid uterus (Pier and Richard 1992). Vascular invasion is common and often results in thrombosis and infarction.

CLINICAL SIGNS AND PATHOLOGY. Clinical signs vary greatly depending on the organ systems involved. In primary respiratory disease, there are miliary nodules in the lungs that histologically are pyogranulomas with intralesional fungal hyphae. In disseminated disease, there may be nodules scattered throughout the body. The fungi are angioinvasive, resulting in massive tissue damage due to vascular invasion and thrombosis. The hyphae proliferate in the necrotic tissue, sometimes in parallel or radial arrays. Typical hyphae of *Aspergillus* spp. are 3–6 μm wide, with parallel walls, evenly distributed septa, and progressive dichotomous branching, with branches at acute angles (Chandler and Watts 1987).

DIAGNOSIS. Typical hyphae in histologic sections suggest a diagnosis, but it is not possible to distinguish them reliably from hyphae of other angioinvasive fungi such as *Fusarium* spp. and *Pseudallescheria boydii* (Chandler and Watts 1987). Diagnosis should be confirmed by culture or immunologic techniques such as fluorescent antibody tests. Because *Aspergillus* spp. are such common contaminants, positive culture should correlate with appropriate histopathologic evidence of tissue invasion (Chandler and Watts 1987).

CONTROL AND TREATMENT. Itraconazole can be used for treatment of captive animals with systemic aspergillosis, but cure is uncommon. Control involves decreasing the number of spores contaminating the environment and controlling other predisposing conditions (Pier and Richard 1992)

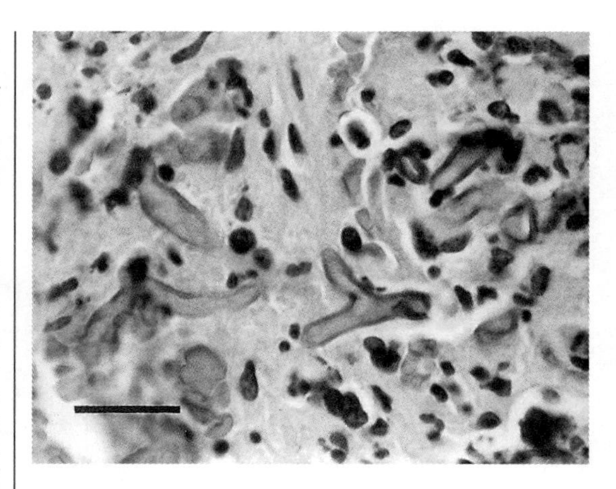

FIG 29.6—Hyphae of a zygomycete in the lung of a Dall porpoise *Phocoenoides dalli.* Hyphae are nonseptate, irregular in width, and branching. They are surrounded by macrophages and neutrophils. H&E, bar = 20 μm.

Candidiasis

Synonyms: Moniliasis, Thrush.

Candida albicans is an ubiquitous yeast considered to be a normal inhabitant of the upper alimentary tract. It also is an opportunistic pathogen of numerous vertebrates and is capable of invading and proliferating in stratified squamous epithelium and, rarely, spreading systemically. Antibiotic therapy and prolonged debilitation are predisposing factors (Medway 1980; Obendorf 1980). Morphologic features characteristic of *Candida* spp. are the combination of oval, 3–6-μm-diameter yeastlike cells, pseudohyphae, and true hyphae 3–5 μm wide (Chandler and Watts 1987).

Gastrointestinal candidiasis, the most common form of disease, has been reported in a wide variety of nonhuman primates (Potkay 1977), captive marine mammals (Dunn et al. 1982), and hand-reared marsupials (Obendorf 1980).

The most common signs of mucocutaneous candidiasis are dysphagia and the presence of ulcers and thick white plaques in the oral cavity and esophagus and less commonly in the colon. Occasionally, if untreated, it can progress to depression and death. Plaques contain large numbers of *Candida* organisms that can be seen in H&E-stained tissue sections but are best studied with PAS or GMS stains. Systemic disease is rare. Oral nystatin and oral or injectable fluconazole have been effective treatments (Finnie 1986; Legendre 1995).

Mucormycoses. The mucormycoses are a group of diseases caused by organisms of the genera *Rhizopus, Mucor, Absidia,* and *Mortierella,* all of which belong to the order Mucorales. The orders Mucorales and the Entomophthorales are within the class Zygomycetes formerly referred to as Phycomycetes. While the names "zygomycosis," "phycomycosis," and "mucormyosis" are often used interchangeably, mucormycosis is the correct term for disease caused by this group of fungi (Chandler and Watts 1987). These are opportunistic infections that occur sporadically in immunocompromised hosts throughout the world. They cause pulmonary, gastrointestinal, and systemic disease. Heavy fungal challenge, disruption of normal flora, a primary local lesion, or lowered host resistance predispose individuals to gastrointestinal infection (Barker et al. 1993). The fungal hyphae typically are tissue and vessel invasive, with resulting infarction and systemic dissemination.

Mucormycosis has been reported in a wide variety of species (Ohbayashi 1971; Best and McCully 1979; Migaki et al. 1982; Obendorf et al. 1993).

Clinical signs and gross and histologic lesions vary depending on the organ system involved. Because of the angioinvasiveness, lesions are often ischemic or hemorrhagic, with marked tissue destruction. Varying degrees of granulomatous and pyogranulomatous inflammation and generally large numbers of fungal hyphae are present. Typical hyphae are broad, thin walled, pleomorphic, varying from 5–20 μm or more, and produce irregular branches that often arise from parent hyphae at right angles. Hyphae are often twisted, folded, or collapsed and have inconspicuous, rare septa (Fig. 29.6). Immunohistochemistry may add confidence to the morphologic diagnosis (Jensen et al. 1994). Definitive diagnosis requires culture, which may be difficult (Barker et al. 1993).

SUBCUTANEOUS MYCOSES

Sporotrichosis

INTRODUCTION AND ETIOLOGY. Sporotrichosis, which is primarily a subcutaneous mycosis caused by *Sporothrix schenckii,* a thermally dimorphic fungus, forms mycelia on decaying vegetation and on Sabouraud's dextrose agar at 25° C–30° C but is yeast-like in tissues and media at 37° C. It is ubiquitous in vegetation and timber and prefers a temperate or tropical climate with high humidity. In tissue, the fungi are yeast-like cells, 2–6 μm in diameter, with elongate "teardrop" or "pipe stem" or, rarely, "cigar-shaped" buds with narrow-based attachments to the parent cells.

DISTRIBUTION AND HOST RANGE. The disease occurs worldwide in temperate as well as tropical climates (Chandler and Watts 1987). In the United States, it is most common in coastal regions and river valleys. Sporotrichosis has been reported in a wide variety of wild and domestic species (Migaki et al. 1978a; Kaplan et al. 1982; Jones and Hunt 1983; Costa et al. 1994; Werner and Werner 1994).

TRANSMISSION AND EPIDEMIOLOGY. Sporotrichosis lesions are usually associated with contamination of a previous wound or scratch. Zoonotic transmission may occur through infected wounds.

CLINICAL SIGNS, PATHOLOGY, AND DIAGNOSIS. Cutaneous, lymphocutaneous, and disseminated disease occurs in people and animals. The lymphocutaneous form is most frequent and is characterized by small, firm, nonpainful, nonpuritic, dermal to subcutaneous nodules that develop at the site of inoculation after an incubation period of 1–3 months. Typically, a single nodule develops that then ulcerates; additional lymphangitic nodules may develop in a linear pattern, and regional lymphadenopathy may occur. Lesions may be ulcerated, papulated, or verrucous or be alopecic or scaly plaques. Ulcers discharge hemorrhagic to serous exudate. The head, extremities, and tail are most commonly affected. Disseminated disease is rare and is indicated by lethargy, depression, anorexia, and fever. Tissue forms of *S. schenckii* usually are scarce and rarely observed by direct examination in histologic preparations (Scott et al. 1995). They occasionally develop a smoothly contoured acidophilic coat or clear PAS-positive capsule that must be differentiated from the *C. neoformans* capsule.

Diagnosis is based on clinical signs, histopathology, fungal cultures, and serologic tests (Goad and Pecquet-Goad 1986). Fluorescent antibody tests are available, but diagnosis is most often established by tissue culture. Suspected samples should be labeled as potentially infectious, due to the risk to laboratory personnel. Serologic tests have been developed (Kaplan et al. 1982), but alone are not diagnostic.

DIFFERENTIAL DIAGNOSES. Differential diagnoses include other causes of chronic ulcers and granulomas, such as foreign-body granulomas, other mycoses, neoplasia, atypical mycobacteriosis, nocardiosis, and other deep bacterial infections.

CONTROL AND TREATMENT. The treatment of choice has been an oral saturated solution of potassium iodide ion continued for 30 days beyond clinical resolution (Scott et al. 1995). Long-term itraconazole and ketoconazole also have been used effectively (Goad and Pecquet-Goad 1986; Werner and Werner 1994; Legendre 1995). Treatments are prolonged, often requiring 6–8 weeks of medication.

PUBLIC HEALTH CONCERNS. Animal-to-person transmission is most often due to inoculation of fungus into bite wounds or other injuries.

Lobomycosis. Lobomycosis, which is also called Lobo's disease and keloidal blastomycosis (Bossart 1984), is a subcutaneous mycosis recognized in humans (Simon-Lopes et al. 1993), Atlantic bottle-nosed dolphins (Migaki et al. 1971; Caldwell et al. 1975), and Guiana dolphin *Sotalia fluviatilis guianensis* (De Vries and Laarmann 1973). *Loboa loboi,* the etiologic agent of lobomycosis, has not been successfully cultivated but can be grown in the feet of experimentally inoculated hamsters *Cricetus cricetus* and mice (Chandler and Watts 1987). Fungal cells are abundant in the cutaneous lesions and can be seen with H&E stain, though they are visualized better with GMS. *Loboa loboi* organisms produce characteristic long chains of budding yeast-like cells occasionally connected by short tubes that have been likened to a string of pearls. They are spherical, oval, elliptical, or crescentic, with an average diameter of 8 μm and a range of 5–12 μm (Fig. 29.7).

DISTRIBUTION. Cases of Lobo's disease in bottlenose dolphins have been described from the waters of Florida; one case was found off the coast of southern Brazil (Simon-Lopes et al. 1993), and another was from the Texas coast of the Gulf of Mexico (Cowan 1993). The affected Guiana dolphin was from the Surinam River.

TRANSMISSION AND EPIDEMIOLOGY. The natural reservoir of *L. loboi* and mechanisms of infection are not known, but exposure to contaminated soil or vegetation through local penetrating injury could explain some human infections. Patients in reported cases have appeared to be otherwise normal and healthy. One Atlantic bottle-nosed dolphin was hypogammaglobulinemic; therefore, immunodeficiency may predispose individuals to infection (Bossart 1984).

CLINICAL SIGNS AND PATHOLOGY. Skin lesions in dolphins are raised white nodules or verrucous crusts on the skin. According to Caldwell et al. (1975), the

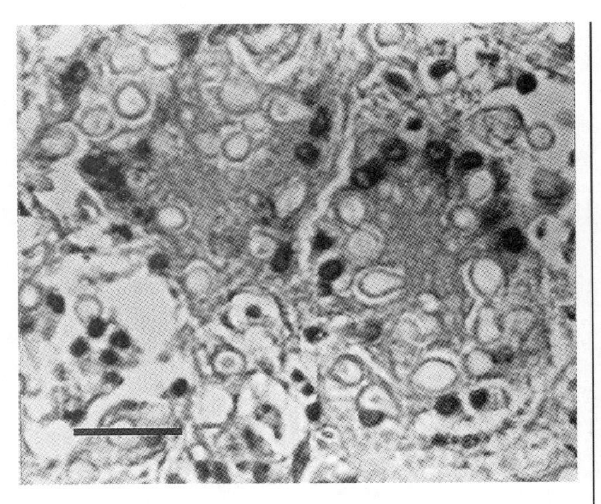

FIG 29.7—*Loboa loboi* organisms within cutaneous lesions in an Atlantic bottlenose dolphin. Organisms are round, to oval, to teardrop shaped and are present free and within multinucleated giant cells. H&E, bar = 20 μm. Glass slide provided by the Charles Louis Davis Foundation.

most serious lesions have been on the areas most often exposed to air: the top of the head, the dorsal fin, and the top of the caudal peduncle and flukes. But lesions may also be present on flanks and ventrum (Simon-Lopes et al. 1993). There may be regional lymphadenopathy, but dissemination does not occur.

Histologically, the papillary dermis contains chronic granulomatous inflammation, with numerous macrophages and multinucleated giant cells; typical fungi are found within the lesions. There may also be marked acanthosis of the overlying epidermis (Simon-Lopes et al. 1993).

DIAGNOSIS. Diagnosis is based on clinical signs and histopathology. Differential diagnoses include cutaneous forms of paracoccidioidomycosis, African histoplasmosis, and blastomycosis. These three can be ruled out by fungal morphology and direct immunofluorescence. A fluorescent antibody test for *L. loboi* is not currently available.

CONTROL AND TREATMENT. An Atlantic bottlenose dolphin was successfully treated with miconazole (Dunn 1990). Treatment of free-ranging wild mammals is not warranted.

SUPERFICIAL FUNGI

Dermatophytosis

Synonyms: Ringworm, tinea.

INTRODUCTION. Dermatophytosis is a complex of diseases caused by any of 31 recognized species of taxonomically related fungi in the genera *Epidermophyton, Microsporum,* and *Trichophyton.* These fungi, called dermatophytes, usually infect the nonviable keratinized portions of the skin, hair, and nails (Chandler and Watts 1987).

ETIOLOGY, DISTRIBUTION, AND HOST RANGE. Dermatophytes are categorized on an ecological basis. The anthrophilic species, mainly *Epidermophyton,* are adapted to humans, only very rarely infect animals, and have lost their saprophytic properties. Zoophilic species principally affect animals. Some species, such as *Microsporum canis,* have become so well adapted to their animal host that inapparent carriers are the rule. Such animals are important sources of infection for humans and other animals. Geophilic dermatophytes, such as *Microsporum gypseum,* normally reside in the soil but are capable of infecting animals and humans. Some species, such as *Trichophyton mentagrophytes,* have both anthrophilic and zoophilic strains (Knudtson et al. 1980). The dermatomycoses occur worldwide and are common in humans and a wide variety of animals.

TRANSMISSION AND EPIDEMIOLOGY. Natural infection is by direct or indirect contact with carrier animals, and experimental transmission can be achieved by rubbing infected hairs, cutaneous scales, or cultivated organisms onto abraded skin (Yager and Scott 1993). Initiation of infection requires alteration of the stratum corneum, either by slight trauma or by continued moisture and maceration. The organisms can survive in the environment for extended periods, but the usual reservoir of infection for the zoophilic species is the inapparent carrier animal. Young animals are more susceptible than adults, and therefore asymptomatic adults can be sources of infection for the young. Severe dermatophyte infection in older animals usually signals underlying systemic disease or reduced immunologic competence (Yager and Scott 1993). Dietary vitamin-A deficiency may predispose individuals to ringworm by debilitating the epithelial tissue (Boever and Rush 1975). Infections are usually self-limiting, with lesions taking a few weeks to several months to regress, depending on the species of fungus, the degree of host adaptation, and individual host response.

CLINICAL SIGNS. Skin lesions begin as focal, round, alopecic areas and may progress to hyperkeratosis, redness, and skin depigmentation. Lesions occur on the head and back and may increase in numbers and size and spread over the body. The lesions generally are not pruritic.

PATHOLOGY. Gross lesions include erythema, follicular papules, scaling, crusting, and alopecia. The alopecia results from breakage of the brittle, parasitized hair shafts. The "classic" ring-shaped lesion caused by central clearing and peripheral expansion is the exception rather than the rule. Histologically, there is no single pattern of inflammation. Folliculitis and furunculosis

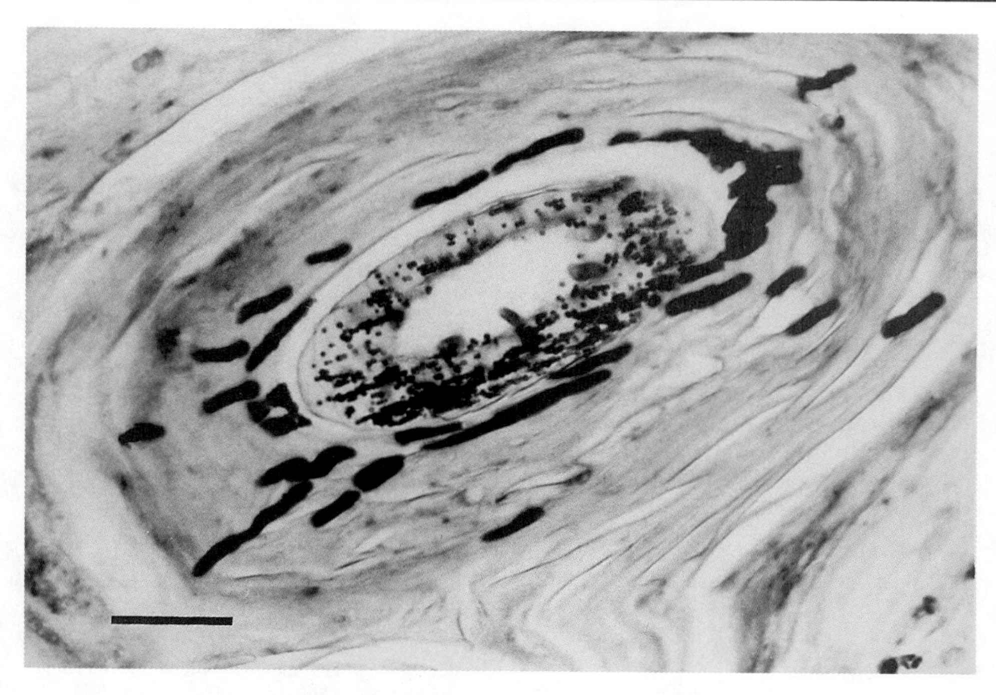

FIG 29.8—Dermatophyte infestation of a hair follicle in a 3-month-old wallaroo *Macropus* spp. *Microsporum gypseum* was isolated from the skin. H&E, bar = 20 μm.

and lesions with marked ortho- and parakeratotic hyperkeratosis, particularly when associated with neutrophilic microabscesses, are typical. Pale blue arthrospores and clear hyphae stand out in relief against the eosinophilic keratin background on H&E-stained sections of skin. Special stains such as PAS or silver stains may be useful for visualization of the organisms (Fig. 29.8).

DIAGNOSIS. Diagnosis is based on history, clinical lesions, microscopic examination of potassium hydroxide (KOH)-digested skin scrapings and plucked hairs, and ultraviolet (UV)-light examination of hairs, skin biopsy, or fungal culture. In skin scrapings digested in 10% KOH, chains of arthrospores may be seen on the surface of the hairs (ectothrix) or within the hairs (endothrix), depending on the species of fungus, and branching hyphae may be seen on the hair shafts. The Wood's lamp, which delivers UV light, causes the metabolites of some dermatophytes, including *M. canis, M. distortum, M. audouinii,* and *T. schoenleinii,* to fluoresce (Chandler and Watts 1987). Not all strains of these species fluoresce, so a negative test does not rule them out. Hair that fluoresces, hair at the margins of lesions, especially broken and misshapen hairs, or scrapings from a toothbrush passed over the animal's coat are the best samples to inoculate onto culture medium. Sabouraud's dextrose agar and dermatophyte test medium are the usual media used (Scott et al. 1995).

DIFFERENTIAL DIAGNOSES. Viral infections such as those caused by pox viruses or caliciviruses, bacterial infections such as *Dermatophilus congolensis* or others that cause folliculitis and furunculosis, and parasitic infections caused by *Demodex* spp. or *Sarcoptes* are differential diagnoses (Janovitz and Long 1984; Phillips et al. 1986).

CONTROL AND TREATMENT. Treatment is not needed in free-ranging wild species. For captive wildlife, there are a wide range of topical and systemic therapy options that are similar to those used in domestic species. Systemic therapies include griseofulvin, ketoconazole, or itraconazole (Knudtson et al. 1980; Mancianti et al. 1998). Species sensitivities to these different medications must be considered. Griseofulvin toxicity is relatively common in cats and was described in an adult cheetah *Acinonyx jubatus* and her four cubs treated for dermatophytosis (Wack et al. 1992).

Zoophilic species may be maintained on carrier animals, and this should be investigated in cases occurring in mixed exhibits and in young animals. Environmental decontamination is important because of the ability of most of the dermatophytes to survive in the environment for extended periods (Scott et al. 1995). Chlorhexidine or sodium hypochlorite (1:10 dilution of household bleach) should be used on all nonporous surfaces. Beddings and other materials that cannot be disinfected should be destroyed. Cages should be disinfected daily.

PUBLIC HEALTH CONCERNS. The zoophilic and geophilic fungi can be transmitted from animals to humans. In many cases, the animal will be asymptomatic (Keymer et al. 1991).

DOMESTIC ANIMAL HEALTH CONCERNS. Dermatomycosis occurs in domestic species and has been reported in various furbearers (Hagen and Gorham 1972; Finley and Long 1978; Knudtson et al. 1980; Janovitz and Long 1984).

ACKNOWLEDGMENTS. Glass slides and gross photos were provided by the University of California, Davis, School of Veterinary Medicine; the Armed Forces Institute of Pathology; and the C.L. Davis Foundation. Thanks to Dr. Julie Schwartz at UC-Davis, who obtained the glass slides, and to Dr. Kimberlee Beckman and Dr. Henry Huntington for their editorial comments.

LITERATURE CITED

Albassam, M.A., R. Bhaunagar, L.J. Lille, and L. Roy. 1986. Adiaspiromycosis in striped skunks in Alberta, Canada. *Journal of Wildlife Diseases* 22:13–18.

Arceneaux, K.A., J. Toboada, and G. Hosgood. 1998. Blastomycosis in dogs: 115 cases (1980–1995). *Journal of the American Veterinary Medical Association* 213:658–64.

Arias, J.R., R.D. Naiff, M.F. Naiff, W.Y. Mok, and M.M.R. Almedia. 1982. Isolation of *Histoplasma capsulatum* from an armadillo (*Dasypus novemcinctus*) in the eastern Amazon of Brazil. *Transactions of the Royal Society of Tropical Medicine and Hygiene* 76:705–706 [cited by Naiff et al. (1985)].

Barker, I.K., A.A. van Dreumel, and N. Palmer. 1993. The alimentary system In *Pathology of domestic animals,* vol. 2, ed. K.V.F. Jubb, P.C. Kennedy, and N. Palmer, 4th ed. San Diego: Academic Press; Harcourt Brace Jovanovich, pp. 255–256.

Barrie, M.T., and C.K. Stadler. 1995. Successful treatment of *Cryptococcus neoformans* infection in an Allen's swamp monkey (*Allenopithecus nigroviridis*) using fluconazole and flucytosine. *Journal of Zoo and Wildlife Medicine* 26:109–114.

Baumgardner, D.J., D.P. Paretsky, and A.C. Yopp. 1995. The epidemiology of blastomycosis in dogs: North central Wisconsin, U.S.A. *Journal of Medical and Veterinary Mycology* 33:171–176.

Beehler, B.A. 1982. Oral therapy for nasal cryptococcosis in a cheetah. *Journal of the American Veterinary Medical Association* 181:1400–1401.

Berry, W.L, J.E. Jardine, and I.W. Espie. 1997. Pulmonary cryptococcoma and cryptococcal meningoencephalomyelitis in a king cheetah (*Acinonyx jubatus*). *Journal of Zoo and Wildlife Medicine* 28:485–490.

Best, P.B., and R.M. McCully. 1979. Zygomycosis (phycomycosis) in a right whale (*Eubalaena australis*). *Journal of Comparative Pathology* 89:341–348.

Blundell, G.P, M.W. Castleberry, E.P. Lowe, and J.L Converse. 1961. The pathology of *Coccidioides immitis* in the *Macaca mulatta. American Journal of Pathology* 39:613–627.

Boever, W.J., and D.M. Rush. 1975. *Microsporum gypseum* infection in a dromedary camel. *Veterinary Medicine and Small Animal Clinician* 70:1190–1192.

Bossart, G.D. 1984. Suspected acquired immunodeficiency in an Atlantic bottlenosed dolphin with chronic-active hepatitis and lobomycosis. *Journal of the American Veterinary Medical Association* 185:1413–1414.

Breznock, A.W., R.V. Henrickson, S. Silverman, and L.W. Schwartz. 1975. Coccidioidomycosis in a rhesus monkey. *Journal of the American Veterinary Medical Association* 167:657–661.

Burton, M., R.J. Morton, E. Ramsay, and E.L. Stair. 1986. Coccidioidomycosis in a ring-tailed lemur. *Journal of the American Veterinary Medical Association* 189:1209–1211.

Butler, T.M., C.A. Gleiser, J.C. Bernal, and L. Ajello. 1988. Case of disseminated African histoplasmosis in a baboon. *Journal of Medical Primatology* 17:153–161.

Caldwell, D.K., M.C. Caldwell, J.C. Woodard, L. Ajello, W. Kaplan, and H.M. McClure. 1975. Lobomycosis as a disease of the Atlantic bottle-nosed dolphin (*Tursiops truncatus* Montague, 1821). *American Journal of Tropical Medicine and Hygiene* 24:105–114.

Campins, H. 1970. Coccidioidomycosis in South America: A review of its epidemiology and geographic distribution. *Mycopathologia et Mycologia Applicata* 40:25–34.

Carmichael, J.W. 1962. *Chrysosporium* and some other aleuriosporic hyphomycetes. *Canadian Journal of Botany* 40:1137–1173.

Cates, M.B., L. Kaufman, J.H. Grabau, J.M. Pletcher, and J.P. Schroeder. 1986. Blastomycosis in an Atlantic bottlenose dolphin. *Journal of the American Veterinary Medical Association* 189:1148–1150.

Chandler, F.W., and J.C. Watts. 1987. *Pathologic diagnosis of fungal infections.* Chicago: American Society of Clinical Pathologists, 302 pp.

Clinkenbeard, K.D., R.L Cowell, and R.D. Tyler. 1988. Disseminated histoplasmosis in dogs: 12 cases (1981–1986). *Journal of the Veterinary Medical Association* 193:1443–1447.

Clyde, V.L., G.V. Kolias Jr., M.E. Roelke, and M.R. Wells. 1990. Disseminated coccidioidomycosis in a western cougar (*Felis concolor*). *Journal of Zoo and Wildlife Medicine* 21:200–205.

Clyde, V.L., E.C. Ramsay, and L. Munson. 1996. A review of blastomycosis in large zoo carnivores. In *Proceedings of the American Association of Zoo Veterinarians annual conference,* ed. C.K. Baer. Puerto Vallarta, Mexico.

Cornell, L.H., K.G. Osborn, J.E. Antrim, and J.G. Simpson. 1979. Coccidioidomycosis in a California sea otter (*Enhydra lutris*). *Journal of Wildlife Diseases* 15:373–378.

Costa, E.O., L.S.M. Diniz, C.F. Netto, C. Arruda, and M.L.A. Dagli. 1994. Epidemiological study of sporotrichosis and histoplasmosis in captive Latin American wild mammals, Sao Paulo, Brazil. *Mycopathologia* 125:19–22.

Cowan, D.F. 1993. Lobo's disease in a bottlenose dolphin (*Tursiops truncatus*) from Matagorda Bay, Texas. *Journal of Wildlife Diseases* 29:488–489.

Darling, S.T. 1906. A protozoon general infection producing pseudotubercles in the lungs and focal necrosis in the liver, spleen and lymph nodes. *Journal of the American Medical Association* 46:1283–1285.

Davis, J.W. 1981. Coccidioidomycosis. In *Infectious diseases of wild mammals,* ed. J.W. Davis, L.H. Karstad, and D.O. Trainer, 2d ed. Ames: Iowa State University Press, pp. 361–365.

De Monbreun, W.A. 1934. The cultivation and cultural characteristics of Darling's *Histoplasma capsulatum. American Journal of Tropical Medicine* 14:93–125.

De Vries, G.A., and J.J. Laarmann. 1973. A case of Lobo's disease in the dolphin *Sotalia guianensis*. *Aquatic Mammals* 1:26–33.

Dillehay, D.L., T.R. Boosinger, and S. MacKenzie. 1985. Coccidioidomycosis in a tapir. *Journal of the American Veterinary Medical Association* 187:1233–1234.

Di Salvo, A.F. 1971. The role of bats in the ecology of *Histoplasma capsulatum*. In *Histoplasmosis,* ed. J. Ajello, E.W. Chick, and M.L. Furcolow. Springfield, IL: Charles C. Thomas, pp 149–161.

Dunbar, M., R.L. Pyle, J.G. Boring, and C.P. McCoy. 1983. Treatment of canine blastomycosis with ketoconazole. *Journal of the American Veterinary Medical Association* 182:156–157.

Dungworth, D.L. 1993. The respiratory system. In *Pathology of domestic animals,* vol. 2, ed. K.V.F. Jubb, P.C. Kennedy, and N. Palmer, 4th ed. San Diego: Academic Press; Harcourt Brace Jovanovich, pp. 665–673.

Dunn, J.L. 1990. Bacterial and mycotic diseases of cetaceans and pinnipeds. In *CRC Handbook of marine mammal medicine: Health, disease, and rehabilitation,* ed. L.A. Dierauf. Boca Raton, FL: CRC, pp. 81–87.

Dunn, J.L., J.D. Buck, and S. Spotte. 1982. Candidiasis in captive cetaceans. *Journal of the American Veterinary Medical Association* 181:1316–1321.

Dvorak, J., M. Otcenasek, and B. Rosicky. 1973. Adiaspiromycosis caused by *Emmonsia crescens,* Emmons and Jellison 1960. *Studies of the Czechoslovak Academy of Sciences* 14:1–120.

El-Khouly, A.B., F.A. Gadir, D.D. Cluer, and G.W. Manefield. 1992. Aspergillosis in camels affected with a specific respiratory and enteric syndrome. *Australian Veterinary Journal* 69:182–186.

Emmons, C.W., and L.L. Ashburn. 1942. The isolation of *Haplosporangium parvum* n. sp. and *Coccidioides immitis* from wild rodents. *Public Health Reports* 57:1715–1727.

Emmons, C.W., J.A. Bell, and B.J. Olson. 1947. Naturally occurring histoplasmosis in *Mus musculus* and *Rattus norvegicus*. *Public Health Reports* 62:1642–1646.

Emmons, C.W., C.H. Binford, U.P. Litz, and K.J. Kwon-chung. 1977. Cryptococcosis. In *Medical mycology.* Philadelphia: Lea and Febiger, p. 206.

Fauquier, D.A., F.M.D. Gulland, J.G. Trupkiewicz, T.R. Spraker, and L.J. Lowenstine. 1996. Coccidioidomycosis in free-living California sea lions (*Zalophus californianus*) in central California. *Journal of Wildlife Diseases* 32:707–710.

Finley, G.G., and J.R. Long. 1978. *Microsporum canis* infection in mink. *Journal of the American Veterinary Medical Association* 173:1226–1227.

Finnie, E.P., 1986. Monotremes and marsupials (Monotremata and Marsupialia). In *Zoo and wild animal medicine,* ed. M.E. Fowler, 2d ed. Philadelphia: W.B. Saunders, pp. 577.

Gales, N., G. Wallace, and J. Dickson. 1985. Pulmonary cryptococcosis in a striped dolphin (*Stenella coeruleoalba*). *Journal of Wildlife Diseases* 21:443–446.

Galgiani, J.N. 1993. Conferences and reviews: Coccidioidomycosis. *Western Journal of Medicine* 159:153–171.

Gatchel, S.L., G. Bjotvedt, and C.R. Leathers. 1986. Experimentally induced *Histoplasma capsulatum* infection in coyotes and a dog. *Journal of the American Veterinary Medical Association* 189:1095–1098.

Goad, D.L., and M.E. Pecquet-Goad. 1986. Osteoarticular sporotrichosis in a dog. *Journal of the American Veterinary Medical Association* 189:1326–1328.

Greer, D.L., and D.N. McMurray. 1981. Pathogenesis of experimental histoplasmosis in the bat, *Artibeus lituratus*. *American Journal of Tropical Medicine and Hygiene* 30:653–659.

Hagen, K.W., and J.R. Gorham. 1972. Dermatomycosis in fur animals: Chinchilla, ferret, mink and rabbit. *Veterinary Medicine and Small Animal Clinician* 67:43–48.

Harwell, G. 1985. Coccidioidomycosis in a river otter, *Lutra canadensis*. In *Proceedings of the annual meeting of the American Association of Zoo Veterinarians,* ed. M.S. Silberman and S.D. Silberman. Scottsdale, AZ: American Association of Zoo Veterinarians, p. 50.

Henrickson, R.V., and E.L. Biberstein. 1972. Coccidioidomycosis accompanying hepatic disease in two Bengal tigers. *Journal of the American Veterinary Medical Association* 161:674–677.

Hoff, G.L., and W.L Bigler. 1981. The role of bats in the propagation and spread of histoplasmosis: A review. *Journal of Wildlife Diseases* 17:191–196.

Hubalek, Z., J. Zejda, J. Nesvadbova, and B. Rychnovsky. 1991. Adiasporomycosis B: A widespread disease of rodents in southern Moravia. *Czechoslovakia Folia Zoologica* 40:107–116.

Hubalek, Z., J. Zejda, S. Svobodova, and J. Kucera. 1993. Seasonality of rodent adiaspiromycosis in a lowland forest. *Journal of Medical and Veterinary Mycology* 31:359–366.

Hutchinson, L.R., F. Duran, C.D. Lane, G.W. Robertstad, and M. Portillo. 1973. Coccidioidomycosis in a giant red kangaroo (*Macropus rufus*). *Journal of Zoo Animal Medicine* 4:22–24.

Ingram, K.A. 1975. Coccidioidomycosis in a colony of chimpanzees. In *Proceedings of the annual meeting of the American Association of Zoo Veterinarians,* ed. M.S. Silberman and S.D. Silberman. Scottsdale, AZ: American Association of Zoo Veterinarians, pp. 127–132.

Jacobs, G.J., L. Medleau, C. Calvert, and J. Brown. 1997. Cryptococcal infection in cats: Factors influencing treatment outcome, and results of sequential serum antigen titers in 35 cats. *Journal of Veterinary Internal Medicine* 11:1–4.

Janovitz, E.B., and G.G. Long. 1984. Dermatomycosis in ranch foxes. *Journal of the American Veterinary Medical Association* 185:1393–1394.

Jellison, W.L. 1981. Adiaspiromycosis. In *Infectious diseases of wild mammals,* ed. J.W. Davis, L.H. Karstad, and D.O. Trainer, 2d ed. Ames: Iowa State University Press, pp. 356–360.

Jensen, H.E., J.B. Jorgensen, and H. Schonheyder. 1989. Pulmonary mycosis in farmed deer: Allergic zygomycosis and invasive aspergillosis. *Journal of Medical and Veterinary Mycology* 27:329–334.

Jensen, H.E., B. Bloch, H.H. Henricksen, H. Schonheyder, and L. Kaufman. 1992. Disseminated histoplasmosis in a badger (*Meles meles*) in Denmark. *APMIS* 100:586–592.

Jensen, H.E., B. Aalbaek, and H. Schonheyder. 1994. Immunohistochemical identification of aetiological agents of systemic bovine zygomycosis. *Journal of Comparative Pathology* 110:65–77.

Jessup, D.A., N. Kock, and M. Berbach. 1989. Coccidioidomycosis in a desert bighorn sheep (*Ovis canadensis nelsoni*) from California. *Journal of Zoo and Wildlife Medicine* 20:471–473.

Johnson, J.H., A.M. Wolf, J.F. Edwards, M.A. Walker, L. Homco, J.M. Jensen, B.R. Simpson, and L. Taliaferro. 1998. Disseminated coccidioidomycosis in a mandril baboon (*Mandrillus sphinx*): A case report. *Journal of Zoo and Wildlife Medicine* 29:208–213.

Johnstone, A.C., H.M. Hussein, and A. Woodgyer. 1993. Adiaspiromycosis in suspected cases of pulmonary tuberculosis in the common brushtail possum (*Trichosurus vulpecula*). *New Zealand Veterinary Journal* 41:175–178.

Jones, T.C., and Hunt, R.D. 1983. Diseases caused by higher bacteria and fungi. In *Veterinary pathology,* 5th ed. Philadelphia: Lea and Febiger, pp. 638–718.

Kano, R., Y. Nakamura, T. Watari, H. Tsujimoto, and A. Hasegawa. 1997. A case of feline cryptococcosis treated with itraconazole. *Mycoses* 40:381–383.

Kaplan, W. 1973. Epidemiology of the principal systemic mycoses of man and lower animals and the ecology of their etiologic agents. *Journal of the American Veterinary Medical Association* 163:1043–1047.

Kaplan, W., J.R. Broderson, and J.N. Pacific. 1982. Spontaneous systemic sporotrichosis in nine-banded armadillos (*Dasypus novemcinctus*). *Sabouraudia* 20:289–294.

Kennedy-Stoskopf, S., and R. Russell. 1983. Blastomycosis in a California sea lion. In *Annual Conference of the International Association of Aquatic Animal Medicine.* Baltimore: International Association of Aquatic Animal Medicine, p. 28. Abstract [cited by Dunn (1990)].

Keymer, I.F., E.A. Gibson, and D.J. Reynolds. 1991. Zoonoses and other findings in hedgehogs (*Erinaceus europaeus*): A survey of mortality and review of the literature. *Veterinary Record* 128:245–249.

Knudtson, W.U., C.E. Gates, and G.R. Ruth. 1980. *Trichophyton mentagrophytes* dermatophytosis in wild fox. *Journal of Wildlife Diseases* 16:465–468.

Kodousek, R., and A.M. Hejtmanek. 1982. Pulmonary adiaspiromycosis of some free living small mammals in the north Moravian region: Pathologic findings and experimental-biological observations. *Acta Universitatis Palackianae Olomucensis Facultatis Medicae* 102:135–138.

Kohn, G.J., S.R. Linne, C.M. Smith, and P.D. Hoeprich. 1992. Acquisition of coccidioidomycosis at necropsy by inhalation of coccidioidal endospores. *Diagnostic Microbiology of Infectious Diseases* 15:527–530.

Krivanec, K., and M. Otcenasek. 1977. Importance of free living mustelid carnivores in circulation of adiasporomycosis. *Mycopathologia* 60:139–144.

Krivanec, K., M. Otcenasek, and B. Rosicky. 1975. The role of polecats of the genus *Putorius* Cuvier, 1817 in natural foci of adiaspiromycosis. *Folia Parasitologica* 22:245–249.

Laakkonen, J.J. Sundell, and T. Soveri. 1998. Lung parasites of least weasels in Finland. *Journal of Wildlife Diseases* 34:816–819.

Lainson R., and J.J. Shaw. 1975. Pneumocystis and histoplasma infections in wild animals from the Amazon region of Brazil. *Transactions of the Royal Society of Tropical Medicine and Hygiene* 69:505–508.

Legendre, A.M. 1995. Antimycotic drug therapy In *Kirk's current veterinary therapy XII: Small animal practice,* ed. J.D. Bonagura and R.W. Kirk. Philadelphia: W.B. Saunders, pp. 327–331.

Leighton, F.A., and G. Wobeser. 1978. The prevalence of adiaspiromycosis in three sympatric species of ground squirrels. *Journal of Wildlife Diseases* 14:362–365.

Lenhard, A. 1985. Blastomycosis in a ferret. *Journal of the American Veterinary Medical Association* 186:70–73.

Lochmiller, R.L., E.C. Hellgren, P.G. Hannan, W.E. Grant, and R.M. Robinson. 1985. Coccidioidomycosis (*Coccidioides immitis*) in the collared peccary (*Tayassu tajacu:* Tayassuidae) in Texas. *Journal of Wildlife Diseases* 21:305–309.

Mancianti, R., F. Pedonese, and C. Zullino. 1998. Efficacy of oral administration of itraconazole to cats with dermatophytosis caused by *Microsporum canis. Journal of the American Veterinary Medical Association* 213:993–995.

Mason R.W., and M. Gauhwin. 1982. Adiaspiromycosis in South Australian hairy-nosed wombats (*Lasiorhinus latifrons*). *Journal of Wildlife Diseases* 18:3–8.

McKenney, F.D., J. Traum, and A.E. Bomestall. 1944. Acute coccidioidomycosis in a mountain gorilla (*Gorilla beringeri*) with anatomical notes. *Journal of the American Veterinary Medical Association* 104:136–140.

Medway, W. 1980. Some bacterial and mycotic diseases of marine mammals. *Journal of the American Veterinary Medical Association* 177:831–834.

Migaki, G., M.G. Valerio, B. Irvine, and F.M. Garner. 1971. Lobo's disease in an Atlantic bottle-nosed dolphin. *Journal of the American Veterinary Medical Association* 159:578–582.

Migaki, G., R.L. Font, W. Kaplan, and E.D. Asper. 1978a. Sporotrichosis in a Pacific white-sided dolphin (*Lagenorhynchus obliquidens*). *American Journal of Veterinary Research* 39:1916–1919.

Migaki, G., R.D. Gunnels, and H.W. Casey. 1978b. Pulmonary cryptococcosis in an Atlantic bottlenosed dolphin (*Tursiops truncatus*). *Laboratory Animal Science* 28:603–606.

Migaki, G., R.E. Schmidt, J.D. Toft, and A.F. Kaufmann. 1982. Mycotic infections of the alimentary tract of nonhuman primates: A review. *Veterinary Pathology* 19(Suppl.7):93–103.

Miller, R.E., and W.J. Boever. 1983. Cryptococcosis in a lion-tailed macaque (*Macaca silenus*). *Journal of Zoo Animal Medicine* 14:110–114.

Morris, P.J., A.M. Legendre, T.L. Bowersock, D.E. Brooks, D.J. Krahwinel, G.M.H. Shires, and M.A. Walker. 1989. Diagnosis and treatment of systemic blastomycosis in a polar bear (*Ursus maritimus*) with itraconazole. *Journal of Zoo and Wildlife Medicine* 20:336–345.

Muir, S., and D. Pappagianis. 1982. Coccidioidomycosis in the llama: Case report and epidemiologic survey. *Journal of the American Veterinary Medical Association* 181:1334–1337.

Naiff, R.D., W.Y. Mok, and M.F. Naiff. 1985. Distribution of *Histoplasma capsulatum* in Amazonian wildlife. *Mycopathologia* 89:165–168.

Nuorva, K.R. Pitkanen, J. Issakainen, N.P. Huttunen, and M. Juhola. 1997. Pulmonary adiaspiromycosis in a two year old girl. *Journal of Clinical Pathology* 50:82–85.

Obendorf, D.L. 1980. Candidiasis in young hand-reared kangaroos. *Journal of Wildlife Diseases* 16:135–140.

Obendorf, D.L., B.F. Peel, and B.L Munday. 1993. *Mucor amphibiorum* infection in platypus (*Ornithorhynchus anatinus*) from Tasmania. *Journal of Wildlife Diseases* 29:485–487.

O'Hara, T.M. 1996. Mycotoxins. In *Noninfectious diseases of wildlife,* ed. A. Fairbrother, L.N. Locke, and G.L. Hoff, 2d ed. Ames: Iowa State University Press, pp. 24–30.

Ohbayashi, M. 1971. Mucormycosis in laboratory-reared rodents. *Journal of Wildlife Diseases* 7:59–62.

Olson, B.J., J.A. Bell, and C.W. Emmons. 1947. Studies on histoplasmosis in a rural community. *American Journal of Public Health* 37:441–449.

Osborn, K.G., L.H. Cornell, J.E. Antrium, and B.E. Joseph. 1985. Disseminated coccidioidomycosis in a stranded California sea lion (*Zalophus californianus*). In *Annual Proceedings of the American Association of Zoo Veterinarians,* ed. M.S. Silberman and S.D. Silberman. Scottsdale, AZ: American Association of Zoo Veterinarians, p. 98.

Otcenasek M., B. Rosicky, K. Krivanec, J. Dvorak, and K. Rasin. 1974. The muskrat as reservoir in natural foci of adiaspiromycosis. *Folia Parasitologica (Praha)* 21:55–57.

Owens, D.R., R.W. Menges, R.F. Sprouse, W. Stewart, and B.E. Hooper. 1975. Naturally occurring histoplasmosis in the chinchilla (*Chinchilla laniger*). *Journal of Clinical Microbiology* 1:486–488.

Pal, M., G.D. Dube, and B.S. Mehrotra. 1984. Pulmonary cryptococcosis in a rhesus monkey (*Macaca mulatta*). *Mykosen* 27:309–312.

Pappagianis, D. 1993. Coccidioidomycosis. *Seminars in Dermatology* 12:301–309.

Pappagianis, D., J. Vanderlip, and B. May. 1973. Coccidioidomycosis naturally acquired by a monkey, *Cercocebus atys,* in Davis, California. *Sabouraudia* 11:52–55.

Peden, W.M., J.L. Richard, D.W. Trampel, and R.E. Brannian. 1985. Mycotic pneumonia and meningoencephalitis due to *Aspergillus terreus* in a neonatal snow leopard (*Panthera uncia*). *Journal of Wildlife Diseases* 21:301–305.

Phillips, P.H., P.K. Davenport, and D.J. Schultz. 1986. *Microsporum gypseum* dermatophytosis in captive Australian sea lions (*Neophoca cinerea*). *Journal of Zoo Animal Medicine* 17:136–138.

Pickett, J.P., C.P. Moore, B.A. Beehler, A. Gendron-Fitzpatrick, and R.R. Dubielzig. 1985. Bilateral chorioretinitis secondary to disseminated aspergillosis in an alpaca. *Journal of the American Veterinary Medical Association* 187:1241–1243.

Pier, A.C., and J.L. Richard. 1992. Mycoses and mycotoxicoses of animals caused by aspergilli. *Biotechnology* 23:233–248.

Potkay, S. 1977. Diseases of marsupials. In *The biology of marsupials,* ed. D. Hunsaker II. New York: Academic, p. 440.

———. 1992. Diseases of the Callitrichidae: A review. *Journal of Medical Primatology* 21:189–236.

Quandt, S.K.F., and J.W. Nesbit. 1992. Histoplasmosis in a 2-toed sloth (*Choloepus didactylus*). *Journal of Zoo and Wildlife Medicine* 23:369–373.

Raju, N.R., R.F. Langham, and R.R. Bennett, 1986. Disseminated histoplasmosis in a fennec fox. *Journal of the American Veterinary Medical Association* 189:1195–1196.

Rapley, W.A., and J.R. Long. 1974. Coccidioidomycosis in a baboon recently imported from California. *Canadian Veterinary Journal* 15:39–41.

Rappelli, P., R. Are, G. Casu, P.L. Fiori, P. Cappuccinelli, and A. Aceti. 1998. Development of a nested PCR for detection of *Cryptococcus neoformans* in cerebrospinal fluid. *Journal of Clinical Microbiology* 36:3438–3440.

Raymond, J.T., M.R. White, T.P. Kilbane, and E.B. Janovitz. 1997. Pulmonary blastomycosis in an Indian fruit bat (*Pteropus giganteus*). *Journal of Veterinary Diagnostic Investigation* 9:85–87.

Reed, R.E., G. Migaki, and J.A. Cummings. 1976. Coccidioidomycosis in a California sea lion (*Zalophus californianus*). *Journal of Wildlife Diseases* 12:372–375.

Reidarson, T.H., L.A. Griner, D. Pappagianis, and J. McBain. 1998. Coccidioidomycosis in a bottlenose dolphin. *Journal of Wildlife Diseases* 34:629–31.

Rosenberg, D.P., C.A. Gleiser, and K.D. Carey. 1984. Spinal coccidioidomycosis in a baboon. *Journal of the American Veterinary Medical Association* 185:1379–1381.

Sanger, V.L. 1981. Histoplasmosis. In *Infectious diseases of wild mammals,* ed. J.W. Davis, L.H. Karstad, and D.O. Trainer, 2d ed. Ames: Iowa State University Press, pp. 356–360.

Scott, D.W., W.H. Miller Jr., and C.E. Griffin. 1995. *Muller and Kirk's small animal dermatology,* 5th ed. Philadelphia: W.B. Saunders, pp. 330–391.

Scrimgeour, E.M., and R.G. Purohit. 1984. Chronic pulmonary cryptococcosis in a *Rattus rattus* from Rabaul, Papua New Guinea. *Transactions of the Royal Society of Tropical Medicine and Hygiene* 78:827–828.

Severo, L.C., J.C. Bohrer, G.R. Geyer, and L. Ferreiro. 1989. Invasive aspergillosis in an alpaca (*Lama pacos*). *Journal of Medical and Veterinary Mycology* 27:193–195.

Sharapov, V.M. 1969. Adiospiromycosis in USSR. *Sibirskoe Otdelenie, Novosibirskii Isvestia Service Biolog. Nauk., Akad. Nauk. SSSR* 1:86–95 [cited by Leighton and Wobeser (1978)].

Simon-Lopes, P.C., G.S. Paula, M.C. Both, F.M. Xavier, and A.C. Scaramello. 1993. First case of lobomycosis in a bottlenose dolphin from southern Brazil. *Marine Mammal Science* 9:329–331.

Spencer, A., C. Ley, P. Canfield, P. Martin, and R. Perry, 1993. Meningoencephalitis in a koala (*Phascolarctos cinereus*) due to *Cryptococcus neoformans* var *gattii* infection. *Journal of Zoo and Wildlife Medicine* 24:519–522.

Splendore, D.A. 1920. Sui parassiti delle arvicole. *Annales Igiene* 30:445–468

Stevens, D.A. 1995. Coccidioidomycosis. *New England Journal of Medicine* 332:1077–1082.

Straub, M., R.J. Troutman, and J.W. Greene. 1961. Coccidioidomycosis in three coyotes. *American Journal of Veterinary Research* 89:811–812.

Stroud, R.K., and B.M. Coles. 1980. Blastomycosis in an African lion. *Journal of the American Veterinary Medical Association* 177:842–844.

Taniyama, H., H. Furuoka, T. Matsui, and T. Ono. 1985. Two cases of adiaspiromycosis in the Japanese pika (*Ochotona hyperborea yesoensis* Kishida). *Japanese Journal of Veterinary Science* 47:139–142.

Tell, L.A., D.K. Nichols, W.P. Fleming, and M. Bush. 1997. Cryptococcosis in tree shrews (*Tupaia tana* and *Tupaia minor*) and elephant shrews (*Macroscelides proboscides*). *Journal of Zoo and Wildlife Medicine* 28:175–181.

Tesh, R.B., and J.D. Schneidau. 1966. Experimental infection of North American insectivorous bats (*Tadarida brasiliensis*) with *Histoplasma capsulatum. American Journal of Tropical Medicine and Hygiene* 15:544–550.

Thiel, R.P., L.D. Mech, G.R. Ruth, J.R. Archer, and L. Kaufman. 1987. Blastomycosis in wild wolves. *Journal of Wildlife Diseases* 23:321–323.

Wack, R.J., L.W. Kramer, and W. Cupps. 1992. Griseofulvin toxicity in four cheetahs (*Acinonyx jubatus*). *Journal of Zoo and Wildlife Medicine* 23:442–446.

Ward, D.A. 1995. Oculomycosis. In *Kirk's current veterinary therapy XII: Small animal practice,* ed. J.D. Bonagura and R.W. Kirk. Philadelphia: W.B. Saunders, pp. 1257–1261.

Weber, M., and R.E. Miller. 1996. Fungal pneumonia in black rhinoceros (*Diceros bicornis*). In *Proceedings of the American Association of Zoo Veterinarians,* ed. C.K. Baer. Puerto Vallarta, Mexico: American Association of Zoo Veterinarians, pp. 34–36.

Weller, R.E., G.E. Dagle, C.A. Malaga, and J.F. Baer. 1990. Hypercalcemia and disseminated histoplasmosis in an owl monkey. *Journal of Medical Primatology* 19:675–680.

Werner, A.H., and B.E. Werner. 1994. Sporotrichosis in man and animals. *International Journal of Dermatology* 33:692–700.

Williamson, W.M., L.S. Lombard, and R.E. Getty. 1959. North American blastomycosis in a northern sea lion. *Journal of the American Veterinary Medical Association* 153:513–515 [cited by Dunn (1990)].

Wilson, T.M., M. Kierstead, and J.R. Long. 1974. Histoplasmosis in a harp seal. *Journal of the American Veterinary Medical Association* 165:815–817.

Woolf, A., C. Gremillion-Smith, J.P. Sundberg, and F.W. Chandler. 1985. Histoplasmosis in a striped skunk (*Mephitis mephitis* Schreber) from southern Illinois. *Journal of Wildlife Diseases* 21:441–443.

Yager, J.A., and D.W. Scott. 1993. The skin and appendages. In *Pathology of domestic animals,* vol. 1, ed. K.V.F. Jubb, P.C. Kennedy, and N. Palmer, 4th ed. New York: Academic, pp. 661–667.

CONTRIBUTORS

Edward Addison (Retired)
Wildlife and Natural Heritage Science
Ontario Ministry of Natural Resources
Peterborough, Ontario, Canada

Ian K. Barker
Canadian Cooperative Wildlife Health Center
Department of Pathobiology
Ontario Veterinary College
University of Guelph
Guelph, Ontario, Canada

David A. Benfield
Veterinary Science Department and Animal Disease Research and Diagnostic Laboratory
Brookings, South Dakota, U.S.A.

Roy G. Bengis
State Veterinarian
Kruger National Park
Skukuza, Republic of South Africa

Corrie C. Brown
Department of Pathology
College of Veterinary Medicine
University of Georgia
Athens, Georgia, U.S.A.

Richard N. Brown
Department of Wildlife
Humboldt State University
Arcata, California, U.S.A.

Kathy Burek
Alaska Veterinary Pathology Services
Eagle River, Alaska, U.S.A.

Elizabeth C. Burgess
Blue Mounds, Wisconsin, U.S.A.

Anthony E. Castro (Retired)
Animal Diagnostic Laboratory
Department of Veterinary Sciences
Pennsylvania State University
University Park, Pennsylvania, U.S.A.

James E. Childs
Special Pathogens Branch, and Viral and Rickettsial Zoonoses Branch
National Center for Infectious Diseases
Centers for Disease Control and Prevention
Atlanta, Georgia, U.S.A.

Richard S. Clifton-Hadley
Epidemiology Department
Veterinary Laboratories Agency
New Haw
Addlestone, Surrey, United Kingdom

William R. Davidson
Southeastern Cooperative Wildlife Disease Study and Warnell School of Forest Resources
University of Georgia
Athens, Georgia, U.S.A.

Jacqueline E. Dawson
Infectious Diseases Pathology Activity
Division of Viral and Rickettsial Diseases
National Center for Infectious Diseases
Centers for Disease Control and Prevention
Atlanta, Georgia, U.S.A.

Dan Dragon
Department of Medical Microbiology and Immunology
University of Alberta Hospitals
Edmonton, Alberta, Canada

Peter A. Durr
Epidemiology Department
Veterinary Laboratories Agency
New Haw
Addlestone, Surrey, United Kingdom

Greg Early
New England Aquarium
Central Warf
Boston, Massachusetts, U.S.A.

Brett Elkin
Department of Resources, Wildlife and Economic Development
Government of the Northwest Territories
Yellowknife, Northwest Territories, Canada

Monika Engels
Institute of Virology
Veterinary Medical Faculty
University of Zürich, Zürich, Switzerland

James F. Evermann
Department of Veterinary Clinical Sciences and Washington Animal Disease Diagnostic Laboratory
College of Veterinary Medicine
Washington State University
Pullman, Washington, U.S.A.

Sidney A. Ewing
Department of Infectious Diseases and Physiology
College of Veterinary Medicine
Oklahoma State University
Stillwater, Oklahoma, U.S.A.

William J. Foreyt
Department of Veterinary Microbiology and Pathology
Washington State University
Pullman, Washington, U.S.A.

Kai Frölich
Institute for Zoo Biology and Wildlife Research
Berlin, Germany

Peter W. Gasper
Avrum Gudelsky Veterinary Center
University of Maryland
College Park, Maryland, U.S.A.

C. Cormack Gates
Faculty of Environmental Design
University of Calgary
Calgary, Alberta, Canada

Dolores Gavier-Widén
Department of Wildlife
National Veterinary Institute
Uppsala, Sweden

Will L. Goff
Agricultural Research Service
United States Department of Agriculture
Pullman, Washington, U.S.A.

Werner P. Heuschele (Deceased)
Center for Reproduction of Endangered Species
Zoological Society of San Diego
San Diego, California, U.S.A.

Martin Hofmann
Institute for Virology and Immunoprophylaxis
Mittelhausern, Switzerland

Elizabeth W. Howerth
Department of Pathology
College of Veterinary Medicine
University of Georgia
Athens, Georgia, U.S.A.

Ronald Jackson
EpiCentre
Massey University
Palmerston North, New Zealand

A. Bennett Jenson
Department of Pathology
Georgetown University School of Medicine
Washington, D.C., U.S.A.

Seamus Kennedy
Veterinary Sciences Division
Stoney Road
Stormont
Belfast, Northern Ireland

Peter J. Kerr
CSIRO Division of Wildlife and Ecology & Vertebrate Biocontrol Cooperative Research Centre
Canberra, Australian Capital Territory, Australia

Norval W. King (Retired)
New England Primate Center
Harvard Medical School
Southborough, Massachusetts, U.S.A.

Peter D. Kirkland
Elizabeth MacArthur Agricultural Institute
New South Wales Agriculture
Camden, New South Wales, Australia

James K. Kirkwood
Universities Federation for Animal Welfare
The Old School
Brewhouse Hill
Wheathampstead, Hertforshire, United Kingdom

Nancy D. Kock
California Marine Wildlife and Veterinary Care and Research Center
California Department of Fish and Game
Santa Cruz, California, U.S.A.

Thijs Kuiken
Institute of Virology
Erasmus University
Rotterdam, The Netherlands

Fredrick A. Leighton
Canadian Cooperative Wildlife Health Center
Department of Veterinary Pathology
Western College of Veterinary Medicine
University of Saskatchewan
Saskatoon, Saskatchewan, Canada

Cor Lenghaus
CSIRO Division of Animal Health
Australian Animal Health Laboratory
Geelong, Victoria, Australia

Ian W. Lugton
Department of Veterinary Clinical Sciences
Massey University
Palmerston North, New Zealand

Courtney Meredith (Retired)
Onderstepoort Veterinary Institute
Onderstepoort, South Africa

Michael W. Miller
Colorado Division of Wildlife
Wildlife Research Center
Fort Collins, Colorado, U.S.A.

James N. Mills
Special Pathogens Branch, and Viral and Rickettsial Zoonoses Branch
National Center for Infectious Diseases
Centers for Disease Control and Prevention
Atlanta, Georgia, U.S.A.

Richard J. Montali
Smithsonian Institution
National Zoological Park
Washington, D.C., U.S.A.

Torsten Mörner
Department of Wildlife
National Veterinary Institute
Uppsala, Sweden

Linda Munson
Department of Pathology, Microbiology, and Immunology
College of Veterinary Medicine
University of California
Davis, California, U.S.A.

Colin R. Parrish
James A. Baker Institute
New York State College of Veterinary Medicine
Cornell University
Ithaca, New York, U.S.A.

Hugh W. Reid
Moredun Research Institute
Penlands Science Park
Midlothian, Scotland, United Kingdom

Laura K. Richman
Johns Hopkins University School of Medicine
Baltimore, Maryland, U.S.A.

Anthony J. Robinson
CSIRO Division of Wildlife and Ecology & Vertebrate Biocontrol Cooperative Research Centre
Canberra, Australian Capital Territory, Australia

Paul Rossiter
Organization of African Unity
Interafrican Bureau for Animal Resources
Nairobi, Kenya

Charles E. Rupprecht
Viral and Rickettsial Zoonoses Branch
Centers for Disease Control and Prevention
Atlanta, Georgia, U.S.A.

Carola M. Sauter-Louis
EpiCentre
Massey University
Palmerston North, New Zealand

Charles Seymour (Deceased)
Department of Animal Health and Biomedical Sciences
School of Veterinary Medicine
University of Wisconsin
Madison, Wisconsin, U.S.A.

David E. Stallknecht
Southeastern Cooperative Wildlife Disease Study
College of Veterinary Medicine
University of Georgia
Athens, Georgia, U.S.A.

Klaus Stöhr
World Health Organization
Geneva, Switzerland

Michael J. Studdert
Faculty of Veterinary Science
University of Melbourne
Parkville, Victoria, Australia

John P. Sundberg
The Jackson Laboratory
Bar Harbor, Maine, U.S.A.

Gavin R. Thomson
ARC-Onderstepoort Institute for Exotic Diseases
Onderstepoort, Republic of South Africa

E. Tom Thorne
Wyoming Game and Fish Department
Cheyenne, Wyoming, U.S.A.

Hana Van Campen
Department of Veterinary Sciences
University of Wyoming
Laramie, Wyoming, U.S.A.

Marc Van Ranst
Rega Institute for Medical Research
Department of Microbiology and Immunology
University of Leuven
Leuven, Belgium

Rowena P. Watson
Avrum Gudelsky Veterinary Center
University of Maryland
College Park, Maryland, U.S.A.

Kevin Whithear
Faculty of Veterinary Science
University of Melbourne
Veterinary Clinical Center
Werribee, Victoria, Australia

Richard Whittington
Microbiology and Immunology Section
Elizabeth Macarthur Agricultural Institute
New South Wales Department of Agriculture
Camden, New South Wales, Australia

John W. Wilesmith
Epidemiology Department
Veterinary Laboratories Agency
New Haw
Addlestone, Surrey, United Kingdom

Elizabeth S. Williams
Department of Veterinary Sciences
University of Wyoming
Laramie, Wyoming, U.S.A.

Gary Wobeser
Canadian Cooperative Wildlife Health Center
Department of Veterinary Pathology
Western College of Veterinary Medicine
University of Saskatchewan
Saskatoon, Saskatchewan, Canada

Leslie W. Woods
California Veterinary Diagnostic Laboratory System
School of Veterinary Medicine
University of California
Davis, California, U.S.A.

Michael Worley
Center for Reproduction of Endangered Species
Zoological Society of San Diego
San Diego, California, U.S.A.

Thomas M. Yuill
Institute for Environmental Studies
University of Wisconsin
Madison, Wisconsin, U.S.A.

INDEX